Biodegradation and Biodeterioration at the Nanoscale

Micro and Nano Technologies

Biodegradation and Biodeterioration at the Nanoscale

Edited by

Hafiz M.N. Iqbal
School of Engineering and Sciences, Tecnologico de Monterrey, Monterrey, Mexico

Muhammad Bilal
School of Life Science and Food Engineering, Huaiyin Institute of Technology, Huai'an, P.R. China

Tuan Anh Nguyen
Microanalysis Department, Institute for Tropical Technology, Vietnam Academy of Science and Technology, Hanoi, Vietnam

Ghulam Yasin
School of Physics and Optical Engineering, Shenzhen University, Shenzhen, P.R. China; Institute for Advanced Study, Shenzhen University, Shenzhen, P.R. China

Elsevier
Radarweg 29, PO Box 211, 1000 AE Amsterdam, Netherlands
The Boulevard, Langford Lane, Kidlington, Oxford OX5 1GB, United Kingdom
50 Hampshire Street, 5th Floor, Cambridge, MA 02139, United States

Copyright © 2022 Elsevier Inc. All rights reserved.

No part of this publication may be reproduced or transmitted in any form or by any means, electronic or mechanical, including photocopying, recording, or any information storage and retrieval system, without permission in writing from the publisher. Details on how to seek permission, further information about the Publisher's permissions policies and our arrangements with organizations such as the Copyright Clearance Center and the Copyright Licensing Agency, can be found at our website: www.elsevier.com/permissions.

This book and the individual contributions contained in it are protected under copyright by the Publisher (other than as may be noted herein).

Notices

Knowledge and best practice in this field are constantly changing. As new research and experience broaden our understanding, changes in research methods, professional practices, or medical treatment may become necessary.

Practitioners and researchers must always rely on their own experience and knowledge in evaluating and using any information, methods, compounds, or experiments described herein. In using such information or methods they should be mindful of their own safety and the safety of others, including parties for whom they have a professional responsibility.

To the fullest extent of the law, neither the Publisher nor the authors, contributors, or editors, assume any liability for any injury and/or damage to persons or property as a matter of products liability, negligence or otherwise, or from any use or operation of any methods, products, instructions, or ideas contained in the material herein.

British Library Cataloguing-in-Publication Data
A catalogue record for this book is available from the British Library

Library of Congress Cataloging-in-Publication Data
A catalog record for this book is available from the Library of Congress

ISBN: 978-0-12-823970-4

For Information on all Elsevier publications
visit our website at https://www.elsevier.com/books-and-journals

Publisher: Matthew Deans
Acquisitions Editor: Simon Holt
Editorial Project Manager: Gabriela D. Capille
Production Project Manager: Debasish Ghosh
Cover Designer: Miles Hitchen

Typeset by MPS Limited, Chennai, India

Contents

List of contributors xix

1. **Biodegradation and biodeterioration at the nanoscale: an introduction** 1
   MUHAMMAD BILAL, PANKAJ BHATT, TUAN ANH NGUYEN AND HAFIZ M.N. IQBAL

   1.1 Introduction 1
   1.2 Nanobioremediation: a step forward to advance traditional bioremediation 2
   1.3 Miscellaneous cases 3
   1.4 Concluding remarks and future outlook 5
   References 5

2. **Biodegradation of materials in presence of nanoparticles** 9
   HIRA MUNIR, KHAJISTA TAHIRA, AHMAD REZA BAGHERI AND MUHAMMAD BILAL

   2.1 Introduction 9
   2.2 Roles of nanoparticles in biodegradation 11
   2.3 Biological synthesis of nanoparticles 12
   2.4 Nanobioremediation 15
   2.5 Role of nanomaterials in bioremediation 16
   2.6 Remediation of pollutants using nanotechnology 16
   2.7 Bioremediation of hydrophobic contaminants 18
   2.8 Biodegradation of synthetic dyes by nanoparticles 20
   2.9 Biodegradation of phenolic compounds 21

	2.10 Biodegradation of antibiotics and personal care products	22
	2.11 Conclusions	23
	References	24
3.	**Interaction of nanomaterials with microbes**	**31**
	MUHAMMAD RIZWAN JAVED, MUHAMMAD HAMID RASHID, ANAM TARIQ, RIFFAT SEEMAB, ANAM IJAZ AND SOHAIL ABBAS	
	3.1 Introduction	31
	3.2 Types of nanoparticles	32
	3.3 Nanomaterial-microbe interaction and their mechanism	33
	3.4 Responses to nanomaterial-microbe interactions	37
	3.5 Microbial mediated synthesis of nanoparticles	37
	3.6 Mechanisms of microbial mediated nanoparticles biosynthesis	41
	3.7 Role of bioreducing agents in nanoparticle biosynthesis	41
	3.8 Biosynthesis pathways	42
	3.9 Applications of nanomaterials	44
	3.10 Challenges pertaining to the applications of nanomaterials	47
	3.11 Conclusion and future perspectives	49
	Acknowledgment	50
	References	50
4.	**Process of biodegradation controlled by nanoparticle-based materials: mechanisms, significance, and applications**	**61**
	ROBERTA ANJOS DE JESUS, JOSÉ ARNALDO SANTANA COSTA, CAIO MARCIO PARANHOS, MUHAMMAD BILAL, RAM NARESH BHARAGAVA, HAFIZ M.N. IQBAL, LUIZ FERNANDO ROMANHOLO FERREIRA AND RENAN TAVARES FIGUEIREDO	
	4.1 Introduction	61
	4.2 Occurrence of emerging pollutants: environmental impact	62
	4.3 Nanoparticle chemistry	67
	4.4 Role of nanoparticles in biodegradation	74
	4.5 Future prospects	76
	Acknowledgments	77
	References	77

5. Nanophotocatalysts for biodegradation of materials — 85
HANIEH FAKHRI AND MAHDI FARZADKIA

5.1 Introduction — 85
5.2 Advanced oxidation processes for water and wastewater treatment — 87
5.3 Methods of improving photocatalytic efficiency — 90
5.4 Effective factors in the advanced oxidation processes — 93
5.5 Photocatalysis mechanisms — 97
5.6 Conclusion — 98
References — 98

6. Effects of nanomaterials on biodegradation of biomaterials — 105
SANAZ SOLEYMANI EIL BAKHTIARI, HAMID REZA BAKHSHESHI-RAD, MAHMOOD RAZZAGHI, AHMAD FAUZI ISMAIL, SAFIAN SHARIF, SEERAM RAMAKRISHNA AND FILIPPO BERTO

Abbreviations — 105
6.1 Introduction — 106
6.2 Effects of different nanomaterials on biodegradation behavior — 109
6.3 Conclusion and future outlooks — 127
Acknowledgments — 127
Funding — 128
Conflicts of Interest — 128
References — 128

7. Enzyme-encapsulated nanoparticles for biodegradation of materials — 137
LEEBA BALAN, JAMUNA SANKER, SRIRAM CHANDRASEKARAN AND SUGUMARI VALLINAYAGAM

7.1 Introduction — 137
7.2 Different types of biodegradable nanoparticles — 137
7.3 What is the need for biodegradation of materials? — 141

	7.4 Enzyme-encapsulated nanoparticles in biodegradation	142
	7.5 Microbes and enzymatic degradation	144
	7.6 Applications of biodegradable nanoparticles	145
	References	148

8. Effects of nanoparticles on the biodegradation of organic materials 153

SABAH BAKHTIARI, ESMAIL DOUSTKHAH, MONA ZAMANI PEDRAM, MASOUD YARMOHAMMADI AND M.ÖZGÜR SEYDIBEYOĞLU

8.1 Introduction	153
8.2 Nanoparticles as enhancers for biodegradation	155
8.3 Inference and future prospects	167
References	167

9. Biodegradation of plastic-based waste materials 175

NIHAN UÇAR, SABAH BAKHTIARI, ESMAIL DOUSTKHAH, MASOUD YARMOHAMMADI, MONA ZAMANI PEDRAM, ELIF ALYAMAÇ AND M. ÖZGÜR SEYDIBEYOĞLU

9.1 Introduction	175
9.2 Waste management of plastics	178
9.3 Plastics identifications and classifications	179
9.4 Plastics biodegradation	181
9.5 Plastics from fossil resources	188
9.6 Plastics from renewable resources	195
9.7 Determination techniques for microbial degradation of plastics	201
9.8 Future prospects	203
9.9 Conclusion	204
References	205

10. Nano-biodegradation of polymers 213

KOMAL RIZWAN, TAHIR RASHEED AND MUHAMMAD BILAL

10.1 Introduction	213

	10.2 Different kinds of degradation of polymers	218
	10.3 Influence of various factors on the polymers biodegradability	226
	10.4 Techniques and methodologies employed for polymer degradation	227
	10.5 Biodegradability of various polymers and impact of nanoparticle	228
	10.6 Different fillers and their influence on biodegradation	231
	10.7 Effect of filler dimension on biodegradation	232
	10.8 Conclusion	233
	References	233
11.	**Nanobiodegradation of plastic waste**	**239**

AYESHA BAIG, MUHAMMAD ZUBAIR, MUHAMMAD NADEEM ZAFAR, MUJAHID FARID, MUHAMMAD FAIZAN NAZAR AND SAJJAD HUSSAIN SUMRRA

	11.1 Introduction	239
	11.2 Plastic waste pollution and environmental impacts	240
	11.3 The impacts of plastic wastes accumulation	241
	11.4 Types of plastics targeted	242
	11.5 Plastic waste disposal methods	243
	11.6 Photooxidative degradation of waste plastics	244
	11.7 Thermal degradation of waste plastics	245
	11.8 Enzymatic degradation of bioplastics	245
	11.9 Reduction strategies for waste management of plastics by using biotechnology	246
	11.10 Biodegradation of plastics	246
	11.11 Biodegradation at the nanoscale level	249
	11.12 Degradation of waste plastics using microbes	251
	11.13 Degradation activity using insects	254
	11.14 Degradation activity using worms	255
	11.15 Conclusion	256
	References	256

12. **Biodegradation of timber industry-based waste materials**   **261**

N. LAVANYA, SUGUMARI VALLINAYAGAM AND KARTHIKEYAN RAJENDRAN

    12.1 Introduction   261
    12.2 Background   262
    12.3 Wood from the tree   262
    12.4 Plantation and cultivation of timber   264
    12.5 Production of timber   268
    12.6 Consumption at wood-based industry   270
    12.7 Annual consumption of wood   273
    12.8 The value chain of timber-based industry   273
    12.9 The waste problem   274
    12.10 Biodegradation   275
    12.11 Role of microorganisms in biodegradation   276
    12.12 Factors affecting microbial degradation   278
    12.13 Degradation by genetically engineered microorganisms   279
    12.14 Conclusion   280

13. **Microbiologically induced deterioration and environmentally friendly protection of wood products**   **283**

OLGA A. SHILOVA, IRINA N. TSVETKOVA, DMITRY YU. VLASOV, YULIA V. RYABUSHEVA, GEORGII S. SOKOLOV, ANATOLY K. KYCHKIN, CHI VăN NGUYÊN AND YULIA V. KHOROSHAVINA

    13.1 Features of wood biodegradation under the influence of wood-destroying fungi   283
    13.2 Chemical means of protecting wood from biodegradation   284
    13.3 Surface hydrophobization   287
    13.4 Nanotechnology for wood protection   289
    13.5 Using sol-gel technology to protect wood from wood-destroying fungi   295
    13.6 Field tests of friendly wood protection coatings in various climatic conditions   307

	13.7 Conclusion	312
	References	315

14. Biodegradable of plastic industrial waste material — 323

A. JAWAHAR NISHA, SUGUMARI VALLINAYAGAM AND KARTHIKEYAN RAJENDRAN

	14.1 Introduction	323
	14.2 Chemical formation of polyester (Apicella, Migliaresi, Nicolais, Iaccarino, & Roccotelli, 1983)	328
	14.3 Structure of polyester	328
	14.4 Nylon and polyethylene	330
	14.5 Biodegradation of natural plastic	330
	14.6 Enzymatic method to degrade polyhydroxyalkanoate	331
	14.7 Biodegradation of blended plastic	332
	14.8 Other degrading polymers	332
	14.9 Biodegradation of thermo set plastics	333
	14.10 Standard testing methods	334
	14.11 Conclusion	336
	References	336

15. Microbiologically induced deterioration and protection of outdoor stone monuments — 339

O.A. SHILOVA, D.Y. VLASOV, T.V. KHAMOVA, M.S. ZELENSKAYA AND O.V. FRANK-KAMENETSKAYA

	15.1 Destruction of a stone under the influence of microorganisms	339
	15.2 Methods to counter stone biodegradation	344
	15.3 Using sol-gel technology to protect marble from biodegradation	350
	15.4 Conclusion	362
	References	363

16. **Microbiologically induced deterioration of cement-based materials** — 369
    N.B. SINGH

    16.1 Introduction — 369
    16.2 Corrosion and deterioration — 371
    16.3 Microbiologically induced corrosion and degradation of cement-based materials — 371
    16.4 Microbiologically-induced corrosion in sewer structures — 376
    16.5 Generation of sulfuric acid — 376
    16.6 Deterioration of concrete materials — 378
    16.7 Measures against concrete biodeterioration — 380
    16.8 Biofilms — 381
    16.9 Minimizing sulfide in sewer environment — 383
    16.10 Antimicrobial agents in concrete — 383
    16.11 Changing redox conditions — 383
    16.12 Inhibiting the activities of sulfate reducing bacteria — 383
    16.13 Chemical removal of sulfide — 384
    16.14 Other measures — 384
    16.15 Biocide — 384
    16.16 Nanomaterials for minimizing microbial attack — 386
    16.17 Conclusion — 386
    References — 388

17. **Microbiologically induced deterioration of concrete** — 389
    ZAIN UL-ABDIN, WAQAS ANWAR AND ANWAR KHITAB

    Abbreviations — 389
    17.1 Introduction — 390
    17.2 Biological deterioration of concrete — 391
    17.3 Classification of biological deterioration — 393
    17.4 Microorganisms and their colonization on concrete — 396

17.5 Microorganisms and their chain on concrete	397
17.6 Microorganism-induced deterioration mechanism process	397
17.7 How to control concrete biodeterioration?	398
References	399

18. Role of nanomaterials in protecting building materials from degradation and deterioration — 405

NAVNEET KAUR DHIMAN, NAVNEET SIDHU, SHEKAR AGNIHOTRI, ABHIJIT MUKHERJEE AND M. SUDHAKARA REDDY

18.1 Scientific background	405
18.2 Nanomaterials for mitigating deterioration	412
18.3 Types of nanomaterials in civil engineering	418
18.4 Microbial biomineralization	450
18.5 Bioconcrete and its limitations	451
18.6 Nanoengineered self-healing concrete	454
18.7 Challenges and future prospects	458
References	458

19. Biodegradation of micropollutants — 477

SARMAD AHMAD QAMAR, ADEEL AHMAD HASSAN, KOMAL RIZWAN, TAHIR RASHEED, MUHAMMAD BILAL, TUAN ANH NGUYEN AND HAFIZ M.N. IQBAL

19.1 Introduction	477
19.2 Advanced physicochemical treatment approaches for pollutants degradation	479
19.3 Photocatalysis	480
19.4 Photocatalytic fuel cells	481
19.5 Sonochemical methods	481
19.6 Nanoremediation	481
19.7 Biosensors for environmental pollutants detection	482
19.8 Biotechnological approaches for micropollutants degradation	483

19.9	Microbial electrochemical system	483
19.10	Enzyme-assisted remediation of micropollutants	485
19.11	Immobilized enzymes for micropollutants degradation	486
19.12	Nanozymes	489
19.13	Metabolic engineering approaches for pollutants degradation	490
19.14	Invention of novel genes involved in bioremediation	492
19.15	Enhanced bioremediation via metabolic engineering processes	492
19.16	Electrochemical and microbial treatment of dye-containing wastewaters	495
19.17	Conclusions	497
	References	498

20. Microbial degradation of environmental pollutants — 509

HAMZA RAFEEQ, SARMAD AHMAD QAMAR, TUAN ANH NGUYEN, MUHAMMAD BILAL AND HAFIZ M.N. IQBAL

20.1	Introduction	509
20.2	Bioremediation: ecological relation between microorganisms	510
20.3	Herbicides, pesticides, and fertilizers as a product of agriculture	511
20.4	Dyestuff-based hazardous pollutants	513
20.5	Potentially toxic heavy metals	514
20.6	Petroleum and aromatic compounds	515
20.7	Polychlorinated biphenyls	516
20.8	Phenazines	519
20.9	Conclusion	521
	Acknowledgments	522
	Conflicts of Interest	522
	References	522
	Further reading	528

21. **Metal oxide nanoparticles for environmental remediation**   529

ROBERTA ANJOS DE JESUS, GEOVÂNIA CORDEIRO DE ASSIS, RODRIGO JOSÉ DE OLIVEIRA, MUHAMMAD BILAL, RAM NARESH BHARAGAVA, HAFIZ M.N. IQBAL, LUIZ FERNANDO ROMANHOLO FERREIRA AND RENAN TAVARES FIGUEIREDO

21.1 Introduction	529
21.2 Fundamentals of biodegradation of organic materials	531
21.3 Performance of metal oxide nanoparticles in the biodegradation of organic matter	537
21.4 Inference and future prospects	546
21.5 Acknowledgments	547
References	547

22. **Metal-organic framework for removal of environmental contaminants**   561

ADNAN KHAN, SUMEET MALIK, NISAR ALI, XIAOYAN GAO, YONG YANG AND MUHAMMAD BILAL

Abbreviations	561
22.1 Introduction	561
22.2 Designing and properties of metal-organic frameworks	564
22.3 Synthetic pathways	565
22.4 Applications of metal-organic framework in environmental remediation	566
22.5 Conclusion	571
References	571

23. **Effects of zeolite-based nanoparticles on the biodegradation of organic materials**   579

FAROOQ SHER, ABU HAZAFA, TAZIEN RASHID, MUHAMMAD BILAL, FATIMA ZAFAR, ZAHID MUSHTAQ AND ZAKA UN NISA

23.1 Introduction	579
23.2 Textile effluent composition	582

23.3	Conventional methods for dye remediation	584
23.4	Advanced oxidation processes for dye remediation	585
23.5	Nanozerovalent iron	587
23.6	Nanoparticle aggregation	589
23.7	Support material for nanozerovalent iron	590
23.8	Future recommendations	593
23.9	Conclusions	594
	References	595

24. Biodegradation of environmental pollutants using horseradish peroxidase — 603

HAMZA RAFEEQ, SARMAD AHMAD QAMAR, SYED ZAKIR HUSSAIN SHAH, SYED SALMAN ASHRAF, MUHAMMAD BILAL, TUAN ANH NGUYEN AND HAFIZ M.N. IQBAL

24.1	Introduction	603
24.2	Carbon nanotubes, carbon nanoonions, and carbon nanodots for horseradish peroxidase immobilization	606
24.3	Graphene and its derivatives for horseradish peroxidase immobilization	608
24.4	Magnetic nanoparticles for horseradish peroxidase immobilization	615
24.5	Magnetic electrospun nanofibers for horseradish peroxidase immobilization	617
24.6	Metal−organic frameworks for horseradish peroxidase immobilization	618
24.7	Mesoporous silica for horseradish peroxidase immobilization	618
24.8	Horseradish peroxidase for environmental applications	619
24.9	Conclusion	623
	Acknowledgment	624
	Conflicts of Interest	624
	References	624

25. Nanobiodegradation of pharmaceutical pollutants — 635
TAHIR RASHEED, KOMAL RIZWAN, SAMEERA SHAFI AND MUHAMMAD BILAL

25.1 Introduction — 635
25.2 Environmental and ecological risks — 636
25.3 Pharmaceutical removal methods — 639
25.4 Agricultural byproducts and biosorbents — 643
25.5 Resins and metal oxide-based adsorbents (metal organic frameworks) — 644
25.6 Nanomaterials — 646
25.7 Conclusion — 647
References — 648

26. Nanobioremediation of insecticides and herbicides — 655
AMMAR ALI, ZAHEER AHMED, RIZWANA MAQBOOL, KHURRAM SHAHZAD, ZAHID HUSSAIN SHAH, MUHAMMAD ZARGHAM ALI, HAMEED ALSAMADANY AND MUHAMMAD BILAL

26.1 Nanobioremediation — 655
26.2 Fundamentals of nanobioremediation technologies — 658
26.3 Nanobioremediation of insecticides and herbicides — 659
26.4 Nanomaterials for remediation and sensing of pesticide — 662
26.5 Nanoparticles — 662
26.6 Bimetallic nanoparticles — 664
26.7 Nanocomposites — 665
26.8 Nanotubes — 666
26.9 Biosensors for pesticide detection — 667
26.10 Enzyme-responsive systems — 667
26.11 Photoresponsive methods — 668
26.12 Conclusion — 670
Conflict of interest — 670
References — 670

27. Microbial-induced corrosion of metals with presence of nanoparticles — 675

MOHAMMAD TABISH, AYESHA ZARIN, MUHAMMAD UZAIR MALIK, MUHAMMAD ABUBAKER KHAN, JINGMAO ZHAO AND GHULAM YASIN

27.1 Introduction — 675
27.2 Corrosive microbes — 677
27.3 Metal-based nanoparticles — 679
27.4 Conclusion — 690
References — 691

Index 701

List of contributors

Sohail Abbas Department of Bioinformatics and Biotechnology, Government College University Faisalabad (GCUF), Faisalabad, Pakistan

Shekar Agnihotri Department of Agriculture and Environmental Sciences, National Institute of Food Technology Entrepreneurship and Management, Sonepat, India

Zaheer Ahmed Department of Plant Breeding and Genetics, University of Agriculture Faisalabad, Faisalabad, Pakistan; Center for Advanced Studies in Agriculture and Food Security, University of Agriculture Faisalabad, Faisalabad, Pakistan

Ammar Ali Department of Plant Breeding and Genetics, University of Agriculture Faisalabad, Faisalabad, Pakistan; Center for Advanced Studies in Agriculture and Food Security, University of Agriculture Faisalabad, Faisalabad, Pakistan

Muhammad Zargham Ali Department of Plant Breeding and Genetics, Pir Mehr Ali Shah Arid Agriculture University, Rawalpindi, Pakistan

Nisar Ali Key Laboratory for Palygorskite Science and Applied Technology of Jiangsu Province, National and Local Joint Engineering Research Center for Deep Utilization Technology of Rock-salt Resource, Faculty of Chemical Engineering, Huaiyin Institute of Technology, Huai'an, P.R. China

Hameed Alsamadany Department of Biological Sciences, King Abdulaziz University, Jeddah, Saudi Arabia

Elif Alyamaç Department of Petroleum and Natural Gas Engineering, Izmir Katip Çelebi University, Izmir, Turkey

Waqas Anwar Department of Civil Engineering, Mirpur University of Science and Technology (MUST), Mirpur, Pakistan

Syed Salman Ashraf Department of Chemistry, College of Arts and Sciences, Khalifa University, Abu Dhabi, United Arab Emirates

Ahmad Reza Bagheri Department of Chemistry, Yasouj University, Yasouj, Iran

Ayesha Baig Department of Chemistry, Faculty of Sciences, University of Gujrat, Gujrat, Pakistan

Hamid Reza Bakhsheshi-Rad Advanced Materials Research Center, Department of Materials Engineering, Najafabad Branch, Islamic Azad University, Najafabad, Iran; Faculty of Engineering, Universiti Teknologi Malaysia, Johor Bahru, Malaysia

Sabah Bakhtiari Faculty of Mechanical Engineering-Energy Division, K.N. Toosi University of Technology, Tehran, Iran

Sanaz Soleymani Eil Bakhtiari Advanced Materials Research Center, Department of Materials Engineering, Najafabad Branch, Islamic Azad University, Najafabad, Iran

Leeba Balan Bionyme Laboratories, Chennai, India

Filippo Berto Department of Mechanical and Industrial Engineering, Norwegian University of Science and Technology, Trondheim, Norway

Ram Naresh Bharagava Laboratory for Bioremediation and Metagenomics Research (LBMR), Department of Microbiology (DM), Babasaheb Bhimrao Ambedkar University (A Central University), Lucknow, India

Pankaj Bhatt State Key Laboratory for Conservation and Utilization of Subtropical Agro-Bioresources, Guangdong Province Key Laboratory of Microbial Signals and Disease Control, Integrative Microbiology Research Centre, South China Agricultural University, Guangzhou, China

Muhammad Bilal School of Life Science and Food Engineering, Huaiyin Institute of Technology, Huai'an, P.R. China

Sriram Chandrasekaran Bionyme Laboratories, Chennai, India

José Arnaldo Santana Costa CDMF, Department of Chemistry, Federal University of São Carlos, São Carlos, Brazil

Geovânia Cordeiro de Assis Chemical Catalysis and Reactivity Group, Institute of Chemistry and Biotechnology, Federal University of Alagoas, Maceió, Brazil

Roberta Anjos de Jesus Institute of Technology and Research (ITP), Tiradentes University (UNIT), Aracaju-Sergipe, Brazil

Rodrigo José de Oliveira Department of Chemistry, State University Paraíba, Campina Grande, Brazil

Navneet Kaur Dhiman Department of Biotechnology, Thapar Institute of Engineering & Technology, Patiala, India

Esmail Doustkhah International Centre for Materials Nanoarchitectonics (MANA), National Institute for Materials Science (NIMS), Ibaraki, Japan

Hanieh Fakhri Research Center for Environmental Health Technology, Iran University of Medical Sciences, Tehran, Iran; Department of Environmental Health Engineering, School of Public Health, Iran University of Medical Science, Tehran, Iran

Mujahid Farid Department of Environmental Sciences, University of Gujrat, Gujrat, Pakistan

Mahdi Farzadkia Research Center for Environmental Health Technology, Iran University of Medical Sciences, Tehran, Iran; Department of Environmental Health Engineering, School of Public Health, Iran University of Medical Science, Tehran, Iran

Luiz Fernando Romanholo Ferreira Institute of Technology and Research (ITP), Tiradentes University (UNIT), Aracaju-Sergipe, Brazil

Renan Tavares Figueiredo Institute of Technology and Research (ITP), Tiradentes University (UNIT), Aracaju-Sergipe, Brazil

O.V. Frank-Kamenetskaya Institute of Silicate Chemistry, Russian Academy of Sciences, Saint Petersburg, Russia; St. Petersburg State University, Saint Petersburg, Russia

Xiaoyan Gao Key Laboratory for Palygorskite Science and Applied Technology of Jiangsu Province, National and Local Joint Engineering Research Center for Deep Utilization Technology of Rock-salt Resource, Faculty of Chemical Engineering, Huaiyin Institute of Technology, Huai'an, P.R. China

Adeel Ahmad Hassan School of Chemistry and Chemical Engineering, Shanghai Jiao Tong University, Shanghai, P.R. China

Abu Hazafa Department of Biochemistry, University of Agriculture, Faisalabad, Pakistan; International Society of Engineering Science and Technology, Coventry, United Kingdom

Anam Ijaz Department of Bioinformatics and Biotechnology, Government College University Faisalabad (GCUF), Faisalabad, Pakistan

Hafiz M.N. Iqbal Tecnologico de Monterrey, School of Engineering and Sciences, Monterrey, Mexico

Ahmad Fauzi Ismail Advanced Membrane Technology Research Center (AMTEC), Universiti Teknologi Malaysia, Johor Bahru, Malaysia

Muhammad Rizwan Javed Department of Bioinformatics and Biotechnology, Government College University Faisalabad (GCUF), Faisalabad, Pakistan

A. Jawahar Nisha Selvam College of Engineering and Technology, Salem, India

T.V. Khamova Institute of Silicate Chemistry, Russian Academy of Sciences, Saint Petersburg, Russia

Adnan Khan Institute of Chemical Sciences, University of Peshawar, Peshawar, Pakistan

Muhammad Abubaker Khan School of Physics and Optical Engineering, Shenzhen University, Shenzhen, P.R. China

Anwar Khitab Department of Civil Engineering, Mirpur University of Science and Technology (MUST), Mirpur, Pakistan

Yulia V. Khoroshavina Institute of Silicate Chemistry of Russian Academy of Sciences, Saint-Petersburg, Russia

Anatoly K. Kychkin Larionov Institute of Physical and Technical Problems of the North of the Siberian Branch of the Russian Academy of Sciences, Yakutsk, Russia

N. Lavanya Department of Botany, Avinashilingam Institute for Home Science and Higher Education for Women, Coimbatore, India

Muhammad Uzair Malik State Key Laboratory of Electrochemical Process and Technology for Materials, College of Materials Science and Engineering, Beijing University of Chemical Technology, Beijing, P.R. China

Sumeet Malik Institute of Chemical Sciences, University of Peshawar, Peshawar, Pakistan

Rizwana Maqbool Department of Plant Breeding and Genetics, University of Agriculture Faisalabad, Faisalabad, Pakistan; Center for Advanced Studies in Agriculture and Food Security, University of Agriculture Faisalabad, Faisalabad, Pakistan

Abhijit Mukherjee　Department of Civil Engineering, Curtin University, Bentley, WA, Australia

Hira Munir　Department of Biochemistry and Biotechnology, University of Gujrat, Gujrat, Pakistan

Zahid Mushtaq　Department of Biochemistry, University of Agriculture, Faisalabad, Pakistan

Muhammad Faizan Nazar　Department of Chemistry, University of Education, Lahore, Multan Campus, Multan, Pakistan

Tuan Anh Nguyen　Institute for Tropical Technology, Vietnam Academy of Science and Technology, Hanoi, Vietnam

Zaka Un Nisa　International Society of Engineering Science and Technology, Coventry, United Kingdom; Faculty of Medicine, Quaid-i-Azam University, Islamabad, Pakistan

Caio Marcio Paranhos　CDMF, Department of Chemistry, Federal University of São Carlos, São Carlos, Brazil

Mona Zamani Pedram　Faculty of Mechanical Engineering-Energy Division, K.N. Toosi University of Technology, Tehran, Iran

Sarmad Ahmad Qamar　Institute of Organic and Polymeric Materials, National Taipei University of Technology, Taipei, Taiwan

Hamza Rafeeq　Department of Biochemistry, Riphah International University, Faisalabad, Pakistan; Department of Biochemistry, University of Agriculture, Faisalabad, Pakistan

Karthikeyan Rajendran　Department of Biotechnology, Mepco Schlenk Engineer College, Sivakasi, India

Seeram Ramakrishna　Nanoscience and Nanotechnology Initiative, National University of Singapore, Singapore

Tahir Rasheed　School of Chemistry and Chemical Engineering, Shanghai Jiao Tong University, Shanghai, P.R. China

Muhammad Hamid Rashid　Industrial Biotechnology Division, National Institute for Biotechnology and Genetic Engineering (NIBGE), Faisalabad, Pakistan

Tazien Rashid Department of Chemical Engineering, NFC Institute of Engineering and Fertilizer Research, Faisalabad, Pakistan

Mahmood Razzaghi Advanced Materials Research Center, Department of Materials Engineering, Najafabad Branch, Islamic Azad University, Najafabad, Iran

M. Sudhakara Reddy Department of Biotechnology, Thapar Institute of Engineering & Technology, Patiala, India

Komal Rizwan Department of Chemistry, University of Sahiwal, Sahiwal, Pakistan

Yulia V. Ryabusheva Saint-Petersburg State University, Saint-Petersburg, Russia

Jamuna Sanker Bionyme Laboratories, Chennai, India

Riffat Seemab Department of Bioinformatics and Biotechnology, Government College University Faisalabad (GCUF), Faisalabad, Pakistan

M. Özgür Seydibeyoğlu Department of Materials Science and Engineering, Izmir Katip Celebi University, Izmir, Turkey; Filinia R&D, Ege University Sciencepark, Izmir, Turkey

Sameera Shafi Institute of Chemistry, The Islamia University of Bahawalpur, Bahawalnagar Campus, Bahawalnagar, Pakistan

Syed Zakir Hussain Shah Department of Zoology, University of Gujrat, Gujrat, Pakistan

Zahid Hussain Shah Department of Plant Breeding and Genetics, Pir Mehr Ali Shah Arid Agriculture University, Rawalpindi, Pakistan

Khurram Shahzad Department of Plant Breeding and Genetics, University of Haripur, Haripur, Pakistan

Safian Sharif Faculty of Engineering, Universiti Teknologi Malaysia, Johor Bahru, Malaysia

Farooq Sher Department of Engineering, School of Science and Technology, Nottingham Trent University, Nottingham, United Kingdom

O.A. Shilova Institute of Silicate Chemistry, Russian Academy of Sciences, Saint Petersburg, Russia

Olga A. Shilova   Institute of Silicate Chemistry of Russian Academy of Sciences, Saint-Petersburg, Russia; Saint-Petersburg State Electrotechnical University "LETI," Saint-Petersburg, Russia

Navneet Sidhu   Department of Biotechnology, Thapar Institute of Engineering & Technology, Patiala, India

N.B. Singh   Chemistry and Biochemistry Department and RTDC, Sharda University, Greater Noida, India

Georgii S. Sokolov   Institute of Silicate Chemistry of Russian Academy of Sciences, Saint-Petersburg, Russia

Sajjad Hussain Sumrra   Department of Chemistry, Faculty of Sciences, University of Gujrat, Gujrat, Pakistan

Mohammad Tabish   State Key Laboratory of Electrochemical Process and Technology for Materials, College of Materials Science and Engineering, Beijing University of Chemical Technology, Beijing, P.R. China

Khajista Tahira   Department of Biochemistry and Biotechnology, University of Gujrat, Gujrat, Pakistan

Anam Tariq   Department of Bioinformatics and Biotechnology, Government College University Faisalabad (GCUF), Faisalabad, Pakistan

Irina N. Tsvetkova   Institute of Silicate Chemistry of Russian Academy of Sciences, Saint-Petersburg, Russia

Nihan Uçar   Department of Materials Science and Engineering, Izmir Katip Celebi University, Izmir, Turkey; Filinia R&D, Ege University Sciencepark, Izmir, Turkey

Zain Ul-Abdin   Department of Civil Engineering, Mirpur University of Science and Technology (MUST), Mirpur, Pakistan

Sugumari Vallinayagam   Department of Biotechnology, Mepco Schlenk Engineering College, Sivakasi, India

D.Y. Vlasov   Institute of Silicate Chemistry, Russian Academy of Sciences, Saint Petersburg, Russia; St. Petersburg State University, Saint Petersburg, Russia

Dmitry Yu. Vlasov   Institute of Silicate Chemistry of Russian Academy of Sciences, Saint-Petersburg, Russia; Saint-Petersburg State University, Saint-Petersburg, Russia

Chi Văn Nguyễn Coastal Branch—Vietnam Russian Tropical Center, Khanh Hoa, Vietnam

Yong Yang Key Laboratory for Palygorskite Science and Applied Technology of Jiangsu Province, National and Local Joint Engineering Research Center for Deep Utilization Technology of Rock-salt Resource, Faculty of Chemical Engineering, Huaiyin Institute of Technology, Huai'an, P.R. China

Masoud Yarmohammadi Faculty of Mechanical Engineering-Energy Division, K.N. Toosi University of Technology, Tehran, Iran

Ghulam Yasin School of Physics and Optical Engineering, Shenzhen University, Shenzhen, P.R. China; Institute for Advanced Study, Shenzhen University, Shenzhen, P.R. China

Fatima Zafar International Society of Engineering Science and Technology, Coventry, United Kingdom; School of Biochemistry and Biotechnology, University of the Punjab, Lahore, Pakistan

Muhammad Nadeem Zafar Department of Chemistry, Faculty of Sciences, University of Gujrat, Gujrat, Pakistan

Ayesha Zarin Institute of Chemical Sciences, Bahauddin Zakariya University, Multan, Pakistan

M.S. Zelenskaya St. Petersburg State University, Saint Petersburg, Russia

Jingmao Zhao State Key Laboratory of Electrochemical Process and Technology for Materials, College of Materials Science and Engineering, Beijing University of Chemical Technology, Beijing, P.R. China

Muhammad Zubair Department of Chemistry, Faculty of Sciences, University of Gujrat, Gujrat, Pakistan

Biodegradation and biodeterioration at the nanoscale: an introduction

Muhammad Bilal[1], Pankaj Bhatt[2], Tuan Anh Nguyen[3], Hafiz M.N. Iqbal[4]

[1]SCHOOL OF LIFE SCIENCE AND FOOD ENGINEERING, HUAIYIN INSTITUTE OF TECHNOLOGY, HUAI'AN, P.R. CHINA [2]STATE KEY LABORATORY FOR CONSERVATION AND UTILIZATION OF SUBTROPICAL AGRO-BIORESOURCES, GUANGDONG PROVINCE KEY LABORATORY OF MICROBIAL SIGNALS AND DISEASE CONTROL, INTEGRATIVE MICROBIOLOGY RESEARCH CENTRE, SOUTH CHINA AGRICULTURAL UNIVERSITY, GUANGZHOU, CHINA [3]INSTITUTE FOR TROPICAL TECHNOLOGY, VIETNAM ACADEMY OF SCIENCE AND TECHNOLOGY, HANOI, VIETNAM [4]TECNOLOGICO DE MONTERREY, SCHOOL OF ENGINEERING AND SCIENCES, MONTERREY, MEXICO

1.1 Introduction

Based on the available literature, biodegradation and biodeterioration at the nanoscale aim to explore two critical issues, among others: (1) exploitation of catalytic cues or adsorbents, which are engineered at the nanoscale, for highly effective and sustainable biodegradation and biodeterioration of thousands of toxic and hazardous chemical compounds of so-called pollutants of serious concern; (2) effects/roles of engineered nanoscale materials in the biodegradation/biodeterioration processes. Regarding the environmental impacts, the biodegradation of polymers, plastics, and environmentally related pollutants of concern has a positive impact. In contrast, biodeterioration/biocorrosion of metals or alloys has a negative impact. In fact, nanoparticles can accelerate or inhibit the biodegradation process, depending on the nature of bacteria, fungi, algae, enzymes, nanomaterials, or bulk materials, especially in the soil environment. In general, biodegradation refers mostly to the microbial-induced degradation of polymers, plastics, and environmentally related pollutants of concern. In this direction, the incorporation of nanomaterials in an organic matrix (nanocomposites) might affect its degradation due to of their nanotoxicity. In the case of metals and their alloys, their biodeterioration/biocorrosion due to microbial biofilms' presence could be reduced by the introduction of nanocrystals/nanophases into the metal matrix (nanoalloys).

This introductory chapter emphasizes the biodegradation and biodeterioration aspects. Various methods are discussed with suitable examples that spotlight the biodegradation of numerous contaminants of emerging issues. Besides, the current research inclinations, positive or negative environmental impacts, legislation, and future directions are also be discussed.

1.2 Nanobioremediation: a step forward to advance traditional bioremediation

Broadly speaking, traditional bioremediation is defined as "a process or method that assist in remediating most of the waste materials under controlled reaction conditions or below the corresponding concentration limits, as set by the regulatory authorities." More specifically, after years of extensive research efforts, it is evident that the implementation of a single technology is inefficient or inappropriate for robust and sustainable remediation of toxic and hazardous chemical compounds. In this context, current advancements in nanotechnology at large, and nanoscale materials in particular, offer new alternatives to traditional bioremediation. Therefore, the combination of nanoscale materials or simply nanomaterials, materials in which at least one relevant length scale is within the range of nanometers, for example, nanoparticles, nanocomposites, and so on, and bioremediation have considerable potential to achieve robust and sustainable remediation (Aguilar-Pérez et al., 2021; Aguilar-Pérez, Heya, Parra-Saldívar, & Iqbal, 2020). Considering this important synergistic action, the appearance of nanobioremediation is found highly useful and being extensively studied. So far, an array of nanoscale materials has been developed and exploited against various contaminants. For instance, Bilal et al. (2021) comprehensively reviewed harnessing the biocatalytic attributes and applied perspectives of nanoengineered laccases for the degradation of emergent pollutants. The work also spotlights several nanostructured materials' unique structural and functional characteristics, including carbon nanotubes (CNTs), graphene and its derivate constructs, nanoparticles, nanoflowers, and metal-organic frameworks as robust matrices for laccase immobilization, which in turn ultimately are used for the degradation of emergent contaminants. Zeb et al. (2020) discussed silica-based nanomaterials as designer adsorbents and used them to mitigate emerging organic pollutants from water matrices. The work also outlines numerous mesoporous silica-based adsorbents and broadly highlight an up to date description, development, and their use for the removal of organic pollutants. In another study by Rasheed et al. (2020), CNTs assisted analytical detection—sensing/delivery cues for environmental monitoring. Several examples are given with particular emphasis to biosensors, and implantations of CNT-based cues to recognize viruses, cancerous cells, glucose, DNA, volatile organic compounds, and various inorganic gases. Nanostructured materials, as an innovative type of support matrices for harnessing the power of horseradish peroxidase (HRP) for tailored environmental applications, have recently been reviewed and discussed (Bilal, Barceló, & Iqbal, 2020). The tailored applications of HRP in different environmentally related sectors are given with suitable examples, such as detection and sensing cues, treatment of wastewater matrices, and mitigation of pollutants (Fig. 1–1; Bilal et al., 2020). Accelerated degradation of environmentally related pollutants from textile industries using selenide-chitosan as high-performance nanophotocatalyst is reported by Ali

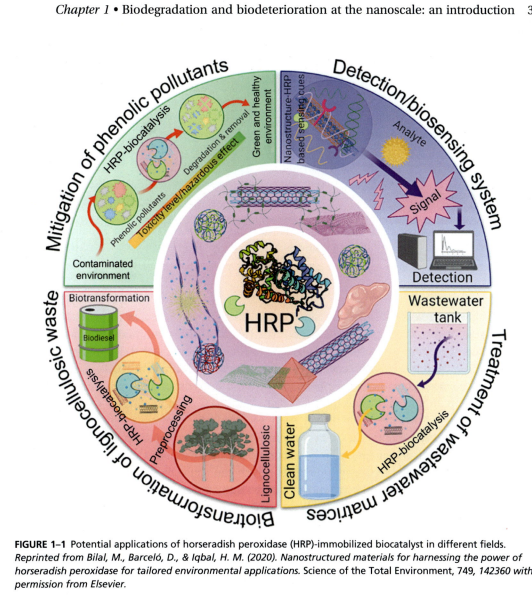

FIGURE 1–1 Potential applications of horseradish peroxidase (HRP)-immobilized biocatalyst in different fields. *Reprinted from Bilal, M., Barceló, D., & Iqbal, H. M. (2020). Nanostructured materials for harnessing the power of horseradish peroxidase for tailored environmental applications.* Science of the Total Environment, 749, 142360 *with permission from Elsevier.*

et al. (2020). Bilal, Adeel, Rasheed, Zhao, and Iqbal (2019) discussed the enzyme-assisted biodegradation of emerging contaminants of high concern (Fig. 1–2).

1.3 Miscellaneous cases

The xenobiotic compounds are increasing into the environment day by day due to various industrial and households' applications (Arora, 2020; Bhatt et al., 2020a). Xenobiotics mainly includes the antibiotics, steroids, pesticides, dyes, nitroaromatics, chlorinated compounds and inorganic compounds (Lin et al., 2020; Mishra et al., 2020). The wide applications of these toxic

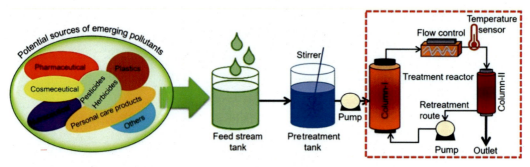

FIGURE 1–2 Schematic illustration of the simplified treatment route for emerging contaminants (ECs) using column-based reactor. *Reprinted from Bilal, M., Adeel, M., Rasheed, T., Zhao, Y., & Iqbal, H. M. (2019). Emerging contaminants of high concern and their enzyme-assisted biodegradation—A review.* Environment International, *124, 336–353 with permission from Elsevier.*

chemicals have increased the residual concentration into the environment (Bhatt et al., 2020c; Bhatt, Huang, Zhang, Sharma, & Chen, 2020b). The cellular toxicity and immune system altered reported into the mammalian cells (Huang, Zhan, Bhatt, & Chen, 2019). The degradation of these chemicals from the environment is an important aspect (Mishra et al., 2020). Both physicochemical and biological methods have been used for the degradation of these xenobiotics from the environment. At present various tools and techniques have been applied for the remediation of the xenobiotics from the environment such as ultrasonic treatment, biochar adsorption, photocatalysis, and nanocomposites. Physiochemical methods are rapid and showed high efficiency in degradation of xenobiotics. It was found that the treatment via the physicochemical methods may produce the secondary pollutants, which can cause higher toxicity. Because of its high efficiency, economical, and ecofriendly characteristics, the removal of xenobiotics from the environmental substrates by indigenous microorganism is an interesting research topic. To date the large number of xenobiotics-degrading microbes have been found and reported, including bacteria, fungi, algae, and genetically engineered bacteria, especially bacteria.

The microbial methods were found to be effective for the treatment of the contaminated sites (Bhatt et al., 2019a; Gangola, Negi, Srivastava, & Sharma, 2016). The indigenous microbial strains are able to mineralized the complex structures into the environmentally accepted form (Bhatt, 2019). Microbes produced the specific enzymes during the interaction with these compounds and able to remediate it from the environment using its metabolic pathway. The bacterial strains belonging to the genera *Pseudomonas, Bacillus, Arthrobacter, Acinetobacter, Xanthomonas, Sphingomonas, Sphingobium,* and other indigenous bacterial strains were reported for the degradation of the xenobiotics from the contaminated environment (Bhatt et al., 2019b; Norris, Gross, Bartholomay, & Coats, 2019) in addition to this fungus belonging to the *Candida, Aspergillus, Fusarium,* and *Trichoderma* documented for their bioremediation potential (Bhatt et al., 2020e). These microbial strains are able to produce the contrasting enzymes that participated in the degradation of a wide group of toxic chemicals. The bioremediation, biostimulation, and bioaugmentation are the common mechanisms by which transformation of the xenobiotics is possible in microbial strains (Cycon & Piotrowska-Seget, 2016). To develop the sustainable environment, there is a need to do research in this area about environmental management of the pollutant (Narancic, Cerrone, Beagan, & O'Connor, 2020). More efforts are needed to make the synergistic interaction between the environmental impact on fate and behavior of environmental contaminants, assortment, and performance of the most suitable bioremediation technique and another relevant technique that can sustain the effective and successful operation and monitoring of a bioremediation process (Kulkarni & Chaudhari, 2007). For recycling of the xenobiotics in a natural system, genetic engineering tools have been used rapidly in recent decades (Bhatt et al., 2020d; Chauhan, Fazlurrahman, Oakeshott, & Jain, 2008). The resource recovery and recycling through the bioremediation-based approach is likely to be beneficial for sustainable development (Álvarez-Barragán et al., 2016). The more future research on the development of the regulations, suitable microbial technology and residue of contaminant might be helpful to reduce the toxic level from the environment.

1.4 Concluding remarks and future outlook

In conclusion, for the past several decades, the growing scientific and technological advancement has directed the evolution of nanotechnology and its successful implementation in various applied sectors of the modern world. This extensive exploitation of nanomaterials has amplified the impact of nanotechnological research programs that have also revolutionized different industrial sectors. More specifically, nanotechnology is being used in the environment as a nanobioremediation that assists the degradation of thousands of toxic and hazardous chemical compounds—so-called pollutants of high concern. Besides its use as nanobioremediation, several nanostructured materials, including CNTs, graphene, and their derivate constructs, nanoparticles, nanoflowers, and metal-organic frameworks have also been fabricated using green synthesis and deployed as sensing and remediation constructs of supreme interest. Nanomaterials that can synergistically work to detect and mitigate pollutants of emerging concern can play a significant role in the on-site remediation of contaminated sites. This will also facilitate the development of a clean and sustainable environment for a better tomorrow. However, the green synthesis of nanoscale materials requires additional consideration for developing smart nanoscale materials for environmental cleanup.

References

Aguilar-Pérez, K. M., Avilés-Castrillo, J. I., Ruiz-Pulido, G., Medina, D. I., Parra-Saldivar, R., & Iqbal, H. M. (2021). Nanoadsorbents in focus for the remediation of environmentally-related contaminants with rising toxicity concerns. *Science of The Total Environment, 779*, 146465.

Aguilar-Pérez, K. M., Heya, M. S., Parra-Saldívar, R., & Iqbal, H. M. (2020). Nano-biomaterials in-focus as sensing/detection cues for environmental pollutants. *Case Studies in Chemical and Environmental Engineering, 2*, 100055.

Ali, N., Ahmad, S., Khan, A., Khan, S., Bilal, M., Ud Din, S., & Khan, H. (2020). Selenide-chitosan as high-performance nanophotocatalyst for accelerated degradation of pollutants. *Chemistry—An Asian Journal, 15*(17), 2660—2673.

Álvarez-Barragán, J., Domínguez-Malfavón, L., Vargas-Suárez, M., González-Hernández, R., Aguilar-Osorio, G., & Loza-Tavera, H. (2016). Biodegradative activities of selected environmental fungi on a polyester polyurethane varnish and polyether polyurethane foams. *Applied and Environmental Microbiology, 82*, 5225—5235. Available from https://doi.org/10.1128/AEM.01344-16.

Arora, P. K. (2020). Bacilli-mediated degradation of xenobiotic compounds and heavy metals. *Frontiers in Bioengineering and Biotechnology, 8*(570307). Available from https://doi.org/10.3389/fbioe.2020.570307.

Bhatt, P. (Ed.), (2019). *Smart bioremediation technologies: Microbial enzymes.* Academic Press.

Bhatt, P., Gangola, S., Bhandari, G., Zhang, W., Maithani, D., Mishra, S., & Chen, S. (2020a). New insights into the degradation of synthetic pollutants in contaminated environments. *Chemosphere, 128827.* Available from https://doi.org/10.1016/j.chemosphere.2020.128827.

Bhatt, P., Gangola, S., Chaudhary, P., Khati, P., Kumar, G., Sharma, A., & Srivastava, A. (2019a). Pesticide induced up-regulation of esterase and aldehyde dehydrogenase in indigenous *Bacillus* spp. *Bioremediation Journal, 23*, 42—52. Available from https://doi.org/10.1080/10889868.2019.1569586.

Bhatt, P., Gangola, S., Chaudhary, P., Khati, P., Kumar, G., Sharma, A., & Srivastava, A. (2019b). Pesticide induced up-regulation of esterase and aldehyde dehydrogenase in indigenous *Bacillus* spp. *Bioremediation Journal.* Available from https://doi.org/10.1080/10889868.2019.1569586.

Bhatt, P., Huang, Y., Zhang, W., Sharma, A., & Chen, S. (2020b). Enhanced cypermethrin degradation kinetics and metabolic pathway in *Bacillus thuringiensis* strain SG4. *Microorganisms, 8*, 1—15. Available from https://doi.org/10.3390/microorganisms8020223.

Bhatt, P., Sethi, K., Gangola, S., Bhandari, G., Verma, A., Adnan, M., . . . Chaube, S. (2020c). Modeling and simulation of atrazine biodegradation in bacteria and its effect in other living systems. *Journal of Biomolecular Structure & Dynamics.* Available from https://doi.org/10.1080/07391102.2020.1846623.

Bhatt, P., Sharma, A., Rene, E. R., Kumar, A. J., Zhang, W., & Chen, S. (2020d). Bioremediation of fipronil using Bacillus sp. FA3: Mechanism, kinetics and resource recovery potential from contaminated environments. *Journal of Water Process Engineering, 39*, 101712. Available from https://doi.org/10.1016/j.jwpe.2020.101712.

Bhatt, P., Zhang, W., Lin, Z., Pang, S., Huang, Y., & Chen, S. (2020e). Biodegradation of allethrin by a novel fungus fusarium proliferatum strain cf2, isolated from contaminated soils. *Microorganisms, 8.* Available from https://doi.org/10.3390/microorganisms8040593.

Bilal, M., Adeel, M., Rasheed, T., Zhao, Y., & Iqbal, H. M. (2019). Emerging contaminants of high concern and their enzyme-assisted biodegradation—A review. *Environment International, 124*, 336—353.

Bilal, M., Ashraf, S. S., Cui, J., Lou, W. Y., Franco, M., Mulla, S. I., & Iqbal, H. M. (2021). Harnessing the biocatalytic attributes and applied perspectives of nanoengineered laccases—A review. *International Journal of Biological Macromolecules, 166*, 352—373.

Bilal, M., Barceló, D., & Iqbal, H. M. (2020). Nanostructured materials for harnessing the power of horseradish peroxidase for tailored environmental applications. *Science of the Total Environment, 749*, 142360.

Chauhan, A., Fazlurrahman., Oakeshott, J. G., & Jain, R. K. (2008). Bacterial metabolism of polycyclic aromatic hydrocarbons: Strategies for bioremediation. *Indian Journal of Microbiology, 48*(1), 95—113. Available from https://doi.org/10.1007/s12088-008-0010-9.

Cycon, M., & Piotrowska-Seget, Z. (2016). Pyrethroid-degrading microorganisms and their potential for the bioremediation of contaminated soils: A review. *Frontiers in Microbiology, 7*(1463). Available from https://doi.org/10.3389/fmicb.2016.01463.

Gangola, S., Negi, G., Srivastava, A., & Sharma, A. (2016). Enhanced biodegradation of endosulfan by *Aspergillus* and *Trichoderma* spp. Isolated from an agricultural field of Tarai Region of Uttarakhand. *Pesticide Research Journal.*

Huang, Y., Zhan, H., Bhatt, P., & Chen, S. (2019). Paraquat degradation from contaminated environments: Current achievements and perspectives. *Frontiers in Microbiology, 10.* Available from https://doi.org/10.3389/fmicb.2019.01754.

Kulkarni, M., & Chaudhari, A. (2007). Microbial remediation of nitro-aromatic compounds: An overview. *Journal of Environmental Management, 85,* 496−512. Available from https://doi.org/10.1016/j.jenvman.2007.06.009.

Lin, Z., Zhang, W., Pang, S., Huang, Y., Mishra, S., Bhatt, P., & Chen, S. (2020). Current approaches to and future perspectives on methomyl degradation in contaminated soil/water environments. *Molecules (Basel, Switzerland), 25,* 1−16. Available from https://doi.org/10.3390/molecules25030738.

Mishra, S., Zhang, W., Lin, Z., Pang, S., Huang, Y., Bhatt, P., & Chen, S. (2020). Carbofuran toxicity and its microbial degradation in contaminated environments. *Chemosphere, 259,* 127419. Available from https://doi.org/10.1016/j.chemosphere.2020.127419.

Narancic, T., Cerrone, F., Beagan, N., & O'Connor, K. E. (2020). Recent advances in bioplastics: Application and biodegradation. *Polymers (Basel), 12.* Available from https://doi.org/10.3390/POLYM12040920.

Norris, E. J., Gross, A. D., Bartholomay, L. C., & Coats, J. R. (2019). Plant essential oils synergize various pyrethroid insecticides and antagonize malathion in *Aedes aegypti. Medical and Veterinary Entomology, 33*(4), 1−14. Available from https://doi.org/10.1111/mve.12380.

Rasheed, T., Hassan, A. A., Kausar, F., Sher, F., Bilal, M., & Iqbal, H. M. (2020). Carbon nanotubes assisted analytical detection−Sensing/delivery cues for environmental and biomedical monitoring. *TrAC Trends in Analytical Chemistry, 132,* 116066.

Zeb, S., Ali, N., Ali, Z., Bilal, M., Adalat, B., Hussain, S., & Iqbal, H. M. (2020). Silica-based nanomaterials as designer adsorbents to mitigate emerging organic contaminants from water matrices. *Journal of Water Process Engineering, 38,* 101675.

2

Biodegradation of materials in presence of nanoparticles

Hira Munir[1], Khajista Tahira[1], Ahmad Reza Bagheri[2], Muhammad Bilal[3]

[1]DEPARTMENT OF BIOCHEMISTRY AND BIOTECHNOLOGY, UNIVERSITY OF GUJRAT, GUJRAT, PAKISTAN [2]DEPARTMENT OF CHEMISTRY, YASOUJ UNIVERSITY, YASOUJ, IRAN [3]SCHOOL OF LIFE SCIENCE AND FOOD ENGINEERING, HUAIYIN INSTITUTE OF TECHNOLOGY, HUAI'AN, P.R. CHINA

2.1 Introduction

Concurrent with population growth, industrialization, and urbanization, environmental pollution has become a main challenge in recent years. This challenge is due to different pollutants and highly undesirable wastes from various artificial sources introduced into the environment, particularly in water sources (Mahmoud, 2020; Rani et al., 2020). These pollutants and wastes not only affect the environment and human health but also are effective on the ecosystem (Liu et al., 2020; Sirajudheen, Karthikeyan, & Meenakshi, 2020). For instance, dye molecules are one of the main groups of pollutants that have complex and rigid structures. These properties make them able to remain in the environment for a long time. Also, their removal using routine approaches is difficult (Bhatti et al., 2020; Li, Gao, Li, & Yang, 2020; Wu et al., 2020). These pollutants can cause different diseases like diverse kinds of cancers, asthma, skin diseases, heart disease, pulmonary diseases, and so on (Hassan, Shahat, El-Daidamony, El-Desouky, & El-Bindary, 2020; Xia et al., 2020). In this regard, the removal of them from the environment is vital. Until now, different methods like precipitation (Kim, Chung, Jeong, & Nam, 2020; Peng & Guo, 2020), filtration (Sahu, Dash, & Pradhan, 2020; Vieira, de Farias, Spaolonzi, da Silva, & Vieira, 2020), membrane (Intrchom, Roy, & Mitra, 2020; Yun et al., 2020), degradation (Chen, Ma, Duan, Lang, & Pan, 2020; Palomino-Ascencio, García-Hernández, Salazar-Villanueva, & Chigo-Anota, 2020), and adsorption (Chen et al., 2020; Tran et al., 2020) have been applied for removal of different pollutants. These methods are based on the application of different materials, which act as adsorbents, catalysts, and supports the for trapping or removal of pollutants.

In recent years, the progress and development of nanotechnology (NT) has been the focus of research, through interconnecting with different forms of life and other branches of science (Baker & Satish, 2012). The notion of NT was first given in 1959 by Richard Feynman (Richard, 1959) and now is among the rapidly emergent branches of science and technology worldwide and being considered as the next industrial revolution (Roco, 2005). It offers a wide number of environmental benefits, which can be classified into three different classes: sensing and detection, treatment and remediation, and prevention of pollution. Particles of smaller size facilitate the production of sensors that can be used in remote locations. The previous decade observed a major focus on nanomaterials (NMs) and nanoparticles (NPs) due to their precise size dependent chemical and physical properties.

NMs are materials which have been prepared and applied at a very small scale, that is, 1–100 nm in at least one dimension (Saleh, 2020). NMs cannot be seen using a simple microscope or naked eyes. They have high reactive atoms and a large surface area, which make them excellent materials with high adsorption and loading capacities (Chen et al., 2020). Generally, the physical properties of NMs are based on their size and shape. The size of NMs can affect different properties of these materials like their chemical, electrical, mechanical, and optical properties. Based on the dimensionality, morphology, state, and chemical composition of NMs, they can be classified to different categories (Olabi et al., 2020). Based on the dimensions, some of the NMs have zero dimension (0D), which can be as spherical NMs, cube, nanorod, polygon, hollow sphere, NPs, metal, and core-shell NMs as well as quantum dots. Besides the dimensionality, NPs also have morphological nature, which can be flatness, sphericity, and aspect ratio (Reina, Peng, Jacquemin, Andrade, & Bianco, 2020). According to uniformity of NPs, they can have different physical properties (Cheng, Wang, Gong, Liu, & Liu, 2020). These amazing features make them excellent candidates for applications in different fields, from daily life to medicine (Arabi, Ostovan, Bagheri, Guo, Li, et al., 2020; Arabi, Ostovan, Bagheri, Guo, Wang, et al., 2020; Bagheri & Aramesh, 2020; Bagheri & Ghaedi, 2020a, 2020b, 2020c; Boyes & van Thriel, 2020; Diouf, El Bari, & Bouchikhi, 2020; Farré & Jha, 2020; Ge, Zhang, Zeng, Gu, & Gao, 2020; N'Dea, Nelson, Gleghorn, & Day, 2020; Xiao, Huang, Wang, Meng, & Yang, 2020). NMs or NPs are substances with at least one dimension less than 100 nm. They are categorized into the following two classes: inorganic and organic. The organic class consist of carbon NPs (fullerenes), however the inorganic class comprises of noble and magnetic metals (such as silver, gold, and palladium) and semiconductor NPs (like TiO_2 and ZnO). The NPs are made up of a top down or bottom-up approach. The conventional technique of NPs formation includes physical or chemical methods. The physical methods are highly expensive and require energy while the chemical methods commonly use some hazardous chemicals. In comparison, the biotechnological methods that emerged in recent years involves different biomolecules present in algae, microbes, vascular, and nonvascular plants to synthesize NPs of the required size and shape. Thus, these NPs formations are significantly clean and green due to the huge relevance with the field of nanobiotechnology. NPs show numerous and significant properties as compared to the bulk materials and mostly have the visible features as they are smaller enough to keep their electrons and create quantum effects. NPs like AuNPs are broadly being used in different fields like catalysis, photonics, biomedicine, and electronics because of their unique features. By the help of different NPs, bioremediation of the radioactive wastes of nuclear weapon production and nuclear power plants has been performed. The S-layer proteins and cells of *Bacillus sphaericus* possesses the ability of cleaning the contaminated wastewater of uranium (Durán, Marcato, De Souza, Alves, & Esposito, 2007). The prepared NPs with larger surface area, high surface energy, perfect dimensions, and spatial confinements were known to have great benefits in the field of magnetic, catalytic, optical, and electronic properties. The area of NT extends with biological NPs, with promising benefits on saving the environment through prevention and treatment of pollution, cleaning up the contaminants. Therefore, environmental nanobiotechnology is one of the rapidly growing fields of research in recent years.

The capacity of NT and especially NPs to reduce pollution is in development and might possibly catalyze many revolutionary changes in the environmental field (Quigg et al., 2013). Remediation of wastewater by using the NPs is one of the applications of NT (Yadav, Singh, Gupta, & Kumar, 2017). In the economic environment, NT is a widespread vector, and still it is essential to search new procedures for the development of better knowledge of NT-based innovation. There are numerous ecofriendly applications of NMs, and these can provide clean H_2O from the polluted H_2O and can also detect different contaminants within the environment, which is known as remediation (Liu et al., 2014; Schrick, Hydutsky, Blough, & Mallouk, 2004). Remediation is known to resolve problems while bioremediation is any method through which different biological agents such as fungi, bacteria, and protists are used for the degradation of contaminants present in the environment into forms that are less toxic (Van Dillewijn et al., 2007). Bioremediation (BRM) have advantages over conventional treatments, such as economics, minimization of biological and chemical sludge, high competency, no requirement of nutrients, possibility of recovery of metal, and biosorbent regeneration (Kratochvil & Volesky, 1998). When BRM is performed without addition of any fertilizer that is known as intrinsic BRM or natural attenuation, whereas in the presence of fertilizers it is known as biostimulated BRM. Bioremediation involves the use of bioreactor, land farming, bioventing, bioaugmentation, bioleaching, composting, rhizofiltration, biostimulation, and phytoremediation (Li & Li, 2011). BRM of a contaminated area is normally performed in two different methods. In the first one, numerous substances like the optimum temperature, concentration of oxygen, and nutrients are utilized to increase the growth of indigenous microorganisms that are usually present at the contaminated area, while in the second method that is less commonly used, some special exogenous microorganisms are supplied for the degradation of contaminants (Prokop, 2000).

2.2 Roles of nanoparticles in biodegradation

Biodegradation is a process based on the breakdown of organic matter by microorganisms such as bacteria and fungi (Shabbir et al., 2020). In the biodegradation process, pollutants and contaminations that are harmful and hazardous to the environment and ecosystem are converted to byproducts that the ecosystem evolved to cycle through (Shi, Zhang, Cheng, & Peng, 2020; Zurier & Goddard, 2021). The biodegradation process has unique properties in terms of simplicity, ease of operation, cheapness, and ready availability, which can be performed for removal and treatment of wastewater pollutants (M-Ridha, Hussein, Alismaeel, Atiya, & Aziz, 2020). In recent years, many efforts have been made to enhance the biodegradation processes. The application of NMs is considered one of the main efforts in biodegradation processes. NPs have exclusive properties such as high chemical and physical stability, high number of reactive atoms, high number of reaction sites, high surface area, high adsorption capacity, and high-loading capacity, which make them excellent candidate for application in biodegradation processes. To this end, many efforts have been done for the synthesis of NPs from microorganisms such as bacteria, fungi, yeast, and so on.

2.3 Biological synthesis of nanoparticles

Conventionally, NPs had been synthesized only via chemical or physical procedures. The demand for biosynthesis of NPs arose because of the higher costs of chemical and physical methods. During the exploration of inexpensive procedures for the formation of NPs, plant extracts and microorganisms were used. The biological synthesis of NPs is the bottom-up method in which the basic reaction occurred is reduction/oxidation. The phytochemicals present in plants or different microbial enzymes with reducing or antioxidant potential are generally responsible for the reduction of metals into the NPs. The plant phytochemicals or the microbial enzymes possessing antioxidants or reducing potential are generally responsible for metal compounds reduction toward their respective NPs. The production of NMs is presently expected to be millions of tons globally, and in the future it is expected to be increased. NMs are known as engineered substances with dimensions of 1–100 nm. For the biological synthesis of NPs, plants, bacteria, algae, yeast, actinomycetes, and fungi can be utilized (Sastry, Ahmad, Khan, & Kumar, 2003).

2.3.1 Nanoparticles synthesized from plants

Green biosynthesis of plant-based NPs is gaining in significance due to the single step during the biosynthesis procedure, lack of toxic materials, and presence of natural capping agents (Gurunathan et al., 2009). The benefits of the usage of plants for biosynthesis of NPs is that they are safe during handling, with a wide variability of metabolites and easy availability. A large number of plant extracts are now being considered for their role during the formation of NPs. Whereas the synthesis from bacteria and fungi needs a quite longer time of incubation for the metallic ion's reduction while the phytochemicals complete the process of synthesis in less time. Hence, in comparison with microorganisms (fungi and bacteria), plant extracts are the better candidates for biosynthesis of NPs (Yadav et al., 2017). In the other work, Kalaiselvi et al. (Kalaiselvi, Mohankumar, Shanmugam, Nivitha, & Sundararaj, 2019) synthesized silver NPs based on the green synthesis approach by application of *Euphorbia tirucalli* as a new method. In this study, the stem of *E. tirucalli* plants was used for the preparation of AgNPs. For synthesis of AgNPs, a specific volume of target species was mixed with AgNO3 solution and kept in the dark. The most important thing about this study is that the concentration of *E. tirucalli* latex extract and precursor solution directly affect the AgNPs synthesis. The final concentration of tirucalli latex extract was 3% while the concentration of $AgNO_3$ was fixed at 5 mM. The synthesis of AgNPs was done at room temperature after 24 h. Based on the Field Emission Scanning Electron Microscopy (FESEM) images, the prepared AgNPs were spherical and cubic and had the size of 20–30 nm. Also, the X-ray diffraction (XRD) pattern of the prepared AgNPs showed their successful synthesis. In the other study, Cr_2O_3 NPs were synthesized using a leaf extract of *Rhamnus virgata* (RV) (Iqbal et al., 2020). The UV-visible spectroscopy has confirmed the formation of Cr_2O_3NPs by the change of color. The leaf extract of RV contains bioactive functional

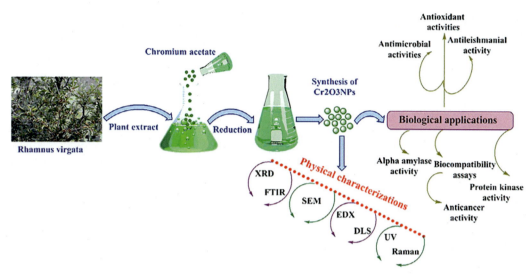

FIGURE 2–1 Synthesis, characterization processes and applications of Cr_2O_3 nanoparticles using *Rhamnus virgate*. Reproduced with permission Iqbal, J., Abbasi, B.A., Munir, A., Uddin, S., Kanwal, S., & Mahmood, T., Facile green synthesis approach for the production of chromium oxide nanoparticles and their different in vitro biological activities. Microscopy Research and Technique, 83(2020), 706−719. Copyright © 2020 Wiley Online Library.

groups, which can act as electron donor groups and reduce the target ions. The RV leaf extract acts as a bioreductant and stabilizing agents. The color change in solutions (light green) indicates the biosynthesis of Cr_2O_3NPs. These processes were evaluated using FTIR analysis. The synthesis and characterization procedure is shown in Fig. 2−1. The prepared NPs had spherical shape with the average size of 28 nm. After synthesis of the proposed Cr_2O_3 NPs, their application for antioxidants propose was also evaluated.

2.3.2 Nanoparticles synthesized from bacteria

The bacteria are able to mobilize and immobilize the metal in a few cases, and the bacteria that can cause the reduction of metals exhibit the capability to precipitate metallic ions at nanometer scale. By performing the formation of different metallic NPs (palladium, platinum, gold, titanium dioxide, silver, and cadmium sulfide), the bacteria are taken into consideration as a potential biofactory. Bacteria are also used as a basic source of enzymes that catalyze certain reactions for the formation of inorganic NPs, which is a new process of a biosynthesis approach (Yadav et al., 2017). Due to the unique metallic binding capabilities of bacterial cells and S-layers, the bacteria have applications in NT and BRM. The features of NPs are controlled by the optimization of some important parameters that manage the cellular activities, enzymatic procedures, and growth conditions of microorganisms. Therefore, more detailed research is still required to determine the precise mechanisms of the reactions and to recognize the proteins and enzymes that are involved in the biosynthesis of NPs. The large-scale production of metallic NPs from bacteria is appealing since this process does not require toxic, costly, and hazardous chemicals for the stabilization and synthesis process (Yadav et al., 2017).

2.3.3 Nanoparticles synthesized from fungi and yeast

Fungi is broadly used as a magnificent source of many extracellular enzymes and biosynthesis of NPs. Sharp and clear-cut dimensions along with monodispersity can be attained using fungi. It can be used in creation of bulk amount of NPs when compared to bacterial species. Fungi secrete some proteins that translate into abundance formation of NPs (Mohanpuria, Rana, & Yadav, 2008). For the biosynthesis of NPs, isolated proteins are being used instead of using the fungi culture, and this method is more hopeful. For the synthesis of metallic NPs by fungi, such enzymes are attained which decreases the salt concentration to metal solid NPs by catalytic reaction (Oksanen, Pere, Paavilainen, Buchert, & Viikari, 2000). Industrial applications of fungi include increased growth rate, more specific enzyme yield, user-friendliness, and cost-effectiveness, which are more beneficial than other fungus procedures (Vahabi, Mansoori, & Karimi, 2011). Fungi have boarder over many biological techniques due to a lot of beneficial applications including easy culture techniques, diversity, less time along with more cost-effectiveness. Moreover, this method is ecofriendly for biosynthesis of NPs (Saxena, Sharma, Gupta, & Singh, 2014). Yeast strains over bacteria contain few advantages because of quick growth and many enzymes biosynthesis by using elementary nutrients. Fig. 2–2 shows the mechanism of synthesis of ZnO NPs using fungi (Ganesan, Hariram, Vivekanandhan, & Muthuramkumar, 2020).

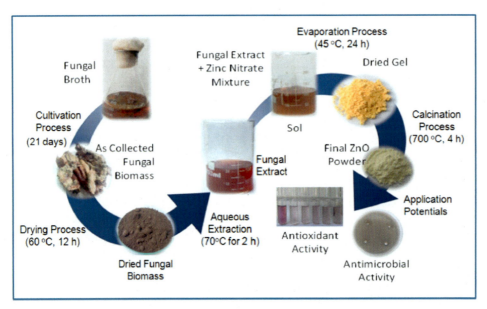

FIGURE 2–2 Mechanism of synthesis of ZnO nanoparticles using fungi. *Reproduced with permission Ganesan, V., Hariram, M., Vivekanandhan, S., & Muthuramkumar, S., Periconium sp. (endophytic fungi) extract mediated sol-gel synthesis of ZnO nanoparticles for antimicrobial and antioxidant applications.* Materials Science in Semiconductor Processing, 105*(2020), 104739. Copyright © 2020 Elsevier.*

2.4 Nanobioremediation

Nanobioremediation (NBR) is the elimination of environmental wastes like heavy metals and pollutants using NPs made by plants and microorganisms by using NT. Environmental reorganization is done by NBR for elimination of wastes. Recent modifications in the NBR technology for the elimination of wastes include physicochemical remediation, carbonization, and biological disposal. For the elimination of environmental wastes, the BRM method is easy and ecofriendly (Singh & Walker, 2006). BRM technology includes three processes: first, using microbes; second, using plants; and third, the use of enzymatic remediation. NT is utilized for retreatment water pollution with heavy metals, soils, and chemical pollutants, and it also enhances phytoremediation efficiency. Many organic pollutants like molinate, chlorpyrifos, and atrazinecan be broken down with nano zero-valent ions (Ghormade, Deshpande, & Paknikar, 2011; Zhang, 2003). Phytoremediation can be combined with enzyme-based BRM that possess NPs (Singh, 2009, 2010). Long-chain hydrocarbons and organochlorines like complex organic compounds are especially impenetrable to plant and microbial mortification. Complex organic compounds are degraded into simpler compounds by nanoencapsulating the enzymes that helps in the breakdown of complex compounds into simple compounds, which helps for easy breakdown by combined activities of plants and microorganisms. NPs of ZnO and CeO2 have enhanced the growth of root and shoot in wheat, corn, and soybean like edible plants (López-Moreno et al., 2010; López-Moreno, de la Rosa, Hernández-Viezcas, Peralta-Videa, & Gardea-Torresdey, 2010); NT can also increase phytoremediation efficiency. Several studies reveal that (Castello, 2013) effect of TiO_2 NPs can be seen on seeds and leaves by showing positive effects with enhanced growth. Nano-TiO_2 enhanced the activity of many enzymes and also increases the nitrate adsorption along with increasing transfer of inorganic into organic nitrogen. NMs appear as a superior alternative to subsist treatment techniques due to its beneficial applications (Dastjerdi & Montazer, 2010). For environmental clean-out, iron NPs found to be the first NPs (Tratnyek & Johnson, 2006). Best advantage of iron-based technology includes treatment of pollutant land. Zn-NPs helps in the breakdown of organic dyes, complete breakdown of many compounds from dye into phenol and pharmaceutical drugs, and used as semiconductor photocatalyst (Tratnyek & Johnson, 2006). Among all NPs, noble metallic NPs like silver and gold metal NPs have extensive uses in various areas but the best application is breakdown of organic dyes. Copper NPs have also helpful in breakdown of organic dyes (Marcelo, Puiatti, Nascimento, Oliveira, & Lopes, 2018). NT also enhances phytoremediation cost-effectiveness (Zhu et al., 2019). A number of remediation technologies are developed continuously with the demand for treating polluted soil, chelate, wastewater, and ground water by using in situ (Abdel-Hamid, Solbiati, & Cann, 2013) and ex situ (Tomei & Daugulis, 2013) protocols. Classical ex situ is common remediation technology, which includes revealing of polluted or toxic material and then treating it using common procedures. Such procedures consume more energy, are more expensive, and leave concentrated harmful waste material, which needs more treatment and discharge. For in situ practices, NPs have more beneficial properties. NPs having small size and original surface fabrication occupy very minute space in subsurface and swing last in the groundwater (Tratnyek & Johnson, 2006). There is no pumping out of ground water and no transportation of soil done for treatment and discharge in the in situ method for nanoremediation (Otto, Floyd, & Bajpai, 2008).

2.5 Role of nanomaterials in bioremediation

Different NMs are utilized in BRM because they enhance the surface area per unit mass of nanosize material, and nanosize material connection with nearby material influence the reactivity. NMs make the chemical reaction easy by using low activation energy and express the quantum effect. NPs are utilized for the determination of toxic stuff and also show surface plasmon resonance. Nanosized metallic and nonmetallic particles of various size and shape are utilized for the environmental clean-out. For example, different sizes, bimetallics, and carbon-based NPs can be utilized because they can be easily dispersed in a polluted zone where microparticles do not stick out and show more reactivity to the redox-amenable pollutants. Oxide-coated Fe^0 can make the outer sphere complexes with pollutants like carbon tetrachloride. The coating with oxides enhances the reactivity by the transfer of electrons, and as a result carbon tetrachloride splits into methane and CO, and a few chlorinated hydrocarbons can split into less harmful materials in batch tests and discipline assessments (Nurmi et al., 2005). TiO_2 nanotubes can be utilized to break down pentachlorophenol (PCP) by photoelectrocatalytic reactions (Quan, Yang, Ruan, & Zhao, 2005). The single metallic NPs can be utilized as biological catalysts to reductive dichlorination (Windt, Aelterman, & Verstraete, 2005). Inactivation of microbial cells, which break down or reform particular chemicals, can be done by utilizing NPs. In the traditional cell inactivation, ammonium oleate make magnetic NP operative and covered on the *Pseudomonas delafieldii* outer layer. Magnetic NP-covered cells become condensed at a particular point, discrete from bulk and reuse for ministration of similar substrate. These microbial cells can be utilized for desulfurizing the organic sulfur from dibenzothiophene, which is an effective fossil fuel.

2.6 Remediation of pollutants using nanotechnology

Remediation has extended and is continually progressing with the following new technologies and refining the process of remediation. The small sized (1–100 nm) nanoiron helps with a remedial utility. Theron, Walker, and Cloete (2008) have completed a detailed review of NT and the art and engineering of manipulating matter at the nanoscale (1–100 nm) by emphasizing the potential of novel NMs for the treatment of contaminated ground, waste, and surface water via toxic metallic ions, microorganisms, and organic and inorganic solutes. Polypyrrole-polyaniline (PPy-PANI) nanofibers are another responsive material utilized for the elimination of Cr (VI) from the solution (Bhaumik, Maity, Srinivasu, & Onyango, 2012). By using a single-stage curing procedure, NT helps in curing the polluted water more effectively (Kamat, Huehn, & Nicolaescu, 2002). NTs incorporate the curing methods that must be safe and avoiding the use of harmful chemicals. The following are a few NMs being used in remediation of pollution.

2.6.1 Nanoiron and its derivatives

The nanoscale zero-valent iron (NZVI) was formed and experimented for the disposal of arsenic (III), which is more toxic and mobile, and major species of As (arsenic) in anoxia ground water (Kanel, Manning, Charlet, & Choi, 2005). NZVI act as a colloidal reactive and is helpful in removing arsenic (V) from ground water (Kanel, Grenèche, & Choi, 2006). Zero-valent iron NPs quickly disperse and inactivate Cr (VI) and Pb (II) from solution and lessen the chromium to Cr (III) and the Pb to Pb (0) (Ponder, Darab, & Mallouk, 2000). Zero-valent Fe-NPs like anionic, hydrophilic, and polyacrylic were active stuff and used for the dehalogenation in groundwater and soils (Schrick et al., 2004). For the degradation of halogenated compounds, the iron form a reactive wall in the way of the polluted ground water (Schrick, Blough, Jones, & Mallouk, 2002). Nickel-iron NPs have more surface area and are used as a reagent for trichloroethylene dehalogenation (Wang & Zhang, 1997). Dechlorination responses of zero-valent metals related superficial results in vanishing of PCP from solution in cooperation with zero-valent metals (Kim & Carraway, 2000).

2.6.2 Dendrimers in bioremediation

Dendrimers is a Greek word that is composed of two words, one is "dendri," which meaning is similar to a branch of a tree, and the other is "meros," which means the component of a tree. They give highly branched and monodispersed structures, which was identified as a member of the polymer field. In the 1980s the first dendrimer was reported by a team of Tomalia et al. (1985), and Newkome, Yao, Baker, and Gupta (1985). It can be considered that dendrimer is a large polymeric structure that is formed by cross-linking of smaller ones. They have shown many potential applications. They are structurally composed of three parts: the central core, interior branch cells, and terminal branch cells (Undre, Singh, & Kale, 2013). The dendrimers have many large pores for the interaction with the other substances (Undre, Singh, Kale, & Rizwan, 2013). The dendrimers-NPs composite can be formed and can be utilized in enhancing the catalytic activity. These composites can also be applied in dye and water treatment industries because of their lower toxicity, high catalytic activity, and large surface area. These have applications in cleaning the water recovery units. The polyamidoamine (PAMAM) dendrimers have unique structure and characteristics, which can be used in the treatment of water. It is a highly effective and innoxious agent for the water treatment. For the removal of different organic pollutants, researchers have designed a filtration unit by using titanium dioxide as a porous ceramic filter, and the pores were incorporated with alkylated poly (propylene imine), poly (ethyleneimine), or β-cyclodextrin; as a result it gives hybrid inorganic or organic module filter, which have a large surface area and mechanical strength.

2.6.3 Nanocrystals and carbon nanotubes in bioremediation

A wide range of environmental practices can be recognized and resolved by carbon-based NMs (Mauter & Elimelech, 2008). For the elimination of ethylbenzene from solution, there are three types: single-walled, multiwalled, and hybrid carbon tubes have been used. The single-walled carbon nanotubes (SWCNTs) carried out preferable ethylbenzene adsorption than the hybrid carbon nanotubes (HCNTs). The SWCNTs are well organized and quick adsorbents for ethylbenzene, which give the best advantage in maintaining high-quality water. It can be utilized for the clean-out of environmental waste to cure ethylbenzene-borne diseases (Bina, Pourzamani, Rashidi, & Amin, 2012). Cyclodextrins (CD) and CNTs have been utilized for the water treatment and waste material observation purposes. For eliminating both organic (p-nitrophenol) and inorganic (Cd^{2+}, Pb^{2+}) pollutants from water, calixarenes, thiacalixarenes, and CNT-based polymeric materials, including these molecules, have been prepared. CNTs is an ecofriendly adsorbent that is inactivated by calcium alginate (CNTs/CA). Copper sorption property via equilibrium studies found copper sorption capacity of CNTs/CA can achieve 67.9 mg/g at a copper equilibrium concentration of 5 mg/L (Li et al., 2010). It helps in eliminating nickel ions from water (Kandah & Meunier, 2007). Magnetic multiple-walled carbon nanotubes (MWCNTs) nanocomposites have been seen to eliminate the cationic dye from the solution (Gong et al., 2009).

2.6.4 Single-enzyme nanoparticles in bioremediation

Enzymes are made up of proteins that are effective and specific in nature. They work as biocatalysts in bioremediation. Their usefulness may be limited due to the minimum stability and relatively low catalytic lifespan of the enzymes as compared to synthetic catalysts. The minimum stability and low lifespan of the enzyme is due to oxidation by making it less efficient. For the efficiency of enzymes, magnetic iron NPs were attached to them to enhance the stability, lifespan, and reusability of enzyme. Due to the attachment of NPs with enzyme, it is easy to detach the product or reactant from enzymes by using magnetic fields. By the help of this mechanism, two catabolic enzymes such as peroxides and trypsin were used in the core-shell magnetic NPs. The published work specifies that magnetic NPs due to high stability, economics, and efficiency are very advantageous in enhancing activity and lifespan of enzyme from a few hours to weeks (Tungittiplakorn, Cohen, & Lion, 2005).

2.7 Bioremediation of hydrophobic contaminants

Polycyclic aromatic hydrocarbons (PAHs) and polychlorinated biphenyls (PCBs) are hydrophobic contaminants that attach firmly with precipitates. The solubility and potency rates are decreased by the absorption of hydrophobic organic contaminant. Decrease in the utilization rate is due to sequestration of pollutants by absorption to clay and by splitting up in the anhydrous liquid phase. Enhances in the "effectual" dispersion of a typical hydrophobic organic pollutants like phenanthrene (PHEN) by the polymer nanonetwork particle helped to increase the delivery of PHEN from the polluted aquifer medium. The progress of poly(ethylene) glycol altered urethane acrylate (PMUA) parent chain helped in increasing the utilization rate of PHEN. PMUA NPs have enhanced the calcification amount of PHEN pellucid in water. PHEN is adsorbed on aquifer medium. PHEN liquefy in a model NAPL (hexadecane) in the existence of aquifer medium. PMUA particles not only increase the discharge of adsorbed and NAPL-isolated PHEN but also enhance its calcination amount. The availability of pollutants in PMUA fragments to bacteria also shows the advantage of increasing in situ decomposition outlay in makeover by innate exhaustion of pollutants. Extracted NPs can be recycled by pump-and-treat and clay-clearing makeover from bioreactor. Stability in the existence of heterogeneous functional bacterial inhabitants have shown PMUA NPs application, allowing them to be repeated after PHEN is attached to the particles and has been reduced by bacteria (Tungittiplakorn et al., 2005).

2.7.1 Soil bioremediation

Polynuclear aromatic hydrocarbons are the hydrophobic living groundwater pollutants that are absorbed firmly to clays and are tough to detach. The manufacturing of amphiphilic polyurethane (APU) NP has been used in makeover of clay polluted with PAHs. The fragments are formed of polyurethane acrylate anionomer (UAA) or PMUA parent chains that can be combined and meshed in water. The particles are formed having a colloidal size (17 – 97 nm as determined by active light dispersion). PAH desorption increased by APU fragment and transfer in such a way similar to surfactant micellars, but contrary to the surface-dynamic element of micellars; the individual meshed parent chains in APU fragments are not free to adsorb to the clay superficial. APU fragments are firm, independent of their agglomeration in the liquid phase. APU fragments have designed to attain the wanted applications. APU particles have hydrophobic interior zones confirmed by experimental analysis have greater affinity for phenanthrene (PHEN) and hydrophilic planes that enhance fragment mobility in clay. Affinity of APU fragment for pollutants such as PHEN can be administered by altering the magnitude of the hydrophobic section utilized in the chain formation. The strength of colloidal APU suspensions in clay is managed by the charge density or the magnitude of the suspended water-soluble chains that occupy on the molecular aspect. The capacity to standard particle applications helped to form distinct NPs optimized for changing pollutant type and clay environment (Tungittiplakorn, Lion, Cohen, & Kim, 2004).

2.7.2 Uranium bioremediation

Biogenic uraninite has a short molecular magnitude and biotic beginning, which made them important for geoscientists to take more interest. Chemical/structural complication of this chief essential NM have been found by modern researchers and have started to light up the chemical/structural complications of this chief essential NMs. Hydrated biogenic uraninite have properties like small size, little scale formation, liveliness, and plane-area suspension rates, which seems alike to the coarser-fragments as well as inert and stoichiometric UO2.

These investigations have essential suggestions for the part of magnitude as a mediator of NP liquid reactivity and for the biological disposal of underground U (VI) pollution (Bargar, Bernier-Latmani, Giammar, & Tebo, 2008).

2.7.3 Bioremediation of heavy metals

Environmental contamination along with heavy metal is faced by most countries. In the previous two decades, hard work has been done to reduce the contamination-spreading history and to arbitrate contaminated clay and water expedient. By field survey, dried decay ponds of guide quarry helped to investigate the domestic collecting plants. Using flame absorption atomic method concentration of heavy metal had been determined for both. In the clay and the plants, heavy metal concentration was investigated by flame absorption atomic methods that were extended in dried decay pond. Concentration of all the poisonous metals (Cu, Zn, Pb, and Ni) was established to be greater than it is found in the innate clay. The exploratory results indicate that six presiding vegetation, namely, *Gundelia tournefortii*, *Centaurea virgata*, *Reseda lutea*, *Scariola orientalis*, *Eleagnum angustifolia*, and *Noaea Mucronata* assembled heavy metals. *Noaea mucronata* belonging to Chenopodiaceae is the good Pb collector and also a best collector for Zn, Cu, and Ni, but the good Fe collector is *Reseda lutea* and the good one for the CD is *Marrubium vulgare*. The biocollecting capacity of NPs formed from *N. mucronata* was estimated in an exploratory water vessel. The quantity of heavy metals reduces manyfold during three days of biological disposal (Mohsenzadeh & Rad, 2012).

2.8 Biodegradation of synthetic dyes by nanoparticles

Synthetic dyes are one of the most major pollutants that have been entered to the environment and water from different sources like in textile industries (Bilal et al., 2016; Nouren et al., 2017). These dyes have rigid and complex structures, which make them stable and can remain in the environment for a long time. Moreover, distribution of these dyes in the water inhibit diffusion of the sunlight to deep waters and subsequently stop photosynthesis process. Most of these synthetic dyes are toxic and harmful and can cause hazardous diseases like carcinogenic, mutagenic, and teratogenic (Bilal et al., 2016; Bilal, Adeel, Rasheed, Zhao, & Iqbal, 2019). To this end, removal of them from environment is vital. Biodegradation of these synthetic dyes using NPs is a popular method that have been extensively applied in recent years. For instance, Li et al. synthesized Fe_3O_4@C-Cu^{2+} NPs and used them for removal of dyes (Li et al., 2020). In this work, after synthesis of proposed NPs, laccase was immobilized on them to enhance biodegradation performance. The applied method not only enhanced catalytic activity but also improved the stability of the immobilized laccase. In this work, at the first step, Fe_3O_4 NPs were prepared and then modified by carbon NPs. Carbon NPs not only increased stability of Fe_3O_4 NPs but also introduce hydroxyl and carboxyl functional groups for interaction with Cu^{2+} ions (Fig. 2–3).

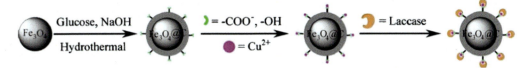

FIGURE 2–3 Synthesis procedure of magnetic nanoparticles and their modification and immobilization. *Reproduced with permission Li, P., Gao, B., Li, A., Yang, H., Evaluation of the selective adsorption of silica-sand/anionized-starch composite for removal of dyes and Cupper (II) from their aqueous mixtures.* International Journal of Biological Macromolecules, 149*(2020), 1285–1293. Copyright © 2020 Elsevier.*

Based on the results, the fabricated Fe_3O_4@C-Cu^{2+} NPs had high ability for loading of bovine serum albumin (436 mg/g) and high activity for laccase (82.3 %) after immobilization. The most important property of the prepared NPs was their high ability for recycling for at least 10 times. In the other work, thermostable and high reusable nanobiocatalyst was constructed and used for removal of dye molecules (Kiran, Rathour, Bhatia, Rana, & Bhatt, 2020). In this work, graphene oxide (GO) was synthesized and then modified with $MnFe_2O_4$ NPs. In the next step, lignin peroxidase (LiP) was immobilized on modified NPs. LiP was achieved from *Pseudomonas fluorescens*. The solgel route was used for synthesis of proposed $MnFe_2O_4$ NPs. According to HR-TEM analysis, $MnFe_2O_4$ NPs had spherical shape, which were well dispersed on the surface of GO monolayer. $MnFe_2O_4$ NPs had average diameter of 10 nm. In the other study, some azo dyes were removed using immobilized laccase on magnetic chitosan NPs (Nadaroglu, Mosber, Gungor, Adıguzel, & Adiguzel, 2019). In this work, laccase was obtained from *Weissella viridescens* LB37. Laccase was covalently interacted on chitosan beads magnetized with Fe_3O_4 NPs via cross-linker L-glutaraldehyde and then used for removal of target dyes molecules.

2.9 Biodegradation of phenolic compounds

Phenolic compounds (PCs) are another hazardous compounds, which have been produced in different industries (Tri et al., 2020). They have become a serious threat for both humans and animals, which can produce many hazardous illnesses. PCs can produce cytoplasmic toxin even at low concentrations. In this regard, they are too toxic and need to be removed from the environment (Sas, Sanchez, Gonzalez, & Dominguez, 2020; Van Tran et al., 2020). Therefore, many affords have been done to remove these materials. Recently, Qiu and coworkers removed PCs using laccase immobilized (Qiu, Wang, Xue, Li, & Hu, 2020). They synthesized magnetic NPs and then modified them by ionic liquid via dialdehyde starch. In the final step, the proposed compound was immobilized with laccase and used for removal of PCs. Fig. 2–4 shows the preparation of the proposed compounds and removal processes of PCs using immobilized laccase. The fabricated compounds had high stability and was applicable for removal of different PCs. The effective parameters on PCs removal were evaluated. The applied method introduced a novel method for enzyme immobilization and its application for removal of different PSs. In the other study, 3D bioprinting was applied for

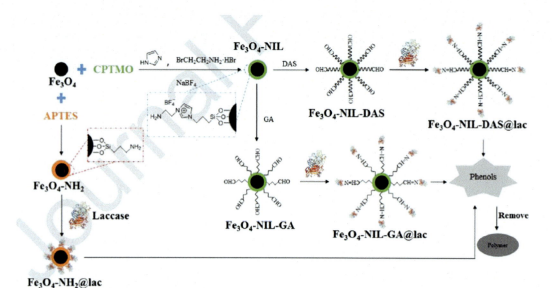

FIGURE 2–4 The preparation of the proposed compounds and removal processes of PCs using immobilized laccase. *Reproduced with permission Qiu, X., Wang, Y., Xue, Y., Li, W., Hu, Y., Laccase immobilized on magnetic nanoparticles modified by amino-functionalized ionic liquid via dialdehyde starch for phenolic compounds biodegradation.* Chemical Engineering Journal, 391*(2020), 123564. Copyright © 2020 Elsevier.*

immobilization of laccases and then applied for the removal of PCs (Liu et al., 2020). In this work, novel immobilization method was used. The features of hydrogel mechanism were investigated and optimized. The morphology pictures and rheology characteristics were used for validation of mechanism properties. The immobilized laccase showed high storage stability and reusability.

2.10 Biodegradation of antibiotics and personal care products

Antibiotics are other important materials that have been entered into the environment and water sources (Reis, Kolvenbach, Nunes, & Corvini, 2020). The residuals of antibiotics can cause many problems. In this regard, their removal is vital. Ciprofloxacin (CIP) is an antibiotic widely used in medicine. High amounts of CIP usually do not metabolize completely and can enter to the environment. Recently, Yang and coworkers removed CIP using biodegradation process in the presence of magnetic NPs (Yang et al., 2017). Based on the results, the addition of magnetic NPs improved degradation of CIP.

The biodegradation of CIP may related to reduction of Fe (III) to Fe (II). Pharmaceuticals and personal care products (PPCPs) are other important pollutants that can produce many problems for humans and animals. In this regard, many efforts have been done for removal

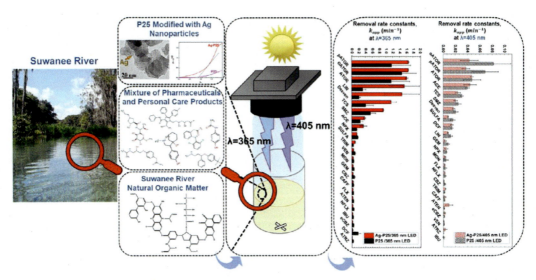

FIGURE 2–5 Removal of PPCPs using P25 modified with Ag nanoparticles. *Reproduced with permission Fattahi, A., Arlos, M. J., Bragg, L. M., Kowalczyk, S., Liang, R., Schneider, O. M. ... Servos, M. R. Photodecomposition of pharmaceuticals and personal care products using P25 modified with Ag nanoparticles in the presence of natural organic matter. Science of the Total Environment, 752(2021), 142000. Copyright © 2021 Elsevier.*

of these pollutants from the environment. For instance, recently, Fattahi used P25 modified with Ag NPs for removal of PPCPs (Fattahi et al., 2021). Fig. 2–5 represents the removal of PPCPs using P25 modified with Ag NPs. In this study, the ability of the synthesized compound for removal of 23 target PPCPs was assessed. The applied method was very applicable and effective for removal of PPCPs.

2.11 Conclusions

Simultaneous with a growing population, the spread of different pollutants in the environment and water sources has become a challenge for researchers—especially environmental researchers. These pollutants can cause and produce various hazardous and harmful diseases for humans and animals. Treatment and removal of these contaminations is important. Until now, many efforts have been performed for the removal of pollutants. One of the methods for encounters with pollutants is the application biodegradation processes. These approaches have amazing properties, which make them good candidates. To improve the performance of biodegradation processes, merging them with NPs can be an ideal choice, which is due to excellent features of NPs in terms of high chemical and physical stability, high surface area, high adsorption capacity, high-loading capacity, and so on. To this end, we have cheered to review recent progress and development in application of biodegradation process-based NPs for removal of various kinds of pollutants. The investigation of the literature proved that the application of different types of NMs and particularly NPs can significantly improve the performance of biodegradation processes. Application of enzymes in biodegradation processes is very popular. But enzymes are also faced with a main drawback, namely, stability. To enhance the stability of enzymes, their immobilization on different supports is an alternative method. One of the best supports is NPs, which have been extensively applied. The other approach for improving the performance of biodegradation processes is the application of selective NPs. For instance, molecularly imprinted NPs can be used as selective materials, which can distinguish target analyte from different compounds. Molecularly imprinted NPs not only are selective, but also their synthesis is easy; they are available and also have high chemical and physical stability. The other materials for application in biodegradation processes are metal-organic frameworks and covalent organic frameworks NPs. These materials have high chemical and physical stability, high surface area, and high adsorption capacity. They can interact with target pollutants via different functional groups and remove them from the environment. We hope that this chapter opens new doors toward researchers and readers for better understanding of properties of NPs and biodegradation processes and the application of them for treatment and removal of different pollutants.

References

Abdel-Hamid, A. M., Solbiati, J. O., & Cann, I. K. (2013). Insights into lignin degradation and its potential industrial applications. *Advances in Applied Microbiology*, 1–28.

Arabi, M., Ostovan, A., Bagheri, A. R., Guo, X., Li, J., Ma, J., & Chen, L. (2020). Hydrophilic molecularly imprinted nanospheres for the extraction of rhodamine B followed by HPLC analysis: A green approach and hazardous waste elimination. *Talanta*, 120933.

Arabi, M., Ostovan, A., Bagheri, A. R., Guo, X., Wang, L., Li, J., ... Chen, L. (2020). Strategies of molecular imprinting-based solid-phase extraction prior to chromatographic analysis. *TrAC Trends in Analytical Chemistry*, 115923.

Bagheri, A. R., & Aramesh, N. (2020). Towards the room-temperature synthesis of covalent organic frameworks: A mini-review. *Journal of Materials Science*, 1–17.

Bagheri, A. R., & Ghaedi, M. (2020a). Green preparation of dual-template chitosan-based magnetic water-compatible molecularly imprinted biopolymer. *Carbohydrate Polymers*, 116102.

Bagheri, A. R., & Ghaedi, M. (2020b). Application of Cu-based metal-organic framework (Cu-BDC) as a sorbent for dispersive solid-phase extraction of gallic acid from orange juice samples using HPLC-UV method. *Arabian Journal of Chemistry*.

Bagheri, A. R., & Ghaedi, M. (2020c). Magnetic metal organic framework for pre-concentration of ampicillin from cow milk samples. *Journal of Pharmaceutical Analysis*.

Baker, S., & Satish, S. (2012). Endophytes: Toward a vision in synthesis of nanoparticle for future therapeutic agents. *International Journal of Bio-Inorganic Hybrid Nanomaterial*, 1, 67–77.

Bargar, J. R., Bernier-Latmani, R., Giammar, D. E., & Tebo, B. M. (2008). Biogenic uraninite nanoparticles and their importance for uranium remediation. *Elements*, 4, 407–412.

Bhatti, H. N., Safa, Y., Yakout, S. M., Shair, O. H., Iqbal, M., & Nazir, A. (2020). Efficient removal of dyes using carboxymethyl cellulose/alginate/polyvinyl alcohol/rice husk composite: Adsorption/desorption, kinetics and recycling studies. *International Journal of Biological Macromolecules*, 150, 861–870.

Bhaumik, M., Maity, A., Srinivasu, V. V., & Onyango, M. S. (2012). Removal of hexavalent chromium from aqueous solution using polypyrrole-polyaniline nanofibers. *Chemical Engineering Journal*, 181–182, 323–333.

Bilal, M., Adeel, M., Rasheed, T., Zhao, Y., & Iqbal, H. M. (2019). Emerging contaminants of high concern and their enzyme-assisted biodegradation—A review. *Environment International, 124*, 336–353.

Bilal, M., Iqbal, M., Hu, H., & Zhang, X. (2016a). Mutagenicity and cytotoxicity assessment of biodegraded textile effluent by Ca-alginate encapsulated manganese peroxidase. *Biochemical Engineering Journal, 109*, 153–161.

Bilal, M., Iqbal, M., Hu, H., & Zhang, X. (2016b). Mutagenicity, cytotoxicity and phytotoxicity evaluation of biodegraded textile effluent by fungal ligninolytic enzymes. *Water Science and Technology, 73*, 2332–2344.

Bina, B., Pourzamani, H., Rashidi, A., & Amin, M. M. (2012). Ethylbenzene removal by carbon nanotubes from aqueous solution. *Journal of Environmental and Public Health, 2012*, 817187.

Boyes, W. K., & van Thriel, C. (2020). Neurotoxicology of nanomaterials, chemical research in toxicology. *Chemical research in toxicology, 33*(5), 1121–1144.

Castello, N. A. (2013). *Targeting structural plasticity to improve Alzheimer disease-related functional deficits*. Irvine: University of California.

Chen, Q., Ma, C., Duan, W., Lang, D., & Pan, B. (2020). Coupling adsorption and degradation in p-nitrophenol removal by biochars. *Journal of Cleaner Production, 271*, 122550.

Chen, Y., Lai, Z., Zhang, X., Fan, Z., He, Q., Tan, C., & Zhang, H. (2020). Phase engineering of nanomaterials. *Nature Reviews Chemistry*, 1–14.

Cheng, L., Wang, X., Gong, F., Liu, T., & Liu, Z. (2020). 2D nanomaterials for cancer theranostic applications. *Advanced Materials, 32*, 1902333.

Dastjerdi, R., & Montazer, M. (2010). A review on the application of inorganic nano-structured materials in the modification of textiles: Focus on anti-microbial properties. *Colloids and Surfaces B: Biointerfaces, 79*, 5–18.

Diouf, A., El Bari, N., & Bouchikhi, B. (2020). A novel electrochemical sensor based on ion imprinted polymer and gold nanomaterials for nitrite ion analysis in exhaled breath condensate. *Talanta, 209*, 120577.

Durán, N., Marcato, P. D., De Souza, G. I., Alves, O. L., & Esposito, E. (2007). Antibacterial effect of silver nanoparticles produced by fungal process on textile fabrics and their effluent treatment. *Journal of Biomedical Nanotechnology, 3*, 203–208.

Farré, M., & Jha, A. N. (2020). Metabolomics effects of nanomaterials: An ecotoxicological perspective. *Environmental Metabolomics*, 259–281.

Fattahi, A., Arlos, M. J., Bragg, L. M., Kowalczyk, S., Liang, R., Schneider, O. M., ... Servos, M. R. (2021). Photodecomposition of pharmaceuticals and personal care products using P25 modified with Ag nanoparticles in the presence of natural organic matter. *Science of the Total Environment, 752*, 142000.

Ganesan, V., Hariram, M., Vivekanandhan, S., & Muthuramkumar, S. (2020). *Periconium* sp. (endophytic fungi) extract mediated sol-gel synthesis of ZnO nanoparticles for antimicrobial and antioxidant applications. *Materials Science in Semiconductor Processing, 105*, 104739.

Ge, J., Zhang, Q., Zeng, J., Gu, Z., & Gao, M. (2020). Radiolabeling nanomaterials for multimodality imaging: New insights into nuclear medicine and cancer diagnosis. *Biomaterials, 228*, 119553.

Ghormade, V., Deshpande, M. V., & Paknikar, K. M. (2011). Perspectives for nano-biotechnology enabled protection and nutrition of plants. *Biotechnology Advances, 29*, 792–803.

Gong, J.-L., Wang, B., Zeng, G.-M., Yang, C.-P., Niu, C.-G., Niu, Q.-Y., ... Liang, Y. (2009). Removal of cationic dyes from aqueous solution using magnetic multi-wall carbon nanotube nanocomposite as adsorbent. *Journal of Hazardous Materials, 164*, 1517–1522.

Gurunathan, S., Kalishwaralal, K., Vaidyanathan, R., Venkataraman, D., Pandian, S. R. K., Muniyandi, J., ... Eom, S. H. (2009). Biosynthesis, purification and characterization of silver nanoparticles using Escherichia coli. *Colloids and Surfaces B: Biointerfaces, 74*, 328–335.

Hassan, N., Shahat, A., El-Daidamony, A., El-Desouky, M., & El-Bindary, A. (2020). Synthesis and characterization of ZnO nanoparticles via zeolitic imidazolate framework-8 and its application for removal of dyes. *Journal of Molecular Structure*, 128029.

Intrchom, W., Roy, S., & Mitra, S. (2020). Functionalized carbon nanotube immobilized membrane for low temperature ammonia removal via membrane distillation. *Separation and Purification Technology, 235*, 116188.

Iqbal, J., Abbasi, B. A., Munir, A., Uddin, S., Kanwal, S., & Mahmood, T. (2020). Facile green synthesis approach for the production of chromium oxide nanoparticles and their different in vitro biological activities. *Microscopy Research and Technique, 83*, 706–719.

Kalaiselvi, D., Mohankumar, A., Shanmugam, G., Nivitha, S., & Sundararaj, P. (2019). Green synthesis of silver nanoparticles using latex extract of *Euphorbia tirucalli*: A novel approach for the management of root knot nematode, Meloidogyne incognita. *Crop Protection, 117*, 108–114.

Kamat, P. V., Huehn, R., & Nicolaescu, R. (2002). A "sense and shoot" approach for photocatalytic degradation of organic contaminants in water. *The Journal of Physical Chemistry. B, 106*, 788–794.

Kandah, M. I., & Meunier, J.-L. (2007). Removal of nickel ions from water by multi-walled carbon nanotubes. *Journal of Hazardous Materials, 146*, 283–288.

Kanel, S. R., Grenèche, J.-M., & Choi, H. (2006). Arsenic(V) removal from groundwater using nano scale zero-valent iron as a colloidal reactive barrier material. *Environmental Science & Technology, 40*, 2045–2050.

Kanel, S. R., Manning, B., Charlet, L., & Choi, H. (2005). Removal of arsenic(III) from groundwater by nanoscale zero-valent iron. *Environmental Science & Technology, 39*, 1291–1298.

Kim, S. H., Chung, H., Jeong, S., & Nam, K. (2020). Identification of pH-dependent removal mechanisms of lead and arsenic by basic oxygen furnace slag: Relative contribution of precipitation and adsorption. *Journal of Cleaner Production, 279*, 123451.

Kim, Y.-H., & Carraway, E. R. (2000). Dechlorination of pentachlorophenol by zero valent iron and modified zero valent irons. *Environmental Science & Technology, 34*, 2014–2017.

Kiran, R. K., Rathour, R. K., Bhatia, D. S., Rana, A. K., & Bhatt, N. T. (2020). Fabrication of thermostable and reusable nanobiocatalyst for dye decolourization by immobilization of lignin peroxidase on graphene oxide functionalized MnFe2O4 superparamagnetic nanoparticles. *Bioresource Technology, 317*, 124020.

Kratochvil, D., & Volesky, B. (1998). Advances in the biosorption of heavy metals. *Trends in Biotechnology, 16*, 291–300.

Li, P., Gao, B., Li, A., & Yang, H. (2020). Evaluation of the selective adsorption of silica-sand/anionized-starch composite for removal of dyes and Cupper (II) from their aqueous mixtures. *International Journal of Biological Macromolecules, 149*, 1285–1293.

Li, Y., Liu, F., Xia, B., Du, Q., Zhang, P., Wang, D., ... Xia, Y. (2010). Removal of copper from aqueous solution by carbon nanotube/calcium alginate composites. *Journal of Hazardous Materials, 177*, 876–880.

Li, Y. Y., & Li, B. (2011). Study on fungi-bacteria consortium bioremediation of petroleum contaminated mangrove sediments amended with mixed biosurfactants. *Advanced Materials Research*, 1163–1167.

Li, Z., Chen, Z., Zhu, Q., Song, J., Li, S., & Liu, X. (2020). Improved performance of immobilized laccase on Fe3O4@ C-Cu2 + nanoparticles and its application for biodegradation of dyes. *Journal of Hazardous Materials*, 123088.

Liu, J., Chen, H., Shi, X., Nawar, S., Werner, J. G., Huang, G., ... Mei, Y. (2020). Hydrogel microcapsules with photocatalytic nanoparticles for removal of organic pollutants. *Environmental Science: Nano, 7*, 656–664.

Liu, J., Shen, X., Zheng, Z., Li, M., Zhu, X., Cao, H., & Cui, C. (2020). Immobilization of laccase by 3D bioprinting and its application in the biodegradation of phenolic compounds. *International Journal of Biological Macromolecules, 164*, 518–525.

Liu, M., Wang, Z., Zong, S., Chen, H., Zhu, D., Wu, L., ... Cui, Y. (2014). SERS detection and removal of mercury (II)/silver (I) using oligonucleotide-functionalized core/shell magnetic silica sphere@ Au nanoparticles. *ACS Applied Materials & Interfaces, 6*, 7371–7379.

López-Moreno, M. L., de la Rosa, G., Hernández-Viezcas, J. Á., Castillo-Michel, H., Botez, C. E., Peralta-Videa, J. R., & Gardea-Torresdey, J. L. (2010). Evidence of the differential biotransformation and genotoxicity of ZnO and CeO2 nanoparticles on soybean (Glycine max) plants. *Environmental Science & Technology, 44*, 7315−7320.

López-Moreno, M. L., de la Rosa, G., Hernández-Viezcas, J. A., Peralta-Videa, J. R., & Gardea-Torresdey, J. L. (2010). X-ray absorption spectroscopy (XAS) corroboration of the uptake and storage of CeO2 nanoparticles and assessment of their differential toxicity in four edible plant species. *Journal of Agricultural and Food Chemistry, 58*, 3689−3693.

Mahmoud, A. E. D. (2020). Graphene-based nanomaterials for the removal of organic pollutants: Insights into linear versus nonlinear mathematical models. *Journal of Environmental Management, 270*, 110911.

Marcelo, C. R., Puiatti, G. A., Nascimento, M. A., Oliveira, A. F., & Lopes, R. P. (2018). Degradation of the reactive blue 4 dye in aqueous solution using zero-valent copper nanoparticles. *Journal of Nanomaterials, 2018*, 4642038.

Mauter, M. S., & Elimelech, M. (2008). Environmental applications of carbon-based nanomaterials. *Environmental Science & Technology, 42*, 5843−5859.

Mohanpuria, P., Rana, N. K., & Yadav, S. K. (2008). Biosynthesis of nanoparticles: Technological concepts and future applications. *Journal of Nanoparticle Research, 10*, 507−517.

Mohsenzadeh, F., & Rad, A. C. (2012). Bioremediation of heavy metal pollution by nano-particles of noaea mucronata. *International Journal of Bioscience, Biochemistry and Bioinformatics, 2*, 85.

M-Ridha, M. J., Hussein, S. I., Alismaeel, Z. T., Atiya, M. A., & Aziz, G. M. (2020). Biodegradation of reactive dyes by some bacteria using response surface methodology as an optimization technique. *Alexandria Engineering Journal, 59*, 3551−3563.

Nadaroglu, H., Mosber, G., Gungor, A. A., Adıguzel, G., & Adiguzel, A. (2019). Biodegradation of some azo dyes from wastewater with laccase from *Weissella viridescens* LB37 immobilized on magnetic chitosan nanoparticles. *Journal of Water Process Engineering, 31*, 100866.

N'Dea, S., Nelson, K. M., Gleghorn, J. P., & Day, E. S. (2020). Design of nanomaterials for applications in maternal/fetal medicine. *Journal of Materials Chemistry B*.

Newkome, G. R., Yao, Z., Baker, G. R., & Gupta, V. K. (1985). Micelles. Part 1. Cascade molecules: A new approach to micelles. A [27]-arborol. *The Journal of Organic Chemistry, 50*, 2003−2004.

Nouren, S., Bhatti, H. N., Iqbal, M., Bibi, I., Nazar, N., Iqbal, D. N., . . . Hussain, F. (2017). Redox mediators assisted-degradation of direct yellow 4. *Polish Journal of Environmental Studies, 26*.

Nurmi, J. T., Tratnyek, P. G., Sarathy, V., Baer, D. R., Amonette, J. E., Pecher, K., . . . Driessen, M. D. (2005). Characterization and properties of metallic iron nanoparticles: Spectroscopy, electrochemistry, and kinetics. *Environmental Science & Technology, 39*, 1221−1230.

Oksanen, T., Pere, J., Paavilainen, L., Buchert, J., & Viikari, L. (2000). Treatment of recycled kraft pulps with *Trichoderma reesei* hemicellulases and cellulases. *Journal of Biotechnology, 78*, 39−48.

Olabi, A., Wilberforce, T., Sayed, E. T., Elsaid, K., Rezk, H., & Abdelkareem, M. A. (2020). Recent progress of graphene based nanomaterials in bioelectrochemical systems. *Science of the Total Environment, 749*, 141225.

Otto, M., Floyd, M., & Bajpai, S. (2008). Nanotechnology for site remediation. *Remediation Journal, 19*, 99−108.

Palomino-Ascencio, L., García-Hernández, E., Salazar-Villanueva, M., & Chigo-Anota, E. (2020). B12N12 nanocages with homonuclear bonds as a promising material in the removal/degradation of the insecticide imidacloprid. *Physica E: Low-Dimensional Systems and Nanostructures*, 114456.

Peng, H., & Guo, J. (2020). Removal of chromium from wastewater by membrane filtration, chemical precipitation, ion exchange, adsorption electrocoagulation, electrochemical reduction, electrodialysis, electrodeionization, photocatalysis and nanotechnology: A review. *Environmental Chemistry Letters*, 1−14.

Ponder, S. M., Darab, J. G., & Mallouk, T. E. (2000). Remediation of Cr(VI) and Pb(II) aqueous solutions using supported, nanoscale zero-valent iron. *Environmental Science & Technology, 34*, 2564–2569.

Prokop, G. (2000). *Management of contaminated sites in Western Europe*. Office for Official Publications of the European Communities.

Qiu, X., Wang, Y., Xue, Y., Li, W., & Hu, Y. (2020). Laccase immobilized on magnetic nanoparticles modified by amino-functionalized ionic liquid via dialdehyde starch for phenolic compounds biodegradation. *Chemical Engineering Journal, 391*, 123564.

Quan, X., Yang, S., Ruan, X., & Zhao, H. (2005). Preparation of titania nanotubes and their environmental applications as electrode. *Environmental Science & Technology, 39*, 3770–3775.

Quigg, A., Chin, W.-C., Chen, C.-S., Zhang, S., Jiang, Y., Miao, A.-J., . . . Santschi, P. H. (2013). Direct and indirect toxic effects of engineered nanoparticles on algae: Role of natural organic matter. *ACS Sustainable Chemistry & Engineering, 1*, 686–702.

Rani, P., Kumar, V., Singh, P. P., Matharu, A. S., Zhang, W., Kim, K.-H., . . . Rawat, M. (2020). Highly stable AgNPs prepared via a novel green approach for catalytic and photocatalytic removal of biological and non-biological pollutants. *Environment International, 143*, 105924.

Reina, G., Peng, S., Jacquemin, L., Andrade, A. F., & Bianco, A. (2020). Hard nanomaterials in time of viral pandemics. *ACS Nano, 14*, 9364–9388.

Reis, A. C., Kolvenbach, B. A., Nunes, O. C., & Corvini, P. F. X. (2020). Biodegradation of antibiotics: The new resistance determinants—Part II. *New Biotechnology, 54*, 13–27.

Richard, P. (1959). There's plenty of room at the bottom. In *Annual Meeting of the American Physical Society*, December 29.

Roco, M. C. (2005). The emergence and policy implications of converging new technologies integrated from the nanoscale. *Journal of Nanoparticle Research, 7*, 129–143.

Sahu, R. L., Dash, R. R., & Pradhan, P. K. (2020). Use of soil conservation service curve number and filtration coefficient approach for simulating Escherichia coli removal during river bank filtration. *Journal of Water Process Engineering, 37*, 101432.

Saleh, T. A. (2020). Nanomaterials: Classification, properties, and environmental toxicities. *Environmental Technology & Innovation, 20*, 101067.

Sas, O. G., Sanchez, P. B., Gonzalez, B., & Dominguez, A. (2020). Removal of phenolic pollutants from wastewater streams using ionic liquids. *Separation and Purification Technology, 236*, 116310.

Sastry, M., Ahmad, A., Khan, M. I., & Kumar, R. (2003). Biosynthesis of metal nanoparticles using fungi and actinomycete. *Current Science, 85*, 162–170.

Saxena, J., Sharma, M. M., Gupta, S., & Singh, A. (2014). Emerging role of fungi in nanoparticle synthesis and their applications. *World Journal of Pharmaceutical Sciences, 3*, 1586–1613.

Schrick, B., Blough, J. L., Jones, A. D., & Mallouk, T. E. (2002). Hydrodechlorination of trichloroethylene to hydrocarbons using bimetallic nickel – iron nanoparticles. *Chemistry of Materials, 14*, 5140–5147.

Schrick, B., Hydutsky, B. W., Blough, J. L., & Mallouk, T. E. (2004). Delivery vehicles for zerovalent metal nanoparticles in soil and groundwater. *Chemistry of Materials, 16*, 2187–2193.

Shabbir, S., Faheem, M., Ali, N., Kerr, P. G., Wang, L.-F., Kuppusamy, S., & Li, Y. (2020). Periphytic biofilm: An innovative approach for biodegradation of microplastics. *Science of the Total Environment, 717*, 137064.

Shi, J., Zhang, B., Cheng, Y., & Peng, K. (2020). Microbial vanadate reduction coupled to co-metabolic phenanthrene biodegradation in groundwater. *Water Research, 186*, 116354.

Singh, B. K. (2009). Organophosphorus-degrading bacteria: Ecology and industrial applications. *Nature Reviews. Microbiology, 7*, 156–164.

Singh, B. K. (2010). Exploring microbial diversity for biotechnology: The way forward. *Trends in Biotechnology, 28*, 111–116.

Singh, B. K., & Walker, A. (2006). Microbial degradation of organophosphorus compounds. *FEMS Microbiology Reviews, 30*, 428–471.

Sirajudheen, P., Karthikeyan, P., & Meenakshi, S. (2020). Mechanistic performance of organic pollutants removal from water using Zn/Al layered double hydroxides imprinted carbon composite. *Surfaces and Interfaces, 20*, 100581.

Theron, J., Walker, J. A., & Cloete, T. E. (2008). Nanotechnology and water treatment: Applications and emerging opportunities. *Critical Reviews in Microbiology, 34*, 43–69.

Tomalia, D. A., Baker, H., Dewald, J., Hall, M., Kallos, G., & Martin, S. (1985). A new class of polymers: Starburst-dendritic macromolecules. *Polymer Journal, 17*, 117–132.

Tomei, M. C., & Daugulis, A. J. (2013). Ex situ bioremediation of contaminated soils: An overview of conventional and innovative technologies. *Critical Reviews in Environmental Science and Technology, 43*, 2107–2139.

Tran, H. N., Tomul, F., Nguyen, H. T. H., Nguyen, D. T., Lima, E. C., Le, G. T., ... Woo, S. H. (2020). Innovative spherical biochar for pharmaceutical removal from water: Insight into adsorption mechanism. *Journal of Hazardous Materials*, 122255.

Tratnyek, P. G., & Johnson, R. L. (2006). Nanotechnologies for environmental cleanup. *Nano Today, 1*, 44–48.

Tri, N. L. M., Thang, P. Q., Van Tan, L., Huong, P. T., Kim, J., Viet, N. M., ... Al, T. (2020). Tahtamouni, Removal of phenolic compounds from wastewaters by using synthesized Fe-nano zeolite. *Journal of Water Process Engineering, 33*, 101070.

Tungittiplakorn, W., Cohen, C., & Lion, L. W. (2005). Engineered polymeric nanoparticles for bioremediation of hydrophobic contaminants. *Environmental Science & Technology, 39*, 1354–1358.

Tungittiplakorn, W., Lion, L. W., Cohen, C., & Kim, J.-Y. (2004). Engineered polymeric nanoparticles for soil remediation. *Environmental Science & Technology, 38*, 1605–1610.

Undre, S. B., Singh, M., & Kale, R. K. (2013). Interaction behaviour of trimesoyl chloride derived 1st tier dendrimers determined with structural and physicochemical properties required for drug designing. *Journal of Molecular Liquids, 182*, 106–120.

Undre, S. B., Singh, M., Kale, R. K., & Rizwan, M. (2013). Silibinin binding and release activities moderated by interstices of trimesoyl, tridimethyl, and tridiethyl malonate first-tier dendrimers. *Journal of Applied Polymer Science, 130*, 3537–3554.

Vahabi, K., Mansoori, G. A., & Karimi, S. (2011). Biosynthesis of silver nanoparticles by fungus Trichoderma reesei (a route for large-scale production of AgNPs). *Insciences Journal, 1*, 65–79.

Van Dillewijn, P., Caballero, A., Paz, J. A., González-Pérez, M. M., Oliva, J. M., & Ramos, J. L. (2007). Bioremediation of 2, 4, 6-trinitrotoluene under field conditions. *Environmental Science & Technology, 41*, 1378–1383.

Van Tran, T., Dai Cao, V., Nguyen, V. H., Hoang, B. N., Vo, D.-V. N., Nguyen, T. D., & Bach, L. G. (2020). MIL-53 (Fe) derived magnetic porous carbon as a robust adsorbent for the removal of phenolic compounds under the optimized conditions. *Journal of Environmental Chemical Engineering, 8*, 102902.

Vieira, W. T., de Farias, M. B., Spaolonzi, M. P., da Silva, M. G. C., & Vieira, M. G. A. (2020). Removal of endocrine disruptors in waters by adsorption, membrane filtration and biodegradation. A review. *Environmental Chemistry Letters, 18*, 1113–1143.

Wang, C.-B., & Zhang, W.-x (1997). Synthesizing nanoscale iron particles for rapid and complete dechlorination of TCE and PCBs. *Environmental Science & Technology, 31*, 2154–2156.

Windt, W. D., Aelterman, P., & Verstraete, W. (2005). Bioreductive deposition of palladium (0) nanoparticles on *Shewanella oneidensis* with catalytic activity towards reductive dechlorination of polychlorinated biphenyls. *Environmental Microbiology, 7*, 314–325.

Wu, J., Yang, J., Feng, P., Huang, G., Xu, C., & Lin, B. (2020). High-efficiency removal of dyes from wastewater by fully recycling litchi peel biochar. *Chemosphere, 246*, 125734.

Xia, L., Zhou, S., Zhang, C., Fu, Z., Wang, A., Zhang, Q., . . . Xu, W. (2020). Environment-friendly Juncus effusus-based adsorbent with a three-dimensional network structure for highly efficient removal of dyes from wastewater. *Journal of Cleaner Production*, 120812.

Xiao, T., Huang, J., Wang, D., Meng, T., & Yang, X. (2020). Au and Au-based nanomaterials: Synthesis and recent progress in electrochemical sensor applications. *Talanta, 206*, 120210.

Yadav, K., Singh, J., Gupta, N., & Kumar, V. (2017). A review of nanobioremediation technologies for environmental cleanup: A novel biological approach. *Journal of Materials and Environmental Sciences, 8*, 740–757.

Yang, Z., Xu, X., Dai, M., Wang, L., Shi, X., & Guo, R. (2017). Accelerated ciprofloxacin biodegradation in the presence of magnetite nanoparticles. *Chemosphere, 188*, 168–173.

Yun, J., Wang, Y., Liu, Z., Li, Y., Yang, H., & Xu, Z.-l (2020). High efficient dye removal with hydrolyzed ethanolamine-Polyacrylonitrile UF membrane: Rejection of anionic dye and selective adsorption of cationic dye. *Chemosphere, 259*, 127390.

Zhang, W.-x (2003). Nanoscale iron particles for environmental remediation: An overview. *Journal of Nanoparticle Research, 5*, 323–332.

Zhu, Y., Xu, F., Liu, Q., Chen, M., Liu, X., Wang, Y., . . . Zhang, L. (2019). Nanomaterials and plants: Positive effects, toxicity and the remediation of metal and metalloid pollution in soil. *Science of the Total Environment, 662*, 414–421.

Zurier, H. S., & Goddard, J. M. (2021). Biodegradation of microplastics in food and agriculture. *Current Opinion in Food Science, 37*, 37–44.

3

Interaction of nanomaterials with microbes

Muhammad Rizwan Javed[1], Muhammad Hamid Rashid[2], Anam Tariq[1], Riffat Seemab[1], Anam Ijaz[1], Sohail Abbas[1]

[1]DEPARTMENT OF BIOINFORMATICS AND BIOTECHNOLOGY, GOVERNMENT COLLEGE UNIVERSITY FAISALABAD (GCUF), FAISALABAD, PAKISTAN [2]INDUSTRIAL BIOTECHNOLOGY DIVISION, NATIONAL INSTITUTE FOR BIOTECHNOLOGY AND GENETIC ENGINEERING (NIBGE), FAISALABAD, PAKISTAN

3.1 Introduction

The term "nanotechnology" was first coined in 1959 by Nobel Physicist Richard P. Feynman in his well-known lecture *There's Plenty of Room at the Bottom* (Feynman, 1960). Nanotechnology is a key enabling technology that involves the managing of substances at the nanolevel (1–100 nm). It is an integration of different fields of science and technology that usually involves processing, isolation, consolidation, and manipulation of materials with desired characteristics that are different from their bulk materials (Invernizzi, 2011). Bionanotechnology (and its counterpart nanobiotechnology) is an interdisciplinary field of biotechnology and nanotechnology that combines the physical, chemical, and biological properties of materials. By combining biological and physical sciences it opens groundbreaking and novel systems with better specificity, sensitivity, and functionality (Fortina, Kricka, Surrey, & Grodzinski, 2005). This interdisciplinary research offers an innovative tool to understand the biological principles guided by nanotechnology. Subsequently, numerous functional nanomachines can be constructed with control switches by integrating biological systems at the molecular level (Amin, Hwang, & Park, 2011). This has not only revolutionary developments in the field of science and technology but also has the possibility to mark an industrial revolution significantly. Nanotechnology-derived nanomaterials have remarkable influence in research and development along with combined impact on business, which directly improves the global economy. More than 1000 commercial products are available based on nanotechnology and expected that they will increase a global business of USD 3 trillion per year in the near future (Roco, 2011).

Bionanotechnology and nanobiotechnology are used to exploit, fabricate, and analyze nano-biosystems to achieve various specific applications and play a vital role in medicine, betterment of diagnosis and treatment, drug delivery, food industry, cosmetics, and cloth finishing.

Nanoparticles (NPs) show various physiochemical properties in the nanorange size, hydrophobic and hydrophilic interactions, composition, and ratio between large surface area and volume (Navya & Daima, 2016). Nanotechnology is currently shaping the global economy through commercialization of beneficial products such as drugs and medicines as well as in environmental and industrial applications (Musee, Thwala, & Nota, 2011). Nanomaterials or NPs are the product of nanotechnology lying in the range of 1–100 nm in size with one dimension. There are various types of NPs having different physiochemical properties depending on the size- and mass-to-volume ratio such as carbon-based, metal-based, and liposome-based NPs. These are not the simplest molecules due to their composition of three layers, that is, (1) surface layer which consists of surfactants, metal ions, polymers, and so forth on which functionality of NPs depends; (2) shell layer that consists of chemically modified material that is different from the core of NPs; and (3) core consists of NP itself. Due to exceptional properties of NPs such as mesoporosity, they have got immense uses in multidisciplinary fields such as CO_2 capturing, drug delivery, gas sensing, and many more (Ganesh, Hemalatha, Peng, & Jang, 2017; Shaalan, Saleh, El-Mahdy, & El-Matbouli, 2016).

3.2 Types of nanoparticles

NPs are classified into different categories based on their chemical composition, morphology, and size. Depending on physiochemical properties, some of the familiar classes of NPs are described in the following sections.

3.2.1 Metallic nanoparticles

Metallic NPs are purely consisting of metal precursors. Their size is less than 100 nm. Alkali and noble metals such as Ag, Cu, and Au have unique opteoelectrical properties due to stability, small size, and large surface area. Silver and gold NPs have a rich history in chemistry. In ancient times gold was used for decorating the steel on glass and for treatment of various diseases. Recently, gold NPs are used for gene and drug delivery, DNA fingerprinting, and protein interactions (Tomar & Garg, 2013). While silver NPs are most effective in antimicrobial effect against viruses, bacteria, and other eukaryotic organisms. They are commonly used as antimicrobial agent in sunscreen lotions, textile industries, and for water treatment (Jha & Prasad, 2010).

3.2.2 Carbon-based nanoparticles

Carbon nanotubes and fullerenes are two major groups of carbon-based NPs. Fullerenes containing nanomaterials are allotropic forms of carbon that consist of globular hollow cage. They have gained much attention at the commercial level due to their structure, high strength, electrical conductivity, versatility, and electron affinity. Some of the renowned fullerenes are C60 and C70 having 7.114 nm and 7.648 nm diameter, respectively (Astefanei, Núñez, & Galceran, 2015). CNTs are tubular structures with the diameter of 1–2 nm. Based on their structure, they resemble a graphite sheet which is rolled around itself. They are usually synthesized by deposition of metal particles with carbon atoms, vaporized from graphite via electrical arc or laser. CNTs have numerous commercial applications such as supporting agents for organic and inorganic catalysts, fillers, and gas absorbents in environmental remediation, penetration into the cell cytoplasm, and carrier for peptide delivery (Saeed & Khan, 2016).

3.2.3 Ceramic nanoparticles

Ceramic NPs consist of nonmetallic inorganic solids such as silica (SiO_2), alumina (Al_2O_3), zirconia (ZrO_2), and titanium oxide (TiO_2), which are produced by successive heating and cooling process. They are small, at approximately 50 nm. They are found in dense, porous, and amorphous, hollow, and polycrystalline forms. These NPs have wide applications in photo degradation of dyes, imaging, catalysis, photocatalysis, and protein absorption (Thomas, Mishra, & Talegaonkar, 2015).

3.2.4 Polymeric nanoparticles

Polymeric NPs are organic in nature including cellulose, chitosan, gelatin, and alginate. They are structurally present in nanocapsule and nanosphere form having size range of 10–100 nm (Rao & Geckeler, 2011). Polymeric NPs are an excellent vehicle for delivery of various genes, drugs, vaccines, and for diagnostic assays. Polymeric NP-based drug delivery system enables target specific activity of particle since polymeric compounds coated with NPs enhances the amount of drug-loading ability (Bhatia, 2016).

3.2.5 Lipid based nanoparticles

Lipids containing NPs are effectively used in biomedical applications. They are generally present in spherical form with 50–100 nm size. Lipid NPs possess a lipid based solid core and a matrix having soluble lipophilic molecules. Emulsifier and surfactants are used to stabilize the outer covering of these NPs. They are versatile, having good absorption ability, biodegradable and biocompatible. Lipid nanotechnology is getting more attention in designing and synthesis of lipid NPs for variety of applications, e.g., RNA release in cancer therapy, cancer diagnosis, drug vehicle and delivery, as a carrier of protein, peptide, and gene (Gujrati et al., 2014).

3.3 Nanomaterial-microbe interaction and their mechanism

The nanomaterial-microbial interaction is highly affected by multiple factors ranging from type of microbes and NPs to environmental factors, for example, temperature, pH, concentration of biomolecules, and ions. Recent scientific knowledge and discoveries about NP mechanistic strategies have revealed their biological actions through structural and functional alterations in bacteria. Advancements have been made to determine the form, surface area, and dimensions of NPs and their impact on biocidal activities of microorganisms, that is, bacteria, algae, and fungi. Interaction of NPs with microbes differs significantly based on shape, surface area, and physiochemical properties of microbes. Physiochemical variations in the surface of microbes such as hydrophobicity of fungal spores and bacterial cell wall composition not only influence the pathobiology but are also important for the type or strength of interaction with nanomaterials (Westmeier et al., 2018). Typically, small-sized NPs have higher toxic activities than larger ones because of higher surface area to volume ratio. This can significantly increase the reactive oxygen species (ROS) production that leads to inactivate or damage the essential cellular biomolecules such as proteins, DNA, or lipids. In addition, shapes of microbes exemplified by rod verses coccus also affect microbial-nanomaterial interaction. It is demonstrated in recent studies that NPs having positive charge can interact efficiently with microbes possessing negative charge bearing cell surface. A comprehensive study of single gene deletion library in *Escherichia coli* shows that mutation in ubiquinone biosynthesis pathway is highly influenced by cationic polystyrene NPs ($PS-NH_2$). Ubiquinone or Q10 coenzyme is a key component of the electron transport chain, which is responsible for aerobic respiration. These positive nanomaterials induce oxidative stress in bacteria, which quench the function of coenzyme Q10 (Ivask et al., 2012). Detailed mechanisms of toxic activity of metallic nanomaterials on bacteria remain doubtful, nevertheless unique attention have been paid to in vitro study of morphological changes that occur in bacteria in the presence of nanomaterials. Fungal infections have gained significant attention due to rising morbidity and mortality in immunosuppressed patients, with the need for severe disease management (Pfaller & Diekema, 2007). Researchers have observed that when *Candida albicans* were treated with silver NPs, membrane depolarization and cell membrane disruption were observed, which led to the release of intracellular trehalose and glucose in suspension thus inhibiting the *C. albicans* growth (Kim et al., 2009).

3.3.1 Mechanism of action

The exact mechanism for antimicrobial activity of NPs is still being studied. Up to now few possibilities have been hypothesized (Fig. 3–1).

3.3.1.1 Translocation and internalization of nanoparticles

NP exposure to microbial cells can cause membrane damage by adsorption of NPs and sometimes followed by penetration into the cell (Pelletier et al., 2010; Stoimenov, Klinger, Marchin, & Klabunde, 2002). Metals have high affinity toward thiol groups present in cysteine residues of transport and respiratory proteins. Cells uptake the metal ions that interact with the NADH dehydrogenase (enzyme of respiratory chain) subsequently inhibiting the ATP synthesis. Furthermore, it is reported that metals also cause interruption in DNA duplication rate (Yang et al., 2009). Silver NPs are widely studied because of their efficient biocidal properties against bacteria, viruses, and fungi. Bacterial cells targeted with silver NPs endure morphological and biochemical changes in plasma membrane stability, resulting in penetration of NPs into the cytoplasm causing cytoplasm retrenchment, inactivating the enzymes

Chapter 3 • Interaction of nanomaterials with microbes 35

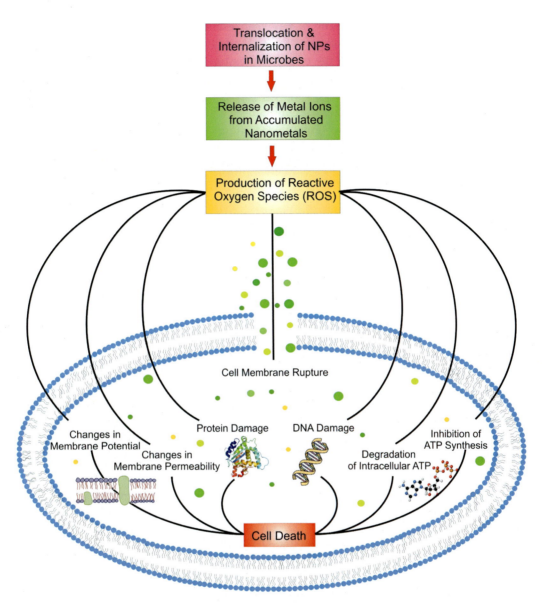

FIGURE 3–1 Mechanistic actions of NPs and response of cells under NPs stress. (1) Translocation and internalization of NPs in microbes; (2) release of metal ions from accumulated nanometals; and (3) production of reactive oxygen species (ROS). These ROS induce damage in cellular structure (proteins and DNA), cause degradation and inhibition of ATP synthesis, alteration in membrane potential, and permeability, which eventually leads to cell death.

and DNA compression, which ultimately is the cause of cell death (Jung et al., 2008). Besides the simple structure and shape of bacteria, they comprise of well-developed cell structure that perform many biological activities. Intracellular delivery of any NPs or solute relies on surface area to volume ratio. Recently, it has been examined that small granular particles either adhere to cell wall or accumulate in the cell. Additionally, accumulation of silver, sulfur, and dense granular particles in the cytoplasm disrupt the membrane of bacteria and aid in the entry of AgNPs into the cell. It also effects the integrity of the cell by intracellular potassium leakage from the cell (Navarro et al., 2008). A possible mechanism of silver NPs interaction might be with phospholipid membrane or thiol groups present in proteins (Lok et al., 2006). Likewise, fungal cell wall inhibits the entrance of AgNPs in cell because it comprises of carbohydrates as major components, which make exclusive and stiff structures. The major constituent of fungal cell wall is chitin, which enables its semipermeability thereby controlling the entry of small particles, whereas they inhibit the transition of larger ones. Due to the presence of AgNPs, pits are created on the surface of cell walls leading to the formation of pores, which subsequently disturb the cell metabolism. Owing to AgNPs, membrane barriers may collapse or alter the membrane potential by the outflow of ions.

3.3.1.2 Generation of reactive oxygen species

Metal ions and NPs generate ROS, which leads to the damaged cellular structures. ROS are reactive species of oxygen that are produced through basic metabolism. Universal intracellular defense mechanisms have developed to cope with this undesired oxygen, to prevent essential biomolecule damage in the cell. A high level of ROS generation under high levels of stress is considered as one of the main mechanisms of action of NPs that inhibit the microbial growth. NPs produce free radicals, that is, peroxide, superoxide (O_2^-), and carbon dioxide radicals (CO_2^-) that lead to cellular damage or cell death (Wang et al., 2014). Presence of higher concentration of oxidative species results in nontoxicity, causing oxidative stress that leads to dysfunction of mitochondria and DNA damage, including single or double DNA strand breakage, sugar and base lesions, protein-DNA cross link (Choi & Hu, 2008; Pathakoti et al., 2013; Valko, Rhodes, Moncol, Izakovic, & Mazur, 2006). Oxidation state of metal in NPs may participate to bactericidal activity, e.g., Cu_2O-NPs have high antibacterial activity than CuO-NPs, representing that oxidation plays an important role in toxicity. When O_2 is utilized to react with Cu_2O and form Cu^{2+}, this may react with superoxide (O_2^-), which leads to continued oxidative stress. This superoxide may reduce Cu^{2+} to Cu^+ and generate H_2O_2, which can react with Cu again making OH^-. High concentration of OH^- have been measured in cells that have been exposed to CuO-NPs than Cu_2O-NPs (Meghana, Kabra, Chakraborty, & Padmavathy, 2015).

3.3.1.3 Leakage of cellular components in response to nanoparticles exposure

NPs have potential to anchor to the cell wall of microbe and penetrate it, which produce pits in the membrane, resulting in structural alterations in the cell membrane such as depolarization, enhance permeability for the liberation of membrane protein, lipopolysaccharides and intracellular components, and degeneration of proteins (McQuillan, Groenaga Infante, Stokes, & Shaw, 2012; Pelletier et al., 2010). Carbon-based NPs have the potential effects of generating mechanical stress on cell membranes. When NPs interact with bacterial cell wall, results in disaggregation of polysaccharide matrix followed by rearrangement into smaller sections. These changes enable the physical association of bacteria and NPs present on bacterial cell surface. Development of deformed or irregular structure of cell wall is accompanied with excretion of intracellular materials (Nair et al., 2009). It has been reported that when bacterial strains *S. aureus* and *E. coli* were treated with Ag^+, both strains underwent lysis. When the damaged cells were observed under transmission electron microscope, it was revealed that cell wall had separated from internal cellular components and electron accumulation were causing the cell lysis (El Badawy et al., 2011; Jung et al., 2008). Similarly, when *Pseudomonas aeruginosa* cells were treated with AgNPs, loss of cell integrity, appearance of cell debris, and cell elongation were also observed that might be possible due to stress environment arresting the cell division (Ramalingam, Parandhaman, & Das, 2016).

3.4 Responses to nanomaterial-microbe interactions

Nanomaterial-microbial interactions play an important role as antimicrobial agents. The responses of NPs against different microorganisms include the release of metal ions, which interact with cell components via different pathways, namely, ROS generation (Simon-Deckers et al., 2009), cell wall damage (McQuillan et al., 2012), formation of pores in cell membrane (Lok et al., 2006), and DNA damage (Devi & Bhimba, 2014), which ultimately inhibit microbial growth. Some responses of nanomaterials against microbes are listed below in Table 3−1.

3.5 Microbial mediated synthesis of nanoparticles

Different methods have been used to synthesize the nanomaterials such as chemical, physical, biological methods, or hybrids of them (Arshad, 2017). Although chemical and physical methods are most common for nonmaterial synthesis but use of toxic chemicals restricts their biomedical applications, mainly in clinical uses (Narayanan & Sakthivel, 2010). Organisms as nanomaterial source could be used as environmentally acceptable nanofactories. NP biosynthesis methods are based on systematic use of organisms like bacteria, yeast, fungi, and algae (Marooufpour, Alizadeh, Hatami, & Lajayer, 2019). Biosynthesis of NPs is considered a cheap, nontoxic method. In biological synthesis no stabilizing agent is required as the biomolecules which are present in microorganisms can not only perform this function but can also regulate properties of NPs like shape and size.

3.5.1 Bacterial and cyanobacterial nanoparticle biosynthesis

Bacteria are present everywhere in the environment and can adjust in various conditions such as oxygenation, temperature, and longer incubation time. Due to their ecofriendly behavior, they can synthesize and reduce the nanomaterials and heavy metals (Arshad, 2017). Bacteria have been employed as nanofactories for synthesis of many NPs, and these nanomaterials have been used in many biomedical applications (Baesman et al., 2007). Silver NPs are synthesized by using *Bacillus brevis* which shows antimicrobial activities against the multidrug resistant *Staphylococcus aureus* and *Salmonella typhi* (Saravanan, Barik, MubarakAli, Prakash, & Pugazhendhi, 2018). *Bacillus* sp. can produce

Table 3–1 Effects/responses of nanoparticle-microbial interaction in bacteria, fungi and algae.

Nanoparticles	Target organism	Responses / effects	References
Responses of NPs on bacteria			
Ag	Escherichia coli Pseudomonas aeruginosa Vibrio cholera Scrub typhus	Membrane disruption and damage of sulfur and phosphorus-containing molecule (DNA)	Morones et al. (2005)
Ag	Escherichia coli	Formation of pores in cell membrane resulting in the cell death	Pal, Tak, and Song (2007)
Ag	Salmonella typhi Pseudomonas aeruginosa Bacillus subtilis Micrococcus luteus	Cell growth inhibition	Bhakya, Muthukrishnan, Sukumaran, and Muthukumar (2016)
Au	Escherichia coli Staphylococcus aureus	Interaction with cell membrane and change in the membrane potential, resulting in the cell lysis	Li et al. (2014)
Au	Escherichia coli	Collapse of membrane potential, inhibition of ATPase activities to decrease the ATP level and hemotaxis	Cui et al. (2012)
CeO_2	Escherichia coli Shewanella oneidensis Bacillus subtilis	Inhibition of cell growth	Pelletier et al. (2010)
Cu	Bacillus subtilis Escherichia coli	Cell membrane disruption	Yoon, Byeon, Park, and Hwang (2007)
ZnO	Escherichia coli	Membrane disruption	Zhang, Jiang, Ding, Povey, and York (2007)
ZnO	Campylobacter jejuni	Disruption of cell membrane, change in the gene expression and development of oxidative stress	Xie, He, Irwin, Jin, and Shi (2011)
ZnO	Staphylococcus aureus Escherichia coli	Cell membrane disruption and generation of reactive oxygen species that leads to cell lysis	Singh, Joyce, Beddow, and Mason (2020)
Responses of NPs on fungi			
Ag	Candida albicans Candida parapsilosis Aspergillus niger	Membrane disruption and inhibition of sulfur and phosphorus-containing compound (DNA)	Devi and Bhimba (2014)
Ag	Candida albicans Saccharomyces cerevisiae	Cell membrane disruption	Nasrollahi, Pourshamsian, and Mansourkiaee (2011)

(Continued)

Table 3–1 (Continued)

Nanoparticles	Target organism	Responses / effects	References
Ag	Cryptococcus neoformans Candida sp.	Disruption of cell wall and cytoplasmic membrane, loss of the cytoplasmic contents	Ishida et al. (2014)
Ag	Aspergillus flavus Aspergillus nomius Aspergillus parasiticus Aspergillus ochraceus Aspergillus melleus	Cell growth inhibition	Bocate et al. (2019)
Al_2O_3	Candida sp. Candida dubliniensis Candida albicans	Penetration in membrane and disruption of morphological as well as physiological activities that lead to cell death	Jalal et al. (2016)
Au-Ag (bimetallic NPs)	Candida parapsilosis Candida krusei Candida glabrata Candida guilliermondii Candida albicans	Growth inhibition	Gutiérrez et al. (2018)
CuO	Aspergillus niger Rhizopus oryzae Aspergillus flavus Cladosporium carrionii Mucor sp. Penicillium notatum Alternaria alternata	Cell growth inhibition	Mageshwari and Sathyamoorthy (2013)
ZnO	Fusarium oxysporum Penicillium expansum	Cell growth inhibition	Yehia and Ahmed (2013)
Responses of NPs on algae			
Ag	Uronema confervicolum	Generation of ROS that leads to cell death	Gonzalez, Fernandez-Rojo, Leflaive, Pokrovsky, and Rols (2016)
Al_2O_3	Chlorella ellipsoidea	Cell wall damage and increase in the ROS production	Pakrashi et al. (2013)
CdSe	Chlorella vulgaris	Changes in the enzymatic activity and cytotoxicity	Movafeghi, Khataee, Rezaee, Kosari-Nasab, and Tarrahi (2019)
ZnO	Chlorococcum sp.	Increase in ROS production, inhibition of the photosynthetic electron transport, and reduction in cell growth	Oukarroum, Halimi, and Siaj (2019)

AgNPs in periplasmic space of bacterial cells (Pugazhenthiran et al., 2009). *Lactobacillus plantarum* (Gram-positive bacteria) can biosynthesize ZnO NPs. *Halomonas elongata* has been used to synthesize triangular CuO-NPs, having antibacterial activity against *S. aureus* and *E. coli* (Rad, Taran, & Alavi, 2018; Selvarajan & Mohanasrinivasan, 2013). *Bacillus cereus* can produce super paramagnetic iron oxide NPs (29 nm), which have anticancer activity against the MCF-7 and 3T3 cell lines (Fatemi, Mollania, Momeni-Moghaddam, & Sadeghifar, 2018). Several studies have shown that cyanobacteria can also be used for the biosynthesis of NPs (Lengke, Fleet, & Southam, 2007). Marine cyanobacterium *Phormidium fragile* can biosynthesize monodispersed AgNPs (5–6.5 nm) (Satapathy & Shukla, 2017). In addition to inorganic NPs, some bacterial genera can also biosynthesize organic nanomaterials, e.g., *Gluconacetobacter*, an aerobic acetic bacteria biosynthesized three-dimensional (3D) network of cellulose nanofibrils (Golmohammadi, Morales-Narvaez, Naghdi, & Merkoci, 2017).

3.5.2 Yeast and mold nanoparticle biosynthesis

"Myconanotechnology" is a term used to describe the biosynthesis of NPs using fungi (Grasso, Zane, & Dragone, 2020). Fungi can effectively secrete the extracellular enzymes while their biomass is another green approach to synthesis nanomaterials (Moghaddam et al., 2015). Molds have characteristic advantages for nanomaterial biosynthesis as compared to bacteria: (1) high extracellular NPs synthesis, (2) uniform growth and easy culturing, (3) high metal tolerance, and (4) higher metal binding and uptake abilities (Guilger-Casagrande & de Lima, 2019). *Aureobasidium pullulans* and *Fusarium oxysporum* have been used to synthesize AuNPs. The biosynthesis of AuNPs was occurred in vacuole and reducing sugars were participating to form tailored spherical AuNPs (Zhang, He, Wang, & Yang, 2011). *Rhizopus stolonifer* extracts mediated biosynthesis of monodispersed AgNPs (2.86 nm) have also been reported (AbdelRahim et al., 2017). Zinc oxide and cobalt oxide NPs are also produced by *Aspergillus niger* and *Aspergillus nidulans* (Kalpana et al., 2018; Vijayanandan & Balakrishnan, 2018). Yeasts also have ability to produce the semiconductor nanomaterials. *Saccharomyces cerevisiae* has been reported to biosynthesize the biocompatible cadmium telluride quantum dots (Luo et al., 2014) and Au-Ag alloy NPs (Wei, 2017; Zheng, Hu, Gan, Dang, & Hu, 2010).

3.5.3 Algae based nanoparticle biosynthesis

Algae are aquatic microorganisms that have great applications in nanotechnology (Dahoumane, Mechouet, Alvarez, Agathos, & Jeffryes, 2016). Algae belonging to *Tetraselmis, Scenedesmus*, and *Desmodesmus* genera have been used for the biosynthesis of metal NPs with antimicrobial activities (Jena et al., 2014; Öztürk, 2019; Senapati, Syed, Moeez, Kumar, & Ahmad, 2012). Mechanisms behind these algae-based nanomaterials include nucleation phenomena, control of dimension, and stabilization of the NP structure mediated by reducing agents (Jena et al., 2014), enzymes of cytoplasmic membrane (Senapati et al., 2012), and biomolecules like polysaccharides, proteins, polyphenols, and phenolic compounds (Öztürk, 2019). *Sargassum muticum* was used for ZnO NPs biosynthesis and has been reported for the reduction of angiogenesis along with apoptotic effects in HepG2 cells (Sanaeimehr, Javadi, & Namvar, 2018). *Gelidium amansii* have been reported to synthesize AgNPs with antimicrobial activities (Ovais et al., 2017). *Sargassum crassifolium, S. ilicifolium* and *Cystoseira trinodis* have also been used for biosynthesis of Au, CuO, and Al_2O_3 NPs, respectively (Gu et al., 2018; Koopi & Buazar, 2018; Maceda et al., 2018).

3.6 Mechanisms of microbial mediated nanoparticles biosynthesis

Microbes like bacteria, yeast, fungi, and algae are considered as biofactories for the reduction of gold, silver, gold-silver alloy, selenium, silica, cadmium, platinum, magnetite, and various other metals to their subsequent nanomaterials for potential applications (Fig. 3–2). Microbes can synthesize NPs extracellularly as well as intracellularly (Li, Xu, Chen, & Chen, 2011).

3.7 Role of bioreducing agents in nanoparticle biosynthesis

Microbial enzymes, proteins, redox mediators, and exopolysaccharides play a significant role as reducing agents for nanomaterial synthesis (Durán et al., 2011; Tyagi, Tyagi, Gola, Chauhan, & Bharti, 2019). Cofactors like nicotinamide adenine dinucleotide (NADH) and reduced form of nicotinamide adenine dinucleotide phosphate (NADPH)-dependent enzymes play an important role as reducing agents via the transfer of the electron from

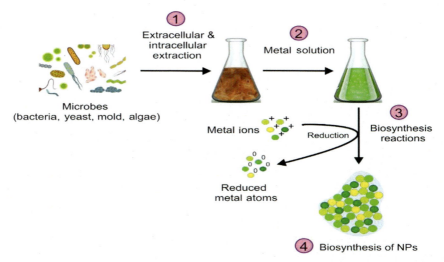

FIGURE 3–2 Mechanistic NP biosynthesis using ecofriendly microbial nanofactories. (1) Growth of microbial strain (bacteria, yeast, fungi, and algae) on a suitable medium followed by NPs extraction either extracellularly (from supernatant) or intracellularly (from pellet). (2) Addition of metal ion solution. (3) Biosynthesis reactions. (4) Change in the color of reaction mixture due to formation of NPs.

NADH by NADH-dependent enzymes, which act as electron carriers (Mukherjee et al., 2018). Cell surface proteins and conductive pili have role in electron transfer thereby resulting in extracellular metal ion reduction (Cologgi, Lampa-Pastirk, Speers, Kelly, & Reguera, 2011; Kitching et al., 2016; Reguera, 2012), e.g., biomineralization of uranium metal ions to uranium NPs through *Geobacter sulfurreducens* (Cologgi et al., 2011). In uranium reduction phenomena of electron transport, pili proteins in *Geobacter* sp. play an important role in uranium immobilization, organization of c-cytochrome, and extracellular uranium reduction (Reguera, 2012). Exopolysaccharides (EPS) have also been considered as NPs producing agents because they have ability to reduce metal ions to synthesize NPs and stabilize them as capping agents (Gahlawat et al., 2016). During silver NPs synthesis in *E. coli* biofilm, hemiacetal and aldehyde groups of EPS acted as bioreducing agents. Nuclear magnetic resonance and Fourier-transform infrared spectroscopy results of metal ions with EPS interaction clearly showed that rhamnose sugar (hemiacetal group) is involved in silver NPs synthesis. In addition, aldehyde groups in pyranose and rhamnose sugars are oxidized to carboxyl groups by sliver ions (Kang, Alvarez, & Zhu, 2014). Redox mediators such as ubiquinol, oxygen/superoxide, and c-type cytochrome redox proteins can also participate in electron transfer and subsequent formation of NPs (Bewley, Ellis, Firer-Sherwood, & Elliott, 2013).

3.8 Biosynthesis pathways

The pathway of biosynthesis are described in the following sections.

3.8.1 Extracellular biosynthesis

Extracellular biosynthesis (Fig. 3–3) involves trapping metal ions on cell surface and reduction of ions in the presence of enzymes (Li et al., 2011). This pathway involves enzyme-mediated biosynthesis which are found on cell membrane or release of enzymes to growth media as extracellular enzymes. Nitrate reductase is an enzyme that catalyze the conversion of nitrate to nitrite. For example, Zn^{2+} bioreduction is initiated by electron transfer from NADH by NADH-dependent reductase that acts as electron carrier. As a result, Zn^{2+} get electron and reduced to Zn^0 and subsequently ZnONPs are synthesized (Shamim, Abid, & Mahmood, 2019). Extracellular synthesis of gold (Au) NPs by the bacterium *Rhodopseudomonas capsulata* is mediated via the secretion of NADH and NADH-dependent enzymes. Au ions accept the electrons and reduced Au^{3+} to Au^0, and thereby, gold NPs are formed (He et al., 2007). Another study reported that NADH-dependent enzymes are involved in Ag^+ ions reduction and formation of Ag NPs (Ingle, Gade, Pierrat, Sonnichsen, & Rai, 2008). Extracellular reductase enzyme produced by *F. oxysporum* causes Au^{3+} and Ag^{1+} reduction to Au-Ag alloy NPs. In an in vitro study, nitrate reductase from *F. oxysporum* in the presences of protein (phytochelatin), electron carrier (4-hydroxyquinoline), and cofactor (NADPH) forms AgNPs (Senapati, Ahmad, Khan, Sastry, & Kumar, 2005). *Aspergillus fumigatus* and *Penicillium fellutanum* can produce AgNPs extracellularly (Bhainsa & D'souza, 2006; Kathiresan, Manivannan, Nabeel, & Dhivya, 2009).

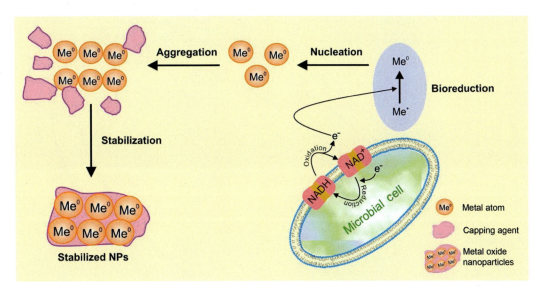

FIGURE 3–3 Schematic diagram of extracellular biosynthesis of NPs. Extracellular biosynthesis involves enzyme-mediated synthesis (nitrate reductase enzyme), which is secreted in growth media. Nitrate reductase reduce metal ions (Me$^+$) to its respective metal atoms (Me^0), leading to nucleation and growth of NPs. Extracellular proteins secreted by microbes act as capping agents for NPs stabilization.

Algae, *Sargassum wightii* has also been reported for its rapid production of AuNPs to reduce Au^{3+} (Singaravelu, Arockiamary, Kumar, & Govindaraju, 2007).

3.8.2 Intracellular biosynthesis

In intracellular biosynthesis (Fig. 3–4), intracellular enzymes and positively charged groups are utilized in taking metallic ions from medium and subsequent enzymatic reduction within the cells (Thakkar, Mhatre, & Parikh, 2010). A microscopic analysis has shown that nanomaterials are accumulated in/on cytoplasmic membrane, periplasmic space, and cell wall due to diffusion of metal ions across membrane (Ovais et al., 2018). Microorganisms can trap or bind ions on cell wall through electrostatic interactions, i.e., metal ions are attracted to negative charges from carboxylate groups (enzymes, polypeptides, and cysteine residues) that are present on cell wall (Selvarajan & Mohanasrinivasan, 2013; Zhang, Yan, Tyagi, & Surampalli, 2011). Trapped ions are reduced to elemental atoms initiated by electron transfer from NADH by NADH-dependent reductase present on cell membrane that act as electron carrier. Peptides, proteins, and amino acids such as cysteine, tryptophan, and tyrosine are present inside the cells, which are responsible for the stabilization of NPs (Yusof, Mohamad, & Zaidan, 2019). Mukherjee et al. (2001) has explained intracellular mechanism of NPs synthesis by using *Verticillium* sp. This is a three-step mechanism that involves trapping, bioreduction, and capping/stabilization. Interaction forces between metal ions and enzymes present on cell wall reduce metal ions in cell wall leading to accumulation of metal atoms and subsequent metallic NPs formation. Transmission electron microscope analysis shows that NPs are formed in cell wall, cytoplasmic membrane, as well as in cytoplasm. AuNPs

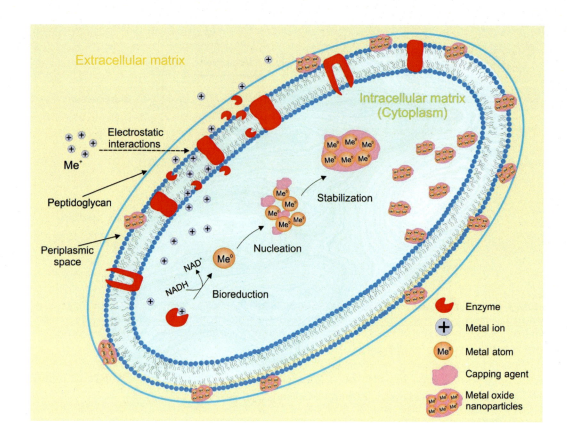

FIGURE 3–4 Intracellular biosynthesis mechanism of metallic (Me^0) NPs. Intracellular synthesis involves metal (Me$^+$) ions transportation across the cell by electrostatic attractions, Me$^+$ are reduced to metal atom (Me^0) by enzymes and generated metallic nuclei grow to form NPs in cytoplasm and may transport to periplasmic space/cell wall after stabilization.

(5–15 nm) synthesized intracellularly by *Rhodococcus* sp. were present on cell wall and on cytoplasmic membrane (Ahmad et al., 2003). Similarly, *Plectonema boryanum*, a filamentous cyanobacterium treated with AuCl$_4$ and Au(S$_2$O$_3$)$_2^{3-}$ solutions resulted in AuNPs biosynthesis at the membrane, while gold sulfide residing intracellularly (Lengke et al., 2007). Au^{3+} and Ag$^+$ aqueous solutions can be reduced by intracellular enzymes of *Brevibacterium casei* and were converted into spherically shaped AuNPs and AgNPs, respectively (Kalishwaralal et al., 2010). Some of the extracellular and intracellular NPs biosynthesis examples are summarized in Table 3–2.

3.9 Applications of nanomaterials

Nanomaterials have contributed significantly to the production and development of products with new and valuable properties in many segments like agriculture, feed, food, health, environment, and so forth (Cushen, Kerry, Morris, Cruz-Romero, & Cummins, 2012; Peters et al., 2014).

Table 3–2 Nanoparticle biosynthesis by using microorganisms.

Microorganism	Nanoparticles	Shape	Size (nm)	Cellular location	References
Bacterial species					
Streptomyces griseoplanus	Ag	Spherical	19.5–20.9	Extracellular	Vijayabharathi, Sathya, and Gopalakrishnan (2018)
Mycobacterium sp.	Au	Spherical	5–55	Extracellular	Camas, Camas, and Kyeremeh (2018)
Bacillus brevis	Ag	Spherical	41–68	Extracellular	Saravanan et al. (2018)
Pantoea ananatis	Ag	Spherical	8.06–91.32	Extracellular	Monowar, Rahman, Bhore, Raju, and Sathasivam (2018)
Alcaligenes sp.	Ag	Spherical	30–50	Extracellular and Intracellular	Divya, Kiran, Hassan, and Selvin (2019)
Bacillus thuringiensis	Ag	Spherical	42	Extracellular and Intracellular	Khaleghi, Khorrami, and Ravan (2019)
Staphylococcus aureus	Se	Spherical	180.1 ± 20.4	Intracellular	Cruz, Mi, and Webster (2018)
Methicillin-resistant Staphylococcus aureus (MRSA)		Spherical/ Rod like	121.4 ± 26.6		
Escherichia coli		Spherical	120.0 ± 15.85		
Pseudomonas aeruginosa		Spherical	171 ± 29.05		
Halomonas elongata	TiO_2 and ZnO	Spherical	104.63 ± 27.75 / 18.11 ± 8.93	Extracellular	Taran, Rad, and Alavi (2018)
Cupriavidus sp.	Ag	Spherical	10–50	Extracellular	Ameen et al. (2020)
Bacillus cereus	Ag	Spherical	18–39	Extracellular	Ahmed et al. (2020)
Bacillus sp.	Se	Spherical	20–50	Extracellular	Bharathi et al. (2020)
Pseudomonas poae	Ag	Spherical	19.8–44.9	Extracellular	Ibrahim et al. (2020)
Lactobacillus sp.	Ag	Spherical	13.84 ± 4.56	Extracellular	Matei, Matei, Matei, Cogălniceanu, and Cornea (2020)
Fungal species					
Aspergillus nidulans	CoO (cobaltous oxide)	Spherical	20.29	Extracellular	Vijayanandan and Balakrishnan (2018)
Trichoderma longibrachiatum	Ag	Monodispersed Spherical	10	Extracellular	Elamawi, Al-Harbi, and Hendi (2018)
Cladosporium cladosporioides	Ag	Spherical	30–60	Extracellular	Hulikere and Joshi (2019)
Phomopsis liquidambaris	Ag	Polydispersed Spherical	18.7	Extracellular	Seetharaman et al. (2018)
Trichoderma asperellum	FeO	Spherical	25 ± 3.94	Extracellular	Mahanty et al. (2019)
Phialemoniopsis ocularis			13.13 ± 4.32		
Fusarium incarnatum			30.56 ± 8.68		
Macrophomina phaseolina	Ag/AgCl	Spherical	5–30	Extracellular	Spagnoletti, Spedalieri, Kronberg, and Giacometti (2019)

(Continued)

Table 3–2 (Continued)

Microorganism	Nanoparticles	Shape	Size (nm)	Cellular location	References
Alternaria tenuissima	ZnO	Spherical	15.45	Extracellular	Abdelhakim, El-Sayed, and Rashidi (2020)
Algal species					
Ulva fasciata	Ag	Spherical	8–60	Extracellular	Negm et al. (2018)
Laurencia catarinensis	Ag	Spherical, Triangular, Rectangle, Polyhedral and Hexagonal	39.41–77.71	Extracellular and intracellular	Abdel-Raouf, Alharbi, Al-Enazi, Alkhulaifi, and Ibraheem (2018)
Padina pavonica	Ag	Spherical, Triangular, Rectangle, Polyhedral and Hexagonal	49.58–86.37	Extracellular and intracellular	Abdel-Raouf, Al-Enazi, Ibraheem, Alharbi, and Alkhulaifi (2019)
Colpomenia sinuosa	Fe_3O_4	Nanospheres	11.24–33.71	Extracellular	Salem, Ismail, and Aly-Eldeen (2019)
Pterocladia capillacea			16.85–22.47		
Chlorella sp.	ZnO	Hexagonal	20 ± 2.2	Extracellular	Khalafi, Buazar, and Ghanemi (2019)
Amphiroa rigida	Ag	Spherical	20–30	Extracellular	Gopu et al. (2020)
Sargassum cymosum	Au	Spherical	7–20	Extracellular	Costa et al. (2020)

From many years, efforts have been made to control plant diseases by using organic and inorganic salts (Talibi, Askarne, Boubaker, Boudyach, & Aoumar, 2011). Use of nanomaterials in plant-disease management is attractive and quite effective approach. Among different ways of using this approach in plant-disease management, one of the simplest ways is the direct application of nanomaterials on foliage or seeds under the soil to protect the plants from the attack of pathogen. Nanomaterials can suppress the pathogens like chemical pesticides (Khan & Rizvi, 2014). Moreover, nanomaterials are also being used as a carrier for herbicides, fungicides, insecticides, and dsRNA for RNAi-mediated protection in plants (Worrall, Hamid, Mody, Mitter, & Pappu, 2018).

Nanomaterials can also be used to reduce toxicity. The working efficiency of microorganisms in degradation of waste can be increased by using nanomaterials as they play an important role in increasing the surface area and lowering the activation energy of microorganisms, thereby reduce the toxicity of pollutants to microorganisms. They also reduce the overall time and cost of remediation (Azubuike, Chikere, & Okpokwasili, 2016).

Nanomedicine is a developing and life-saving field of research. It is playing an important role in the advanced diagnosis and treatment of human diseases (Fadeel & Garcia-Bennett, 2010). Nanomaterials are being used for fluorescent biological labels, biodetection of pathogens, image contrast enhancements, as a gene or drug delivery agent, for bone treatments, in cancer treatments, to study phagokinetics, for hyperthermia (destruction of tumor through heating), tissue engineering, and more (Li et al., 2011).

The detection of cancer at its initial stage is the critical step in the treatment of cancer. For such purposes, researchers are discovering innovative tools at the nanoscale, which are performing very effective roles in cancer treatment (Wee et al., 2005), for example, identification of those properties of cancerous cells that differentiate them from closely related nonpathogenic cells (Gourley et al., 2005). At cellular level, the cancer is being detected by quantum dots and nanoscale cantilever. If the cancer is not detected at the initial stage, separation of cancerous cells from the healthy cells has to be carried out to prevent its spread (Zeenia et al., 2003). Drug delivery systems based on NPs can be characterized for their localization in tumor cells by coating them with tumor specific antibodies, hormones, peptides, sugars, anticarcinogenic drugs, and so on (Subramani, 2006), thereby making possible the effective identification of cancerous cells. Nanobiosensors are also being used for the identification of a particular type of cells or areas in a body. This identification is based on factors like difference in size, concentration, temperature, volume, displacement, and so forth of a particular cell (like cancer cells) and the normal cells. At molecular level, nanobiosensors are used to deliver medicines as well as monitor the development of the disease management (Fan et al., 2005).

E. coli, *B. cereus*, and *S. aureus* are the most commonly known food borne pathogens. Different assays are being used to identify the toxins, but they are mostly unable to detect the activities of the toxins (Singh et al., 2019). Rapid detection techniques for toxins using nanomaterials have been developed, e.g., for the enterotoxins of *S. aureus*. The techniques are rapid, more efficient, easy, and have great potential as compared to traditional methods for the detection of pathogens (Wu et al., 2016).

In food storage and packaging, gold nanomaterials are being used (Ahmed, Khan, Anwar, Ali, & Shah, 2016). They are also being used as a biosensor for the detection of contaminants in the food products (Bajpai et al., 2018). Silver nanomaterials are commonly used as a health supplement, antiodorant, and antimicrobial agent in food storage applications. the US Food and Drug Administration allowed the use of silver as silver nitrate (at concentration <17 g/kg) for killing and preventing the growth of bacteria in bottled water in 2009 (Duncan, 2011). It has also been reported that reduction of bacterial growth and increase of food shelf life can be achieved by using them in minced meat (Mahdi, Vadood, & Nourdahr, 2012). A few additional industrial applications of nanomaterials have been summarized in Table 3–3.

3.10 Challenges pertaining to the applications of nanomaterials

Though the ultrasmall size of nanomaterials makes them useful in many fields of life, the same feature is the cause of many adverse effects on humans, plants, animals, and the environment. When the airborne nanofertilizers, nanopesticides, or nanoformulations deposit on leaf or flower, it may clog the stomata opening. This deposition may also form a toxic barrier layer on the surface of the stigma, which results in the prevention of pollen germination and tube germination into the stigma. The translocation of minerals, water, and photosynthates may also be affected by nanomaterials if they enter the vascular tissues. The nanomaterials

Table 3–3 Industrial applications of nanomaterials.

Industry	Application	Nanomaterial used	Beneficial effects	References
Food	Packaging	Silver	Inhibition of the growth of aerobic bacteria	Toker, Kayaman-Apohan, and Kahraman (2013)
		Silver, Zinc oxide	Inhibition of the growth of yeast, and bacteria	Emamifar, Kadivar, Shahedi, and Soleimanian-Zad (2010)
	Food additive	Titanium dioxide	Used to enhance the white color of dairy products, candies, etc.	Weir, Westerhoff, Fabricius, Hristovski, and Von Goetz (2012)
		Iron oxide	Used as a food colorant (in the form of iron oxide)	Zimmermann and Hilty (2011)
Agriculture	Plant protection and production	Silver	Control of wilting and lentil pathogens	Ashrafi, Rastegar, Jafarpour, and Kumar (2010)
		Porous hollow silica nanoparticles	Controls delivery system for water-soluble pesticide	Khot, Sankaran, Maja, Ehsani, and Schuster (2012)
	Food	Gold	Detection of heavy metal contamination in sea-food; detection of methyl paraoxon, carbofuran and phoxim pesticides	Yin, Ai, Xu, Shi, and Zhu (2009)
	Storage	Silver	Antimicrobial agent for the storage of food and beverage	Dasgupta et al. (2016)
		Silver	Improving the shelf life of vegetables and fruits	Fernández, Picouet, and Lloret (2010)
Health and medicines	Tracking stem cells	Super paramagnetic iron oxide	Used as a probe for noninvasive cell tracking	Li et al. (2013)
	Diagnosis	Gold	Drug delivery and biomarking of drug-resistant leukemia cells	Li et al. (2007)
		Silver	Diagnosis and detection of cancer	Wu et al. (2008)
	Antimicrobial activity	Silver	Inhibits the growth of *E. coli* and *S. aureus*	Jiang, Qin, Guo, and Zhang (2010)
		Zinc oxide (ZnO)	Inhibits the growth of *E. coli* and *M. luteus*	Farouk, Moussa, Ulbricht, Schollmeyer, and Textor (2014)
Environment	Water treatment	Titanium dioxide (TiO_2)	Removal of organic contaminants from different media	Ali Mansoori, Bastami, Ahmadpour, and Eshaghi (2008)
	Degradation	Iron, zinc, and tin	Degradation of halogenated organic compound (HOCs)	Ali Mansoori et al. (2008)

(Continued)

Table 3–3 (Continued)

Industry	Application	Nanomaterial used	Beneficial effects	References
Textile	Fiber strength	Carbon nanofibers	Increase the tensile strength of composite fibers	Patra and Gouda (2013)
		Carbon black nanoparticles	Improve the abrasion resistance and toughness of composite fibers	Patra and Gouda (2013)
		Clay nanoparticles (nanoflakes)	Provide electrical, heat and chemical resistance; anti-UV and anti-corrosive effect to composite fibers	Patra and Gouda (2013)

may also get inhaled by human beings and animals so deep into the lungs that they may cause severe illness. These small particles may enter the bloodstream and may assemble in kidneys or lungs resulting in proinflammatory effects, which lead to protein fibrillation, inflammation, and genotoxicity induction. Because of longer persistence, greater reactivity, and magnified transport of nanopesticides, they may generate new types of contaminations in soil and water bodies (Khan & Rizvi, 2014). Another challenge pertaining to nanomaterials is agglomeration, i.e., ability of nanomaterials to bind together both in wet as well as dry state through van der Waal's forces (Powers, Palazuelos, Moudgil, & Roberts, 2007). The process of agglomeration results in bringing the NPs out of the nanorange by increasing the size and decreasing the surface area, hence making the safety of novel nanomaterials questionable as they are no longer in the nanorange (Balbus et al., 2007).

3.11 Conclusion and future perspectives

Bionanotechnology has received great interest in recent years owing to the synthesis of particles that are different from their bulk materials with required features. Microorganisms (bacteria, yeast, fungi, and algae) have been used for the biosynthesis of NPs. Microbes provide the best platform for biosynthesis of NPs in a cost-effective and ecofriendly way. Primary tasks for involvement of microorganisms in NP synthesis is the selection of strains, optimization of intrinsic characters of microbes such as replication, growth rate, and biochemical activities. Control of size, shape, and monodispersity of synthesized NPs is another important challenge for obtaining the desired effects of NPs. Moreover, maintaining the optimal conditions necessary for growth and enzymatic activities of microbes is a crucial challenge for NPs biosynthesis. Interaction between microorganisms and NPs are generally related to be toxic and have been exploited for antimicrobial applications. Inhibition mechanism of NPs against microorganisms include release of metal ions that interact with components of cell through different pathways such as formation of pores in plasma membrane, formation of ROS, cell wall damage, damage of DNA, and cell cycle arrest, which ultimately stop the cell growth.

However, microorganisms can reduce the toxic effects of NPs and their beneficial interactions can be exploited for useful applications in food, medicine, agriculture, and textile industries. NP-microbial interactions hold importance in many ways: (1) discharge of NPs in soil and water affect microbial variety; and (2) NP antimicrobial activity could be used for applications in agriculture, food, and medicine. However, NP interactions with microbial DNA, membrane proteins, and inside the microbial cells need to be deeply studied. The biosynthesis of NPs is still in the development stage and need more efforts to understand the characteristic mechanisms, behavior, biosynthesis, and interaction of NPs with microbes.

Acknowledgment

All the authors are gratefully acknowledged for their contribution.

References

Abdelhakim, H. K., El-Sayed, E., & Rashidi, F. B. (2020). Biosynthesis of zinc oxide nanoparticles with antimicrobial, anticancer, antioxidant and photocatalytic activities by the endophytic *Alternaria tenuissima*. *Journal of Applied Microbiology, 128*(6), 1634–1646.

AbdelRahim, K., Mahmoud, S. Y., Ali, A. M., Almaary, K. S., Mustafa, A. E.-Z. M. A., & Husseiny, S. M. (2017). Extracellular biosynthesis of silver nanoparticles using *Rhizopus stolonifer*. *Saudi Journal of Biological Sciences, 24*(1), 208–216.

Abdel-Raouf, N., Al-Enazi, N. M., Ibraheem, I. B. M., Alharbi, R. M., & Alkhulaifi, M. M. (2019). Biosynthesis of silver nanoparticles by using of the marine brown alga *Padina pavonia* and their characterization. *Saudi Journal of Biological Sciences, 26*(6), 1207–1215.

Abdel-Raouf, N., Alharbi, R. M., Al-Enazi, N. M., Alkhulaifi, M. M., & Ibraheem, I. B. M. (2018). Rapid biosynthesis of silver nanoparticles using the marine red alga *Laurencia catarinensis* and their characterization. *BENI-SEUF University Journal of Basic and Applied Sciences, 7*(1), 150–157.

Ahmad, A., Senapati, S., Khan, M. I., Kumar, R., Ramani, R., Srinivas, V., & Sastry, M. (2003). Intracellular synthesis of gold nanoparticles by a novel alkalotolerant actinomycete, *Rhodococcus* species. *Nanotechnology, 14*(7), 824.

Ahmed, A., Khan, A. K., Anwar, A., Ali, S. A., & Shah, M. R. (2016). Biofilm inhibitory effect of chlorhexidine conjugated gold nanoparticles against *Klebsiella pneumoniae*. *Microbial Pathogenesis, 98*, 50–56.

Ahmed, T., Shahid, M., Noman, M., Niazi, M. B. K., Mahmood, F., Manzoor, I., ... Yan, C. (2020). Silver nanoparticles synthesized by using *Bacillus cereus* SZT1 ameliorated the damage of bacterial leaf blight pathogen in rice. *Pathogens, 9*(3), 160.

Ali Mansoori, G., Bastami, T. R., Ahmadpour, A., & Eshaghi, Z. (2008). Environmental application of nanotechnology. *Annual review of nano research* (Vol. 2, pp. 439–493). World Scientific.

Ameen, F., AlYahya, S., Govarthanan, M., ALjahdali, N., Al-Enazi, N., Alsamhary, K., ... Alharbi, S. (2020). Soil bacteria *Cupriavidus* sp. mediates the extracellular synthesis of antibacterial silver nanoparticles. *Journal of Molecular Structure, 1202*, 127233.

Amin, R., Hwang, S., & Park, S. H. (2011). Nanobiotechnology: An interface between nanotechnology and biotechnology. *Nanotechnology, 6*(02), 101–111.

Arshad, A. (2017). Bacterial synthesis and applications of nanoparticles. *Nanoscience and Nanotechnology, 11*(2), 119.

Ashrafi, S., Rastegar, M., Jafarpour, B., Kumar, S. (2010). Possibility use of silver nano particle for controlling Fusarium wilting in plant pathology. In *Paper presented at the symposium of international conference on food and agricultural applications of nanotechnologies*, São Pedro SP, Brazil.

Astefanei, A., Núñez, O., & Galceran, M. T. (2015). Characterisation and determination of fullerenes: A critical review. *Analytica Chimica Acta, 882*, 1–21.

Azubuike, C. C., Chikere, C. B., & Okpokwasili, G. C. (2016). Bioremediation techniques-classification based on site of application: Principles, advantages, limitations and prospects. *World Journal of Microbiology and Biotechnology, 32*(11), 180.

Baesman, S. M., Bullen, T. D., Dewald, J., Zhang, D., Curran, S., Islam, F. S., ... Oremland, R. S. (2007). Formation of tellurium nanocrystals during anaerobic growth of bacteria that use Te oxyanions as respiratory electron acceptors. *Applied and Environmental Microbiology, 73*(7), 2135–2143.

Bajpai, V. K., Kamle, M., Shukla, S., Mahato, D. K., Chandra, P., Hwang, S. K., ... Han, Y.-K. (2018). Prospects of using nanotechnology for food preservation, safety, and security. *Journal of Food and Drug Analysis, 26*(4), 1201–1214.

Balbus, J. M., Maynard, A. D., Colvin, V. L., Castranova, V., Daston, G. P., Denison, R. A., ... Kulinowski, K. M. (2007). Meeting report: Hazard assessment for nanoparticles-report from an interdisciplinary workshop. *Environmental Health Perspectives, 115*(11), 1654–1659.

Bewley, K. D., Ellis, K. E., Firer-Sherwood, M. A., & Elliott, S. J. (2013). Multi-heme proteins: Nature's electronic multi-purpose tool. *Biochimica et Biophysica Acta (BBA)-Bioenergetics, 1827*(8–9), 938–948.

Bhainsa, K. C., & D'souza, S. (2006). Extracellular biosynthesis of silver nanoparticles using the fungus *Aspergillus fumigatus*. *Colloids and Surfaces, B, 47*(2), 160–164.

Bhakya, S., Muthukrishnan, S., Sukumaran, M., & Muthukumar, M. (2016). Biogenic synthesis of silver nanoparticles and their antioxidant and antibacterial activity. *Applied Nanoscience, 6*(5), 755–766.

Bharathi, S., Kumaran, S., Suresh, G., Ramesh, M., Thangamani, V., & Pughazhventhan, S. (2020). Extracellular synthesis of nanoselenium from fresh water bacteria *Bacillus* sp., and its validation of antibacterial and cytotoxic potential. *Biocatalysis and Agricultural Biotechnology*, 101655.

Bhatia, S. (2016). *Nanoparticles types, classification, characterization, fabrication methods and drug delivery applications. Natural polymer drug delivery systems* (pp. 33–93). Springer.

Bocate, K. P., Reis, G. F., de Souza, P. C., Junior, A. G. O., Durán, N., Nakazato, G., ... Panagio, L. A. (2019). Antifungal activity of silver nanoparticles and simvastatin against toxigenic species of *Aspergillus*. *International Journal of Food Microbiology, 291*, 79–86.

Camas, M., Camas, A. S., & Kyeremeh, K. (2018). Extracellular synthesis and characterization of gold nanoparticles using *Mycobacterium* sp. BRS2A-AR2 isolated from the aerial roots of the Ghanaian mangrove plant, *Rhizophora racemosa*. *Indian Journal of Microbiology, 58*(2), 214–221.

Choi, O., & Hu, Z. (2008). Size dependent and reactive oxygen species related nanosilver toxicity to nitrifying bacteria. *Environmental Science & Technology, 42*(12), 4583–4588.

Cologgi, D. L., Lampa-Pastirk, S., Speers, A. M., Kelly, S. D., & Reguera, G. (2011). Extracellular reduction of uranium via *Geobacter* conductive pili as a protective cellular mechanism. *Proceedings of the National Academy of Sciences, 108*(37), 15248–15252.

Costa, L., Hemmer, J., Wanderlind, E., Gerlach, O., Santos, A., Tamanaha, M., ... Radetski, C. (2020). Green synthesis of gold nanoparticles obtained from algae *Sargassum cymosum*: Optimization, characterization and stability. *BioNanoScience, 10*(4), 1049–1062.

Cruz, D. M., Mi, G., & Webster, T. J. (2018). Synthesis and characterization of biogenic selenium nanoparticles with antimicrobial properties made by *Staphylococcus aureus*, methicillin-resistant *Staphylococcus aureus* (MRSA), *Escherichia coli*, and *Pseudomonas aeruginosa*. *Journal of Biomedical Materials Research, 106*(5), 1400–1412.

Cui, Y., Zhao, Y., Tian, Y., Zhang, W., Lü, X., & Jiang, X. (2012). The molecular mechanism of action of bactericidal gold nanoparticles on *Escherichia coli*. *Biomaterials*, *33*(7), 2327–2333.

Cushen, M., Kerry, J., Morris, M., Cruz-Romero, M., & Cummins, E. (2012). Nanotechnologies in the food industry-recent developments, risks and regulation. *Trends in Food Science & Technology*, *24*(1), 30–46.

Dahoumane, S. A., Mechouet, M., Alvarez, F. J., Agathos, S. N., & Jeffryes, C. (2016). Microalgae: An outstanding tool in nanotechnology. *Bionatura*, *1*(4), 196–201.

Dasgupta, N., Ranjan, S., Rajendran, B., Manickam, V., Ramalingam, C., Avadhani, G. S., & Kumar, A. (2016). Thermal co-reduction approach to vary size of silver nanoparticle: Its microbial and cellular toxicology. *Environmental Science and Pollution Research*, *23*(5), 4149–4163.

Devi, J. S., & Bhimba, B. V. (2014). Antibacterial and antifungal activity of silver nanoparticles synthesized using *Hypnea muciformis*. *Biosciences Biotechnology Research Asia*, *11*, 235–238.

Divya, M., Kiran, G. S., Hassan, S., & Selvin, J. (2019). Biogenic synthesis and effect of silver nanoparticles (AgNPs) to combat catheter-related urinary tract infections. *Biocatalysis and Agricultural Biotechnology*, *18*, 101037.

Duncan, T. V. (2011). Applications of nanotechnology in food packaging and food safety: Barrier materials, antimicrobials and sensors. *Journal of Colloid and Interface Science*, *363*(1), 1–24.

Durán, N., Marcato, P. D., Durán, M., Yadav, A., Gade, A., & Rai, M. (2011). Mechanistic aspects in the biogenic synthesis of extracellular metal nanoparticles by peptides, bacteria, fungi, and plants. *Applied Microbiology and Biotechnology*, *90*(5), 1609–1624.

Elamawi, R. M., Al-Harbi, R. E., & Hendi, A. A. (2018). Biosynthesis and characterization of silver nanoparticles using *Trichoderma longibrachiatum* and their effect on phytopathogenic fungi. *Egyptian Journal of Biological Pest Control*, *28*(1), 28.

El Badawy, A. M., Silva, R. G., Morris, B., Scheckel, K. G., Suidan, M. T., & Tolaymat, T. M. (2011). Surface charge-dependent toxicity of silver nanoparticles. *Environmental Science & Technology*, *45*(1), 283–287.

Emamifar, A., Kadivar, M., Shahedi, M., & Soleimanian-Zad, S. (2010). Evaluation of nanocomposite packaging containing Ag and ZnO on shelf life of fresh orange juice. *Innovative Food Science & Emerging Technologies*, *11*(4), 742–748.

Fadeel, B., & Garcia-Bennett, A. E. (2010). Better safe than sorry: Understanding the toxicological properties of inorganic nanoparticles manufactured for biomedical applications. *Advanced Drug Delivery Reviews*, *62*(3), 362–374.

Fan, R., Karnik, R., Yue, M., Li, D., Majumdar, A., & Yang, P. (2005). DNA translocation in inorganic nanotubes. *Nano Letters*, *5*(9), 1633–1637.

Farouk, A., Moussa, S., Ulbricht, M., Schollmeyer, E., & Textor, T. (2014). ZnO-modified hybrid polymers as an antibacterial finish for textiles. *Textile Research Journal*, *84*(1), 40–51.

Fatemi, M., Mollania, N., Momeni-Moghaddam, M., & Sadeghifar, F. (2018). Extracellular biosynthesis of magnetic iron oxide nanoparticles by *Bacillus cereus* strain HMH1: Characterization and in vitro cytotoxicity analysis on MCF-7 and 3T3 cell lines. *Journal of Biotechnology*, *270*, 1–11.

Fernández, A., Picouet, P., & Lloret, E. (2010). Cellulose-silver nanoparticle hybrid materials to control spoilage-related microflora in absorbent pads located in trays of fresh-cut melon. *International Journal of Food Microbiology*, *142*(1–2), 222–228.

Feynman, R. P. (1960). There's plenty of room at the bottom. California Institute of Technology, Engineering Science Magazine.

Fortina, P., Kricka, L. J., Surrey, S., & Grodzinski, P. (2005). Nanobiotechnology: The promise and reality of new approaches to molecular recognition. *Trends in Biotechnology*, *23*(4), 168–173.

Gahlawat, G., Shikha, S., Chaddha, B. S., Chaudhuri, S. R., Mayilraj, S., & Choudhury, A. R. (2016). Microbial glycolipoprotein-capped silver nanoparticles as emerging antibacterial agents against cholera. *Microbial Cell Factories*, *15*(1), 25.

Ganesh, M., Hemalatha, P., Peng, M. M., & Jang, H. T. (2017). One pot synthesized Li, Zr doped porous silica nanoparticle for low temperature CO_2 adsorption. *Arabian Journal of Chemistry, 10*, S1501–S1505.

Golmohammadi, H., Morales-Narvaez, E., Naghdi, T., & Merkoci, A. (2017). Nanocellulose in sensing and biosensing. *Chemistry of Materials: A Publication of the American Chemical Society, 29*(13), 5426–5446.

Gonzalez, A. G., Fernandez-Rojo, L., Leflaive, J., Pokrovsky, O. S., & Rols, J.-L. (2016). Response of three biofilm-forming benthic microorganisms to Ag nanoparticles and Ag^+: The diatom *Nitzschia palea*, the green alga *Uronema confervicolum* and the cyanobacteria *Leptolyngbya* sp. *Environmental Science and Pollution Research, 23*(21), 22136–22150.

Gopu, M., Kumar, P., Selvankumar, T., Senthilkumar, B., Sudhakar, C., Govarthanan, M., ... Selvam, K. (2020). Green biomimetic silver nanoparticles utilizing the red algae *Amphiroa rigida* and its potent antibacterial, cytotoxicity and larvicidal efficiency. *Bioprocess and Biosystems Engineering*, 1–7.

Gourley, P. L., Hendricks, J. K., McDonald, A. E., Copeland, R. G., Barrett, K. E., Gourley, C. R., ... Naviaux, R. K. (2005). Mitochondrial correlation microscopy and nanolaser spectroscopy-new tools for biophotonic detection of cancer in single cells. *Technology in Cancer Research & Treatment, 4*(6), 585–592.

Grasso, G., Zane, D., & Dragone, R. (2020). Microbial nanotechnology: Challenges and prospects for green biocatalytic synthesis of nanoscale materials for sensoristic and biomedical applications. *Nanomaterials, 10*(1), 11.

Gu, H., Liu, C., Zhu, J., Gu, J., Wujcik, E. K., Shao, L., ... Zhang, J. (2018). *Introducing advanced composites and hybrid materials* (Vol. 1, pp. 1–5). Springer.

Guilger-Casagrande, M., & de Lima, R. (2019). Synthesis of silver nanoparticles mediated by fungi: A review. *Frontiers in Bioengineering and Biotechnology, 7*, 287.

Gujrati, M., Malamas, A., Shin, T., Jin, E., Sun, Y., & Lu, Z.-R. (2014). Multifunctional cationic lipid-based nanoparticles facilitate endosomal escape and reduction-triggered cytosolic siRNA release. *Molecular Pharmaceutics, 11*(8), 2734–2744.

Gutiérrez, J. A., Caballero, S., Díaz, L. A., Guerrero, M. A., Ruiz, J., & Ortiz, C. C. (2018). High antifungal activity against candida species of monometallic and bimetallic nanoparticles synthesized in nanoreactors. *ACS Biomaterials Science & Engineering, 4*(2), 647–653.

He, S., Guo, Z., Zhang, Y., Zhang, S., Wang, J., & Gu, N. (2007). Biosynthesis of gold nanoparticles using the bacteria *Rhodopseudomonas capsulata*. *Materials Letters, 61*(18), 3984–3987.

Hulikere, M. M., & Joshi, C. G. (2019). Characterization, antioxidant and antimicrobial activity of silver nanoparticles synthesized using marine endophytic fungus-*Cladosporium cladosporioides*. *Process Biochemistry, 82*, 199–204.

Ibrahim, E., Zhang, M., Zhang, Y., Hossain, A., Qiu, W., Chen, Y., ... Li, B. (2020). Green-synthesization of silver nanoparticles using endophytic bacteria isolated from garlic and its antifungal activity against wheat Fusarium head blight pathogen *Fusarium graminearum*. *Nanomaterials, 10*(2), 219.

Ingle, A., Gade, A., Pierrat, S., Sonnichsen, C., & Rai, M. (2008). Mycosynthesis of silver nanoparticles using the fungus *Fusarium acuminatum* and its activity against some human pathogenic bacteria. *Current Nanoscience, 4*(2), 141–144.

Invernizzi, N. (2011). Nanotechnology between the lab and the shop floor: What are the effects on labor? *Journal of Nanoparticle Research, 13*(6), 2249–2268.

Ishida, K., Cipriano, T. F., Rocha, G. M., Weissmüller, G., Gomes, F., Miranda, K., & Rozental, S. (2014). Silver nanoparticle production by the fungus *Fusarium oxysporum*: Nanoparticle characterisation and analysis of antifungal activity against pathogenic yeasts. *Memorias do Instituto Oswaldo Cruz, 109*(2), 220–228.

Ivask, A., Suarez, E., Patel, T., Boren, D., Ji, Z., Holden, P., ... Godwin, H. (2012). Genome-wide bacterial toxicity screening uncovers the mechanisms of toxicity of a cationic polystyrene nanomaterial.

Environmental Sciences: An International Journal of Environmental Physiology and Toxicology, 46(4), 2398–2405.

Jalal, M., Ansari, M. A., Shukla, A. K., Ali, S. G., Khan, H. M., Pal, R., . . . Cameotra, S. S. (2016). Green synthesis and antifungal activity of Al2O3 NPs against fluconazole-resistant *Candida* spp isolated from a tertiary care hospital. *RSC Advances, 6*(109), 107577–107590.

Jena, J., Pradhan, N., Nayak, R. R., Dash, B. P., Sukla, L. B., Panda, P. K., & Mishra, B. K. (2014). Microalga *Scenedesmus* sp.: A potential low-cost green machine for silver nanoparticle synthesis. *Journal of Microbiology and Biotechnology, 24*(4), 522–533.

Jha, A. K., & Prasad, K. (2010). Green synthesis of silver nanoparticles using Cycas leaf. *International Journal of Green Nanotechnology: Physics and Chemistry, 1*(2), P110–P117.

Jiang, S., Qin, W., Guo, R., & Zhang, L. (2010). Surface functionalization of nanostructured silver-coated polyester fabric by magnetron sputtering. *Surface and Coatings Technology, 204*(21–22), 3662–3667.

Jung, W. K., Koo, H. C., Kim, K. W., Shin, S., Kim, S. H., & Park, Y. H. (2008). Antibacterial activity and mechanism of action of the silver ion in *Staphylococcus aureus* and *Escherichia coli*. *Applied and Environmental Microbiology, 74*(7), 2171–2178.

Kalishwaralal, K., Deepak, V., Pandian, S. R. K., Kottaisamy, M., BarathManiKanth, S., Kartikeyan, B., & Gurunathan, S. (2010). Biosynthesis of silver and gold nanoparticles using Brevibacterium casei. *Colloids and Surfaces B, 77*(2), 257–262.

Kalpana, V., Kataru, B. A. S., Sravani, N., Vigneshwari, T., Panneerselvam, A., & Rajeswari, V. D. (2018). Biosynthesis of zinc oxide nanoparticles using culture filtrates of Aspergillus niger: Antimicrobial textiles and dye degradation studies. *OpenNano, 3*, 48–55.

Kang, F., Alvarez, P. J., & Zhu, D. (2014). Microbial extracellular polymeric substances reduce Ag^+ to silver nanoparticles and antagonize bactericidal activity. *Environmental Science & Technology, 48*(1), 316–322.

Kathiresan, K., Manivannan, S., Nabeel, M., & Dhivya, B. (2009). Studies on silver nanoparticles synthesized by a marine fungus, *Penicillium fellutanum* isolated from coastal mangrove sediment. *Colloids and Surfaces B, 71*(1), 133–137.

Khalafi, T., Buazar, F., & Ghanemi, K. (2019). Phycosynthesis and enhanced photocatalytic activity of zinc oxide nanoparticles toward organosulfur pollutants. *Scientific Reports, 9*(1), 1–10.

Khaleghi, M., Khorrami, S., & Ravan, H. (2019). Identification of *Bacillus thuringiensis* bacterial strain isolated from the mine soil as a robust agent in the biosynthesis of silver nanoparticles with strong antibacterial and anti-biofilm activities. *Biocatalysis and Agricultural Biotechnology, 18*, 101047.

Khan, M. R., & Rizvi, T. F. (2014). Nanotechnology: Scope and application in plant disease management. *Plant Pathology Journal, 13*(3), 214–231.

Khot, L. R., Sankaran, S., Maja, J. M., Ehsani, R., & Schuster, E. W. (2012). Applications of nanomaterials in agricultural production and crop protection: A review. *Crop Protection (Guildford, Surrey), 35*, 64–70.

Kim, K.-J., Sung, W. S., Suh, B. K., Moon, S.-K., Choi, J.-S., Kim, J. G., & Lee, D. G. (2009). Antifungal activity and mode of action of silver nano-particles on *Candida albicans*. *Biometals: An International Journal on the Role of Metal Ions in Biology, Biochemistry, and Medicine, 22*(2), 235–242.

Kitching, M., Choudhary, P., Inguva, S., Guo, Y., Ramani, M., Das, S. K., & Marsili, E. (2016). Fungal surface protein mediated one-pot synthesis of stable and hemocompatible gold nanoparticles. *Enzyme and Microbial Technology, 95*, 76–84.

Koopi, H., & Buazar, F. (2018). A novel one-pot biosynthesis of pure alpha aluminum oxide nanoparticles using the macroalgae *Sargassum ilicifolium*: A green marine approach. *Ceramics International, 44*(8), 8940–8945.

Lengke, M. F., Fleet, M. E., & Southam, G. (2007). Biosynthesis of silver nanoparticles by filamentous cyanobacteria from a silver (I) nitrate complex. *Langmuir: The ACS Journal of Surfaces and Colloids, 23*(5), 2694–2699.

Li, J., Wang, X., Wang, C., Chen, B., Dai, Y., Zhang, R., . . . Fu, D. (2007). The enhancement effect of gold nanoparticles in drug delivery and as biomarkers of drug-resistant cancer cells. *ChemMedChem: Chemistry Enabling Drug Discovery, 2*(3), 374−378.

Li, L., Jiang, W., Luo, K., Song, H., Lan, F., Wu, Y., & Gu, Z. (2013). Superparamagnetic iron oxide nanoparticles as MRI contrast agents for non-invasive stem cell labeling and tracking. *Theranostics, 3*(8), 595−615.

Li, X., Robinson, S. M., Gupta, A., Saha, K., Jiang, Z., Moyano, D. F., . . . Rotello, V. M. (2014). Functional gold nanoparticles as potent antimicrobial agents against multi-drug-resistant bacteria. *ACS Nano, 8*(10), 10682−10686.

Li, X., Xu, H., Chen, Z.-S., & Chen, G. (2011). Biosynthesis of nanoparticles by microorganisms and their applications. *Journal of Nanomaterials*, 2011.

Lok, C.-N., Ho, C.-M., Chen, R., He, Q.-Y., Yu, W.-Y., Sun, H., . . . Che, C.-M. (2006). Proteomic analysis of the mode of antibacterial action of silver nanoparticles. *Journal of Proteome Research, 5*(4), 916−924.

Luo, Q. Y., Lin, Y., Li, Y., Xiong, L. H., Cui, R., Xie, Z. X., & Pang, D. W. (2014). Nanomechanical analysis of yeast cells in CdSe quantum dot biosynthesis. *Small (Weinheim an der Bergstrasse, Germany), 10*(4), 699−704.

Maceda, A. F., Ouano, J. J. S., Que, M. C. O., Basilia, B. A., Potestas, M. J., Alguno, A. C. (2018). Controlling the absorption of gold nanoparticles via green synthesis using *Sargassum crassifolium* extract. In *Paper presented at the Key Engineering Materials*.

Mageshwari, K., & Sathyamoorthy, R. (2013). Flower-shaped CuO nanostructures: Synthesis, characterization and antimicrobial activity. *Journal of Materials Science & Technology, 29*(10), 909−914.

Mahanty, S., Bakshi, M., Ghosh, S., Chatterjee, S., Bhattacharyya, S., Das, P., . . . Chaudhuri, P. (2019). Green synthesis of iron oxide nanoparticles mediated by filamentous fungi isolated from Sundarban mangrove ecosystem, India. *BioNanoScience, 9*(3), 637−651.

Mahdi, S., Vadood, R., & Nourdahr, R. (2012). Study on the antimicrobial effect of nanosilver tray packaging of minced beef at refrigerator temperature. *Global Veterinaria, 9*(3), 284−289.

Marooufpour, N., Alizadeh, M., Hatami, M., & Lajayer, B. A. (2019). Biological synthesis of nanoparticles by different groups of bacteria. *Microbial nanobionics* (pp. 63−85). Springer.

Matei, A., Matei, S., Matei, G.-M., Cogălniceanu, G., & Cornea, C. P. (2020). Biosynthesis of silver nanoparticles mediated by culture filtrate of lactic acid bacteria, characterization and antifungal activity. *The EuroBiotech Journal, 4*(2), 97−103.

McQuillan, J. S., Groenaga Infante, H., Stokes, E., & Shaw, A. M. (2012). Silver nanoparticle enhanced silver ion stress response in *Escherichia coli* K12. *Nanotoxicology, 6*(8), 857−866.

Meghana, S., Kabra, P., Chakraborty, S., & Padmavathy, N. (2015). Understanding the pathway of antibacterial activity of copper oxide nanoparticles. *RSC Advances, 5*(16), 12293−12299.

Moghaddam, A. B., Namvar, F., Moniri, M., Tahir, P. M., Azizi, S., & Mohamad, R. (2015). Nanoparticles biosynthesized by fungi and yeast: A review of their preparation, properties, and medical applications. *Molecules (Basel, Switzerland), 20*(9), 16540−16565.

Monowar, T., Rahman, M., Bhore, S. J., Raju, G., & Sathasivam, K. V. (2018). Silver nanoparticles synthesized by using the endophytic bacterium *Pantoea ananatis* are promising antimicrobial agents against multidrug resistant bacteria. *Molecules (Basel, Switzerland), 23*(12), 3220.

Morones, J. R., Elechiguerra, J. L., Camacho, A., Holt, K., Kouri, J. B., Ramírez, J. T., & Yacaman, M. J. (2005). The bactericidal effect of silver nanoparticles. *Nanotechnology, 16*(10), 2346.

Movafeghi, A., Khataee, A., Rezaee, A., Kosari-Nasab, M., & Tarrahi, R. (2019). Toxicity of cadmium selenide nanoparticles on the green microalga *Chlorella vulgaris*: Inducing antioxidative defense response. *Environmental Science and Pollution Research, 26*(36), 36380−36387.

Mukherjee, K., Gupta, R., Kumar, G., Kumari, S., Biswas, S., & Padmanabhan, P. (2018). Synthesis of silver nanoparticles by *Bacillus clausii* and computational profiling of nitrate reductase enzyme involved in production. *Journal of Genetic Engineering and Biotechnology, 16*(2), 527−536.

Mukherjee, P., Ahmad, A., Mandal, D., Senapati, S., Sainkar, S. R., Khan, M. I., ... Alam, M. (2001). Bioreduction of AuCl4 − ions by the fungus, *Verticillium* sp. and surface trapping of the gold nanoparticles formed. *Angewandte Chemie International Edition, 40*(19), 3585−3588.

Musee, N., Thwala, M., & Nota, N. (2011). The antibacterial effects of engineered nanomaterials: Implications for wastewater treatment plants. *Journal of Environmental Monitoring: JEM, 13*(5), 1164−1183.

Nair, S., Sasidharan, A., Rani, V. D., Menon, D., Nair, S., Manzoor, K., & Raina, S. (2009). Role of size scale of ZnO nanoparticles and microparticles on toxicity toward bacteria and osteoblast cancer cells. *Journal of Materials Science. Materials in Medicine, 20*(1), 235.

Narayanan, K. B., & Sakthivel, N. (2010). Biological synthesis of metal nanoparticles by microbes. *Advances in Colloid and Interface Science, 156*(1−2), 1−13.

Nasrollahi, A., Pourshamsian, K., & Mansourkiaee, P. (2011). Antifungal activity of silver nanoparticles on some of fungi. *International Journal of Nano Dimension, 1*(3), 233.

Navarro, E., Baun, A., Behra, R., Hartmann, N. B., Filser, J., Miao, A.-J., ... Sigg, L. (2008). Environmental behavior and ecotoxicity of engineered nanoparticles to algae, plants, and fungi. *Ecotoxicology (London, England), 17*(5), 372−386.

Navya, P., & Daima, H. K. (2016). Rational engineering of physicochemical properties of nanomaterials for biomedical applications with nanotoxicological perspectives. *Nano Convergence, 3*(1), 1.

Negm, M. A., Ibrahim, H. A., Shaltout, N. A., Shawky, H. A., Abdel-mottaleb, M., & Hamdona, S. (2018). Green synthesis of silver nanoparticles using marine algae extract and their antibacterial activity. *Sciences, 8*(03), 957−970.

Oukarroum, A., Halimi, I., & Siaj, M. (2019). Cellular responses of *Chlorococcum* Sp. algae exposed to Zinc oxide nanoparticles by using flow cytometry. *Water, Air, and Soil Pollution, 230*(1), 1.

Ovais, M., Khalil, A. T., Ayaz, M., Ahmad, I., Nethi, S. K., & Mukherjee, S. (2018). Biosynthesis of metal nanoparticles via microbial enzymes: A mechanistic approach. *International Journal of Molecular Sciences, 19*(12), 4100.

Ovais, M., Raza, A., Naz, S., Islam, N. U., Khalil, A. T., Ali, S., ... Shinwari, Z. K. (2017). Current state and prospects of the phytosynthesized colloidal gold nanoparticles and their applications in cancer theranostics. *Applied Microbiology and Biotechnology, 101*(9), 3551−3565.

Öztürk, B. Y. (2019). Intracellular and extracellular green synthesis of silver nanoparticles using *Desmodesmus* sp.: Their antibacterial and antifungal effects. *Caryologia International Journal of Cytology, Cytosystematics and Cytogenetics, 72*(1), 29−43.

Pakrashi, S., Dalai, S., Prathna, T., Trivedi, S., Myneni, R., Raichur, A. M., ... Mukherjee, A. (2013). Cytotoxicity of aluminium oxide nanoparticles towards fresh water algal isolate at low exposure concentrations. *Aquatic Toxicology (Amsterdam, Netherlands), 132*, 34−45.

Pal, S., Tak, Y. K., & Song, J. M. (2007). Does the antibacterial activity of silver nanoparticles depend on the shape of the nanoparticle? A study of the gram-negative bacterium *Escherichia coli*. *Applied and Environmental Microbiology, 73*(6), 1712−1720.

Pathakoti, K., Morrow, S., Han, C., Pelaez, M., He, X., Dionysiou, D. D., & Hwang, H.-M. (2013). Photoinactivation of *Escherichia coli* by sulfur-doped and nitrogen−fluorine-codoped TiO2 nanoparticles under solar simulated light and visible light irradiation. *Environmental Science & Technology, 47*(17), 9988−9996.

Patra, J. K., & Gouda, S. (2013). Application of nanotechnology in textile engineering: An overview. *Journal of Engineering and Technology Research, 5*(5), 104−111.

Pelletier, D. A., Suresh, A. K., Holton, G. A., McKeown, C. K., Wang, W., Gu, B., ... Allison, M. R. (2010). Effects of engineered cerium oxide nanoparticles on bacterial growth and viability. *Applied and Environmental Microbiology, 76*(24), 7981−7989.

Peters, R., Brandhoff, P., Weigel, S., Marvin, H., Bouwmeester, H., Aschberger, K., ... Botelho Moniz, F. (2014). Inventory of nanotechnology applications in the agricultural, feed and food sector. *EFSA Supporting Publications*, *11*(7), 621E.

Pfaller, M., & Diekema, D. (2007). Epidemiology of invasive candidiasis: A persistent public health problem. *Clinical Microbiology Reviews*, *20*(1), 133–163.

Powers, K. W., Palazuelos, M., Moudgil, B. M., & Roberts, S. M. (2007). Characterization of the size, shape, and state of dispersion of nanoparticles for toxicological studies. *Nanotoxicology*, *1*(1), 42–51.

Pugazhenthiran, N., Anandan, S., Kathiravan, G., Prakash, N. K. U., Crawford, S., & Ashokkumar, M. (2009). Microbial synthesis of silver nanoparticles by *Bacillus* sp. *Journal of Nanoparticle Research*, *11*(7), 1811.

Rad, M., Taran, M., & Alavi, M. (2018). Effect of incubation time, $CuSO_4$ and glucose concentrations on biosynthesis of copper oxide (CuO) nanoparticles with rectangular shape and antibacterial activity: Taguchi method approach. *Nano Biomedicine and Engineering*, *10*(1), 25–33.

Ramalingam, B., Parandhaman, T., & Das, S. K. (2016). Antibacterial effects of biosynthesized silver nanoparticles on surface ultrastructure and nanomechanical properties of gram-negative bacteria viz. *Escherichia coli* and *Pseudomonas aeruginosa*. *ACS Applied Materials & Interfaces*, *8*(7), 4963–4976.

Rao, J. P., & Geckeler, K. E. (2011). Polymer nanoparticles: Preparation techniques and size-control parameters. *Progress in Polymer Science*, *36*(7), 887–913.

Reguera, G. (2012). *Electron transfer at the cell-uranium interface in* Geobacter *spp.* Portland Press Ltd.

Roco, M. C. (2011). *The long view of nanotechnology development: The national nanotechnology initiative at 10 years* (pp. 427–445). Springer.

Saeed, K., & Khan, I. (2016). Preparation and characterization of single-walled carbon nanotube/nylon 6, 6 nanocomposites. *Instrumentation Science &. Technology*, *44*(4), 435–444.

Salem, D. M., Ismail, M. M., & Aly-Eldeen, M. A. (2019). Biogenic synthesis and antimicrobial potency of iron oxide (Fe_3O_4) nanoparticles using algae harvested from the Mediterranean sea, Egypt. *The Egyptian. Journal of Aquatic Research*, *45*(3), 197–204.

Sanaeimehr, Z., Javadi, I., & Namvar, F. (2018). Antiangiogenic and antiapoptotic effects of green-synthesized zinc oxide nanoparticles using *Sargassum muticum* algae extraction. *Cancer Nanotechnology*, *9*(1), 3.

Saravanan, M., Barik, S. K., MubarakAli, D., Prakash, P., & Pugazhendhi, A. (2018). Synthesis of silver nanoparticles from *Bacillus brevis* (NCIM 2533) and their antibacterial activity against pathogenic bacteria. *Microbial Pathogenesis*, *116*, 221–226.

Satapathy, S., & Shukla, S. P. (2017). Application of a marine cyanobacterium *Phormidium fragile* for green synthesis of silver nanoparticles. *Indian Journal of Biotechnology*, *16*(1), 110–113.

Seetharaman, P. K., Chandrasekaran, R., Gnanasekar, S., Chandrakasan, G., Gupta, M., Manikandan, D. B., & Sivaperumal, S. (2018). Antimicrobial and larvicidal activity of eco-friendly silver nanoparticles synthesized from endophytic fungi *Phomopsis liquidambaris*. *Biocatalysis and Agricultural Biotechnology*, *16*, 22–30.

Selvarajan, E., & Mohanasrinivasan, V. (2013). Biosynthesis and characterization of ZnO nanoparticles using *Lactobacillus plantarum* VITES07. *Materials Letters*, *112*, 180–182.

Senapati, S., Ahmad, A., Khan, M. I., Sastry, M., & Kumar, R. (2005). Extracellular biosynthesis of bimetallic Au-Ag alloy nanoparticles. *Small (Weinheim an der Bergstrasse, Germany)*, *1*(5), 517–520.

Senapati, S., Syed, A., Moeez, S., Kumar, A., & Ahmad, A. (2012). Intracellular synthesis of gold nanoparticles using alga *Tetraselmis kochinensis*. *Materials Letters*, *79*, 116–118.

Shaalan, M., Saleh, M., El-Mahdy, M., & El-Matbouli, M. (2016). Recent progress in applications of nanoparticles in fish medicine: A review. *Nanomedicine: Nanotechnology, Biology*, *12*(3), 701–710.

Shamim, A., Abid, M. B., & Mahmood, T. (2019). Biogenic synthesis of zinc oxide (ZnO) nanoparticles using a fungus (*Aspargillus niger*) and their characterization. *International Journal of Chemistry*, *11*(2), 119–126.

Simon-Deckers, A., Loo, S., Mayne-L'hermite, M., Herlin-Boime, N., Menguy, N., Reynaud, C., ... Carriere, M. (2009). Size-, composition-and shape-dependent toxicological impact of metal oxide nanoparticles and carbon nanotubes toward bacteria. *Environmental Science & Technology*, *43*(21), 8423–8429.

Singaravelu, G., Arockiamary, J., Kumar, V. G., & Govindaraju, K. (2007). A novel extracellular synthesis of monodisperse gold nanoparticles using marine alga, Sargassum wightii Greville. *Colloids and Surfaces B*, *57*(1), 97–101.

Singh, G., Joyce, E. M., Beddow, J., & Mason, T. J. (2020). Evaluation of antibacterial activity of ZnO nanoparticles coated sonochemically onto textile fabrics. *Journal of Microbiology Biotechnology and Food Sciences*, *9*(4), 106–120.

Singh, J., Vishwakarma, K., Ramawat, N., Rai, P., Singh, V. K., Mishra, R. K., ... Sharma, S. (2019). Nanomaterials and microbes' interactions: A contemporary overview. *3 Biotech*, *9*(3), 68.

Spagnoletti, F. N., Spedalieri, C., Kronberg, F., & Giacometti, R. (2019). Extracellular biosynthesis of bactericidal Ag/AgCl nanoparticles for crop protection using the fungus *Macrophomina phaseolina*. *Journal of Environmental Management*, *231*, 457–466.

Stoimenov, P. K., Klinger, R. L., Marchin, G. L., & Klabunde, K. J. (2002). Metal oxide nanoparticles as bactericidal agents. *Langmuir: The ACS Journal of Surfaces and Colloids*, *18*(17), 6679–6686.

Subramani, K. (2006). Applications of nanotechnology in drug delivery systems for the treatment of cancer and diabetes. *International Journal of Nanotechnology*, *3*(4), 557–580.

Talibi, I., Askarne, L., Boubaker, H., Boudyach, E. H., & Aoumar, A. A. B. (2011). *In vitro* and *in vivo* antifungal activities of organic and inorganic salts against citrus sour rot agent *Geotrichum candidum*. *Plant Pathology Journal*, *10*, 138–145.

Taran, M., Rad, M., & Alavi, M. (2018). Biosynthesis of TiO_2 and ZnO nanoparticles by *Halomonas elongata* IBRC-M 10214 in different conditions of medium. *BioImpacts*, *8*(2), 81–89.

Thakkar, K. N., Mhatre, S. S., & Parikh, R. Y. (2010). Biological synthesis of metallic nanoparticles. *Nanomedicine: Nanotechnology, Biology and Medicine*, *6*(2), 257–262.

Thomas, S. C., Mishra, P. K., & Talegaonkar, S. (2015). Ceramic nanoparticles: Fabrication methods and applications in drug delivery. *Current Pharmaceutical Design*, *21*(42), 6165–6188.

Toker, R., Kayaman-Apohan, N., & Kahraman, M. (2013). UV-curable nano-silver containing polyurethane based organic-inorganic hybrid coatings. *Progress in Organic Coatings*, *76*(9), 1243–1250.

Tomar, A., & Garg, G. (2013). Short review on application of gold nanoparticles. *Global Journal of Pharmacology*, *7*(1), 34–38.

Tyagi, S., Tyagi, P. K., Gola, D., Chauhan, N., & Bharti, R. K. (2019). Extracellular synthesis of silver nanoparticles using entomopathogenic fungus: Characterization and antibacterial potential. *SN Applied Sciences*, *1*(12), 1545.

Valko, M., Rhodes, C., Moncol, J., Izakovic, M., & Mazur, M. (2006). Free radicals, metals and antioxidants in oxidative stress-induced cancer. *Chemico-Biological Interactions*, *160*(1), 1–40.

Vijayabharathi, R., Sathya, A., & Gopalakrishnan, S. (2018). Extracellular biosynthesis of silver nanoparticles using *Streptomyces griseoplanus* SAI-25 and its antifungal activity against *Macrophomina phaseolina*, the charcoal rot pathogen of sorghum. *Biocatalysis and Agricultural Biotechnology*, *14*, 166–171.

Vijayanandan, A. S., & Balakrishnan, R. M. (2018). Biosynthesis of cobalt oxide nanoparticles using endophytic fungus *Aspergillus nidulans*. *Journal of Environmental Management*, *218*, 442–450.

Wang, L., He, H., Yu, Y., Sun, L., Liu, S., Zhang, C., & He, L. (2014). Morphology-dependent bactericidal activities of Ag/CeO_2 catalysts against *Escherichia coli*. *Journal of Inorganic Biochemistry*, *135*, 45–53.

Wee, K. W., Kang, G. Y., Park, J., Kang, J. Y., Yoon, D. S., Park, J. H., & Kim, T. S. (2005). Novel electrical detection of label-free disease marker proteins using piezoresistive self-sensing micro-cantilevers. *Biosensors & Bioelectronics, 20*(10), 1932–1938.

Wei, R. (2017). Biosynthesis of Au-Ag alloy nanoparticles for sensitive electrochemical determination of paracetamol. *International Journal of Electrochemical Science, 12,* 9131–9140.

Weir, A., Westerhoff, P., Fabricius, L., Hristovski, K., & Von Goetz, N. (2012). Titanium dioxide nanoparticles in food and personal care products. *Environmental Science & Technology, 46*(4), 2242–2250.

Westmeier, D., Posselt, G., Hahlbrock, A., Bartfeld, S., Vallet, C., Abfalter, C., ... Stauber, R. H. (2018). Nanoparticle binding attenuates the pathobiology of gastric cancer-associated *Helicobacter pylori*. *Nanoscale, 10*(3), 1453–1463.

Worrall, E. A., Hamid, A., Mody, K. T., Mitter, N., & Pappu, H. R. (2018). Nanotechnology for plant disease management. *Agronomy, 8*(12), 285.

Wu, Q., Cao, H., Luan, Q., Zhang, J., Wang, Z., Warner, J. H., & Watt, A. (2008). Biomolecule-assisted synthesis of water-soluble silver nanoparticles and their biomedical applications. *Inorganic Chemistry, 47*(13), 5882–5888.

Wu, S., Duan, N., Gu, H., Hao, L., Ye, H., Gong, W., & Wang, Z. (2016). A review of the methods for detection of *Staphylococcus aureus* enterotoxins. *Toxins, 8*(7), 176.

Xie, Y., He, Y., Irwin, P. L., Jin, T., & Shi, X. (2011). Antibacterial activity and mechanism of action of zinc oxide nanoparticles against *Campylobacter jejuni*. *Applied and Environmental Microbiology, 77*(7), 2325–2331.

Yang, W., Shen, C., Ji, Q., An, H., Wang, J., Liu, Q., & Zhang, Z. (2009). Food storage material silver nanoparticles interfere with DNA replication fidelity and bind with DNA. *Nanotechnology, 20*(8), 085102.

Yehia, R. S., & Ahmed, O. F. (2013). In vitro study of the antifungal efficacy of zinc oxide nanoparticles against *Fusarium oxysporum* and *Penicillium expansum*. *African Journal of Microbiological Research, 7*(19), 1917–1923.

Yin, H., Ai, S., Xu, J., Shi, W., & Zhu, L. (2009). Amperometric biosensor based on immobilized acetylcholinesterase on gold nanoparticles and silk fibroin modified platinum electrode for detection of methyl paraoxon, carbofuran and phoxim. *Journal of Electroanalytical Chemistry, 637*(1–2), 21–27.

Yoon, K.-Y., Byeon, J. H., Park, J.-H., & Hwang, J. (2007). Susceptibility constants of *Escherichia coli* and *Bacillus subtilis* to silver and copper nanoparticles. *The Science of the Total Environment, 373*(2–3), 572–575.

Yusof, H. M., Mohamad, R., & Zaidan, U. H. (2019). Microbial synthesis of zinc oxide nanoparticles and their potential application as an antimicrobial agent and a feed supplement in animal industry: A review. *Journal of Animal Science and Biotechnology, 10*(1), 57.

Zeenia, K., Yaguchi, T., Sunil, C. K., Hirano, T., Wadhwa, R., & Taira, K. (2003). Mortalin imaging in normal and cancer cells with quantum dot immuno-conjugates. *Cell Research, 13*(6), 503–507.

Zhang, L., Jiang, Y., Ding, Y., Povey, M., & York, D. (2007). Investigation into the antibacterial behaviour of suspensions of ZnO nanoparticles (ZnO nanofluids). *Journal of Nanoparticle Research, 9*(3), 479–489.

Zhang, X., He, X., Wang, K., & Yang, X. (2011). Different active biomolecules involved in biosynthesis of gold nanoparticles by three fungus species. *Journal of Biomedical Nanotechnology, 7*(2), 245–254.

Zhang, X., Yan, S., Tyagi, R., & Surampalli, R. (2011). Synthesis of nanoparticles by microorganisms and their application in enhancing microbiological reaction rates. *Chemosphere, 82*(4), 489–494.

Zheng, D., Hu, C., Gan, T., Dang, X., & Hu, S. (2010). Preparation and application of a novel vanillin sensor based on biosynthesis of Au-Ag alloy nanoparticles. *Sensors and Actuators. B, Chemical, 148*(1), 247–252.

Zimmermann, M. B., & Hilty, F. M. (2011). Nanocompounds of iron and zinc: Their potential in nutrition. *Nanoscale, 3*(6), 2390–2398.

Process of biodegradation controlled by nanoparticle-based materials: mechanisms, significance, and applications

Roberta Anjos de Jesus[1], José Arnaldo Santana Costa[2], Caio Marcio Paranhos[2], Muhammad Bilal[3], Ram Naresh Bharagava[4], Hafiz M.N. Iqbal[5], Luiz Fernando Romanholo Ferreira[1], Renan Tavares Figueiredo[1]

[1]INSTITUTE OF TECHNOLOGY AND RESEARCH (ITP), TIRADENTES UNIVERSITY (UNIT), ARACAJU-SERGIPE, BRAZIL [2]CDMF, DEPARTMENT OF CHEMISTRY, FEDERAL UNIVERSITY OF SÃO CARLOS, SÃO CARLOS, BRAZIL [3]SCHOOL OF LIFE SCIENCE AND FOOD ENGINEERING, HUAIYIN INSTITUTE OF TECHNOLOGY, HUAI'AN, P.R. CHINA [4]LABORATORY FOR BIOREMEDIATION AND METAGENOMICS RESEARCH (LBMR), DEPARTMENT OF MICROBIOLOGY (DM), BABASAHEB BHIMRAO AMBEDKAR UNIVERSITY (A CENTRAL UNIVERSITY), LUCKNOW, INDIA [5]TECNOLOGICO DE MONTERREY, SCHOOL OF ENGINEERING AND SCIENCES, MONTERREY, MEXICO

4.1 Introduction

In the last decades, environmental problems related to anthropogenic causes have been constantly increasing, mainly due to the inadequate disposal of waste, whether by industrial activities or simple disposal of products used in our daily lives. This problem has been further aggravated by population growth and the increasingly recurring need to generate energy, fuels, food, consumption of freshwater, plant protection, personal care, and pharmaceutical products, among others. Thus, it is increasingly common to find pollutants from these activities in aquatic bodies, soil, air, flora, and fauna, which can pose serious risks to the environment, human health, and living species (Wang, Jing, et al., 2020; Wang, Yang, et al., 2020; Zhou et al., 2020).

The main problem with the entry of these pollutants into the environment is due to the high toxicity of these compounds to living species because even in low concentrations, they can have several harmful properties (Costa et al., 2020; Yadav et al., 2019). Many of these pollutants that have been increasingly found in the environment are the so-called emerging pollutants (EPs), which can be derived from industrial activities until cyanotoxins are produced by cyanobacteria; thus the main categories of EPs will be discussed in this chapter.

Given the aforementioned, scientific methodologies for the environmental remediation need to be developed to mitigate the environmental impacts caused by the uncontrolled release of EPs into the environment. Given this, nanostructured materials with multifunctional properties are an interesting alternative to be used in the remediation of these pollutants. In general, nanoparticles (NPs) mainly include particles based on metals, clay minerals, graphene, activated carbon, carbon nanotubes, among others (Chen et al., 2019). These materials have some physical-chemical characteristics to be applied to environmental remediation, such as high surface area, nanoscale size, optical properties, reactivity, resistance, among others (Liu, Jiang, et al., 2020; Liu, Rowe, et al., 2020).

Our approach aims to present the main EPs found in the environment and their possible sources and environmental implications, as well as the recurrent methods of synthesis of NPs that can be used as an alternative for biodegradation of EPs.

4.2 Occurrence of emerging pollutants: environmental impact

Nowadays, environmental contamination with EPs has increased enormously, mainly due to the uncontrolled population growth and industrial activities. Environmental contamination with organic and inorganic compounds from daily industrial activities is one of the main polluting sources of natural waters, soils, flora, and fauna (EL-Sheshtawy, El-Hosainy, Shoueir, El-Mehasseb, & El-Kemary, 2019; Feizpoor, Habibi-Yangjeh, Yubuta, & Vadivel, 2019; Sodré, 2012). Since the uncontrolled release of these pollutants, even in low concentrations, can cause serious environmental and public health problems for humans, as well as posing a constant threat to aquatic and terrestrial organisms. Also, an impressive number of water treatment plants around the world, especially in developing countries, still employ classical methods that do not take into account the presence of EPs.

EPs are well-known as toxic, allergenic, pathogenic, carcinogenic, mutagenic, recalcitrant, bioaccumulative, nonbiodegradable, teratogenetic, and other (Costa et al., 2020; Yadav et al., 2019). To monitor EPs worldwide, the network of reference laboratories, research centers, and related organizations for monitoring of emerging environmental substances (NORMAN Network) was created in 2005 (NORMAN, 2020a; Peña-Guzmán et al., 2019). The NORMAN network has identified a list of several EPs, which are classified within some categories, such as biocides, disinfection byproducts, drugs of abuse, flame retardants, food additives, gasoline additives, industrial chemicals, personal care products, pharmaceuticals, plant protection products, surfactants, trace metals, and others (NORMAN, 2020b). Table 4−1 presents the main categories and some representatives of the EPs prepared by The NORMAN network.

Table 4–1 NORMAN list of emerging substances (NORMAN, 2020b).

Use category I	Use category II	Acronym	Individual substance
Algal toxins	—	—	Microcystin-LA / Cyanoginosin-LA
			Microcystin-LR
			Microcystin-RR
			Microcystin-YR
Antimicrobial agent/ Moth repellent	Fragrances	—	Camphor
	—		Isoborneol
Bioterrorism/Sabotage agents	—	—	Trichloronitromethane (Chloropicrin)
Biocide transformation products	—	—	Methyl triclosan
Biocides	—	—	(Benzyloxy)methanol
			(Ethylenedioxy)dimethanol
			1,3-dichloro-5,5-dimethylhydantoin
			2-Butanone, peroxide
		CMI	5-Chloro-2-methyl-3(2H)-isothiazolone
		ADAO	Amines, C10—16-alkyldimethyl, N-oxides
		—	Bromochloro-5,5-dimethylimidazolidine-2,4-dione
			Bronopol / Bronosol
			Diclosan / 5-chloro-2-(4-chlorphenoxy)phenol
			Methylene dithiocyanate
Disinfection byproducts (drinking water)	—	—	1,1-Dibromopropanone
			1,3-Dichloroketone
			2-Chlorophenol
		NMOR	4-Nitrosomorpholine
			Bromoacetonitrile
		HBCDD	Hexabromocyclododecane
		NDPA	N-nitrosodiphenylamine
Drugs of abuse	—	—	Amphetamine
			Cocaine
			Heroin
			MDMA (Ecstasy)
			Methamphetamine
			Morphine
			THC-COOH (Cannabis)
Flame retardants	Brominated flame retardants	DBP-TAZTO	1-(2,3-Dibromopropyl)-3,5-diallyl-1,3,5-Triazine-2,4,6(1H,3H,5H)-trione
	Chlorine containing flame retardants	DDC-Ant	1,2,3,4,5,6,7,8,12,12,13,13-Dodecachloro-1,4,4a,5,8,8a,9,9a,10,10a-decahydro-1,4:5,8:9,10-Trimethanoanthracene
	Brominated flame retardants	HCTBPH	1,2,3,4,7,7-hexachloro-5-(2,3,4,5-tetrabromophenyl)-Bicyclo[2.2.1]hept-2-ene
		BCMP-BCMEP	

(Continued)

Table 4–1 (Continued)

Use category I	Use category II	Acronym	Individual substance
	Phosphorous containing flame retardants		2,2-Bis(chloromethyl)-1,3-propanediol bis[bis (2-chloro1-methylethyl) phosphate]
	Brominated flame retardants	EBTEBPI	N,N'-Ethylenebis(tetrabromophthalimide)
	Brominated flame retardants	—	Octabromodiphenyl ethers
Food additives	Artificial sweeteners	—	Sucralose
	Humectants		Triacetin
Gasoline additives	—	MTBE	Methyl-tert-butyl ether
Industrial chemicals	—	4,4'-DDMS	1-Chloro-2,2-bis(p-chlorophenyl)ethane
		—	1,3-Dinitropyrene
			2-(2-Naphthalenyl)benzothiophene
			2-Bromoanisole
			2-Methylanthraquinone
	PPP transformation products		3-Chloroaniline
	—		3-Nitrobenzanthrone
	—		Benzenesulfonamide
	Complexing agents	NTA	Nitrilotriacetic acid
	—	—	p-Cresol
Lubricants	Chlorine containing flame retardants	LCCP	Long-chained chloroalkanes (C18–30)
	Chlorine containing flame retardants	MCCP	Medium-chained chloroalkanes, (C14–17)
Moth repellent	Fragrances	—	Camphor
			Isoborneol
Personal care products	Fragrances	—	4-Oxoisophorone
	Sunscreen agents		Benzaldehyde
	Fragrances		Benzylacetate
	Antioxidants	BHA	Butylated hydroxyanisole
	—	—	d-Limonene
	Siloxanes		Decamethylcyclopentasiloxane (D5)
	Parabens		Ethyl paraben
			Methyl paraben
	Fragrances		Musk xylene
	Sunscreen agents		Octocrylene
			Oxybenzone
	Antioxidants		tert-Butylhydroquinone
	—		Triclosan
	Foam stabilizer		Triethylcitrate
Pharmaceuticals	Pharmaceuticals transformation products	—	1-Hydroxy Ibuprofen
	Antiinflammatory		Aceclofenac

(Continued)

Table 4−1 (Continued)

Use category I	Use category II	Acronym	Individual substance
	Analgesic		Acetaminophen (Paracetamol)
	Antibacterial		Amoxicillin
	Beta-Blockers		Atenolol
	Antibacterial		Azithromycin
	Steroids and hormones		Betamethasone
	Diuretic		Caffeine
	Steroids and hormones		Dexamethasone
	Antiinflammatory		Ibuprofen
	Antibacterial		Penicillin G
	Beta-blockers	—	Propranolol
Plant protection products	Pesticide metabolites	—	2-Aminobenzimidazole
	Herbicides		Aclonifen
	—		Bentazone
	Herbicides		Bromacil
	—	—	Cypermethrin
			Deltamethrin
			Iodofenphos
			Malathion
	Insecticides		Metrifonate (Trichlorfon)
			Parathion
	—		Permethrin
			Thiamethoxam
Plasticizers	Household products	BBP	Benzylbutylphthalate
	—	—	Bisphenol A
		—	Diethyl phthalate
	—	NBBS	N-butyl-benzenesulfonamide
	Technical additives	TIBP	Tris(2-methylpropyl) phosphate
Surfactants	Detergents	NPE2C	2-(2-(4-Nonylphenoxy)ethoxy)acetic acid
		NPE2O group	4-Nonylphenol di-ethoxylate / 2-(2-(4-Nonylphenoxy)ethoxy)ethanol
		—	Naphthalene sulfonic acid
Trace metals and their compounds	—	—	Tetraethyl lead
Other	—	—	Tetramethyl lead
			1-Decanol
		TRCP	1,2,3-Trichloropropene
		—	2-(Methylthio)benzothiazol
			2-Bromophenol
			Androstenone
			Chlorate
			Formylpiperidine
			Nitrobenzene

Human contamination with EPs, through exposure to contaminated water or air, as well as activities related to the handling of these compounds, can lead to several human health problems. Thus, mechanisms for attenuation and environmental remediation of these compounds increasingly need to be developed to minimize the possible impacts caused by EPs. Some remediation methods such as adsorption, biological treatment, coagulation, photocatalytic method, ozonation, ion exchange, and membrane separation have been developed in the literature (Barrocas, Chiavassa, Conceição Oliveira, & Monteiro, 2020; Benjedim et al., 2020). Among these remediation techniques, the biodegradation process is very attractive to be used in the wastewater treatment process containing EPs, especially considering the toxicity of EPs.

Among the main EPs, we can highlight pharmaceutical products, which can be a major threat to human health and wildlife around the world (Tan et al., 2020). As can be seen in the classification of the NORMAN network (NORMAN, 2020a) and summarized in Table 4–1, the number of pharmaceutical products classified as EPs is enormous because of the increasing need of society for pharmaceutical products. Different pharmaceutical products are found as residues in natural waters worldwide due to the inadequate disposal by the human population as well as the absence of an efficient wastewater treatment mechanism, mainly antidepressant (Fursdon, Martin, Bertram, Lehtonen, & Wong, 2019; Tan et al., 2020), antibacterial (Ojemaye & Petrik, 2019; Orimolade, Koiki, Peleyeju, & Arotiba, 2019; Xia et al., 2020), antiinflammatory (Li, Lian, Wu, Zou, & Tan, 2020; Zhou et al., 2019), analgesic (Lladó, Solé-Sardans, Lao-Luque, Fuente, & Ruiz, 2016; Orimolade et al., 2019), anticonvulsant (Guégan, De Oliveira, Le Gleuher, & Sugahara, 2020; Ojemaye & Petrik, 2019), steroids and hormones (Acheampong et al., 2019; Zhou et al., 2019), diuretic (Zhou et al., 2019), antiulcerative (Park, Moon, Shin, & Kim, 2018; Zhou et al., 2019), beta-blockers (Guégan et al., 2020; Zhou et al., 2019), and others. Thus, the continuous inadequate disposal of pharmaceutical compounds in natural waters can represent serious environmental problems for aquatic species, as well as for the ecosystem, due to the bioaccumulation ability of these compounds.

Plant protection products are another category of EPs that can represent serious environmental pollution problems, this category is represented by pesticides, which according to the European Commission (European Parliament, 2009) the term includes, among others, herbicides, insecticides, acaricides, fungicides, nematicides, molluscicides, rodenticides, growth regulators, repellents, rodenticides, and biocides. However, despite several advantages of using pesticides in the agricultural and industrial sectors, such as increased productivity and pest control, the indiscriminate use of these pesticides worldwide has caused several environmental problems, especially in the direct contamination of the environment, arable soil, aquatic bodies, agricultural products, as well as in the indirect contamination of fruits, vegetables, and food for human and animal consumption (Wang, Jing, et al., 2020; Wang, Yang, et al., 2020; Zhang et al., 2020), mainly due to the ecotoxicity, teratogenicity, carcinogenicity, and/or mutagenicity of the pesticide residues that may remain in the crop (Li, 2018; Ock, Kim, & Choi, 2020; Rahman, Islam, Haque, & Shahjahan, 2019; Santos et al., 2019; Wang, Jing, et al., 2020; Wang, Yang, et al., 2020; Zhang et al., 2020).

Personal care products are also classified by the NORMAN network (NORMAN, 2020a) as EPs, especially products used as sunscreens, body lotions, perfumes, lipstick, and others, as well as household chemicals. Generally, these EPs enter the environment, especially in aquatic bodies, through inappropriate disposal of domestic sewage, wastewater, leaching of contaminated solid products, and so on (Rachel, Kranert, & Philip, 2020; Wang, Jing, et al., 2020; Zhou et al., 2020). On the other side, these pollutants can also be found in soils, sediments, and sludge with different concentrations (Ramprasad & Philip, 2018). Thus, effective wastewater treatment techniques containing personal care products, as well as EPs, need to be developed to mitigate the impacts caused to the environment, human health, and living species, due to the release of these compounds in water bodies. Since, in most countries, traditional water treatment methods do not efficiently remove EPs from treated water and sewage (Ramprasad & Philip, 2018; Wang, Jing, et al., 2020).

The eutrophication of water bodies, especially freshwater intended for human consumption, is a global environmental problem, and this process is mainly due to human activities (Chen, Ding, & Zhou, 2020; Modley et al., 2019). In this way, human eutrophication considerably raises the concentrations of pollutants in aquatic ecosystems, such as rivers, ponds, lakes, or slow-flowing waters, which can considerably increase the concentration of nutrients such as phosphorus, nitrogen, and others (Chen et al., 2020). Thus, in the face of this problem, freshwater bodies can lead to the proliferation of cyanotoxins that are generated by cyanobacteria, among cyanotoxins we can highlight microcystins (MCs), which have their growth linked to human activities and climate change (Shahmohamadloo, Simmons, & Sibley, 2020; Shan et al., 2020; Woller-Skar, Russell, Gaskill, & Luttenton, 2020; Zaki, Merican, Muangmai, Convey, & Broady, 2020). MCs are classified as hepatotoxic cyclic heptapeptides and according to Zaki et al. (2020) they can cause hepatocellular apoptosis, as well as induce DNA oxidative destruction. According to the same authors, these hepatotoxins are responsible globally for the poisoning of domestic animals, pets, and wildlife, as well as they may represent a threat to human health (Liu, Jiang, et al., 2020; Liu, Rowe, et al., 2020; Woller-Skar et al., 2020; Zaki et al., 2020). In this way, MCs are listed as EPs according to the classification of the NORMAN network because of their deleterious effects on the environment (Shahmohamadloo, Poirier, Ortiz Almirall, Bhavsar, & Sibley, 2020; Shahmohamadloo, Simmons, et al., 2020; Shan et al., 2020). Given the deleterious properties of the EPs mentioned here, we can suggest that the biodegradation process can be an efficient method for the remediation of these pollutants found in the environment.

4.3 Nanoparticle chemistry

With the advent of nanotechnology since the last century, several materials on a nanoscale have been produced. NPs include particulate substances in 0D, 1D, 2D, or 3D with dimensions less than 100 nm, which have their physical and chemical properties influenced by the size of their dimensions (Jeevanandam, Barhoum, Chan, Dufresne, & Danquah, 2018; Khan, Saeed, & Khan, 2019).

NPs are composed of three layers: (1) the surface layer, which can be functionalized with a wide variety of molecules, metal ions, surfactants, and polymers; (2) the shell layer, which is a material that is chemically different from the core in all aspects; and (3) the core, which is essentially the central portion of the NPs (Shin, Cho, Kannan, Lee, & Kim, 2016). Due to these particularities, NPs aroused immense interest from researchers in multidisciplinary fields (Alsaba, Al Dushaishi, & Abbas, 2020; Ealia & Saravanakumar, 2017; Kaur & Gupta, 2009).

With an emphasis on morphology, size of its dimensions, and chemical properties, NPs are divided into several classes, namely, metallic, semiconductor, polymeric, ceramic, lipid-based, carbon-based, among others (Khan et al., 2019).

Metallic NPs have unique optoelectrical properties and are prepared solely from the use of metallic precursors. The synthesis controlled by facet, size, and shape of metallic NPs is important in materials with advanced properties currently prepared due to their advanced multifunctional optical properties (Dreaden, Alkilany, Huang, Murphy, & El-Sayed, 2012). For example, coating the surface of materials with gold NPs is a common procedure in scanning electron microscopy analyses to obtain high-resolution microscopic images.

Ceramic NPs are obtained from nonmetallic solids and can be found in amorphous, polycrystalline, dense, porous, or hollow forms (Sigmund et al., 2006). The robustness and other innate properties of ceramics offer technology for catalysis, biomedical sciences, and optics. Jiang et al. (2017) synthesized chiral ceramic NPs dispersed in ethanol, WO_3-x·H_2O, and observed that the proximity of amino acids to the mineral surface was associated with the catalytic abilities of NP that facilitated the formation of peptide bonds, leading to dipeptides Asp-Asp and Asp-Pro (Jiang et al., 2017).

Polymeric NPs are organic materials presented in the form of nanospheres or nanocapsules (Mansha, Khan, Ullah, & Qurashi, 2017). Nanospheres are matrix particles that present mostly solid mass and other molecules that are absorbed in the outer limit of the spherical surface. The nanocapsules have solid mass encapsulated completely inside the particle (Rao & Geckeler, 2011). Polymeric NPs are readily functionalized and, therefore, used for different drug formulations, including adsorption, dissolution, entrapment, encapsulation, or chemical bonding of drug molecules. Feng et al. (2010) prepared NPs by an electrostatic assembly of cationic-conjugated polymer (PFO) and poly(l-glutamic acid) anionic drug combined with anticancer drug doxorubicin (PFO/PG-Dox). They observed that after NPs are exposed to carboxypeptidase or taken up by cancer cells, the poly(l-glutamic acid) was hydrolyzed and released Dox in cancer cells.

Lipid-based NPs contain lipid moieties and have a solid core made of lipids and a matrix that contains soluble lipophilic molecules and are therefore effectively used in many biomedical applications (Khan et al., 2019).

Semiconductor NPs have properties between metals and non-metals, have wide-bandgap intervals, therefore, undergo significant changes in their properties with bandgap interval adjustment. Due to these characteristics, they are very important materials in photocatalysis, photo optics, and electronics (Sun, 2000). The most efficient application of these NPs is in photocatalytic water splitting due to the proper bandgap position of certain semiconductors.

Li et al. (2019) doped TiO_2 NPs during water photolysis at elevated temperatures, and the production of H_2 was observed with an evolution rate greater than 11,000 μmol/g/h without any sacrificial reagent at 270°C.

Carbon-based NPs are represented by two classes: fullerenes and carbon nanotubes. Fullerenes are nanomaterials made of the hollow globular cage, as allotropic forms of carbon and have several properties such as electrical conductivity, high strength, structure, electronic affinity, and versatility (Astefanei, Núñez, & Galceran, 2015). Nanotubes are elongated tubular structures, 1−2 nm in diameter (Ibrahim, 2013). Due to their unique characteristics, these NPs are used in many commercial applications, such as fillers, efficient gas adsorbents for environmental remediation and as a support medium for different inorganic and organic catalysts, among others (Khan et al., 2019; Mabena, Sinha Ray, Mhlanga, & Coville, 2011; Ngoy, Wagner, Riboldi, & Bolland, 2014; Saeed & Khan, 2014).

4.3.1 Preparation of NPs

The development of fast and reliable experimental protocols for the synthesis of NPs is one of the main aspects of current research in nanotechnology. The methods for the synthesis of NPs are divided into two approaches: (1) bottom-up approach and (2) top-down approach (see Fig. 4−1) (Wang & Xia, 2004).

In the top-down method, NPs are synthesized by breaking, molding, and slicing bulk materials. This approach has the advantage of synthesizing nanomaterials in large quantities and in a short time. In the bottom-up method, atoms or molecules are assembled step by

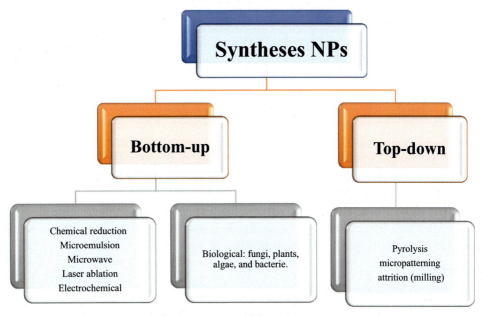

FIGURE 4–1 NPs preparation approach: (A) bottom-up and (B) top-down.

step to build NPs. This approach has the advantage of synthesizing homogeneous nanostructures with perfect crystallographic and surface structures. However, the methods have some disadvantages. Thus, the search for the development of new synthetic nanoparticle routes with controlled and adjustable properties constitutes a wide and challenging field of investigation.

4.3.1.1 Bottom-up approach

In this approach NPs that have appreciable stability, highly desirable for the performance of their applications are formed from relatively simpler substances and use the chemical properties of atoms or molecules that are assembled step by step to manufacture NPs using the concepts of molecular self-assembly and/or molecular recognition. The most common methods are presented below (Pareek, Bhargava, Gupta, Jain, & Panwar, 2017).

Chemical reduction

This method is the most used in the preparation of metallic NPs in solutions. It involves reducing salt in the presence of various reducing agents (e.g., hydrogen, hydrazine, alcohols, carbon monoxide, $LiAlH_4$, $NaBH_4$) in an appropriate medium containing a stabilizing agent (Mishra, Chandran, & Khan, 2014). By this method, gold NPs are conventionally prepared by reducing Au^+ ions in aqueous solutions. Turkevich, Stevenson, and Hillier (1951) demonstrated the citrate-mediated reduction of the aqueous $HAuCl_4$ solution for the synthesis of Au NPs with a size of 20 nm.

Microemulsion

NPs are synthesized by dispersing the reagents (salts and reducing agent) in two separate emulsions (water in oil or water in water) and mixing them in the presence of surfactants. The Brownian movement resulting from the developed micelles leads to intermicellar collisions. As a result, the reactants are mixed to form nucleation followed by NPs (Zieliska-Jurek, Reszczyska, Grabowska, & Zalesk, 2012). Ganguli, Ahmad, Vaidya, and Ahmed (2008) synthesized, characterized and discussed their dielectric properties of various NPs ($SrTiO_3$, $BaTiO_3$, Sr_2TiO_4, Ba_2TiO_4, $SrZrO_3$, (Ba, Pb)ZrO_3, CeO_2, ZrO_2, SnO_2, and $CaCO_3$) by the reverse micellar route.

Microwave

The use of microwave irradiations is considered a fast and simple method that allows high selectivity of shape and size and has been successfully implemented for the synthesis of Au, Ag, and Pt NPs (Walters & Parkin, 2009). The principle of the technique consists of irradiations use for the synthesis of "one-pot" of metallic NPs using the solution of metal salts and polymer surfactants. Henam, Ahmad, Shah, Parveen, and Wani (2019) prepared NPs of copper(II) oxide and iron(III) oxide by this method.

Laser ablation

The laser ablation process involves removing material from a solid surface by irradiation with a laser beam. This method emerged as a potential and reliable alternative for the synthesis of noble metals NPs in comparison to the chemical reduction method. It is considered a simple and versatile method because it allows the preparation of NPs with control of the dimensions. The main advantage of the method refers to the possibility of synthesizing metallic NPs in aqueous and organic solvents (Amendola, Polizzi, & Meneghetti, 2006). Jiménez, Abderrafi, Abargues, Valdés, and Martínez-Pastor (2010) reported a simple and powerful method for the rapid and scalable synthesis of colloidal metal-silica NPs. It was based on laser ablation of a solid target submerged in an aqueous solution of metal salts, the reduction of which gave rise to NPs. The technique was applied for the manufacture of several inert silica metal NPs (silver, gold, and silver-gold).

Electrochemical method
In this method, NPs are prepared by dissolving a metal plate at the anode and then reducing the intermediate metal salt that is formed at the cathode. The main advantage of this method is that there is no need to use templates (Yanilkin, Nastapova, Nasretdinova, & Osin, 2017).

Green synthesis (biological)
This process is based on the principles of green chemistry and has attracted many researchers due to the viability and less toxic nature of the processes. One of the main attributes of this method is to imitate nature's approaches to the synthesis of NPs, that is, they are economic and ecological processes where the synthesis of NPs is carried out through biological systems using water as a solvent. Bacteria, fungi, aloe vera, human cells, and others are used for the synthesis of NPs (Khan et al., 2019).

4.3.1.2 Top-down approach
In this approach, NPs are prepared essentially by the cleavage of a larger entity to generate small subset entities with the aid of tools with the aim of cutting, milling, molding, and sculpting the bulk materials to the desired size and shape (Pareek et al., 2017). The most common methods are presented subsequently.

Pyrolysis
In this method, aggregates and agglomerates are formed that have a wide particle size distribution resulting from the burning of a precursor in the form of steam. The main limitation is that the method requires an enormous amount of energy (Pareek et al., 2017).

Micropatterning
This technique is widely used to prepare NPs in electronics, photolithography being the most common micropattern method. The basic principle is the use of light (X-rays or UV), electrons, ions, strongly concentrated electron beam, or electrostatic forces that selectively remove nanosized structures from a precursor and develop ordered arrays of NPs (Pareek et al., 2017).

Attrition (milling)
Industrially, this method is important for the synthesis of materials and consists of grinding macro and microscale materials in a ball mill to generate nanometer-sized particles. In this process, the researchers developed a variety of grinding devices for different purposes since

several parameters influence the size, shape, and associated physical properties of the NPs. In short, for the large-scale production of NPs, the friction process is highly advantageous. This technique offers the opportunity to prepare alloys and nanocomposites, which cannot be synthesized by conventional routes. However, the friction process has some limitations, such as internal stresses, defects, and surface contamination NPs (Pareek et al., 2017).

4.3.2 Applications of nanoparticles

Due to their versatile properties, NPs have many potential applications in the field of electronics, information technology, energy, catalysis, defense and security, cosmetics, environment, and others (Martis, Badve, & Degwekar, 2012; Todescato et al., 2016; Weiss, Takhistov, & McClements, 2006).

4.3.2.1 Catalysis

NPs are commonly used as catalysts due to their remarkable characteristics in terms of effectiveness in the enzymatic load, high surface area, and resistance to mass transfer. Metallic NPs are used as catalysts for various chemical reactions such as hydrogenation, hydroformylation, carbonylation, among others. The catalytic applications of tin oxide NPs have been well reported in oxidation reactions (Bai et al., 2017; Inomata, Albrecht, & Yamamoto, 2018). Manjunathan et al. (2018) reported the application of mesoporous tin oxide NPs as an acid catalyst for the synthesis of 5-hydroxy-2-phenyl-1,3-dioxane and ethanol from acetalization and glycerol ketalization, respectively, and epoxidation cyclohexene oxide cyclohexene.

4.3.2.2 Biomedical applications

NPs are important in the health field because of their ability to administer drugs in the ideal dosage range, usually resulting in greater therapeutic efficiency, fewer side effects, and better patient compliance (Alexis, Pridgen, Molnar, & Farokhzad, 2008). Iron oxide NPs are commonly used in in vivo biomedical applications, such as enhancing contrast by magnetic resonance, tissue repair, detoxification of biological fluids, hyperthermia, drug delivery, and cell separation (Ali et al., 2016). Rajendran, Karunagaran, Mahanty, and Sen (2015) synthesized iron NPs and assessed their possible anticancer activity against HepG2 liver cancer cells. They observed that NPs exhibited significant cytotoxicity in HepG2 cells (Rajendran et al., 2015). Ag NPs are being used more and more in dressings, catheters, and products from various families due to their antimicrobial activity (AshaRani, Low Kah Mun, Hande, & Valiyaveettil, 2009).

4.3.2.3 Energy capture

The concern with the preservation of the environment and the scarcity of oil in the future intensify the search for new sources of energy (Hussein, 2015). Therefore, scientists are changing their research strategies to generate renewable energy from easily available resources at a cheap and environmentally clean cost. In this context, NPs are the best candidates due to their large surface area, optical behavior, and catalytic nature. Especially in photocatalytic applications, NPs are widely used for the generation of H_2 from the electrochemical water splitting (Basheer & Ali, 2019; Tada, Naya, & Fujishima, 2018; Tian et al., 2018). Méndez, González-Millán, and García-Macedo (2019) modified TiO_2 NPs with Pd, Pt, Ag, and Au (0.5% by weight) to evaluate hydrogen production via water splitting using methanol as a hole eliminator at room temperature, atmospheric pressure in UV-vis light. In this work, it was observed that the NPs of TiO_2 based on Au and Pt showed a higher amount and rate of hydrogen production than the respective ones based on Ag and Pd.

4.3.3 Toxicity

Despite various industrial applications, NPs have a degree of toxicity to the environment and living organisms. NPs are expected to be introduced into the environment in increasing quantities in the coming decades, as their use in industrial processes and consumer products increases.

In industrial effluents, the use of NPs for environmental remediation contaminates the environment through water, soil, and air through production processes (emissions to air and water, waste) or, even more likely, as a result of the routine use of products that contain NPs (for example, TiO_2 in sunscreens). Researchers are currently facing the problem of establishing a balance between the positive therapeutic effect of NPs and the side effects related to their toxicity. This has attracted growing concern from a number of interested parties (Sukhanova et al., 2018).

The toxicity of NPs in living organisms is largely determined by their physical and chemical characteristics, such as size, shape, area, surface charge, catalytic activity, and the presence or absence of a shell and active groups on the surface.

NPs are characterized by having a very large surface area resulting in a high catalytic activity and reaction capacity, thus facilitating diffusion processes and gas exchange. NPs easily enter cells and cell organelles due to their sizes that are comparable to the size of protein globules (2–10 nm), DNA helix diameter (2 nm), and thickness of cell membranes (10 nm) (Sukhanova et al., 2018). Pan et al. (2007) evaluated the toxicity of gold NPs with varying sizes (0.8–15 nm). The authors noted that 15 nm NPs are 60 times less toxic than 1.4 nm NPs. These data suggest that 1.4 nm NPs can enter the nucleus.

The toxicity of NPs is strongly dependent on its shape (spheres, ellipsoids, cylinders, leaves, cubes, and stems). For example, Zhao et al. (2013) compared the effects of hydroxyapatite NPs in different ways (needle, plate, stem, and spherical) in cultures of BEAS-2B cells and observed that the plate and needle shapes cause the death of a greater proportion of cells than spherical and rod shapes. This is partly accounted for by the capacity of plate-like and needle-like NPs for damaging cells and tissue upon direct contact.

The chemical composition and crystalline structure of NPs must also be considered to toxicity. Some metal ions, such as Ag and Cd are toxic and therefore cause damage to cells, whereas metal ions, such as Fe and Zn, are biologically useful, but in high concentrations, they can damage cellular pathways and therefore cause high toxicity. Gurr, Wang, Chen, and Jan (2005) demonstrated that NPs with a rutile-like crystal structure (prism-shaped TiO_2

crystals) cause oxidative damage of DNA, lipid peroxidation, and formation of micronuclei, which indicates abnormal chromosome segregation during mitosis, whereas NPs with the anatase-like crystal structure (octahedral TiO_2 crystals) of the same size are nontoxic.

The surface functionality of NPs plays an important role in their toxicity because it largely determines the interactions of NPs with biological systems (Sukhanova et al., 2018). Hühn et al. (2013) observed that positively charged gold NPs were absorbed by cells in large quantities and more quickly than negatively charged NPs. Therefore, to reduce hazardous effects, NPs must be recycled and used throughout the life cycle. Thus, changes in the properties of NPs can reduce the related risks.

4.4 Role of nanoparticles in biodegradation

Nowadays, NPs are receiving tremendous attention from industry and academia due to their potential in developing new highly active materials. Recent studies reveal a new role for NPs as enhancers of biodegradation. NPs influence the growth profile of degrading microorganisms by increasing the rate of biodegradation. Thus, a new paradigm in research consists of supplementing microbial degradation with NPs (See Fig. 4–2).

4.4.1 General considerations of biodegradation

Biodegradation is defined as the chemical degradation caused by the action of naturally occurring microorganisms, such as bacteria and fungi, by enzymatic action, in metabolic products of microorganisms (Poornima et al., 2015).

Recently, the use of microorganisms to degrade emerging contaminants has gained notable importance due to the inefficiency of the physical and chemical disposal methods used for pollutants, as they cause many environmental problems.

Microorganisms play a substantial role in the biological decomposition of emerging contaminants (Shah, Hasan, Hameed, & Ahmed, 2008). Several studies have been published using fungi and bacteria in the process of biodegradation of polyethylene (Artham & Doble, 2010; Esmaeili, Pourbabaee, Alikhani, Shabani, & Kumar, 2014; Orhan & Büyükgüngör, 2000).

The actions of microorganisms in biodegradation are influenced by two different processes: (1) direct action: the deterioration of materials that serve as a nutritive substance for the growth of microorganisms; and (2) indirect action: the influence of the metabolic products of microorganisms, for example, discoloration or additional deterioration (Poornima et al., 2015).

4.4.2 Nanoparticles as enhancers of biodegradation

The harmful effects caused by contaminants emerging in recent years on the environment and human health have focused on the need for attention and awareness for their degradation, especially those that are characteristically inert and resistant to microbial attack and

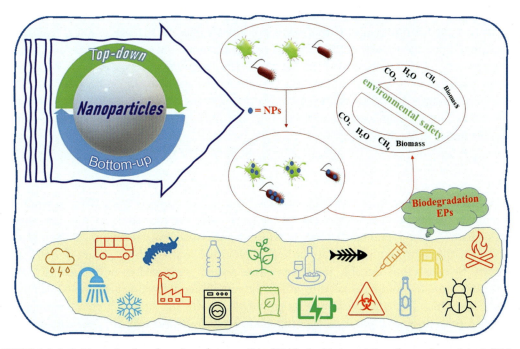

FIGURE 4–2 Probable mechanism of coaction of nanoparticles (NPs) in the biodegradation of emerging pollutants (EPs).

therefore remain in nature without any deformation for a long time, such as petroleum-derived plastics, dye, and insecticides.

4.4.2.1 Iron nanoparticles

Fe NPs ranging in size from 10.6 to 37.8 nm were synthesized and evaluated for the growth profile of the degrading microbial consortium of low-density polyethylene. It was observed that NPs improved the durability of the exponential phase in 36 h, accelerating bacterial growth. Furthermore, NPs with 10.6 nm significantly increased the efficiency of the consortium's biodegradation. The researchers pointed out that bacterial interactions with NPs drastically influence the main metabolic processes, that is, NPs and consortia help to capture O_2 from the organic chain, further facilitating degradation by bioactive hydrolysis (Kapri, Zaidi, Satlewal, & Goel, 2010).

The biodegradation of ciprofloxacin (CIP) was investigated using magnetite NPs (Fe_3O_4). It was observed that in the presence of 2-bromoethanesulfonate (BES) the biodegradation of CIP in enrichments supplemented with Fe_3O_4 was concomitant with an inhibition in the production of methane and 67% higher than in enrichments not corrected by Fe_3O_4. Fe(II) concentrations were also significantly increased in BES and enrichments supplemented with Fe_3O_4. This indicated that there may be a positive relationship between CIP biodegradation and the microbial reduction from Fe(III) to Fe(II) (Yang et al., 2017).

The strain of *Bacillus fusiformis* (BFN) was used for the biodegradation of phenol with the help of Ni-doped iron-based NPs that were present at different pH values (3, 6, and 8). It was observed that the growth of BFN and the rate of phenol biodegradation were accelerated in the presence of NPs both at pH 8 and at pH 6. The H_2 generated by the corrosion of iron can be used as an electron donor and an energy source for the cultivation of BFN. Microscopic images with X-ray energy mapping demonstrated that these NPs adhered to the BFN surface, but no significant changes in BFN morphology were observed (Kuang, Zhou, Chen, Megharaj, & Naidu, 2013).

4.4.2.2 Titanium nanoparticles

Polyethylene-doped TiO_2 and TiO_2 NPs were synthesized and evaluated for photolytic and photocatalytic degradation. It has been observed that direct photons from UV rays irradiate the nanocomposite surface, which causes TiO_2 to enter an excited level and, therefore, causes chain scission, branching, and cross-linking by oxidation reactions. Polyethylene-doped TiO_2 NPs showed a higher rate of degradation due to their ability to act as holes and electrons, by which they are able to absorb more oxygen compared to nondoped TiO_2 NPs and release more hydrogen peroxide to attack the remaining matrix; despite the NPs of non-doped TiO_2, which degrade the sample by photolytic activity and show a greater degradation than pure polyethylene. Therefore, according to the degradation capacity, we can deduce: Doped TiO_2 > Undoped TiO_2 > Pure LDPE (Asghar et al., 2011).

4.4.2.3 Silver nanoparticles

Silver NPs were used for the first time in the biodegradation of di (2-Ethylhexyl) phthalate by a bacterial strain. In the absence of NPs the rate of biodegradation was 30% to 66%, while biodegradation in the presence of NPs was 100% in the 72 h period. The formation of a self-assembled monolayer of Ag NPs in the bacterial strain was observed, which in turn improved the biodegradation of di(2-ethylhexyl) phthalate, improving the bioavailability and easy assimilation of di(2-ethylhexyl) phthalate as a carbon source by bacterial cells. In addition, *the Rhodococcus jostii* strain showed good tolerance to Ag toxicity, and the extracellular esterase enzyme in the medium of mineral salts was able to reduce Ag ions and stabilize NPs. This biodegradation approach has technological and ecological efficiency that can be applied to various cleaning micropollutants (Annamalai & Vasudevan, 2020).

4.5 Future prospects

The accelerated population growth has been accompanied by several problems, among them the environmental impacts. The emergence of several classes of emerging contaminants has caused concern about the potential for contamination. Microbial degradation has proven to be effective in minimizing impacts to the environment due to the improper disposal of EPs. Thus, the researchers discovered NPs as enhancers of microbial degradation capacity and measures were then designed to reduce environmental risks and human health with the use of NPs in the biodegradation process.

In the literature, the use of iron-based NPs for biodegradation studies is frequently reported. Therefore, this field of study has the potential to be explored with other documented NPs, such as Ag, Ti, and Si, among others, that influence bacterial growth profiles. The degradation mechanisms are also not known in detail and, consequently, open a new direction for studies related to biodegradation to make the area solid and useful at a commercial level.

Acknowledgments

The authors would like to thank the Institute of Technology and Research at Tiradentes University (ITP-UNIT), Fundação de Amparo à Pesquisa do Estado de São Paulo (FAPESP, Grant 2017/06775-5), the Foundation of Support to Research and Technological Innovation of the State of Sergipe [FAPITEC/SE], the Coordination for the Improvement of Higher Education Personnel (CAPES) under the Finance Code 001 and the National Council for Scientific and Technological Development (CNPq) for financial support (process no 315405/2018-0; 421147/2016-4).

References

Acheampong, E., Dryden, I. L., Wattis, J. A. D., Twycross, J., Scrimshaw, M. D., & Gomes, R. L. (2019). Modelling emerging pollutants in wastewater treatment: A case study using the pharmaceutical 17A − ethinylestradiol. *Computers & Chemical Engineering, 128*, 477−487. Available from https://doi.org/10.1016/j.compchemeng.2019.06.020.

Alexis, F., Pridgen, E., Molnar, L. K., & Farokhzad, O. C. (2008). Factors affecting the clearance and biodistribution of polymeric nanoparticles. *Molecular Pharmaceutics, 5*, 505−515. Available from https://doi.org/10.1021/mp800051m.

Ali, A., Zafar, H., Zia, M., ul Haq, I., Phull, A. R., Ali, J. S., & Hussain, A. (2016). Synthesis, characterization, applications, and challenges of iron oxide nanoparticles. *Nanotechnology, Science and Appllications, 9*, 49−67. Available from https://doi.org/10.2147/NSA.S99986.

Alsaba, M. T., Al Dushaishi, M. F., & Abbas, A. K. (2020). A comprehensive review of nanoparticles applications in the oil and gas industry. *Journal of Petroleum Exploration and Production Technology, 10*, 1389−1399. Available from https://doi.org/10.1007/s13202-019-00825-z.

Amendola, V., Polizzi, S., & Meneghetti, M. (2006). Laser ablation synthesis of gold nanoparticles in organic solvents. *The Journal of Physical Chemistry: B, 110*, 7232−7237. Available from https://doi.org/10.1021/jp0605092.

Annamalai, J., & Vasudevan, N. (2020). Enhanced biodegradation of an endocrine disrupting micro-pollutant: Di (2-ethylhexyl) phthalate using biogenic self-assembled monolayer of silver nanoparticles. *The Science of the Total Environment, 719*, 137115. Available from https://doi.org/10.1016/j.scitotenv.2020.137115.

Artham, T., & Doble, M. (2010). Biodegradation of physicochemically treated polycarbonate by fungi. *Biomacromolecules, 11*, 20−28. Available from https://doi.org/10.1021/bm9008099.

Asghar, W., Qazi, I. A., Ilyas, H., Khan, A. A., Awan, M. A., & Rizwan Aslam, M. (2011). Comparative solid phase photocatalytic degradation of polythene films with doped and undoped TiO 2 nanoparticles. *Journal of Nanomaterials, 2011*, 1−8. Available from https://doi.org/10.1155/2011/461930.

AshaRani, P. V., Low Kah Mun, G., Hande, M. P., & Valiyaveettil, S. (2009). Cytotoxicity and genotoxicity of silver nanoparticles in human cells. *ACS Nano, 3*, 279−290. Available from https://doi.org/10.1021/nn800596w.

Astefanei, A., Núñez, O., & Galceran, M. T. (2015). Characterisation and determination of fullerenes: A critical review. *Analytica Chimica Acta, 882*, 1–21. Available from https://doi.org/10.1016/j.aca.2015.03.025.

Bai, X., Chai, S., Liu, C., Ma, K., Cheng, Q., Tian, Y., . . . Li, X. (2017). Insight into copper oxide-tin oxide catalysts for the catalytic oxidation of carbon monoxide: Identification of active copper species and a reaction mechanism. *ChemCatChem, 9*, 3226–3235. Available from https://doi.org/10.1002/cctc.201700460.

Barrocas, B., Chiavassa, L. D., Conceição Oliveira, M., & Monteiro, O. C. (2020). Impact of Fe, Mn co-doping in titanate nanowires photocatalytic performance for emergent organic pollutants removal. *Chemosphere, 250*, 126240. Available from https://doi.org/10.1016/j.chemosphere.2020.126240.

Basheer, A. A., & Ali, I. (2019). Water photo splitting for green hydrogen energy by green nanoparticles. *International Journal of Hydrogen Energy, 44*, 11564–11573. Available from https://doi.org/10.1016/j.ijhydene.2019.03.040.

Benjedim, S., Romero-Cano, L. A., Pérez-Cadenas, A. F., Bautista-Toledo, M. I., Lotfi, E. M., & Carrasco-Marín, F. (2020). Removal of emerging pollutants present in water using an *E-coli* biofilm supported onto activated carbons prepared from argan wastes: Adsorption studies in batch and fixed bed. *The Science of the Total Environment, 720*, 137491. Available from https://doi.org/10.1016/j.scitotenv.2020.137491.

Chen, G., Ding, X., & Zhou, W. (2020). Study on ultrasonic treatment for degradation of Microcystins (MCs). *Ultrasonics Sonochemistry, 63*, 104900. Available from https://doi.org/10.1016/j.ultsonch.2019.104900.

Chen, Y., Liang, W., Li, Y., Wu, Y., Chen, Y., Xiao, W., . . . Li, H. (2019). Modification, application and reaction mechanisms of nano-sized iron sulfide particles for pollutant removal from soil and water: A review. *Chemical Engineering Journal, 362*, 144–159. Available from https://doi.org/10.1016/j.cej.2018.12.175.

Costa, J. A. S., de Jesus, R. A., Santos, D. O., Mano, J. F., Romão, L. P. C., & Paranhos, C. M. (2020). Recent progresses in the adsorption of organic, inorganic, and gas compounds by MCM-41-based mesoporous materials. *Microporous Mesoporous Materials, 291*, 109698. Available from https://doi.org/10.1016/j.micromeso.2019.109698.

Dreaden, E. C., Alkilany, A. M., Huang, X., Murphy, C. J., & El-Sayed, M. A. (2012). The golden age: Gold nanoparticles for biomedicine. *Chemical Society Reviews, 41*, 2740–2779. Available from https://doi.org/10.1039/C1CS15237H.

Ealia, A. M., & Saravanakumar, M. P. (2017). A review on the classification, characterisation, synthesis of nanoparticles and their application. *IOP Conference Series: Materials Science and Engineering, 263*, 032019. Available from https://doi.org/10.1088/1757-899X/263/3/032019.

EL-Sheshtawy, H. S., El-Hosainy, H. M., Shoueir, K. R., El-Mehasseb, I. M., & El-Kemary, M. (2019). Facile immobilization of Ag nanoparticles on g-C3N4 /V2O5 surface for enhancement of post-illumination, catalytic, and photocatalytic activity removal of organic and inorganic pollutants. *Applied Surface Science, 467–468*, 268–276. Available from https://doi.org/10.1016/j.apsusc.2018.10.109.

Esmaeili, A., Pourbabaee, A. A., Alikhani, H. A., Shabani, F., & Kumar, L. (2014). Colonization and biodegradation of photo-oxidized low-density polyethylene (LDPE) by new strains of *Aspergillus* sp. and *Lysinibacillus* sp. *Bioremediation Journal, 18*, 213–226. Available from https://doi.org/10.1080/10889868.2014.917269.

European Parliament. (2009). Directive 2009/128/EC of the European Parliament and the Council of 21 October 2009 establishing a framework for Community action to achieve the sustainable use of pesticides. *Official Journal of the European Union, 309*, 71–86. Available from https://doi.org/10.3000/17252555.L_2009.309.

Feizpoor, S., Habibi-Yangjeh, A., Yubuta, K., & Vadivel, S. (2019). Fabrication of TiO2/CoMoO4/PANI nanocomposites with enhanced photocatalytic performances for removal of organic and inorganic pollutants under visible light. *Materials Chemistry and Physics, 224*, 10–21. Available from https://doi.org/10.1016/j.matchemphys.2018.11.076.

Feng, X., Lv, F., Liu, L., Tang, H., Xing, C., Yang, Q., & Wang, S. (2010). Conjugated polymer nanoparticles for drug delivery and imaging. *ACS Applied Materials & Interfaces, 2*, 2429−2435. Available from https://doi.org/10.1021/am100435k.

Fursdon, J. B., Martin, J. M., Bertram, M. G., Lehtonen, T. K., & Wong, B. B. M. (2019). The pharmaceutical pollutant fluoxetine alters reproductive behaviour in a fish independent of predation risk. *The Science of the Total Environment, 650*, 642−652. Available from https://doi.org/10.1016/j.scitotenv.2018.09.046.

Ganguli, A. K., Ahmad, T., Vaidya, S., & Ahmed, J. (2008). Microemulsion route to the synthesis of nanoparticles. *Pure and Applied Chemistry. Chimie Pure et Appliquee, 80*, 2451−2477. Available from https://doi.org/10.1351/pac200880112451.

Guégan, R., De Oliveira, T., Le Gleuher, J., & Sugahara, Y. (2020). Tuning down the environmental interests of organoclays for emerging pollutants: Pharmaceuticals in presence of electrolytes. *Chemosphere, 239*. Available from https://doi.org/10.1016/j.chemosphere.2019.124730.

Gurr, J.-R., Wang, A. S. S., Chen, C.-H., & Jan, K.-Y. (2005). Ultrafine titanium dioxide particles in the absence of photoactivation can induce oxidative damage to human bronchial epithelial cells. *Toxicology, 213*, 66−73. Available from https://doi.org/10.1016/j.tox.2005.05.007.

Henam, S. D., Ahmad, F., Shah, M. A., Parveen, S., & Wani, A. H. (2019). Microwave synthesis of nanoparticles and their antifungal activities. *Spectrochimica Acta Part A: Molecular and Biomolecular Spectroscopy, 213*, 337−341. Available from https://doi.org/10.1016/j.saa.2019.01.071.

Hühn, D., Kantner, K., Geidel, C., Brandholt, S., De Cock, I., Soenen, S. J. H., ... Parak, W. J. (2013). Polymer-coated nanoparticles interacting with proteins and cells: Focusing on the sign of the net charge. *ACS Nano, 7*, 3253−3263. Available from https://doi.org/10.1021/nn3059295.

Hussein, A. K. (2015). Applications of nanotechnology in renewable energies—A comprehensive overview and understanding. *Renewable and Sustainable Energy Reviews, 42*, 460−476. Available from https://doi.org/10.1016/j.rser.2014.10.027.

Ibrahim, K. S. (2013). Carbon nanotubes-properties and applications: A review. *Carbon Letters, 14*, 131−144. Available from https://doi.org/10.5714/CL.2013.14.3.131.

Inomata, Y., Albrecht, K., & Yamamoto, K. (2018). Size-dependent oxidation state and CO oxidation activity of tin oxide clusters. *ACS Catalysis, 8*, 451−456. Available from https://doi.org/10.1021/acscatal.7b02981.

Jeevanandam, J., Barhoum, A., Chan, Y. S., Dufresne, A., & Danquah, M. K. (2018). Review on nanoparticles and nanostructured materials: History, sources, toxicity and regulations. *Beilstein Journal of Nanotechnology, 9*, 1050−1074. Available from https://doi.org/10.3762/bjnano.9.98.

Jiang, S., Chekini, M., Qu, Z.-B., Wang, Y., Yeltik, A., Liu, Y., ... Kotov, N. A. (2017). Chiral ceramic nanoparticles and peptide catalysis. *Journal of the American Chemical Society, 139*, 13701−13712. Available from https://doi.org/10.1021/jacs.7b01445.

Jiménez, E., Abderrafi, K., Abargues, R., Valdés, J. L., & Martínez-Pastor, J. P. (2010). Laser-ablation-induced Synthesis of SiO 2-Capped noble metal nanoparticles in a single step. *Langmuir: The ACS Journal of Surfaces and Colloids, 26*, 7458−7463. Available from https://doi.org/10.1021/la904179x.

Kapri, A., Zaidi, M. G. H., Satlewal, A., & Goel, R. (2010). SPION-accelerated biodegradation of low-density polyethylene by indigenous microbial consortium. *International Biodeterioration & Biodegradation, 64*, 238−244. Available from https://doi.org/10.1016/j.ibiod.2010.02.002.

Kaur, A., & Gupta, U. (2009). A review on applications of nanoparticles for the preconcentration of environmental pollutants. *Journal of Materials Chemistry, 19*, 8279. Available from https://doi.org/10.1039/b901933b.

Khan, I., Saeed, K., & Khan, I. (2019). Nanoparticles: Properties, applications and toxicities. *Arabian Journal of Chemistry, 12*, 908−931. Available from https://doi.org/10.1016/j.arabjc.2017.05.011.

Kuang, Y., Zhou, Y., Chen, Z., Megharaj, M., & Naidu, R. (2013). Impact of Fe and Ni/Fe nanoparticles on biodegradation of phenol by the strain Bacillus fusiformis (BFN) at various pH values. *Bioresource Technology*, *136*, 588–594. Available from https://doi.org/10.1016/j.biortech.2013.03.018.

Li, Y., Lian, J., Wu, B., Zou, H., & Tan, S. K. (2020). Phytoremediation of pharmaceutical-contaminated wastewater: Insights into rhizobacterial dynamics related to pollutant degradation mechanisms during plant life cycle. *Chemosphere*, *253*, 126681. Available from https://doi.org/10.1016/j.chemosphere.2020.126681.

Li, Y., Peng, Y.-K., Hu, L., Zheng, J., Prabhakaran, D., Wu, S., ... Tsang, S. C. E. (2019). Photocatalytic water splitting by N-TiO2 on MgO (111) with exceptional quantum efficiencies at elevated temperatures. *Nature Communications*, *10*, 4421. Available from https://doi.org/10.1038/s41467-019-12385-1.

Li, Z. (2018). Health risk characterization of maximum legal exposures for persistent organic pollutant (POP) pesticides in residential soil: An analysis. *Journal of Environmental Management*, *205*, 163–173. Available from https://doi.org/10.1016/j.jenvman.2017.09.070.

Liu, J., Jiang, J., Meng, Y., Aihemaiti, A., Xu, Y., Xiang, H., ... Chen, X. (2020). Preparation, environmental application and prospect of biochar-supported metal nanoparticles: A review. *Journal of Hazardous Materials*, *388*, 122026. Available from https://doi.org/10.1016/j.jhazmat.2020.122026.

Liu, Q., Rowe, M. D., Anderson, E. J., Stow, C. A., Stumpf, R. P., & Johengen, T. H. (2020). Probabilistic forecast of microcystin toxin using satellite remote sensing, in situ observations and numerical modeling. *Environmental Modelling & Software*, *128*, 104705. Available from https://doi.org/10.1016/j.envsoft.2020.104705.

Lladó, J., Solé-Sardans, M., Lao-Luque, C., Fuente, E., & Ruiz, B. (2016). Removal of pharmaceutical industry pollutants by coal-based activated carbons. *Process Safety and Environmental Protection*, *104*, 294–303. Available from https://doi.org/10.1016/j.psep.2016.09.009.

Mabena, L. F., Sinha Ray, S., Mhlanga, S. D., & Coville, N. J. (2011). Nitrogen-doped carbon nanotubes as a metal catalyst support. *Applied Nanosciences*, *1*, 67–77. Available from https://doi.org/10.1007/s13204-011-0013-4.

Manjunathan, P., Marakatti, V. S., Chandra, P., Kulal, A. B., Umbarkar, S. B., Ravishankar, R., & Shanbhag, G. V. (2018). Mesoporous tin oxide: An efficient catalyst with versatile applications in acid and oxidation catalysis. *Catalysis Today*, *309*, 61–76. Available from https://doi.org/10.1016/j.cattod.2017.10.009.

Mansha, M., Khan, I., Ullah, N., & Qurashi, A. (2017). Synthesis, characterization and visible-light-driven photoelectrochemical hydrogen evolution reaction of carbazole-containing conjugated polymers. *International Journal of Hydrogen Energy*, *42*, 10952–10961. Available from https://doi.org/10.1016/j.ijhydene.2017.02.053.

Martis, E., Badve, R., & Degwekar, M. (2012). Nanotechnology based devices and applications in medicine: An overview. *Chronicles Young Science*, *3*, 68. Available from https://doi.org/10.4103/2229-5186.94320.

Méndez, F. J., González-Millán, A., & García-Macedo, J. A. (2019). Surface modification of titanium oxide as efficient support of metal nanoparticles for hydrogen production via water splitting. *Materials Chemistry and Physics*, *232*, 331–338. Available from https://doi.org/10.1016/j.matchemphys.2019.04.057.

Mishra, R. R., Chandran, P., & Khan, S. S. (2014). Equilibrium and kinetic studies on adsorptive removal of malachite green by the citrate-stabilized magnetite nanoparticles. *RSC Advances*, *4*, 51787–51793. Available from https://doi.org/10.1039/C4RA07651F.

Modley, L. A. S., Rampedi, I. T., Avenant-Oldewage, A., Mhuka, V., Nindi, M., & Van Dyk, C. (2019). Microcystin concentrations and liver histopathology in *Clarias gariepinus* and *Oreochromis mossambicus* from three impacted rivers flowing into a hyper-eutrophic freshwater system: A pilot study. *Environmental Toxicology and Pharmacology*, *71*, 103222. Available from https://doi.org/10.1016/j.etap.2019.103222.

Ngoy, J. M., Wagner, N., Riboldi, L., & Bolland, O. (2014). A CO2 capture technology using multi-walled carbon nanotubes with polyaspartamide surfactant. *Energy Procedia, 63*, 2230–2248. Available from https://doi.org/10.1016/j.egypro.2014.11.242.

NORMAN (2020a). *Network of reference laboratories, research centres and related organisations for monitoring of emerging environmental substances.* NORMAN. <https://www.norman-network.net/> Accessed 5.18.20.

NORMAN (2020b). *List of emerging substance.* NORMAN. <https://www.norman-network.net/?q = node/81#sub29> Accessed 5.18.20.

Ock, J., Kim, J., & Choi, Y. H. (2020). Organophosphate insecticide exposure and telomere length in U.S. adults. *The Science of the Total Environment, 709*, 135990. Available from https://doi.org/10.1016/j.scitotenv.2019.135990.

Ojemaye, C. Y., & Petrik, L. (2019). Occurrences, levels and risk assessment studies of emerging pollutants (pharmaceuticals, perfluoroalkyl and endocrine disrupting compounds) in fish samples from Kalk Bay harbour, South Africa. *Environmental Pollution (Barking, Essex: 1987), 252*, 562–572. Available from https://doi.org/10.1016/j.envpol.2019.05.091.

Orhan, Y., & Büyükgüngör, H. (2000). Enhancement of biodegradability of disposable polyethylene in controlled biological soil. *International Biodeterioration & Biodegradation, 45*, 49–55. Available from https://doi.org/10.1016/S0964-8305(00)00048-2.

Orimolade, B. O., Koiki, B. A., Peleyeju, G. M., & Arotiba, O. A. (2019). Visible light driven photoelectrocatalysis on a FTO/BiVO 4/BiOI anode for water treatment involving emerging pharmaceutical pollutants. *Electrochimica Acta, 307*, 285–292. Available from https://doi.org/10.1016/j.electacta.2019.03.217.

Pan, Y., Neuss, S., Leifert, A., Fischler, M., Wen, F., Simon, U., . . . Jahnen-Dechent, W. (2007). Size-dependent cytotoxicity of gold nanoparticles. *Small (Weinheim an der Bergstrasse, Germany), 3*, 1941–1949. Available from https://doi.org/10.1002/smll.200700378.

Pareek, V., Bhargava, A., Gupta, R., Jain, N., & Panwar, J. (2017). Synthesis and applications of noble metal nanoparticles: A review. *Advanced Science, Engineering and Medicine, 9*, 527–544. Available from https://doi.org/10.1166/asem.2017.2027.

Park, J., Moon, Gh, Shin, K. O., & Kim, J. (2018). Oxalate-TiO2 complex-mediated oxidation of pharmaceutical pollutants through ligand-to-metal charge transfer under visible light. *Chemical Engineering Journal, 343*, 689–698. Available from https://doi.org/10.1016/j.cej.2018.01.078.

Peña-Guzmán, C., Ulloa-Sánchez, S., Mora, K., Helena-Bustos, R., Lopez-Barrera, E., Alvarez, J., & Rodriguez-Pinzón, M. (2019). Emerging pollutants in the urban water cycle in Latin America: A review of the current literature. *Journal of Environmental Management, 237*, 408–423. Available from https://doi.org/10.1016/j.jenvman.2019.02.100.

Poornima, P., Ey., Swati, P., Harshita., Manimita., Shraddha., . . . Tiwari, A. (2015). Nanoparticles accelerated in-vitro biodegradation of LDPE: A review. *Advances in Applied Science Research, 6*.

Rachel, A., Kranert, M., & Philip, L. (2020). Fate and impact of pharmaceuticals and personal care products during septage co-composting using an in-vessel composter. *Waste Management (New York, N.Y.), 109*, 109–118. Available from https://doi.org/10.1016/j.wasman.2020.04.053.

Rahman, M. S., Islam, S. M. M., Haque, A., & Shahjahan, M. (2020). Toxicity of the organophosphate insecticide sumithion to embryo and larvae of zebrafish. *Toxicology Reports, 7*, 317–323. Available from https://doi.org/10.1016/j.toxrep.2020.02.004.

Rajendran, K., Karunagaran, V., Mahanty, B., & Sen, S. (2015). Biosynthesis of hematite nanoparticles and its cytotoxic effect on HepG2 cancer cells. *International Journal of Biological Macromolecules, 74*, 376–381. Available from https://doi.org/10.1016/j.ijbiomac.2014.12.028.

Ramprasad, C., & Philip, L. (2018). Contributions of various processes to the removal of surfactants and personal care products in constructed wetland. *Chemical Engineering Journal, 334*, 322–333. Available from https://doi.org/10.1016/j.cej.2017.09.106.

Rao, J. P., & Geckeler, K. E. (2011). Polymer nanoparticles: Preparation techniques and size-control parameters. *Progress in Polymer Science, 36*, 887−913. Available from https://doi.org/10.1016/j.progpolymsci.2011.01.001.

Saeed, K., & Khan, I. (2014). Preparation and properties of single-walled carbon nanotubes/poly(butylene terephthalate) nanocomposites. *Iranian Polymer Journal, 23*, 53−58. Available from https://doi.org/10.1007/s13726-013-0199-2.

Santos, L. F. S., de Jesus, R. A., Costa, J. A. S., Gouveia, L. G. T., de Mesquita, M. E., & Navickiene, S. (2019). Evaluation of MCM-41 and MCM-48 mesoporous materials as sorbents in matrix solid phase dispersion method for the determination of pesticides in soursop fruit (*Annona muricata*). *Inorganic Chemistry Communications, 101*, 45−51. Available from https://doi.org/10.1016/j.inoche.2019.01.013.

Shah, A. A., Hasan, F., Hameed, A., & Ahmed, S. (2008). Biological degradation of plastics: A comprehensive review. *Biotechnology Advances, 26*, 246−265. Available from https://doi.org/10.1016/j.biotechadv.2007.12.005.

Shahmohamadloo, R. S., Poirier, D. G., Ortiz Almirall, X., Bhavsar, S. P., & Sibley, P. K. (2020). Assessing the toxicity of cell-bound microcystins on freshwater pelagic and benthic invertebrates. *Ecotoxicology and Environmental Safety, 188*, 109945. Available from https://doi.org/10.1016/j.ecoenv.2019.109945.

Shahmohamadloo, R. S., Simmons, D. B. D., & Sibley, P. K. (2020). Shotgun proteomics analysis reveals sublethal effects in Daphnia magna exposed to cell-bound microcystins produced by Microcystis aeruginosa. *Comparative Biochemistry and Physiology Part D: Genomics Proteomics, 33*, 100656. Available from https://doi.org/10.1016/j.cbd.2020.100656.

Shan, K., Wang, X., Yang, H., Zhou, B., Song, L., & Shang, M. (2020). Use statistical machine learning to detect nutrient thresholds in Microcystis blooms and microcystin management. *Harmful Algae, 94*, 101807. Available from https://doi.org/10.1016/j.hal.2020.101807.

Shin, W.-K., Cho, J., Kannan, A. G., Lee, Y.-S., & Kim, D.-W. (2016). Cross-linked composite gel polymer electrolyte using mesoporous methacrylate-functionalized SiO_2 nanoparticles for lithium-ion polymer batteries. *Scientific Reports, 6*, 26332. Available from https://doi.org/10.1038/srep26332.

Sigmund, W., Yuh, J., Park, H., Maneeratana, V., Pyrgiotakis, G., Daga, A., . . . Nino, J. C. (2006). Processing and structure relationships in electrospinning of ceramic fiber systems. *Journal of the American Ceramic Society, 89*, 395−407. Available from https://doi.org/10.1111/j.1551-2916.2005.00807.x.

Sodré, F. F. (2012). Fontes Difusas de Poluição da Água: Características e métodos de controle. *Artig. Temáticos do AQQUA, 1*, 9−16.

Sukhanova, A., Bozrova, S., Sokolov, P., Berestovoy, M., Karaulov, A., & Nabiev, I. (2018). Dependence of nanoparticle toxicity on their physical and chemical properties. *Nanoscale Research Letters, 13*, 44. Available from https://doi.org/10.1186/s11671-018-2457-x.

Sun, S. (2000). Monodisperse FePt nanoparticles and ferromagnetic FePt nanocrystal superlattices. *Science, 287*, 1989−1992. Available from https://doi.org/10.1126/science.287.5460.1989.

Tada, H., Naya, S., & Fujishima, M. (2018). Water splitting by plasmonic photocatalysts with a gold nanoparticle/cadmium sulfide heteroepitaxial junction: A mini review. *Electrochemistry Communications, 97*, 22−26. Available from https://doi.org/10.1016/j.elecom.2018.10.005.

Tan, H., Polverino, G., Martin, J. M., Bertram, M. G., Wiles, S. C., Palacios, M. M., . . . Wong, B. B. M. (2020). Chronic exposure to a pervasive pharmaceutical pollutant erodes among-individual phenotypic variation in a fish. *Environmental Pollution (Barking, Essex: 1987), 263*, 114450. Available from https://doi.org/10.1016/j.envpol.2020.114450.

Tian, B., Lei, Q., Tian, B., Zhang, W., Cui, Y., & Tian, Y. (2018). UV-driven overall water splitting using unsupported gold nanoparticles as photocatalysts. *Chemical Communications, 54*, 1845−1848. Available from https://doi.org/10.1039/C7CC09770K.

Todescato, F., Fortunati, I., Minotto, A., Signorini, R., Jasieniak, J., & Bozio, R. (2016). Engineering of semiconductor nanocrystals for light emitting applications. *Materials (Basel)* (9, p. 672). Available from https://doi.org/10.3390/ma9080672.

Turkevich, J., Stevenson, P. C., & Hillier, J. (1951). A study of the nucleation and growth processes in the synthesis of colloidal gold. *Discussions of the Faraday Society, 11*, 55. Available from https://doi.org/10.1039/df9511100055.

Walters, G., & Parkin, I. P. (2009). The incorporation of noble metal nanoparticles into host matrix thin films: Synthesis, characterisation and applications. *Journal of Materials Chemistry, 19*, 574–590. Available from https://doi.org/10.1039/B809646E.

Wang, Q.-Y., Yang, J., Dong, X., Chen, Y., Ye, L.-H., Hu, Y.-H., ... Cao, J. (2020). Zirconium metal-organic framework assisted miniaturized solid phase extraction of phenylurea herbicides in natural products by ultra-high-performance liquid chromatography coupled with quadrupole time-of-flight mass spectrometry. *Journal of Pharmaceutical and Biomedical Analysis, 180*, 113071. Available from https://doi.org/10.1016/j.jpba.2019.113071.

Wang, Y., Jing, B., Wang, F., Wang, S., Liu, X., Ao, Z., & Li, C. (2020). Mechanism Insight into enhanced photodegradation of pharmaceuticals and personal care products in natural water matrix over crystalline graphitic carbon nitrides. *Water Research, 180*, 115925. Available from https://doi.org/10.1016/j.watres.2020.115925.

Wang, Y., & Xia, Y. (2004). Bottom-up and top-down approaches to the synthesis of monodispersed spherical colloids of low melting-point metals. *Nano Letters, 4*, 2047–2050. Available from https://doi.org/10.1021/nl048689j.

Weiss, J., Takhistov, P., & McClements, D. J. (2006). Functional materials in food nanotechnology. *Journal of Food Science, 71*, R107–R116. Available from https://doi.org/10.1111/j.1750-3841.2006.00195.x.

Woller-Skar, M. M., Russell, A. L., Gaskill, J. A., & Luttenton, M. R. (2020). Microcystin in multiple life stages of Hexagenia limbata, with implications for toxin transfer. *Journal of Great Lakes Research, 46*, 666–671. Available from https://doi.org/10.1016/j.jglr.2020.03.007.

Xia, B., Deng, F., Zhang, S., Hua, L., Luo, X., & Ao, M. (2020). Design and synthesis of robust Z-scheme ZnS-SnS2 n-n heterojunctions for highly efficient degradation of pharmaceutical pollutants: Performance, valence/conduction band offset photocatalytic mechanisms and toxicity evaluation. *Journal of Hazardous Materials, 392*, 122345. Available from https://doi.org/10.1016/j.jhazmat.2020.122345.

Yadav, A., Raj, A., Purchase, D., Ferreira, L. F. R., Saratale, G. D., & Bharagava, R. N. (2019). Phytotoxicity, cytotoxicity and genotoxicity evaluation of organic and inorganic pollutants rich tannery wastewater from a common effluent treatment plant (CETP) in Unnao district, India using *Vigna radiata* and *Allium cepa*. *Chemosphere, 224*, 324–332. Available from https://doi.org/10.1016/j.chemosphere.2019.02.124.

Yang, Z., Xu, X., Dai, M., Wang, L., Shi, X., & Guo, R. (2017). Accelerated ciprofloxacin biodegradation in the presence of magnetite nanoparticles. *Chemosphere, 188*, 168–173. Available from https://doi.org/10.1016/j.chemosphere.2017.08.159.

Yanilkin, V. V., Nastapova, N. V., Nasretdinova, G. R., & Osin, Y. N. (2017). Electrosynthesis of gold nanoparticles mediated by methylviologen using a gold anode in single compartment cell. *Mendeleev Communications, 27*, 274–277. Available from https://doi.org/10.1016/j.mencom.2017.05.019.

Zaki, S., Merican, F., Muangmai, N., Convey, P., & Broady, P. (2020). Discovery of microcystin-producing Anagnostidinema pseudacutissimum from cryopreserved Antarctic cyanobacterial mats. *Harmful Algae, 93*, 101800. Available from https://doi.org/10.1016/j.hal.2020.101800.

Zhang, Y., Calabrese, E. J., Zhang, J., Gao, D., Qin, M., & Lin, Z. (2020). A trigger mechanism of herbicides to phytoplankton blooms: From the standpoint of hormesis involving cytochrome b559, reactive oxygen species and nitric oxide. *Water Research, 173*, 115584. Available from https://doi.org/10.1016/j.watres.2020.115584.

Zhao, X., Ng, S., Heng, B. C., Guo, J., Ma, L., Tan, T. T. Y., ... Loo, S. C. J. (2013). Cytotoxicity of hydroxyapatite nanoparticles is shape and cell dependent. *Archives of Toxicology, 87*, 1037−1052. Available from https://doi.org/10.1007/s00204-012-0827-1.

Zhou, R., Lu, G., Yan, Z., Jiang, R., Bao, X., & Lu, P. (2020). A review of the influences of microplastics on toxicity and transgenerational effects of pharmaceutical and personal care products in aquatic environment. *The Science of the Total Environment, 732*, 139222. Available from https://doi.org/10.1016/j.scitotenv.2020.139222.

Zhou, S., Di Paolo, C., Wu, X., Shao, Y., Seiler, T. B., & Hollert, H. (2019). Optimization of screening-level risk assessment and priority selection of emerging pollutants—The case of pharmaceuticals in European surface waters. *Environment International, 128*, 1−10. Available from https://doi.org/10.1016/j.envint.2019.04.034.

Zieliska-Jurek, A., Reszczyska, J., Grabowska, E., & Zalesk, A. (2012). *Nanoparticles preparation using microemulsion systems. Microemulsions—An introduction to properties and applications* (p. 55) InTech. Available from https://doi.org/10.5772/36183.

5

Nanophotocatalysts for biodegradation of materials

Hanieh Fakhri[1,2], Mahdi Farzadkia[1,2]

[1]RESEARCH CENTER FOR ENVIRONMENTAL HEALTH TECHNOLOGY, IRAN UNIVERSITY OF MEDICAL SCIENCES, TEHRAN, IRAN [2]DEPARTMENT OF ENVIRONMENTAL HEALTH ENGINEERING, SCHOOL OF PUBLIC HEALTH, IRAN UNIVERSITY OF MEDICAL SCIENCE, TEHRAN, IRAN

5.1 Introduction

Treatment processes are a practical approach to solve environmental challenges in the world due to population growth and unbalanced climate pattern. Among various contaminants, stable organic compounds such as hormones, pharmaceuticals, personal care products, surfactants, and pesticides have attracted more attention. Nevertheless, development of various treatment technologies is essential to obtain acceptable removal of common contaminants and especially stable organic pollutants. To have a clean future, various technological approaches were investigated like adsorption, biological treatment, UV treatment, ozone treatment, membrane separation technologies, and so forth. However, the need for complex and expensive systems, low efficiency, long operating times, and production of hazardous byproducts have restricted practical usage.

With the advent of nanotechnology in the 1990s, much of the research was focused on nanophotocatalysis, which originated from the unique properties of nanomaterials. Nanophotocatalysis is an ideal choice to treat polluted systems due to its low cost, high efficiency, and no/low toxicity (Li & Shi, 2016). Photocatalysis can use free energy (solar energy) to oxidize or decrease pollutants via the produced active radical ions under light illumination (Deng et al., 2020; Motahari, Mozdianfard, Soofivand, & Salavati-Niasari, 2014). The technology will have significant implications for treatment systems, including the modification of the electronic structure and the increase in surface area, which will eventually lead to extensive changes in efficiency. In this process, solar energy is converted into chemical energy, and then this chemical energy destroys the pollutant during oxidation-reduction reactions.

Fig. 5–1 shows an overview of the photocatalytic process. After adsorbing of the light with more energy or equal to the photocatalytic band gap, electron-hole pairs will form. Here, the electrons are excited onto the conduction band (CB), and holes remain on the valance band (VB). These carriers with enough time to transfer to the surface of the photocatalyst initiate the redox-

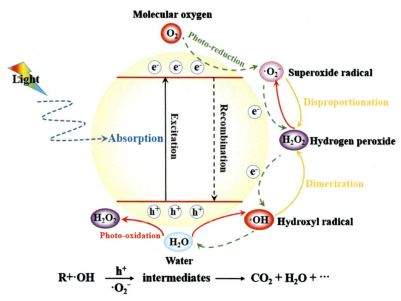

FIGURE 5–1 Schematic of the photocatalysis process (Long, Li, Wei, Zhanga, & Ren, 2020).

reduce reactions. On the surface of photocatalysts, the electron acceptors such as O_2 and H_2O can be reduced by electrons to generate reactive oxygen radicals. On the other hand, the holes are interacted with the donor types and resulted in the mineralization of the contaminants (Rostami-Vartooni, Nasrollahzadeh, Salavati-Niasari, & Atarod, 2016; Li, Li, Ai, Jia, & Zhang, 2016).

Particularly in water treatment processes, nanomaterials are divided into three categories: (1) nanosorbents, (2) nanocatalysts, and (3) nanoparticles.

Nanoadsorbents: Nanosorbents are compounds with high adsorption capacity that have active functional groups. These compounds generally have a high surface area and have mineral functional groups. Examples of such materials include: silica, activated carbon, clay materials, metal oxides, and metal (Rao, 2014). The main advantage of nanotechnology in the adsorption process is the increase in the surface area of nanomaterials compared to the balk sample.

Nanomembranes: Nanomembranes are another efficient class of nanomaterials used in large-scale refining. The efficiency of these materials can be enhanced by changing the size of the holes and the presence of various functional groups (Rao, 2014; Petrinic, Andersen, Sostar-Turk, & Le Marechal, 2007).

Nanophotocatalysts: Nanophotocatalysts are semiconductor compounds that have at least one nanodimension. Numerous studies have shown that nanomaterials behave quite differently from attributing balk material in chemical reactions. Suitable photocatalytic characteristics include stability, cheapness, nontoxicity, and active visible light. Of course, the size of the band gap and the position of the CB and VB are very important. The speed of light absorption and the production of electron-hole pairs due to modification of surface to volume ratio have been reported that have dramatic positive effects on the efficiency.

5.2 Advanced oxidation processes for water and wastewater treatment

Traditional methods of wastewater treatment, such as adsorption, coagulation, and filtration, have inherent limitations in practical applications, such as the use of complex systems, low efficiency, high energy consumption, and the production of higher toxicity byproducts. In this context, it is important to introduce an advanced oxidation processes (AOPs) that are able to use sunlight as an energy source and has the ability to completely destroy pollutants to harmless compounds.

The main actor in the AOPs is the active oxidizing radicals, which along with various processes like sonolysis, ozonation, UV, and Fenton processes, lead to destruction of organic and inorganic pollutants. The point to consider is that the AOP is environmentally friendly and prevents the transmission of the polluting phase (unlike chemical precipitation and adsorption), and on the other hand, it is able to destroy the pollutants and turn it into harmless materials. The following categories are the most common used AOPs to destroy according to types of pollutant and photocatalysts.

- H_2O_2 + UV (direct photolysis) (Krupa & Wierzejewska, 2016)
- H_2O_2 + $Fe^{2+/3+}$ (classic, homogeneous Fenton) (Pereira, Oliveira, & Murad, 2012)
- H_2O_2 + Fe/support (heterogeneous Fenton) (Zhu et al., 2019)
- H_2O_2 + $Fe^{2+/3+}$ + UV (photo-Fenton) (Rahim Pouran, Abdul Aziz, & Wan Daud, 2015)
- O_3 (direct ozone feeding) (Rodríguez et al., 2008)
- O_3 + UV (photoozone feeding) (Gomes, Matos, Gmurek, Quinta-Ferreira, & Martins, 2019)
- O_3 + catalysts (catalytic ozone feeding) (Emam, 2012)
- H_2O_2 + O_3 (Li, He, Wang, Meng, & Zeng, 2015)
- TiO_2 + UV (photocatalysis) (Thiruvenkatachari, Vigneswaran, & Moon, 2008)

Important factors in photocatalytic systems are the rate of production of active oxidizing radicals and the degree of association of these active radicals with the pollutant.

In recent years, active photocatalysts in visible light have attracted a lot of attention. In fact, the use of the sun as a free source of energy promises practical applications of photocatalytic processes. Activation under visible light is valuable because more than 46% of sunlight is visible light. Indeed, the basic step in designing active photocatalysts is improvement of solar energy exploitation. In the following sections, three categories of materials that have attracted more attention will be introduced.

5.2.1 Metal oxide

So far, the most popular type of photocatalyst used has been metal oxide-based material (Khan, Adil, & Al-Mayouf, 2015). Several factors including crystallinity phase and size, effective surface area, pore size, and adsorption capacity affect the photocatalytic ability of these materials (Mamaghani, Haghighat, & Lee, 2017), which depending on the class usage, the effect of them can vary. TiO_2, ZnO, SnO_2, and CeO_2 are most commonly used photocatalysts

(Song, Wu, Tang, Qi, & Yan, 2008). The energy levels of VB and CB belong to some metal oxide were indicated in Fig. 5–2 (Reddy, Hassan, & Gomes, 2015; Saison, 2013; Schreck & Niederberger, 2019). It is to be noted that the CB of metal oxide must be higher than the potential of O_2/O_2^- to give an electron and change oxygen to superoxide; while the VB requires to be more lower than H_2O/H^+ to receive an electron and oxidize to the OH radical (Saison, 2013). The main problems of the metal oxides are the wide band gaps and low separation of charge carriers (Choi, Termin, & Hoffmann, 1994). Metal oxide photocatalysts at nano size often have a low band gap and reduce the transfer way of charge carriers to the surface. In addition, these nanophotocatalysts have large surface to volume ratio, which can adsorb more pollutants (Reddy et al., 2015).

5.2.2 Carbon-based structure

In recent years, much research has been done on the photocatalytic properties of metal oxides such as TiO_2 and ZnO. However, low surface area and low light absorption caused their low photocatalytic efficiency. These photocatalysts are often active in UV light, and their active types in visible light suffer from the high the electron-hole recombination. One of the efficient strategies is their combination with low dimensional materials (i.e., graphene) via bandgap alteration (Faraji et al., 2019). Various organic carbon-based materials have been utilized to dominate the mentioned confines. Fig. 5–3 indicates function of carbon-based materials in both energy and environmental usage in three various scopes, namely pollutant photodegradation, photocatalytic hydrogen production, and CO_2 reduction.

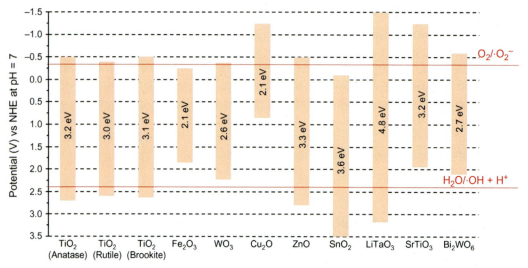

FIGURE 5–2 The band gaps, CB energy levels, and VB energy levels of some metal oxide photocatalyst (Long, Li, Wei, Zhanga, & Ren, 2020).

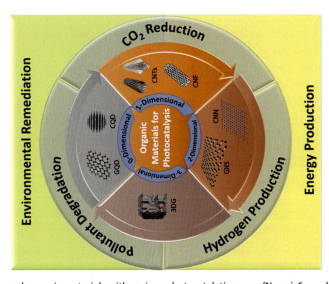

FIGURE 5–3 Carbon-based organic materials with various photocatalytic usage (Naseri, Samadi, Ebrahimi, Kheirabadi, & Moshfegh, 2020).

These compounds can have a variety of functions in heterogeneous photocatalysis. For example, graphitic carbon nitride (g-C_3N_4) nanosheets are semiconductors with relatively good separation efficiency of electron-hole pairs and are an excellent candidate in photocatalytic applications. The use of graphene and carbon as substrates has also shown satisfactory results. Graphene-based photocatalysts supply a numerous number of active sites and indicate admirable electron mobility. Moreover, carbon nanotubes and carbon-based quantum dots enhance the optical properties of the photocatalyst and also improve the ability to separate charge carriers. Another special feature of carbon compounds is the high surface area of these compounds, which helps to absorb more pollutants and ultimately increases the number of possible photocatalytic reactions.

5.2.3 Metal organic framework based structure

Metal organic frameworks (MOFs) are porous crystals that are formed from metal clusters (e.g., secondary building units or SBUs) and organic linkers (e.g., multidentate bridging ligands) via strong coordination bondings (Cheng et al., 2018; Kukkar, Vellingiri, Kim, & Deep, 2018). Since the late 1990s, MOFs have been used in various fields such as photocatalysis (Shen, Liang, & Wu, 2015; Fakhri & Bagheri, 2020a, b), sensing (Kumar, Deep, & Kim, 2015; Amini, Kazemi, & Safarifard, 2020), electro-conductivity (Sun, Campbell, & Dincă, 2016), and so forth. the theoretical bandgaps of MOFs was in the range from 1.0 to 5.5 eV depending to type of the metal cluster and organic linkers (Shen et al., 2015). It was reported that optical property of MOFs considerably enhanced by modulating the structure of MOFs (Zeng, Guo, He, & Duan, 2016). In the MOFs, both the metal clusters and organic ligands can act to enhance photon absorption and catalytic activity. However, their wide usage is limited due to their complex and expensive synthesis procedure, low stability in polar media (Kumar, Kim, Kim, Szulejko, & Brown, 2016), wide bandgap, and nonreusability. In order to optimize their photocatalytic activity, metal and metal-free semiconductors like graphene and g-C_3N_4 were used to dope in MOFs structure. Doping of MOFs with magnetic metal Fe, Co, and Ni improves their photocatalytic efficiency but are not proper to the separation and recovery of materials. However, recent advances in the design, synthesis, and application of MOFs are significant and will shed light on the future of these materials.

5.3 Methods of improving photocatalytic efficiency

In order to examine the importance of modifying surface properties and band gaps, it is best to explain the mechanism of optical adsorption by a photocatalyst. Active photocatalysts in visible light must have a band gap of 1.2–2.3 eV to cover the oxidation and reduction potentials of the water. As mentioned earlier, after light irradiation, electron-hole pairs are formed according to the electronic structure of the semiconductor. In fact, with the formation of a pair of free electron-hole, the reduction oxidation reactions begin. For example, Martha reported (Martha, Sahoo, & Parida, 2015) that the position of the band gap for water splitting processes should be such that the CB is more negative than the reduction potential of water. On the other hand, for oxidation reactions, the VB should be placed at more positive than the oxidation potential of water. In following, the most common ways to modify a photocatalyst present.

5.3.1 Metal doping

The doping method refers to the introduction of external elements in the structure of the host material and is used today as one of the efficient methods in modifying the structure of the photocatalyst. Metal/nonmetal doping of semiconductors lead to reduce their bandgap, resulted in the improvement of the visible-light ability, enhancement of the electrical conductivity, and proper separation of photoinduced electron-hole pairs (Martha et al., 2015; Wetchakun, Wetchakun, & Sakulsermsuk, 2019). Fig. 5–4 indicated the image of the suggested charge separation procedure where the electrons of metals can pass to the CB of the n-type semiconductor after light absorption. Then, the oxidizing of metal particle from M0 to M1 occurred (Fig. 5–4A), resulting in the indirect response of the n-type semiconductor to the visible light from the metal surface plasmon resonance.

5.3.2 Surface sensitization by organic ligands

One of the most common recent methods for photocatalyst modification is surface modification with organic matter. For example, after sensitizing a semiconductor with dye under light illumination, electrons move from dye to CB of semiconductor, and then the fast electron transfer initiates oxidative reactions (Martha et al., 2015). It is worth noting

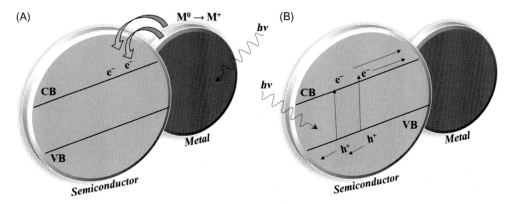

FIGURE 5–4 Illustrations of electron transitions between metal and semiconductor: (A) the plasmon-induced charge separation and (B) charge separation via Schottky barrier from electron excitation (Hadnadjev-Kostic, Vulic, Dostanic, & Loncarevic, 2020).

that dye-sensitization strategy can appear of electron injection in femtoseconds and the start of the charge carriers recombination in nanoseconds to milliseconds. Therefore, it can be said that dye sensitization can be a promising method. Several works about the surface synthesizing of TiO_2 with organic dyes like erythrosin B (Kamat & Fox, 1983), eosin (Puangpetch, Sommakettarin, Chavadej, & Sreethawong, 2010), and other organic compounds (Milicevi et al., 2015; Vukoje et al., 2016) were reported. In the other research, the surface modification of Al_2O_3 by 5-aminosalicylic acid indicated enhanced photocatalytic ability in the methylene blue degradation reaction under both simulated solar light and visible-light irradiation. It is be noted that Al_2O_3 has a larger band gap than 6 eV (Đorđevic et al., 2017; Đorđevic et al., 2019).

5.3.3 Semiconductor coupling

One of the successful strategies in improving the efficiency of optical absorption as well as reduction of electron-hole recombination is the coupling of semiconductors with each other. The mechanism of action of this method is similar to the modification of the surface by dye, with the difference that in this method, a semiconductor with a small band gap is used. In designing an effective coupling system, it should be noted that the energy level of the CB and the VB (belonging to the second component) should be more positive and negative than attributing bands of the base semiconductor, respectively. According to the charge separation mechanism, the electron migrates from a higher CB potential level to the lower CB potential level. On the other hand, holes can be moved from the lower VB level to the semiconductor with higher VB level (Dong et al., 2015; Wetchakun et al., 2019; Fig. 5–5). However, the final efficiency is influenced by other parameters such as the amount of visible-light absorption and the optimal amount of band gap (Ho & Yu, 2006; Wu, Yu, & Fu, 2006; Rehman, Ullah, Butt, & Gohar, 2009).

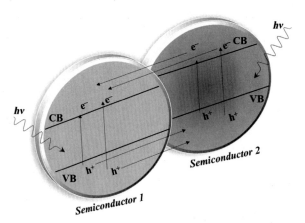

FIGURE 5–5 Illustration of the semiconductor coupling (Hadnadjev-Kostic, Vulic, Dostanic, & Loncarevic, 2020).

5.3.4 Co-doping

Co-doping strategies with double metal/nonmetal dopants widely utilize the modification of photocatalyst. This attention is owing to having synergic effects to enhance the visible-light absorption efficiency. Furthermore, it can reduce the recombination rate of the charge carriers (Thota, Tirukkovalluri, & Bojja, 2014). Some co-doped works were tabulated in Table 5–1.

5.3.5 Morphology control

One interesting way to improve the performance of a photocatalyst is control the morphology of the compounds. For example, for different morphologies of TiO_2, the rate of separation of charge carriers is different. In addition, each of crystal faces has a specific photocatalytic activity. Previous research has shown the (001) facets in TiO_2 are presented more oxidative power than the (101) facets in degradation of pollutants (Johansson et al., 2010; Wanbayor & Ruangpornvisuti, 2010). Nevertheless, the activity of the (101) facets is more proper than the (001) facets for the reduction of heavy metals or decomposition of acetaldehyde (Omote et al., 2005). Therefore, various facets of TiO_2 have various ability in the photocatalytic reactions; (001) is the facet contributing in photocatalytic oxidation and (101) is the facet contributing in photocatalytic reduction (Peng et al., 2008; Zheng, Meng, & Tang, 2009). Furthermore, it is worth noting that crystalline phases also have different capabilities in separating charge carriers (Song et al., 2008; Djurišić, Leung, & Ching, 2014). Moreover, controlling the crystal structure can also affect the photocatalytic behavior. Particularly, perovskite compounds with a layered and nanoscale structure (e.g., La-$NaTaO_3$) have a separated reduction and oxidation sites (Torres-Martínez et al., 2010). This unique structure can reduce the recombination rate of charge carriers, resulting in superior photocatalytic efficiency (Wei et al., 2010).

Table 5–1 Some of co-doped photocatalytic systems.

Photocatalyst	Dopant	Pollutant model	References
BiOCl	Yb^{3+}/Er^{3+}	Rhodamine B	Li, Li, Wang, Liu, and Ren (2017)
TiO_2	N/S	Toluene	Lee, Shie, Yang, and Chang (2016)
$CsTaWO_6$	S/N	Water splitting	Marschall et al. (2011)
$BiVO_4$	B/Eu	Methyl orange	Wang, Che, Niu, Dang, & Dong (2013)
$g-C_3N_4$	Fe/P	Rhodamine B	Hu et al. (2014)
ZnO	In/Sn	Direct red 31	Bhatia, Verma, & Bedi (2017)

It is found that hollow structures can enhance the light-capturing capability of a photocatalyst by supplying substantial scattering and reflecting lanes via inner surface shell (Li, Yu, & Jaroniec, 2016). Wang et al. (Sun et al., 2012) have synthesized hollow spheres of polymeric carbon nitrides (CN) as the support. The observation indicated CN hollow spheres with size smaller than 430 nm were formed that have shells with thickness from 56 to 85 nm. Some of the hollow structure is fairly capable of harvesting light and finally to enhance quantum yield of 7.5% in 10 vol.% triethanolamine (TEOA) scavenger aqueous solution by using 3 wt.% platinum (Pt) as the cocatalyst.

Other research has presented that the photocatalytic ability can be enhanced via copolymerization (Zheng, Pang, Liu, & Wang, 2015), postannealing (Zheng, Huang, & Wang, 2015), and molybdenum disulfide (MoS_2) incorporation (Zheng, Zhang, Hou, & Wang, 2016). In other study (Tong et al., 2017), a type of CN hollow spheres with triple-shelled was synthesized. As seen in Fig. 5–6A and B, the shell thickness were 20, 20, and 40 nm from outside to inside, respectively, and for the outer sphere, the total size was 400 nm. These triple-shelled CN (TSCN) hollow spheres were more efficient than single-shelled CN (SSCN) and double-shelled CN (DSCN) counterparts and its adsorption edge have more shift to visible region in compared with SSCN and DSCN (Fig. 5–6C) that can be corresponded to the consecutive scattering and reflection. Fig. 5–6D shows that the TSCN hollow spheres have the best performance in photocatalytic H_2-production (Fig. 5–6D), under the visible light illumination (l > 420 nm).

5.4 Effective factors in the advanced oxidation processes

In order to increase photocatalytic efficiency, it is necessary to examine the factors affecting the photocatalytic process, such as pH, catalyst loading, and the presence of oxidant. In the following, the mentioned parameters will be examined.

5.4.1 pH of solution

Solution pH has various effects on the rate of adsorption, which occurs in the first phase of the photocatalytic procedure. Various species of degradation products generated are affected by the amount of deprotonation and electrical charge of surface functional groups. In some states, a highest value for the degradation efficiency is achieved, which can attribute to the point of zero charge of the photocatalyst. Determining this point leads to control

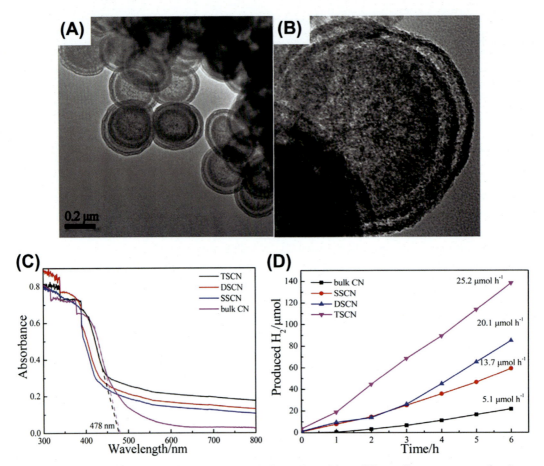

FIGURE 5–6 (A and B) TEM images of triple-shelled CN hollow sphere, (C) UV-diffuse reflectance spectra of various CN samples, and (D) Photocatalytic H_2 production over various CN samples under visible light irradiation. SSCN, DSCN, and TSCN represent single-shelled, double-shelled, and triple-shelled CN, respectively (Tong et al., 2017).

photocatalyst-pollutant electrostatic absorption or repulsion. For example, Jamali et al.'s and Jallouli's research have reported that the pH of 4.5 and 5 were as optimum values for the degradation of phenol and ibuprofen, respectively (Jallouli et al., 2018; Jamali, Vanraes, Hanselaer, & Van Gerven, 2013). Since the point of zero charge of used photocatalyst was nearly 6.6, at the pH 4.5 and 5 the ibuprofen and phenol molecules are negatively charged while surface of photocatalyst was positive. Hence, the maximum electrostatic interaction between the pollutant molecules and the photocatalyst led to the enhanced removal rate. As seen in Fig. 5–7A, alkaline pH is proper to remove of phenols (Xiong & Xu, 2016). This work reported that loading of Pt in TiO_2 structure and adding borate in the solution changed the production and decomposition of H_2O_2 where the production of OH radicals is due to H_2O_2. The highest adsorption of borate on TiO_2 was placed at pH 9.0. Addition of borate improved

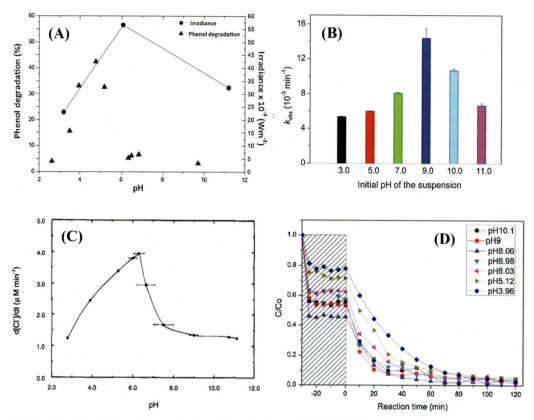

FIGURE 5–7 Effect of pH on (A) phenol degradation (Jamali et al., 2013), (B) phenol degradation (Xiong & Xu, 2016), (C) CCl$_4$ dechlorination rates (Choi & Hoffmann, 1995), (D) degradation of norfloxacin (Tang et al., 2016).

the value of H$_2$O$_2$ generation of Pt/TiO$_2$. In addition, fluorescence determination indicated a maximum intensity for Pt/TiO$_2$ + borate in pH 9, which finally lead to maximum phenol degradation in pH 9. Also, it is noteworthy to mention, the pH of solution has no impact on the degradation yield of nitrotoluenes and dinitrotoluenes in Fig. 5–7B (Kumar & Davis, 1997). This fact attributes to lack of any pH-dependent variation in nitrotoluene proton donating/accepting over the pH 3–11. Choi et al. also indicates that oxic and anoxic states are important to degrade of CCl$_4$ Fig. 5–7C (Choi & Hoffmann, 1995). They stated that a highest degradation at pH = 6 is due to pH$_{zpc}$ of TiO$_2$ which intensely affects the electrostatic interactions between TiO$_2$ and the pollutant.

5.4.2 The effect of H$_2$O$_2$

Since H$_2$O$_2$ is one of the sources of the OH radical production, its presence affects the final efficiency (Mohabansi, Patil, & Yenkie, 2011). H$_2$O$_2$ accelerates and improves the photocatalytic process via OH radical generation in some paths: (1) via interaction with holes (h$^+$)

generated by semiconductor, (2) via interaction with transition metal like iron, and (3) via self-decomposition. Mahmoodi et al. reported that H_2O_2 has an optimum point (450 mg/L) in the decolorization procedure [Eqs. (5.1)–(5.3)] (Mahmoodi, Arami, & Limaee, 2006).

$$H_2O_2 + 2h^+ \rightarrow O_2 + 2H^+ \tag{5.1}$$

$$H_2O_2 + OH\% \rightarrow HO_2\% + H_2O \tag{5.2}$$

$$HO_2\% + OH\% \rightarrow H_2O + O_2 \tag{5.3}$$

Furthermore, Kourdali's review (Kourdali, Badis, & Boucherit, 2014) and Bautista et al., presented that the maximum removal efficiency of humic acid was obtained at the optimum value of about 0.01% of H_2O_2 concentration. Nevertheless, the optimal amount is not always easily determined in the reactions, and in some cases, by increasing the amount of H_2O_2, we will only increase the efficiency.

5.4.3 Direct photolysis

UV lamps wavelength of 254 nm with a relative high photoelectric conversion efficiency are one way to remove contaminants, which are much less efficient than photocatalytic methods. These low-pressure mercury lamps can produce mercury resonance lines at 254 and 185 nm. Direct photolysis is an easy way to remove of pollutants. A typical photolysis reaction which is carried out without a photocatalyst, indicates a lower yield than photocatalytic reaction. In this reaction, UV lamps, xenon lamps, or solar simulators can be utilized. It is noteworthy that final photolysis efficiency is effected by some parameters, for example, pH of solution and pollutants concentration (Shemer & Linden, 2007; Truppi et al., 2019).

5.4.4 Dosage of photocatalyst

In order to prevent overconsumption of the photocatalyst, it is necessary to determine the optimal amount of photocatalyst, which of course depends on the size of the reactor, the concentration of the pollutant, and the inherent characteristics of the photocatalyst. Studies have shown that different photocatalytic systems exhibit a variety of behaviors by increasing the amount of photocatalyst. In some cases, an increase in photocatalysts will be accompanied by an increase in aggregation, which will eventually reduce the number of active photocatalytic centers. In some cases, increasing the photocatalyst causes a turbidity of the solution and increases the amount of light scattering, which will eventually reduce the photocatalytic efficiency. In the investigation of dibenzo-p-dioxins by using TiO_2 films under UV light, it is found that maximum efficiency was achieved with TiO_2 dosage of 200 μg (Choi, Hong, Chang, & Cho, 2009). In the other case, Tseng et al. presented that the oxidation efficiency of chlorophenol increased with increasing dosage of TiO_2 and reached a maximum efficiency in the presence of 3 g/L of catalyst (Tseng & Huang, 1991). Chaudhary and Thakur (Chaudhary & Thakur, 2012) reported that TiO_2 with 0.1 wt.% was the optimum dosage in pilot-scale studies to the decontamination of organics and heavy metals. In other study,

Bansal et al., assayed the photocatalytic efficiency in the concentration range from 0.25 to 2.0 g/L. Fig. 5–8 shows the removal efficiency improves with an increase in catalyst dosage. They also demonstrate when photocatalyst dosage increases, the amount of photons absorbed on the photocatalyst reduced due to the radiation scattering, therefore more increase in catalyst dosage reduced the decolorization efficiency. In Bansal, Singh, and Sud's research (Bansal, Singh, & Sud, 2010), optimum dosage of photocatalyst was obtained 1 g/L for ZnO and 1.5 g/L for TiO_2.

5.5 Photocatalysis mechanisms

According to reported works, the main path of photocatalysis consists of several steps: (1) produce of electron-hole pairs; (2) charge separation; (3) redox reactions between the charge carriers and the adsorbents; and (4) desorption of reaction fragments. Initially, the electron-hole pairs are formed by irradiating light with energy equal to or greater than its bandgap. Then the electrons move from the CB to the VB, leading to the creation of positive holes. It is reported that many of the photoinduced electron/hole pairs (>90%) recombine rapidly after excitation. The residual

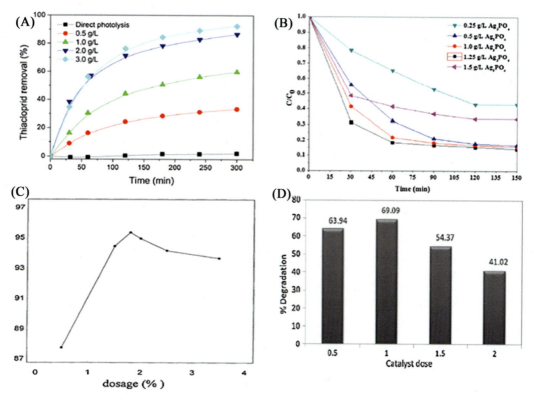

FIGURE 5–8 Effect of catalyst dosage on (A) thiacloprid (Abramović, Banić, & Krstić, 2013) (B) azo dyes (Li et al., 2016), (C) azo dyes (Zhu et al., 2000), and (D) azo dye (Bansal et al., 2010).

charge carriers are able to initiate a broad range of chemical reactions. Here, the photoinduced holes could either directly oxidize the pollutant molecules or react with OH ions or H_2O to form OH radicals. The OH radicals are strong oxidizing agents that can oxidize many varieties of pollutant models. On the other hand, the photogenerated electrons reduce the pollutant molecule or react with electron receivers such as O_2 and reducing it to a superoxide radical anion. Finally, sum of these active radicals degrade pollutants.

5.6 Conclusion

Today, one of the chief efficient technologies to remove stable and hazardous contaminants is via the photocatalytic process. Especially, over the past 10 years, this technology has been used increasingly due to the easy procedure, low cost, and high efficiency. These findings verified the its high potential for having a clean future. According to the latest investigations, the photocatalytic method can be used as an efficient way to remove both organic and nonorganic pollutants. In this regard, in order to achieve the highest efficiency, structural and optical modifications can be very helpful, and the study of these methods has opened numerous windows for researchers in the field of environmental health. In this chapter, we focused on introducing famous photocatalyst types, modification of approaches of photocatalysts, and effective parameters on efficiency. Despite significant and rapid advances in recent years, the technical challenges of this technology need to be addressed to eventually bring efficient photocatalytic reactors into the industry. For example, reactors equipped with pollutant detection sensors. Moreover, novelty in the regulation of light sources and photocatalysts create the novel systems to more qualification and efficiency. However, their industrial applications depend on the final cost of manufacturing the photocatalytic reactor, the final efficiency, and energy consumption.

References

Abramović, B. F., Banić, N. D., & Krstić, J. B. (2013). Degradation of thiacloprid by ZnO in a laminar falling film slurry photocatalytic reactor. *Industrial & Engineering Chemistry Research, 52*, 5040–5047.

Amini, A., Kazemi, S., & Safarifard, V. (2020). Metal-organic framework-based nanocomposites for sensing applications—A review. *Polyhedron, 177*, 114260.

Bansal, P., Singh, D., & Sud, D. (2010). Photocatalytic degradation of azo dye in aqueous TiO 2 suspension: Reaction pathway and identification of intermediates products by LC/MS. *Separation and Purification Technology, 72*, 357–365.

Bhatia, S., Verma, N., & Bedi, R. (2017). Effect of aging time on gas sensing properties and photocatalytic efficiency of dye on In-Sn co-doped ZnO nanoparticles. *Materials Research Bulletin, 88*, 14–22.

Chaudhary, R., & Thakur, R. S. (2012). Photocatalytic treatment of industrial wastewater containing chromium as a model pollutant-effect on process parameters and kinetically studies. *Journal of Renewable and Sustainable Energy, 4*, 053121.

Cheng, M., Lai, C., Liu, Y., Zeng, G., Huang, D., & Zhang, C. (2018). Metal-organic frameworks for highly efficient heterogeneous Fenton-like catalysis. *Coordination Chemistry Reviews, 368*, 80–92.

Choi, W., & Hoffmann, M. R. (1995). Photoreductive mechanism of CCl_4 degradation on TiO_2 particles and effects of electron donors. *Environmental Science & Technology, 29*, 1646–1654.

Choi, W., Hong, S. J., Chang, Y. S., & Cho, Y. (2009). Photocatalytic degradation of polychlorinated dibenzo-p-dioxins on TiO_2 film under UV or solar light irradiation. *Environmental Science & Technology, 34*, 4810–4815.

Choi, W., Termin, A., & Hoffmann, M. R. (1994). The role of metal ion dopants in quantum-sized TiO_2: Correlation between photoreactivity and charge carrier recombination dynamics. *The Journal of Physical Chemistry, 98*(51), 13669–13679.

Deng, F., Luo, Y., Li, H., Xia, B., Luo, X., Luo, S., & Dionysiou, D. D. (2020). Efficient toxicity elimination of aqueous Cr(VI) by positively-charged BiOClxI1-x, BiOBrxI1-x and BiOClxBr1-x solid solution with internal hole-scavenging capacity via the synergy of adsorption and photocatalytic reduction. *Journal of Hazardous Materials, 383*, 121–127.

Djurišić, A. B., Leung, Y. H., & Ching, Ng. A. M. (2014). Strategies for improving the efficiency of semiconductor metal oxide photocatalysis. *Materials Horizons, 1:*, 400–410.

Dong, H., Zeng, G., Tang, L., Fan, C., Zhang, C., He, X., & He, Y. (2015). An overview on limitations of TiO_2-based particles for photocatalytic degradation of organic pollutants and the corresponding countermeasures. *Water Research, 79*, 128–146.

Đorđević, V., Dostanić, J., Lončarević, D., Ahrenkiel, S. P., Sredojević, D. N., Švrakić, N., Belic, M., et al. (2017). Hybrid visible-light responsive Al_2O_3 particles. *Chemical Physics Letters, 685*, 416–421.

Đorđevic, V., Sredojević, D. N., Dostanić, J., Lončarević, D., Ahrenkiel, S. P., Svrakić, N., Brothers, E., et al. (2019). Visible light absorption of surface-modified Al2O3 powders: A comparative DFT and experimental study. *Microporous and Mesoporous Materials, 273*, 41–49.

Emam, E. A. (2012). Effect of ozonation combined with heterogeneous catalysts and ultraviolet radiation on recycling of gas-station wastewater. *Egyptian Journal of Petroleum, 21*, 55–60.

Fakhri, H., & Bagheri, H. (2020a). Highly efficient Zr-MOF@WO3/graphene oxide photocatalyst: Synthesis, characterization and photodegradation of tetracycline and malathion. *Materials Science in Semiconductor Processing, 107*, 104815.

Fakhri, H., & Bagheri, H. (2020b). Two novel sets of UiO-66@ metal oxide/graphene oxide Z-scheme heterojunction: Insight into tetracycline and malathionphotodegradation. *Journal of Environmental Science, 91*, 222–236.

Faraji, M., Yousefi, M., Yousefzadeh, M., Zirak, M., Naseri, N., Jeon, T., Choi, W., et al. (2019). Two-dimensional materials in semiconductor photoelectrocatalytic systems for water splitting. *Energy & Environmental Science, 12*(1), 59–95.

Gomes, J., Matos, A., Gmurek, M., Quinta-Ferreira, R., & Martins, R. (2019). Ozone and photocatalytic processes for pathogens removal from water: A review. *Catalysts, 9*, 46.

Hadnadjev-Kostic, M., Vulic, T., Dostanic, J., & Loncarevic, D. (2020). *Design and application of various visible light responsive metal oxide photocatalysts. Handbook of Smart Photocatalytic Materials* (pp. 65–99). Elsevier

Ho, W., & Yu, J. C. (2006). Sonochemical synthesis and visible light photocatalytic behavior of CdSe and CdSe/TiO_2 nanoparticles. *Journal of Molecular Catalysis A: Chemical, 247*, 268274.

Hu, S., Ma, L., You, J., Li, F., Fan, Z., Lu, G., . . . Gui, J. (2014). Enhanced visible light photocatalytic performance of g-C 3 N 4 photocatalysts co-doped with iron and phosphorus. *Applied Surface Science, 311*, 164–171.

Jallouli, N., Pastrana-Martínez, L. M., Ribeiro, A. R., Moreira, N. F. F., Faria, J. L., Hentati, O., . . . Ksibi, M. (2018). Heterogeneous photocatalytic degradation of ibuprofen in ultrapure water, municipal and pharmaceutical industry wastewaters using a TiO_2/UV-LED system. *Chemical Engineering Journal, 334*, 976–984.

Jamali, A., Vanraes, R., Hanselaer, P., & Van Gerven, T. (2013). A batch LED reactor for the photocatalytic degradation of phenol. *Chemical Engineering and Processing: Process Intensification, 71*, 43–50.

Johansson, E. M. J., Plogmaker, S., Walle, L. E., Schölin, R., Borg, A., Sandell, A., & Rensmo, H. (2010). Comparing surface binding of the maleic anhydride anchor group on single crystalline anatase TiO_2 (101), (100), and (001) surfaces. *Journal of Physical Chemistry C, 114*, 15015−15020.

Kamat, P. V., & Fox, M. A. (1983). Photo-sensitization of TiO_2 colloids by Erythrosin-B in acetonitrile. *Chemical Physics Letters, 102*, 379384.

Khan, M. M., Adil, S. F., & Al-Mayouf, A. (2015). *Metal oxides as photocatalysts*. Elsevier.

Kourdali, S., Badis, A., & Boucherit, A. (2014). Degradation of direct yellow 9 by electroFenton: Process study and optimization and, monitoring of treated water toxicity using catalase. *Ecotoxicology and Environmental Safety, 110*, 110−120.

Krupa, J., & Wierzejewska, M. (2016). New aspects of UV photolysis of hydrogen peroxide Nitrogen matrix isolation FTIR and theoretical studies. *Journal of Photochemistry and Photobiology A: Chemistry, 330*, 134−139.

Kukkar, D., Vellingiri, K., Kim, K.-H., & Deep, A. (2018). Recent progress in biological and chemical sensing by luminescent metal-organic frameworks. *Sensors and Actuators B, Chemical, 273*, 1346−1370.

Kumar, P., Deep, A., & Kim, K.-H. (2015). Metal organic frameworks for sensing applications. *Trends in Analytical Chemistry, 73*, 39−53.

Kumar, P., Kim, K. H., Kim, Y. H., Szulejko, J. E., & Brown, R. J. (2016). A review of metal organic resins for environmental applications. *Journal of Hazardous Materials, 320*, 234−240.

Kumar, S., & Davis, A. P. (1997). Heterogeneous photocatalytic oxidation of nitrotoluenes. *Water Environment Research, 69*, 1238−1245.

Lee, C. H., Shie, J. L., Yang, Y. T., & Chang, C. Y. (2016). Photoelectrochemical characteristics, photodegradation and kinetics of metal and non-metal elements co-doped photocatalyst for pollution removal. *Chemical Engineering Journal, 303*, 477−488.

Li, D., & Shi, W. (2016). Recent developments in visible-light photocatalytic degradation of antibiotics. CuihuaXuebao/Chinese. *Journal of Catalysis, 37*, 792−799.

Li, G., He, J., Wang, D., Meng, P., & Zeng, M. (2015). Optimization and interpretation of O3 and O3/H2O2 oxidation processes to pretreat hydrocortisone pharmaceutical wastewater. *Environmental Technology, 36*, 1026−1034.

Li, H., Li, J., Ai, Z., Jia, F., & Zhang, L. (2016). Oxygen vacancy-mediated photocatalysis of BiOCl: Reactivity, selectivity, and perspectives. *Angewandte Chemie - International Edition, 57*, 122−138.

Li, H., Li, W., Wang, F., Liu, X., & Ren, C. (2017). Fabrication of two lanthanides co-doped Bi 2 MoO 6 photocatalyst: Selection, design and mechanism of Ln 1/Ln 2 redox couple for enhancing photocatalytic activity. *Applied Catalysis B, 217*, 378−387.

Li, R., Song, X., Huang, Y., Fang, Y., Jia, M., & Ma, W. (2016). Visible-light photocatalytic degradation of azo dyes in water by Ag_3PO_4: An unusual dependency between adsorption and the degradation rate on pH value. *Journal of Molecular Catalysis A: Chemistry, 421*, 57−65.

Li, X., Yu, J., & Jaroniec, M. (2016). Hierarchical photocatalysts. *Chemical Society Reviews, 45*, 2603−2636.

Long, Z., Li, Q., Wei, T., Zhanga, G., & Ren, Z. (2020). Historical development and prospects of photocatalysts for pollutant removal in water. *Journal of Hazardous Materials, 395*, 122599.

Mahmoodi, N. M., Arami, M., & Limaee, N. Y. (2006). Photocatalytic degradation of triazinic ring-containing azo dye (Reactive Red 198) by using immobilized TiO_2 photoreactor: Bench scale study. *Journal of Hazardous Materials, 133*, 113−118.

Mamaghani, A. H., Haghighat, F., & Lee, C. S. (2017). Photocatalytic oxidation technology for indoor environment air purification: The state-of-the-art, *Applied Catalysis B: Environmental* (203, p. 247269).

Marschall, R., Mukherji, A., Tanksale, A., Sun, C., Smith, S. C., Wang, L., & Lu, G. Q. M. (2011). Preparation of new sulfur-doped and sulfur/nitrogen co-doped CsTaWO 6 photocatalysts for hydrogen production from water under visible light. *Journal of Materials Chemistry, 21*, 8871−8879.

Martha, S., Sahoo, P. C., & Parida, K. M. (2015). An overview on visible light responsive metal oxide based photocatalysts for hydrogen energy production. *RSC Advances, 5*, 61535−61553.

Milićević, B., Đorđević, V., Lončarević, D., Ahrenkiel, S. P., Dramićanin, M. D., & Nedeljković, J. M. (2015). Visible light absorption of surface modified TiO_2 powders with bidentate benzene derivatives. *Microporous and Mesoporous Materials, 217*, 184189.

Mohabansi, N., Patil, V., & Yenkie, N. (2011). A comparative study on photo degradation of methylene blue dye effluent by advanced oxidation process by using TiO_2/ZnO photo catalyst. *Rasayan Journal of Chemistry, 4*, 814−819.

Motahari, F., Mozdianfard, M. R., Soofivand, F., & Salavati-Niasari, M. (2014). NiO nanostructures: Synthesis, characterization and photocatalyst application in dye wastewater treatment. *RSC Advances, 4*, 27654−27660.

Naseri, A., Samadi, M., Ebrahimi, M. M., Kheirabadi, M., & Moshfegh, A. Z. (2020). Heterogeneous photocatalysis by organic materials: from fundamental to applications. *Current Developments in Photocatalysis and Photocatalytic Materials*, 457−473.

Omote, M., Kitaoka, H., Kobayashi, E., Suzuki, O., Aratake, K., Sano, H., . . . Podloucky, R. (2005). Spectral, tensor, and ab initio theoretical analysis of optical second harmonic generation from the rutile TiO_2 (110) and (001) faces. *Journal of Physics. Condensed Matter: An Institute of Physics Journal, 17*, S175−S200.

Peng, C. W., Ke, T. Y., Brohan, L., Richard-Plouet, M., Huang, J. C., Puzenat, E., . . . Lee, C. Y. (2008). (101)-exposed anatase TiO_2 nanosheets. *Chemistry of Materials, 20*, 2426−2428.

Pereira, M., Oliveira, L., & Murad, E. (2012). Iron oxide catalysts: Fenton and Fentonlike reactions−A review. *Clay Minerals, 47*, 285−302.

Petrinic, I., Andersen, N. P. R., Sostar-Turk, S., & Le Marechal, A. M. (2007). The removal of reactive dye printing compounds using nanofiltration. *Dyes and Pigments, 74*, 512−518.

Puangpetch, T., Sommakettarin, P., Chavadej, S., & Sreethawong, T. (2010). Hydrogen production from water splitting over Eosin Y-sensitized mesoporous-assembled perovskitetitanatenanocrystalphotocatalysts under visible light irradiation. *International Journal of Hydrogen Energy, 35*, 12428−12442.

Rahim Pouran, S., Abdul Aziz, A. R., & Wan Daud, W. M. A. (2015). Review on the main advances in photo-Fenton oxidation system for recalcitrant wastewaters. *Journal of Industrial and Engineering Chemistry, 21*, 53−69.

Rao, L. N. (2014). Nanotechnological methodology for treatment of waste water. *International Journal of ChemTech Research, 6*, 2529−2533.

Reddy, K. R., Hassan, M., & Gomes, V. G. (2015). Hybrid nanostructures based on titanium dioxide for enhanced photocatalysis. *Applied Catalysis A: General, 489*, 1−16.

Rehman, S., Ullah, R., Butt, A. M., & Gohar, N. D. (2009). Strategies of making TiO_2 and ZnO visible light active. *Journal of Hazardous Materials, 170*, 560−569.

Rodríguez, A., Rosal, R., Perdigón-Melón, J., Mezcua, M., Agüera, A., Hernando, M., Letón, P., Fernandez-Alba, A., & García-Calvo, E. (2008). Ozone-based technologies in water and wastewater treatment. *Emerging contaminants from industrial and municipal waste*, 127−175.

Rostami-Vartooni, A., Nasrollahzadeh, M., Salavati-Niasari, M., & Atarod, M. (2016). Photocatalytic degradation of azo dyes by titanium dioxide supported silver nanoparticles prepared by a green method using Carpobrotusacinaciformis extract. *Journal of Alloys and Compounds, 689*, 15−20.

Saison, T., Gras, P., Chemin, N., Chaneac, M., Durupthy, O., Brezova, D., Colbeua-Justin, C., et al. (2013). New insights into Bi_2WO_6 properties as a visible-light photocatalyst. *Journal of Physical Chemistry C, 117*(44), 22656−22666.

Schreck, M., & Niederberger, M. (2019). Photocatalytic gas phase reactions. *Chemistry of Materials, 31*(3), 597618.

Shemer, H., & Linden, K. G. (2007). Aqueous photodegradation and toxicity of the polycyclic aromatic hydrocarbons fluorene, dibenzofuran, and dibenzothiophene. *Water Research, 41*, 853−861.

Shen, L., Liang, R., & Wu, L. (2015). Strategies for engineering metal-organic frameworks as efficient photocatalysts. *Chinese Journal of Catalysis, 36*, 2071–2088.

Song, X. M., Wu, J. M., Tang, M. Z., Qi, B., & Yan, M. (2008). Enhanced photoelectrochemical response of a composite titania thin film with single-crystalline rutile nanorods embedded in anatase aggregates. *Journal of Physical Chemistry C, 112*, 19484–19493.

Sun, J., Zhang, J., Zhang, M., Antonietti, M., Fu, X., & Wang, X. (2012). Bioinspired hollow semiconductor nanospheres as photosynthetic nanoparticles. *Nature Communications, 3*, 1139.

Sun, L., Campbell, M. G., & Dincǎ, M. (2016). Electrically conductive porous metal–organic frameworks. *Angewandte Chemie International Edition, 55*, 3566–3579.

Tang, L., Wang, J., Zeng, G., Liu, Y., Deng, Y., Zhou, Y., . . . Guo, Z. (2016). Enhanced photocatalytic degradation of norfloxacin in aqueous Bi_2WO_6 dispersions containing nonionic surfactant under visible light irradiation. *Journal of Hazardous Materials, 306*, 295–304.

Thiruvenkatachari, R., Vigneswaran, S., & Moon, I. S. (2008). A review on UV/TiO_2 photocatalytic oxidation process. *Korean Journal of Chemical Engineering, 25*, 64–72.

Thota, S., Tirukkovalluri, S. R., & Bojja, S. (2014). Effective catalytic performance of manganese and phosphorus co-doped titaniananocatalyst for Orange-II dye degradation under visible light irradiation. *Journal of Environmental Chemical Engineering, 2*, 1506–1513.

Tong, Z., Yang, D., Li, Z., Nan, Y., Ding, F., Shen, Y., & Jiang, Z. (2017). Thylakoid-inspired multishell g-C3N4 nanocapsules with enhanced visible-light harvesting and electron transfer properties for high-efficiency photocatalysis. *ACS Nano, 11*, 1103–1112.

Torres-Martínez, L. M., Gómez, R., Vázquez-Cuchillo, O., Juárez-Ramírez, I., Cruz-López, A., & Alejandre-Sandoval, F. J. (2010). Enhanced photocatalytic water splitting hydrogen production on RuO_2/La:$NaTaO_3$ prepared by sol-gel method. *Catalysis Communications, 12*, 268–272.

Truppi, A., Petronella, F., Placido, T., Margiotta, V., Lasorella, G., Giotta, L., . . . Comparelli, R. (2019). Gram-scale synthesis of UV–vis light active plasmonicphotocatalytic nanocomposite based on TiO_2/Au nanorods for degradation of pollutants in water. *Applied Catalysis B, 243*, 604–613.

Tseng, J., & Huang, C. (1991). Removal of chlorophenols from water by photocatalytic oxidation. *Water Science and Technology, 23*, 377–387.

Vukoje, I., Kovač, T., Džunuzović, J., Džunuzović, E., Lončarević, D., Ahrenkiel, S. P., & Nedeljković, J. (2016). Photocatalytic ability of visible-light-responsive TiO_2 nanoparticles. *Journal of Physical Chemistry C, 120*(33), 18560–18569.

Wanbayor, R., & Ruangpornvisuti, V. (2010). Adsorption of CO, H_2, N_2O, NH_3 and CH_4 on the anatase TiO_2 (0 0 1) and (1 0 1) surfaces and their competitive adsorption predicted by periodic DFT calculations. *Materials Chemistry and Physics, 124*, 720–725.

Wang, M., Che, Y., Niu, C., Dang, M., & Dong, D. (2013). Effective visible light-active boron andeuropium co-doped $BiVO_4$ synthesized by sol–gel method for photodegradion of methyl orange. *Journal of Hazardous Materials, 262*, 447–455.

Wei, L., Shifu, C., Sujuan, Z., Wei, Z., Huaye, Z., & Xiaoling, Y. (2010). Preparation and characterization of p-n heterojunctionphotocatalyst p-$CuBi_2O_4$/n-TiO_2 with high photocatalytic activity under visible and UV light irradiation. *Journal of Nanoparticle Research, 12*, 1355–1366.

Wetchakun, K., Wetchakun, N., & Sakulsermsuk, S. (2019). An overview of solar/visible light-driven heterogeneous photocatalysis for water purification: TiO_2- and ZnO-based photocatalysts used in suspension photoreactors. *Journal of Industrial and Engineering Chemistry, 71*, 1949.

Wu, L., Yu, J. C., & Fu, X. (2006). Characterization and photocatalytic mechanism of nanosized CdS coupled TiO_2 nanocrystals under visible light irradiation. *Journal of Molecular Catalysis A: Chemistry, 244*, 2532.

Xiong, X., & Xu, Y. (2016). Synergetic effect of Pt and borate on the TiO_2-photocatalyzed degradation of phenol in water. *Journal of Physical Chemistry C, 120*, 3906–3912.

Zeng, L., Guo, X., He, C., & Duan, C. (2016). Metal−organic frameworks: Versatile materials for heterogeneous photocatalysis. *ACS Catalysis, 6*, 7935−7947.

Zheng, D., Huang, C., & Wang, X. (2015). Post-annealing reinforced hollow carbon nitride nanospheres for hydrogen photosynthesis. *Nanoscale, 7*, 465−470.

Zheng, D., Pang, C., Liu, Y., & Wang, X. (2015). Shell-engineering of hollow g-C3N4 nanospheres via copolymerization for photocatalytic hydrogen evolution. *Chemical Communications, 51*, 9706−9709.

Zheng, D., Zhang, G., Hou, Y., & Wang, X. (2016). Layering MoS2 on soft hollow g-C3N4 nanostructures for photocatalytic hydrogen evolution. *Applied Catalysis A-General, 521*, 2−8.

Zheng, R., Meng, X., & Tang, F. (2009). Synthesis, characterization and photodegradation study of mixed-phase titania hollow submicrospheres with rough surface. *Applied Surface Science, 255*, 5989−5994.

Zhu, C., Wang, L., Kong, L., Yang, X., Wang, L., Zheng, S., . . . Zong, H. H. (2000). Photocatalytic degradation of AZO dyes by supported TiO_2 + UV in aqueous solution. *Chemosphere, 41*, 303−309.

Zhu, Y., Zhu, R., Xi, Y., Zhu, J., Zhu, G., & He, H. (2019). Strategies for enhancing the heterogeneous Fenton catalytic reactivity: A review. *Applied Catalysis B Environmental, 255*, 117739.

6

Effects of nanomaterials on biodegradation of biomaterials

Sanaz Soleymani Eil Bakhtiari[1], Hamid Reza Bakhsheshi-Rad[1,2], Mahmood Razzaghi[1], Ahmad Fauzi Ismail[3], Safian Sharif[2], Seeram Ramakrishna[4], Filippo Berto[5]

[1]ADVANCED MATERIALS RESEARCH CENTER, DEPARTMENT OF MATERIALS ENGINEERING, NAJAFABAD BRANCH, ISLAMIC AZAD UNIVERSITY, NAJAFABAD, IRAN [2]FACULTY OF ENGINEERING, UNIVERSITI TEKNOLOGI MALAYSIA, JOHOR BAHRU, MALAYSIA [3]ADVANCED MEMBRANE TECHNOLOGY RESEARCH CENTER (AMTEC), UNIVERSITI TEKNOLOGI MALAYSIA, JOHOR BAHRU, MALAYSIA [4]NANOSCIENCE AND NANOTECHNOLOGY INITIATIVE, NATIONAL UNIVERSITY OF SINGAPORE, SINGAPORE [5]DEPARTMENT OF MECHANICAL AND INDUSTRIAL ENGINEERING, NORWEGIAN UNIVERSITY OF SCIENCE AND TECHNOLOGY, TRONDHEIM, NORWAY

Abbreviations

3D	Three-dimensional
ATR	Attenuated total reflectance
BCP	Biphasic calcium phosphate
BG	Bioactive glass
BTR	Bone tissue regeneration
Ca-P	Calcium phosphate
CBN	Carbon-based nanomaterial
β-CD	β-Cyclodextrin
CNT	Carbon nanotube
CS	Chitosan
DNA	Deoxyribonucleic acid
ECM	Extracellular matrix
EDC	1-ethyl-3- (3-dimethylaminopropyl) carbodiimide
FESEM	Field emission scanning electron microscopy
FTIR	Fourier transform infrared
GelMA	Gelatin methacryloyl
GO	Graphene oxide
GQD	Graphene quantum dot
HA	Hyaluronic acid
HAp	Hydroxyapatite
MSC	Mesenchymal stem cell

MSN	Mesoporous silica nanoparticle
MWCNT	Multiwall carbon nanotube
nAg	Nanosilver
nAu	Nanogold
nHAp	Nanohydroxyapatite
NM	Nanomaterial
NP	Nanoparticle
$nTiO_2$	Nanotitanium dioxide
nZnO	Nanozinc oxide
PAAG	Poly(acrylamide-co-acrylamidoglycolic acid)/guar gum
PBS	Phosphate-buffered saline
PCL	Polycaprolactone
PGA	Polyglycolic acid
PLA	Polylactic acid
PLAGA	Polylactide-co-glycolide
PLGA	Poly(lactate-co-glycolate)
PLLA	Poly-L-lactide
PPF	Poly(propylene fumarate)
PPP	Poly(para-phenylene)
PU	Polyurethane
PVA	Poly(vinyl alcohol)
SBF	Simulated body fluid
TCP	Tricalcium phosphate
TGA	Thermogravimetric analysis

6.1 Introduction

In tissue engineering, the scaffolds' rate of degradation must be controlled to meet the newly allocated extracellular matrix (ECM) and maintain mechanical properties (Chang, Ahuja, Ma, & Liu, 2017; Kikuchi et al., 2004; Klammert et al., 2010; Yuan, Shen, Chua, & Zhou, 2019). Biodegradable materials used in tissue engineering, drug delivery systems, and biomedical applications must degrade gradually to release their contents and create space for newly formed tissue. The ideal rate of degradation is when it equals the formation of new tissue (Gopinathan et al., 2020; Sopyan, Gunawan, Shah, & Mel, 2016; Wang et al., 2016; Xavier et al., 2015). Rapid degradation causes the scaffold to lose its function as a carrier for cell growth, while slow degradation rates reduce available space and prevent new tissue formation (Hassan, Dave, Chandrawati, Dehghani, & Gomes, 2019; Unagolla & Jayasuriya, 2020; Zhu, 2010). In general, the breaking-down mechanism of polymer-based materials, including hydrogels, could be classified in three ways of simple dissolution, hydrolysis, and enzymatic cleavage. The hydrogels that are crosslinked physically lose their shape and dissolve in a solution with changes in environmental conditions such as pH and temperature. The hydrogels that are ionic crosslinked react to changes in the solution's ionic strength, and hydrogels dissolve as ions diffuse out of the hydrogel (Calori, Braga, de Jesus, Bi, & Tedesco, 2020; Dorishetty, Dutta, & Choudhury, 2020; Zinge & Kandasubramanian, 2020). Simple dissolution is not a degradation process because hydrogel materials do not break down into small molecules. Hydrolysis of unstable ester bonds in the polymer backbone, for example, lactide or glycolide segments in poly(lactate-co-glycolate) (PLGA), is the most common mechanism for the degradation of hydrogel (Hennink & van Nostrum, 2012; Liu, Zhuang, Shuai, & Peng, 2013; Wiltfang et al., 2002). Parameters that determine the hydrolysis rate include molecular weight, porosity, crosslinking density, morphology, and residual monomers of the polymer (Moseke & Gbureck, 2010; Shi, Wang, Chen, & Huang, 2008; Yeo et al., 2020). Increasing the molecular weight, crosslinking density, or hydrophobicity of the polymer lowers the hydrogel's degradation rate. Like local pH and fillers' inclusion, some other parameters may also play a role in the degradation rate (Timmer, Ambrose, & Mikos, 2003; Yeo et al., 2020). For example, the incorporation of β-tricalcium phosphate (TCP) ceramic filler into the poly(propylene fumarate) (PPF) matrix delays degradation and acts as an internal buffer for acidic decomposition products resulting from ester groups' hydrolysis (Calori et al., 2020; Chang et al., 2017; Dorishetty et al., 2020; Gopinathan et al., 2020; Hassan, Dave et al., 2019; Hennink & van Nostrum, 2012; Kikuchi et al., 2004; Klammert et al., 2010; Liu et al., 2013; Moseke & Gbureck, 2010; Shi et al., 2008; Sopyan et al., 2016; Timmer et al., 2003; Unagolla & Jayasuriya, 2020; Wang et al., 2016; Wiltfang et al., 2002; Xavier et al., 2015; Yeo et al., 2020; Zhu, 2010; Zinge & Kandasubramanian, 2020). Although the degradation mechanisms are different, the hydrogel degradation products should not be toxic and should not activate an immune response. The degradation of the hydrogel in the body forms natural molecules that are known as biocompatible. For example, hydrolysis of poly(α-hydroxy acid) produces lactic/glycolic acid, which enters the tricarboxylic acid cycle and is finally excreted from the body (Calori et al., 2020; Chang et al., 2017; Dorishetty et al., 2020; Gopinathan et al., 2020; Hassan, Dave et al., 2019; Hennink & van Nostrum, 2012; Kikuchi et al., 2004; Klammert et al., 2010; Li & Chang, 2005; Liu et al., 2013; Moseke & Gbureck, 2010; Shi et al., 2008; Sopyan et al., 2016; Timmer et al., 2003; Unagolla & Jayasuriya, 2020; Wang et al., 2016; Wiltfang et al., 2002; Xavier et al., 2015; Yeo et al., 2020; Zhu, 2010; Zinge & Kandasubramanian, 2020). Several research groups have studied various bioactive ceramic materials with homogeneous composite structures and different weight losses (Li & Chang, 2005). Studies have shown that the ideal degradation rate is consistent with the rate at which new tissue grow. Biomaterials' characteristics, including the degradation rate of biomaterials, particularly polymer-based scaffolds, must be maintained until the newly generated tissue has adequate mechanical properties for replacing this supporting function (Díaz, Puerto, Ribeiro, Lanceros-Mendez, & Barandiarán, 2017; Hassan, Yassin et al., 2019; Kim, Lim, Naren, Yun, & Park, 2016; Polo-Corrales, Latorre-Esteves, & Ramirez-Vick, 2014). As previously discussed, technological strategies are expected to control the rate of degradation of biomaterials by the incorporation of ceramic and metal nanoparticles (NPs) in biomaterials paving the way for such degradation systems (Alge et al., 2012; Leijten et al., 2017; Qi et al., 2017; Schantz et al., 2003; Shin et al., 2013). In this study, while discussing the degradation behavior of biomaterials, especially those made from polymers containing ceramics and metallic fillers, the challenges and perspectives in providing a roadmap for further advancement in this field are highlighted.

6.1.1 Nanomaterials

Richard Feynman, a Nobel Prize winner, proposed nanotechnology's basic idea by proposing the development of molecular machines in 1959. Since the scientific community has studied this role, nanotechnology can play a role in all aspects of society. Nanotechnology's intrigue lies in controlling materials' properties by assembling such materials on the nanoscale (Liu & Webster, 2007; Zhang & Webster, 2009). Nanotechnology has made tremendous advances in recent decades. More recently, nanomaterials (NMs), a substance that is a basic structural unit, particles, grains, fibers, or other compounds with at least one dimension less than 100 nm (Siegel & Fougere, 1995), have been addressed to improve disease prevention, diagnosis, and treatment. NMs could be nanostructured powders, gels, films, coatings, or components in the range of nanoscale. They are unique as they display quantum mechanical properties that lead to unique properties compared to their bulk forms. Some of NMs properties are dimensional quantification, novel crystal phases, single charge effects, light scattering interference, charge depletion, ballistic electron transfer, and so forth (Hodes, 2007; Kumar, 2000). Since their discovery, NMs have been extensively investigated for industrial and medical applications. An important aspect of gaining more recognition in the health care sector is the safety issues associated with NMs' treatment and use (Andronescu et al., 2016; Zhang & Webster, 2009). NMs represent some unique properties, unlike their bulk forms, because of decreasing the material dimensions, changing the physicochemical properties, exponentially increasing the stiffness, surface area, and surface-to-volume ratio. A wide range of NPs, nanoscale surfaces, nanofibers, nanoporous scaffolds, and carbon nanotubes (CNTs) could be made with advanced manufacturing and processing technologies. These substances are increasingly being introduced into tissue-engineering biomaterials, particularly scaffolds for developing new biomimetic substitutes for replacing damaged organs and tissues. Another common application of NMs is their use in wound healing (Padmanabhan & Kyriakides, 2015).

6.1.2 Application of nanomaterials in tissue engineering

Using artificial organs/tissues or organ transplant is the first choice for repairing damaged tissues and organs when severely damaged or lost due to cancer, congenital disabilities, or trauma, and conventional medications are no longer applicable. However, organ transplant practices currently face several problems. About three decades ago, tissue engineering, as a new paradigm, emerged as an alternative tissue and organ reconstruction method (Ikada, 2006). Tissue engineering has shown great promise as a solution to the tremendous demand for tissue and organ transplantation. Tissue engineering mainly revolves around three main pillars: biomaterial scaffolds, cells, and growth factors/bioactive molecules (Singla, Abidi, Dar, & Acharya, 2019). Biomaterial scaffolds locate cells, direct them for growth, and expose them to sufficient nutrients, oxygen, metabolites, and appropriate growth factors to enhance differentiation and function (Zhang, Zhou, & Zhang, 2014). The characteristics of NMs considered in tissue engineering consist of several factors, including: (1) nanoscale provides cells/tissues with the correct structure for adhesion, differentiation, and proliferation; (2) NMs act in the same structure as the

ECM in terms of chemical composition and physical structure (Singla et al., 2019; Yi, Rehman, Zhao, Liu, & He, 2016; Zhang & Webster, 2009; Zhang et al., 2014); (3) they have outstanding mechanical properties and can reinforce various organic/synthetic scaffolds (Guo, Lei, Li, & Ma, 2015; Singla et al., 2019); (4) the high conductivity of carbon NPs provides local electrical stimulation for bone tissue engineering (Edwards, Werkmeister, & Ramshaw, 2009; Shi, Votruba, Farokhzad, & Langer, 2010; Singla et al., 2019); (5) micro/nanoencapsulation of essential growth factors helps them to release in a controlled and smooth manner at the target location (De Witte, Fratila-Apachitei, Zadpoor, & Peppas, 2018; Singla et al., 2019; van Rijt & Habibovic, 2017); and (6) NMs provide enhanced biocompatibility, bioactivity, and improve local interactions with cells or proteins. The degradation of the biomaterial scaffold is a crucial factor for successful tissue regeneration (Zhang et al., 2014). Various NMs or biomaterials integrated with NPs can be used to control scaffolds' biodegradation, which can further alter the degradation process. Recently, various NMs have been used as scaffolds for various organ and tissue engineering such as skin, bone, cartilage, heart, liver, corneal, nervous system, pancreas, and so on.

6.1.3 Application of nanomaterials in wound healing

A wound is a kind of injury that occurs for various reasons, such as surgery, trauma, burns, and diabetes. Among them, diabetic foot ulcers and burns are becoming life-endangering issues. Based on available data, approximately ten million people worldwide suffer from chronic burns and wounds (Albertini, Di Sabatino, Calonghi, Rodriguez, & Passerini, 2013; Kumar et al., 2013). This type of wound requires timely and correct treatment. The use of an ideal dressing with appropriate properties can improve healing the wound and cause patient well-being. A perfect wound dressing should have sufficient flexibility, mechanical strength, tear strength, and should not stick to the wound (Jayakumar, Prabaharan, Kumar, Nair, & Tamura, 2011). Wound healing has been studied in detail to develop an ideal process that enables rapid healing and reduces scarring to remain functional. The conventional wound care method is topical treatments such as antibacterial or colloidal drugs that prevent infection and promote wound healing. Based on nanotechnology that studies ultrafine particles (up to 100 nm in diameter) and related phenomena, metallic NPs such as silver, gold, and zinc are increasingly applied in dermatology because of their positive effect to promote wound healing, treatment, and prevention of bacterial infections (Mihai, Dima, Dima, & Holban, 2019). The effects of common NMs on the rate of degradation of tissue-engineering scaffolds and wound dressings have been reported.

6.2 Effects of different nanomaterials on biodegradation behavior

6.2.1 Nano zinc-oxide (nZnO) effects on biodegradation behavior

Zinc is often required for various cellular and enzymatic processes and plays an important role in wound healing, especially burns. Zinc-oxide nanoparticles (nZnO) have antimicrobial properties and are often used in cosmetics. Zinc-oxide is often used in skin creams due to its excellent astringent and gentle antiinflammatory, dry, and antiseptic properties. The size and concentration of NPs also affect wound healing properties. The nZnO with the ultrafine particle size and its concentration shows strong antibacterial activity and does not cause side effects. Zn^{2+} ions released from ZnO may promote wound healing by increasing the migration of keratinocytes (Deepachitra, Lakshmi, Sivaranjani, Chandra, & Sastry, 2015). The effects of nZnO on the biodegradation of tissue-engineering scaffolds and wound dressing have been reported. In a research, Kumar et al. (2013) evaluated the potential of wound healing of β-chitin hydrogel/nZnO composite bandage. This study synthesized β-chitin hydrogel via dissolving the β-chitin powder in methanol/$CaCl_2$ solvent and adding distilled water. Subsequently, the nZnO powder was added to the β-chitin hydrogel and blended for having a uniform suspension. The resulting suspension was frozen at 0°C for 12 h. The frozen specimens were then freeze-dried for 24 h to attain a permeable nanocomposite bandage. The outcomes from the composite bandage's biodegradation evaluation showed similar degradation rates for the samples with and without nZnO. All bandages degraded by about 40%–50% after soaking for one week in phosphate-buffered saline (PBS)-containing lysozyme. In this study, the bandage's degradation rate was not affected by nZnO particles, and the bandage showed controlled degradation (Kumar et al., 2013). In another study, Cleetus et al. (2020) investigated alginate hydrogels with integrated ZnO NPs for wound healing. In this in vitro study, they developed a three-dimensional (3D) printing compound of nZnO with photocatalytic activity encapsulated into a hydrogel (alginate) structure for antibacterial purposes applicable in wound healing. The results indicated that the degradation of bioprinted samples immersed for 28 days in PBS has differences in the nZnO incorporated samples compared to the alginate alone ones. Within seven days, the alginate gel samples began to lose structural integrity. Conversely, 0.5% and 1% ZnO incorporated specimens retained their structure for a long period of time and remained unchanged for 28 days. The long life of nZnO incorporated 3D printed gels was attributed to the crosslinking of both calcium and zinc ions into alginate (Chan, Jin, & Heng, 2002; Cleetus et al., 2020). Besides the fact that the zinc/alginate binding site is different from the calcium binding site, zinc ions are also less selective, which appears to lead to better crosslinking of alginate gels with nZnO (Chan et al., 2002; Cleetus et al., 2020). In another study, Mohandas, Sudheesh Kumar, Raja, Lakshmanan, and Jayakumar (2015) investigated the alginate hydrogel/nZnO composite bandage on infected wounds. In this study, a mixture of nZnO and alginate hydrogel was freeze-dried to develop a composite alginate hydrogel/nZnO dressing. In vitro biodegradation assay results showed that the nZnO-containing composite bandage exhibited a controlled degradation profile compared to the control dressing, which did not have nZnO (Mohandas et al., 2015). Sudheesh Kumar et al. (2012) evaluated the flexible and microporous nZnO incorporated chitosan (CS) hydrogel composite bandage in vitro and in vivo. The biodegradation test outcomes showed that the nZnO content did not change the dressing's degradation rate because of its low concentration (Sudheesh Kumar et al., 2012). In other research, Vedhanayagam, Unni Nair, and Sreeram (2018) evaluated the functional role and morphology of dendrimers of collagen/nZnO scaffolds to apply as a wound dressing. In this study, the specimens' weight loss upon exposure to a degradation condition was considered an indication of biodegradation. The outcomes showed that the collagen scaffold's weight loss was about 98% on day 28, and the weight loss of the nZnO-containing scaffold was 72%−80% over the same period. These results suggest that the nZnO incorporated scaffold is chemically more stable than the monolithic collagen one (Vedhanayagam et al., 2018). Rakhshaei, Namazi, Hamishehkar, Kafil, and Salehi (2019) evaluated in situ synthetic CS-gelatin/ZnO nanocomposite scaffold for drug delivery application. The results of the degradation tests obtained in this study showed that in situ formed nZnO, did not significantly affect nanocomposite biodegradation due to its low concentration (Rakhshaei et al., 2019). In another study, Ullah, Zainol, and Idrus (2017) studied morphology, mechanical properties, and cellular suitability of the nZnO incorporated CS/collagen 3D porous scaffold. As a result of 28-day biodegradation tests, the average biodegradation rate of CS/collagen scaffold on day 10 was approximately equal to the biodegradation rates of 0.5%, 2.0%, and 4.0% nZnO reinforced CS/collagen scaffolds. The biodegradation rate of the 1.0% nZnO reinforced CS/collagen composite was higher than that of the CS/collagen scaffold (Ullah et al., 2017). In another study by Ahmadzadeh, Babaei, and Goudarzi (2018) the localization and degradation of nZnO in biocompatible polylactic acid/polycaprolactone (PLA/PCL) mixtures were evaluated through extensive rheological characterization. Attenuated total reflectance−Fourier transform infrared spectroscopy (ATR-FTIR), and hydrolysis analysis outcomes showed that the addition of nZnO caused faster degradation of the PLA/PCL mixture. The results showed that biodegradation accelerated with increasing nZnO content and that the highest weight loss was associated with the composite containing 6 wt.% of nZnO. This suggests that the synergistic effect of higher nZnO and acidic compounds contents leads to higher degradation rates, resulting in the faster formation of water-soluble oligomers (Ahmadzadeh et al., 2018). In another study, Augustine, Kalarikkal, and Thomas (2016) evaluated the effect of nZnO on the in vitro biodegradation of electrospun PCL membranes in simulated body fluid (SBF). Based on the morphological and crystallinity assessments, it was seen that ZnO NPs accelerate the degradation of PCL (Augustine et al., 2016). As the duration of PCL homopolymer biodegradation was more than two years, ε-caprolactone and dl-lactide copolymer were synthesized for accelerating the biodegradation rate of the polymeric scaffold. The mechanical properties and degradation rate could be adjusted by changing the PCL-PLA copolymer composition and adding additives, including ceramic materials (Yuan et al., 2019). Moreover, the degradation processes of poly-L-lactide (PLLA), polylactide-co-glycolide (PLAGA), and PLGA exhibit random and bulky hydrolysis of ester bonds in the polymer chain and may lead to premature destruction of the scaffold (Kikuchi et al., 2004). Besides, the release of acidic biodegradation products can lead to highly inflammable reactions in vivo. The kinetics of biodegradation is affected by several parameters, including chemical composition, geometry, scaffold's material phase (crystalline, semicrystalline, or amorphous), polymer porosity, hydrophobicity, and the existence of additives (Yuan et al., 2019).

In another study, Lizundia et al. (2017) evaluated tunable hydrolysis of nZnO incorporated PLLA scaffolds. The results showed that the scaffold degradation profile could be successfully adjusted by altering the amount of nZnO and changing the scaffold's porosity. The results revealed that the separated water on the nZnO particles' surface begins a hydrolysis reaction with a decrease in the chemical bond strength of the adjacent PLLA chain, which causes further division into water-soluble oligomers. As can be seen in Fig. 6–1, it can be observed that the addition of nZnO to the PLLA matrix speeds up the hydrolysis of the scaffold (Lizundia et al., 2017). In another study by Vicentini, Smania, and Laranjeira (2010), a CS/poly(vinyl alcohol) (PVA) film incorporated with nZnO and plasticizer was made. In this study, nZnO was prepared via Pechini method from polyester by a reaction between citric acid and ethylene glycol in which the ions of metal were dissolved and incorporated into CS and PVA mixed films with various concentrations of polyoxyethylene sorbitan monooleate, Tween 80 (T80). The outcomes of degradation tests showed that CS/PVA films produced lower degradation and swelling than films containing nZnO and T80, probably due to the

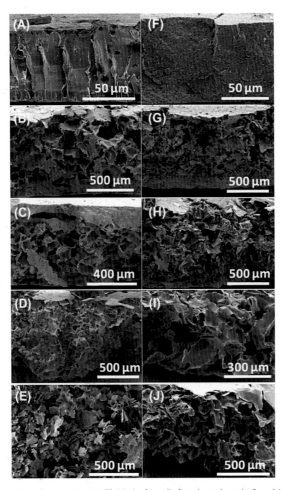

FIGURE 6–1 FE-SEM micrographs of PLLA/ZnO scaffolds before (left column) and after (right column) being hydrolytically degraded for 20 days (right column) with various porosities: (A) 0%; (B) 50%; (C) 70%; (D) 75%; and (E) 90% porosities (Lizundia, Mateos, & Vilas, 2017).

lack of pores and plasticizers. When nZnO and T80 were added to the film, an increase in degradation and swelling was observed (Vicentini et al., 2010). In another study, Felice et al. (2018) controlled PCL-ZnO nanofiber scaffolds' degradation for bone tissue engineering and assessed their antibacterial activity. In this study, degradation analysis showed that the ZnO dispersion in fibers plays an essential role in material degradation. The obtained results showed that the nZnO powder significantly accelerates the degradation of the sample. In this regard, the rate of degradation could be controlled by changing the concentration of the ZnO nanopowder and the ZnO dispersion in the fiber (Felice et al., 2018). The results obtained from the revised article suggest that adding a certain amount of nZnO to tissue-engineered scaffold results in controlled and rapid biodegradation, and a small amount of nZnO does not change the rate of scaffold degradation.

6.2.2 Nanosilver effects on biodegradation behavior

Nanosilver (nAg) particles typically have at least one dimension in a size range between 1 and 100 nm (Chen, Ouyang, Kong, Zhong, & Xing, 2013; Mohamed & Xing, 2012). As the particle size decreases, the surface-to-volume ratio of the nAg particles increases dramatically, resulting in significant changes in physical, chemical, and biological properties. For hundreds of years, silver has been among the most widely used materials in medical applications (Ge et al., 2014). Recently, nAg particles have increased interest in biomedical applications due to their antimicrobial, antifungal, antiviral, and antiinflammatory activities (Varner, Sanford, El-Badawy, Feldhake, & Venkatapathy, 2010; Zhong, Xing, & Maibach, 2010). Also, the effect of nAg on the biodegradation of wound dressing and tissue-engineered scaffolds have been reported. In the study of Anisha, Biswas, Chennazhi, and Jayakumar (2013), CS-hyaluronic acid (HA)/nAg composite sponges for drug-resistant bacteria-infected diabetic wounds were fabricated. The nanocomposites were fabricated through homogeneously mixing CS, HA, and nAg particles and then a lyophilization process for obtaining a porous and flexible construction. As a result of in vitro biodegradation analysis, it was found that the degradation rate of nAg reinforced CS/HA composite sponge was higher compared to that of CS/HA control one. The higher degradation rate could be attributed to the comparatively higher swelling and porosity of the nanocomposite sponges, which makes the N-acetyl glucosamine group available for the action of lysozyme (Anisha et al., 2013). In another study by Song et al. (2015), nAg in situ hybridized collagen scaffold was designed to regenerate full-thickness burned skin. The outcomes revealed that the addition of nAg resulted in significant influence on the degradation, and the complete degradation time of the composite scaffold with 1% and 2% nAg particles increased to 46 and 70 h, correspondingly. Based on the outcomes, two parameters can affect the degradation process. First, collagenase diffusion decreases with increasing crosslinking strength. Second, nAg particles or Ag^+ ions can affect the activity of the enzyme (Song et al., 2015). In another study, Lu et al. (2018) produced antimicrobial and biodegradable nAg reinforced mesoporous silica nanoparticles (MSNs) tissue nanoadhesives to heal the wound faster. In this study, nAg reinforced MSN nanoadhesives were produced via reducing ultrafine nAg particles on both surface and large pores of biodegradable MSNs in a controlled manner. The results showed that nAg-MSNs exhibit the same degradation pathway as MSNs, indicating that nAg integration has a limited effect on degradation (Lu et al., 2018). In another study, Yazdimamaghani et al. (2014) evaluated a macroporous hybrid gelatin/bioactive glass (BG)/nAg scaffolds with controlled biodegradation property and antibacterial activity for bone tissue engineering. In this study, nAg was successfully synthesized in situ in gelatin solution by reducing heat treatment in a simple, "green" method in which gelatin acts as a natural reducing agent and stabilizer. The results of in vitro biodegradation assessment of the enzyme showed that the addition of nAg increased the in vitro biodegradation in samples without BG; however, the biodegradation was significantly reduced by the addition of BG (Yazdimamaghani et al., 2014). In another study, Palem, Shimoga, Kang, and Lee (2020) investigated the production of nAg incorporated guar gum composite hydrogels for biomedical and environmental applications. The results obtained for the degradation behavior showed that poly(acrylamide-co-acrylamidoglycolic acid)/guar gum (PAAG) hydrogel showed higher weight loss (%) than PAAG/nAg hydrogel. This was due to the releasing of nAg particles from the hydrogel network in an aqueous medium which adheres to or penetrated the negatively charged bacterial cell wall. This can negatively affect cell activity by interacting with the phosphate and sulfur functions resulting from cell death. As a result, the cells' metabolic activity was decreased (biodegradation was reduced), and a strong interaction between the nAg particles and the guar gum hydrogel network was observed (Palem et al., 2020). In another study, Liu et al. (2015) evaluated the biodegradability, antimicrobial activity, bioactivity, and cytotoxicity of PLA composite hybrids produced by fusion with hydroxyapatite (HAp) nanorods and nAg particles. Degradation tests showed that a large surface-to-volume ratio of nAg particles spread over the PLA matrix's surface that absorbs more water, which significantly contributes to the hydrolytic cleavage of the PLA ester bonds (Liu et al., 2015). In another study, Le Thi, Lee, Thi, Park, and Park (2018) evaluated the catechol-rich gelatin hydrogel hybridized in situ with incorporated nAg particles to enhance antibacterial activity. In vitro degradation results showed that higher $AgNO_3$ concentrations corresponding to higher nAg particle content resulted in a lower biodegradation rate (Le Thi et al., 2018). In another study, Balakrishnan and Thambusamy (2020) evaluated the preparation of a riboflavin-containing PVA/β-cyclodextrin (β-CD) polymeric nanofiber matrix for in vivo wound dressing application. In vitro biodegradation of various PVA/β-CD nanofiber formulations were performed in lysozyme solution at 37°C for 1, 5, 10, 15, and 20 days. As the incubation time increased, the rate of biodegradation of each nanofiber scaffold increased. The obtained results revealed that the composite nanofiber scaffolds show a higher biodegradation rate. This may be due to the large average fiber diameter and the nanofiber scaffold's high antimicrobial activity, which allows a larger number of microorganisms to attach to the polymeric nanofiber. The obtained results indicated that the nAg particles and riboflavin components in nanofiber scaffolds had influenced the composite scaffold's biodegradation rate, which protects the nanofiber scaffold from water penetration, which may be useful for the wound dressing (Balakrishnan & Thambusamy, 2020). In another study by Saravanan et al. (2011), a biocomposite scaffold containing CS/ Nanohydroxyapatite (nHAp)/nAg was prepared and evaluated for bone tissue engineering. Results obtained from in vitro biodegradation studies showed that the introduction of nAg to the CS/nHAp scaffold reduces biodegradation, suggesting that the CS/nHAp/nAg scaffold will be available longer until new bone tissue growth begins (Saravanan et al., 2011). From the results obtained from peer-reviewed articles, it can be said that adding Ag NPs to different combinations has different effects and does not follow the same behavior.

6.2.3 Nanogold effects on biodegradation behavior

Among the various types of NMs, metal NPs, especially gold (Au) NPs, are attracting great interest in various scientific fields due to their special properties, including high coefficient of X-ray absorption, simplicity of synthesis, and easy control of the particle's physical and chemical properties (Elahi, Kamali, & Baghersad, 2018; Zhang, Wang et al., 2014; Zhang, Chu et al., 2014). Gold NPs are used in tissue-engineering usages, and the effect of nAu particles on wound dressing biodegradation and tissue-engineering scaffolds has been reported. In the study of Tamayo et al. (2018), porous nAu/polyurethane (PU) scaffolds were designed, and their mechanical and thermal characteristics were evaluated. The influence of Au NPs content on biodegradation and the thermal degradation of a porous PU matrix containing different amounts of Au NPs were also assessed using thermogravimetric analysis (TGA). The degradation temperature generally increased with the incorporation of nAu particles. Besides, the obtained results showed that the heat resistance increases with the incorporation of nAu particles. The obtained in vitro degradation results revealed that the degradation rate of PU/0.32 Au was higher than that of the other samples. The improved hydrophilicity due to the presence of NPs has the potential to accelerate polymer biodegradation. This is because the increase in hydrophilicity improves hydrolysis by promoting the integration of water molecules into the polymer's internal structure (Tamayo et al., 2018). Baei et al. (2016) produced electrically conductive Au NPs reinforced CS thermosensitive hydrogels for cardiac tissue-engineering application. In this study, a biodegradation assessment was performed to measure hydrogel's stability under cell culture conditions. During the first week of incubation, the CS degradation rate was less than 20%, but for CS-2Au NPs this rate was less than 10%. After 2 weeks of culture, the stability of CS-2Au NPs was 2.3 times higher than that of the CS sample. Au NPs content improved the biostability of CS hydrogels (Baei et al., 2016). Aly, Eldesouky, and Eid (2012) produced gold NPs reinforced ceramic scaffolds for bone tissue engineering and evaluated cells' properties and viability. The results showed that the monolithic ceramic scaffold's degradation rate was higher than that of Au NPs reinforced ceramic scaffold as the gold NPs delay the degradation due to the scaffold's reduced solubility (Aly et al., 2012). In another study, Grant et al. (2014) fabricated crosslinking collagen utilizing Au NPs for improving the mechanical properties and degradation resistance while also keeping its natural biocompatibility and microstructure. In this study, rat tail collagen type I was bound to Au NPs using 1-ethyl-3- (3-dimethylaminopropyl) carbodiimide (EDC), a zero-length crosslinker. The results obtained from the biodegradation assessment indicated that the Au NPs reinforced collagen scaffolds are more resistant to degradation than the monolithic collagen scaffold while still keeping an open microstructure

(Grant et al., 2014). Based on the results obtained from peer-reviewed articles, it can be said that the addition of Au NPs can improve the scaffold's biostability.

6.2.4 Carbon-based nanomaterials effects on biodegradation behavior

In science and technology, carbon-based nanomaterials (CBNs) are becoming an attractive NM due to the existence in a wide variety of carbon allotropes, ranging from known allotrope phases such as amorphous carbon, graphite, and diamond to advantageous CNTs, graphene oxide (GO), graphene quantum dots (GQDs), and carbon-based fullerene materials that have become precious recently (Maiti, Tong, Mou, & Yang, 2019; Mostofizadeh, Li, Song, & Huang, 2011). In addition, the effects of CBNs on the biodegradation of bone substitutes and tissue-engineered scaffolds have been reported. In a study by Bakhtiari, Karbasi, Tabrizi, and Ebrahimi-Kahrizsangi (2019), CS/multiwalled carbon nanotubes (MWCNTs) composite as a bone substitute was fabricated, and its physical, mechanical, bioactivity, and biodegradation were evaluated. Biodegradability assessment showed that incorporating MWCNTs to CS film reduces the biodegradation rate of CS film (Bakhtiari et al., 2019). In another study, Mirmusavi et al. (2019) designed poly 3-hydroxybutyrate (P3HB)-CS-MWCNTs/silk nano/microcomposite scaffolds for cartilage tissue-engineering usage and evaluated their physical, mechanical, and biological properties. According to the biodegradation analysis results, it could be said that the composite containing MWCNTs had a moderate rate of degradation compared to the scaffold without MWCNTs, and the slower degradation rate of P3HB-CS-MWCNTs/silk is due to the interaction between the carboxyl groups of the MWCNTs and amine groups of CS as a polycationic biopolymer in polar solutions. Ionic bonds between the amine and carboxyl groups present in MWCNTs and P3HB may increase chemical stability and reduce the rate of degradation of the P3HB-CS-MWCNTs/silk scaffold (Mirmusavi et al., 2019). Lan et al. (2019) fabricated biphasic calcium phosphate (BCP)/PVA scaffold reinforced with CNTs for bone tissue-engineering application. Degradation analysis showed that by changing BCP and CNTs powder concentrations, the scaffold's degradation could be altered. According to the obtained results, it could be said that the biodegradation rate is steadily decreased with increasing the concentration of CNTs from 0.05% to 0.25%. With further increasing the CNTs content by 0.5%, the weight loss was increased because of the change in the scaffold's internal structure and the increasing the mean size of pores (Lan et al., 2019). In another study, Tavakoli, Karbasi, and Soleymani Eil Bakhtiari (2020) investigated the physical, mechanical, and biological degradation of CS/GO composites for bone substitution. The results obtained from the biodegradation test showed that the addition of a slight amount of GO to the CS film and the interaction between the COOH functional group of GO and the CS matrix improved the crystallinity, thereby reducing the biodegradation rate of the film (Tavakoli et al., 2020). Aidun et al. (2019) fabricated GO reinforced PCL/CS/collagen electrospun scaffold to study in vitro evaluation of biological and physicochemical activities. Biodegradation analysis showed that the GO incorporation in the PCL/CS/collagen scaffolds increases the water absorption rate by adding hydrophilic groups, resulting in increasing the degradation rate. The researchers reported the scaffold's biodegradation after

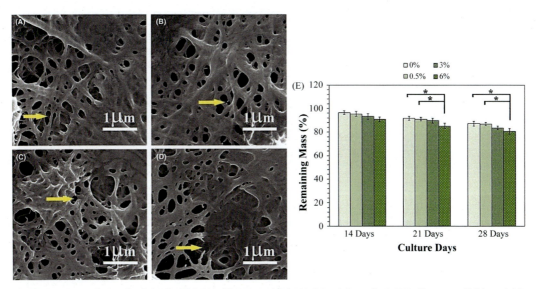

FIGURE 6–2 Degradation for (A) 0%, (B) 0.5%, (C) 3%, and (D) 6% GO reinforced PCL/CS/collagen scaffolds and (E) changes in remaining mass for the scaffolds with different ratios of GO (*$P < .05$) (Aidun et al., 2019).

8 weeks, as can be seen in Fig. 6–2 (Aidun et al., 2019). Some pores were broken, and some cracks in the nanoscale were seen on the fibers (Aidun et al., 2019; Ezati, Safavipour, Houshmand, & Faghihi, 2018; Saravanan et al., 2017). Their results also showed that the degradation rate of scaffolds was increased with raising the GO because of scaffold's saturation with GO that makes them brittle, as shown in Fig. 6–2E (Aidun et al., 2019). In another study, Saravanan et al. (2017) fabricated a scaffold containing CS, gelatin, and GO for in vitro and in vivo evaluations for bone tissue regeneration (BTR). Biodegradation studies showed that the CS/gelatin scaffolds containing 0.1% and 0.25% GO maintain structural integrity with 50% and 38% degradation, respectively. The results revealed that if the GO concentration exceeds 0.25%, the rate of degradation increases to almost 65%, which may be due to the GO sheet's saturation in the CS/gelatin scaffolds that can make them brittle. Hence, the 0.25% GO/CS/gelatin scaffold exhibited better structural integrity than other GO concentrations added to the CS/gelatin scaffold (Saravanan et al., 2017). In another study, Mohammadi, Shafiei, Asadi-Eydivand, Ardeshir, and Solati-Hashjin (2017) produced the electrospinning GO reinforced PCL nanocomposite scaffold for bone tissue-engineering application. Biodegradation evaluation of the fabricated scaffolds showed that the addition of GO nanosheets could increase the scaffolds' rate of degradation. The narrow diameter and the high ratio of surface-to-volume fiber in the composite can provide more space for water molecules and promote water absorption (Zhang et al., 2012). The GO sheet has multiple functional groups on its surface, increasing the density of hydrophilic groups in the polymer chain and increasing the system's overall hydrophilicity. Given this, increasing the amount of GO nanosheets in the PCL matrix could increase the PCL chain's moisture adsorption and hydration, resulting in faster biodegradation (Mohammadi et al., 2017).

Shin et al. (2013) fabricated a bioactive CNTs-based ink for 3D printing of flexible electronic components. Deoxyribonucleic acid (DNA) was used for dispersing CNTs in gelatin methacryloyl (GelMA) and HA. The produced inks were used to make hybrid films. They reported enzyme-mediated membrane degradation during the incubation of constructs in a solution of type II collagenase. The strong binding between the CNT and the DNA chains resulted in a highly stable composite ink construct limiting ink degradation (Shin et al., 2013). The Pseudomonas enzyme degree of degradation depends on the effective area and pores that can increase the possible locations for the enzyme's binding activity. Besides, enzymatic biodegradation of PCL and nanocomposites could be changed by altering the content of CNTs. Thermal analysis displayed that complete thermal biodegradation of PCL happens at 500°C. According to the results obtained from the reviewed papers, it can be said that the biodegradation rate decreases when a certain amount of MWCNTs is added, and the biodegradation rate increases when a certain amount of GO is added.

6.2.5 Nanotitanium dioxide effects on biodegradation behavior

TiO_2 NMs are used in various applications, from conventional products such as sunscreens to modern devices like photovoltaic cells, and are applied in various environmental and biomedical applications like water treatment, biosensing, and drug delivery. These applications' importance and diversity have led to tremendous interest and significant advances in the production, characterization, and basic understanding of TiO_2 NMs over the past decades (Chen & Mao, 2007; Chen & Selloni, 2014; Chen, Nanayakkara, & Grassian, 2012; Fujishima & Honda, 1972; Hoffmann et al., 1995; Linsebigler, Lu, & Yates, 1995; Thompson & Yates, 2006). Hereafter, the effects of nanotitanium dioxide ($nTiO_2$) on the biodegradation of tissue-engineered scaffolds have been reported. In Kumar's study (Kumar, 2018), $nTiO_2$ doped CS scaffolds were designed for bone tissue engineering and fabricated employing a freeze-drying method. For evaluating the CS/TiO_2 scaffolds' in vitro degradation, the samples were incubated in PBS containing lysozyme at pH 7.4. Results obtained from biodegradation tests showed that the addition of TiO_2 reduces the rapid degradation of the scaffold. The TiO_2 NPs limited the formation of strong bonds during the preparation of the specimens. Thus, it could be concluded that the NPs-doped scaffold can be used to produce biocompatible implants with lighter weight (Kumar, 2018). In another study, Pelaseyed, Madaah Hosseini, and Samadikuchaksaraei (2020) constructed $PLGA/TiO_2$ nanocomposite scaffolds for tissue engineering. The assessment of the biodegradation of the scaffolds showed a significant drop in pH at the beginning of degradation, as the molecular chains were small enough to be released from the scaffold into the PBS solution (Fig. 6–3A) (Pelaseyed et al., 2020). As shown in Fig. 6–3B, the outcomes revealed that in $nTiO_2$ reinforced PLGA nanocomposite, the degradation rate increases with raising TiO_2 content and incubation time (Pelaseyed et al., 2020). In another study, Khoshroo et al. (2017) fabricated a 3D PCL microspheres/TiO_2 nanotube composite scaffold for bone tissue engineering. The obtained biodegradation test results showed that the biodegradation rate was comparatively higher in the 0.5 wt.% TiO_2 nanotube-reinforced PCL nanocomposite comparing with the monolithic PCL when the

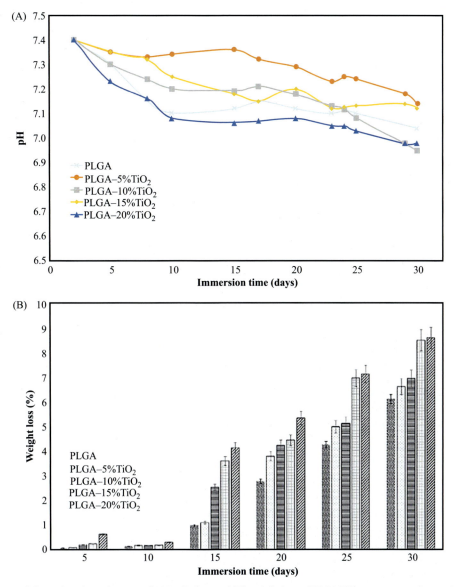

FIGURE 6–3 (A) Results of pH changes of PBS solution and (B) weight loss of PLGA/TiO$_2$ nanocomposites and pure PLGA (Pelaseyed et al., 2020).

specimens were placed in SBF for 6 weeks. It is worth noting that PCL is less hydrophilic and highly crystalline than its different composites. Also, the rapid penetration of water into PCL is unacceptable because of its natural hydrophobicity. In addition, the PCL/TiO$_2$ nanotube composite has a larger surface area and better water adsorption capacity. As a result, the nanocomposites showed a higher biodegradation rate compared to monolithic PCL.

The degradation and mineralization of PCL/TiO$_2$ nanotube nanocomposites are faster and higher than the monolithic PCL. Hence, PCL/TiO$_2$ nanotube composites are outstanding materials for bioimplants and other biomedical applications, where the smaller size and larger surface area of TiO$_2$ nanotubes can improve the composite performance (Khoshroo et al., 2017). In another study, Eslami et al. (2018) developed a novel PLGA/TiO$_2$ nanotubes bioactive composite for bone tissue engineering. They applied weight loss measurements to assess the biodegradation of the scaffold. As a result of evaluating the degradation rate, the weight of the PLGA and PLGA/TiO$_2$ nanotube composite decreased slightly during the first 4 weeks after incubation in SBF solution. Besides, the weight loss ratio of the PLGA matrix was higher than that of the PLGA/TiO$_2$ nanotube composite. The weight loss of pure PLGA was 50% in 3 weeks of incubation and reached about 90% after 35 days of incubation. The TiO$_2$ nanotube-reinforced PLGA composite reached 40% and 75% in the 3rd and 5th weeks, respectively. It should be noted that the pure PLGA weight loss was 10% greater than that of TiO$_2$ nanotubes-reinforced PLGA after 21 days of incubation. This shows that the dispersion of TiO$_2$ nanotubes in the PLGA matrix plays an important role in reducing biodegradation. The reason is that even though TiO$_2$ nanotubes with a high surface area/volume ratio reduce the contact angle to make the polymer surface more hydrophilic, the TiO$_2$ nanotubes in the polymer chains can decrease the interaction between water molecules with ester bonds (Valenzuela, Michniak, & Kohn, 2011). Alternatively, it is argued that the decrease in the PLGA degradation rate may be related to the slow degradation of TiO$_2$ nanotubes (Eslami et al., 2018; Jayakumar et al., 2011). In another study, Jayakumar et al. (2011), chitin-CS/nTiO$_2$ scaffolds were prepared for tissue engineering. In vitro biodegradation evaluations revealed that the biodegradation increases with time in both chitin/CS and chitin/CS/nTiO$_2$ scaffolds. However, the rate of degradation had been reduced by the addition of nTiO$_2$, which may be due to the slow degradation of TiO$_2$ NPs that existed in the composite scaffold. It can be concluded that the composite scaffold's degradation can be modified by the incorporation of nTiO$_2$ (Jayakumar et al., 2011).

In another study by Johari, Hosseini, and Samadikuchaksaraei (2018), a new fluorinated silk/nTiO$_2$ composite scaffold was fabricated for bone tissue engineering. The outcomes of biodegradation evaluations of the fabricated specimens revealed that with increasing the number of fluorinated nTiO$_2$ particles in the scaffold, the scaffold's porosity was increased, and the structure became less compact. As a result of the study, the nanocomposite scaffold's degradation rate increased with increasing the concentration of fluorinated nTiO$_2$ particles (Johari et al., 2018). Fan et al. (2016) fabricated nTiO$_2$ reinforced collagen/CS porous scaffold for wound-healing application. For the evaluation of scaffold biodegradation, the produced scaffolds were put into PBS solution. Obtained results indicated that the degradation rate of scaffolds with different loadings of nTiO$_2$ (1%, 3%, 5%, and 7%) had decreased because of the small size of nTiO$_2$ particles, which can quickly enter into the collagen and CS molecules, and form comparatively stable hydrogen bonds with hydroxyl, amino, and carboxyl of collagen and CS. Overall it can be said that the addition of nTiO$_2$ can improve the resistance to degradation of the scaffolds (Fan et al., 2016). In another study by Shafaghi et al. (2020) influence of TiO$_2$ doping on biodegradation rate, strength, and microstructure of borate BG scaffolds was evaluated. A titanium-containing borate glass series with different TiO_2 loadings of 0, 5, and 15 mol.% were used in this study. Obtained results from in vitro biodegradation assessment of the scaffolds showed that TiO_2 is an intermediate oxide (Brauer, Karpukhina, Law, & Hill, 2010); it could be a network modifier, which depolymerizes the glass network and increases the concentration of nonbridging oxygen species within the glass (increasing solubility); or a network former, which is involved in the structure of the glass network and raises the concentration of bridging oxygens; thus, declining the solubility (Shafaghi et al., 2020). Based on the outcomes, it can be concluded that $nTiO_2$ has different effects on biodegradation rate in different combinations (Table 6–1).

Table 6–1 Biodegradation results of samples containing different NMs.

Sample composition	Used NM	Effect of NMs on degradation results	Application	Reference
β-Chitin hydrogel/nZnO composite wound dressing	nZnO	The composite bandages containing nZnO displayed controlled biodegradation with improved hemostatic potential compared to the control bandages	Wound healing	Kumar et al. (2013)
Alginate hydrogel/nZnO composite	nZnO	Samples containing nZnO maintained their structures for a more extended period of time and remained intact for 28 days	Wound healing	Cleetus et al. (2020)
Alginate hydrogel/nZnO composite bandage	nZnO	The incorporation of nZnO particles controlled the degradation profile	Wound healing	Mohandas et al. (2015)
CS hydrogel/nZnO composite bandage	nZnO	The incorporation of nZnO particles did not change the degradation of bandages	Wound healing	Sudheesh Kumar et al. (2012)
Collagen/ZnO scaffold	nZnO	The composite scaffold containing nZnO particles was chemically more stable than the collagen scaffold	Wound healing	Vedhanayagam et al. (2018)
CS-gelatin/ZnO nanocomposite hydrogel scaffold	nZnO	Low concentration of the nZnO had not any significant impact on the nanocomposite hydrogel biodegradation	Drug delivery	Rakhshaei et al. (2019)
CS-collagen 3D Porous scaffold	nZnO	The composite scaffold containing 1.0% of nZnO had a high biodegradation rate	Tissue engineering	Ullah et al. (2017)
PLA/PCL/nZnO nanocomposite	nZnO	The degradation was accelerated in nanocomposite containing 6.0 wt. % of nZnO	Biocompatible polymer	Ahmadzadeh et al. (2018)
PCL membrane	nZnO	The nZnO particles accelerated the degradation of PCL membrane	Biomaterials application	Augustine et al. (2016)

(Continued)

Table 6-1 (Continued)

Sample composition	Used NM	Effect of NMs on degradation results	Application	Reference
PLLA scaffold	nZnO	The nZnO particles accelerated the hydrolytic degradation kinetics of the PLLA scaffold	Tissue engineering	Lizundia et al. (2017)
CS/PVA/nZnO films	nZnO	The nZnO particles enhanced the degradation rate	Hydrophilic wound and burn dressings	Vicentini et al. (2010)
PCL-ZnO nanofibrous scaffolds	nZnO	The nZnO particles enhanced and regulated the degradation rate	Bone tissue engineering	Felice et al. (2018)
nAg reinforced CS/HA composite sponge	nAg	The nAg particles enhanced the degradation percentage	Wound healing	Anisha et al. (2013)
Collagen scaffolds	nAg	The nAg incorporation prolonged the total degradation times	Wound healing	Song et al. (2015)
nAg-MSNs tissue nanoadhesive	nAg	The nAg reinforced MSNs showed the same manner of degradation as MSN ones	Wound healing	Lu et al. (2018)
Gelatin/BG/nAg scaffold	nAg	The nAg particles increased the in vitro enzyme degradation	Bone tissue engineering	Yazdimamaghani et al. (2014)
PAAG/nAg nanocomposite hydrogel	nAg	Degradability was decreased in nanocomposite containing nAg	Biomedical and environmental application	Palem et al. (2020)
PLA/HAp-nanorods/nAg particles	nAg	nAg promoted hydrolytic scission of ester bonds of PLA	Bone tissue engineering	Liu et al. (2015)
Gelatin/AgNP composite hydrogels	nAg	nAg incorporation decreased degradation rate	Biomedical applications, such as wound management and surface coating	Le Thi et al. (2018)
Ag NPs-riboflavin/PVA/β-CD	nAg and riboflavin	Composite nanofibrous scaffolds showed the highest degradation percentage	Wound dressing application	Balakrishnan and Thambusamy (2020)
CS/nHAp/nAg	nAg	The nAg incorporation decreased the biodegradation of CS/nHAp scaffold	Bone tissue engineering	Saravanan et al. (2011)
nAu/PU scaffold	nAu	The degradation rate for PU-Au 0.32 was higher than the other samples	Soft tissue applications	Tamayo et al. (2018)
nAu/CS hydrogel	nAu	The incorporation of AuNPs enhanced the biostability of the CS hydrogel	Cardiac tissue engineering	Baei et al. (2016)
Ceramic/nAu scaffold	nAu	Au NPs delayed degradation	Bone Tissue Engineering	Aly et al. (2012)
AuNPs/collagen bioscaffold	nAu	Resistance to degradation was increased in AuNPs/collagen scaffold	Soft tissue applications	Grant et al. (2014)
CS/MWCNTs composite film	MWCNTs	Adding MWCNTs to CS film caused a decrease in biodegradation rate	Bone substitute	Bakhtiari et al. (2019)

(Continued)

Table 6–1 (Continued)

Sample composition	Used NM	Effect of NMs on degradation results	Application	Reference
P3HB/CS/MWCNTs/silk scaffold	MWCNTs	The incorporation of MWCNTs reduced the biodegradation rate	Cartilage tissue-engineering applications	Mirmusavi et al. (2019)
PVA/BCP/CNTs	CNTs	The degradation ratio of scaffolds could be changed by altering the concentration of BCP and CNTs powders	Bone tissue engineering	Lan et al. (2019)
CS/GO composite film	GO	Adding GO to CS films increased the biodegradation time	Bone substitute	Tavakoli et al. (2020)
PCL/CS/collagen/GO	GO	GO increased the degradation rate	Bone tissue engineering	Aidun et al. (2019)
CS/gelatin/GO	GO	The degradation rate was increased for the loading beyond 0.25% of GO	Bone tissue engineering	Saravanan et al. (2017)
PCL/GO	GO	The incorporation of GO enhanced the degradation rate	Bone tissue-engineering applications	Mohammadi et al. (2017)
CS/TiO_2 scaffold	$nTiO_2$	The incorporation of $nTiO_2$ reduced the fast degradation of the CS scaffold	Bone tissue-engineering applications	Kumar (2018)
PLGA/$nTiO_2$ nanocomposite scaffold	$nTiO_2$	The weight loss was increased with raising TiO_2 content and incubation time	Tissue engineering	Pelaseyed et al. (2020)
3D PCL microsphere/TiO_2 nanotube composite scaffold	$nTiO_2$	The incorporation of $nTiO_2$ nanotube caused faster degradation in composite scaffold	Bone tissue engineering	Khoshroo et al. (2017)
PLGA/TiO_2 nanotube	$nTiO_2$	Dispersion of TiO_2 nanotube caused a decline in degradation rate	Bone tissue engineering	Eslami et al. (2018)
Chitin-CS/nano TiO_2-composite scaffold	$nTiO_2$	By adding $nTiO_2$ the degradation rate was decreased	Tissue-engineering applications	Jayakumar et al. (2011)
Fluoridated silk fibroin/TiO_2 nanocomposite scaffold	Fluoridated TiO_2 NPs	By increasing the concentration of fluoridated TiO_2 NPs, the degradation rates of nanocomposite scaffolds were increased	Bone tissue engineering	Johari et al. (2018)
$nTiO_2$/collagen-CS porous scaffold	$nTiO_2$	The incorporation of $nTiO_2$ improved the degradation resistance	Wound repairing	Fan et al. (2016)
Borate BG scaffold	$nTiO_2$	The degradation rates of borate BG scaffolds could be controlled by changing the amount of $nTiO_2$ loading	Total knee arthroplasty and revision total knee arthroplasty	Shafaghi et al. (2020)

6.2.6 Calcium phosphates effects on biodegradation behavior

Bioceramic bonding with polymer matrices has been studied to enhance the performance of bone substitutes by modifying the biodegradation kinetics and increasing bone conductivity (Chang et al., 2017; Kikuchi et al., 2004; Yuan et al., 2019). Bioceramics like calcium phosphate (Ca-P) and bioglasses like SrO, CaO, and P_2O_5 show high biocompatibility and bioactivity in both in vivo and in vitro environments, improving the density, phase stability, and biodegradability of composite scaffolds (Sopyan et al., 2016). Ca-P is a group of ceramic materials with remarkable differences in physical, mechanical, and biological properties and is often used to replace bone and tooth tissue due to its natural bone and tooth-like composition (Chang et al., 2017; Wang et al., 2016; Yuan et al., 2019). For example, HAp (Ca_{10} $(PO_4)_6$ $(OH)_2$) contains a favorable Ca/P ratio (close to 1.67) found in natural tissues such as bone and only affects the rate of degradation of biomaterials. Another popular Ca-P material is β-TCP that can be used to improve biodegradability kinetics and scaffold dissolution because of its higher biodegradation rate compared to HAp (Klammert et al., 2010). Wiltfang et al. (2002) studied TCP as a biodegradable filler for piglet bone marrow defects. Their results showed that osteoclasts dissolve the bioceramic particles to redecorate the material and that the newly formed collagen replaces the destroyed organic matrix. PLLA compounds such as β-TCP/PLLA exhibit higher modulus, fracture toughness, and strength-to-weight ratio compared to pure β-TCP components (Liu et al., 2013). Thus, biochemically enhanced polymers are promising compounds for the development of biodegradable implant and bone tissue engineering (Moseke & Gbureck, 2010). In this regard, the weight loss (degradation rate) and microstructure of the sample correspond to the degree of biodegradation of the polyglycolic acid (PGA)/HAp support in PBS solution (Fig. 6–4A and B) (Yeo et al., 2020). Scaffolding collapses as new tissue grows in the original

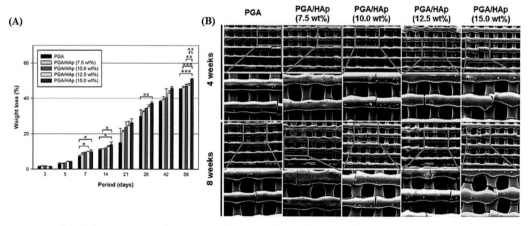

FIGURE 6–4 (A) Biodegradation profile of PGA and PGA/HAp composite scaffolds depends on the HAp ratios (*$P < .05$, **$P < .01$, ***$P < .001$), and (B) Photomicrographs of PGA/HAp composite scaffolds during biodegradation in PBS solution. Scale bars = 500 mm (Yeo et al., 2020).

bone tissue. The PGA degradation process is corrosive and appears to be a two-step process. The degradation rate of all scaffolds increased with the amount of HAp after one week. In the first step, the water diffuses into the amorphous regions of the polymer matrix, breaking the ester bonds. The second stage begins after the amorphous region is destroyed, as a result of which the crystalline portion of the polymer is exposed to hydrolytic attack. The polymer chains dissolve after the crystal regions collapse (Shi et al., 2008; Yeo et al., 2020). The degradation of HAp-based hydrogels has been shown to affect the extent of bone healing and the newly formed collagen arrangement in critical bone defects in rat calvarial (Gopinathan et al., 2020). Degradable printable HAp/TCP provided this secondary space and exhibited greater BTR than templates with lower degradation rates than printed HAp templates (Moseke & Gbureck, 2010). On the other hand, in the slowly degraded printed PPF-reinforced Ca-P templates, the loaded mesenchymal stem cells (MSCs) did not promote more BTR than unloaded templates. This could be attributed to the nonexistence of any secondary space for loaded MSCs to play their supposed role (Alge et al., 2012). Lately, Kim et al. (2016) discovered the influence of the biodegradation rate of the fast-degrading printed magnesium phosphate (MgP) templates with and without additional microporosity for BTR. The MgP templates completely degraded in 4 weeks, where the added micropore construction within the template struts resulted in better BTR (Hassan, Yassin et al., 2019). It has been reported that with the increase of BGs ratio, BTR increased relative to the increase of in vivo biodegradation of the printed template (Qi et al., 2017). On the other hand, compared with the commercially available porous templates, most 3D-printed templates displayed higher biodegradation and BTR (Cleetus et al., 2020). This can be due to interconnected porosity in 3D printed, which leads to faster biodegradation and improved BTR and remodeling activities and BTR (Alge et al., 2012; Hassan, Yassin et al., 2019; Kim et al., 2016; Qi et al., 2017; Schantz et al., 2003). In another investigation, it was shown that hydrogels' porous nature allows in vivo biodegradation. However, gelatin and alginate hydrogels with nHAp particles change the mechanism of degradation. The solubility of nHAp is low in the biological environment, but its uptake is quick compared with gelatin and alginate hydrogels because of its high biocompatibility with osteoblasts. As a result, biodegradation happens along with pores' formation, which results in nutrient transfer and effective cell differentiation to improve bone formation (Wang et al., 2016; Xavier et al., 2015). Nanosilicates possess a high degree of anisotropy, which accounts for their improved surface interactions with biological moieties. The existence of both negative and positive charges on the nanosilicate surface helps it to interact with anionic, cationic, and neutral polymers. Complete biodegradation of subcutaneous nanosilicate-gelatin implant in vivo confirmed the potential biomedical applications of such hydrogels (Gopinathan et al., 2020; Hassan, Dave et al., 2019; Xavier et al., 2015; Zhu, 2010). However, some other studies (Bakhsheshi-Rad, Idris, & Abdul-Kadir, 2013; Bakhsheshi-Rad, Hamzah, Daroonparvar, Ebrahimi-Kahrizsangi, & Medraj, 2014; Bakhsheshi-Rad, Hamzah, Daroonparvar, Kasiri-Asgarani, & Medraj, 2014; Bakhsheshi-Rad, Hamzah, Daroonparvar, Saud, & Abdul-Kadir, 2014; Bakhsheshi-Rad, Hamzah, Daroonparvar, Yajid, Kasiri-Asgarani et al., 2014; Bakhsheshi-Rad, Hamzah, Daroonparvar, Yajid, & Medraj, 2014; Bakhsheshi-Rad, Hamzah, Saud, & Medraj, 2015; Bakhsheshi-Rad, Hamzah et al., 2016b; Bakhsheshi-Rad et al.,

2018; Bakhsheshi-Rad, Akbari et al., 2019; Bakhsheshi-Rad, Ismail et al., 2019; Iqbal, Iqbal, Iqbal, Bakhsheshi-Rad, & Alsakkaf, 2020; Khalajabadi, Kadir, Izman, Bakhsheshi-Rad, & Farahany, 2014; Saberi, Bakhsheshi-Rad, Karamian, Kasiri-Asgarani, & Ghomi, 2020) were evaluated the degradation behavior of biodegradable metallic-based materials. Among them, magnesium received significant attention since it is naturally found in bone tissue and is essential for human metabolism. Mg^{2+} ions formed during biomaterial degradation also contribute to tissue healing and growth (Bakhsheshi-Rad et al., 2013; Bakhsheshi-Rad, Hamzah, Daroonparvar, Yajid, Medraj, 2014; Bakhsheshi-Rad et al., 2016; Bakhsheshi-Rad et al., 2018; Bakhsheshi-Rad, Akbari et al., 2019). Magnesium-based fracture fixation products have received significant consideration because of their low density (1.78 g/cm^3) and their high specific strength, significant mechanical characteristics, including Young's modulus (40–45 GPa), and compressive yield strength (65–100 MPa) matching the natural bone (density 1.8 g/cm^3; elastic modulus: 3–20 GPa; compressive yield strength:130–180 MPa). Thus, magnesium and its alloy have become good candidates for fabricating degradable bone implants such as screws, bone plates, and pins (Bakhsheshi-Rad, Hamzah, Daroonparvar, Kasiri-Asgarani et al., 2014; Bakhsheshi-Rad, Hamzah, Daroonparvar, Ebrahimi-Kahrizsangi et al., 2014; Bakhsheshi-Rad, Hamzah, Daroonparvar, Saud et al., 2014; Bakhsheshi-Rad, Hamzah, & Kasiri-Asgarani, 2016a). Implants made of magnesium and its alloys can eliminate the need for further surgery to remove the implant, as it will degrade during the growing back of the damaged tissue (Bakhsheshi-Rad, Hamzah, Daroonparvar, Yajid, Kasiri-Asgarani et al., 2014). However, magnesium degraded too rapidly in vivo and produces a large amount of hydrogen gas beneath the skin, limiting their wide applications (Khalajabadi et al., 2014). To deal with this problem, surface modification with ceramic Ca-P and Ca-silicate coating, the hybrid bioceramic-biopolymer coating is implemented to decrease biodegradable magnesium's initial degradation rate to allow the bone tissue is entirely healed. Overall, the preparation of HA (Bakhsheshi-Rad et al., 2013; Bakhsheshi-Rad, Hamzah, Daroonparvar, Yajid, Medraj, 2014; Bakhsheshi-Rad et al., 2018), fluorine-doped hydroxyapatite (FHAp) (Bakhsheshi-Rad, Hamzah, Daroonparvar, Ebrahimi-Kahrizsangi et al., 2014; Bakhsheshi-Rad, Hamzah, Daroonparvar, Saud et al., 2014; Bakhsheshi-Rad, Hamzah, Daroonparvar, Yajid, Kasiri-Asgarani et al., 2014; Bakhsheshi-Rad et al., 2016a), brushite (DCPD) (Bakhsheshi-Rad et al., 2015), åkermanite (Bakhsheshi-Rad, Akbari et al., 2019), hardystonite (Bakhsheshi-Rad et al., 2016b), and bredigite (Saberi et al., 2020) as ceramic-based materials composite coatings could control long-term degradation behavior of biodegradable magnesium. The incorporation of ceramic-based NMs with special characteristics demonstrates great promise as filler to reduce biodegradable magnesium degradation rate (Khalajabadi et al., 2014). Since a great degradation rate considerably amplified pH value and hydrogen gas evolution rate, which may lead to some toxicity and inhibition of cells to attach the composite surface. Moreover, a higher degradation rate increases significantly more corrosion products that could reduce the specimens' cell attachment (Iqbal et al., 2020). The results are of significance in the ceramic-based coating, hybrid Ca-P ceramic-biopolymer layer design as coating materials is able to reduce degradation rate of biodegradable magnesium for orthopedic applications.

6.3 Conclusion and future outlooks

In addition to desirable physical and biological properties, the rate of degradation of a material is an important factor that can be customized based on the application. Regarding this issue, it has been reported that the degradation mostly depends on the dissolution of bulk based on some mechanisms such as photolysis, hydrolysis (ester or enzyme), disentanglement, and a combination of the mentioned mechanisms (Unagolla & Jayasuriya, 2020). Typically, a combination of hydrophobic and hydrophilic polymers can be used to control degradation to the desired level. The rate of biodegradation must be equal to the rate of cell growth to replace the degraded material with the ECM produced by the cell. For example, biodegradation of a printed structure changes the physiological environment of cells with decreased mechanical strength (Dorishetty et al., 2020). In addition, the breakdown of substances creates more space for proliferation, cell migration, and vascular invasion. Developing the spatiotemporal aspect of this type of hydrogel is still a challenge (Unagolla & Jayasuriya, 2020). Biodegradable materials for synthesis, implantation, or scaffolding must have the mechanical strength to withstand in vivo stress and time-dependent degradation to allow new tissue could be replaced. Therefore, bionanocomposites must meet biocompatibility and biodegradability requirements to promote and sustain tissue growth and degrade at the rate at which new tissues are formed. The rate of substance degradation affects tissue/scaffold interactions as it can affect host response, cell growth, and tissue regeneration (Zinge & Kandasubramanian, 2020). Alternatives such as nanoadditives are active substances that significantly affect the manufacturing method, processing factors, degradation properties, cellular responses, and biological activity. From this point of view, the size, shape, and properties of nanoadditive particles greatly influence the rate of materials' degradation, including scaffolds, 3D structures, and implants (Calori et al., 2020; Zinge & Kandasubramanian, 2020). Although the desired rate of degradation and mechanism can be specifically designed, the loss in mechanical properties is still inevitable. The modulating of the material's biodegradation rate is considered a breakthrough in materials science and design (Chang et al., 2017; Unagolla & Jayasuriya, 2020; Yuan et al., 2019; Zinge & Kandasubramanian, 2020). In this regard, existing strategies and recently improved designs for controlling the rate of biomaterial degradation should be pursued as they significantly impact mechanical properties, cell adhesion, proliferation, differentiation, tissue deposition, and host response. In summary, controlling the rate of degradation of biomaterials, especially polymers, has opened up a new paradigm in biomedical, tissue engineering, and regenerative medicine fields. Success in this rapidly growing field of research requires an interdisciplinary approach and sustainable funding.

Acknowledgments

The authors would like to thank the Norwegian University of Science and Technology, Universiti Teknologi Malaysia and Islamic Azad University, Najafabad branch for providing the facilities for this research.

Funding

This research received no external funding.

Conflicts of Interest

The authors declare that they have no competing/financial conflict of interests in this paper.

References

Ahmadzadeh, Y., Babaei, A., & Goudarzi, A. (2018). Assessment of localization and degradation of ZnO nanoparticles in the PLA/PCL biocompatible blend through a comprehensive rheological characterization. *Polymer Degradation and Stability, 158*, 136–147.

Aidun, A., Safaei Firoozabady, A., Moharrami, M., Ahmadi, A., Haghighipour, N., Bonakdar, S., & Faghihi, S. (2019). Graphene oxide incorporated polycaprolactone/chitosan/collagen electrospun scaffold: Enhanced osteogenic properties for bone tissue engineering. *Artificial Organs, 43*(10), E264–E281.

Albertini, B., Di Sabatino, M., Calonghi, N., Rodriguez, L., & Passerini, N. (2013). Novel multifunctional platforms for potential treatment of cutaneous wounds: Development and in vitro characterization. *International Journal of Pharmaceutics, 440*(2), 238–249.

Alge, D. L., Bennett, J., Treasure, T., Voytik-Harbin, S., Goebel, W. S., & Chu, T. M. (2012). Poly (propylene fumarate) reinforced dicalcium phosphate dihydrate cement composites for bone tissue engineering. *Journal of Biomedical Materials Research. Part A, 100*(7), 1792–1802.

Aly, A. F., Eldesouky, A. S., & Eid, K. A. (2012). Evaluation, characterization and cell viability of ceramic scaffold and nano-gold loaded ceramic scaffold for bone tissue engineering. *American Journal of Biomedical Science, 4*(4), 316–326.

Andronescu, E., Brown, J. M., Oktar, F. N., Agathopoulos, S., Chou, J., & Obata, A. (2016). Nanomaterials for medical applications: Benefits and risks. *Journal of Nanomaterials, 2016*, 1–2.

Anisha, B. S., Biswas, R., Chennazhi, K. P., & Jayakumar, R. (2013). Chitosan–hyaluronic acid/nano silver composite sponges for drug resistant bacteria infected diabetic wounds. *International Journal of Biological Macromolecules, 62*, 310–320.

Augustine, R., Kalarikkal, N., & Thomas, S. (2016). Effect of zinc oxide nanoparticles on the in vitro degradation of electrospun polycaprolactone membranes in simulated body fluid. *International Journal of Polymeric Materials and Polymeric Biomaterials, 65*(1), 28–37.

Baei, P., Jalili-Firoozinezhad, S., Rajabi-Zeleti, S., Tafazzoli-Shadpour, M., Baharvand, H., & Aghdami, N. (2016). Electrically conductive gold nanoparticle-chitosan thermosensitive hydrogels for cardiac tissue engineering. *Materials Science and Engineering: C, 63*, 131–141.

Bakhsheshi-Rad, H. R., Akbari, M., Ismail, A. F., Aziz, M., Hadisi, Z., Pagan, E., ... Chen, X. (2019). Coating biodegradable magnesium alloys with electrospun poly-L-lactic acid-åkermanite-doxycycline nanofibers for enhanced biocompatibility, antibacterial activity, and corrosion resistance. *Surface and Coatings Technology, 377*, 124898.

Bakhsheshi-Rad, H. R., Hamzah, E., Daroonparvar, M., Ebrahimi-Kahrizsangi, R., & Medraj, M. (2014). In-vitro corrosion inhibition mechanism of fluorine-doped hydroxyapatite and brushite coated Mg–Ca alloys for biomedical applications. *Ceramics International, 40*(6), 7971–7982.

Bakhsheshi-Rad, H. R., Hamzah, E., Daroonparvar, M., Kasiri-Asgarani, M., & Medraj, M. (2014). Synthesis and biodegradation evaluation of nano-Si and nano-Si/TiO_2 coatings on biodegradable Mg-Ca alloy in simulated body fluid. *Ceramics International, 40*(9), 14009–14018.

Bakhsheshi-Rad, H. R., Hamzah, E., Daroonparvar, M., Saud, S. N., & Abdul-Kadir, M. R. (2014). Bi-layer nano-TiO$_2$/FHA composite coatings on Mg-Zn-Ce alloy prepared by combined physical vapour deposition and electrochemical deposition methods. *Vacuum, 110*, 127−135.

Bakhsheshi-Rad, H. R., Hamzah, E., Daroonparvar, M., Yajid, M. A., Kasiri-Asgarani, M., Abdul-Kadir, M. R., & Medraj, M. (2014). In-vitro degradation behavior of Mg alloy coated by fluorine doped hydroxyapatite and calcium deficient hydroxyapatite. *Transactions of Nonferrous Metals Society of China, 24*(8), 2516−2528.

Bakhsheshi-Rad, H. R., Hamzah, E., Daroonparvar, M., Yajid, M. A., & Medraj, M. (2014). Fabrication and corrosion behavior of Si/HA nano-composite coatings on biodegradable Mg-Zn-Mn-Ca alloy. *Surface and Coatings Technology, 258*, 1090−1099.

Bakhsheshi-Rad, H. R., Hamzah, E., Ismail, A. F., Sharer, Z., et al. (2015). Synthesis and corrosion behavior of a hybrid bioceramic-biopolymer coating on biodegradable Mg alloy for orthopedic implants. *Journal of Alloys and Compounds, 648*, 1067−1071.

Bakhsheshi-Rad, H. R., Hamzah, E., Ismail, A. F., Aziz, M., Daroonparvar, M., Saebnoori, E., & Chami, A. (2018). In vitro degradation behavior, antibacterial activity and cytotoxicity of TiO$_2$-MAO/ZnHA composite coating on Mg alloy for orthopedic implants. *Surface and Coatings Technology, 334*, 450−460.

Bakhsheshi-Rad, H. R., Hamzah, E., Ismail, A. F., Kasiri-Asgarani, M., Daroonparvar, M., Parham, S., ... Medraj, M. (2016). Novel bi-layered nanostructured SiO$_2$/Ag-FHAp coating on biodegradable magnesium alloy for biomedical applications. *Ceramics International, 42*(10), 11941−11950.

Bakhsheshi-Rad, H. R., Hamzah, E., Kasiri-Asgarani, M., et al. (2016a). Deposition of nanostructured fluorine-doped hydroxyapatite−polycaprolactone duplex coating to enhance the mechanical properties and corrosion resistance of Mg alloy for biomedical applications. *Materials Science and Engineering: C, 60*, 526−537.

Bakhsheshi-Rad, H. R., Hamzah, E., Kasiri-Asgarani, M., et al. (2016b). Fabrication, degradation behavior and cytotoxicity of nanostructured hardystonite and titania/hardystonite coatings on Mg alloys. *Vacuum, 129*, 9−12.

Bakhsheshi-Rad, H. R., Hamzah, E., Saud, S. N., & Medraj, M. (2015). Effect of electrodeposition parameters on the microstructure and corrosion behavior of DCPD coatings on biodegradable Mg-Ca-Zn alloy. *International Journal of Applied Ceramic Technology, 12*(5), 1054−1064.

Bakhsheshi-Rad, H. R., Idris, M. H., & Abdul-Kadir, M. R. (2013). Synthesis and in vitro degradation evaluation of the nano-HA/MgF$_2$ and DCPD/MgF$_2$ composite coating on biodegradable Mg-Ca-Zn alloy. *Surface and Coatings Technology, 222*, 79−89.

Bakhsheshi-Rad, H. R., Ismail, A. F., Aziz, M., Hadisi, Z., Omidi, M., & Chen, X. (2019). Antibacterial activity and corrosion resistance of Ta$_2$O$_5$ thin film and electrospun PCL/MgO-Ag nanofiber coatings on biodegradable Mg alloy implants. *Ceramics International, 45*(9), 11883−11892.

Bakhtiari, S. S., Karbasi, S., Tabrizi, S. A., & Ebrahimi-Kahrizsangi, R. (2019). Chitosan/MWCNTs composite as bone substitute: Physical, mechanical, bioactivity, and biodegradation evaluation. *Polymer Composites, 40*, E1622−E1632.

Balakrishnan, S. B., & Thambusamy, S. (2020). Preparation of silver nanoparticles and riboflavin embedded electrospun polymer nanofibrous scaffolds for in vivo wound dressing application. *Process Biochemistry, 88*, 148−158.

Brauer, D. S., Karpukhina, N., Law, R. V., & Hill, R. G. (2010). Effect of TiO$_2$ addition on structure, solubility and crystallisation of phosphate invert glasses for biomedical applications. *Journal of Non-Crystalline Solids, 356*(44−49), 2626−2633.

Calori, I. R., Braga, G., de Jesus, P. D., Bi, H., & Tedesco, A. C. (2020). Polymer scaffolds as drug delivery systems. *European Polymer Journal, 129*, 109621.

Chan, L. W., Jin, Y., & Heng, P. W. (2002). Cross-linking mechanisms of calcium and zinc in production of alginate microspheres. *International Journal of Pharmaceutics, 242*(1−2), 255−258.

Chang, B., Ahuja, N., Ma, C., & Liu, X. (2017). Injectable scaffolds: Preparation and application in dental and craniofacial regeneration. *Materials Science and Engineering: R: Reports, 111*, 1−26.

Chen, H., Nanayakkara, C. E., & Grassian, V. H. (2012). Titanium dioxide photocatalysis in atmospheric chemistry. *Chemical Reviews, 112*(11), 5919–5948.

Chen, J., Ouyang, J., Kong, J., Zhong, W., & Xing, M. M. (2013). Photo-cross-linked and pH-sensitive biodegradable micelles for doxorubicin delivery. *ACS Applied Materials & Interfaces, 5*(8), 3108–3117.

Chen, X., & Mao, S. S. (2007). Titanium dioxide nanomaterials: Synthesis, properties, modifications, and applications. *Chemical Reviews, 107*(7), 2891–2959.

Chen, X., & Selloni, A. (2014). Introduction: Titanium dioxide (TiO_2) nanomaterials. *Chemical Reviews, 114*(19), 9281–9282.

Cleetus, C. M., Primo, F. A., Fregoso, G., Raveendran, N. L., Noveron, J. C., Spencer, C. T., ... Joddar, B. (2020). Alginate hydrogels with embedded ZnO nanoparticles for wound healing therapy. *International Journal of Nanomedicine, 15*, 5097.

De Witte, T. M., Fratila-Apachitei, L. E., Zadpoor, A. A., & Peppas, N. A. (2018). Bone tissue engineering via growth factor delivery: From scaffolds to complex matrices. *Regenerative Biomaterials, 5*(4), 197–211.

Deepachitra, R., Lakshmi, R. P., Sivaranjani, K., Chandra, J. H., & Sastry, T. P. (2015). Nanoparticles embedded biomaterials in wound treatment: A review. *Journal of Chemical and Pharmaceutical Sciences, 8*, 324–329.

Díaz, E., Puerto, I., Ribeiro, S., Lanceros-Mendez, S., & Barandiarán, J. M. (2017). The influence of copolymer composition on PLGA/nHA scaffolds' cytotoxicity and in vitro degradation. *Nanomaterials, 7*(7), 173.

Dorishetty, P., Dutta, N. K., & Choudhury, N. R. (2020). Bioprintable tough hydrogels for tissue engineering applications. *Advances in Colloid and Interface Science, 281*, 102163.

Edwards, S. L., Werkmeister, J. A., & Ramshaw, J. A. (2009). Carbon nanotubes in scaffolds for tissue engineering. *Expert Review of Medical Devices, 6*(5), 499–505.

Elahi, N., Kamali, M., & Baghersad, M. H. (2018). Recent biomedical applications of gold nanoparticles: A review. *Talanta, 184*, 537–556.

Eslami, H., Lisar, H. A., Kashi, T. S., Tahriri, M., Ansari, M., Rafiei, T., ... Tayebi, L. (2018). Poly (lactic-co-glycolic acid)(PLGA)/TiO_2 nanotube bioactive composite as a novel scaffold for bone tissue engineering: In vitro and in vivo studies. *Biologicals: Journal of the International Association of Biological Standardization, 53*, 51–62.

Ezati, M., Safavipour, H., Houshmand, B., & Faghihi, S. (2018). Development of a PCL/gelatin/chitosan/ β-TCP electrospun composite for guided bone regeneration. *Progress in Biomaterials, 7*(3), 225–237.

Fan, X., Chen, K., He, X., Li, N., Huang, J., Tang, K., ... Wang, F. (2016). Nano-TiO_2/collagen-chitosan porous scaffold for wound repairing. *International Journal of Biological Macromolecules, 91*, 15–22.

Felice, B., Sánchez, M. A., Socci, M. C., Sappia, L. D., Gómez, M. I., Cruz, M. K., ... Rodríguez, A. P. (2018). Controlled degradability of PCL-ZnO nanofibrous scaffolds for bone tissue engineering and their antibacterial activity. *Materials Science and Engineering: C, 93*, 724–738.

Fujishima, A., & Honda, K. (1972). Electrochemical photolysis of water at a semiconductor electrode. *Nature, 238*(5358), 37–38.

Ge, L., Li, Q., Wang, M., Ouyang, J., Li, X., & Xing, M. M. (2014). Nanosilver particles in medical applications: Synthesis, performance, and toxicity. *International Journal of Nanomedicine, 9*, 2399.

Gopinathan, J., Hao, T. N., Cha, E., Lee, C., Das, D., & Noh, I. (2020). 3D printable and injectable lactoferrin-loaded carboxymethyl cellulose-glycol chitosan hydrogels for tissue engineering applications. *Materials Science and Engineering: C, 113*, 111008.

Grant, S. A., Spradling, C. S., Grant, D. N., Fox, D. B., Jimenez, L., Grant, D. A., & Rone, R. J. (2014). Assessment of the biocompatibility and stability of a gold nanoparticle collagen bioscaffold. *Journal of Biomedical Materials Research Part A: An Official Journal of The Society for Biomaterials, The Japanese Society for Biomaterials, and The Australian Society for Biomaterials and the Korean Society for Biomaterials, 102*(2), 332–339.

Guo, B., Lei, B., Li, P., & Ma, P. X. (2015). Functionalized scaffolds to enhance tissue regeneration. *Regenerative Biomaterials, 2*(1), 47−57.

Hassan, M., Dave, K., Chandrawati, R., Dehghani, F., & Gomes, V. G. (2019). 3D printing of biopolymer nanocomposites for tissue engineering: Nanomaterials, processing and structure-function relation. *European Polymer Journal, 121*, 109340.

Hassan, M. N., Yassin, M. A., Suliman, S., Lie, S. A., Gjengedal, H., & Mustafa, K. (2019). The bone regeneration capacity of 3D-printed templates in calvarial defect models: A systematic review and meta-analysis. *Acta Biomaterialia, 91*, 1−23.

Hennink, W. E., & van Nostrum, C. F. (2012). Novel crosslinking methods to design hydrogels. *Advanced Drug Delivery Reviews, 64*, 223−236.

Hodes, G. (2007). When small is different: Some recent advances in concepts and applications of nanoscale phenomena. *Advanced Materials, 19*(5), 639−655.

Hoffmann, M. R., et al. (1995). Environmental applications of semiconductor photocatalysis. *Chemical Reviews, 95*(1), 69−96.

Ikada, Y. (2006). Challenges in tissue engineering. *Journal of the Royal Society Interface, 3*(10), 589−601.

Iqbal, N., Iqbal, S., Iqbal, T., Bakhsheshi-Rad, H. R., Alsakkaf, A., et al. (2020). Zinc-doped hydroxyapatite-zeolite/polycaprolactone composites coating on magnesium substrate for enhancing in-vitro corrosion and antibacterial performance. *Transactions of Nonferrous Metals Society of China, 30*(1), 123−133.

Jayakumar, R., Prabaharan, M., Kumar, P. S., Nair, S. V., & Tamura, H. (2011). Biomaterials based on chitin and chitosan in wound dressing applications. *Biotechnology Advances, 29*(3), 322−337.

Jayakumar, R., et al. (2011). Fabrication of chitin−chitosan/nano TiO_2-composite scaffolds for tissue engineering applications. *International Journal of Biological Macromolecules, 48*(2), 336−344.

Johari, N., Hosseini, H. R., & Samadikuchaksaraei, A. (2018). Novel fluoridated silk fibroin/TiO_2 nanocomposite scaffolds for bone tissue engineering. *Materials Science and Engineering: C, 82*, 265−276.

Khalajabadi, S. Z., Kadir, M. R., Izman, S., Bakhsheshi-Rad, H. R., & Farahany, S. (2014). Effect of mechanical alloying on the phase evolution, microstructure and bio-corrosion properties of a Mg/HA/TiO_2/MgO nanocomposite. *Ceramics International, 40*(10), 16743−16759.

Khoshroo, K., Kashi, T. S., Moztarzadeh, F., Tahriri, M., Jazayeri, H. E., & Tayebi, L. (2017). Development of 3D PCL microsphere/TiO_2 nanotube composite scaffolds for bone tissue engineering. *Materials Science and Engineering: C, 70*, 586−598.

Kikuchi, M., Koyama, Y., Yamada, T., Imamura, Y., Okada, T., Shirahama, N., . . . Tanaka, J. (2004). Development of guided bone regeneration membrane composed of β-tricalcium phosphate and poly (L-lactide-co-glycolide-co-ε-caprolactone) composites. *Biomaterials, 25*(28), 5979−5986.

Kim, J. A., Lim, J., Naren, R., Yun, H. S., & Park, E. K. (2016). Effect of the biodegradation rate controlled by pore structures in magnesium phosphate ceramic scaffolds on bone tissue regeneration in vivo. *Acta Biomaterialia, 44*, 155−167.

Klammert, U., Gbureck, U., Vorndran, E., Rödiger, J., Meyer-Marcotty, P., & Kübler, A. C. (2010). 3D powder printed calcium phosphate implants for reconstruction of cranial and maxillofacial defects. *Journal of Cranio-Maxillofacial Surgery, 38*(8), 565−570.

Kumar, C. S. (2000). Nanomaterials for medical applications. *Kirk-Othmer Encyclopedia of Chemical Technology*, Dec 4.

Kumar, P. (2018). Nano-TiO_2 doped chitosan scaffold for the bone tissue engineering applications. *International Journal of Biomaterials, 2018*.

Kumar, S., Lakshmanan, V. K., Raj, M., Biswas, R., Hiroshi, T., Nair, S. V., & Jayakumar, R. (2013). Evaluation of wound healing potential of β-chitin hydrogel/nano zinc oxide composite bandage. *Pharmaceutical Research, 30*(2), 523−537.

Lan, W., Zhang, X., Xu, M., Zhao, L., Huang, D., Wei, X., & Chen, W. (2019). Carbon nanotube reinforced polyvinyl alcohol/biphasic calcium phosphate scaffold for bone tissue engineering. *RSC Advances, 9*(67), 38998–39010.

Le Thi, P., Lee, Y., Thi, T. T., Park, K. M., & Park, K. D. (2018). Catechol-rich gelatin hydrogels in situ hybridizations with silver nanoparticle for enhanced antibacterial activity. *Materials Science and Engineering: C, 92*, 52–60.

Leijten, J., Seo, J., Yue, K., Trujillo-de Santiago, G., Tamayol, A., Ruiz-Esparza, G. U., ... Zhang, Y. S. (2017). Spatially and temporally controlled hydrogels for tissue engineering. *Materials Science and Engineering: R: Reports, 119*, 1–35.

Li, H., & Chang, J. (2005). In vitro degradation of porous degradable and bioactive PHBV/wollastonite composite scaffolds. *Polymer Degradation and Stability, 87*(2), 301–307.

Linsebigler, A. L., Lu, G., & Yates, J. T., Jr (1995). Photocatalysis on TiO_2 surfaces: Principles, mechanisms, and selected results. *Chemical Reviews, 95*(3), 735–758.

Liu, C., Chan, K. W., Shen, J., Wong, H. M., Yeung, K. W., & Tjong, S. C. (2015). Melt-compounded polylactic acid composite hybrids with hydroxyapatite nanorods and silver nanoparticles: Biodegradation, antibacterial ability, bioactivity and cytotoxicity. *RSC Advances, 5*(88), 72288–72299.

Liu, D., Zhuang, J., Shuai, C., & Peng, S. (2013). Mechanical properties' improvement of a tricalcium phosphate scaffold with poly-l-lactic acid in selective laser sintering. *Biofabrication, 5*(2), 025005.

Liu, H., & Webster, T. J. (2007). Nanomedicine for implants: A review of studies and necessary experimental tools. *Biomaterials, 28*(2), 354–369.

Lizundia, E., Mateos, P., & Vilas, J. L. (2017). Tuneable hydrolytic degradation of poly (l-lactide) scaffolds triggered by ZnO nanoparticles. *Materials Science and Engineering: C, 75*, 714–720.

Lu, M. M., Bai, J., Shao, D., Qiu, J., Li, M., Zheng, X., ... Dong, W. F. (2018). Antibacterial and biodegradable tissue nano-adhesives for rapid wound closure. *International Journal of Nanomedicine, 13*, 5849.

Maiti, D., Tong, X., Mou, X., & Yang, K. (2019). Carbon-based nanomaterials for biomedical applications: A recent study. *Frontiers in Pharmacology, 9*, 1401.

Mihai, M. M., Dima, M. B., Dima, B., & Holban, A. M. (2019). Nanomaterials for wound healing and infection control. *Materials, 12*(13), 2176.

Mirmusavi, M. H., Zadehnajar, P., Semnani, D., Karbasi, S., Fekrat, F., & Heidari, F. (2019). Evaluation of physical, mechanical and biological properties of poly 3-hydroxybutyrate-chitosan-multiwalled carbon nanotube/silk nano-micro composite scaffold for cartilage tissue engineering applications. *International Journal of Biological Macromolecules, 132*, 822–835.

Mohamed, A., & Xing, M. M. (2012). Nanomaterials and nanotechnology for skin tissue engineering. *International Journal of Burns and Trauma, 2*(1), 29.

Mohammadi, S., Shafiei, S. S., Asadi-Eydivand, M., Ardeshir, M., & Solati-Hashjin, M. (2017). Graphene oxide-enriched poly (ε-caprolactone) electrospun nanocomposite scaffold for bone tissue engineering applications. *Journal of Bioactive and Compatible Polymers, 32*(3), 325–342.

Mohandas, A., Sudheesh Kumar, P. T., Raja, B., Lakshmanan, V. K., & Jayakumar, R. (2015). Exploration of alginate hydrogel/nano zinc oxide composite bandages for infected wounds. *International Journal of Nanomedicine, 10*, 53.

Moseke, C., & Gbureck, U. (2010). Tetracalcium phosphate: Synthesis, properties and biomedical applications. *Acta Biomaterialia, 6*(10), 3815–3823.

Mostofizadeh, A., Li, Y., Song, B., & Huang, Y. (2011). Synthesis, properties, and applications of low-dimensional carbon-related nanomaterials. *Journal of Nanomaterials, 2011*, 1–21.

Padmanabhan, J., & Kyriakides, T. R. (2015). Nanomaterials, inflammation, and tissue engineering. *Wiley Interdisciplinary Reviews: Nanomedicine and Nanobiotechnology, 7*(3), 355–370.

Palem, R. R., Shimoga, G., Kang, T. J., & Lee, S. H. (2020). Fabrication of multifunctional Guar gum-silver nanocomposite hydrogels for biomedical and environmental applications. *International Journal of Biological Macromolecules, 159*, 474–486.

Pelaseyed, S. S., Madaah Hosseini, H. R., & Samadikuchaksaraei, A. (2020). A novel pathway to produce biodegradable and bioactive PLGA/TiO$_2$ nanocomposite scaffolds for tissue engineering: Air–liquid foaming. *Journal of Biomedical Materials Research, Part A, 108*(6), 1390–1407.

Polo-Corrales, L., Latorre-Esteves, M., & Ramirez-Vick, J. E. (2014). Scaffold design for bone regeneration. *Journal of Nanoscience and Nanotechnology, 14*(1), 15–56.

Qi, X., Pei, P., Zhu, M., Du, X., Xin, C., Zhao, S., ... Zhu, Y. (2017). Three dimensional printing of calcium sulfate and mesoporous bioactive glass scaffolds for improving bone regeneration in vitro and in vivo. *Scientific Reports, 7*, 42556.

Rakhshaei, R., Namazi, H., Hamishehkar, H., Kafil, H. S., & Salehi, R. (2019). In situ synthesized chitosan–gelatin/ZnO nanocomposite scaffold with drug delivery properties: Higher antibacterial and lower cytotoxicity effects. *Journal of Applied Polymer Science, 136*(22), 47590.

Saberi, A., Bakhsheshi-Rad, H. R., Karamian, E., Kasiri-Asgarani, M., & Ghomi, H. (2020). A study on the corrosion behavior and biological properties of polycaprolactone/bredigite composite coating on biodegradable Mg-Zn-Ca-GNP nanocomposite. *Progress in Organic Coatings, 147*, 105822.

Saravanan, S., Chawla, A., Vairamani, M., Sastry, T. P., Subramanian, K. S., & Selvamurugan, N. (2017). Scaffolds containing chitosan, gelatin and graphene oxide for bone tissue regeneration in vitro and in vivo. *International Journal of Biological Macromolecules, 104*, 1975–1985.

Saravanan, S., Nethala, S., Pattnaik, S., Tripathi, A., Moorthi, A., & Selvamurugan, N. (2011). Preparation, characterization and antimicrobial activity of a bio-composite scaffold containing chitosan/nano-hydroxyapatite/nano-silver for bone tissue engineering. *International Journal of Biological Macromolecules, 49*(2), 188–193.

Schantz, J. T., Teoh, S. H., Lim, T. C., Endres, M., Lam, C. X., & Hutmacher, D. W. (2003). Repair of calvarial defects with customized tissue-engineered bone grafts I. Evaluation of osteogenesis in a three-dimensional culture system. *Tissue Engineering, 9*, 113–126.

Shafaghi, R., Rodriguez, O., Phull, S., Schemitsch, E. H., Zalzal, P., Waldman, S. D., ... Towler, M. R. (2020). Effect of TiO$_2$ doping on degradation rate, microstructure and strength of borate bioactive glass scaffolds. *Materials Science and Engineering: C, 107*, 110351.

Shi, J., Votruba, A. R., Farokhzad, O. C., & Langer, R. (2010). Nanotechnology in drug delivery and tissue engineering: From discovery to applications. *Nano Letters, 10*(9), 3223–3230.

Shi, Y., Wang, Y., Chen, J., & Huang, S. (2008). Experimental investigation into the selective laser sintering of high-impact polystyrene. *Journal of Applied Polymer Science, 108*(1), 535–540.

Shin, S. R., Jung, S. M., Zalabany, M., Kim, K., Zorlutuna, P., Kim, S. B., ... Khademhosseini, A. (2013). Carbon-nanotube-embedded hydrogel sheets for engineering cardiac constructs and bioactuators. *ACS Nano, 7*(3), 2369–2380.

Siegel, R. W., & Fougere, G. E. (1995). Mechanical properties of nanophase metals. *Nanostructured Materials, 6*, 205–216.

Singla, R., Abidi, S. M., Dar, A. I., & Acharya, A. (2019). Nanomaterials as potential and versatile platform for next generation tissue engineering applications. *Journal of Biomedical Materials Research, Part B: Applied Biomaterials, 107*(7), 2433–2449.

Song, J., et al. (2015). Nano-silver in situ hybridized collagen scaffolds for regeneration of infected full-thickness burn skin. *Journal of Materials Chemistry B, 3*(20), 4231–4241.

Sopyan, I., Gunawan., Shah, Q. H., & Mel, M. (2016). Fabrication and sintering behavior of zinc-doped biphasic calcium phosphate bioceramics. *Materials and Manufacturing Processes, 31*(6), 713–718.

Sudheesh Kumar, P. T., Lakshmanan, V. K., Anilkumar, T. V., Ramya, C., Reshmi, P., Unnikrishnan, A. G., ... Jayakumar, R. (2012). Flexible and microporous chitosan hydrogel/nano ZnO composite bandages for wound dressing: In vitro and in vivo evaluation. *ACS Applied Materials & Interfaces, 4*(5), 2618–2629.

Tamayo, L., Acuña, D., Riveros, A. L., Kogan, M. J., Azocar, M. I., Páez, M., ... Cerda, E. (2018). Porous nano-gold/polyurethane scaffolds with improved antibiofilm, mechanical, and thermal properties and with reduced effects on cell viability: A suitable material for soft tissue applications. *ACS Applied Materials & Interfaces, 10*(16), 13361–13372.

Tavakoli, M., Karbasi, S., & Soleymani Eil Bakhtiari, S. (2020). Evaluation of physical, mechanical, and biodegradation of chitosan/graphene oxide composite as bone substitutes. *Polymer-Plastics Technology and Materials, 59*(4), 430–440.

Thompson, T. L., & Yates, J. T. (2006). Surface science studies of the photoactivation of TiO_2 new photochemical processes. *Chemical Reviews, 106*(10), 4428–4453.

Timmer, M. D., Ambrose, C. G., & Mikos, A. G. (2003). In vitro degradation of polymeric networks of poly (propylene fumarate) and the crosslinking macromer poly (propylene fumarate)-diacrylate. *Biomaterials, 24*(4), 571–577.

Ullah, S., Zainol, I., & Idrus, R. H. (2017). Incorporation of zinc oxide nanoparticles into chitosan-collagen 3D porous scaffolds: Effect on morphology, mechanical properties and cytocompatibility of 3D porous scaffolds. *International Journal of Biological Macromolecules, 104*, 1020–1029.

Unagolla, J. M., & Jayasuriya, A. C. (2020). Hydrogel-based 3D bioprinting: A comprehensive review on cell-laden hydrogels, bioink formulations, and future perspectives. *Applied Materials Today, 18*, 100479.

Valenzuela, L. M., Michniak, B., & Kohn, J. (2011). Variability of water uptake studies of biomedical polymers. *Journal of Applied Polymer Science, 121*(3), 1311–1320.

van Rijt, S., & Habibovic, P. (2017). Enhancing regenerative approaches with nanoparticles. *Journal of the Royal Society Interface, 14*(129), 20170093.

Varner, K., Sanford, J., El-Badawy, A., Feldhake, D., & Venkatapathy, R. (2010). *State of the science literature review: Everything nanosilver and more* (p. 363) Washington DC: US Environmental Protection Agency.

Vedhanayagam, M., Unni Nair, B., & Sreeram, K. J. (2018). Collagen-ZnO scaffolds for wound healing applications: Role of dendrimer functionalization and nanoparticle morphology. *ACS Applied Bio Materials, 1*(6), 1942–1958.

Vicentini, D. S., Smania, A., Jr, & Laranjeira, M. C. (2010). Chitosan/poly (vinyl alcohol) films containing ZnO nanoparticles and plasticizers. *Materials Science and Engineering: C, 30*(4), 503–508.

Wang, X. F., Lu, P. J., Song, Y., Sun, Y. C., Wang, Y. G., & Wang, Y. (2016). Nano hydroxyapatite particles promote osteogenesis in a three-dimensional bio-printing construct consisting of alginate/gelatin/hASCs. *RSC Advances, 6*(8), 6832–6842.

Wiltfang, J., Merten, H. A., Schlegel, K. A., Schultze-Mosgau, S., Kloss, F. R., Rupprecht, S., & Kessler, P. (2002). Degradation characteristics of α and β tri-calcium-phosphate (TCP) in minipigs. *Journal of Biomedical Materials Research: An Official Journal of The Society for Biomaterials, The Japanese Society for Biomaterials, and The Australian Society for Biomaterials and the Korean Society for Biomaterials, 63*(2), 115–121.

Xavier, J. R., Thakur, T., Desai, P., Jaiswal, M. K., Sears, N., Cosgriff-Hernandez, E., ... Gaharwar, A. K. (2015). Bioactive nanoengineered hydrogels for bone tissue engineering: A growth-factor-free approach. *ACS Nano, 9*(3), 3109–3118.

Yazdimamaghani, M., Vashaee, D., Assefa, S., Walker, K. J., Madihally, S. V., Köhler, G. A., & Tayebi, L. (2014). Hybrid macroporous gelatin/bioactive-glass/nanosilver scaffolds with controlled degradation behavior and antimicrobial activity for bone tissue engineering. *Journal of Biomedical Nanotechnology, 10*(6), 911–931.

Yeo, T., Ko, Y. G., Kim, E. J., Kwon, O. K., Chung, H. Y., & Kwon, O. H. (2020). Promoting bone regeneration by 3D-printed poly (glycolic acid)/hydroxyapatite composite scaffolds. *Journal of Industrial and Engineering Chemistry*, *94*, 343–351.

Yi, H., Rehman, F. U., Zhao, C., Liu, B., & He, N. (2016). Recent advances in nano scaffolds for bone repair. *Bone Research*, *4*(1), 1.

Yuan, S., Shen, F., Chua, C. K., & Zhou, K. (2019). Polymeric composites for powder-based additive manufacturing: Materials and applications. *Progress in Polymer Science*, *91*, 141–168.

Zhang, H., Zhou, L., & Zhang, W. (2014). Control of scaffold degradation in tissue engineering: A review. *Tissue Engineering Part B: Reviews*, *20*(5), 492–502.

Zhang, L., & Webster, T. J. (2009). Nanotechnology and nanomaterials: Promises for improved tissue regeneration. *Nano Today*, *4*(1), 66–80.

Zhang, Y., Chu, W., Foroushani, A. D., Wang, H., Li, D., Liu, J., . . . Yang, W. (2014). New gold nanostructures for sensor applications: A review. *Materials*, *7*(7), 5169–5201.

Zhang, Y., et al. (2012). Graphene: A versatile nanoplatform for biomedical applications. *Nanoscale*, *4*(13), 3833–3842.

Zhang, Z., Wang, J., Nie, X., Wen, T., Ji, Y., Wu, X., . . . Chen, C. (2014). Near infrared laser-induced targeted cancer therapy using thermoresponsive polymer encapsulated gold nanorods. *Journal of the American Chemical Society*, *136*(20), 7317–7326.

Zhong, W., Xing, M. M., & Maibach, H. I. (2010). Nanofibrous materials for wound care. *Cutaneous and Ocular Toxicology*, *29*(3), 143–152.

Zhu, J. (2010). Bioactive modification of poly (ethylene glycol) hydrogels for tissue engineering. *Biomaterials*, *31*(17), 4639–4656.

Zinge, C., & Kandasubramanian, B. (2020). Nanocellulose based biodegradable polymers. *European Polymer Journal*, *133*, 109758.

7

Enzyme-encapsulated nanoparticles for biodegradation of materials

Leeba Balan[1], Jamuna Sanker[1], Sriram Chandrasekaran[1], Sugumari Vallinayagam[2]

[1]BIONYME LABORATORIES, CHENNAI, INDIA [2]DEPARTMENT OF BIOTECHNOLOGY, MEPCO SCHLENK ENGINEERING COLLEGE, SIVAKASI, INDIA

7.1 Introduction

The field of nanotechnology has attained much growth and have gained much importance in recent years especially from last decades, because of their attractive applications in biology and medicine. Nanotechnology is concentrated in many areas like agriculture, electronics, transportation, food industry, cosmetics, medicines, bioimaging, and drug delivery (Vasir & Labhasetwar, 2007). Mostly due to their size, which ranges from 1–100 nm is an added advantage in drug delivery and mostly they are used widely in medicinal fields. Due to their significant size and shape, they are used for site-specific drug delivery (Mahapatro & Singh, 2011).

Polymer-based nanomaterials serve as excellent vehicles in delivering genes, vaccines, drugs, and biomolecules because of their submicron-sized polymeric colloidal particles. Therapeutic material of interest can be loaded in these polymers and delivered at their sites (Labhasetwar, Song, & Levy, 1997). Polymer-based nanoparticles are chosen for drug delivery because they have the tendency to move across the biological barriers and into the cell membranes. These nanoparticles help in the delivery of hydrophobic drugs, hydrophilic drugs, vaccines, and proteins in the body. They can be specifically designed according to the parts to which the drug is to be delivered (Park et al., 2009). In this chapter, we have discussed about the biodegradable nanoparticles that are currently used, biodegradation of nanoparticles and their types, applications, and their upcoming challenges are discussed briefly.

7.2 Different types of biodegradable nanoparticles

Biodegradable nanoparticles show a high compatibility in biosafety, and they are based on polymers with an average size of 10–500 nm. Mostly they are used to carry out therapeutic agents by encapsulation or embedding method with the matrix. The size and surface of the polymers can be adjusted according to the dosage of drugs to be released (Han et al., 2018; Mahapatro & Singh, 2011). Biodegradable nanoparticles are mainly constituted with polymer-based nanomaterials which are classified as poly-e-caprolactone (PCL), nanomaterials such as chitosan, and poly-D-L-lactide-co-glycolide (PLGA).

The rate of degradation of these polymer-based materials depends on the pH, temperature, size, and structure of the nanomaterials, which is processed safely from the body without inducing any complications. Some of the types of biodegradable nanoparticles are explained below.

7.2.1 Poly-e-caprolactone

PCL is used as a source in drug delivery because these substances degrade by hydrolysis of the linkages (esters) in humans under normal physiological conditions found to be less toxic or nontoxic for making long-term implantable devices. PCL are mostly prepared by nanoprecipitation, solvent displacement, and solvent evaporation. These biodegradable nanomaterials can be degraded by microorganisms, and it is a semicrystalline polymer with a melting point of 55°C–60°C (Song et al., 2018; Fig. 7–1).

7.2.2 Poly-D-L-lactide-co-glycolide

PLGA is a biodegradable material consists of hydrophobic polylactic acid (PLA) and hydrophilic polyglycolic acid (PGA), which can be degraded into glycolic acid with natural metabolites (Gunatillake, Adhikari, & Gadegaard, 2003; Fig. 7–2). PLGA are considered to be the best biodegradable polymers and they are frequently used in the drug delivery process because they can be easily hydrolyzed and degraded into nontoxic products such as CO_2 and H_2O, which gets easily eliminated from the body (Makadia & Siegel, 2011; Rezvantalab et al., 2018). Generally, PLGA is used in conjugation with other biomaterials to enhance the ability of bone regeneration by low osteoinductivity and suboptimal mechanical properties (Félix Lanao et al., 2013; Pan, Ding, & Poly, 2012).

7.2.3 Chitosan

Chitosan is a modified copolymer prepared of D-glucosamine and N-acetyl glucosamine bonded with β (1–4) linkages (Islam, Dmour, & Taha, 2019; Fig. 7–3). Chitosans are biodegraded by enzymes such as lysozymes which breaks the bonding between acetylated units

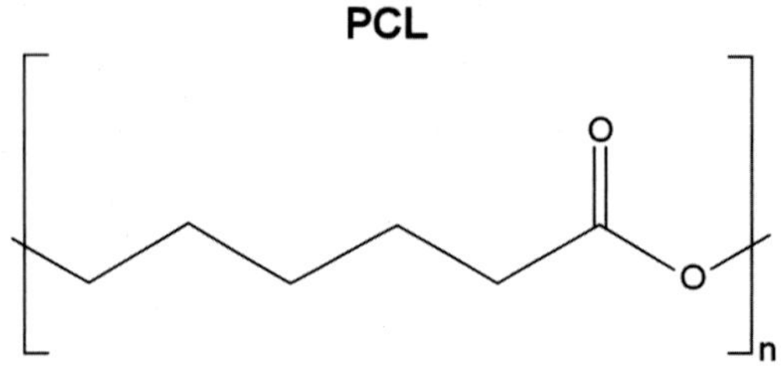

FIGURE 7–1 Structure of poly-e-caprolactone (PCL) (Song et al., 2018).

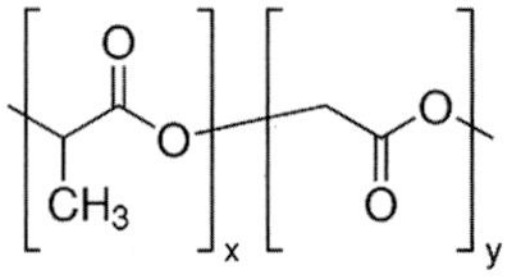

FIGURE 7–2 Structure of poly-D-L-lactide-co-glycolide (PLGA) (Rezvantalab et al., 2018).

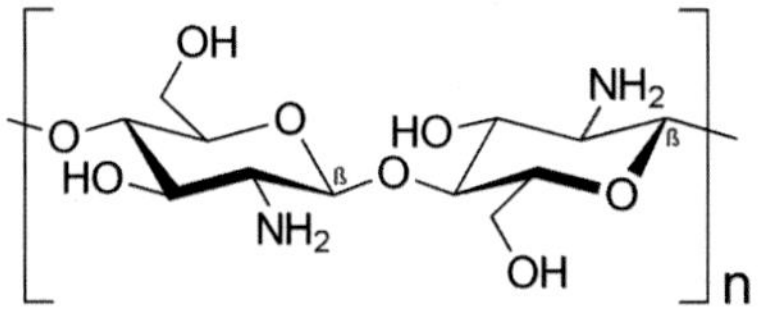

FIGURE 7–3 Structure of chitosan (Islam et al., 2019).

and degrades to oligosaccharides (Muzzarelli, 1997). Chitosan has the capability of degrading Mg^{2+}/Ca^{2+} present in the bacterial cell wall and reacts with the anionic groups of phospholipids present in the bacterial cell wall, which leads to the changes in the cell wall permeability and the exudation of the cellular contents (Chung & Chen, 2008; Kim, Seo, & Moon, 2008).

Chitosan-based materials have tissue-engineered applications with relation to adipose, vascular, nerve, and connective tissues (Akman, Tigli, Gumusderelioglu, & Nohutcu, 2010; Gobin, Butler, & Mathur, 2006; Wu, Black, Santacana-Laffitte, & Patrick, 2007). Chitosan is used widely in the field of bone regeneration which is remarkably demonstrated in the case of saos-2 osteosarcoma cell lines showing biocompatibility. Chitosan, combined with a natural silica material—diatomite a scaffold—is produced for bone tissue regeneration (Cicco, Vona, & deGiglio, 2015; Levengood & Zhang, 2014). Carboxymethyl chitin nanoparticles incorporated with CaCl2 and FeCl3, exhibited nontoxic property against L929 mouse models and antibacterial effects against Staphylococcus strains (Dev, Binulal, & Anitha, 2010a). Studies on drug delivery mechanism toward HIV are found to be exciting (Gu, Al-Bayati, Ho, & Ea, 2017). Studies on antibradykinin B2 antibody conjugated chitosan nanoparticles and antitransferrin have the potential to penetrate across the blood-brain barrier, thus inhibiting replication of HIV in the neural system (Gu et al., 2017). Moreover, the drug release pattern under in vitro condition studies have reported that PLA/chitosan nanoparticles loaded with lamivudine (type 1 and type 2 HIV selective inhibitor) decreased when the pH of medium changed from alkaline to acidic condition and from acidic to neutral condition (Dev, Mohan, & Sreeja, 2010b). The results show that the chitosan-based nanoparticles have a potential carrier system for HIV-controlled drug delivery. Cancer- and HIV-based nanodrug delivery vehicles are chitosan-based. An ultrasound-based insulin delivery system was incorporated for insulin release into the blood stream to regulate blood glucose levels, which can provide both long-term sustainable and fast on-demand response (Di, Yu, & Wang, 2017). Chitosan-based nanomaterials have their own unique qualities in different areas like tissue engineering, wound-healing, regenerative mechanisms, drug-delivery processes, and anticancer properties (Hillyard, Doczi, & Kiernan, 1964; Kozen, Kircher, Henao, Godinez, & Johnson, 2008; Sugano, Watanabe, Kishi, Izume, & Ohtakara, 1988; Ueno, Mori, & Fujinaga, 2001). One of the important property of chitosan is the mucoadhesive nature and drug-delivery mechanism (Sashiwa & Shigemasa, 1999).

7.2.4 Dendrimers

Dendrimers are smallest sphere-shaped nanocarriers with 1–100 nm in size with amphicilic copolymers consisting of both hydrophilic and hydrophobic units to carry drugs with poor solubility (Singh, Garg, Goyal, & Rath, 2016). These dendrimers are amphicilic in nature with copolymers like hydrophilic and hydrophobic monomers hence, drugs can be integrated by means of covalent bonding, encapsulation, or electrostatic interaction methods (Santos, Veiga, & Figueiras, 2019; Singh et al., 2016; Fig. 7–4).

Dendrimers are convenient in their usage as drug carriers because they possess high surface volume with much branching, biocompatibility, polyvalency, low immunogenicity, and high water compatibility (Duncan & Izzo, 2005; Santos et al., 2019). Poly(amidoamine) dendrimers are most commonly used in the field of nanomedicines for drug and gene delivering systems, since they can be transported into the cells or across the cells by endocytic pathways, and moreover, as modifications and charge on the surface of the dendrimers can be made, cytotoxicity takes place during the interactions with the cell membranes (Duncan & Izzo, 2005).

7.2.5 Lipid-based nanoparticles

Lipid-based nanoparticles mostly consist of liposomes, which are spherical-shaped vesicles in size ranging from 10 to 100 nm; this type of nanoparticle was in usage for decades for drug delivery systems (Pattni, Chupin, & Torchilin, 2015; Tapeinos, Battaglini, & Ciofani, 2017; Fig. 7–5). Using lipid-based nanoparticles, the aqueous part of liposomes are used for loading hydrophilic agents, while the lipid ends are used for loading hydrophobic agents during the drug-delivery process (Li, Tan, Li, Shen, & Wang, 2017). Mostly, liposomes have the ability of self-assembling and enable easy drug-loading mechanisms, and due to similar composition between liposomes and cell membranes, liposomes are considered to be more compatible than the synthetic ones (Li et al., 2017; Sercombe et al., 2015). Lipid-based nanoparticles offer new approaches toward cancer therapies as the polymers have the tendency

Linear   Branched   Dendron   Dendrimer

FIGURE 7–4 Structure of dendrimer (Singh et al., 2016).

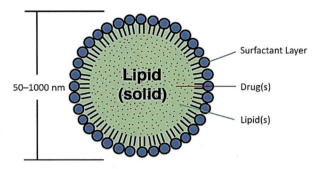

FIGURE 7–5 Structure of solid lipid nanoparticle (Pratibha & Conway, 2014).

to break down under acidic conditions, which is similar to that of tumor microenvironments to release the payloads but found to be stable under physiological conditions (Chiang, Lyu, Wen, & Lo, 2018).

7.3 What is the need for biodegradation of materials?

Biodegradation is a process of degrading materials by the use of biological methods such as microorganisms (bacteria and fungi). Generally plastic materials are highly focused to degrade since they are harmful to the environment. Apart from microorganisms used for the degradation process, certain nanomaterials are now focused to remove the toxic materials from the surroundings (Gottschalk & Nowack, 2011). Some of the nanomaterials like CNTs and graphene and their derivatives are found to inert in nature, stable, and difficult to degrade as they are used in drug-delivery processes; many studies have reported their presence in the environment (Li et al., 2014). Studies are being carried out to eliminate the synthesized nanoparticles from the environment by using both microbes and enzymes. When these nanoparticles are treated, one of the final products obtained is found to be CO_2 evolution (Chen, Qin, & Zeng, 2016; Fig. 7–6).

Various technologies are being involved in the degradation of the nanomaterials, and among these, Raman spectroscopy, Visible/Near-Infrared spectroscopy (Vis-NIR) spectrum, and Transmission Electron Microscope (TEM) are more often used. Among molecular methods, molecular docking, followed by molecular dynamics and homology modeling, are used (Rose et al., 2015). It has been widely recognized that the synthesis of nanomaterials are widely used in the fields of drug delivery, biosensors, bioimaging, and enzyme immobilization; hence, their usage and mass production have increased rapidly. Polymer nanoparticles like PLA, PCL, PLGA, and chitosan can be degraded and safely processed in the body. The degradation of these polymer nanoparticles are greatly affected by certain internal factors such as structure of the nanoparticles, size, and their molecular weight and some of the external factors like pH and temperature (Panyam & Labhasetwar, 2003; Fig. 7–7).

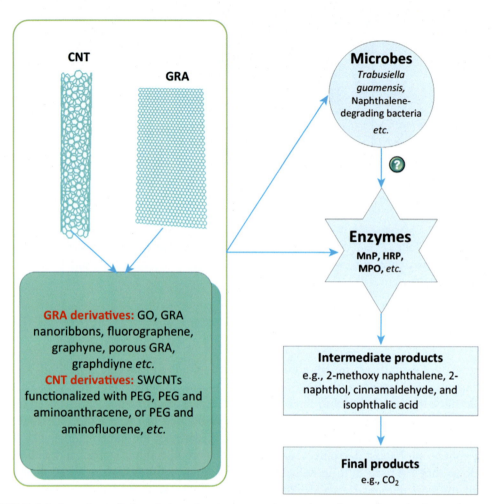

FIGURE 7–6 A flow chart of degradation of carbon nanotubes and graphene using microbes and enzymes (Chen et al., 2016).

7.4 Enzyme-encapsulated nanoparticles in biodegradation

Enzyme encapsulation therapy is currently used in cancer treatment as a specific enzyme, and the interaction site forms the enzyme therapy rather than radiation and chemotherapy for the treatment. Encapsulation of specific or unspecified enzymes nanoparticles has given out an assured production of enzyme-encapsulated nanoparticles at low cost and their increased homogeneity (Slowing, Vivero-Escoto, Wu, & Lin, 2008). The enzymes, which are encapsulated on the nanoparticles, act as a delivery vehicle where the enzymes can be either released or remain at the site, and allow further for the enzyme-substrate interaction (Volodkin, Petrov, Prevot, & Sukhorukov, 2004). Fig. 7–8 represents the possible two methods of enzyme encapsulation with polymer nanoparticles.

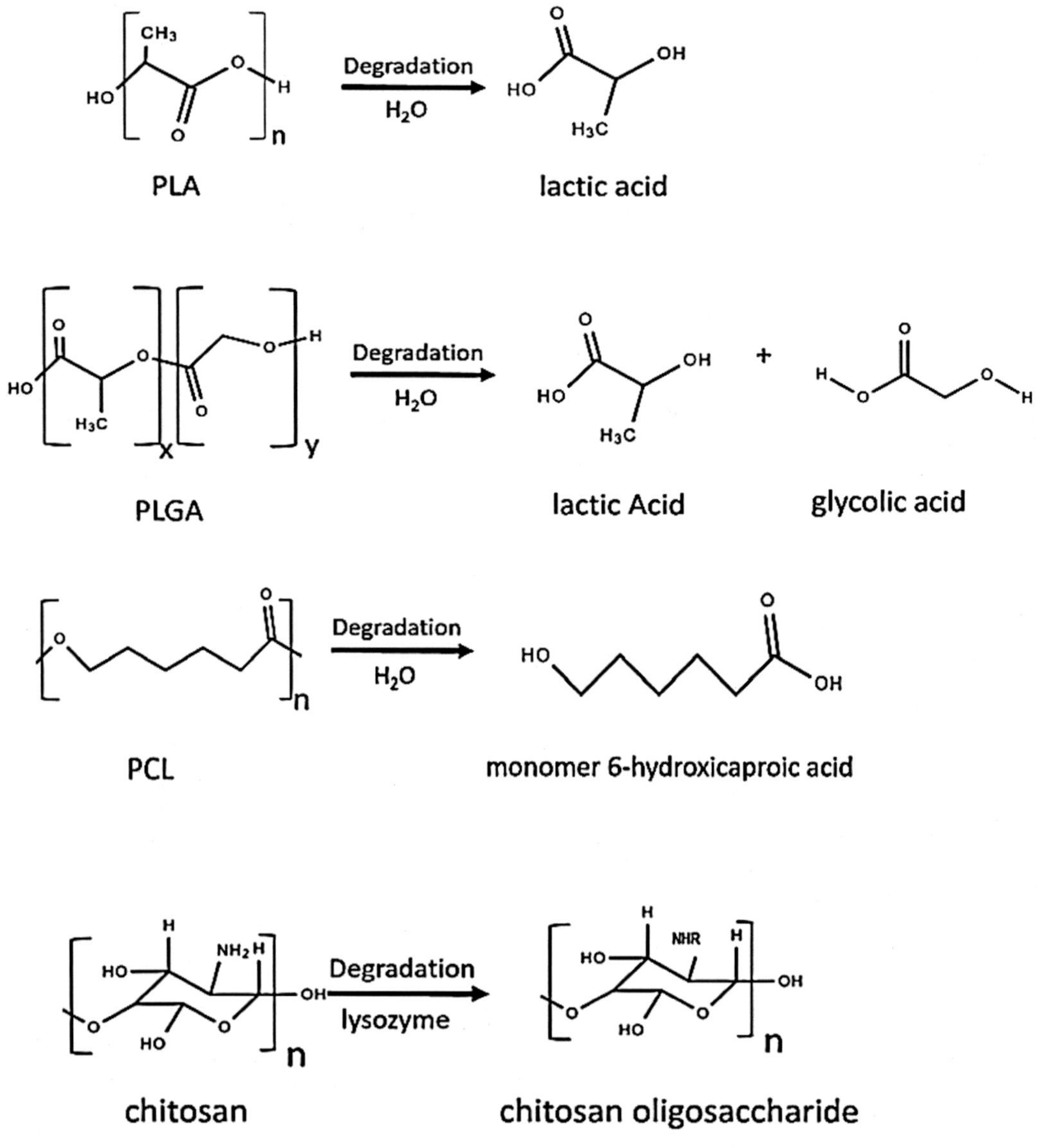

FIGURE 7–7 Biodegradation of commonly used polymer nanoparticles (Panyam & Labhasetwar, 2003).

The schematic flow chart A represents the volumetric expansion and contraction of the polymer to encapsulate the enzymes and B represents coating of enzymes absorbed polymer nanoparticles for enzyme encapsulation. These kinds of enzyme-encapsulated nanoparticles can be used locally or specifically targeted (Volodkin et al., 2004).

Enzyme-facilitated degradation of natural or synthetic polymers can be on a random or terminal mode pattern. Apart from the purpose of enzyme encapsulation, all enzymes are

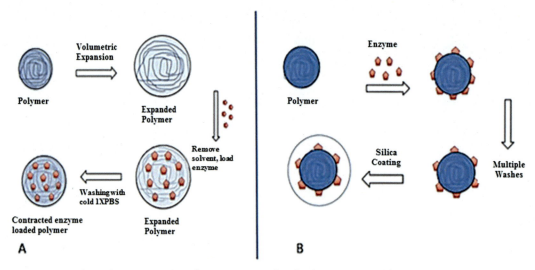

FIGURE 7–8 A schematic representation of enzyme-encapsulated polymer nanoparticles.

found to be specific to substrate and catalytic to the confirmation of chain formation (Chu, 2005). Polymer degradation is greatly influenced by drug loading in PLA/PGA matrix by acceleration of acidific catalysis when the drug is acidic. Degradation rate decreases due to the interactions between carboxyl chain ends and drug molecules. Degradation of PLA/PGA matrix depends on parameters like porosity, size, base catalysis, dimension of devices, load, and morphology of incorporated compounds (Li, 2017).

7.5 Microbes and enzymatic degradation

Soil bacteria like *Trabusiella guamensis* has the ability to survive in the soil contaminated with nanomaterials and it was found out that they able to adapt themselves in those contaminated areas. The survivability was clearly observed and found that the bacteria was able to biotransform carbon nanotubes (CNTs) by oxidation process (Chen et al., 2016). Apart from bacteria, fungus also has the potential to degrade nanomaterials. *Sparassis latifolia* is a type of mushroom secrets lignin peroxidase (LiP) that has the capacity to degrade raw-grade carboxylated CNTs and thermally heated nanomaterials. White-rot fungus, *Phanerochaete chrysosporium*, is widely used to degrade lignin, dyes, aromatic hydrocarbons, and other pollutants (Chen et al., 2016). In addition to microorganism degradation, there are enzyme-linked degradation, which also takes place where horseradish peroxidase, myeloperoxidase, and H_2O_2 degrades nanomaterials. Hexagonal boron nitrile carbon sheets are found to be degraded by myeloperoxidase enzyme (Li et al., 2014). Some of the enzymes, which have the property of degrading CNTs, graphenes, and their derivatives, are listed below in Table 7–1.

A large number of microbes and enzymes are found to be involved in the biodegradation of these nanomaterials, but only a very few studies are being postulated and

Table 7–1 List of enzymatic biodegradation of carbon nanotubes, graphene and their derivatives (Chen et al., 2016).

Enzymes	Nanomaterial substrates
Lactoperoxidase	SWCNTs
Horseradish peroxidase	SWCNTs, MWCNTs, and GO
Myeloperoxidase	SWCNTs and GO
Xanthine oxidase	MWCNTs
Eosinophil peroxidase	SWCNTs
LiP	SWCNTs, oxidized, and reduced Graphenylene-based nanoribbons (GRA) nanoribbons
MnP	SWCNTs
Tyrosinase	MWCNTs
Laccase	SWCNTs, MWCNTs

GO, Graphene oxide; MWCNTs, multi-walled carbon nanotubes; SWCNTs, single-walled carbon nanotubes.

recorded in this case, and hence field applications are more complex than laboratory conditions to work in this area to learn about the microbial action and enzymatic activity (Chen et al., 2016).

In recent findings, the mechanism of microbial and enzymatic interaction with nanomaterials are studied extensively using molecular docking methods, where the structure and properties of the enzymes are obtained from protein data bank (PDB) and by means of homology modeling, the interaction sites between a ligand and a receptor is studied (Yuriev & Ramsland, 2013). Molecular-based studies are found to be helpful in picturing the molecular modeling and the mechanism involved in biodegradation and oxidation of nanomaterials, which can be clearly depicted by their structure orientation and docking results (Liu et al., 2014) (Fig. 7–9).

7.6 Applications of biodegradable nanoparticles

Recent research emphasized the promising role of biodegradable nanoparticles in the field of biology, especially in drug delivery and as bioresorbable surgical devices. Biodegradable nanoparticles have been widely used for its drug-delivery application due to its site-specific transportation of drugs, vaccines, and biomolecules-based drugs (Patra, Das, & Fraceto, 2018). Based on the size, shape, and structure of the specific nanoparticles, they possess different pharmacokinetic properties, encapsulation efficiency, and drug-releasing mechanisms. The biodegradable nanoparticles are mostly used in the field of biomedical applications due to its safety profile with improved biodegradability and releasing capacity. Hence, these nanoparticles have been used in the biomedical field for target-based drug delivery, bioimaging, and diagnostics. Various biodegradable matrices have been used for the medical applications. In the following sections we discuss the most essential applications of biodegradable nanoparticles.

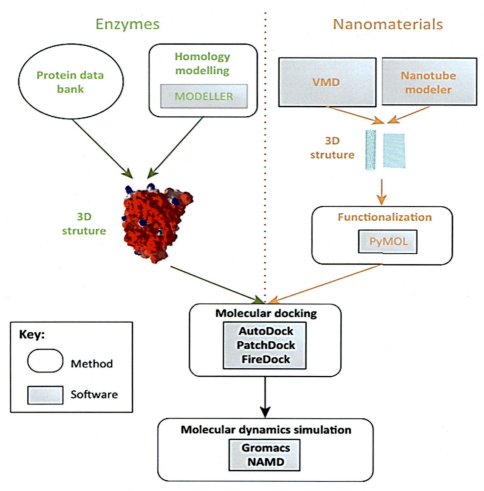

FIGURE 7–9 Molecular studies on biodegradation of carbon nanotubes (Chen et al., 2016).

7.6.1 Biodegradable nanoparticles in imaging technology

Recently, synthesized nanoparticles have been used to produce progressive multifunctional drug-delivery materials. Therefore, these biodegradable nanoparticles have been employed in various therapeutic needs consequently as real time imaging, targeting, therapeutic cargo and regulated release of drug (Patra et al., 2018; Senapati, Mahanta, Kumar, & Maiti, 2018; Sohail et al., 2020). Biodegradable nanoparticles has been accepted in the medical field because of its unique properties like regulated and sustained drug delivery, enhanced pharmacokinetics properties, without causing side effects and efficient biodegradability which high light these materials as an ideal carrier systems and widely employed in the field of medicine (Swierczewska, Crist, & McNeil, 2018). Recently, biodegradable nanomaterials have been used as contrast agents as biomedical imaging purpose as it easy to locate the target tissue with higher resolution (Siddique & Chow, 2020). Therefore, nanoparticles are employed in several bioimaging instruments like computed tomography, photoacoustic imaging, positron emission tomography (PET), magnetic resonance imaging (MRI), and fluorescence imaging (Hahn, Singh, & Sharma, 2011). For instance, metal-based nanoparticles have been emphasized as radiosensitizers in the chemotherapy to treat cancer (Liu, Zhang, & Li, 2018). Metal nanoparticles synthesized from gold and silver produces array of colors in the visible light based on plasmon resonance in owing to combined oscillations generated at the surface of nanoparticles (Huang & El-Sayed, 2010). Due to this unique plasmon absorbance features of these metal nanoparticles have been utilized in various medical field as chemical sensors and biosensors (Sadrolhosseini, Noor, & Moksin, 2012). Gold nanoparticles play a vital role in the development of contrast agents in bioimaging. Gold nanoparticles have attracted much interest in the field of bioimaging due to their new and unique characters of gold nanoparticles like less cytotoxicity, biocompatibility, optical properties, bioimaging, biosensing, drug delivery, and cancer therapies (Chien, 2012; Chhour, 2017). Gold nanoparticles are utilized in radiotherapy and photothermal therapy in the primary diagnosis and cancer therapy (Yao et al., 2016). Studies conducted by Bobyk and his colleagues on glioma mice models showed increased efficacy and enhanced survival rate with gold nanoparticles in combination with low energy radiation therapy (Bobyk et al., 2013). Currently, cancer theranostic application includes the recognition of novel biomarkers of cancer therapy, molecular diagnostics tools, and molecular imaging tools and aids the early detection of cancer. These cancer nanotheranostics are related with external stimuli such as light, temperature, magnetism, and sound (i.e., photodynamic and photothermal). Recently, nanomaterials have been used as PET and SPECT systems to improve the image resolution and proper diagnosis. These cancer nanotheranostics could be categorized as conventional (liposomes, micelles and nanogels) and biomimetics. In addition to this, polyethylene glycols (PEG) are used as functionalized agents due to their increased bioavailability (Suk et al., 2015). Silicon-based naphthalocyanine PEG nanoparticles were developed, which was encapsulated with PEG-b-poly (ε-Caprolactone) for fluorescence-based surgery. Biomimetics nanotheranostics have been focused by several researchers because of the preparation of these nanoparticles with the combination of proteins, phospholipids, cholesterol, microbes, apatite and exosomes. Magnetic nanotheranostics comprised of magnetic substances like magnetite and iron oxide with the external magnetic field could emphasize the role of nanocarrier at the site of a tumor. The most common magnetic nanotheranostics diagnostic tool is MRI, which is a noninvasive tool to get a clear picture of anatomy and physiological processes of the body (Wallyn, Anton, & Vandamme, 2019).

7.6.2 Biodegradable nanoparticles in the industries, material sciences, and electronics

Research efforts in the past 20 years have resulted in advancing the promising applications of nanoparticle-based materials in industrial and construction-based fields (Mohajerani et al., 2019; Piccinno et al., 2012). Nanocrystals could provide very interesting base materials for material science by their size-dependent manner. The overwhelmed benefits of nanoparticles have been recognized in the manufacture and marketable materials like pharmaceuticals products, microelectronics, and aerospace materials (Khan, Saeed, & Khan, 2019). Recently, in the food processing, packaging, and preservation industries, these biodegradable nanoparticles have been utilized (Pathakoti, Manubolu, & Hwang, 2017; Pradhan et al., 2015; Thiruvengadam, Rajakumar, & Chung, 2018). An emerging resonant energy transfer system comprised of biodegradable nanoparticle-based organic dyes and noble metals have been utilized in both biophotonics as well as in material science (Grel, Ratajczak, Jakiela, & Stobiecka, 2020). Nowadays, nanoparticles have been utilized in the production of electronic equipment at low cost with more efficacies due to their unique structural, optical, and electrical properties of metal-based nanoparticles used in the new generation of electronics, sensors, and photonic materials (Tuantranont, 2013). To conclude that biodegradable nanoparticles have gained much attention due to their unique properties and potential biomedical application in almost all the field of science and technology.

References

Akman, A. C., Tigli, R. S., Gumusderelioglu, M., & Nohutcu, R. M. (2010). bFGF-loaded HA-chitosan: A promising scaffold for periodontal tissue engineering. *Journal of Biomedical Materials Research. Part A*, *92*(3), 953–962.

Bobyk, L., Magali, E., Pierre, D., & Helene, E. (2013). Photoactivation of Gold Nanoparticles for Glioma Treatment. *Nanomedicine: Nanotechnology, Biology, and Medicine*, *9*(7), 1089–1097.

Chen, M., Qin, X., & Zeng, G. (2016). Biodegradation of carbon nanotubes, graphene and their derivatives. *Trends in Biotechnology*, *1449*, 11.

Chhour, P. (2017). Design and synthesis of gold nanoparticle contrast agents for atherosclerosis imaging with computed tomography.Publicly Accessible Penn Dissertations. 2221.

Chiang, Y. T., Lyu, S. Y., Wen, Y. H., & Lo, C. L. (2018). Preparation and characterization of electrostatically cross-linked polymer–liposomes in anticancer Therapy. *International Journal of Molecular Sciences*, *19*, 1615.

Chien, C.-C., et al. (2012). Gold nanoparticles as high-resolution X-ray imaging contrast agents for the analysis of tumor-related micro-vasculature. *Journal of Nanobiotechnology*, *10*(1), 1–12.

Chu, C. C. (2005). Surface degradation and microenvironmental outcomes. *Surfaces and Interfaces for Biomaterials*, 585–618.

Chung, Y.-C., & Chen, C.-Y. (2008). Antibacterial characteristics and activity of acid-soluble chitosan. *Bioresource Technology*, *99*(8), 2806–2814.

Cicco, S. R., Vona, D., & deGiglio, E. (2015). Chemically modified diatoms biosilica for bone cell growth with combined drug-delivery and antioxidant properties. *Chempluschem*, *80*(7), 1104–1112.

Dev, A., Binulal, N. S., & Anitha, A. (2010a). Preparation of poly(lactic acid)/ chitosan nanoparticles for anti-HIV drug delivery applications. *Carbohydrate Polymers*, *80*(3), 833–838.

Dev, A., Mohan, J. C., & Sreeja, V. (2010b). Novel carboxymethyl chitin nanoparticles for cancer drug delivery applications. *Carbohydrate Polymers*, *79*(4), 1073–1079.

Di, J., Yu, J., & Wang, Q. (2017). Ultrasound-triggered noninvasive regulation of blood glucose levels using microgels integrated with insulin nanocapsules. *Nano Research*, *10*(4), 1393–1402.

Duncan, R., & Izzo, L. (2005). Dendrimer biocompatibility and toxicity. *Advanced Drug Delivery Reviews*, *57*, 2215–2237.

Félix Lanao, R. P., Jonker, A. M., Wolke, J. G., Jansen, J. A., van Hest, J. C., & Leeuwenburgh, S. C. (2013). Physicochemical properties and applications of poly (lactic-co-glycolic acid) for use in bone regeneration. *Tissue Engineering. Part B, Reviews, 19*(4), 380–390.

Gobin, A. S., Butler, C. E., & Mathur, A. B. (2006). Repair and regeneration of the abdominal wall musculofascial defect using silk fibroin-chitosan blend. *Tissue Engineering, 12*(12), 3383–3394.

Gottschalk, F., & Nowack, B. (2011). The release of engineered nanomaterials to the environment. *Journal of Environmental Monitoring, 13*, 1145–1155.

Grel, H., Ratajczak, K., Jakiela, S., & Stobiecka, M. (2020). Gated resonance energy transfer (gRET) controlled by programmed death protein ligand 1. *Nanomaterials, 10*(8), 1592.

Gu, J., Al-Bayati, K., Ho, E. A., & Ea, H. (2017). Development of antibody-modified chitosan nanoparticles for the targeted delivery of siRNA across the blood-brain barrier as a strategy for inhibiting HIV replication in astrocytes. *Drug Delivery and Translational Research, 7*(4), 497–506.

Gunatillake, P. A., Adhikari, R., & Gadegaard, N. (2003). Biodegradable synthetic polymers for tissue engineering. *European Cells and Materials, 5*, 1–16.

Hahn, M. A., Singh, A. K., Sharma, P., et al. (2011). Nanoparticles as contrast agents for in-vivo bioimaging: Current status and future perspectives. *Analytical and Bioanalytical Chemistry, 399*, 3–27.

Han, J., Zhao, D., Li, D., Wang, X., Jin, Z., & Zhao, K. (2018). Polymer-based nanomaterials and applications for vaccines and drugs. *Polymers (Basel)* (10, p. 31).

Hillyard, I. W., Doczi, J., & Kiernan, P. B. (1964). Antacid and antiulcer properties of the polysaccharide chitosan in the rat. *Proceedings of the Society for Experimental Biology and Medicine. Society for Experimental Biology and Medicine (New York, N.Y.), 115*, 1108–1112.

Huang, X., & El-Sayed, M. A. (2010). Gold nanoparticles: Optical properties and implementations in cancer diagnosis and photothermal therapy. *Journal of Advanced Research, 1*(1), 13–28.

Islam, N., Dmour, I., & Taha, M. O. (2019). Degradability of chitosan micro/nanoparticles for pulmonary drug delivery. *Heliyon, 5*, e01684.

Khan, I., Saeed, K., & Khan, I. (2019). Nanoparticles: Properties, applications and toxicities. *Arabian Journal of Chemistry, 12*(7), 908–931.

Kim, I. Y., Seo, S. J., & Moon, H. S. (2008). Chitosan and its derivatives for tissue engineering applications. *Biotechnology Advances, 26*(1), 1–21.

Kozen, B. G., Kircher, S. J., Henao, J., Godinez, F. S., & Johnson, A. S. (2008). An alternative hemostatic dressing: Comparison of CELOX, HemCon, and QuikClot. *Academic Emergency Medicine: Official Journal of the Society for Academic Emergency Medicine, 15*(1), 74–81.

Labhasetwar, V., Song, C., & Levy, R. J. (1997). Nanoparticle drug delivery system for restenosis. *Advanced Drug Delivery Reviews, 24*(1), 63–85.

Levengood, S. L., & Zhang, M. (2014). Chitosan-based scaffolds for bone tissue engineering. *Journal of Materials Chemistry B., 2*(21), 3161–3184.

Li, S. (2017). 2—Synthetic biodegradable medical polyesters. *Science and Principles of Biodegradable and Bioresorbable Medical Polymers*, 37–78.

Li, Y., et al. (2014). Surface coating-dependent cytotoxicity and degradation of graphene derivatives: Towards the design of nontoxic, degradable nano-graphene. *Small (Weinheim an der Bergstrasse, Germany), 10*, 1544–1554.

Li, Z., Tan, S., Li, S., Shen, Q., & Wang, K. (2017). Cancer drug delivery in the nano era: An overview and perspectives (review). *Oncology Reports, 38*, 611–624.

Liu, W.-W., et al. (2014). Synthesis and characterization of graphene and carbon nanotubes: A review on the past and recent developments. *Journal of Industrial and Engineering Chemistry, 20*, 1171–1185.

Liu, Y., Zhang, P., Li, F., et al. (2018). Metal-based nanoenhancers for future radiotherapy: Radiosensitizing and synergistic effects on tumor cells. *Theranostics, 8*(7), 1824–1849.

Mahapatro, A., & Singh, D. K. (2011). Biodegradable nanoparticles are excellent vehicle for site directed invivo delivery of drugs and vaccines. *Journal of Nanobiotechnology, 9*, 55.

Makadia, H. K., & Siegel, S. J. (2011). Poly lactic-co-glycolic acid (PLGA) as biodegradable controlled drug delivery carrier. *Polymers (Basel), 3*, 1377–1397.

Mohajerani, A., et al. (2019). Nanoparticles in construction materials and other applications, and implications of nanoparticle use. *Materials (Basel, Switzerland), 12*(19), 3052.

Muzzarelli, R. A. A. (1997). Human enzymatic activities related to the therapeutic administration of chitin derivatives. *Cellular and Molecular Life Sciences: CMLS, 53*(2), 131–140.

Pan, Z., Ding, J., & Poly, D. J. (2012). Poly (lactide-co-glycolide) porous scaffolds for tissue engineering and regenerative medicine. *Interface Focus, 2*(3), 366–377.

Panyam, J., & Labhasetwar, V. (2003). Biodegradable Nanoparticles for Drug and Gene Delivery to Cells and Tissue. *Advanced Drug Delivery Reviews, 55*, 329–347.

Park, K., Lee, S., Kang, E., Kim, K., Choi, K., & Kwon, I. C. (2009). New generation of multifunctional nanoparticles for cancer imaging and therapy. *Advanced Functional Materials, 19*(10), 1553–1566.

Pathakoti, K., Manubolu, M., & Hwang, H.-M. (2017). Nanostructures: Current uses and future applications in food science. *Journal of Food and Drug Analysis, 25*(2), 245–253.

Patra, J. K., Das, G., Fraceto, L. F., et al. (2018). Nano based drug delivery systems: Recent developments and future prospects. *Journal of Nanobiotechnology, 16*(1), 71.

Pattni, B. S., Chupin, V. V., & Torchilin, V. P. (2015). New developments in liposomal drug delivery. *Chemical Reviews, 115*, 10938–10966.

Piccinno, F., et al. (2012). Industrial production quantities and uses of ten engineered nanomaterials in Europe and the world. *Journal of Nanoparticle Research, 14*(9), 1109.

Pradhan, N., Singh, S., Ojha, N., Shrivastava, A., Barla, A., Rai, V., & Bose, S. (2015). Facets of nanotechnology as seen in food processing, packaging, and preservation industry. *BioMed Research International, 2015*.

Pratibha, G. K., & Conway, B. R. (2014). Solid lipid nanoparticles: A potential approach for dermal drug delivery. *American Journal of Pharmacological Sciences, 2*(5A), 1–7.

Rezvantalab, S., Drude, N. I., Moraveji, M. K., Güvener, N., Koons, E. K., Shi, Y., … Kiessling, F. (2018). PLGA-based nanoparticles in cancer treatment. *Frontiers in Pharmacology, 9*.

Rose, P. W., et al. (2015). The RCSB protein data bank: Views of structural biology for basic and applied research and education. *Nucleic Acids Research, 43*, D345–D356.

Sadrolhosseini, A. R., Noor, A. S. M., & Moksin, M. M. (2012). Application of surface plasmon resonance based on a metal nanoparticle. *Plasmonics-Principles and Applications*, 253–282.

Santos, A., Veiga, F., & Figueiras, A. (2019). Dendrimers as pharmaceutical excipients: Synthesis, properties, toxicity and biomedical applications. *Materials (Basel)* (13, p. 65).

Sashiwa, H., & Shigemasa, Y. (1999). Chemical modification of chitin and chitosan 2: Preparation and water soluble property of N-acylated or N-alkylated partially deacetylated chitins. *Carbohydrate Polymers, 39*(2), 127–138.

Senapati, S., Mahanta, A. K., Kumar, S., & Maiti, P. (2018). Controlled drug delivery vehicles for cancer treatment and their performance. *Signal Transduction and Targeted Therapy, 3*, 7, Published 2018 Mar 16.

Sercombe, L., Veerati, T., Moheimani, F., Wu, S. Y., Sood, A. K., & Hua, S. (2015). Advances and challenges of liposome assisted drug delivery. *Frontiers in Pharmacology, 6*, 286.

Siddique, S., & Chow, J. C. L. (2020). Application of nanomaterials in biomedical imaging and cancer therapy. *Nanomaterials, 10*(9), 1700.

Singh, B., Garg, T., Goyal, A. K., & Rath, G. (2016). Recent advancements in the cardiovascular drug carriers. *Artificial Cells, Nanomedicine and Biotechnology, 44*, 216−225.

Slowing, I. I., Vivero-Escoto, J. L., Wu, C. W., & Lin, V. S. (2008). Mesoporous silica nanoparticles as controlled release drug delivery and gene transfection carriers. *Advanced Drug Delivery Reviews, 60*(11), 1278−1288.

Sohail, M., Guo, W., Li, Z., Xu, H., Zhao, F., Chen, D., & Fu, F. (2020). Nanocarrier-based drug delivery system for cancer therapeutics: A review of the last decade. *Current Medicinal Chemistry*, Oct 5.

Song, R., Murphy, M., Li, C., Ting, K., Soo, C., & Zheng, Z. (2018). Current development of biodegradable polymeric materials for biomedical applications. *Drug Design, Development and Therapy, 12*, 3117−3145.

Sugano, M., Watanabe, S., Kishi, A., Izume, M., & Ohtakara, A. (1988). Hypocholesterolemic action of chitosans with different viscosity in rats. *Lipids, 23*(3), 187−191.

Suk, H-II, Seong-Whan Lee, S. W., & Shen, D. (2014). Hierarchical Feature Representation and Multimodal Fusion with Deep Learning for AD/MCI Diagnosis. *Neuroimage, 101*, 569−582.

Swierczewska, M., Crist, R. M., & McNeil, S. E. (2018). Evaluating nanomedicines: Obstacles and advancements. *Methods in Molecular Biology, 1682*, 3−16.

Tapeinos, C., Battaglini, M., & Ciofani, G. (2017). Advances in the design of solid lipid nanoparticles and nanostructured lipid carriers for targeting brain diseases. *Journal of Controlled Release: Official Journal of the Controlled Release Society, 264*, 306−332.

Thiruvengadam, M., Rajakumar, G., & Chung, I.-M. (2018). Nanotechnology: Current uses and future applications in the food industry. *3 Biotechnology, 8*(1), 74.

Tuantranont, A. (2013). *Applications of nanomaterials in sensors and diagnostics. Springer series on chemical sensors and biosensors*. Berlin Heidelberg: Springer.

Ueno, H., Mori, T., & Fujinaga, T. (2001). Topical formulations and wound healing applications of chitosan. *Advanced Drug Delivery Reviews, 52*(2), 105−115.

Vasir, J. K., & Labhasetwar, V. (2007). Biodegradable nanoparticles for cytosolic delivery of therapeutics. *Advanced Drug Delivery Reviews, 10*(8), 718−728, 59.

Volodkin, D. V., Petrov, A. I., Prevot, M., & Sukhorukov, G. B. (2004). Matrix polyelectrolyte microcapsules: New system for macromolecule encapsulation. *Langmuir: The ACS Journal of Surfaces and Colloids, 20*(8), 3398−3406.

Wallyn, J., Anton, N., & Vandamme, T. F. (2019). Synthesis, principles, and properties of magnetite nanoparticles for in vivo imaging applications-A review. *Pharmaceutics, 11*(11), 601.

Wu, X., Black, L., Santacana-Laffitte, G., & Patrick, C. W. (2007). Preparation and assessment of glutaraldehyde-crosslinked collagen−chitosan hydrogels for adipose tissue engineering. *Journal of Biomedical Materials Research. Part A, 81A*(1), 59−65.

Yao, C., et al. (2016). Gold nanoparticle mediated phototherapy for cancer. *Journal of Nanomaterials*.

Yuriev, E., & Ramsland, P. A. (2013). Latest developments in molecular docking: 2010−2011 in review. *Journal of Molecular Recognition: JMR, 26*, 215−239.

8

Effects of nanoparticles on the biodegradation of organic materials

Sabah Bakhtiari[1], Esmail Doustkhah[2], Mona Zamani Pedram[1], Masoud Yarmohammadi[1], M.Özgür Seydibeyoğlu[3,4]

[1]FACULTY OF MECHANICAL ENGINEERING-ENERGY DIVISION, K.N. TOOSI UNIVERSITY OF TECHNOLOGY, TEHRAN, IRAN [2]INTERNATIONAL CENTRE FOR MATERIALS NANOARCHITECTONICS (MANA), NATIONAL INSTITUTE FOR MATERIALS SCIENCE (NIMS), IBARAKI, JAPAN [3]DEPARTMENT OF MATERIALS SCIENCE AND ENGINEERING, IZMIR KATIP CELEBI UNIVERSITY, IZMIR, TURKEY [4]FILINIA R&D, EGE UNIVERSITY SCIENCEPARK, IZMIR, TURKEY

8.1 Introduction

During past two centuries, the range of synthesized organic compounds have been increased excessively including raw materials, plastics, fuels, detergents, and other useful substances. However, the pollution and hazards resulted from the production of these compounds as well as remaining in the environment after usage seems to be a serious concern. There are plenty categories of organic compounds that could be considered as the main issues in the current era, such as polychlorinated biphenyls, dichlorodiphenyltrichloroethane, polybrominated diphenyl ethers, dechlorane plus, and decabromodiphenyl ethane (Xing et al., 2005). Polychlorinated biphenyls can be used in the capacitors, transformers, and electric fluids. Polybrominated diphenyl ethers, dechlorane plus, and dichlorodiphenyltrichloroethane are known as fire retardant in many industries such as in the production of fabrics and polymers (Alaee et al., 2003). In fact, three various sources are known to be responsible as the most pollutant industries including military wastes, industrial activities, and agriculture chemical materials. On the other hand, petroleum product, polycyclic aromatic hydrocarbons, aromatic compounds, dioxins, most of the chloro-derivatives of acetic acids, phosphates derivatives, carbamates, organometallic compounds, and the most highlighted one, plastics or other degradable resistant polymers (Connell et al., 2009). Furthermore, the importance of the pesticides and fertilizers in modern agriculture are well-known for everyone. Agrochemical compounds save the crop production from being lost for 70% during production and storage steps. The other sources of the organic pollutant compounds such as fuels in vehicles as well as urban and industrial wastes are more familiar than any demands for definition (Connell et al., 2009) (Fig. 8–1).

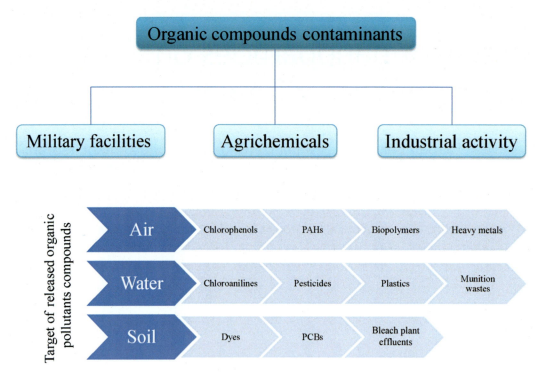

FIGURE 8–1 Sources of organic pollutants.

Basically, chemical compounds with environmental hazardous are the compounds containing carbon with at least one negative side effect in nature. Organic pollutant compounds are categorized into three various classes including hydrocarbons, compounds containing heteroatoms such as oxygen, nitrogen and phosphorus, and organometallic materials. Most of the organic pollutant compounds have some similar properties such as low polarity, lipophilicity, poor solubility in water, and persistence in the environment. The group containing heteroatoms owe some properties such as relatively high solubility in water, low fat solubility, and relatively low persistence in the environment (Connell et al., 2009).

Accordingly, removing of the organic compounds pollutants have been interested in many studies during recent decades. Among various methods for removing of the organic materials, biodegradation of the contaminants has been highlighted due to its ability in elimination of the organic compounds by ecofriendliness, green, and facilitated processes. Furthermore, nanoparticles (NPs) have shown incredible effects on the biodegradation of organic compounds. Their extraordinary properties in adsorption, trapping, releasing electrons, and catalytic activity confirmed their ability in enhancement of the contaminants elimination. In this regard, the effect of the NPs on the biodegradation of the various organic compounds has been discussed in this chapter by focusing on the most important applied NPs.

8.2 Nanoparticles as enhancers for biodegradation
8.2.1 Carbon-based nanoparticles
8.2.1.1 Carbon nanotubes

Carbon nanotubes (CNTs) are some of the most attractive nanomaterials because CNTs have excellent performance (Tasis et al., 2006). It should be mentioned that CNTs are known as green compounds with low hazards. Furthermore, CNTs have proved to be able to remove heavy metals, organic compounds, and a range of biological contaminants (Upadhyayula et al., 2009). Noteworthy, the effect of the CNTs on the enhancement of the enzymes activities has been established before (Zhang et al., 2020). Therefore it could be understood that the bioavailability of the organic contaminants are most affected by CNTs. Biodegradation of the atrazine could be considered as one of the most highlighted examples of the CNT effect on the biodegradation of the organic compounds. Whether the biodegradation of atrazine by Acinetobacter lwoffii DNS32 can make a distinction in the presence of two representative carbon materials (CMs), namely, biochars (BCs), and CNTs is reconnoitered by inspecting the effect of CMs on the biodegradation rate, the feasibility of bacteria, and the appearance of atrazine genes in aqueous medium (Yang et al., 2017a). The disturbance of CNTs on the fate and transport of organic pollutants in the environment have subjected to more concerns (Petersen et al., 2011). Reportedly, sequestering organic contaminants in soils subjects CNTs to function like black carbon (Towell et al., 2011). The following process illustrates that biodegradation processes might be affected by CNTs. Primarily, they hinder biodegradation through plummeting bacterial activity because of the cytotoxicity nature of CNTs. The bactericidal activity of CNTs on model Gram-positive or Gram-negative bacteria has been demonstrated by several studies (Zhao & Liu, 2012). This antibacterial activity was connected to the aqueous phase properties of the CNTs, comprising their length (Yang et al., 2010), functionalization (Rodrigues, Jaisi, & Elimelech, 2013), electronic structure (Vecitis et al., 2010), and dispersity (Liu et al., 2009). Next, their bioavailability is specified by the freely dissolved concentrations of pollutants. Notably, their low freely dissolved concentrations end in slow degradation. Down to their outstanding adsorption capacity, CNTs may decline biodegradation rates and efficacy through decreasing the bioavailability and bioaccessibility of the organic contaminants. Another ubiquitous pollutant is atrazine, which can be detected through air, soil, and aquatic systems. Yan et al. and Chen et al. reported the adsorption and desorption of atrazine on CNTs (Yan et al., 2008). It is assumed that its biodegradation rate and efficiency is influenced by the adsorption and desorption of atrazine on CNTs (Chen et al., 2008).

8.2.1.2 Graphene based

Graphene holds a flat two-dimensional layer of carbon atoms filled in a honeycomb-like lattice (Geim & Novoselov, 2010). The unique features of graphene have made it so spectacular that many researchers have heeded it. As graphene and its derivatives are able to remove pollutants from water, they have been remarkably discovered. One of the derivatives of the graphene is graphene oxide which has been used as adsorbent for RhB and malachite green

156 Biodegradation and Biodeterioration at the Nanoscale

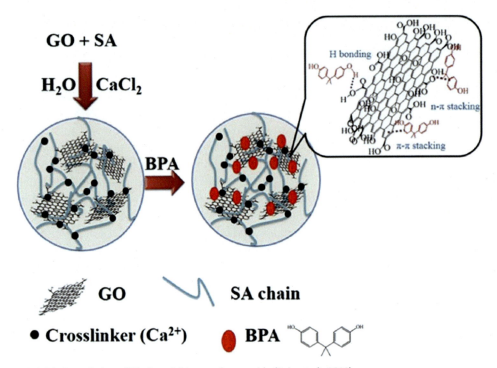

FIGURE 8–2 Biodegradation of Bisphenol A by graphene oxide (Baig et al., 2019).

dyes from aqueous media (Vadodaria, Onyianta, & Sun, 2018). Moreover, graphene has been used as an enhancer in biodegradation of the endocrine disruptors and personal care products in the environment that can be found in wastewater (Kasprzyk-Hordern, Dinsdale, & Guwy, 2009). Endocrine disruptors might disturb or moderate the endocrine system of human and other living organisms; the two distinctive endocrine disruptors are Bisphenol A and nonylphenol (Cajthaml, 2015). Many studies have been performed on the biodegradation of Bisphenol A and nonylphenol (Cabana et al., 2007). The findings clarify the proofs that the binding conformations of Bisphenol A and nonylphenol, would be changed in the presence of the graphene. As a consequence, their biodegradation processes would be influenced (Fig. 8–2).

8.2.1.3 *Fullerene nanoparticles*

Fullerenes are known as the molecular allotropes of carbon. The presence of conjugated π-electron on this hollow spheres NP makes it possible to perform many functionalization reactions on this NP. It should be mentioned that the C60 is the most studied derivation of the fullerene category (Klupp, Margadonna, & Prassides, 2016). Recently, fullerenes have been highlighted as a catalyst for biodegradation of the organic compounds, contaminants, microorganisms, etc. (Auwerter et al., 2017; Wang et al., 2012). Kümmerer et al. studied the biodegradation of the starch and other polysaccharide compounds using C60 and C70 fullerenes (Kümmerer et al., 2011). A similar degradation study performed on the biodegradation by fullerene NPs once it was practiced in minimal broth Davis without dextrose so that the growth cycle of low-density polyethylene (LDPE)-degrading bacterial consortia was affected (Kapri et al., 2010; Sah et al., 2010).

8.2.2 Iron oxide magnetic nanoparticles

Fe_3O_4 magnetic NP is interested in many studies due to its outstanding properties such as low toxicity, mature synthetic technology, rapid separation from substrates, and predominantly cyclic catalytic property (Rossi et al., 2014). Moreover, a new generation of environmental remediation technologies is provided by nanoscale iron particles. They present solutions that are not costly to some of the most inspiring environmental crackdown difficulties (Zhang, 2003). Owing to biocompatibility, large surface areas, high surface reactivity, and superparamagnetic properties, magnetic Fe_3O_4 NPs are extensively practiced to confiscate a wide variety of environmental contaminants, such as gases (Rodriguez, 2006), contaminated chemicals (Ponder, Darab, & Mallouk, 2000), organic pollutants (Lien & Zhang, 2001), and biological substances (Bosetti et al., 2002). Recently, the immobilization of microorganisms which utilize Fe_3O_4 NPs is identified as a novel dimension of the industrialization of microbial cell immobilization in environmental remediation (Guobin et al., 2005). One of the most highlighted research on the biodegradation of organic compounds using iron oxide magnetic NPs was performed on ciprofloxacin. Ciprofloxacin is identified as a third-generation fluoroquinolone frequently practiced in human and veterinary medicine. Usually, high proportions of ciprofloxacin are partly metabolized in humans and livestock, and defecated as the parent substance (Daughton & Ternes, 1999). The soil–water ecosystem with wastewater treatment plant, manure, and sludge can be extended by ciprofloxacin remains (Golet et al., 2003). The possible growth and spread of antibiotic resistance causes potential threats to the ecosystem and human health. Thus, it is essential to deliberate ciprofloxacin elimination before introducing it into the environment. Accordingly, Yang et al. established that the Fe_3O_4 enrichments can enhance ciprofloxacin biodegradation (Yang et al., 2017b). Biodegradation of carbazole was evaluated using encapsulated *Sphingobium yanoikuyae* XLDN2−5 cells in the mixture of Fe_3O_4 NPs and gellan gum (Sun et al., 2017). Moreover, it should be mentioned that the proper biodegradation activity and reusability at low Fe_3O_4 NPs concentration was divulged by the final immobilized *S. yanoikuyae* XLDN2−5 cells when the concentration activity is low at high Fe_3O_4 NPs concentration (Wang et al., 2007). The carbazole biodegradation activities of *S. yanoikuyae* XLDN2−5 cell/Fe_3O_4 NP biocomposites are identified as an essential representative of environmentally microorganisms used for carbazole bioremediation.

Organisms and environments through cytoplasmic toxins even at low concentration are endangered by phenolic compounds in industrial wastewater discharged from many industries such as metallurgy, petrochemicals, chemical organic synthesis, and pharmaceuticals (Lin & Juang, 2009). Primarily, laccase is classified as a blue copper of oxidoreductases extensively detected in nature, particularly in fungi. Laccase has been demonstrated to catalyze the degradation of an extensive variety of potential substrates such as phenols, aromatic amines. At last, the only byproduct is water (Torres, Bustos-Jaimes, & Le Borgne, 2003). In the field of treating industrial wastewater, scientists have dissected laccase (Zainudin, Abdullah, & Mohamed, 2010). To compensate or the deficiency of free enzymes, enzyme immobilization technology seems to be a perfect method (Dong et al., 2019). Accordingly, a great use possible for the treatment of phenol-containing wastewater was achieved by functionalizing of dialdehyde starch with Fe_3O_4 magnetic NPs modified with amino-functionalized ionic liquid. To simultaneously immobilize laccase, this organic macromolecule could be regarded as a crosslinking agent. With the contribution of the amino-functionalized ionic liquid, magnetic NPs restrained the organic polymer compound DAS. By means of DAS as the crosslinking agent, laccase was covalently immobilized onto the carrier. In reverse to the other measured immobilized laccases, the highest enzyme loading, activity holding, and excellent elimination efficacy for phenolic compounds was scored by Fe_3O_4-NIL-DAS@lac at a wider pH and temperature scope (Qiu et al., 2020).

On the other hand, the textile industry generated the release of synthetic dyes wastewater that is measured to be the main sources of water contamination (Bilal et al., 2016b). It is projected that 2%–20% of the entire yearly production of textile dyes used in industry are directly cleared as aqueous wastes to natural water bodies (Bilal et al., 2016a). These synthetic dyes are very steady, and their accrual always subjects to highly opaque watercourses that avert an acceptable light entrance for the photosynthetic organisms to perform their function usually, resulting in poor oxygenation of the environment (Iqbal & Khera, 2015; Nouren et al., 2017). Consequently, to prevent ecosystem deterioration, the in effect treatment of dyeing wastewater without producing any secondary pollution is indispensable. To obstruct synthetic dyes from release and contamination, various treatment technologies including adsorption, coagulation, and oxidation have been advanced by scientists to eliminate of synthetic dyes from wastewater, in which biodegradation using laccase has progressively expected more attention as it is low-cost, effective, and less-damaging to the environment (Saber-Samandari et al., 2017). Magnetic NPs are promising carriers, which affords an efficient recycling of biocatalysts (Kadam, Jang, & Lee, 2017). Direct immobilization of enzymes on Fe_3O_4 NPs is typically restricted, because bare Fe_3O_4 NPs are highly susceptible to acidic and oxidative conditions (Chandra et al., 2010). As aforementioned, Fe_3O_4@C-Cu^{2+} NPs are fabricated to immobilize laccase. Joining the facts of easy parting of the Fe_3O_4 core and chelated Cu^{2+} of carbon shell, the Fe_3O_4@C-Cu^{2+} NPs were set up to hold a simple immobilization procedure, high loading capacity, and high enzyme activities, stabilities and reusability of the immobilized laccase. To degrade several synthetic dyes, the immobilized laccase was proved to be a valid biocatalyst (Li et al., 2020).

Huang, Fulton, and Keller (2016) presented a novel method for removal of polycyclic aromatic hydrocarbons using modified magnetic iron oxide NPs (Fig. 8–3). Polycyclic aromatic hydrocarbons and possibly toxic metal ions such as cadmium are identified as determined organic pollutants. Owing to their high harmfulness and long perseverance, they meaningfully endanger the public health and the environment (Hübner, Astin, & Herbert, 2010). E-waste processing sites (Luo et al., 2011) manufactured gas plant sites (Thavamani, Megharaj, & Naidu, 2012), and river sediments (Feng et al., 2012; Thavamani et al., 2012) are among

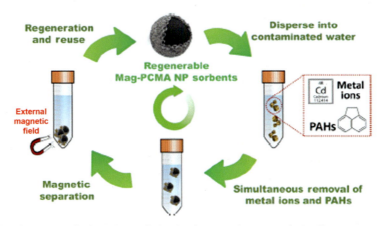

FIGURE 8–3 Polycyclic aromatic hydrocarbons elimination by Fe_3O_4 (Huang et al., 2016).

the several sites contaminated by both polycyclic aromatic hydrocarbons and heavy metals. It has been known that modified magnetic NPs are effective on elimination of such severe hazardous contaminants.

To recycle agricultural waste, it is suitable to utilize composting as an ecofriendly, economical, and practical waste system (Zeng et al., 2018). Composting microbes utilized the readily obtainable fractions of organic materials in agricultural straw, whereas the enzymatic activities affect polymeric organic compounds to be degraded into directly accessible carbon and nitrogen sources (Jurado et al., 2015). Among the polymers, lignin, and cellulose are identified as the foremost constituents which it is difficult to degrade them. Accordingly, the composting process is reserved and the final compost quality is declined. Exogenous microorganisms or amendment of BC, zeolite, and so forth are among the methods that scientists have practiced different methods to improve the organic materials degradation during composting (Hagemann et al., 2018). The exclusive physical, electronic, optical, and biological feature of engineered NPs hold exclusive physical, electronic, optical, and biological feature, their application is swiftly improved such as architecture, aerospace and airplanes, computer memory, catalysts, chemicals, environmental protection, and so forth (Wang et al., 2018). The growth of the plant, such as mung beans, could be endorsed by Fe_2O_3 NPs (He et al., 2011). Reportedly, dehydrogenase and urease activity in soil positively replied to the actions of iron oxide NPs, including Fe_2O_3 and Fe_3O_4 NPs. Wherein Fe_2O_3 NPs displayed more important impacts, as variations in enzyme activities could be prompted by the alterations of microbial community triggered by iron oxide NPs (He et al., 2011). During composting, in terms of organic material degradation and the nitrogen cycle, the actual considerations of performance of biological process are dehydrogenase and urease (Ren et al., 2018). On the other hand, enzyme activities and organic materials degradation through composting process with the alteration of iron oxide NPs have not been remarkably scrutinized. It is worthy to mention that Ag or Ag-based NPs have been the main subject of most of the studies ever done in regard to nanomaterials in composting systems (Gitipour et al., 2013) (Fig. 8–4).

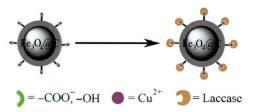

FIGURE 8–4 Dye biodegradation by Fe_3O_4 functionalized nanoparticles (Li et al., 2020).

8.2.3 Titanium oxide–based nanoparticles

Titanium oxide (TiO_2) is identified as a semiconductor with incredible photocatalytic activity (Asghar et al., 2011). This NP has been widely used for removing gaseous and water contamination. It is able to eliminate volatile organic compounds from indoor air, for example, dichloromethane (Kowit et al., 2012). Regarding to its photocatalytic activity, it has been implicated on LDPE to achieve effective biodegradation by Asghar et al. (2011).

2,4,5-Trichlorophenol as one the most important chlorinated organic compounds with high toxicity has been biodegraded by TiO_2 NP. This compound is widely used in solvents, herbicides, pesticides, plastics, or their precursors (Häggblom, 1992; Holladay, Bevelhimer, & Brandt, 1994). It should be mentioned that the kinetics of some chlorinated compounds are often slower than usual and their biodegradation occurs only at low concentration (Dahlen & Rittmann, 2002). By comparison, chemical oxidation, including aromatic compounds, is a robust method of breaking down large organic molecules (Scott & Ollis, 1995). Complete mineralization by chemical oxidation, however, is economically prohibitive and toxicity and oxygen demand are maintained by the products of partial chemical oxidation (Belháčová et al., 1999). To address the drawbacks of both approaches, an emerging approach incorporates chemical oxidation and biodegradation (Scott & Ollis, 1996). Chemical oxidation partially transforms biorecalcitrant organic compounds into intermediates in such a combined treatment that can be completely mineralized by biodegradation (Bandara et al., 1997). The advanced oxidation method and biological treatment separately, or sequential pairing, were used in most previous attempts at integrated treatments (Marco, Esplugas, & Saum, 1997). Advanced oxidation processes rely on indiscriminate free-radical reactions, creating a variety of materials that can be overoxidized or not readily biodegradable. Excessive oxidation loses oxidant and increases the cost of treatment and the objective of downstream biodegradation is defeated by low biodegradability. When the advanced oxidation phase and biodegradation occur together, the issues of concurrent treatment can be overcome; this is called intimate coupling. In a novel photocatalytic circulating-bed biofilm reactor exploiting macroporous cellulose carriers, Marsolek et al. (2008) successfully demonstrated the principle of intimate coupling of photocatalysis and biodegradation for 2,4,5-trichlorophenol. To eliminate this particle, TiO_2 NPs were either fixed to or distributed in slurry form on the outer surface of biofilm carriers. The TiO_2 photocatalyzed 2,4,5-trichlorophenol when illuminated with UV light into intermediates that could biodegrade bacteria in the biofilm inside the microporous carriers, where the bacteria were protected from UV light and free radicals generated as part of advanced photocatalytic oxidation. It should be mentioned that the applied carriers by Marsolek et al. was cellulose based. They were charred by UV irradiation and hydroxyl radical attack, although they had good wet density for circulation and perfect macropores for biofilm accumulation. This charring ultimately resulted in a degradation of the physical structure of the carriers. The biofilm-carrier must therefore be made of a material that is ideally suited for intimate coupling with an advanced oxidation processes. An ideal carrier for intimate photocatalysis and biodegradation coupling should have porosity suitable for inside biofilm accumulation, density that allows good circulation in the reactor, capacity to retain a large amount of TiO_2, and long-term longevity in the photocatalytic circulating-bed biofilm reactor.

Solar TiO_2 photocatalytic degradation of the insecticide UltracidTM, a commercial formulation containing methidathion as the active ingredient, has been investigated by García-Ripoll et al. (2007). In less than 2 h of irradiation, total removal of methidathion can be achieved, although longer solar exposures are needed for full solution mineralization (7–8 h). Enabled respirometry of the sludge indicates that the solution is detoxified when methidathion is removed, so further irradiation does not seem necessary. A biodegradable nanocomposite of polymer TiO_2 with antimicrobial activity arising from photooxidation of composite TiO_2 was investigated by Hitoshi Ando et al. By the solvent cast process, poly(ε-caprolactone-co-L-lactide)-visible-light-sensitive TiO_2 nanocomposite films were prepared. Since they consist of TiO_2 (which is nontoxic and environmentally friendly) and a biodegradable polymer, these nanocomposites are expected to be applicable in a wide range of fields. In addition, a simple method that requires only mixing a small amount of TiO_2 into the polymer can prepare the nanocomposites; The method can therefore be used not only for P(CL/LA) but also for other biodegradable polymers, thus imparting to others the role of on/off biodegradation (Ando et al., 2015). Biodegradation of polymer nanocomposites has recently been suppressed by obstructing the diffusion of microorganisms and enzymes into polymers and introducing particles of clay and metal (Bikiaris, 2013). However, biodegradation suppression was not controllable, and after being discarded, the nanocomposites were not rapidly degraded. Using TiO_2 in composites can be a very effective way for biodegradable polymers to provide an on/off biodegradation mechanism. Compared to pure polymer films under UV irradiation, Miyauchi, Li, and Shimizu (2008) reported that the rate of enzymatic hydrolysis of nanocomposite films containing 5 wt.% of TiO_2 was increased as the incorporated TiO_2 decomposed the base polymer by photooxidation. Fukuda and Tsuji (2005) and Buzarovska and Grozdanov (2012) have also evaluated the biodegradability of polymer TiO_2 nanocomposite films in enzymatic hydrolysis experiments. Yew et al. reported that biodegradation of polymer TiO_2 nanocomposite films was suppressed in soil and tropical mangrove sediment regardless of whether the films were exposed to sunlight, and also that the films were photolyzed upon exposure to high intensity sunlight. They proposed that even without exposure to sunlight, the suppression of biodegradation of films may have been caused by the introduction of high TiO_2 concentrations (up to 57% by weight), resulting in less surface area for microorganisms to function on. These findings indicate that it is possible to impart an antimicrobial property that can regulate polymer biodegradation by compositing low TiO_2 concentrations that do not induce polymer photolysis. In addition, UV-sensitive TiO_2 was used in all of the previous studies.

A hybrid material (conversion phosphors-TiO$_2$) was coupled with a biofilm to achieve enhanced removal of tetracycline to establish a more green and effective process for tetracycline removal. Wang et al. showed that while tetracycline is a form of refractory contaminant, tetracycline can be extracted from wastewater by coupling biofilms with hybrid material. The removal of tetracycline by biofilm hybrid material was due to low biofilm extracellular polymeric substances (EPS), which can increase the generation of free radicals by acting in the coupling system as a transient medium. By controlling its antibiotic resistance genes and superoxide dismutase to ensure the existence of EPS, biofilm tolerated tetracycline stress in the coupled system. They offered a new way of combining the microbial community with hybrid materials that are visible-light-driven to degrade organic recalcitrant contaminants in water. In the future, improved tetracycline removal rates are required for further investigation in different scenarios (Wang et al., 2020). Tetracycline is a commonly used antibiotic, but it is difficult for tetracycline to more effectively biodegrade (Prado, Ochoa, & Amrane, 2009). Means for the biodegradation of tetracycline will be of great benefit. Biofilms and new usable materials (like Ag/TiO$_2$) are two methods for improving tetracycline biodegradation (Xiong et al., 2018). Microbial aggregates are biofilms that are typically used to extract organic contaminants, nitrogen, phosphorus, and heavy metals from wastewater (Wu, Liu, & Rene, 2018). Due to its moderate cost and ability to maintain slow-growing microorganisms, biofilm technology is a promising approach to biodegrading refractory pollutants in wastewater (Su et al., 2020). However, for highly refractory compounds such as tetracycline (Su et al., 2020), a biofilm process also needs to be combined with other technologies to improve degradation performance. When excited by light at a certain wavelength, some functional materials such as TiO$_2$ can be used to break down TC by producing electron–hole pairs (Fan et al., 2020). The electron–hole pairs produce free radicals that could react quickly with tetracycline, such as the hydroxyl radical (•OH) (Zhu et al., 2018).

8.2.4 Metal-based nanoparticles

8.2.4.1 Nickel/iron nanoparticles

As mentioned before, phenolic compounds are recognized as the determined organic contaminants, which can be normally detected in the soil and ground water (Du et al., 2011). These compounds frequently exist in industrial effluents and they are often released through the environment. According to the fact that phenols are toxic, it brings about great environmental hurdles (Lin & Juang, 2009). As time dragged on, researchers have regarded the expansion of procedures for the remediation of phenolic compounds from industrial wastewater (Lin & Juang, 2009). Notable the microorganisms, which are utilized to biologically degrade phenols in wastewater, are advantageous because this degradation can be possibly completed by them. In fact, such microorganisms contribute to develop nontoxic products (Zuo et al., 2009). Henceforth, a hopeful approach is the biodegradation of phenols by microorganisms (Zainudin et al., 2010). According to studies, it is possible to aerobically biodegrade phenols even though they inadequately biodegradable and determined in the environment (Olaniran & Igbinosa, 2011). To use microorganisms to aerobically biodegrade phenols, it is essential to practice enzyme oxygenases so that atmospheric oxygen integrate oxygen into their substrate. Nonetheless, there are several technical experiments in the biodegradation such as low degradation rates, long clean-up times, and in effectual supply of appropriate electron donors (H_2, acetate) to the bacteria (Xiu et al., 2010). To scrutinize these subjects, effective biodegradation requires increasing the biodegradation rates and providing electron donors to the microorganisms. Lately, iron-based NPs combined with biodegradation to decrease nitrate. Reportedly, H_2 is formed because of Nanoscale Zerovalent Iron (nZVI) corrosion. Nonetheless, the formation of H_2 completely depends on the medium pH (An et al., 2010). The *Bacillus fusiformis* strain can be practiced to biologically degrade phenols (Lin, Gan, & Chen, 2010). Ye Kuang et al. has indicated that, the impacts of different medium pH values on the biodegradation of phenol by *B. fusiformis* in the presence of iron-based NPs such as nZVI and Ni/Fe were examined to upsurge biodegradation rates. The addition of Ni/Fe to *B. fusiformis* was evaluated to understand the difference between Ni/Fe and nZVI impacting on the growth of *B. fusiformis* and the biodegradation of phenol. Evidently, the growth of *B. fusiformis* or biodegradation of phenol in the presence of Ni/Fe was similar to that of nZVI (Lin et al., 2010).

8.2.4.2 Zinc oxide nanoparticles

The enormous presence of organics, sulfate, and synthetic dyes detect industrial effluents arising from textile, pulp, and paper industries (Blázquez, Gabriel, Baeza, & Guisasola, 2017). Among the highest volume/mass manufactured nanomaterials, Zinc oxide NPs are extensively utilized in pigments, textile, semiconductors, sunscreens, and food additives (Seow et al., 2008). Consequently, it is anticipated that NPs will find their way into aquatic, terrestrial, and atmospheric environment. For the sake of the distinctive physical and chemical features the probable impacts these nanomaterials should be recognized (Ma, Williams, & Diamond, 2013). ZnO-NPs toxicity and inhibition to several microorganisms has been detected (Feris et al., 2010). Reportedly, a portion of ZnO-NPs may find their way into wastewater treatment plants (Brar et al., 2010). As the necessary decomposers of the pollutants are microorganisms in wastewaters, the effectiveness of pollutants removal could be reduced by released ZnO-NPs. Within the past few years, few studies have examined the effects of ZnO-NPs on microbial communities in biodegradation of the organic pollutants and set up their inhibitory effects on organic elimination, methane production, nitrification, denitrification, and phosphorus exclusion (Hou et al., 2013). Rasool et al. pioneered the providence biotransformation and biodegradation of azo dye and cosubstrate (glucose) in presence of ZnO-NPs. In fact, azo dye decolorization and sulfate decreases are both oxidation-reduction reactions, in which both sulfate and azo dye serve as electron acceptor. Eight reducing equivalents are required to biologically diminish to sulfide. To efficiently remove sulfate and azo dyes, it is essential to access suitable and sufficient electron donors (Rasool & Lee, 2016).

Notably, benzo[ghi]perylene (BghiP) has been identified as priority pollutant owing to its toxicity, mutagenic and carcinogenic behavior. It is noteworthy that the molecular weight of BghiP polycyclic aromatic hydrocarbon is high since it holds six fused benzene rings (Habs, Jahn, & Schmähl, 1984). In case they include more than four rings, they are detected to be remarkably hydrophobic and minimally bioavailable. Additionally, they are intractable to be degraded (Seo et al., 2007). Considering its biodegradative abilities and environmental sustainability, scientists have deliberated the remediation of polycyclic aromatic hydrocarbon contaminated sites by microorganisms (Su et al., 2018). Few papers have been issued on degradation of BghiP by means of fungi (Su et al., 2018) and yeast (Hesham et al., 2006). Moreover, they have been highly considered for numerous industrial, medical and environmental applications as well as remediation of number of pollutants (Ibrahim, 2018). According to the high surface areas and other features of NPs, they have been utilized as reductants or catalysts so that different reactions are applied. As ZnO-NPs are recyclable, easy to handle, inexpensive, nonvolatile, nonexplosive, and can serve as ecofriendly catalyst for many organic transformations, they are widely practiced (Safaei-Ghomi & Ghasemzadeh, 2017). Furthermore, researched have regarded the influence of NPs as the activity of microbes are simulated by their exclusive influence on microbiological responses placed (Shin & Cha, 2008). Nevertheless, particularly restricted studies have been reported on consequence of NPs on biodegradation of pollutants in presence of produced biosurfactants in the development mediu1m by means of microbes (El-Sheshtawy & Ahmed, 2017).

8.2.4.3 Manganese peroxidase

Manganese peroxidase, which has been isolated from several fungi (Baborová et al., 2006), has been previously applied in environmental bioremediation due to its ability to mediate oxidation and removal of a broad range of contaminants including polycyclic aromatic hydrocarbons, phenolic compounds (Tuor et al., 1992), and azo dyes (Mielgo et al., 2003) using H_2O_2 as the terminal electron acceptor. This NP has also been applied in environmental bioremediation because it is able to mediate oxidation and remove a wide assortment of contaminants as well as polycyclic aromatic hydrocarbons (Baborová et al., 2006), phenolic compounds (Hirano et al., 2000), and azo dyes by means of H_2O_2 as the terminal electron acceptor. The restricted constancy of free enzymes under natural environments restrained their usage for remediation although they have several benefits. Furthermore, macrosized enzyme immobilization, including surface binding and encapsulation, has been practiced to improve enzymatic stability on manganese peroxidase NPs (Olsson & Ögren, 1983). As immobilized enzymes covalently bound to solid surfaces or physically filled in solid matrices, they have indicated to be more stable against different inhibitors including organic solvents and thermal inactivation. All the aforesaid features are caused by their tough covalent binding and further substrate dispersal resistance from solid matrices (Wei et al., 2013). To improve enzymatic stability on the manganese peroxidase NPs without considerably influencing catalytic efficacy, some nanosized encapsulation approaches have been established (Patterson, Prevelige, & Douglas, 2012).

8.2.4.4 Silver nanoparticles

A considerable enactment of nanomaterials is achieved by the fast-stepping development in NP design. Indeed, complex architecture and improved chemical and physical properties are illustrated. This paves the way for hopeful opportunities in the theranostic field

(Bogart et al., 2014). Nonetheless, while an unrivaled theranostic potential in a test tube may be displayed by NPs, when the nanoobjects pass in the cell or the organism, they influence differently their effect might be different (and quickly dismantled) (Feliu et al., 2016). Once in vivo, the surface of the NPs is quickly covered by the biomolecules and the cell affects the NP recognition. The main question of the past decade was what the cells see (Lynch, Salvati, & Dawson, 2009) considering the interaction between the bodily proteins and NPs and a protein corona cover them; this is vital to the NPs − cell interaction (Monopoli et al., 2011). When there is a remarkable level of intricacy, on the cell (Caballero-Díaz et al., 2013) or organism level (Su et al., 2014), silver (Ag) NP degradation is evaluated and enactment becomes progressively intricate. According to the fact that cell division and following dilution of NPs interlocks with NP bioprocessing and dissolution, they firmly harm the biodegradation measures on the cellular level. It is hard to recover a reliable bioprocessing fingerprint and quantitative degradation index, and it is difficult to distinguish AgNPs and ionic silver using common analytical methods along with Ag NP biodegradation, biodistribution, and elimination. To identify the biodegradation of AgNPs within the organisms, improvised analytical methods are occasionally practiced (Jiang et al., 2015). The high plasmonic potential of AgNps pave the way to develop strategies so that Ag nanostructures from degradation is preserved (Baida et al., 2009). Some of the strategies contain surface alteration of Ag nanostructures with an extra protecting layer of inorganic or organic material, such as silica, silica@titania double shells, and self-assembled monolayers of organic thiols (Jiang, Zeng, & Yu, 2007). Although the complexity of the chemistry stresses on the development of NPs, it is possible to obtain only incomplete and temporary protection of silver.

8.2.5 Nanobarium titanate

Primarily, nanobarium titanate is added to the marginal broth to effect the development cycle of LDPE-degrading bacterial consortia. The phase, exponential phase, and stationary phase are affected by it. Its function mostly relies on plummeting the length of lag phase and growing the extent of exponential and inactive phase. Nanobarium titanate acts as a supportive nutritional component to hasten the development of bacterial consortia. Therefore it contributes the conglomerates in plastic waste biodegradation. According to the published paper by Scherer et al. (1999), this NP is known to be effective on the biodegradation of the plastics.

8.2.6 Metal−organic framework nanoparticles

Metal−organic frameworks (MOFs) are identified as porous crystalline hybrid materials, of which inorganic and organic units connect them by strong bonds. They are recognized for their large pore volume, high surface area, and tunable pore size (Furukawa et al., 2013). The restriction in materials recycle can be controlled by adding magnetism. To remove drugs (Bayazit et al., 2017) and organic dyes (Wang et al., 2016) from water, several magnetic MOFs are put into practice. Bayazit et al. prepared Fe_3O_4/MIL-101(Cr) to adsorb ciprofloxacin in water and magnetic separation recycled the adsorbent (Wang et al., 2016). Organics including phenols, anilines, and dyes can be oxidized by laccase, which is a multicopper oxidoreductase enzyme. It is normally set up in plants and fungi. Further, its catalytic reaction structures slight catalytic condition, low energy consumption, little pollution and high selectivity. Commercial applications in water treatment are remark these specifications (Majeau, Brar, & Tyagi, 2010). Enzyme immobilization is regarded as a hotspot, and supports immobilize enzymes. Thus, they can be simply practiced, stored and recycled (Chen et al., 2018). Attributable to high specific surface area, tunable porosity, excellent chemical/thermal stability and little dispersal restriction, MOFs are new candidates so that enzymes are immobilized (Cui et al., 2018). Pang et al. synthesized a micromesoporous Zr-MOF so as to immobilize laccase. After immobilization, the stability and reusability of laccase were both improved. The biomacromolecules and MOFs are bridged to one another by functional groups. Wei et al. (2015) synthesized Fe_3O_4-COOH@MIL-101 so that protein biomarkers from bacterial cell lysates are selectively improved.

8.2.7 Clay nanoparticles

Nanoclays with an ultralarge interfacial area per volume to form nanocomposites are identified as excessive NPs have paved the way to enhance the biodegradation and other enactments of materials. One of the main gains of clay nanocomposites is their enhanced barrier properties when their flexibility and optical clarity of the pure polymer are still preserved. The features of the matrix polymer are developed with the contribution of a larger interfacial region along with proper organic modification. Other thermoplastic/biodegradable polymers, such as polycarbonate (Balsamo et al., 2001), polysaccharides (Ciardelli et al., 2005), and chitosan (Wu, 2005) are grafted or blended with each other to improve the mechanical and thermal features of polycaprolacton (PCL). Furthermore, complete details on the crystal structure and structural motifs of melt crystallized PCL in the presence of nanoclay (Homminga et al., 2006) have been provided. The limited weight percentage of different NPs comprising organically modified layered silicates have developed the mechanical (Lepoittevin et al., 2003), thermal (Chrissafis et al., 2007), and gas barrier properties (Gorrasi et al., 2003) of the matrix PCL. The thermal (Chrissafis et al., 2007), environmental (Rosa, Filho, Chui, Calil, & Guedes, 2003), and enzymatic (Sanchez, Tsuchii, & Tokiwa, 2000) degradation of pure PCL have also been considered throughout the last two decades. The enzymatic degradation rates of pure PCL are not as fast as those of nanocomposites. In the presence of organically modified nanoclay, the matrix PCL degrades at a measured higher rate. It is also important to ponder the controlled biodegradation. To put it in the nutshell, it is essential to reduce or boost the degree of biodegradation of polymer aside from its natural degradation. Considering the fact that inorganic layered silicate minerals, or nanoclays are commercially available, they are not costly and they include other benefits such as significant property enhancement and relatively simple processability, they are identified as one of the suitable groups of nanofillers used to produce BNCs (De Azeredo, 2009). Another clay is Laponite, which probably subject to unique features even though not much research has been done to mull over it as it is able to develop PLA-based nanocomposites. Laponite is identified as a hectorite clay, which is completely synthesized. It is one of the subdivisions of smectite phyllosilicate minerals. Laponite is capable of swelling and exfoliating (Aouada, Mattoso, & Longo, 2013). The applications of synthetic clays like Laponite have several benefits including high structural regularity, single layer dispersions of NPs, and low level of impurities.

8.3 Inference and future prospects

Due to the heavy mass of pollutant wastes of organic materials and a reduced amount of influential procedures for its management, organic compounds degradation has become an essential concern that needs considering. Henceforth, scientists have concluded NPs the ones which are able to accelerate degradation. It has been demonstrated that the biodegradation enhancement could be accessible using NPs like nanobarium titanate, fullerene 60, and supermagnetic iron oxide. Moreover, the doped TiO_2 and undoped TiO_2 NPs are known as two of the most common NPs in biodegradation of the organic compounds by photocatalytic activity in the presence of visible light. Although TiO_2 has been indicated to be hazardous for the environment, it is economic and efficient to enhance the degradation. Therefore some processes have been designed to decline environmental perils. They have been demonstrated to be somewhat applied in antioxidants, altering surface properties, larger-sized particles, and a better fixation in the matrix. Although NPs have been identified as a new field of science, they have not been profoundly scrutinized. Henceforth, this issue has been heeded by several researchers. Since biodegradation enhancers of organic compounds can serve as a nonthreatening solution, the effects of NPs have been extensively applied across various fields.

References

Alaee, M., et al. (2003). An overview of commercially used brominated flame retardants, their applications, their use patterns in different countries/regions and possible modes of release. *Environment International, 29*(6), 683–689.

An, Y., et al. (2010). Effect of bimetallic and polymer-coated Fe nanoparticles on biological denitrification. *Bioresource Technology, 101*(24), 9825–9828.

Ando, H., et al. (2015). Biodegradation of a poly(ε-caprolactone-*co*-L-lactide)–visible-light-sensitive TiO_2 composite with an on/off biodegradation function. *Polymer Degradation and Stability, 114*, 65–71.

Aouada, F. A., Mattoso, L. H., & Longo, E. (2013). A simple procedure for the preparation of laponite and thermoplastic starch nanocomposites: Structural, mechanical, and thermal characterizations. *Journal of Thermoplastic Composite Materials, 26*(1), 109–124.

Asghar, W., et al. (2011). Comparative solid phase photocatalytic degradation of polythene films with doped and undoped TiO_2 nanoparticles. *Journal of Nanomaterials*, 2011.

Auwerter, L. C. C., et al. (2017). Effects of nanosized titanium dioxide (TiO_2) and fullerene (C60) on wastewater microorganisms activity. *Journal of Water Process Engineering, 16*, 35–40.

Baborová, P., et al. (2006). Purification of a new manganese peroxidase of the white-rot fungus *Irpex lacteus*, and degradation of polycyclic aromatic hydrocarbons by the enzyme. *Research in Microbiology, 157*(3), 248–253.

Baida, H., et al. (2009). Quantitative determination of the size dependence of surface plasmon resonance damping in single Ag@SiO$_2$ nanoparticles. *Nano Letters, 9*(10), 3463–3469.

Baig, N., et al. (2019). Graphene-based adsorbents for the removal of toxic organic pollutants: A review. *Journal of Environmental Management, 244*, 370–382.

Balsamo, V., et al. (2001). Thermal characterization of polycarbonate/polycaprolactone blends. *Journal of Polymer Science Part B: Polymer Physics, 39*(7), 771–785.

Bandara, J., et al. (1997). Chemical (photo-activated) coupled biological homogeneous degradation of *p*-nitro-*o*-toluene-sulfonic acid in a flow reactor. *Journal of Photochemistry and Photobiology A: Chemistry, 111*(1–3), 253–263.

Bayazit, Ş. S., et al. (2017). Preparation of magnetic MIL-101 (Cr) for efficient removal of ciprofloxacin. *Environmental Science and Pollution Research, 24*(32), 25452–25461.

Belháčová, L., et al. (1999). Inactivation of microorganisms in a flow-through photoreactor with an immobilized TiO$_2$ layer. *Journal of Chemical Technology & Biotechnology: International Research in Process, Environmental & Clean Technology, 74*(2), 149–154.

Bikiaris, D. N. (2013). Nanocomposites of aliphatic polyesters: An overview of the effect of different nanofillers on enzymatic hydrolysis and biodegradation of polyesters. *Polymer Degradation and Stability, 98*(9), 1908–1928.

Bilal, M., et al. (2016a). Chitosan beads immobilized manganese peroxidase catalytic potential for detoxification and decolorization of textile effluent. *International Journal of Biological Macromolecules, 89*, 181–189.

Bilal, M., et al. (2016b). Mutagenicity and cytotoxicity assessment of biodegraded textile effluent by Ca-alginate encapsulated manganese peroxidase. *Biochemical Engineering Journal, 109*, 153–161.

Blázquez, E., Gabriel, D., Baeza, J. A., & Guisasola, A. (2017). Evaluation of key parameters on simultaneous sulfate reduction and sulfide oxidation in an autotrophic biocathode. *Water Research, 123*, 301–310. Available from https://doi.org/10.1016/j.watres.2017.06.050.

Bogart, L. K., et al. (2014). *Nanoparticles for imaging, sensing, and therapeutic intervention*. ACS Publications.

Bosetti, M., et al. (2002). Silver coated materials for external fixation devices: In vitro biocompatibility and genotoxicity. *Biomaterials, 23*(3), 887–892.

Brar, S. K., et al. (2010). Engineered nanoparticles in wastewater and wastewater sludge–Evidence and impacts. *Waste Management, 30*(3), 504–520.

Buzarovska, A., & Grozdanov, A. (2012). Biodegradable poly(L-lactic acid)/TiO$_2$ nanocomposites: Thermal properties and degradation. *Journal of Applied Polymer Science, 123*(4), 2187–2193.

Caballero-Díaz, E., et al. (2013). The toxicity of silver nanoparticles depends on their uptake by cells and thus on their surface chemistry. *Particle & Particle Systems Characterization, 30*(12), 1079–1085.

Cabana, H., et al. (2007). Elimination of endocrine disrupting chemicals nonylphenol and bisphenol A and personal care product ingredient triclosan using enzyme preparation from the white rot fungus *Coriolopsis polyzona*. *Chemosphere, 67*(4), 770–778.

Cajthaml, T. (2015). Biodegradation of endocrine-disrupting compounds by ligninolytic fungi: Mechanisms involved in the degradation. *Environmental Microbiology, 17*(12), 4822–4834.

Chandra, V., et al. (2010). Water-dispersible magnetite-reduced graphene oxide composites for arsenic removal. *ACS Nano, 4*(7), 3979–3986.

Chen, C., et al. (2018). Spacer arm-facilitated tethering of laccase on magnetic polydopamine nanoparticles for efficient biocatalytic water treatment. *Chemical Engineering Journal, 350*, 949–959.

Chen, G.-C., et al. (2008). Effects of copper, lead, and cadmium on the sorption and desorption of atrazine onto and from carbon nanotubes. *Environmental Science & Technology, 42*(22), 8297–8302.

Chrissafis, K., et al. (2007). Comparative study of the effect of different nanoparticles on the mechanical properties and thermal degradation mechanism of in situ prepared poly(ε-caprolactone) nanocomposites. *Composites Science and Technology, 67*(10), 2165–2174.

Ciardelli, G., et al. (2005). Blends of poly-(ε-caprolactone) and polysaccharides in tissue engineering applications. *Biomacromolecules, 6*(4), 1961–1976.

Connell, D. W., et al. (2009). Chemistry of organic pollutants, including agrochemicals. *Environmental and Ecological Chemistry, 6*, 181.

Cui, J., et al. (2018). Optimization protocols and improved strategies for metal-organic frameworks for immobilizing enzymes: Current development and future challenges. *Coordination Chemistry Reviews, 370*, 22–41.

Dahlen, E. P., & Rittmann, B. E. (2002). Two-tank suspended growth process for accelerating the detoxification kinetics of hydrocarbons requiring initial monooxygenation reactions. *Biodegradation, 13*(2), 101–116.

Daughton, C. G., & Ternes, T. A. (1999). Pharmaceuticals and personal care products in the environment: Agents of subtle change? *Environmental Health Perspectives, 107*(Suppl. 6), 907–938.

De Azeredo, H. M. (2009). Nanocomposites for food packaging applications. *Food Research International, 42*(9), 1240–1253.

Dong, Z., et al. (2019). Carbon nanoparticle-stabilized pickering emulsion as a sustainable and high-performance interfacial catalysis platform for enzymatic esterification/transesterification. *ACS Sustainable Chemistry & Engineering, 7*(8), 7619–7629.

Du, Y., et al. (2011). Photodecomposition of 4-chlorophenol by reactive oxygen species in UV/air system. *Journal of Hazardous Materials, 186*(1), 491–496.

El-Sheshtawy, H., & Ahmed, W. (2017). Bioremediation of crude oil by *Bacillus licheniformis* in the presence of different concentration nanoparticles and produced biosurfactant. *International Journal of Environmental Science and Technology, 14*(8), 1603–1614.

Fan, G., et al. (2020). TiO_2-graphene 3D hydrogel supported on Ni foam for photoelectrocatalysis removal of organic contaminants. *Journal of Nanoscience and Nanotechnology, 20*(4), 2645–2649.

Feliu, N., et al. (2016). In vivo degeneration and the fate of inorganic nanoparticles. *Chemical Society Reviews, 45*(9), 2440–2457.

Feng, S., et al. (2012). Genotoxicity of the sediments collected from Pearl River in China and their polycyclic aromatic hydrocarbons (PAHs) and heavy metals. *Environmental Monitoring and Assessment, 184*(9), 5651–5661.

Feris, K., et al. (2010). Electrostatic interactions affect nanoparticle-mediated toxicity to gram-negative bacterium *Pseudomonas aeruginosa* PAO1. *Langmuir: The ACS Journal of Surfaces and Colloids, 26*(6), 4429–4436.

Fukuda, N., & Tsuji, H. (2005). Physical properties and enzymatic hydrolysis of poly(L-lactide)–TiO_2 composites. *Journal of Applied Polymer Science, 96*(1), 190–199.

Furukawa, H., et al. (2013). The chemistry and applications of metal-organic frameworks. *Science (New York, NY), 341*(6149).

García-Ripoll, A., et al. (2007). Increased biodegradability of UltracidTM in aqueous solutions with solar TiO_2 photocatalysis. *Chemosphere, 68*(2), 293–300.

Geim, A., & Novoselov, K. (2010). The rise of graphene, 2007. *Nature Materials, 6*, 183.

Gitipour, A., et al. (2013). The impact of silver nanoparticles on the composting of municipal solid waste. *Environmental Science & Technology, 47*(24), 14385–14393.

Golet, E. M., et al. (2003). Environmental exposure assessment of fluoroquinolone antibacterial agents from sewage to soil. *Environmental Science & Technology, 37*(15), 3243–3249.

Gorrasi, G., et al. (2003). Vapor barrier properties of polycaprolactone montmorillonite nanocomposites: Effect of clay dispersion. *Polymer, 44*(8), 2271–2279.

Guobin, S., et al. (2005). Improvement of biodesulfurization rate by assembling nanosorbents on the surfaces of microbial cells. *Biophysical Journal, 89*(6), L58–L60.

Habs, M., Jahn, S., & Schmähl, D. (1984). Carcinogenic activity of condensate from coloquint seeds (*Citrullus colocynthis*) after chronic epicutaneous administration to mice. *Journal of Cancer Research and Clinical Oncology, 108*(1), 154–156.

Hagemann, N., et al. (2018). Effect of biochar amendment on compost organic matter composition following aerobic composting of manure. *Science of the Total Environment, 613*, 20–29.

Häggblom, M. M. (1992). Microbial breakdown of halogenated aromatic pesticides and related compounds. *FEMS Microbiology Reviews, 9*(1), 29–71.

He, S., et al. (2011). The impact of iron oxide magnetic nanoparticles on the soil bacterial community. *Journal of Soils and Sediments, 11*(8), 1408–1417.

Hesham, A. E.-L., et al. (2006). Isolation and identification of a yeast strain capable of degrading four and five ring aromatic hydrocarbons. *Annals of Microbiology, 56*(2), 109.

Hirano, T., et al. (2000). Degradation of Bisphenol A by the lignin-degrading enzyme, manganese peroxidase, produced by the white-rot basidiomycete, *Pleurotus ostreatus. Bioscience, Biotechnology, and Biochemistry, 64*(9), 1958–1962.

Holladay, S., Bevelhimer, M., & Brandt, C. (1994). Quality assurance/quality control summary report for Phase 1 of the Clinch River remedial investigation. In: *Environmental restoration program*. Oak Ridge National Lab.

Homminga, D., et al. (2006). Crystallization behavior of polymer/montmorillonite nanocomposites. Part II. Intercalated poly(ε-caprolactone)/montmorillonite nanocomposites. *Polymer, 47*(5), 1620–1629.

Hou, L., et al. (2013). Removal of ZnO nanoparticles in simulated wastewater treatment processes and its effects on COD and NH_4^+-N reduction. *Water Science and Technology, 67*(2), 254–260.

Huang, Y., Fulton, A. N., & Keller, A. A. (2016). Simultaneous removal of PAHs and metal contaminants from water using magnetic nanoparticle adsorbents. *Science of the Total Environment, 571*, 1029–1036.

Hübner, R., Astin, K. B., & Herbert, R. J. (2010). 'Heavy metal'—Time to move on from semantics to pragmatics? *Journal of Environmental Monitoring, 12*(8), 1511–1514.

Ibrahim, H. M. (2018). Characterization of biosurfactants produced by novel strains of *Ochrobactrum anthropi* HM-1 and *Citrobacter freundii* HM-2 from used engine oil-contaminated soil. *Egyptian Journal of Petroleum, 27*(1), 21–29.

Iqbal, M., & Khera, R. A. (2015). Adsorption of copper and lead in single and binary metal system onto *Fumaria indica* biomass. *Chemistry International, 1*(3), 157b–163b.

Jiang, X., et al. (2015). Fast intracellular dissolution and persistent cellular uptake of silver nanoparticles in CHO-K1 cells: Implication for cytotoxicity. *Nanotoxicology, 9*(2), 181–189.

Jiang, X., Zeng, Q., & Yu, A. (2007). Thiol-frozen shape evolution of triangular silver nanoplates. *Langmuir: The ACS Journal of Surfaces and Colloids, 23*(4), 2218–2223.

Jurado, M., et al. (2015). Enhanced turnover of organic matter fractions by microbial stimulation during lignocellulosic waste composting. *Bioresource Technology, 186*, 15–24.

Kadam, A. A., Jang, J., & Lee, D. S. (2017). Supermagnetically tuned halloysite nanotubes functionalized with aminosilane for covalent laccase immobilization. *ACS Applied Materials & Interfaces, 9*(18), 15492–15501.

Kapri, A., et al. (2010). SPION-accelerated biodegradation of low-density polyethylene by indigenous microbial consortium. *International Biodeterioration & Biodegradation, 64*(3), 238–244.

Kasprzyk-Hordern, B., Dinsdale, R. M., & Guwy, A. J. (2009). The removal of pharmaceuticals, personal care products, endocrine disruptors and illicit drugs during wastewater treatment and its impact on the quality of receiving waters. *Water Research, 43*(2), 363–380.

Klupp, G., Margadonna, S., & Prassides, K. (2016). *Fullerenes. Reference module in materials science and materials engineering.* Elsevier.

Kowit, S., et al. (2012). Photo catalytic oxidation activity of carbon supported nano TiO_2-LDPE film. *Journal of Environmental Research and Development, 7*, 45–50.

Kümmerer, K., et al. (2011). Biodegradability of organic nanoparticles in the aqueous environment. *Chemosphere, 82*(10), 1387–1392.

Lepoittevin, B., et al. (2003). Polymer/layered silicate nanocomposites by combined intercalative polymerization and melt intercalation: A masterbatch process. *Polymer, 44*(7), 2033–2040.

Li, Z., et al. (2020). Improved performance of immobilized laccase on Fe_3O_4@C-Cu^{2+} nanoparticles and its application for biodegradation of dyes. *Journal of Hazardous Materials, 399*, 123088.

Lien, H.-L., & Zhang, W.-x (2001). Nanoscale iron particles for complete reduction of chlorinated ethenes. *Colloids and Surfaces A: Physicochemical and Engineering Aspects, 191*(1–2), 97–105.

Lin, C., Gan, L., & Chen, Z.-L. (2010). Biodegradation of naphthalene by strain *Bacillus fusiformis* (BFN). *Journal of Hazardous Materials, 182*(1–3), 771–777.

Lin, S.-H., & Juang, R.-S. (2009). Adsorption of phenol and its derivatives from water using synthetic resins and low-cost natural adsorbents: A review. *Journal of Environmental Management, 90*(3), 1336–1349.

Liu, S., et al. (2009). Sharper and faster "nano darts" kill more bacteria: A study of antibacterial activity of individually dispersed pristine single-walled carbon nanotube. *ACS Nano, 3*(12), 3891–3902.

Luo, C., et al. (2011). Heavy metal contamination in soils and vegetables near an e-waste processing site, South China. *Journal of Hazardous Materials, 186*(1), 481–490.

Lynch, I., Salvati, A., & Dawson, K. A. (2009). What does the cell see? *Nature Nanotechnology, 4*(9), 546–547.

Ma, H., Williams, P. L., & Diamond, S. A. (2013). Ecotoxicity of manufactured ZnO nanoparticles—A review. *Environmental Pollution, 172*, 76–85.

Majeau, J.-A., Brar, S. K., & Tyagi, R. D. (2010). Laccases for removal of recalcitrant and emerging pollutants. *Bioresource Technology, 101*(7), 2331–2350.

Marco, A., Esplugas, S., & Saum, G. (1997). How and why combine chemical and biological processes for wastewater treatment. *Water Science and Technology, 35*(4), 321–327.

Marsolek, M. D., et al. (2008). Intimate coupling of photocatalysis and biodegradation in a photocatalytic circulating-bed biofilm reactor. *Biotechnology and Bioengineering, 101*(1), 83–92.

Mielgo, I., et al. (2003). Oxidative degradation of azo dyes by manganese peroxidase under optimized conditions. *Biotechnology Progress, 19*(2), 325–331.

Miyauchi, M., Li, Y., & Shimizu, H. (2008). Enhanced degradation in nanocomposites of TiO_2 and biodegradable polymer. *Environmental Science & Technology, 42*(12), 4551–4554.

Monopoli, M. P., et al. (2011). Physical – chemical aspects of protein corona: Relevance to in vitro and in vivo biological impacts of nanoparticles. *Journal of the American Chemical Society, 133*(8), 2525–2534.

Nouren, S., et al. (2017). By-product identification and phytotoxicity of biodegraded Direct Yellow 4 dye. *Chemosphere, 169*, 474–484.

Olaniran, A. O., & Igbinosa, E. O. (2011). Chlorophenols and other related derivatives of environmental concern: Properties, distribution and microbial degradation processes. *Chemosphere, 83*(10), 1297–1306.

Olsson, B., & Ögren, L. (1983). Optimization of peroxidase immobilization and of the design of packed-bed enzyme reactors for flow injection analysis. *Analytica Chimica Acta, 145*, 87–99.

Patterson, D. P., Prevelige, P. E., & Douglas, T. (2012). Nanoreactors by programmed enzyme encapsulation inside the capsid of the bacteriophage P22. *ACS Nano, 6*(6), 5000−5009.

Petersen, E. J., et al. (2011). Potential release pathways, environmental fate, and ecological risks of carbon nanotubes. *Environmental Science & Technology, 45*(23), 9837−9856.

Ponder, S. M., Darab, J. G., & Mallouk, T. E. (2000). Remediation of Cr (VI) and Pb (II) aqueous solutions using supported, nanoscale zero-valent iron. *Environmental Science & Technology, 34*(12), 2564−2569.

Prado, N., Ochoa, J., & Amrane, A. (2009). Biodegradation by activated sludge and toxicity of tetracycline into a semi-industrial membrane bioreactor. *Bioresource Technology, 100*(15), 3769−3774.

Qiu, X., et al. (2020). Laccase immobilized on magnetic nanoparticles modified by amino-functionalized ionic liquid via dialdehyde starch for phenolic compounds biodegradation. *Chemical Engineering Journal, 391*, 123564.

Rasool, K., & Lee, D. S. (2016). Effect of ZnO nanoparticles on biodegradation and biotransformation of co-substrate and sulphonated azo dye in anaerobic biological sulfate reduction processes. *International Biodeterioration & Biodegradation, 109*, 150−156.

Ren, X., et al. (2018). Effect of exogenous carbonaceous materials on the bioavailability of organic pollutants and their ecological risks. *Soil Biology and Biochemistry, 116*, 70−81.

Rodrigues, D. F., Jaisi, D. P., & Elimelech, M. (2013). Toxicity of functionalized single-walled carbon nanotubes on soil microbial communities: Implications for nutrient cycling in soil. *Environmental Science & Technology, 47*(1), 625−633.

Rodriguez, J. A. (2006). The chemical properties of bimetallic surfaces: Importance of ensemble and electronic effects in the adsorption of sulfur and SO_2. *Progress in Surface Science, 81*(4), 141−189.

Rosa, D., Filho, R. P., Chui, Q. S. H., Calil, M. R., & Guedes, C. G. F. (2003). *European Polymer Journal, 39*(2), 233−237.

Rossi, L. M., et al. (2014). Magnetic nanomaterials in catalysis: Advanced catalysts for magnetic separation and beyond. *Green Chemistry, 16*(6), 2906−2933.

Saber-Samandari, S., et al. (2017). Adsorption of anionic and cationic dyes from aqueous solution using gelatin-based magnetic nanocomposite beads comprising carboxylic acid functionalized carbon nanotube. *Chemical Engineering Journal, 308*, 1133−1144.

Safaei-Ghomi, J., & Ghasemzadeh, M. A. (2017). Zinc oxide nanoparticle promoted highly efficient one pot three-component synthesis of 2,3-disubstituted benzofurans. *Arabian Journal of Chemistry, 10*, S1774−S1780.

Sah, A., et al. (2010). Implications of fullerene-60 upon in-vitro LDPE biodegradation. *Journal of Microbiology and Biotechnology, 20*(5), 908.

Sanchez, J. G., Tsuchii, A., & Tokiwa, Y. (2000). Degradation of polycaprolactone at 50°C by a thermotolerant *Aspergillus* sp. *Biotechnology Letters, 22*(10), 849−853.

Scherer, T. M., et al. (1999). Hydrolase activity of an extracellular depolymerase from *Aspergillus fumigatus* with bacterial and synthetic polyesters. *Polymer Degradation and Stability, 64*(2), 267−275.

Scott, J. P., & Ollis, D. F. (1996). Engineering models of combined chemical and biological processes. *Journal of Environmental Engineering, 122*(12), 1110−1114.

Scott, J. P., & Ollis, D. F. (1995). Integration of chemical and biological oxidation processes for water treatment: Review and recommendations. *Environmental Progress, 14*(2), 88−103.

Seo, J.-S., et al. (2007). Isolation and characterization of bacteria capable of degrading polycyclic aromatic hydrocarbons (PAHs) and organophosphorus pesticides from PAH-contaminated soil in Hilo, Hawaii. *Journal of Agricultural and Food Chemistry, 55*(14), 5383−5389.

Seow, Z., et al. (2008). Controlled synthesis and application of ZnO nanoparticles, nanorods and nanospheres in dye-sensitized solar cells. *Nanotechnology, 20*(4), 045604.

Shin, K.-H., & Cha, D. K. (2008). Microbial reduction of nitrate in the presence of nanoscale zero-valent iron. *Chemosphere, 72*(2), 257–262.

Su, C.-K., et al. (2014). Quantitatively profiling the dissolution and redistribution of silver nanoparticles in living rats using a knotted reactor-based differentiation scheme. *Analytical Chemistry, 86*(16), 8267–8274.

Su, X., et al. (2018). Revealing potential functions of VBNC bacteria in polycyclic aromatic hydrocarbons biodegradation. *Letters in Applied Microbiology, 66*(4), 277–283.

Su, Y., et al. (2020). Towards a simultaneous combination of ozonation and biodegradation for enhancing tetracycline decomposition and toxicity elimination. *Bioresource Technology, 304*, 123009.

Sun, H., et al. (2017). Effect of Fe_3O_4 nanoparticles on *Sphingobium yanoikuyae* XLDN2-5 cells in carbazole biodegradation. *Nanotechnology for Environmental Engineering, 2*(1), 5.

Tasis, D., et al. (2006). Chemistry of carbon nanotubes. *Chemical Reviews, 106*(3), 1105–1136.

Thavamani, P., Megharaj, M., & Naidu, R. (2012). Multivariate analysis of mixed contaminants (PAHs and heavy metals) at manufactured gas plant site soils. *Environmental Monitoring and Assessment, 184*(6), 3875–3885.

Torres, E., Bustos-Jaimes, I., & Le Borgne, S. (2003). Potential use of oxidative enzymes for the detoxification of organic pollutants. *Applied Catalysis B: Environmental, 46*(1), 1–15.

Towell, M. G., et al. (2011). Impact of carbon nanomaterials on the behaviour of 14C-phenanthrene and 14C-benzo-[a] pyrene in soil. *Environmental Pollution, 159*(3), 706–715.

Tuor, U., et al. (1992). Oxidation of phenolic arylglycerol β-aryl ether lignin model compounds by manganese peroxidase from *Phanerochaete chrysosporium*: Oxidative cleavage of an α-carbonyl model compound. *Biochemistry, 31*(21), 4986–4995.

Upadhyayula, V. K., et al. (2009). Application of carbon nanotube technology for removal of contaminants in drinking water: A review. *Science of the Total Environment, 408*(1), 1–13.

Vadodaria, S. S., Onyianta, A. J., & Sun, D. (2018). High-shear rate rheometry of micro-nanofibrillated cellulose (CMF/CNF) suspensions using rotational rheometer. *Cellulose, 25*(10), 5535–5552.

Vecitis, C. D., et al. (2010). Electronic-structure-dependent bacterial cytotoxicity of single-walled carbon nanotubes. *ACS Nano, 4*(9), 5471–5479.

Wang, T., et al. (2016). Facile fabrication of Fe_3O_4/MIL-101 (Cr) for effective removal of acid red 1 and orange G from aqueous solution. *Chemical Engineering Journal, 295*, 403–413.

Wang, X., et al. (2007). Degradation of carbazole by microbial cells immobilized in magnetic gellan gum gel beads. *Applied and Environmental Microbiology, 73*(20), 6421–6428.

Wang, Y., et al. (2020). Biodegradation of tetracycline using hybrid material (UCPs-TiO_2) coupled with biofilms under visible light. *Bioresource Technology*, 124638.

Wang, Y., et al. (2018). How to construct DNA hydrogels for environmental applications: Advanced water treatment and environmental analysis. *Small (Weinheim an der Bergstrasse, Germany), 14*(17), 1703305.

Wang, Y., Westerhoff, P., & Hristovski, K. D. (2012). Fate and biological effects of silver, titanium dioxide, and C60 (fullerene) nanomaterials during simulated wastewater treatment processes. *Journal of Hazardous Materials, 201–202*, 16–22.

Wei, J.-P., et al. (2015). Synthesis of magnetic framework composites for the discrimination of *Escherichia coli* at the strain level. *Analytica Chimica Acta, 868*, 36–44.

Wei, W., et al. (2013). Construction of robust enzyme nanocapsules for effective organophosphate decontamination, detoxification, and protection. *Advanced Materials, 25*(15), 2212–2218.

Wu, C.-S. (2005). A comparison of the structure, thermal properties, and biodegradability of polycaprolactone/chitosan and acrylic acid grafted polycaprolactone/chitosan. *Polymer, 46*(1), 147–155.

Wu, Y., Liu, J., & Rene, E. R. (2018). Periphytic biofilms: A promising nutrient utilization regulator in wetlands. *Bioresource Technology, 248*, 44–48.

Xing, Y., et al. (2005). A spatial temporal assessment of pollution from PCBs in China. *Chemosphere, 60*(6), 731–739.

Xiong, H., et al. (2018). Roles of an easily biodegradable co-substrate in enhancing tetracycline treatment in an intimately coupled photocatalytic-biological reactor. *Water Research, 136*, 75–83.

Xiu, Z.-m, et al. (2010). Effects of nano-scale zero-valent iron particles on a mixed culture dechlorinating trichloroethylene. *Bioresource Technology, 101*(4), 1141–1146.

Yan, X., et al. (2008). Adsorption and desorption of atrazine on carbon nanotubes. *Journal of Colloid and Interface Science, 321*(1), 30–38.

Yang, C., et al. (2010). Antimicrobial activity of single-walled carbon nanotubes: Length effect. *Langmuir: The ACS Journal of Surfaces and Colloids, 26*(20), 16013–16019.

Yang, F., et al. (2017a). Effects of biochars and MWNTs on biodegradation behavior of atrazine by *Acinetobacter lwoffii* DNS32. *Science of the Total Environment, 577*, 54–60.

Yang, Z., et al. (2017b). Accelerated ciprofloxacin biodegradation in the presence of magnetite nanoparticles. *Chemosphere, 188*, 168–173.

Zainudin, N. F., Abdullah, A. Z., & Mohamed, A. R. (2010). Characteristics of supported nano-TiO_2/ZSM-5/silica gel (SNTZS): Photocatalytic degradation of phenol. *Journal of Hazardous Materials, 174*(1–3), 299–306.

Zeng, G., et al. (2018). Pathway and mechanism of nitrogen transformation during composting: Functional enzymes and genes under different concentrations of PVP-AgNPs. *Bioresource Technology, 253*, 112–120.

Zhang, W., et al. (2020). Effects of carbon nanotubes on biodegradation of pollutants: Positive or negative? *Ecotoxicology and Environmental Safety, 189*, 109914.

Zhang, W.-x (2003). Nanoscale iron particles for environmental remediation: An overview. *Journal of Nanoparticle Research, 5*(3–4), 323–332.

Zhao, X., & Liu, R. (2012). Recent progress and perspectives on the toxicity of carbon nanotubes at organism, organ, cell, and biomacromolecule levels. *Environment International, 40*, 244–255.

Zhu, Y., et al. (2018). Phosphorus and Cu^{2+} removal by periphytic biofilm stimulated by upconversion phosphors doped with Pr^{3+}-Li^+. *Bioresource Technology, 248*, 68–74.

Zuo, X., et al. (2009). Design of a carbon nanotube/magnetic nanoparticle-based peroxidase-like nanocomplex and its application for highly efficient catalytic oxidation of phenols. *Nano Research, 2*(8), 617–623.

Biodegradation of plastic-based waste materials

Nihan Uçar[1,2], Sabah Bakhtiari[3], Esmail Doustkhah[4], Masoud Yarmohammadi[3], Mona Zamani Pedram[3], Elif Alyamaç[5], M. Özgür Seydibeyoğlu[1,2]

[1]DEPARTMENT OF MATERIALS SCIENCE AND ENGINEERING, IZMIR KATIP CELEBI UNIVERSITY, IZMIR, TURKEY [2]FILINIA R&D, EGE UNIVERSITY SCIENCEPARK, IZMIR, TURKEY [3]FACULTY OF MECHANICAL ENGINEERING-ENERGY DIVISION, K.N. TOOSI UNIVERSITY OF TECHNOLOGY, TEHRAN, IRAN [4]INTERNATIONAL CENTRE FOR MATERIALS NANOARCHITECTONICS (MANA), NATIONAL INSTITUTE FOR MATERIALS SCIENCE (NIMS), IBARAKI, JAPAN [5]DEPARTMENT OF PETROLEUM AND NATURAL GAS ENGINEERING, IZMIR KATIP ÇELEBI UNIVERSITY, IZMIR, TURKEY

9.1 Introduction

Inevitably, the terms "materials" and "products" differ from one another. To comprehend their potentiality in terms of mechanical, chemical, physical, and biological properties, the specifications of basic materials are verified in the laboratory. Then, materials are changed or contrived into products. As a definite form of a product, its thickness, shape, its inclination to migrate from one section to another, and so forth are regarded as the main extrinsic features; they are so essential in view of littering. However, the fundamental features of materials should be also pondered. Notably, they are supposed to be preserved while changing into final product procedures (Vert et al., 2012). The fact is that plastics are usually organic compounds that are synthetic or semisynthetic. They hold high molecular mass and along chain polymeric molecules is included in them (Shah, Hasan, Hameed, & Ahmed, 2008). Holistically, plastics embrace side-linked molecular groups and some organic and inorganic blends added as additives, plasticizers, fillers, and a main chain organic link (Zheng, Yanful, & Bassi, 2005). Elements such as carbon, hydrogen, nitrogen, oxygen, chlorine, and bromine are the building blocks of plastics. Plastics subject to important gains from the viewpoints of weight, durability, and low cost in terms of manufacture comparison with other materials. Since plastics can be molded into various size and shapes, they can be extensively practiced (Zheng et al., 2005). Plastics have been exceedingly applied in every aspect of life and technology. Polyethylene (PE), polystyrene (PS), polyvinyl chloride (PVC), polypropylene (PP), and PE terephthalatete are among the plastics that are mostly put into practice (Mohee & Unmar, 2007). Plastics are vastly applied in every field such as health care industries, building constructions, transportations, households, and for disposals items, such as wraps, cups, spoons, trays plates, and so forth (Sudesh & Iwata, 2008). Notably, all the mostly used plastics cannot be recyclable. As a result, plastics are accumulated in landfills. Then, they are discharged in water. Henceforth, they do not decompose in the environment. Thus, environment and its varied ecosystems are endangered. One of the main global concerns is plastic pollution in the sea. They are found everywhere. In every square kilometer, almost 580,000 plastic pieces can be concentrated (Willis, Maureaud, Wilcox, & Hardesty, 2018). Between 1000 and 3000 t of plastic floating in the Mediterranean Sea has been reported. This contamination is due to the physical/mechanical degradation of macroplastics in landfills that produce microplastics that are transferred directly to rivers, seas, and oceans via leachate. Also, additives found in plastic formulations are not normally covalently bonded to polymer chains and can therefore leak from plastics and enter the marine environment. Plastics that enter the marine environment can also absorb persistent organic pollutants due to the hydrophobic properties of these compounds, or they can be a vector agent for the spread of harmful organisms living on plastic surfaces (Avio, Gorbi, & Regoli, 2017).

Although plastics are not inactive, they comprise several toxic chemicals. They are capable of dispersing in the form of microplastic contaminants. Plastics consist of chemicals which bring about chronic respiratory illnesses and other health difficulties. Dioxins bring about cancer and neurological harm and damage the progress of reproductive systems because the dioxins freed from plastic polymers are dreadful organic pollutants (Kavlock et al., 2002). Several procedures and methods are put into practice so as to overcome this problem. Consequently, many local industries have benefited the additives to manufacture various types of degradable plastic bags or somewhat degradable bags. Numerous industries have used carbohydrates to synthesize plastics, which can be degraded under UV radiation. Furthermore, microorganisms degrade them (Mohee & Unmar, 2007). The high CH linkage causes the strength of plastic polymers. The biological degradation of plastics is counted as one of the ways to degrade such regenerated plastics (Ludevese-Pascual, Laranja, Amar, Bossier, & De Schryver, 2019). To scrutinize the degradation of plastic film, it is essential to consider both the polymer strength and density. A common system has been designed by the Plastic Bottle Institute so that the disposable products can be diminished and recycled (Sen & Raut, 2015). With the contribution of the technological progress and swift global economic development, plastic supplies stay persistent. Hitherto, lots of plastics like PE and PVC are nonbiodegradable and accordingly, they become accumulation pollutants. For the sake of properties such as exclusive chemical structures, molecular weight, melting temperature, elasticity, modality, and crystalline structures, the degradation procedure is influenced by the plastic polymers (Wang, Liu, Zhao, & Jin, 2020). One of the significant dimensions of the biodegradable processes is the molecular size (Haider, Völker, Kramm, Landfester, & Wurm, 2019). Biodegradable plastics (BPs) are types of plastics which can be disintegrated on account of the activity of microbes such as bacteria and algae. Aliphatic polyesters are biodegradable due to their perchance hydrolysable ester bonds, whereas aromatic polyesters are mainly vulnerable to microbial attacks. The degradation degree of some BPs is remarkable in that they can be completely metabolized by microorganisms into carbon and water (Jain, Jain, Singhai, & Jain, 2017). When being mixed with organic wastes and liquids, conventional plastics can change organic wastes and liquids into hard and impractical ones. Polymers can be recycled without the necessity of costly treatment procedures. Disposing insufficient plastic waste brings about challenges through which flora and fauna are influenced. In comparison with chemical treatments in the Indian context, biological methods are more significant even though the greatest challenge is the cost (Wang & Wan, 2015). Incineration is considered as a combustion procedure that is mainly practiced to manage soil waste. This procedure takes in inactive materials and high organic, which their degrading on burning is rather difficult (Rist, Almroth, Hartmann, & Karlsson, 2018). Microbes are included in the polymer degradation in the course of degradation of plastics. Microbes contribute to degrade natural and synthetic plastics. When the microplastics that contain remains are burned, they subject to through discharging noxious gases (Adrados et al., 2012). A vast load on the degradation process is created for the sake the recalcitrant nature of plastics. Plastic waste disposal becomes an insurmountable challenge when disposal techniques and inefficient regulations are not adequate enough (Vesilind, Peirce, & Weiner, 2013). It is possible to interchangeably use the terms "biodegradation" and "biodegradability." Nevertheless, this is improper. Basically, biodegradation is dealt with a procedure occurring in specific conditions. It can also be measured. As a whole, the term "biodegradability" denotes a property, the probability of a material to undertake a biodegradation procedure. More clearly, it refers to the capability of being degraded by means of biological mediators. It is noteworthy that abrasion is the main cause of degradation. Moreover, other factors, such as light, heat, and so forth can cause it. In such cases, one is dealt with photodegradation, thermodegradation, and so on. The possibility for biodegradation is well-defined by dint of a standardization process. Similar theoretical and standardization methods are applied in all divisions. Hence, it is not specific to the field of BPs. Conclusively, materials can be biodegradable based on a standardized evaluation procedure. The traceability of consequences to an accepted reference measurement system and upsurges transparency of statements can be confirmed by the usage of standardization procedures. A product can be called biodegradable or recyclable just in case the product does not finish up in a disposal infrastructure that uses the biodegradability or recyclability features after being used by the customer. If the recyclable product can be simply composed and transferred to a recycling facility to be distorted into the same or new product, recycling can be efficient. There are materials that cannot be recycled or biodegraded, but recovery of energy (waste to energy) can burn them. It is not recommended to choose landfills as a repository of plastic and organic waste. Nowadays, it is suitable to use landfills as plastic-lined tombs because they can hinder biodegradation due to slight or no moisture and insignificant microbial activity. It should be pointed out that it is not efficient to bury organic waste such as lawn and yard waste, paper, food, BPs, and other inert materials in such landfills. The integration of BPs with disposal infrastructures is divulged in Fig. 9–1. The biodegradable function of the plastic products is used by them. Amid disposal choices, composting is an environmentally

FIGURE 9–1 Different methods in biodegradation of plastics.

sound method to transmit biodegradable waste, such as the new BPs, into useful soil alteration products (Majid, Hori, Akiyama, & Doi, 1994).

9.2 Waste management of plastics

Plastic waste management includes methods and strategies used to recycle plastics that are disposed of in landfills or water bodies or pollute the environment in a different way. With this method, it can both reduce production costs and help protect the environment. Plastic waste management is carried out in accordance with regulations set by local, state, and federal governments and may differ from one region to another.

Plastics play an important role in human life because of their low cost and versatility. However, it is one of the biggest environmental pollutants and takes a long time to break down. When plastics are burned, chemicals that are dangerous to the ecosystem are released. These chemicals mix into the food chain and water as microplastics, causing diseases or damage to human health. Therefore waste management is very important (Rajmohan, Ramya, Viswanathan, & Varjani, 2019).

The physical method aims to reduce the volume of waste such as squeezing, pulverizing, and incineration. Commonly used are UV degradation and photooxidation. However, these methods also create byproducts that harm the environment (Gilpin, Wagel, & Solch, 2003).

In the chemical method, chemicals are used that break the polymeric bond of plastics and make them harmless. But these chemicals also create great waste (Lin & Yang, 2009). The only advantage of incineration is that energy can be supplied in the second process from plastic. However, with the burning of plastics, harmful gases that cause air pollution are released (Singh et al., 2017). Pyrolysis, another thermochemical method, is the burning of solid waste without an oxidizer. Two types are fast pyrolysis for biooil production and slow pyrolysis for charcoal production (Becidan, 2007). In the third method, gasification, the waste is converted into commonly used gases such as hydrogen and carbon dioxide in the presence of oxygen or air at a controlled rate (Nikolaidis & Poullikkas, 2017).

In the biological method, plastics degrade without producing harmful byproducts. It is the transformation of solid wastes into methane and fertilizer by microbial intervention in the absence of oxygen (Shah et al., 2008).

The increase in the world population causes an increase in the amount of plastic waste. In developing countries, plastic consumption is above the world average due to economic development and increasing urbanization rate. Even though landfill is prohibited by the

European Union, 50% of waste is still dumped in landfills. In countries such as Sweden, Germany, Austria, Germany, Denmark, and the Netherlands, the recycling rate of plastic waste is around 80%. Due to the biodegradation of renewable resources such as agricultural waste, it enables more bioplastics to be used unlike conventional plastics. The production of polymers that are sustainable and can be obtained from renewable sources can be recycled, reused, and most importantly, do not harm the environment, is very important for the future of humanity (Sharma & Jain, 2020).

The most important issue in integrated urban solid waste systems is the management of plastic waste. Both material and energy recovery can be achieved with different plans. There are five scenarios defined, P0, P1, P2, P3, and P4. In P0, plastic waste is partially incinerated with energy recovery and partially directed to mechanical biological treatment. In P1, clean plastics are separated for recycling. Mixed plastics are separated for mechanical improvement and separation into certain polymer types in P2. In P3, the mixed plastic part is welded together with metals. Plastic is mechanically separated from waste before it is burned in P4. In the study the most appropriate strategy for the management of plastic waste has been researched. It turned out that neither scenario alone is the best option (Rigamonti et al., 2014).

The large amount of plastic waste dumped into the environment worldwide is causing pollution. Waste plastics that accumulate in the environment can be broken down into small pieces such as microplastics and nanoplastics depending on the weather conditions, which will harm the environment and people more than large plastics. Therefore plastic production and disposal should be considered. BPs have become the focus of recent research due to their potential biodegradability and harmlessness, and this will be the most effective approach to managing the plastic waste environmental deposition problem. However, in the long run, it is unclear whether BPs will be a promising solution for waste disposal and global plastic pollution. Most conventional plastics cannot be replaced with these BPs. Biodegradation of BPs requires certain environmental conditions in the environment that are not always reliable. Also, changes in human behavioral awareness will affect the development and implementation of BPs. BPs should not be viewed as a technical solution, because with the promotion of an effective technology, the garbage doesn't change. Therefore BPs can only be part of the solution. Efficiency in providing environmental solutions for plastic waste management depends on a combination of affordable waste classification technologies and investments in organic waste treatment facilities. Therefore there is still a long way to go to solve global plastic pollution through BPs (Shen et al., 2020).

9.3 Plastics identifications and classifications

Considering the features and chemical structures, different types of plastics can be developed.

9.3.1 Thermal properties

Thermoplastics and thermosetting polymers are the two main classifications of plastics according to the thermal properties of plastics.

9.3.1.1 Thermoplastics properties
When thermoplastics are heated, their chemical composition cannot be changed. However, its polymers can be molded several times. PE, PP, PS, PVC, and polytetrafluoroethylene are among the different types of these polymers. They are remarked as usual plastics and their molecular weight ranges from 20,000 to 500,000 AMU. They include various numbers of repeating units taken from a simple monomer unit (Alshehrei, 2017).

9.3.1.2 Thermosetting polymers
Another classification of plastics is thermosetting polymers. Thermosetting polymers are solid in nature. So, it is not possible to melt or transform them. As a matter of fact, the chemical change is not flexible. Since the structure of the aforementioned plastics is a cross-linked one, while the thermoplastic has a linear structure. In fact, thermosetting polymers cannot be recycled such as phenol–formaldehyde, polyurethanes (PU), and so forth (Gnanavel, Valli, & Thirumarimurugan, 2012).

9.3.2 Design properties
Likewise, plastics are categorized according to their significance to the manufacturing process and design. Many other categorizations have been devised to illustrate the manufacturing process and design of plastics. Electrical conductivity, durability, tensile strength, degradability, and thermal stability are considered as the main parameters in this respect.

9.3.3 Degradability properties
To distinguish degradable and nondegradable polymers from one another, the chemical properties of plastics can be applied as suitable measures (Ghosh, Pal, & Ray, 2013). Nonbiodegradable plastics are usually identified as synthetic ones taken from petrochemicals. Numerous repetitions of small monomer units are detected in them. In fact, their molecular weight is high. Similar to BP, the components of living plants, animals, and algae as source of cellulose, starches, protein, and algal materials are biodegradable because they are produced by renewable resources. Various microorganisms can produce them (Imre & Pukánszky, 2013). Interaction with UV, water, enzymes, and gradual changes in pH are the parameters breaking down the BPs. Photodegradable bioplastics, compostable bioplastics, biobased bioplastics, and biodegradable bioplastics are identified as the four types of biodegradable bioplastics (Arikan & Ozsoy, 2014). The light sensitive groups of photodegradable bioplastic are directly connected to the backbone of the polymer. Their polymer structure can be disintegrated by being exposed to the ultraviolet radiation. Then, they are simply degraded by bacteria. It is noteworthy that sunlight influence landfill and it prevents the degradation of plastics (Singh et al., 2017). Primarily, bioplastics are identified as "plastics." Their carbon content is 100% taken from renewable agricultural and forestry resources, such as corn starch, soybean protein, and cellulose. From a biological point of view, a composting process contributes to decompose the compostable bioplastics. As a result, it takes place at the same rate as other compostable materials. Considerably, no visible toxic remainders are

left. Standardized test is practiced to specify its total biodegradability, its disintegration degree, and the possible ecological toxicity of its degraded materials so that a plastic as biocompostable is designated. Microorganisms entirely degrade the biodegradable bioplastics. Significantly, no visible toxic remainders are left. The term "biogradable" is attributed to materials which can be split or broken down into biogases and biomass (typically carbon dioxide and water). This is resulted by being exposed to humidity and a microbial environment (Jain, Kosta, & Tiwari, 2010).

9.4 Plastics biodegradation

Any physical or chemical change in a material affected by biological activity is well-defined by biodegradation. The degradation of both natural and synthetic plastics holds microorganisms such as bacteria, fungi, and actinomycetes. Plastics are typically biodegraded aerobically in nature, anaerobically in residues and landfills and somewhat aerobically in dung and soil. Within aerobic biodegradation, carbon dioxide, and water are produced, whereas carbon dioxide, water, and methane are produced by anaerobic biodegradation (Ishigaki et al., 2004). When microorganisms change biochemical into compounds, the result of this conversion is biodegradation. Characteristic of plastics, such as its mobility, crystalline, molecular weight, the kind of functional groups, and additives added to the polymers are the main factors providing the conditions of the degradation of plastics. The presence of either microbes or enzymes is essential to carry out this phenomenon. Two phases intervene the biodegradation of high molecular weight polymers (Fig. 9–2). They include (1) depolymerization: to be diffuse across the cell membrane, and (2) cellular metabolism: to integrate the monomeric components (Singh et al., 2011). Microorganisms biologically degrade plastic colonization on the plastic surface. Henceforth, the molecular weight of plastics is decreased. After that, the polymer is changed to its monomers. Then, the monomers analyzed to carbon dioxide and water and methane labeled as mineralization (Shah et al., 2008; Zheng et al., 2005). As some polymers are too large, it is hard for the cellular membrane to pass through. Consequently, it should be depolymerized into smaller monomers. Then, microorganisms absorb it. Researchers have heeded the biogradable plastics because they may include hydrolyzable ester bonds changing into H_2O and CO_2 under aerobic condition. Furthermore, under anaerobic condition in addition to H_2O and CO_2, CH_4 is produced (Fig. 9–2). Compost, microbial, and enzymatic are detected as the three different ways for the degradation of biopolymer. Singh et al. Sharma and Jain (2020) carried out a study and used compost media, microorganism (*Pseudomonas stutzeri*), and pure enzyme to mull over the comparative biodegradation study of poly(hydroxybutyrate-*co*-valerate) and their nanohybrids with layered silicate in terms of percentage weight loss. In comparison with compost for a similar set of samples, it was specified that the biodegradation rate is expressively higher in enzymatic media (proteinase-k and lipase). Additionally, various nanoclays can tune the rate of degradation, that is, the degradation rate can be enhanced by 15A nanoclays, whereas the rate of degradation is diminished by the 30B nanoclay in comparison with the rate of biodegradation of pure

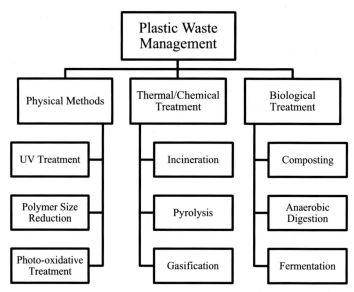

FIGURE 9–2 Technologies for solid waste management (Rajmohan et al., 2019).

PHBV (Polyhydroxybutyrate-co-valerate). This relative rate remains alike in in compost, microorganism and enzymatic media. Lactic acid and lactoyl lactic acid into ethyl ester of lactoyl lactic acid in the biotic environment are quickly changed by microorganisms, whereas it was not noticed in abiotic medium (Hakkarainen, Karlsson, & Albertsson, 2000). As observed by means of gas chromatography mass spectrometry, to produce new degradation products such as ethyl ester of lactoyl lactic acid, acetic acid, and propanoic acid in a biotic medium, lactic acid, and lactoyl lactic acids are quickly integrated from the polymer films. In the presence of *Fusarium moniliforme* and *Pseudomonas putida*, degradation of L- and DL-lactic acid oligomers was inspected as lactic acid, lactyl lactic acid dimers, and higher oligomers in liquid culture. HPLC was applied to scrutinize the outcome of these compounds. Notwithstanding the enantiomer composition, two microorganisms entirely used DL- and L-lactic acids under the particular conditions (Torres, Li, Roussos, & Vert, 1996). The procedure of the HPLC calibration was practiced by Codari, Moscatelli, Storti, and Morbidelli (2010) to measure both the production of poly(lactic acid) (PLA) by polycondensation and its matching degradation. Moreover, they proposed the mechanism of hydrolytic degradation of cyclic lactide into the linear dimer hydrolyzed into monomer.

9.4.1 Composting

Primarily, composting is considered as a biological decomposition and conversion procedure to treat and recycle biodegradable organic waste into a humus-like substance. The necessary requirements for decomposition include source of microbial inoculum such as bacteria, fungi, algae, and so forth. In addition, the type of compost and the parameters such as temperature, moisture, and pH of the media are controlled by the actual composting conditions. According to these studies, it is demonstrated that the most appropriate environment for biodegradation of PLA embrace the high temperature, high moisture content, and plenty of microorganisms in compost (Kale, Auras, & Singh, 2006). 1H NMR and ESI-MS analysis of the compost extracts as a function of time confirm the effect of microorganisms present in the compost throughout biodegradation procedure of PLA and PLA/a-PHB (poly(hydroxy butyrate)) (Sikorska et al., 2015). The plastic waste is regarded plausible when 90% of its original dry weight of the sample pass through a 2.0 mm sieves. Presently accessible compostable plastics are frequently qualified by several agencies like European standard EN 13432 (2000) and their counterparts ASTM D 6400-04, ASTM D 6868 (US), AS 4736 (Australia), ISO 17088 and ISO 18606 (worldwide) and BNQ-9011-911/2007 (Canada) (Sikorska et al., 2015). Noteworthy improvement of biodegradation rate in presence of clay or hydroxyapatite is indicated by the biodegradation of poly(hydroxybutyrate-*co*-hydroxyvalerate) in compost media (Maiti & Prakash Yadav, 2008). During biodegradation, steady upsurge of surface roughening is divulged by the surface morphology of pure Biopol and its nanocomposites. Surface roughening is resulted from biodegradation in compost media. The removal of polymers is responsible for this phenomenon. Also, its scope relies on the nanoparticle used after the degradation kinetics (Singh et al., 2013). Two steps are required for the occurrence of the biodegradation in compost media (1) hydrolysis of high molecular weight to lower molecular weight oligomers by using acids, bases, or moisture and (2) consumption of the oligomers by the microorganisms in compost producing CO_2, H_2O, and humus (Maiti & Prakash Yadav, 2008).

9.4.2 Microbial degradation

The action of several microorganisms is necessary to do the biodegradation of polymers. The range of microbial degradation of polymers relies on its dissemination, plenty, variety, activity, and adaptation of microbiota besides the chemical and physical features of polymers. Plate count process and the clear zone technique on combined polyester agar plates are the methods practiced for showing of polyester-degrading microbes. It is identified as a suitable technique to assess the distinction and influence of polymer-degrading microorganisms in the environment (Mao, Liu, Gao, Su, & Wang, 2015). The order of PHB = PCL (polycaprolactone) > PBS (polybutylene succinate) > PLA reveals the reduction of the population of aliphatic polyester-degrading microorganisms (Pranamuda & Tokiwa, 1999). The zone method enriches the Gram-negative rod-shaped mesophilic bacteria. Through attacking both crystalline and amorphous zone, these Gram-negative rod-shaped mesophilic bacteria are active for the degradation of low as well as high molecular weight PLA (Kim & Park, 2010). Nevertheless, different microbial species need either intracellular enzymes or extracellular enzymes through which assimilated the degradation products as sole carbon source are integrated. Consequently, understanding of the enzymatic mechanism is necessary to assess the microbial degradation procedure. This will be provided in enzymatic section. A thermophilic bacterium, *Ralstonia* sp. (MRL-TL), secluded from hot spring water, is able to degrade more than 50 wt.% of PCL film in 10 days screening an esterase activity in the culture broth and the same strain is operative for the degradation of other polyesters such as PES (Polyethersulfone), PLA, PHB, and PHBV even under harsh conditions (Shah et al., 2015).

9.4.3 Enzymatic degradation and mechanism

Enzymes are identified as biological catalyst which presents well-balanced option to manage plastic waste and thus grow sustainability. Two stages are required to do the enzymatic degradation of aliphatic polyesters. Dealing with adsorbing enzyme on the surface of the substrate through surface-binding domain is remarked as the first step. It causes the growth of enzyme-substrate complex. The second step is concerned with the hydrolysis of the ester bonds is the second step. This results in the formation of soluble degradation products. The extensively premeditated enzymes used for biopolymer degradation are PHB depolymerase, PCL-depolymerase, proteinase-k, cutinase, serine proteases, trypsin, elastase esterase, and lipases. The stereochemistry of the polymer are important for its degradation that is some stereoisomers are degraded in the succeeding direction of L-L > L-D > D-L > D-D at 37°C and pH 8.6 (Li, Tenon, Garreau, Braud, & Vert, 2000). In the same way, PLA50-mes degrades at a faster rate than PLA50-rac caused by higher water uptake ratio which facilitates enzymatic attack (Li et al., 2000). Diverse kind of biodegradable polyesters such as PLA, PBS, PBSA, PES, and PCL can be catalyzed by the recombinant enzyme DL-PLA depolymerase (Akutsu-Shigeno et al., 2003). The hydrolysis of polyester by lipase is separated into two successive phases (Fig. 9–3); initially, nucleophilic attack of serine on ester bond that form tetrahedral acyl-serine intermediate, which eventually hydrolyzed by the another attack of nucleophile, that is, water molecule (Röttig & Steinbüchel, 2013). Fig. 9–4a features out that activity of lipase relies on the giving ability of hydroxyl proton toward the cleavage of ester linkage. Reportedly, it is relevant that pH of media powerfully influences the activity of enzyme for a specific substrate through possible change of active site at dissimilar pH (Sugihara et al., 1992). Obviously, the conformation of enzyme is changed by pH of the medium and altered the conformations are altered by this subject to differ hydrolysis rate. By these means, the biodegradation rate of polymers is detected in enzyme media (Singh et al., 2010). The depolymerase activity of lipase *Pseudomonas cepacia* toward PCL, PLA, and PU are very diverse and displays highest activity at acidic (pH ~ 6.5), nearly neutral (pH ~ 7.2), and simple condition (pH ~ 7.7) for PU, PLA, and PCL, correspondingly.

9.4.4 Factors influencing the biodegradability of polymers

Reviewing of the parameters which influence the degree of biodegradability for different polymers are provided in this section. The biodegradability relies on the chemical and physical features of the polymers along with biotic and abiotic issues. Abiotic parameters such as pH, temperature, humidity, salinity, presence or absence of oxygen, and the supply of essential nutrients not only influence the polymers to be degraded but also affect the biotic factors such as microbial diversity, population, and action potential of enzyme activity. Furthermore, the range of degradation relies on the features of polymer including morphology, crosslinking, purity,

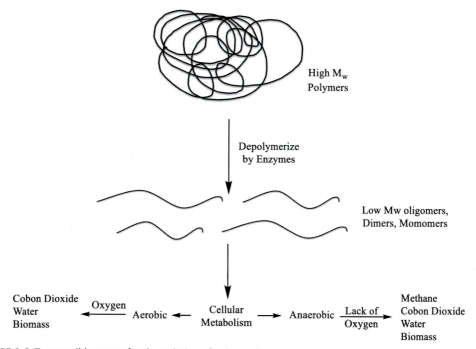

FIGURE 9–3 Two possible routes for degradation of polymers by enzymes.

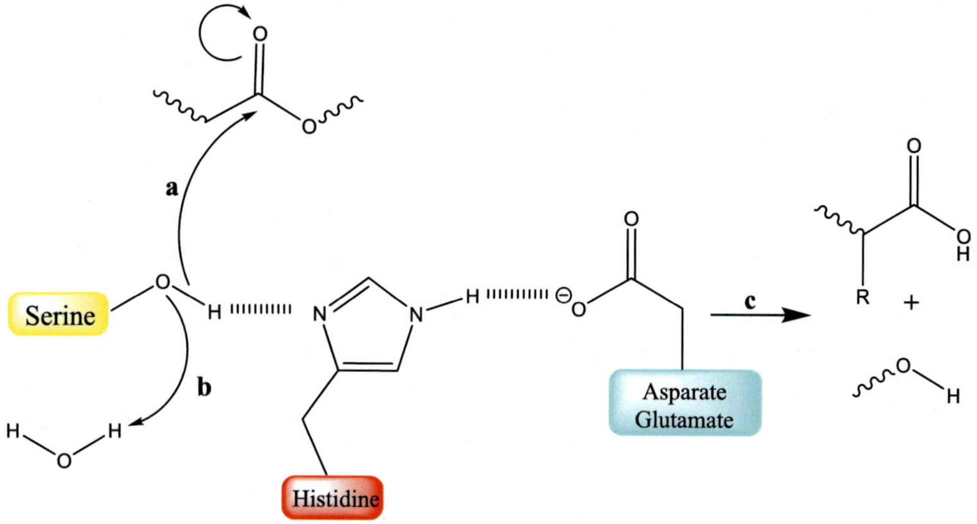

FIGURE 9–4 The hydrolysis of polyester by lipase.

diffusivity, porosity, chemical reactivity, mechanical strength, thermal tolerance, and resistance to electromagnetic radiation (Nampoothiri, Nair, & John, 2010). The surface morphology (surface area, shape, size, hydrophilicity, water permeability, and hydrophobic characteristics), intrinsic properties (monomeric composition, presence of terminal carboxyl or hydroxyl groups, type of enantiomer, alignment of group and its crosslinking, molecular weight and its distribution), high order configurations (glass transition temperature, melting temperature, modulus of elasticity, etc.), and processing conditions (ratio of crystallinity and amorphous region, copolymerization, choice of additives and organic/inorganic fillers) of polymers play important roles in determining the degradation kinetic and influence the mechanism of biodegradation (Park & Xanthos, 2009). More vulnerability to lipolytic enzymes and microbial degradation are indicated by the aliphatic polyesters in comparison to aromatic polyesters and PURs due to their monomeric composition. The increase of molecular weight causes the reduction of biodegradation, for example, low molecular weight PCL degrades more rapidly in comparison with high molecular weight PCL by means of lipase. Notwithstanding of enantiomeric composition against the faster degradation of high molecular weight L-PLA as compared to D-PLA, the low molecular weight oligomers completely degrade (Nampoothiri et al., 2010). Furthermore, their rate of biodegradation is remarkably affected by the morphology of the polymers (Paul et al., 2005). The fraction of crystallinity and amorphous domain are essential to specify the biodegradability. The polymer chains are orderly placed within the crystal structure. They are more resistant against degradation, whereas the amorphous areas are roughly filled. Thus, they are vulnerable toward microbial attack and enzyme catalysis. The biodegradation rate has significantly been reduced for the samples crystallized at 52°C as compared to room temperature satisfied samples for pure PCL due to greater amount of crystallinity as obvious from the higher heats of fusion and bigger size of spherules for crystallized sample regarding the satiated one (Singh et al., 2010). An upsurge in crystallinity of the polymer subjects to reduce the rate of degradation of PLA. On the whole, higher melting temperature (Tm) takes down the biodegradation of polymer for the sake of connections among polymer chains influencing the heat of fusion (ΔH) value and internal rotational energies matching to the rigidity (or the flexibility) of the polymer molecules (Jeon & Kim, 2016). lipase but D-PLA, PGA hydrolyze low molecular weight L-PLA and their copolymers such as DL-PLA, poly(L-lactide-*co*-glycolide), and poly(D-lactide-coglycolide), and it is not easy to hydrolyze high molecular weight L-PLA. They are resilient to enzymatic degradation because of its high crystallinity and high melting temperature (Tm) (Tokiwa & Calabia, 2006). Henceforth, copolymerization effects on the physical properties and biodegradation behavior by lowering the crystallinity. Fillers embedded in polymer matrix are also to significant switch the rate of biodegradation of polymer and good dispersion of filler (exfoliated nanostructure) degrades the polymer at quicker rate in comparison with composite taking interposed nanostructure or pure polymer (Lee, Seong, & Youn, 2005).

9.4.5 Synthetic and other approaches to biodegradable polymers

Many years ago, the overall fundamental features facilitating the biodegradation of polymers were well-defined in primary work by means of unsophisticated testing protocols. However, it demonstrated to be suitable enough to establish strategies for future synthesis plans. Indications were that biodegradation was improved by features such as linear polymer chain structures, low molecular weight, hydrophilicity, noncrystallinity, heteroatom backbone polymers (condensation polymers) and low levels of functional groups such as amines and carboxylic acids. These strategies have been extensively practiced in the growth of biodegradable polymers, chiefly plastics, which are developing today (Vroman & Tighzert, 2009). It is necessary to point out that the guidelines are similarly appropriate to polymers based on fossil or renewable resources (Fig. 9−5). Condensation polymers are susceptible to hydrolytic degradation for the sake of their structures with hydrophilic support linkages such as esters and amides; while additional polymers with hydrocarbon backbones (carbon chain) are vulnerable to oxidative degradation. Commercial polymers are extensively practiced today to develop commodity, this intrinsic faintness was documented and diminished through making condensation polymers as hydrophobic as possibly reliable with presentation, and in addition, polymers were boosted against oxidation through compounding with antioxidants. It was expected that all the selected approaches may develop modifications of existing polymers, but new, more hydrophilic condensation polymers are supposed to be developed by them. Considering the learning profile of new progresses, this approach has led to comparatively costly biodegradable polymers. Following the tack of modifying the very inexpensive commodity addition polymers, polyolefins, a few approaches have been developed to check and upsurge the oxidation rate. Therefore degradation and biodegradation rates are increased by hydrophilicity of the polymer. Within the last few years, scientists have heeded polymers from renewable resources and lower cost feedstocks to fossil fuels. They have been considerable for the sake of their energy and chemicals. To increase the available and widely used starches and celluloses, new natural polymers have emerged such as poly(hydroxyalkanoates) (PHAs)

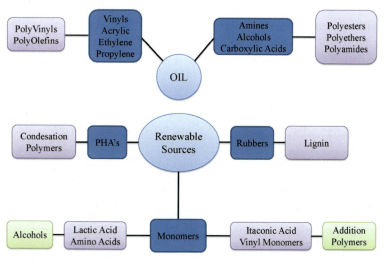

FIGURE 9–5 Pathways to condensation and addition polymer from fossil and renewable sources.

(Sun et al., 2018). The outcome of monomers from natural resources is like the productions of fossil resources from the view of structure, to some extent. Notably, there is an exceeding interest to them. The new biodegradable polymers have been done by dint of outstanding polymerization technologies. Such phenomenon has been carried out by new monomers from renewable resources, amines, alcohols, acids, vinyls, and so forth for the sake of synthetic tractability already established industrially along with condensation polymers. Lactic acid monomer as a precursor to its polymer, PLA is a great instance of this method founded by NatureWorks, LLC (formerly Cargill-Dow Corporation) and others with restricted commercialization recognized (Maharana, Mohanty, & Negi, 2009). Starch and cellulose are natural polymers which are chemically altered and mixed with other degradable and biodegradable synthetic and natural polymers so that they can be applied in different fields such as packaging, films, detergents, and so forth. Further, their biodegradation is preserved by natural polymers.

9.5 Plastics from fossil resources

9.5.1 Polyurethane

PUR is identified as a polymer included organic units, which urethane links stick them to each other. Thermosetting polymers hold most of the PUR. Ester type PU (ES-PU) and ether type PU (ET-PU) are the main types of PUR. PU is widely applied in the production of cushions, rubber products, leathers, adhesives, and so forth. In comparison with the polyester-type PUR, polyether-type PUR is less probable to be affected by fungal attack (Darby & Kaplan, 1968). It was indicated that polydiethylene adipate (M_n 2500 and 2690) produces ES-PU which is degraded by *Comamonas acidovorans* starin TB-35. It also hydrolyzes TDI (Toluene diisocyanate) ES-PU. Diethylene glycol and adipic acid are released, too (Nakajima-Kambe, Onuma, Kimpara, & Nakahara, 1995). It is demonstrated that cholesterol esterase degrades ES-PU produced from TDI, PCL-diol (M_n 1250), and ethylenediamine. Furthermore, the hard segment components are also released (Fukushima et al., 2010).

9.5.2 Polycaprolactone

It is considered that PCL, which is petrochemical polymer, can be biologically graded. PLC is mostly recognized by its biodegradability and nontoxicity although it comes from a synthetic origin. Basically, it is identified as polyester including methylene units and ester groups. Further, bacteria and fungi can degrade it. PLC is produced by the ring opening polymerization of ecaprolactone (Mohee & Unmar, 2007). Since graphene/PCL composite has been proved to be nontoxic nature through cell proliferation, it is identified as hopeful biodegradable material (Murray, Thompson, Sayyar, & Wallace, 2015). The existence of hydrolytically labile aliphatic ester linkages has made PLC endure degradation although the rate of degradation is somewhat slow (3–4 years). Regarding the natural degradation rate of pure polymer, any increase or decrease in the biodegradation rate of nanocomposite is considered fine-tuned. With the contribution of different nanoparticles, this is viable by directing the pH of the medium. In the presence of nanoparticles, higher biodegradation is made by the greater enzyme (PCL-depolymerase) activity in slight alkaline nanocomposite. Notably, the pH of PCL solution is estimated to be ~7.0. In comparison with PCL-C18 nanocomposite, the biodegradation rate is defined by higher basic pH in PCL-30B nanocomposite. In comparison with pure PCL, alkaline pH together with higher amorphous contents subject to enhance enzymatic activity in nanocomposites. By means of scanning confocal images, the relative morphology illustrates the highest enzymatic degradation in PCL-30B and the lowest rate in pure PCL as evident from the relative roughened surface (Singh et al., 2010). As observed, the aerobic and anaerobic microorganisms extensively existing in the ecosystems degrades the PCL. To investigate the degradation of high molecular weight PCL, *Penicillium* sp. strain 26−1 isolated from soil was practiced. Within 12 days, the PLC was entirely degraded (Palmisano & Pettigrew, 1992). Enzymes degrade or absorb unsaturated aliphatic and alicyclic polyester, while aromatic polyesters do not. It should be mentioned that soil is the source of *Aspergllus* sp. strain ST-01, which causes the degradation of PCL (Fukushima et al., 2010). In the amorphous region, very rapid degradation of PCL is confirmed by *Aspergillus flavus* and *Penicillium funiculosum* (Geyer, Jambeck, & Law, 2017). PCL can be degraded by some fungal phytopathogens, too (Shah et al., 2008). Under anaerobic conditions, PCL can be degraded by a new microbial species. It is a subdivision of the genus Clostridium (Geyer et al., 2017). Pseudomonas, *Bacillus licheniformis* (Paul et al., 2005), and *Mucor miehei* are among other microorganisms that are able to degrade PLC. Allı, Aydın, Allı, and Hazer (2015) proved the enzymatic degradation of various attached PCL (PLina-g-PCL and PLina-g-PSt-g-PCL copolymers) presenting controlled degradation in existence of *Pseudomonas lipase* relying on the nature of comonomer and its extent. Multicomponent composites of PCL (filled with the fibers of PHB) with cationic (Cloisite) or anionic (Perkalite) nanoclay is described with changing degree of biodegradation. In the matrix PCL the rate of biodegradation has been controlled by the type and amount of clay and fibers. It is permitted that the filled PHB fibers are able to modulate the properties of thermal stability of PCL. Furthermore, the biodegradation process is banned (Marega & Marigo, 2015). Pawar et al. (2015) scrutinized the influence of two different multiwall carbon nanotubes, namely pristine (*p*-MWNTs) and amine functionalized (a-MWNTs) on the biodegradation of PCL. They used lipase to divulge inactive rate in presence of *p*-MWNTs composite, whereas the biodegradation is alike in presence of a-MWNTs with rebased on natural degradation of pure PCL. The type of soil is significant because it influences PLC degradation.

9.5.3 Polypropiolactone

Evidently, polypropiolactone (PPL) is identified to be biodegradable aliphatic polyester. Its mechanical properties are significant. The structural units of PHB and PCL are similar to those of PPL. Thus, PHB depolymerase and lipase can degrade it (Sivan, Szanto, & Pavlov, 2006).

Bacillus sp. is responsible for several degrading microorganisms of PPL (Nishida, Suzuki, & Tokiwa, 1998), as well as *R. delemar* can degrade PPL (Tokiwa & Suzuki, 1977) or *Streptomyces* sp. may degrade PPL (Geyer et al., 2017).

9.5.4 Polybutylene succinate

Basically, PBS is identified as an aliphatic synthetic polyester. Its features are similar to the properties of PP. The microorganisms degrading PBS are very extensively dispersed through the environment. The degradation of PBS is caused by *Amycolatopsis* sp. HT-6 (Geyer et al., 2017).

9.5.5 Polyvinyl chloride

PVC is identified as a water-soluble synthetic polymer synthesized by hydrolysis of polyvinyl acetate. Although it is not as biodegradable as plastics, it could be biologically degraded by hydrolysis of polyvinyl acetate (Shah et al., 2008). *Pseudomonas chlororaphis, Pseudomonas putida* AJ, and *Ochrobactrum* TD are the microorganisms used for biodegradation of PVCs (Danko, Luo, Bagwell, Brigmon, & Freedman, 2004). Limited reports have been registered on the biodegradation of PVC polymers and plastics. As reported by Webb et al., fungal strains are detected as the only source of carbon and degrade the plasticizer dioctyl adipate are degraded by them. Primarily, the tested fungal strains acquired from the surface of plasticized PVC (pPVC) films exposed to the atmosphere could grow with the complete pPVC (0.5 mm thick films). In 2014 Ali et al. indicated that sewage sludge soak the isolation of four fungal strains within 10 months after thin pure PVC films are buried in sterilized soil. *Phanerochaete chrysosporium, Lentinus tigrinus, Aspergillus niger,* and *Aspergillus sydowii* are identified as isolators which are capable of growing in mineral salt medium. After 7 weeks of incubation, the PVC polymer is degraded. Particularly, it is detected that the polymer molecular weight is decreased in the presence of *Phanerochaete chrysosporium*. Reportedly, the pPVC by *Pseudomonas aeruginosa* and *Achromobacter* sp. bacterial strains are biodegraded. Then, they are secluded from hydrocarbon contaminated soil samples (Ali et al., 2014). Pure strains of *Bacillus* and *Pseudomonas* were able to degrade PVC films. Eventually, Giacomucci et al. carried out a research and indicated that PVC films could be degraded by pure strains of *Bacillus* and *Pseudomonas*. After 90 days of PVC film incubation with the bacterial strains, the average molecular weight M_n (%) was diminished almost 10% by means of GPS measurements. Thus it is implied that the PVC polymer could be attacked by both strains. In fact, smaller fragments were developed by some chains scission. Exactly at this incubation time, the films were entirely split. Furthermore, the PVC wastes could be degraded by *P. citronellolis*. After 30 days incubation, these types of wastes were utilized to pack fruits and vegetables. By means of gravimetric, it is revealed that weight losses between 13.07% ± 0.36% and 18.58% ± 0.01% for PVC films protected with the bacterium in comparison with maximum loss of 8.39% ± 1.10% in the case of the abiotic controls. Giacomucci et al. illustrated that various anaerobic marine consortia were capable of degrading plasticized pretreated PVC. These anaerobic marine consortia were improved and their connected microbial communities categorized. The films seemingly acted against both the additives and the PVC polymer chains could be degraded by 3/16 enriched consortia after 7 months of incubation. The fact is that these consortia subjected the gravimetric weight loss of up to 11.67% ± 0.58%. Notably a remarkable reduction in the M_n% from 100% up to ∼ 91% was detected by gel permeation chromatography (GPC). Henceforth, it is suggested the formation of some polymer chain scission and the development of several fragments, whereas thermogravimetrical analysis (TGA) discovered a reduction in thermal steadiness. As the incubation lasted up to 24 months, it paved the way to select other three consortia through which the PVC polymer chains could be broken down. The PVC film-associated communities of the biodegrading microcosms was categorized and included microbial phyla carefully connected to those previously reported from laboratory/field enrichments as able to degrade halogenated organic compounds and hydrocarbons. In fact, it was permitted to hypothesize that dechlorination of PVC film. Hereafter, the study presented visions on the possible destiny of PVC plastics made known to into marine environment (Giacomucci, Raddadi, Soccio, Lotti, & Fava, 2020).

9.5.6 Polyethylene

Primarily, PE includes short reiterating monomer units with strong interchain hydrogen bonding. In consequence, a highly hydrophobic polymer is developed and the enzyme susceptibility is restricted. Therefore microbial action can only affect the polymer surface with a restricted number of free chain ends. Moreover, extensive studies have been done to mull over the biodegradability of PE. Examples are summarized as (1) a 10-year soil incubation (25°C) study under controlled conditions using 14C-labeled low-density PE film samples (20 mm thick), where the degree of biodegradation was estimated by the yield of 14 CO_2 and showed 0.5% CO_2 evolution for pre-UV-irradiated samples and 0.2% CO_2 evolution for nonirradiated samples (Albertsson & Karlsson, 1990); (2) PE samples buried for an estimated 32 years under a garden soil in Japan that showed only partial degradation using Fourier transform infrared (FTIR) microscopy and differential scanning calorimetry (DSC) techniques (Otake, Kobayashi, Asabe, Murakami, & Ono, 1995); (3) a 120-day laboratory soil burial study in which PE was incubated at 28 1C and presented minimal biodegradation (César, Mariani, Innocentini-Mei, & Cardoso, 2009); (4) even though undergoing a continuous slow abiotic degradation in soil, linear low-density PE mulching films with prooxidants buried under field conditions for 8.5 years were recovered intact (Briassoulis, Babou, Hiskakis, & Kyrikou, 2015b); (5) LLDPE mulching films (20 mm thick) with prooxidants that were degraded affectedly and then buried in soil for 6 years and 10 months were slowly distorted into microfragments in the soil (Briassoulis, Babou, Hiskakis, & Kyrikou, 2015a); (6) Vieyra, Aguilar-Méndez, and San (2013) labeled a unchanged PE samples buried in compost for 125 days; (7) maximum weight loss of 1.5 mm think samples was 1.5%–2.5% and 0.5%–0.8% for low-density PE and high-density PE, correspondingly, after 6 months wrapped up in the bay of Bengal, India. With the contribution of copolymerization, biodegradable additives, photoinitiators can enhance the biofragmentation of PE. Eventually, PE is changed into progressively smaller fragments until nonvisible particles through the combination of abiotic and biotic degradation process until the formation of nonvisible particles.

9.5.7 Polyethylene

The environmental degradation of PE consists of a two-step process. The first step is dealt with rate-determining step being the abiotic oxidation. Bioassimilation or mineralization of the oxidized products follows it. Its high molecular mass and existence of nonpolar C−C and C-H bonds are counted as the main barriers of PE degradation. Semicrystalline regions are formed by the macromolecular chains, which are closely lined up. The motionlessness of PE was affected by a combination of all these factors. To avert degradation during processing, antioxidant package is typically introduced to the polymer. Morphology and surface area of the material are counted as other physical features influencing the degradation rate. As a whole, the degradation of PE is started at weak sites and continues via the development of hydroperoxide intermediates so that carbonyl compounds are formed. These can undertake Norrish I and II type reactions. This subjects to chain scission. Finally, the loss of mechanical features will be obtained. When prodegradants is exposed to light/heat, they decompose so that free radicals are developed and PE macroradicals are produced. Moreover, hydroperoxides are produced as the result of the subsequent reactions of PE macroradicals with molecular oxygen. The autooxidation and chain cleavage of PE occurs by the use of classical free radical chain reactions, the autooxidation and chain cleavage of PE are detected in the last procedure. Interchanging between two oxidation states differing by a single unit is considered as the distinctive feature of the transition metals. It leads to the catalytic decomposition of hydroperoxides. By this means pushing forward the whole reaction is pushed forward. When the light initiates the oxidative degradation, it lingers under dark thermal conditions. According to the fact that the crystalline regions are unreachable to oxygen, they nearly stay uninfluenced. The oxidative degradation of PE brings about the formation of some low molecular weight degradation products detected by chromatographic analysis. Indeed, chromatographic fingerprinting has been effectively practiced as a technique so that aging in diverse environments can be distinguished. Alkanes, alkenes, aldehydes, ketones, alcohols, mono- and dicarboxylic acids, lactones, keto acids, and esters are among the mostly recognized. As the biodegradable fraction could be properly assessed as the percentage of polymer extractable in an appropriate solvent, it is only the oxidation products that can be biologically degraded. The molecular weight of PE can be swiftly diminished to as low as 5000 Da by the oxidative degradation especially when enhanced oven aging is practiced. A large portion of PE degradation products could still have plenty of a molecular mass to be simply used up by microorganisms, although polar compounds with rather higher molecular bulk can be biodegraded. Initially, larger molecules need oxidizing by the extracellular or cell wall−associated enzymes to pass through the cell wall. Then, they can indirectly act through production of diffusible radicals several times. The water-insoluble degradation products are mobilized by the biosurfactants aid adhesion of cells to the material to pass through the cells. Then, in the cytoplasmic and/or periplasmic space convert them (Lemaire, Arnaud, & Gardette, 1991). The β-oxidation pathway uses the molecules with smaller size. This trail also transports such molecules across the cytoplasmic membrane. Roy, Hakkarainen, Varma, and Albertsson (2011) have stated that some research on the biodegradation of synthetic polymers and plastics has been done on PE films through terrestrial deriving pure and mixed aerobic cultures. The formation of functional groups was illustrated by means of microscopic observations indicating surface unevenness and incidence of deep cavities. Moreover, a 46.7% decrease in the viscous area of biodegraded PE was revealed with the contribution of phase imaging. Also, a loss was detected in crystalline content of the biodegraded PE by dint of Raman spectroscopy. Skariyachan et al. (2016) informed on very high weight loss percentages getting up to 81% ± 4% and up to 75% ± 2% of LDPE strips after 120 days of incubation with tailored associations composed of bacterial isolates gained from plastic garbage meting out areas or plastic-contaminated cow muck, respectively (Skariyachan et al., 2017). This research also revealed a very high degradation of LDPE pellet, High Density Polyethylene (HDPE) pellet, and strips. Additionally, there is another report presented by Sudhakar et al. on a determined gravimetric weight loss of 2.5% for low LDPE and 0.8% for HDPE afterward 6 months of incubation in situ in ocean waters. Balasubramanian et al. reported that HDPE incubated at 30°C within 30 days in shaken flasks comprising synthetic medium in the presence of *Arthrobacter* sp. or *Pseudomonas* sp. secluded from the plastic waste dumped sites of the Gulf of Mannar region (India) divulged approximately 12% and 15% of gravimetric weight loss, correspondingly. Harshvardhan and Jha stated that a gravimetric weight loss of up to 1.75% of LDPE incubated for 30 days has been detected in the presence of marine bacterial isolates in Bushnell−Haas medium under shaken flasks.

9.5.8 Polypropylene

So far, not much research has been done to realize more about the biodegradation of PP polymers and plastics. Cacciari et al. (1993) pioneered to report that microbial communities were capable of degrading isotactic PP. Cacciari et al. stated that the polymer had 40% methylene chloride extractable compounds and a mixture of hydrocarbons (between $C_{10}H_{22}$ and $C_{31}H_{64}$) were perceived and recognized in the extract after 175 days incubation in the presence of the bacterial consortia gained from different soil samples that were rich in plastic wastes. It is noteworthy that only a limited number of papers have been issued on the biodegradation of pretreated and/or prooxidant additive-containing PPs. It lasted 1 year to present the report on biodegradation of thermally pretreated 0.05 mm thick PP films by means of a soil microbial consortium under aerobic conditions. In this case, a gravimetric weight was reduced up to 10.7% for the thermally pretreated PP matched to merely 0.4% for the untreated polymer (Arkatkar, Arutchelvi, Bhaduri, Uppara, & Doble, 2009). They used four bacterial types, namely, *Pseudomonas azotoformans* MTCC 7616 and three isolates enriched from one soil sample recovered from a plastic dumping site (identified as *P. stutzeri*, *Bacillus subtilis*, and *Bacillus flexus*). They deliberated the biodegradation of pretreated (PPUT), chemically (aquaregia pretreated or Fenton's pretreated) and physically pretreated (thermal and UV-pretreated) PP films After 1 year of observing, biodegradation was mainly detected on the sole UV-pretreated PP inoculated with *B. flexus* where a 2.5% weight loss of the polymeric film was verified even though the three strains were capable of developing biofilm on the surface of all polymeric films. Jeyakumar, Chirsteen, and Doble (2013) indicated that

UV-pretreated prooxidant blended-PP incubated for 1 year with *Phanerochaete chrysosporium* NCIM 1170 and *Engyodontium album* MTP091 fungal strains in shake flasks presented 18.8% and 9.42% gravimetric weight losses, correspondingly. Fontanella et al. (2013) put the *Rhodococcus rhodochrous* ATCC 29672 into practice and compared the biodegradability of diverse PP films (51-μm) comprising several prooxidant additives (centered on Mn/Fe, Co, and Mn) and the additive-free films. It lasted 6 months to come into the conclusion that no biodegradation was detected in the case of additive-free polymers, whereas the biodegradability of PP films encompassing Mn + Fe or Mn additive that were capable of supporting bacterial growth. Jeon and Kim (2016) secluded a bacterium from a soil sample recovered from an open storage yard for municipal solid waste. This bacterium was detected as *Stenotrophomonas panacihumi* PA3−2. Thus the capability of degrading PP was demonstrated. Under composting conditions at 37°C, it took 90 days to carry out the experiment. Regarding the KS M3100-1:2002; MOD ISO 14855:1999 method and GPC analysis, the biodegradation activity was assessed. Furthermore, Aravinthan, Arkatkar, Juwarkar, and Doble (2016) reported the biodegradation of UV and thermally pretreated PP by dint of a coculture of *Bacillus* and *Pseudomonas*. As the presence of the pretreated PP is regarded as a chief carbon source, high bacterial growth rate was verified in its presence during 12 months of incubation. In the last part of the experiment, 22.7% thermogravimetric (TG) weight loss and a gravimetric weight loss percentage of 1.95% ± 0.18% were displayed by UV-treated PP open to the coculture. Skariyachan et al. (2018) used aerobic thermophilic artificial bacterial consortia made due to *Brevibacillus* spp. and *Aneurinibacillus* sp. isolates gained from waste management landfills and sewage treatment plants. They finally reported on the biodegradation of PP pellet and strips. On the surface of the plastic, biofilm formation was observed after 140 days incubation. Also, there was a record on a gravimetric weight loss of up to 56.3% ± 2%. There have been the developments of a high-density biofilm even though Sudhakar et al. (2007) testified on a 0.5% gravimetric weight loss with unblended PP (1.5 mm thickness) after 6 months of incubation in situ in ocean waters. Auta, Emenike, Jayanthi, and Fauziah (2018) used *Rhodococcus* sp. strain 36% and 4.0% via *Bacillus* sp. strain 27, isolated from mangrove systems (Peninsular Malaysia), of UV-pretreated PP after 40 days incubation. Finally, they presented their complete report on a gravimetric weight loss of 6.4%. Furthermore, PP biodegradation was established by means of FTIR spectroscopy and scanning electron microscopy (SEM) analyses, which structural and morphological variations were illustrated in the PP microplastics with microbial treatment. Auta et al. (2018) practiced *Rhodococcus* sp. strain 36% and 4.0% through *Bacillus* sp. strain 27, isolated from mangrove systems (Peninsular Malaysia), of UV-pretreated PP after 40 days incubation and they testified that PP biodegradation was further confirmed using FTIR spectroscopy and SEM analyses, which revealed structural and morphological changes in the PP microplastics with microbial treatment. Mohanrasu et al. (2018) used *Brevibacillus borstelensis* isolated from marine sediment and presented reports on the biodegradation of HDPE plastic bags. A weight loss of 11.4% was verified and pits and cavities development on the HDPE film surface were displayed by SEM observation after 30 days of incubation in shaken flasks.

9.5.9 Polystyrene

It is noteworthy that pure microbial cultures have been practiced in several studies focus on assessing the biodegradation of PS. Mor and Sivan (2008) used the actinomycete *Rhodococcus ruber* isolate C208 and considered biofilm formation on pure PS flakes. 0.8% of the gravimetric weight was decreased as the result of incubation of the biofilm under shaken flask for up to 8 weeks. Tian et al. (2017) employed the fungal strain *Penicillium variabile* and stated the mineralization of ozone-pretreated PS. Ho, Roberts, and Lucas (2018) conducted a broad study on biodegradation of PS and modified/blended PS. Currently, they have published the results of their review. Johnston et al. (2018) have currently conducted a suitable approach to couple biodegradation of prodegraded high impact and general-purpose PS flakes to produce PHAs. The bacterial strain *Cupriavidus necator* H16 was indeed shown to be able to grow in the presence of oxidized (by thermal and ozone pretreatment). Myristicate and costearate containing prodegraded PS as carbon source and produce PHAs. Syranidou et al. (2017) stated that PS films that are naturally worn and composed from two coastal sites in Chania (Greece) were exposed to be degraded in seawater microcosms according to Mn reduction of up to 32%, a gravimetric weight diminution of 4.7% and the observation of cracks/fissures on the surface of the films after 6 months of incubation. Auta, Emenike, and Fauziah (2017) stated that biodegradation of UV-pretreated PS by *B. cereus* and *B. gottheilii* strains were secluded from mangrove.

9.6 Plastics from renewable resources

9.6.1 Poly(hydroxyalkanoate)

Since PHAs are naturally biodegradable and biocompatible polymers, they become striking biomaterials in clinical use because of their microbial source. The microbial-source biodegradable polymers are necessary, particularly for bioimplants to elude the necessity for another surgery to eliminate the nondegradable polymers. PHAs are polyesters that bacterial fermentation of sugars and lipids produce them. PHAs are accrued intracellularly and operate as carbon and energy reserves. PHB homopolymer is not proper to be extensively applied since it is stiff and brittle. To progress the physical properties, copolymer of 3-hydroxyvalerate with 3-hydroxybutyrate (HB), commercially identified as Biopol, has been planned. The genera *Alcaligenes*, *Bacillus*, and *Pseudomonas* are encompassed by microorganisms, which are able to accumulate PHAs. Koller et al. have presented more details in this regard. Moreover, a new research has indicated the PHA synthesis by means of marine purple bacteria. In comparison with its copolyester PHBV, PHB degrades more quickly in anaerobic setting. However, the degradation rate is almost slow in aerobic condition (Abou-Zeid, Müller, & Deckwer, 2001). An adequately high molecular mass is included in the PHA molecules taken out from bacterial cells display features alike to PE and PP. The most of PHAs are chiefly linear polyesters made up of 3-hydroxy fatty acid monomers. An ester bond is formed by the carboxyl group and the hydroxyl group of adjacent monomers in these PHAs. It is regarded that neat PHA polymers are directly biodegradable microorganisms in the environment that can colonize under conditions of soil burial by means of thin-film samples, PHA biodegradation research has been done. For example: (1) based on polyesters of HB, Kunioka, Kawaguchi, and Doi (1989) utilized PHB film samples (0.07 mm thick), and demonstrated that increasing the content of 4-HB led to the enhancement the rate of biodegradation; (2) Mergaert, Webb, Anderson, Wouters, and Swings (1993) practiced 2 mm thick PHB samples stated a maximum weight loss of 0.64% per day for samples hatched at 40°C; (3) Woolnough, Yee, Charlton, and Foster (2010) employed 0.1 mm thick films informed 50% weight loss in 50 days at temperatures of up to 30°C;(4) it was indicated that the water inorganic composition, water temperature, and PHA chemical structure affects the scope of PHA biodegradation in aquatic reservoirs (Voinova, Gladyshev, & Volova, 2008); (5) studies have established that when samples have undergone pre-UV exposure, the biodegradability of PHAs is improved under conditions of soil burial (Saad, Khalil, & Sabaa, 2010); and (6) the soil experiments demonstrated that the microbial communities were found to differ from the microbial communities of the setting soil on the surface of PHA polymers (Volova, Prudnikova, Vinogradova, Syrvacheva, & Shishatskaya, 2017); (7) the daily mass loss of PHA films and pellets was 0.04%−0.33% and 0.02%−0.18%, correspondingly, was inclined by polymer chemical composition, specimen shape, and microbial community in tropical Vietnamese soils (Boyandin et al., 2013). The capability degrading bacteria to secrete specific extracellular PHA depolymers signifies the biodegradability of PHA. As homopolymers are not able to degrade as fast as PHA copolymers, the biodegradability is also affected by the physicochemical properties of the polymer. For example, PHB/polyhydroxyvalerate (PHV) copolymers degrade to a higher degree compared to PHB under aqueous conditions, because PHB has a more crystalline structure with a high melting point. For instance, in comparison with PHB under aqueous conditions, PHB/PHV copolymers degrade to a higher degree. Within an 8-week sediment burial study directed in a tropical mangrove ecosystem at Sungai Pinang, the same outcome was achieved. Inclusively, the processing of PHAs is considered to be a challenging one because it needs low temperatures for their decomposition. Indeed, thermal stability is restricted. Adding plasticizers can improve properties like toughness and processability. This addition permits processing and occurs at lower temperatures. Thus thermal degradation is eluded.

In this regards and as most famous type of this group, PHB can be highlighted which bacteria develop it so that carbon and energy are stored (Ludevese-Pascual et al., 2019). Regardless the development of any toxic products, PHB can be biologically degraded both in the aerobic and anaerobic settings. It is determined that PHB is degraded by microorganisms from *Bacillus, Pseudomonas,* and *Streptomyces* species (DeMoss, 2019). It is assessed that PHB-degrading microorganisms hold 0.5%−9.6% of the whole microorganisms (Shah et al., 2008). The ambient temperature paves the way to separate most of the PHB-degrading microorganisms. When the temperature gets higher, PHB could be degraded by some of them. In fact, composting at high temperature is regarded as one of the most suitable technologies to recycle BP. The composting process is considerably affected by the thermophilic microorganisms (Willis et al., 2018). This specifies that significant microorganisms are the ones which are capable of degrading polyesters at high temperatures. In comparison with the value of the Streptomyces strains, the actinomycete takes higher value of PHB-degrading activity (DeMoss, 2019). 90% of PHB film was degraded by *Aspergillus* sp. After 5 days of being cultivated at 50°C (Sivan et al., 2006). Microorganisms like *Alcaligenes faecalis, Schlegelella*, and thermodepolymerans are also responsible for the degradation of plastics (Shimao, 2001).

9.6.2 Poly(lactic acid) and its copolymers

Member aliphatic polyesters are PLA composed of lactic acid (2-hydroxy propionic acid). PLA is biologically produced by animal, plants, and microbes via fermentation under oxygen restraining circumstances. It can be synthetically obtained from renewable materials such as ethanol, acetylene, or acetaldehyde. PLA is basically biothermoplastic aliphatic polyester which its properties are the same as PS and PET (Polyethylene Terephthalate). The polymerization of lactic acid produces it. Since PLA is regarded as a compostable polymer, it lessens the solid waste disposal problems. PLGA is another type of copolymer which synthesizes through the ROP (ring-opening polymerization) of lactide and glycolide. The degradation rate of PLGA is enhanced by the higher ratios of hydrophilic PGA (Steinbach, Seo, & Saltzman, 2016). This part also presents the relative studies on biodegradable microspheres of poly(DL-lactide-*co*-glycolide) with poly(ethylene glycol) derivates (Garcia, 1999). As PLA is typically remarked as a biodegradable one, its choice might sound controversial in this section. PLAGA [Poly(lactic-co-glycolic) Acid] is another type of polymer which is highly applied in biomedical arena because it is biologically compatible and it can control the degradation behavior. Hitherto, the terrestrial and aquatic environments are not suitable enough for degradation, whereas it can be composted under considerably certain situations. The PLA monomer is regarded as a chiral molecule presenting D-lactic acid and L-lactic acid, which are the two isometric forms. Further, it is considered that the L-isoform of PLA is biodegradable; nonetheless the microbial groups measured able to do this are not extensively be dispersed in soil. Also, it is reflected nonbiodegradable even though the D-PLA is hydrolyzable in water. The biodegradability of PLA has been premeditated under soil burial, composting, and in aquatic conditions. Particular instances comprise as below: (1) soil burial considers using L-PLA over a 6 week time period caused in no weight loss (Ohkita & Lee, 2006); (2) under composting conditions rigid film samples (0.3 mm thickness) and reported the disintegration of the samples and the development of acidic water-soluble degradation products after 70 days (temperature up to 59°C) (Sikorska et al., 2015); (3) PLA food utensils lost 34% weight during composting and degraded to small fragments after 7 weeks exposure at 65°C (Mulbry, Reeves, & Millner, 2012); (4) PLA weight loss of 2.5% was observed in a simulated marine environment over 600 days (Pelegrini et al., 2016); (5) in simulated home composting experiment PLA packaging material presented no visual indication of microbial breakdown and 5% weight loss after 180 days. Higher temperatures of composting exposures are responsible for the distinction in degradation rates between soil and aqueous media in comparison with composting. Water is absorbed by PLA and it goes through important abiotic hydrolysis, while these processes are slow or insignificant in soil and water when the temperatures are increased. Nevertheless, if the compost is to suppose as an end-product, it is essential to evaluate the occurrence of fragmented PLA particles. The present concerns are considered for the sake of the accrual of microplastics in environmental systems. Since neat PLA is very brittle, some additives should be added to achieve the elongation more than 10%. It is obvious that PLA might not be convenient for usages a high mechanical performance is needed except it is changed. To form blends and composites, the probable modifications include plasticization of PLA and its copolymerization. In this way, the features of PLA like stiffness, permeability, crystallinity, and thermal stability can be improved. It is considered that the hydrolysis process of PLA is slow via the support of ester groups. Nonetheless, exposing PLA to temperatures above 50°C can enhance the process. In comparison with organic waste in commercial facilities, it is still remarked that PLA is gradually degraded. Sangeetha et al. presented a remarkable review of PLA chemistry and various blends and composites based on PLA, whereas Karamanlioglu et al. analyzed the steadiness and degradation of PLA in a range of differing environments. According to predominant degradation of the amorphous regions of the polymer chain, a significant increase in crystallinity is detected after degradation at higher temperatures (Felfel, Hossain, Parsons, Rudd, & Ahmed, 2015). In several fields, its application is limited by poor stiffness and hydrophobicity of PLA which is typically enhanced by preparing its nanocomposite by means of nanometer dimension particle surrounded in the polymer matrix. The hydrolysis of its ester bonds degrades it. It includes nontoxic lactic acid and glycolic acid that normal metabolic ways can remove it from the body (Kamaly, Yameen, Wu, & Farokhzad, 2016). At room temperature, a free-flowing sol is formed by the synthesis of PEG–PLGA–PEG triblock copolymers, whereas it is changed into a gel at body temperature. Hydrolysis makes the gel degraded. From the in situ gel, preferential mass loss of PEG-rich segment is indicated (Jeong, Bae, & Kim, 2000). To improve the gene delivery efficacy in rat skeletal muscle, Chang et al. (2007) practiced the nonionic amphiphilic biodegradable PEG–PLGA–PEG copolymer.

9.6.3 Others polymers

9.6.3.1 Cellulose

Primarily, cellulose is linear polymer made up of D-glucose subunits, which β-1,4glycoside bonds linked them. Apparently, either in organized or nonorganized chain form is identified as crystalline or amorphous cellulose, correspondingly. Cellulose is inclined to enzyme or microbes (typically eubacteria and fungi) with the formation of synergistic interface with noncellulolytic species directing to complete degradation of cellulose (Pérez, Munoz-Dorado, De la Rubia, & Martinez, 2002). The enzyme necessary for the biodegradation of cellulose is secreted by fungi like *Trichoderma reesei* and *Phanerochaete chrysosporium* to biologically degrade cellulose after the backbone linkages, β-1,4-glycosidic bond of cellulose are hydrolyzed (Singhania, 2009). Crystalline and amorphous cellulose are efficiently degraded by enzymes. The internal bonds of amorphous cellulose are hydrolyzed by out of two classes of cellulose, endoglucanases, and as a consequence, cellobiose molecules are released by CBHs (cellobiohydrolase) and EGs. β-Glucosidases enzyme hydrolyzes these molecules to glucose ones (Garcia, 1999).

9.6.3.2 Starch

Amylose and amylopectin are identified as the two different macromolecules which compose starch. Amulose consists of a linear α-D-glucopyranosyl unit linked with α-1,4 bonds, while amylopectin is a highly branched fraction of α-1,4 linked glucan chains connected by α-1,6 linkages. Starch is remarked as an important source of energy for a wide range of green plants (Miao, Li, Huang, Jiang, & Zhang, 2015). One of the most cost-effective, plentiful, and renewable biomaterials is starch-based BP which is broadly applied in disposal and biomedical practices. Scientists have done extensive research on starch-based materials which have been either chemically or physically altered to figure out its drug delivery and degradation kinetics. To realize the influence of distinct structures on enzymatic degradation, fungal-amylase is employed to hydrolyze starch films with different molecular, crystalline, and granular types (Li et al., 2015). Researchers have put the indoor soil burial method into practice to mull over the degradation of cassava starch-based composite films. The entry of soil microorganism is upheld by the enhanced water sorption, and the starch film is used as a source of energy to develop them. Weight decrease is practiced to observe the loss of matrix constituents of the films. With the contribution of SEM, the morphology is premeditated (Maran, Sivakumar, Thirugnanasambandham, & Sridhar, 2014).

9.6.3.3 Gelatin

Collagen is a fibrous insoluble protein, of which gelatin is made from through chemical denaturation. It includes three polypeptide chains holding reiterating Gly-X-Y structure (glycine-proline-hydroxyprolinem) organized in a triple helix and hydrogen and hydrophobic bonds stabilize it. Distraction of the gelatin structure rising from the loss of triple helix conformation is brought about by alterations in pH, temperature, or presence of denaturing chemicals. To obtain a high gelling strength avoiding the extensive degradation of the peptide structure, it is essential to carefully control the processing of gelatin (Hanani, Roos, & Kerry, 2014). In comparison with CPX−MMT (montmorillonite)−gel or MMT−gel, much faster rate in CPX−gel (ciprofloxacin loaded) is indicated by the influence of lysozyme and proteinase-k on the degradation behavior of gelatin and its composites with MMT (Berto et al., 2017). For the sake of the large number of hydrophilic amino and carboxyl groups, The CPX−gel is rapidly degraded. The physical structure of CPX−gel scaffold is including higher porosity and leaner pore walls. Therefore the two main parameters influencing the cellular behavior and tissue regeneration are the gelation and biodegradation (Li, Rodrigues, & Tomas, 2012).

9.6.3.4 Alginate

Brown alga produces alginate, a water-soluble linear polymer. Alginate consists of (1−4)-β-D-mannuronic acid (M) and (1−4)-α-L-guluronic acid (G) units in the form of homopolymeric (MM- or GG-blocks) and hetero-polymeric sequences (MG- or GM-blocks) (Gao, Liu, Chen, & Zhang, 2009). Changing molecular weight, chemical structure, and crosslinking controls the degradation kinetics of alginate. With the contribution of an oxidation reaction using uronic acid residues by inducing hydrolytically labile acetal-like groups within the alginates, it is paved to modulate the degradation rate of alginate gels (Kong, Kaigler, Kim, & Mooney, 2004). Nevertheless, the gel is transformed by the oxidation and it becomes more malleable. Then, it subjects to a reverse relation between degradation rate and gel rigidity.

9.6.3.5 Chitosan

Chitosan is regarded as one of the significant biodegradables and biocompatible polymer, consisted of β-(1,4)-linked 2-deoxy-2-amino-D-glucopyranose and somewhat of β-(1,4)-connected 2-deoxy-2-acetamido-D-glucopyranose. It is achieved by the alkaline deacetylation of chitin which is the main constituent of the exoskeletal in crustaceans (Mahanta et al., 2015). Xia reported that definite enzymes (chitinase/chitosanase) and nonspecific enzymes (cellulases, pectinases, papains, and lipases) hydrolyze chitosan (Xia, Liu, & Liu, 2008). The chitinase is isolated by Brzezinska et al. that produces actinomycete *Streptomyces rimosus* for biodegradation of chitinous substances. After 4 weeks of implantation in rats with respect to the control of only 7% mass loss, ~21% mass loss can be displayed by the lysozyme loaded chitosan films. Capillary electrophoresis—mass spectroscopy has contributed to detect the degraded products of monomer and oligomer of glucosamine and *N*-acetyl-glucosamine (Mawad et al., 2015).

9.6.3.6 Guar gum

Another polygalactomannan achieved from the seeds of *Cyamopsis tetragonalobus* is derived from guar gum (GG). It is taken from the seeds of *C. tetragonalobus*, which is native to the northwestern region of India. GG is identified as a nonionic, water-soluble, and biodegradable hetero-polysaccharide, made up of a β(1→4) D-mannose backbone which is accidentally connected to α(1→6) D-galactose units (Mudgil, Barak, & Khatkar, 2012). The physicochemical and rheological characteristics of GG are significantly influenced by enzymatic hydrolysis of GG. It is effectually combined in food products as dietary fiber; it is identified as a beneficial one (Maran et al., 2014).

9.6.3.7 Agar

Primarily, Agar is made up of a combination of assorted galactans, mostly consisted of 3,6-anhydro-L-galactoses (or L-galactose-6-sulfates) D-galactoses and L-galactoses (routinely in the forms of 3,6-anhydro-L-galactoses or L-galactose-6-sulfates) alternately linked by β-(1,4) and α-(1,3) linkages (Chi, Chang, & Hong, 2012). Seawater and marine sediments are the habitats of several microorganisms hydrolyzing and metabolizing as the main sources of carbon and energy. The hydrolysis of agar is catalyzed by Agarolytic microorganisms such as *Saccharophagus degradans* and *Streptomyces coelicolor* normally produce α-agarase, β-agarase and β-porphyranase enzyme (Hanani et al., 2014). A reduction in agar mechanical properties triggered through a decrease in molecular size and a diminution in the number of sulfate groups are upheld by the photo degradation process, temperature and humidity fluctuations. Crystallinity is changed by the aforementioned variation. Basically, it subjects to develop microfractures and embrittlement and stimulate microbial attack (Freile-Pelegrín et al., 2007).

9.7 Determination techniques for microbial degradation of plastics

There is a variety of plastic degrading microorganism relying on several parameters including environmental conditions, temperature, and pH. Among environmental factors, the moisture content performs a significant role in the development of microorganisms. The microbes can be activated by the water content of soil (Tokiwa, Calabia, Ugwu, & Aiba, 2009). The increase of moisture content may leads to enhance the hydrolytic cleavage of microbes. Temperature and pH are considered as the main parameters influencing degradation (Wilcox, Van Sebille, & Hardesty, 2015). The increment of temperature leads to reduce the degradation capacity of enzymes. Thus polymers with high melting point are less probable to be degraded. The changes in pH influence the rate of hydrolytic reaction. The microbial growth rate is affected by changes in pH. Also, the degradation rate is influenced (Fukushima et al., 2010). To quantify the biodegradation activities toward oil-based synthetic polymeric materials, various techniques can be practiced altogether.

9.7.1 Gravimetric determination of weight loss

Assessing the gravimetric weight loss is considered as one of the commonly methods to analyze the polymer degradation assays. Definitely, this method is not deliberated as evidence to confirm the degradation particularly about plastics which their formulations are dealt with a remarkable amount of additives. In consequence, the release of additives or their soluble components can be responsible for the weight loss. Since weight losses rely on chemical hydrolysis and fragmentation/disintegration of plastics, they should be conservatively and carefully deduced particularly while interpreting weathered polymeric materials or to a loss of the material. Moreover, when the substrate is powder polymer, it might subject to overestimate the weight loss percentage. According to the fact that biodegradation extents are slow and restricted, the weight loss is a gradual process. To evaluate and describe the assumed biodegradation activities, it is necessary to integrated the gravimetric technique with other ones (Giacomucci, Raddadi, Soccio, Lotti, & Fava, 2019).

9.7.2 Thermogravimetrical analysis

The thermal stability of a polymer is measured by TGA. Therefore any decline in the stability in this polymer is recognized as a sign of polymer degradation. In case the plastic directed to TGA with high amount of additives in its formulation, more attention should be paid. It is noteworthy that heat stabilizers are also the additive constituents contributing the last thermal stability of the plastic. The same is detected in PVC plastic film including up to 35% (w/w) of additive constituents. Indeed, it is suitable to consider biodegradation of the polymer chains as the plastic film directed to biodegradation assay. In comparison with its matching abiotic control film, a lower thermal stability is revealed (Giacomucci et al., 2020).

9.7.3 Differential scanning calorimetric analysis

DSC is applied as a suitable technique to evaluate various thermal features of materials and ponder the thermal transition of synthetic polymers, such as glass transition (Tg). Lucas et al. (2008) demonstrated that a reduction in the steadiness of the polymer resulted from its degradation is frequently caused by changing Tg to lower temperatures.

9.7.4 Gel permeation chromatography

The inputs on the number average molecular weight (M_n) and the molecular weight distribution of the polymer are provided by GPC. Chain scission is proved by a reduction in the M_n of the polymer. As the analysis is done on the bulk polymer, this technique cannot be very sensitive. It is counted as one of its downsides because it hinders finding the biodegradation of the polymer in the early stages. The reason is that this type of biodegradation primarily takes place on the polymer surface. Henceforth, if this technique is mingled with other ones, it will be efficient enough to evaluate the synthetic plastics biodegradation (Ali et al., 2014).

9.7.5 Fourier transform infrared spectroscopy

FTIR is usually applied to divulge the chemical alterations of the polymer structure and observe chemical changes in polymeric film. Lucas et al. indicated that microbial attack helps detect the development of functional groups. This technique makes it hard to identify the functional groups formed because of microbial attack to the polymer chains. Indeed, if the plastic film formulation includes plenty of additives, this technique is not helpful enough. Moreover, it is essential to completely remove the biofilm developed on the plastic surface. In consequence, the functional groups are not erroneously interpreted or identified. In fact, the structural changes of the polymer and not to the cell debris are epitomized by the observed functional groups (Lucas et al., 2008).

9.7.6 Microscopy observations of the surface

The formation of cracks and holes or the formations of biofilm are the physical features, which their evaluation contributes to assess polymer biodegradation. This evaluation indicates the microbial colonization. Initially, the stereomicroscopy observations are deliberated and these observations are followed by dint of higher magnification microscopes like SEM. The use of SEM observations paves the way to discover the microbial colonization of the polymer/plastic surface. As the polymer surface could be used by the microbe as a support for biofilm formation, the colonization of the polymer cannot be a reliable evidence to prove the biodegradation ability of the microorganism. To verify the biodegradation activity, more studies are necessary (Esmaeili, Pourbabaee, Alikhani, Shabani, & Esmaeili, 2013).

9.7.7 Radiolabeling

The classification of the carbon in the polymer is included in this technique to be practiced as substrate for microbial development with carbon isotope ^{14}C. Assessing the radioactive gas produced ($^{14}CO_2$, $^{14}CH_4$) verifies the mineralization. The hitches and cost of preparing the radioactive polymer as well as the necessity of specific measures for management and disposal of the radiolabeled samples limit the usage of this technique even though it is not critical and very accurate for the assessment of polymer biodegradation (Federle et al., 2002).

9.7.8 Standard methods

ISO (International Organization for Standardization), CEN (European Committee for Standardization), and ASTM (American Society for Testing and Materials) are among the organizations which have developed various standard tests to assess the biodegradation of plastics/polymers under aerobic or anaerobic conditions. The evaluation of gas produced after a specific incubation period is the foundation of these tests. Nonetheless, a negligible amount of gas could be released in the case of plastics derived from petroleum. It seems that this method is not trustworthy because its sensitivity is very low. Additionally, it is essential to be used along with others. Furthermore, the gas production might be related to the degradation of other compounds existing in the matrix in which the test is done (compost, soil, etc.). Eventually, Harrison et al. stated that no specific technique has been presented to biologically assess the degradability of plastic in aquatic systems (Ho et al., 2018).

9.7.9 Other recently reported analytical techniques

Along with the aforementioned analytical methods typically practiced for the assessment of conventional plastics/polymers biodegradation, they have not been optimized or standardized so that the biodegradation of fossil-based conventional plastics/polymers can be regularly checked. For example, these contain: (1) reflectometric interference spectroscopy, a convenient method for the assessment of difference in the physical thickness of a biodegradable polymer and which has been used for observing enzymatic biodegradation of thin PCL polymer film but not yet applied in the case of conventional plastics/polymers (Ooya, Sakata, Choi, & Takeuchi, 2016); and (2) elemental analyzer/isotope ratio mass spectrometry, a method founded on the approximation of carbon stable isotopes (^{13}C) which could reveal (bio)degradation of plastic material as shown by an escalation of ^{13}C values (Berto et al., 2017).

9.8 Future prospects

Using BP in every field is the most creative way to meet the problems associated with the accrual plastics waste from different sources Specific applications are excessively expanding as the result of the production of BP including packaging stuff, disposable medical items, agricultural films, fishery materials, surgical framework, and so forth. Forthcoming studies and research should aim to improve the use of biobased BP in definite usages to develop a safer environment. Several microbial research groups are able to transform certain plastics into simpler products. It is highly needed to synthesize BP. Plastics can be used again in case they are recycled in a balanced way. Scientists who have done extensive studies in this issue should strive to develop ecofriendly plastics or materials to benefit a more sustainable environment.

9.9 Conclusion

The environment and ecosystem are threatened by the destructive effects of burning plastics or the accrual of huge amount of plastics in landfills. Furans and dioxins are examples of gases produced by burning plastic. These are harmful greenhouse gases which reduce ozone layer. Severe health problems and soil pollution are caused by dioxin. Henceforth, several research has been conducted to dissect the biodegradation of plastic waste to achieve a safer environment and diminish the perilous effects of plastics on environment. Reportedly, several fungal genera (e.g., *Acremonium, Cladosporium, Debaryomcyces, Emericellopsis, Eupenicillium, Aspergillus, Aureobasidium, Penicillium,* and *Thermoascus*) and bacterial genera (e.g., *Schlegelella, Amycolatopsis, Clostridium,* and *Pseudomonas*) have been able to degrade different types of plastics like PHB, PPL, PCL, PEA (Polyesteracetals), and PVC. Polyester and PUR are among plastics which microorganisms can easily biodegrade through making hydrolytic cleavage. It is better to utilize BP more in our daily life so as to benefit safer aquatic life. It is supposed that researchers figure out more microbial strains for the potential of biodegradation. Reportedly, the biodegradation rates and extents are frequently restricted and the accessible data demonstrate that microbial reworking to conventional plastics arising in the environment is gradually happening in both terrestrial and aquatic compressed settings. Considerably, plastic wastes have a great contribution to marine trash and its undesirable consequences on the marine ecosystem are being verified. Under terrestrial environments, the biodegradability of major types of plastics and plastic wastes has been mostly assessed. This is done on the postulation that plastics are expected to be finally composed or landfilled. As a matter of fact, it is essential to measure better alternatives to specify the real in situ biodegradation of wastes of conventional plastics in marine habitats. Holistically, it is concluded that biodegradation of microplastics is able to somewhat help decrease in situ marine litter or alleviate environmental influence of plastic pollution. Hence, it is essential to achieve strategies which aim at improving the recycling rates of distinctly collected plastics and efficiently managing the plasmix. Remarkably, marine litter problem can be abandoned by (1) forbidding unnecessary plastic products and limit the usage of plastic microgranules in profitable products, (2) merging discerning and effectual collection and recycling of the dissimilar plastics utilized in our everyday life, (3) slowly accepting biodegradable (bio)plastics, through beginning from those used in marine habitats for making ready fishing gears, tubular net for marine aquaculture, mussel-culture socks or additives for painting and upkeep of ships and vacation boats; and (4) removing the rubbish dump and landfill discharge of microplastics.

References

Abou-Zeid, D.-M., Müller, R.-J., & Deckwer, W.-D. (2001). Degradation of natural and synthetic polyesters under anaerobic conditions. *Journal of Biotechnology, 86*, 113−126.

Adrados, A., De Marco, I., Caballero, B. M., López, A., Laresgoiti, M. F., & Torres, A. (2012). Pyrolysis of plastic packaging waste: A comparison of plastic residuals from material recovery facilities with simulated plastic waste. *Waste Management, 32*, 826−832.

Akutsu-Shigeno, Y., Teeraphatpornchai, T., Teamtisong, K., Nomura, N., Uchiyama, H., Nakahara, T., & Nakajima-Kambe, T. (2003). Cloning and sequencing of a poly(DL-lactic acid) depolymerase gene from *Paenibacillus amylolyticus* strain TB-13 and its functional expression in *Escherichia coli*. *Applied and Environmental Microbiology, 69*, 2498−2504.

Albertsson, A.-C., & Karlsson, S. (1990). The influence of biotic and abiotic environments on the degradation of polyethylene. *Progress in Polymer Science, 15*, 177−192.

Ali, M. I., Ahmed, S., Robson, G., Javed, I., Ali, N., Atiq, N., & Hameed, A. (2014). Isolation and molecular characterization of polyvinyl chloride (PVC) plastic degrading fungal isolates. *Journal of Basic Microbiology, 54*, 18−27.

Allı, S., Aydın, R. S. T., Allı, A., & Hazer, B. (2015). Biodegradable poly(ε-caprolactone)-based graft copolymers via poly(linoleic acid): In vitro enzymatic evaluation. *Journal of the American Oil Chemists' Society, 92*, 449−458.

Alshehrei, F. (2017). Biodegradation of synthetic and natural plastic by microorganisms. *Journal of Applied & Environmental Microbiology, 5*, 8−19.

Aravinthan, A., Arkatkar, A., Juwarkar, A. A., & Doble, M. (2016). Synergistic growth of *Bacillus* and *Pseudomonas* and its degradation potential on pretreated polypropylene. *Preparative Biochemistry & Biotechnology, 46*, 109−115.

Arikan, E. B., & Ozsoy, H. D. (2014). Time to bioplastics. In *Akademik platform* (pp. 743−750). Istanbul, Turkey.

Arkatkar, A., Arutchelvi, J., Bhaduri, S., Uppara, P. V., & Doble, M. (2009). Degradation of unpretreated and thermally pretreated polypropylene by soil consortia. *International Biodeterioration & Biodegradation, 63*, 106−111.

Auta, H., Emenike, C., & Fauziah, S. (2017). Screening of *Bacillus strains* isolated from mangrove ecosystems in Peninsular Malaysia for microplastic degradation. *Environmental Pollution, 231*, 1552−1559.

Auta, H. S., Emenike, C. U., Jayanthi, B., & Fauziah, S. H. (2018). Growth kinetics and biodeterioration of polypropylene microplastics by *Bacillus* sp. and *Rhodococcus* sp. isolated from mangrove sediment. *Marine Pollution Bulletin, 127*, 15−21.

Avio, C. G., Gorbi, S., & Regoli, F. (2017). Plastics and microplastics in the oceans: From emerging pollutants to emerged threat. *Marine Environmental Research, 128*, 2−11.

Becidan, M. (2007). *Experimental studies on municipal solid waste and biomass pyrolysis*.

Berto, D., Rampazzo, F., Gion, C., Noventa, S., Ronchi, F., Traldi, U., ... Giovanardi, O. (2017). Preliminary study to characterize plastic polymers using elemental analyser/isotope ratio mass spectrometry (EA/IRMS). *Chemosphere, 176*, 47−56.

Boyandin, A. N., Prudnikova, S. V., Karpov, V. A., Ivonin, V. N., Đỗ, N. L., Nguyễn, T. H., ... Filipenko, M. L. (2013). Microbial degradation of polyhydroxyalkanoates in tropical soils. *International Biodeterioration & Biodegradation, 83*, 77−84.

Briassoulis, D., Babou, E., Hiskakis, M., & Kyrikou, I. (2015a). Degradation in soil behavior of artificially aged polyethylene films with pro-oxidants. *Journal of Applied Polymer Science, 132*.

Briassoulis, D., Babou, E., Hiskakis, M., & Kyrikou, I. (2015b). Analysis of long-term degradation behaviour of polyethylene mulching films with pro-oxidants under real cultivation and soil burial conditions. *Environmental Science and Pollution Research, 22*, 2584–2598.

Cacciari, I., Quatrini, P., Zirletta, G., Mincione, E., Vinciguerra, V., Lupattelli, P., & Sermanni, G. G. (1993). Isotactic polypropylene biodegradation by a microbial community: Physicochemical characterization of metabolites produced. *Applied and Environmental Microbiology, 59*, 3695–3700.

César, M., Mariani, P., Innocentini-Mei, L., & Cardoso, E. (2009). Particle size and concentration of poly (ε-caprolactone) and adipate modified starch blend on mineralization in soils with differing textures. *Polymer Testing, 28*, 680–687.

Chang, C.-W., Choi, D., Kim, W. J., Yockman, J. W., Christensen, L. V., Kim, Y.-H., & Kim, S. W. (2007). Non-ionic amphiphilic biodegradable PEG–PLGA–PEG copolymer enhances gene delivery efficiency in rat skeletal muscle. *Journal of Controlled Release, 118*, 245–253.

Chi, W.-J., Chang, Y.-K., & Hong, S.-K. (2012). Agar degradation by microorganisms and agar-degrading enzymes. *Applied Microbiology and Biotechnology, 94*, 917–930.

Codari, F., Moscatelli, D., Storti, G., & Morbidelli, M. (2010). Characterization of low-molecular-weight PLA using HPLC. *Macromolecular Materials and Engineering, 295*, 58–66.

Danko, A. S., Luo, M., Bagwell, C. E., Brigmon, R. L., & Freedman, D. L. (2004). Involvement of linear plasmids in aerobic biodegradation of vinyl chloride. *Applied and Environmental Microbiology, 70*, 6092–6097.

Darby, R. T., & Kaplan, A. M. (1968). Fungal susceptibility of polyurethanes. *Applied Microbiology, 16*, 900–905.

DeMoss, J. A. (2019). *Nature is our teacher: More-than-human communities within the deep nature connection movement*. University of Georgia.

Esmaeili, A., Pourbabaee, A. A., Alikhani, H. A., Shabani, F., & Esmaeili, E. (2013). Biodegradation of low-density polyethylene (LDPE) by mixed culture of *Lysinibacillus xylanilyticus* and *Aspergillus* niger in soil. *PLoS One, 8*, e71720.

Federle, T. W., Barlaz, M. A., Pettigrew, C. A., Kerr, K. M., Kemper, J. J., Nuck, B. A., & Schechtman, L. A. (2002). Anaerobic biodegradation of aliphatic polyesters: Poly(3-hydroxybutyrate-*co*-3-hydroxyoctanoate) and poly(ε-caprolactone). *Biomacromolecules, 3*, 813–822.

Felfel, R. M., Hossain, K. M. Z., Parsons, A. J., Rudd, C. D., & Ahmed, I. (2015). Accelerated in vitro degradation properties of polylactic acid/phosphate glass fibre composites. *Journal of Materials Science, 50*, 3942–3955.

Fontanella, S., Bonhomme, S., Brusson, J.-M., Pitteri, S., Samuel, G., Pichon, G., . . . Delort, A.-M. (2013). Comparison of biodegradability of various polypropylene films containing pro-oxidant additives based on Mn, Mn/Fe or Co. *Polymer Degradation and Stability, 98*, 875–884.

Freile-Pelegrín, Y., Madera-Santana, T., Robledo, D., Veleva, L., Quintana, P., & Azamar, J. (2007). Degradation of agar films in a humid tropical climate: Thermal, mechanical, morphological and structural changes. *Polymer Degradation and Stability, 92*, 244–252.

Fukushima, K., Abbate, C., Tabuani, D., Gennari, M., Rizzarelli, P., & Camino, G. (2010). Biodegradation trend of poly(ε-caprolactone) and nanocomposites. *Materials Science and Engineering: C, 30*, 566–574.

Gao, C., Liu, M., Chen, J., & Zhang, X. (2009). Preparation and controlled degradation of oxidized sodium alginate hydrogel. *Polymer Degradation and Stability, 94*, 1405–1410.

Garcia, J. (1999). Comparative degradation study of biodegradable microspheres of poly(DL-lactide-*co*-glycolide) with poly(ethyleneglycol) derivates. *Journal of Microencapsulation, 16*, 83–94.

Geyer, R., Jambeck, J., & Law, K. (2017). Producción, uso y destino de todos los plásticos jamás fabricados. *Science Advances, 3*, 1207–1221.

Ghosh, S. K., Pal, S., & Ray, S. (2013). Study of microbes having potentiality for biodegradation of plastics. *Environmental Science and Pollution Research, 20*, 4339–4355.

Giacomucci, L., Raddadi, N., Soccio, M., Lotti, N., & Fava, F. (2019). Polyvinyl chloride biodegradation by *Pseudomonas citronellolis* and *Bacillus flexus*. *New Biotechnology, 52*, 35–41.

Giacomucci, L., Raddadi, N., Soccio, M., Lotti, N., & Fava, F. (2020). Biodegradation of polyvinyl chloride plastic films by enriched anaerobic marine consortia. *Marine Environmental Research*, 104949.

Gilpin, R. K., Wagel, D. J., & Solch, J. G. (2003). Production, distribution, and fate of polychlorinated dibenzo-p-dioxins, dibenzofurans and related organohalogens in the environment. *Dioxins and Health*, 55–87.

Gnanavel, G., Valli, V., & Thirumarimurugan, M. (2012). A review of biodegradation of plastics waste. *International Journal of Pharmaceutical and Chemical Sciences, 1*, 670–673.

Haider, T. P., Völker, C., Kramm, J., Landfester, K., & Wurm, F. R. (2019). Plastics of the future? The impact of biodegradable polymers on the environment and on society. *Angewandte Chemie International Edition, 58*, 50–62.

Hakkarainen, M., Karlsson, S., & Albertsson, A. C. (2000). Influence of low molecular weight lactic acid derivatives on degradability of polylactide. *Journal of Applied Polymer Science, 76*, 228–239.

Hanani, Z. N., Roos, Y., & Kerry, J. (2014). Use and application of gelatin as potential biodegradable packaging materials for food products. *International Journal of Biological Macromolecules, 71*, 94–102.

Ho, B. T., Roberts, T. K., & Lucas, S. (2018). An overview on biodegradation of polystyrene and modified polystyrene: The microbial approach. *Critical Reviews in Biotechnology, 38*, 308–320.

Imre, B., & Pukánszky, B. (2013). Compatibilization in bio-based and biodegradable polymer blends. *European Polymer Journal, 49*, 1215–1233.

Ishigaki, T., Sugano, W., Nakanishi, A., Tateda, M., Ike, M., & Fujita, M. (2004). The degradability of biodegradable plastics in aerobic and anaerobic waste landfill model reactors. *Chemosphere, 54*, 225–233.

Jain, P., Jain, A., Singhai, R., & Jain, S. (2017). Effect of bio-degradation and non degradable substances in environment. *International Journal of Life Sciences, 1*, 58–64.

Jain, R., Kosta, S., & Tiwari, A. (2010). Polyhydroxyalkanoates: A way to sustainable development of bioplastics. *Chronicles of Young Scientists, 1*, 10.

Jeon, H. J., & Kim, M. N. (2016). Isolation of mesophilic bacterium for biodegradation of polypropylene. *International Biodeterioration & Biodegradation, 115*, 244–249.

Jeong, B., Bae, Y. H., & Kim, S. W. (2000). In situ gelation of PEG-PLGA-PEG triblock copolymer aqueous solutions and degradation thereof. *Journal of Biomedical Materials Research: An Official Journal of the Society for Biomaterials, the Japanese Society for Biomaterials, and the Australian Society for Biomaterials and the Korean Society for Biomaterials, 50*, 171–177.

Jeyakumar, D., Chirsteen, J., & Doble, M. (2013). Synergistic effects of pretreatment and blending on fungi mediated biodegradation of polypropylenes. *Bioresource Technology, 148*, 78–85.

Johnston, B., Radecka, I., Hill, D., Chiellini, E., Ilieva, V. I., Sikorska, W., ... Keddie, D. (2018). The microbial production of polyhydroxyalkanoates from waste polystyrene fragments attained using oxidative degradation. *Polymers, 10*, 957.

Kale, G., Auras, R., & Singh, S. P. (2006). Degradation of commercial biodegradable packages under real composting and ambient exposure conditions. *Journal of Polymers and the Environment, 14*, 317–334.

Kamaly, N., Yameen, B., Wu, J., & Farokhzad, O. C. (2016). Degradable controlled-release polymers and polymeric nanoparticles: mechanisms of controlling drug release. *Chemical Reviews, 116*, 2602–2663.

Kavlock, R., Boekelheide, K., Chapin, R., Cunningham, M., Faustman, E., Foster, P., & Zacharewski, T. (2002). NTP Center for the Evaluation of Risks to Human Reproduction: phthalates expert panel report on the reproductive and developmental toxicity of di(2-ethylhexyl) phthalate. *Reproductive Toxicology (Elmsford, N.Y.), 16*, 529–653.

Kim, M. N., & Park, S. T. (2010). Degradation of poly(L-lactide) by a mesophilic bacterium. *Journal of Applied Polymer Science, 117*, 67–74.

Kong, H. J., Kaigler, D., Kim, K., & Mooney, D. J. (2004). Controlling rigidity and degradation of alginate hydrogels via molecular weight distribution. *Biomacromolecules, 5*, 1720–1727.

Kunioka, M., Kawaguchi, Y., & Doi, Y. (1989). Production of biodegradable copolyesters of 3-hydroxybutyrate and 4-hydroxybutyrate by *Alcaligenes eutrophus*. *Applied Microbiology and Biotechnology, 30*, 569–573.

Lee, S. K., Seong, D. G., & Youn, J. R. (2005). Degradation and rheological properties of biodegradable nanocomposites prepared by melt intercalation method. *Fibers and Polymers, 6*, 289–296.

Lemaire, J., Arnaud, R., & Gardette, J.-L. (1991). Low temperature thermo-oxidation of thermoplastics in the solid state. *Polymer Degradation and Stability, 33*, 277–294.

Li, M., Witt, T., Xie, F., Warren, F. J., Halley, P. J., & Gilbert, R. G. (2015). Biodegradation of starch films: The roles of molecular and crystalline structure. *Carbohydrate Polymers, 122*, 115–122.

Li, S., Tenon, M., Garreau, H., Braud, C., & Vert, M. (2000). Enzymatic degradation of stereocopolymers derived from L-, DL- and meso-lactides. *Polymer Degradation and Stability, 67*, 85–90.

Li, Y., Rodrigues, J., & Tomas, H. (2012). Injectable and biodegradable hydrogels: Gelation, biodegradation and biomedical applications. *Chemical Society Reviews, 41*, 2193–2221.

Lin, Y.-H., & Yang, M.-H. (2009). Tertiary recycling of commingled polymer waste over commercial FCC equilibrium catalysts for producing hydrocarbons. *Polymer Degradation and Stability, 94*, 25–33.

Lucas, N., Bienaime, C., Belloy, C., Queneudec, M., Silvestre, F., & Nava-Saucedo, J.-E. (2008). Polymer biodegradation: Mechanisms and estimation techniques—A review. *Chemosphere, 73*, 429–442.

Ludevese-Pascual, G., Laranja, J. L., Amar, E., Bossier, P., & De Schryver, P. (2019). Artificial substratum consisting of poly-β-hydroxybutyrate-based biodegradable plastic improved the survival and overall performance of postlarval tiger shrimp *Penaeus monodon*. *Aquaculture Research, 50*, 1269–1276.

Mahanta, A. K., Mittal, V., Singh, N., Dash, D., Malik, S., Kumar, M., & Maiti, P. (2015). Polyurethane-grafted chitosan as new biomaterials for controlled drug delivery. *Macromolecules, 48*, 2654–2666.

Maharana, T., Mohanty, B., & Negi, Y. (2009). Melt−solid polycondensation of lactic acid and its biodegradability. *Progress in Polymer Science, 34*, 99–124.

Maiti, P., & Prakash Yadav, J. P. (2008). Biodegradable nanocomposites of poly(hydroxybutyrate-*co*-hydroxyvalerate): The effect of nanoparticles. *Journal of Nanoscience and Nanotechnology, 8*, 1858–1866.

Majid, M., Hori, K., Akiyama, M., & Doi, Y. (1994). In Y. Doi, & K. Fukuda (Eds.), *Biodegradable plastics and polymers*. New York: Elsevier, BV.

Mao, H., Liu, H., Gao, Z., Su, T., & Wang, Z. (2015). Biodegradation of poly(butylene succinate) by *Fusarium* sp. FS1301 and purification and characterization of poly(butylene succinate) depolymerase. *Polymer Degradation and Stability, 114*, 1–7.

Maran, J. P., Sivakumar, V., Thirugnanasambandham, K., & Sridhar, R. (2014). Degradation behavior of biocomposites based on cassava starch buried under indoor soil conditions. *Carbohydrate Polymers, 101*, 20–28.

Marega, C., & Marigo, A. (2015). Effect of electrospun fibers of polyhydroxybutyrate filled with different organoclays on morphology, biodegradation, and thermal stability of poly(ε-caprolattone). *Journal of Applied Polymer Science, 132*.

Mawad, D., Warren, C., Barton, M., Mahns, D., Morley, J., Pham, B. T., ... Lauto, A. (2015). Lysozyme depolymerization of photo-activated chitosan adhesive films. *Carbohydrate Polymers, 121*, 56–63.

Mergaert, J., Webb, A., Anderson, C., Wouters, A., & Swings, J. (1993). Microbial degradation of poly(3-hydroxybutyrate) and poly(3-hydroxybutyrate-*co*-3-hydroxyvalerate) in soils. *Applied and Environmental Microbiology, 59*, 3233–3238.

Miao, M., Li, R., Huang, C., Jiang, B., & Zhang, T. (2015). Impact of β-amylase degradation on properties of sugary maize soluble starch particles. *Food Chemistry, 177*, 1–7.

Mohanrasu, K., Premnath, N., Prakash, G. S., Sudhakar, M., Boobalan, T., & Arun, A. (2018). Exploring multi potential uses of marine bacteria; an integrated approach for PHB production, PAHs and polyethylene biodegradation. *Journal of Photochemistry and Photobiology B: Biology, 185*, 55–65.

Mohee, R., & Unmar, G. (2007). Determining biodegradability of plastic materials under controlled and natural composting environments. *Waste Management, 27*, 1486−1493.

Mor, R., & Sivan, A. (2008). Biofilm formation and partial biodegradation of polystyrene by the actinomycete *Rhodococcus ruber*. *Biodegradation, 19*, 851−858.

Mudgil, D., Barak, S., & Khatkar, B. (2012). Effect of enzymatic depolymerization on physicochemical and rheological properties of guar gum. *Carbohydrate Polymers, 90*, 224−228.

Mulbry, W., Reeves, J. B., & Millner, P. (2012). Use of mid-and near-infrared spectroscopy to track degradation of bio-based eating utensils during composting. *Bioresource Technology, 109*, 93−97.

Murray, E., Thompson, B. C., Sayyar, S., & Wallace, G. G. (2015). Enzymatic degradation of graphene/polycaprolactone materials for tissue engineering. *Polymer Degradation and Stability, 111*, 71−77.

Nakajima-Kambe, T., Onuma, F., Kimpara, N., & Nakahara, T. (1995). Isolation and characterization of a bacterium which utilizes polyester polyurethane as a sole carbon and nitrogen source. *FEMS Microbiology Letters, 129*, 39−42.

Nampoothiri, K. M., Nair, N. R., & John, R. P. (2010). An overview of the recent developments in polylactide (PLA) research. *Bioresource Technology, 101*, 8493−8501.

Nikolaidis, P., & Poullikkas, A. (2017). A comparative overview of hydrogen production processes. *Renewable and Sustainable Energy Reviews, 67*, 597−611.

Nishida, H., Suzuki, S., & Tokiwa, Y. (1998). Distribution of poly(β-propiolactone) aerobic degrading microorganisms in different environments. *Journal of Environmental Polymer Degradation, 6*, 43−58.

Ohkita, T., & Lee, S. H. (2006). Thermal degradation and biodegradability of poly(lactic acid)/corn starch biocomposites. *Journal of Applied Polymer Science, 100*, 3009−3017.

Ooya, T., Sakata, Y., Choi, H. W., & Takeuchi, T. (2016). Reflectometric interference spectroscopy-based sensing for evaluating biodegradability of polymeric thin films. *Acta Biomaterialia, 38*, 163−167.

Otake, Y., Kobayashi, T., Asabe, H., Murakami, N., & Ono, K. (1995). Biodegradation of low-density polyethylene, polystyrene, polyvinyl chloride, and urea formaldehyde resin buried under soil for over 32 years. *Journal of Applied Polymer Science, 56*, 1789−1796.

Palmisano, A. C., & Pettigrew, C. A. (1992). Biodegradability of plastics. *Bioscience, 42*, 680−685.

Park, K., & Xanthos, M. (2009). A study on the degradation of polylactic acid in the presence of phosphonium ionic liquids. *Polymer Degradation and Stability, 94*, 834−844.

Paul, M.-A., Delcourt, C., Alexandre, M., Degée, P., Monteverde, F., & Dubois, P. (2005). Polylactide/montmorillonite nanocomposites: Study of the hydrolytic degradation. *Polymer Degradation and Stability, 87*, 535−542.

Pawar, S. P., Kumar, S., Misra, A., Deshmukh, S., Chatterjee, K., & Bose, S. (2015). Enzymatically degradable EMI shielding materials derived from PCL based nanocomposites. *RSC Advances, 5*, 17716−17725.

Pelegrini, K., Donazzolo, I., Brambilla, V., Coulon Grisa, A. M., Piazza, D., Zattera, A. J., & Brandalise, R. N. (2016). Degradation of PLA and PLA in composites with triacetin and buriti fiber after 600 days in a simulated marine environment. *Journal of Applied Polymer Science, 133*.

Pérez, J., Munoz-Dorado, J., De la Rubia, T., & Martinez, J. (2002). Biodegradation and biological treatments of cellulose, hemicellulose and lignin: An overview. *International Microbiology, 5*, 53−63.

Pranamuda, H., & Tokiwa, Y. (1999). Degradation of poly(L-lactide) by strains belonging to genus *Amycolatopsis*. *Biotechnology Letters, 21*, 901−905.

Rajmohan, K. V. S., Ramya, C., Viswanathan, M. R., & Varjani, S. (2019). Plastic pollutants: Effective waste management for pollution control and abatement. *Current Opinion in Environmental Science & Health, 12*, 72−84.

Rigamonti, L., Grosso, M., Møller, J., Sanchez, V. M., Magnani, S., & Christensen, T. H. (2014). Environmental evaluation of plastic waste management scenarios. *Resources, Conservation and Recycling, 85*, 42−53.

Rist, S., Almroth, B. C., Hartmann, N. B., & Karlsson, T. M. (2018). A critical perspective on early communications concerning human health aspects of microplastics. *Science of the Total Environment, 626*, 720−726.

Röttig, A., & Steinbüchel, A. (2013). Acyltransferases in bacteria. *Microbiology and Molecular Biology Reviews, 77*, 277–321.

Roy, P. K., Hakkarainen, M., Varma, I. K., & Albertsson, A.-C. (2011). Degradable polyethylene: Fantasy or reality. *Environmental Science & Technology, 45*, 4217–4227.

Saad, G. R., Khalil, T. M., & Sabaa, M. W. (2010). Photo-and bio-degradation of poly(ester-urethane) s films based on poly [(R)-3-hydroxybutyrate] and poly(ε-caprolactone) blocks. *Journal of Polymer Research, 17*, 33.

Sen, S. K., & Raut, S. (2015). Microbial degradation of low density polyethylene (LDPE): A review. *Journal of Environmental Chemical Engineering, 3*, 462–473.

Shah, A. A., Hasan, F., Hameed, A., & Ahmed, S. (2008). Biological degradation of plastics: A comprehensive review. *Biotechnology Advances, 26*, 246–265.

Shah, A. A., Nawaz, A., Kanwal, L., Hasan, F., Khan, S., & Badshah, M. (2015). Degradation of poly(ε-caprolactone) by a thermophilic bacterium *Ralstonia* sp. strain MRL-TL isolated from hot spring. *International Biodeterioration & Biodegradation, 98*, 35–42.

Sharma, B., & Jain, P. (2020). Deciphering the advances in bioaugmentation of plastic wastes. *Journal of Cleaner Production*, 123241.

Shen, M., Song, B., Zeng, G., Zhang, Y., Huang, W., Wen, X., & Tang, W. (2020). Are biodegradable plastics a promising solution to solve the global plastic pollution? *Environmental Pollution (Barking, Essex: 1987), 263*, 114469.

Shimao, M. (2001). Biodegradation of plastics. *Current Opinion in Biotechnology, 12*, 242–247.

Sikorska, W., Musiol, M., Nowak, B., Pajak, J., Labuzek, S., Kowalczuk, M., & Adamus, G. (2015). Degradability of polylactide and its blend with poly [(R, S)-3-hydroxybutyrate] in industrial composting and compost extract. *International Biodeterioration & Biodegradation, 101*, 32–41.

Singh, N., Hui, D., Singh, R., Ahuja, I., Feo, L., & Fraternali, F. (2017). Recycling of plastic solid waste: A state of art review and future applications. *Composites Part B: Engineering, 115*, 409–422.

Singh, N. K., Purkayastha, B. D., Roy, J. K., Banik, R. M., Yashpal, M., Singh, G., ... Maiti, P. (2010). Nanoparticle-induced controlled biodegradation and its mechanism in poly(ε-caprolactone). *ACS Applied Materials & Interfaces, 2*, 69–81.

Singh, N. K., Purkayastha, B. P. D., Panigrahi, M., Gautam, R. K., Banik, R. M., & Maiti, P. (2013). Enzymatic degradation of polylactide/layered silicate nanocomposites: Effect of organic modifiers. *Journal of Applied Polymer Science, 127*, 2465–2474.

Singh, N. K., Purkayastha, B. P. D., Roy, J. K., Banik, R. M., Gonugunta, P., Misra, M., & Maiti, P. (2011). Tuned biodegradation using poly(hydroxybutyrate-*co*-valerate) nanobiohybrids: Emerging biomaterials for tissue engineering and drug delivery. *Journal of Materials Chemistry, 21*, 15919–15927.

Singhania, R. R. (2009). *Cellulolytic enzymes. Biotechnology for agro-industrial residues utilisation* (pp. 371–381). Springer.

Sivan, A., Szanto, M., & Pavlov, V. (2006). Biofilm development of the polyethylene-degrading bacterium *Rhodococcus ruber*. *Applied Microbiology and Biotechnology, 72*, 346–352.

Skariyachan, S., Manjunatha, V., Sultana, S., Jois, C., Bai, V., & Vasist, K. S. (2016). Novel bacterial consortia isolated from plastic garbage processing areas demonstrated enhanced degradation for low density polyethylene. *Environmental Science and Pollution Research, 23*, 18307–18319.

Skariyachan, S., Patil, A. A., Shankar, A., Manjunath, M., Bachappanavar, N., & Kiran, S. (2018). Enhanced polymer degradation of polyethylene and polypropylene by novel thermophilic consortia of *Brevibacillus* sps. and *Aneurinibacillus* sp. screened from waste management landfills and sewage treatment plants. *Polymer Degradation and Stability, 149*, 52–68.

Skariyachan, S., Setlur, A. S., Naik, S. Y., Naik, A. A., Usharani, M., & Vasist, K. S. (2017). Enhanced biodegradation of low and high-density polyethylene by novel bacterial consortia formulated from plastic-contaminated cow dung under thermophilic conditions. *Environmental Science and Pollution Research, 24*, 8443–8457.

Steinbach, J. M., Seo, Y.-E., & Saltzman, W. M. (2016). Cell penetrating peptide-modified poly(lactic-*co*-glycolic acid) nanoparticles with enhanced cell internalization. *Acta Biomaterialia, 30*, 49–61.

Sudesh, K., & Iwata, T. (2008). Sustainability of biobased and biodegradable plastics. *Clean—Soil, Air, Water, 36*, 433–442.

Sudhakar, M., Trishul, A., Doble, M., Kumar, K. S., Jahan, S. S., Inbakandan, D., ... Venkatesan, R. (2007). Biofouling and biodegradation of polyolefins in ocean waters. *Polymer Degradation and Stability, 92*, 1743–1752.

Sugihara, A., Ueshima, M., Shimada, Y., Tsunasawa, S., & Tominaga, Y. (1992). Purification and characterization of a novel thermostable lipase from *Pseudomonas cepacia*. *The Journal of Biochemistry, 112*, 598–603.

Sun, J., Shen, J., Chen, S., Cooper, M. A., Fu, H., Wu, D., & Yang, Z. (2018). Nanofiller reinforced biodegradable PLA/PHA composites: Current status and future trends. *Polymers, 10*, 505.

Syranidou, E., Karkanorachaki, K., Amorotti, F., Franchini, M., Repouskou, E., Kaliva, M., ... Corvini, P. F.-X. (2017). Biodegradation of weathered polystyrene films in seawater microcosms. *Scientific Reports, 7*, 1–12.

Tian, L., Kolvenbach, B., Corvini, N., Wang, S., Tavanaie, N., Wang, L., ... Ji, R. (2017). Mineralisation of 14C-labelled polystyrene plastics by *Penicillium variabile* after ozonation pre-treatment. *New Biotechnology, 38*, 101–105.

Tokiwa, Y., & Calabia, B. P. (2006). Biodegradability and biodegradation of poly(lactide). *Applied Microbiology and Biotechnology, 72*, 244–251.

Tokiwa, Y., Calabia, B. P., Ugwu, C. U., & Aiba, S. (2009). Biodegradability of plastics. *International Journal of Molecular Sciences, 10*, 3722–3742.

Tokiwa, Y., & Suzuki, T. (1977). Hydrolysis of polyesters by lipases. *Nature, 270*, 76–78.

Torres, A., Li, S., Roussos, S., & Vert, M. (1996). Degradation of L- and DL-lactic acid oligomers in the presence of *Pseudomonas putida* and *Fusarium moniliforme*. *Journal of Environmental Polymer Degradation, 4*, 213–223.

Vert, M., Doi, Y., Hellwich, K.-H., Hess, M., Hodge, P., Kubisa, P., ... Schué, F. (2012). Terminology for biorelated polymers and applications (IUPAC Recommendations 2012). *Pure and Applied Chemistry, 84*, 377–410.

Vesilind, P. A., Peirce, J. J., & Weiner, R. F. (2013). *Environmental pollution and control*. Elsevier.

Vieyra, H., Aguilar-Méndez, M. A., & San Martín-Martínez, E. (2013). Study of biodegradation evolution during composting of polyethylene–starch blends using scanning electron microscopy. *Journal of Applied Polymer Science, 127*, 845–853.

Voinova, O., Gladyshev, M., & Volova, T. G. (2008). *Comparative study of PHA degradation in natural reservoirs having various types of ecosystems. Macromolecular symposia* (pp. 34–37). Wiley Online Library.

Volova, T. G., Prudnikova, S. V., Vinogradova, O. N., Syrvacheva, D. A., & Shishatskaya, E. I. (2017). Microbial degradation of polyhydroxyalkanoates with different chemical compositions and their biodegradability. *Microbial Ecology, 73*, 353–367.

Vroman, I., & Tighzert, L. (2009). Biodegradable polymers. *Materials, 2*, 307–344.

Wang, H., Liu, Q., Zhao, X., & Jin, Z. (2020). Synthesis of reactive DOPO-based flame retardant and its application in polyurethane elastomers. *Polymer Degradation and Stability, 183*, 109440.

Wang, J., & Wan, Z. (2015). Treatment and disposal of spent radioactive ion-exchange resins produced in the nuclear industry. *Progress in Nuclear Energy, 78*, 47–55.

Wilcox, C., Van Sebille, E., & Hardesty, B. D. (2015). Threat of plastic pollution to seabirds is global, pervasive, and increasing. *Proceedings of the National Academy of Sciences, 112*, 11899–11904.

Willis, K., Maureaud, C., Wilcox, C., & Hardesty, B. D. (2018). How successful are waste abatement campaigns and government policies at reducing plastic waste into the marine environment? *Marine Policy, 96*, 243–249.

Woolnough, C. A., Yee, L. H., Charlton, T., & Foster, L. J. R. (2010). Environmental degradation and biofouling of "green" plastics including short and medium chain length polyhydroxyalkanoates. *Polymer International, 59*, 658–667.

Xia, W., Liu, P., & Liu, J. (2008). Advance in chitosan hydrolysis by non-specific cellulases. *Bioresource Technology, 99*, 6751–6762.

Zheng,, Y., Yanful,, E. K., & Bassi,, A. S. (2005). A review of plastic waste biodegradation. *Critical Reviews in Biotechnology, 25*, 243–250.

10
Nano-biodegradation of polymers

Komal Rizwan[1], Tahir Rasheed[2], Muhammad Bilal[3]

[1]DEPARTMENT OF CHEMISTRY, UNIVERSITY OF SAHIWAL, SAHIWAL, PAKISTAN
[2]SCHOOL OF CHEMISTRY AND CHEMICAL ENGINEERING, SHANGHAI JIAO TONG UNIVERSITY, SHANGHAI, P.R. CHINA [3]SCHOOL OF LIFE SCIENCE AND FOOD ENGINEERING, HUAIYIN INSTITUTE OF TECHNOLOGY, HUAI'AN, P.R. CHINA

10.1 Introduction

Due to the marvelous performance of poly(ethylene terephthalate), poly(vinyl chloride) (PVC), polystyrene (PS), polypropylene, and polyethylene (PE) are extensively used as nonbiodegradable plastics for the last few decades. For many years after the time of expiration, most of these plastics persist and are producing plastic waste in a massive amount. Plastic production has been increased drastically from 1.5 million tons to 245 million tons from 1950 to 2008 (Chanprateep, 2010). China, Japan, and the United States produced PVC-based waste approximately 1,686,000, 214,000, and 435,000 tons, respectively (Wu et al., 2014). Every year many kinds of petroleum-derived synthetic plastics produce 140 million tons, and the industrial waste of these plastics are introduced into the environment in significant amounts (Nampoothiri, Nair, & John, 2010). The European Bioplastics Association and the US Environmental Protection Agency (EPA) reported on the production and recycling of plastics (US EPA, 2007). After all, these plastics are nonbiodegradable so unaffected to microbial attack, and after use, these are used to fill land in burial sites for many years, which causes scarcity of space and producing bad effects onto the animals by eating and enticement.

One of the prominent causes of change in climatic conditions results to cause air pollution is the nonscientific mechanism of burning that is adopted due to lack of landfill space. For the installation of plastics, the oceans recognize as the main sink and plastics are changed into microscopic fiber-like pieces, which distinguished as "microplastic" and collect in the ground of the ocean. Though food chain biomagnification occurs at a higher level by engulfing these microplastics by aquatic living organisms, and it causes cancer, impairs reproduction, and also affects the endocrine systems (Bang et al., 2012). For example, styrene, which is considered a neurotoxin, is a chemical that is used for the production of PS used for food-item packaging (damage the peripheral and central nervous systems), the source for cancer, and chromosomal abnormalities (Dare et al., 2004). PVC is produced mainly from an industrial solvent the vinyl chloride (VC). At life-history cellular and gene level the toxicity of VC has determined in marine living organisms *Daphnia magna*.

The action of juvenile hormone the esterase in *D. magna* increased prominently by the exposure of VC and cases metamorphosis and premature reproduction (Houde, Douville, Gagnon, Sproull, & Cloutier, 2015). Alternation in cytokine expression and oxidative stress leads to asthma and allergies are caused by phthalates, which is measured as a well-known plasticizer, is an endocrine-disrupting chemical compound (North, Takaro, Diamond, & Ellis, 2014). Imbalance of ecological hierarchy is a great threat due to the accumulation of untreated wastes. Consequently, to control these environmental issues due to nonbiodegradable plastic disposal, the production of biodegradable plastics has been increased; through this action environmental pollution can also be controlled.

10.1.1 Broad-spectrum perspectives and status of biodegradable polymers

The synthetic plastics, which are not degradable, can be substituted by biodegradable polymers as a sustainable alternative. Globally, every year biodegradable plastic accomplished less than 1% of the 181 million metric tons of synthetic plastics produced. However, every year the market of bioplastic is grown by 20%–30%, but this growth rate is not sufficient to meet the global need (Nampoothiri et al., 2010). In 2013 the 1.6 million tons of bioplastics global production need is accounted for according to European Bioplastic Association, and by 2018 the 6.7 million tons of bioplastics were expected to increase (I. F. B. A. B. European Bioplastics, Nova-Institut, 2014). To replace the synthetic plastics, commercially the biodegradable plastics such as poly(lactic acid) (PLA), poly(ε-caprolactone) (PCL), poly(butylene succinate-co-adipate) (PBSA), polyhydroxyvalerate, polyhydroxyalkanoates (PHA), and poly(butylene succinate) (PBS) have synthesized.

Hydrolyzable ester bonds of biodegradable plastics, especially polyesters, have gained attention because they can be changed into oligomer or monomer respective units in bioactive environments. Under ordinary conditions like conventional plastics, biodegradable plastics can be used in a likewise manner, and after disposal biodegradable plastics could be degraded in the existence of microorganisms. According to ISO 14855-2 by using a microbial oxidative degradation analyzer, the biodegradability of PLA and PCL measured under controlled conditions and according to TC61/SC5/WG22, in which biodegradation of plastics standardized by nine tests. Information on biodegradable plastics and biomass-based plastic is provided by Japan who is a leader in biotechnology among Asian countries.

Bioplastics are divided into two categories by Japan Bioplastics Association (JBPA) (J. B. Association, 2021) as: (1) biodegradable plastics: synthetic polymer with an at least average mol. weight of 1000 Da, referred to biodegradable synthetic plastics. Biodegradable high polymers may include chemically improved poly(amino acid) and starch. In the natural environment the first biodegradable plastic was green PLA, which converted into carbon dioxide and water microbiologically. (2) Biomass-based plastics produced by using raw material of biomass. Biomass-based plastics are produced by using renewable organic ingredients either biochemically or chemically. Further, biomass-based plastics divided into two categories: (1) partially biomass-based plastics and (2) totally biomass-based plastics. Twenty percent of petrochemical plastics have been aimed to replace biomass-based plastics by JBPA. The same kind of bioplastics production ideas and European Bioplastics Organization has reported degradation of bioplastics, for example, either from the plastics or from renewable resources (biobased) derived bioplastics are compostable and biodegradable. It is proclaimed by the European Bioplastics Association that biobased plastic may be 100% nonbiodegradable; on the other hand, fossil-fuel based plastics might be 100% biodegradable.

Lim and his colleagues (Lim, Raku, & Tokiwa, 2005) studied the hydrolysis of different polyesters as PCL, PHB (polyhydroxybutyrate), PLA, and respective copolymers, and suggested that biodegradation is not dependent on the resource of a specific polymer. Polyesters have features similar to conventional plastics. Natural polymers include PHAs and PLA, while the petroleum-based polymers include PBS and PCL, and are biodegradable in nature. Biocompatible and biodegradable polyester like PLA is formed in the microbial fermentation process by using corn as resources. The great intention has been received by PLA due to its effective properties and its low cost. Synthesis of PLA can be carried out by the ring-opening polymerization of lactide or through the condensation of lactic acid by using a catalyst. PLA polymer present in three forms of isomers (1) poly (DL-lactide), (2) poly (D-lactide), and (3) poly (L-lactide) according to enantiomers ratio.

Bacterial polymer like PHAs produces as a byproduct intracellulary in a biochemical process in many bacteria and stores as energy backup and carbon. Depending upon different growth conditions and different bacterial species, about 120 different variations exist in the structure of PHAs. The PHA bacterial polymer consists of poly(3-hydroxybutyrate-co-3-hydroxy valerate), pol (3-hydroxybutyrate), and other related copolymers. Depending upon the structure of poly(3-hydroxybutyrate-co-3-hydroxy valerate) it has a melting point between $120°C-180°C$ and sold under the label of Biopol. Through the ring-opening polymerization process, the PCL is a synthetic biodegradable polymer prepared by using ε-caprolactone as raw material and has a melting temperature between $60°C-65°C$. The depletion rate of PCL depends on the degree of crystallinity and molecular weight. PBS is a biodegradable polyester synthetic in nature that has some outstanding properties like better toughness, much chemical resistance, and thermal properties, which make it important in the biodegradable polymers family. PBS is prepared from glycols (ethylene glycol and/or 1, 4-butanediol) and aliphatic dicarboxylic acids (adipic and succinic acid) by the polycondensation process.

10.1.2 Challenges and benefits of biodegradable polymers

Biodegradable plastics can bear different environmental influences while they are in use, which is a positive factor, and in the bioactive agents they undergo the degradation process easily. Biodegradable plastics have a very important role due to their important applications in the agricultural field, and with packaging, applications in biomedical implementation, and environmental safety. Due to low production cost and extremely hydrolyzable ester bonds, PLA, which is a biodegradable polymer, is frequently and commercially used (Fukushima et al., 2011). Rasal and his colleagues (Rasal, Janorkar, & Hirt, 2010) revealed that petroleum-based plastic (PBP) formation requires more energy while the PLA formation requires less production energy, which is

25%−55% lower than PBP. Moreover, in the future, this may be developed more. So, in the place of petrochemicals as a resource, the lactic acid obtained from the fermentation process used as a renewable source has increased. The physicochemical parameters can be used to control the formation of active (optically) molecules either D-(−)-lactic acid or L-(+)-lactic acid and obtained stereospecific copolymers. Therefore, in comparison with the chemical synthesis of lactic acid, the biotechnological production formation gives many advantages (John, Nampoothiri, & Pandey, 2006; Rasal et al., 2010).

As biodegradable polyesters are obtained from sustainable resources and are compostable, in this way solid waste disposal issues can be controlled, since biodegradable polyesters are a very auspicious material. These biodegradable polyesters are frequently used for food packaging and in different other consumer products because they have very little immunogenic characteristics (Lim, Auras, & Rubino, 2008). In many different areas of interest, the biodegradable polymers proved to be very profitable. Many biodegradable polymers characteristics like slow degradation rate, restricted gas barrier properties, mechanical resistance, poor thermal, low-melt viscosity, low-heat distortion temperature, brittleness make its access limited to industrial sectors (Fukushima, Tabuani, & Camino, 2009; Kumar, Depan, Tomer, & Singh, 2009). To overcome these disadvantages the process of copolymerization, filled system, and blending may use to increase biodegradable polymers thermomechanical characteristics.

To meet the functional and structural characteristics of biodegradable polymers in a very reasonable way, the nanometer dimension particles can be used in the formation of nanocomposites. Many properties of biodegradable polymers can be changed by the introduction of nanoparticles into the biopolymers. In comparison to the unfilled biopolymers the 1−5 wt.% layered silicate nanocomposites showed enhanced flame retardance, barrier, mechanical, thermal, and degradation behavior characteristics (Kumar et al., 2009). The economical production of biodegradable polymer is also important (Hamad, Kaseem, Ko, & Deri, 2014), and the regaining of biopolymer through fermentation is another factor that limits its production. The production of polymers may also lower due to the lower growth rate of microorganisms in agrarian's byproducts, and as a result, it has eventually enhanced the downstream processing cost (Hamad et al., 2014). In this respect, confinement and screening of strong and productive microorganisms (particularly bacterial and fungal strains) from the common environments are fundamental to utilize less rate feedstock. PHB crude material accounts for about 30%−40% of the whole cost (Kim & Chang, 1998). Subsequently, the investigation has concentrated, to develop the growth media by employing low-cost carbon substrates to decrease the developmental rates and to enhance the productivity for controlled conditions. The fermentations physicochemical parameters depend on the temperature, pH, and medium incubation time which effect the metabolism process of microbes which enhance the production of bioploymer.

Bora (2013) investigated that PHB production from *Bacillus megaterium* bacteria varied from 24%−48% (w/w) (dry cell weight). On PHB production and cell development the Box Bohn plan used to study the four factors which have interactive effect. For optimization of fermenting media the integrator approach of measurable strategies and response surface methodology was utilized. This made a difference in understanding the intuitive among different physicochemical conditions in the fabrication of PHB. This research opened the way to new research and also decreased the many experiments for multiple factors and also increased the generation of PHB. In any case with the latter progress in the recombinant techniques and hereditary engineering, to advance PHA generation, the PHA genes are expressed in different bacteria through the cloning process. To get an understanding of the biochemical pathways, which includes the production of biodegradable polymers, a wider study is required with genetic profiles of microbes.

10.1.3 Bioplastics, biodegradable polymers, and biodegradation

Biodegradation (biochemical process) includes hydrolysis of bonds in different conditions through the microbes or their various excreted molecules as surfactants and enzymes taking after the exact path. Biodegradation might be anaerobic or aerobic, which depends on the type of degradation (Nair & Laurencin, 2007). The material is referred to as "biodegradable" if it decayed into extreme end products of water and CO_2 in the presence of a certain microbial environment. The degradation of biodegradable polymers emphatically depends on the chemical structure of biopolymers rather than their crude materials (http://www.bpf.co.uk/plastipedia/polymers/polymer-bio-based-degradables.aspx). Renewable carbon resources (particularly biomass) are used fully or partially to obtain biobased plastics, which satisfying the concept of "carbon neutrality" to decreases the outflow of carbon dioxide and reliance on fossil assets. The carbon dioxide produced after the burning of biobased plastic again changed to biomass by the process of photosynthesis is an example of carbon neutrality.

Bioplastics are nonbiodegradable or biodegradable it is not a matter of concern. Therefore, the usage of bio or renewable carbon for the production of bioplastic reduces the carbon footprints as opposite to fossil and petro carbon (Iwata, 2015; Lagaron & Lopez-Rubio, 2011). According to the European bioplastic association, the biodegradable could be 100% fossil-based plastic and nonbiodegradable might be 100% biobased plastics. Like polymer of PE which is obtained from ethanol in the chemical dehydration process is not biodegradable while on the other hand biobased PHA is biodegradable (http://www.plastice.org/fileadmin/files/biobased.pdf; available at http://www.sciencedirect.com/science/article/pii/S0040403998005036). In addition, products are recognized as biodegradable whereas more data about level of biodegradation, time frame and essential environmental conditions should be given. Subsequently, plastic fabric products are characterized as a bioplastic on the off chance that it is either biodegradable, biobased, or highlights both characteristics.

There are three main groups are present in bioplastic products/material:

1. Bioplastics are specifically extricated from biomass, such as lipids, proteins like soy, and gluten, and can be polysaccharides (cellulose, starch, and chitosan). Through destructurization and in the particular sums of plasticizers like polyalcohols and water starch can be converted into thermoplastic (Mohanty, Misra, & Hinrichsen, 2000). Some others are cellulose-based bioplastics, pea protein-based bioplastics, and soy-based

bioplastics (Fernández-Espada, Bengoechea, Cordobés, & Guerrero, 2016; Perez-Puyana, Felix, Romero, & Guerrero, 2016).

2. Bioplastic gained from biomass-based monomers some chemical action for their transformation to bioplastic, for example, production of PLA from the monomer of lactic acid intermediate acquired from the fermentation of corn-starch required some changes chemically.
3. Genetically or natural microbes that can collect in between the cells in the form of storage granules can be used to produce the microbial bioplastics such as PBS, PHB, and PHA. Bioplastics may be different from others in their downstream processing, fermentation methods, fermentation conditions, media ingredients, type of microorganisms, stress conditions, survival rate, and microbial origin in physical characteristics, macromolecular structure, and monomer composition (Keshavarz & Roy, 2010).

More PHA collection appeared in acidic (pH, 6; 39%), monitored by basic (pH, 8; 46%) and higher in neutral (pH, 7; 56%). As of now, the most impediments for bulk bioplastics production, recovery, and cost-effectiveness.

10.2 Different kinds of degradation of polymers

Degradation is not a reversible process, in degradation, the biological agents, thermal, chemical erosion, and photooxidation reduce the molecular weight and cause a mechanical failure. As a result of the decaying process, the half-life of products gets decreased and becomes brittle and delicate (Pandey, Reddy, Kumar, & Singh, 2005). The process of degradation happens primarily through the splitting of the backbone or side chains of the polymer. In natural conditions, the degradation or decaying of polymers is provoked by the cohabitation of biotic and abiotic components. The biotic system incorporates microorganisms, protein, or the presence of both that execute organic action for their development and utilizing polymers. While abiotic factor like presence or absence of oxygen, salinity, humidity, temperature, and pH initiates the benign development of microbial species and also create impacts on their metabolic action. In this manner, bioreactors are outlined in this conduct that imitates natural conditions amid degradation. The biodegradability of distinctive plastics in the presence of aerobic conditions prominently depends on the sort of biodegradable plastics (Adamcová & Vaverková, 2014). The degradation of biopolymeric products is also affected by physical and chemical factors such as radiolysis or photolysis, hydrolysis, oxidation, and thermal activation.

10.2.1 Weathering

The natural and environmental conditions such as rainfall, radiation, salinity, oxygen, soil texture, pH, and temperature also regularly affect the aging behavior of biopolymers (polyesters). Chain splitting of polymers and crosslinking reactions that occur in a continuous manner can cause structural changes during the maturing process of polyesters. Both responses cause the misfortune of the physical characteristics of the matured polymers. Hence, aging characteristics are vital variables to choose polymeric products expiry date (Fang et al., 2014). Poly(3-hydroxybutyrate-co-3-hydroxyvalerate)/poly(butylene adipate coterephthalate) mix with halloysite nanoclay film and then exposed to photooxidative weathering under UV appearing with more prominent color alter and lessening in transparency in nanocomposite film in comparison of the unfilled polymer because of structure alterations (Scarfato, Acierno, & Russo, 2015). Photoaging involves recombination and chain splitting and this phenomenon is considered an oxidative process. Fang et al. (2014) synthesized the antiaging polyester elastomer (BEE) nanocomposites by utilizing carbon-blend to study the glass transition temperature at various temperatures, the mechanism through tensile testing, and aging behavior. Under these factors, BEE appeared with the dominance of chain splitting at the advanced stage and with initial crosslinking phenomena. Stloukal et al. (2012) showed that photooxidation can develop a similar insoluble polymeric gel (crosslinking). During photooxidation, the copolyester with bigger aromatic content is also pronounced. Hence, against crosslinking, PLA displayed chain sessions significantly (Stloukal et al., 2012). The loss of tensile quality/ tensile strength within the tune of 53.7% and 78% (in the xenon chamber) for 72 h appeared in weathering of sugar palm starch biocomposite and sugar palm fibers (Sahari, Sapuan, Zainudin, & Maleque, 2014). PLA, PBSA, PBS, PCL, and PHBV polymers were examined through weight loss strategy under the effect of soil factors and climatic conditions and degradation order was found to be PBSA = PHBV = PCL > PBS > PLA (Hoshino et al., 2001).

10.2.2 Photodegradation

This is a type of degradation in which gamma-ray and ultraviolet rays are used as a source of high-frequency electromagnetic radiation to irradiate and check the degradation of the polymer. Mass of polymeric molecules decreases after absorption of a photon and this occurs due to chain scission of the polymeric molecule. Chromophoric motifs are necessary to initiate the photoresponse because they are important to the absorbance of light. To explain the photooxidation of polymers, the singlet oxygen mechanism has been proposed and certain sensitizers/inducer is mixed to polymer framework, which actuates the exchange of energy in the polymer framework (Yousif & Haddad, 2013). Combination of a polymeric molecule with various inorganic and organic compounds considered to assist the free radicals to produce hydroperoxide (Rabek & Ranby, 1975).

10.2.3 Thermal degradation

Thermal energy of the system increased with the elevation of temperature and subsequently at high-temperature process of degradation also increased. Hyon and colleagues (Hyon, Jamshidi, & Ikada, 1984) studied PLA filaments were 50% degraded at 100°C after 30 h and it remained undisturbed at 37°C for six months in phosphate buffer. Fukushima and his colleagues (Fukushima et al., 2011) examined the impact of temperature on degradation of PLA in phosphate buffer exhibiting a significantly quicker debasement at 58°C against the control at 37°C. In the presence of both aerobic conditions and anaerobic conditions the poly(L-lactide) (PLLA) biodegradation observed to elevate at 52°C than 37°C (Itävaara, Karjomaa, & Selin, 2002). PLLA degradation rate increased significantly when the temperature is near to glass transition (55°C) (Itävaara et al., 2002). The same mechanism is followed for degradation of PLLA, while activation energies for poly(lactic-co-glycolic acid) (PLGA) degradation at below and over Tg temperatures was found to be quite variable (Agrawal, Huang, Schmitz, & Athanasiou, 1997; Weir, Buchanan, Orr, Farrar, & Dickson, 2004). A number of reactions occur during thermal degradation of oligo-L-lactide at 200°C in the presence of nitrogen atmosphere which includes the production of various carbon- and oxygen-centered macroradicals, selective Sn-catalyzed depolymerization, concerted nonradical reactions, radical reactions, cis elimination, intermolecular ester exchange, and intramolecular ester exchange (Kopinke, Remmler, Mackenzie, Möder, & Wachsen, 1996). At high temperature, the weakening of molecules occurs due to hydrolysis and overheating polyesters, on other hand, the process of degradation of the polyesters is not so simple it includes a number of different processes (Carrasco, Pagès, Gámez-Pérez, Santana, & Maspoch, 2010).

10.2.4 Hydrolysis

An alteration in the refractive index of the test sample as a result of H_2O uptake is induced by the method of hydrolytic degradation of polymer matrix (Li & McCarthy, 1999). The ester motifs in the PLA framework are hydrolyzed by the presence of hydroxyl (OH) motifs in clays or silicate layers. In the amorphous zone the process of hydrolytic degradation of PLA occurs very quickly in water medium while the production of lactic acid oligomer occurs in the crystalline region, which leads to enhanced carboxylic acid motifs concentration in the medium. Intrinsically, the hydrolytic degradation is catalyzed by the carboxylic acid groups. Hence, it is studied that PLA hydrolytic degradation occurred in the bulk of the materials it does not takes place at its surface. The productions of degradation items and lessen of lactic acid oligomer are found to follow the two factors such as dielectric constant and pH of the media. Chain-end splitting occurs in the hydrolysis process taking place in acidic pH. On the other hand, lactoyl lactate scission proceeds in basic media (De Jong et al., 2001). PLA depolyemrization occurs by the attack of alkaline species on esteric group carbon which leads to hydrolysis of the polymer (Lucas et al., 2008). An intramolecular transesterification process under the basic conditions explained the PLA chain-end degradation, whereas the protonation of OH motif creates an intramolecular hydrogen bonding in the acidic media. Basic hydrolysis occurs quickly in comparison to basic hydrolysis especially in the case of polyester. Under acidic and basic conditions the mechanism of hydrolysis is explained in Fig. 10−1B and C. Dopico-García et al. (2013) has reported the alkali decomposition products, enhancing concentration of the lactic acid, and hydrolyzed oligomers.

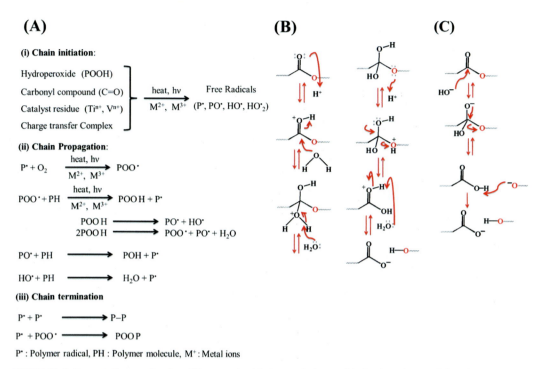

FIGURE 10–1 Degradation mechanism: (A) general oxidation and photooxidation in polymers; (B) acidic hydrolysis; and (C) basic hydrolysis. *Reproduced from Yousif, E., & Haddad, R. (2013). Photodegradation and photostabilization of polymers, especially polystyrene. SpringerPlus, 2, 398, with permission from Springer Open.*

10.2.5 Biodegradation

Biodegradation is a process in which the original chemical compound or inorganic molecules formed from organic wastes. In presence of microorganism or enzymes this biochemical process happens. Two phases come out in the biodegradation of higher molecular weight polymers

1. Depolymerization: polymer diffuse through the cell membrane; and
2. Cellular metabolism: monomeric constituents assimilate.

In the primary phase the depolymerization process is encouraged by extracellular enzymes, secreted by the microbial cells and this occurs by the means of hydrolysis mechanism and during this the bigger and complex molecules break down to simpler ones such as dimer or oligomers and eventually into single monomeric constituents. In the second phase to produce the carbon dioxide gas and water molecules from the low molecular weight monomeric molecules, the low molecular weight monomers pass through the cell membranes and the various kind of biochemical reactions to make the original molecule. Due to conversion of biodegradable plastics particularly polyester into water and carbon dioxide gas under anaerobic conditions along with the formation of methane, these biodegradable polymers gained a lot of attention for their potential hydrolysable ester linkages.

Biodegradable plastic can be degraded with three different processes named as (1) enzymatic, (2) microbial, and (3) compost. Singh et al. (2011) studied the biodegradation of poly (hydroxybutyrateco-valerate) and poly(hydroxybutyrateco-valerate) nanohybrids with layered silicate for their percentage weight loss by utilizing pure enzyme and microorganism (*Pseudomonas stutzeri*) as compost medium. In comparison to compost the biodegradation of plastic found considerably more in enzymatic media (lipase and proteinase-k) for same set of samples. Degradation rate can be influenced by employment of different nanoclays, as 15A nanoclay enhanced the degradation, whereas 30B nanoclay decreases the degradation in comparison to the pure PHBV, and this degradation rate remained intact in microorganism, enzyme media, and also in compost. Ethyl ester of lactoyl lactic acid obtained by role of microorganism for conversion of lactoyl lactic acid and lactic acid in biotic media, and it had not been found in the abiotic media (Hakkarainen, Karlsson, & Albertsson, 2000).

The new degradation products as propanoic acid, acetic acid, and ethyl ester of lactoyl lactic acid produced, when lactoyl lactic acids, and lactic acid quickly assimilated from polymer films in the biotic medium determined through gas chromatography—mass spectrometry (GC-MS). Higher oligomers, lactyllactic acid dimers and lactic acid were found in liquid culture during degradation of DL-Lactic acid oligomers and L-Lactic acid oligomers in the existence of *Pseudomonas putida* and *Fusarium moniliforme*. All this process and compounds are observed by utilizing HPLC. In the presence of controlled conditions L-lactic acids and DL-lactic acids was completely utilized by two microorganisms irrespective of enantiomers (Torres, Li, Roussos, & Vert, 1996). Codari and coworkers (Codari, Moscatelli, Storti, & Morbidelli, 2010) quantified the degradation and the formation of PLA (by polycondensation process).

10.2.5.1 Composting

Compositing could be a natural process of decomposition and a transformation, in which a humus-like matter is produced from biodegradable organic waste by the recycling and treating. Microbial inoculum like algae, fungi, bacteria, and mineral salts are used for the organic waste degradation in presence of aerobic or aerobic conditions. Actual conditions for composting are based on various factors as type of compost, temperature, medium pH, and moisture. The increased temperature, more moisture, and plenitude of microbes in compost is demonstrated good and reasonable atmosphere for PLA biodegradation (Kale, Auras, & Singh, 2006, 2007). The effect of microbes existing in the compost during biodegradation of the PLA and PLA/a were checked by NMR and EI/MS investigation of compost extracts (Sikorska et al., 2015). If 90% of original polymers dried weight sample passed by 2.0 millimeter sieves was considered as reasonable deterioration of plastic waste (Shah, Hasan, Hameed, & Ahmed, 2008). The BNQ - 9011-911/2007 (Canada), ISO 18606 (worldwide), AS 4736 (Australia), ISO 17088, ASTM D 6868 (US), ASTM D 6400-04, and European standard EN 13432 (2000) are the different agencies of the world which certified the present compostable plastics in market (Sikorska et al., 2015). Enhancement in poly(hydroxybutyrate-co-hydroxyvalerate) biodegradation occurred in compost medium in presence of clay or hydroxyapatite (Maiti & Prakash Yadav, 2008). Subsequent enhancement in roughness of surface of original biopolymer and biopolymer nanocomposite during biodegradation was observed. Surface roughening happened due to removal of biopolymer molecules during degradation process which completely dependent on two factors like degradation kinetics and nanoparticles used in the preparation of the bioplastics (Singh, Purkayastha et al., 2013). Two steps followed in compost media for the biodegradation of bioplastics (1) hydrolysis of higher weight molecules to low molecular weight dimers or oligomers by utilizing moisture—base or acid; (2) production of carbon dioxide, humus, and water by the microorganisms by the oligomers or dimers digestion (Maiti & Prakash Yadav, 2008).

10.2.5.2 Microbial degradation

By the action of different types of microorganisms, the polymer biodegradation occurs. The degree of microbial biodegradation of the biopolymers depends completely on the physical characteristic and chemical characteristics of polymers along with adaptation of microbiotic, abundance, diversity, and distribution of the polymers. The clear-zone technique and plate count method are the important processes utilized for the screening of polyester degrading microbes. This technique is very efficient for the investigation of potency and diversity of polymer-degrading microorganisms in the atmosphere (Mao, Liu, Gao, Su, & Wang, 2015). In the order of PHB = PCL > PBS > PLA, the aliphatic polyester-degrading microorganisms decreased (Pranamuda & Tokiwa, 1999; Tokiwa & Calabia, 2006). By attacking both amorphous and crystalline zone of the polymer, higher and low molecular weight polymer PLA degraded by mesophilic bacteria which has been enhanced with clear-zone technique (Kim & Park, 2010). Fundamental understanding of the enzymatic mechanism is required for the investigation of microbial biodegradation method. The bacteria spp. *Ralstonia sp.* (MRL-TL) within 10 days can degrade the 50 wt.% of film of PCL and in culture broth displaying an esterase activity. Same bacterium also showed the biodegradation of PHBV, PHB, PLA, and PES in the presence of harsh environment (Shah et al., 2015).

10.2.5.3 Mechanistic pathway of enzymatic degradation

Enzymes are natural catalyst and are vital choice for plastic waste administration. There are two steps involved in the enzymatic degradation of aliphatic esters polymers: (1) development of enzyme substrate complex and this happens through adsorption of enzyme on substrate surface; and (2) the soluble degraded products form in second step by the hydrolysis of esters linkages. For the biopolymer degradation the lipases, esterase, elastase, trypsin, serine proteases, cutinase, proteinase-k, PCL depolymerase, and PHB depolymerase enzymes are broadly utilized. For the biodegradation process the polymer stereochemistry played a very significant role; for example, different stereomers of PLA is degraded in the subsequent manner L-L > L-D > D-L > D-D at pH 8.6°C and 37°C (Li, Tenon, Garreau, Braud, & Vert, 2000). Enzymatic attack is facilitated by the larger amount of water uptake thereof the PLA50-rac is degraded at lower speed than PLA50-mes (Li et al., 2000). Various types of biodegradable polyesters like PCL, PES, PBSA, PBS, and PLA are degraded by the use of the DL-PLA depolymerase recombinant enzyme (Akutsu-Shigeno et al., 2003). Moreover, bacterial protease (proteinase-k) enzyme adsorb more on the surface of the nanoparticles (onto single-walled carbon nanotubes) having high surface consequently, low mass transfer resistance, more enzyme loading and higher catalytic activity obtained (Eker, Asuri, Murugesan, Linhardt, & Dordick, 2007). In the future, further modification of the surface of nanoparticles through functionalization and amendments in biopolymer are important things for polymers degradations through ester bond hydrolysis.

Various lipases share exceedingly moderated α/β hydrolase overlay comprising of the catalytic group of three serine, aspartate/glutamate, and histidine as nucleophilic elbow. Interaction of histidine with acid residues (aspartate/glutamate) occurs to readjustment of imidazole motif by shifting of hydrogens in way which increase the serine nucleophilicity. The hydrolysis of biopolymer, the polyester by lipase is separated into two successive stages; firstly, nucleophilic assault of serine on ester bond that forms tetrahedral acyl-serine intermediate which inevitably hydrolyzed by the second nucleophilic attack that is water molecule (Röttig & Steinbüchel, 2013).

The lipase activity mainly depends on the ability of donation of OH proton to the ester linkage cleavage as delineated within the Fig. 10–2. Medium pH highly affects the enzymatic

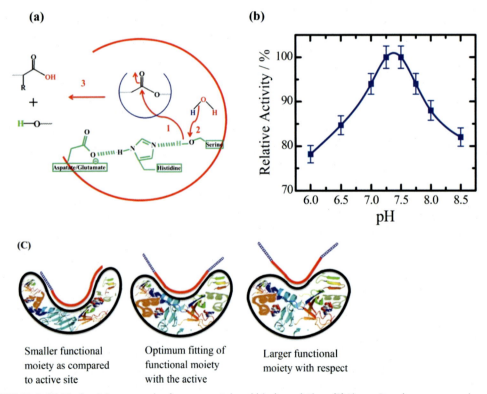

FIGURE 10–2 (A) Mechanistic approach of enzyme catalyzed biodegradation, (B) Lipase Pseudomonas cepacia activity as a function of pH showing the highest activity at pH ~7.4, (C) Models depicting the conformational changes of enzyme (Lipase) at various pH. *Reproduced from Kumar, S., & Maiti, P. (2015). Understanding the controlled biodegradation of polymers using nanoclays. Polymer, 76, 25–33 with permission from Elsevier.*

action for specific substrate through changes of active sites at variable pH (Sugihara, Ueshima, Shimada, Tsunasawa, & Tominaga, 1992). With the alteration of media pH, enzyme conformation also gets changed and change in conformations can varied the speed of hydrolysis and degradation of the biopolymers in enzyme medium (Kumar & Maiti, 2015). The action of depolymerase lipase *Pseudomonas cepacia* toward PCL, PLA, and PU was exceptionally distinctive and showed the most elevated action at acidic (pH ~ 6.5), neutral (pH ~ 7.2) and basic condition (pH ~ 7.7) PU, PLA, and PCL, respectively. Fascinatingly, various altered nanoclays, specifically 30B, 15A, and C18, scattering in H_2O have comparable pH near peak in curve of pH against action of depolymerase (Kumar & Maiti, 2015). The biodegradation notably decreased utilizing other two nanoclays containing variable pH. A demonstration of changed configuration of enzyme was reported at variable pH for a particular polymer chain (Kumar & Maiti, 2015). The hydrolytic action of *Fusarium solani* cutinase shows more prominent biodegradation at pH 8, whereas it drops its ability at pH 3 driving to scant loss of weight in the same time. Mostly surface charge around active sites gets to be more positive at profoundly acidic condition responsible in a loss of cutinase ability for PCL biodegradation. Contrastingly, the *Humicola insolens* derived cutinase exhibited 60% loss of weight at pH 3 (in 6 h) to observe the enzyme ability loss as pH function. The surface potentials (electrostatic) of *F. solani* at pH 5 and 3 is considered 4.4 and 13.3, whereas *H. insolens* shows 1.1, 5.6, 15.8 values at pH 8, 5.0, and 3.0 (Baker, Poultney, Liu, Gross, & Montclare, 2012). By adding proper disulfide bonds and surface charge engineering the enzyme stability and functions can be enhanced over variable pH values.

The enzymatic biodegradation of PBS-co-butylene adipate is lower in comparison of biodegradation of PBS-co-hexane succinate (PBS-co-HS), emerging from stereoisomerism influence of the backbone of the copolymer. By utilizing MALDI-TOF-MS the enzymatic biodegradation has examined for PBS-based copolymers. In expansion, mechanism of enzymatic biodegradation and the arrangement of protein substrate complex has evaluated with molecular-docking phenomena. Substrates docking having the hexane succinate (HS) within the *P. cepacia* lipase active site is steadier in comparison of other different substrates. Catalytic residues of Ser-87 interacts with ester linkages, the carbonyl motifs linked with oxygen residues of Leu-17 and Gln-88 that assists the proper pathway (catalytic). Addition of linear hexa monomers enhance the PBS copolymer hydrophobicity and this led to increase the catalytic degradation potential. PBS-co-HS may reach the down side of active site to tie the active-residue amid coupling to PC lipase enzyme, Subsequently enhance the catalytic effectiveness by enhancing degradation potential (Li, Zhang, Qin, Zhang, & Qiu, 2015).

By immobilizing lipase within the blended organic solvent having a little water, the biodegradation of 1,4-cyclohexane dimethanol (PBS-co-CL-co-CHDM), PCL, ternary-copolymer of PBS, 1,4-cyclohexane dimethanol (PBS-co-CHDM), parallel copolymer of PBS, ε-caprolactone copolymer (PBS-co-CL), and PBS are catalyzed. The degradation rate observed were 40, 43, 53 and 54% in PBS, P(BS-co-CL), P(BS-co-CHDM), P(BS-co-CL-co-CHDM). By using NMR examination the biodegraded products of PBS-co-CL are confirmed (Ding, Zhang, Yang, & Qiu, 2012).

10.3 Influence of various factors on the polymers biodegradability

The biodegradation is dependent on physical and chemical properties of polymeric molecules beside abiotic and biotic variables. Abiotic variables like salinity, humidity, temperature, pH, existence, or nonappearance of oxygen affect the degradation of polymers, the biotic variables such as microorganism's variety differences, populace, and activity of the enzyme. In addition, the degree of biodegradation depends on the characteristics of polymer counting morphology, cross connecting, resistance to electromagnetic radiation, thermal tolerance, mechanical strength, chemical reactivity, porosity, diffusivity, and purity (Nampoothiri et al., 2010). The morphology of surface (as hydrophobic characteristics, water permeability, hydrophilicity, size, shape, and surface area), processing conditions (organic/inorganic fillers, choice of additives, copolymerization and ratio of crystallinity and amorphous region), high order configurations (modulus of elasticity, melting temperature, glass transition temperature), and intrinsic properties (its distribution, molecular weight, its crosslinking, alignment of group, type of enantiomer, existence of terminal carboxyl -COOH or OH motifs, and monomeric composition) of the polymers significantly play role to find the degradation rate and its mechanistic process (Park & Xanthos, 2009). The aliphatic polyesters appear more susceptibilities to enzymes (lipolytic) and microorganisms-based degradation as compared to polyurethanes and polyesters.

Biodegradation rate decreases with increment of molecular mass as lower molecular mass PCL degrades very quickly in comparison of higher molecular mass PCL utilizing lipase. Moreover, lower molecular mass oligomers degrade completely notwithstanding of enantiomeric composition against the quicker biodegradation of higher molecular weight L-PLA in comparison of D-PLA (Tokiwa & Calabia, 2006). The biodegradation rate of polymer is affected greatly with the morphology of the polymers (Paul et al., 2005). In determining the biodegradability, the degree of crystallinity and amorphous regions plays a vital part. Within crystal configurations the polymeric chains are arranged orderly therefore making it resilient to degradation while in amorphous regions the loosely packed chains can be easily attacked by microbes and enzymes. The rate of degradation was found decreased for the samples crystallized at 52°C in comparison of those samples which were quenched at room temperature for pure PCL because of high crystallinity (Singh et al., 2010). With an enhancement in crystallinity of the polymer the degradation rate of PLA has decreased. Commonly, the rate of polymer biodegradation decreased with the higher melting temperature due to interaction among polymeric chains (Kale, Auras, Singh, & Narayan, 2007).

The lipase can hydrolyze low molecular weight L-PLA and its copolymers such as poly(D-lactide-coglycolide), poly(L-lactide-co-glycolide) and DL-PLA, whereas, higher molecular weight L-PLA, PGA (polyglycolic acid) and D-PLA are not effectively hydrolysable and resilient to enzymatic biodegradation because of its higher melting temperature (Tm) and higher crystallinity (Tokiwa & Calabia, 2006). Consequently, through copolymerization, biodegradation behavior and physical properties can be modified and make them more vulnerable toward biodegradation by bringing down the Tm and crystallinity in comparison of the homopolymers. Fillers inserted in polymer framework also play a role to control the rate of polymer biodegradation, and fillers dispersed in a better way quickly degrade the polymers in comparison of composite having intercalated nanostructure or pure polymer (Lee, Seong, & Youn, 2005).

10.4 Techniques and methodologies employed for polymer degradation

The degree of biodegradation is determined by measuring the weight loss of the sample under study. For molecular characterization almost all products obtained after biodegradation of the high-throughput analytical strategies are used which includes chromatography as HPLC, GPC (gel permeation chromatography), and GC, and that can be coupled with NMR and mass-spectroscopy for better results.

The surface analysis of materials degraded is done by scanning electron microscopy (SEM) or atomic force microscopy of optical microscopy. As a result, weight loss of polymer the degree of crystallinity examined through X-ray diffraction and differential filtering calorimetric. The different level of polymer degradation accounted by melting temperature (Tm) and glass transition (Tg) temperatures. Comparative examinations of tensile strength of polymer before and after the process of biodegradation gave useful data regarding mechanical stability of biodegraded materials (Carraher & Seymour, 2003; Grassie & Scott, 1988). To elucidate disposal and emission of carbon dioxide, the radio-labeling technique utilizing ^{14}C-labeled testing material which release $^{14}CO_2$ (carbon dioxide) through biodegradation. The ^{14}C-labeled ethylene diamine with polyester urea-urethane was vulnerable to enzymatic biodegradation observed by radiolabel released (Santerre, Labow, & Adams, 1993).

In pretargeting concepts, the short-lived isotopes such as 13N, 68Ga and 18F have utilized, which offer an incredible adaptability with respect to the time-point imaging and illustrate basic significance within starting hours (Stockhofe, Postema, Schieferstein, & Ross, 2014). Waste disposal administrations are major disadvantage of this strategy. To account the degree of biodegradation, diverse standards have been created by authorized organizations such as CEN (Europe), ISO (international standards), JIS (Japan), DIN (Germany), and ASTM (USA).

By utilizing standard methods NF X 41−514, ASTM G 22−76, ASTM G 21−90, ISO 14853 (for anaerobic condition), and ISO 14851 (for aerobic condition) various biopolymers have been studied for their biodegradation (Allı, Aydın, Allı, & Hazer, 2015). Quantitative parameters such as composition, biogas generation, pressure variations, oxygen uptake, and mass variations used to measure the efficiency of the biodegradation and percentage of degradation was assessed after 28 days of biodegradation (Massardier-Nageotte, Pestre, Cruard-Pradet, & Bayard, 2006). The oxygen is required for the process of polymer biodegradation in microbial oxidation. The utilization of the oxygen, keeping in view the biochemical oxygen demand and carbon dioxide exhaust, provides indication of degradation of the polymer and this assessed through the respirometric test and Sturm test methodologies. The biogas volume created through the biodegradation of starch/poly-caprolactone mix found 9.4% carbon dioxide and 23% methane at finishing of incubation (28 days) was measured in the anaerobic condition within the test bottles (Massardier-Nageotte et al., 2006). The biochemical oxygen demand and weight loss of biosynthetic poly (3-hydroxybutyrate-co-14% 3-hydroxyvalerate) found to be 100% and 78% for 28 days (Pawar et al., 2015).

Comparison of biodegradability of PCL, cellulose acetate (CA), and their respective mixtures was carried out as a purpose of production of CO_2 (Calil, Gaboardi, Guedes, & Rosa, 2006). Quicker biodegradation is observed in 40PCL/60CA blend in comparison of other blends because of higher carbon dioxide formation in an aerobic environment (Sturm test). Mixture of CL with PCL decreases the melting temperature of PCL but enhance the melting temperature of CA, which shows the immiscibility of polymers and assists the reduction of crystallinity and enhance the amorphous domains which favors the biodegradability (Calil et al., 2006).

During biodegradation, the production carbon dioxide was determined by the Sturm test, and its further confirmation was done by gas chromatography and infrared spectroscopy (Calmon, Dusserre-Bresson, Bellon-Maurel, Feuilloley, & Silvestre, 2000). In term of molecular weight of PLGA and drug discharge, the PLGA biodegradation at two diverse pH conditions through GPC and HPLC was assessed. Under different pH of the medium SEM imaginings uncover surface setting and formation of pores. Conceivable dispersion of the repeating units and the end groups was distinguished through Tandem MS. Reversed stage HPLC empowered the division of the homologous arrangement of polymeric molecules individually on the premise of expanding repeating nonpolar units (Dimzon et al., 2015). By utilizing HPLC, lactide linkages and glycolide biodegradation profile was determined which uncovering fact biodegradation of PLGA microspheres in comparison of PDLLA particles because of higher hydrophilicity (Giunchedi, Conti, Scalia, & Conte, 1998). Degree of crystallinity and degradation rate was examined in films having oligomer and found starting mass loss and H_2O assimilation (Schliecker, Schmidt, Fuchs, Wombacher, & Kissel, 2003).

10.5 Biodegradability of various polymers and impact of nanoparticle

Lactic acid (2-hydroxy propionic acid) is used as a precursor for the preparation of PLA, and it is part of the aliphatic polyesters. Through fermentation under oxygen-restricting conditions, PLA is produced by organisms, plants, and animals. Artificially, it may be inferred from renewable material intermediates such as acetaldehyde, ethylene, and ethanol (Södergård & Stolt, 2002). PLA is a compostable polymer and in this manner reduces disposal issue of solid waste. There are primarily two simulated composting strategies: (1) gravimetric estimation respirometric frameworks, (2) aggregate estimation respirometric framework utilized to survey the PLA bottles biodegradation (Kale, Auras, Singh, & Narayan, 2007). PCL-coated PLLA showed sluggish rate of degradation as compared to a sample, which is uncoated.

By utilizing, proteinase-k and *Rhizopus arrhizus* lipase, PLLA fiber-reinforced PCL and PCL/PLLA mixture film examined during enzymatic degradation (Tsuji, Kidokoro, & Mochizuki, 2006). A bimodal pattern has been showed in degradation (in vitro) of

PLA/phosphate glass fiber nanocomposites at 37°C and at higher temperature the crests moved toward low molecular weight. Enhancement in crystallinity was noted at higher temperature degradation because of major degradation in amorphous zones of polymeric chains (Felfel, Hossain, Parsons, Rudd, & Ahmed, 2015). The nanoparticles are nucleating agents; subsequently they regulate spherulite dimension of the network that assist to controls the biodegradation. Amorphous domains are enhanced with better interactions, and this enhances the polymer degradation.

As compared to neat polymer network, layered polylactide silicate nanocomposites exhibited significant advancement in biodegradability. Within the organic modifier of 30B nanoclay, the presence of excess OH groups quickens the hydrolytic decay (Singh, Purkayastha et al., 2013). High interaction between the functionalized MWCNT (PLA-FMWCNT) and PLA comes about littlest spherulitic dimension in PLA-FMWCNT and become vulnerable to hydrolysis carried out by the enzymes because of bigger amorphous regions. By utilizing SEM the pure PLA and PLA nanohybrids morphology are observed afore and later biodegradation for comparing the biodegradation relative rates appearing more noteworthy loss of materials from 30B nanocomposite in comparison of pure PCL and the nanocomposite having 15 A nanoclay (Singh, Purkayastha et al., 2013). In the presence of different nanoparticles such as MEE nanoclays and MAE (Dimethyl ditallow ammonium), similar trends of biodegradation showed quicker rate through the control of the amorphous substance or spherulite size of PLA (Panigrahi, Singh, Gautam, Banik, & Maiti, 2010). Polymerization (Ring-opening) of lactide and glycolide prepared PLGA-type copolymer. By changing the monomers ratio, tuning of properties of polymers is done. Increased amount of hydrophilic PGA enhances the PLGA degradation.

Through ring-opening polymerization, the PCL is prepared. Indeed in spite of the fact of its synthetic origin, PCL is not harmful and conveniently biodegradable. In four days *Penicillium oxalicum* degrades PCL creating H_2O-soluble 6-hydroxyhexanoic acid examined by the mass spectrometry (Li, Yu et al., 2012). Graphene/polycaprolactone nanocomposite is promising substance which is biodegradable exhibiting its nontoxic harmful nature (Murray, Thompson, Sayyar, & Wallace, 2015). Hence in presence of nanoparticles the more noteworthy enzyme (PCL depolymerase) action in slight alkaline nanocomposite exhibits higher biodegradation. This is to specify here that pH of PCL solution is approximately 7.0. As compared to PCL-C18 nanocomposite, the higher basic pH in PCL-30B nanocomposite explained the greater rate of biodegradation. The increase in enzymatic action in nanocomposites is because of alkaline pH in conjunction of higher amorphous substance in comparison to pure PCL.

Allı and coworkers (Allı et al., 2015) explained the enzymatic biodegradation of different grafted PCL such as PLina-g-PStg- PCL copolymer and PLina-g-PCL copolymer exhibiting controlled degradation in the presence of *Pseudomonas lipase* depending on its extent and comonomer nature. Within the framework of PCL, the sort and amount of clay and fibers have controlled the rate of biodegradation. The filled PHB strands permit to balance the characteristics of thermal steadiness of PCL and inhibit degradation. Pawar et al. (2015) has considered the impact of two distinctive multiwall carbon nanotubes as pristine (p-MWNTs) and amine-functionalized (a-MWNTs) on the PCL biodegradation (Pawar et al., 2015). The drowsy rate appeared in the presence of p-MWNTs composite utilizing lipase enzymes whereas the biodegradation is comparable in existence of a-MWNTs with regard to characteristic biodegradation of pure PCL.

One of the most promising biodegradable aliphatic polyesters are PBS, invented in the early 1990s. Against other biodegradable plastics, PBS is a competitive material due to its predominant process, high resistance to chemicals, great thermal and mechanical properties. By utilizing maleic anhydride grafted PBS (PBS-g-MA) as compatibilizer, the biodegradable nanocomposites of PBS with organo montmorillonite (OMMT) have been synthesized, and because of improved barrier characteristics on adding of OMMT it showed slower weight loss (biodegradation) as compared to neat PBS. Biodegradation in compost-soil burial measurement of PBS/jute composites exhibits approximately 62% weight loss after 180 days, which is higher in comparison to pure PBS. With expanding jute rate within the composites, the rate of biodegradation of PBS increases. As filler, the cotton fiber is employed in the PBS for enhancing biodegradability of composite and to improve polymer strength as well. Moreover, increased the rate of biodegradation observed when carbon-fiber treated with silane (a coupling agent) (Calabia et al., 2013).

PHAs are biocompatible and biodegradable polymer. Biocompatibility and natural biodegradability make them appealing substances in medical application due to their microbe-based origin. The microbial origin polymers which are easily biodegradable are alluring particularly for bioimplantations to evade the prerequisite for another operation to evacuate the polymers which are nondegradable in nature. It is not reasonable for wide applications as PHB homopolymer is hard and fragile. Biopol has deliberated to enhance the physical characteristics and this is copolymer of 3-hydroxybutyrate (HB) with 3-hydroxyvalerate (HV). Only some poly (hydroxy alkanoates) are accessible for therapeutic research including copolymer of poly 3-hydroxyoctanoate, copolymer of 3-HB and 3-hydroxyhexanoate (PHBHHx), poly(4-HB) (P4HB), 3-hydroxyvalerate (PHBV), and copolymers of 3-HB (Singh et al., 2011). In 30 days, the blend film of PCL/PHBV and poly(HB-co-valerate) (PHBV) is entirely biodegrade and the PHBV film is highly liable attack of microbes as compared to film of PCL (Gonçalves & Martins-Franchetti, 2010). In two anaerobic sludge, the biodegradation of the natural polyesters poly (β-HB-cohydroxyvalerate) (PHBV) and poly (β-HB) (PHB) is examined showing critical degradation in a time scale of 6–10 weeks. As compared to its copolyester PHBV, PHB degrades more quickly in anaerobic environment, and in oxygen consuming conditions the biodegradation rate is generally moderate (Abou-Zeid, Müller, & Deckwer, 2001).

To create nanocomposites the two-dimensional nanofiller in PHA has introduced and appeared to be a PHA-based bio-nanocomposite with progressed properties. The nanoparticle directs a controlled biodegradation and improves the physical properties within an alluring time frame under specific physiological conditions. By utilizing two diverse naturally adjusted nanoclays (15A and 30B) the rate of biodegradation of PHBV was tuned by Singh et al. (2011). While comparing the rate of degradation of pure PHBV it has significantly been sluggish in 30B containing composite while greater degradation was observed in 15A nanoclay.

10.6 Different fillers and their influence on biodegradation

Loss of weight of silica-filled-polylactide nanocomposites (PLA) synthesized has exhibited variation during the enzyme biodegradation. PLA (neat) the rate of biodegradation is approximately 0.055 mg/cm^2/h while the rates are expanded to 0.36 mg/cm^2/h for the 9 wt.% of nanosilica test sample (PLA9). Fig. 10–3 presented the quicker PLA biodegradation in comparison of the neat polymer. In phosphate buffer arrangement of pH 7.4 at 37°C the hydrolytic degradation of PLA/TiO$_2$ nanocomposites determined, illustrate the impact of nanofillers on PLA degradation (Luo, Wang, & Wang, 2012). By utilizing twin-screw extruder and self-reinforcing procedure, the composite having the 20 wt.% of osteoconductive β-tricalcium-phosphate and bioabsorbable poly-L, D-lactide (80 wt.%) synthesized. The degradation of the lattice polymer altered in the presence of the β-tricalcium phosphate. The degradation rate of the self-reinforced lattice polymer was slower than the degradation rate of the self-reinforced composites due to the neutralization of acidic degradation products of the polymer by alkaline degradation products of β-TCP (tricalcium phosphate) particles (Niemelä, 2005). By utilizing two natural friendly fillers, carbon nanospheres inferred from cellulose and calcium sulfate anhydrite synthesis of clean and green bioplastic composites of PLA has been reported (Sobkowicz, Feaver, & Dorgan, 2008). PLA composites with CaCO$_3$ coated with the stearic acid (PCC) and halloysite nanotubes compared to the morphology of

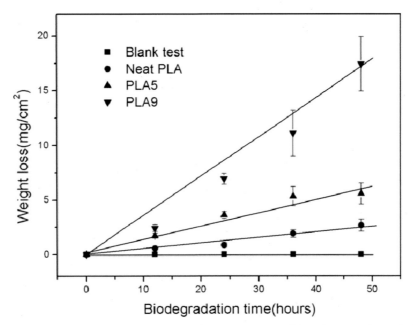

FIGURE 10–3 Weight loss profile of neat PLA and PLA-silica nanocomposite film as function of time during enzymatic degradation. *Reproduced from Li, Y., Han, C., Bian, J., Han, L., Dong, L., & Gao, G. (2012). Rheology and biodegradation of polylactide/silica nanocomposites.* Polymer Composites, 33, *1719–1727, with permission from John Wiley and Sons.*

composites, mechanical characteristics, and crystallization behavior. Cold crystallization temperatures and the glass transition were decreased by PCC nanoparticles which showed a better nucleating effect. In presence of PCC particles the morphological examination of fractures surface of PLA nanocomposites appeared great dispersion of fillers, formation of smaller scale voids, and larger plastic distortion (Shi, Zhang, Siligardi, Ori, & Lazzeri, 2015).

To form composites of biodegradable polymers, the employments of plant fiber as filler are very attractive due to their adequate accessibility at low cost. Moreover, natural filaments in composites improved the thermal properties and increase mechanical strength of biopolymers. To explore biodegradability, the composites of aliphatic polyesters such as acidic anhydride-treated (AA-) abaca filaments, poly(lactic corrosive) (PLA) with 10 wt.% untreated, PBS, PHBV, and PCL were synthesized. The weight loss of network polymers was found to be in array of PCL > PHBV > PBS > PLA (Teramoto, Urata, Ozawa, & Shibata, 2004). With different natural fibers such as hemp, kenaf, banana fiber, bamboo, jute, and flax, the PLA appeared with great mechanical and thermal properties with great biodegradability capacity. To observe the disintegration rate of PLA, okra filaments (5–10 mm) were also presented in PLA network (Fortunati et al., 2013).

10.7 Effect of filler dimension on biodegradation

Scientists are focusing on making the composites biodegradable by inserting variable dimensions of nanofillers in the polymer framework. It is vital to note that over its counterpart of small scale- or macrodimension, the nanofillers are favored due to its huge specific surface zone (Liu, Yu, Cheng, & Yang, 2009). Addition of the magnesia nanoparticles in PLLA (pure), the biodegradation rate of the g-MgO/PLLA nanocomposites was found to be expanded continuously. The mechanical property of polymer enabling adsorption of protein was improved by adding the altered MgO-nanoparticles. The enhanced mechanical characteristics, protein attachment behavior of g-MgO/PLLA nanocomposites, and quicker biodegradation rate were invaluable for biomedical application (Bikiaris, 2013; Ma et al., 2011). In the poly(3-HB-co-4-HB) [P(3HB-co-4HB)]/silica-based nanocomposites the rate of enzymatic degradation of P(3HB-co-4HB) has upgraded. Hydrophilic nature of silica enhanced the degradation by enzymes of P(3HB-co-4HB)/silica nanocomposites (Han et al., 2012). Moreover, the investigation of one dimension and two dimension nanofillers has been expended by the development of nanoscience and nanotechnology. On basis of one and two axis, the one dimension carbon nanotubes (CNT) and the two dimension nanofillers are characterized showing intercalated or exfoliated pattern in nanocomposites. CNT may lead to great strengthening filler, substituting routine nanofillers in the fabrication of advanced multifunctional polymer nanohybrids and its allotropes of carbon categorized under one dimension nanoparticles. There are two sub domains which includes single-wall-nanotubes, multiwall-nanotubes (MWNT), double-wall-nanotubes, and functionalized MWCNT (FMWCNT). Mechanical behavior, crystallization, and biodegradation of the PLA and CNT induced were detailed with great properties (Singh, Singh et al., 2013). After the enzymatic biodegradation (96 h) the least weight loss which was approximately 25% found in pure PLA. The loss of weight of PLA-MWCNT and PLA-FMWCNT found 30 and 60%, individually. Dipolar interaction between organic modifier OH motif and esteric motif of PLA, is detailed that two-dimensional silicates (layered) contains layered thicknesses (1 nm) and exceptionally high proportions (50–1000). Crystallization behavior and improved materials properties shown by the PLA/organically altered layered silicate nanocomposite with a synchronous enhancement in biodegradability as compared to flawless PLA (Ray, Yamada, Ogami, Okamoto, & Ueda, 2002). In compost peat the changes in loss of weight due to biodegradation are very moderate for the perfect PLA (greatest 17% weight loss in 240 days), whereas the rates of biodegradation were upgraded essentially to 27 wt.% and 31 wt.% loss, separately within the 15A and 30B nanoclays. In expansion to the truth that organically adjusted surface can assist initiate the compatibility with polymer lattice (Singh, Purkayastha et al., 2013).

10.8 Conclusion

In this chapter, biodegradable polymers and the role of individual biodegradable polymer with their characteristic applications have been emphasized. Diverse types of degradation have been talked about with suitable reasons and with their particular mechanism. In various media (compost, organisms, and chemical) rate of biodegradation has been compared and tuning of biodegradation rate (both increase and decrease with regard to pure polymer biodegradation) has been outlined, utilizing organically modified layered silicate or carbon nanosheets, nanoparticles, and nanocomposites.

References

Abou-Zeid, D.-M., Müller, R.-J., & Deckwer, W.-D. (2001). Degradation of natural and synthetic polyesters under anaerobic conditions. *Journal of Biotechnology, 86*, 113–126.

Adamcová, D., & Vaverková, M. (2014). Biodegradation of degradable/biodegradable plastic material in controlled composting environment. *Polish Journal of Environmental Studies, 23*, 1465–1474.

Agrawal, C. M., Huang, D., Schmitz, J., & Athanasiou, K. (1997). Elevated temperature degradation of a 50:50 copolymer of PLA-PGA. *Tissue Engineering, 3*, 345–352.

Akutsu-Shigeno, Y., Teeraphatpornchai, T., Teamtisong, K., Nomura, N., Uchiyama, H., Nakahara, T., et al. (2003). Cloning and sequencing of a poly (DL-lactic acid) depolymerase gene from *Paenibacillus amylolyticus* strain TB-13 and its functional expression in *Escherichia coli*. *Applied and Environmental Microbiology, 69*, 2498–2504.

Allı, S., Aydın, R. S. T., Allı, A., & Hazer, B. (2015). Biodegradable poly (ε-caprolactone)-based graft copolymers via poly (linoleic acid): In vitro enzymatic evaluation. *Journal of the American Oil Chemists' Society, 92*, 449–458.

Baker, P. J., Poultney, C., Liu, Z., Gross, R., & Montclare, J. K. (2012). Identification and comparison of cutinases for synthetic polyester degradation. *Applied Microbiology and Biotechnology, 93*, 229–240.

Bang, D. Y., Kyung, M., Kim, M. J., Jung, B. Y., Cho, M. C., Choi, S. M., et al. (2012). Human risk assessment of endocrine-disrupting chemicals derived from plastic food containers. *Comprehensive Reviews in Food Science and Food Safety, 11*, 453–470.

Bikiaris, D. N. (2013). Nanocomposites of aliphatic polyesters: An overview of the effect of different nanofillers on enzymatic hydrolysis and biodegradation of polyesters. *Polymer Degradation and Stability, 98,* 1908−1928.

Bora, L. (2013). Polyhydroxybutyrate accumulation in *Bacillusmegaterium* and optimization of process parameters using response surface methodology. *Journal of Polymers and the Environment, 21,* 415−420.

Calabia, B. P., Ninomiya, F., Yagi, H., Oishi, A., Taguchi, K., Kunioka, M., et al. (2013). Biodegradable poly (butylene succinate) composites reinforced by cotton fiber with silane coupling agent. *Polymers, 5,* 128−141.

Calil, M., Gaboardi, F., Guedes, C., & Rosa, D. (2006). Comparison of the biodegradation of poly (ε-caprolactone), cellulose acetate and their blends by the Sturm test and selected cultured fungi. *Polymer Testing, 25,* 597−604.

Calmon, A., Dusserre-Bresson, L., Bellon-Maurel, V., Feuilloley, P., & Silvestre, F. (2000). An automated test for measuring polymer biodegradation. *Chemosphere, 41,* 645−651.

Carraher, C., & Seymour, R. (2003). *Polymer chemistry: Testing and spectrometric characterization of polymers.* (ed.). New York: Marcel Dekker.

Carrasco, F., Pagès, P., Gámez-Pérez, J., Santana, O., & Maspoch, M. L. (2010). Processing of poly (lactic acid): Characterization of chemical structure, thermal stability and mechanical properties. *Polymer Degradation and Stability, 95,* 116−125.

Chanprateep, S. (2010). Current trends in biodegradable polyhydroxyalkanoates. *Journal of Bioscience and Bioengineering, 110,* 621−632.

Codari, F., Moscatelli, D., Storti, G., & Morbidelli, M. (2010). Characterization of low-molecular-weight PLA using HPLC. *Macromolecular Materials and Engineering, 295,* 58−66.

Dare, E., Tofighi, R., Nutt, L., Vettori, M., Emgård, M., Mutti, A., et al. (2004). Styrene 7, 8-oxide induces mitochondrial damage and oxidative stress in neurons. *Toxicology, 201,* 125−132.

De Jong, S., Arias, E. R., Rijkers, D., Van Nostrum, C., Kettenes-Van den Bosch, J., & Hennink, W. (2001). New insights into the hydrolytic degradation of poly (lactic acid): Participation of the alcohol terminus. *Polymer, 42,* 2795−2802.

Dimzon, I. K., Trier, X., Frömel, T., Helmus, R., Knepper, T. P., & de Voogt, P. (2015). High resolution mass spectrometry of polyfluorinated polyether-based formulation. *Journal of the American Society for Mass Spectrometry, 27,* 309−318.

Ding, M., Zhang, M., Yang, J., & Qiu, J. H. (2012). Study on the enzymatic degradation of aliphatic polyester−PBS and its copolymers. *Journal of Applied Polymer Science, 124,* 2902−2907.

Dopico-García, S., Ares-Pernas, A., Otero-Canabal, J., Castro-López, M., López-Vilariño, J. M., González-Rodríguez, V., et al. (2013). Insight into industrial PLA aging process by complementary use of rheology, HPLC, and MALDI. *Polymers for Advanced Technologies, 24,* 723−731.

Eker, B., Asuri, P., Murugesan, S., Linhardt, R. J., & Dordick, J. S. (2007). Enzyme−carbon nanotube conjugates in room-temperature ionic liquids. *Applied Biochemistry and Biotechnology, 143,* 153−163.

Fang, B., Kang, H., Wang, R., Wang, Z., Wang, W., & Zhang, L. (2014). Aging behavior and mechanism of bio-based engineering polyester elastomer nanocomposites. *Journal of Applied Polymer Science, 131.*

Felfel, R. M., Hossain, K. M. Z., Parsons, A. J., Rudd, C. D., & Ahmed, I. (2015). Accelerated in vitro degradation properties of polylactic acid/phosphate glass fibre composites. *Journal of Materials Science, 50,* 3942−3955.

Fernández-Espada, L., Bengoechea, C., Cordobés, F., & Guerrero, A. (2016). Thermomechanical properties and water uptake capacity of soy protein-based bioplastics processed by injection molding. *Journal of Applied Polymer Science, 133.*

Fortunati, E., Puglia, D., Monti, M., Santulli, C., Maniruzzaman, M., Foresti, M. L., et al. (2013). Okra (*Abelmoschus esculentus*) fibre based PLA composites: Mechanical behaviour and biodegradation. *Journal of Polymers and the Environment, 21,* 726−737.

Fukushima, K., Tabuani, D., & Camino, G. (2009). Nanocomposites of PLA and PCL based on montmorillonite and sepiolite. *Materials Science and Engineering: C, 29*, 1433–1441.

Fukushima, K., Tabuani, D., Dottori, M., Armentano, I., Kenny, J., & Camino, G. (2011). Effect of temperature and nanoparticle type on hydrolytic degradation of poly (lactic acid) nanocomposites. *Polymer Degradation and Stability, 96*, 2120–2129.

Giunchedi, P., Conti, B., Scalia, S., & Conte, U. (1998). In vitro degradation study of polyester microspheres by a new HPLC method for monomer release determination. *Journal of Controlled Release, 56*, 53–62.

Gonçalves, S., & Martins-Franchetti, S. (2010). Action of soil microorganisms on PCL and PHBV blend and films. *Journal of Polymers and the Environment, 18*, 714–719.

Grassie, N., & Scott, G. (1988). *Polymer degradation and stabilisation*. CUP Archive.

Hakkarainen, M., Karlsson, S., & Albertsson, A. C. (2000). Influence of low molecular weight lactic acid derivatives on degradability of polylactide. *Journal of Applied Polymer Science, 76*, 228–239.

Hamad, K., Kaseem, M., Ko, Y. G., & Deri, F. (2014). Biodegradable polymer blends and composites: An overview. *Polymer Science Series A, 56*, 812–829.

Han, L., Han, C., Cao, W., Wang, X., Bian, J., & Dong, L. (2012). Preparation and characterization of biodegradable poly (3-hydroxybutyrate-co-4-hydroxybutyrate)/silica nanocomposites. *Polymer Engineering & Science, 52*, 250–258.

Hoshino, A., Sawada, H., Yokota, M., Tsuji, M., Fukuda, K., & Kimura, M. (2001). Influence of weather conditions and soil properties on degradation of biodegradable plastics in soil. *Soil Science and Plant Nutrition, 47*, 35–43.

Houde, M., Douville, M., Gagnon, P., Sproull, J., & Cloutier, F. (2015). Exposure of *Daphnia magna* to trichloroethylene (TCE) and vinyl chloride (VC): Evaluation of gene transcription, cellular activity, and life-history parameters. *Ecotoxicology and Environmental Safety, 116*, 10–18.

Hyon, S. H., Jamshidi, K., & Ikada, Y. (1984). *Polymers as biomaterials* (pp. 51–65). New York: Plenum.

I.F.B.A.B. European Bioplastics, Nova-Institut. (2014). <http://www.corbion.com/media/203221/eubp_factsfigures_bioplastics_2013.pdf>.

Itävaara, M., Karjomaa, S., & Selin, J.-F. (2002). Biodegradation of polylactide in aerobic and anaerobic thermophilic conditions. *Chemosphere, 46*, 879–885.

Iwata, T. (2015). Biodegradable and bio-based polymers: Future prospects of eco-friendly plastics. *Angewandte Chemie International Edition, 54*, 3210–3215.

J. B. Association. (2021). <http://www.jbpaweb.net/english/english.htm>.

John, R. P., Nampoothiri, K. M., & Pandey, A. (2006). Solid-state fermentation for L-lactic acid production from agro wastes using *Lactobacillus delbrueckii*. *Process Biochemistry, 41*, 759–763.

Kale, G., Auras, R., & Singh, S. P. (2006). Degradation of commercial biodegradable packages under real composting and ambient exposure conditions. *Journal of Polymers and the Environment, 14*, 317–334.

Kale, G., Auras, R., & Singh, S. P. (2007). Comparison of the degradability of poly (lactide) packages in composting and ambient exposure conditions. *Packaging Technology and Science: An International Journal, 20*, 49–70.

Kale, G., Auras, R., Singh, S. P., & Narayan, R. (2007). Biodegradability of polylactide bottles in real and simulated composting conditions. *Polymer Testing, 26*, 1049–1061.

Keshavarz, T., & Roy, I. (2010). Polyhydroxyalkanoates: Bioplastics with a green agenda. *Current Opinion in Microbiology, 13*, 321–326.

Kim, B. S., & Chang, H. N. (1998). Production of poly (3-hydroxybutyrate) from starch by *Azotobacter chroococcum*. *Biotechnology Letters, 20*, 109–112.

Kim, M. N., & Park, S. T. (2010). Degradation of poly (L-lactide) by a mesophilic bacterium. *Journal of Applied Polymer Science, 117*, 67–74.

Kopinke, F.-D., Remmler, M., Mackenzie, K., Möder, M., & Wachsen, O. (1996). Thermal decomposition of biodegradable polyesters—II. Poly (lactic acid). *Polymer Degradation and Stability, 53*, 329–342.

Kumar, A. P., Depan, D., Tomer, N. S., & Singh, R. P. (2009). Nanoscale particles for polymer degradation and stabilization—Trends and future perspectives. *Progress in Polymer Science, 34*, 479–515.

Kumar, S., & Maiti, P. (2015). Understanding the controlled biodegradation of polymers using nanoclays. *Polymer, 76*, 25–33.

Lagaron, J. M., & Lopez-Rubio, A. (2011). Nanotechnology for bioplastics: Opportunities, challenges and strategies. *Trends in Food Science & Technology, 22*, 611–617.

Lee, S. K., Seong, D. G., & Youn, J. R. (2005). Degradation and rheological properties of biodegradable nanocomposites prepared by melt intercalation method. *Fibers and Polymers, 6*, 289–296.

Li, C.-T., Zhang, M., Qin, J.-X., Zhang, Y., & Qiu, J.-H. (2015). Study on molecular modeling and the difference of PC lipase-catalyzed degradation of poly (butylene succinate) copolymers modified by linear monomers. *Polymer Degradation and Stability, 116*, 75–80.

Li, F., Yu, D., Lin, X., Liu, D., Xia, H., & Chen, S. (2012). Biodegradation of poly (ε-caprolactone)(PCL) by a new *Penicillium oxalicum* strain DSYD05-1. *World Journal of Microbiology and Biotechnology, 28*, 2929–2935.

Li, S., & McCarthy, S. (1999). Further investigations on the hydrolytic degradation of poly (DL-lactide). *Biomaterials, 20*, 35–44.

Li, S., Tenon, M., Garreau, H., Braud, C., & Vert, M. (2000). Enzymatic degradation of stereocopolymers derived from L-, DL-and meso-lactides. *Polymer Degradation and Stability, 67*, 85–90.

Li, Y., Han, C., Bian, J., Han, L., Dong, L., & Gao, G. (2012). Rheology and biodegradation of polylactide/silica nanocomposites. *Polymer Composites, 33*, 1719–1727.

Lim, H.-A., Raku, T., & Tokiwa, Y. (2005). Hydrolysis of polyesters by serine proteases. *Biotechnology Letters, 27*, 459–464.

Lim, L.-T., Auras, R., & Rubino, M. (2008). Processing technologies for poly (lactic acid). *Progress in Polymer Science, 33*, 820–852.

Liu, L., Yu, J., Cheng, L., & Yang, X. (2009). Biodegradability of poly (butylene succinate)(PBS) composite reinforced with jute fibre. *Polymer Degradation and Stability, 94*, 90–94.

Lucas, N., Bienaime, C., Belloy, C., Queneudec, M., Silvestre, F., & Nava-Saucedo, J.-E. (2008). Polymer biodegradation: Mechanisms and estimation techniques—a review. *Chemosphere, 73*, 429–442.

Luo, Y.-B., Wang, X.-L., & Wang, Y.-Z. (2012). Effect of TiO2 nanoparticles on the long-term hydrolytic degradation behavior of PLA. *Polymer Degradation and Stability, 97*, 721–728.

Ma, F., Lu, X., Wang, Z., Sun, Z., Zhang, F., & Zheng, Y. (2011). Nanocomposites of poly (L-lactide) and surface modified magnesia nanoparticles: Fabrication, mechanical property and biodegradability. *Journal of Physics and Chemistry of Solids, 72*, 111–116.

Maiti, P., & Prakash Yadav, J. P. (2008). Biodegradable nanocomposites of poly (hydroxybutyrate-co-hydroxyvalerate): The effect of nanoparticles. *Journal of Nanoscience and Nanotechnology, 8*, 1858–1866.

Mao, H., Liu, H., Gao, Z., Su, T., & Wang, Z. (2015). Biodegradation of poly (butylene succinate) by *Fusarium* sp. FS1301 and purification and characterization of poly (butylene succinate) depolymerase. *Polymer Degradation and Stability, 114*, 1–7.

Massardier-Nageotte, V., Pestre, C., Cruard-Pradet, T., & Bayard, R. (2006). Aerobic and anaerobic biodegradability of polymer films and physico-chemical characterization. *Polymer Degradation and Stability, 91*, 620–627.

Mohanty, A., Misra, Ma, & Hinrichsen, G. (2000). Biofibres, biodegradable polymers and biocomposites: An overview. *Macromolecular Materials and Engineering, 276*, 1–24.

Murray, E., Thompson, B. C., Sayyar, S., & Wallace, G. G. (2015). Enzymatic degradation of graphene/polycaprolactone materials for tissue engineering. *Polymer Degradation and Stability, 111*, 71–77.

Nair, L. S., & Laurencin, C. T. (2007). Biodegradable polymers as biomaterials. *Progress in Polymer Science, 32*, 762−798.

Nampoothiri, K. M., Nair, N. R., & John, R. P. (2010). An overview of the recent developments in polylactide (PLA) research. *Bioresource Technology, 101*, 8493−8501.

Niemelä, T. (2005). Effect of β-tricalcium phosphate addition on the in vitro degradation of self-reinforced poly-l, d-lactide. *Polymer Degradation and Stability, 89*, 492−500.

North, M. L., Takaro, T. K., Diamond, M. L., & Ellis, A. K. (2014). Effects of phthalates on the development and expression of allergic disease and asthma. *Annals of Allergy, Asthma & Immunology, 112*, 496−502.

Pandey, J. K., Reddy, K. R., Kumar, A. P., & Singh, R. (2005). An overview on the degradability of polymer nanocomposites. *Polymer Degradation and Stability, 88*, 234−250.

Panigrahi, M., Singh, N. K., Gautam, R. K., Banik, R. M., & Maiti, P. (2010). Improved biodegradation and thermal properties of poly (lactic acid)/layered silicate nanocomposites. *Composite Interfaces, 17*, 143−158.

Park, K., & Xanthos, M. (2009). A study on the degradation of polylactic acid in the presence of phosphonium ionic liquids. *Polymer Degradation and Stability, 94*, 834−844.

Paul, M.-A., Delcourt, C., Alexandre, M., Degée, P., Monteverde, F., & Dubois, P. (2005). Polylactide/montmorillonite nanocomposites: Study of the hydrolytic degradation. *Polymer Degradation and Stability, 87*, 535−542.

Pawar, S. P., Kumar, S., Misra, A., Deshmukh, S., Chatterjee, K., & Bose, S. (2015). Enzymatically degradable EMI shielding materials derived from PCL based nanocomposites. *RSC Advances, 5*, 17716−17725.

Perez-Puyana, V., Felix, M., Romero, A., & Guerrero, A. (2016). Effect of the injection moulding processing conditions on the development of pea protein-based bioplastics. *Journal of Applied Polymer Science, 133*.

Pranamuda, H., & Tokiwa, Y. (1999). Degradation of poly (L-lactide) by strains belonging to genus *Amycolatopsis*. *Biotechnology Letters, 21*, 901−905.

Rabek, J., & Ranby, B. (1975). Role of singlet oxygen in photo-oxidative degradation and photostabilization of polymers. *Polymer Engineering & Science, 15*, 40−43.

Rasal, R. M., Janorkar, A. V., & Hirt, D. E. (2010). Poly (lactic acid) modifications. *Progress in Polymer Science, 35*, 338−356.

Ray, S. S., Yamada, K., Ogami, A., Okamoto, M., & Ueda, K. (2002). New polylactide/layered silicate nanocomposite: Nanoscale control over multiple properties. *Macromolecular Rapid Communications, 23*, 943−947.

Röttig, A., & Steinbüchel, A. (2013). Acyltransferases in bacteria. *Microbiology and Molecular Biology Reviews, 77*, 277−321.

Sahari, J., Sapuan, S. M., Zainudin, E. S., & Maleque, M. A. (2014). Degradation characteristics of SPF/SPS biocomposites. *Fibres & Textiles in Eastern Europe, 22*, 94−96.

Santerre, J., Labow, R., & Adams, G. (1993). Enzyme−biomaterial interactions: Effect of biosystems on degradation of polyurethanes. *Journal of Biomedical Materials Research, 27*, 97−109.

Scarfato, P., Acierno, D., & Russo, P. (2015). Photooxidative weathering of biodegradable nanocomposite films containing halloysite. *Polymer Composites, 36*, 1169−1175.

Schliecker, G., Schmidt, C., Fuchs, S., Wombacher, R., & Kissel, T. (2003). Hydrolytic degradation of poly (lactide-co-glycolide) films: Effect of oligomers on degradation rate and crystallinity. *International Journal of Pharmaceutics, 266*, 39−49.

Shah, A. A., Hasan, F., Hameed, A., & Ahmed, S. (2008). Biological degradation of plastics: A comprehensive review. *Biotechnology Advances, 26*, 246−265.

Shah, A. A., Nawaz, A., Kanwal, L., Hasan, F., Khan, S., & Badshah, M. (2015). Degradation of poly (ε-caprolactone) by a thermophilic bacterium *Ralstonia* sp. strain MRL-TL isolated from hot spring. *International Biodeterioration & Biodegradation, 98*, 35−42.

Shi, X., Zhang, G., Siligardi, C., Ori, G., & Lazzeri, A. (2015). Comparison of precipitated calcium carbonate/polylactic acid and halloysite/polylactic acid nanocomposites. *Journal of Nanomaterials, 2015*.

Sikorska, W., Musiol, M., Nowak, B., Pajak, J., Labuzek, S., Kowalczuk, M., et al. (2015). Degradability of polylactide and its blend with poly [(R, S)-3-hydroxybutyrate] in industrial composting and compost extract. *International Biodeterioration & Biodegradation, 101*, 32–41.

Singh, N. K., Purkayastha, B. D., Roy, J. K., Banik, R. M., Yashpal, M., Singh, G., et al. (2010). Nanoparticle-induced controlled biodegradation and its mechanism in poly (ε-caprolactone). *ACS Applied Materials & Interfaces, 2*, 69–81.

Singh, N. K., Purkayastha, B. P. D., Panigrahi, M., Gautam, R. K., Banik, R. M., & Maiti, P. (2013). Enzymatic degradation of polylactide/layered silicate nanocomposites: Effect of organic modifiers. *Journal of Applied Polymer Science, 127*, 2465–2474.

Singh, N. K., Purkayastha, B. P. D., Roy, J. K., Banik, R. M., Gonugunta, P., Misra, M., et al. (2011). Tuned biodegradation using poly (hydroxybutyrate-co-valerate) nanobiohybrids: Emerging biomaterials for tissue engineering and drug delivery. *Journal of Materials Chemistry, 21*, 15919–15927.

Singh, N. K., Singh, S. K., Dash, D., Gonugunta, P., Misra, M., & Maiti, P. (2013). CNT Induced β-phase in polylactide: Unique crystallization, biodegradation, and biocompatibility. *The Journal of Physical Chemistry C, 117*, 10163–10174.

Sobkowicz, M. J., Feaver, J. L., & Dorgan, J. R. (2008). Clean and green bioplastic composites: Comparison of calcium sulfate and carbon nanospheres in polylactide composites. *CLEAN–Soil, Air, Water, 36*, 706–713.

Södergård, A., & Stolt, M. (2002). Properties of lactic acid based polymers and their correlation with composition. *Progress in Polymer Science, 27*, 1123–1163.

Stloukal, P., Verney, V., Commereuc, S., Rychly, J., Matisova-Rychlá, L., Pis, V., et al. (2012). Assessment of the interrelation between photooxidation and biodegradation of selected polyesters after artificial weathering. *Chemosphere, 88*, 1214–1219.

Stockhofe, K., Postema, J. M., Schieferstein, H., & Ross, T. L. (2014). Radiolabeling of nanoparticles and polymers for PET imaging. *Pharmaceuticals, 7*, 392–418.

Sugihara, A., Ueshima, M., Shimada, Y., Tsunasawa, S., & Tominaga, Y. (1992). Purification and characterization of a novel thermostable lipase from *Pseudomonas cepacia*. *The Journal of Biochemistry, 112*, 598–603.

Teramoto, N., Urata, K., Ozawa, K., & Shibata, M. (2004). Biodegradation of aliphatic polyester composites reinforced by abaca fiber. *Polymer Degradation and Stability, 86*, 401–409.

Tokiwa, Y., & Calabia, B. P. (2006). Biodegradability and biodegradation of poly (lactide). *Applied Microbiology and Biotechnology, 72*, 244–251.

Torres, A., Li, S., Roussos, S., & Vert, M. (1996). Degradation of L-and DL-lactic acid oligomers in the presence of *Pseudomonas putida* and *Fusarium moniliforme*. *Journal of Environmental Polymer Degradation, 4*, 213–223.

Tsuji, H., Kidokoro, Y., & Mochizuki, M. (2006). Enzymatic degradation of biodegradable polyester composites of poly (L-lactic acid) and poly (ε-caprolactone). *Macromolecular Materials and Engineering, 291*, 1245–1254.

US EPA. (2007). <http://www.sustainableplastics.org/problems/plastics-recyclingremains->.

Weir, N., Buchanan, F., Orr, J., Farrar, D., & Dickson, G. (2004). Degradation of poly-L-lactide. Part 2: Increased temperature accelerated degradation. *Proceedings of the Institution of Mechanical Engineers, Part H: Journal of Engineering in Medicine, 218*, 321–330.

Wu, Y., Wang, G., Wang, Z., Liu, Y., Gu, P., & Sun, D. (2014). Comparative study on the efficiency and environmental impact of two methods of utilizing polyvinyl chloride waste based on life cycle assessments. *Frontiers of Environmental Science & Engineering, 8*, 451–462.

Yousif, E., & Haddad, R. (2013). Photodegradation and photostabilization of polymers, especially polystyrene. *SpringerPlus, 2*, 398.

11

Nanobiodegradation of plastic waste

Ayesha Baig[1], Muhammad Zubair[1], Muhammad Nadeem Zafar[1], Mujahid Farid[2], Muhammad Faizan Nazar[3], Sajjad Hussain Sumrra[1]

[1]DEPARTMENT OF CHEMISTRY, FACULTY OF SCIENCES, UNIVERSITY OF GUJRAT, GUJRAT, PAKISTAN [2]DEPARTMENT OF ENVIRONMENTAL SCIENCES, UNIVERSITY OF GUJRAT, GUJRAT, PAKISTAN [3]DEPARTMENT OF CHEMISTRY, UNIVERSITY OF EDUCATION, LAHORE, MULTAN CAMPUS, MULTAN, PAKISTAN

11.1 Introduction

Plastic is a term commonly used for a range of high molecular mass organic polymers which are obtained mostly from various hydrocarbon and the petroleum derivatives. There is generally an ever-increasing trend toward production and the consumption of plastic materials due to their increased domestic and industrial applications. However, wide spectrum of such polymers is basically nonbiodegradable with only few exceptions. The extensive applications of plastics, community behavior toward proper disposal of their wastes, and lack of waste management pose significant threats to the environment. This has raised a growing concern among the stakeholders to devise innovative strategies and policies for the management of plastic wastes, application of biologically degradable polymers especially in their packaging and educating the people about proper waste disposal of plastics. Current degradation strategies of polymers rely on photo, thermal, biological, and chemical procedures (Ru, Huo, & Yang, 2020).

Over past 50 years, the plastic materials have been used as a substitution the materials like metal, wood, and paper for a wide range of different applications, due to their excellent characteristics in terms of stability, lightness, low cost, and durability. Ironically, properties like toughness and durability of plastics are in turn resulting in a serious threat to environment, and causing them to become resistant to the biological degradation. This resistance has now become a challenge to waste management processes, particularly those involving sustainable waste management. This is where actual problems begin (Luyt & Malik, 2019).

In 2011, over 230 million tons of plastic materials were produced, with Europe accounting for about 25%, Asia 37%, and NAFTA countries 23% of total global production. High demand for durable and cheap plastics led to a global waste problem (Nikodinovic-Runic et al., 2013). The current methods used for plastic waste disposal include incineration, landfilling, chemical, and mechanical recycling. In most countries, especially developing countries, landfilling is major method for disposing waste of plastics due to its low cost and operability (Ahmed et al., 2018).

Biodegradation is a biochemical process in which a particular compound ultimately breaks into very simpler molecules by an enzymatic activity, for example, degradation of the hydrocarbons by fungi and bacteria (Ramírez-García, Gohil, & Singh, 2019), and biodegradable bioplastics are those that are completely degraded by microorganisms, without any toxic residues (Alshehrei, 2017). Biodegradable polymers (BDPs) can be defined as the polymers derived from various renewable resources, especially the materials, which can be decomposed to simpler substances (CH_4, other inorganic compounds, CO_2, biomass, and water) under microbial enzymatic action. In the European Union and the United States these polymers are mostly convenient bottles for beverages and water packing due to their special properties: transparency, lightness, and softness. These BDPs can make a significant contribution to utilization of the renewable resources, reduction of landfill, and material recovery (Gheorghe, Anastasiu, Mihaescu, & Ditu, 2019).

There is a continuous negligence by public sectors and government as well as the general public toward the impact of waste of plastics on environmental adversity. Plastics in the environment adversely affect soil fertility, releasing toxic and volatile chemicals in environment, which affect lives by the ingestion of the plastic particles and also the entanglement of animals in different ecosystems (Kumari & Chaudhary, 2020). Animals are dying because of plastic waste both by swallowing waste plastic debris for exerting catastrophic outcomes on surroundings and by being stuck in waste plastic traps. Some plastic products affect human health when we recognize that they mimic human hormones. Among these, vinyl chloride is considered carcinogenic to human beings and the mammary cursive molecular mass. While degradable plastics are usually fabricated from starch and have comparatively lower molecular weights, these degradable plastics are getting broken down when exposed to UV radiations, water, enzymes, and slight changes in pH (Fesseha & Abebe, 2019).

11.2 Plastic waste pollution and environmental impacts

Waste plastics pollute marine, freshwater, terrestrial, and groundwater environments. Plastics are prepared from petroleum products and many other additives, which are highly toxic even in very small concentrations when these are dissolved in soil or water, land and contaminating water, and, when burned, result in releasing some carcinogenic gases and ultimately causing air pollution. Some of constituents of plastic like phenol, benzene, and chloride are carcinogenic, and the gases emitted during conversion of plastic as well as the liquid hydrocarbons spoil air and earth (Vanapalli, Samal, Dubey, & Bhattacharya, 2019). The efficient decomposition or degradation of plastic bags usually takes about 1000 years. Plastic waste pollution has recently been considered a threat to the biosphere including human beings due to its persistent nature and widespread distribution. Plastic causes global warming and pollution not only because of increasing the problem of landfilling and waste disposal but also because of the release of dioxins and CO_2 due to burning

(Kale, Deshmukh, Dudhare, & Patil, 2015). Although the use of plastics has brought much convenience and comfort to the modern daily life, much more attention is being paid to plastic pollution in the environment. A common solution to this problem is to replace the traditional plastic items with bioplastics that are biodegradable and/or biobased plastics. The biodegradability of biobased plastics is highly influenced by their chemical and physical structure. The utilization of renewable resources such as agricultural wastes and their biodegradation in different environments has enabled these plastics to be more easily acceptable as compared to conventional plastics (Emadian, Onay, & Demirel, 2017).

11.2.1 Leaching of highly toxic chemicals from plastic wastes

Plastics contain mixtures of various additives to enhance their particular physical characteristics, which can leach into the surrounding environment. This leaching process occurs primarily at the surface of plastic particle, with the possibility of the constant diffusion of toxic chemicals from the core of particle to surface. Thus, leaching from the plastic particles could provide a long-term source of different chemicals into body fluids and tissues, despite the fact that most of these chemicals have short half-lives ($t_{1/2}$) in the body and are not persistent. Plastic additives of concern to the health of human beings include bisphenol A, organotin, triclosan, phthalates, and brominated flame retardants (Galloway, 2015).

11.3 The impacts of plastic wastes accumulation

Plastics are undoubtedly an essential pillar material in the present society due to their moldable, cost-effective, and resistant nature. Plastic materials are easily dispersed by ocean, wind, and river currents across the globe (Oliveira, Ameixa, & Soares, 2019). Due to the combination of direct litter and poor waste management, plastic is now very ubiquitous on the whole surface of earth (and has also accumulated in both marine and terrestrial deposits) that there is suggestion it should also be recognized as the geological indicators for the Anthropocene epoch. The controlled or uncontrolled accumulation of plastic waste adversely impacts the environmental matrices either by entering the ocean or accumulating at the landfills. (Dilkes-Hoffman, Pratt, Lant, & Laycock, 2019). However, concerns about use and waste disposal include plastic waste accumulation in natural habitats and in landfills, physical troubles for wildlife resulting from entanglement in plastic, leaching of toxic chemicals from plastics and potential for plastic products to transfer toxic chemicals to humans and wildlife (Thompson, Moore, Vom Saal, & Swan, 2009). Furthermore, additional problems or concerns that occur with plastics (i.e., negative influences on the human health and degradation-resistance under natural conditions) also remain intractable.

11.3.1 Plastic wastes on land

The vast majority of different plastics are manufactured, utilized, and then discarded on the land, and it is well known that plastic waste can accumulate in terrestrial environments.

In rural areas, plastic use and disposal are dominated by agricultural practices ("plasticulture"). The macroplastics are highly concentrated in convergence points of the anthropogenic activity (e.g., road verges or urban environments) and regions of the higher population density, as well as in proximity to various waste-processing sites. Anthropogenic activities related to reuse and waste disposal have the effect of moving plastic particles and ultimately initiating environmental release, for example, use of sewage sludge as soil fertilizer or conditioner to the agricultural soil leads to release of small-sized particles of plastics present within sludge. Therefore, repeated sludge treatment leads to microplastic accumulation within receiving soils. In addition, illegal dumping of waste serves as a direct input to many plastic wastes on land (Hurley, Horton, Lusher, & Nizzetto, 2020).

11.3.2 Plastic wastes in freshwater and marine ecosystems

The occurrence of plastic litter in the freshwater systems has also been documented for many decades. Plastics have been found across all the freshwater systems including urban drainage networks, dams, rivers, and lakes (Fig. 11−1). Within river systems, transport of macroplastics using plastic tracer items has been demonstrated. For smaller plastic particles, such as nano and microplastics, transport processes are comparatively harder for experimental assessment due to their heterogeneity, associated analytical challenges, and small size. Freshwater sediments represent potential sinks for the plastic particles (Hurley et al., 2020). Floating plastic wastes also have a negative impact on marine ecosystems. However, there is a lack of precise information and knowledge about the sources, quantity, role, accumulation, and transport of plastic products in the oceans (Urbanek, Rymowicz, & Mirończuk, 2018).

11.4 Types of plastics targeted

Plastic waste consists of various kinds of plastic, such as polypropylene (PP), low-density polyethylene (LDPE), high-density polyethylene (HDPE) polystyrene (PS), polyethylene tetrachloride (PET), and polyvinyl chloride (PVC). Plastics are mainly classified into two types, thermoplastics and the thermosetting plastics (Zheng, Yanful, & Bassi, 2005). Thermoplastics are linear shaped,

FIGURE 11–1 Plastics disposed in water bodies. *Retrieved from https://www.britannica.com/science/plastic-pollution.*

less rigid than thermosets, and can be easily molded into different shapes by heating, for example, PP, polyethylene (PE), and so forth; in contrast the thermoset plastics cannot be molded due to their three dimensional cross-linked structure, and they become comparatively harder after heating, for example, melamine formaldehyde and Bakelite. Thermoset plastics, such as polyurethane and polyester, are biodegradable due to amide bonds or ester bonds that are potentially susceptible to the hydrolytic cleavage (Anjana, Hinduja, Sujitha, & Dharani, 2020). Many other chemicals and polymers are also added to plastic to enhance its elasticity, durability, and other properties, for example, the addition of vinyl chloride in the preparation of PVC plastic (Shahnawaz, Sangale, & Ade, 2019).

11.5 Plastic waste disposal methods

Currently, three main methods used for handling plastic waste are burying in the landfill, incineration, and recycling of plastic waste. Each of these methods has its own inherent drawbacks.

11.5.1 Landfilling

The major drawback associated with plastic waste disposal by the landfill facility occupies space that could be used for many other productive means, like agriculture. Plastic components of the landfill waste have been considered to persist for more than 20 years. This is due to limited oxygen availability in landfills, and because the surrounding environment is anaerobic. The limited degradation experienced by plastics is mainly due to thermooxidative degradation, and anaerobic conditions in the landfills only act to further limit the degradation rates. The plastic debris in the landfills also serves as a source for different types of secondary environmental pollutants. The pollutants of note include a number of volatile organics, such as xylenes, benzene, toluene, trimethyl benzenes, and ethyl benzenes released both as contained in leachate and as gases and endocrine disrupting compounds (Webb, Arnott, Crawford, & Ivanova, 2013). Lack of capacity of the waste landfills has now become a serious issue, especially in urban areas, and effective decomposition or reduction of waste is essential to secure adequate capacity of landfill. Among many compounds in the waste landfill sites, plastic materials are estimated to make up about 20%–30% of volume municipal solid waste sites of landfills (Adamcová and Vaverková, 2014). In landfill sites, plastic waste can release toxic chemicals and other carcinogens that pollute groundwater. Additionally, these toxic chemicals can also disrupt soil fertility (Erdogan, 2020).

11.5.2 Incineration

Another method routinely used for plastic waste disposal is incineration. The plastic incineration process overcomes some of drawbacks placed on landfill in that no significant space is required, and there is a capability for the energy recovery in form of heat. Disposal of plastic waste such as PVC, PS, and PE has been indicated as a worldwide environmental problem. However, due to their nonbiodegradability and chemical stability, waste PE items are being mostly decomposed by the process of incineration, which will release toxic byproducts (Li et al., 2010).

11.5.3 Recycling

The recycling of waste plastic is an important matter nowadays. Plastic products are composed mainly of nonrenewable resources, natural, or petroleum gas. The burning of petroleum to make new plastic products releases harmful gases in the environment. Recycling old plastic items will therefore use far less energy than making it from new. Moreover, recycling plastic reduces the amount of space required for landfilling of plastics. The high quality and low cost of recycled products encourages industries, especially the construction industry, to use recycled plastic products, with a more specific focus on plastic waste, the second major waste component. Recycling is a preferred solution for plastic waste management because it falls into the category of less environmental impact as indicated by the Global Warming Potential and Human Toxicity Potential (Al-Maaded, Madi, Kahraman, Hodzic, & Ozerkan, 2012).

Recycling is an environmentally attractive solution, but only a small part of waste plastics can be successfully recycled, and the remaining goes to burial sites. Landfills are usually lower in number but are rarely satisfactory, whereas incineration causes the generation of highly toxic fumes that are released into the environment and ultimately cause air pollution (Bhatia, Girdhar, Chandrakar, & Tiwari, 2013).

11.6 Photooxidative degradation of waste plastics

Light-induced degradation of plastics is a physical process caused by irradiation of the high-frequency electromagnetic radiations, especially gamma and ultraviolet rays. The absorption of photons causes degradation of polymer main chain leading to reduction in the molar mass. Prerequisite for initiation of a photochemical reaction is presence of the chromophoric groups within molecules that absorb quantum of light (Kumar & Maiti, 2016). Photodegradable bioplastics have light sensitive groups directly connected into the backbone of polymer. In landfills, however, there is a lack of sunlight, so these plastics remain nondegraded (Alshehrei, 2017).

In an experiment, fragmentation of the oxo-biodegradable plastic bags was observed after about 21 months of sunlight exposure. The control samples of plastics were cut with scissors, but after sunlight exposure, it was not possible to cut bags. These plastic bags were easily fragmented by using hands, resulting in a powder form. Thus, results conclude that in order to completely degrade the plastics, it takes more than 18 months of exposure to sunlight. The oxo-biodegradable degradation of PE was observed through sunlight exposure up to 90 days in the soil with pH and moisture control. It was concluded that PE films with pro-oxidant additives had comparatively less superficial and structural modifications, than films without additives. Thus, action of pro-oxidants by effect of sunlight usually depends on time and conditions of the exposure to sunlight (da Luz, da Silva, dos Santos, & Kasuya, 2019).

11.7 Thermal degradation of waste plastics

Thermal degradation (cracking and/or pyrolysis) is considered a feasible means of waste plastic reuse. Indeed, viability of the thermal cracking has found uses in a number of various processing plants. The coprocessing of coal and waste plastics has also been recognized as a pathway to fuel production. The drawback to the thermal degradation is requisite high temperatures, which results in a broad range of various products (Keane, 2010).

11.8 Enzymatic degradation of bioplastics

The increasing waste and water pollution due to available degradation methods of plastics have led to emergence of bioplastics and biological degradation with the microbial extracellular enzymes. The microbes use BDPs as substrates under starvation and in the unavailability of nutrients (Fig. 11−2). Microbial enzymatic degradation is an appropriate method from the bioremediation point of view as no waste accumulates (Bano et al., 2017). Under nutrient-limited conditions, microbes that produce and store polyhydroxyalkanoates (PHA) can metabolite and degrade it when the limit is removed. However, the ability of microbes to store PHA does not guarantee their ability to degrade/deteriorate it in the environment. Due to their large size, individual polymers cannot be transported directly across the cell wall in bacteria; therefore, the bacteria need to secrete hydrolases capable of converting polymers into their hydroxyl acid monomers (Alshehrei, 2017). Similarly, the synthetic aliphatic polyesters, for example, polycaprolactone and PE adipate, are hydrolyzed by hog liver esterase and various lipases (Iwamoto & Tokiwa, 1994). Müller et al. demonstrated that PET, which is usually considered a "nonbiodegradable" polymer, can be efficiently depolymerized

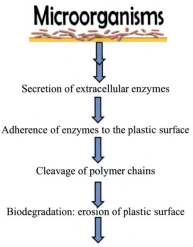

FIGURE 11–2 Schematic illustration of enzymatic degradation of plastics. *From Das & kumar. 2014.*

by a hydrolase from *Thermobifida fusca*. However, under comparable conditions, Lipases from *Candida Antarctica* and *Pseudomonas* sp. did not degrade PET (Müller, Schrader, Profe, Dresler, & Deckwer, 2005).

11.9 Reduction strategies for waste management of plastics by using biotechnology

In general, biodegradability of the plastics in environment is extremely slow and poor. Several reasons advanced for this include (1) lack of suitable culture conditions or nutrients, (2) low bioavailability of solid substrates, and (3) hydrophobic nature of surface. However, studies concluded that significant potentials may exist to discover or develop organisms or microorganisms with appropriate types of metabolic machinery for biodegradation of plastics (Bassi, 2017).

11.9.1 Combined approaches

As plastics are quite recalcitrant types of materials, the combined approach by using a photochemical or a thermochemical step followed by the bioprocessing and upgrading is basically an excellent example for describing the role of biotechnology in waste management of plastics. The production of the biopolymers has always been of interest. Photodegradation has been combined with the biodegradation. Bassi investigated degradation potential of the LDPE films in biotic and abiotic environments and confirmed the oxidative degradation under abiotic conditions after exposure of films to the UV radiation for periods ranged from 0 to 42 days. The authors indicated the presence of some carbonyl groups and also reported that the biotic activity was subsequent to photodegradation (Bassi, 2017).

In another study, films of PP and PE were first exposed to gamma radiations and then inoculated endophytic fungi (lactase-enzyme producing) from two endemic plants, *Humboldtia brunonis* and *Psychotria flavida*. Biodegradation of films was then evaluated over a period of 90 days. Scanning electron microscopy (SEM), Fourier transform infrared (FTIR) spectroscopy, and differential scanning calorimetric information were collected to evaluate (Bassi, 2017).

11.10 Biodegradation of plastics

Biodegradation is the process by which microbes (mainly fungi and bacteria) alter or transform (through enzymatic and metabolic action) structure of various chemicals introduced into environment (Urbanek et al., 2018). Several polymers that are considered to be "biodegradable" are basically hydrobiodegradable, photodegradable, or bioerodible. These different classes of polymers all come under a broader category, that is, environmentally degradable polymers (Nair, Sekhar, Nampoothiri, & Pandey, 2017).

Microorganisms, such as fungi and bacteria, are involved in biodegradation of both synthetic and natural plastics in the same way as various photosynthetic bacteria, archaebacteria, lower eukaryotes, aerobes, and anaerobes. All of these microorganisms are found extensively in soil and compost materials, while some are in the aquatic environments (Gewert, Plassmann, & MacLeod, 2015). Among many other microorganisms, certain microalgae strains (i.e., *Scenedesmus* sp. *Chlorella* sp., and *Ulva* sp.) and cyanobacterial strains (e.g., *Nostoc* sp., *Phormidium* sp., and *Chlamydomonas* sp.) have been recommended as potential candidates for production of bioplastics. The degradation products and processes must be understood to evaluate and detect potential environmental hazards (Syrpas & Venskutonis, 2020). Plastics are usually biodegraded partly aerobically in soil and compost, aerobically in nature, and anaerobically in landfills and sediments. Aerobic biodegradation produces water and carbon dioxide, while water, carbon dioxide, and methane are produced during anaerobic biodegradation.

11.10.1 Aerobic biodegradation

Aerobic biodegradation, also known as aerobic respiration, is an important component of the natural attenuation of pollutants or contaminants in many hazardous waste areas. Aerobic microbes use oxygen (O_2) as an electron acceptor, and synthesize small organic compounds by breaking down organic chemicals. Water and CO_2 are produced as byproducts of this process (Ahmed et al., 2018).

$$C\ plastic + O_2 \rightarrow H_2O + CO_2 + Biomass + C\ residual$$

11.10.2 Anaerobic biodegradation

In anaerobic conditions, polymers or organic contaminants are broken down by microorganisms. Some anaerobic bacteria use manganese, sulfate, carbon dioxide, nitrate, and iron as their electron acceptors to synthesize small organic compounds by breaking down organic chemicals. New microbial pathways and enzymes need to be discovered to optimize the conditions under which the polymers can be effectively degraded (Grima, Bellon-Maurel, Feuilloley, & Silvestre, 2000).

$$C\ plastic \rightarrow CH_4 + H_2O + CO_2 + Biomass + C\ residual$$

Aerobic and anaerobic biodegradation mechanism pathways are demonstrated in Fig. 11–3. Intracellular and extracellular enzymes are involved in the biological degradation of polymers. In degradation process, microbial exoenzymes break down the complex polymers into smaller molecules like monomers, dimers, and oligomers. These small and water-soluble molecules can pass through the outer semipermeable bacterial membranes to be utilized as energy and carbon sources. This initial step of polymer degradation is called depolymerization. When the end products of anaerobic degradation, that is, water, carbon dioxide, and methane, are formed, degradation is called mineralization (Alshehrei, 2017).

Biodegradable plastics are considered by many as promising solution to plastic waste pollution because they are environmentally friendly and can be derived from the renewable feedstock, thereby decreasing greenhouse gas emissions. For example, lactic acid and PHA

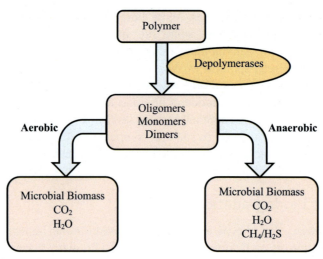

FIGURE 11–3 Plastic degradation under aerobic and anaerobic conditions. *From Pathak, V.M., & Kumar, N. (2017). Implications of SiO2 nanoparticles for in vitro biodegradation of low-density polyethylene with potential isolates of Bacillus, Pseudomonas, and their synergistic effect on Vigna mungo growth. Energy, Ecology and Environment, 2(6), 418–427.*

can be prepared by the fermentative biotechnological methods using microorganisms and agricultural products. Biodegradable plastics offer many advantages such as enhanced soil fertility, reduction in cost of plastic waste management, and low accumulation of large plastic products in the environment (which will minimize major injuries to wild animals). In addition, biodegradable plastics can also be recycled to the useful metabolites (oligomers and monomers) by enzymes and microorganisms. A second method involves degradation of petroleum-derived plastic products by biological processes, for example, the degradation of some aliphatic polyesters (Tokiwa, Calabia, Ugwu, & Aiba, 2009).

11.10.3 Biodegradation standards for plastics

Early work on standardization of test methods of biodegradation was carried out in 1980s, for instance by the Organization for Economic Cooperation and Development. In 1996, the European Commission mandated the European Committee for Standardization to develop the standards for packaging, also including those for the compostability (Harrison, Boardman, O'Callaghan, Delort, & Song, 2018). The biodegradation standards for plastics are established in following two important categories for biodegradation.

One standard is for biodegradation testing methodology, and the other is for biodegradation performance specifications. The first standard is basically a test method that specifies a methodology for measurement of biodegradation and accurately stimulates an intended environment. Second standard is a specification standard, which assigns minimum value for establishing biodegradation. Both of these standards are sufficient and necessary to adequately establish biodegradation performance of the plastic materials. For plastic materials, waste disposal environments can include litter, anaerobic digestion, home compost, industrial compost, ocean water, and landfill (Greene, 2014). The commonly used methods for the plastic waste disposal are considered to be inadequate for management of plastic waste, and hence there is a growing concern for utilization of effective microorganisms meant for the biodegradation of nondegradable synthetic polymers (Kale et al., 2015).

11.10.4 Factors affecting biodegradation of plastics

The degradation process of plastics is affected by the following factors:

1. Plastics present on the seafloor are less prone to degradation compared to those in surface waters for which the decomposition takes less time due to hot water temperature and increased sunlight penetration.
2. The surface conditions (such as surface area, hydrophobic, and hydrophilic properties), high order structures (melting temperature, glass transition temperature, crystal structure, crystallinity and modulus of elasticity), and the first-order structures (chemical structure, molecular mass, and molecular mass distribution) of polymers play an important role in biodegradation processes (Tokiwa et al., 2009).
3. The polyesters without side chains are more assimilated than those with side chains.
4. Location of waste plastics, type of organism, environment status, moisture content, pH, light (UV), and polymer characteristics (such as molecular weight, polymer chains, functional groups, crystallinity, chemical structure, tacticity and mobility) are the most essential factors for the successful growth of microorganisms and a need to be considered (Rojas-Parrales, Orantes-Sibaja, Redondo-Gómez, & Vega-Baudrit, 2018).

11.11 Biodegradation at the nanoscale level

The degradation or deterioration of plastic debris/waste materials may result in the formation of nanoplastics. Their specific surface area for sorption of toxic heavy metals and organic pollutants make the study of nanoplastics an urgent priority. There is a limited understanding on the fate, occurrence, abundance, and distribution of nanoplastics in the environment due to lack of suitable methods for the identification and separation of nanoplastics from various complex environmental matrices. One of the key reasons for limited reports on nanoplastics is the lack of practically applicable methods for the evaluation of nanoplastics in the environment, which hinders the act of accurate prediction of potential health risks of nanoplastics on humans (Li, Li, Hao, Yu, & Liu, 2020). It is experimentally evident that about three weeks of ultraviolet irradiation generates particles of plastics in the nanometer (nm) size range. These nanoparticles (1–100 nm) exhibit different physical and chemical properties than the macroscopic objects based on same material. At micro and nanoscales, PE, PVC, PS, and PET were observed. Recently nanoplastics have been collected in North Atlantic, while oligomers and styrene monomers have been found in water and sand from shorelines worldwide. These nanoplastics reveal that the fingerprint of colloidal fraction of the seawater was chiefly of pollutants and environmental pollution and attributed to the combination of plastics. The composition of polymer varied among size classes (Bratovcic, 2019). The sources of nanoplastics include

- Engineered nanomaterials:
  - Agricultural nanotechnology
  - Pharmaceutical products
  - Personal care products
- Nanosized plastic fragments

In Fig. 11−4 is shown the process of gradual degradation of the macroplastics into nano and microplastics, their interactions with the medium and impacts on human health, leading to different diseases. It has been reported that native microbes, of plastic dumpsite, have more potential for the conversion of polymers into water and carbon dioxide. Similarly, presence of nanomaterials like 0.01% nanobarium titanate nanosized particles has been reported to increase LDPE biodegradation. Super magnetic iron-oxide nanoparticles and fullerene-60 have been reported to support growth trend of microbes in a progressive way. *Pseudomonas*

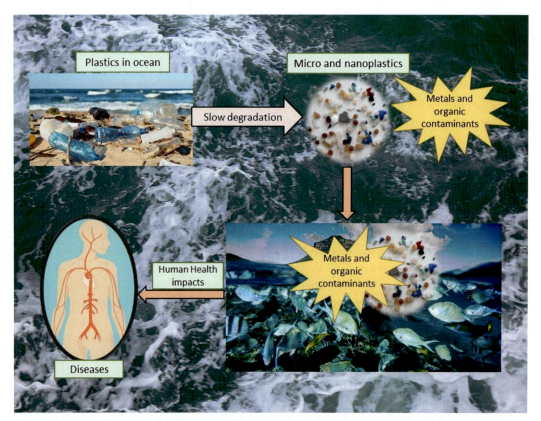

FIGURE 11–4 Schematic representation of plastic degradation into micro and nanoplastics and their impacts on human health. *From Bratovcic, A. (2019). Degradation of micro-and nano-plastics by photocatalytic methods. Journal of Nanoscience and Nanotechnology Applications, 3, 304.*

aeruginosa, isolated from a local dumpsite, effectively degrades PE at 1% concentration of titania nanoparticles (Alvi, Qazi, Baig, Andleeb, & Yaqoob, 2016). Although overlooked for several years, amount of plastic waste accumulated in environment has been rapidly increasing as a result of the material's lightweight and durable nature. Once discarded on the land, particles of plastic debris make their way to water bodies which act as sinks for the low-density litter (Brunner, Fischer, Rüthi, Stierli, & Frey, 2018).

11.12 Degradation of waste plastics using microbes

The increasing waste and water pollution due to the available decomposition methods for plastic degradation have ultimately lead to biological degradation with the microbial (fungi and bacteria) extracellular enzymes and emergence of the biodegradable plastics. The microbes use BDPs as substrate in unavailability of the microbial nutrients and under starvation. From bioremediation point of view, microbial enzymatic degradation of plastics is found suitable as no waste plastic accumulation occurs (Roohi et al., 2017). The BDPs are usually designed to degrade or deteriorate fast by microbes due to their ability to degrade most of the inorganic or organic materials, including hemicelluloses, starch, cellulose, and lignin (Kale et al., 2015). Biodegradation uses the functions of various microbial species to convert polymers (organic substrates) to smaller molecular weight fragments. Biodegradation efficiency obtained by microorganisms is directly related to key properties such as crystallinity and molecular weight of the polymers. The chemical and physical properties of polymers are very important for their biodegradation (Glaser, 2019). Recently, a number of microorganisms have been reported and investigated for producing polyester-degrading enzymes. The microbial biodegradation is broadly accepted and is still carried out for its enhanced efficiency (Usha, Sangeetha, & Palaniswamy, 2011). Polymer biodegradation can take place in a series of stages (Fig. 11–5):

1. Biodeterioration (altering the physical and chemical properties of the polymer).
2. Biofragmentation (breakdown of polymer in monomers and oligomers via enzymatic cleavage).
3. Assimilation: (uptake of molecules by the microorganisms).
4. Mineralization: (production of oxidized metabolites such as CH_4, H_2O, and CO_2).

11.12.1 Degradation activity by bacteria

A total of 10 bacterial cultures were isolated from the waste disposal areas contaminated with PE (synthetic polymer) and examined for the polymer degradation. For further characterization and isolation of polymer-degrading microorganisms, these selected bacterial strains were transferred to polymer-containing medium. On the basis of growth of these isolated cultures in polymer-containing medium, five out of ten, potential bacterial strains were found suitable for degradation of LDPE and screened by LDPE utilization as carbon source. Biochemical tests identified the isolates C 2 5, B1 4-, V8, V1, and V4, and as *Pseudomonas*,

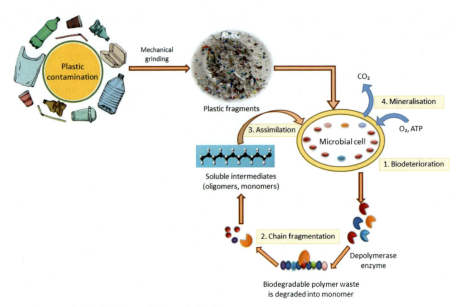

FIGURE 11–5 Stages of microbial degradation of plastics.

Paracoccus, *Bacillus*, *Pseudomonas*, and *Acinetobacter* species. 16S ribosomal DNA sequences of all the isolates were then successfully amplified through polymerase chain reaction and the size of amplified bands was approximately 950 bp (Pathak & Kumar, 2017).

Raddadi and Fava demonstrated that in a recent research study, pure strains of *Pseudomonas* and *Bacillus* were able to successfully degrade PVC films. The gel permeation chromatographic (GPC) measurements indicated 10% reduction of average molecular weight after 90 days of the PVC films incubation with bacterial strains, indicating that both of these strains were able to cause degradation and chain scission of PVC polymer leading to formation of its smaller fragments. At this incubation time, PVC films were completely fragmented (Raddadi & Fava, 2019). The *Streptomyces* species have a greater potential to degrade plastics and PE when compared with other fungi and bacteria (Usha et al., 2011).

11.12.2 Degradation activity by fungi

Aspergillus species are mostly aerobic and are present in oxygen-rich environments, where these species usually grow as mold on surface of substrate, due to higher oxygen tension. Several *Aspergillus* species have been qualified for degrading waste plastic materials due to their particular enzymes, and PE is not an exception. Some research studies have concluded that formation of the biofilms of *A. fumigatus* and *A. terreus* was recognized as a result of surface moistness and was observed on surface of low-density PE (without additives). Other species of this genus, such as *A. flavus* and *A. tubingensis*, have shown highest rate of degradation of HDPE without any pro-oxidant additive and pretreatment (Rojas-Parrales et al., 2018).

Rhizopus oryzae NS5, lab isolate fungal strain, is capable of not only utilizing surface of LDPE as source of carbon but also adhering to LDPE surface efficiently. The degradation process has been confirmed by weight loss, change in functional groups, morphological changes and mechanical properties changes. Even though it is relatively a slow method, the prevailing analysis gives an insight to evidences of the biodegradation of LDPE. It indicates a remarkable possibility of discovering the existence of microorganisms in surroundings that may have the ability to degrade artificial plastics (Awasthi, Srivastava, Singh, Tiwary, & Mishra, 2017).

Two strains of *Lysinibacillus* sp. and *Aspergillus* sp. with their extraordinary abilities to degrade LDPE were separated from the landfill soils of Tehran by using screening and enrichment culture procedures. The ultraviolet (UV) and nonUV-irradiated pure films of LDPE without the pro-oxidant additives in presence and absence of specific mixed cultures of the selected microorganism was done for 126 days. The CO_2 measurements in soil indicated that biodegradation in uninoculated treatments or experiments were slow and were nearly 8.6% and 7.6% of mineralization measured for UV and nonUV-irradiated LDPE, respectively, after treatment of 126 days. In contrast, in presence of selected microorganisms, the biodegradation process was much more efficient and percent biodegradations were 15.8% and 29.5% for nonUV-irradiated and the UV-irradiated films, respectively. Percentage decrease in carbonyl index was comparatively higher for UV-irradiated LDPE when biodegradation process was done in the soil inoculated with selected microorganisms (Pandey, Swati, Harshita, Shraddha, & Tiwari, 2015).

11.12.3 Degradation activity by algae

Algae is one of the high potential organisms which is very useful for humanity. Algae are also often used as an effective biodegradation or bioremediation agent. A term related to biodegradation is biomineralization, in which the living organisms produce minerals such as CH_4, H_2O, and CO_2 (Fig. 11–6). Biosorption and biodegradation capacity of several cyanobacterial species, namely, *Synechococcus* sp., *Cyanothece* sp., *Oscillatoria* sp., *Nostoc* sp., and *Nodularia* sp. are found and tested to be efficient in degrading and removing the contaminant.

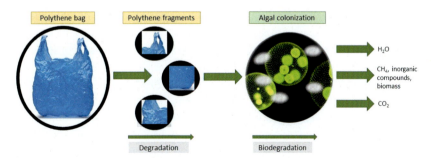

FIGURE 11–6 Algal colonization and degradation. *Adapted From Sarmah, P., & Rout, J. (2020). Role of algae and cyanobacteria in bioremediation: Prospects in polyethylene biodegradation. In Advances in cyanobacterial biology (pp. 333–349): Elsevier.*

Due to their rapid growth and ability to produce lignolytic and oxidative enzymes, Cyanobacteria have been commonly used in biodegradation and bioremediation of waste plastics. Several algal species, *N. pupula*, *A. spiroides*, and *S. dimorphus*, are found to be effective in PE degradation. The algal species are found to grow in large numbers on PE sheets (Sarmah & Rout, 2020). Sharma et al. demonstrated that during summer season, when water bodies are mostly dry, the partially degraded/decomposed polythenes exposed to the sunlight, break into small pieces with algal and bacterial attachment and are released into the atmosphere. Biofilm microorganisms increase the degradation of PE. The microorganisms, especially blue green algae with extracellular polymeric mucilaginous substances, are the primary colonizers of construction materials, rocks, walls, and so forth, and play an important role in their biodeterioration. It has also been reported that the extracellular substances of Cyanobacteria can chelate the surrounding medium, resulting in increased availability and solubility of nutrients (Sharma, Dubey, & Pareek, 2014).

11.13 Degradation activity using insects

Observation of ingestion, penetration, and damage to plastic materials by various insects and their larvae ultimately leads to research on the biodegradation of plastics using insects. The larvae of darkling beetles, especially *Tenebrio obscurus* and *Tenebrio molitor* larvae, showed capacity of rapid degradation of PS. *T. molitor* larva also degrades LDPE. The biodegradation is usually assessed on the basis of modification of ingested polymers, ^{13}C isotopic tracer tests, plastic mass balance, and formation of the biodegraded intermediates. Ingested LDPE or PS polymers can be depolymerized up to 60% to 70% within about 12–24 h after one or two weeks' adaptation. Ingested PE or PS supports larvae with energy for various life activities but not the growth. The cofeeding normal diet enhances PE and PS consumption rate significantly (Yang & Wu, 2020). PS which accounts for an essential fraction of the plastic wastes, is very difficult to biodegrade because of its unique molecular structure. Thus, the chemical modification and biodegradation of PS is very limited.

PS biodegradation can be done by another darkling beetle, *Plesiophthalmus davidis*. Woo et al. demonstrated that, in 14 days, *P. davidis* ingested about 34.27 ± 4.04 mg of Styrofoam per larva, and it survived by only feeding on PS foam (Styrofoam). FTIR spectroscopy verified that the ingested PS foam was oxidized. Gel permeation chromatography analysis showed decrease in the average molecular weight of residual PS in frass compared with feed Styrofoam. In extracted gut flora cultured for 20-days with PS films, cavities and biofilms were observed by atomic force microscopy (AFM) and SEM. X-ray photoelectron spectroscopy studies also revealed that the C-O bonding was introduced to biodegraded PS films. Microbial community analysis indicated that *Serratia* was present in significant amounts in the gut flora, and increased in approximately sixfold when larvae were Styrofoam-fed for two weeks. This concludes that *P. davidis* larvae and *Serratia*, that is, its gut bacteria could be utilized to rapidly degrade and chemically modify PS (Woo, Song, & Cha, 2020). The larvae of *Tribolium castaneum* (red flour beetle) are also observed for chewing and eating the extruded polystyrene foam. The analysis of gut microbiome of bran- and bran-fed *T. castaneum* larvae indicated that *Acinetobacter* sp. is strongly associated with the PS ingestion (Wang, Xin, Shi, & Zhang, 2020).

11.14 Degradation activity using worms

Tenebrio molitor (mealworm) is capable of mineralizing and degrading Styrofoam (PS foam). This fact arises our curiosity to explore many other insect species which have same ability as mealworms. An insect larva, *Zophobas atratus* (superworm), is newly proved to be capable of degrading, eating and mineralizing PS. Superworms and Styrofoam can live together as sole diet and those fed with bran (normal diet) over a 28-day time period. The average rate of Styrofoam consumption for each superworm is estimated at 0.58 mg/d that is four times greater than that of mealworm. The analyses of frass, using thermogravimetric interfaced FTIR, solid-state ^{13}C cross-polarization/magic angle spinning nuclear magnetic resonance spectroscopy, and GPC, demonstrated that the formation of low molecular-weight byproducts and depolymerization of the long-chain PS molecules, also occurs in the larval gut. The PS-degrading capability of superworm is inhibited by antibiotic suppression of the gut microbiota, showing that gut microbiota contributes to PS degradation (Yang, Wang, & Xia, 2020).

Multiple recent reports demonstrated accelerated biodegradation of PE, most common petroleum-based plastic, by employing different macroorganisms such as larvae of *Galleria mellonella* (greater wax moth) and *Tenebrio molitor* (mealworms), which seemingly chew, ingest, and then potentially biodegrade the plastics. Both larvae make holes by chewing plastic items, and indications conclude that in immediate surroundings of these holes in the PE films, the morphology of plastic product seems to have changed. We cannot detect any significant changes of the PE film substrates, either by infrared spectroscopy or gravimetric analysis after contact with homogenate biomass-paste alone, which may indicate that either specimens are different to those present in previous work, or in turn prove that the living macroorganisms are essential for degradation. Additionally, the larvae can also convert macroplastics to microplastic or fragments in the environment (Billen, Khalifa, Van Gerven, Tavernier, & Spatari, 2020). It is found that Indian mealworms (larvae of *Plodia interpunctella*), or waxworms are capable of chewing and then eating PE films. Two bacterial strains, *Bacillus* species YP1 and *Enterobacter asburiae* YT1, capable of degrading PE can be isolated from gut of this worm. Over 28-day incubation period of these two bacterial strains on PE films, viable biofilms are formed, and PE film's hydrophobicity decreases. The obvious damage including cavities and pits (approximately 0.3–0.4 μm in depth) is also observed on surfaces of PE films using AFM and SEM. Thus, presence of the PE-degrading bacteria in waxworm's gut, provide a promising evidence for biodegradation of PE in environment (Yang, Yang, Wu, Zhao, & Jiang, 2014).

11.15 Conclusion

Fortunately, the public and scientific awareness of plastic as global threat is now rising. For reducing the impacts of waste plastics on environment, two major efforts have always been pursued: one is the isolation of selected microorganisms for biodegradation of plastic wastes, and the other is synthesis of biodegradable plastics. The use of BDPs can reduce the cost of labor used to remove conventional plastics from the environment as they degrade naturally. In addition, the degradation and decomposition of BDPs enhances the longevity of landfills by reducing the garbage volume and consequently stabilizes the environment. Different techniques such as biotechnology, thermal, enzymatic, and photodegradation have unequivocally described the decisive role of microorganisms in plastic degradation. While various techniques for plastic degradation are documented, biodegradation has gained much more attention across the globe. The microbes on the PE surfaces in dumping sites were proved to be very effective in degrading PE. Besides other microorganisms, the colonizing algae on PE surfaces were recognized as less hazardous and nontoxic.

References

Adamcová, D., & Vaverková, M. (2014). Degradation of biodegradable/degradable plastics in municipal solid-waste landfill. *Polish Journal of Environmental Studies, 23*(4).

Ahmed, T., Shahid, M., Azeem, F., Rasul, I., Shah, A. A., Noman, M., . . . Muhammad, S. (2018). Biodegradation of plastics: Current scenario and future prospects for environmental safety. *Environmental Science and Pollution Research, 25*(8), 7287–7298.

Al-Maaded, M., Madi, N., Kahraman, R., Hodzic, A., & Ozerkan, N. (2012). An overview of solid waste management and plastic recycling in Qatar. *Journal of Polymers and the Environment, 20*(1), 186–194.

Alshehrei, F. (2017). Biodegradation of synthetic and natural plastic by microorganisms. *Journal of Applied & Environmental Microbiology, 5*(1), 8–19.

Alvi, S., Qazi, I., Baig, A., Andleeb, S., & Yaqoob, A. (2016). Survivability of polyethylene degrading microbes in the presence of titania nanoparticles, 5: 3. *Journal of Nanomaterials and Molecular Nanotechnology, 7*, 2.

Anjana, K., Hinduja, M., Sujitha, K., & Dharani, G. (2020). Review on plastic wastes in marine environment—Biodegradation and biotechnological solutions. *Marine Pollution Bulletin, 150*, 110733.

Awasthi, S., Srivastava, N., Singh, T., Tiwary, D., & Mishra, P. K. (2017). Biodegradation of thermally treated low density polyethylene by fungus *Rhizopus oryzae* NS 5. *3 Biotech, 7*(1), 73.

Bano, K., Kuddus, M., Zaheer, M., Zia, Q., Khan, M., Gupta, A., & Aliev, G. (2017). Microbial enzymatic degradation of biodegradable plastics. *Current Pharmaceutical Biotechnology, 18*(5), 429–440.

Bassi, A. (2017). Biotechnology for the management of plastic wastes. In: *Current developments in biotechnology and bioengineering* (pp. 293–310). Elsevier.

Bhatia, M., Girdhar, A., Chandrakar, B., & Tiwari, A. (2013). Implicating nanoparticles as potential biodegradation enhancers: A review. *Journal of Nanomedicine and Nanotechnology, 4*(175), 2.

Billen, P., Khalifa, L., Van Gerven, F., Tavernier, S., & Spatari, S. (2020). Technological application potential of polyethylene and polystyrene biodegradation by macro-organisms such as mealworms and wax moth larvae. *Science of The Total Environment*, 139521.

Bratovcic, A. (2019). Degradation of micro-and nano-plastics by photocatalytic methods. *Journal of Nanoscience and Nanotechnology Applications, 3*, 304.

Brunner, I., Fischer, M., Rüthi, J., Stierli, B., & Frey, B. (2018). Ability of fungi isolated from plastic debris floating in the shoreline of a lake to degrade plastics. *PLoS One, 13*(8), e0202047.

da Luz, J. M. R., da Silva, Md. C. S., dos Santos, L. F., & Kasuya, M. C. M. (2019). Plastics polymers degradation by fungi. In: *Microorganisms*. IntechOpen.

Dilkes-Hoffman, L., Pratt, S., Lant, P., & Laycock, B. (2019). The role of biodegradable plastic in solving plastic solid waste accumulation. In: *Plastics to energy* (pp. 469–505). Elsevier.

Emadian, S. M., Onay, T. T., & Demirel, B. (2017). Biodegradation of bioplastics in natural environments. *Waste Management, 59*, 526–536.

Erdogan, S. (2020). Recycling of waste plastics into pyrolytic fuels and their use in IC engines. In: *Sustainable mobility* (pp. 1–23). IntechOpen.

Fesseha, H., & Abebe, F. (2019). Degradation of plastic materials using microorganisms: A review. *Public Health Open Journal, 4*, 57–63. Available from: https://doi.org/10.17140/PHOJ-4-136.

Galloway, T. S. (2015). Micro-and nano-plastics and human health. In: *Marine anthropogenic litter* (pp. 343–366). Cham: Springer.

Gewert, B., Plassmann, M. M., & MacLeod, M. (2015). Pathways for degradation of plastic polymers floating in the marine environment. *Environmental Science: Processes & Impacts, 17*(9), 1513–1521.

Gheorghe, I., Anastasiu, P., Mihaescu, G., & Ditu, L.-M. (2019). Advanced biodegradable materials for water and beverages packaging. In: *Bottled and packaged water* (pp. 227–239). Elsevier.

Glaser, J. A. (2019). Biological degradation of polymers in the environment. In: *Plastics in the environment* (pp. 1–22). IntechOpen.

Greene, J. (2014). Degradation and biodegradation standards for starch-based and other polymeric materials. In: *Starch polymers* (pp. 321–356). Elsevier.

Grima, S., Bellon-Maurel, V., Feuilloley, P., & Silvestre, F. (2000). Aerobic biodegradation of polymers in solid-state conditions: A review of environmental and physicochemical parameter settings in laboratory simulations. *Journal of Polymers and the Environment, 8*(4), 183–195.

Harrison, J. P., Boardman, C., O'Callaghan, K., Delort, A.-M., & Song, J. (2018). Biodegradability standards for carrier bags and plastic films in aquatic environments: A critical review. *Royal Society Open Science, 5*(5), 171792.

Hurley, R., Horton, A., Lusher, A., & Nizzetto, L. (2020). Plastic waste in the terrestrial environment. In: *Plastic waste and recycling* (pp. 163–193). Elsevier.

Iwamoto, A., & Tokiwa, Y. (1994). Enzymatic degradation of plastics containing polycaprolactone. *Polymer Degradation and Stability, 45*(2), 205–213.

Kale, S. K., Deshmukh, A. G., Dudhare, M. S., & Patil, V. B. (2015). Microbial degradation of plastic: A review. *Journal of Biochemical Technology, 6*(2), 952–961.

Keane, M. (2010). Catalytic processing of waste polymer composites. In: *Management, recycling and reuse of waste composites* (pp. 122–151). Elsevier.

Kumar, S., & Maiti, P. (2016). Controlled biodegradation of polymers using nanoparticles and its application. *RSC Advances, 6*(72), 67449–67480.

Kumari, A., & Chaudhary, D. R. (2020). Engineered microbes and evolving plastic bioremediation technology. In: *Bioremediation of pollutants* (pp. 417–443). Elsevier.

Li, P., Li, Q., Hao, Z., Yu, S., & Liu, J. (2020). Analytical methods and environmental processes of nanoplastics. *Journal of Environmental Sciences, 94*, 88–99. Available from https://doi.org/10.1016/j.jes.2020.03.057.

Li, S., Xu, S., He, L., Xu, F., Wang, Y., & Zhang, L. (2010). Photocatalytic degradation of polyethylene plastic with polypyrrole/TiO2 nanocomposite as photocatalyst. *Polymer-Plastics Technology and Engineering, 49*(4), 400–406.

Luyt, A. S., & Malik, S. S. (2019). Can biodegradable plastics solve plastic solid waste accumulation? In: *Plastics to energy* (pp. 403−423). Elsevier.

Müller, R. J., Schrader, H., Profe, J., Dresler, K., & Deckwer, W. D. (2005). Enzymatic degradation of poly (ethylene terephthalate): Rapid hydrolyse using a hydrolase from *T. fusca*. *Macromolecular Rapid Communications, 26*(17), 1400−1405.

Nair, N., Sekhar, V., Nampoothiri, K., & Pandey, A. (2017). Biodegradation of biopolymers. In: *Current developments in biotechnology and bioengineering* (pp. 739−755). Elsevier.

Nikodinovic-Runic, J., Guzik, M., Kenny, S. T., Babu, R., Werker, A., & Connor, K. E. (2013). Carbon-rich wastes as feedstocks for biodegradable polymer (polyhydroxyalkanoate) production using bacteria. In: *Advances in applied microbiology* (Vol. 84, pp. 139−200). Elsevier.

Oliveira, M., Ameixa, O. M., & Soares, A. M. (2019). Are ecosystem services provided by insects "bugged" by micro (nano) plastics? *Trends in Analytical Chemistry, 113,* 317−320.

Pandey, P., Swati, P., Harshita, M., Shraddha, M. Y., & Tiwari, A. (2015). Nanoparticles accelerated in vitro biodegradation of LDPE: A review. *Advances in Applied Science Research, 6*(4), 17−22.

Pathak, V. M., & Kumar, N. (2017). Implications of SiO_2 nanoparticles for in vitro biodegradation of low-density polyethylene with potential isolates of *Bacillus, Pseudomonas*, and their synergistic effect on *Vigna mungo* growth. *Energy, Ecology and Environment, 2*(6), 418−427.

Raddadi, N., & Fava, F. (2019). Biodegradation of oil-based plastics in the environment: Existing knowledge and needs of research and innovation. *Science of the Total Environment, 679,* 148−158.

Ramírez-García, R., Gohil, N., & Singh, V. (2019). Recent advances, challenges, and opportunities in bioremediation of hazardous materials. In: *Phytomanagement of polluted sites* (pp. 517−568). Elsevier.

Rojas-Parrales, A., Orantes-Sibaja, T., Redondo-Gómez, C., & Vega-Baudrit, J. (2018). Biological degradation of plastics: Polyethylene biodegradation by Aspergillus and Streptomyces species—A review. In: *Integrated and sustainable environmental remediation* (pp. 69−79). ACS Publications.

Roohi., Kulsoom, B., Mohammed, K., Mohammed, R. Z., Qamar, Z., Mohammed, F. K., ... Gjumrakch, A. (2017). Microbial enzymatic degradation of biodegradable plastics. *Current Pharmaceutical Biotechnology, 18*(5), 429−440.

Ru, J., Huo, Y., & Yang, Y. (2020). Microbial degradation and valorization of plastic wastes. *Frontiers in Microbiology, 11.* Available from https://doi.org/10.3389/fmicb.2020.00442.

Sarmah, P., & Rout, J. (2020). Role of algae and cyanobacteria in bioremediation: Prospects in polyethylene biodegradation. In: *Advances in cyanobacterial biology* (pp. 333−349). Elsevier.

Shahnawaz, M., Sangale, M. K., & Ade, A. B. (2019). Case studies and recent update of plastic waste degradation. In: *Bioremediation technology for plastic waste* (pp. 31−43). Springer.

Sharma, M., Dubey, A., & Pareek, A. (2014). Algal flora on degrading polythene waste. *CIBTech Journal of Microbiology, 3,* 43−47.

Syrpas, M., & Venskutonis, P. R. (2020). Algae for the production of bio-based products. In: *Biobased products and industries* (pp. 203−243). Elsevier.

Thompson, R. C., Moore, C. J., Vom Saal, F. S., & Swan, S. H. (2009). Plastics, the environment and human health: Current consensus and future trends. *Philosophical Transactions of the Royal Society B: Biological Sciences, 364*(1526), 2153−2166.

Tokiwa, Y., Calabia, B. P., Ugwu, C. U., & Aiba, S. (2009). Biodegradability of plastics. *International Journal of Molecular Sciences, 10*(9), 3722−3742.

Urbanek, A. K., Rymowicz, W., & Mirończuk, A. M. (2018). Degradation of plastics and plastic-degrading bacteria in cold marine habitats. *Applied Microbiology and Biotechnology, 102*(18), 7669−7678.

Usha, R., Sangeetha, T., & Palaniswamy, M. (2011). Screening of polyethylene degrading microorganisms from garbage soil. *Libyan Agriculture Research Center Journal International, 2*(4), 200−204.

Vanapalli, K. R., Samal, B., Dubey, B. K., & Bhattacharya, J. (2019). Emissions and environmental burdens associated with plastic solid waste management. In: *Plastics to energy* (pp. 313–342). Elsevier.

Wang, Z., Xin, X., Shi, X., & Zhang, Y. (2020). A polystyrene-degrading Acinetobacter bacterium isolated from the larvae of *Tribolium castaneum*. *Science of the Total Environment, 726*, 138564. Available from https://doi.org/10.1016/j.scitotenv.2020.138564.

Webb, H. K., Arnott, J., Crawford, R. J., & Ivanova, E. P. (2013). Plastic degradation and its environmental implications with special reference to poly (ethylene terephthalate). *Polymers, 5*(1), 1–18.

Woo, S., Song, I., & Cha, H. J. (2020). Fast and facile biodegradation of polystyrene by the gut microbial flora of *Plesiophthalmus davidis* larvae. *Applied and Environmental Microbiology, 86*. Available from https://doi.org/10.1128/AEM.01361-20.

Yang, J., Yang, Y., Wu, W.-M., Zhao, J., & Jiang, L. (2014). Evidence of polyethylene biodegradation by bacterial strains from the guts of plastic-eating waxworms. *Environmental Science & Technology, 48*(23), 13776–13784.

Yang, S.-S., & Wu, W.-M. (2020). Biodegradation of Plastics in Tenebrio Genus (Mealworms). In: *Microplastics in terrestrial environments: Emerging contaminants and major challenges* (pp. 385–422). Springer.

Yang, Y., Wang, J., & Xia, M. (2020). Biodegradation and mineralization of polystyrene by plastic-eating superworms *Zophobas atratus*. *Science of the Total Environment, 708*, 135233. Available from https://doi.org/10.1016/j.scitotenv.2019.135233.

Zheng, Y., Yanful, E. K., & Bassi, A. S. (2005). A review of plastic waste biodegradation. *Critical Reviews in Biotechnology, 25*(4), 243–250.

12

Biodegradation of timber industry-based waste materials

N. Lavanya[1], Sugumari Vallinayagam[2], Karthikeyan Rajendran[2]

[1]DEPARTMENT OF BOTANY, AVINASHILINGAM INSTITUTE FOR HOME SCIENCE AND HIGHER EDUCATION FOR WOMEN, COIMBATORE, INDIA [2]DEPARTMENT OF BIOTECHNOLOGY, MEPCO SCHLENK ENGINEERING COLLEGE, SIVAKASI, INDIA

12.1 Introduction

Timber are of two types such as hardwood timber and softwood timber. Hardwood timber species include oaks, gums, maples, hickories, walnut, and so on. Softwood timber species are Douglas-fir, Hemlock, true Firs, Pines, Spruces, Cedars, Redwood, and so on.

In India, based on the 2019 numbers, land surface is about 297.3 mha. The India State of Forest Report (2019), a biennial assessment report of India's forests, elaborates that India has around 80.72 mha of forested land (forest and tree cover), which constitutes 24.56% of the total land area. India's annual change rate is positive and varies between 0.4% and 1.0% per year, indicating a constant expansion in forest area by about 0.66−1 mha/year through regeneration and afforestation projects. About 86% of forest area is publicly owned, administered by the government, and those public lands are reserved for communities and indigenous groups; the other 14% is privately owned. Public lands can be classified as protected, reserved, unclassed, and village forests. FAO has also released the State of World's Forests 2020 stating India has 72 mha of forested land, constituting 2% of the total land area.

India is one of 17 mega-biodiversity countries, hosting 7% of the world's biodiversity. Many plant species are endemic to India.

Indian forest types include tropical wet evergreens, semievergreens, deciduous, tropical dry deciduous, subtropical pine forests, alpine, dry alpine scrub, temperate montane forests, swamps, mangroves, thorn. Champion and Seth in 1968 published a seminal classification system of forest types of India. It provides a description of 200 forest types and its hierarchical levels (ISFR, 2019).

According to the Desertification and Land Degradation Atlas of India, prepared by the Space Applications Centre, Indian Space Research Organization, Department of Space, an area equivalent to 96.4 mha, or about 29.32% of the total geographical area of the country, is undergoing the process of degradation. From this, about 10.67% of forests are under degradation. This is due to heavy land use pressure. Forests are under pressure due to several reasons, ranging from fuel wood collection, extraction of resources, fodder, wood, non-timber forest products, grazing, and so forth. There are about 170,000 villages in and around forests with approximately 275 million people living in them.

Commonly harvested species from natural forests in India include Teak (*Tectona grandis*), both from natural and planted forests, Sal (*Shorea robusta*), and Khair (*Acacia catechu*). Common planted species include, among others, fast-growing (and short rotation) species such as Teak (*T. grandis*), widely planted timber species, and most of it is harvested from planted forests, Eucalyptus (*Eucalyptus* spp.), Poplar (*Poplus* spp), Acacia (*Acacia* spp), and Subabul *(Leucaena leucocephala)*.

India cannot meet its own demand for wood products with domestic supply, and hence is one of the top most importing country of tropical woods. However, India's tropical log consumption levels has seen a decline in 2017 with previous year. India's imports have continued to decline steadily, dropping to 2.6 million m^3 in 2017 and 2.4 million m^3 in 2018. The decline in imports reflects demand levels in India's plywood industry, which is a significant end-user of tropical logs.

12.2 Background

Historical buildings are very significant for the cultural identity of the people so it is important to conserve it. Amid growing environmental consciousness and increasing demand for timber products, the importance of fulfilling growing demand for these products on the one hand, and at the same minimizing environmental impacts, is increasingly recognized. Wood is decomposed by many biological agents such as bacteria, fungi, and insects. The environmental benefits of using timber are not straightforward; although it is a natural product, a large amount of energy is used to dry and process it. Much of this can come from the biomass of the tree itself, but that requires investment in plant, which is not always possible in an industry that is widely distributed among many small producers. This chapter composes the biodegradation of timber industry based on waste materials.

12.3 Wood from the tree

Timber is one of the significant resource of forest product used for construction work around the world. Majorly timber is used for building construction that leads to small and large wood modification that enable much larger buildings. Deforestation take place normally due to agriculture rather than logging of timber. One should use timber for various reasons, such as unlimited potential design; construction is based on both engineering and culture practice and high strength-to-weight ratio. Due to the evolution of plants, trees have been formed and have great variability by means of angiosperms (flowering plants) and almost all gymnosperms (uncovered seeded plants). Industrially, wood obtained from angiosperm is called hardwood, and wood obtained from gymnosperms are softwood. *Trema micrantha*, Royal

Empress Trees, Eucalyptus, and Willow are some of the fast-growing popular tree species. Based on the apical meristem, the growth of the plant takes place. But in trees the secondary growth is obtained by means of vascular cambium. Due to the mechanism of secondary growth oldest part of the wood were formed in the center of the tree and young xylem (sapwood) form the outerlayer of the tree. when the cell died and leads to form a hollow, which is said to be heartwood. Heartwood and sapwood have various different properties in construction. Resin and polyphenols protect the dead cell from fungal attack. The annual rings of tree are produced based on the heartwood and sapwood, which is nonuniform and consists of early and late wood. Wood is made up of cellulose, hemicellulose, lignin, and extractives. Mechanical properties of the wood are mostly affected by knots, twists, and spiral grain. In wood water absorption and dehydration are take place by hydroxyl groups that leads to swelling and shrinkage respectively. When the wood contains 20% of moisture, it leads to attack fungi and bacteria that affect mechanical properties. Impregnation, chemical modification, and thermal modification are some of the modification methods in wood processing that increase wood stability and durability.

Timber is significantly used for construction of building, in which large amounts of usage of timber causes deforestation. Forest is one of the major sources for the production of supply chain of softwood and hardwood timbers. Timber is the unique material for construction of building compare to other materials. Compared to the human time scale, forests are long lived, ranging from 35 to 70 years depending upon the species and habitat. Changes in forest causes societal impacts, as well as ecology and environmental changes, which leads to sustainability of timber woods that affect Earth's natural resources.

The worldwide supply of wood is a multifaceted network of harvesters, processors, and distributors. After harvesting the timber (Fig. 12–1). It is termed as a round wood and further processing takes place, transported into a sawmill to remove the bark and surface defects.

Softwood also posses structurally optimized construction materials known as "engineered timber." The benefits of this wood consists of laminated timber, adhesives, dimensional stability, and stronger durability. The coating surface of the wood products can be painted, which is against physical barriers such as weathering and degradation. The strength of the timber is similar to that of reinforced concrete. Heartwood is slightly stronger and softwood is a little weaker, even though timber cannot match modern high-strength concrete.

Forests cover most of our land surface in which one-third of our world is covered by the total land area. Forests can be classified as

1. Pristine primary forests
2. Modified natural forests and
3. Planted forest

Not all forests are used for production; half of the forests are protected by environment and heritage as a primary objective. Sustainable quantity of forest wood can be harvested without depleting or humiliating the forest resource. The forest is one of the major benefits for the reduction of greenhouse gases and to mitigate climate change.

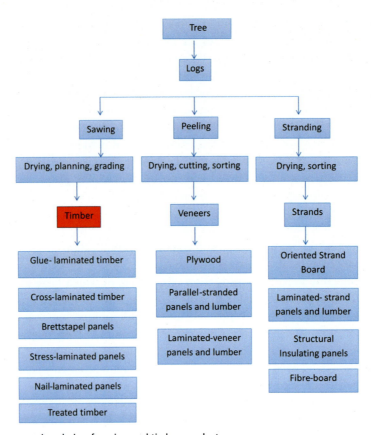

FIGURE 12–1 The processing chain of engineered timber products.

Trees and their derived product are significantly helpful for construction of buildings in society around the world for thousands of years. Timber is mainly used for the construction of contemporary tall buildings. Wood has a natural preference, and building with wood is good for construction, although a huge amount of energy is used to dry and progress it. The previous reports shows that forests could maximize timber production as a boon to both people and planet.

12.4 Plantation and cultivation of timber

Plantation of timber species has been widely cultivated in tropical regions that produce high quality timber. Important timber tree species include Teak, Rosewood, Illuppai, Sal, Sempgam, Red sander, Gmar, Neem, *Melia dubia*, and Indian kino tree. Other timber species are Ailannthus, Eucalyptus, *Casuarina equisatifolia*, *Casuarina junghuhniana*, and Thornless bamboo.

12.4.1 Teak

Teak is indigenous to India. In natural forests, rotation is 100–120 years; in artificial regeneration it is about 70–80 years; and in coppice regeneration it is 40–60 years. Teak mostly occurs in monsoon climates, under tropical and subtropical conditions, with sandy loam soil, 6–7 pH not exceeding 8.5, and well drained. Rainfall of about 1000–1500 mm is adequate and even less than 750 mm per year. Spacing for cultivation is about 1.8 × 1.8 m, 2 × 2 m is generally initially applied (subsequently thinned in stages); 50%–60% germination percentage and 2.5–3 kg seeds are required to prepare one mother bed and derived 1000–1500 seedlings. Stumps are prepared out of seedlings that are 2.5 cm long, collar (2–3 cm), and 22–23 cm of tap root. It is done in the premonsoon period, which has high success. Square or line planting is done in 45–60 cm^3 size pits for seedlings and for stump planting, 15 cm diameter holes, and 30 cm depth. Special features of this species contain the presence of "Tectol" phenol in the sap, which gives high resistance to sap wood rot and termites. It is mainly used for timber, its value is superior, so-called "King of Trees." It can be used for all purposes. It yields a volume of 1.58 cum of timber per year per tree (Fig. 12–2).

12.4.2 Rosewood

Rosewood is indigenous to India and distributed in the sub-Himalayan tract of Eastern U.P. to Sikkim (Fig. 12–3). It also occurs in central, western, and southern India. Maximum

FIGURE 12–2 Various useful parts of Teak tree.

FIGURE 12–3 Picture of Rosewood tree.

temperature ranges from 38°C to 50°C and minimum temperature ranges from 0°C to 15°C and the annual rainfall ranges from 750 to 5000 mm. It occurs in a variety of soils and geological formations but it attains its best development in well-drained deep, moist soils. Rosewood is not as fire hardy as Teak. It does withstand annual fires, however, but benefits from protection. Seedlings and saplings are readily browsed by cattle and goats. Light grazing helps to put down weeds and benefit its growth, but heavy, uncontrolled grazing is injurious. The wood is very hard, close-grained, strong, durable, ornamental, and is used for a large number of purposes such as for furniture, paneling, ornamental work, ordinance work, agricultural implements, and so forth. It is exported to Europe under the name of rosewood or Bombay blockwood. Wood is also used for making cart wheels and gun carriages.

12.4.3 Illupai

Botanical name: *Bassia latifolia*

The tree is an indigenous species and common in Madhya Pradesh, Maharashtra, Gujarat, Peninsular India, Chhota Nagpur, Bihar, and Orissa. It occurs almost all over India. It is a characteristic tree of moist, dry, mixed deciduous forests (Fig. 12−4). The spacing should be 3 m × 3 m or 5 m × 5 m. It performs well in the areas with rainfall between 800−1800 mm. It is a tree of dry tropical and subtropical climates. The tree grows on a wide variety of soils but prefers sandy soil. It also performs in alluvial soil. The ripe fruits are collected by shaking the branches during the months of July and August. The fruits are then rubbed and washed to separate the seeds. Fresh seeds can be sown in nursery beds or polyethylene bags during the months of July and August. The seeds are sown at a depth of 1.5−2.5 cm and should be covered with soil. In the nursery beds, the spacing should be 30 cm × 15 cm. Germination starts after 10−15 days. The seedlings should be watered daily during the first year of growth. Seedlings should be protected from direct sunlight during the first year of growth. Nursery seedlings can be transplanted into the fields after 12 months during the rainy season. For field plantation, pits (0.5 m × 0.5 m) are made in May and June. The pits should be filled with soil and Farm Yard Manure (FYM) after the second shower of rains. Seedling transplantation should be carried out in the evenings, and

FIGURE 12–4 Various parts of Illupai tree.

the taproots should be handled carefully. The soil around the seedlings should be loosened every three months and weeding should be done simultaneously during the first two years. The wood is coarse to even-textured, very strong, hard and heavy. The seed oil is used in the manufacture of soap, glycerin, and margarine. The flower corollas are eaten raw or cooked. The flowers yield alcohol and the leaves are used as fodder. The seed cake can be used as biofertilizer. Decoction of bark is used in curing bleeding gums and ulcers. Its flowers are used in curing coughs and bronchitis.

12.4.4 Sal

Scientific name: *Shorea robusta* (Fig. 12–5)
　　Nativity: India
　　Climate

- Maximum temperature: 360 cm to 440°C
- Minimum temperature: 110 cm and 170°C
- The average annual precipitation is 1000–3500 mm

　　Soil

- Well-drained, loamy soil
- Mountains, rivers, fertile soil rich in areas

　　Cultivation

- Nursery preparation: 1:1:1 (Sand: Soil: Farm Yard Manure)
- Planting Spacing: 2 × 2.5 m
- Weeding, cleaning, and pruning operations should be done up to the age of three-year-old saplings
- LLow demand species
- It can be grown in partially shaded areas

　　Plant protection

- Young seedlings are susceptible to reactions with animals
- Deer and cows cause damage for tubers

FIGURE 12–5 Picture of Sal wood.

12.4.5 Red sander

Scientific name: *Pterocarpus santalinus* (Fig. 12–6)
  Life span: 20–25 years
  Planting spacing: 4 × 4 m
  Nurseries technologies

- Seeds are viable up to one year
- Nursery bed size: 1.2 × 1.2 × 0.03 m (Length × width × height)
- Seeds are mixed with sand and broadcasted over the mother bed
- The mother bed should be covered with a thin layer of soil
- Keep a layer of paddy straw for mulching

  Plantation techniques

- Land should be plowed thoroughly
- Planting spacing: 45 × 45 × 45 cm (Length × width × height)
- 12-month-old saplings should be planted in pits
- Weeding, cleaning, pruning, and irrigation should be done up to three years old

  Utility values

- Wood can be used for the perfume industry
- Wood can be used for medicinal purposes

12.5 Production of timber

Production of timber is the primary source for revenue and to maximize the output of wood. Production is not a linear process and also includes market and nonmarket products, financial benefits, management practices, environmental implication, and some management practices.

According to the International Tropical Timber Organization [2017] the production of timber is almost 50 million m^3 of logs, of which only a minor portion was exported (Fig. 12–7).

FIGURE 12–6 Picture of Red sander tree.

	Production quantity (x 1000 m³)	Imports quantity (x 1000 m³)	Domestic consumption (x 1000 m³)	Exports quantity (x 1000 m³)
Logs (Ind. Roundwood)	49 517	4 383	53 881	19
Sawnwood	6 889	869	7 744	14
Veneer	295	415	702	8
Plywood	2 537	141	2 627	51

ITTO (2019), data 2017

FIGURE 12–7 Production and Utilization of timber species.

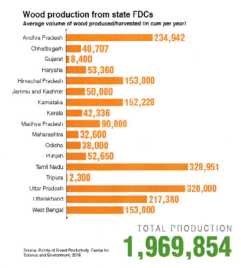

FIGURE 12–8 Wood Production from variouse states in India.

For the production of timber species, the forest must contain higher temperature with sufficient soil moisture for plant growth. The warmer condition leads to affect the wood by insects and increases weed population, which negatively impacts development and increases mortality rate.

In Asia and other developing countries, the demand of timber wood has been increased, which is more productive and sustainable. The alternative of wood products, such as cement, steel, and coal, affects the global environment by CO_2 emission. Recent research suggested that emission of CO_2 is increased by drops in tree growth, and it can be declined if the overall growth of trees increases. In future, the production of timber will be increased in the tropic and subtropic regions, and rotation of fast-growing timber species can be between 10–20 years. The value of timber is increased from a few dollars to several dollars (Fig. 12–8).

The average annual production of timber by Forest Development Corporations (FDCs) is 1.97 z million cum, and makes up a significant part of the total from forests (3.175 million cum). FDCs have not been able to fulfill the dream of raising the productivity of Indian forests. At 0.77 cum per ha, their annual productivity does not compare with that of Trees outside the **forest** (TOF) (3.06 cum per ha).

Timber sources

- Find out the different sources of legal timber
- Determine which source type your timber comes from various source

Timber sources type	Description of source type
Government reserved forests/ unclassed forests	Timber from Government reserved forests/unclassed forests (can be natural forests, plantations, degraded areas, or barren land). They are managed either solely by the State Forest Department or jointly by the State Forest Department and local communities through Joint Forest Management Committees. They may only be harvested by the Forest Department. The following documents are required: • Forest working plans (Forest management prescriptions per the National Working Plan Code for a period of 10 years) • Transit passes which details the origin and destination of all consignments
Private plantations	Timber from private plantations, including block plantations, agro-forestry plantations, farm forestry plantations, and industrial plantations. The following documents are required: • Permits from the Forest Department or local panchayats (local elected representative bodies) per harvesting rules for specific species • Harvest permission letter from the Forest Department/local government (panchayat) head • Transit pass is optional as per transit rules. In cases where the transit pass requirement is waived, substitute documents such as Agricultural Produce Market Committee tax receipts are considered legal documents

12.6 Consumption at wood-based industry

Currently, the best estimates of state-wise wood consumption under nonfuelwood categories are available only for three industrial sectors: housing, furniture and agricultural implements. The combined annual consumption of timber in the three categories, is 33.61 million cum, the round wood equivalent of which is 48 million cum. These calculations are made assuming a wood lifespan of 20 years in construction, 15 years in household furniture, 10 years in commercial furniture, and 5 years in agricultural implements. (Table 12−1).

12.6.1 Furniture

Booming real estate and housing, growth in tourism and hospitality sectors, and changing demographics have been the key drivers of the furniture market in India. With a market size of approximately 165,000 crore, the industry employs around 500,000 workers. There are 1420 registered furniture factories in the country, out of which 1157 are in operation. Unorganized sector constitutes approximately 85% of the market, but the share of the organized sector is increasing. Wood-based furniture dominates the market with a

Table 12–1 Consumption of timber under three major categories[a].

State/ Union Territory	House construction	Furniture	Agricultural implements	Total	Percentage
Andhra Pradesh	23.4	3.2	2.1	28.6	6.8
Arunachal Pradesh	0.7	0.1	0	0.8	0.2
Assam	7.7	2.3	0.5	11	2.6
Bihar	9.2	3.0	0.3	12,5	3
Chhattisgarh	12.0	0.9	0.9	13.8	3.3
Gujarat	27.5	2.4	0.3	30.2	7.2
Haryana	5.9	1.5	0.5	7.9	1.9
Himaohal Pradesh	5.7	0.5	0.1	6.3	1.5
Jammu and Kashmir	5.3	1.1	0.1	6.4	1.5
Jharkhand	4.9	0.9	0.3	6.1	1.5
Karnataka	17.9	1.9	2.6	22.5	5.3
Kerala	15.2	3.3	NA	18.5	4.4
Madhya Pradesh	26.3	1.8	1.8	29.8	7.1
Maharashtra	55.6	4.6	5.1	65.3	15.5
Odisha	7.1	1.3	0.4	8.8	2.1
Punjab	9.2	3.4	0.8	13.4	3.2
Rajasthan	10.4	1.6	0.7	12.6	3.0
Tamil Nadu	18.5	2.5	0.2	21.1	5.0
Uttar Pradesh	42	15.1	4.3	61.4	14.6
Uttarakhand	4.8	0.5	0.1	5.4	1.3
West Bengal	14.6	4.1	0.5	19.2	4.6
Northeastern states	5	1.1	0.0	6.1	1.5
Union Territories	11.6	0.9	0.0	12.6	3.0
Total	340.2	58.5	21.6	420.3	100

[a]Figures in this table do not represent annual consumption.
Source: India State of Forest Report. **Forest** Survey of India(FSI), 2011.

65% share worth 1468 billion, with teakwood being the most popular raw material, particularly in western and southern regions (Fig. 12–9). However, cheaper cane and bamboo is becoming a favorite as well. The future of the furniture industry of India is bright, with a huge growth in demand, but poor business models within the unorganized sector discourage competiveness. The procurement of raw material presents its own set of problems (Table 12–2).

12.6.2 Fuelwood

Almost nine-tenths of all wood produced in India is consumed as fuelwood. Wood is a major source of energy for cooking in Indian households, mainly in rural areas, and meets around 60% of all domestic energy needs of the country. The annual consumption of fuelwood in India is 332.95 million cum (216.42 million tons; per capita about 17.7 kg in rural and 6.3 kg in urban areas). About 853 million Indians make use of this source of energy. Out of this, about 200 million people source nearly 90.4 million cum (58.75 million tons) of fuelwood from forests alone

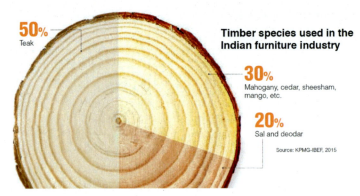

FIGURE 12-9 Timber species use in furniture industry.

Table 12-2 Timber used for Furniture industry.

Furniture material	2010−11 (Rs. Crores)	2014−15 (Rs. Crores)	2019−20 (Rs. Crores)	CARG (projected) 2014−20 (%)
Wood	281	468	936	14.9
Plastic	33	65	136	16
Bamboo and cane	13	29	91	25.9
Metal	72	122	272	17.3
Others	21	36	75	15.8
Total	420	720	1510	16

Based on the: MGC estimate, 2016.

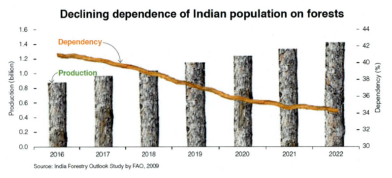

FIGURE 12-10 Declining of forest population with respect to human population.

(Fig. 12−10). In recent years, the number of people dependent on fuelwood has been increasing at a rate slower than the rate of population growth. This provides a golden opportunity to the government to promote alternative sources of energy in the rural landscape (Table 12−3). Fig. 12−10. Declining of forest population with respect to human population.

Table 12–3 State-wise annual fuelwood consumption.

State/ Union territory	Total fuelwood (million tonnes)	Fuelwood from (million tonnes)	Fuelwood from TOF (million tonnes)
Andhra Pradesh + Telangana	24.3	3	21.3
Arunachal Pradesh	0.4	0.3	0.1
Assam	11.4	2.5	3.9
Bihar	11.5	0.5	11
Chhatisgarh	4.4	1.4	3
Gujarat	9.7	2.2	7.5
Haryana	1.5	0	1.5
Himachal Pradesh	1.2	1.2	0.1
Jammu and Kashmir	1.4	1	0.4
Jharkhand	4.8	2.9	2
Karnataka	21	5.8	15.2
Kerala	14.5	2.2	12.4
Madhya Pradesh	13.7	7.2	6.5
Maharashtra	9.5	4.5	5
Odisha	8.9	3	5.9
Punjab	3.4	0	3.3
Rajasthan	18.8	3.7	15.1
Tamil Nadu	12.4	2.6	9.8
Uttar Pradesh	19.1	1.3	17.8
Uttarakhand	2.6	2.1	0.4
West Bengal	14.2	6.4	7.8
Northeastern states	5.3	3.8	1.5
Union Territories	2.6	1.3	1.4

Source: India Stats of Forest Report, FSI, 2011.

12.7 Annual consumption of wood

In its 2011 assessment, **Forest** Survey of India (FSI) has provided the estimates of consumption of timber in only three sectors (in nonfuel wood categories)—housing, furniture, and agricultural implements.

Adding all categories, the total estimated wood consumption (excluding fuelwood) in India comes to about 69 million cum per year. This may be a gross underestimation considering that a large share of wood markets, especially panel and plywood, and furniture markets are fairly unorganized, and no official estimates are available for the same. Fuelwood alone amounts to approximately 90% total wood production in India (Fig. 12–11). It is still the dominant energy source in rural India, which indicates the paucity of alternatives in the countryside.

12.8 The value chain of timber-based industry

In the past few years, the wood supply chain has undergone dramatic changes, which have led to increases in forest productivity and management. The mechanism of logging and their

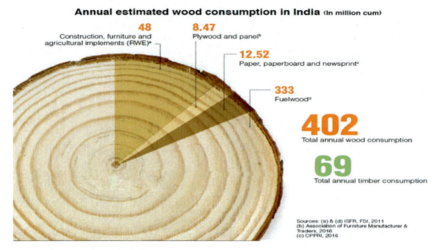

FIGURE 12–11 Wood consumption in India.

equipment has changed. From 1995 to 2005, Asia is only the continent that shows a negative regional production growth rate in the world, and China has become the largest round wood importer in the world. Independent timber harvesting companies play significant roles in the wood supply chain that yield timber on public and private forest land and distribute it to forest manufactured goods mills.

The supply chain study of softwood and hardwood timbers from the forest, used for construction of buildings in industry. The supply chain of wood is unique when compared to other construction materials such as steel, concrete, brick, timber, cement, aggregate, and sand. Timber can be harvested in which top soil remains integral, seedlings are allowed to sprout, and forests are nurtured before yield but other construction materials required rock, ores, or soil that are mechanically detached from the ground.

Forests are long lived in terms of human time scales, with recommended rotations for forestry harvests ranging from 35 to 70 years depending on species and location (one or two rotations per year for most cereal crops). As such, changes to our forests can impact society, ecology, and the environment. Yet, one must keep in mind that even the longest forestry rotations are just a blink on any geological time scale, that is, the time scale for the replenishment of the Earth's resources (rocks, ores, and soils) required in the supply chain of other construction materials. In that regard, timber is the only widely used building material that can be considered to be truly sustainable.

12.9 The waste problem

Waste wood is one of the major problem in timber industry and comprises the most important part of waste materials. The utilization of primary wood leads to prevent wood waste,

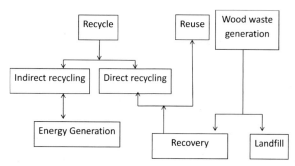

FIGURE 12–12 Schematic flow of recycling, reuse, and recovery of wood products.

which can help to reduce environmental impacts. Out of 1 m^3 of cutting the tree from forest, of which 50% goes to waste in the form of damaged residuals, followed by abandoned logs (3.75%), stumps (10%), tops and branches (33.75%), and butt trimmings (2.5%) are obtained. In Germany, 401 million tons of waste were produced in 2015, in which wood waste was around 11.9 million tones. The major primary sources of wood were wood packaging (21%), wood processing industry (14%), demolition and construction (26.7%), municipal waste (20.7%), imported wood (9.7%), and private households and railway construction (8%). Until 2007, 1,781,000 tons of waste wood were generated in Australia per annum. These large amounts wood waste were disposed of by means of steam production boiler for drying wood products or dumped in a specific region. It causes the environmental impact and mitigation, and also leads to the reduction of forest resources. The main source for the wood waste are low-quality logs, which causes defects, bark, off-cuts, sawdust, slabs, and edged trimmings from sawn timber. Due to new technologies, these low-quality logs can be utilized, and they reduce the wood wastage and recovery (Fig. 12–12).

Instead of reuse, recycled and refurbished waste wood has been disposed of in specific sites that create pollution in the environment and emission of greenhouse gases in many ways, such as transport from the source to landfill site, disposal of synthetic materials contributes to toxic waste, leach from the landfill sites, and such materials take up a large amount of space in landfill sites that need for new waste disposal sites (Box 12–1).

12.10 Biodegradation

Biodegradation is defined as the biologically catalyzed reduction in density of chemical compounds. Wood is decomposed by various biological agents such as bacteria, fungi, and insects if sufficient environment conditions are obtained. The biodegradation process attacks wood and uses cell wall components to cause mechanical disintegration. In a natural environment, microbes and other insects process the colonies and start to degrade the wood. By means of biodegradation, the structure polymer and degradation of wood leads to final products such as CO_2 and water.

> **BOX 12–1** The wood product carbon impact equation.
>
> $A - B - C - D = E$
>
> A. *Manufacturing carbon*: Manufacturing uses energy, and most energy production results in carbon dioxide release.
> B. *Biofuel*: Wood residues are often burned for energy during the manufacture of wood products.
> C. *Carbon storage*: Carbon dioxide is absorbed from the atmosphere during photosynthesis by the growing tree. This carbon is converted to wood, bark, and other parts of the tree.
> D. *Substitution*: There are alternatives to wood products for most applications. However, almost all of these nonwood alternatives require more energy for their manufacture, and the energy used is almost entirely fossil carbon.
> E. *Total carbon footprint or carbon credit*: The biofuel (B), carbon storage (C), and substitution (D) effects reduce the carbon footprint of wood products. In fact, these effects together are almost always greater than the manufacturing carbon (A), so the overall carbon effect of using wood products is a negative carbon footprint (i.e., carbon credit or storage). Thus using wood products can help us to reduce contributions to climate change and conserve energy resources.
>
> *Source*: Bergman et al. (2014)

Table 12–4 Biodegradation is described as an environmental bioremediation.

Type of organism	Character of wood
Bacteria	No, or less-intense, decomposition of wood cells.
Fungi	Rot of wood cells caused by wood-decaying fungi: white-, brown-, soft-rot, and moderate impairment of wood by wood-staining fungi and molds.
Insects	Damage of wood by feeding marks produced by larvae or imagoes.
Marine borers	Damage of wood by bore holes and tunnels.
Parasitic spermatophytes	Growth of plant roots in the wood of a living tree.
Synanthropic vertebrate	Damage of wood by biting it out, hiding places in panels, and excrements.

Table 12–5 Lignin and wood sugars in fungi and type of degradation

Fungi and type of degradation	Weight loss	Lignin	Glucose	Xylose	Mannose
Coriolus versicolor	0	19.0	44.9	24.3	2.1
Phellinus pini	65.3	18.8	38.1	18.7	1.6
Fomitopsis pinicola	17.3	9.7	45.5	23.2	2.3

12.11 Role of microorganisms in biodegradation

(Tables 12–4 and 12–5)

12.11.1 Marine borers

Various kinds of marine borers will attack wood in saltwater or in brackish water. They are found in all of the seas and oceans, but are most active at temperatures of about 10°C.

12.11.2 Shipworms

Shipworms are wormlike mollusks, related to clams and oysters, the most destructive of which are the teredos. They are a particular problem in wharves and harbors, where they attack pilings, docks and so on; in extreme cases, they can destroy structural timbers within a year. They tunnel into the wood, creating tunnels up to 12 mm in diameter and up to 1 m in length. They live on the wood borings and the organic material contained in seawater.

12.11.3 Wood lice

Wood lice are crustaceans related to crabs and lobsters. They too bore into the wood, but unlike teredos, they confine themselves to the region just below the wood surface. The damage is greatest in the intertidal zone, particularly as the regions weakened by the small burrows are further damaged by wave action, and by abrasion from floating debris.

12.11.4 Mold and stain fungi

These live on the carbohydrates stored mostly in the sapwood, and thus have little or no effect on the strength properties of the wood. They thus cause mostly esthetic damage to the wood. (It should be noted, however, that some molds may pose a health hazard in some cases to humans, particularly if there is a high level of mold present in a home). The stains produced by these fungi can take on a variety of colors; black, blue, and gray are most common, but brighter colors such as various shades of yellow, red, green, and orange also occur. On softwoods, the resulting stains can often be brushed off or surfaced off. On hardwoods, however, the stains generally penetrate too deeply to be removed easily.

12.11.5 Bacteria

Bacterial growth may occur in wood that has a high moisture content—freshly cut, in contact with damp soil, or stored in water or under a water spray. While bacterial decay is not generally a major problem, it may cause a significant loss in strength over long periods of time (decades). Some bacteria may increase the permeability or absorptivity of wood.

12.11.6 Insects

There are a number of different wood-destroying insects that may infest wood, including termites and some species of beetles and ants. Collectively, they are responsible for a great deal of damage.

12.11.7 Termites

Termites are social insects that live in large colonies. There are three main classes of termites: subterranean termites, damp wood termites, and dry wood termites. Subterranean termites are responsible for most of the serious wood damage. They live in underground nests in damp soil and will infest wood either directly in contact with the nest, or which they can reach through mud tubes (up to 100 m long) that they construct. The "worker" termites excavate tunnels and chambers through the wood; it is the formation of these galleries that weakens the wood.

12.12 Factors affecting microbial degradation

Microorganisms can humiliate various organic pollutants suitable to their metabolic machinery and to their ability to acclimatize to inhospitable environments. Thus, microorganisms are foremost players in site remediation. However, their effectiveness depends on many factors, including the chemical nature and the concentration of pollutants, their availability to microorganisms, and then the physicochemical characteristics of the environment. So, factors that influence the speed of pollutants degradation by microorganisms are either associated with the microorganisms and their nutritional requirements (biological factors) or associated to the environment (environmental factors).

12.12.1 Biological factors

A biotic factor is that of the metabolic ability of microorganisms. The biotic factors that affect the microbial degradation of organic compounds include direct inhibition of enzymatic activities and therefore the proliferation processes of degrading microorganisms. This inhibition can occur, for instance, if there's a contest between microorganisms for limited carbon sources, antagonistic interactions between microorganisms or the predation of microorganisms by protozoa and bacteriophages. the speed of contaminant degradation is usually hooked in to the concentration of the contaminant and therefore the amount of "catalyst" present. during this context, the quantity of "catalyst" represents the amount of organisms ready to metabolize the contaminant also because the amount of enzymes(s) produced by each cell. Furthermore, the extent to which contaminants are metabolized is essentially a function of the precise enzymes involved and therefore the "affinity" for the contaminant and the availability of the contaminant. additionally, sufficient amounts of nutrients and oxygen must be available during a usable form and in proper proportions for unrestricted microbial growth to occur. Other factors that influence the speed of biodegradation by controlling the rates of enzyme-catalyzed reactions are temperature, pH, and moisture. Biological enzymes involved within the degradation pathway have an optimum temperature and cannot have an equivalent metabolic turnover for each temperature. Indeed, the speed of biodegradation is decreased by roughly one-half for every 10°C decrease in temperature. Biodegradation can occur under a wide range of pH; however, a pH of 6.5–8.5 is usually optimal for biodegradation in most aquatic and terrestrial systems. Moisture influences the speed of contaminant metabolism because it influences the type and amount of soluble materials that are available also because of the pressure and pH of terrestrial and aquatic systems.

12.12.2 Environmental factors

Soil type and soil organic matter content affect the potential for adsorption of an compound to the surface of a solid. Absorption is a similar process wherein a contaminant penetrates into the majority mass of the soil matrix. Both adsorption and absorption reduce the supply of the contaminant to most microorganisms and therefore the rate at which the chemical is metabolized is proportionately reduced. Variations in porosity of the unsaturated and saturated zones of the aquifer matrix may control the movement of fluids and pollutant migration in groundwater. The power of the matrix to transmit gases, like oxygen, methane, and CO_2, is reduced in fine-grained sediments and also when soils become more saturated with water. This will affect the speed and type of biodegradation happening.

12.13 Degradation by genetically engineered microorganisms

Molecular biology offers the utensils to optimize the biodegradative capacity of microbes, accelerate the evolution of "novel" activities, and build completely "new" pathways through the group of catabolic segments from different microbes.

Genes liable for degradation of environmental pollutants, for instance, toluene, chlorobenzene acids, and other halogenated pesticides and toxic wastes are identified. For each compound, one separate plasmid is required—it is not that one plasmid can degrade all the toxic compounds of various groups.

The plasmids are grouped into four categories:

1. OCT plasmid that degrades, octane, hexane, and decane;
2. XYL plasmid that degrades xylene and toluenes;
3. CAM plasmid that decompose camphor; and
4. NAH plasmid that degrades naphthalene [130].

The potential for creating, through genetic manipulation, microbial strains ready to degrade a spread of various sorts of hydrocarbons has been demonstrated. They successfully formed a multiplasmid-containing *Pseudomonas* strain accomplished of oxidizing aliphatic, aromatic, terpenic, and polyaromatic hydrocarbons.

Pseudomonas putida, which contained the XYL and NAH plasmid also as a hybrid plasmid derived by recombinating parts of CAM and OCT developed by conjugation, could degrade camphor, octane, salicylate, and naphthalene, and will grow rapidly on petroleum because it has been capable of metabolizing hydrocarbons more efficiently than the other single plasmid.

This product of gene-splicing was called a superbug (oil eating bug). The plasmids of *P. putida* degrading various chemical compounds are TOL (for toluene and xylene), RA500 (for 3, 5-xylene) pAC 25 (for 3-cne chlorobenxoate), and pKF439 (for salicylate toluene).

Plasmid WWO of *P. putida* is one member of a group of plasmids now termed as TOL plasmid. It had been the primary living being to be the topic of a property case. At that time, it seemed that molecular techniques, either through plasmid breeding or sheer gene-splicing, could rapidly produce microbes with higher catalytic abilities, ready to basically degrade any environmental pollutant. Reports on the degradation of environmental pollutants by genetically engineered microorganisms are focused on genetically engineered bacteria using different gene-splicing technologies: Pathway modification, modification of substrate specificity by *Comamonas testosteroni* VP44.

The application of gene-splicing for heavy metals removal has aroused great interest. For instance, *Ralstonia eutropha* AE104 (pEBZ141) was used for chromium removal from industrial wastewater, and therefore the recombinant photosynthetic bacterium, *Rhodopseudomonas palustris*, was constructed to simultaneously express mercury transport system and metallothionein for Hg2+ removal from heavy metal wastewater.

For polychlorinated biphenyls degradation, chromosomally located Polychlorinated biphenyls (PCB) catabolic genes of *R. eutropha* A5, *Achromobacter* sp. LBS1C1, and *A. denitrificans* JB1 were transferred into an important metal resistant strain *R. eutropha* CH34 through natural conjugation.

Genetic engineering of endophytic and rhizospheric bacteria to be used in plant-associated degradation of toxic compounds in soil is taken into account one among the foremost promising new technologies for remediation of contaminated environmental sites. Many bacteria within the rhizosphere show only limited ability in degrading organic pollutants. With the occurrence of biology, the genetically engineered rhizobacteria with the contaminant-degrading gene are constructed to carry out the rhizoremediation. Examples about the molecular mechanisms involved within the degradation of some pollutants like trichloroethylene and PCBs has been studied.

For heavy metals, introduced *Arabidopsis thaliana* gene for phytochelatin synthase (PCS; PCSAt) into *Mesorhizobium huakuii* subsp. rengei strain B3 then established the symbiosis between *M. huakuii* subsp. *rengei* strain B3 and *Astragalus sinicus*. The gene was expressed to supply phytochelatins and accumulate Cd2+, under the control of bacteroid-specific promoter, the nifH gene.

Finally, the utilization of genetically engineered microorganisms (GEM) strains as an inoculum during seeding would preclude the issues related to competition between strains during a mixed culture. However, there's considerable controversy surrounding the discharge of such genetically engineered microorganisms into the environment, and field testing of those organisms must therefore be delayed until the problems of safety, and therefore the potential for ecological damage are resolved.

12.14 Conclusion

Wood is the oldest and one of the most widely used construction materials. It combines a high strength-to-weight ratio, appealing esthetic properties, ease of construction, the ability to be repaired, and cost effectiveness. It is also a renewable resource, and its production is much less energy intensive than that of other construction materials, such as steel, aluminum, or concrete. The pattern of microbial degradation caused by different microorganisms are distinct and can be used to identify the type of degradation that may be present in wood. The microorganisms are responsible for degradation as well as provide important insights concerning the condition of wood.

13

Microbiologically induced deterioration and environmentally friendly protection of wood products

Olga A. Shilova[1,2], Irina N. Tsvetkova[1], Dmitry Yu. Vlasov[1,3], Yulia V. Ryabusheva[3], Georgii S. Sokolov[1], Anatoly K. Kychkin[4], Chi Văn Nguyễn[5], Yulia V. Khoroshavina[1]

[1]INSTITUTE OF SILICATE CHEMISTRY OF RUSSIAN ACADEMY OF SCIENCES, SAINT-PETERSBURG, RUSSIA [2]SAINT-PETERSBURG STATE ELECTROTECHNICAL UNIVERSITY "LETI," SAINT-PETERSBURG, RUSSIA [3]SAINT-PETERSBURG STATE UNIVERSITY, SAINT-PETERSBURG, RUSSIA [4]LARIONOV INSTITUTE OF PHYSICAL AND TECHNICAL PROBLEMS OF THE NORTH OF THE SIBERIAN BRANCH OF THE RUSSIAN ACADEMY OF SCIENCES, YAKUTSK, RUSSIA [5]COASTAL BRANCH—VIETNAM RUSSIAN TROPICAL CENTER, KHANH HOA, VIETNAM

13.1 Features of wood biodegradation under the influence of wood-destroying fungi

Many papers confirm a significant variety of living organisms capable of destroying wood. However, the main role in this process is played by wood-destroying basidiomycetes, which cause various types of wood rot in the natural environment, as well as in buildings and structures (Goodell, Qian, & Jellison, 2008; Kirk & Cowling, 1985; Pánek, Reinprecht, & Hulla, 2014; Schmidt, 2007; Verevkin, Kononov, Serdyukova, & Zaytsev, 2019). The main wood components that are destroyed by basidiomycetes are cellulose, hemicellulose, and lignin. The enzymes secreted by fungi convert these wood components into more readily available water-soluble compounds. This transformation is possible only if there is a certain amount of free moisture in the wood. Organic substances must be in aqueous solutions, which ensure their diffusion into the fungal cell through the cell wall.

Among the basidiomycetes that destroy wood, a special place belongs to domestic fungi that develop on industrial wood. They are capable of damaging wooden structures in residential buildings, engineering structures, mines, and historical buildings (Schmidt, 2007).

This group includes the genera: *Serpula, Tapinella, Coniophora, Antrodia, Neolentinus, Gloeophyllum, Cylindrobasidium,* and *Phlebiopsis*. Domestic fungi destroy cellulose in wood, causing destructive brown rot (Worrall & Wang, 1991). The high rate of destructive processes is due to the activity of cellulolytic enzymes of wood-destroying fungi, secreted into the substrate (Baldrian & Valášková, 2008; Selvam et al., 2017). When affected by domestic fungi, the wood takes on a brown color, which is associated with an increase in the specific content of lignin in it. It becomes brittle, easily breaks and crumbles, cracks, disintegrates into prismatic fragments and loses in volume and weight.

Another type of wood destruction is called "white rot." It is mainly associated with lignin utilization, with very low cellulose utilization by fungi (Abbott & Wicklow, 1984; Ander, Eriksson, & Yu, 1984; Blanchette, 1995; Blondeau, 1989). The main role in this process belongs to the ligninolytic enzymes of wood-destroying fungi (Glasser, 2019; Verevkin et al., 2019), which they release into the environment.

The development of the decay process begins in wood with a moisture content of at least 18%–20% in the presence of air and at a positive temperature (in the range from 5°C to 45°C). However, in wood with very high moisture content, fungi develop more slowly due to the lack of free air. Wood-destroying fungi easily tolerate prolonged exposure to low temperatures, but heating above 100°C, especially in the presence of water vapor, ensures wood sterilization (Dickinson & Pugh, 1974).

Plants are known to produce various compounds that have a protective function (Clausen & Yang, 2007). Many types of wood contain substances that counteract biodegradation. Moreover, as a rule, heartwood is more resistant than sapwood. There are two large groups of substances that provide plant resistance to pathogenic fungi. One group of substances is constantly present in the plant, regardless of whether the pathogen has attacked it or not. These substances are called preinfectious compounds. Another group of substances is absent in a healthy plant, but appears in it after infection with a fungus. These substances are called postinfectious compounds. Examples of natural biocides are pinosylvin found in pine heartwood or thujaplicin in thuja (Coombs & Trust, 1973; Stirling, Ruddick, Xue, Morris, & Kennepohl, 2015). Pinosylvin is an aromatic substance, a phenolic compound. It is found in the wood of conifers and protects the wood from fungi and insects (Plumed-Ferrer et al., 2013). Other compounds in woody plants are known to have antibacterial, antifungal, and antioxidant properties (Baya, Soulounganga, Gelhaye, & Grardin, 2001; Chedgy, Lim, & Breuil, 2009; Plumed-Ferrer et al., 2013; Chedgy, 2010). However, despite the presence of defense mechanisms in woody plants, the wood used by humans requires additional protection. For this purpose, various chemical methods are most often used, for example, impregnation with antiseptics.

13.2 Chemical means of protecting wood from biodegradation

Over the past 20 years, a wide variety of formulations of protective coatings and biocides have been proposed to counteract wood biodeterioration agents (Morrell, 2017). The growing interest in the search and study of the properties of new biocides for wood worldwide is associated, first of all, with a significant tightening of requirements for the safety of protective agents {European Directive—the Biocidal Products Regulation [Regulation (EU) 528/2012]}. Numerous studies devoted to the analysis of the effect of biocides on human health and the environment have led to the abandonment of the use of the most toxic substances (chromium and arsenic-containing fungicides), which for a long time led the market for wood preservatives.

Antiseptics for the protection of wood materials must be highly toxic in relation to the inhabiting microorganisms. However, it is desirable that they meet a number of requirements, namely, to be harmless to humans and animals; maintain protective activity for a given period; have low water solubility (i.e., low leachability); easy to penetrate into wood without deteriorating its physical and mechanical properties; do not have an unpleasant odor; and be resistant to high temperatures and wood processing.

For antiseptic processing of wood, water-, organo-, and oil-soluble antiseptics are used (Mazela, Polus-Ratajczak, Hoffmann, & Goslar, 2005). The following groups of antiseptics meet modern environmental requirements to the greatest extent: boron-containing substances, carbamates, isothiozolones, quaternary ammonium compounds (QACs), sulfamides, triazoles, inorganic copper compounds, copper naphthenates and citrates, modified creazote oils (not containing benzo-α-pyrene), and metal dioxides (Reinprecht, 2010).

Let's describe some of them in more detail. Boric acid H_3BO_3 and boron-containing wood preservatives are widely used (Cai, Lim, Fitzkee, Cosovic, & Jeremic, 2020; Neklyudov & Ivankin, 2005; Tsvetkova et al., 2019). Boron-containing inorganic fungicides, which include boric acid, sodium tetraborate, sodium octaborate, zinc borates, and several others, are traditionally used to protect wood used indoors. They provide protection against damage of wood-destroying fungi and insects, and also significantly reduce the flammability of wood. Typically, borates are used as aqueous solutions. To achieve effective protection, the absorption of the antiseptic in terms of dry boron-containing substance must be at least 3 kg/m^3 (Mazanik, 2011). The main advantage of inorganic boron-containing agents is their low toxicity in relation to humans and relative harmlessness to the environment. The disadvantage lies in their ease of washing out from wood, as well as in insufficiently high efficiency against mold fungi. Inorganic borates used to protect wood outdoors are modified by introducing fixatives, water-repellent additives, or polymerizing monomers into their composition, which reduce the leaching of the antiseptic. The most famous fixative that forms a stable complex with boron ions is polyvinyl alcohol. Tannins, silicone gel, and animal proteins can be used for this purpose (Mohareb, Thévenon, Wozniak, & Gérardin, 2011). A very promising substance, the fungicidal properties of which are just beginning to be investigated, is didecyldimethylammonium tetrafluoroborate (DBF). This new antiseptic demonstrates high efficacy against cellulose and lignin-destructive fungi, as well as low leachability even without the use of acryl-silicone fixing additives.

An effective representative of the class of fungicidal carbamates is 3-iodo-2-propynyl-butyl carbamate (IPBC) (Badreshia & Marks, 2002). This organic fungicide can be used in the form of solutions in organic solvents (acetone and xylene), as well as in the form of aqueous emulsions. IPBC has a broad spectrum of activity and is effective against a wide variety of fungal groups. However, it shows the greatest efficiency in protecting raw wood from damage by mold and wood-staining fungi. In the presence of borates, the inhibitory effect of IPBC increases dramatically. There is evidence of an almost 50% increase in the effectiveness of IPBC against cellulose-destroying fungi when the antiseptic is added to the solution of α-amino-isobutyric acid. IPBC is moderately toxic, noncarcinogenic, and has no gene-modifying effect.

Isothiazolones such as DCOIT (4,5-dichloro-2-n-octyl-4-isothiazol-3-one) and Kathon (5-chloro-2-methyl-4-isothiazol-3-one) are used as solutions in organic solvents, or in the form of emulsions in water for preventive wood treatment in an amount of 0.15–1.28 kg/m^3 (Steen, Ariese, Hattum, Jacobsen, & Jacobson, 2004). In this case, it is necessary to use two-component formulations including a fixing polymer.

QACs have been used for wood protection for the past 30 years. Generally, QACs show their maximum efficiency in relation to wood-staining and mold fungi and significantly less—in relation to wood-destroying fungi. The exception is didecyldimethylammonium chloride, which has a fairly high inhibitory ability against micro- and macromycetes. QACs are water-soluble and miscible with alcohol (Mazanik, 2011). One of the main advantages of long-chain ammonium compounds is anchoring in wood by ion exchange (Jin & Preston, 1991). This is due to the characteristics of the quaternary ammonium salts $[R_4N]^+X^-$, where four organic radicals are covalently bonded to the nitrogen atom. In the wood, these compounds are fixed by reactions with the carbonyl groups of lignin and hemicellulose (Mazela et al., 2005). QACs are not recommended for the protection of wood that will be used in contact with the ground due to their low stability and ability to increase the water absorption of wood. Currently, the main area of application for QACs is the treatment of wood, which is used both indoors and outdoors, but without contact with the ground. Generally, combined protective agents are used, which, along with QACs, include various cobiocides: copper- and boron-containing substances, triazoles, and so forth.

Sulfonamides are also used as biocidal additives. Like QACs, sulfonamides are most effective in protecting against mold and wood-staining fungi. They are used in the form of organic solutions and aqueous emulsions (Mazanik, 2011).

Another class of organic fungicides that are widely used in modern antiseptics are 1,2,4-triazoles, which are stable in the environment and have low toxicity for warm-blooded animals. They are used as solutions in polar and nonpolar organic solvents, as well as in the form of aqueous emulsions. These substances inhibit the growth of all types of wood-damaging fungi. However, as independent antiseptics, they are not sufficiently effective against basidiomycetes.

Preservatives based on copper compounds are the undisputed leaders among the products intended for the long-term protection of wood during its operation. Marketing research from North America (Marzi, 2015) showed that copper salt-based wood preservatives represent 50% of the global wood preservative market. New preservatives based on copper nanomaterials have taken over at least 50% of this sector in the past few years. The high efficiency of copper-containing protective agents in relation to wood destructor fungi has been known for a long time. In recent decades, traditional recipes have been revised. In the 20th century, inorganic copper-containing compounds were most often used ($CuSO_4 \times 5H_2O$, CuO, $CuCO_3 \times Cu(OH)_2$, salts based on Cu/Cr, Cu/Cr/B, Cu/Cr/As, ammonia/Cu/As, Cu/Cr/B/Zn. $CuCO_3 \times Cu(OH)_2$, salts based on Cu/Cr, Cu/Cr/B, Cu/Cr/As, ammonia/Cu/As, Cu/Cr/B/Zn). Currently, the most interesting are organic copper-containing reagents such as bis/N-cyclohexyl diacenium copper oxide (Cu-HDO), copper 8-quinolinolate, various naphthenates, and ammonium carboxylates (e.g., ammonia copper citrate) (Green & Clausen, 2005).

The main problem with the use of copper-based protective agents is the existence of a rather large group of copper-tolerant fungi, for example, *C. versicolor* and *Poria placenta* (Goodell, Daniel, Jellison, & Qian, 2006; Green & Clausen, 2005) attribute this to their ability to release oxalic acid, under the influence of which crystalline copper oxalate, insoluble in water, is formed, which fungi can no longer absorb. To protect the wood from copper-resistant fungi, biocide formulations are being developed in which copper is replaced by other active ingredients (Beaudelaire et al., 2020).

An alternative to copper-containing compositions for wood preservation can be inorganic sol-gel compositions based on TiO_2 and SiO_2, in combination with other reagents (Miyafuji & Saka, 2001). This approach can improve the protection of wood from white and brown rot fungi and termites. In recent years, wood processing using sol-gel technology has attracted the increasing interest of the world community (Hübert & Mahr, 2017; Lu, Feng, & Zhan, 2014; Marzi, 2015; Tsvetkova et al., 2019), which will be described in more detail in the following sections.

Despite the long history of the study of domestic fungi, effective methods of protecting wood from these destructors have not yet been found. The protective drugs available on the market provide protection for no more than five years, while they contain quite dangerous biocides. Therefore, the search for new options for wood protection continues.

Methods for hydrophobization of building materials, including wood, have been used for a long time. However, studies on imparting high hydrophobicity to the surface of wood, up to superhydrophobicity, are among the promising developments for protecting the facades of wooden buildings from weathering. These areas of research will be discussed in the next section.

13.3 Surface hydrophobization

Traditionally, wood hydrophobization is carried out using polymers and organosilicon compounds, while hygroscopicity decreases due to blocking of hydroxyl groups of cellulose and lignin, filling of the capillary-porous structure of wood, as well as changes in the density of crosslinking of the ligno-carbohydrate matrix. With complex fire and biological protection, wood is silylated with an insignificant degree of chemical modification. In this case, the content of chemically bound silicon should be about 1% (Cappelletto et al., 2013; De Vetter, Van Den Bulcke, & Van Acker, 2010).

Due to its hydroxyl groups, wood is hydrophilic and has a high ability to interact with moisture. Both the absorption of water and its desorption lead to undesirable swelling and shrinkage of the wood. Cracks should also be avoided, which are the preferred sites for fungal attack. Controlling the moisture content of wood through various treatments is a promising strategy for increasing wood durability.

A relatively new method of chemical hydrophobization of wood, which significantly increases its strength and durability, is the method of acetylation. It consists in the impregnation of wood to its full depth with acetic anhydride to replace hydroxyl groups with acetyl ones, followed by processing with furan polymers.

Coatings with high water repellency on wood surfaces were also prepared using zinc oxide and stearic acid (Wang, Piao, & Lucas, 2011) or based on hydrothermally grown ZnO nanorods (Fu, Yu, Sun, Li, & Liu, 2012).

Unger (Unger, Bücker, Reinsch, & Hübert, 2013) reduced the moisture absorption of pine sapwood from 25 to 5%–10% as a result of impregnation with silica sols. The moisture absorption of samples from pine sapwood after their impregnation with a sol based on titanium dioxide decreased from 24% to 4%, and the volume of treated samples due to moisture absorption increased by only 2% instead of 13% for untreated samples (Hübert & Mahr, 2017; Hübert, Unger, & Bücker, 2010). These facts led to the conclusion that the use of sols based on silicon dioxide, titanium dioxide, or their mixtures significantly reduces the water absorption of the experimental samples in comparison with the untreated control, that is, increases the water resistance of wood. For example, after immersion in water for 6 h, the water absorption of all samples treated with sols was 47%–52% lower in comparison with untreated control samples (from pine sapwood). This trend remained unchanged for the tested samples for a long time—after 96 h.

The use of sols for wood impregnation is aimed, among other things, at improving the hydrophobicity of the wood surface. It is known (Hübert & Mahr, 2017), that silanes functionalized with organic spacers, in combination with alkoxysilanes, increase water repellency and reduce water absorption and leaching of chemicals from wood. A measure of water repellency can be the surface tension of water or the wetting angle of the surface with water droplets, which is usually in the range of 20–80 degrees for untreated wood. The application of water-repellent coatings using sols based on vinyltriethoxysilane and SiO_2 nanoparticles led to an increase in the contact angle with water up to 150 degrees (Chang, Tu, Wang, & Liu, 2015), that is, to the formation of a superhydrophobic surface. The phenomenon of superhydrophobicity is based on the so-called "lotus effect," as a result of which drops roll off the surface, it acquires good water-repellent properties. For example, (Marzi, 2015) proposes nanocoatings for this purpose, which have a nanorough surface due to filling with silica nanoparticles. At the same time, the good air permeability of such coatings is maintained. Superhydrophobic coatings can be obtained using silicon dioxide nanoparticles as a structure-forming agent, and alkoxysilanes, for example, tetraethoxysilane, methyltriethoxysilane, and so forth, as precursors of the polymer matrix (Kanokwijitsilp, Traiperm, Osotchan, & Srikhirin, 2016; Tsvetkova et al., 2019) or polyvinyl (Liu et al., 2013).

Impregnation of wood with sols can be carried out using ultrasound. This approach significantly reduces the hygroscopicity of wood and, as a result, improves its mechanical characteristics (Lu et al., 2014).

In addition to the use of sols, silanization is traditionally used as one of the options for increasing the hydrophobicity of the wood surface. For example, wood is treated with alkyltrichloro/alkyltrialkoxysilanes. Such processing allows obtaining a surface with contact angles of more than 100 degrees (Mohammed-Ziegler, Tánczos, Hórvölgyi, & Agoston, 2008).

Impregnation of wood with compositions based on commercial SiO_2 nanoparticles with an average size of 20 nm, mixed with a perfluoroalkyl methacrylic copolymer, makes it possible to obtain contact angles of more than 160 degrees (Hsieh, Chang, & Lin, 2011).

However, the decisive factor is the long-term protection of the impregnation or coating. This is especially true for superhydrophobic surfaces, which tend to lose the superhydrophobic effect.

13.4 Nanotechnology for wood protection

Nanotechnology is becoming more and more part of everyday life. At the beginning of the 21st century, it began to be used in compositions for protecting wood from biodegradation (Elvin, 2007; Evans, Matsunaga, & Kiguchi, 2008). Authors (Mishra et al., 2018) in their review article note that wood science and nanomaterials science interact with each other in two different aspects: (1) the manufacture of lignocellulosic nanomaterials obtained from wood and plant sources, and (2) surface or volumetric modification of wood with nanoparticles.

The current trend is the use of metal nanoparticles as a functional component in wood modification. Metal nanoparticles impart bactericidal, fungicidal, and other useful properties to wood-processing preparations. At the same time, unlike drugs based on organic biocides, the emission of volatile organic compounds into the atmosphere is minimized (Mishra et al., 2018). It has been experimentally proven that the fungicidal properties of nanobiocides with an average particle size of $\sim 100-200$ nm differ significantly from the properties of the corresponding metals. Nanoparticles easily penetrate into the wood through the pores of the cell walls, which allows for full impregnation, as well as high uniformity of their distribution over the volume of the protected material. This contributes to the mechanical hardening of the wood, protection from moisture, and from biodegradation.

To obtain bactericidal coatings and to impregnate wood, metal nanoparticles are included in compositions, both water-based and organic-based. The addition of only a few percent of nanoparticles can significantly change the chemical, physical-mechanical, bactericidal, fungicidal, and other properties of protective coatings (Ansell, 2013; Shilova et al., 2019). The presence of nanoparticles in paint, varnish, and sol-gel compositions can impart additional functions to protective coatings, such as self-cleaning ability, photocatalytic activity, water resistance, fire resistance, and provide protection against graffiti (Ansell, 2013; Elvin, 2017; Evans et al., 2008; Mishra et al., 2018; Mukhopadhyay, 2011).

In recent years, the so-called green synthesis has become widespread—the production of metal nanoparticles by the reduction of their salts using plant extracts without the use of toxic chemicals (Korkmaz, 2020; Korkmaz et al., 2020; Nesrin et al., 2020). Metal nanoparticles (Cu, Ag, Au, Cu-Ag, and Ag-Au) are obtained by reduction with plant extracts obtained, for example, from tree bark (Burlacu, Tanase, Coman, & Berta, 2019). Based on a review of literature data, Burlacu et al. (2019) state that metal nanoparticles obtained by the green synthesis method are active reducing agents, have antioxidant, antibacterial, anticarcinogenic activity, and are effective catalysts for purifying water from toxic organic contaminants.

Mishra et al. (2018) believe that the use of nanomaterials will make it possible to more effectively solve problems of wood protection, including from biodegradation. Thanks to the use of preparations with nanoscale particles, it is possible to significantly reduce the cost of the process of creating modified wood with the desired properties and a longer service life. At the same time, less harm will be done to the environment. However, the commercialization of any wood-processing product still requires extensive testing and regulatory documentation.

There are several ready-to-use commercial wood protection nanobiocides on the market today. Compositions have been developed based on copper microparticles ranging in size from 10 to 700 nm (Freeman & McIntyre, 2008). Authors (Civardi et al., 2015) investigated the effectiveness of recently marketed commercial formulations containing a nanosized powder fraction obtained from basic copper carbonate $CuCO_4 \cdot Cu(OH)_2$. The aim of this study was to assess whether these wood preservatives are more effective against Cu-tolerant (brown rot) fungi than conventional preservatives containing copper ions or bulk $CuCO_3 \times Cu(OH)_2$. The idea was that in copper-resistant fungi (brown rot), the concentration threshold for nanoparticles could be lower than the threshold for Cu^{2+} ions. Therefore, Cu-tolerant wood-destroying fungi may not recognize copper nanoparticles, which can penetrate into fungal cells and have an inhibitory effect on them. However, the first results obtained did not reveal any special effects of using micronized copper against the brown rot of *Rhodonia placenta*. The most effective were compositions based on micronized copper mixed with tebuconazole. The low efficiency of $CuCO_3 \cdot Cu(OH)_2$ is mainly due to poor wood penetration. The wood samples did not change color toward green/blue due to the presence of copper, and unreacted $CuCO_3 \cdot Cu(OH)_2$ appeared only as fine dust on the surface of the samples. Civardi et al. (2015) believe that further studies of the interaction of copper nanoparticles with other wood-degrading copper-tolerant fungi are necessary to obtain a more complete picture of the effect of nanoparticles on wood-degrading fungi. In addition, field studies are needed to confirm laboratory findings and evaluate the long-term effects of copper nanoparticles.

Preparations based on silver nanoparticles aimed for protecting wood deserve special attention. Silver is a well-known biocidal agent and has been available in the form of colloidal nanoparticles for over 100 years (Nowack, Krug, & Height, 2011; Rai, Yadav, & Gade, 2009). Silver nanoparticles are introduced into paint and varnish and sol-gel-derived polymer matrices as a biocidal additive, in some cases instead of biocidal organic compounds (Kaegi et al., 2010).

The effectiveness of the bactericidal action of colloidal silver is explained by the ability to suppress the work of the enzyme, which ensures oxygen exchange in the simplest organisms, which leads to their death. Spherical silver nanoparticles and no more than 20 nm in size have the best antimicrobial properties. The use of silver in the form of nanoparticles makes it possible to reduce the concentration of silver by hundreds of times while maintaining all its bactericidal properties for a very long time. Preparations with colloidal silver fix well and are difficult to remove (Stenina & Chesnokova, 2017).

The fungicidal activity of preparations based on silver nanoparticles was studied on strains of molds *Aspergillus niger Teigh*, *Aspergillus flavus Link*, *Penicillium chrysogenum Thom*, *Ulocladium ilicis Thom*, both in laboratory and in the field. Studies have shown that the drugs used inhibit the growth of test cultures at a concentration of nanosized metal particles of the order of 10^{-2}%... 10^{-4}% (in laboratory conditions, when applied to paper disks); and at a concentration of 10^{-3}%... 10^{-2}% (in natural conditions, on the wall). At the same time, the solvents used (water or isooctane) do not affect the fungicidal properties (Dmitrieva, Chmutin, Yarovaya, & Linnik, 2009).

Thus, in Pica & Ficai (2016) the antibacterial properties of some film-forming compositions used for long-term protection of wood, concrete, drywall, and other building materials were investigated. The film-forming composition has antibacterial properties due to the presence of nanosilver in its composition at an optimal concentration of 100−200 ppm. It has been shown that the resulting coatings are not destroyed even when exposed to 200 wet friction cycles, and the antibacterial activity, both against Gram-negative and Gram-positive bacteria, remains after 200 washing cycles.

Silver nanoparticles not only prevent wood biodegradation, but are also used for antitermite treatment of wood building structures (Kartal, Green, & Clausen, 2009).

At the same time, studies on the effectiveness of the use of silver nanoparticles in protective coatings for outdoor use showed that about 30% of nanosilver was washed out within one year as a result of natural weathering (Kaegi et al., 2010).

Authors (Pařil et al., 2017) studied the fungicidal efficacy of copper and silver nanoparticles against two woody fungi on samples of European beech sapwood (*Fagus sylvatica L.*) and Scots pine (*Pinus sylvestris L.*), which were impregnated under vacuum with dispersions of copper and silver nanoparticles in two concentrations — 1 and 3 g/L. Beechwood samples were tested against white-rot fungus (*Trametes versicolor*) and pine wood against brown rot fungus (*Poria placenta*) according to EN113 European beech (*F. sylvatica L.*) and Scots pine (*P. sylvestris L.*) sapwood specimens were vacuum impregnated using dispersions of copper and silver nanoparticles within two concentrations, that is, 1 and 3 g/L.

It was found that the amount of nanoparticles in the wood did not increase proportionally with increasing concentration of nanoparticles in dispersions, but an increase of only 1.5−2 times was achieved. Average leaching of 15%−35% was observed for copper nanoparticles, depending on the wood species used and the copper concentration. The smallest leaching (maximum 15%) was observed for pine sapwood impregnated with a dispersion of silver nanoparticles with a concentration of 3 g/L. The greatest antifungal effect against both tested fungi was found when treated with a dispersion of copper nanoparticles at a concentration of 3 g/L. However, as the authors point out, this treatment effect seems almost negligible after the leaching test.

Thus, more research is needed to evaluate the benefits and harms of using silver nanoparticles in wood protection coatings (Künniger, Heeb, & Arnold, 2014; Pařil et al., 2017).

In addition to metal nanoparticles, metal oxide nanoparticles are used to protect wood. Authors (De Peres, Delucis, Amico, & Gatto, 2019) synthesized zinc oxide nanorods by the microwave solvothermal method. It was shown that the resulting nanorods, agglomerated in the form of hedgehog structures, have high crystallinity, high purity, and high photocatalytic activity. Pine samples were impregnated with compositions containing 1%–5% zinc oxide. It has been found that such an impregnation is effective in increasing the resistance of pine wood to white rot. The fungicidal properties of TiO_2 and ZnO nanoparticles in the form of suspensions in polyvinyl butyral (PVB) were studied in (Harandi, Ahmadi, & Mohammadi Achachluei, 2016). Poplar wood samples were treated with the test compositions in vacuum. Two groups were incubated with rainbow fungus (*T. versicolor*) and then tested in both light and dark for seven weeks. The third group was studied under conditions of accelerated aging under the influence of temperature, humidity, and ultraviolet radiation variation. Suspensions of TiO_2 (1%) and ZnO (5%) nanoparticles in PVB did not exhibit fungicidal properties under dark conditions. Under dark conditions, only suspensions with a concentration of 2% nanoparticles turned out to be effective. We can state the weak fungicidal activity of all compositions. Tests under conditions of accelerated aging made it possible to reveal the protective effect of the use of nanoparticles against aging factors in comparison with samples treated with a consolidating composition without nanoparticles.

Nanoparticles of metal fluorides MgF_2 and CaF_2 were used to treat wood against brown putrefactive fungi *Coniophora puteana* and *R. placenta*. It was found that untreated wood samples had higher weight loss ($\sim$30%) compared to treated samples, which had an average weight loss of 2% (versus *C. puteana*) and 14% (versus *R. placenta*), respectively. Authors (Usmani, Stephan, Hübert, & Kemnitz, 2018) believe that metal fluoride nanoparticles could provide a viable alternative to modern wood preservatives.

The interest of researchers in natural bioactive materials is enduring. By now, the role of carbohydrate-containing biopolymers in intercellular interactions, in cell differentiation, and in the formation of multicellular systems has been proven and widely studied. It should be emphasized that the practical use of natural polysaccharides is growing steadily. Of particular note is the simplest chitin derivative—chitosan, as well as its various modifications and composites. The review (Varlamov, Il'ina, Shagdarova, Lunkov, & Mysyakina, 2020; Varlamov & Mysyakina, 2018) considers some aspects of the distribution of chitin and chitosan in nature, methods for modifying chitosan, and the prospects for using chitosan and its derivatives with bactericidal, fungicidal, and antioxidant activity. The study (Khademibami, Jereniic, Slunulsky, & Barnes, 2020) evaluated the effectiveness of chitosan oligomers and related nanoparticles as environmentally friendly wood preservatives. Commercial low-molecular-weight chitosan was depolymerized using sodium nitrite. Then the oligomers were modified to form quaternized chitosan oligomers. Commercial low-molecular-weight chitosan was depolymerized using sodium nitrite. Quaternized chitosan oligomers formed after original oligomer modification. "Chitosan-TPP" nanoparticles formed by ionic gelation after mixing both quaternized and nonquaternized oligomers with tripolyphosphate (TPP). Southern pine wood samples were treated with various chitosan-based solutions and suspensions in a vacuum impregnation process. The results of the weight gain and volume increase showed that the samples treated with the "quaternized chitosan-TPP" formulation had greater weight and volume gains after treatment compared to the samples treated with the unquaternized chitosan-TPP formulation and control samples. The data obtained showed that the weight loss increases with the use of quaternized nanoparticles' chitosan-TPP. Although the quaternized nanoparticles were positively charged, they could not attach to the cell walls and were washed out. Therefore, these nanoparticles are likely to be used as wood preservatives in nonleaching applications.

Authors Janesch, Czabany, et al. (2020) proposed a simple procedure for the preparation of coatings to protect spruce wood from ultraviolet radiation. They used a layer-by-layer coating method, alternately immersing the samples in a suspension of cerium dioxide nanoparticles and in solutions of biopolymers (chitosan or cationic starch containing groups capable of imparting a positive charge in an aqueous medium at an appropriate pH value). CeO_2 nanoparticles dispersed in citric acid had a negative charge; chitosan and cationic starch formed a positively charged layer. The coatings created were transparent and resulted in only slight discoloration of the surface. At the same time, they reduced the amount of color change caused by UV light across all CIELAB parameters. The good performance of the coatings was achieved already after three coating cycles.

As discussed earlier, the problem of ensuring the resistance of wood to biodegradation is closely related to the need to increase the water resistance of wood. To solve this problem, the authors (Janesch, Arminger, Gindl-Altmutter, & Hansmann, 2020) proposed a simple method for producing superhydrophobic coatings on wood using environmentally friendly materials. Spruce wood samples were coated by immersion in a mixture of tung oil and natural beeswax, followed by the deposition of micronized sodium chloride particles. Sodium chloride particles served as a template for the formation of a microstructured relief. As a result, the surface acquired superhydrophobic properties (contact angle over 160 degrees). After rinsing with water, no traces of sodium chloride remained on the surface, but superhydrophobic properties were retained. As noted in Section 13–3, the superhydrophobicity of the surface will tend to roll off water droplets, thereby reducing water absorption.

The idea of creating coatings with a superhydrophobic surface was also worked out by the authors (Shah et al., 2017). Alumina nanoparticles were used to form a tough, rough surface, which was then treated with polydimethylsiloxane (PDMS). Several alternating layers of nanoparticles and PDMS were applied to provide a durable superhydrophobic coating.

Another effective technique for environmentally friendly wood protection using nanotechnology is the encapsulation of the biocide in nanostructured matrices (Teng et al., 2018). This can be achieved using paint and varnish or sol-gel technology (Shilova et al., 2019). The authors (Pantano et al., 2018) used acrylic paint to fix the nanoparticles of copper oxide CuO on the wood surface. They evaluated the in vitro effect of CuO nanoparticles encapsulated in this paint against three species of micromycetes: *C. puteana*, *Gloeophyllum trabeum*, and *T. versicolor*. The fungicidal activity of these CuO nanoparticles encapsulated in an acrylic matrix was compared with the action of both a suspension of micronized basic copper carbonate $CuCO_3 \cdot Cu(OH)_2$ (for pressure treatment of wood) and the European standard for wood treatment with aqueous copper amine (when the amine is dissolved in water to the state of molecular dispersion). A thorough study allowed the authors (Pantano et al., 2018) to conclude that the use of acrylic paint is not promising for encapsulating biocide nanoparticles intended to protect wood from biodegradation under the influence of wood-destroying fungi.

An alternative method for fixing biocides on the surface of wood can be the sol-gel technology, the promising use of which was already indicated earlier. Sol-gel synthesis is based on the transition of true solutions to sol and then to gel. Various hydrolyzable compounds, including silicon and titanium alkoxides, alkali silicates, various salts, and boric acid can be used as precursors for the preparation of sols (Brinker & Scherer, 1990). The use of sols containing nontoxic and friendly compounds is a promising approach to multifunctional protection and improvement of wood properties (Hübert & Mahr, 2017; Subasri, Reethika, & Soma Raju, 2013; Tsvetkova et al., 2019).

The first attempts to improve the fire resistance of wood using sol-gel technology (although it was not yet called that) were made back in the 19th century with the use of potassium water glass. Systematic research in this area was carried out by Furuno in the 1980s of the last century, who investigated the silica mineralization of wood and carried out treatments with aqueous solutions of alkali silicates. In the 20th century, examples of wood impregnation with a mixture of alkoxysilanes are known (Saka & Ueno, 1997; Schneider & Brebner, 1985). This approach is inspired by the natural process of the formation of silicon-rich wood, which is formed over millions of years by the penetration of silicic acid into the wood tissue. Silica gel is formed as a result of polycondensation, which further reacts to form quartz (chalcedony) and opal (wood opal).

Some researchers have used sol-gel technology to impregnate wood with organic or hybrid (organic-inorganic) materials. However, their technical use and commercialization are still pending. This is largely due to the lengthy process of entering the market. It can be said that such sol-gel-derived impregnations create wood composites because a relatively large number of new ingredients are introduced into the wood and, as a result, its properties change (Hübert & Mahr, 2017). However, they note that it should be borne in mind that as a result of these impregnations, chemical interaction between modifiers and wood does not always occur; basically, the structure of the tree is preserved.

Interestingly, the impregnation of wood with a sol based on silicon dioxide and titanium dioxide with biocidal additives increases the resistance to leaching of these biocides, since a water-insoluble gel is formed and fixed in the wood matrix. In this way, boron and copper leaching can be reduced (Zhang, 2015). In addition to this, gels fixed in the wood structure can also significantly reduce the release of hazardous active agents (e.g., $CuCl_2$ or boron), wood preservatives into the environment (Altun, Ozcifci, Şenel, Baysal, & Toker, 2010; Feci, Palanti, Predieri, & Vignali, 2009; Kartal, Yoshimura, & Imamura, 2009; Palanti, Feci, Predieri, & Francesca, 2012). It is assumed that the aforementioned antileaching effect is associated with a decrease in the mobility of ions of active biocidal compounds inside the wood due to their encapsulation in a gel matrix, due to a decrease in water penetration into the bulk of the wood. Unlike sol-gel compositions, various combinations of solutions, both water-based and based on organic solvents and organic compounds (organofunctionalized silanes and organic biocides) do not reduce the effect of leaching of biocides (De Vetter, Depraetere, Stevens, Janssen, & van Acke, 2009).

In a review article (Hübert & Mahr, 2017), the following strategies were described for the further introduction of sol-gel technology in wood processing: (1) the use of less effective, but environmentally friendly sol-gel-derived impregnations as an alternative to highly toxic impregnation; (2) the introduction of well-known effective biocidal additives in sol-gel-derived SiO_2, TiO_2, SiO_2-TiO_2 matrices to increase their biocidal activity; and (3) inoculation of classical biocidal wood preservatives (such as boron and copper) in a sol-gel matrix to prevent or reduce leaching. Despite the fact that sol-gel impregnations do not provide the same high characteristics as traditional commercial preparations for improving bioresistance and fire resistance, they can significantly improve the characteristics of wood in a number of parameters and be an alternative to traditional agents. In the future, it is promising to conduct research to study the long-term effect of sol-gel-derived impregnations, as well as to test them on wood composites such as plywood and veneers.

The next section will be devoted to the use of sol-gel technology to protect wood from the action of wood-destroying fungi using examples from the authors' personal experience.

13.5 Using sol-gel technology to protect wood from wood-destroying fungi

13.5.1 Hybrid coatings based on wax and silicones

As described in the previous sections of this chapter (see Section 13–4), sol-gel coatings are promising for protecting wood from biodegradation, in particular from wood-destroying fungi action.

As can be seen from Table 13–1, the use of hybrid tetraethoxysilane (TEOS)-derived sol-gel compositions with beeswax and nanodispersed hydrophobized SiO_2 filler (aerosil—colloidal pyrogenic silica or fumed silica) increases the degree of hydrophobicity and significantly reduces water absorption. The degree of hydrophobicity also depends on the amount of aerosil introduced. The addition of boric acid, on the contrary, increases the hydrophilicity. The addition of organosilicon varnishes to the composition significantly reduces water absorption and improves adhesion (Tsvetkova et al., 2019). After inoculation of wood, protected with coatings of compositions 1, 2, and 3 (see Table 13–1) with two wood-destroying fungi *Serpula lacrymans* and *Pleurotus ostreatus*, followed by their cultivation for four months, a decrease in their weight was observed in all three variants of coatings. The most effective inhibition of the development of wood-destroying fungi was provided by the coating with the biocidal additive H_3BO_3.

As a promising area of research, the wide opportunities that the sol-gel technology creates for the hydrophobization of wood surface and the creation of both superhydrophobic and superhydrophilic self-cleaning coatings are notable. A similar effect can be achieved by

Table 13–1 Properties and results of evaluating the fungicidal activity of the TEOS and beeswax derived coatings applied to the wood of a pine.

Sign: 1st layer// 2nd layer	Composition of hybrid sols, wt. %				Parameters				Weight loss, Dm.%	
	Mixture: TEOS/Wax	Modifier/ amount	Aerosil R 972	Biocidal additive	Contact angle (± 1°)	Roll-off angle (± 2°)	Adhesion criteria, in points	H$_2$O absorption, % after 8 days	*Serpula lacrymans*	*Pleurotus ostreatus*
Ethyl silicate// TEOS-wax-R^{972}	99.0	–	1.0	–	152	40	3	35	5.1	9.1
TEOS-derive sol & H$_3$BO$_3$// TEOS-wax-R^{972}—H$_3$BO$_3$	56.3	–	0.2	43.5	20	90	3	26	4.1	4.2
TEOS-wax-Block-R-972	63.7	Block-copolymer/ 36.1	0.2	–	145	50	2	21	4.9	9.0
Without coatings					92	90	–	98	4.0	5.2

[a]*Note:* [a]*TEOS-derived sol & H$_3$BO$_3$*—TEOS:C$_3$H$_5$(OH)$_3$:C$_2$H$_5$OH:H$_2$O:H$_3$BO$_3$ = 1:6:1:0,05:3 (mole ratio); *TEOS-wax*—beeswax was dissolved in TEOS in a ratio of 1:5, and stirred for 20 min R-972—hydrophobized silica nanopowder; *Bloc*—polyphenylsilsesquioxane-polydimethylsiloxane block-copolymer, and silicone oxime—vinyl-tris-(acetoxim)silane CH$_2$ = CH–Si[O–N = C(CH$_3$)$_2$]$_3$ was used as the hardener for the curing of composites at room temperature; *Biocidal additive*—boric acid solution in glycerol.

introducing nanodispersed fillers into sols—silicon oxides, titanium oxides, fluoroplastic nanoparticles, and so forth. Hybrid sol-gel compositions, obtained by mixing silica sols with various modifying additives, such as—organosilicon varnishes, block copolymers, and even with natural materials, for example, with beeswax, are also promising for hydrophobizing wood surfaces. Table 13−1 shows examples of hybrid sol-gel compositions that we used to change the surface state of pine sapwood,—from hydrophilic to hydrophobic and superhydrophobic (Tsvetkova et al., 2019).

At the same time, the use of silicon dioxide nanoparticles in sol-gel compositions as a structure-forming agent should be treated with caution. Many types of fungi actively use it for their vital activity (Breene, 1990; Voronkov, Zelchan, & Lukevits, 1975), which can adversely affect the protective properties of coatings. For example, after 4 months of incubation of pinewood samples infected with *S. lacrymans* micromycetes, a coating based on a mixture of TEOS with wax with an increased aerosil content (1 wt.%) lost more weight than other coatings of similar compositions, but with a lower aerosil content (0.2 wt.%) (Table 13−1).

Our experience in the development and research of various protective coatings shows that, despite the hydrophobic surface, biocidal additives are the most important for the suppression of the growth and development of biodeterioration agents, including wood-destroying fungi. We will delve in more detail on the study of the effectiveness of various biocidal preparations of mild action in comparison with the effectiveness of a nonbiocidal protective coating in the next section.

13.5.2 The use of low-toxic biocides to protect wood from wood-destroying fungi

13.5.2.1 Evaluation of the effectiveness of biocidal compositions for protecting wood from xylotrophic fungi on a nutrient medium

Characterization of the formulations included in the nutrient medium tests. We have carried out a study of the effect of green biocidal compounds, including in the nano state, on the growth and development of wood-destroying fungi. In the first series of experiments, promising wood protection formulations based on biocides that have a minimum impact on the environment were included in the tests (Table 13−2).

In the present work, the reducing ability of hydridesiloxane was used to obtain nanoparticles of metals —copper and silver. It is known that in organic chemistry, hydridesiloxanes are used as reducing agents of imines to amines, aldehydes to alcohols, and so forth, but there is currently no information on their use as reducing agents for metal salts. In the presented method, low-molecular-weight polymethylhydridesiloxane was used as a reducing agent, and silver nitrate and copper (II) acetate were used as a source of silver and copper ions. Water, linseed oil, and tributyl phosphate (TBP) (hereinafter referred to as the medium) were chosen as the medium for the nanoparticles. The stabilization and hydrophobization of the resulting nanoparticles in water occurred as a result of their treatment with polyvinyl alcohol (PVA) dissolved in water. No particle stabilization was required in the linseed oil/siloxane and TBP/siloxane two-phase system.

Table 13–2 Characteristics of low-toxic biocides included in the tests.

Active substance	Denotement	Biocide composition
Quaternary ammonium compound (QAC)	#1	A mixture of benzalkonium chloride compounds
	#2	Tetrabutylammonium bromide
Boric acid H_3BO_3	#3	Silica sol doped by H_3BO_3 "TEOS-derived sol & H_3BO_3"
Ag nanoparticles	#4	Linseed oil nanosilver colloid: *"Silver Nano Oil 500,"* LLC «Mikron» (Saint-Petersburg, Russia); 500 ppm
	#5	Aqueous nanosilver colloid: *"Silver Nano Aqua 500,"* LLC «Mikron» (Saint-Petersburg, Russia); 500 ppm
	#6	Tributyl phosphate nanosilver colloid: *"Silver Nano TBP 500,"* LLC «Mikron» (Saint-Petersburg, Russia); 500 ppm
	#7	Tributyl phosphate nanosilver colloid: *"Silver Nano TBP 3000,"* LLC «Mikron» (Saint-Petersburg, Russia); 3000 ppm
Ag/Cu nanoparticles	#8	Linseed oil nanosilver and nanocopper colloid: *"Silver & Cooper Nano Oil,"* LLC «Mikron» (Saint-Petersburg, Russia); 2.5 mg Ag0 and 5 mg Cu0 in 1 g of colloidal solutions

Preparation of a colloidal solution of silver nanoparticles in water "Nano Aqua 500" (# 5, Table 13–2). 1.5–2 g of polymethylhydridesiloxane $HO\text{-}(CH_3SiHO)_n\text{-}H$, where n = 30, was added to 1000 mL of an aqueous solution of silver nitrate (1 g/L) containing 3%–5% wt. % PVA. The reduction was carried out at room temperature at constant stirring for two days.

Preparation of colloidal solutions of Ag and Ag/Cu nanoparticles in linseed oil —"Silver Nano oil 500" and "Silver & Cooper Nano Oil" (# 4, # 8, Table 13–2, respectively). For 1–2 h, 1 g of silver nitrate (for # 4) or 0.5 g of silver nitrate with 0.5 g of copper acetate (for # 8) in 1000 mL of linseed oil was stirred at a temperature of 40°C–50°C, then the heating was turned off and 1.5–2 g of polymethylhydridesiloxane $HO\text{-}(CH_3SiHO)_n\text{-}H$, where n = 30 was added.

Preparation of colloidal solution of Ag nanoparticles in tributyl phosphate (TBP) "Silver Nano TBP 500" (# 6, Table 13–2) *and "Silver Nano 3000"* (# 7, Table 13–2) 1 g of silver nitrate was dissolved in 1000 mL TBP by stirring for 1–2 h at a temperature of 40°C–50°C. Then the heating was turned off and 1.5–2 g of polymethylhydridesiloxane $HO\text{-}(CH_3SiHO)_n$-H, where $n = 30$, was added.

For all the above colloidal solutions of silver nanoparticles, the reduction was carried out at room temperature and with constant stirring for one day. As the reduction reaction proceeded, the color of the reaction mixture changed from colorless to yellow and then to dark red. This indicated the appearance of silver nanoparticles in the reaction mixture. The colloidal solution of copper nanoparticles had a cherry-red color and after aging for five to seven days changed to green, and a small dark precipitate also appears. The green color of the colloidal solution of copper nanoparticles is associated with their oxidation by atmospheric oxygen in the solution and the formation of copper ions Cu^{+2}. A colloidal solution of Ag/Cu nanoparticles was prepared in a similar way at an Ag/Cu ratio of 1 to 10, it had a red-orange color.

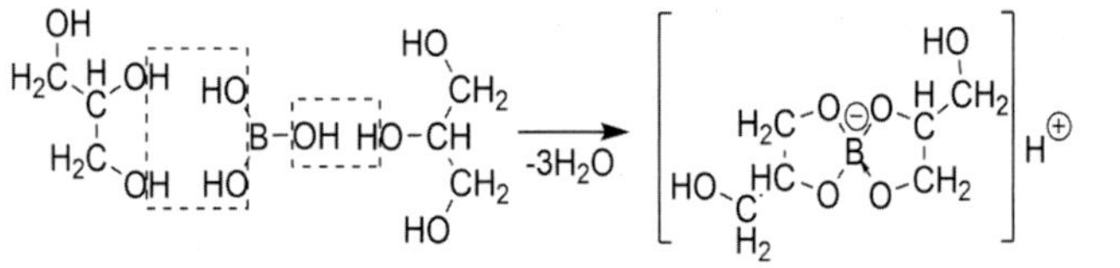

FIGURE 13–1 Scheme illustrating the formation of a boron complex compound as a result of the interaction of glycerol and boric acid.

Synthesis of silica sol doped by H_3BO_3 "TEOS-derived sol & H_3BO_3" (#3, Table 13–2). Silica sol with boric acid was obtained for wood impregnation by successive mixing of the initial components: tetraethoxysilane (TEOS), ethanol, and a solution of boric acid in glycerol, in a molar ratio: TEOS:$C_3H_5(OH)_3$:C_2H_5OH:H_2O:H_3BO_3 = 1:6:1:1:0.05:2.

A mixture of glycerol with boric acid was stirred while heating until complete dissolution. The sol matured for one day (Tsvetkova et al., 2006). This method allows adding more boric acid in dissolved form. As a result of the donor-acceptor interaction of boric acid with glycerol, a complex compound is formed (Currie, Byrom, Bruce, & Schmidt, 2020), see Fig. 13–1; it is a much stronger acid than the original weak boric acid (Shvarts, Ignash, & Belousova, 2005). It is known that complex acids have surface-active properties (Kuvshinov, Altunina, Stasieva, & Kuvshinov, 2019). As a result, silica sol containing a complex boron compound penetrates deep into the pores and is firmly fixed in the solid wood.

Preparation of a water-alcohol solution of quaternary ammonium salts (QAS). For the studies, we used aqueous solutions of two types of QAS: tetrabutylammonium bromide (10 wt.%) and alkyldimethylbenzylammonium chloride (20 wt.%).

The presence of a cationic surfactant improves the wettability of wood with an antiseptic agent and promotes the adhesion of biologically active cations on its surface due to electrostatic interaction (Varfolomeev et al., 1991).

Test objects. The experimental work included six strains of domestic-fungi from the collection of basidiomycetes of Komarov Botanical Institute of Russian Academy of Sciences (LE-BIN), the characteristics of which are shown in Table 13–3. The strains for research were selected on the basis of the results of an earlier study of their micro- and macromorphological characteristics, molecular-genetic identification, as well as physiological and biochemical characteristics.

The strains were maintained by a traditional method of subculture on wort-agar bevels at a temperature of 4°C–5°C. Duplicate strains were stored in the form of disks in sterile 2 mL cryovials with distilled water at room temperature or at 4°C–5°C (depending on the type of fungus), as well as in cryovials at −80° C with 10% glycerol as a cryoprotectant (freezing speed 1°C/min).

Testing procedure. Tests were carried out in laboratory conditions in several stages. Initially, the intensity of the effect of biocidal compositions on the vital activity of domestic fungi was determined by adding them to the composition of a solid nutrient medium. Two methods were used:

1. The biocidal composition was added directly to the nutrient medium (2% agarized beer wort) in the amount of 5% of the medium volume;

Table 13-3 Characterization of strains of house-fungi from the LE-BIN collection included in the tests.

Strain # LE-BIN	Verified data	Origin	Collecting year	# In the NCBI genebank
1029*	*Antrodia xantha* (Fr.) Ryvarden.	Russia	1996	KY433983
1370*	*Coniophora puteana* (Schumach.) P. Karst.	Cuba, Topes de Collantes National Park	1984	KY433978
2058*	*Gloeophyllum sepiarium* (Wulfen) P. Karst.	Finland, Helsinki region	2005	KY433980
0963*	*Neolentinus lepideus* (Fr.) Redhead & Ginns.	Russia, Pskov region	1994	KY433992
2278*	*Neolentinus lepideus* (Fr.) Redhead & Ginns.	Russia, Republic of Altai	2008	KY433985
1192*	*Serpula lacrymans* (Wulfen) J. Schröt.	Russia, Saint-Petersburg	2000	KY352493

2. The biocidal composition was introduced into a hole cut in a nutrient medium (2% agar beer wort) at the center of a Petri dish. This method made it possible to assess not only the biocidal properties of a substance, but also its ability to penetrate into the substrate.

Variants with the same nutrient medium, but with the addition of distilled water to the medium or into the hole without biocide, served as control.

Then domestic fungi were sown on the surface of the prepared media and their growth was monitored. Evaluation of the effect of biocides on the growth and development of domestic fungi was carried out by comparison with the control (medium without the addition of biocide) for 30 days. In all experiments, the replication was fivefold.

Test results. Most of the biocidal compositions provided for testing (see Table 13−2) were very well absorbed by the culture medium, except for compositions based on linseed oil, which were not completely absorbed. The test results obtained are presented in Table 13−4.

As a result of the experiments, three substances showed the best biocidal effect: #1 (QAC, benzalkonium chloride), #3 (silica sol doped by H_3BO_3 "*TEOS-derive sol & H_3BO_3*"), and #7 (tributyl phosphate nanosilver colloid: "*Silver Nano TBP 3000*"). Biocidal compositions ##1, 3, and 7 in all replicates exhibited a persistent biocidal effect against all tested species of domestic fungi and were found to be the most promising for further work. At the same time, composition #5 (Aqueous nanosilver colloid: "*Silver Nano Aqua 500*") showed no biocidal effect on all fungal strains. Composition 6 (Tributyl phosphate nanosilver colloid: "*Silver Nano TBP 500*") initially inhibited the growth of the studied species (during the first six to eight days), after which the fungi developed almost as in the control. This fact indicates the ability of the tested fungi to quickly adapt to this composition, or the enzymes and organic acids secreted by the fungus neutralize the biocidal effect of the composition. The effect of

Table 13–4 Assessment of the impact of biocidal compounds on the development of domestic fungi.

##, see Table 13–2	Antrodia xanta 1029	Coniophora puteana 1370	Neolentinus lepideus 0963	Neolentinus lepideus 2278	Gleophillum sepiarium 2058	Serpula lacrimans 1192
1	−	−	−	−	−	−
2	+	+++	+++	+++	++	++
3	−	−	−	−	−	−
4	+	+++	++	++	+	++
5	+++	+++	+++	+++	+++	+++
6	−+++	−+++	−++	+	−+++	−+++
7	−	−	−	−	−	−
8	+	++	++	++	++	++

compositions #2 QAC (tetrabutylammonium bromide), #4 (linseed oil nanosilver colloid "*Silver Nano Oil 500*"), and #8 (linseed oil nanosilver and nanocopper colloid: "*Silver & Cooper Nano Oil*") turned out to be different (selective) in relation to different types and strains of fungi included in the test. For example, composition #4 suppressed the development of the fungus *Gloeophyllum sepiarium* 2058, but practically had no effect on the *C. puteana* 1370 strain. Composition #8 significantly suppressed the development of *Antrodia xantha* 1029 and had a mild biocidal effect on other fungi species. The effect of composition #2 (QAC) varied most markedly in relation to the studied strains (from a significant suppression of fungal growth to the absence of a biocidal effect).

The sizes of the obtained Ag and Ag/Cu nanoparticles for colloids ##4−8 were measured via dynamic light scattering method. It turned out that the size of nanoparticles was not the same everywhere and varied from 50 to 2000 nm in different colloidal solutions. It was found that biocidal activity decreased with increasing agglomerate size (Table 13–5). For example, the size of nanoparticle aggregates in aqueous colloids #5 ("*Silver Nano Aqua 500*") reached 1400 nm. Perhaps this is due to the use of PVA as a stabilizing agent. Interestingly, at a lower concentration of silver nanoparticles, larger aggomerates formed. Thus, aggregates in TBP colloidal solutions were significantly higher at a concentration of Ag nanoparticles equal to 500 ppm ("*Silver Nano TBP 500*") than at 3000 ppm ("*Silver Nano TBP 3000*") — 2000 nm versus 50−80 nm. The biocidal properties of colloidal solutions of silver nanoparticles with a size of ∼50−80 nm in terms of their biological activity can be compared with compositions based on already known biocidal additives—boric acid and benzalkonium chloride.

In general, the obtained data indicate that a number of tested compositions are promising for suppressing the development of wood-destroying fungi. Their assignment to low-toxic biocides (not containing potent toxic substances) largely explains the incomplete suppression of the growth of domestic fungi, as well as the selectivity of action on individual strains. In addition, the domestic fungi themselves differ quite significantly in adaptive properties, the manifestation of protective reactions, growth rate, and other indicators. The most sensitive to

Table 13-5 The efficiency of the effect of biocidal compositions on the development of domestic fungi, depending on the size of nanoparticles.

##, see Table 13-2	Biocide composition	Suppression of growth of house-fungi (*Coniophora puteana* and *Antrodia xanta*) by biocide			Average particle or agglomerate size, nm
		Complete	Partial	No growth suppression	
1	QAC—a mixture of benzalkonium chloride compounds	+		−	−
3	Silica sol doped by H_3BO_3 "TEOS-derive sol & H_3BO_3"	+		−	−
7	Tributyl phosphate nanosilver colloid: "Silver Nano TBP 3000," LLC «Mikron» (Saint-Petersburg, Russia); 3000 ppm	+			50−80
8	Linseed oil nanosilver and nanocopper colloid: "Silver & Cooper Nano Oil"		+		200
4	Linseed oil nanosilver colloid "Silver Nano Oil 500"		+		300−400
2	QAC—tetrabutylammonium bromide		+		−
5	Aqueous nanosilver colloid "Silver Nano Aqua 500"			+	1300−1400
6	Tributyl phosphate nanosilver colloid "Silver Nano TBP 500"			+	2000

the action of biocides was the *A. xantha* 1029 strain, while the most resistant was *C. puteana* 1370. It was found that the biocidal activity of silver and copper nanoparticles decreases with an increase in the size of nanoparticle agglomerates in colloidal solutions. Apparently, to maintain the biocidal activity of nanoparticles, it is necessary to select the type and concentration of the flocculating agent more carefully, avoiding the use of long-chain polymers.

13.5.2.2 Evaluation of the effectiveness of biocidal compositions for protecting wood from xylotrophic fungi in a moist chamber

Characterization of the formulations included in the nutrient medium tests. In another series of experiments, boric acid, and QAC—alkyldimethylbenzylammonium chloride were studied as mild biocides. For environmental reasons, due to their relatively low toxicity to mammals, there is an increase in interest for using these compounds for wood preservation, even though boron compounds are susceptible to leaching if they are not chemically fixed in the wood (Stirling & Morris, 2012; Terzi, Kartal, White, Shinoda, & Imamura, 2011).

Three different protective coatings were selected as objects of study (Table 13−6). Uncoated pine woodblocks were used as a control. The commercial coating was used as a reference. All tests were carried out in triplicate.

Table 13–6 Brief description of experimental samples.

Composition name	Description of coatings composition
"Organosilicon & wax"	Composition based on weather- and water-resistant organosilicon varnish with the addition of beeswax
"Organosilicon&wax&sol-H_3BO_3"	Based on weather- and water-resistant silicone varnish with the addition of beeswax and TEOS-derived sol doped by boric acid
"Organosilicon&wax&QAC"	Based on weather- and water-resistant silicone varnish with the addition of beeswax and quaternary ammonium salt (alkyldimethylbenzylammonium chloride)
"Commercial"	Water-borne antiseptic varnish for interior work and elements of wooden facades
Control	Comparison sample (uncoated)

Sol-gel synthesis of film-forming compositions and coatings. When obtaining the "Organosilicon&wax" coating, a base composition was initially prepared based on organosilicon varnish and beeswax, which was previously dissolved in TEOS, at a ratio of 1/5. As a film-forming base, an atmospheric and moisture resistant organosilicon varnish was used, which is a solution of polymethylphenylsiloxane resin in aromatic solvent in a weight ratio of 1/1 (KO-08, JSC "Khimprom," Novocheboksarsk, Russia). The hardener was aminopropyltriethoxysilane (1 wt.% of the varnish mass). For comparison and to prevent the appearance of wood-destroying fungi or a significant reduction in their number, a low-toxic biocidal additive was also introduced into the other two compositions—"*Organosilicon&wax&sol-H_3BO_3*" or "*Organosilicon&wax&QAC*," respectively (see Table 13–6).

The synthesis of the "*TEOS-derive sol&H_3BO_3*" sol was described above in Section 13–5.2.1. The resulting silica sol was mixed with the "*Organosilicon&wax*" siliceous composition for 1 h.

The "*Organosilicon&wax&QAC*" composition was prepared by mixing an aqueous solution of alkyldimethylbenzylammonium chloride (50 wt.%) and a siliceous composition (20 wt.%) (Table 13–6).

The obtained compositions were impregnated with coniferous wood bars 10 cm × 2 cm × 2.5 cm in size, without their preliminary grinding or polishing. After applying the coatings, the wood was dried in air at room temperature for one day and then again impregnated with the same compositions.

Test objects. To study the ability of protective coatings to increase the bioresistance of wood to wood-destroying fungi, the following domestic fungi were selected: *A. xantha* 1029 and *C. puteana* 1370 from the collection of basidiomycetes Komarov Botanical Institute of Russian Academy of Sciences (LE-BIN). The choice for their inclusion in the tests was justified by the fact that pure cultures of these fungi have a high growth rate and form abundant mycelium in laboratory conditions, release a large amount of organic acids, and also cause wood rotting like brown rot and destroy mainly coniferous wood. These types differ in their requirements for humidity. The domestic fungi *A. xantha* 1029 is more demanding on the moisture content of the substrate, for its growth it is necessary that the substrate has a moisture content of at least 60%, and *C. puteana* 1370 requires lower substrate moisture values (from 40%) for its growth and development.

Test procedure. Before infecting the wood with domestic fungi, all bars (coated and control) were soaked in water to create the required moisture content of the test substrate and to check the resistance of the coating to moisture. Then all the bars were placed in containers, in which the mycelium of the domestic fungi developed on a small surface covered with agar nutrient medium. Each bar had contact with this mycelium at the beginning of the experiment. To maintain the required moisture content of the substrate and simulate the environment close to the operating conditions of wood materials with increased moisture, moist chambers were used (where the temperature and humidity values that were optimal for the growth of domestic fungi were maintained). The bars inoculated with *A. xantha* 1029 and *C. puteana* 1370 mycelium were placed in a moist chamber, where the optimum humidity and temperature were maintained for the development of each type of domestic-fungus. The inoculated bars were kept in a moist chamber for four months. Then they were removed from the containers and subjected to inspection and analysis.

Experiment Results. As a result of the carried out tests, it can be noted that three out of four experimental coatings did not change the color of the wood, and one of them, —"*Commercial*," has a dark brown color, and the wood under its influence also becomes dark brown. Coating with H_3BO_3 ("*Organosilicon&wax&sol-H_3BO_3*") is unstable in water. When a bar is soaked, the water becomes cloudy. This is probably due to the leaching process. Microscopic studies using a binocular microscope revealed details of the development of fungi on the experimental coated bars and in the control (Figs. 13–2–13–4).

Initially, upon contact of all the bars (both experimental and control samples) with the mycelium of domestic fungi, local growth of mycelium on the wood was observed. Subsequently, the situation changed depending on the type of coating. In the control (wood-blocks were not treated with protective compounds), the formation of fruiting bodies of domestic fungi was observed. *A. xantha* 1029 was characterized by the formation of open, yellowish fruiting bodies with characteristic pores (Fig. 13–2).

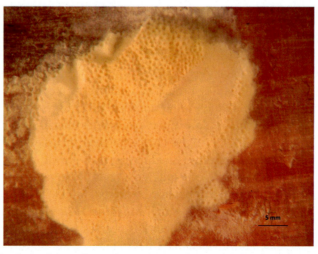

FIGURE 13–2 The fruiting body of *Antrodia xantha* 1029. Control-wood is not treated with protective compounds.

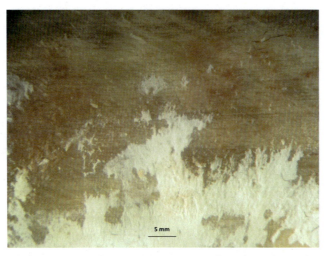

FIGURE 13–3 Dying off of *Antrodia xantha* 1029 mycelium on the surface of wood coated with "Organosilicon&wax&QAC."

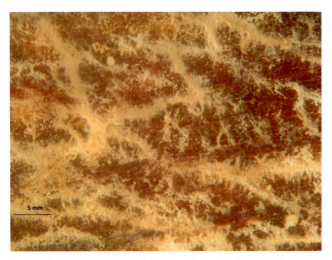

FIGURE 13–4 Growth of *Coniophora puteana* 1370 over the surface of the "Commercial" coated bar. The growth is not observed over the entire surface, the mycelium is a thin cobweb, the fruiting body did not develop.

On blocks of wood coated with the *"Organosilicon&wax&QAC"* composition, there was a rapid decrease of mycelium growth and even its gradual death (Fig. 13–3). When such a bar with mycelium was moved into a moist chamber without a nutrient medium, the fungus stopped developing (i.e., it could not use wood treated with this substance as a nutrient source).

On wood bars coated with *"Commercial,"* mycelium growth was observed over the entire surface of the bar, but it had a cobweb consistency uncharacteristic for domestic fungi, and did not form strands and clusters. The fungus *C. puteana* 1370 only locally formed a thin cobweb mycelium (Fig. 13−4), which gradually died off and the fruiting body did not develop. The fungus *A. xantha* 1029 developed poorly mainly along the edges of the bar, where the coating of the bar with a protective compound was uneven.

The other two protective compounds, "*Organosilicon&wax*" and "*Organosilicon&wax&sol-H_3BO_3*," did not have a significant effect on the development of domestic fungi under test conditions. Moreover, the "*Organosilicon&wax*" coating was preferable for the development of fungi, because stimulation of their growth was observed when compared to control. Domestic fungi formed fluffy fibrous growths of mycelium over the entire surface of the treated bar. The buds of fruiting bodies on these bars formed more often and faster in comparison with other test variants. This can probably be explained by the possible availability of silicon, which is an essential component of all plants (Voronkov et al., 1975). Fungi show significant growth, and this was noticed by us earlier for nonbiocidal compositions (Tsvetkova et al., 2019), where amorphous silicon dioxide with a size of 10−40 nm was used. Apparently, the enzymatic activity of fungi makes it possible to easily destroy C-Si-C, Si-O-C bonds. The results are shown in Table 13−7.

Experiments showed that the "*Commercial*" coating has a fungistatic effect on domestic fungi, but is inferior in this parameter to the "*Organosilicon&wax&QAC*" coating, which showed the best result. The coating with H_3BO_3 ("*Organosilicon&wax&sol-H_3BO_3*") has almost no biocidal properties (or has a weak fungistatic effect), and the rate of the development of brownies and fungi on bars with this composition was close to control. Boric acid inhibits mycelium growth (as seen in the examples above) and is widely used in protective coatings for wood (Baysal & Yalinkilic, 2005; Yalinkilic et al., 1998). However, it is well-known main disadvantage is its tendency to leach (Baysal & Yalinkilic, 2005; Yalinkilic et al., 1998), however, it is not suitable as a biocide if the wood is in constant contact with water. In this case, boric acid leached out even though it was encapsulated in silica sol.

Thus, of the four coatings tested, the "*Organosilicon&wax&QAC*" coating has the greatest biocidal effect against domestic fungi—based on a mixture of organosilicon varnish and wax containing a QAC (alkyldimethylbenzylammonium chloride). The revealed effect can be called fungistatic, since the biocide included in the coating does not completely kill the fungus, but inhibits its development. In the future, there is a gradual dying off of the mycelium of the domestic fungi. This coating proved to be effective against two types of domestic fungi included in the tests. QACs have a strong affinity for carboxylic acid and phenolic groups found in wood (Jin & Preston, 1991). As a result, they could interact with extractive substances that contain carboxylic acids, phenolic hydroxyl groups, and with the aromatic part of lignin.

The "*Organosilicon&wax*" coating did not show biocidal properties, but on the contrary, promoted the growth of mycelium and the formation of fruit bodies of two types of domestic fungi. The development of fungi was even higher here than in the control. This can be explained by the fact that the composition contains easily assimilated biogenic silicon, which stimulates the growth and development of domestic fungi in the absence of an inhibiting biocide.

Table 13–7 Results of tests of protective coatings for wood against fungi in a moist chamber.

Domestic fungi specie	Organosilicon & wax & QAC	Commercial	Organosilicon&wax & sol-H_3BO_3	Organosilicon & wax	Control
Evaluation of the growth of domestic fungi on treated wood blocks, in points					
Antrodia xantha 1029	1	2	4	4	3
Coniophora puteana 1370	1	2	3	5	4
The change of wood color					
Antrodia xantha 1029	No changes	Dark brown initially	Brown rays	Spotted light browning	Brown rays
Coniophora puteana 1370	Light browning along cracks	Dark brown initially	The entire surface darkened	The entire surface turned brown	The entire surface turned brown
Formation of rudiments of fruiting bodies					
Antrodia xantha 1029	Did not develop	Locally around the edges	Did not develop	Abundantly over the entire surface	Locally around the edges
Coniophora puteana 1370	Did not develop	Did not develop	Small crusts	Small crusts	Do not develop

Note: 0—death of the mycelium was observed after initial weak growth; 1—growth was observed locally (not over the entire surface of the bar), thin cobweb mycelium; 2—growth over the entire surface of the bar, the mycelium was slightly denser than the cobweb; 3—growth over the entire surface of the bar, dense mycelium in places, strands formed; 4—growth over the entire surface of the bar, the mycelium was very dense with the formation of crusts and films.

13.6 Field tests of friendly wood protection coatings in various climatic conditions

Full-scale tests of wood samples with various protective coatings in natural conditions were carried out in several different climatic zones, namely: in a harsh, sharply continental climate in the Far North of Russia, in the area of Yakutsk (Republic of Sakha (Yakutia)) (Tsvetkova et al., 2019); as well as in two completely different climatic zones in Vietnam, at the weather stations of the Vietnam Russian Tropical Center (Vietnam) in the area of Hanoi, which is characterized by a subequatorial monsoon climate, and in the area of Nha Trang with a tropical savanna climate. The sapwood part of pine wood was used for the tests; the size of the wooden sticks was: 100 mm × 25 mm × 20 mm. The samples were not previously subjected to any grinding or polishing.

Far North, Republic of Sakha (Yakutia). Weather station of the Larionov Institute of Physical and Technical Problems of the North of the Siberian Branch of the Russian Academy of Sciences. The climate of Yakutsk, located beyond the Arctic Circle in the permafrost zone, is characterized by the world's maximum and very sharp seasonal temperature drops: a

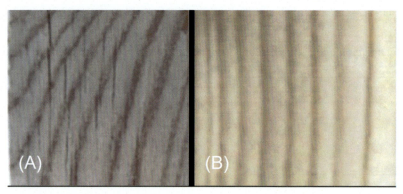

FIGURE 13–5 The surface of pine blocks with coatings obtained from compositions based on a mixture of wax and TEOS (Table 13–1, # 1) (A), and a mixture of wax and TEOS with modifiers H_3BO_3 and glycerin (Table 13–1, # 2) (B) after full-scale climatic tests in a sharply continental climate (within 11 months).

unique combination of summer heat and severe frosts in winter. During the 11-month test of the experimental samples, the air temperature ranged from $-47°C$ to $+28.6°C$; relative humidity from 15% to 97%; solar radiation from 0.66 to 240 MJ/m^2; ultraviolet radiation: from 1.7438 to 43.4925 MJ/m^2 (Tsvetkova et al., 2019).

The exposure of samples to harsh conditions of Yakutia revealed a tendency of wood to crack (Fig. 13–5). Wood surface with coatings based on a mixture of wax and TEOS, including modified with a block copolymer (see Table 13–1, ## 1, 3), retained hydrophobicity (contact angle >100°C, roll-off angle 50 degrees). However, these coatings did not protect the wood from moisture loss and cracking. The appearance of individual cracks was detected (Fig. 13–5A). In contrast, the presence of boric acid and glycerol in the coating prevented cracking (Fig. 13–5A). Wood with these coatings retained a smooth and even surface.

The exposure of samples to harsh conditions of Yakutia revealed a tendency of wood to crack (Fig. 13–5). Wood surface with coatings based on a mixture of wax and TEOS, including modified with a block copolymer (see Table 13–1, ## 1, 3), retained hydrophobicity (contact angle >100°C, roll-off angle 50 degrees). However, these coatings did not protect the wood from moisture loss and cracking. The appearance of individual cracks was detected (Fig. 13–5A). In contrast, the presence of boric acid and glycerol in the coating prevented cracking (Fig. 13–5B). Wood with these coatings retained a smooth and even surface.

Vietnam-Russia Tropical Center. Experimental samples were tested from mid-March 2020 to December 2020 at two climate stations located in the Hanoi, Hoa Lac, and in the Coastal Branch of the Tropical Center (Nha Trang, Dam Bay)—see Table 13–8. The samples were attached to racks installed on the grass or on a concrete platform in the open air. Another batch of samples was exposed under a canopy protecting them from direct sunlight; in this case, the samples were *suspended in a free state.*

Dam Bay, Coastal Branch—Vietnam Russian Tropical Center, Nha Trang, Vietnam. Test conditions. The average daytime temperature for March–December was $26.7°C–32.8°C$; precipitation in March–August 284 mm, September–October 268 mm, November–December 56 mm.

Table 13–8 Test conditions for wood samples coated with drying linseed oil and organosilicon varnish (solution of polyphenylsiloxane resin in toluene) with various biocidal additives.

#	Description of the biocidal additive	Test location	
		Dam Bay	Hoa Lac
1	Quaternary ammonium compound (tetrabutylammonium bromide), QAC, see Table 13–2, #2.	on grass under the canopy on a concrete site	mycological site[a] on a concrete site
2	Silica sol doped by H_3BO_3: "TEOS-derive sol & H_3BO_3," see Table 13–2, # 3	on grass under the canopy on a concrete site	mycological site[a] on a concrete site
3	Linseed oil nanosilver and nanocopper colloids: "Silver & Cooper Nano Oil," see Table 13–2, # 8	on grass under the canopy on a concrete site	mycological site[a] on a concrete site
4	Tributyl phosphate nanosilver colloid: "Silver Nano TBP 3000," see Table 13–2, # 7	on grass under the canopy on a concrete site	mycological site[a] on a concrete site
Control	on a concrete site	on a concrete site	

[a]*On the mycological site*—the samples are suspended in a free state under a canopy.

All experimental specimens located in the open air, both on grassy and on concrete sites, changed color—they acquired a light gray color. The appearance of the samples changed under the influence of climatic factors, which indicates the initial stage of infection with wood-coloring fungi. The color change and appearance of defects are more pronounced after exposure on a grassy area (Fig. 13–6). Following the deterioration of the surfaces, the coatings can be arranged in the following order: # 4, # 2, # 3, and # 1 (see Table 13–8).

Thus, the best protective properties were revealed for the biocide, which is a colloidal solution of silver nanoparticles in tributyl phosphate, followed by silica sol with boric acid additives in glycerol. The worst of all was the coating where QAC—tetrabutylammonium bromide—was used as a biocide. It is possible that this phenomenon is associated with leaching of QAC from the coating under high temperature and humidity conditions.

The surface condition of the samples under a canopy, protected from direct sunlight and precipitation, turned out to be slightly better; all of them did not change color to gray (Fig. 13–7). However, the effects of biocides turned out to be completely different.

The smallest number of dark spots appeared on the sample with coating #3 (see Table 13–8), containing both silver and copper nanoparticles; the worst was the coating where only silver nanoparticles were used as a biocide. Coatings with QAC and boric acid with glycerin also performed well.

Hoa Lac Research Station, Hanoi, Vietnam. Test conditions. The average daytime temperature for March–December was 21°C–35°C; precipitation in April–June 190 mm, in August–September 1010 mm, in October–December 113 mm.

The effect of biocidal additives on changing the appearance of samples tested in the subequatorial monsoon climate of Hoa Lac over a given time period (10-and-a-half months) is more clearly traced than in tests in the tropical Dam Bay climate (Figs. 13–8 and 13–9). In the absence of direct sunlight and precipitation (on a mycological site under a canopy), QACs, as well as a colloidal solution of silver and copper nanoparticles in linseed oil in the composition of

FIGURE 13–6 The appearance of a stand with experimental samples installed on a grass site (Dam Bay), test duration—10.5 months. Coating compositions correspond to the description and numbers in Table 13–8. From left to right: ## 4, 3, 2, 1.

FIGURE 13–7 The appearance of the stand with experimental samples, installed under a canopy (Dam Bay), test duration:—10.5 months. Coating compositions correspond to the description and numbers in Table 13–8. From left to right: ## 1, 2, 3, 4.

protective coatings, showed the greatest protective effect. The protective effect of silica ash with boric acid and glycerin, as well as silver nanoparticles (in the absence of copper nanoparticles) was significantly weaker. On the surface of samples with these coatings, a dark coating and sound spots appeared, which indicates the beginning of a fungal infection.

The appearance of signs of fungal infection is more pronounced on samples that were exposed on an open concrete area (Fig. 13–10).

As seen from Figs. 13–8–13–10 field tests of protective coatings with various biocidal additives based on drying oil and organosilicon glue at the climate station in Hoa Lac, located in the subequatorial monsoon climate zone, showed similar results, both when tested in the open air and when protected from direct sunlight rays and rain. The best results were found

Chapter 13 • Deterioration and protection of wood products 311

FIGURE 13–8 View of the stand for atmospheric tests at the mycological site.

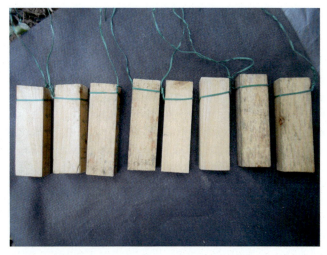

FIGURE 13–9 The appearance of experimental samples after testing at the mycological site from mid-March to December 2020, Hoa Lac. From left to right, the numbers of pairs of samples: ## 1, 1; 2, 2; 3, 3; 4, 4 (see Table 13–8).

for coating with a biocidal additive of a colloidal solution of silver and copper nanoparticles in linseed oil, as well as for QAC (to a much lesser extent in the open air). As demonstrated earlier, under tropical savanna climate, the same tendency for the strongest biocidal action of a colloidal solution of silver and copper nanoparticles, as well as QAC, was observed only when tested under a canopy. In the case of outdoor tests, these biocides performed worse, especially QAC, which can be associated with the phenomenon of biocide leaching.

312 Biodegradation and Biodeterioration at the Nanoscale

FIGURE 13–10 The appearance of experimental samples on an open concrete area after testing from mid-March to December 2020, Hoa Lac. From left to right, sample numbers are ## 1, 2, 3, 4, and an uncoated control.

13.7 Conclusion

Wood is one of the most beautiful natural materials widely demanded as a building and ornamental material. In this regard, the problem of preserving its beautiful natural appearance and preventing its destruction under the influence of unfavorable climatic factors and biodeterioration agents does not cease to be relevant. The main role in the biodegradation of wood is played by wood-destroying fungi, primarily basidiomycetes, which cause various types of wood decay in the natural environment, as well as in buildings and structures. The main wood components that are destroyed by basidiomycetes are cellulose, hemicellulose, and lignin. The enzymes secreted by fungi convert these wood components into more readily available water-soluble compounds. Among the basidiomycetes that destroy wood, a special place belongs to domestic fungi that develop on industrial wood. They are capable of damaging wooden structures in residential buildings, engineering structures, mines, and historical buildings. This group includes the genera: *Serpula, Tapinella, Coniophora, Antrodia, Neolentinus, Gloeophyllum, Cylindrobasidium, Phlebiopsis*. Domestic fungi destroy cellulose in wood, causing destructive brown rot. The high rate of destructive processes is due to the activity of cellulolytic enzymes of wood-destroying fungi, secreted into the substrate. Another type of wood destruction is called white rot. It is mainly associated with lignin utilization. The main role in this process belongs to the ligninolytic enzymes of wood-destroying, which they release into the environment.

Despite a large amount of research in this area, there are still no universal techniques, which would prevent wood degradation and at the same time would be friendly for the environment. Commercially available protective biocides provide protection for no more than five years, while they contain quite dangerous substances. Therefore, studies on the development of new effective and environmentally friendly (green) compositions for wood protection are still relevant and arouse the interest of the world scientific community.

In this review, modern methods and approaches for the chemical protection of wood using paint and varnish compositions containing various antiseptic and fungicidal additives, as environmentally friendly as possible, were considered. For antiseptic processing of wood, water-, organo-, and oil-soluble antiseptics are used. The following groups of antiseptics meet modern environmental requirements to the greatest extent: boron-containing substances, carbamates, isothiozolones, QACs, sulfamides, triazoles, inorganic copper compounds, copper naphthenates and citrates, modified creazote oils (not containing benzo-α-pyrene), and metal oxides.

The main problem with the use of copper-based protective agents is the existence of a rather large group of copper-tolerant fungi. An alternative to copper-containing compositions for wood preservation can be inorganic sol-gel compositions based on TiO_2 and SiO_2, in combination with other reagents. This approach can improve the protection of wood from white and brown rot fungi.

The technique of hydrophobization is still used to reduce water absorption. Wood hydrophobization is carried out using polymers and organosilicon compounds, while hygroscopicity decreases due to blocking of hydroxyl groups of cellulose and lignin, filling of the capillary-porous structure of wood, as well as changes in the density of crosslinking of the ligno-carbohydrate matrix. A relatively new method of chemical hydrophobization of wood, which significantly increases its strength and durability, is the method of acetylation.

Nanotechnology is becoming more and more part of everyday life. At the beginning of the 21st century, it began to be used in compositions for protecting wood from biodegradation. Impregnation of wood or protective coatings application, using suspensions of metal nanoparticles (Cu, Cu-Ag, Ag, Au, Ag-Au, Zn, and Al), micronized copper obtained, for example, on the basis of $CuCO_3 \times Cu(OH)_2$, metal oxides (ZnO, TiO_2, CeO_2), metal fluorides MgF_2 and CaF^2, as well as suspensions and solutions of natural bioactive materials (chitosan, cationic starch), can improve the characteristics of wood preservatives, thereby increasing the service life of wood. Nanoparticles of various natures are also used to create superhydrophobic coatings that would reduce water absorption, thereby inhibiting the biodegradation process of wood.

In recent years, wood processing using sol-gel technology has attracted the increasing interest of the world community. The use of sols for wood impregnation is aimed, among other things, at improving the hydrophobicity of the wood surface. Impregnation of wood with sols can be carried out using ultrasound. The application of water-repellent coatings using sols based and SiO_2 nanoparticles led to an increase in the contact angle with water up to 150 degrees, that is, to the formation of a superhydrophobic surface. The phenomenon of superhydrophobicity is based on the so-called "lotus effect," as a result of which drops roll off the surface, it acquires good water-repellent properties. However, the decisive factor is the long-term protection of the impregnation or coating. This is especially true for superhydrophobic surfaces, which tend to lose the superhydrophobic effect. Despite the fact that sol-gel impregnations do not provide the same high characteristics as traditional commercial preparations for improving bioresistance and fire resistance, they can significantly improve the characteristics of wood in a number of parameters and be an alternative to traditional agents. In the future, it is promising to conduct research to study the long-term effect of sol-gel-derived impregnations, as well as to test them on wood composites such as plywood and veneers.

It was shown in laboratory conditions that hybrid sol-gel compositions, obtained by mixing silica sols with various modifying additives—organosilicon varnishes, block copolymers, nanodispersed hydrophobized SiO_2 filler (aerosil—colloidal pyrogenic silica or fumed silica), and even with natural materials (e.g., beeswax), are also promising for hydrophobizing wood surfaces. The degree of hydrophobicity significantly depends on the amount of aerosil introduced. At the same time, the use of silicon dioxide nanoparticles in sol-gel compositions as a structure-forming agent should be treated with caution. Many types of fungi actively use it for their vital activity, which can adversely affect the protective properties of coatings. The addition of organosilicon varnishes to the composition significantly reduces water absorption and improves adhesion. Our experience in the development and research of various protective coatings shows that, despite the hydrophobic surface, biocidal additives are the most important for the suppression of the growth and development of biodeterioration agents, including wood-destroying fungi.

In general, the obtained data indicate that a number of tested compositions based on mixtures of wax with TEOS, containing QACs, H_3BO_3, and Ag, Ag/Cu nanoparticles in water, oil, or tributyl phosphate, are promising for suppressing the development of wood-destroying fungi.

Their assignment to low-toxic biocides (not containing potent toxic substances) largely explains the incomplete suppression of the growth of domestic fungi, as well as the selectivity of action on individual strains. In addition, the domestic fungi themselves differ quite significantly in adaptive properties, the manifestation of protective reactions, growth rate, and other indicators. It was found that the biocidal activity of silver and copper nanoparticles decreases with an increase in the size of nanoparticle agglomerates in colloidal solutions (from ~50 to 2000 nm). Apparently, in order to maintain the biocidal activity of nanoparticles, it is necessary to select the type and concentration of the flocculating agent more carefully, avoiding the use of long-chain polymers.

Full-scale tests in the Far North in a very cold climate (11 months with temperature drops from $-47°C$ to $+28.6°C$ and humidity jumps from 15% to 97%) coatings based on a mixture of wax and TEOS with a polyphenylsiloxane resin and glyphthalic varnish or with H_3BO_3 and glycerol, both with a highly hydrophobic surface, and with a highly hydrophilic surface, showed high crack resistance. The addition of boric acid with glycerol helped to preserve the original color of the coatings, that is, prevented the colonization of wood by biodeterioration agents.

Full-scale tests of alkyd coatings based on linseed oil and organosilicon varnish for weather resistance in a hot tropical climate with a large amount of precipitation ($21°C-35°C$, precipitation 608–1313 mm) turned out to be more extreme for wood. Better results were found for coatings that were exposed under a canopy (protected from direct sunlight and rainfall). The addition of a QAC and Ag/Cu nanoparticles as biocides was found to be the most effective. In the case of outdoor tests, these biocides showed worse results, especially a QAC, which may be due to the phenomenon of leaching of biocides.

References

Abbott, T. P., & Wicklow, D. T. (1984). Degradation of lignin by *Cyathus* species. *Applied and Environmental Microbiology*, *47*(3), 585–587. Available from https://doi.org/10.1128/aem.47.3.585-587.1984.

Altun, S., Ozcifci, A., Şenel, A., Baysal, E., & Toker, H. (2010). Effects of silica gel on leaching resistance and thermal properties of impregnated wood. *Wood Research*, *55*(4), 101–112.

Ander, P., Eriksson, K. E., & Yu, H. S. (1984). Metabolism of lignin-derived aromatic acids by wood-rotting fungi. *Journal of General Microbiology*, *130*(1), 63–68. Available from https://doi.org/10.1099/00221287-130-1-63.

Ansell, M. P. (2013). Multi-functional nano-materials for timber in construction. *Proceedings of Institution of Civil Engineers: Construction Materials*, *166*(4), 248–256. Available from https://doi.org/10.1680/coma.12.00035.

Badreshia, S., & Marks, J. G. (2002). Iodopropynyl butylcarbamate. *American Journal of Contact Dermatitis*, *13*(2), 77–79. Available from https://doi.org/10.1053/ajcd.2002.30728.

Baldrian, P., & Valášková, V. (2008). Degradation of cellulose by basidiomycetous fungi. *FEMS Microbiology Reviews*, *32*(3), 501–521. Available from https://doi.org/10.1111/j.1574-6976.2008.00106.x.

Baya, M., Soulounganga, P., Gelhaye, E., & Grardin, P. (2001). Fungicidal activity of β-thujaplicin analogues. *Pest Management Science*, *57*(9), 833–838. Available from https://doi.org/10.1002/ps.379.

Baysal, E., & Yalinkilic, M. K. (2005). A new boron impregnation technique of wood by vapor boron of boric acid to reduce leaching boron from wood. *Wood Science and Technology*, *39*(3), 187–198. Available from https://doi.org/10.1007/s00226-005-0289-1.

Beaudelaire, K. G. Y., Zhuang, B., Aladejana, J. T., Li, D., Hou, X., & Xie, Y. (2020). Influence of mesoporous inorganic Al-B-P amphiprotic surfactant material resistances of wood against brown and white-rot fungi (part 1). *Coatings*, *10*(2). Available from https://doi.org/10.3390/coatings10020108.

Blanchette, R. A. (1995). Degradation of lignocellulose complex in wood. *Canadian Journal of Botany*, *73*(1), 999–1010. Available from https://doi.org/10.1139/b95-350.

Blondeau, R. (1989). Biodegradation of natural and synthetic humic acids by the white rot fungus *Phanerochaete chrysosporium*. *Applied and Environmental Microbiology*, *55*(5), 1282–1285. Available from https://doi.org/10.1128/AEM.55.5.1282-1285.198.

Breene, W. M. (1990). Nutritional and medicinal value of specialty mushrooms. *Journal of Food Protection*, *53*(10), 883–899. Available from https://doi.org/10.4315/0362-028x-53.10.883.

Brinker, C. J., & Scherer, G. W. (1990). *Sol-gel science: The physics and chemistry of sol-gel processing*. Academic Press.

Burlacu, E., Tanase, C., Coman, N. A., & Berta, L. (2019). A review of bark-extract-mediated green synthesis of metallic nanoparticles and their applications. *Molecules (Basel, Switzerland)*, *24*(23). Available from https://doi.org/10.3390/molecules24234354.

Cai, L., Lim, H., Fitzkee, N. C., Cosovic, B., & Jeremic, D. (2020). Feasibility of manufacturing strand-based wood composite treated with β-Cyclodextrin-Boric acid for fungal decay resistance. *Polymers*, *12*(2). Available from https://doi.org/10.3390/polym12020274.

Cappelletto, E., Maggini, S., Girardi, F., Bochicchio, G., Tessadri, B., & Di Maggio, R. (2013). Wood surface protection with different alkoxysilanes: A hydrophobic barrier. *Cellulose*, *20*(6), 3131–3141. Available from https://doi.org/10.1007/s10570-013-0038-9.

Chang, H., Tu, K., Wang, X., & Liu, J. (2015). Fabrication of mechanically durable superhydrophobic wood surfaces using polydimethylsiloxane and silica nanoparticles. *RSC Advances, 5*(39), 30647–30653. Available from https://doi.org/10.1039/c5ra03070f.

Chedgy, R. J. (2010). Secondary metabolites of Western red cedar (*Thuja plicata*): their biotechnological applications and role in conferring natural durability. LAP Lambert Academic Publishing, Saarbrücken, Germany. ISBN-10: 3838346610, ISBN-13: 978-3838346618.

Chedgy, R. J., Lim, Y. W., & Breuil, C. (2009). Effects of leaching on fungal growth and decay of western redcedar. *Canadian Journal of Microbiology, 55*(5), 578–586. Available from https://doi.org/10.1139/W08-161.

Civardi, C., Schubert, M., Fey, A., Wick, P., Schwarze, F. W. M. R., & Sarrocco, S. (2015). Micronized copper wood preservatives: Efficacy of ion, nano, and bulk copper against the brown rot fungus *Rhodonia placenta*. *PLoS One, 10*(11), e0142578. Available from https://doi.org/10.1371/journal.pone.0142578.

Clausen, C. A., & Yang, V. (2007). Protecting wood from mould, decay, and termites with multi-component biocide systems. *International Biodeterioration and Biodegradation, 59*(1), 20–24. Available from https://doi.org/10.1016/j.ibiod.2005.07.005.

Coombs, R. W., & Trust, T. J. (1973). The effect of light on the antibacterial activity of β thujaplicin. *Canadian Journal of Microbiology, 19*(9), 1177–1180. Available from https://doi.org/10.1139/m73-190.

Currie, R. B., Byrom, G. B., Bruce, A., & Schmidt, R. G. (2020). Incorporation of stable glycerol-borate complex into phenolic resin as a glue line treatment in engineered wood products. *Wood Material Science and Engineering, 15*(4), 190–197. Available from https://doi.org/10.1080/17480272.2018.1551931.

De Peres, M. L., Delucis, R. d A., Amico, S. C., & Gatto, D. A. (2019). Zinc oxide nanoparticles from microwave-assisted solvothermal process: Photocatalytic performance and use for wood protection against xylophagous fungus. *Nanomaterials and Nanotechnology, 9*. Available from https://doi.org/10.1177/1847980419876201.

De Vetter, L., Depraetere, G., Stevens, M., Janssen, C., & van Acke, J. (2009). Potential contribution of organosilicon compounds to reduced leaching of biocides in wood protection. *Annals of Forest Science*, 209. Available from https://doi.org/10.1051/forest/2008091, 209.

De Vetter, L., Van Den Bulcke, J., & Van Acker, J. (2010). Impact of organosilicon treatments on the wood-water relationship of solid wood. *Holzforschung, 64*(4), 463–468. Available from https://doi.org/10.1515/HF.2010.069.

Dickinson, C. H., & Pugh, G. J. F. (Eds.), (1974). *Biology of plant litter decomposition* (Vol. 147). Academic Press.

Dmitrieva, M. B., Chmutin, L. A., Yarovaya, M. S., & Linnik, M. A. (2009). Determination of fungicidal activity of preparations based on silver nanoparticles. *Nanotechnika, 20*(4), 45–53. Available from http://www.nanotech.ru/journal/word/cont09-4.pdf.

Elvin, G. (2007). The nano revolution. A science that works on the molecular scale is set to transform the way we build, Architect Magazine. *The Journal of the American Institute of Architect*. https://www.architectmagazine.com/technology/the-nano-revolution_o

Elvin, G. (2017). Nanotechnology in architecture. In B. Raj, M. Van de Voorde, & Y. Mahajan (Eds.), *Nanotechnology for energy sustainability* (pp. 967–996). Wiley-VCH Verlag GmbH & Co. KGaA, Chapter 39. Available from https://doi.org/10.1002/9783527696109.ch39.

Evans, P., Matsunaga, H., & Kiguchi, M. (2008). Large-scale application of nanotechnology for wood protection. *Nature Nanotechnology, 3*(10), 577. Available from https://doi.org/10.1038/nnano.2008.286.

Feci, N., Palanti, D., Predieri, & Vignali, F. (2009). Effectiveness of sol-gel treatments coupled with copper and boron against subterranean termites. In *Proceedings 40th of the annual meeting* (pp. 9–30493). http://repositorio.lnec.pt:8080/jspui/handle/123456789/16964

Freeman, M. H., & McIntyre, C. R. (2008). A comprehensive review of copper-based wood preservatives with a focus on new micronized or dispersed copper systems. *Forest Products Journal, 58*(11), 6–27.

Fu, Y., Yu, H., Sun, Q., Li, G., & Liu, Y. (2012). Testing of the superhydrophobicity of a zinc oxide nanorod array coating on wood surface prepared by hydrothermal treatment. *Holzforschung, 66*(6), 739–744. Available from https://doi.org/10.1515/hf-2011-0261.

Glasser, W. G. (2019). About making lignin great again—some lessons from the past. *Frontiers in Chemistry, 7*. Available from https://doi.org/10.3389/fchem.2019.00565.

Goodell, B., Daniel, G., Jellison, J., & Qian, Y. (2006). Iron-reducing capacity of low-molecular-weight compounds produced in wood by fungi. *Holzforschung, 60*(6), 630–636. Available from https://doi.org/10.1515/HF.2006.106.

Goodell, B., Qian, Y., & Jellison, J. (2008). *Fungal decay of wood: Soft rot-brown rot-white rot,* . ACS symposium series (Vol. 982, pp. 9–31). American Chemical Society. Available from https://doi.org/10.1021/bk-2008-0982.ch002.

Green, F., & Clausen, C. A. (2005). Copper tolerance of brown-rot fungi: Oxalic acid production in southern pine treated with arsenic-free preservatives. *International Biodeterioration and Biodegradation, 56*(2), 75–79. Available from https://doi.org/10.1016/j.ibiod.2005.04.003.

Harandi, D., Ahmadi, H., & Mohammadi Achachluei, M. (2016). Comparison of TiO2 and ZnO nanoparticles for the improvement of consolidated wood with polyvinyl butyral against white rot. *International Biodeterioration and Biodegradation, 108*, 142–148. Available from https://doi.org/10.1016/j.ibiod.2015.12.017.

Hsieh, C. T., Chang, B. S., & Lin, J. Y. (2011). Improvement of water and oil repellency on wood substrates by using fluorinated silica nanocoating. *Applied Surface Science, 257*(18), 7997–8002. Available from https://doi.org/10.1016/j.apsusc.2011.04.071.

Hübert, T., & Mahr, M. S. (2017). *Handbook of sol-gel science and technology.* https://doi.org/10.1007/978-3-319-19454-7_106-2

Hübert, T., Unger, B., & Bücker, M. (2010). Sol-gel derived TiO2 wood composites. *Journal of Sol-Gel Science and Technology, 53*(2), 384–389. Available from https://doi.org/10.1007/s10971-009-2107-y.

Janesch, J., Arminger, B., Gindl-Altmutter, W., & Hansmann, C. (2020). Superhydrophobic coatings on wood made of plant oil and natural wax. *Progress in Organic Coatings, 148*. Available from https://doi.org/10.1016/j.porgcoat.2020.105891.

Janesch, J., Czabany, I., Hansmann, C., Mautner, A., Rosenau, T., & Gindl-Altmutter, W. (2020). Transparent layer-by-layer coatings based on biopolymers and CeO2 to protect wood from UV light. *Progress in Organic Coatings, 138*. Available from https://doi.org/10.1016/j.porgcoat.2019.105409.

Jin, L., & Preston, F. (1991). The interaction of wood preservatives with lignocellulosic substrates: I. Quaternary ammonium compounds. *Holzforschung, 45*(6), 455–459. Available from https://doi.org/10.1515/hfsg.1991.45.6.455.

Kaegi, R., Sinnet, B., Zuleeg, S., Hagendorfer, H., Mueller, E., Vonbank, R., ... Burkhardt, M. (2010). Release of silver nanoparticles from outdoor facades. *Environmental Pollution, 158*(9), 2900–2905. Available from https://doi.org/10.1016/j.envpol.2010.06.009.

Kanokwijitsilp, T., Traiperm, P., Osotchan, T., & Srikhirin, T. (2016). Development of abrasion resistance SiO2 nanocomposite coating for teak wood. *Progress in Organic Coatings, 93*, 118–126. Available from https://doi.org/10.1016/j.porgcoat.2015.12.004.

Kartal, S. N., Green, F., & Clausen, C. A. (2009). Do the unique properties of nanometals affect leachability or efficacy against fungi and termites? *International Biodeterioration and Biodegradation, 63*(4), 490–495. Available from https://doi.org/10.1016/j.ibiod.2009.01.007.

Kartal, S. N., Yoshimura, T., & Imamura, Y. (2009). Modification of wood with Si compounds to limit boron leaching from treated wood and to increase termite and decay resistance. *International Biodeterioration and Biodegradation, 63*(2), 187–190. Available from https://doi.org/10.1016/j.ibiod.2008.08.006.

Khademibami, L., Jereniic, D., Slunulsky, R., & Barnes, H. M. (2020). Chitosan oligomers and related nanoparticles as environmentally friendly wood preservatives. *BioResources*, *15*(2), 2800−2817. Available from https://doi.org/10.15376/biores.15.2.2800-2817.

Kirk, T. K., & Cowling, E. B. (1985). *Biological decomposition of solid wood. The chemistry of solid wood* (Vol. 207, pp. 455−487). ASC Publications Chapter 12. Available from https://doi.org/10.1021/ba-1984-0207.ch012.

Korkmaz, N. (2020). Bioreduction: The biological activity, characterization, and synthesis of silver nanoparticles. *Turkish J. Chem*, *44*, 325−334. Available from http://hdl.handle.net/11772/4119.

Korkmaz, N., Ceylan, Y., Hamid, A., Karadağ, A., Bülbül, A. S., Aftab, M. N., ... Şen, F. (2020). Biogenic silver nanoparticles synthesized via Mimusops elengi fruit extract, a study on antibiofilm, antibacterial, and anticancer activities. *Journal of Drug Delivery Science and Technology*, *59*, 101864. Available from https://doi.org/10.1016/j.jddst.2020.101864.

Künniger, T., Heeb, M., & Arnold, M. (2014). Antimicrobial efficacy of silver nanoparticles in transparent wood coatings. *European Journal of Wood and Wood Products*, *72*(2), 285−288. Available from https://doi.org/10.1007/s00107-013-0776-2.

Kuvshinov, V. A., Altunina, L. K., Stasieva, L. A., & Kuvshinov, I. V. (2019). Acidity study of donor-acceptor complexes of boric acid with polyols for oil displacing compositions. *Journal of Siberian Federal University. Chemistry*, *12*(3), 364−373. Available from https://doi.org/10.17516/1998-2836-0133.

Liu, F., Wang, S., Zhang, M., Ma, M., Wang, C., & Li, J. (2013). Improvement of mechanical robustness of the superhydrophobic wood surface by coating PVA/SiO 2 composite polymer. *Applied Surface Science*, *280*, 686−692. Available from https://doi.org/10.1016/j.apsusc.2013.05.043.

Lu, Y., Feng, M., & Zhan, H. (2014). Preparation of SiO2−wood composites by an ultrasonic-assisted sol−gel technique. *Cellulose*, *21*(6), 4393−4403. Available from https://doi.org/10.1007/s10570-014-0437-6.

Marzi, T. (2015). Nanostructured materials for protection and reinforcement of timber structures: A review and future challenges. *Construction and Building Materials*, *97*, 119−130. Available from https://doi.org/10.1016/j.conbuildmat.2015.07.016.

Mazanik, N. V. (2011). Modern bioprotectors for wood. *Proceedings of BSTU. Scientific Journal*, *140*(2), 181−184. Available from https://elib.belstu.by/handle/123456789/2214.

Mazela, B., Polus-Ratajczak, I., Hoffmann, S. K., & Goslar, J. (2005). Copper monoethanolamine complexes with quaternary ammonium compounds in wood preservation. Biological testing and EPR study. *Wood Research*, *50*(2), 1−17.

Mishra, P. K., Giagli, K., Tsalagkas, D., Mishra, H., Talegaonkar, S., Gryc, V., & Wimmer, R. (2018). Changing face of wood science in modern era: Contribution of nanotechnology. *Recent Patents on Nanotechnology*, *12*(1), 13−21. Available from https://doi.org/10.2174/1872210511666170808111512.

Miyafuji, H., & Saka, S. (2001). Na2O-SiO2 wood-inorganic composites prepared by the sol-gel process and their fire-resistant properties. *Journal of Wood Science*, *47*(6), 483−489. Available from https://doi.org/10.1007/BF00767902.

Mohammed-Ziegler, I., Tánczos, I., Hórvölgyi, Z., & Agoston, B. (2008). Water-repellent acylated and silylated wood samples and their surface analytical characterization. *Colloids and Surfaces A: Physicochemical and Engineering Aspects*, *319*(1−3), 204−212. Available from https://doi.org/10.1016/j.colsurfa.2007.06.063.

Mohareb, A., Thévenon, M. F., Wozniak, E., & Gérardin, P. (2011). Effects of polyvinyl alcohol on leachability and efficacy of boron wood preservatives against fungal decay and termite attack. *Wood Science and Technology*, *45*(3), 407−417. Available from https://doi.org/10.1007/s00226-010-0344-4.

Morrell, J. J. (2017). *Protection of wood: A global perspective on the future. Wood is good: Current trends and future prospects in wood utilization* (pp. 213−226). Singapore: Springer. Available from https://doi.org/10.1007/978-981-10-3115-1_20.

Mukhopadhyay, A. K. (2011). Next-generation nano-based concrete construction products: A review. In K. Gopalakrishnan, B. Birgisson, P. Taylor, & N. O. Attoh-Okine (Eds.), *Nanotechnology in civil infrastructure* (pp. 207–223). Springer Science and Business Media LLC. Available from https://doi.org/10.1007/978-3-642-16657-0_7.

Neklyudov, A. D., & Ivankin, A. N. (2005). Treatment of a wood—one of methods of preservation of forest riches of country. *Vestnik Moskovskogo Gosudarstvennogo Universiteta Lesa—Lesnoj Vestnik (Bulletin of the Moscow State University of Forests—Forest Bulletin)*. 2. 77–89. <https://www.elibrary.ru/download/elibrary_18106785_54321267.pdf>

Nesrin, K., Yusuf, C., Ahmet, K., Ali, S. B., Muhammad, N. A., Suna, S., & Fatih, Ş. (2020). Biogenic silver nanoparticles synthesized from *Rhododendron ponticum* and their antibacterial, antibiofilm and cytotoxic activities. *Journal of Pharmaceutical and Biomedical Analysis, 179*. Available from https://doi.org/10.1016/j.jpba.2019.112993.

Nowack, B., Krug, H. F., & Height, M. (2011). 120 years of nanosilver history: Implications for policy makers. *Environmental Science and Technology, 45*(4), 1177–1183. Available from https://doi.org/10.1021/es103316q.

Palanti, S., Feci, E., Predieri, G., & Francesca, V. (2012). Copper complexes grafted to amino-functionalized silica gel as wood preservatives against fungal decay: Mini-blocks and standard test. *BioResources, 7*(4), 5611–5621. Available from https://doi.org/10.15376/biores.7.4.5611-5621.

Pánek, M., Reinprecht, L., & Hulla, M. (2014). Ten essential oils for beech wood protection - efficacy against wood-destroying fungi and moulds, and effect on wood discoloration. *BioResources, 9*(3), 5588–5603. Available from https://doi.org/10.15376/biores.9.3.5588-5603.

Pantano, D., Neubauer, N., Navratilova, J., Scifo, L., Civardi, C., Stone, V., . . . Wohlleben, W. (2018). Transformations of nanoenabled copper formulations govern release, antifungal effectiveness, and sustainability throughout the wood protection lifecycle. *Environmental Science and Technology, 52*(3), 1128–1138. Available from https://doi.org/10.1021/acs.est.7b04130.

Pařil, P., Baar, J., Čermák, P., Rademacher, P., Prucek, R., Sivera, M., & Panáček, A. (2017). Antifungal effects of copper and silver nanoparticles against white and brown-rot fungi. *Journal of Materials Science, 52*(5), 2720–2729. Available from https://doi.org/10.1007/s10853-016-0565-5.

Pica, A., & Ficai, A. (2016). A new generation of antibacterial film forming materials. *Revista de Chimie, 67*(1), 34–37. Available from http://www.revistadechimie.ro/archive.asp.

Plumed-Ferrer, C., Väkeväinen, K., Komulainen, H., Rautiainen, M., Smeds, A., Raitanen, J. E., . . . Von Wright, A. (2013). The antimicrobial effects of wood-associated polyphenols on food pathogens and spoilage organisms. *International Journal of Food Microbiology, 164*(1), 99–107. Available from https://doi.org/10.1016/j.ijfoodmicro.2013.04.001.

Rai, M., Yadav, A., & Gade, A. (2009). Silver nanoparticles as a new generation of antimicrobials. *Biotechnology Advances, 27*(1), 76–83. Available from https://doi.org/10.1016/j.biotechadv.2008.09.002.

Reinprecht, L. (2010). Fungicides for wood protection—world viewpoint and evaluation/testing in Slovakia. In O. Carisse (Ed.), *Fungicides*. IntechOpen. Available from https://doi.org/10.5772/13233.

Saka, S., & Ueno, T. (1997). Several SiO2 wood-inorganic composites and their fire-resisting properties. *Wood Science and Technology, 31*(6), 457–466. Available from https://doi.org/10.1007/bf00702568.

Schmidt, O. (2007). Indoor wood-decay basidiomycetes: Damage, causal fungi, physiology, identification and characterization, prevention and control. *Mycological Progress, 281*. Available from https://doi.org/10.1007/s11557-007-0545-x, 281.

Schneider, M. H., & Brebner, K. I. (1985). Wood-polymer combinations: The chemical modification of wood by alkoxysilane coupling agents. *Wood Science and Technology, 19*(1), 67–73. Available from https://doi.org/10.1007/BF00354754.

Selvam, K., Senbagam, D., Selvankumar, T., Sudhakar, C., Kamala-Kannan, S., Senthilkumar, B., & Govarthanan, M. (2017). Cellulase enzyme: Homology modeling, binding site identification and molecular docking. *Journal of Molecular Structure, 1150*, 61–67. Available from https://doi.org/10.1016/j.molstruc.2017.08.067.

Shah, S. M., Zulfiqar, U., Hussain, S. Z., Ahmad, I., Habib-ur-Rehman., Hussain, I., & Subhani, T. (2017). A durable superhydrophobic coating for the protection of wood materials. *Materials Letters, 203*, 17−20. Available from https://doi.org/10.1016/j.matlet.2017.05.126.

Shilova, O. A., Yu., Zelenskaya, M. S., Ryabusheva, Y. V., Khamova, T. V., Glebova, I. B., ... Marugin, A. M. (2019). Structure, properties and effects of biomineralization for epoxy-titanate sol-gel coatings used to protect materials from biodegradation. In O. Frank-Kamenetskaya, D. Vlasov, E. Panova, & S. Lessovaia (Eds.), *Processes and phenomena on the boundary between biogenic and abiogenic nature* (pp. 619−638). Cham: Springer. Available from https://doi.org/10.1007/978-3-030-21614-6_33, Print ISBN 978-3-030-21613-9, Chapter 33.

Shvarts, E. M., Ignash, R. T., & Belousova, R. G. (2005). Reactions of polyols with boric acid and sodium monoborate. *Russian Journal of General Chemistry, 75*(11), 1687−1692. Available from https://doi.org/10.1007/s11176-005-0492-7.

Steen, R. J. C. A., Ariese, F., Hattum, B. V., Jacobsen, J., & Jacobson, A. (2004). Monitoring and evaluation of the environmental dissipation of the marine antifoulant 4,5-dichloro-2-n-octyl-4-isothiazolin-3-one (DCOIT) in a Danish Harbor. *Chemosphere, 57*(6), 513−521. Available from https://doi.org/10.1016/j.chemosphere.2004.06.043.

Stenina, E. I., & Chesnokova, T. Yu (2017). Research of opportunities of use of colloidal solution of nanodimensional particles of silver as the biocide for wood in severe conditions of exploitation. *Proceedings of BSTU. Scientific Journal. Series, 1*(192), 152−155. Available from https://elib.belstu.by/handle/123456789/2214.

Stirling, R., & Morris, P. I. (2012). Treatments to minimize extractives stain in western red cedar. *BioResources, 7*(2), 2461−2468. Available from https://doi.org/10.15376/biores.7.2.2461-2468.

Stirling, R., Ruddick, J. N. R., Xue, W., Morris, P. I., & Kennepohl, P. (2015). Characterization of copper in leachates from ACQ- and MCQ-treated wood and its effect on basidiospore germination. *Wood and Fiber Science, 47*(3), 209−216. Available from http://wfs.swst.org/index.php/wfs/article/view/2329/2250.

Subasri, R., Reethika, G., & Soma Raju, K. R. C. (2013). Multifunctional sol-gel coatings for protection of wood. *Wood Material Science and Engineering, 8*(4), 226−233. Available from https://doi.org/10.1080/17480272.2013.834967.

Teng, T.-J., Mat Arip, M. N., Sudesh, K., Nemoikina, A., Jalaludin, J., Ng, E.-P., & Lee, H.-L. (2018). Conventional technology and nanotechnology in wood preservation: A review. *BioResources, 13*(4), 9220−9252. Available from https://doi.org/10.15376/biores.13.4.teng.

Terzi, E., Kartal, S. N., White, R. H., Shinoda, K., & Imamura, Y. (2011). Fire performance and decay resistance of solid wood and plywood treated with quaternary ammonia compounds and common fire retardants. *European Journal of Wood and Wood Products, 69*(1), 41−51. Available from https://doi.org/10.1007/s00107-009-0395-0.

Tsvetkova, I. N., Krasil'nikova, L. N., Khoroshavina, Y. V., Galushko, A. S., Frantsuzova Yu, V., Kychkin, A. K., & Shilova, O. A. (2019). Sol-gel preparation of protective and decorative coatings on wood. *Journal of Sol-Gel Science and Technology, 92*(2), 474−483. Available from https://doi.org/10.1007/s10971-019-04996-3.

Tsvetkova, I. N., Shilova, O. A., Shilov, V. V., Shaulov, A. Y., Gomza, Y. P., & Khashkovskii, S. V. (2006). Sol-gel synthesis and investigation of hybrid organic-inorganic borosilicate nanocomposites. *Glass Physics and Chemistry, 32*, 218−227. Available from https://doi.org/10.1134/s1087659606020155.

Unger, B., Bücker, M., Reinsch, S., & Hübert, T. (2013). Chemical aspects of wood modification by sol-gel-derived silica. *Wood Science and Technology, 47*(1), 83−104. Available from https://doi.org/10.1007/s00226-012-0486-7.

Usmani, S. M., Stephan, I., Hübert, T., & Kemnitz, E. (2018). Nano metal fluorides for wood protection against fungi. *ACS Applied Nano Materials, 1*(4), 1444−1449. Available from https://doi.org/10.1021/acsanm.8b00144.

Varfolomeev, Ju. A., Shchegolev, A. T., Kurbatova, N. A., Poromova, T. M., Klobukova, N. N., Smirnov, S. F., ... Lezhenin, V. V. (1991). Preparation for wood protection against injury by wood-staining and mold fungi. https://patents.s3.yandex.net/RU2048288C1_19951120.pdf

Varlamov, V. P., Il'ina, A. V., Shagdarova, B. T., Lunkov, A. P., & Mysyakina, I. S. (2020). Chitin/chitosan and its derivatives: Fundamental problems and practical approaches. *Biochemistry (Moscow)*, *85*, 154–176. Available from https://doi.org/10.1134/S0006297920140084.

Varlamov, V. P., & Mysyakina, I. S. (2018). Chitosan in biology, microbiology, medicine, and agriculture. *Microbiology (Reading, England)*, *87*(5), 712–715. Available from https://doi.org/10.1134/s0026261718050168.

Verevkin, A. N., Kononov, G. N., Serdyukova, J. V., & Zaytsev, V. D. (2019). Biodegradation of wood by wood-destroying fungi enzyme complexes. *Lesnoi Vestnik/Forest Bulletin*, *23*(5), 95–100. Available from https://doi.org/10.18698/2542-1468-2019-5-95-100.

Voronkov, G. I., Zelchan., & Lukevits. (1975). In K. Ruhlmann (Ed.), *Silizium und Leben. Biochemie, Toxikologie und Pharmakologie der Verbindungen des Siliziums*. Akademie Verlag. Available from https://doi.org/10.1002/food.19760200620.

Wang, C., Piao, C., & Lucas, C. (2011). Synthesis and characterization of superhydrophobic wood surfaces. *Journal of Applied Polymer Science*, *119*(3), 1667–1672. Available from https://doi.org/10.1002/app.32844.

Worrall, J. J., & Wang, C. J. K. (1991). Importance and mobilization of nutrients in soft rot of wood. *Canadian Journal of Microbiology*, *37*(11), 864–868. Available from https://doi.org/10.1139/m91-148.

Yalinkilic, M. K., Tsunoda, K., Takahashi, M., Gezer, E. D., Dwianto, W., & Nemoto, H. (1998). Enhancement of biological and physical properties of wood by boric acid-vinyl monomer combination treatment. *Holzforschung*, *52*(6), 667–672. Available from https://doi.org/10.1515/hfsg.1998.52.6.667.

Zhang, W. (2015). Leachability of boron from trimethyl borate (TMB)/poplar wood composites prepared by sol-gel process. *Wood Research*, *60*(3), 471–476. Available from http://www.centrumdp.sk/wr/201503/13.pdf.

14

Biodegradable of plastic industrial waste material

A. Jawahar Nisha[1], Sugumari Vallinayagam[2], Karthikeyan Rajendran[2]

[1]SELVAM COLLEGE OF ENGINEERING AND TECHNOLOGY, SALEM, INDIA [2]DEPARTMENT OF BIOTECHNOLOGY, MEPCO SCHLENK ENGINEERING COLLEGE, SIVAKASI, INDIA

14.1 Introduction

14.1.1 Biodegradation

Biodegradation is the natural process to break down chemical chains of materials by using microorganisms such as bacteria, fungi, and so forth that are naturally growing, biodegrading substances in the environment (Alexander, 1999).

The term "composting" is used to describe the biodegradation of packaging materials. Four criteria are offered by the European Union for composting.

1. *Chemical composition*: Usage of heavy materials and easily volatile materials should be limited.
2. *Biodegradability*: More than 90% of materials converted into organic substances such as CO_2, water, and minerals, which is used for plant growth and to improve soil strength (Tokiwa, Calabia, Ugwu, & Aiba, 2009).
3. *Disintegrability*: Original mass can be converted to particles that pass through a sieve 2 mm $\times$ 2 mm.
4. *Quality*: Substance that inference composition and toxic substances should be absent.

14.1.2 Plastic

Prehistoric people used wood and soil to make utensils and other equipment for their use. Later they used glass utensils (Brydson, 1999). In 1839 discovery of plastic was started. In 1907 Bakelite from Belgium was discovered—a plastic from vulcanized rubber and polystyrene. Huge production of plastic began in the 1940s (Brydson, 1999). Twenty types of plastics now are used from all over the world. Plastics are very useful in our day-to-day life. All types of industries, factories, and shops use plastics, even though small shops also use plastics. Today 99% of people use plastics in their daily life (Momani, 2009).

Plastics are polymers. Polymer is nothing but a combination of small chains (Bartman & Swift, 1983). After usage of plastic throughout the environment. Plastics are mostly nonbiodegradable, and they remain in the soil for several years. They will cause plants to stop growing. Plastics are poisonous to living things (Chanda, 2017). If we throw plastics in open surroundings, they will be eaten by many animals. Plastic enters an animal's digestive tract and causes some digestive problems, leading to death (Singh, 2005). Many food items are packed inside plastics and are supplied to many countries. After several days, the food becomes poisonous to our health (Lament, 1993).

14.1.3 Polymer production

Plastics are polymers produced from fossil fuels (Belboom & Leonard, 2016). Fossil fuels are dangerous in several stages. The formation of fossil fuels take many million years (Garcia-Garibay & Marshall, 1991). Plastic is produced from petroleum, and their products are heated under certain conditions (Belboom & Leonard, 2016). They become split in to several monomers having different characteristics. The monomers are combined by link each link of chain made up of carbon, hydrogen, oxygen, and so forth (Garcia-Garibay & Marshall, 1991). Different combinations of monomers produce plastic resin.

Resin can be developed based on its stability (Garcia-Garibay & Marshall, 1991). Resin of plastic is divided into two categories, based on their stability: thermosets and thermoplastics.

Thermosets: thermosets are polymers that do not regain their original state when exposed to heat. It is irreversible (e.g., primary parts of automobiles; Kim & Robertson, 1992).

Thermoplastic: thermoplastic is another type that can be easily shaped (e.g., cards, bottles, and so forth; Georgopoulos, Tarantili, Avgerinos, Andreopoulos, & Koukios, 2005).

14.1.4 Commonly used plastics (Hylton, 2004)

- Polyvinylchloride
- Polyethylene terephthalate
- High density polyethylene
- Low density polyethylene
- Polypropylene
- Polystyrene

14.1.5 Industries

Industry is a place where different processes take place to produce a finished product. In this world several types of industries are present, such as food industry, chemical industry, steel industry, cement industry, and so forth. Every industry produces a finished product (Shreve and Brink, 1977). It plays a major role in this world. Based on production, they are classified into four types.

- Primary
- Secondary

- Tertiary
- Quaternary

Secondary industries are further classified into two types: light and heavy

These four types of industries use plastic in their daily production. They use large amounts of plastic but they convert finished products in small quantities; the remaining will be released as waste (Fuglie et al., 2011).

14.1.6 Industrial plastic waste

The plastic wastes from industries are throughout the environment with proper disposal, but plastics are harmful to our environment. More than 1 million plastics come from industrial processes per month (Arena & Di Gregorio, 2014). From these plastics, only a few degrade with microorganisms present in soil, and the remaining amount of plastics are buried under the soil due to environmental changes. These plastics remain under the soil for several million years and spoil soil fertility (Scott, Czernik, Piskorz, & Radlein, 1990). They emit harmful rays, which react with chemicals. Industrial plastic waste materials such as rapping materials, waste pipes, packaging materials, and so forth (Aguado & Serrano, 2007) based on their degradation of plastic are classified into two types: (1) biodegradable plastic and (2) nonbiodegradable plastic (Shent, Pugh, & Forssberg, 1999).

14.1.7 Biodegradable plastic

Biodegradable plastic is generally made up of natural materials that occur in our environment. Plastic is dumped under the soil due to environmental processes after dumping some microorganisms in the soil to degrade plastic. Not all organisms are able to degrade plastics (Nonato, Mantelatto, & Rossell, 2001). Those organisms have plastic-degrading capacities, but only degrade plastic into biomass, which is useful to plants to utilize biomass and for growth (e.g., disposable tableware, biowastage bags, and carrier bags).

$$\text{Biodegradable plastic} \rightarrow CO_2 + O_2 + \text{biomass}$$

14.1.8 Nonbiodegradable plastics

Nonbiodegradable plastics are generally made up of fossil fuels such as petroleum. These types of plastic are also dumped due to natural processes but they remain stable under soil. Soil microorganisms are not able to degrade these types of plastics. They still stay under the soil for several years (e.g., paints, plastics, batteries, etc.; Steinbüchel, 2005).

$$\text{Nonbiodegradable plastic} \rightarrow \text{No degradation to remain}$$

14.1.9 Biodegradation of plastics

Biodegradation processes are carried out by microorganisms such as bacteria and fungi. They degrade natural and synthetic polymers under controlled conditions because different microorganisms are involved in the degradation process (Shah, Hasan, Hameed, & Ahmed, 2008). Each microorganism involves processes in different climatic conditions. Degrading processes also depend upon factors of polymers, such as polymer type, character, nature, stability, and so forth. Polymer characteristics such as mobility, tactility, crystallinity, molecular weight, and types of functional groups are attached (Bhardwaj, Gupta, & Tiwari, 2013).

Polymers are first converted to monomer, and after they enter cellular membranes of microorganisms. Generally, polymers are large chains that do not involve inside cellular membranes. The breakdown of polymers happen physically and biologically, and physical processes such as altering heating/cooling, freezing/thawing, wetting/drying, which causes mechanical damage and creates cracks in polymers. In the bacterial degradation process, some bacteria that are not able to degrade the plastics. At this time fungi are involved to degradation process. Fungi first breaks the polymer chain into smaller molecules such as monomers, dimers, timers, and oligomers. After breaking up small molecules involved in the process, abiotic hydrolysis also degrades plastic under environmental conditions. Higher molecular-weight polymers do not directly involve the degradation process. First they reduce their molecular weight by breaking linkage after they enter in the degradation process. Molecular weight increases while solubility decrease (Shimao, 2001). This causes unfavorable conditions for microorganisms that attack and do not enter inside the semipermeable membrane. These conditions of cellular enzymes are involved in degrading polymers. Two categories of enzymes are involved: (1) extracellular and (2) intracellular depolymerase. Extracellular depolymerase break down large chains into small chains. These small chains enter inside the semipermeable membranes and utilize carbon end-product produce CO_2, H_2O, and some minerals. Some dominant groups are degradable activity determined by the environment based on their character.

14.1.10 Types of biodegradation (Ono et al., 1995)

Three types:

1. Biodeterioration
2. Biofragmentation
3. Assimilation

14.1.10.1 Biodeteriotion
Biodeterioration or surface-level degradation is one of the processes of biodegradation. It alters the mechanical, chemical, and physical properties of the substance when the substance is exposed to abiotic factors in the outer environment. It weakens the structure and makes it biodegradable. Some abiotic factors, such as temperature and pressure affect biodeterioration. It is similar to biofragmentation (e.g., corrosion of metals by microorganisms; Allsopp, Seal, & Gaylarde, 2004).

14.1.10.2 Biofragmentation
Biofragmentation of polymer is a lytic process which breaks down polymers into oligomers and monomers. Breakdown of polymers occur with or without the presence of oxygen

(aerobic and anaerobic digestion). In anaerobic digestion volume and mass of the plastic materials become reduce. Natural gas will produce during this process. The result of biofragmentation is assimilation (Coma, Copinet, Couturier, & Prudhomme, 1995).

14.1.10.3 Assimilation

The assimilation stage is when some of the products enter easily within the cell by the membrane carrier. Some other products that do not enter directly involve biotransformation produce product after they enter inside the cell. After entering, cell products enter catabolic pathways to produce adenosine triphosphate (Alba & Nee, 1997).

Aerobic biodegradation formula (Clemente, MacKinnon, & Fedorak, 2004):

$$C_{polymer} + O_2 \rightarrow C_{residue} + C_{biomass} + CO_2 + H_2O$$

Anaerobic biodegradation formula

$$C_{polymer} + O_2 \rightarrow C_{residue} + C_{biomass} + CO_2 + CH_4 + H_2$$

14.1.11 Biodegradable plastics

Approximately 160 million tons of plastics come from industries worldwide. These plastics are more stable and they do not enter readily into the degradation process. But with the help of microorganisms, it can be possible to degrade plastic. The major four types of plastic biodegradation methods can be determined with the help of microorganisms (Lenz & Marchessault, 2005).

1. Biodegradation of synthetic plastic
2. Biodegradation of natural plastic
3. Biodegradation of polymer blends
4. Other degradable polymer

14.1.12 Biodegradation of synthetic plastic

Degradation of plastics in the natural process is very slow by using microorganisms. Generally, degradation of plastics, initially oxidation or hydrolysis, takes place to reduce their molecular weight by using enzymes. It causes high-molecular weight into low-molecular weight compounds (Alshehrei, 2017).

- Polylactic acid
- Polycaprolactone
- Polyester
- Polyvinyl alcohol
- Polyurethane
- Nylon

Physical properties: crystallinity, orientation

Morphological property: surface area

Synthetic plastics, their physical properties, and morphological properties affect their degradation rate.

14.1.13 Polyester

Polyester is synthetic fiber derived from coal, air, waste, and petroleum. Polyester is a fiber from chemical reaction, which occurs between acid and alcohol (Shimao, 2001). Polyesters are polymers that are monomer with bonding of ester linkages. Several types of polymers are present in nature that can be degrading by several enzymes such as esterase which are ubiquitous living organism (Shimao, 2001). Generally ester bonds can be easily hydrolyzed and several synthetic esters are used as biodegradable plastic in industries. Since large, semicommercial industries use polyester in the place of plastic. In industries 18% of plastics are made up of polyester (33.5%; Rehm, 2003).

14.2 Chemical formation of polyester (Apicella, Migliaresi, Nicolais, Iaccarino, & Roccotelli, 1983)

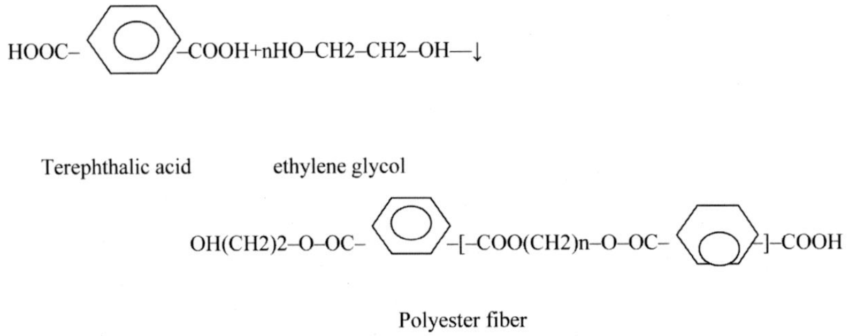

HOOC–⟨◯⟩–COOH+nHO–CH2–CH2–OH—↓

Terephthalic acid   ethylene glycol

OH(CH2)2–O–OC–⟨◯⟩–[–COO(CH2)n–O–OC–⟨◯⟩–]–COOH

Polyester fiber

14.3 Structure of polyester

$$R - OH + R - COOH \rightarrow Ester$$

(R is any hydrocarbon group)

14.3.1 Polycaprolactone

Polycaprolactone (PCL) is a synthetic polymer that degrades from natural microorganisms. Polyhydroxybutyrate (PHB)-degrading bacteria are naturally present in environments; likewise, PCL-degrading bacteria also naturally occurs in this environment. PCL is degrading by lipase and esterase. One type of PCL-degrading bacteria includes lipase, which degrades glycerides. PHB depolymerase does not hydrolyze PCL. Another researcher discovered some fungi phytopathogens have plastic-degrading abilities. These degrade and act as PCL depolymerase (Labet & Thielemans, 2009), for example, Fusarium.

The *Fusarium solani* strain naturally lack PCL depolymerase activity. Wild-type strains induce cutin and PCL hydrocylate, which induce cutinase activity in *F. solani*. The chemical structures of PCL trimer is similar in two monomers that induce cutinase activity. This becomes PCL depolymerase.

Chemical structure of polycaprolactone:

$$-[-O-COO-(CH_2)_5-]n$$

Chemical formula:

$$[C_6H_{10}O_2]n$$

Polylactic acid:

Many companies use polylactic acid (PLA) as a biodegradable plastic at a semicommercial scale. PLA is taken from animals and plants and used in polymers for medical purposes. Several enzymes are degrading PLA in animals and plants, such as proteinase K, and so forth. Only a few organisms are available to degrade PLA, but they are not widespread in the natural environment.

Case study:

Pranamuda et al.(1999) tested 45 soil samples and detected only one sample containing PCL-degrading microorganisms (Pranamuda & Tokiwa, 1999). The microorganism can be plated in an agar-contained PLA powder which form a clear zone in the agar plate (Shimao, 2001). This ratio is much lower than PCL degradation. A PLA-degrading actinomycete, an *Amycolatotsis* sp. strain isolated from the sample, reduced 100 mg of PLA film by 60% after 14 days in liquid culture at 30°C.

Chemical structure of polylactic acid:

$$-[-CHCH_3 - COO - O-]n$$

Chemical formula:

$$[C_3H_4O_2]n$$

14.3.2 Polyurethane

Polyurethane (PUR) is produced by diisocyanate polyaddition process. In many processes PUR link with polyalkylene ether and polyester sequence with molecular mass 200 and 6000. This process is incomplete, and their growth is not supported by PUR. With the addition of carbon sources and nutrients, most of the degradation is carried by esterase. TB-35 strain does not utilize polyester PUR but it does utilize polyester PUR, so it is suitable for degradation (Akutsu, Nakajima-Kambe,

Nomura, & Nakahara, 1998). The PUR-degrading enzyme can be isolated by *Comamonas acidovorans*. It converts the PUR chain into diethylene glycol and adipic acid. The degradation of PUR is inhibiting the presence of detergents. It does not inhibit water-ester solubility.

Chemical structure of polyurethane:

$$R - N = C = O + H_2O \rightarrow R1 - NH - COO - O - H \rightarrow R - NH_2 + CO_2(g)$$

$$R - N = C = O + R - NH_2 \rightarrow -R - NH - COO - NH - R-$$

Chemical formula:

$$C_{27}H_{36}N_2O_{10}$$

14.4 Nylon and polyethylene

A high-molecular nylon 66 membrane was found to be degraded by lignin-degrading white rot fungi under ligninolytic conditions. This enzyme was purified from a culture supernatant of white rot fungi similar to manganese peroxidase (Kudva, Keskkula, & Paul, 1999). In some cases high-molecular polyethylene degrade from lignin-degradation under limited carbon and nitrogen conditions. Manganese peroxidase are purified by strains such as phanerocheate chrysosporium.

Types of nylon:

- Nylon 66
- Nylon 6/6-6
- Nylon 6/9
- Nylon 6/10
- Nylon 6/12
- Nylon 11
- Nylon 12
- Nylon 6

Chemical structure of nylon:

$$-NH - (CH_2)_5 - COO - NH - (CH_2)_5 - COO - NH - (CH_2)_5 - COO - NH - (CH_2)_5 - COO$$

14.5 Biodegradation of natural plastic

14.5.1 Polyhydroxyalkanoates

Polyhydroxybutyrate (PHB) is a naturally occurring polyester that accumulates bacterial cells into carbon and energy-storing compounds (Mergaert, Anderson, Wouters, Swings, & Kersters, 1992). PHB contains polymers of polyhydroxyalkanoate (PHA) such as 3-hydroxyvalerate. It is used to manufacture biodegradable plastics, which are mostly used in large-scale industries. PHA and PHB are metabolized by several microorganisms and produce no PHA depolymerase, and their genes have been isolated (Mergaert et al., 1992). PHA is serenine hydrolases. Their protein sequences have four regions: signal sequence, substrate-binding domain, and catalytic domain with lipase box, catalyst, and substrate-linking domain. PHA depolymerase contain one substrate-binding domain and PHB contains two substrate-binding domains. It contains a high-level enzyme specificity and absorption of enzymes. In terms of substrate specificity, PHB degrade R-chain and cyclic R oligomers. The enzyme is an endoesterase that determine the spin of chain and its active site. Binding site contains four subsites. Three of the sites occupied HB site to cleavage occur, the center two occupy (R) HB unit. Other terminal sites occupy the (S) HB unit. All four subsites occupied must be at a maximum for cleavage to take place. They can be either thermoplastic or elastomeric, and their melting points are from 40°C to 180°C (Mergaert et al., 1992). PHA copolymer is called poly-3-butyrate-co-hydroxyvaleate is less stiff and tough, thus used in packaging materials (Alshehrei, 2017).

14.6 Enzymatic method to degrade polyhydroxyalkanoate

Generally, enzymatic methods include the involvement of two enzyme intracellular and extracellular enzymes. Basically, PHA is very large in nature. It does not easily enter the cellular membrane of microorganisms. Extracellular enzymes are produced in various microorganisms grown and purified. PHB depolymerase enzymes are produced by various microorganisms such as alcaligens, comamonas, pseudomonas, and so forth (Kanmani, Kumaresan, Aravind, Karthikeyan, & Balan, 2016). It contains lipase box peptidase which contains Gly-x-ser-x-Gly hydrolases. The primary structure of PHB depolymerase microorganisms contains three regions.

1. Subdomain
2. Catalytic domain
3. Linker region

14.6.1 Subdomain

Subdomains generally bind with two solid PHB plastics.

14.6.2 Catalytic domain

Catalytic domains are composed of catalytic trials.

14.6.3 Linker region

Linker regions connect two domains and play a major role in connecting.
Biochemical property:

- Shares below 100 KDA
- PHA shares between 50 KDA
- Highly stable under limited pH, temperature, and ionic stability
- Some bonds are covalently active in active site regions

Some organism in PHA and their pH

- *Pseudomonas pickett*: 7.5–9.8
- *Penicillium funiculosum*: 5.5–7
- PHBV: 7

Seven *PHA depolymerase of extracellular enzymes are generally the same but differ slightly in their character* (Yamada, Mukai, & Doi, 1993).

- PHB depolymerase A
- PHB depolymerase B
- PHB depolymerase D
- P (3HB-CO-3HV)
- PHB depolymerase C
- Poly (D-3-hydroxyvalverate depolymerase)
- PHBV

Chemical structure of polyhydroxyalkanoate:

$$PH3B : -[-O - CHCH_3 - CH_2COO-]n$$
$$PHBV : -[-O - CHCH_3 - CH_2COO - O - CHCH_2CH_3 - CH_2COO-]n$$
$$PHV : -[-O - CHCH_2CH_3 - CH_2COO-]n$$

14.7 Biodegradation of blended plastic

Initially the rate of degradation controlled by microorganisms readily degrade. Its interfaces with structural ability, surface area, and enzyme attacks in less surface area. It results in plastic exposure to degrade by thedegenerative enzymes from the microorganisms. Several biodegradation processes of starch-biobased polymers are available, but we generally use four blended polymers (Raghavan, 1995):

1. Thermoplastic starch product
2. Starch synthetic aliphatic polyester blends
3. Starch Polybutylene succinate (PBS)/ Poly butylene succinate-co-butylene adipate (PBSA) polyester blends
4. Starch PVOH blends

14.8 Other degrading polymers

Some other biodegrading polymers are also available, but research about their degradation and mechanisms are not proven. Only a few degrading polymers are proven (Rajagopal, Srinivasa, & Wineman, 2007). They are

- Ethylene vinyl alcohol
- Photobiodegradable plastics

- Biodegradation of thermoset plastics

14.8.1 Ethylene vinyl alcohol

Ethylene vinyl alcohol is water-soluble plastic. It has a high cost to degrade, so it is not used highly, except in a few industries that use these plastics for manufacturing. It is used in oxygen barriers and in multifilm layer barriers.

14.8.2 Photobiodegradation plastic

Photobiodegradation plastic is light sensitive. It easily breaks down their bonds when passing through light. It will design based on their light. When light enters, bonds contained in plastic become weaker and brittle. Some time photosensitizers are used. Photosensitizers include diketones, ferrocene derivative, and carbonyl sensitizer. In this type, there are two stages (Scott, 1990).

First stage:

At the first stage UV light passes through molecules and breaks down into small molecules. In the microbial degradation process, they convert plastic into carbon dioxide and water. This process depends upon exposure of light intensity, season, geography, climatic condition, and so forth.

Second stage:

At the second stage, plastics have low molecular weight, and light enters, ready to degrade. Because there is a small molecular weight materilas in the sample, so degradation will happen due to physical stress. It directly coverts plastic into carbon dioxide and water. There is no intermediate step to degrade plastic.

14.9 Biodegradation of thermo set plastics

This plastic degradation depends upon temperature. Some plastic-degrading microorganisms degrade plastic at a high temperature, and some others degrade microorganisms at normal and low temperatures. These types of plastics are already see on biodegradation in synthetic plastic (e.g., PUR and PCL).

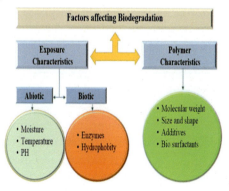

14.10 Standard testing methods

14.10.1 Experimental observation

Each and every observation change will take place. Effects include surface roughening, holes or crack formation, and color changes, which will take place while degradation of microorganisms occurs. But this result is not proven that biodegradation takes place. Only we see the changes take place in scanning electron microscope and atomic force microscope. After degradation crystalline appears on the surface, along with amorphous polymer fraction and etching from slower degradation. Another method uses an atomic force microscopy to determine their degradation activity. Many other methods are also available to determine biodegradation changes (Schartel, Wachtendorf, & Hennecke, 1999). They include Fourier transform infrared spectroscopy, differential scanning colorimetry, nuclear magnetic resonance, x-ray photoelectron spectroscopy, and x-ray diffraction. These are the other techniques used to determine biodegradation changes happen and will be seen clearly.

14.10.2 Carbon dioxide evolution and oxygen consumption

During degradation, microorganisms involved in this process utilize oxygen present in the air. Microorganisms contain carbon sour to grow. Utilized oxygen and present carbon react and give out carbon dioxide gas (Turner & Carlile, 1983). At this time, measure the amount of oxygen utilized by microorganisms and measure amount of carbon dioxide eliminated from the degradation process is measured. Using measurement of O_2 consumption and CO_2 evolution proven degradation takes place.

14.10.3 Forming clear zone

It is an easy test to determine if biodegradation happens or not. Microorganisms, which degrade plastic, can be isolated from their environment. Prepare an agar plate containing plastic, which are finely mixed (Feijó, Sainhas, Hackett, Kunkel, & Hepler, 1999). The microorganisms are introduced in the agar plate. Microorganisms degrade the finely mixed plastic present in the agar plate, which forms a clear zone around microorganisms present in the plate. This shows that particular organisms degraded plastics is proven.

14.10.4 Radiolabeling

CO_2 and C_{14} measurement is simple, nondestructive, and measurements are accurate. Approximate C_{14} testing materials are used in radiolabeling. Microorganisms can be isolated and grown in artificially formed environments (Eichenberger, Patra, & Holland, 2019). When the radiolabeling C_{14} passes through this environment, some amount of carbon dioxide evolution can take place. Using scintillation, countermeasure the amount of carbon dioxide evolved. K value can be determined by using a formula and measurement taken by altering time. Obtained values will be plotted in the following graph.

Graph:[x- axis contain K value]
[y- axis contain time]

14.10.5 Control composting test

In this test, plastics are buried under the ground. It contains both biodegrading and nonbiodegrading plastic. Thermostatic organisms introduce inside under high temperature (58°C). Water content is 50% of their total weight for the recommended time period. After completion of time period microorganisms (Weng, Wang, Wang, & Wang, 2010), which we introduce, decompose the plastic present inside. Finally measure the amount of carbon dioxide evolved from the mixture of polymer minus the amount of carbon dioxide unamended in the mixture of polymer. The difference is measured, and proven biodegradation takes place.

Flow chart:

Test material

↓

ASTM D 6400

↓

Biodegradation

↓

Controlled laboratory scale composting test

ASTM D 5338

↓

Introduce required microorganisms for degrading

Process

↓

>60 % OR 90% carbon conversion into carbon dioxide

↓

Formation of carbon dioxide more than 60%

↓

Check their stability and toxicity

↓

identify as biodegradable plastic materials

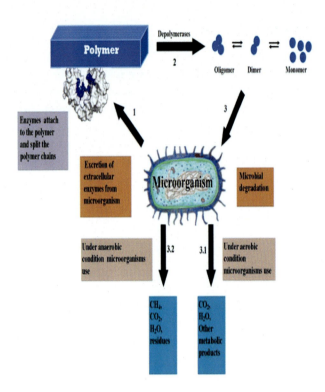

14.11 Conclusion

In this chapter, we explore how industrial plastic waste and microorganisms are involved in the mechanisms to degrade plastics under their environmental conditions, including industrial plastic waste and their discharging mechanisms, as well as types of plastic waste, which microorganisms are able to degrade. This increases our knowledge about the mechanisms that detect microorganisms to degrade plastics in biological processes.

References

Aguado, J., & Serrano, D. P. (2007). *Feedstock recycling of plastic wastes*. Royal society of chemistry.

Akutsu, Y., Nakajima-Kambe, T., Nomura, N., & Nakahara, T. (1998). Purification and properties of a polyester polyurethane-degrading enzyme from Comamonas acidovorans TB-35. *Applied and Environmental Microbiology, 64*(1), 62–67.

Alba, R., & Nee, V. (1997). Rethinking assimilation theory for a new era of immigration. *International Migration Review, 31*(4), 826–874.

Alexander, M. (1999). *Biodegradation and bioremediation*. Gulf Professional Publishing.

Allsopp, D., Seal, K. J., & Gaylarde, C. C. (2004). *Introduction to biodeterioration*. Cambridge University Press.

Alshehrei, F. (2017). Biodegradation of synthetic and natural plastic by microorganisms. *Journal of Applied & Environmental Microbiology, 5*(1), 8−19.

Apicella, A., Migliaresi, C., Nicolais, L., Iaccarino, L., & Roccotelli, S. (1983). The water ageing of unsaturated polyester-based composites: Influence of resin chemical structure. *Composites, 14*(4), 387−392.

Arena, U., & Di Gregorio, F. (2014). Energy generation by air gasification of two industrial plastic wastes in a pilot scale fluidized bed reactor. *Energy, 68,* 735−743.

Bartman, B., & Swift, G. (1983). Acetoacetate functionalized polymers and monomers useful for crosslinking formulations. In: Google Patents.

Belboom, S., & Leonard, A. (2016). Does biobased polymer achieve better environmental impacts than fossil polymer? Comparison of fossil HDPE and biobased HDPE produced from sugar beet and wheat. *Biomass and Bioenergy, 85,* 159−167.

Bhardwaj, H., Gupta, R., & Tiwari, A. (2013). Communities of microbial enzymes associated with biodegradation of plastics. *Journal of Polymers and the Environment, 21*(2), 575−579.

Brydson, J. A. (1999). *Plastics materials.* Elsevier.

Chanda, M. (2017). *Plastics technology handbook.* CRC Press.

Clemente, J. S., MacKinnon, M. D., & Fedorak, P. M. (2004). Aerobic biodegradation of two commercial naphthenic acids preparations. *Environmental Science & Technology, 38*(4), 1009−1016.

Coma, V., Copinet, A., Couturier, Y., & Prudhomme, J. C. (1995). Biofragmentation of acetylated starch by the α-amylase of *Aspergillus oryzae. Starch-Stärke, 47*(3), 100−107.

Eichenberger, L. S., Patra, M., & Holland, J. P. (2019). Photoactive chelates for radiolabelling proteins. *Chemical Communications, 55*(16), 2257−2260.

Feijó, J., Sainhas, J., Hackett, G., Kunkel, J., & Hepler, P. (1999). Growing pollen tubes possess a constitutive alkaline band in the clear zone and a growth-dependent acidic tip. *The Journal of Cell Biology, 144*(3), 483−496.

Fuglie, K., Heisey, P., King, J. L., Day-Rubenstein, K., Schimmelpfennig, D., Wang, S. L., ... Karmarkar-Deshmukh, R. (2011). Research investments and market structure in the food processing, agricultural input, and biofuel industries worldwide. *USDA-ERS Economic Research Report* (130).

Garcia-Garibay, M., & Marshall, V. (1991). Polymer production by Lactobacillus delbrueckii ssp. bulgaricus. *Journal of Applied Bacteriology, 70*(4), 325−328.

Georgopoulos, S. T., Tarantili, P., Avgerinos, E., Andreopoulos, A., & Koukios, E. (2005). Thermoplastic polymers reinforced with fibrous agricultural residues. *Polymer Degradation and Stability, 90*(2), 303−312.

Hylton, D. C. (2004). *Understanding plastics testing* (Vol. 27). Hanser Publishers.

Kanmani, P., Kumaresan, K., Aravind, J., Karthikeyan, S., & Balan, R. (2016). Enzymatic degradation of polyhydroxyalkanoate using lipase from *Bacillus subtilis. International journal of Environmental Science and Technology, 13*(6), 1541−1552.

Kim, J. K., & Robertson, R. E. (1992). Toughening of thermoset polymers by rigid crystalline particles. *Journal of Materials Science, 27*(1), 161−174.

Kudva, R., Keskkula, H., & Paul, D. (1999). Morphology and mechanical properties of compatibilized nylon 6/polyethylene blends. *Polymer, 40*(22), 6003−6021.

Labet, M., & Thielemans, W. (2009). Synthesis of polycaprolactone: A review. *Chemical Society Reviews, 38*(12), 3484−3504.

Lament, W. J. (1993). Plastic mulches for the production of vegetable crops. *Horttechnology, 3*(1), 35−39.

Lenz, R. W., & Marchessault, R. H. (2005). Bacterial polyesters: Biosynthesis, biodegradable plastics and biotechnology. *Biomacromolecules, 6*(1), 1−8.

Mergaert, J., Anderson, C., Wouters, A., Swings, J., & Kersters, K. (1992). Biodegradation of polyhydroxyalkanoates. *FEMS Microbiology Reviews*, *9*(2–4), 317–321. Available from https://doi.org/10.1111/j.1574-6968.1992.tb05853.x.

Momani, B. (2009). Assessment of the Impacts of Bioplastics. Energy Usage (Doctoral dissertation, Worcester Polytechnic Institute.

Nonato, R., Mantelatto, P., & Rossell, C. (2001). Integrated production of biodegradable plastic, sugar and ethanol. *Applied Microbiology and Biotechnology*, *57*(1–2), 1–5.

Ono, D., Yamamura, S., Nakamura, M., Takeda, T., Masuyama, A., & Nakatsuji, Y. (1995). Biodegradation of different carboxylate types of cleavable surfactants bearing a 1, 3-dioxolane ring. *Journal of the American Oil Chemists' Society*, *72*(7), 853–856.

Pranamuda, H., & Tokiwa, Y. (1999). Degradation of poly (L-lactide) by strains belonging to genus Amycolatopsis. *Biotechnology Letters*, *21*(10), 901–905.Chicago.

Raghavan, D. (1995). Characterization of biodegradable plastics. *Polymer-Plastics Technology and Engineering*, *34*(1), 41–63.

Rajagopal, K., Srinivasa, A., & Wineman, A. (2007). On the shear and bending of a degrading polymer beam. *International Journal of Plasticity*, *23*(9), 1618–1636.

Rehm, B. H. (2003). Polyester synthases: Natural catalysts for plastics. *Biochemical Journal*, *376*(1), 15–33.

Schartel, B., Wachtendorf, V., & Hennecke, M. (1999). Chemiluminescence: A promising new testing method for plastic optical fibers. *Journal of Lightwave Technology*, *17*(11), 2291.

Scott, D., Czernik, S., Piskorz, J., & Radlein, D. S. A. (1990). Fast pyrolysis of plastic wastes. *Energy & Fuels*, *4*(4), 407–411.

Scott, G. (1990). Photo-biodegradable plastics: Their role in the protection of the environment. *Polymer Degradation and Stability*, *29*(1), 135–154.

Shah, A. A., Hasan, F., Hameed, A., & Ahmed, S. (2008). Biological degradation of plastics: A comprehensive review. *Biotechnology Advances*, *26*(3), 246–265.

Shent, H., Pugh, R., & Forssberg, E. (1999). A review of plastics waste recycling and the flotation of plastics. *Resources, Conservation and Recycling*, *25*(2), 85–109.

Shimao, M. (2001). Biodegradation of plastics. *Current Opinion in Biotechnology*, *12*(3), 242–247.

Shreve, R. N., & Brink, J. A., Jr (1977). *Chemical process industries*. McGraw-Hill Book Co.

Singh, B. (2005). Harmful effect of plastic in animals. *The Indian Cow: The Scientific and Economic Journal*, *2*(6), 10–18.

Steinbüchel, A. (2005). Non-biodegradable biopolymers from renewable resources: Perspectives and impacts. *Current Opinion in Biotechnology*, *16*(6), 607–613.

Tokiwa, Y., Calabia, B. P., Ugwu, C. U., & Aiba, S. (2009). Biodegradability of plastics. *International Journal of Molecular Sciences*, *10*(9), 3722–3742.

Turner, C., & Carlile, W. (1983). Microbial activity in blocking composts. 1. Measurement of CO_2 evolution and O_2 consumption. In *Paper presented at the International Symposium on Substrates in Horticulture other than Soils* In Situ 150.

Weng, Y.-X., Wang, Y., Wang, X.-L., & Wang, Y.-Z. (2010). Biodegradation behavior of PHBV films in a pilot-scale composting condition. *Polymer Testing*, *29*(5), 579–587.

Yamada, K., Mukai, K., & Doi, Y. (1993). Enzymatic degradation of poly (hydroxyalkanoates) by Pseudomonas pickettii. *International Journal of Biological Macromolecules*, *15*(4), 215–220.

15

Microbiologically induced deterioration and protection of outdoor stone monuments

O.A. Shilova[1], D.Y. Vlasov[1,2], T.V. Khamova[1], M.S. Zelenskaya[2], O.V. Frank-Kamenetskaya[1,2]

[1]INSTITUTE OF SILICATE CHEMISTRY, RUSSIAN ACADEMY OF SCIENCES, SAINT PETERSBURG, RUSSIA [2]ST. PETERSBURG STATE UNIVERSITY, SAINT PETERSBURG, RUSSIA

15.1 Destruction of a stone under the influence of microorganisms

Microbiological damage to a natural or artificial stone is usually understood as a noticeable change in the properties of a material due to the effect of microorganisms on it in the process of their vital activity (a residual effect on the material by metabolic products of microbes after their death is possible as well). A significant number of studies have been devoted to the study of biological colonization of rocks and minerals in various environmental conditions. Most of them are associated with biological damage to cultural heritage sites made of natural stone and exhibited under the open air (Sterflinger et al., 2018; Toreno et al., 2018; Sazanova, Zelenskaya, Bobir, & Vlasov, 2020). In urban environments, carbonate rocks (limestones and marbles) are most susceptible to biological damage.

The data accumulated in the literature make it possible to identify the main types of microorganisms impact on stone material. Biophysical (mechanical) destruction occurs due to the growth of microorganisms directly in the material. The main precondition for the mechanical (biophysical) effect of microorganisms on a stony substrate is their ability to penetrate through the uneven stone surface relief with subsequent growth (Fig. 15–1). Some lithobiotic microorganisms are able to penetrate into stone to a considerable depth. This can lead to surface disintegration and loss of stone elements, and, accordingly, have a noticeable effect on the rock structure. Fouling of the stone surface can also lead to a change in porosity, water absorption, vapor permeability, dampness, surface temperature, etc.

Along with mechanical effects, microorganisms have an aggressive biochemical effect on the stone. It is due to the ability to use individual components of the material as sources of energy for growth and development. In addition, in the process of vital activity, microbes produce various substances that have a destructive effect on the material. Among them, there are organic and inorganic acids, pigments, polysaccharides, enzymes, and so forth (Sazanova, Shchiparev, & Vlasov, 2014; Sazanova et al., 2015). Their impact leads to biochemical weathering (intense corrosion) of carbonate rocks (marbles and limestones). Corrosion processes occurring on the surface of carbonate rocks under the action of microscopic fungi and bacteria lead to the crystallization of secondary minerals, oxalates of various metals (Rusakov et al., 2014), primarily calcium oxalates (monohydrate whewellite and dihydrous weddellite) (Fig. 15–2).

This phenomenon, called the induced biomineralization, can lead to the formation of an oxalate patina on the stone surface (Gadd et al., 2014). The intensive dissolution of calcite preceding the crystallization under the influence of aggressive metabolic products (primarily organic acids) can lead to the formation of microkarst (Fig. 15–3), which is very typical for rocks containing mineral components with different solubility (Frank-Kamenetskaya et al., 2019).

The intensity of biochemical weathering directly depends on the species composition of the lithobiotic community and on the number of microorganisms involved in it, as well as on external conditions. One of the main factors of the biochemical effect on the stone is the release of organic acids by microorganisms, which are capable of extracting Ca, Al, Si, Fe, Mn, and Mg cations from minerals, forming stable organometallic complexes, or chelates (Sazanova et al., 2015). It should be emphasized that organic acids have a selective effect on the metal ions contained in the rocky substrate. For example, propionic or lactic acids promote the mobilization of calcium and potassium

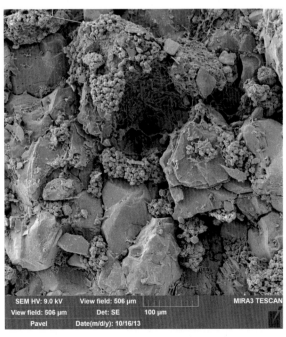

FIGURE 15–1 Scanning electron microscopy image of the weathered surface of inhomogeneous Ruskeala marble colonized by microorganisms.

FIGURE 15–2 Crystals of calcium oxalates, formed on the marble surface by microorganisms.

FIGURE 15–3 Formation of microkarst on Ruskeala marble.

ions, while they have a weak effect on the ions of aluminum, manganese, silicon, and iron. At the same time, oxalic, silicic, and citric acids are much more active with respect to aluminum ions.

Oxalic acid is the most dangerous for the stone. It has the broadest spectrum of action on polyvalent cations and initiates metasomatic crystallization processes on the surface of carbonate substrate.

One of the most dangerous forms of destruction of carbonate rocks is sulfatization, which leads to the formation of a gypsum-rich patina, or "gypsum crust" (Fig. 15–4). As a result of a chemical process, calcite is transformed into gypsum (Fig. 15–5).

As studies of gypsum crusts have shown, all stages of sulfatization of carbonate rock take place with the participation of microorganisms, primarily microscopic fungi and bacteria, which contribute to the accumulation of moisture and atmospheric pollution in the gypsum crust during its formation (Frank-Kamenetskaya et al., 2019).

FIGURE 15–4 Gypsum crust on white statuary marble.

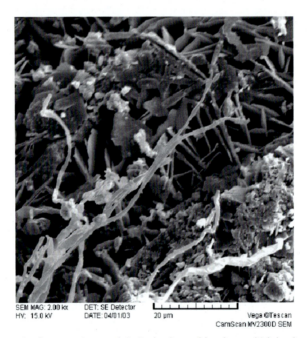

FIGURE 15–5 Lamellar crystals of gypsum in a crack in Ruskeala marble, along which hyphae of fungi grow.

Biofilm formation can be considered as a successful strategy for the existence and survival of microorganisms on a stony substrate in various environmental conditions, including the extreme ones. A biofilm is a structure formed by a complex of microorganisms immersed in an extracellular polymeric substance secreted by them, which functions as an organic matrix, such as integration, protection, and attachment to a substrate (adhesion) (Gulotta, Villa, Cappitelli, & Toniolo, 2018; Morton, 1996). It was found that the organic matrix is composed of polysaccharides, lipopolysaccharides, proteins, lipids, fatty acids, and enzymes (Sazanova (nee Barinova) et al., 2020). Due to the pigmentation of microorganisms (chlorophyll, carotenoids, melanins, etc.), a change in the color of the stone surface is observed during the formation of the primary biofilm (Grbić, Vukojević, Simić, Krizmanić, & Stupar, 2010). Gradually, microorganisms completely or partially overgrow the stone surface, changing its appearance. The accumulation of nutrients in biofilms allows microorganisms to occupy oligotrophic habitats. In general, the existence of microorganisms in the form of biofilms helps to survive in adverse conditions such as after drying out, at extreme temperatures, under osmotic stress, UV radiation, and toxic chemicals.

Biofilms formed by a complex of microorganisms can be divided by color into the following groups:

- dark-colored;
- with a predominance of green color;
- with a predominance of colors from yellow to brown; and
- bright orange, pink, and red.

The presence of the green hue in biofilms indicates the presence of photosynthetic pigments of algae and cyanobacteria in them. Yellow, orange, brown biofilms are characterized by the presence of carotenes, carotenoids, and products formed during the breakdown of chloroplasts and the destruction of chlorophyll (e.g., phycobilliproteins). In biofilms of bright orange, pink, and reddish hue, the presence of bacterial pigments or degradation products of iron-enriched algae, and cyanobacteria is noted. The dark color of biofilms is due to the presence of melanins and melanoids, products formed during the breakdown of chloroplasts, and the destruction of chlorophyll, as well as in the presence of various minerals (with iron or manganese). Dark-colored biofilms can contain all of the aforementioned components.

Phototrophic organisms—algae and cyanobacteria—are usually the first to settle on the surface of the stone; lichens can also appear. Following phototrophs, heterotrophs (fungi and organotrophic bacteria) appear, using the accumulated biomass of phototrophic organisms as their nutrients. As organic matter accumulates on the surface of the stony substrate, the proportion of heterotrophic organisms begins to prevail over phototrophic ones. In particular, the role of microscopic fungi, which can form the basis of biofilms on rocky substrates, increases significantly, especially in urban environments (De Leo & Urzì, 2015; Sazanova et al., 2020). They can develop by consuming the waste products of algae and bacteria, as well as substances deposited from the atmosphere.

The properties of the stone surface play an important role in its colonization by microorganisms. An uneven, rough surface is preferable for the adhesion of microorganisms (Prieto & Silva, 2005). In temperate climates, microorganisms settle mainly on the surface of the stone, while in tropical and subtropical climates, as well as in polar regions, microorganisms more often penetrate the rock mass to protect themselves from sunlight, temperature changes, and drying (Warscheid & Braams, 2000).

It should be emphasized that biofilms composed of different groups of microorganisms often exhibit a synergistic destructive effect on the underlying stone material, which can be significantly greater than that corresponding to individual groups of microorganisms. This fact must be taken into account when searching for effective means of protecting stone monuments from biological damage, including when testing protective compositions aimed at suppressing biodeterioration agents.

15.2 Methods to counter stone biodegradation

Biodegradation damages stone buildings and structures, deteriorates their appearance, and there may be a danger of collapse. Therefore a scientific approach to protecting stone buildings and structures from destruction, including biodegradation, presupposes scientifically grounded monitoring of the state of these objects, for which catalogs and databases are created in many countries (Doehne & Price, 2010; Price, 1998). The most famous organization in the world is the Getty Conservation Institute, Los Angeles, United States. They are actively involved in the problems of protecting buildings and structures in Europe, especially in Italy, Spain, France, as well as in South America—in Brazil and Argentina. Recently, these problems have been closely tackled in China, the work of Japanese scientists is also well known. Scientific centers are concentrated in the largest universities in the world. In St. Petersburg (Russia), monitoring of the state of outdoor monuments and their environment has been carried out for many years by scientists from St. Petersburg State University in collaboration with the Museum of Urban Sculpture, State Pedagogical University named after A.I. Herzen, and other scientific and industrial organizations of St. Petersburg. Students and postgraduates are actively involved in the work. A specialized database was created to store and structure the received material (Frank-Kamenetskaya et al., 2019). A special service at the State Museum of Urban Sculpture carries out constant monitoring of the state of the monuments (Vlasov, Zelenskaya, & Frank-Kamenetskaya, 2002).

An important step in protecting cultural heritage sites is the choice of methods for their preservation. Modern methods for the preservation of stone monuments from the effects of natural and anthropogenic factors can be reduced to three complementary areas, namely, (1) engineering-constructive methods of protection; (2) thermophysical methods; and (3) chemical methods of protection (Sizov, 2020).

The engineering-constructive methods of protecting cultural monuments include creating drainages, ensuring good waterproofing, and thermal insulation. Often, special protective structures are built, for example, protective pavilions of various designs. The modern original protective structure was erected over the temple of Apollo Epicurean, 5th century BC, located in the mountains of the western Peloponnese (Greece) at an altitude of 1130 m above sea level (Sizov, 2020). In some cases, the originals are replaced with copies. For example, statues of white Carrara marble in the Summer Garden in St. Petersburg have been replaced with copies.

Thermophysical methods are aimed at maintaining the optimal temperature and humidity of the air and are applicable mainly for stone structures, monuments, and products located indoors.

Chemical methods of stone protection include surface or deep cleaning, structural strengthening, and protective treatment (antiseptic treatment, and hydrophobization).

Stone cleaning includes technically and esthetically challenging independent tasks. Recently, controlled methods of cleaning that are neutral to stone are increasingly used along with the old washing of stone structures and monuments with water and chemical reagents, as well as the mechanical removal of biofouling. These include traditional steam blasting; an improved "sandblasting" method that uses particles of various sizes and hardness as abrasives, from corundum to nutshells, and allow regulating the energy of the cleaning "jet" (Sizov, 2020). In addition, methods of ultrasonic and laser cleaning of stone from biodeterioration agents are widely used.

Mechanical tools often damage the surface of the stone, and the use of chemicals often leads to discoloration. As a result, part of the author's surface may be lost and the esthetic properties of the monuments may deteriorate. In addition, the use of mechanical and chemical methods is undesirable from an environmental point of view, as it leads to environmental pollution.

An example of gentle cleaning is laser cleaning, which is becoming more and more popular in restoration and conservation work (Cooper, 1998; Verges-Belmin, 1997). Lasers have been successfully used, for example, in the restoration of such world-famous monuments as Notre Dame Cathedral (Paris, France) (Verges-Belmin, 1997) and the Parthenon Temple (Athens, Greece) (Pouli, Frantzikinaki, Papakonstantinou, Zafiropulos, & Fotakis, 2005). The experience of using laser technology to remove biological films from the surface of stone monuments is described in Leavengood, Twilley, and Asmus (2000) and Vlasov et al. (2019).

The physical principles of laser technology are described in detail by several researchers (Cooper, 1998; Luk'yanchuk, 2002; Parfenov, Gerashenko, Gerashenko, & Grigor'eva, 2010). In laser cleaning, particles of pollutants are removed from the surface of the processed object as a result of the absorption of high-intensity laser radiation. With the correct choice of the type and output parameters of the laser, the cleaning process can work selectively as the laser removes pollution only in the place on the surface of the monument where the beam of its radiation is directed. This favorably distinguishes laser cleaning from, for example, chemical treatment, in which not only the treatment area is exposed to the uncontrolled effect of a chemical reagent, but also the entire adjacent part of the object's surface (which most often does not need cleaning). Another advantage of laser cleaning is the ability to remove dirt from hard-to-reach areas of the surface of monuments, which is a serious problem when using traditional technologies. To clean the surface of stone monuments, pulsed Nd:YAG lasers operating at a wavelength of 1.06 mm are usually used (Cooper, 1998; Siano, Fabiani, Caruso, Pini, & Salimbeni, 2000; Siano et al., 2007). To reduce the risk of damage to monuments when using them, laser radiation with low energy of individual pulses and a high repetition rate is used. This allows for effectively removing surface contamination and at the same time excludes the possibility of damage to the stone substrate due to a significant reduction of heat absorbed by it (Rode et al., 2008).

The choice of processing methods for the object is determined taking into account the preliminary analysis of the material, the determination of the forms of destruction of the stone, as well as the identification of the main biodeterioration agents. Currently, in the practices of restoration and conservation works, the most widely used method of protecting monuments from biodegradation is the chemical method (Fidanza & Caneva, 2019; Silva et al., 2016; Tezel & Pavlostathis, 2015). This method of protection is mainly based on the use of biocides—chemicals that inhibit the growth of biodeterioration agents biodeterioration agents biodeterioration agents or completely destroy them on a specific material. A preliminary analysis of the composition of destructors on damaged materials or structures is a necessary step for effective conservation treatment of monuments using biocides. This stage of analysis is necessary to obtain the most effective result, since, in the case of revealing a wide range of biodeterioration agents, it seems rational to use several biocides or their mixture in succession to enhance the protective effect. The use of biocides leads not only to an increase in the biological resistance of the material, but also to a decrease in its moisture capacity. However, despite the high efficiency of biocides in protecting materials from biodeterioration, there are many objections to their widespread use. First, most biocides are highly toxic, not only to biodeterioration agents, but also to the environment, and can also hazardous to human health. In addition, the widespread use of biocides can provoke the emergence of resistance of microorganisms to the components of biocides or lead to the emergence of new forms of microorganisms with increased resistance to biocides. Such microorganisms can pose a serious threat not only as biodeterioration agents, but also as human conditional pathogens. Secondly, individual components or combinations of biocides can cause serious changes in the physicochemical characteristics of the material. Biocides are often added as fillers to paints, primers, electrolytes, and conversion fillers. However, the use of biocides can cause a change in color, structure of the material, and also accelerate the destruction processes. In the case of the most valuable cultural objects, this is one of the main limitations when using chemical methods of material protection. Thirdly, the short duration of the biocidal action leads to the fact that the treated surfaces can be re-colonized by both old and new, more aggressive microorganisms, occupying the vacated ecological niches. In the restoration, there were cases when biocidal compositions, sometime after their application, themselves became a source of nutrition for microorganisms (Fig. 15–6).

Taking these effects into account, it is worth selecting biocides more carefully and individually. Despite the existence of a wide range of biocides for destroying microorganisms, the effectiveness of the antimicrobial action depends on the correct choice of the protective composition, method, and processing conditions. The main requirements for biocidal formulations are a wide spectrum of action, duration of exposure, no negative effect on the material, a fairly high penetrating ability, and low toxicity. One of the important requirements for the newest biocidal preparations is their directed effect on the cellular structures of biodegradation and the processes taking place in them.

Since the end of the 20th century, nanotechnology has started being used for conservation using nanoparticles with biocidal and photocatalytic properties (Doehne & Price, 2010; Goffredo, Accoroni, & Totti, 2019; Ruffolo & La Russa, 2019). Nanopowders are used as an independent substance in the form of suspensions, or as additives for building solutions and

FIGURE 15–6 Micromycetes colonies on Carrara marble sculpture (Saint Petersburg) a year after treatment by organotin and quaternary ammonium biocides.

building masses. These additives include titanium dioxide and commercial photosenses offered by various companies (Crupi et al., 2018; Ruffolo & La Russa, 2019), as well as other substances with mild antibacterial action, such as detonation nanodiamond (DND) (Khamova et al., 2012; Khamova, Shilova, Kopitsa, Almásy, & Rosta, 2014; Schrand, Hens, & Shenderova, 2009).

In modern technologies for protecting stone surfaces, nanodispersed titanium dioxide (preferably in the crystalline modification of anatase), which is capable of generating reactive oxygen forms, is used quite successfully. However, the biocidal activity of nanodispersed titanium dioxide depends on photoactivation by UV radiation, since it is not a biocide in the dark (Goffredo et al., 2019; Seven et al., 2004). Its mechanisms of action are still being investigated. To increase the photo- and biocidal activity of titanium dioxide, many researchers recommend using additives of metal nanoparticles (e.g., Ag and Cu) (Arreche & Vázquez, 2020; Barberia-Roque, Obidi, Gámez-Espinosa, Viera, & Bellotti, 2019; Goffredo et al., 2019; Pinna, Salvadori, & Galeotti, 2012), metal oxides (e.g., CuO, ZnO, etc.) (Zarzuela, Moreno-Garrido, Blasco, Gil, & Mosquera, 2018), as well as doping TiO_2 in situ during the synthesis with Ag, Sr, and Fe (III) compounds (La Russa et al., 2014).

In addition to the ability of titanium dioxide under the influence of UV irradiation to release reactive oxygen species that inhibit the growth of biodeterioration agents, coatings based on titanium dioxide (usually in the form of anatase) after treatment with water or steam (e.g., after rains) acquire superhydrophilic properties and help to clean the surface from contamination (Aflori et al., 2013). At the same time, the size of the anatase particles does not have to be nanosized to achieve the self-cleaning effect, while micron-sized particles are not suitable for providing the antibacterial effect. Unfortunately, the superhydrophilic surface in the air quickly loses its properties as it becomes contaminated with organic matter. To restore superhydrophilicity, constant UV irradiation is necessary. Interestingly, the surface topography, which provides both superhydrophobic and superhydrophilic properties, also affects its biocidal activity (Kaminskii et al., 2019).

Nanodispersed titanium dioxide (usually anatase) is widely used for protective biologically active coatings. For example, the authors of La Russa et al. (2012), based on anatase nanopowder with a particle size of ~ 20 nm (0.3 wt.%) and an acrylic polymer (4 wt.%), prepared an aqueous suspension that was applied to marble and limestone. The resulting composition impregnated porous limestone well, but did not adhere well enough to the marble surface. In addition, it is known that there is a problem of degradation of the polymer matrix, which is relevant for all organic polymers (Aflori et al., 2013). To improve the adhesion of such coatings to form a matrix specially synthesized hybrid nanocomposites are used instead of an aqueous suspension of an acrylic polymer, for example, based on 3-(trimethoxysilyl) propyl methacrylate and titanium alkoxide (Allen et al., 2004).

Nanoparticles (e.g., SiO_2, ZnO, Al_2O_3, and TiO_2) are used to improve the mechanical properties of bioactive polymer coatings (Arreche & Vázquez, 2020; Sbardella, Bracciale, & Santarelli, 2020).

Nanoparticles are used in paint and varnish and sol-gel compositions to form coatings with a superhydrophobic surface (Shevchenko, Shilova, Kochina, Barinova, & Belyi, 2019; Xu, Li, Zhang, Zhang, & Xu, 2012). Many researchers consider coatings with a superhydrophobic surface to be very promising for protecting stone surfaces from biodegradation, since the surface structure ensures minimal contact with water and, possibly, with microorganisms. However, as the authors rightly note (Ruffolo & La Russa, 2019), the use of superhydrophobic coatings with a nanostructured surface in the field of preserving cultural heritage is still very limited. This is due to the lack of long-term tests, as well as the degradation of superhydrophobic properties under the influence of negative environmental factors.

Despite the advances in the use of nanoparticles to protect stone surfaces from biodegradation, the scientific community is still debating the issue of the safety of their use and their impact on the environment and human health.

As we move from the 1990s of the 20th century to modern times in the 21st century, more and more attention is paid to ecology when choosing methods for protecting the stone surfaces of buildings and structures. More and more attention is paid to environmentally friendly methods of protection. At the same time, in almost all works, the question is raised about the need for long-term observations in order to select reagents and methods that will not harm either the environment or the protected stone material.

The market for chemical reagents produced for cleaning, protecting, and consolidating stone surfaces of buildings and structures is expanding every year. These are hydrophobic, biocidal, and fungicidal impregnations, both based on organic solvents and on a water basis. Aqueous emulsions based on acrylic polymers, fluoropolymers, and alkoxysilanes, which are often preferred in hot climates over compositions based on volatile organic solvents (La Russa et al., 2012; Pinna et al., 2012). Researchers are attracted by the possibility to obtain synergistic effects due to the simultaneous use of consolidating reagents in mixtures based on alkoxysilanes and organic solvents and polymers (epoxy resins, acrylic polymers, etc.) and protective coatings (Cardiano et al., 2003; La Russa et al., 2012; Shilova et al., 2020).

The so-called "lime" technology is also known (Doehne & Price, 2010; Price, 1998). This is an innovative technology to combat the gypsum-enriched patina formed on the surface of natural and synthetic carbonate materials, in which not just a lime solution, but calcium hydroxide based on nanosized lime particles are used to carry out the replacement reaction for the gypsum formed on the surface of carbonate rocks.

Environmentally friendly efficient biotechnology conservation (Zhu & Dittrich, 2016) and bioremediation (Doehne & Price, 2010; Price, 1998) using microorganisms, primarily bacteria, are being actively developed. The studies of bacterial carbonate crystallization and the ability of bacteria or microbial communities for calcite recrystallization and cementation formed the basis for practical developments for the conservation of marble and limestone monuments ("healing" of stone) (Achal & Mukherjee, 2015; DeJong, Fritzges, & Nüsslein, 2006; Wang & Qian, 2014). To date, research in the applied field is carried out in two main directions: the search for new species of bacteria capable of calcite redepositing and the selection of conditions for this process (Zhu & Dittrich, 2016).

In biotechnologies utilizing *Bacillus subtilis* the time of the crystallization, as well as the thickness of carbonate crust can be controlled by variation of the amount of glucose in the medium (Sazanova (nee Barinova) et al., 2020). At the same time, it is necessary to take precautions to prevent contamination of the marble surface by a different airborne microbe. The regular routine care associated with the chemical cleaning of the surface of marble monuments can significantly reduce the number of microorganisms (first of all, microscopic fungi) on the rock surface (Vlasov et al., 2019). After such cleaning, it is possible to use "beneficial" bacteria to restore the surface of the monument, taking into account the specific parameters of the crystallization medium.

There were proposals to use microscopic fungi that produce oxalic acid to protect marble from environmental influences, under the influence of which poorly soluble calcium oxalates are formed on the marble surface (Rampazzi, 2019). Protective oxalate patina on the surface of carbonate rocks was also tried to be obtained chemically (without the participation of microorganisms) (Burgos-Cara, Ruiz-Agudo, & Rodriguez-Navarro, 2017).

Environmentally friendly biotechnology for the protection of marble and limestone monuments with the help of microorganisms undoubtedly belongs to the future. They are very effective for healing cracks and reducing porosity in marble and other carbonate materials. However, such technologies are still used very limitedly. Their widespread use is slowed not only by the lack of interest from the industry, but also by the significant dependence of the biological activity of the applied strains of microorganisms on the variability of environmental conditions.

15.3 Using sol-gel technology to protect marble from biodegradation

Sol-gel technology is one of the most promising technologies for making protective coatings. The sol-gel process is based on the reactions of hydrolysis and polycondensation of hydrolysis products. Hydrolyzable compounds (precursors) can be metal alkoxides (e.g., $Si(OEt)_4$ and $Ti(OPr^i)_4$), as well as metal salts (e.g., $TiCl_4$) and inorganic acids (e.g., H_3PO_4 and H_3BO_3). The resulting sols have film-forming properties, that is, when applied to stone surfaces, they can form thin nanosized and nanostructured coatings. This method is technologically simple and economical enough, and does not require complex and power-consuming equipment. It is possible to expand the functional properties of coatings and improve their performance characteristics by modifying the sols with inorganic and organic additives that have bactericidal, photocatalytic, and fungicidal properties.

Summarized subsequently are the results of our many years of experimental research on the development, study, and application of long-term self-cleaning environmentally friendly sol-gel-derived epoxysiloxane coatings with antimicrobial properties to protect marble monuments from biodegradation under the influence of mold fungi and unfavorable climatic factors (Frank-Kamenetskaya, Vlasov, & Shilova, 2011; Khamova et al., 2012; Shilova et al., 2020). The main idea is to avoid the use of additives of highly active toxic substances, such as biocides (e.g., organophosphorus compounds and quaternary ammonium compounds, tin and copper-containing reagents), which are widely used in practice for introduction into paint and varnish, organosilicon, and sol-gel compositions. The main disadvantages of their use include short durability, destruction of marble surface, and toxicity (after biocide treatment). Instead of highly toxic biocidal additives, we introduce biocide compositions into the sol-gel that have a minimal environmental impact, they do not destroy biodeterioration agents (bacteria, fungi), but only inhibit their development. Such biocides, which can be called "mild" biocides or even green biocides. They do not provoke the emergence of resistance in aggressive forms of biodeterioration agents (see, Fig. 15–6).

Various photosensitizing substances can be used as green biocidal additives; they are capable of producing ozone and other reactive oxygen species under the action of ultraviolet radiation. Such additives include titanium dioxide and commercial photosensitizers, offered by various companies (Crupi et al., 2018; Ruffolo & La Russa, 2019). Other substances that do not have photocatalytic properties but are characterized by a mild antimicrobial effect, for example, DND (detonation nanodimond), can be used as well (Khamova et al., 2012, 2014; Schrand et al., 2009).

For the fixation of biologically active substances on the stone surface, they are enclosed into film-forming organic-inorganic or sol-gel compositions based on tetraethoxisilane Si(OEt)$_4$ (TEOS), tetrabutoxytitanium Ti(OBu)$_4$, polydimethylsiloxane, acrylate hybrid nanocomposites, and so forth (Crupi et al., 2018; Sbardella et al., 2020; Shilova et al., 2020). However, acrylics and siloxanes have proven to be the most suitable for this purpose, although they still cannot meet all the requirements for an ideal protective coating for stone monuments.

Organic-inorganic epoxysiloxane and epoxy-titanate sols (Frank-Kamenetskaya et al., 2011; Khamova et al., 2012; Shilova et al., 2020) performed well among the sol-gel compositions studied by us (Fig. 15−7). When applied to the stone surface, they form a thin, transparent film that does not spoil the decorative properties of the stone, but has good adhesion to the surface, protecting it from adverse environmental influences. It is the organic-inorganic film-forming base that was chosen because, on one hand, the organic (epoxy) component ensures good adhesion of the coating to the surface even during cold solidification. This eliminates the need for the heat treatment of coatings. On the other hand, the inorganic, polysiloxane, or titanate component promotes faster curing of the coating and improves its hardness.

Previously, we studied coatings based on epoxy-titanate sols for protecting stone, both undoped and doped with titanium dioxide nanoparticles, as well as DND (Shilova et al., 2020). It was shown that the in situ formation of the crystalline anatase phase in epoxy-titanate coatings significantly increases the efficiency of inhibition of the growth and development of mold and yeast-like fungi (*Aspergillus niger*, *Cladosporium cladosporioides*, and *Hormonema dematioides*). The ordering of the mesostructure of epoxy-titanate coatings at the nanolevel, the formation of a fractal structure of nanocomposites also contributes to an increase in their antifungicidal activity (Shilova et al., 2020). The addition of titanium dioxide (commercial nanopowder P25, Degussa, 0.1−0.5 wt.%) to both epoxy-titanate and epoxy-siloxane sols enhance their fungicidal activity. However, during field tests in the atmosphere of a large metropolis for a long time (1 year or more), a whitish coating appears on the marble surface. Therefore we rejected the addition of titanium dioxide nanopowders to the sol-gel composition.

Below we will describe in more detail on epoxysiloxane coatings doped with DND (green biocide).

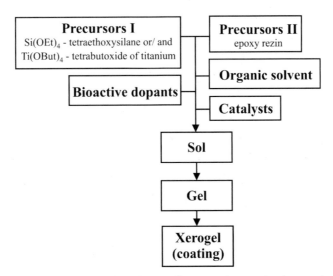

FIGURE 15–7 Scheme of preparation sol-gel-derived epoxysiloxane and epoxy titanium coatings.

Produced coatings satisfy the following important demands: low-temperature synthesis, high vapor penetration, transparency (invisible), and an absence of shine.

15.3.1 Materials and methods

15.3.1.1 Synthesis of epoxysiloxane sols

Two versions of sol-gel epoxysiloxane film-forming compositions doped by DND have been developed.

Technique No 1. Initial components: TEOS $Si(OEt)_4$, aliphatic epoxy resin DEG-1, uncured epoxy resin ED-20, 15% boron trifluoride BF_3 solution in diethylene glycol, distilled water, isopropyl alcohol.

To create the sol, a 50% TEOS solution in isopropyl alcohol and a mixture of ED-20 and DEG-1 resins in isopropyl alcohol were first prepared, after which the resulting components were mixed in a 1:1 ratio. Then water and BF_3 hardener in diethylene glycol were added to the resulting mixture which was intensively stirred. Then, a DND suspension in an amount of 0.2 wt.% was introduced into the formed system which was intensively mixed after that (Khamova et al., 2012).

Technique No 2. Initial components: TEOS, EPONEX 1510 epoxy resin, which is a hydrogenated analog of ED-20 epoxy uncured resin, acetone, 15% boron trifluoride BF_3 solution in diethylene glycol, 1N HNO_3 solution (Shilova, 2013).

To prepare the epoxysiloxane sol, TEOS was mixed with acetone and a solution of nitric acid, then EPONEX 1510 and the hardener BF_3 in diethylene glycol were added to the resulting solution and the resulting mixture was intensively stirred. The ratio TEOS: EPONEX 1510 = 1:1. The resulting solution was kept for several hours, after which a DND suspension in an amount of 0.2 wt.% was introduced and the resulting composition was intensively stirred.

As a result of the above operations, sols were obtained, which were stored in closed containers at room temperature for at least 1 day and then used to form coatings.

Epoxysiloxane coatings were applied with a brush to glass slides, pieces of marble, as well as to certain areas of cultural monuments made of Carrara marble (Necropolis of the 18th century, St. Alexander Nevsky Lavra of the Holy Trinity, Saint Petersburg, Russia). The drying of coatings was carried out in air under ambient conditions.

15.3.1.2 Environmentally friendly fungicidal additive: detonation nanodiamond

DND is produced by Special Design Bureau "Technolog", Saint Petersburg, Russia. DND was added to epoxysiloxane sols as a 3 wt.% aqueous suspension.

DND is a biologically active material (Khamova et al., 2012, 2014; Schrand et al., 2009). In laboratory tests, it was found that, opposite to photosensitizers, including titanium dioxide, DND inhibited the development of a number of molds (e.g., *Aspergillus versicolor, Cladosporium herbarum, Penicillium spinulosum,* and *Ulocladium chartarum*) both under illumination and in the dark. At the same time, DND did not suppress the development of *A. niger* and *Chaetomium globosum* micromycetes, and the development of *Scopulariopsis brevicaulis* was only slightly suppressed. Thus like most "mild" biocides, DND's fungicidal activity is selective in relation to different strains of molds, which are biodeterioration agents of stone.

15.3.1.3 Methods for studying the microstructure of coatings

The morphology of the surface of epoxysiloxane coatings was studied via scanning electron microscopy (SEM) by means of Carl Zeiss NVision 40 workstation at accelerating voltages of 1–30 kV using secondary and backscattered electron detectors. The study was carried out without preliminary deposition of conducting materials on the surface of the samples. X-Ray Spectrum Microanalysis and mapping of the distribution of elements over the coating surface were performed using an Oxford Instruments X-Max detector.

The structure of the coatings at the mesoscopic level was studied by the method of small-angle X-ray scattering (SAXS) on xerogels obtained from epoxysiloxane sols used for coating. SAXS measurements were carried out by means of a small-angle scattering unit Molecular Metrology SAXS System (Institute of Macromolecular Chemistry, Prague, Czech Republic) operating in axial geometry and using a CuK_α microfocus X-ray generator ($\lambda = 0.154$ nm) Osmic MicroMax 002 operating in the 45 kV mode and 0.66 mA (30 W). The spectrometer is equipped with a gas-filled detector with an active area of 20 cm in diameter (Gabriel design). The use of two sample positions made it possible to measure the intensity of X-ray scattering in the range of scattering vectors $4.5 \times 10^{-3} < q < 1.1$ A^{-1}. All samples were measured in vacuum at room temperature. To obtain the differential small-angle scattering cross section $d\Sigma(q)/d\Omega$ in absolute units, we used the standard procedure of normalization to the scattering cross-section by amorphous carbon (Glassy Carbon), which, when scattered at small angles in the region $0.02 < q < 0.09$ A^{-1}, gives a plateau with an intensity of 3.805 cm^{-1}. This value is almost 250 times higher than the corresponding value for water H_2O, and thus allows for more accurate and quicker obtaining of the intensity in absolute values.

15.3.1.4 Laboratory testing of epoxysiloxane coatings for biostability

To test epoxysiloxane coatings with and without DND for biostability, three species of micromycetes were selected: *C. cladosporioides*, *H. dematioides*, and *A. niger*. Strains of these micromycetes were obtained by isolating them into a pure culture from the surface of a damaged stone substrate (marble of ancient monuments from the Necropolises of the State Museum of Urban Sculpture, Saint Petersburg, Russia). All strains are characterized as active biodeterioration agents of materials in various climatic and environmental conditions. They are capable of developing and causing damage to solid materials, including rock substrates. The selected strains have different strategies for colonization of the hard surface, differ in the nature of sporulation and mycelium development.

For the preparation of a spore suspension fungi cultures with an age of 14 to 28 days were used. The suspension was prepared in sterile distilled water according to GOST 9.048–89 (National state standard of Russia; consists of a list of prescriptions regarding the quality of goods in any industry). The spore concentration in 1 mL of the suspension was: 4.3105 spores for *A. niger*, 1.8105 spores for *C. cladosporioides*, 3104 spores for *H. dematioides*. Cover glass slides coated with epoxysiloxane coatings were inoculated with a freshly prepared spore suspension. Then the contaminated slides were placed in moist chambers. The results of the experiment were studied after 4 weeks (28 days) from the moment of inoculation (according to GOST 9.048−89) according to the special scale characterizing the degree of development of molds on experimental samples (Table 15−1).

15.3.1.5 Climatic tests of protective coating experimental samples in Antarctica
Epoxy-siloxane coatings with DND and without it were applied to the slides. These experimental samples were exposed for 3 months at Bellingshausen station in Antarctica (Fig. 15−8). The climatic conditions in which the samples were located can be called extreme, because the temperature changed from plus values (during the day) to minus values (at night) daily.

Full-scale tests of protective epoxysiloxane coatings on marble monuments
To carry out full-scale tests of coatings in an urban environment gravestones made of white Carrara marble were selected in the Necropolis of the 18th century by a group of specialists from the State Museum of Urban Sculpture and from Saint Petersburg State University: Princess S.D. Viljegorskaya, Batyushkova A.G., Ramburg A.I. (Fig. 15−9A), and to the Unknown (Fig. 15−9B).

In order to assess their effectiveness, the applied coatings were being observed for 1−3 years, two times a year. To assess the condition of biological samples they were studied by nondamaging methods: taking photographs, special prints from the surface onto a nutrient medium, as well as a smear from the surface (1 dm^2). Each sample was taken in triplicate.

15.3.2 Results and discussion

15.3.2.1 Structure and composition of protective epoxysiloxane coatings
Checking the presence of a protective layer on the marble surface (measuring changes in its continuity and structure by the SEM method) showed that the marble has a sufficiently developed surface (Fig. 15−10A). It is possible to form continuous organic-inorganic coatings

Table 15–1 Ordinal scale for assessing the effectiveness of suppressing the growth and development of micromycetes.

Score	Score characteristics
0	Germination of spores and conidia not detected under the microscope.
1	Swollen conidia are visible under the microscope, some of them germinate, but there is no further growth. Mycelium does not develop.
2	Swollen conidia are visible under the microscope; a significant part of them germinates. A slight growth of mycelium is observed (locally).
3	The germination of most conidia is seen under the microscope. Short or slightly branching hyphae are visible, hyphal intertwines are locally formed.
4	Well-developed fungi are seen under the microscope, mycelium is being formed, and sporulation can form locally.
5	The development of fungi is noticeable to the naked eye.

FIGURE 15–8 Exposition of protective coatings at the experimental site at Bellingshausen station, coast of West Antarctica.

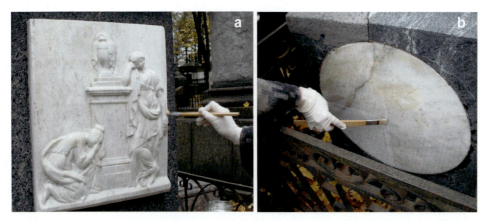

FIGURE 15–9 The process of applying a protective epoxysiloxane coating doped with detonation nanodiamond: (A) on a marble board with a relief (monument to I.O. Ramburg); and (B) on a round marble insert of the monument to the Unknown. Autumn 2016.

on it, which differ in their structure depending on the method of preparing the sol (first and second options; see Section 15–3.1). On the coatings obtained by technique 1, there are rather large areas of phase separation: without DND additive—from ~ 5 to $50\,\mu m$ (Fig. 15–10B-1); with DND additive—from ~ 5 to $90\,\mu m$ (Fig. 15–10C-1). Numerous small

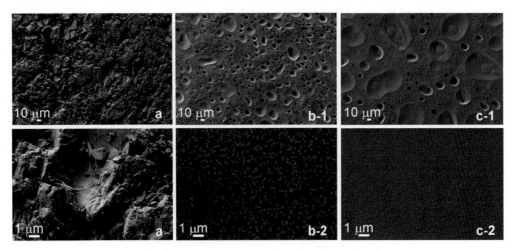

FIGURE 15–10 Scanning electron microscopy images of clean marble surface (A) and with epoxy-siloxane coatings: no detonation nanodiamond (DND) (B-1, B-2) and with 0.20 wt.% DND (C-1, C-2) obtained by technique 1 (*above*) and technique 2 (*below*) 10.

objects resembling pores are also visible. In this case, the doping of the epoxysiloxane coating with DND leads to a certain ordering of the surface structure; not only the enlargement of the delamination regions occurs, but also their orientation in a certain direction. Since the coating is organic-inorganic, it can be assumed that microseparation into organic (epoxy) and inorganic (siloxane) regions occurs, which is typical for hybrid organic-inorganic materials (Khamova et al., 2014; Shilova, 2013).

The results of the quantitative determination of C, O, and Si content (Table 15–2) confirm this assumption. Large areas contain O, C, and a minimum amount of Si. A porous phase is distributed around them, containing O, Si, and a minimum amount of C. It is quite possible that the resulting structure consists of two interpenetrating network (IPN) or semi-interpenetrating network (semi-IPN).

The surface morphology of coatings obtained by technique 2 differs from that described above (Fig. 15–10B-2 and C-2, *bottom row*). SEM images show numerous small objects less than 1 μm in size. At the same time, the addition of DND helps to reduce the size of these formations (Fig. 15–10C-2, *bottom row*). In this case, apparently, coatings with microphase separation into epoxy and siloxane components are also formed, but the size of the areas of the separation is an order of magnitude smaller. Thus DND is a structuring agent, the presence of which contributes to the improvement of the uniformity of the coating structure, which should have a positive effect on its physical and mechanical properties.

The results of the small-angle X-ray scattering (SAXS) study of the structure formation of epoxysiloxane xerogels obtained from sols synthesized by technique 2, both without and with DND, confirm this hypothesis (Khamova et al., 2014). Based on SAXS data, it was revealed that epoxysiloxane materials at the mesoscopic scale are characterized by a two-level fractal structure, in the formation of which the dominant role belongs to the siloxane component. At the first

Table 15–2 Quantitative elemental composition of the epoxysiloxane coating modified with 0.20 wt.% detonation nanodiamond obtained by technique 1.

Measurement area	Element content, wt.%		
	C	O	Si
Large individual objects	64	12	24
(at three different points on the surface)	64	12	24
	64	12	24
Porous phase (at three different points on the surface)	38	22	40
	39	21	40
	38	22	40

(bottom) level, inside the epoxy phase, the primary SiO_2 nanoparticles form. Their characteristic size (d_{c1}) is ~35 Å. Then, they aggregate into mass-fractal clusters with an anisodiametric, branched structure and a characteristic size d_{c2} ~ 400 Å. These clusters are structural units (bricks) of the spatial structural network of epoxysiloxane xerogel, as well as epoxysiloxane coatings.

The introduction of DND into epoxysiloxane compositions leads to a change in the mesostructure of the primary SiO_2 particles. Nanoparticles with a structure of the surface fractal are formed. Brinker and Scherer (1990) showed that the fractal dimension of objects corresponding to the structure of surface fractals can vary from two to three, that is, from a smooth surface to a three-dimensional object. In this case, under the influence of the structuring agent DND, the surface of primary SiO_2 nanoparticles changes from almost smooth ($D_{S1} = 2.04 \pm 0.04$, D_{S1} is the fractal dimension of the surface fractal formed at the first structural level) to a developed fractal surface ($D_{S1} = 2.28 \pm 0.04$), with a simultaneous increase in their characteristic size from $d_{c1} \approx 35$ Å to $d_{c1} \approx 44$ Å.

In addition, it was revealed that in the presence of DND in epoxysiloxane compositions at the second structural level, mass-fractal clusters are formed, the dimension of which increases (from $D_{M2} = 1.86 \pm 0.02$ to $D_{M2} = 1.88 \pm 0.02$; D_{M2} is the fractal dimension of the mass fractal, formed at the second structural level), with a simultaneous decrease in their characteristic size from $d_{c2} \approx 400$ Å to d_{c2} ~250 Å. This clearly indicates an increase in the density of the epoxysiloxane xerogel structure. Consequently, the introduction of DND into epoxysiloxane sols provides strengthening of both the formed xerogel structure and coating structure due to the formation of denser mass-fractal clusters of smaller sizes. The strengthening effect of DND on the coating structure is also evidenced by the field tests data carried out in Antarctica at Bellingshausen station. As can be seen from AFM (atomic force microscopy) data (Fig. 15–11), doping of coatings with DND increases their cracking resistance.

The cracking of the coatings occurred in the unfavorable conditions of Antarctica, when during the day the air temperature constantly changed from negative (at night) to positive values (during the day). Breaks appeared along the boundaries of the phases (epoxy and siloxane). It is not surprising that the smaller the phase separation regions were, the mechanically stronger the coatings were.

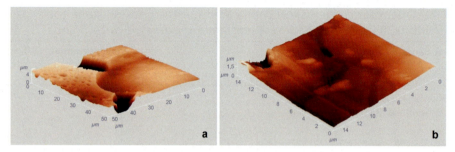

FIGURE 15-11 3D AFM images of epoxysiloxane coatings obtained according to technique 1, without detonation nanodiamond (DND) (A) and with DND (B) after testing for 3 months in Antarctica.

Thus on the basis of the obtained results, it can be concluded that coatings with microphase separation into organic (epoxy) and inorganic (siloxane) components are formed from epoxysiloxane sols, which is typical for organic-inorganic coatings. The resulting structure consists of two IPN or semiIPN. However, despite the phase separation at the micro-level, the coatings meet the requirements for biostable protective coatings to protect stone structures, including objects of cultural heritage. They do not impair the decorative properties of the stone, have a mild biocidal effect; they inhibit the development of biodeterioration agents (molds) and are ecologically safe for the environment.

15.3.2.2 Biostability of epoxysiloxane coatings
Laboratory test results

Coatings based on epoxysiloxane sols obtained by the techniques described above, both without and with DND, were tested for biostability. The influence of the investigated coatings on the growth and development of micromycetes of three types of molds can be traced according to the data presented in Figs. 15–12 and 15–13.

Epoxysiloxane matrices in general have an inhibitory effect on the development of micromycetes. In particular, suppression of conidial germination and mycelium development was observed in *C. cladosporioides* and *A. niger*. At the same time, the prevailing yeast-like growth was recorded in *H. dematioides*. DND additive improved the fungicidal activity for epoxysiloxane coatings prepared according to technique 1, which is characterized by a pronounced microphase separation. The greatest effect of suppressing aggressive biodeterioration agents was obtained for coatings obtained according to technique 2, both with DND and without DND. Thus a tendency toward an increase in the inhibitory effect for coatings with a more uniform structure can be noted.

Field test results

To confirm the presence of antimicrobial activity of the synthesized coatings, they were additionally tested in an urban environment, the results of which are presented in Tables 15–3 and 15–4 and Fig. 15–16.

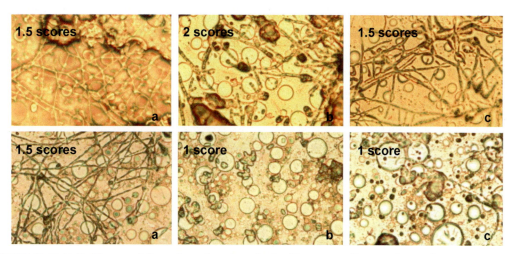

FIGURE 15–12 Optical images of the surface of coatings obtained from epoxysiloxane sols according to technique 1 without detonation nanodiamond (DND) (*top row*) and with DND (*bottom row*) after inoculation with microscopic fungi Cladosporium cladosporioides (A), Hormonema dematioides (B), and Aspergillus niger (C). Microphase separation is clearly visible. Scores indicate the fungicidal activity of the coatings.

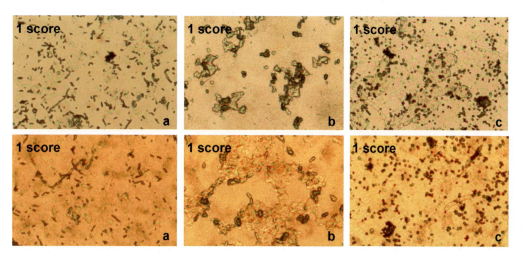

FIGURE 15–13 Optical images of the surface of coatings obtained from epoxysiloxane sols according to technique 2 without detonation nanodiamond (DND) (*top row*) and with DND (*bottom row*), after inoculation with microscopic fungi Cladosporium cladosporioides (A), Hormonema dematioides (B), and Aspergillus niger (C). Scores characterize the fungicidal activity of the coatings.

In samples taken 1 year after treatment with a nondamaging method (an imprint from the surface on an artificial nutrient medium and a smear from a surface of 1 dm^2), in variants with DND, only a weak development of several types of micromycetes was recorded with a low

Table 15–3 Characteristics of the state of the monument to the Unknown, which round marble inserts were protected with epoxysiloxane coatings prepared by technique 2; and without any protective coating (3 years after application).

DND content in the coating	Microorganisms	Species composition of fungi, CFU, per 1 dm² of surface
0 (control)	Aureobasidium pullulans—dominant; Cladosporium cladosporioides; Cladosporium herbarum; Coniosporium sp.	3600
0.20 wt.%	A. pullulans—dominant; C. cladosporioides; C. herbarum	2300

DND, detonation nanodiamond.

Table 15–4 Changes in the surface state of marble monuments with epoxysiloxane coatings doped with detonation nanodiamond (0.2 wt.%) prepared by technique 2 (1 and 3 years after coatings application).

Description of the state of the marble	Microorganisms	
	Species composition	CFU, per 1 dm² of surface
Monument to A.I. Ramborg		
1 year after application		
The marble is in good condition. The relief shows no signs of the development of fungi and algae.	Aureobasidium pullulans; C. cladosporioides	600
3 years after application		
The marble is in good condition, there are no fungi or algae, only small mud layering.	Aurebasidium pullulans; Cladosporium cladosporioides; Cladosporium herbarum; nonspore-bearing light-colored fungi	1500
Monument to A.G. Batyushkova		
1 year after application		
The marble is in good condition. On the body of the vase, mud layers are insignificant; there are no colonies of fungi and no algae deposits. On the rim of the vase, where water accumulates, the beginning of the development of colonies of dark-colored fungi (a little), algae scurf (locally), mud layers are very few.	A. pullulans; C. herbarum; Hormonema dematioides; Paecilomyces variotii	700
3 years after application		
There are fungi on the rim and on the body of the vase. On the rim, there are mud and algae deposits; biofilm development.	Alternaria alternata; A. pullulans—dominant; C. cladosporioides—dominant; C. herbarum; Coniosporium sp.; Epicoccum nigrum; H. dematioides; nonspore-bearing light-colored fungi	7600

number of colony-forming units. At the same time, these fungi developed only on the surface and it could only occur due to the deposition of atmospheric pollution on the treated surface, which served as a source of nutrition for the fungi. The revealed fungi did not change the properties of the marble surface. The greatest biocidal effect was obtained for an epoxysiloxane coating containing 0.20 wt.% DND. This result is comparable to the results of highly effective organotin compound-based biocides. At the same time, on the untreated side of the monument, an active development of biodeterioration agents was noted: several dark and light-colored micromycetes at once. On the basis of this and previous laboratory experiments (Khamova et al., 2012), a DND concentration of 0.2 wt.% was chosen for doping epoxy-siloxane coatings.

As follows from Table 15−3 data and Fig. 15−14, the epoxy coating itself has an inhibitory effect on biodeterioration agents (fungi). This effect is further enhanced by doping the coating with a small amount of DND.

The protective effect of the coating is influenced not only by the surface structure of both the stone itself and the protective coating, but also by the location of the experimental site. As follows from the data given in Table 15−4 the vertically located surface of the monument is less susceptible to biodegradation (Table 15−4, the monument to A.I. Ramburg, Fig. 15−15). On the contrary, the neck of the vase on the monument to A.G. Batyushkova, which is constantly filled with rainwater, is substantially colonized by microorganisms. Monuments under trees are also more susceptible to biodegradation (Table 15−4, the monument to A.G. Batyushkova, Fig. 15−16).

Thus the tests performed showed that epoxysiloxane coatings based on the two investigated types of compositions, with a more or less homogeneous microstructure, both with and without DND doping, exhibit an inhibitory effect on biodeterioration agents (fungi). At the same time, the greatest biocidal effect was obtained in coatings containing DND, regardless of the composition of the epoxysiloxane matrix, with the duration of field tests from 1 to 3 years.

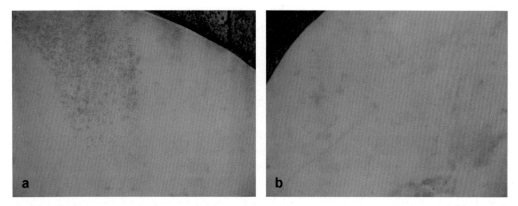

FIGURE 15–14 The exterior of the marble fragments of the monument to the Unknown after 3 years of outdoor exposure: (A) uncoated marble (control); (B) with an epoxysiloxane coating prepared by technique 2 containing 0.2 wt.% detonation nanodiamond.

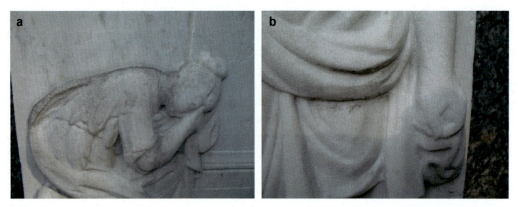

FIGURE 15–15 The exterior of the marble fragments of the monument to A.I. Ramborg with an epoxysiloxane coating prepared by technique 2 containing 0.2 wt.% detonation nanodiamond after outdoor exposure during 1 year (A) and 3 years (B).

FIGURE 15–16 The exterior of the marble fragments of the monument to A.G. Batyushkova with an epoxysiloxane coating prepared by technique 2 containing 0.2 wt.% detonation nanodiamond after outdoor exposure during 1 year (A) and 3 years (B, C).

15.4 Conclusion

The features of the biophysical (mechanical) and biochemical effects of microorganisms on the stone material of outdoor monuments, leading to their destruction, as well as factors contributing to the processes of stone biodegradation are reviewed. Special attention is paid to the process of biofilm formation, which can be considered as a successful strategy for the existence and survival of microorganisms on a rocky substrate in various environmental conditions, including extreme ones. The evolution of the colonization of the stony substrate by microorganisms is discussed; the composition of the biological matrix formed on the stone is analyzed. It was emphasized that biofilms formed by different groups of microorganisms exhibit a synergistic destructive effect on the stone substrate, which can be significantly greater than that corresponding to individual groups of microorganisms. This fact must be taken into account when searching for effective means of protecting stone monuments from biological damage, including when testing protective compositions aimed at suppressing biodeterioration agents biodeterioration agents.

The analysis of modern methods of cleaning sculptural monuments, as well as stone surfaces of buildings and structures from biodeterioration agents is presented. The advantages and disadvantages of mechanical and chemical cleaning methods are described. Laser technology is preferred as a gentle modern method of stone surface treatment (especially on cultural heritage sites).

It is shown that the use of protective coatings remains the most popular way to protect the stone from biodegradation. In addition to traditional organic and organosilicon film formers (epoxy resins, silanes, acrylates), suspensions and water-based paint and varnish compositions are recommended to form a protective layer on the stone surface.

Nanotechnology is increasingly used to protect the stone from biodegradation. This primarily applies to the use of biologically active nanoparticles (Ag, Cu, carbon nanoparticles, etc.), as well as metal oxides (TiO_2, CuO, ZnO, etc.), which are introduced in small amounts into varnishes, sol-gel compositions, or suspension, and have an inhibitory effect on the development of microorganisms. Much attention in the scientific community is paid to the use of photosensitizer additives and most notably nanodispersed titanium dioxide to suppress the growth of microorganisms and the appearance of biofilms. To increase its photocatalytic and bioactivity, titanium dioxide is doped with various compounds, for example, Fe (III) compounds, at the production stage. The attention of researchers is attracted by the possibilities of creating coatings with both superhydrophilic and superhydrophobic surfaces.

It was suggested that the future of conservation of unique marble monuments with cultural heritage is connected to the use of biotechnology, using the possibility of secondary crystallization of calcite on the surface of the monument under the action of bacteria, which leads to the healing of cracks and a decrease in the porosity of marble and other carbonate materials.

Particular attention is paid to the sol-gel technologies, which are successfully used for the formation of long-lived protective coatings on marble that inhibit the development of biodeterioration agents. The long-term experience of obtaining and using sol-gel-derived epoxysiloxane coatings for the protection of marble outdoor sculptural monuments in the urban environment is summarized. Their ability to inhibit the growth of a number of micromycetes, the most aggressive biodeterioration agents (fungi) of stone has been shown. The possibility of enhancing the inhibitory effect of the coatings by introducing small amounts of DND powder, a green biocide which is bioactive both under the light and in the dark, is discussed.

References

Achal, V., & Mukherjee, A. (2015). A review of microbial precipitation for sustainable construction. *Construction and Building Materials, 93*, 1224–1235. Available from https://doi.org/10.1016/j.conbuildmat.2015.04.051.

Aflori, M., Simionescu, B., Bordianu, I.-E., Sacarescu, L., Varganici, C.-D., Doroftei, F., ... Olaru, M. (2013). Silsesquioxane-based hybrid nanocomposites with methacrylate units containing titania and/or silver nanoparticles as antibacterial/antifungal coatings for monumental stones. *Materials Science and Engineering B*, *178*(19), 1339–1346. Available from https://doi.org/10.1016/j.mseb.2013.04.004.

Allen, N. S., Edge, M., Ortega, A., Sandoval, G., Liauw, C. M., Verran, J., ... McIntyre, R. B. (2004). Degradation and stabilisation of polymers and coatings: Nano versus pigmentary titania particles. *Polymer Degradation and Stability*, *85*(3), 927–946. Available from https://doi.org/10.1016/j.polymdegradstab.2003.09.024.

Arreche, R., & Vázquez, P. (2020). Green biocides to control biodeterioration in materials science and the example of preserving World Heritage Monuments. *Current Opinion in Green and Sustainable Chemistry*, *25*. Available from https://doi.org/10.1016/j.cogsc.2020.100359.

Barberia-Roque, L., Obidi, O. F., Gámez-Espinosa, E., Viera, M., & Bellotti, N. (2019). Hygienic coatings with bioactive nano-additives from *Senna occidentalis*-mediated green synthesis. *NanoImpact*, *16*. Available from https://doi.org/10.1016/j.impact.2019.100184.

Brinker, C. J., & Scherer, G. W. (1990). *Sol-gel science: The physics and chemistry of sol-gel processing*. San-Diego: Academic Press Inc.

Burgos-Cara, A., Ruiz-Agudo, E., & Rodriguez-Navarro, C. (2017). Effectiveness of oxalic acid treatments for the protection of marble surfaces. *Materials and Design*, *115*, 82–92. Available from https://doi.org/10.1016/j.matdes.2016.11.037.

Cardiano, P., Mineo, P., Sergi, S., Ponterio, R. C., Triscari, M., & Piraino, P. (2003). Epoxy-silica polymers as restoration materials. Part II. *Polymer*, *44*(16), 4435–4441. Available from https://doi.org/10.1016/S0032-3861(03)00432-4.

Cooper, M. (1998). *Laser cleaning in conservation: An introduction*. Oxford: Butterworth Heinemann.

Crupi, V., Fazio, B., Gessini, A., Kis, Z., La Russa, M. F., Majolino, D., ... Venuti, V. (2018). TiO_2-SiO_2-PDMS nanocomposite coating with self-cleaning effect for stone material: Finding the optimal amount of TiO_2. *Construction and Building Materials*, *166*, 464–471. Available from https://doi.org/10.1016/j.conbuildmat.2018.01.172.

De Leo, F., & Urzì, C. (2015). Microfungi from deteriorated materials of cultural heritage. In J. K. Mistra, J. P. Tewari, S. K. Deshmukh, & C. Vagvolgui (Eds.), *Fungi from different substrates* (pp. 144–158). Science Publishers, Press.

DeJong, J. T., Fritzges, M. B., & Nüsslein, K. (2006). Microbially induced cementation to control sand response to undrained shear. *Journal of Geotechnical and Geoenvironmental Engineering*, *132*(11), 1381–1392. Available from https://doi.org/10.1061/(ASCE)1090-0241(2006)132:11(1381).

Doehne, E. F., & Price, C. A. (2010). *Stone conservation: An overview of current research* (2nd ed.). Los Angeles: Getty Conservation Institute Getty Publications.

Fidanza, M. R., & Caneva, G. (2019). Natural biocides for the conservation of stone cultural heritage: A review. *Journal of Cultural Heritage*, *38*, 271–286. Available from https://doi.org/10.1016/j.culher.2019.01.005.

Frank-Kamenetskaya, O. V., Vlasov, D. Y., Manurtdinova, V. V., Zelenskay, M. S., Parfenov, V. A., Rytikova, V. V., ... Grishkin, V. M. (2019). Monitoring of the state of stone and bronze monuments. In O. V. Frank-Kamenetskaya, D. Y. Vlasov, & V. V. Rytikova (Eds.), *The effect of the environment on Saint Petersburg's cultural heritage: Results of monitoring the historical necropolis monuments* (pp. 145–159). Springer Nature Switzerland AG.

Frank-Kamenetskaya, O. V., Vlasov, D. Y., & Shilova, O. A. (2011). Biogenic crystal genesis on a carbonate rock monument surface: The main factors and mechanisms, the development of nanotechnological ways of inhibition. In S. V. Krivovichev (Ed.), *Minerals as advanced materials II* (pp. 401–413). Berlin Heidelberg: Springer-Verlag.

Gadd, G. M., Bahri-Esfahani, J., Li, Q., Rhee, Y. J., Wei, Z., Fomina, M., & Liang, X. (2014). Oxalate production by fungi: Significance in geomycology, biodeterioration and bioremediation. *Fungal Biology Reviews*, *28*(2–3), 36–55. Available from https://doi.org/10.1016/j.fbr.2014.05.001.

Goffredo, G. B., Accoroni, S., & Totti, C. (2019). Chapter 25. Nanotreatments to inhibit microalgal fouling on building stone surfaces. In: *Nanotechnology in eco-efficient construction. Materials, processes and applications. Woodhead publishing series in civil and structural engineering* (2nd ed., pp. 619–647). Woodhead Publishing. Available from https://doi.org/10.1016/B978-0-08-102641-0.00025-6.

Grbić, M. L., Vukojević, J., Simić, G. S., Krizmanić, J., & Stupar, M. (2010). Biofilm forming cyanobacteria, algae and fungi on two historic monuments in Belgrade, Serbia. *Archives of Biological Sciences, 62*(3), 625–631. Available from https://doi.org/10.2298/ABS1003625L.

Gulotta, D., Villa, F., Cappitelli, F., & Toniolo, L. (2018). Biofilm colonization of metamorphic lithotypes of a renaissance cathedral exposed to urban atmosphere. *Science of the Total Environment, 639*, 1480–1490. Available from https://doi.org/10.1016/j.scitotenv.2018.05.277.

Kaminskii, V. V., Aleshkin, A. V., Zul'karneev, E. R., Zatevalov, A. M., Kiseleva, I. A., Efimova, O. G., ... Boinovich, L. B. (2019). Development of a bacteriophage complex with superhydrophilic and superhydrophobic nanotextured surfaces of metals preventing healthcare-associated infections (HAI). *Bulletin of Experimental Biology and Medicine, 167*(4), 500–503. Available from https://doi.org/10.1007/s10517-019-04559-0.

Khamova, T. V., Shilova, O. A., Kopitsa, G. P., Almásy, L., & Rosta, L. (2014). Small-angle neutron scattering study of the mesostructure of bioactive coatings for stone materials based on nanodiamond-modified epoxy siloxane sols. *Physics of the Solid State, 56*(1), 105–113. Available from https://doi.org/10.1134/S1063783414010156.

Khamova, T. V., Shilova, O. A., Vlasov, D. Y., Ryabusheva, Y. V., Mikhal'chuk, V. M., Ivanov, V. K., ... Dolmatov, V. Y. (2012). Bioactive coatings based on nanodiamond-modified epoxy siloxane sols for stone materials. *Inorganic Materials, 48*(7), 702–708. Available from https://doi.org/10.1134/S0020168512060052.

La Russa, M. F., Macchia, A., Ruffolo, S. A., De Leo, F., Barberio, M., Barone, P., ... Urzì, C. (2014). Testing the antibacterial activity of doped TiO_2 for preventing biodeterioration of cultural heritage building materials. *International Biodeterioration and Biodegradation, 96*, 87–96. Available from https://doi.org/10.1016/j.ibiod.2014.10.002.

La Russa, M. F., Ruffolo, S. A., Rovella, N., Belfiore, C. M., Palermo, A. M., Guzzi, M. T., & Crisci, G. M. (2012). Multifunctional TiO_2 coatings for cultural heritage. *Progress in Organic Coatings, 74*(1), 186–191. Available from https://doi.org/10.1016/j.porgcoat.2011.12.008.

Leavengood, P., Twilley, J., & Asmus, J. F. (2000). Lichen removal from Chinese Spirit Path figures of marble. *Journal of Cultural Heritage, 1*(2), S71–S74. Available from https://doi.org/10.1016/S1296-2074(00)00191-6.

Luk'yanchuk, B. S. (2002). *Laser cleaning*. Singapore: World Scientific Publishing.

Morton, L. H. G. (1996). The Involvement of biofilms in biodeterioration processes. *International Biodeterioration and Biodegredation, 37*(2), 126–135.

Parfenov, V. A., Gerashenko, A. N., Gerashenko, M. D., & Grigor'eva, I. D. (2010). Laser cleaning of historical monuments. *Scientific and Technical Journal of Information Technologies, Mechanics and Optics, 10*(2), 11–17.

Pinna, D., Salvadori, B., & Galeotti, M. (2012). Monitoring the performance of innovative and traditional biocides mixed with consolidants and water-repellents for the prevention of biological growth on stone. *Science of the Total Environment, 423*, 132–141. Available from https://doi.org/10.1016/j.scitotenv.2012.02.012.

Pouli, P., Frantzikinaki, K., Papakonstantinou, E., Zafiropulos, V., & Fotakis, C. (2005). Pollution encrustation removal by means of combined ultraviolet and infrared laser radiation: The application of this innovative methodology on the surface of the Parthenon West Frieze. In K. Dickman, C. Fotakis, & J. F. Asmus (Eds.), *Lasers in the conservation of artworks* (Vol. 100, pp. 333–340). Berlin, Heidelberg: Springer. Available from https://doi.org/10.1007/3-540-27176-7_41.

Price, C. (1998). Stone conservation: An overview of current research. *Journal of the American Institute for Conservation, 37*(2), 223–224. Available from https://doi.org/10.2307/3179804.

Prieto, B., & Silva, B. (2005). Estimation of the potential bioreceptivity of granitic rocks from their intrinsic properties. *International Biodeterioration and Biodegradation, 56*(4), 206−215. Available from https://doi.org/10.1016/j.ibiod.2005.08.001.

Rampazzi, L. (2019). Calcium oxalate films on works of art: A review. *Journal of Cultural Heritage, 40,* 195−214. Available from https://doi.org/10.1016/j.culher.2019.03.002.

Rode, A. V., Baldwin, K. G. H., Wain, A., Madsen, N. R., Freeman, D., Delaporte, P., & Luther-Davies, B. (2008). Ultrafast laser ablation for restoration of heritage objects. *Applied Surface Science, 254*(10), 3137−3146. Available from https://doi.org/10.1016/j.apsusc.2007.10.106.

Ruffolo, S. A., & La Russa, M. F. (2019). Nanostructured coatings for stone protection: An overview. *Frontiers in Materials, 6,* 147. Available from https://doi.org/10.3389/fmats.2019.00147.

Rusakov, A. V., Frank-Kamenetskaya, O. V., Gurzhiy, V. V., Zelenskaya, M. S., Izatulina, A. R., & Sazanova, K. V. (2014). Refinement of the crystal structures of biomimetic weddellites produced by microscopic fungus *Aspergillus niger*. *Crystallography Reports, 59*(3), 362−368. Available from https://doi.org/10.1134/S1063774514030146.

Sazanova, K., Osmolovskaya, N., Schiparev, S., Yakkonen, K., Kuchaeva, L., & Vlasov, D. (2015). Organic acids induce tolerance to zinc- and copper-exposed fungi under various growth conditions. *Current Microbiology, 70*(4), 520−527. Available from https://doi.org/10.1007/s00284-014-0751-0.

Sazanova, K. V., Shchiparev, S. M., & Vlasov, D. Y. (2014). Formation of organic acids by fungi isolated from the surface of stone monuments. *Microbiology (Reading, England), 83*(5), 516−522. Available from https://doi.org/10.1134/S002626171405021X.

Sazanova, K. V., Zelenskaya, M. S., Bobir, S., & Vlasov, D. Y. (2020). Micromycetes in the biofilms on stone monuments of Saint Petersburg. *Mycology and Phytopathology, 54*(5), 329−339. Available from https://doi.org/10.31857/S0026364820050104.

Sazanova (nee Barinova), K. V., Frank-Kamenetskaya, O. V., Vlasov, D. Y., Zelenskaya, M. S., Vlasov, A. D., Rusakov, A. V., & Petrova, M. A. (2020). Carbonate and oxalate crystallization by interaction of calcite marble with *Bacillus subtilis* and *Bacillus subtilis−Aspergillus niger* association. *Crystals, 10*(9), 756. Available from https://doi.org/10.3390/cryst10090756.

Sbardella, M. P., Bracciale, M. L., & Santarelli, J. M. A. (2020). Waterborne modified-silica/acrylates hybrid nanocomposites as surface protective coatings for stone monuments. *Progress in Organic Coatings, 149,* 105897. Available from https://doi.org/10.1016/j.porgcoat.2020.105897.

Schrand, A. M., Hens, S. A. C., & Shenderova, O. A. (2009). Nanodiamond particles: Properties and perspectives for bioapplications. *Critical Reviews in Solid State and Materials Sciences, 34*(1−2), 18−74. Available from https://doi.org/10.1080/10408430902831987.

Seven, O., Dindar, B., Aydemir, S., Metin, D., Ozinel, M. A., & Icli, S. (2004). Solar photocalytic disinfection of a group of bacteria and fungi aqueous suspensions with TiO_2, ZnO and sahara desert dust. *Journal of Photochemistry and Photobiology A: Chemistry, 165*(1−3), 103−107. Available from https://doi.org/10.1016/j.jphotochem.2004.03.005.

Shevchenko, V. Y., Shilova, O. A., Kochina, T. A., Barinova, L. D., & Belyi, O. V. (2019). Environmentally friendly protective coatings for transport. *Herald of the Russian Academy of Sciences, 89*(3), 279−286. Available from https://doi.org/10.1134/S1019331619030080.

Shilova, O. A. (2013). Synthesis and structure features of composite silicate and hybrid TEOS-derived thin films doped by inorganic and organic additives. *Journal of Sol-Gel Science and Technology, 68*(3), 387−410. Available from https://doi.org/10.1007/s10971-013-3026-5.

Shilova, O. A., Vlasov, D. Y., Zelenskaya, M. S., Ryabusheva, Y. V., Khamova, T. V., Glebova, I. B., ... Frank-Kamenetskaya, O. V. (2020). Sol-gel derived TiO_2 and epoxy-titanate protective coatings: Structure, property, fungicidal activity and biomineralization effects. In O. V. Frank-Kamenetskaya, D. Vlasov, E. Panova, & S. Lessovaia (Eds.), *Processes and phenomena on the boundary between biogenic and abiogenic nature. Lecture notes in earth system sciences* (pp. 619−638). Cham: Springer. Available from https://doi.org/10.1007/978-3-030-21614-6_33.

Siano, S., Fabiani, F., Caruso, D., Pini, R., & Salimbeni, R. (2000). Laser cleaning of stones: Assessment of operative parameters, damage thresholds and associated optical diagnostics. In *Proc. SPIE, ALT '99 international conference on advanced laser technologies (25 February 2000)* (Vol. 4070, pp. 27–35). Society of Photo-Optical Instrumentation Engineers. https://doi.org/10.1117/12.378171.

Siano, S., Giamello, M., Bartoli, L., Mencaglia, A., Parfenov, V., & Salimbeni, R. (2007). Phenomenological characterisation of stone cleaning by different laser pulse duration and wavelength. In J. Nimmrichter, W. Kautek, & M. Schreiner (Eds.), *Lasers in the conservation of artworks. Springer proceedings in physics* (Vol. 116, pp. 87–96). Berlin Heidelberg: Springer-Verlag. Available from https://doi.org/10.1007/978-3-540-72310-7_11.

Silva, M., Salvador, C., Candeias, M. F., Teixeira, D., Candeias, A., & Caldeira, A. T. (2016). Toxicological assessment of novel green biocides for cultural heritage. *International Journal of Conservation Science, 7*(1), 265–272. Available from http://www.ijcs.uaic.ro/public/IJCS-16-SI09_Silva.pdf.

Sizov, B. T. (2020). Methods for preserving stone monuments from external atmospheric influences. In *ART Konservatsiya: A specialized social resource for information assistance in the field of preservation, conservation and restoration of monuments of material culture*. http://art-con.ru/node/7031.

Sterflinger, K., Little, B., Pinar, G., Pinzari, F., de los Rios, A., & Gu, J. D. (2018). Future directions and challenges in biodeterioration research on historic materials and cultural properties. *International Biodeterioration and Biodegradation, 129*, 10–12. Available from https://doi.org/10.1016/j.ibiod.2017.12.007.

Tezel, U., & Pavlostathis, S. G. (2015). Quaternary ammonium disinfectants: Microbial adaptation, degradation and ecology. *Current Opinion in Biotechnology, 33*, 296–304. Available from https://doi.org/10.1016/j.copbio.2015.03.018.

Toreno, G., Isola, D., Meloni, P., Carcangiu, G., Selbmann, L., Onofri, S., … Zucconi, L. (2018). Biological colonization on stone monuments: A new low impact cleaning method. *Journal of Cultural Heritage, 30*, 100–109. Available from https://doi.org/10.1016/j.culher.2017.09.004.

Verges-Belmin, V. (1997). Comparison of three cleaning methods—micro-sandblasting, chemical and Q-switched Nd:YAG laser on a portal of the cathedral Notre-Dame in Paris. In W. Kautek, & E. Konig (Eds.), *Lasers in conservation of artworks* (pp. 1–24). Vienna: Restaratorendblatter, Mayer & Comp.

Vlasov, D. Y., Parfenov, V. A., Zelenskaya, M. S., Plotkina, Y. V., Geludova, V. M., Frank-Kamenetskaya, O. V., & Marugin, A. M. (2019). Methods of monument protection from damage and their performance. In O. Frank-Kamenetskaya, D. Vlasov, & V. Rytikova (Eds.), *The effect of the environment on Saint Petersburg's cultural heritage. Geoheritage, geoparks and geotourism (Conservation and management series)*. Cham: Springer. Available from https://doi.org/10.1007/978-3-319-79072-5_7.

Vlasov, D. Y., Zelenskaya, M. S., & Frank-Kamenetskaya, O. V. (2002). Micromycetes on marble monuments of Alexander-Nevskaya Lavra Museum Necropolis (St. Petersburg). *Mycology and Phytopathology, 36*(3), 7–10.

Wang, R., & Qian, C. (2014). In situ restoration of the surface defects on cement-based materials by bacteria mineralization with spraying method. *Journal Wuhan University of Technology, Materials Science Edition, 29*(3), 518–526. Available from https://doi.org/10.1007/s11595-014-0951-2.

Warscheid, T., & Braams, J. (2000). Biodeterioration of stone: A review. *International Biodeterioration and Biodegradation, 46*(4), 343–368. Available from https://doi.org/10.1016/S0964-8305(00)00109-8.

Xu, F., Li, D., Zhang, Q., Zhang, H., & Xu, J. (2012). Effects of addition of colloidal silica particles on TEOS-based stone protection using n-octylamine as a catalyst. *Progress in Organic Coatings, 75*(4), 429–434. Available from https://doi.org/10.1016/j.porgcoat.2012.07.001.

Zarzuela, R., Moreno-Garrido, I., Blasco, J., Gil, M. L. A., & Mosquera, M. J. (2018). Evaluation of the effectiveness of CuONPs/SiO_2-based treatments for building stones against the growth of phototrophic microorganisms. *Construction and Building Materials, 187*, 501–509. Available from https://doi.org/10.1016/j.conbuildmat.2018.07.116.

Zhu, T., & Dittrich, M. (2016). Carbonate precipitation through microbial activities in natural environment, and their potential in biotechnology: A review. *Frontiers in Bioengineering and Biotechnology, 4*, 4. Available from https://doi.org/10.3389/fbioe.2016.00004.

16

Microbiologically induced deterioration of cement-based materials

N.B. Singh

CHEMISTRY AND BIOCHEMISTRY DEPARTMENT AND RTDC, SHARDA UNIVERSITY, GREATER NOIDA, INDIA

16.1 Introduction

Ordinary Portland cement (OPC) is a very important binding material used in concretes. OPC consists of tricalcium silicate, dicalcium silicate, tricalcium aluminate, tetracalcium alumino ferrite, and gypsum. When this is mixed with sand, aggregate, and water, concrete is formed. Formation of paste, mortars, and concrete is represented by Fig. 16–1.

When water is added, different phases in OPC hydrate with different rates and number of hydration products are formed. Following chemical reactions occur.

$$2(3CaO.SiO_2) + 6H_2O \rightarrow 3CaO.2SiO_2.3H_2O + 3Ca(OH)_2 \quad (16.1)$$

$$2(2CaO.SiO_2) + 4H_2O \rightarrow 3CaO.2SiO_2.3H_2O + Ca(OH)_2 \quad (16.2)$$

$$3CaO.Al_2O_3 + 3CaSO_4.2H_2O + 26H_2O \rightarrow 3CaO.Al_2O_3.3CaSO_4.32H_2O \quad (16.3)$$

$$4CaO.Al_2O_3.Fe_2O_3 + 2Ca(OH)_2 + 2CaSO_4.2H_2O + 18H_2O \rightarrow \quad (16.4)$$

$$6CaO.Al_2O_3.Fe_2O_3.2CaSO_4.24H_2O)$$

In reactions 1 and 2, product $CaO.2SiO_2.3H_2O$ (C-S-H) formed is the main glue. It is responsible for hardness and cohesion of the material and is well established that C-S-H has nanostructure. Concretes, blended cements, admixtures, and nanomaterials (NMs) are also being used to improve the quality. When we talk of quality, the major emphasis is given on strength and durability. Durability is an important property of concrete and determines the service life of concrete structures and therefore is of great concern for practical applications. Because of interactions between environment and the concrete and different hydrates mentioned, mechanical and physical properties change (deteriorate). Out of different factors, which deteriorate the concrete structures carbonation, chloride, microbe, sulfate, alkali attacks, freezing and thawing, abrasion, and steel corrosion are the most important (CYR, 2013). Deterioration is linked with porosity and permeability as shown in Fig. 16–2.

Biodegradation and Biodeterioration at the Nanoscale. DOI: https://doi.org/10.1016/B978-0-12-823970-4.00016-6

FIGURE 16–1 Representation of paste, mortar and concrete made from Portland cement.

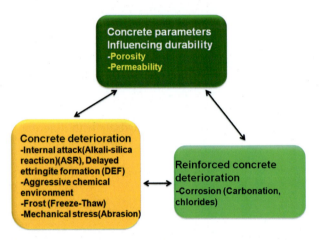

FIGURE 16–2 Parameters and concrete deterioration.

Concretes are also used for sewage waste disposal pipelines, where corrosion is a big problem. In sewage environment, various types of chemical and biological aggressive ions exist. The presence of number of bacteria in the sewage wastes corrodes the concrete and other cementitious materials. Corrosion occurs in many concrete structures associated with water and wastewater treatment. A typical corrosion and degradation of cement based materials in sewer systems particularly by microbial-induced corrosion (MIC) is shown in Fig. 16–3 (Grengg et al., 2018). Under these conditions deterioration occurs in less than a decade.

Many wastewater collection and treatment plants have the problem of concrete deterioration with time (Kong, Fang, Zhou, Han, & Lu, 2018). The mechanism of deterioration is very complex. Apart from physical and chemical effects, microbial attack is a major source for deterioration, due to the harmful acids produced by the metabolic activity of many bacteria. Microbial-influenced concrete deterioration (MICD) is very harmful and effective at

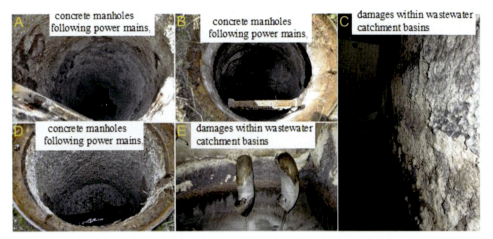

FIGURE 16–3 Typical deterioration affected by microbial-induced corrosion under different conditions.

deterioration. It shortens the service life of structure. In this chapter MICD of cement-based materials will be discussed in detail.

16.2 Corrosion and deterioration

There are a number of ways the cementitious materials in structures are deteriorated. The type of attack, effect, and chemical reaction during the process of corrosion and deterioration of cement based materials are given in Table 16–1 (Lokesh, Satish Reddy, & Kotaiah, 2014).

16.3 Microbiologically induced corrosion and degradation of cement-based materials

Microorganisms are present everywhere in nature and grow at a high rate in water, air, and soil. Microorganisms can influence corrosion depending on the type of microorganism, material, and electrolytes present in the system. The overall corrosion, combining all the mechanisms, is called MIC (Little et al., 2020). These corrosions when occur in cementitious systems particularly in concrete structures lead to degradation. This type of degradation is more common in sewer pipes because of different types of microbes. MIC or biocorrosion can be understood as

1. an electrochemical process
2. microorganisms affect corrosion
3. nutrients and water must be available in addition to microorganisms

Table 16–1 Different corrosion and degradation of cement-based materials.

Chemical substance	Effect on cement-based materials	Chemical reactions and models
Sulfuric acid	When cement-based structures particularly concretes are in contact with sulfuric acid, deterioration occurs	Chemical reactions of cement-based materials in presence of H_2SO_4 $Ca(OH)_2 + H_2SO_4 \rightarrow CaSO_4 \cdot 2H_2O$ $3CaO \cdot 2SiO_2 \cdot 3H_2O + H_2SO_4 \rightarrow CaSO_4 \cdot 2H_2O + Si(OH)_4$
Sodium sulfate	0.5% or more concentrated sodium sulfate solutions strongly attack, cement-based materials causing deterioration and expansion	Free lime obtained during Portland cement hydration in concrete reacts with sulfates dissolved in the water: "$Ca(OH)_2 + Na_2SO_4 + 2H_2O \rightarrow CaSO_4, 2H_2O + NaOH$" Conversion of aluminates into ettringite, expands considerably causing deterioration of the structure $3CaOAl_2O_3 \cdot 12H_2O + 3CaSO_4 \cdot 2H_2O + 13H_2O \rightarrow$ $3CaO \cdot Al_2O_3 \cdot 3CaSO_4 \cdot 31H_2O$ (Ettringite) Ettringite formation due to external sulfate attack and internal sulfate attack are shown by the following models. 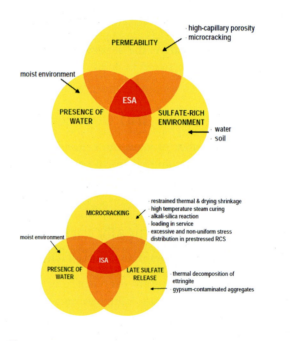
NaOH	NaOH is harmful if the concrete is made with alkali reactive aggregates	—

(Continued)

Table 16–1 (Continued)

Chemical substance	Effect on cement-based materials	Chemical reactions and models
HNO_3	Destroys concrete	Reaction of Nitric acid with hydration products is given below $2HNO_3 + Ca(OH)_2 \rightarrow Ca(NO_3)_2 \cdot 2H_2O$ $Ca(NO_3)_2 \cdot 2H_2O + 3CaO \cdot Al_2O_3 \cdot 8H_2O \rightarrow 3CaO \cdot Al_2O_3 \cdot Ca(NO_3)_2 \cdot 10H_2O$ This is responsible for shrinkage
Acetic acid	Deteriorates slowly	Cement hydration products react with acetic acid $2CH_3COOH + Ca(OH)_2 \rightarrow Ca(CH_3COO)_2 + 2H_2O$ $2CH_3COOH + C\text{-}S\text{-}H \rightarrow SiO_2 + Ca(CH_3COO)_2 + 2H_2O$
Carbonic acid	Corrosive and permeable in concrete; causes slow degradation	Carbon dioxide from atmosphere in presence of water is converted to carbonic acid, enters the concrete structure and deteriorates slowly
HCl	Destroys concrete	$Ca(OH)_2 + 2HCl \rightarrow CaCl_2 + 2H_2O$ $Ca_6Al_2(SO_4)_3(OH)_{12} \cdot 26H_2O \rightarrow 3Ca^{2+} + 2[Al(OH)_4]^- + 4OH^- + 26H_2O$ $3Ca^{2+} + 2[Al(OH)_4]^- + 4OH^- + 12HCl \rightarrow 3CaCl_2 + 2AlCl_3 + 12H_2O$
H_2S	Harmful in moist, oxidizing environments	—
SO_2	Gas in water forms acids and deteriorates	—
Magnesium sulfate	0.5% or above strongly attack concretes	—
Sewage	In many cases damage the concrete after long exposure	Complex reactions occur and are discussed in the next section
Alkali-Silica Reaction	Water with high alkali content degrades the structure due to alkali-silica reaction	The model can be represented

- alkali-rich portland cement
- environmental alkali salts (sea water and de-icing agents)
- moist environment
- ALKALI CONTENT
- PRESENCE OF WATER
- ALKALI REACTIVE SILICA — amorphous silica or strained quartzs in aggregate particles
- ASR

(Continued)

Table 16–1 (Continued)

Chemical substance	Effect on cement-based materials	Chemical reactions and models
Microbial attack	Microbes degrades cement-based structures	Microorganisms such as bacteria, fungi, microalgae, etc. degrade cement-based materials known as microbially-induced corrosion (MIC). Number of parameters affect MIC as represented (Little et al., 2020).

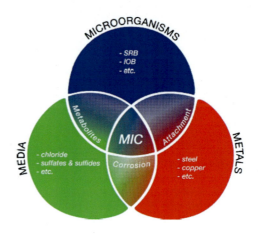

(SRB: sulfate-reducing bacteria: IOB: iron-oxidizing bacteria)

Microorganisms might be acting as catalysts for accelerating the reactions. Such processes may involve number of aggressive chemical compounds. Only certain type of microorganisms deteriorates the structure and depends on temperature, pH, and nutrient availability. Some of the microbes responsible for corrosion and degradation of cementitious structures are discussed. The microorganisms traditionally comprise fungi, bacteria, algae, and small unicellular animal-like protists (Table 16–2) (Natarajan, 2018; Stott, 2010).

Concrete because of high alkalinity [due to formation of $Ca(OH)_2$], is usually less susceptible to biological attack. Surface erosion facilitates the colonization of microbes. The colonization of sulfur-oxidizing and sulfur-reducing bacteria on concrete is associated with the sulfur cycle in their environment. The sulfur-reducing bacteria (anaerobic) convert sulfate into sulfide, which combines with hydrogen to form hydrogen sulfide. Hydrogen sulfide formed oxidizes into sulfuric acid by sulfur-oxidizing bacteria. Due to formation of sulfuric acid, alkali neutralization reaction occurs at the surface of concrete and pH is decreased, facilitating the concrete surface for further microbial colonization by neutraphilic and/or acidophilic organisms. Surface pH of concrete further reduces due to microbial growth. Due to different microorganisms, biogenic organic acids and carbon dioxide are produced, which are extremely corrosive toward concrete (Wei, Jiang, Liu, Zhou, & Sanchez-Silva, 2013).

Table 16–2 Classification of microbes in microbial-induced corrosion with properties.

Microbes	Features	Properties
Bacteria	Bacteria are a single walled biological cell	• Sulfur-oxidizing bacteria (*Acidithiobacillus* spp.) • Acid-producing bacteria (inorganic sulfuric acid by *Acidithiobacillus*, organic acids by *Bacillus* spp.) • Sulfate-reducing bacteria (*Desulfovibrio, Desulfotomaculum, Desulfobacter*) • Iron-oxidizing bacteria or metal-oxidizing bacteria and metal depositing bacteria. (*Gallionella, Crenothrix, Leptothrix, Sphaerotilus*) • Metal-reducing bacteria (*Pseudomonas, Shewanella*). • Slime producing bacteria (*Bacillus, Flavobacterium, Aerobacter, Pseudomonas*)
Fungi	Member of the group of eukaryotic organisms	*Cladosporium resinae* • *Paecilomyces varioti* • *Aspergillus niger*. All the fungi produce organic acids • *Penicillium cyclospium*
Algae	Typically aquatic plant	Blue green algae
Microbial consortia	Two or more microbial groups living symbiotically	Mutualism among different groups of microorganisms

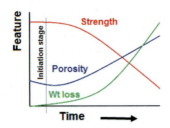

FIGURE 16–4 Change of properties of concretes on microbial attack.

When microbes attack the cementitious structures, strength decreases, porosity and weight losses increases as shown in Fig. 16–4 (Márquez, Sánchez-Silva, & Husserl, 2013).

16.4 Microbiologically-induced corrosion in sewer structures

MIC is a very complex phenomena in concrete structures exposed to sewer environments and reduces the lifetime significantly. Biological, chemical, physicochemical, and electrochemical processes are involved. Microbiologically induced corrosion is considered to be the most important one. Parker in 1945 found that bacteria plays an important role in deteriorating sewer concrete and proposed that the process was due to sulfur cycle. By biological and chemical oxidation of H_2S, variety of sulfur species is formed. One of the species formed is H_2SO_4 which reacts with concrete leading to deterioration. Also by microbial metabolism, HNO_3 and organic acids are formed, which cause concrete corrosion in sewage. Sewage provides nutrients, oxygen, and moisture responsible for biofilm formation and growth, which causes concrete corrosion. Sulfur and nitrogen compounds are present in fresh sewage entering into wastewater. These compounds give energy to bacteria in sewers (Kiliswaa, Scrivener, & Alexander, 2019). N- and S-compounds are oxidized by bacteria into nitrate (NO_3^-) and sulfate (SO_4^{2-}). To begin with, aerobic microorganisms grow in the sewage but later on anaerobic-anoxic microorganisms dominate due to depletion of oxygen and organic compounds. Anaerobic-anoxic microorganisms [proteobacteria that include sulfate-reducing bacteria (SRB) and nitrate-reducing bacteria (NRB)] are responsible for MIC or biogenic corrosions in sewers.

Sulfide (S^{2-}) is obtained due to reduction of SO_4^{2-} by SRB whereas nitrite (NO_2^-) is obtained due to reduction of NO_3^- by NRB. Sulfide the most important component for MIC in sewer exists as S^{2-}, HS^-, and H_2S depending on pH (Fig. 16–5) (Kiliswaa et al., 2019).

MIC of concrete exposed to sewer environments can be divided into number of steps/stages (Fig. 16–6) (Wu, Wang, Wu, & Kan, 2020).

16.5 Generation of sulfuric acid

In sewer systems, gaseous $H_2S_{(g)}$ can partition into moisture films on the surfaces above the waterline. Abiotic (chemical) and biotic (biological) reactions will take place, with the

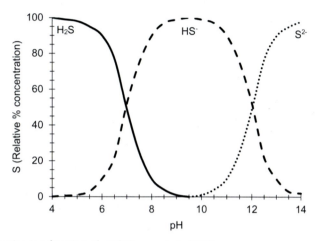

FIGURE 16–5 Sulfide species as a function of pH (Kiliswaa et al., 2019).

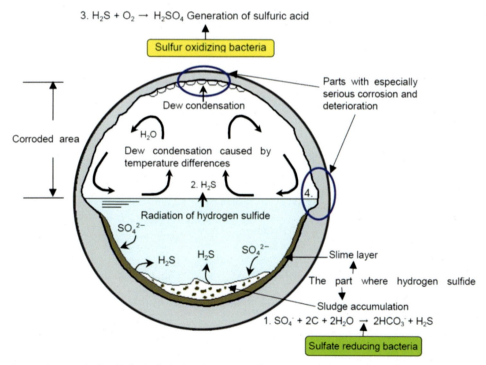

FIGURE 16–6 Events of microbiologically-induced corrosion of concrete in sewer environment (Wu et al., 2020).

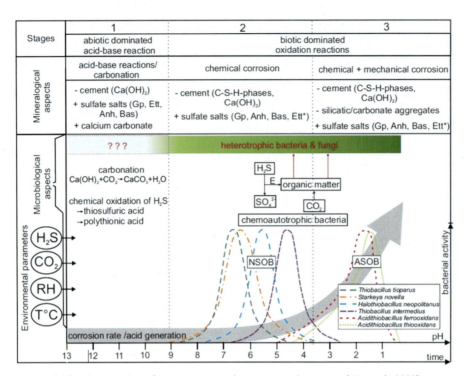

FIGURE 16–7 Model for the corrosion of concrete exposed to sewer environments (Wu et al., 2020).

formation of sulfuric acid leading to concrete corrosion (Fig. 16–7) (Wu et al., 2020). SOB (Sulfur-Oxidizing Bacteria) plays a critical role in the process.

16.6 Deterioration of concrete materials

H_2SO_4 formed by SOB will react with cement hydration products and following reactions leading to deterioration of cementitious system occur.

$$H_2SO_4 + CaCO_3' \rightarrow CaSO_4 + H_2CO_3$$
$$H_2SO_4 + Ca(OH)_2 \rightarrow CaSO_4 + 2H_2O$$
$$H_2SO_4 + CaO \cdot SiO_2 \cdot 2H_2O \rightarrow CaSO_4 + Si(OH)_4 + H_2O$$
$$3CaSO_4 + 3CaO \cdot Al_2O_3 \cdot 6H_2O + 26H_2O' \rightarrow 3CaO \cdot Al_2O_3 \cdot 3CaSO_432H_2O$$

Different type of cements and aggregates deteriorate with different rates and to different extents.

In calcium aluminate cement (CAC) based binder, the hydrates that are neutralized by the acids generated in sewers are hydrogarnet (CAH_{10}, C_2AH_8, and C_3AH_6) and gibbsite (alumina gel—AH_3). When these hydrates react with biogenic sulfuric acid, the following reactions occur (Kiliswaa et al., 2019).

$$C_3AH_6 + 3H_2SO_4 \rightarrow 3C\bar{S} + AH_3 9H$$
$$3.\ C\bar{S}H_2 + C_3AH_6 + H_{20} \rightarrow C_6A\bar{S}_3H_{32}$$
$$2AH_3 + 3H_2SO_4 \rightarrow Al_2(SO_4)_3 + 12H$$

CAC-based binder gives various cement hydrates responsible for a longer chain of dissolution and neutralization and thus more resistant to deterioration.

A number of factors responsible for deterioration is given in Table 16–3 (Wu et al., 2020).

Biodeterioration of concrete depends on pH. High pH restricts the deterioration of freshly prepared concrete against microbial activity although bacteria gradually lowers the pH of concrete. Deterioration is higher at lower pH. pH-dependent microbial succession of

Table 16–3 Factors influencing microbiologically-induced corrosion rate (Wu et al., 2020).

Factor	Description	Effect
Sulfate content	More sulfates gives more sulfide formation	+
DO	High DO reduce SRB activity, reducing sulfate conversion to sulfide	−
	DO (high) increases the oxidation of $H_2S_{(aq)}$, reducing $H_2S_{(g)}$ in sewers	−
BOD	High BOD promote sulfide generation	+
pH	High pH discourage $H_2S_{(g)}$ buildup	−
Temperature	(1) Microbial activities increase at high temperature	+
	(2) More $H_2S_{(aq)}$ is released at high temperature in sewers	+
Relative humidity	H_2SO_4 formation is increased	+
Alkalinity	High alkalinity slows down the MIC rate	−
High aluminate cements	High aluminate cements show better performance as compared to Portland cements in MIC	−
Geopolymer	Geopolymer performs better than Portland cements under acid attack	−
Water/cement ratio, SCMs, porosity	Mixed results are obtained	N.A.
SRB	More SRB increase higher MIC rates	+
SOB	SOB increase MIC rates	+
Others	Other microorganisms or bacteria are not known	N.A.
Hydraulics and design	(1) Liquid-gas surface area for H_2S increases at high turbulence	+
Turbulence	(2) Re-aeration encouraged at high velocity leading SOB growth	+
	(3) The thickness of the slime layer is affected at high velocity	NA
Retention time	(1) DO is consumed at a long retention time increasing H_2S to sewer head space	+
	(2) Formation of H_2S(aq) is increased at a long retention time leading to a loose slime layer	+
Ventilation conditions	(1) Release of H_2S(aq) to the gas phase is favored at adequate ventilation	+
	(2) Concentration of H_2S(g) is limited at adequate ventilation leading to decrease MIC rates	−

BOD, Biological Oxygen Demand; *DO*, Dissolved oxygen; *MIC*, microbial-induced corrosion; *SRB*, sulfate-reducing bacteria; *SCM*, Supplementry Cementitious Materials.

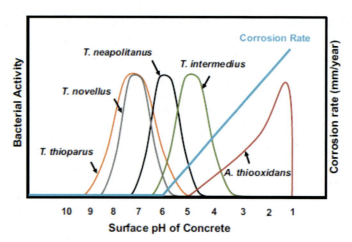

FIGURE 16–8 Variation of bacterial activity and corrosion rate with pH (Noeiaghaei et al., 2017).

Thiobacillus species on the surface of fresh concrete exposed to hydrogen sulfide is presented in Fig. 16–8 (Noeiaghaei, Mukherjee, Dhami, & Chae, 2017).

A pH drop stimulates the growth of SOB on the surface of concrete, increasing the formation of H_2SO_4, leading to enhancement of biodegradation of concrete. As a result of reaction between biologically produced H_2SO_4, gypsum, ettringite, or thaumosite are formed. Due to formation of these compounds, internal cracking occurs, reducing the compressive strength by 20% (Noeiaghaei et al., 2017).

Carbonation of concrete contributes to the neutralization of concrete surface and is due to diffusion of carbon dioxide into the structure (Noeiaghaei et al., 2017). The reaction occurs with calcium hydroxide and calcium silicate hydrates. As the hydrates react with CO_2, the pH of surface layer of concrete reduces to ~9 and becomes less alkaline. Carbonation may be advantageous by decreasing porosity but can be detrimental as it initiates microorganism's growth on the surface of concrete. Carbonation and bacterial colonization can even damage plain concrete. The rate of carbonation depends on number of factors such as calcium hydroxide content, porosity of concrete, moisture content of concrete, and environmental parameters such as humidity and atmospheric concentrations of CO_2.

Relative humidity (RH) is another important parameter that influences biodeterioration of concrete. Higher RH deteriorates faster (Noeiaghaei et al., 2017). Temperature also affects the rate of biodeterioration because of changes of abiotic and biotic reaction rates. Environmental factors, type of materials present in concretes and water-solid ratio affect biogenic deterioration.

16.7 Measures against concrete biodeterioration

Most of the concrete pipes for sewage disposal are underground and subjected for a long period of time to erosion, acid, microbial and other corrosive attack. As a consequence concrete deteriorates resulting into huge economic losses. Repair and replacement of sewer pipes is very expensive, but in addition, sewer pipes corrosion and deterioration cause extensive damage to roads and pavements. Number of methods has been developed to protect the deterioration.

Biodeteriation can be minimized by controlling the environment responsible for corrosion. In sewer systems, if the rate of generation of sulfide and emission of H_2S can be minimized, deterioration can be controlled to some extent. This can be done by adding, NaOH, oxygen, $Mg(OH)_2$, iron salts, and nitrates in sewer systems (Noeiaghaei et al., 2017).

Biodeterioration of concrete occurs when water and nutrients are available and biocorrosive attack depends on a number of factors, such as the type of cementitious materials, porosity, environmental conditions, architectural conditions, and so forth. By using biocides and protective coatings, biodegradation can be minimized. Coating with epoxy resins, an embedded antimicrobial agent, and silver bearing zeolites, prevent deterioration.

Biodeterioration can be controlled by appropriate coating (Table 16–4) (Noeiaghaei et al., 2017).

16.8 Biofilms

In order to develop biofilms normal sewage and sewage with different compositions (AS1, AS2, AS3, AS4) (Table 16–5) (Kong, Liu, Cao, & Fang, 2018) have been used.

Table 16–4 Coating systems to protect biodeterioration of concretes (Noeiaghaei et al., 2017).

Coating material	Remarks
Polyester-based polymers	Bond strength with concrete surface is high and resist the reaction of acidic solution
Epoxy	Excellent durability against *Thiobacillus ferrooxidans*
Silane/siloxanes	Due to sulfate attack, low reduction of compressive
Glass fiber mat reinforced epoxy	In acidic environments, no failure for reinforced up to 20 months
Polyurethane	Reduced solution penetration after three year exposure to 3% H_2SO_4
Epoxy and polyurethane	Good protection even after 60-day immersion in 2.5% H_2SO_4
Epoxy containing silver functionalized zeolites	Reduces microbial population
Epoxy mixed with Cu_2O and Ag_2O	92% and 99.9%, reduction of sulfide formation for cuprous and silver oxide coated concrete pipes

Table 16–5 Nutrient solution (g/20 kg mother liquid) (Kong, Liu, et al., 2018).

Sewage samples	COD (mg/L)	Glucose	NaCl	Starch	Peptone	Urea	$(NH_4)_2HPO_4$	$MgSO_4$
AS1	300	6.0	0.6	12.0	6.0	4.0	1.8	1.3
AS2	3000	62.0	1.0	125.0	14.0	8.0	4.5	2.4
AS3	6000	110.0	1.2	200.0	28.5	12.0	6.7	3.6
AS4	9000	167.7	1.5	307.2	46.4	20.0	5.6	3.0

Fig. 16–9 indicates that mass losses increases with time but the values were lower in presence of biofilms. However, with the increase of COD, mass losses increased. *COD, Chemical Oxygen Demand.*

Further higher COD (Chemical Oxygen Demand) decreased pH considerably, making the concrete to deteriorate fast (Fig. 16–10) (Kong, Liu, et al., 2018). Thus in any sewage by any means, if the pH is high, the concrete is more durable.

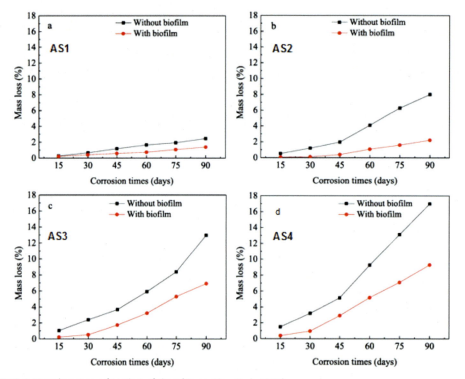

FIGURE 16–9 Mass losses as a function of time (Kong, Liu, et al., 2018).

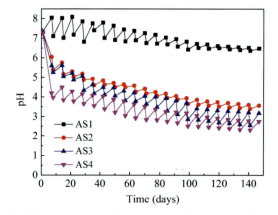

FIGURE 16–10 pH changes with different concentrations in sewage (Kong, Liu, et al., 2018).

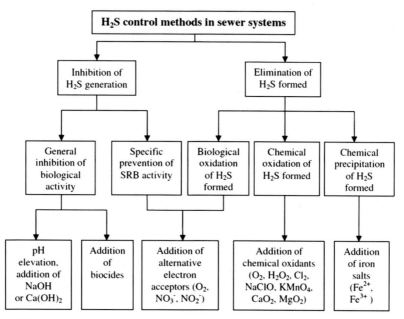

FIGURE 16–11 Sulfide control in sewer systems (Wang et al., 2020).

16.9 Minimizing sulfide in sewer environment

MIC can be minimized by reducing sulfide as given in Fig. 16–11 (Wang, Wu, Kan, & Wu, 2020).

16.10 Antimicrobial agents in concrete

Chemicals which have antimicrobial (bactericidal) property or can inhibit MIC can be added in sewers to protect concrete from deterioration. Heavy metal ions are generally added to reduce MIC, where they rupture the walls of bacteria and kill them. Other than heavy metal ions, chemicals such as nitrofuran, calcium formate, isothiazoline/carbamate, etc. are also used. A few antimicrobial agents are given in Table 16–6 (Wang et al., 2020).

16.11 Changing redox conditions

By adding nitrate to sewers, sulfide production is prevented. In the presence of nitrate the redox potential of wastewater is increased, and MIC is reduced (Wang et al., 2020).

16.12 Inhibiting the activities of sulfate reducing bacteria

If biofilms, where SRB reside and are destroyed by any means, MIC can be reduced. By adding NaOH, pH of wastewater is enhanced and if pH goes above 9; $H_2S_{(aq)}$ in wastewater is almost negligible, inhibiting the activity of SRB for at least 2 weeks.

Table 16–6 Antimicrobial agents to minimize microbiologically-induced corrosion of concrete (Wang et al., 2020).

Antimicrobial agents	Effects
Sodium tungstate	At 50 µM concentration inhibits five strains of ASOB
Ammonium chloride, copper slag, zinc oxide, cetyl-methyl-ammonium bromide, and sodium bromide	Bacteriostatic properties are found similar to commercial biocides
Nickel, calcium tungstate	By adding nickel and calcium tungstate (0.075%) in sewer containing 28 ppm H_2S, microbial-induced corrosion is reduced
Antimicrobial fibers	In 2–4 mm of the area, no growth of *Escherichia coli* and *Staphylococcus aureus* bacteria
Titanium dioxide	On white cements, Algal growth was not seen in presence of TiO_2 nanoparticles
Nickel, tungsten and fluosilicates	SOB growths are inhibited
Copper and copper oxide	After 4 h contact, 95% of *E. coli* is killed
Copper oxide	Nano copper oxide film inhibits ASOB growth
Silver–silica composite	99% bacteria (both gram positive and negative) was killed
Isothiazoline/carbamate	Isothiazoline/carbamate has outstanding antifungal effects
Aluminum ions	Calcium aluminate cements have high durability in presence of Al
Calcium formate	Calcium formate (50 mM) inhibits the growth of the SOB (ASOB)
Copper, silver, and zinc ions contained in zeolites	Cu/Zn is found to show antimicrobial effect similar to Ag

16.13 Chemical removal of sulfide

When some oxidizing agents like $KMnO_4$ and H_2O_2 are added, sulfides are oxidized and MIC is prevented.

16.14 Other measures

Oxidation of sulfide by biological methods has also been tried to control sulfide levels in sewers. Formaldehyde or para formaldehyde, nitrate, etc. are used to oxidize the H_2S generated by SRB. Use of MgO_2 and CaO_2 in controlling the formation of sulfides has also been studied.

Studies have shown that sulfide can be converted to elemental sulfur by microbial fuel cells (Fig. 16–12) (Wang et al., 2020). The advantage of this method is that electricity is generated in the process. The electricity generated can be used for other works.

16.15 Biocide

Bacteria, lichen, fungi, algae, archaea, animals, and higher plants cause biodeterioration of cultural heritage (Fig. 16–13) (Kakakhel et al., 2019). Biocides can be used for preventing deteriorations of cultural heritage by microorganisms. Some of the biocides used for controlling biodeterioration of heritage buildings are given in Table 16–7.

Chapter 16 • Microbiologically induced deterioration of cement-based materials 385

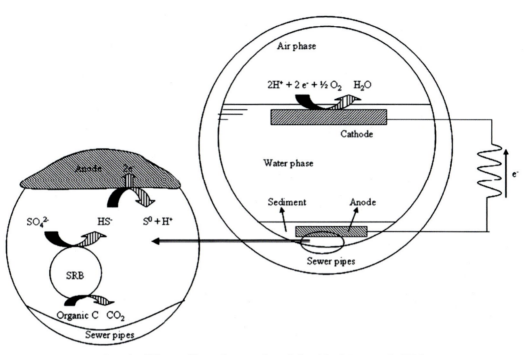

FIGURE 16–12 Conversion of sulfide to sulfur and generation of electricity (Wang et al., 2020).

FIGURE 16–13 Biodeterioration of the world heritage (Kakakhel et al., 2019).

Table 16–7 Protection from biological deterioration of cultural heritage by biocides.

Name of biocide	Active components
Des Novo R	Essential oil and benzalkonium chloride
Anatase	Titanium dioxide (TiO_2)
Biotine R	N-octyl-isothiazolinone (3%−5%) þ 3-iodoprop-2-ynyl N-butylcarbamate (10%−25%)
Algophase R	2,3,5,6-tetrachloro-4-methylsulfonyl-pyridine
Clove Extract	Eugenol
Devor Mousse R	Quaternary ammonium and benzalkonium chloride
Colloidal silver	Nanosilver
Lichenicida 464	4,5-dichloro-2-octyl-4-isothiazolin-3-one þ 3-iodo-2-propynyl N butyl carbamate þ 2-octyl-4-isothiazolin-3-one þ benzyl alcohol
Polymyxins R	Nonribosomal synthesized secondary metabolites
DES	2-phenylethoxy-quinazoline
Glifene SL	N-(phosphonomethyl) glycine; 30%−40%
Kimistone Biocide	Didecyldimethylammonium chloride
Polybor TR	Borates/Boric acid
Metatin N58-	Dimethyl benzyl ammonium bromide
Mora Poultice	Ammonium/Sodium Bicarbonates
Panacide	Dichlorophene
Parmetol DF12	Isothiazoline Derivatives
Preventol R180	(80%), isopropyl alcohol (2%), Alkyl dimethyl benzyl ammonium chloride
Prenentol R	Ammonium quaternary compound
Nanoparticles Prepared	ZnO
Rocima 103	Alkyl-benzyl-dimethylammonium
Sanosil S003	Hydrogen peroxide and Silver nitrate
Secondary Metabolites	Lipopeptide
Sinoctan PS	Iodopropynylbutyl, 4,5-Dichloro-2-octylisothiazolinone,
Ucarcide R	di-aldehyde product

16.16 Nanomaterials for minimizing microbial attack

NMs either mixed in the concrete or coated on the surface, minimizes corrosion. NMs rupture the cell of bacteria and thus kills. As a result microbial corrosion can be hindered. The overall action of NMs in killing the bacteria is given in Fig. 16−14 (Ogunsona, Muthuraj, Ojogbo, Valerio, & Mekonnen, 2020).

16.17 Conclusion

Microbial corrosions, including other type of corrosions of cementitious materials with reference to concrete, have been discussed. Microbial corrosion is more pronounced for concrete structure during sewage disposals. Different types of protective measures including use of

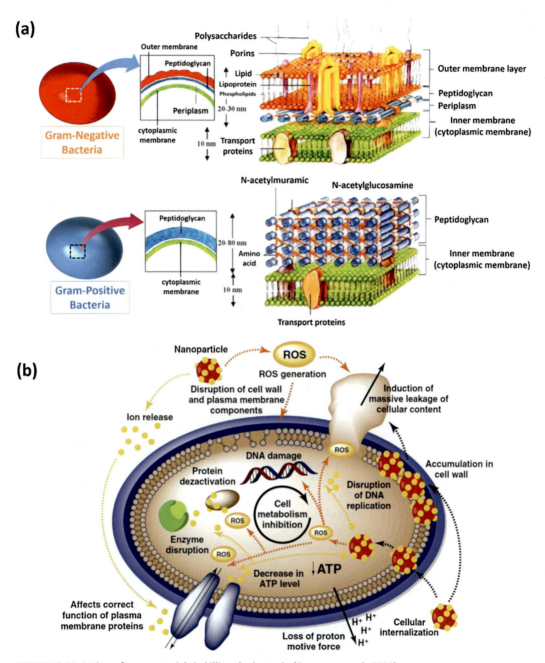

FIGURE 16–14 Action of nanomaterials in killing the bacteria (Ogunsona et al., 2020).

NMs has been discussed. Attempts have been made to understand the mechanism of microbial deterioration of concretes and protection.

References

Cyr, M. (2013). *Influence of supplementary cementitious materials (SCMs) on concrete durability.* Woodhead Publishing Limited, 2013, 153-197.

Grengg, C., Mittermayr, F., Ukrainczyk, N., Koraimann, G., Kienesberger, S., & Dietzel, M. (2018). Advances in concrete materials for sewer systems affected by microbial induced concrete corrosion: A review. *Water Research, 134*, 341–352.

Kakakhel, M. A., Wu, F., Gu, J.-D., Feng, H., Shah, K., & Wang, W. (2019). Controlling biodeterioration of cultural heritage objects with biocides: A review. *International Biodeterioration & Biodegradation, 143*, 104721.

Kiliswaa, M. W., Scrivener, K. L., & Alexander, M. G. (2019). The corrosion rate and microstructure of Portland cement and calcium aluminate cement-based concrete mixtures in outfall sewers: A comparative study. *Cement and Concrete Research, 124*, 105818.

Kong, L., Liu, C., Cao, M., & Fang, J. (2018). Mechanism study of the role of biofilm played in sewage corrosion of mortar. *Construction and Building Materials, 164*, 44–56.

Kong, L., Fang, J., Zhou, X., Han, M., & Lu, H. (2018). Assessment of coatings for protection of cement paste against microbial induced deterioration through image analysis. *Construction and Building Materials, 191*, 342–353.

Little, B. J., Blackwood, D. J., Hinks, J., Lauro, F. M., Marsili, E., Okamoto, A., . . . Flemming, H.-C. (2020). Microbially influenced corrosion—Any progress? *Corrosion Science, 170*, 108641.

Lokesh, K., Satish Reddy, M., & Kotaiah, B. (2014). Degradation of concrete structures and protective measures. *International Journal of Engineering Research & Technology (IJERT), 3*(5), 1415–1420.

Márquez, M., Sánchez-Silva, & Husserl, J. 2013 Review of reinforced concrete deterioration mechanisms. In VIII international conference on fracture mechanics of concrete and concrete structures FraMCoS-8 J. F, pp. 1–9.

Natarajan, K.A. (2018). Biofouling and microbially influenced corrosion, 355–393.

Noeiaghaei, T., Mukherjee, A., Dhami, N., & Chae, S.-R. (2017). Biogenic deterioration of concrete and its mitigation technologies. *Construction and Building Materials, 149*, 575–586.

Ogunsona, E. O., Muthuraj, R., Ojogbo, E., Valerio, O., & Mekonnen, T. H. (2020). Engineered nanomaterials for antimicrobial applications: A review. *Applied Materials Today, 18*, 100473.

Stott, J.F.D. (2010). Corrosion in microbial environments, 1169–1190.

Wang, T., Wu, K., Kan, L., & Wu, M. (2020). Current understanding on microbiologically induced corrosion of concrete in sewer structures: A review of the evaluation methods and mitigation measures. *Construction and Building Materials, 247*, 118539.

Wei, S., Jiang, Z., Liu, H., Zhou, D., & Sanchez-Silva, M. (2013). Microbiologically induced deterioration of concrete - a review. *Brazilian Journal of Microbiology, 44*(4), 1001–1007.

Wu, M., Wang, T., Wu, K., & Kan, L. (2020). Microbiologically induced corrosion of concrete in sewer structures: A review of the mechanisms and phenomena. *Construction and Building Materials, 239*, 117813.

17

Microbiologically induced deterioration of concrete

Zain Ul-Abdin, Waqas Anwar, Anwar Khitab

DEPARTMENT OF CIVIL ENGINEERING, MIRPUR UNIVERSITY OF SCIENCE AND TECHNOLOGY (MUST), MIRPUR, PAKISTAN

Abbreviations

Ag	Silver
BIC	Biological-Induced Corrosion
BID	Biological-Induced Deterioration
Br^-	Bromide
$Ca(OH)_2$	Calcium Hydroxide
$CaSO_4$	Calcium Sulfate
$CaWO_4$	Calcium Tungstate
CH_4	Methane
CO_2	Carbon dioxide
Cu	Copper
Cu-S	Copper Slag
EPI	Extracellular Polymeric Ingredients
FA	Fly Ash
GGBFS	Ground Granulated Blast Furnace Slag
H_2S	Hydrogen Sulfide
H_2SO_4	Sulfuric Acid
Me^+	Heavy Metal Ion
MIC	Microbiological-Induced Corrosion
MICD	Microbiologically Influenced Concrete Deterioration
MID	Microbiological-Induced Deterioration
MIMD	Microbiologically Influenced Metal Deterioration
NH_3	Ammonia
NH_4^+	Ammonium
Ni	Nickel
SCM's	Supplementary Cementitious Materials
SF	Silica Fume
SOB	Sulfur-oxidizing bacteria
VSOCs	Volatile Organic Compounds
W	Tungsten
ZnO	Zinc oxide
$ZnSiF_6$	Zinc Silicon Fluoride Hexahydrate

17.1 Introduction

17.1.1 Conventional concrete

Concrete is a multiphase, porous, and strong basic composite material (Alexander, Bertron, & De Belie, 2013). It is mainly a blend of aggregates (gravel and rock) and paste (cement, sand, and water) that include numerous hydrated composites of calcium silicates, calcium aluminates, alumino ferrites, and gypsum. The structural properties of concrete are mainly due to a series of hydration chemical reactions that result in the formation of a stable matrix, which binds all ingredients together into a stone-like hard mass (Khitab & Anwar, 2016; Khitab, Anwar, Mansouri, Tariq, & Mehmood, 2015). To date, there exists multiple ranges of concretes, with Portland cement concretes being the most common. To achieve diverse physiochemical properties a wide range of minerals and chemical admixtures such as plasticizers, superplasticizers (e.g., polycarboxylate ether), and/or supplementary cementitious materials, such as fly ash (FA), volcanic ash, silica fume (SF), and blast furnace slag can be incorporated to admixtures (Khitab, Arshad, Awan, & Khan, 2013; Parthasarathi, Saraf, Prakash, & Satyanarayanan, 2019).

Concrete is normally alkaline and its pH varies from 11–13 (Khitab, Lorente, & Ollivier, 2005; Khitab, 2005; Wang, Wu, Kan, & Wu, 2020). The alkaline character of concrete is due to the creation of calcium hydroxide ($CaOH_2$) because it is the byproduct of hydrated cement. The atmospheric carbon dioxide (CO_2) is a mean to reduce the concrete pH up to 9.5 due to carbonation of concrete. During this process; the CO_2 is dissolved in water and carbonic acid (H_2CO_3) is formed, this acid reacts with alkaline component of the cement and results in the lowering of pH which may provide a medium for microbes' adhesion in concrete (Baltrusaitis & Grassian, 2010).

17.1.2 Deterioration in concrete

For the life cycle of structure and the components of concrete, deterioration plays a very important role. Main causes of deterioration include sulfate attack, chloride ingress, aging, and live organism actions. Microorganisms have complex structures, and they always stay in contact with water in sanitary and sewage systems, substructures, and marine structures. To degrade existing structures, two processes are the most important: (1) microbiological-induced corrosion (MIC) and (2) microbiological-induced deterioration (Khitab, Anwar, Ul-Abdin, Tayyab, & Ibrahim, 2019).

17.1.2.1 Microbiological-induced deterioration

The corrosion or cracking action by microorganism as produced by inorganic and organic acids conciliate integrity of concrete generating substantial problems. Mostly, deterioration means the reduction of structural capacity with time as a result of "Weakened Material and External Agencies Actions" (Bastidas-Arteaga, Sánchez-Silva, Chateauneuf, & Silva, 2008).

Biodeterioration is also termed as microbiologically induced deterioration. It is an unwanted change in the material characteristics caused by an external agent termed as living organisms, resulting in lowering of the structural quality. There are two major boundaries of biodeterioration, namely, biology and engineering, known as microbiology in combined form. It is necessary to study the chemical and physical mechanisms of microbes activity. Mechanical characteristics can be affected through the microbes outbreak. In concrete structures, biodeterioration disturbs the solid concrete mass by enhancing porosity and develops cracks. The exposure to living microorganisms speeds up damage processes that may eventually lead to undesirable performance or cause catastrophic failure (Sanchez-Silva & Rosowsky, 2008).

Deterioration is a multidimensional process that depends on many factors like structural type, constitutive material, operational characteristics, and environmental circumstances. The consequences of deterioration are destructive and impact the performance of the infrastructure systems, reduce the structural capacity, and ultimately pose potential threat to the economy (Bastidas-Arteaga & Sa, 2008). In a harsh environment, the deterioration of concrete structure arising from biological sources, namely, living organisms is very significant. These organisms grow on concrete surfaces in environments that are favorable for their survival—available water and lower pH. The activity of living organisms is very prominent in conductive environments with high carbon dioxide concentrations (like carbonation), relative humidity (e.g., from 60% to 98%), chloride ions (marine conditions), salt concentrations (deicing salts), a small amount of acids (sanitary and sewage system), and sulfates (contact soils). Fig. 17-1 shows the MIC deterioration of concrete manholes. It is very important to note that biodeterioration and biodegradation are two different terms, which are used mistakenly as same. Biodegradation is a naturally occurring breakdown of materials by microorganisms such as bacteria and fungi. It occurs in certain special circumstances. According to investigators, biodegradation is intended to utilize the microorganisms for improving the material properties with a beneficial or positive purpose while biodeterioration is related to the negative impact of microorganism activity (Grengg et al., 2018).

17.1.2.2 Microbiological-influenced concrete deterioration

The microbiological-influenced concrete deterioration (MICD) process can be observed on concrete surfaces exposed to the marine environment. This process is different from microbiological-influenced metal deterioration (MIMD). The mechanism of MIMD is an electrochemical process while the MICD includes the chemical deterioration of concrete hydration products such as $Ca(OH)_2$ by organic ingredients especially bacteria (Hernandez, Kucera, Thierry, Pedersen, & Hermansson, 1994; Örnek, Wood, Hsu, & Mansfeld, 2002; Örnek et al. 2002; Zuo & Wood, 2004). The similarity between MIMD and MICD is that the deterioration process is initiated by the activity of living organisms such as bacteria, algae, and fungi (Scheerer, Ortega-Morales, & Gaylarde, 2009).

17.2 Biological deterioration of concrete

17.2.1 Mechanism

In the deterioration of concrete, living organisms, especially microorganisms, play a very important role. Recently, the development in experimental techniques has provided a way to

FIGURE 17–1 Microbiologically induced concrete deterioration of concrete manholes. *Adapted with permission from Grengg, C., Mittermayr, F., Ukrainczyk, N., Koraimann, G., Kienesberger, S., & Dietzel, M. (2018). Advances in concrete materials for sewer systems affected by microbial induced concrete corrosion: A review.* Water Research, *134, 341–352.*

understand how physical, mechanical, and chemical characteristics of concrete as well as its ingredients in terms of strength and durability were altered because of biological-induced deterioration (Hudon, Mirza, & Frigon, 2011; Sanchez-Silva & Rosowsky, 2008; Wei, Jiang, Liu, Zhou, & Sanchez-Silva, 2013). The concrete structures, which come in contact with an aggressive environment such as chloride ion penetration and sulfate attack are often more significantly deteriorated (Khitab, Lorente, & Ollivier, 2006). These aggressive environmental conditions lead to an initial level of deterioration and reduce the concrete structural capacity of carrying the load. Biological-induced corrosion (BIC) of concrete can be found in building foundations and walls and is shown in Fig. 17–2 (Cwalina, 2008). Furthermore, the process of BIC can be detected in dams, harbor structures, tanks, pipelines, silos, cooling structures, and bridges.

Progressive deterioration is mainly due to chloride ingress, which has been proved by previous studies (Hussain, Rizwan, & Khitab, 2015; Khitab, Lorente, & Ollivier, 2004; Khitab, Arshad, Ali, Kazmi, & Munir, 2014). Moreover, the deterioration process is accelerated by the biological process and it alters physicochemical characteristics of concrete. Act of microorganisms disturbs the concrete primarily by the erosive action on the exposed surface of the

FIGURE 17–2 Lichens in concrete foundation due to Algae. *Adapted with permission from Cwalina, B. (2008). Biodeterioration of concrete.* Architecture Civil Engineering Environment (ACEE) 4, 1, 133 140. Copyright 2008.

concrete by leading to a decrease in the defense cover depth. Concrete porosity can grow more, which increases the passage of degrading ingredients into the concrete, and hence accelerates the spalling, cracking, and other defects. It can also decrease the structural service life (Khitab, Anwar, & Arshad, 2017). Fungi, bacteria, algae, live organisms and lichens are the causes of biodeterioration of concrete, as shown in Fig. 17–2.

In addition to it, drilling and erosion of the concrete also occurs (Bastidas-Arteaga et al., 2008). Living organisms, owing to their wide variety, can accelerate deterioration in concrete: The summary of the main organisms, responsible for the deterioration of concrete is given in Table 17–1.

17.3 Classification of biological deterioration

Gaylarde et al. classified the biodeterioration into three types which are explained subsequently (Gaylarde, Ribas Silva, & Warscheid, 2003).

17.3.1 Physical or mechanical classification

In physical or mechanical biodeterioration, the activity of living organisms is very harmful. It affects mechanical properties and material components. It is mostly associated with the live microorganism process. This can have adverse effects on the material structure by movement or by growth but it does not utilize the material as a meal source, namely, gnawing by rodents and root damage. The microorganisms involved in this type of deterioration process

Table 17–1 Organisms participated in deterioration of concrete table.

Sr. No.	Live Organisms	Category	References
1	*Candelariella* ssp	Algae	Edwards et al. (1999)
2	*Acarospora cervina*		
3	*Ulva fasciata*		Khitab (2005)
4	*Chaetomorpha antennina*		
5	*Exophiala* sp.	Fungus	Khitab et al. (2005)
6	*Alternaria alternata*		
7	*Aspergillus niger*		Lajili, Devillers, Grambin-Lapeyre, and Bournazel (2008)
8	*Alternaria* sp		
9	*Penicillium oxalicum*		Khitab et al. (2019)
10	*Pestalotiopsis maculans*		
11	*Trichoderma asperellum*		
12	*Thiobacillus thioparus*	Bacteria	Bastidas-Arteaga et al. (2008), Sanchez-Silva and Rosowsky (2008)
13	*Thiobacillus neapolitanus*		
14	*Thiobacillus novellus*		
15	*Thiomonas perometablis*		Vollertsen, Nielsen, Jensen, Wium-Andersen, and Hvitved-Jacobsen (2008)
16	*Acidithiobacillus thiooxidans*		Bastidas-Arteaga et al. (2008), Sanchez-Silva and Rosowsky (2008), Bastidas-arteaga and Sa (2008), Grengg et al. (2018)

FIGURE 17–3 Physical deterioration of farm (Scheerer et al., 2009).

are autotroph bacteria. These bacteria yield complex organic compounds gaining energy either by photosynthesis (photoautotrophic) or due to inorganic chemical reactions (chemoautotrophic). The physical deterioration of a farm is shown in Fig. 17–3.

17.3.2 Soiling (esthetic)/fouling classification

In fouling or soiling (esthetic) deterioration, a biofilm or microbial layer is formed on the surface of the material structure due to the existence of microorganisms, their dead bodies, metabolic products and excreta. In this type of deterioration, the performance of the material structure is not disturbed but causes an unacceptable appearance. Formation of biofilm on the window of the Convent of Christ in Tomar, Portugal, and the stone surfaces of the façade of the Cathedral of Monza (Italy) are shown in Fig. 17–4 (Rosado et al., 2020; Gulotta, Villa, Cappitelli, & Toniolo, 2018). Biofilms are composed of groups comprising microbial cells that are rooted in a self-produced matrix of extracellular polymeric ingredients and are attracted to each other and/or to a surface (Flemming et al., 2016). They depict a strong spatial and chronological heterogeneous ecology organized by internal procedures and ecological circumstances. Mixed culture biofilms, such as environmental biofilms, can be composed of a wide range of microorganisms including autotrophic and heterotrophic bacteria, fungi, algae, and archaea. Life important compounds such as nutrients are provided by either gaseous diffusion or throughout interstitial lancing the biofilm. The microorganisms involved in this type of deterioration process are Heterotroph bacteria. These bacteria utilize organic carbon as an energy source to synthesize the life-essential modules (Ichhpujani, Bhatia, & Ichhpujani, 2008).

17.3.3 Chemical classification

The chemical classification of biodeterioration is further divided into two types, namely, assimilatory and dissimilatory, as discussed in the following sections.

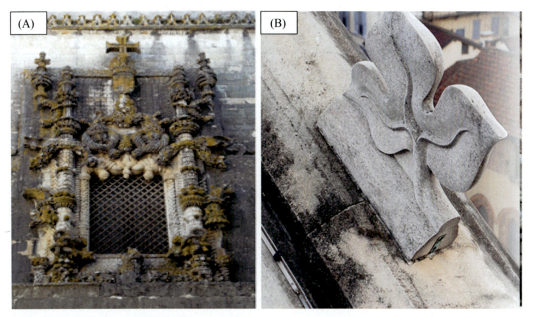

FIGURE 17–4 Formation of biofilm (a) window of the convent of Christ, Tomar, Portugal (b) façade of the Cathedral of Monza, Italy (Gulotta et al., 2018; Rosado et al., 2020). *Adapted with permission from Elsevier Copyright Clearance Center's RightsLink.*

17.3.3.1 Assimilatory

In assimilatory chemical biodeterioration, the live microorganisms utilize the material structural components as food source that are carbon and energy sources. It modifies the material properties such as degradation of metals and fuels (Southgate, 2005).

17.3.3.2 Dissimilatory

In dissimilatory chemical biodeterioration, the live organisms severely affect the material properties and structure by harmful substances and excrete waste products such as hydrogen sulfide (H_2S) and iron sulfide (FeS) (Obieze et al., 2020).

17.4 Microorganisms and their colonization on concrete

After the construction work is completed, concrete usually shows less resistance to biological attack. It is just because of its high alkaline nature due to calcium hydroxide formation, for example, $Ca(OH)_2$ as a cement byproduct. Microorganism activity is very least effective at high pH of concrete (Sand, 2014). Usually, the roughness on the concrete surface is because of the frictional effect of structural elements with the collaboration of other materials and water's erosive action. The colonization process of microbes on the concrete surface is started due to the presence of moisture and nutrients. In the case of aquatic or marine environment especially for the case of wastewater treatment plants and sewer line build-up from concrete, sulfate is presented in abundant. It is very necessary to introduce the sulfur cycle, which is very prominent microbial activity. Under anaerobic conditions, organic matter especially sulfur-reducing bacteria can be reduced by the sulfur, and it may alter sulfate into sulfide, namely, ultimately combine with hydrogen (H_2) to form hydrogen sulfide gas (H_2S) via the following chemical reactions (Hernandez et al., 1994).

Overall Reaction:

$$\text{SRB (Organic ingredients)} + SO_4^{-2} \text{ (Sulfate ions)} \rightarrow S^{-2} + H_2O + CO_2 \tag{17.1}$$

$$S^{-2} + 2H^+ \text{(Hydrogen Ions)} \rightarrow H_2S \tag{17.2}$$

$$H_2S \text{ (Hydrogen Sulfide)} + 2O_2 \xrightarrow{\text{(SOB)}} H_2SO_4 \text{ (Sulfuric Acid)} \tag{17.3}$$

Reaction mechanism involving the deterioration of concrete by sulfuric acid is shown in Fig. 17–5.

With the passage of time, pH or alkaline nature of the concrete surface is step by step reduced by neutralization of H_2S and carbonation (Lahav, Lu, Shavit, & Loewenthal, 2004; Matos & Aires, 1995; Nielsen, Yongsiri, Hvitved-Jacobsen, & Vollertsen, 2005; Zhang et al., 2008). Meanwhile, the sulfur-oxidizing bacteria convert the H_2S into sulfuric acid (H_2SO_4); that further reduces the pH value. The formation of H_2SO_4 is very corrosive to the concrete surfaces in terms of deterioration. When the pH of the concrete surface is further lowered, the acidophilic and neutrophilic living organisms create more microbial colonization. The key organisms that participate during the colonization, include *Thiobacillus* species (Mori et al., 1992). It is also noted that

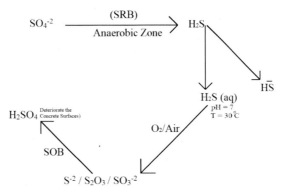

FIGURE 17–5 Complete reaction mechanism of deterioration of concrete.

the production of hazardous and odorless gases, such as hydrogen sulfide (H_2S), carbon dioxide (CO_2), methane (CH_4), ammonia (NH_3), and volatile organic compounds during the deterioration process of concrete by microbes, is a subject of major concern. These harmful gases symbolize a substantial health risk for public workers, sewer machinists, as well as for private occupants of affected societies (Gutierrez et al., 2008; Gutierrez et al., 2014). Particularly, H_2S is well known for its collective odor and health-hazardous potential.

17.5 Microorganisms and their chain on concrete

The pH of concrete reduces because of continuous growth of microorganisms that results in the release of biogenic material, that is, sulfuric acid and polythionic (Diercks, Sand, & Bock, 2000; Islander et al., 1992; Botanik, Mikrobiologie, Hamburg, & Republic, 2000; Sand, 2014). As soon as the pH of the concrete surface further reduces below 9, some sulfur bacteria species that are *Thiobacillus* start reproducing in presences of oxygen nutrients and moisture attaches in the chain form of the concrete surface (Sand, 2014; Diercks et al., 2000). When pH is below 5 then *Thiobacillus thiooxidans* start their growth and release a huge amount of sulfuric acid that further drops pH near or equal to 1.5 (Sand & Bock, 2008). When the pH value is close to 3, the growth of *T. thiooxidans* is very rapid. The lower value of pH favors the composition of sulfur element and the oxidation process as carried by *T. thiooxidans*, which converts sulfur into sulfate ion. When the pH is further decreased close to 1, this condition is inhibitory for the bacterial species even for *T. thiooxidans*. The microbial chain on the concrete surface as produced by *T. thiooxidans* is dominated and the organic products' waste as excreted by bacteria is utilized by acidophilic heterotrophs (Peters, Friedel, & McTavish, 1992).

17.6 Microorganism-induced deterioration mechanism process

Microscopic fungi, algae, and bacteria are the main microorganisms initiating concrete biodeterioration. The carbon dioxide (CO_2), butyric acid, acetic acid, and lactic acid (biogenic organic acid) are produced by various living microorganisms. It can be severely affected by the surface of the concrete by corrosive action (Cwalina, 2008; Bertron, Escadeillas, & Duchesne, 2004; Siripong & Rittmann, 2007). The concrete surfaces like concrete pipelines, concrete floors that are directly exposed or in contact with sewer water, or wastewater can be deteriorated by a corrosive agent named as hydrogen sulfide (H_2S). SOB in wastewater are aerobic bacteria that oxidize the hydrogen sulfide (H_2S) into sulfuric acid (H_2SO_4). This biogenic acid is highly corrosive to concrete surfaces as it degrades the cementitious material in concrete, generating different hydrogen states of $CaSO_4$ (Mori et al., 1992) and ettringite ($3CaO \cdot Al_2O_3 \cdot CaSO_4 \cdot 12H_2O$ or $3CaO \cdot Al_2O_3 \cdot 3CaSO_4 \cdot 31H_2O$). Gypsum gives a protective layer to the concrete surface and basic corrosion protects the reinforcement bars in concrete. When the protective layer of gypsum is removed then the concrete surface can be intensely affected by the attack of acid. In addition, the production of ettringite induces cracks in the concrete because of its large volume and high internal pressure as generated by internal stresses (Aviam et al., 2004). Finally, the deteriorated material is removed off from the concrete surface and process of concrete corrosion is accelerated. Its internal concrete surfaces are exposed to the deleterious processes (Mori et al., 1992). The live microorganisms can penetrate in the concrete surface without any observable crack (Sanchez-Silva & Rosowsky, 2008). The most common process of their entrance in the concrete is via microcracks or by concrete capillaries. Laboratory analysis of some research work depicts that many species of microorganisms like fungi (*Cladosporium*, mycelia, yeasts, hypha, etc.), *Thiobacillus*, *actinomycetes* (bacteria), algae, and even protozoa can be found between the matrix of sand and cement. Microorganisms presence in concrete increases the concrete porosity and hence alters the concrete diffusivity and severely affects the concrete durability (Sanchez-Silva & Rosowsky, 2008). Higher porosity values of concrete result greater surface wear and hence minimize the depth of shielded concrete cover, over the reinforcement. When concrete covers are reduced, the high diffusivity can enhance the process of deterioration and even corrode the steel due to reaction between the H_2S gas and concrete reinforcement (Abdelmseeh, Jofriet, Negi, & Hayward, 2005; Abdelmseeh, Jofriet, Negi, & Hayward, 2006; Parande, Ramsamy, Ethirajan, Rao, & Palanisamy, 2006).

17.7 How to control concrete biodeterioration?

The biodeterioration of concrete is basically due to the presence of nutrients and water. The intensity of biocorrosive attacks by living microorganisms depends upon the material properties, porosity, and permeability of exposed surface, and site-specific environmental conditions. The word control means to take suitable measures to prevent the growth of damaging microorganisms (Warscheid & Braams, 2000). The concrete surface can be protected from the attack of microorganisms by utilizing concrete modification in mix design, a protective coating of biocides, water layer consolidants/repellants, and additives (Dai, Akira, Wittmann, Yokota, & Zhang, 2010). Concrete mix modification results in an increase in the alkalinity. Alkalinity and corrosion rate have an inverse relation. Silica content in the concrete mix reacts with $[Ca(OH)_2]$ in the presence of water to form calcium silicate hydrate cementitious compound. In the case of SF concrete, it modifies the mechanical and durability properties of concrete and makes it less prone to corrosion (Yilmaz, 2010). In the case of polymer-modified concrete, the polymers can create a film; cement hydration products and polymers arrange themselves in the form of complex networks, and aggregates are embedded in them (Ohama, 1995). In addition, the biocides provide good results to protect the concrete against microbial attack (Alum, Rashid, Mobasher, & Abbaszadegan, 2008). The selection of biocides depends on the activities of living microorganism that settle stone concrete (Allan, 1999). Polypropylene or other fibers may be organized as bacteriostatic composites systems. The antibacteriostatic agents are introduced as an effective instrument to control microbial growth on concrete surfaces in intense sewer environments and hence protect the concrete surface for the long duration. Moreover, geopolymer concretes are introduced as highly resistant in acid environments, thus representing a possible green alternative to conventional cement-based construction materials. The utilization of antimicrobial agents has been documented in many applications that included the study of the tolerance of heavy metals ion (Me^+) on iron-reducing microbial communities in soils and antimicrobial action of copper coatings in the public sector (Wei, Yang, Wang, Tay, & Gao, 2014; Burkhardt, Bischoff, Akob, Büchel, & Küsel, 2011).

In conventional concretes, antimicrobial additives have also been utilized to control the microbial action. These additives are available in the form of supplementary cementitious materials such as FA, SF, ground granulated blast furnace slag, copper slag and metal ions such as zinc oxide (ZnO), ammonium (NH_4^+) and bromide (Br^-), Copper (Cu), Nickel (Ni), Silver (Ag), Calcium Tungstate $(CaWO_4)$, Tungsten (W), and Zinc Silicon Fluoride Hexahydrate $(ZnSiF_6)$ (Alum et al., 2008; Hashimoto et al., 2015; Negishi et al., 2005; Psarras, 2011). Concrete mixtures including 10% ZnO have acceptable antimicrobial characteristics that are similar to commercial biocides. The heavy metal ions (Me^+) start reacting with the negatively charged cell wall of bacteria to create a toxic effect by the formation of complexes within the bacterial membrane. This reaction acts as a retarder for the activity of sensitive enzymes, destroying the osmotic stability of the cell and finally resulting in the leakage of intracellular constituents (Nies, 1999).

References

Abdelmseeh, V. A., Jofriet, J. C., Negi, S. C., & Hayward, G. L. (2005). Corrosion of reinforced concrete specimens exposed to hydrogen sulfide and sodium sulfate. *CIGR Ejournal*, 1–15.

Abdelmseeh, A. V., Jofriet, J. C., Negi, S. C., & Hayward, G. (2006). Sulphide, sulphate and sulphuric acid corrosion of concrete in laboratory tests. *Solid Mechanics and Its Applications*, 140, 55–65, no. Hewayde 2005.

Alexander, M., Bertron, A., & De Belie, N. (2013). Performance of cement-based materials in aggressive aqueous environments, *10*.

Allan, M.L. (1999). Evaluation of coatings and mortars for protection of concrete cooling towers structures from microbiologically influenced corrosion in geothermal power plants. Brookhaven National Laboratory Report BNL-66980, New York.

Alum, A., Rashid, A., Mobasher, B., & Abbaszadegan, M. (2008). Cement & concrete composites cement-based biocide coatings for controlling algal growth in water distribution canals. *Cement and Concrete Composites, 30*, 839–847.

Aviam, O., Bar-Nes, G., Zeiri, Y., & Sivan, A. (2004). Accelerated biodegradation of cement by sulfur-oxidizing bacteria as a bioassay for evaluating immobilization of low-level radioactive waste. *Applied and Environmental Microbiology, 70*(10), 6031–6036.

Baltrusaitis, J., & Grassian, V. H. (2010). Carbonic acid formation from reaction of carbon dioxide and water coordinated to Al(OH)3: A quantum chemical study. *The Journal of Physical Chemistry. A*.

Bastidas-Arteaga, E., & Sa, M. (2008). Coupled reliability model of biodeterioration, chloride ingress and cracking for reinforced concrete structures. *Structural Safety, 30*, 110–129.

Bastidas-Arteaga, E., Sánchez-Silva, M., Chateauneuf, A., & Silva, M. R. (2008). Coupled reliability model of biodeterioration, chloride ingress and cracking for reinforced concrete structures. *Structural Safety*.

Bertron, A., Escadeillas, G., & Duchesne, J. (2004). Cement pastes alteration by liquid manure organic acids: Chemical and mineralogical characterization. *Cement and Concrete Research, 34*, 1823–1835.

Botanik, A., Mikrobiologie, A., Hamburg, U., & Republic, F. (2000). Thiobacilli of the corroded concrete walls of the Hamburg sewer system, 1983.

Burkhardt, E. M., Bischoff, S., Akob, D. M., Büchel, G., & Küsel, K. (2011). Heavy metal tolerance of Fe(III)-reducing microbial communities in contaminated creek bank soils. *Applied and Environmental Microbiology, 77*(9), 3132–3136.

Cwalina, B. (2008). Biodeterioration of concrete, pp. 133–140.

Dai, J., Akira, Y., Wittmann, F. H., Yokota, H., & Zhang, P. (2010). Water repellent surface impregnation for extension of service life of reinforced concrete structures in marine environments: The role of cracks. *Cement & Concrete Composites, 32*, 101–109.

Diercks, M., Sand, W., & Bock, E. (2000). Universitiit Hamburg, Institut Jfir Allgemeine Botanik, Mikrobiologie, Ohnhorststr. 18, D-2000 Hamburg 52 (Federal Republik of Germany) Degradation of concrete by biogenic acids Biogenic sulphuric acid corrosion, 47, 514–516, 1991.

Edwards, K. J., Goebel, B. M., Rodgers, T. M., Schrenk, M. O., Gihring, T. M., Cardona, M. M., ... Banfield, J. F. (1999). Geomicrobiology of pyrite (Fes2) dissolution: Case study at iron mountain, California. *Geomicrobiology Journal, 16*(2), 155–179.

Flemming, H. C., Wingender, J., Szewzyk, U., Steinberg, P., Rice, S. A., & Kjelleberg, S. (2016). Biofilms: An emergent form of bacterial life. *Nature Reviews. Microbiology, 14*(9), 563–575.

Gaylarde, C., Ribas Silva, M., & Warscheid, T. (2003). Microbial impact on building materials: An overview. *Materials and Structures, 36*(259), 342–352.

Grengg, C., Mittermayr, F., Ukrainczyk, N., Koraimann, G., Kienesberger, S., & Dietzel, M. (2018). Advances in concrete materials for sewer systems affected by microbial induced concrete corrosion: A review. *Water Research, 134*, 341–352.

Gulotta, D., Villa, F., Cappitelli, F., & Toniolo, L. (2018). Biofilm colonization of metamorphic lithotypes of a renaissance cathedral exposed to urban atmosphere. *The Science of the Total Environment, 639*, 1480–1490, Oct.

Gutierrez, O., Mohanakrishnan, J., Sharma, K. R., Meyer, R. L., Keller, J., & Yuan, Z. (2008). Evaluation of oxygen injection as a means of controlling sulfide production in a sewer system. *Water Research, 42*(17), 4549–4561.

Gutierrez, O., Sudarjanto, G., Ren, G., Ganigué, R., Jiang, G., & Yuan, Z. (2014). Assessment of pH shock as a method for controlling sulfide and methane formation in pressure main sewer systems. *Water Research, 48*(1), 569–578.

Hashimoto, S., MacHino, T., Takeda, H., Daiko, Y., Honda, S., & Iwamoto, Y. (2015). Antimicrobial activity of geopolymers ion-exchanged with copper ions. *Ceramics International, 41*(10), 13788–13792.

Hernandez, G., Kucera, V., Thierry, D., Pedersen, A., & Hermansson, M. (1994). Corrosion inhibition of steel by bacteria. *Corrosion, 50*(10), COV3.

Hudon, E., Mirza, S., & Frigon, D. (2011). Biodeterioration of concrete sewer pipes: State of the art and research needs. *Journal of Pipeline Systems Engineering and Practice, 2*(2), 42−52.

Hussain, N., Rizwan, M., & Khitab, A. (2015). Modeling of chloride ion penetration in concrete with empirical approach. In *2nd International conference on engineering sciences*.

Ichhpujani, R., Bhatia, R., & Ichhpujani, R. (2008). *Nutrition and growth of bacteria. Essentials of Medical Microbiology*.

Islander, B. R. L., Devinny, J. S., Member, A., Mansfeld, F., Postyn, A., & Shih, H. (1992). Microbial ecology of crown corrosion in sewers. *Journal of Environmental Engineering, 117*(6), 751−770.

Khitab, A. (2005). *Modélisation des transferts ioniques dans les milieux poreux saturés: Application à la pénétration des chlorures à travers les matériaux cimentaires*. France: Institut National Des Science Appliquees de Toulouse.

Khitab, A., & Anwar, W. (2016). *Classical Building Materials. Advanced research on nanotechnology for civil engineering applications* (pp. 1−27).

Khitab, A., Anwar, W., & Arshad, M. T. (2017). Predictive models of chloride penetration in concrete: An overview. *Journal of Engineering and Applied Sciences, 1*(1), 1−14.

Khitab, A., Anwar, W., Mansouri, I., Tariq, M. K., & Mehmood, I. (2015). Future of civil engineering materials: A review from recent developments. *Reviews on Advanced Materials Science, 42*(1).

Khitab, A., Anwar, W., Ul-Abdin, Z., Tayyab, S., & Ibrahim, O.A. (2019). *Applications of self healing nano concretes*.

Khitab, A., Arshad, M. T., Ali, S. A., Kazmi, S. M. S., & Munir, M. J. (2014). Modeling of chloride ingress in concrete using Fick's laws: Review and historical perspective. *Science International, 26*(4), 1519−1521.

Khitab, A., Arshad, M. T., Awan, F. M., & Khan, I. (2013). Development of an acid resistant concrete: A review. *International Journal of Sustainable Construction Engineering, 4*(2), 33−38.

Khitab, A., Lorente, S., & Ollivier, J. P. (2006). Chloride attack through concrete: Time effects. *ACI Special Publications, 234*, 191−204.

Khitab, A., Lorente, S., & Ollivier, J. P. (2004). *Chloride diffusion through saturated concrete: Numerical and experimental results. Advances in Concrete through Science and Engineering*.

Khitab, A., Lorente, S., & Ollivier, J. P. (2005). Predictive model for chloride penetration through concrete. *Magazine of Concrete Research, 57*(9).

Lahav, O., Lu, Y., Shavit, U., & Loewenthal, R.E. (2004). Modeling hydrogen sulfide emission rates in gravity sewage collection systems. 1382−1389.

Lajili, H., Devillers, P., Grambin-Lapeyre, C., & Bournazel, J. P. (2008). Alteration of a cement matrix subjected to biolixiviation test. *Materials and Structures, 41*(10), 1633−1645.

Matos, J. S., & Aires, C. M. (1995). Mathematical modelling of sulphides and hydrogen sulphide gas build-up in the Costa do Estoril sewerage system. *Water Science and Technology: A Journal of the International Association on Water Pollution Research, 31*(7), 255−261.

Mori, T., Nonaka, T., Tazaki, K., Koga, M., Hikosaka, Y., & Noda, S. (1992). Interactions of nutrients, moisture and pH on microbial corrosion of concrete sewer pipes. *Water Research, 26*(1), 29−37. Available from http://dx.doi.org/10.1016/0043-1354(92)90107-f.

Negishi, A., Muraoka, T., Maeda, T., Takeuchi, F., Kanao, T., Kamimura, K., & Sugio, T. (2005). Growth inhibition by tungsten in the sulfur-oxidizing bacterium *Acidithiobacillus thiooxidans*. *Bioscience, Biotechnology, and Biochemistry, 69*(11), 2073−2080.

Nielsen, A. H., Yongsiri, C., Hvitved-Jacobsen, T., & Vollertsen, J. (2005). Simulation of sulfide buildup in wastewater and atmosphere of sewer networks. *Water Science and Technology: A Journal of the International Association on Water Pollution Research, 52*(3), 201−208.

Nies, D. H. (1999). Microbial heavy-metal resistance. *Applied Microbiology and Biotechnology, 51*(6), 730–750.

Obieze, C. C., Chikere, C. B., Selvarajan, R., Adeleke, R., Ntushelo, K., & Akaranta, O. (2020). Functional attributes and response of bacterial communities to nature-based fertilization during hydrocarbon remediation. *International Biodeterioration & Biodegradation*.

Ohama, Y. (1995). *Handbook of polymer-modified concrete and mortars: Properties and process technology* (1st ed.). William Andrew.

Örnek, D., Jayaraman, A., Syrett, B., Hsu, C. H., Mansfeld, F., & Wood, T. (2002). Pitting corrosion inhibition of aluminum 2024 by Bacillus biofilms secreting polyaspartate or γ-polyglutamate. *Applied Microbiology and Biotechnology, 58*(5), 651–657.

Örnek, D., Wood, T. K., Hsu, C. H., & Mansfeld, F. (2002). Corrosion control using regenerative biofilms (CCURB) on brass in different media. *Corrosion Science, 44*(10), 2291–2302.

Parande, A. K., Ramsamy, P. L., Ethirajan, S., Rao, C. R. K., & Palanisamy, N. (2006). Deterioration of reinforced concrete in sewer environments. *Proceedings of the Institution of Civil Engineers: Municipal Engineer, 159*(1), 11–20.

Parthasarathi, N., Saraf, D.S., Prakash, M., & Satyanarayanan, K.S. (2019). Analytical and experimental study of the reinforced concrete specimen under elevated temperature. In *Materials today: Proceedings*.

Peters, D. H., Friedel, H. A., & McTavish, D. (1992). Azithromycin: A review of its antimicrobial activity, pharmacokinetic properties and clinical efficacy. *Drugs*.

Psarras, G. C. (2011). Smart polymer systems: A journey from imagination to applications. *Express Polymer Letters, 5*(12), 1027.

Rosado, T., Dias, L., Lança, M., Nogueira, C., Santos, R., Martins, M. R., ... Caldeira, A. T. (2020). Assessment of microbiota present on a Portuguese historical stone convent using high-throughput sequencing approaches. *Microbiologyopen, 9*(6), 1067–1084.

Sanchez-Silva, M., & Rosowsky, D. V. (2008). Biodeterioration of construction materials: State of the art and future challenges. *Journal of Materials in Civil Engineering, 20*(5), 352–365.

Sand, W. (2014). Importance of hydrogen sulfide, thiosulfate, and methylmercaptan for growth of thiobacilli during simulation of concrete corrosion, August 1987.

Sand, W., & Bock, E. (2008). Concrete corrosion in the Hamburg Sewer system concrete corrosion in the Hamburg, April 2014, 37–41.

Scheerer, S., Ortega-Morales, O., & Gaylarde, C. (2009). Chapter 5 Microbial deterioration of stone monuments-an updated overview. *Advances in Applied Microbiology*.

Siripong, S., & Rittmann, B. E. (2007). Diversity study of nitrifying bacteria in full-scale municipal wastewater treatment plants. *Water Research, 41*, 1110–1120.

Southgate, D. A. T. (2005). *Dietary fibre and health. Dietary Fibre*.

Vollertsen, J., Nielsen, A. H., Jensen, H. S., Wium-Andersen, T., & Hvitved-Jacobsen, T. (2008). Corrosion of concrete sewers-The kinetics of hydrogen sulfide oxidation. *The Science of the Total Environment, 394*(1), 162–170.

Wang, T., Wu, K., Kan, L., & Wu, M. (2020). Current understanding on microbiologically induced corrosion of concrete in sewer structures: A review of the evaluation methods and mitigation measures. *Construction and Building Materials, 247*, 118539.

Warscheid, T., & Braams, J. (2000). Biodeterioration of stone: A review. *International Biodeterioration & Biodegradation, 46*(4), 343–368.

Wei, S., Jiang, Z., Liu, H., Zhou, D., & Sanchez-Silva, M. (2013). Microbiologically induced deterioration of concrete: A review. *Brazilian Journal of Microbiology, 44*(4), 1001–1007.

Wei, X., Yang, Z., Wang, Y., Tay, S. L., & Gao, W. (2014). Polymer antimicrobial coatings with embedded fine Cu and Cu salt particles. *Applied Microbiology and Biotechnology, 98*(14), 6265–6274.

Yilmaz, K. (2010). A study on the effect of fly ash and silica fume substituted cement paste and mortars. *Scientific Research and Essays, 5*(9), 990–998.

Zhang, L., De Schryver, P., De Gusseme, B., De Muynck, W., Boon, N., & Verstraete, W. (2008). Chemical and biological technologies for hydrogen sulfide emission control in sewer systems: A review. *Water Research, 42*(1–2), 1–12.

Zuo, R., & Wood, T. K. (2004). Inhibiting mild steel corrosion from sulfate-reducing and iron-oxidizing bacteria using gramicidin-S-producing biofilms. *Applied Microbiology and Biotechnology, 65*(6), 747–753.

18

Role of nanomaterials in protecting building materials from degradation and deterioration

Navneet Kaur Dhiman[1,*], Navneet Sidhu[1,*], Shekar Agnihotri[2], Abhijit Mukherjee[3], M. Sudhakara Reddy[1]

[1]DEPARTMENT OF BIOTECHNOLOGY, THAPAR INSTITUTE OF ENGINEERING & TECHNOLOGY, PATIALA, INDIA [2]DEPARTMENT OF AGRICULTURE AND ENVIRONMENTAL SCIENCES, NATIONAL INSTITUTE OF FOOD TECHNOLOGY ENTREPRENEURSHIP AND MANAGEMENT, SONEPAT, INDIA [3]DEPARTMENT OF CIVIL ENGINEERING, CURTIN UNIVERSITY, BENTLEY, WA, AUSTRALIA

18.1 Scientific background

18.1.1 Introduction

The development of science and technology and the worldwide expansion of the construction industry have put forward interdisciplinary fields to meet the rising demand for more strong and smart civil structures. With nanotechnology, the innovation of new construction materials and self-healing efficiency, the face of the construction industry has improved significantly. Buildings and other construction materials frequently suffer from physical, chemical, and biological damage, which results in a severe deterioration of mechanical and structural properties of concrete. Every year, the world spends millions of dollars on the production, maintenance, and repair of damaged structures (Cailleux & Pollet, 2009). Different repair mechanisms are implemented to restore and enhance the strength of buildings and other cementitious material. Soudki (2001) classified the repair mechanisms into five different categories namely surface treatment, strength enhancement, stabilization, waterproofing, and protection. From past few years, different repair mechanisms viz. organic and inorganic treatments have been adopted all over the world. Organic treatments consist of various surface coating (acrylic, polymers, epoxy resin, oleo resinous, polyester resin, polyethylene, polyurethane, vinyl, and polymer-modified mortar) and hydrophobic impregnation (oils, waxes, silane, and siloxane-based sealers) (Sánchez et al., 2018; Seifan, Samani, & Berenjian, 2016). Inorganic treatment

---

[*] Both authors have made equal contributions to this chapter.

includes pore-blocking products, water glass (Thompson, Silsbee, Gill, & Scheetz, 1997), calcium silicate (Moon, Shin, & Choi, 2007), magnesium fluorosilicate, and sodium silicate (Pan, Shi, Zhang, Jia, & Chong, 2018).

However, there are few limitations viz. moisture sensitivity, low heat resistance, unsustainability, susceptibility to degradation and delamination, variation in thermal expansion coefficients between concrete, and the sealers associated with these conventional repair mechanisms, which has demanded some innovative approaches to address them (De Muynck, Cox, De Belie, & Verstraete, 2008). To understand the defects and restoration techniques, there is a need to recognize the type and cause of the damage to select the suitable repair mechanism and products for stable service. Researchers are looking to design ecofriendly and cost-effective concrete composites that can elicit a more durable and smarter repair mechanism in terms of its strength and durability.

Recently, a breakthrough of nanotechnology in the construction industry has greatly influenced the performance of building materials. The term *nano* comes from the Greek word, which means dwarf. One nanometer is one-billionth of a meter (Adams & Barbante, 2013; Ganesh, 2012). Zhu, Bartos, and Porro (2004) described nanotechnology as the innovation to create material at a small scale by using different techniques that allow understanding and manipulation of materials at the nanoscale from 0.1 to 100 nm (nanometer). It has brought a tremendous boost in the research and development of the civil industry (Norhasri, Hamidah, & Fadzil, 2017). An application such as nanoengineered concrete and nanocoatings are one of the emerging technologies for the prevention and control of concrete deterioration. Nanoparticles improve the self-sensing efficiency of any concrete structure by providing electrical resistance. Concrete has low conductivity which can be enhanced by adding conductive microparticles and nanoparticles (D'Alessandro, Ubertini, Laflamme, & Materazzi, 2015).

This chapter introduces the new domain, nanotechnology and its application along with microbial-induced calcium carbonate precipitation (MICCP) to address the prevailing issue of various deterioration factors affecting buildings and other construction materials. In the past few years, nanotechnology has achieved a tremendous amount of success in varied other fields such as medicine, microelectronics and material science. Although the application of the nanomaterials in civil structures is still under expansion, few studies have been reported demonstrating the proficient role of nanomaterials in developing ecoefficient and self-sensing repair systems that will be covered in this chapter.

18.1.2 Contemporary defects and biogenic deterioration in buildings and other cementitious materials

One of the most extensively used artificial construction materials is concrete, and its consumption has been growing exponentially with increases in population (Miller & Moore, 2020). Concrete is artificially created, hard, composite material consisting of cement, sand, aggregates, and water. The global construction of concrete infrastructure has built an unprecedented demand for cement worldwide. Annual production of concrete is more than

10 billion tons, which is estimated to grow twofold by the year 2050 (Castro-Alonso et al., 2019). Properties such as mechanical strength and durability make concrete the best hardening agent for buildings, bridges, dams, runways, tunnels, and other infrastructures. Proper maintenance and repair of concrete structured buildings are essential to ensure durability, strength, and design steadiness since there can be many reasons for concrete deterioration (Wang, Ersan, Boon, & De Belie, 2016). However, concrete possesses low tensile strength and less ductility due to which it validates fragility toward the formation of cracks. Poor construction practices during the fabrication of concrete material can lead to damage that will resurface years after the construction. The deterioration mechanism can be initiated due to any single factor, but thereafter it can continue as a combination of different mechanisms. It occurs when the material is exposed to unfavorable conditions viz. weather, water, or chemicals over an extended period. Different concrete damages can be categorized in accordance to the cause of the damage, type of damage, mechanisms of attack, kinds of deficient structures, financial loss due to different defects, amount, and extent of repair measures (Kovler & Chernov, 2009).

18.1.2.1 Factors responsible for the deterioration of the building

Many structural defects are the result of inefficient design, specification, execution, and scanty materials. Various authors have diversified the cause of constructional damages, which are principally divided into three categories:

1. *Physical induced deterioration*: concrete and other building material are attributable to physical damage due to mechanical enfeeblement and fatigue stress. Mostly caused by freeze–thaw, shrinkage, salt assemblage, crystallization, thermal effect, fire, and so forth. Freeze–thaw is the cumulative effect of the repetitive freeze–thaw process resulting in hydrophobic layer formation, which may cause significant expansion leading to crack, scaling, and crumbling of the material (Moncmanová, 2007). Freeze–thaw causes internal tension and weakens the building structure. Heidari et al. (2016) reported deterioration of Anahita Temple Stone, Kangavar, West of Iran due to continuous free-thaw cycles. In addition, concrete buildings are also vulnerable to slowly deform with time even in the absence of externally applied loads. Such deformations are termed as shrinkage commonly observed as plastic shrinkage and dry shrinkage in buildings, which could be a result of both intrinsic and extensive factors. The intrinsic factors such as design strength, the elastic modulus of the fraction of aggregate in the concrete mix, and the aggregate size are material characteristics of concrete. On the other hand, extensive factors may vary after the casting; they include temperature, pore water content, age at loading, a period of curing, and so forth (Bazant & Wittmann, 1982). If the tensile stress that forms exceed the tensile strength of the concrete due to water loss by evaporation or by suction to dry concrete causes cracking (Hobbs, 2001).
2. *Mechanical induced deterioration*: Overload, explosion, and loadings can generate stress higher than the strength of the concrete, resulting in cracking or failure of the structure. Concrete and other building materials are strong enough to support loads. However, due

to poor construction practices, concrete buildings are unable to reach the desired design strength. Damage caused by an earthquake is a classic example of the overloading of concrete structures. Apart from that, loss of support, soil movement beneath the building can cause cracking and structure failure.

3. *Chemical induced deterioration*: The chemical causes for the deterioration of concrete involve hydrolysis, dissolution, and oxidation of the constituents or reactions between aggressive fluids and concrete (Moncmanová, 2007). The reaction results in the formation of expansive products, such as ettringite in sulfate attack, alkali-aggregate attack, and iron hydroxide in the corrosion of reinforcing steel in concrete (Anisuddin & Khaleeq, 2005). These reactions can prompt enough constrain to disrupt the cement matrix, resulting in a loss of cohesion and strength. Reddy, Krishna, Tadepalli, and Kumar (2020) studied deterioration to quantify the effect of corrosive conditions because of chemical attack due to acid (HCl, H_2SO_4) chloride (NaCl), and sulfate ($MgSO_4$) at 5% concentration for a period of 3, 7, 14, and 28 days. Mass loss was identified due to exposures to different acids in the order of H_2SO_4, HCl, NaCl, and $MgSO_4$. The sulfate attack is considered a serious challenge for historical monuments and building stones. According to Scherer (2004) permeable materials like stone, mortar, and concrete can be at risk of damage when salt crystals grow from a supersaturated solution in its pores. Surface scaling is caused by physical sulfate attacks, unlike the chemical sulfate attack that involves chemical interactions between sulfate ions and the concrete hydration products (Haynes, O'Neill, & Mehta, 1996; Nehdi, Suleiman, & Soliman, 2014). It induces adequate pressure to wreck the cement paste, resulting in a loss of cohesion and strength. Concrete can also be severally affected by exposure to the acids. Acids react with the calcium hydroxide of the hydrated Portland cement, eroding concrete surface by dissolving the hydraulic binders and some aggregates. Industrial waste, fuel combustions contain sulfurous gases which combine with moisture to form sulfuric acid, is a deteriorating agent for concrete surface (Hobbs, 2001). Alkali-aggregates present in the concrete are more or less chemically inert. Despite that, aggregates react with the hydroxyl ions (OH^-) in the pore solution causing expansion and cracking over a couple of years (Thomas, 2011). This alkali-aggregate reactivity deterioration has two forms, alkali-silica reaction (ASR) and alkali-carbonate reaction (ACR). ACR is a relatively rare phenomenon. ASR is one of the harshest conditions that can damage concrete. Aggregates contain a certain amount of silica in their composition, which reacts with OH^- ions in concrete to form a gel. Gel swells in the presence of moisture, causing expansion in the concrete (Fournier & Bérubé, 2000; Mo et al., 2020). Corrosion is the most significant factor for the deterioration of structures. The degradation of the protective layer formed on the metal surface leads to esthetic as well as structural damage. Concrete has low tensile strength and porosity, therefore concrete is usually reinforced with embedded steel bars. However, reinforced concrete (RC) structures are deteriorated by chloride ingress and carbonation (Christou, Tantele, & Votsis, 2014). Corrosion of reinforced steel rebar is initiated by the degradation of the protective concrete layer around the steel rebar. One of the fragilities associated with carbonation and chloride ions is the lowering of the pH causing the

anodic dissolution of iron and the cathodic reduction of oxygen as represented in the following equations (El-Reedy, 2017).

$$2Fe \rightarrow 2Fe^{2+} + 4e^- \tag{18.1}$$

$$2H_2O + O_2 + 4e^- \rightarrow 4OH^- \tag{18.2}$$

Thus water and oxygen are the main cause of concrete corrosion. To maintain electrical neutrality, the Fe^{2+} migrate through the concrete pore water to these cathodic sites where they combine to form iron hydroxides or rust.

$$2Fe^{2+} + 4OH^- \rightarrow 2Fe(OH)_2 \tag{18.3}$$

$$2Fe(OH)_2 + 2H_2O + O_2 \rightarrow 2Fe(OH)_3 \tag{18.4}$$

$$2Fe(OH)_3 \rightarrow 2Fe_2O_3 \cdot H_2O \tag{18.5}$$

Accumulation of ferric hydroxide causes an increase in size and volume of original steel rebar making it brittle, affecting the workability of RC structures. It has immense consequences on concrete deterioration and causes the spalling of cementitious materials such as bridges or buildings. The resulting corrosion assists the appearance of cracks to an unacceptable level (Rodrigues, Gaboreau, Gance, Ignatiadis, & Betelu, 2020). Once the rebars are exposed to an external environment, the rate of corrosion occurs at a faster pace, leading to loss of surface area of rebars (Chen, Yu, & Leung, 2018).

4. *Biogenic deterioration of building materials*: In addition to mechanical and chemical processes, biological agents (i.e., living microorganisms) can also influence the durability of cementitious building structures. Hueck (2001) described biodeterioration as an unwanted modification in the material due to the metabolic activity of microorganisms. In other words, biodeterioration is generally defined as microbiologically induced deterioration that causes undesirable physical, chemical, and esthetic alterations and damages (Wei et al., 2015). Although concrete buildings are not susceptible to the growth of microorganisms because of their high pH, microorganisms (bacteria, algae, lichens, fungi, and cyanobacteria) often grow on concrete and stone surfaces with favorable conditions (i.e., available water, low pH, etc.). Buildings and stone monuments are always exposed to severe weather conditions and climatic change. Different environmental factors, viz. temperatures, rainfall, relative humidity, pollution, and biological colonization can accelerate the growth of wide diversity of microorganisms on various monuments, buildings, paintings, and so forth (Biswas, Sharma, Harris, & Rajput, 2013; Webster & May, 2006). Carbonation and neutralization of hydrogen sulfide that build upon the concrete building structure lowers the pH of the alkaline concrete surface which creates surroundings for further microbial growth (Zhang et al., 2008). Microbial aggregations over the surface of the building structure, stone monuments, and works of art can directly cause physical and chemical degradation (Warscheid & Braams, 2000). Microorganisms can pass through the

concrete matrix even if there are no noticeable cracks in the concrete. The most prevailing method for their ingress is via microcracks or through the capillaries in the building structure. The presence of water either on surfaces or in cavities enhances the growth of microorganisms (Sanchez-Silva & Rosowsky, 2008). Water also takes up soluble salts (sulfates, nitrates) from the soil, and atmosphere which damage the cementitious and stone structures leading to degradation. Microbial assemblage due to favorable conditions leads to the secretion of corrosive metabolites such as acids and enzymes from the metabolic activity of microorganisms that can react with the binding material of concrete structure (Kip & Van Veen, 2015). This microbial activity of microorganisms can be stratified as per their effects on concrete and stone surfaces, concrete matrices, and crack development.

Biodeterioration of building structures can be categorized into (1) physical deterioration where the material structure is damaged by microbial growth (e.g., physical or mechanical breaking); (2) esthetic deterioration due to fouling and visual damage (formation of biofilm); and (3) chemical deterioration due to excretion of metabolites or other substances, such as hydrogen sulfide and acids which adversely affect the structural properties of material, that is, increased porosity and weakening of the mineral matrix in the concrete structure (Cwalina, 2008). Sulfur-oxidizing, nitrogen-metabolizing, and acid-producing bacteria, such as *Thiobacillus, Nitrosomonas*, and *Nitrobacter* as well as fungi, which produce organic acids, are the most active microorganisms affecting stone degradation and concrete structure leading to metal corrosion and blistering of paint. These microorganisms utilize inorganic salts as electron acceptors during cellular metabolic energy production, triggering the formation of inorganic acids (sulfuric, nitric). This biogenic release of acid by sulfur oxidizing bacteria *Thiobacillus thioparus, Thiobacillus novellus, Thiobacillus neapolitanus, Thiobacillus intermedius*, and *Acidithiobacillus thiooxidans* such as sulphuric acid (H_2SO_4) degrades the cementitious material in concrete and causes a corroding layer on the surface of the concrete. Yousefi, Allahverdi, and Hejazi (2013) concluded that sulfur-oxidizing bacteria, including *A. thiooxidans*, play a key role in this biodegradation due to sulfuric acid formation. After 90 days of exposure to a semicontinuous culture of *A. thiooxidans*, 96% decrease in compressive strength, 11% in length, and 43% reduction in mass were observed in concrete samples. Algae and cyanobacteria are claimed to be an important family in the formation of a biofilm that utilizes carbon dioxide as their carbon source and sunlight as their energy source and function as a nutritive layer for more deteriorating organisms (Bravery, 1988; Warscheid & Braams, 2000). Biofilm maintains and accelerates the salt and water concentrations. Most of these biofilm-producing microorganisms can produce pigments to shield themselves from harmful ultraviolet (UV) light, which may leave an imprint on the exposed areas of buildings causing both esthetic and structural deterioration of the building (Ahmed, Usman, & Scholz, 2018; Bertron, 2014). Deterioration of concrete structures by bacteria and fungus build-up is not only limited to outdoor damage but indoor too, due to a humid environment. Such buildings are at risk of various health issues for residents because of the production of airborne particles such as spores, allergens, toxins, and other metabolites (Ahmed et al., 2018;

Hughes et al., 2014). *Cladosporium, Aspergillus, Penicillium, Alternaria, Stachybotrys,* and *Helminthosporium* are the most common outdoor molds degrading esthetic and structural frameworks of the building, while *Cladosporium, Alternaria, Aspergillus,* and *Penicillium* are more common indoor molds (Etzel et al., 1998). Bacterial Gram-positive (*Clostridium* and *Streptomyces*) and Gram-negative (*Bacillus*) cause indoor proliferation and damage to the building. These organisms manifest cellulolytic, proteolytic, and amylolytic activity, which further produces organic acids, assisting in the degradation of materials (Borrego, Lavin, Perdomo, Gómez de Saravia, & Guiamet, 2012).

Climatic conditions (humidity, temperature, and wind velocity) can adversely affect the mechanical, chemical, and biological processes of decay. The impact of these factors varies and can accelerate the desiccation, scaling, and cracking process (Moncmanová, 2007). Humidity or water retention plays a significant role in metal corrosion and damaging of the cementitious buildings. Due to its bipolar nature, it acts as a good solvent for ionic components, liquids, and gases. Water channels the transportation of salts in pores and cracks, which further expedite the chemical deterioration of the building. All these deterioration factors have immense effects on buildings and other cementitious structures specifically, deflection, scaling, spalling, cracking, leakage, disintegration, wear, delamination, and settlement over the period (Hobbs, 2001; Richardson, 2002).

The development of cracks in concrete buildings, pavements, roads, or bridges is a major failure for the civil industry. The crack formation can be a result of one or a combination of many factors, such as external loads, buckling, fatigue, drying shrinkage, thermal contraction, freeze−thaw cycles, ASR, and corrosion (Achal, Mukherjee, & Reddy, 2013; Samani & Attard, 2012). Mechanisms of deterioration and crack formation by these factors have been explained earlier. Cracks can be categorized into two different groups: structural and nonstructural cracks. Structural cracks are due to inefficient design, poor construction practice, overloading, settlement of the foundation, and temperature variation on the other hand nonstructural cracks are the result of internal stress. Nonstructural cracks are induced due to chemical attack (Salgiya, Jain, & Tongiya, 2019). All these miscellaneous factors hinder the mechanical and durability properties of concrete such as compressive strength and permeability, consequently reducing the service life of the concrete and affecting structural safety (Achal & Mukherjee, 2015). The surface layer of concrete flakes off when exposed to repetitive freezing and thawing environments. Another major damage is scaling, one of the common problems in winter-region countries. The exact reason for scaling is still unknown but it is characterized by sudden thermal change and builds up of osmotic pressure (Kovler & Chernov, 2009; Larosche, 2009). Scaling can be severe if aggregates are exposed to the external environment, which may further degrade the structural properties of concrete. Besides that, spalling is another damage characterized by falling off a large area from the concrete surface, exposing reinforced bars to the environment which can further enhance corrosion of reinforced bars. Spalling can occur due to extreme temperature variation and fire, the thermal effect and weathering (Safiuddin, 2017). Fire induced spalling is a major concern as the concrete layer can break off. Spalling is usually observed under the concrete bridge deck, roofs, floor, and building walls.

18.1.3 Current practices to overcome degradation

In construction materials, degradation phenomenon such as crack formation is almost implausible to prevent given the exposure to harsh environmental conditions. In early stages of decay, concrete structure and strength are not compromised however, undisputedly, it can become a serious threat to building materials' lifespan in the long term. Therefore to impede and terminate crack formation at an early-stage preventive strategy are crucial. There are two main types of methods, passive and active for treatment of cracks and pores in concrete. Surface cracks can be healed through passive treatments, while active methods can cure both internal and external cracks. Conventional passive treatment methods include exterior coatings of polymers and chemical mixtures such as polyurethane, epoxy resins, acrylics, waxes, and so forth. These sealants can increase the durability, avoid infiltration of aggressive substances, and are applicable to many existing building structures (Seifan et al., 2016). However, these methods have various limitations such as low heat resistance, unsustainability, inferior weather resistance, moisture sensitivity, and vulnerability to deterioration (Dhami, Reddy, & Mukherjee, 2012; Van Tittelboom, De Belie, De Muynck, & Verstraete, 2010). On the contrary, active treatment techniques work independently and have the ability of immediate activation upon crack formation thereby, sealing the crack regardless of its position. This treatment comprises of techniques such as autogenous healing, polymeric encapsulation and microbial production of calcium carbonate (Wu, Johannesson, & Geiker, 2012). Autogenous healing in the presence of moisture fills cracks through hydration of unhydrated cement particles or carbonation of dissolved calcium hydroxide (Edvardsen, 1999). Calcium carbonate thus produced by reaction of calcium hydroxide with atmospheric carbon dioxide is one of the most valuable and versatile fillers to plug the voids, porosities and cracks in concrete. However, autogenous healing is highly dependent upon presence of humidity in the surrounding environment, concrete matrix composition, and amount of unhydrated cement (Wang, Xing, Zhang, Han, & Qian, 2013). Besides, autogenous healing can only fill cracks ranging from 0.1 to 0.3 mm (Ahn & Kishi, 2008; Şahmaran, Keskin, Ozerkan, & Yaman, 2008). Another type of active treatment involves encapsulation of polymeric material, which contributes toward curing of cracks through conversion of healing agent to foam in the presence of moisture. However, in some cases extension of existing cracks has been reported owing to the different chemistry of healing agent than cement/concrete. Additionally, this practice requires capsules, which may further influence the concrete workability and mechanical properties. The aforementioned drawbacks render existing treatments ineffective and difficult to commercialize therefore alternative innovative passive and active treatment methods are of greater interest nowadays.

18.2 Nanomaterials for mitigating deterioration

Construction materials technology is being continuously upgraded to withstand severe environmental conditions, such as temperature fluctuation, biogenic deterioration, and abrasion wear. The employment of nanomaterials is gaining colossal attention as one of the effective methods to improve the performance of cementitious composites. Nanomaterials are usually materials such as fine powders, liquids, or solids with particle sizes ranging between 0.1–100 nm (Sierra-Fernandez et al., 2018). These materials possess unique physical and chemical properties such as their increased surface-to-mass ratio, diffusivity, and electrical, optical, antimicrobial, and thermal properties. Therefore nanomaterials are being utilized in various fields of application such as food technology (Kumar, 2015), biomedical, biomaterials (Agnihotri & Dhiman, 2017; Agnihotri, Dhiman, & Tripathi, 2018), materials technology, and civil engineering (Singh, 2014) etc.

The capability to resist the detrimental environmental forces to which the building material is subjected throughout its lifespan is known as durability. However, the amount of penetration of deteriorating ions and fluids into the composite directly affects the durability performance of cementitious composites (Adesina, 2020). The ultrasmall size of nanomaterials enables them to penetrate deep and work as pore fillers between the grains of the cement, thereby resulting in a denser structure, which provides higher resistance to the penetration of deleterious materials (David, Ion, Grigorescu, Iancu, & Andrei, 2020). Moreover, owing to the pozzolanic properties of nanomaterials the reaction mixture yields more product and provide consistent densification to the microstructure (Adesina, 2019; Hou et al., 2013). For instance, the calcium hydroxide in the pore solution of cementitious composite reacts with nanomaterials to form surplus calcium silicate hydrate (CSH) (Land & Stephan, 2012; Madani, Bagheri, & Parhizkar, 2012). The durability enhancement because of increase in hydration products can be attributed to the high surface area of nanomaterials serving as a nucleation area for the formation of hydration products. The overall increase in durability also keeps in check the water absorption as most harmful ions infiltrate the composite using water as a medium. It ultimately lowers the susceptibility of composites to various durability threats. Jalal, Pouladkhan, Harandi, and Jafari (2015) reported a 35% decrease in the water sorption of self-compacting concrete with application of 2% nanosilica. Sikora et al. (2016) found out that the use of about 1.5% nano-Fe_3O_4 in cement pastes resulted in a significant reduction in the chloride ion penetration. Abrasion resistance, which is also a durability property is vital to ensure well performance of certain types of infrastructures. Li, Wang, He, Lu, and Wang (2006) showed that concrete mixtures altered with nanotitanium have a higher abrasion resistance compared to nanosilica when used in same dosage. However, the researchers also concluded that abrasion resistance subsided with higher dosage of the nanomaterials and increase in the water to binder ratio of the composite. Abrasion performance is also dependent upon composite compositions, type of finishing and curing method employed in the process. The susceptibility to the acid attack of building composites buried under the ground is quite high where the acidic ions penetrate these structures through the surrounding soil or moisture in the environment. These composites may be exposed to various types of acidic ions but every attack results in deterioration by weakening the matrix through dissolution of hydration products. Diab and coworkers explored the prospect of using nanomaterials to inculcate acid resistance properties to concrete. It was inferenced that with the incorporation of nanomaterials such as nanosilica and nanometakaolin the resistance of concrete to acid attack can be amplified significantly. The samples subjected to nitric acid and sulfuric acid displayed increased acid attack resistance which was evident through lower loss in compressive strength and mass. This enhancement was attributed majorly to pozzolanic contribution, formation of silicate chain length and pore-filling ability of the nanomaterials which inhibits the penetration of the acidic solution into the composite (Diab, Elyamany, Abd Elmoaty, & Sreh, 2019). The additional formation of hydration products, less water permeability, and densified concrete structure as a result of addition of nanomaterials also attributed to augmented freeze and thaw resistance. There has been reports of incorporation of nanosilica into concrete to enhance its resistance to deterioration from freeze–thaw cycles. Kalhori, Bagherzadeh, Bagherpour, and Akhlaghi (2020) deduced that introduction of nanosilica and nanoclay into shotcrete mixtures exhibited a lower reduction in the compressive strength when subjected to freeze–thaw cycles. Although, use of nanosilica resulted in higher freeze–thaw resistance compared to nanoclay nevertheless, the refinement of the composite's microstructure by these nanomaterials was the cause of the enhancement in both cases. Temperature elevation is a prominent factor in degradation of building materials. However, studies showing elevated temperature resistance through incorporation of nanomaterials are rather limited. Nonetheless, nanosilica-modified cement paste has been reported to have elevated temperature resistance even at 500°C. A study described improvement in mechanical strength when cement paste samples were subjected to elevated temperature in the range of 400°C–700°C (Ibrahim, Hamid, & Taha, 2012). Similarly, Guler, Türkmenoğlu, and Ashour (2020) also concluded that at optimum concentration of 1.5% use of nanosilica coupled with nanoalumina enhanced the concrete stability at elevated temperatures. Although, the results from these studies are pretty conclusive but more research and development in this field is required in imminent future. In addition to the abovementioned enhancements accompanied with application of nanomaterials, its use also has been associated with cement content reduction. For instance, approximately 1 kg of nanomaterials reduces cement content by fourfold (~4 kg) in order to achieve desired effect and properties (Qing, Zenan, Deyu, & Rongshen, 2007; Senff, Labrincha, Ferreira, Hotza, & Repette, 2009). More studies with respect to enhancement in strength antimicrobial activity, physical, and chemical attack resistance in building materials on account of nanoparticles addition are mentioned in Tables 18–1–18–3, respectively.

As discussed earlier, biodeterioration caused by the growth of microorganisms such as bacteria, fungi, algae, and lichens on the surface of building structures can result in undesirable changes in material properties. The release of corrosive metabolites and formation of biological stains creates an environment that instigates corrosion, detriments, durability, and/or can cause damage to esthetic quality (Graziani et al., 2013; Park, Kim, Nam, Phan, & Kim, 2009). Surface coatings and integration of antimicrobial agents, within the coating system or matrix, are generally used for microbial growth and metabolism inhibition in order to protect against biodeterioration (Alum, Rashid, Mobasher, & Abbaszadegan, 2008; Dhiman, Agnihotri, & Shukla, 2019). For instance, a study reported that concrete doped with calcium formate, a growth inhibitory compound, protected concrete pipes against bacterial degradation in sewer systems (Yamanaka et al., 2002). However, these agents

Table 18-1 Nanomaterials in conjunction with building composites for strength enhancement.

Type of material	Nanoparticles, size, shape, density, conc.	Application area	Findings	Reference
Carbon nanotubes (CNTs)/concrete mix	CNTs density varied between 0.15–0.35 g/cm^3	Durable and highly stress resistant concrete	Overall water absorption reduction varying from 10.22% to 17.76%; Compressive strength increased by 26.69% while split-tensile strength increased to 66.3% with 0.045% multiwall CNTs	Madhavi, Pavithra, Singh, Vamsi Raj, and Paul (2013)
TiO$_2$/cement concrete	—	High compressive strength concrete	2% TiO$_2$ addition increased compressive strength by 12%, 22.71% and 27% at 7, 28 and 120 days respectively	Salemi et al. (2014)
Colloidal nano-TiO$_2$/concrete	Length 25 and 30 nm	High compressive strength concrete	Addition of nano-TiO$_2$ in the range 1%–5% increased compressive strength of cementitious material	Zhang et al. (2015)
Nanosilica/concrete		Durable concrete structure	Less chloride permeability and better durability than normal concrete	Gopinath et al. (2012)
Nano-SiO$_2$/binary-blended concrete	15 nm; 80 nm	Stiffer compressive strength concrete	Concrete specimens with SiO$_2$ particles of size 15 nm were stiffer compared to specimens containing SiO$_2$ particles of diameter 80 nm during curing	Givi et al. (2010)
Nano-Al$_2$O$_3$/concrete	15 nm; varied percentage between 0.5%–2%	Optimized setting time concrete with durability	Initial and final setting time of concrete decreased with increased Al$_2$O$_3$ NPs; sped up the hydration process	Nazari, Riahi, Riahi, Shamekhi, and Khademno (2010b)
Nano-Fe$_2$O$_3$/concrete	15 nm	High-strength self-compacting concrete	Strength and water permeability improved by addition of Fe$_2$O$_3$ NPs in cement paste up to 4% by weight	Khoshakhlagh, Nazari, and Khalaj (2012)
Nanometakaolin (clay)/ordinary Portland cement (OPC) mortar	Surface area-48 m^2/g; average dimensions 200 × 100 × 20 nm;	High compressive and tensile strength OPC mortar	Compressive and tensile strength of cement mortars with nanometakaolin was higher than control	Morsy, Alsayed, and Aqel (2010)
Nanoclay/concrete		Ultrahigh performance concrete; cement replacement	At 3% nanoclay in concrete highest compressive strength value recorded as 129.8 MPa; low chloride penetration	Faizal, Hamidah, Norhasri, Noorli, and Hafez (2015)
Nanoclay/earth bricks		Sustainable construction material significant compressive strength	5% inclusion of nanoclay developed compressive strength 4.8 times that of normal clay bricks	Niroumand, Zain, and Alhosseini (2013)

Table 18-2 Nanomaterials in construction compounds for antimicrobial activity.

Type of material	Nanoparticles (NPs), size, shape, density, conc.	Application area	Findings	Reference
Nano-TiO$_2$/wood	—	Antifungal coatings	Prevented colonization of *Hypocrea lixii* and *Mucor circinelloides* (wood degrading fungi)	De Filpo et al. (2013)
Nano-TiO$_2$/limestone	—	Antifungal hard to remove coatings	Antifungal activity against *Penicillium chysogenum* and *Cladosporium cladosporioides* after accelerated spraying water equivalent to 5000 mm rainfalls	Veltri et al. (2019)
p-nitrophenol-nanosilica (PNP-nSiO$_2$)	2%	Antimicrobial waterborne paint	Improved antifungal activity by 100% and antibacterial activity by 50% against *Aspergillus niger* and *Escherichia coli*	Dileep et al. (2020)
AgNPs/Gypsum-Grout	643 ppm	Antimicrobial building composite	Antibacterial against *Staphylococcus aureus*, *Pseudomonas aeruginosa*, and *E. coli*; MIC-26.8 μg/mL; antifungal against *Aspergillus niger*, MIC- 428 μg/mL	da Silva et al. (2019)
AgNPs/silicone	5–15 ppm; 1%–3%	Antifungal silicone paints	2% or 3% AgNPs antifungal against *C. cladosporioides*, *Alternaria alternate*, *Rhodotorula mucilaginosa*, *A. niger*, *Geotrichium candidum*	Banach et al. (2014)
Nano-CuO/concrete	—	Antibacterial concrete	Declination in live bacterial cell number at maximum specific growth rate of 1.1×10^2/day and decay rates ranging from 1.4–2.6×10^2/day.	Haile et al. (2010)
ZnO NPs/acrylic polymer (Paraloid B44)	—	Antifungal nanocomposite coatings	Biocidal effect against fungal attack of *A. niger*, *Penicillium sp.* on historic marble columns	Aldosari et al. (2019)
Mortar/ZnO NPs/palygorskite	—	Antimicrobial commercial mortar	Antibacterial against *E. coli*, *S. aureus* bacteria in darkness and visible light	Rosendo et al. (2020)

Table 18-3 Nanomaterial building composites for physical and chemical resistance.

Type of material	Nanoparticles, size, shape, density, conc.	Application area	Findings	Reference
Nano-TiO_2/stone	0.05 g/100 mL	Protective coating	1.75 times acid resistance higher and four times higher weather resistance of treated stone than untreated	Shu et al. (2020)
Nano-TiO_2/magnetite/concrete	2,4,6wt.%	High temperature resistant concrete	At 6wt.% nano-TiO_2 concrete retained higher compressive strength even at higher temperatures (600°C)	Nikbin et al. (2020)
Nano-SiO_2/concrete	0–5wt.%, 10 nm	Chemical (sulfate) resistant concrete	Cement mortars by partially immersed in 5% Na_2SO_4 solution displayed higher sulfate resistance with increased nanomaterial conc.	Huang et al. (2020)
Nano-SiO_2/cement	3wt.%	Thermally resistant cement mortar	Nano-SiO_2 improved thermal resistance and prevented the crack extension in the range of 200°C–400°C	Horszczaruk et al. (2017)
APTS-nanozinc oxide/bitumen	—	UV aging resistant pavement material	Modified bitumen displayed lowest viscosity aging index and carbonyl index after UV aging	Liu et al. (2015)
Nano-ZnO/Trinidad Lake asphalt/SBS-modified asphalt	3%	Antiaging pavement asphalt	3% nano-ZnO and 25% Trinidad Lake asphalt improved the comprehensive aging resistance	Zhan et al. (2020)
Nano-Al_2O_3/cement	0.5, 1, 3, 5wt.%	Chloride resistant cement	At 5wt.% nano-Al_2O_3 increased by bound chloride content 37.2%	Yang et al. (2019)
Nano-Al_2O_3/Rice husk ash/concrete	1–4wt.%	Degradation resistant and ecofriendly concrete	resistance to hydrochloric acid attack and chloride permeation	Meddah, Praveenkumar, Vijayalakshmi, Manigandan, and Arunachalam (2020)
Nano-Fe_2O_3/mortar/fly ash	2wt.%	Chloride resistant self-compacting mortar	Modified mortar displayed decreased chloride permeability values of 44%	Madandoust, Mohseni, Mousavi, and Namnevis (2015)
Nanokaolinite clay/concrete	0%, 1%, 3%, 5%	Freeze–thaw and chloride resistant concrete	Compressive strength improvement of 34% after 125 freeze–thaw cycles; 59% reduction and 64% increment in the chloride diffusion coefficient and electrical resistivity in the concrete samples with 5% nanoclay after 75 freeze–thaw cycles	Fan, Zhang, Wang, and Shah (2015)
Nanoclay/concrete/minerals	3%	Freeze–thaw and frost resistant self-compacting concrete	Overall characteristics improved in temperature range of −18°C to +4°C.	Langaroudi and Mohammadi (2018)

work in high dosages and their effectiveness is usually temporary moreover, they might affect the structural properties of material rather than improving it (Sun, Jiang, Bond, Keller, & Yuan, 2015). Coatings of epoxy resin work as protective materials against chemical and biological degradation but only initially afterwards their performance dramatically reduces from delamination and poor bonding of coatings to the concrete substrate (Berndt, 2011). Current advances suggest reducing the antimicrobial agent size, which results in larger surface area to volume ratios and consequently superior properties. Ultimately, which leads to greater interaction with the surrounding microorganisms thus killing them with toxic ions or reactive oxygen species (ROS) released from the surface (Seil & Webster, 2012). In comparison to their bulk configuration, nanoantimicrobial compounds display multiple actions to destroy microbes. These mechanisms include, damage of the cell membrane by either direct contact with nanoparticles or photocatalytic production of ROS, release of toxic ions, interruption of electron transport, protein oxidation, modification of membrane charges, degradation of DNA, RNA, and proteins by ROS, lowering the production of adenosine triphosphate (ATP) due to acidification and ROS production (Li et al., 2008). The development of antimicrobial building materials with application of nanomaterials can be achieved through two different approaches. The first method includes incorporation of nanomaterials within the concrete mix, as new building material; second, surface treatment over partially deteriorated concrete structures (Haile, Nakhla, Allouche, & Vaidya, 2010). The latter approach, that is, surface coated material, provides larger nanomaterial to microorganism contact area than with embedded nanoparticles. Depending upon the type of nanomaterial used, its influence on building composite can be variable. For instance, titanium dioxide (TiO_2) is currently used as a cement admixture and protective coatings that provides the self-cleaning and bactericidal properties, whereas copper oxide (CuO) or zinc oxide (ZnO) provides only bactericidal properties. Silica nanoparticles can be used in self-cleaning composites, which are designed to be more resistant to microbial overgrowth than conventional concrete.

Conclusively for reasons mentioned above, nanomaterials are proposed as an admixture to construction material composites. These substances can play multiple roles in building materials from durability enhancement to imparting antimicrobial characteristics. To summarize, nanomaterials with their ultrafine nature and ability to work even at very low concentrations enable the modifications of building composites in direction that is unreachable with conventional admixtures, fillers and additives. There is plethora of nanomaterials which are currently being studied and employed in construction materials and elaborate details of which are discussed in the next section of this chapter.

18.3 Types of nanomaterials in civil engineering

The extraordinary property of materials at the nanometer scale enable novel applications. Consequently, various nanomaterials have been used in construction compounds to make it a factual smart material that resists degradation and deterioration.

18.3.1 Inorganic nanomaterials

18.3.1.1 Nanotitanium dioxide (TiO$_2$)

Titanium dioxide (TiO$_2$) are small inorganic material with typical size of 15 nm and high purity of 99.9% that can be found in nanocrystals/nanogranules form. Depending upon the preparation method, TiO$_2$ can also be obtained in various others forms, including nanoparticles, nanofibers and nanotubes (Chen & Mao, 2007). Properties such as the high surface area offered by the small size of the TiO$_2$ particles and the increased antimicrobial activity make it ideal for different applications. In recent decades, this nanomaterial due to its unique characteristics has been used widely in ointments, paints, lacquers, and UV protection. These nanoparticles work by optimizing the rheological or mechanical properties of pristine materials and endow them with self-cleaning features through dirt repellent, photocatalytic, and superhydrophylic properties (Chauhan, Sillu, & Agnihotri, 2019). These nanosized elements also display different photoactivities when incorporated into commercial polymers (Allen et al., 2005). Thus with new and improved properties of building materials that contain nano-TiO$_2$ are now employed particularly for the concrete pavements used in road surfaces, parking lots, bridge decks, and airfield runways (Kwalramani & Syed, 2018).

Research has been conducted to study the effect of nano TiO$_2$ in construction composites. It has been demonstrated that nano-TiO$_2$ accelerates the early-age hydration of cement, increase compressive and flexural strengths and improves abrasion resistance of concrete, which makes it suitable for construction composites. Zhang, Cheng, Hou, and Ye (2015) reported an increase in the compressive strength of cementitious material using colloidal nano-TiO$_2$ (25–30 nm) with varied percentage of 1, 3, and 5wt.%. Nano-TiO$_2$ as an additive accelerated hydration and filled the concrete pores. Mohseni, Naseri, Amjadi, Khotbehsara, and Ranjbar (2016) integrated rice husk ash (RHA) and nano-TiO$_2$ as supplementary cementitious materials in cement mortars in various proportions. The authors used ultrasonic pulse velocity (UPV) tests to estimate the effectiveness of crack repairs and observed greater values of UPV with increased percentage of nano-TiO$_2$. It was manifested higher UPV values indicated higher quality and durability of cement-based materials and 5wt.% content of nano-TiO$_2$ was found to be optimal proportion of cement replacement. Similarly, Praveenkumar, Vijayalakshmi, and Meddah (2019) designed concrete with (0% to 5%) nano-TiO$_2$ and RHA. Concrete mixes with a combination of 10% RHA and 3% nano-TiO$_2$ showed the highest strengths and durability performances (Fig. 18–1). This blended ternary cementitious system revealed, not only an appreciable strength improvement but also reduction in permeability to chloride ingress and enhanced resistance to deterioration by hydrochloric acid attack. Sastry, Sahitya, and Ravitheja (2020) employed nano-TiO$_2$ between 1% and 5% as a partial replacement in fly ash based geopolymer concrete. Strength properties increased as the percentage of nano-TiO$_2$ increased and at 5% nano-TiO$_2$ concentration, maximum enhancement among all mixes was observed. The enhancement of compressive strength, split tensile and flexural strengths with respect to control was 54.96%, 32.63%, and 22.22% respectively. Additionally, nano-TiO$_2$ enhanced the chloride resistance and sulfate resistance and reduced the water absorption. Gopalakrishnan, Vignesh, and Jeyalakshmi (2020) studied the inclusion

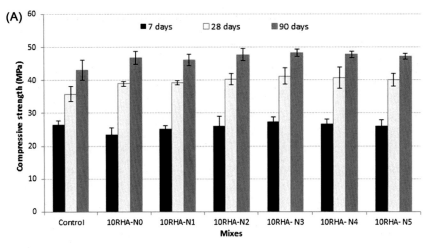
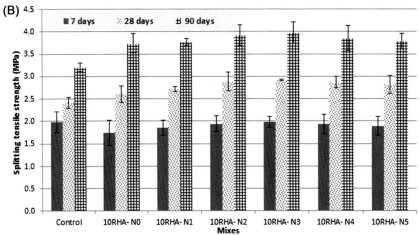
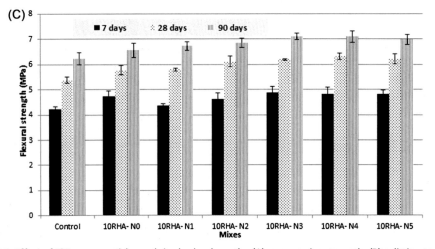

FIGURE 18–1 Effect of TiO$_2$ nanoparticles and rice husk ash on the (A) compressive strength, (B) splitting tensile strength, (C) flexural strength of concrete where mixes containing 10% RHA and TiO$_2$ nanoparticles at proportions of 1%, 2%, 3%, 4%, and 5% were labeled as 10RHA-N1, 10RHA-N2, 10RHA-N3, 10RHA-N4, and 10RHA-N5, respectively. *RHA*, rice husk ash. *Reprinted with permission from Praveenkumar, T. R., Vijayalakshmi, M. M., & Meddah, M. S. (2019). Strengths and durability performances of blended cement concrete with TiO$_2$ nanoparticles and rice husk ash.* Construction and Building Materials, 217, *343–351, Elsevier.*

of nano-TiO$_2$ as an admixture (2–10wt.%). The combination of nano-TiO$_2$ and wastewater cure improved early-age hydration, reduced setting time, decreased miniaturized scale porosity, and size for microcracks and defects. Consequently, 10%wt. of nano-TiO$_2$ resulted in development of mortar strength and durability for economic efficiency by improving its overall microstructure. Moreover, specific properties of nano-TiO$_2$, such as photosynthetic properties and organic pollutants disinfection, permitted it to effectively absorb various contaminants from mixed water; thereby reduced the negative effects of these components on the compressive strength of cement mortar. The electrical resistance to all possible forms of degradation was markedly enhanced by using higher nano-TiO$_2$ materials.

A study by Salemi, Behfarnia, and Zaree (2014) depicted that 2wt.% of TiO$_2$ increased the compressive strength by 12%, 22.71%, and 27% at 7, 28, and 120 days, respectively, compared to plain cement concrete specimens. The increase in compressive strength of concrete was due to promoted cement hydration as nano-TiO$_2$ with high reactivity acted as nucleus for cement phases and therefore worked as nanoreinforcer. As a result, it densified the microstructure and interferential transition zone, reduced porosity, and filled pores to increase compressive strength.

Li, Cui, Zhou, and Hu (2019) modified three types of polymer coatings polyurethane, epoxy resin, and chlorinated rubber with nano-TiO$_2$ and concrete specimens fabricated with these coatings were tested for long-term chloride resistance. Nano-TiO$_2$ effectively reduced the microdefects in coating films and alleviated damages due to periodical UV-accelerated aging. In addition, after UV aging, nano-TiO$_2$ reduced Coulomb fluxes of concrete with chlorinated rubber, epoxy resin, and polyurethane coatings by 32.2%, 40.2%, and 59.7%, respectively. Moreover, nano-TiO$_2$ increased the chloride resistance of concrete with chlorinated rubber, epoxy resin, and polyurethane coatings by 76.8%, 61.8%, and 60%, respectively. Liu, Li, and Xu (2019) compared the effects of SiO$_2$ and TiO$_2$ nanoparticles addition on the durability and internal deterioration of concrete subjected to 75 freeze–thaw cycles. It was evaluated through 3D pore distribution that, nano-TiO$_2$ with size 30 nm effectively prevented pore and crack expansion. Moreover, nano-TiO$_2$ concrete had a smaller number of harmful pores and less total internal porosity than nano-SiO$_2$ concrete with the increase of freezing and thawing cycles. Shu and coworkers studied the protective effects of nano-TiO$_2$ modified sol coatings on stone-built cultural heritage. The acid resistance was found to be 1.75 times higher and weather resistance cycle number of treated stone was four times higher than untreated or stone treated with organic protective materials. Further, with 0.05 g/100 mL concentration of nano-TiO$_2$ and four layers of coating material did not affect the appearance of stone-built cultural heritage (Shu, Yang, Liu, & Luo, 2020). Moreover, treated stone also showed excellent water absorption and water vapor permeability than untreated stone. Fernandes, Ferreira, Bernardo, Avelino, and Bertini (2020) investigated the degradation of sulfur dioxide (SO$_2$) by cement mortars where nano-TiO$_2$ was incorporated as cement replacement at various concentrations of 2.5, 5, 7.5 and 10wt.%. Nano-TiO$_2$ modified mortars due to the filler effect of nanoparticles contributed to the enhancement in open porosity and favored the increase in dry bulk density of mortars. Furthermore, increase in compressive and flexural strength of mortars was observed. Therefore it was concluded that overall improvements made in physical, mechanical and photocatalytic properties of mortars by nano-TiO$_2$ enabled the decontamination of mortars due to the action of the SO$_2$ pollutant.

Nikbin et al. (2020) described the shielding performance of heavy concrete containing magnetite aggregates and different contents (2%, 4%, and 6%) of nano-TiO_2, exposed to various temperatures cycles of 25°C, 200°C, 400°C, and 600°C. The results showed that specimens containing 6% nano-TiO_2 met the building design requirements because of retainment of higher compressive strength even at higher temperatures.

Moreover, it also to be noted that nano-TiO_2 is used in concrete building material as a coating material because of its special properties such as self-cleaning, sterilization, and related environmental benefits like purification of air on sunny days and infiltration of water on rainy days. Nano-TiO_2 in concrete pavement also adsorbs the pollution caused by the vehicles; adsorbed pollutant oxidizes under UV light and afterwards the water washes pollutants away. Nano-TiO_2 imparts self-cleaning attributes to pristine matrices in which these are incorporated. Thus inculcation of TiO_2 nanoparticles defends the building composites from degrading effects of organic and inorganic pollutants through powerful photocatalytic reactions. Lettieri, Colangiuli, Masieri, and Calia (2019) investigated the self-cleaning efficiency of nano-TiO_2 coated limestone surfaces exposed to an urban environment. At the end of the exposure period (1 year), TiO_2 coating preserved the surface color and protected limestone from pollutants at early ages. However, self-cleaning efficiency was reduced to negligible final rates after 8 months because of deactivation of the photocatalysts by the soluble salt ions either adsorbed from the environment or produced by the photocatalytic abatement of pollutants. Another group of researchers tested the hydrophobicity and self-cleaning effectiveness (Colangiuli, Lettieri, Masieri, & Calia, 2019) of nano-TiO_2/fluoropolymer coatings applied on limestone in a field study. Nanocomposite coated stone surfaces, which were exposed for a year were protected against dirt and water penetration meanwhile, TiO_2 nanoparticles did not affect the protective ability of the polymer. However, polymer modification by aging resulted in progressive decline in photocatalytic efficiency due to the loss of the photocatalyst as polymer failed to hold the nanoparticles on coating surface. Additionally, embedment of nanosized titania within the polymer limited the adsorption, accumulation of soluble salt ions and reduced soot deposition on the coated surface. Moreover, titania-based coatings preserved the color properties of the coated surfaces better than the fluoropolymer. Wang, Lim, and Tan (2020) concluded the self-cleaning behavior of acidic anatase TiO_2 hydrosol modified hardened Portland cement paste. The presence of TiO_2 hydrosol resulted in better self-cleaning performance owing to the retardation of hydration at early age. The increment in additional surface defects of TiO_2 dominated the enhancement of self-cleaning performance of hardened cement paste at lower TiO_2 hydrosol concentration of <0.05wt.% while the contribution of the surface electron capture effect of hydration products in self-cleaning performance enhancement became prevalent at higher TiO_2 hydrosol concentration of >0.05wt.%.

Bacteria and fungi proliferation are one of the main causes responsible for construction materials degradation and for health problems (Bolashikov & Melikov, 2009; Santucci et al., 2007), as these microorganisms are responsible for mycotoxins growth. One of the most important applications of materials with photo catalytical properties concerns the destruction of such microbes. In recent times, nanotitanium dioxide has been used as a biocide as it can adsorb UV-light with wavelength 380 nm and photocatalytically produce hydroxyl radicals in the presence of water (Salthammer & Fuhrmann, 2007). It has been exhibited in several in vitro studies that these hydroxyl radicals being strong oxidizers can attack and compromise organic compounds, such as proteins, DNA and therefore kill growing microorganisms (Shukla et al., 2011). Presently, TiO_2-based nanomaterials have been tested on different materials in order to consolidate or restore the construction composites from degradation. For example, La Russa and coworkers (La Russa et al., 2014) studied the efficiency of TiO_2 nanoparticles dispersed in an acrylic polymer solution on limestone and marble samples. The biocidal efficacy of TiO_2 against *Aspergillus niger* fungi demonstrated a high efficiency of growth inhibition on both types of samples. In addition, nanoparticle integration increased the oxidation speed of the methylene blue stains as revealed by photodegradation tests. Moreover, the composition of the sample played an important role, as the treated limestone surfaces apparently were not affected by solar radiation, on the contrary, the coating on marble was almost inefficient after aging. In a study by De Filpo, Palermo, Rachiele, and Nicoletta (2013) efficiency of nano-TiO_2 as antifungal and biocidal agents for wood working was tested against two species of fungi, *Hypocrea lixii* and *Mucor circinelloides*, which are known for the rapid degradation of the wood. The results concluded that the photocatalytic activity of TiO_2 nanoparticles prevented fungal colonization on wood samples for a longer period of time then compared to untreated samples (Fig. 18–2). Goffredo et al. (2017) applied different nano treatments that consisted of photocatalytic titanium dioxide (TiO_2) with silver and copper on softwood and hardwood surfaces. Treated surfaces displayed partial biocidal ability against the development of soft-rot fungus *A. niger*. Nonetheless, study concluded them to be a promising tool to reduce harmful mold development and for better preservation of wooden artifacts.

Others (Chen, Yang, & Wu, 2009) used wood specimens coated with a TiO_2 thin film (1.5 mg/cm^2) and noticed that the photocatalytic reaction prevented fungi growth. Building

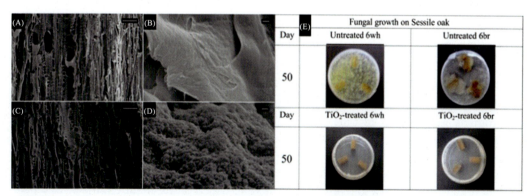

FIGURE 18–2 SEM images (a, b) of untreated Sessile oak at low (100×) & high (17k×) magnification; (c, d) TiO_2-treated Sessile oak at low (100×) & high (17k×) magnification. (e) Day fifty fungal growth on untreated and TiO_2-treated Sessile oak samples where, 6wh and 6br stands for Sessile oak sample contaminated by white-rot (*Hypocrea lixii*) and brown-rot (*Mucor circinelloides*) fungi, respectively. No fungal attack evident in TiO_2-treated Sessile oak samples (Reprinted with permission from De Filpo et al. (2013), Elsevier).

composites made with clay bricks are subjected to weathering for a long time, and this causes their biodeterioration particularly upon exposure to algae and cyanobacteria. Consequently, at a later stage esthetics and elemental integrity is compromised. Nowadays, the use of photocatalytic products for the prevention of organic contamination of building façades is increasing. Graziani and D'Orazio (2015) studied the efficiency of nanocoatings against biofouling was evaluated by sprinkling suspension of algae and cyanobacteria on the bricks' surface for a duration of 12 weeks. Photocatalytic nanocoatings were able to inhibit biofouling on bricks' surfaces and parameters of substrata such as porosity and roughness influenced the algal adhesion. Calabria et al. (2010) demonstrated that application of TiO_2 thin films with thickness of 20–50 nm in adobe blocks managed to increase the bactericidal capacity. Those authors claimed that TiO_2 thin films could be more cost-efficient than conventional paints. Veltri and coworkers reported a new subsurface deposition method where limestone samples were semi immersed in a TiO_2 solution so that nanoparticles could be transported via capillary diffusion into pores and grains of stone surfaces. The authors claimed that nanoparticles loaded with this method were more difficult to remove than in the case of direct surface coating. It was justified when treated sample surface still preserved decent antifungal activity against *Penicillium chrysogenum* and *Cladosporium cladosporioides* even after accelerated washing and drying cycles of spraying water equivalent to 5000 mm rainfalls (Veltri, Palermo, De Filpo, & Xu, 2019).

18.3.1.2 Nanosilica

Silica (SiO_2) constitutes over 50% of earth's crust and is a colorless, crystalline solid substance that is inert and acid resistant. It has been used as basic material for manufacturing of absorbents, plastics fillers, sensors, and anticorrosive agents. Besides this, nanodispersed silica has taken over the world's nanomaterials market (Pivinskii, 2007). In sustainable construction sector, nanosilica can replace microsilica and silica fume because of its superior performance in creation of ultrahigh performance concrete (UHPC) (Ibrahim, Ramyar, Hamid, & Raihan Tah, 2011). Strong reactions produced by nanosilica in UHPC are comparable to conventional materials involving their strength, performance and durability improvement (Qing et al., 2007). The incorporation of nanosilica brings about both physical and chemical change in binder matrix. Physically, it fills the abysses in hydrated cement paste because of its very small specific surface thus leading to denser structure and higher compressive strength. Chemically, nanosilica through pozzolanic reaction produces more hydration products in the form of CSH, which is identified as the source of strength gain in concrete. This was demonstrated by Vera-Agullo et al. (2009) where nanoparticles increased the hydration process, thus enhanced the mechanical properties at only three days. Additionally, nanosilica helps in reduction of CO_2 emissions and subsequent greenhouse effect by minimizing the use of cement in concrete, which is the main culprit in this phenomenon (Maheswaran, Bhuvaneshwari, Palani, Nagesh, & Kalaiselvam, 2013). However, the researchers are confronted with a problem while using nanosilica is the dispersion of the nanoparticles within the cement paste. The distribution of nanosilica particles within the cement paste plays an essential role and governs the overall performance. It has been found that colloidal nanosilica disperses better than powdered form of silica nanoparticles. Moreover, an excessive number of nanoparticles can cause agglomeration due to their high surface energy, which will provide a nonuniform dispersion. Therefore caution must be taken in employment of nanosilica in context of its quantity and morphological state.

Ecoconcretes such as self-compacting concrete and high performance concrete are mixtures where cement is replaced by waste materials or supplementary waste materials and are often associated with problems of segregation, loss of strength, degradation, and so forth. Addition of nanosilica in the corresponding mixtures accelerates setting time and increases their compressive strength. For instance, Givi, Rashid, Aziz, and Salleh (2010) studied the effects of nanosilica (15 and 80 nm) on compressive, flexural, and tensile strength of binary-blended concrete. This experimental study concluded that concrete specimens containing nanosilica with average diameter of 15 nm were stiffer compared to specimens containing nanosilica of diameter 80 nm during initial days of curing. Rojas, Pineda-Gómez, and Guapacha (2020) investigated the effect of the silica nanoparticles as an additive at different concentrations of 3, 5, and 7wt.% upon properties of fiber cement boards. The mechanical analysis showed that there was an increment in bending resistance up to 16.25% for the cement board made with 5wt.%, nanosilica due to formation of hydration products such as hydrated calcium silicate or tobermorite gel. It was inferred the nanoparticles had high pozzolanic reactivity that promoted the reaction with portlandite to form tobermorite, which in turn favored the strength of the fiber cement boards. Similarly, Snehal, Das, and Akanksha (2020) revealed that presence of 3wt.% of colloidal nanosilica in binary-blended cement composites amplified the hydration and pozzolanic activity, thereby promoted the densification of cement microstructure. Baloch, Usman, Rizwan, and Hanif (2019) incorporated nanosilica (1%, 2%, and 3%) in self-compacting cement pastes containing superabsorbent polymer with an attempt to increase strength. Nanosilica at 2wt.% was proven to be beneficial in improving the mechanical strength by increasing hydration attributes and refining the microstructure. Although, polymer particles reduced fluidity, increased the air content and reduced strength slightly due to the formation of macropores. However, denser microstructure of the nanosilica-cement more than compensated for the macro pore formation and thereby, enhanced cement strength. Li, Huang, Cao, Sun, and Shah (2015) used nanosilica (0.5%, 1.0%, and 1.5%) in conjunction with nanolimestone (nano-$CaCO_3$-2.0%, 3.0%, and 4.0%) to alter the properties of UHPC. The authors found that upon incorporation of nanomaterials compressive and tensile strength of concretes were improved with respect to concretes without any additives. The microstructure with highest values of density and mechanical strength was obtained with content levels of 1% nanosilica and 3% nanolimestone, respectively. Super plasticized mortars with 0.25% of selected nanosilica showed a 16% increase of one-day compressive strength, reaching 63.9 MPa; the 28-day strength of these mortars was 95.9 MPa and the 28-day flexural strength increase of nanosilica mortars was 18%, reaching 27.1 MPa. Zhang, Li, Chen, Shi, and Ling (2019) investigated the durability of concrete specimens containing nanosilica (1%, 3%, 5%, 7%, and 9%) and steel fiber (0.5%, 1%, 1.5%, 2%, and 2.5%). Fig. 18–3A and B show the total cracking area per unit area and cracks number of concrete samples containing 15% fly ash and five different nanosilica

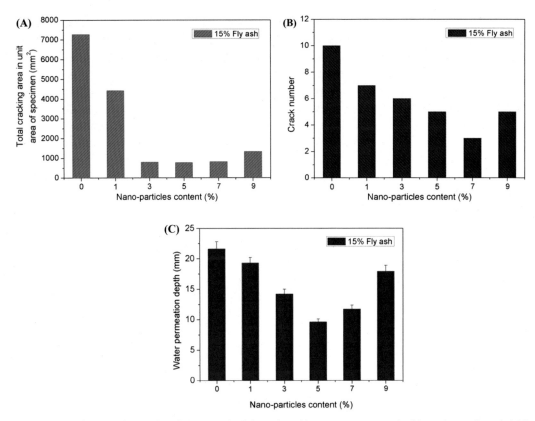

FIGURE 18–3 Influence of nanosilica dosage on the (A) total cracking area per area unit, (B) cracks number, and (C) water permeation depth of concrete specimens. *Reprinted with permission from Zhang, P., Li, Q., Chen, Y., Shi, Y., & Ling, Y. F. (2019). Durability of steel fiber-reinforced concrete containing SiO$_2$ nano-particles. Materials, 12(13), 2184, MDPI Licensed under the Creative Commons Attribution License.*

dosages. It was observed that when nanosilica dosages increased from 1% to 7%, crack numbers in the concrete specimens decreased and the least number of cracks were achieved with 7% nanosilica. Moreover, the total cracking area significantly decreased for nanosilica contents between 3% and 5%. However, when nanosilica reached a maximum of 9%, increment of 71.8% in the total cracking area and number of cracks were detected. On the other hand, water permeation depth (Fig. 18–3C) of the concrete specimens significantly reduced when the nanosilica content increased between 1% and 5%. Therefore within a certain limit nanosilica enhanced the durability of concrete specimens and this improvement level decreased for nanosilica dosages of 7% and 9%. Conventionally, in construction composites silica fume has become the gold standard for achieving strength and durability. However, there are few studies where nanosilica have either shown improvements on the same level of silica fume or have outperformed it. For instance, the compressive strength evaluation of cement mortar with nanosilica and silica fume was discussed for different water-to-cement ratio (Jo, Kim, Tae, & Park, 2007). From the experimental results it was confirmed that the compressive strength of mortars with nanosilica were higher than those of mortars containing silica fume at 7 and 28 days. Qing et al. (2007) demonstrated that nanosilica concrete gained early strength than silica-fume concrete. Moreover, nanosilica particles acted as ultrafiller for microvoids inside the concrete, which lead to dense and refined smart microstructures.

Jalal et al. (2015) found that the use of 2% nanosilica resulted in an approximately 35% decrease in the water sorption of self-compacting concrete. Ji (2005) reported a significant reduction in the water permeability of concrete incorporating nanosilica as partial replacement of Portland cement. The decrease in the water permeability of concrete incorporating nanosilica was attributed to the significant production of additional CSH, which results in refinement of the microstructure and a corresponding reduction in permeability. Mohammed et al. (2018) evaluated that nanosilica inclusion caused a reduction of 13% in the pore amount of the cementitious paste. However, the workability was modified negatively and based on scanning electron microscope (SEM) results nanosilica concentration beyond the optimum amount reduced permeability and infiltration rates. Fu and coworkers treated the wood surface by SiO_2 nanoparticles and concluded that water absorption, hygroscopic expansion rate of the treated wood was lower than control sample. Additionally, in treated wood, discoloration resistance improved by 1.5 times, aging resistance and hydrophobicity increased as depicted by contact angle test (Fu, Liu, Cheng, Sun, & Qin, 2016). These studies confirmed that nanosilica could be considered a potential material for a consolidant in wood preservation and restoration from degrading conditions.

Cement-based materials when exposed to sulfate environment suffer more severe damage under partial immersion condition than when fully immersed (Haynes, O'Neill, Neff, & Mehta, 2008; Najjar, Nehdi, Soliman, & Azabi, 2017) owing to the dual chemical and physical effects of the sulfate attack process. The mechanism of deterioration by sulfate species is still controversial and the opinions put emphasis on whether the damage to the exposed part of building materials, that comes in air contact, is caused by physical or chemical processes (Nehdi et al., 2014). Chemical sulfate attack involves the penetration of sulfate into cementitious materials and their consequent reaction with the cement hydration products to form harmful products such as ettringite and gypsum. This complex process results in the softening, expansion, cracking, scaling, and disintegration of these materials. On the other hand, physical sulfate attack involves the crystallization of sulfates in the pores of cement-based materials in the absence of chemical reactions, which results in the progressive surface scaling and flaking of the materials. Huang et al. (2020) tested the sulfate resistance of nanosilica containing cement mortars by partially immersing them in 5% Na_2SO_4 solution. The study inferenced that as the nanosilica content increased from 0 to 5wt.%, the sulfate resistance of cement mortars under partial immersion condition also increased. Moreover, the mechanism of deterioration was deduced to be chemical attack and coarse nanosilica (50 nm) displayed more potential than fine nanosilica (10 nm) (Fig. 18–4). Lately, the same author (Huang et al., 2020) conducted a study to observe microstructure changes to nanosilica-cement composites when partially exposed to sulfate attack. Distribution characteristics of

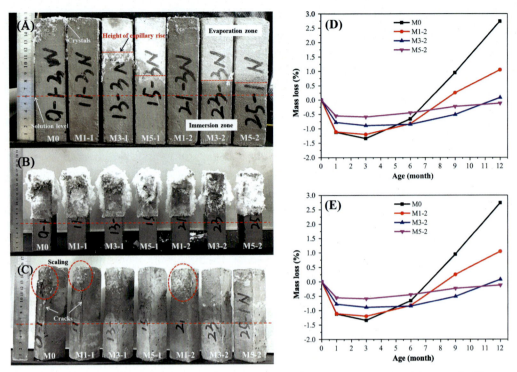

FIGURE 18–4 Visual appearance of specimens partially exposed to the 5% Na_2SO_4 solution: (A) 1 month, (B) 12 months with white crystals, (C) 12 months after removal of white crystals; mass loss of specimens partially exposed to the 5% Na_2SO_4 solution: (D) nanosilica particle size = 10 nm; (E) nanosilica particle size = 50 nm. (M0 = control; nanosilica-free mixture), and the M1-1, M3-1, and M5-1 = cement with 1, 3, and 5wt.% nanosilica (10 nm), and M1-2, M3-2, and M5-2 = cement with 1, 3, and 5wt.% nanosilica (50 nm) respectively. *Reprinted with permission Huang, Q., Zhu, X., Zhao, L., Zhao, M., Liu, Y., & Zeng, X. (2020). Effect of nanosilica on sulfate resistance of cement mortar under partial immersion.* Construction and Building Materials, 231, *117180, Elsevier.*

the sulfate products within the composites did not change with the incorporation of nanosilica. However, the addition of nanosilica reduced the amount of sulfate products in both the immersed and evaporation portions, and their amount decreased with the increase of nanosilica content. Evaluation of the sulfate resistance of concrete mixtures incorporating nanosilica and nanometakaolin was carried out by Diab et al. (2019). The study showed that concrete mixture incorporated with nanosilica exhibited higher resistance to sulfate attack than nanoclay which was evident in the lower strength loss, expansion and weight loss of the samples used. The effect of nanosilica on the sulfate resistance of concrete mixtures in terms of the loss in compressive strength and expansions. Nasution, Imran, and Abdullah (2015) also reported a significant enhancement in the sulfate attack resistance of concrete mixtures incorporating nanosilica as a 10% replacement of Portland cement. The enhancement in the sulfate resistance of concrete mixtures incorporating nanosilica was attributed to the formation of secondary products in the matrix due to the reaction of the nanosilica with calcium hydroxide. The formation of these secondary products results in the densification of the microstructure and reduction in the calcium aluminate content of the concrete. Normally, concrete begins to lose its mechanical strength during acid rain or similar harsh conditions that further leads to cracking, weight loss and finally, destruction of structures. Mahdikhani, Bamshad, and Shirvani (2018) used an artificial acid rain simulator to test the performance of different concrete mixtures modified with 0−6wt.% nanosilica. The results concluded that nanosilica had positive influence on mechanical properties and durability of concrete specimens. Additionally, mechanical properties and durability of concrete specimens are improved upon increasing the pH value of artificial acid rain.

The study by Zhang and Li (2011) showed that the chloride ion penetration of cementitious composites can be reduced with the incorporation of nanomaterials such as nanosilica and nanotitanium up to a dosage of 1%. Sakr and coworkers composed superficial coatings of colloidal nanosilica (5%−50%) for protection of surface layer of concrete (Sakr, Bassuoni, & Ghazy, 2020). Coatings with 50% loading ratio of nanosilica achieved the least penetration depth, absorption/desorption percentage, and mass loss of concrete under aggravated physical salt attack. Gopinath, Mouli, Murthy, Iyer, and Maheswaran (2012) performed rapid chloride permeability tests (RCPT) on nanosilica-coated concrete to determine chloride permeability and electrical conductance. Nanocoated concrete showed less chloride permeability and hence better durability. This study shed light on the possibility of using nanosilica as a coating material for normal concrete to increase durability. Supit and Shaikh (2015) determined the durability properties of high-volume fly ash concrete with addition of nanosilica. The incorporation of nanosilica into ordinary concrete increased the compressive strength up to 150% more at early ages and at 28, 56, and 90 curing days, the compressive strength showed increments that ranged between 45%−75%. The nanosilica. The 4% nanosilica-modified concrete decreased its water absorption (2−3 times lower) in comparison to concrete without nanosilica because of accelerated hydration process and denser microstructure. The resistance of chloride penetration was studied at ages of 28 and 90 days, in which the mixture with 2% nanosilica registered the lowest penetration value. Thus nanosilica-modified concretes could be classified as low permeability concretes.

Kalhori et al. (2020) reported a significant enhancement in the freeze−thaw resistance of shotcrete mixtures with the incorporation of nanosilica and nanoclay at optimum concentrations of 6 and 4wt.% respectively. The findings from the study showed that shotcrete mixtures incorporating nanomaterials esp. nanosilica performed better than nanoclay and exhibited a lower reduction in the compressive strength after subjecting to freeze−thaw cycles. Bastami and coworkers prepared high-strength concrete (HSC) with nanosilica and subjected these samples to varying degrees of temperature (400°C−800°C) to check its thermal stability and performance (Bastami, Baghbadrani, & Aslani, 2014). There was an indication of enhancement in mechanical properties at elevated temperature up to 400°C, after that great mass loss occurred at 800°C and residual strength only remained about 25%. Horszczaruk, Sikora, Cendrowski, and Mijowska (2017) investigated the effect of nanosilica on the behavior of cement mortars containing quartz aggregates, magnetite and barite exposed to elevated temperature. The results demonstrated that optimal 3wt.% of nanosilica content, in the cement mortar improved the thermal resistance and prevented the crack extension especially in the range of 200°C—400°C. Kumar, Singh, and Singh (2017) evaluated the performance of silica nanoparticles incorporated HSC under elevated temperature conditions. It was reported that nanosilica in concrete decreased the rate of degradation at high temperature by delaying the heat transfer by 11%, 18%, 22%, and 15% at 200°C, 400°C, 600°C, and 800°C, respectively. The compressive strength and split-tensile strength increased by 40% and 13% respectively, for nanosilica concrete upon exposure to 400°C, compared to no increase in control specimen's strength after 200°C. Moreover, exposure to 800°C left higher residual compressive (7%) and split-tensile strength (8%) in nanosilica specimens and stress—strain curves revealed that test samples exhibited brittle failure up to 600°C while control samples only up to 400°C. It was assessed that nanosilica altered microstructure and lead to enhanced mechanical and thermal stability by the formation of higher C—S—H content. Dileep, Jacob, and Narayanankutty (2020) used chemically modified nanosilica, that is, p-nitrophenol-nanosilica (PNP-NS) to develop a waterborne paint with improved antimicrobial properties. These modified paints with optimum loading of 2% PNP-NS improved the antifungal activity by 100% and antibacterial activity by 50% against *A. niger* and *Escherichia coli*, respectively. In vitro, modified silica inhibited the growth of *E. coli*, *A. niger*, *Chlorella vulgaris*, and marine water *Chlorella*. Moreover, on immersion in water the paint with modified silica did not show any blisters.

18.3.1.3 Nanosilver

Nanosilver has been extremely popular because of its extraordinary size and shape dependent optical characteristics, efficient surface plasmon excitation, high electrical, and thermal conductivity. The most relevant feature of silver nanoparticles (AgNPs) is their ability to kill various bacteria, virus, and fungi via mechanisms that work at the cellular and molecular level. Therefore nanosilver has been widely recognized as a strong broad-spectrum antimicrobial, which prevents microbial resistance and has proven to be an effective biocide against drug-resistant strains. Silver nanoparticles with dimensions less than 100 nm can interact with the carbon matrix of the bacteria and bind to cell proteins which leads to dysfunction and cell death. These properties of AgNPs along with environmental safety, efficiency at low dosages, and low toxicity to mammalian cells have formed the basis for its preferential use. Thus nanosilver has been successfully employed for biomedical applications, water disinfection, food technology, cosmetics formulations, and bactericidal coatings on surfaces of diverse materials (Carrillo-González, Martínez-Gómez, González-Chávez, & Hernández, 2016).

Significant number of studies on biodeterioration control have been performed to determine the antimicrobial activity of AgNPs on specific microbes, which however may or may not be involved in the biodeterioration process. Microorganisms that are tested for biodeterioration include clinically isolated multidrug resistant bacteria model microorganisms, such as *E. coli* (Maiti, Krishnan, Barman, Ghosh, & Laha, 2014), *Staphylococcus aureus* (Agnihotri, Mukherji, & Mukherji, 2014), and *Bacillus subtilis* (Bellissima et al., 2014). Therefore few authors contemplate that before any restoration and conservation treatments of buildings can be established, recognition, and characterization of biodeteriorants must be done initially. Consequently, only limited information is existing on the AgNPs role in microorganism inhibition that takes into account the microbial flora that participates directly in deterioration of building materials.

Nevertheless, there are significant amount of studies quoting the antibacterial benefits of silver nanoparticles in building materials. For instance, grout and gypsum are commonly used finishing materials in constructions however; hygroscopic and porous nature makes them susceptible to microorganism infestation. The fungi proliferation in grout is prevented conventionally through waterproofing via polymeric additives though, it is not fully efficient. Thus addition of nanoantimicrobial agent to the grout mixture would prevent the passage of water between ceramics and proliferation of microorganisms. da Silva et al. (2019) incorporated 643 ppm of AgNPs in gypsum-grout mixture and found that the blend exhibited antibacterial activity against *S. aureus*, *Pseudomonas aeruginosa*, and *E. coli* at MIC of 26.8 µg/mL. Additionally, antifungal efficacy was also demonstrated at higher MIC of 428 µg/mL against *A. niger*. Moreover, in biofilm test SEM images revealed crenation of *S. aureus* bacteria in contact with specimens containing AgNPs. Carrillo-González et al. (2016) suggested preventive/corrective treatment on walls of pre-Hispanic city, Teotihuacan using green AgNPs synthesized from plant leaf extracts of *Foeniculum vulgare*, and *Tecoma stans*. The potential of AgNPs to protect three types of stone materials stucco, basalt and calcite was analyzed against bacterial and fungal species isolated from colored stains, patinas and biofilms that cause biodeterioration of archeological walls surfaces. It was deduced that AgNPs from *F. vulgare* were more effective for in vitro microbial growth inhibition than those from *T. stans*. AgNPs treated calcite blocks, inhibited *Pectobacterium carotovorum* growth by 98% and *Alternaria alternata* by 99%. On stucco and basaltic blocks, it reduced microbial growth by 96% and 98%, respectively, which were highly significant relative to that on untreated blocks.

Paints and lacquers are used as bactericidal coatings to protect the building materials from surface damage by deteriorants. There are quite a number of articles, which report about the antimicrobial effects of nanosilver in paints. It has been established that low concentrations of silver ions released from the paints were very effective in killing resistant microorganisms, such as gram-positive and Gram-negative bacteria, fungi and yeasts. Authors have demonstrated the effectiveness of silver nanoparticles at a concentration of 1%–3%, corresponding to 5–15 ppm nanosilver. The use of silver nanoparticles at a concentration of 2% or 3% in the silicone paint is outperformed the known biocide against various fungal spores such as *C. cladosporioides*, *Alternaria alternate*, *Rhodotorula mucilaginosa*, *A. Niger*, and *Geotrichium candidum* (Banach, Szczygłowska, Pulit, & Bryk, 2014). In another study, the incorporation of silver, copper, and zinc oxide nanoparticles in indoor waterborne paints was assessed for antifungal activity (Bellotti et al., 2015). Results evaluated through SEM indicated better bioresistance in acrylic paints that contained with silver nanoparticles with small size of 10 nm. However, silver is not always added as nanosilver to the coatings or paints. In fact, many nanoparticles containing coatings are supported on zeolites which are ceramic solid materials with a three-dimensional grid-like structure resulting in a network of orthogonal pores running throughout each crystal. It has been demonstrated in the past that paints containing silver zeolite showed good antibacterial effects (Galeano, Korff, & Nicholson, 2003). Recently, Machado, Pereyra, Rosato, Moreno, and Basaldella (2019) used zeolite supported silver nanoparticles as additive for outdoor waterborne coating formulations. The final composite displayed antifungal efficiency against *Trichoderma harzianum Rifai*, the biocidal action of which was found comparable to that of traditional isothiazolinone-based biocides. Thus the replacement of conventional biocides, which usually show toxicity and have poor efficiency to reset on the coating surface was suggested in the study.

18.3.1.4 Nanocopper

Copper is an inorganic semiconductor metal with exceptional mechanical, optical, electrical and magnetic properties which makes it lucrative for many applications such as in supercapacitors, semiconductors, magnetic storage media, and sensors (David et al., 2020). Additionally, the oxides of copper at nanoscale have intrinsic antimicrobial properties similar to that of nanosilver although less intense. Nanocopper has been used as pesticides in agricultural practices and possess biocidal tendency against a variety of organisms such as algae, bacteria. These nanoparticles can work even at low concentration range of 40–60 μg/mL (Griffitt, Luo, Gao, Bonzongo, & Barber, 2008) with toxicity mechanism involving ROS generation thus causing damage to most biological molecules.

Haile et al. (2010) demonstrated the bactericidal characteristics of nanocopper oxide coated concrete pipes against *A. thiooxidans*. These electrochemically coated concrete pipes showed reduction in oxygen uptake rate and cellular ATP as a result of decline in live bacterial cell number at maximum bacterial specific growth rate of 1.1×10^2/day and decay rates ranging from $1.4–2.6 \times 10^2$/day. Additionally, during bacterial dry cell weight measurement it was found that for the bacterium inhibition in the nanocopper oxide coated concrete pipes minimum concentration limits ranged from 2.3–2.6 mg Cu/mg dry cell weight. Similarly, Zarzuela and coworkers depicted the effectiveness of nanostructured $CuONPs/SiO_2$-based multifunctional treatment on building stones colonized by phototrophic microorganisms such as algae (Zarzuela, Moreno-Garrido, Blasco, Gil, & Mosquera, 2018). The treated surfaces showed delay in algal colonization on account of viability loss of microorganisms at longer exposure times by CuO nanoparticles and initial setting time change due to modifications by SiO_2 matrix in roughness and surface tension. Copper nanoparticles have also been used in antibacterial paints for instance, Zielecka, Bujnowska, Kępska, Wenda, and Piotrowska (2011) studied nanocopper as antimicrobial additives for architectural paints. Copper nanoparticles (CuNPs) that were immobilized in silica nanospheres displayed complete eradication of algal growth however, only in paint samples containing 0.5 ppm of CuNPs. In contrast, algae intensively covered control samples and so were the 10% of the tested surfaces with the paint containing copper NPs with concentration of 0.1 ppm.

Besides concrete and building stones, construction materials such as wood are also susceptible to various physical, chemical and biological degradation. In few studies, use of copper oxide (CuO) nanoparticles have been mentioned. It has been studied that synthesis method and parameters involved during it shape the properties of nanoparticles. For instance, in green synthesis of metallic nanoparticles from sustainable plant extracts allows the prospect of combining the intrinsic property of the plant extract such as biocidal efficiency for potential applications in wood protection. Shiny, Sundararaj, Mamatha, and Lingappa (2019) engineered CuO nanoparticles (CuONPs) from plant extracts which are acknowledged for wood preservation characteristics such as *Azadirachta indica, Pongamia pinnata, Lantana camara,* and *Citrus reticulata*. CuO nanoparticles thus produced were tested for their effectiveness against wood termites. Results from this study stated that in comparison to control, CuONPs obtained from all plant extracts provided termite protection over the period of six months. It was inferenced that such formulations can protect the wood from biodeterioration more efficiently without harming the environment. Similarly, Akhtari and Nicholas (2013) tested and compared the effectiveness of CuO and ZnO nanoparticles as potential wood consolidants. CuO nanoparticles have been found to be much more effective in protecting wood against termites, compared to ZnO nanoparticles. Besides its antimicrobial activity, another important aspect of using the CuONPs as a wood consolidant is the fact that the effective removal of the CuO layer from the wood surface was achieved with the help of chelated agents, without causing damage to the treated wood initially (Kartal, Terzi, Woodward, Clausen, & Lebow, 2013).

18.3.1.5 Nanozinc

Zinc oxide (ZnO) nanoparticles along with its different modifications are among the most frequently employed nanomaterials, with widespread areas of application such as ceramics, pharmaceuticals, cosmetics, paints and lacquers. ZnO nanoparticles show significant antimicrobial activity against various microorganisms, which is significantly dependent on the chosen concentration and particle size (Beyth, Houri-Haddad, Domb, Khan, & Hazan, 2015). In a way similar to other metallic nanoparticles, toxicity of nanozinc oxide is a result of the dissolution and subsequent release of zinc ions cause extensive damage to species of bacteria, crustaceans, algae, and ciliates (Xia et al., 2008).

Utilization of zinc oxide-based nanomaterials as a potential enhancer in conservation and restoration applications of cultural heritage buildings has been intensively studied, which demonstrates that it can be used successfully in alleviating building degradation. Particularly, antimicrobial aspect of zinc nanoparticles has been exploited in various studies concerning the protection of building materials. Aldosari, Darwish, Adam, Elmarzugi, and Ahmed (2019) devised a nanocomposite coating using ZnO nanoparticles and acrylic polymer (Paraloid B44) with intention of inculcating combined biocidal effect against fungal attack on historic marble columns. The coating layer of ZnO nanoparticles protected the stone surface from fungal species such as *A. niger* and *Penicillium* sp. in comparison to control and samples coated only with acrylic polymer (Fig. 18–5). Moreover, the said coating also imparted self-protection properties such as resistance to UV aging, relative humidity, and thermal effect. With similar application area, Schifano and coworkers tested the efficacy of ZnO-nanorods (ZnNRs) and graphene nanoplatelets (GNPs) decked with ZnNRs (ZNGs) against environmental biodeteriogens Gram-positive *Arthrobacter aurescens* and Gram-negative isolates of *Achromobacter spanius*. ZNGs with their synergistic characteristics functioned as

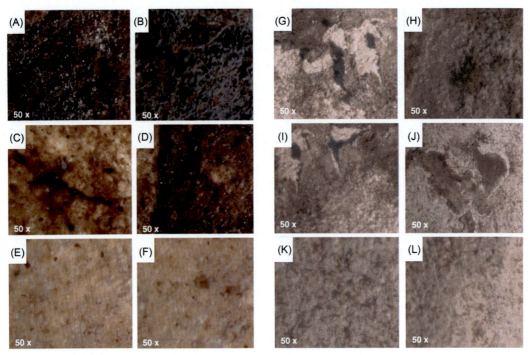

FIGURE 18–5 Stereo microscopy of fungal colonization growth on experimental marble samples after 4 weeks since inoculation by *Aspergillus niger*. (A), (B) Untreated marble samples; (C), (D) experimental samples treated with synthesized B44; (E), (F) samples treated with synthesized ZnO/polymer nanocomposites; by *Penicillium sp*. fungi; (G), (H) untreated marble samples; (I), (J) experimental samples treated with synthesized B44; and (K), (L) samples treated with synthesized ZnO/polymer nanocomposites. *Reprinted with permission from Aldosari, M. A., Darwish, S. S., Adam, M. A., Elmarzugi, N. A., & Ahmed, S. M. (2019). Using ZnO nanoparticles in fungal inhibition and self-protection of exposed marble columns in historic sites. Archaeological and Anthropological Sciences, 11(7), 3407–3422, Springer.*

nanoneedles, which pierced the bacterial wall and demonstrated high antibacterial and antibiofilm activities. The study was concluded on three types of stones in historical buildings in Sicily and Malta. Treated Noto stones revealed viability reduction up to 60% for both bacterial strains and treated common yellow brick induced a strong reduction of *A. aurescens* and *A. spanius* cell viability at 90%. Meanwhile, reduction of about 70% in bacterial viability was obtained in the case of treated Carrara marble (Schifano et al., 2020). Even more studies have employed ZnO NPs in conjunction with beneficial materials and compounds. For instance, Gómez-Ortíz et al. (2013) prepared antimicrobial surfaces for two limestone lithotypes using $Ca(OH)_2$-ZnO and $Ca(OH)_2$-TiO_2 suspensions. ZnO-based systems demonstrated superior antifungal properties against *Penicillium oxalicum*, *A. niger*, and maintained such responsiveness in dark and under illumination. Additionally, coating systems that contained 50% of nano-ZnO and pure zincite nanoparticulate films showed promising performance on low porosity limestone.

In another study, researchers (Rosendo et al., 2020) investigated the potential of ZnO NPs immobilized in palygorskite clay mineral (Pal) as an antimicrobial agent in a commercial mortar with aim to control microorganisms and improve stability for building applications. The mortar/ZnO/Pal composites worked exceptionally well against *E. coli* and *S. aureus* bacteria in darkness and visible light. Thus this composite could be an alternative antimicrobial material for building surfaces.

Further, Nazari and Riahi (2011) stated improvements in mechanical and physical properties such as flexural strength, of self-compacting concrete doped with ZnO_2 nanoparticles of size 15 nm. The authors suggested to use ZnO_2 nanoparticles up to concentration of 4wt.% which accelerated C–S–H gel formation, however, more content than this reduced the flexural strength because of unsuitable dispersion of nanoparticles in the concrete matrix. Additionally, ZnO_2 nanoparticles improved the pore structure of concrete by shifting the distributed pores to harmless and few-harm pores. On similar grounds, another group specified maximum flexural and split-tensile strength enhancement in self-compacted concrete after the addition of 0.5% ZnO nanoparticles with average size 30 nm (Arefi & Rezaei-Zarchi, 2012).

Aging and weathering are two other phenomena that lead to deterioration of construction materials such as bitumen pavements, asphalt concrete, and few wood composites by the action of heat, sunlight, oxygen or combination of these factors (Cong, Wang, Li, & Chen, 2012). Researchers have encountered two types of aging in building composites, thermal aging and UV radiation aging. These are two very different types of aging however, when UV aged structures are further exposed to temperature fluctuation, hefty load and other external environmental factors, it leads to formation of crack. Furthermore, the damage is even more intense in the geographical regions where there is high solar radiation intensity and high relative air humidity (Lee, Amirkhanian, & Kim, 2009). Consequently, more attention is being paid to mitigate UV radiation induced aging. Such kind of study has been done by Liu, Zhang, Hao, and Zhu (2015), where they used nanozinc oxide modified with three kinds of surface groups, γ-methacryloxypropyl trimethoxy silane (MTS), 3-aminopropyltriethoxysilane (APTS), and γ-(2,3-epoxypropoxy) propytrimethoxysilane (EPTMS), to improve the UV radiation aging resistance of bitumen. These modifiers improved the dispersion stability of nanozinc oxide in target matrix. It was concluded that APTS-nanozinc oxide modified bitumen showed the lowest viscosity aging index values and carbonyl index after UV aging, indicating its good UV aging resistance. Similarly, a group of researchers utilized nano-ZnO and organic expanded vermiculite to improve antiaging performance of different types of bitumen (Zhu, Zhang, Xu, & Wu, 2018). These modified bitumens displayed effective aging resistance because of retardation of deterioration in physical and rheological properties and the carbonyl formation during thermal-oxidative aging or during photo-oxidative aging. Zhan, Xie, Wu, and Wang (2020) provided an effective method to solve the aging problem of Styrene-butadiene-styrene copolymer (SBS)-modified asphalt in pavement industry, which eventually causes it to lose its function prematurely and affect its service life. Therefore the authors suggested incorporation of at least 3% nano-ZnO and 25% Trinidad Lake asphalt (TLA) into SBS-modified asphalt to improve its antiaging ability. Nano-ZnO and TLA worked synergistically to improve the comprehensive aging resistance by effectively reducing the breakage of the C=C double bond in butadiene and the increase of carbonyl groups. In recent years, in outdoor environments employment of wood fillers such as wood plastic composites in thermoplastic polymers has become the new norm for applications such as decking and fencing. However, in such harsh conditions they are exposed to weathering and biological attacks. For protection of such structures, Rasouli et al. (2016) investigated the ZnO NPs (1%–4%) effect on degradation resistance enhancement of wood–high density polyethylene composites through simulation of artificial weathering. ZnO nanoparticles employment at higher mixing ratio decreased surface degradation during weathering, which was justified by less crack formation, minimal tensile strength loss, UV absorption and contact angle changes. ZnO particles. All these factors limited the deterioration of polymer chains located in the boundary of ZnO particles, especially at the surface of the composites.

18.3.1.6 Nanoalumina

Nanoalumina (Al_2O_3) is seldomly reported in construction materials in contrast to its parent alumina. The employment of nanoalumina in concrete leads to formation of C-A-S (calcium-aluminum-silicate) gel in response to reaction with calcium hydroxide produced from hydration of calcium silicates. The amount of surface area available dictates the rate of the pozzolanic reaction. In addition, nano-Al_2O_3 as nanofiller enhances the hydration-gel product porosity and without this enhancement silica mediated hydration becomes weaker on account of silica unable to penetrate the interior assembly of the gel. Therefore nanoalumina paves the way for easy amalgamation of binding materials into the interior microstructures of the hydration-gel to start the enhancement (Richardson, 1999). The incorporation of nano-Al_2O_3 improves concrete characteristics, in terms of higher split tensile and flexural strength. Nanoalumina reduces the segregation and flocculation in high performance concrete by accelerating its early setting time (Sikora et al., 2018). The greenhouse effect caused by cement use can also be tamed by using nano-Al_2O_3 as partial replacement of cement. As studied by Nazari, Riahi, Riahi, Shamekhi, and Khademno (2010c) cement could be favorably substituted in the concrete mixture with nano-Al_2O_3 particles of average size 15 nm. The maximum replacement limit of 2% was imposed, as the optimal level of performance at only 1% replacement with nano-Al_2O_3 nanoparticles was achieved. The biocidal tendency of aluminum oxide (Al_2O_3) nanoparticles is not clearly established for antimicrobial treatment. Their bactericidal effect is relatively mild and they work only at high concentrations unless in combination with other nanomaterials such as AgNPs (Beyth et al., 2015). Nevertheless, all these merits stated above make nanoalumina one of the most important materials to produce smart and sustainable building materials in technologies using self-healing.

Karthikeya and Senthil (2016) studied effect of nanoalumina particles (15 nm) as partial replacement of cement in different percentages such as 0.5%, 1.0%, 1.5%, and 2% on strength parameters such as compressive strength, split-tensile strength and flexural strength experimentally. The study observed compressive strength increase of about 18.42%, split-tensile strength increase of about 32.6%, and flexural strength increase of about 44.9% for mix containing 1.5% of nanoalumina compared to control mix. The use of nano-Al_2O_3 can accelerate the formation process of C–S–H gel, especially at early ages, which enhances the strength of composites. For instance, Muzenski, Flores-Vivian, and Sobolev (2019) fabricated ultrahigh strength cement-based materials using Al_2O_3 nanofibers with a content of 0.25% by weight of cementitious materials, which improved the compressive strength up to 200 MPa. This represents an increment of 30%, which was the result of nanofibers acting as a seed to generate hydration products and contribute the reinforcement for the C−S−H formations, which decrease the number of microcracks. Moreover, longer dispersion time reduced the fibers agglomeration and the compressive strength at 28 days age achieved higher values for specimens with Al_2O_3 nanofibers dispersed for 3 hours. Ahmed and Alkhafaji (2020) recorded an increase of 36% in compressive strength of concrete doped with 1% nanoalumina as a replacement of cement after 28 days of curing. Additionally, the study concluded 1% nanoalumina to be the optimal content for better abrasion resistance and durability abrasion performance among various concentration of 0.5%, 1%, 1.5%, and 2% used during the experiment. Gowda, Narendra, Nagabushan, Rangappa, and Prabhakara (2017) reported the influence of nano-Al_2O_3 at concentrations of 1%, 3%, and 5% in the water absorption and electrical resistivity of cement mortars. The water absorption had a small reduction with the addition of 1% and 3% nano-Al_2O_3. However, the water absorption registered a small increment with the addition of 5% nano-Al_2O_3. The highest electrical resistivity of the cement mortar is achieved with 5% nano-Al_2O_3. More recently, Zhan and coworkers incorporated γ-nano-Al_2O_3 particles with size of 30 nm into Portland cement pastes and found that nanoalumina cement exhibited improved compressive strength at all ages due to the reduced macropores and the accelerated cement hydration (Zhan, Xuan, & Poon, 2019). Muzenski et al. (2019) reported that even at a very small dosage of 0.25wt.% aluminum oxide nanofibers up to 30% increase in compressive strength (200 MPa) was achieved in oil well cement-based mortars due to a "shish kebab" effect. Furthermore, the experimental results illustrated that nanofibers provided the CSH nanoreinforcement, reduction of shrinkage; nano-Al_2O_3 could potentially reduce initiation of cracks and replace conventional additives such as silica fume or metakaolin in future. Another insight in use of nanoalumina was given by Shao, Zheng, Zhou, Zhou, and Zeng (2019) who examined Portland cement blend with 5% nanoalumina and concluded that it reduced sulfur species concentration, increased solid volume, reduced porosity and refined pore structure in the cement paste which consequently led to an enhancement of strength at later stages.

Along with nanoparticles, waste and byproducts (in macro- and microdimensions) are also gaining widespread recognition and consideration for their use as a partial substitute of the pristine concrete constituents used in various construction applications. For instance, RHA which is an agricultural waste material, can be recycled to obtain economic and environmental benefits. Mohseni, Khotbehsara, Naseri, Monazami, and Sarker (2016) studied the effects of nanoalumina and RHA in polypropylene fiber (PPF)-reinforced cement mortars. The compressive strength of the mortar samples increased up to 18% and 20% due to the addition of 3% nano-Al_2O_3 with 20% RHA at 28 and 90 days, respectively. Meanwhile, the flexural strength of the mortar samples increased up to 34% and 41% by adding 3% nano-Al_2O_3 with 10% RHA at 28 and 90 days respectively. These augmentations were the result of nano-Al_2O_3 addition, which in turn generated a denser microstructure in the mortar.

Similarly, Meddah et al. (2020) evaluated mechanical and durability properties, including resistance to hydrochloric acid attack and chloride permeation of modified concrete containing 10% RHA and nanoalumina (1—4wt.%) as partial replacement of Portland cement. It was deduced that enhancement of strengths and durability properties were an aspect of nano-Al_2O_3 as a filling material which increased the volume of calcium silicate hydrates (C—S—H) formed. Moreover, 3% content of Al_2O_3 nanoparticles was found to be the optimum cement substitution that leads to production of efficient, degradation resistant and ecofriendly concrete material.

Chloride ingress is one of the main causes for the degradation of RC structures. Increasing the chloride-binding capacity and decreasing chloride ion permeability of concrete is generally thought as a feasible way to restrain the chloride ingress. Li et al. (2006) reported chloride ion permeability reduction of cementitious composites incorporating nanoalumina. The change was attributed to the densification capability of the nanoalumina. Similarly, significant improvement in the chloride ion permeability was observed by Joshaghani, Balapour, Mashhadian, and Ozbakkaloglu (2020), when nanomaterials such as nano-TiO_2, nano-Al_2O_3, and nanomaghemite were incorporated at different dosages into self-compacting concrete. In comparison to nanotitanium and nanomaghemite, the use of nanoalumina resulted in higher improvement in the chloride ion penetration resistance of the composites. Moreover, it also decreased the water absorption and chloride penetration and improved the overall durability of concrete. Yang et al. (2019) investigated the effect of nano-Al_2O_3 (0.5%, 1.0%, 3.0%, and 5wt.% dosages) on the chloride-binding capacity of cement samples. The samples were exposed to NaCl solution at 0.05, 0.1, 0.3, 0.5, and 1.0 M, respectively in conventional equilibrium tests for examination of chloride-binding capacity. The experimental results suggested an increase of 37.2% in the bound chloride content at 0.05 M NaCl, when 5.0% of nano-Al_2O_3 was used as additive. Thus an appropriate addition of nano-Al_2O_3 improved the chloride binding of cement paste samples. Salemi et al. (2014) found that the cement could be advantageously replaced with nanoalumina particles up to maximum limit of 2% with average particle size of 15 nm. It was found that water absorption increased from 13.4% to 117% in pristine concrete while, for concrete containing nano-Al_2O_3 it increased from 6.88% to 32.11%. Enhancement of frost resistance in terms of water absorption was attributed to consumption of $Ca(OH)_2$ formed during hydration process due to high reactivity of nanoparticles. Behfarnia and Salemi (2013) compared the efficiency of nanosilica and nanoalumina to modify the frost resistance and mechanical properties of concrete. Experimental results from this research demonstrated that although, compressive strength of nanoalumina concrete was lower than that of nanosilica. However, the frost resistance of concrete containing nano-Al_2O_3 was better than that containing the same amount of nano-SiO_2 (Fig. 18—6). The study by Farzadnia, Ali, and Demirboga (2013) demonstrated the effect of nano-Al_2O_3 (1,3,5wt.%) incorporation in cement mortars in terms of residual compressive strength, relative elastic modulus and gas permeability coefficient. Nano-Al_2O_3 modified mortar samples displayed low mass loss up to temperature of 200°C. Further at 1wt.% nano-Al_2O_3 addition, increment of 16%, and reduction of up to 38% were observed in compressive strength and gas permeability respectively. Similarly, Guler et al. (2020) also concluded that

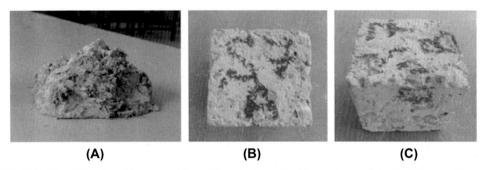

FIGURE 18–6 After 300 cycles of freeze and thaw (A) control sample, (B) sample containing 5wt.% nanosilica, and (C) sample containing 3wt.% nanoalumina. *Reprinted with permission from Behfarnia, K., & Salemi, N. (2013). The effects of nano-silica and nano-alumina on frost resistance of normal concrete.* Construction and Building Materials, 48, 580–584, Elsevier.

the binary use of nanosilica and nanoalumina at a dosage of 1.5% is optimum to enhance the stability of concrete at elevated temperatures.

18.3.1.7 Nanoiron oxide

Nano-Fe_2O_3 as nanofiller is tremendously reactive that improves the performance characteristics of composite cement. Nanoferric oxide generates nanoreinforcement by reacting with $Ca(OH)_2$ and filling hollow pores thereby, densifying the microstructure with the hydration products formed such as CFH, CSH, and CFSH. Moreover, it enhances durability, aggressive attack and thermal resistance, compressive and flexural strength, reaction kinetics, rheological characteristics, and microstructural properties of matrices as well as reduces the total cost of cementitious building construction and porosity (Heikal, Zaki, & Ibrahim, 2020). Concrete specimens with nano-Fe_2O_3 are also proven to have self-diagnostic ability of stress and this extraordinary feature can be used for health monitoring of building structures in real time without use of any embedded sensors (Kwalramani & Syed, 2018). Such peculiar characteristics of nano-Fe_2O_3 help in construction of smart and sustainable building materials. Studies pertaining to benefits of incorporation of iron-oxide nanoparticles (ION) are summarized subsequently.

Nanoparticle-modified smart cement composites with better sensing properties have become popular, so that the progress can be monitored at various stages of construction. In such application area, Vipulanandan and Mohammed (2015) altered cement with 0.38 water-to-cement ratio with iron-oxide nanoparticles (ION) and evaluated the piezoresistive smart cement behavior. It was reported that with addition of 1% nano-Fe_2O_3 compressive strength and modulus of elasticity increased by 26%, 40%; 29%, 28% after 1 day; and 28 days of curing, respectively. Additionally, depending on the curing time and nanoparticle content piezoresistivity of smart cement with nano-Fe_2O_3 was enhanced by magnitude of 750 times in comparison to unmodified cement. Authors also presented a nonlinear curing model to predict the changes in electrical resistivity with curing time. Nazari, Riahi, Riahi, Shamekhi, and Khademno (2010a) conducted studies on the compressive strength and workability of concrete by partial replacement of cement with nanophase Fe_2O_3 (i.e., 0%, 0.5%, 1%, 1.5%, and 2% by cement weight) and with 0.4 water-binder ratio. A significant achievement in high compressive strength was obtained in comparison to control specimen. It was also inferenced that with increased nano-Fe_2O_3 dosage, the workability of concrete decreased. Earlier investigations by the same author suggested the inclusion of Fe_2O_3 nanoparticles up to maximum replacement level of 2.0% by volume of cement. Consequently, nano-Fe_2O_3 particles blended concrete displayed significantly higher compressive strength, improved split-tensile strength and reduced setting time of fresh concrete compared to conventional concrete. Fe_2O_3 nanoparticles modified the pore structure of cement by significantly reducing the total porosity of the concrete and permeability. Furthermore, these improved novel characteristics reduced the influence of harmful ions that leads to deterioration of the concrete mix. Yet another group of researchers concluded increased compressive strength in cement mortar samples with different additions of nano-Fe_2O_3 (3%, 5%, and 10% by cement weight). The surface morphology got denser with increased nano-Fe_2O_3 content and maximum mechanical strength was achieved using 10% of nano-Fe_2O_3 with values 66.81%, 69.76%, and 25.20% at curing ages of 7, 14, and 28 days, respectively (Fang, Kumari, Zhu, & Achal, 2018).

Khoshakhlagh and coworkers worked on high-strength self-compacting concrete and perceived the concrete properties change after incorporation of different weight percentages (1%–5% of cement) of nano-Fe_2O_3 with average particle size of 15 nm (Khoshakhlagh et al., 2012). The flexural, compressive, and tensile strength, and the water permeability of the concrete specimens could be improved with the incorporation of nano-Fe_2O_3 up to 4wt.%. Hydration heat and workability of concrete specimens were also improved. Empirical relations for prediction of flexural and split-tensile strength based upon corresponding compressive strength at certain age of curing were also proposed in the study. Heikal et al. (2020) besides reporting the compressive strength and bulk density enhancement also, informed about the durability resistance of cement matrix with nano-Fe_2O_3 against SO_4^{2-} or Cl^- anions attack. It was stated that samples, which contained 0.5%–1.0% nano-Fe_2O_3, lead to formation of denser and finer-matrix with lower porosity, which blocked the empty pores for the diffusion of sulfate and chloride ions inside. Madandoust et al. (2015) partially replaced Portland cement in self-compacting mortar (SCM) with 25wt.% fly ash and three different nanoparticles such as nano-SiO_2, nano-Fe_2O_3 and nano-CuO up to 5wt.%. It was observed that compressive strength of the SCM specimens increased by using up to 4wt.% nano-SiO_2, 2wt.% nano-Fe_2O_3, and 3wt.% nano-CuO nanoparticles; however, the workability increased only slightly. The chloride permeability values decreased 60%, 44%, and 44% by addition of nano-SiO_2, nano-Fe_2O_3 and nano-CuO, respectively. Rashad (2013) presented a study and compared the effects of nano-Fe_2O_3, nano-Al_2O_3, nano-Fe_3O_4, and nanoclay on properties of cement composites such as mechanical strength, hydration heat, water absorption, workability, setting time, and durability. The inclusion of nano-Fe_2O_3 in the cementitious matrix decreased the water absorption and heat rate values as well as accelerated the peak times. In addition, nano-Fe_2O_3 (0.5%–5% in concretes and 0.5%–10% in mortars) added into the cementitious matrix improved the compressive strength. More recently, a comparative study involving nano-SiO_2, nano-TiO_2, and nano-Fe_2O_3 (1%, 3%, and 5% wt.%) and fly ash

(30% wt.%) in cement mortar by Ng et al. (2020) explored the effects on mechanical properties and microstructure. While the other nanomaterials had negative effect on the workability of the fresh mortar, nano-Fe_2O_3 created an exception with positive effect supported by an increase of 10.9% at excessive dosage of 5wt.% due to their intense agglomeration effect.

18.3.2 Carbon-based nanomaterials

18.3.2.1 Carbon nanotubes

Carbon nanotubes (CNTs) are microscopic hollow tubular channels with average diameter 20–40 nm, length 1–10 μm, and surface area 35 m^2/g with >90% carbon purity. These tubes are attributed with peculiar characteristics such as exceptionally high tensile strength (200 GPa) & Young's modulus (1 TPa) (Kwalramani & Syed, 2018), ultrahigh aspect ratio, desirable electrical and thermal conductivity which makes them indispensable for material science applications. CNTs can exist in the form of one wall (single-walled CNTs) or several walls [multiwalled CNTs (MWCNT)], of rolled graphene sheets (Musso, Tulliani, Ferro, & Tagliaferro, 2009), which are held by strong bonds. The mechanical properties of CNTs make them typically 100 times stronger than the steel therefore; their addition in any concrete mix can reduce or eliminate the usage of the steel reinforcement (Sasmal, Bhuvaneshwari, & Iyer, 2013).

CNTs are considered helpful in improving engineering properties of construction materials and can be used as potential additives for diverse structural applications. CNTs as reinforcing material transfers the macroscopic behavior of reinforcement to nanoscopic level by acting as a filler producing denser concrete, prevent crack growth and propagation at early age, reduce final setting time of concrete and improve quality of bond interaction between cement paste and aggregates. Other significant merits of CNTs include environmental sustainability and flexibility to produce sustainable concrete, which may be changed into distinctive or rigid type. However, employability of CNTs as multifunctional composites faces two major issues, that is, strong agglomeration among tubes and bond linkage between CNTs and the binder matrix (Kowald & Trettin, 2009). These two factors put a limitation on the use of maximum amount of the CNTs (set at 1%–1.5% by the weight of cement) in cementitious composites as higher quantities severely reduce workability. However, these problems can be minimized by use of surfactant such as polycarboxylate super plasticizers, which enhances the dispersion process. Moreover, chemically functionalized CNTs with polar end groups for example, −OH, −COOH improve the overall workability (Cwirzen et al., 2009). In the building materials sector, researchers have made several experimental studies to benefit from the advantageous effects of CNTs. For instance, CNTs can decrease the formation and growth of microcracks in concrete and generate stronger cement composites. The CNTs cement-based composites have also been found to display strain-sensing behavior that can measure their electrical parameters under applied loads (García-Macías, D'Alessandro, Castro-Triguero, Pérez-Mira, & Ubertini, 2017; Singh, Kalra, & Saxena, 2017). This behavior can allow the development of strain-sensing systems of concrete structures for potential applications of damage detection and structural health monitoring. The researches regarding the effect of CNTs in enhancing the mechanical properties of cementitious materials are summarized below.

Luo (2009) studied five different contents (0.1%, 0.2%, 0.5%, 1.0%, and 2.0%) of MWCNT for cementitious composites. The flexural and compressive strength was reported to be enhanced by 34.58%, and 21.7% in 0.1 and 0.2wt.% MWCNTs, respectively. Similarly, Kaur and his coworkers scrutinized the effect of MWCNT in concrete. According to results, 0.075% CNTs progressively increased by 23.35% and 40% in compressive strength and flexural strength respectively. Abrasion resistance was lowered by 40.22% in concrete specimens (Kaur, Dhami, Goyal, Mukherjee, & Reddy, 2016). Coherent findings were observed in UHPC (Suchorzewski, Prieto, Mueller, & Malaga, 2019), asphalt concrete (Yoo, Kim, Kim, Kim, & Shin, 2019), mortar (Sedaghatdoost & Behfarnia, 2018), and RC (Hawreen & Bogas, 2019). In another case, multiple types and content of MWCNTs on the mechanical properties of RPC were examined under the water and heat curing method (Ruan, Han, Yu, Zhang, & Wang, 2018). The four sets of MWCNTs used in the study were categorized as T1 (functionalized MWCNTs with carboxyl groups), T2 (functionalized MWCNTs with hydroxyl groups), T3 (helical MWCNTs through catalytic cracking), and T4 (nickel-coated MWCNTs). Experimental results showed that with the incorporation of 0.25% and 0.50% MWCNTs in T3 sample, enhanced flexural/compressive strength of the RPC incorporated specimens. While, on the other hand, specimen T2 of 0.50% dosage of MWCNTs recorded a 3.8% decrease and increase of 27.2% of the flexural strength with 0.25% dosage. Moreover, all the four types with 0.25% and 0.50% content showed amplified compressive strength of the RPC specimens under water curing. The maximum compressive strength of 39.2% increase was calculated in the specimen T2 with 0.25% MWCNTs content. Metaxa, Seo, Konsta-Gdoutos, Hersam, and Shah (2012) prepared a multiwalled purified carbon nanotube (MWCNT) at a concentration of 0.26wt.% by chemical vapor deposition method to accelerate mechanical properties. Dispersion of the MWCNTs was investigated by different methods (ultrasonication, ultracentrifugation, decantation, and ultrasonication) in an aqueous/surfactant solution. As reported by mechanical test, the ultracentrifugation method preserved the solubility of the concentrated MWCNT suspensions, without affecting the reinforcing properties of the admixture. Moreover, compared to the plain cement paste samples, the MWCNT reinforced nanocomposite samples had higher flexural strength and Young's modulus.

CNTs can function as nanofiller in the concrete matrix and has the potency to improve the strength and durability of concrete. Despite this fact, CNT also has a major challenge because of very low surface friction, which makes it tough to bind with the cement matrix and proclivity to agglomerate due to the van der Waals attraction force. Recently alternative approaches, such as the use of silica fume (Kim, Nam, & Lee, 2014; Scrivener & Kirkpatrick, 2008), sonification (Hassan, Fattah, & Tamimi, 2015), and surfactant addition (Makar, Margeson, & Luh, 2005) have been tested to propagate CNT in the cement mix. Silica fume has a very small size, therefore gets easily dispersed with agglomerated CNTs. Kim et al. (2014) reported enhanced compressive strength and hydration activity of silica fumes with CNTs by Ca^{2+} ions from the cement. Similarly, Tamimi, Hassan, Fattah, and Talachi (2016) also communicated improved compressive and flexural strengths for the functionalized

CNTs with 30% silica fumes. Results were confirmed using a SEM and energy-dispersive spectroscopy. Homogenous work was reported by Konsta-Gdoutos and Aza (2014), Stynoski, Mondal, and Marsh (2015) in the development of a firm and stable concrete composite with CNTs.

Madhavi et al. (2013) investigated to study strength and durability properties such as compressive strength, split-tensile strength, and water absorption of concrete with varying proportions (0.015%, 0.03%, and 0.045% by weight of cement) of MWCNT. The density of CNT used in this experiment was 0.15–0.35 g/cm^3 with 3–15 CNTs walls. As a result, water absorption reduced from 10.22% to 17.76% with an increased number of CNTs. Also, compressive strength was observed to increase by 26.69% while split-tensile strength escalated to 66.3% by adding 0.045% multiwall CNT. A significant amount of work has been reported by researchers for the dispersion of carbon nanomaterials in cementitious matrices to boost the mechanical strength and durability characteristics of concrete. MacLeod et al. (2020) reported a remarkable increase of 40%, 21% with OPC and 46%, 37% with HE (High early strength) binders in strength at 1 and 28 days curing the effect of mixing time, respectively. Much appreciation is underlined for extended mixing time for improved CNT homogenization, compressive strength, and reduced bleed quantity. Carbon nanofibers (CNFs) are a quasi-one-dimensional carbon material between CNT and carbon fiber grown in the chemical vapor-phase (Kim, Hayashi, Endo, & Dresselhaus, 2013). CNFs are different from CNT. CNFs are diversified into two different classes, hollow and solid CNFs according to their structural property. CNF possess low density and maintains electrical as well as thermal conductivity due to its higher degree of crystalline orientation. Such properties can be enhanced for producing fiber-RC.

It has been found that both CNT can effectively serve as nanoreinforcements in cementitious nanocomposites. Preliminary work carried out by for CNT reinforced composites highlighted the successful improvement in compressive and flexure strength of specimens (Chen & Chung, 1996; Li, Zhang, & Ou, 2007). Gao, Sturm, and Mo (2009) conducted an experimental study to determine the mechanical and electrical properties of concrete containing CNF. The study results indicated higher compressive strength and improved electrical conductivity for damage evaluation and self-health monitoring of concrete. Similar study was carried out by Galao, Baeza, Zornoza, and Garcés (2014) on the strain and damage sensing properties on CNF cement composites. The objective of this investigation was to prove the sensitivity of these CNF composites to sense their damage. Different variables (cement paste, curing age, current density, loading rate, or maximum stress) were analyzed using the gage factor. CNF dosage was kept in the range of 0%–2%. All specimens with varied CNF content showed displayed good strain-sensing capacities for curing periods of 28 days. Nevertheless, a 2% reinforced CNF cement paste was sensitive to its structural damage. Peyvandi, Sbia, Soroushian, and Sobolev (2013) evaluated performance efficiencies of CNFs with HSC and UHPC. Generally, two different fibers, poly vinyl alcohol (PVA) and steel, were investigated in link with CNF in UHPC. In UHPC, the amalgamation of CNF and steel fiber contained a significantly higher fiber volume fraction than the combination of CNF and PVA fiber owing to the high elastic modulus of steel fibers and flexural strength. Wang et al.

(2020) composed uniform microfoam bubbles to produce a lightweight cementitious composite (CNF-LCC) with 1500 ± 50 kg/m^3 density. CNFs were attributed to increase the compressive strength for cylinders (150 × 300 mm cylinders) at 1, 7, and 28 days by 18.5%, 16.5%, and 12.7%, respectively. The flexural strength was significantly enhanced in CNF and CNFLCC by 37.1% and 50.8%, respectively. Apart from CNF compost concrete, carbon nanomaterials have been implied in bio cleaning of marble surfaces. Barbhuiya and Chow (2017) found increase in compressive strength of cement composites fabricated with CNFs. CNFs were also proved effective in removing microbial degradation from the stone surface. Valentini, Diamanti, Carbone, Bauer, and Palleschi (2012) reported effective removal of black crust using CNFs (CNF-COOH), which were dispersed in Tween 20 medium. Results were compared with SWCNTs and the enzyme-based cleaning treatments. CNFs are advantageous over carbon fibers as they fill up the voids in the cement matrix up to the nanoscale.

18.3.2.2 Graphene-based nanomaterials

Another excellent nanomaterial, namely graphene is considered significant nanosized additive for cementitious materials. In contrast to other nanomaterials, graphene displays unique atom-thick sp^2 bonded 2D structure which gives rise to extraordinary properties such as high surface area, ultrahigh tensile strength and desirable chemical, thermal and optical properties (Tong et al., 2016). Therefore these materials have garnered augmented interests in upcoming engineering applications. GNPs and graphene oxide nanoplatelets (GONPs) are a new generation of nanoparticles comprising of graphene stacks. These nanoparticles inherit many aspects and attributes of its parent compound. Moreover, these are quite economical nanoparticles among the nanomaterials used as additives in construction materials. Recently, the performance of cement-based materials incorporating graphene family nanomaterials (GFN) such as graphene, graphene oxide (GO), reduced graphene oxide (rGO), and graphene nanosheets (GNS) have been examined (Gao et al., 2019; Tragazikis, Dassios, Dalla, Exarchos, & Matikas, 2019). Thus GFN reinforced cement-based materials can improve their structural strength and durability, as well as allow self-cleaning surfaces and self-sensing abilities (Belli, Mobili, Bellezze, Tittarelli, & Cachim, 2018; Liu et al., 2019). The rapid development of graphene-based nanoparticles such as GO and graphene platelets (GNP) have provided a new opportunity and direction toward cement modification. According to research by Hu et al. (2019b) compressive strength of mortar increased by 120.6%, 124.1%, 126.7%, and 133% as compared to plain mortar at curing period of 1, 3, 7, and 28 days, respectively in nanosilica-coated GO reinforced cement composites. However, Hu, Guo, Fan, and Chen (2019a) operated GO via triethanolamine (TEA), to enhance its mechanical behavior of cement composite. Compressive strength was observed to escalate in TEA-GO reinforced cement samples ranging 9.4%–31% while it showed a comparatively less increase in GO reinforced cement (4.1%–17.2%). Pan et al. (2015) observed 15%–33% and 42%–59% enhancement in compressive and flexural strength by incorporation of 0.05wt.% GO in cement specimens. Later, M Wang et al. (2016) worked upon Hummers' method to prepare the GO nanosheets and 0.05wt.% GO aqueous dispersion in the cement mixture. Along with his coworkers, Wang developed a 3D network structure for GO nanosheet cement with hydration properties. The implementation of GO nanosheets had a favorable effect on the tensile properties of cement. Compressive and flexure strength of GO nanosheets modified cement increased by 25.28% and 56.62% respectively. Wang and Yao (2020) also reported a significant impact of CFGO (Chemical functionalization of GO) on the flowability and mechanical properties of cement.

In another study, influence of GO and rGO on the performance of cement-based composite was characterized by (Qureshi & Panesar, 2019). Uniformity and efficient mixture of both GO and rGO with the content of 0.02%, 0.04%, and 0.06% of cement weight was established. A significant decrease in final setting time and workability of GO reinforced cement specimens with 0.06% content was observed due to the hydrophobic behavior of rGO. The remarkable increase of 10.2%, 7.8%, and 10.6%, and 9.6%, 13.3%, and 14.9%, in compressive strength of GO and rGO based specimens, was recorded, respectively. Li et al. (2017) also reported improvement in compressive strength by adding 0.04wt.% GO in the silicate cement matrix. Li doubted the functionality of GO but proposed GO agglomerate to enhance the tensile property of cement-based materials. Further, Li et al. (2018) investigated the flexure, compressive, and tensile strength of mortar with GO at a dosage of 0.02 and 0.04wt.% at the age of 3, 7, and 28 days. According to the results, 0.02% and 0.04% based GO increased compressive strength as compared to the control sample. About 18% and 14.8% increase in tensile and flexure strength was recorded at 28 days respectively. Similar results were reported by Gong et al. (2015), who demonstrated a 40% increase in compressive strength with 0.03% by weight GO in mortar. Further, elaborated the improvement in durability properties of mortar by a decrease in porosity by 13.5% as compared to the control samples. Capillary pores in the GO-cement samples were observed to be 0.173 mL/g, which is 27.7% lower than that of the plain cement.

To address the limitation of previous studies to protect erosion of the protective layer on a metal surface by application of chemicals and existing polymers, graphene and its derivatives are being explored. Due to distinctive nanostructures of nanobased coating, superior ion-diffusion barrier properties, large specific surface area, and good compatibility with polymer (Zhu et al., 2010; Zohhadi, Aich, Matta, Saleh, & Ziehl, 2015). Zhang et al. (2015) found that erosion resistance for GNS based epoxy resin coating with (0, 0.1, 0.4, and 0.7wt.%) dosage was much better than using only epoxy resin on the metal surface. Young modulus and thermal stability were observed to increase by 213% and 73°C in 0.7wt.% polyvinylpyrrolidone/reduced GO (PVP/rGO) samples loading. GO can be easily dispersed in water-based polymer due to hydroxyl groups on the basal planes and carboxylic acid groups at their edges. Zheng et al. (2020) reported significantly lower water absorption and chloride diffusion coefficient for (100 × 100 × 50) mm concrete samples sprayed with epoxy resin nanocomposite modified with GOs.

GNPs and GONPs portray 2D sheet-like structure with graphene at nanoscale (less than 10 nm) to consummate reinforcement for high and smart structural materials. Varied types of GNPs and GONPs can be characterized according to their size and oxidation agent, to influences the durability related properties of cementitious materials. Lv et al. (2013) stated 78.6%, 60.7%, and 38.9% surge in the tensile, flexural and compressive strength of concrete by adding 0.03% GONPs (concerning the weight of cement) respectively. GNPs dosage range has been continuously optimized by researchers from 0.01% to 0.1% by weight of cement. Tong and his group carried out an experimental investigation to study the effect GNPs and GONPs on compressive strength of mortar cubes (50 × 50 × 50 mm), cylinders (101.6 × 202.3 mm), and beams (70 × 70 × 350 mm) upon deterioration by acids, freeze−thaw cycles. As confirmed by the AFM and SEM images GNPs, GONPs, and C-S-H gels in the cement matrix have develops better interfacial bond which has the potential to minimize the chemical attack induced by an acidic solution (Tong et al., 2016). It was found that during a freeze−thaw activity of graphene-based composites, 18.4%, 27.1%, and 8.39% of water move from nanopore to the neighboring capillary pores in the C-S-H gels, GNPs reinforced C-S-H gels and GONPs-reinforced C-S-H gels respectively. In another study conducted by Du, Gao, and Dai Pang (2016) reported fall in water penetration depth, chloride diffusion, and migration coefficients in concrete by 80%, 80%, and 37% upon using 1.5% of GNP. Similarly, Iqbal, Khushnood, Baloch, Nawaz, and Tufail (2020) scrutinized the comparison of graphite nano/micro platelets (GNMPs 0.1, 0.3, and 0.5% by weight of cement) and HSCs based on mechanical and durability properties. GNMPs were observed to be performing better, as the compressive strength of GNMP modified mixes increased by 20%, 28%, and 24% by adding 0.1%, 0.3%, and 0.5% GNMPs, respectively, as compared to HSC mix at room temperature. Loss in compressive strength at elevated temperature (100°C−600°C) was recorded to be higher in HSC specimens than GNMPS modified samples. GNMPs modified concrete samples with 0.3% were selected to be performing better based on mechanical and durability properties. Implementation of nanoparticles with modified concrete mixture has proven to be accelerating the flexure and compressive strength of concrete with increased durability properties. However, in-field use of such nanoparticles is limited due to the low production and low cost-effectiveness of GO/GNPs.

18.3.3 Nanoclay

Nanoclay are nanoparticles of layered mineral silicates that possess many novel properties and provide ample scope for nanotechnology maneuvering. Nanoclay endow composites with improved properties that are notably attributed to its stability, swelling capacity, interlayer spacing, elevated hydration and robust chemical reactivity. There are various forms of nanoclays, which are categorized into montmorillonite, bentonite, kaolinite, hectorite, and halloysite based on their chemical compositions and morphologies (Sillu, Kaushik, & Agnihotri, 2020). These mostly are deposits of mineral having platelet structures of average thickness of 1 nm and width of 70−150 nm (Huseien, Shah, & Sam, 2019). The added advantage of clay as a source material is that it is inexpensive and easily accessible. Although these materials have been diversely employed in the polymeric systems and immobilization strategies (Sillu & Agnihotri, 2019) however, nanoclay as a potential constructional supplement needs to be further explored.

There are few studies, which describe use of nanoclay as an additive to improve the mechanical and binding concrete characteristics. Morsy et al. (2010) investigated effect of nanoclay on the mechanical properties such as compressive and tensile strength, phase composition and microstructure of OPC mortar. In this study, nanometakaolin (surface area of

48 m²/g and average dimensions 200 × 100 × 20 nm) as 0%, 2%, 4%, 6%, and 8% by weight of cement partially substituted OPC. The enhancement in tensile strength clocked at 49%, whereas the enhancement in compressive strength was 7% even at 8% replacement of cement with same water-binder ratio. Qian, Zhou, and Schlangen (2010) studied the inclusion of nanoclay with water (worked as the inner water provider to promote hydration alongside the microcracks). It was shown that the recovery level could be considerably improved with the incorporation of nanoclay in the mixtures. The mechanism of healing depended on the reaction between calcium hydroxide and nanoclay to formulate CSH gel led to healing the crack. Hamed, El-Feky, Kohail, and Nasr (2018) investigated the inclusion and dispersion of nanoclay as cement partial replacements at various percentages of 5%, 7.5%, and 10%. It was deduced that nanoclay particles at 7.5wt.% dispersed through sonication brought maximum enhancement in all aspects of cements' mechanical properties. The claims were justified by the magnitude of gains in the compressive, split tensile, flexural, and bond strengths ranged from 1.42–3.74x with sonicated nanoclay mixes. In addition, filling effect, pozzolanic reactivity, nucleation effect, and needle effect were among the major alterations observed in concrete under SEM on account of nanoclay. Dejaeghere, Sonebi, and De Schutter (2019) wanted to optimize the fresh mix properties, rheological parameters, and compressive strength of cement mortars containing nanoclay (0.5%–2.5%), superplasticizer (0.6%–3%), and fly ash (5%–20%). The formed card-house like microstructure increased the resistance against break down which thereby, had a positive influence on rheological and mechanical properties. Norhasri, Hamidah, and Fadzil (2019) reported a gradual increase in compressive strength of UHPC upon introduction of nanometaclay as an additive which ranged from 1, 3%, 5%, 7%, and 9% of cement weight. However, owing to the large surface area of metaclay nanoparticles retardation effect was observed on hardening period and workability of concrete pastes. More recent work on the use of nanoclay include a study by, Ghadikolaee and coworkers, who demonstrated that enhancement in compressive and flexural strength of plain mortar up to 25% and 20%, respectively was achieved with inculcation of 3wt.% halloysite nanotubes (HNTs). Furthermore, introduction of HNTs lead to more calcium hydroxide consumption, denser microstructure and reduced the capillary water absorption by 25% and electrical resistivity by 28%. Ultimately, HNTs inclusion produced sustainable concrete structures with reduced pore size and porosity in plain mortar (Ghadikolaee, Korayem, Sharif, & Liu, 2020).

Besides modifying mechanical properties, nanoclay particles are also used for enhancing resistance to chloride penetration, temperature fluctuation, freeze–thaw resistance, and reducing permeability. It has been established earlier in the text that ingress of chlorides into concrete initiate corrosion of steel reinforcement and deteriorate concrete member affecting the durability of RC. Faizal et al. (2015) experimentally studied ability of UHPC to withstand the action of chloride penetration by incorporating nanoclay as cement replacement as 1%, 3%, and 5% by weight into UHPC. The results showed that incorporation of nanoclay in concrete mixes caused reduction in workability of mix but marginally increased compressive strength of UHPC at 28-days compared to mixes without nanoclays with optimum value of cement replacement as 3% by weight. The study recorded highest value of compressive

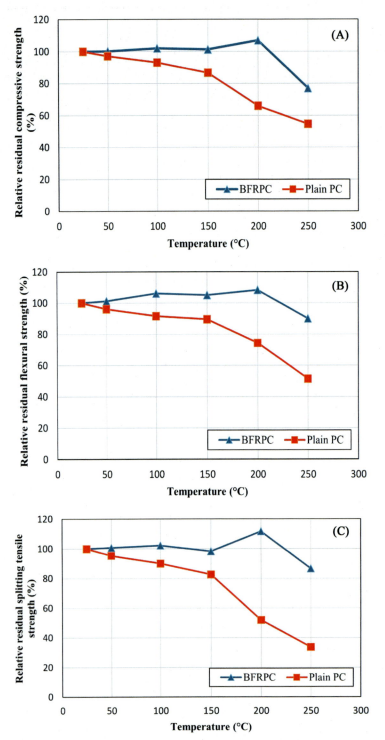

FIGURE 18–7 (A) Relative residual compressive strength, (B) relative residual flexural strength, (C) relative residual splitting tensile strength at different temperatures, and (D) effect of nanoclay on the mechanical properties of basalt fiber-reinforced polymer concrete after exposure to high temperatures. *Reprinted with permission from Niaki, M. H., Fereidoon, A., & Ahangari, M. G. (2018). Experimental study on the mechanical and thermal properties of basalt fiber and nanoclay reinforced polymer concrete.* Composite Structures, 191, 231–238, Elsevier.

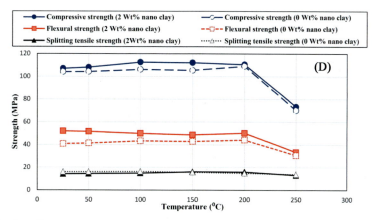

FIGURE 18–7 (Continued).

strength as 129.8 MPa with UHPC made using 3% nanoclay due to creation of nanopozzolanic reaction and development of more hydration-gel. Moreover, increase in percentage of nanoclay resulted in low chloride penetration with increase in age of specimens as nanoclays act as ultrafiller material and reduce diffusion coefficient of chloride ion. Niaki, Fereidoon, and Ahangari (2018) improved the mechanical properties compressive, flexural, splitting tensile and impact strengths (Fig. 18–7A–C) of quaternary epoxy-based polymer concrete via inclusion of basalt fiber and clay nanoparticles. In addition, thermal stability of reinforced polymer composite concrete exposed to high temperatures up to 250°C was enhanced (Fig. 18–7D). Similarly, Wang (2017) observed a correlation between compressive strength and thermal conductivity coefficients of concrete modified with nanoclay (0.1%–0.5wt.%). It was inferenced that the strength of nanoclay concrete increased at temperature <300°C and was lowered by 10% when temperature reached 1000°C. Another study evaluating thermal performance and fire resistance of montmorillonite nanoclay (0–2wt.%) modified cement mortar was performed by Irshidat and Al-Saleh (2018). At higher temperatures, nanoclay significantly reduced the degradation in the tensile and flexural strengths of cement mortar that is associated to elevated temperatures exposure and maximum relative strength improvement due to 2% nanoclay addition at 400°C. Moreover, it was observed that elevated temperatures caused hairline cracks along the cement matrix and the presence of nanoclay decreased the density and the width of detected microcracks.

Fan et al. (2015) substituted ordinary Portland cement in concrete with nanokaolinite clay at varied weight percentages of 0%, 1%, 3%, and 5%. Compressive strength improvement of 34% was demonstrated in nanoclay-altered concrete even after 125 freeze–thaw cycles. Moreover, a 59% reduction and 64% increment were observed in the chloride diffusion coefficient and electrical resistivity in the concrete samples with 5% nanoclay after 75 freeze–thaw cycles. Another group of researchers, fabricated self-compacting concrete with different portions of nanoclay and mineral admixtures and subjected the specimens to freeze–thaw cycles in temperature range of −18°C to +4°C (Langaroudi & Mohammadi, 2018).

The incorporation of 3% nanoclay improved the overall characteristics of specimens which was evident by amplification in freeze—thaw and frost resistance. Hakamy, Shaikh, and Low (2014) showed that the use of nanoclay in hemp fabric reinforced composites reduced the porosity significantly. The decrease in the porosity of the composites was associated with the formation of more hydration products which results in the densification of the microstructure. However, the optimum dosage to achieve lower porosity was deemed to be 1% nanoclay as there is a consequential increase in the porosity at dosages higher than 1%.

18.4 Microbial biomineralization

Recently, biomineralization has received more attention as an ecofriendly method of protecting and remediating building materials. It is a biogeochemical phenomenon occurring in nature caused by prokaryotic microorganisms by altering their metabolic rates with the surrounding environment, leading to the formation of an inorganic matter (Dhami, Reddy, & Mukherjee, 2013). More than 60 biominerals are currently known, most of which are calcium carbonates, silicates, and iron oxides or sulfides (Weiner & Dove, 2003). Moreover, biominerals vary distinctly in shape, size, crystallinity, isotopic, and trace element composition (Bazylinski, Frankel, & Konhauser, 2007; Takahashi, Hirata, Shimizu, Ozaki, & Fortin, 2007). Mineral precipitation by prokaryotes can be characterized into three different categories based upon their modes of biomineralization. The biologically induced mineralization (BIM; Lowenstam, 1981), biologically controlled mineralization (BCM: Mann, 1983), and biologically influenced mineralization (BIFM; Fortin & Beveridge, 2000). The latter is also known as organic matrix-mediated mineralization or sometimes biologically passive biomineralization (Dupraz et al., 2009).

In the case of BIM, minerals are formed due to the interaction between the biological activity of microorganisms and the environment. Biominerals are formed as a result of the metabolic activity of bacteria and the changes occurring in the chemical equilibrium of the surrounding. Biominerals like carbonates, oxides, phosphates, sulfates, and sulfides all fall under this list. Whereas in another case, BCM that is, biological controlled mineralization microorganism controls the rate of nucleation, growth, and location of precipitates formed (Bazylinski & Frankel, 2003; Krajewska, 2018). Weiner and Dove (2003) categorized mineral precipitates as inter-, intracellularly, and extracellularly. Bacteria employ control over the composition, size, habit, and location of the minerals formed. BCM depends on the bacteria and the external environment does not affect mineral precipitation. The metabolic ability of microorganisms to precipitate biominerals motivated researchers globally for exploring the miscellaneous benefits of biomineralization for bioengineering and biomedical applications.

Calcium carbonate is considered to be most remarkable due to its efficient bonding capacity and compatibility with concrete compositions. Carbonate minerals are precipitated via biologically induced biomineralization. Calcium forms strong bonding with carbonate than any other mineral. Carbonate is a polymorph that has three anhydrous crystalline phases; calcite, aragonite, and vaterite. The shape, form, and size of a crystal are highly dependent on the source of the environment. Diverse calcium sources form a crystal with different shapes. The most stable form of calcium carbonate precipitated crystal is rhombohedral shape (calcite) induced by calcium chloride (Favre, Christ, & Pierre, 2009; Gorospe et al., 2013). Other forms of crystals are lettuce like or lamellar shape (vaterite) prompted by calcium acetate, while calcium lactate and calcium gluconate instigate a more complex shape that leads to the growth of vaterite with a spherical shape (Tai & Chen, 1998).

18.5 Bioconcrete and its limitations

The adverse effect of microorganisms on building materials has been acknowledged for centuries. The majority of microorganisms such as bacteria, cyanobacteria, fungi, algae, lichens, and so forth have prodigious effect on the surroundings and are considered great source of biodeterioration (Bundeleva et al., 2014; Ferrari et al., 2015; Verdier, Coutand, Bertron, & Roques, 2014; Liu et al., 2020). In spite of all this, microorganisms can also have a positive influence on the cementitious materials. From the last two decades, the valuable role of microorganisms in enhancing the restoration and durability of building materials, stones, and concrete are under extensive research. The major benefit of microbial biomineralization is MICCP due to its efficient bonding capacity and compatibility with concrete compositions. It is considered the noblest, environment friendly and economical material for engineering applications. The MICP for restoration and refining the surfaces of ornamental stone was introduced by Adolphe, Loubiere, Paradas, and Soleilhavoup (1990) for which they received the patent known as "bioconcept of calcite or biodeposition,". Currently, this biomaterial is known as bio concrete or MICCP based concrete. Bio concrete in other words is described as the concrete prepared by admixing bacteria with the capability of bacteria for precipitation of carbonate that aids in improving the overall properties of concrete and crack remediation (Jonkers, 2011). Self-healing concrete is another new technology that is gaining importance later. In which concrete is cast by immobilizing bacterial spores along with calcium source, in which when water enters the cracks, it opens up the pellets, causing the bacteria to germinate and produce calcite (Jonkers & Schlangen, 2009; Wiktor & Jonkers, 2015). Many microorganisms can precipitate calcium carbonate through different mechanisms or pathways; photosynthesis, ureolysis, ammonification, denitrification, sulfate reduction, anaerobic sulfide oxidation, and methane oxidation, by increasing pH or dissolved inorganic carbon. Both autotrophic and heterotrophic pathways are involved in forming an alkaline environment for bacterial-induced precipitation. The cell wall of the bacteria acts as a nucleation site for precipitation having Ca^{2+} ions attached to the surface (Dhami et al., 2013).

Microbes enhance the self-repair ability of cementitious and other building materials. The major interest of MICCP in concrete and limestone marble is the restoration of damaged materials, strength enhancement, resistance to freeze–thaw, high carbonation, low permeability, resistance to chloride attack, and have low maintenance cost. It improves the quality of bioconcrete by providing resistance toward deterioration. The metabolic activity of microbes has been employed through different metabolic pathways to restore the damage caused to concrete and stone infrastructure. Different diversity of microbes such as *Bacillus sphaericus, Bacillus licheniformis, Bacillus pasteurii, Bacillus pseudofirmus, Bacillus magaterium, Bacillus cohnii* utilizes organic compound such as urea and calcium source to proliferate the carbonate precipitation (Achal et al., 2013; Dhami et al., 2013; Jonkers & Schlangen, 2009; Joshi, Goyal, Mukherjee, & Reddy, 2019; Wang, Van Tittelboom, De Belie, & Verstraete, 2012). Some halophilic and halotolerant bacteria in particularly *Exiguobacterium mexicanum, Halomonas, Flavobacterium,* and *Acinetobacter* are reported to show carbonate precipitation (Bansal, Dhami, Mukherjee, & Reddy, 2016; Ferrer et al., 1988; Rivadeneyra, Delgado, Párraga, Ramos-Cormenzana, & Delgado, 2006; Stabnikov, Chu, Myo, & Ivanov, 2013). Achal et al. (2013) investigated the calcite precipitation by *Bacillus* sp. CT5 bacteria for enhancing the durability and heal crack of 27.2 mm depth in cement structures. Over 50% reduction of the porosity, low chloride ion permeability as determined by the RCPT, and 40% improved compressive strength in their mortar specimens. Nain et al. (2019) reported 22.5%, 14.3%, and 15.8% increase in compressive strength and 18.49%, 25.3%, and 19.58% split-tensile strength of upon treatment with *B. subtilis*, *Bacillus megaterium* and their consortia respectively. Recently, according to the results of the microscopic observation, after the cracks were healed, Jafarnia, Saryazdi, and Moshtaghioun (2020) found *Sporosarcina pasteurii* bacteria were able to heal crack of 0.58 mm width completely at 30 days by using bacteria in curing environment, while cracks up to 0.16 mm width were completely healed in bacteria admixed concrete samples. Further increase of 23% compressive strength and 64% in electrical resistivity of concrete treated with bacteria containing 10% percentage zeolite was observed. *Bacillus* is the most extensively used bacteria in bio concrete research studies as a bioagent for carbonate precipitation because of its high tolerance to the alkaline conditions and moisture. Moreover, it can form spores upon unfavorable conditions and is harmless to man, which makes it suitable to use as self-healing agent in concrete (Gupta, Dai Pang, & Kua, 2017). Similarly, photosynthetic microorganism's viz. algae and cyanobacteria assimilate the calcium carbonate precipitation by oxygenic photosynthetic pathway. Species such as *Dichothrix* can precipitate calcite and aragonite in seawater. *Lyngbya* and *Gloeocapsa* species can precipitate hydro-magnesite in alkaline lakes. These bacteria can easily survive in alkaline environment conditions. Badger, Hanson, and Price (2002) reported calcification as an extracellular process in photoautotrophic bacteria (e.g., cyanobacteria) which occurs in the exopolysaccharide sheath or proteinaceous surface layer (S-layer) that surrounds the cells. Recently, *Synechococcus* PCC8806 was used to prepare microbial concrete and it provided 38% greater calcite precipitation as compared to the controlled sample (Zhu, Paulo, Merroun, & Dittrich, 2015). Further, Zhu, Lu, and Dittrich (2017) investigated the calcification potential of *Gloeocapsa* PCC73106 in mortar samples, and demonstrated that UV-killed cells were able to enhance the compressive strength, decrease the water absorption, and the porosity.

Comprehensive research has been put into effect to address the applications of phototrophic and heterotrophic bacteria in enhancing the strength of concrete and stone restoration (Table 18–4). However, there are a few limitations to this technology. One of the drawbacks of infusing bacteria in a concrete matrix as a means of creating self-healing concrete is its highly alkaline conditions. Secondly, low survival rate of microorganisms due to

Table 18–4 Applications of MICCP to prevent deterioration of monumental stones and cementitious materials.

Microorganisms	Application	Evaluation of specimens	Reference
Desulfovibrio vulgaris spp. *vulgaris* ATCC 29579 coupled with a nonionic detergent	Removal of sulfate based black crust, gray deposit	Environmental scanning electron microscope (**ESEM**), X-ray energy dispersive system (EDS), Fourier-transform infrared spectroscopy (FTIR), X-ray powder diffraction (XRD)	Troiano et al. (2013)
Bacillus sphaericus LMG 22557 melamine-based microencapsulation	Crack healing in cementitious material	Water permeability	Wang et al. (2014)
Bacillus pumilus ACA-DC 4061	Biomineralization on marble stone	Chromatic analysis, XRD, SEM, FTIR	Daskalakis et al. (2015)
Bacillus subtilis LMG 3589	Improved mechanical properties and resistance to salt deterioration in limestone	Water absorption, optical microscopy, SEM	Micallef, Vella, Sinagra, and Zammit (2016)
Gloeocapsa PCC73106	Strength enhancement of cementitious material	Compressive strength, water absorption, porosity, SEM-EDS, XRD	Zhu et al. (2017)
Penicillium chrysogenum	Bio cementation of sand stone	Compressive strength, SEM, XRD	Fang et al. (2018)
Sporosarcina pasteurii	Crack healing in marble	X-ray CT scan, permeability	Minto et al. (2018)
Bacillus cereus	Crack healing in cementitious material	Water permeability, rapid chloride permeability test. SEM, FTIR, XRD, DSC	Wu, Hu, Zhang, Xue, and Zhao (2019)
Bacillus pasteurii (DSM 33)	Antierosion of tiles	Capillary water absorption test, durability test, acid resistance test, water vapor permeability test	Liu et al. (2020)
B. subtilis using cellulose fiber as a bacteria-carrier	Crack healing in cementitious material	Compressive strength, UPV	Singh and Gupta, (2020)
B. subtilis	Strength enhancement	water absorption, electrical resistance, compressive strength, chloride ion penetration, carbonation depth, water penetration depth	Salmasi and Mostofinejad (2020)

DSC, differential scanning calorimetry; FTIR, Furrier transformed infrared spectroscopy; SEM, scanning electron microscopy.

shear stresses to the bacteria during the concrete mixing process and the dry condition of hardened concrete (Seifan & Berenjian, 2020). Moreover, the size of pores in most of the cases is smaller than 0.5 μm, while the size of bacteria is in the range of 1−3 μm. It is challenging for bacteria to endure such harsh conditions. Jonkers and Schlangen (2007) observed a decrease in the population of bacteria from 10^9 spores/cm^3 to 1.15×10^6 spores/cm^3 (0.12% of the original inoculum) of *B. cohnii* spores, 0.01% *Bacillus halodurans* spores (10^9 spores/cm^3 to 1.07×10^5 spores/cm^3 of the original inoculum) and 0.06% *B. pseudofirmus* spores (10^9 spores/cm^3 to 5.62×10^5 spores/cm^3 of the original inoculum) survived 10 days after inoculation to cement paste. Similarly, Achal, Mukherjee, and Reddy (2011) reported a rapid decline in the concentration of *B. megaterium* from 10^7 to 10^5 cfu/mL after 3 days, and 10^4 cfu/mL after 28 days in mortar specimens (survival rate 0.1%). Basaran (2013) outlined 9×10^3 MPN/mL (0.4% of the original inoculums) viable cells after 28 days curing period. Out of all the viable cells, about 50% were observed to be spores. Principally, unprotected bacterial cells and spores have poor survival ability inside the cementitious matrix. Studies have been reported in which different implementation techniques are used such as the addition of nutrient sources and other materials for example, hollow fibers, microencapsulation, superabsorbent polymers, polyurethane, sol-gel ceramics, calcium sulphoaluminate cement (González et al., 2020; Lors, Ducasse-Lapeyrusse, Gagné, & Damidot, 2017; Snoeck & De Belie, 2019). Apart from these an alternative approach has been examined for immobilizing bacteria by using carrier material for instance silica gel (Wang et al., 2012), GNPs (Khaliq & Ehsan, 2016), hydrogel (Wang et al., 2015); zeolite (Bhaskar, Hossain, Lachemi, Wolfaardt, & Kroukamp, 2017) and polymer microcapsules (Wang, Soens, Verstraete, & De Belie, 2014). Immobilization of bacterial strain safeguards the bacteria from harsh concrete conditions such as a high pH (12.5), low nutrient and water access, thus enhancing the performance of bacterial admixed or bacterial treated concrete. Wang et al. (2014) successfully used the hydrogel in healing a crack of 0.5 mm width. Owning to the hydrophilic nature of Hydrogels, it stores sufficient water to promote bacterial growth and $CaCO_3$ precipitation. Similarly, Bhaskar et al. (2017) showed improvement in self-healing efficiency of mortar upon using zeolite as carrier material for immobilizing bacteria. Chen, Qian, and Huang (2016) used ceramsite as carrier material to immobilize bacteria *Bacillus mucilaginous* and Brewer's yeast in prism specimens ($40 \times 40 \times 160$ mm) and cylinders specimens (110×45 mm). Due to the precipitation of calcite in the specimens, the flexural strength of specimens was successfully escalated from 56% to 87.5%. Because of self-healing efficiency, specimens healed crack up to 0.3−0.4 mm width and showed decreased water penetration.

18.6 Nanoengineered self-healing concrete

The construction industry has undergone significant development over time. Many researchers and scientists have employed the metabolic role of bacteria to improve the strength and self-healing ability of concrete and other building materials (Jafarnia et al., 2020; Mondal &

Ghosh, 2018; Van Tittelboom et al., 2010). With innovation in the synthesis of nanoparticles, nanotechnology has brought tremendous possibilities of immobilizing bacteria and offering improved self-healing. It has magnificently brought development in every domain including medicine, engineering, manufacturing, electronics, and computer sciences. However, compared to other areas, application in construction engineering is still in progress.

Nanotechnology can enhance the longevity of construction materials by delivering smarter and stronger structural composites, low-cost coatings, and better cementitious materials. The construction industry has undergone significant development over time. Various properties of nanoparticles such as particle size, surface area, and specific site action can be utilized to reduce microdefects in the hardened cement (Abulmagd & Etman, 2018). It improves the intrinsic bond between the cement and aggregate mixture. Moreover, nanomaterials offer high chemical reactivity, corrosion resistance, strength and durability, which are of particular interest to civil industry. Nanoscience offers resistance against biodegradation of concrete structures by embedding nanomaterials into a concrete frame or in coating compositions (Peyvandi et al., 2013; Shi, Liu, Yang, & Han, 2008). Mechanical performance (density, tensile and compressive strength) of concrete structures can be enhanced as nanoparticles act as an excellent filling medium through cement pores or voids.

Limited studies are available on the potential use of nanoscience with immobilized bacteria in cementitious material (Table 18–5). Due to the ongoing research of manufacturing smart cement composite and cost-efficient repair mechanisms, researchers are now focusing on using a bacterial strain with nanoparticles to enhance the self-healing ability of the nanoengineered concrete structures. Recently, Gupta and his coworkers have successfully immobilized bacterial spores using biochar combined with superabsorbent polymer and a polypropylene microfiber. These carrier materials ensured the availability of water to bacteria and helped in controlling crack propagation during the damage of mortar. As a result, mechanical strength was reported to enhance by 38%. Durability and water penetration properties of cement were improved (Gupta, Kua, & Dai Pang, 2018). Rajasegar and Kumaar (2020) reported a hybrid effect of nanosilica, rice husk, and PVA on self-healing efficiency of cementitious materials. Improvement in compressive strength and crack remediation was reported in concrete specimens composited using nanosilica and RHA. A continuous investigation is carried out by researchers to withdraw the self-sufficient healing of nanoparticles in enhancing the properties of concrete and mortar. As micro cracks and other damage are inevitable, owning to deterioration by physical and chemical factors. Cracks make concrete unresisting to various detrimental deterioration factors due to ingress of harmful compounds. Recently, Shaheen et al. (2019) examined the structural properties of concrete by immobilizing *B. subtilis'* bacteria in iron-oxide nano/microparticles (INMPs) and bentonite nano/microparticles (BNMPs). Authors observed that INMPs successfully healed cracks up to 1.2 mm width while BNMPs immobilized samples healed cracks up to only 0.1 and 0.45 mm width. Further, concrete cylindrical specimens of (150 mm × 300 mm) were tested for strength enhancement, showing 85% escalation in compressive strength in INMPs specimens. Whereas, BNMPs prepared specimens were observed to improve by 45% and 65%, respectively.

Table 18-5 Applications of nanoengineered bio concrete.

Bacterial strain	Immobilizing material	Specimens used	Application	Key findings	Reference
Sporosarcina pasteurii	Glass beads, Siran	(25.4 × 25.4 × 152) mm beam and (50.8 × 50.8 × 50.8) mm mortar cube 76.2 × 76.2 × 285.8 mm beams	Crack healing, Stiffness, Compressive strength enhancement, freeze–thaw	12% ↑ in stiffness 24% ↑ in compressive strength 98% retain in original weight after freeze–thaw cycle than only 69% in control 3.175 mm wide and 12.7 mm deep in beams, and 3.175 mm wide and 25.4 mm deep in cubes crack	Bang et al. (2010)
Bacillus subtilis	Lightweight aggregates (LWA), graphene nanoplatelets (GNP)	150 × 100 mm 150 × 300 mm	Crack healing, Compressive strength	9.8% ↑ in compressive strength 12% ↑ in compressive strength	Khaliq and Ehsan (2016)
Bacillus sphaericus	Biochar, superabsorbent polymer (SAP), and a polypropylene microfiber (PP)	Cylinder 200 mm (h) × 100 mm (d) cube (50 mm).	Crack remediation, Durability and Strength enhancement	700 μm width Crack healed 38% ↑ in compressive strength 65% ↓ in water penetration 70% ↓ water absorption	Gupta et al. (2018)
B. sphaericus and Bacillus licheniformis	3-aminopropyltriethoxy silane (APTES) coated Fe_3O_4	—	$CaCO_3$ yield viability of bacteria	$CaCO_3$-specific yield range 8.3 × 10^4 to 1.3 × 10^5 mg/OD600 in all the concentration Specific yield higher in 50 μg/mL of APTES coated IONs. highest viability in 100 μg/mL APTES coated IONs	Seifan et al. (2018a)
B. sphaericus and B. licheniformis	Iron-oxide nanoparticle (ION)	Cylindrical disk (100 × 50 mm)	Crack remediation, Water absorption	26% ↓ initial water absorption 22% ↓ secondary water absorption	Seifan et al. (2018b)
B. subtilis	Oxide nano/microparticles (INMPs) and bentonite nano/microparticles (BNMPs)	Concrete cylindrical specimens (150 × 300) mm	Crack remediation, Durability and Strength enhancement	INMPs healed cracks up to 1.2 mm width with 85% ↑ in compressive strength BNMPs healed cracks up to 0.15 and 0.45 mm with 65% ↑ in compressive strength	Shaheen et al. (2019)
B. sphaericus and B. licheniformis	Magnetic ION	Cylinder (100 × 50) mm	Durability enhancements	↓ water absorption	Seifan et al. (2019)

Earlier, Seifan, Ebrahiminezhad, Ghasemi, Samani, and Berenjian (2018b) immobilized magnetic ION with *Bacillus licheniformis* (ATCC 9789) and *B. sphaericus* (NZRM 4381) by incubating at 150 rpm (35°C) for 15 min. The maximum bacterial growth was observed in the 150 μg/mL IONs. The highest concentration for induced $CaCO_3$ precipitation (34.54 g/L) was achieved by the bacterial cells immobilized with 300 μg/mL IONs. Out of all the nanoparticles, the application of magnetic ION is in demand, as they possess exceptional physical and chemical properties because of their high intrinsic anisotropy and surface activity (Wu et al., 2010). The nanoscience approach ensures the survival of bacteria from severe conditions of concrete. Studies have been conducted by Seifan et al. (2018b), Seifan, Ebrahiminezhad, Ghasemi, Samani, and Berenjian (2018a) in composing two types of nanoparticles notably naked Fe_3O_4 and 3-aminopropyltriethoxy silane (APTES) coated Fe_3O_4 to study the role of ION in the survival of bacterial and its ability to induce $CaCO_3$ precipitation in the concrete framework. Experimental results showed amino coated ION were effective in promoting bacterial growth and naked iron-oxide particles were efficient in $CaCO_3$ precipitation. The morphology of precipitated $CaCO_3$ crystals and interrelationship of IONs with the bacterial cell surface was confirmed using SEM and XRD diffraction. The efficiency of the immobilization technique of bacteria is affected by three different factors such as particle size, magnetization force, and dispersion. Later, Seifan and his colleagues carried out an experimental study to determine the effect of immobilized bacterial cells with IONs on the durability of concrete (Seifan, Ebrahiminezhad, Ghasemi, & Berenjian, 2019). In this study, the negatively charged bacterial strain was immobilized with nanoparticles by adsorption of magnetic IONs. This technique was productive in decreasing the water absorption and permeable pore space of concrete specimens.

Immobilization of bacteria and nanoparticles in carrier materials prevents microbes from crushing during mixing and help in inducing self-healing cementitious property. Nanoparticles also provide the additional benefit of protecting structures from corrosion and enhancing the strength and durability property. Khaliq and Ehsan (2016) found that using lightweight aggregate and GNPs as additional material along with *B. subtilis* gave a better result in strength improvement as compared to control. Khaliq along with his coworker designed four different mix proportions (Mix 1–Mix 4) for incorporating *B. subtilis* bacteria in the concrete mixture. In "Mix 1" no bacterial spore specimens were added whereas, in "Mix 2" specimens, bacteria were added precisely by mixing the bacterial broth in water during mixing of concrete. In "Mix 3" lightweight aggregates (LWA), were used as carrier material by soaking them in bacterial broth for 24 hours. In "Mix 4" GNP were introduced with superplasticizer (Sikament-520) to ensure the uniform distribution of GNP. GNPs were soaked with the bacterial solution before mixing in concrete. Precracked samples (150 × 300 mm) were tested for compressive strength and self-healing efficiency at 3, 7, 14, and 28 days. GNP incorporated specimens were reported maximum crack healing at the age of 3 and 7 days than 28 days. Specimens with LWA as carrier material displayed consistent crack healing but were not as efficient as GNP. Compressive strength was observed to slightly higher in all the samples with bacterial consolidation.

Therefore the fusion of microbes with nanoparticles may significantly modify the microstructural properties of cementitious materials. It can also produce physical effects such as $CaCO_3$ precipitation, improved mechanical strength and reduced water absorption. In the case of self-healing systems based on nanoparticles, Evangelia and Maria (2020) successfully used nano-SiO$_2$ and nano-CaO as filler nanoparticles agent in a cement matrix. To test the Self-healing efficiency in specimens $40 \times 40 \times 40$ mm^3, compressive strength was tested to increase by 13.5% and 23.4%, at the age of 7 and 28 days, respectively. The crack (width <100 μm) healed partially for the nanomodified compositions within the first 14 days of curing and healed completely at the age of 28 days. The recent development in nanotechnology science could lead to a new era of the construction industry that is much stronger, more durable, and attractive as well as smarter in damage control.

18.7 Challenges and future prospects

The demands and needs of the general masses to protect degrading construction composites call for new challenges in the exploration of some advanced, reliable, smart but effective building materials that should be safe for both humans and the environment. In this context, nanotechnology is playing major role in some high priorities' areas such as provision of biocidal efficacy, chemical and physical attack resistance, and strength enhancement to protect construction compounds from degradation and deterioration. However, the incorporation of nanomaterials in building materials should be monitored carefully to verify their potential in particular applications. As nanomaterials come with their own set of problems such as toxicity, agglomeration, colloidal suspension leaching to surrounding environment. For these reasons, it is important to continue research on finalizing the nanoparticle concentration, morphology, dimensions, and interaction between microorganisms, materials, and nanomaterials. Furthermore, in provision of biocidal efficacy to building materials, future experiments should focus not only on type of microorganisms for testing but also on the relevancy and metabolic potential of microorganisms. To ensure the microbial survivability and accomplishment of crack healing without affecting mechanical characteristics of concrete is a major concern. Nanomaterials functions as additional surface area for bacterial nucleation and crystal growth by filling up the voids and provide C-S-H for hydration. However, there is still need to work on immobilization techniques and carrier materials for formulating long-lasting, sturdy, and cost-efficient bioconcrete. Moreover, most of the investigations have been done at the laboratory scale. Therefore this task requires detailed analysis of field study before commercialization. The application of studied nanocomposites and materials should also be taken into consideration, so that the outcome would give a sufficient measure of the consequences arising from the given use.

References

Abulmagd, S., & Etman, Z. A. (2018). Nanotechnology in repair and protection of structures state-of-the-art. *Journal of Civil & Environmental Engineering, 8*, 306. Available from https://doi.org/10.4172/2165-784X.1000306.

Achal, V., & Mukherjee, A. (2015). A review of microbial precipitation for sustainable construction. *Construction and Building Materials, 93*, 1224–1235.

Achal, V., Mukherjee, A., & Reddy, M. S. (2011). Microbial concrete: Way to enhance the durability of building structures. *Journal of Materials in Civil Engineering, 23*(6), 730–734.

Achal, V., Mukherjee, A., & Reddy, M. S. (2013). Biogenic treatment improves the durability and remediates the cracks of concrete structures. *Construction and Building Materials, 48*, 1–5.

Adams, F. C., & Barbante, C. (2013). Nanoscience, nanotechnology and spectrometry. In *Spectrochimica Acta. Part B: Atomic Spectroscopy, 86*, 3–13.

Adesina, A. (2019). Durability enhancement of concrete using nanomaterials: An overview . *Materials science forum* (Vol. 967, pp. 221–227). Trans Tech Publications Ltd.

Adesina, A. (2020). Nanomaterials in cementitious composites: Review of durability performance. *Journal of Building Pathology and Rehabilitation, 5*(1), 1–9.

Adolphe, J. P., Loubiere, J. F., Paradas, J., Soleilhavoup, F., (1990). Proc_ed_e de traitement biologique d' une surface artificielle. European Patent No. 90400697.0 (after French Patent No. 8903517, 1989).

Agnihotri, S., Dhiman, N., & Tripathi, A. (2018). *Antimicrobial surface modification of polymeric biomaterials* (pp. 435–486). Elsevier.

Agnihotri, S., & Dhiman, N. K. (2017). Development of nano-antimicrobial biomaterials for biomedical applications. In *Advances in biomaterials for biomedical applications* (pp. 479–545). Singapore: Springer.

Agnihotri, S., Mukherji, S., & Mukherji, S. (2014). Size-controlled silver nanoparticles synthesized over the range 5–100 nm using the same protocol and their antibacterial efficacy. *RSC Advances, 4*(8), 3974–3983.

Ahmed, N. Y. & Alkhafaji, F. F. (2020). Enhancements and mechanisms of nano alumina (Al_2O_3) on wear resistance and microstructure characteristics of concrete pavement. In *IOP Conference Series: Materials Science and Engineering*, 871, p. 012001. IOP Publishing.

Ahmed, T., Usman, M., & Scholz, M. (2018). Biodeterioration of buildings and public health implications caused by indoor air pollution. *Indoor and Built Environment, 27*(6), 752–765.

Ahn, T.H. & Kishi, T. (2008). The effect of geo-materials on the autogenous healing behavior of cracked concrete. Concrete repair, rehabilitation and retrofitting II, Cape Town, South Africa, pp. 125–126.

Akhtari, M., & Nicholas, D. (2013). Evaluation of particulate zinc and copper as wood preservatives for termite control. *European Journal of Wood and Wood Products, 71*(3), 395–396.

Aldosari, M. A., Darwish, S. S., Adam, M. A., Elmarzugi, N. A., & Ahmed, S. M. (2019). Using ZnO nanoparticles in fungal inhibition and self-protection of exposed marble columns in historic sites. *Archaeological and Anthropological Sciences, 11*(7), 3407–3422.

Allen, N. S., Edge, M., Sandoval, G., Verran, J., Stratton, J., & Maltby, J. (2005). Photocatalytic coatings for environmental applications. *Photochemistry and Photobiology, 81*(2), 279–290.

Alum, A., Rashid, A., Mobasher, B., & Abbaszadegan, M. (2008). Cement-based biocide coatings for controlling algal growth in water distribution canals. *Cement and Concrete Composites, 30*(9), 839–847.

Anisuddin, S. & Khaleeq, S. (2005). Deterioration and rehabilitation of concrete structures in hot and arid regions. In *Proceedings of the twenty-first annual ARCOM conference*. Vol. 7, No. 9.

Arefi, M. R., & Rezaei-Zarchi, S. (2012). Synthesis of zinc oxide nanoparticles and their effect on the compressive strength and setting time of self-compacted concrete paste as cementitious composites. *International Journal of Molecular Sciences, 13*(4), 4340–4350.

Badger, M. R., Hanson, D., & Price, G. D. (2002). Evolution and diversity of CO_2 concentrating mechanisms in cyanobacteria. *Functional Plant Biology, 29*(3), 161–173.

Baloch, H., Usman, M., Rizwan, S. A., & Hanif, A. (2019). Properties enhancement of super absorbent polymer (SAP) incorporated self-compacting cement pastes modified by nano silica (NS) addition. *Construction and Building Materials, 203*, 18–26.

Banach, M., Szczygłowska, R., Pulit, J., & Bryk, M. (2014). Building materials with antifungal efficacy enriched with silver nanoparticles. *Journal of Chemical Sciences, 5*, 085.

Bansal, R., Dhami, N. K., Mukherjee, A., & Reddy, M. S. (2016). Biocalcification by halophilic bacteria for remediation of concrete structures in marine environment. *Journal of Industrial Microbiology & Biotechnology*, *43*(11), 1497−1505.

Barbhuiya, S., & Chow, P. (2017). Nanoscaled mechanical properties of cement composites reinforced with carbon nanofibers. *Materials*, *10*(6), 662.

Basaran, Z. (2013). Biomineralization in cement-based materials: inoculation of vegetative cells.

Bastami, M., Baghbadrani, M., & Aslani, F. (2014). Performance of nano-silica modified high-strength concrete at elevated temperatures. *Construction and Building Materials*, *68*, 402−408.

Bazant, Z. P. & Wittmann, F. H. (1982). Creep and shrinkage in concrete structures.

Bazylinski, D. A., & Frankel, R. B. (2003). Biologically controlled mineralization in prokaryotes. *Reviews in Mineralogy and Geochemistry*, *54*(1), 217−247.

Bazylinski, D. A., Frankel, R. B., & Konhauser, K. O. (2007). Modes of biomineralization of magnetite by microbes. *Geomicrobiology Journal*, *24*(6), 465−475.

Behfarnia, K., & Salemi, N. (2013). The effects of nano-silica and nano-alumina on frost resistance of normal concrete. *Construction and Building Materials*, *48*, 580−584.

Belli, A., Mobili, A., Bellezze, T., Tittarelli, F., & Cachim, P. (2018). Evaluating the self-sensing ability of cement mortars manufactured with graphene nanoplatelets, virgin or recycled carbon fibers through piezoresistivity tests. *Sustainability*, *10*(11), 4013.

Bellissima, F., Bonini, M., Giorgi, R., Baglioni, P., Barresi, G., Mastromei, G., & Perito, B. (2014). Antibacterial activity of silver nanoparticles grafted on stone surface. *Environmental Science and Pollution Research*, *21*(23), 13278−13286.

Bellotti, N., Romagnoli, R., Quintero, C., Domínguez-Wong, C., Ruiz, F., & Deyá, C. (2015). Nanoparticles as antifungal additives for indoor water borne paints. *Progress in Organic Coatings*, *86*, 33−40.

Berndt, M. L. (2011). Evaluation of coatings, mortars and mix design for protection of concrete against sulphur oxidising bacteria. *Construction and Building Materials*, *25*(10), 3893−3902.

Bertron, A. (2014). Understanding interactions between cementitious materials and microorganisms: A key to sustainable and safe concrete structures in various contexts. *Materials and Structures*, *47*(11), 1787−1806.

Beyth, N., Houri-Haddad, Y., Domb, A., Khan, W., & Hazan, R. (2015). Alternative antimicrobial approach: Nano-antimicrobial materials. *Evidence-Based Complementary and Alternative Medicine*, *2015*.

Bhaskar, S., Hossain, K. M. A., Lachemi, M., Wolfaardt, G., & Kroukamp, M. O. (2017). Effect of self-healing on strength and durability of zeolite-immobilized bacterial cementitious mortar composites. *Cement and Concrete Composites*, *82*, 23−33.

Biswas, J., Sharma, K., Harris, K. K., & Rajput, Y. (2013). Biodeterioration agents: Bacterial and fungal diversity dwelling in or on the pre-historic rock-paints of Kabra-pahad, India. *Iranian Journal of Microbiology*, *5*(3), 309.

Bolashikov, Z. D., & Melikov, A. K. (2009). Methods for air cleaning and protection of building occupants from airborne pathogens. *Building and Environment*, *44*(7), 1378−1385.

Borrego, S., Lavin, P., Perdomo, I., Gómez de Saravia, S., & Guiamet, P. (2012). Determination of indoor air quality in archives and biodeterioration of the documentary heritage. *International Scholarly Research Notices*, *2012*.

Bravery, A. F. (1988). Biodeterioration of paint-a state-of-the-art comment. In *Biodeterioration* (7, pp. 466−485). Dordrecht: Springer.

Bundeleva, I. A., Shirokova, L. S., Pokrovsky, O. S., Bénézeth, P., Ménez, B., Gérard, E., & Balor, S. (2014). Experimental modeling of calcium carbonate precipitation by cyanobacterium *Gloeocapsa* sp. *Chemical Geology*, *374*, 44−60.

Cailleux, E. & Pollet, V. (2009). Investigations on the development of self-healing properties in protective coatings for concrete and repair mortars. In *Proceedings of the second international conference on self-healing materials*, Chicago, USA.

Calabria, J., Vasconcelos, W. L., Daniel, D. J., Chater, R., McPhail, D., & Boccaccini, A. R. (2010). Synthesis of sol–gel titania bactericide coatings on adobe brick. *Construction and Building Materials, 24*(3), 384–389.

Carrillo-González, R., Martínez-Gómez, M. A., González-Chávez, M. D. C. A., & Hernández, J. C. M. (2016). Inhibition of microorganisms involved in deterioration of an archaeological site by silver nanoparticles produced by a green synthesis method. *Science of the Total Environment, 565*, 872–881.

Castro-Alonso, M. J., Montañez-Hernandez, L. E., Sanchez-Muñoz, M. A., Macias Franco, M. R., Narayanasamy, R., & Balagurusamy, N. (2019). Microbially Induced Calcium carbonate Precipitation (MICP) and its potential in bioconcrete: Microbiological and molecular concepts. *Frontiers in Materials, 6*, 126.

Chauhan, A., Sillu, D., & Agnihotri, S. (2019). Removal of pharmaceutical contaminants in wastewater using nanomaterials: A comprehensive review. *Current Drug Metabolism, 20*(6), 483–505.

Chen, F., Yang, X., & Wu, Q. (2009). Antifungal capability of TiO_2 coated film on moist wood. *Building and Environment, 44*(5), 1088–1093.

Chen, H., Qian, C., & Huang, H. (2016). Self-healing cementitious materials based on bacteria and nutrients immobilized respectively. *Construction and Building Materials, 126*, 297–303.

Chen, P. W., & Chung, D. D. L. (1996). Concrete as a new strain/stress sensor. *Composites Part B: Engineering, 27*(1), 11–23.

Chen, X., & Mao, S. S. (2007). Titanium dioxide nanomaterials: Synthesis, properties, modifications, and applications. *Chemical Reviews, 107*(7), 2891–2959.

Chen, Y., Yu, J., & Leung, C. K. (2018). Use of high-strength strain-hardening cementitious composites for flexural repair of concrete structures with significant steel corrosion. *Construction and Building Materials, 167*, 325–337.

Christou, G., Tantele, E. A., & Votsis, R.A. (2014). Effect of environmental deterioration on buildings: A condition assessment case study. In *Proceedings of the second international conference on remote sensing and geoinformation of the environment*. Vol. 9229, (p. 92290Y). International Society for Optics and Photonics.

Colangiuli, D., Lettieri, M., Masieri, M., & Calia, A. (2019). Field study in an urban environment of simultaneous self-cleaning and hydrophobic nanosized TiO_2-based coatings on stone for the protection of building surface. *Science of the Total Environment, 650*, 2919–2930.

Cong, P., Wang, J., Li, K., & Chen, S. (2012). Physical and rheological properties of asphalt binders containing various antiaging agents. *Fuel, 97*, 678–684.

Cwalina, B. (2008). Biodeterioration of concrete. *Archives of Civil and Mechanical Engineering, 4*, 133–140.

Cwirzen, A., Habermehl-Cwirzen, K., Nasibulina, L. I., Shandakov, S. D., Nasibulin, A. G., Kauppinen, E. I., . . . Penttala, V. (2009). CHH cement composite. In *Nanotechnology in construction* (3, pp. 181–185). Berlin, Heidelberg: Springer.

D'Alessandro, A., Ubertini, F., Laflamme, S., & Materazzi, A. L. (2015). Towards smart concrete for smart cities: Recent results and future application of strain-sensing nanocomposites. *Journal of Smart Cities, 1*(1), 3.

da Silva, G. D., Guidelli, E. J., de Queiroz-Fernandes, G. M., Chaves, M. R. M., Baffa, O., & Kinoshita, A. (2019). Silver nanoparticles in building materials for environment protection against microorganisms. *International Journal of Environmental Science and Technology, 16*(3), 1239–1248.

Daskalakis, M. I., Rigas, F., Bakolas, A., Magoulas, A., Kotoulas, G., Katsikis, I., . . . Mavridou, A. (2015). Vaterite bio-precipitation induced by *Bacillus pumilus* isolated from a solutional cave in Paiania, Athens, Greece. *International Biodeterioration & Biodegradation, 99*, 73–84.

David, M. E., Ion, R. M., Grigorescu, R. M., Iancu, L., & Andrei, E. R. (2020). Nanomaterials used in conservation and restoration of cultural heritage: An up-to-date overview. *Materials, 13*(9), 2064.

De Filpo, G., Palermo, A. M., Rachiele, F., & Nicoletta, F. P. (2013). Preventing fungal growth in wood by titanium dioxide nanoparticles. *International Biodeterioration & Biodegradation, 85*, 217–222.

Dejaeghere, I., Sonebi, M., & De Schutter, G. (2019). Influence of nano-clay on rheology, fresh properties, heat of hydration and strength of cement-based mortars. *Construction and Building Materials, 222*, 73–85.

De Muynck, W., Cox, K., De Belie, N., & Verstraete, W. (2008). Bacterial carbonate precipitation as an alternative surface treatment for concrete. *Construction and Building Materials, 22*(5), 875–885.

Dhami, N. K., Reddy, S. M., & Mukherjee, A. (2012). Biofilm and microbial applications in biomineralized concrete. *Advanced Topics in Biomineralization*, 137–164.

Dhami, N. K., Reddy, M. S., & Mukherjee, A. (2013). Biomineralization of calcium carbonate polymorphs by the bacterial strains isolated from calcareous sites. *Journal of Microbiology and Biotechnology, 23*(5), 707–714.

Dhiman, N. K., Agnihotri, S., & Shukla, R. (2019). Silver-based polymeric nanocomposites as antimicrobial coatings for biomedical applications. In *Nanotechnology in modern animal biotechnology* (pp. 115–171). Singapore: Springer.

Diab, A. M., Elyamany, H. E., Abd Elmoaty, M., & Sreh, M. M. (2019). Effect of nanomaterials additives on performance of concrete resistance against magnesium sulfate and acids. *Construction and Building Materials, 210*, 210–231.

Dileep, P., Jacob, S., & Narayanankutty, S. K. (2020). Functionalized nanosilica as an antimicrobial additive for waterborne paints. *Progress in Organic Coatings, 142*, 105574.

Du, H., Gao, H. J., & Dai Pang, S. (2016). Improvement in concrete resistance against water and chloride ingress by adding graphene nanoplatelet. *Cement and Concrete Research, 83*, 114–123.

Dupraz, C., Reid, R. P., Braissant, O., Decho, A. W., Norman, R. S., & Visscher, P. T. (2009). Processes of carbonate precipitation in modern microbial mats. *Earth-Science Reviews, 96*(3), 141–162.

Edvardsen, C. (1999). Water permeability and autogenous healing of cracks in concrete. In *Innovation in concrete structures: Design and construction* (pp. 473–487). Thomas Telford Publishing.

El-Reedy, M. A. (2017). *Steel-reinforced concrete structures: Assessment and repair of corrosion*. CRC Press.

Etzel, R. A., Balk, S. J., Bearer, C. F., Miller, M. D., Shannon, M. W., & Shea, K. M. (1998). American Academy of Pediatrics: Toxic effects of indoor moulds. *Pediatrics, 101*, 712–714.

Evangelia, T., & Maria, S. (2020). Effect of nano-SiO_2 and nano-CaO in autogenous self-healing efficiency. *Materials Today: Proceedings*.

Faizal, M. M., Hamidah, M. S., Norhasri, M. M., Noorli, I., & Hafez, M. M. E. (2015). Chloride permeability of nanoclayed ultra-high performance concrete. In *Proceedings of the international civil and infrastructure engineering conference* (pp. 613–623). Singapore: Springer.

Fan, Y., Zhang, S., Wang, Q., & Shah, S. P. (2015). Effects of nano-kaolinite clay on the freeze–thaw resistance of concrete. *Cement and Concrete Composites, 62*, 1–12.

Fang, C., Kumari, D., Zhu, X., & Achal, V. (2018). Role of fungal-mediated mineralization in biocementation of sand and its improved compressive strength. *International Biodeterioration & Biodegradation, 133*, 216–220.

Farzadnia, N., Ali, A. A. A., & Demirboga, R. (2013). Characterization of high strength mortars with nano alumina at elevated temperatures. *Cement and Concrete Research, 54*, 43–54.

Favre, N., Christ, M. L., & Pierre, A. C. (2009). Biocatalytic capture of CO_2 with carbonic anhydrase and its transformation to solid carbonate. *Journal of Molecular Catalysis B: Enzymatic, 60*(3–4), 163–170.

Fernandes, C. N., Ferreira, R. L., Bernardo, R. D., Avelino, F., & Bertini, A. A. (2020). Using TiO_2 nanoparticles as a SO_2 catalyst in cement mortars. *Construction and Building Materials, 257*, 119542.

Ferrari, C., Santunione, G., Libbra, A., Muscio, A., Sgarbi, E., Siligardi, C., & Barozzi, G. S. (2015). Review on the influence of biological deterioration on the surface properties of building materials: Organisms, materials, and methods. *International Journal of Design & Nature and Ecodynamics, 10*(1), 21–39.

Ferrer, M. R., Quevedo-Sarmiento, J., Rivadeneyra, M. A., Bejar, V., Delgado, R., & Ramos-Cormenzana, A. (1988). Calcium carbonate precipitation by two groups of moderately halophilic microorganisms at different temperatures and salt concentrations. *Current Microbiology, 17*(4), 221–227.

Fortin, D., & Beveridge, T. J. (2000). Mechanistic routes towards biomineral surface development. In E. Baeuerlein (Ed.), *Biomineralisation: From biology to biotechnology and medical application*. Verlag, Germany: Wiley-VCH.

Fournier, B., & Bérubé, M. A. (2000). Alkali-aggregate reaction in concrete: A review of basic concepts and engineering implications. *Canadian Journal of Civil Engineering, 27*(2), 167–191.

Fu, Y., Liu, X., Cheng, F., Sun, J., & Qin, Z. (2016). Modification of the wood surface properties of *Tsoongiodendron odorum* chun with silicon dioxide by a sol-gel method. *BioResources, 11*(4), 10273–10285.

Galao, O., Baeza, F. J., Zornoza, E., & Garcés, P. (2014). Strain and damage sensing properties on multifunctional cement composites with CNF admixture. *Cement and Concrete Composites, 46*, 90–98.

Galeano, B., Korff, E., & Nicholson, W. L. (2003). Inactivation of vegetative cells, but not spores, of *Bacillus anthracis, B. cereus*, and *B. subtilis* on stainless steel surfaces coated with an antimicrobial silver-and zinc-containing zeolite formulation. *Applied and Environmental Microbiology, 69*(7), 4329–4331.

Ganesh, V. K. (2012). Nanotechnology in civil engineering. *European Scientific Journal, 8*(27).

Gao, D., Sturm, M., & Mo, Y. L. (2009). Electrical resistance of carbon-nanofiber concrete. *Smart Materials and Structures, 18*(9), 095039.

Gao, Y., Jing, H., Zhou, Z., Chen, W., Du, M., & Du, Y. (2019). Reinforced impermeability of cementitious composites using graphene oxide-carbon nanotube hybrid under different water-to-cement ratios. *Construction and Building Materials, 222*, 610–621.

García-Macías, E., D'Alessandro, A., Castro-Triguero, R., Pérez-Mira, D., & Ubertini, F. (2017). Micromechanics modeling of the uniaxial strain-sensing property of carbon nanotube cement-matrix composites for SHM applications. *Composite Structures, 163*, 195–215.

Ghadikolaee, M. R., Korayem, A. H., Sharif, A., & Liu, Y. M. (2020). The halloysite nanotube effects on workability, mechanical properties, permeability and microstructure of cementitious mortar. *Construction and Building Materials, 267*, 120873.

Givi, A. N., Rashid, S. A., Aziz, F. N. A., & Salleh, M. A. M. (2010). Experimental investigation of the size effects of SiO_2 nano-particles on the mechanical properties of binary-blended concrete. *Composites Part B: Engineering, 41*(8), 673–677.

Goffredo, G. B., Citterio, B., Biavasco, F., Stazi, F., Barcelli, S., & Munafò, P. (2017). Nanotechnology on wood: The effect of photocatalytic nanocoatings against *Aspergillus niger*. *Journal of Cultural Heritage, 27*, 125–136.

Gómez-Ortíz, N., De la Rosa-García, S., González-Gómez, W., Soria-Castro, M., Quintana, P., Oskam, G., & Ortega-Morales, B. (2013). Antifungal coatings based on $Ca(OH)_2$ mixed with ZnO/TiO_2 nanomaterials for protection of limestone monuments. *ACS Applied Materials & Interfaces, 5*(5), 1556–1565.

Gong, K., Pan, Z., Korayem, A. H., Qiu, L., Li, D., Collins, F., ... Duan, W. H. (2015). Reinforcing effects of graphene oxide on Portland cement paste. *Journal of Materials in Civil Engineering, 27*(2), A4014010.

González, Á., Parraguez, A., Corvalán, L., Correa, N., Castro, J., Stuckrath, C., & González, M. (2020). Evaluation of Portland and Pozzolanic cement on the self-healing of mortars with calcium lactate and bacteria. *Construction and Building Materials, 257*, 119558.

Gopalakrishnan, R., Vignesh, B., & Jeyalakshmi, R. (2020). Mechanical, electrical and microstructural studies on nano-TiO_2 admixtured cement mortar cured with industrial wastewater. *Engineering Research Express, 2*(2), 025010.

Gopinath, S., Mouli, P. C., Murthy, A. R., Iyer, N. R., & Maheswaran, S. (2012). Effect of nano silica on mechanical properties and durability of normal strength concrete. *Archives of Civil Engineering, 58*(4), 433–444.

Gorospe, C. M., Han, S. H., Kim, S. G., Park, J. Y., Kang, C. H., Jeong, J. H., & So, J. S. (2013). Effects of different calcium salts on calcium carbonate crystal formation by *Sporosarcina pasteurii* KCTC 3558. *Biotechnology and Bioprocess Engineering, 18*(5), 903–908.

Gowda, R., Narendra, H., Nagabushan, B. M., Rangappa, D., & Prabhakara, R. (2017). Investigation of nano-alumina on the effect of durability and micro-structural properties of the cement mortar. *Materials Today: Proceedings, 4*(11), 12191–12197.

Graziani, L., & D'Orazio, M. (2015). Biofouling prevention of ancient brick surfaces by TiO_2-based nanocoatings. *Coatings, 5*(3), 357–365.

Graziani, L., Quagliarini, E., Osimani, A., Aquilanti, L., Clementi, F., Yéprémian, C., & D'Orazio, M. (2013). Evaluation of inhibitory effect of TiO_2 nanocoatings against microalgal growth on clay brick façades under weak UV exposure conditions. *Building and Environment, 64*, 38–45.

Griffitt, R. J., Luo, J., Gao, J., Bonzongo, J. C., & Barber, D. S. (2008). Effects of particle composition and species on toxicity of metallic nanomaterials in aquatic organisms. *Environmental Toxicology and Chemistry: An International Journal, 27*(9), 1972–1978.

Guler, S., Türkmenoğlu, Z. F., & Ashour, A. (2020). Performance of single and hybrid nanoparticles added concrete at ambient and elevated temperatures. *Construction and Building Materials, 250*, 118847.

Gupta, S., Dai Pang, S., & Kua, H. W. (2017). Autonomous healing in concrete by bio-based healing agents: A review. *Construction and Building Materials, 146*, 419–428.

Gupta, S., Kua, H. W., & Dai Pang, S. (2018). Healing cement mortar by immobilization of bacteria in biochar: An integrated approach of self-healing and carbon sequestration. *Cement and Concrete Composites, 86*, 238–254.

Haile, T., Nakhla, G., Allouche, E., & Vaidya, S. (2010). Evaluation of the bactericidal characteristics of nano-copper oxide or functionalized zeolite coating for bio-corrosion control in concrete sewer pipes. *Corrosion Science, 52*(1), 45–53.

Hakamy, A., Shaikh, F. U. A., & Low, I. M. (2014). Characteristics of hemp fabric reinforced nanoclay–cement nanocomposites. *Cement and Concrete Composites, 50*, 27–35.

Hamed, N., El-Feky, M. S., Kohail, M., & Nasr, E. A. R. (2018). Investigating the effect of nano clay on concreterebars bond strength modes. *International Journal of Scientific and Engineering Research, 8*(12), 1621–1628.

Hassan, N. M., Fattah, K., & Tamimi, A. (2015). Standardizing protocol for incorporating cnts in concrete. In *Proceedings of the world congress on new technologies* (pp. 15–17). Barcelona, Spain.

Hawreen, A., & Bogas, J. A. (2019). Creep, shrinkage and mechanical properties of concrete reinforced with different types of carbon nanotubes. *Construction and Building Materials, 198*, 70–81.

Haynes, H., O'Neill, R., & Mehta, P. K. (1996). Concrete deterioration from physical attack by salts. *Concrete International, 18*(1), 63–68.

Haynes, H., O'Neill, R., Neff, M., & Mehta, P. K. (2008). Salt weathering distress on concrete exposed to sodium sulfate environment. *ACI Materials Journal, 105*(1), 35.

Heikal, M., Zaki, M. E., & Ibrahim, S. M. (2020). Characterization, hydration, durability of nano-Fe_2O_3-composite cements subjected to sulphates and chlorides media. *Construction and Building Materials, 269*, 121310.

Hobbs, D. W. (2001). Concrete deterioration: Causes, diagnosis, and minimising risk. *International Materials Reviews, 46*(3), 117–144.

Horszczaruk, E., Sikora, P., Cendrowski, K., & Mijowska, E. (2017). The effect of elevated temperature on the properties of cement mortars containing nanosilica and heavyweight aggregates. *Construction and Building Materials, 137*, 420–431.

Hou, P., Kawashima, S., Kong, D., Corr, D. J., Qian, J., & Shah, S. P. (2013). Modification effects of colloidal nano SiO_2 on cement hydration and its gel property. *Composites Part B: Engineering, 45*(1), 440–448.

Hu, M., Guo, J., Fan, J., & Chen, D. (2019a). Dispersion of triethanolamine-functionalized graphene oxide (TEA-GO) in pore solution and its influence on hydration, mechanical behavior of cement composite. *Construction and Building Materials, 216*, 128–136.

Hu, M., Guo, J., Li, P., Chen, D., Xu, Y., Feng, Y., ... Zhang, H. (2019b). Effect of characteristics of chemical combined of graphene oxide-nanosilica nanocomposite fillers on properties of cement-based materials. *Construction and Building Materials, 225*, 745–753.

Huang, Q., Zhu, X., Zhao, L., Zhao, M., Liu, Y., & Zeng, X. (2020). Effect of nanosilica on sulfate resistance of cement mortar under partial immersion. *Construction and Building Materials, 231*, 117180.

Hueck, H. J. (2001). The biodeterioration of materials—An appraisal. *International Biodeterioration & Biodegradation, 48*(1–4), 5–11.

Hughes, P., Fairhurst, D., Sherrington, I., Renevier, N., Morton, L. H. C., Robery, P. C., & Cunningham, L. (2014). Microbial degradation of synthetic fibre-reinforced marine concrete. *International Biodeterioration & Biodegradation, 86*, 2–5. Available from https://doi.org/10.1016/j.ibiod.2013.02.015.

Huseien, G. F., Shah, K. W., & Sam, A. R. M. (2019). Sustainability of nanomaterials based self-healing concrete: An all-inclusive insight. *Journal of Building Engineering, 23*, 155–171.

Ibrahim, R. K., Hamid, R., & Taha, M. R. (2012). Fire resistance of high-volume fly ash mortars with nanosilica addition. *Construction and Building Materials, 36*, 779–786.

Ibrahim, R. K., Ramyar, K., Hamid, R., & Raihan Tah, M. (2011). The effect of high temperature on mortars containing silica fume. *Journal of Applied Sciences, 11*(14), 2666–2669.

Iqbal, H. W., Khushnood, R. A., Baloch, W. L., Nawaz, A., & Tufail, R. F. (2020). Influence of graphite nano/micro platelets on the residual performance of high strength concrete exposed to elevated temperature. *Construction and Building Materials, 253*, 119029.

Irshidat, M. R., & Al-Saleh, M. H. (2018). Thermal performance and fire resistance of nanoclay modified cementitious materials. *Construction and Building Materials, 159*, 213–219.

Jafarnia, M. S., Saryazdi, M. K., & Moshtaghioun, S. M. (2020). Use of bacteria for repairing cracks and improving properties of concrete containing limestone powder and natural zeolite. *Construction and Building Materials, 242*, 118059.

Jalal, M., Pouladkhan, A., Harandi, O. F., & Jafari, D. (2015). Comparative study on effects of Class F fly ash, nano silica and silica fume on properties of high performance self-compacting concrete. *Construction and Building Materials, 94*, 90–104.

Ji, T. (2005). Preliminary study on the water permeability and microstructure of concrete incorporating nano-SiO_2. *Cement and Concrete Research, 35*(10), 1943–1947.

Jo, B. W., Kim, C. H., Tae, G. H., & Park, J. B. (2007). Characteristics of cement mortar with nano-SiO_2 particles. *Construction and Building Materials, 21*(6), 1351–1355.

Jonkers, H. M. (2011). Bacteria-based self-healing concrete. *Heron, 56*(1/2).

Jonkers, H. M. & Schlangen, E. (2007). Crack repair by concrete-immobilized bacteria. In *Proceedings of the first international conference on self-healing materials*, 18 (p. 20).

Jonkers, H. M. & Schlangen, E. (2009). Towards a sustainable bacterially-mediated self-healing concrete. In *Proceedings of second international conference on self-healing materials*, Chicago.

Joshaghani, A., Balapour, M., Mashhadian, M., & Ozbakkaloglu, T. (2020). Effects of nano-TiO_2, nano-Al_2O_3, and nano-Fe_2O_3 on rheology, mechanical and durability properties of self-consolidating concrete (SCC): An experimental study. *Construction and Building Materials, 245*, 118444.

Joshi, S., Goyal, S., Mukherjee, A., & Reddy, M. S. (2019). Protection of concrete structures under sulfate environments by using calcifying bacteria. *Construction and Building Materials, 209*, 156–166.

Kalhori, H., Bagherzadeh, B., Bagherpour, R., & Akhlaghi, M. A. (2020). Experimental study on the influence of the different percentage of nanoparticles on strength and freeze−thaw durability of shotcrete. *Construction and Building Materials, 256*, 119470.

Kartal, S. N., Terzi, E., Woodward, B., Clausen, C. A., & Lebow, S. T. (2013). Removal of nano-and micronized-copper from treated wood by chelating agents. In *Proceedings IRG annual meeting; the international research group on wood protection, section 5, sustainability and environment.* IRG/WP 13−50294, 15 p. (pp. 1−15).

Karthikeya, R. U., & Senthil, G. K. (2016). An experimental study on strength parameters of nano alumina and GGBS on concrete. *International Journal for Research in Emerging Science and Technology, 3*(4).

Kaur, G., Dhami, N. K., Goyal, S., Mukherjee, A., & Reddy, M. S. (2016). Utilization of carbon dioxide as an alternative to urea in bio cementation. *Construction and Building Materials, 123*, 527−533.

Khaliq, W., & Ehsan, M. B. (2016). Crack healing in concrete using various bio influenced self-healing techniques. *Construction and Building Materials, 102*, 349−357.

Khoshakhlagh, A., Nazari, A., & Khalaj, G. (2012). Effects of Fe_2O_3 nanoparticles on water permeability and strength assessments of high strength self-compacting concrete. *Journal of Materials Science & Technology, 28*(1), 73−82.

Kim, H. K., Nam, I. W., & Lee, H. K. (2014). Enhanced effect of carbon nanotube on mechanical and electrical properties of cement composites by incorporation of silica fume. *Composite Structures, 107*, 60−69.

Kim, Y. A., Hayashi, T., Endo, M., & Dresselhaus, M. S. (2013). Carbon nanofibers. In *Springer handbook of nanomaterials* (pp. 233−262). Berlin, Heidelberg: Springer.

Kip, N., & Van Veen, J. A. (2015). The dual role of microbes in corrosion. *The ISME Journal, 9*(3), 542−551.

Konsta-Gdoutos, M. S., & Aza, C. A. (2014). Self-sensing carbon nanotube (CNT) and nanofiber (CNF) cementitious composites for real time damage assessment in smart structures. *Cement and Concrete Composites, 53*, 162−169.

Kovler, K., & Chernov, V. (2009). Types of damage in concrete structures. In *Failure, distress and repair of concrete structures* (pp. 32−56). Woodhead Publishing.

Kowald, T., & Trettin, R. (2009). Improvement of cementitious binders by multi-walled carbon nanotubes. In *Nanotechnology in construction* (3, pp. 261−266). Berlin, Heidelberg: Springer.

Krajewska, B. (2018). Urease-aided calcium carbonate mineralization for engineering applications: A review. *Journal of Advanced Research, 13*, 59−67.

Kumar, L. Y. (2015). Role and adverse effects of nanomaterials in food technology. *Journal of Toxicology and Environmental Health, 2*(2).

Kumar, R., Singh, S., & Singh, L. (2017). Studies on enhanced thermally stable high strength concrete incorporating silica nanoparticles. *Construction and Building Materials, 153*, 506−513.

Kwalramani, M. A., & Syed, Z. I. (2018). Application of nanomaterials to enhance microstructure and mechanical properties of concrete. *International Journal of Integrated Engineering, 10*(2).

Land, G., & Stephan, D. (2012). The influence of nano-silica on the hydration of ordinary Portland cement. *Journal of Materials Science, 47*(2), 1011−1017.

Langaroudi, M. A. M., & Mohammadi, Y. (2018). Effect of nano-clay on workability, mechanical, and durability properties of self-consolidating concrete containing mineral admixtures. *Construction and Building Materials, 191*, 619−634.

Larosche, C. J. (2009). Types and causes of cracking in concrete structures. In *Failure, distress and repair of concrete structures* (pp. 57−83). Woodhead Publishing.

La Russa, M. F., Macchia, A., Ruffolo, S. A., De Leo, F., Barberio, M., Barone, P., & Urzì, C. (2014). Testing the antibacterial activity of doped TiO_2 for preventing biodeterioration of cultural heritage building materials. *International Biodeterioration & Biodegradation, 96*, 87−96.

Lee, S. J., Amirkhanian, S. N., & Kim, K. W. (2009). Laboratory evaluation of the effects of short-term oven aging on asphalt binders in asphalt mixtures using HP-GPC. *Construction and Building Materials, 23*(9), 3087–3093.

Lettieri, M., Colangiuli, D., Masieri, M., & Calia, A. (2019). Field performances of nanosized TiO_2 coated limestone for a self-cleaning building surface in an urban environment. *Building and Environment, 147*, 506–516.

Li, G., Cui, H., Zhou, J., & Hu, W. (2019). Improvements of nano-TiO_2 on the long-term chloride resistance of concrete with polymer coatings. *Coatings, 9*(5), 323.

Li, H., Zhang, M. H., & Ou, J. P. (2007). Flexural fatigue performance of concrete containing nano-particles for pavement. *International Journal of Fatigue, 29*(7), 1292–1301.

Li, Q., Mahendra, S., Lyon, D. Y., Brunet, L., Liga, M. V., Li, D., & Alvarez, P. J. (2008). Antimicrobial nanomaterials for water disinfection and microbial control: Potential applications and implications. *Water Research, 42*(18), 4591–4602.

Li, W., Huang, Z., Cao, F., Sun, Z., & Shah, S. P. (2015). Effects of nano-silica and nano-limestone on flowability and mechanical properties of ultra-high-performance concrete matrix. *Construction and Building Materials, 95*, 366–374.

Li, W., Li, X., Chen, S. J., Liu, Y. M., Duan, W. H., & Shah, S. P. (2017). Effects of graphene oxide on early-age hydration and electrical resistivity of Portland cement paste. *Construction and Building Materials, 136*, 506–514.

Li, X., Li, C., Liu, Y., Chen, S. J., Wang, C. M., Sanjayan, J. G., & Duan, W. H. (2018). Improvement of mechanical properties by incorporating graphene oxide into cement mortar. *Mechanics of Advanced Materials and Structures, 25*(15–16), 1313–1322.

Li, Z., Wang, H., He, S., Lu, Y., & Wang, M. (2006). Investigations on the preparation and mechanical properties of the nano-alumina reinforced cement composite. *Materials Letters, 60*(3), 356–359.

Liu, H. Y., Zhang, H. L., Hao, P. W., & Zhu, C. Z. (2015). The effect of surface modifiers on ultraviolet aging properties of nano-zinc oxide modified bitumen. *Petroleum Science and Technology, 33*(1), 72–78.

Liu, J., Li, Q., & Xu, S. (2019). Reinforcing mechanism of graphene and graphene oxide sheets on cement-based materials. *Journal of Materials in Civil Engineering, 31*(4), 04019014.

Liu, S., Wang, R., Yu, J., Peng, X., Cai, Y., & Tu, B. (2020). Effectiveness of the anti-erosion of an MICP coating on the surfaces of ancient clay roof tiles. *Construction and Building Materials, 243*, 118202.

Lors, C., Ducasse-Lapeyrusse, J., Gagné, R., & Damidot, D. (2017). Microbiologically induced calcium carbonate precipitation to repair microcracks remaining after autogenous healing of mortars. *Construction and Building Materials, 100*(141), 461–469.

Lowenstam, H. A. (1981). Minerals formed by organisms. *Science (New York, N.Y.), 211*(4487), 1126–1131.

Luo, J. L. (2009). Fabrication and functional properties of multi-walled carbon nanotube/cement composites (Dissertation for the Doctoral Degree in Engineering). Harbin: Harbin Institute of Technology.

Lv, S., Ma, Y., Qiu, C., Sun, T., Liu, J., & Zhou, Q. (2013). Effect of graphene oxide nanosheets of microstructure and mechanical properties of cement composites. *Construction and Building Materials, 49*, 121–127.

Machado, G. E., Pereyra, A. M., Rosato, V. G., Moreno, M. S., & Basaldella, E. I. (2019). Improving the biocidal activity of outdoor coating formulations by using zeolite-supported silver nanoparticles. *Materials Science and Engineering: C, 98*, 789–799.

MacLeod, A. J., Fehervari, A., Gates, W. P., Garcez, E. O., Aldridge, L. P., & Collins, F. (2020). Enhancing fresh properties and strength of concrete with a pre-dispersed carbon nanotube liquid admixture. *Construction and Building Materials, 247*, 118524.

Madandoust, R., Mohseni, E., Mousavi, S. Y., & Namnevis, M. (2015). An experimental investigation on the durability of self-compacting mortar containing nano-SiO_2, nano-Fe_2O_3 and nano-CuO. *Construction and Building Materials, 86*, 44–50.

Madani, H., Bagheri, A., & Parhizkar, T. (2012). The pozzolanic reactivity of monodispersed nanosilica hydrosols and their influence on the hydration characteristics of Portland cement. *Cement and Concrete Research, 42*(12), 1563–1570.

Madhavi, T. C., Pavithra, P., Singh, S. B., Vamsi Raj, S. B., & Paul, S. (2013). Effect of multiwalled carbon nanotubes on mechanical properties of concrete. *International Journal of Scientific Research, 2*(6), 166–168.

Mahdikhani, M., Bamshad, O., & Shirvani, M. F. (2018). Mechanical properties and durability of concrete specimens containing nano silica in sulfuric acid rain condition. *Construction and Building Materials, 167*, 929–935.

Maheswaran, S., Bhuvaneshwari, B., Palani, G. S., Nagesh, R., & Kalaiselvam, S. (2013). An overview on the influence of nano silica in concrete and a research initiative. *Research Journal of Recent Sciences, 3*, 2502, ISSN2277.

Maiti, S., Krishnan, D., Barman, G., Ghosh, S. K., & Laha, J. K. (2014). Antimicrobial activities of silver nanoparticles synthesized from *Lycopersicon esculentum* extract. *Journal of Analytical Science and Technology, 5*(1), 1–7.

Makar, J., Margeson, J., & Luh, J. (2005). Carbon nanotube/cement composites-early results and potential applications. In *Proceedings of the third international conference on construction materials: performance, innovations and structural implications* (pp. 1–10). Vancouver Canada.

Mann, S. (1983). Mineralization in biological systems. In *Inorganic elements in biochemistry* (pp. 125–174). Berlin, Heidelberg: Springer.

Meddah, M. S., Praveenkumar, T. R., Vijayalakshmi, M. M., Manigandan, S., & Arunachalam, R. (2020). Mechanical and microstructural characterization of rice husk ash and Al_2O_3 nanoparticles modified cement concrete. *Construction and Building Materials, 255*, 119358.

Metaxa, Z. S., Seo, J. W. T., Konsta-Gdoutos, M. S., Hersam, M. C., & Shah, S. P. (2012). Highly concentrated carbon nanotube admixture for nano-fiber reinforced cementitious materials. *Cement and Concrete Composites, 34*(5), 612–617.

Micallef, R., Vella, D., Sinagra, E., & Zammit, G. (2016). Bio calcifying *Bacillus subtilis* cells effectively consolidate deteriorated Globigerina limestone. *Journal of Industrial Microbiology & Biotechnology, 43*(7), 941–952.

Miller, S. A., & Moore, F. C. (2020). Climate and health damages from global concrete production. *Nature Climate Change, 10*, 1–5.

Minto, J. M., Tan, Q., Lunn, R. J., El Mountassir, G., Guo, H., & Cheng, X. (2018). 'Microbial mortar'-restoration of degraded marble structures with microbially induced carbonate precipitation. *Construction and Building Materials, 180*, 44–54.

Mo, K. H., Ling, T. C., Tan, T. H., Leong, G. W., Yuen, C. W., & Shah, S. N. (2020). Alkali-silica reactivity of lightweight aggregate: A brief overview. *Construction and Building Materials, 270*, 121444.

Mohammed, B. S., Liew, M. S., Alaloul, W. S., Khed, V. C., Hoong, C. Y., & Adamu, M. (2018). Properties of nano-silica modified pervious concrete. *Case Studies in Construction Materials, 8*, 409–422.

Mohseni, E., Khotbehsara, M. M., Naseri, F., Monazami, M., & Sarker, P. (2016). Polypropylene fiber reinforced cement mortars containing rice husk ash and nano-alumina. *Construction and Building Materials, 111*, 429–439.

Mohseni, E., Naseri, F., Amjadi, R., Khotbehsara, M. M., & Ranjbar, M. M. (2016). Microstructure and durability properties of cement mortars containing nano-TiO_2 and rice husk ash. *Construction and Building Materials, 114*, 656–664.

Moncmanová, A. (Ed.), (2007). *Environmental deterioration of materials* (vol. 21). Wit Press.

Mondal, S., & Ghosh, A. D. (2018). Investigation into the optimal bacterial concentration for compressive strength enhancement of microbial concrete. *Construction and Building Materials, 183*, 202–214.

Moon, H. Y., Shin, D. G., & Choi, D. S. (2007). Evaluation of the durability of mortar and concrete applied with inorganic coating material and surface treatment system. *Construction and Building Materials, 21*(2), 362–369.

Morsy, M. S., Alsayed, S. H., & Aqel, M. (2010). Effect of nano-clay on mechanical properties and microstructure of ordinary Portland cement mortar. *International Journal of Civil & Environmental Engineering IJCEE-IJENS, 10*(01), 23–27.

Musso, S., Tulliani, J.-M., Ferro, G., & Tagliaferro, A. (2009). Influence of carbon nanotubes structure on the mechanical behavior of cement composites. *Composites Science and Technology, 69*(11–12), 1985–1990.

Muzenski, S., Flores-Vivian, I., & Sobolev, K. (2019). Ultra-high strength cement-based composites designed with aluminium oxide nano-fibers. *Construction and Building Materials, 220*, 177–186.

Nain, N., Surabhi, R., Yathish, N. V., Krishnamurthy, V., Deepa, T., & Tharannum, S. (2019). Enhancement in strength parameters of concrete by application of *Bacillus bacteria*. *Construction and Building Materials, 202*, 904–908.

Najjar, M. F., Nehdi, M. L., Soliman, A. M., & Azabi, T. M. (2017). Damage mechanisms of two-stage concrete exposed to chemical and physical sulfate attack. *Construction and Building Materials, 137*, 141–152.

Nasution, A., Imran, I., & Abdullah, M. (2015). Improvement of concrete durability by nanomaterials. *Procedia Engineering, 125*, 608–612.

Nazari, A., & Riahi, S. (2011). The effects of zinc dioxide nanoparticles on flexural strength of self-compacting concrete. *Composites Part B: Engineering, 42*(2), 167–175.

Nazari, A., Riahi, S., Riahi, S., Shamekhi, S. F., & Khademno, A. (2010a). Benefits of Fe_2O_3 nanoparticles in concrete mixing matrix. *Journal of American Science, 6*(4), 102–106.

Nazari, A., Riahi, S., Riahi, S., Shamekhi, S. F., & Khademno, A. (2010b). The effects of incorporation Fe_2O_3 nanoparticles on tensile and flexural strength of concrete. *Journal of American Science, 6*(4), 90–93.

Nazari, A., Riahi, S., Riahi, S., Shamekhi, S. F., & Khademno, A. (2010c). Mechanical properties of cement mortar with Al_2O_3 nanoparticles. *Journal of American Science, 6*(4), 94–97.

Nehdi, M. L., Suleiman, A. R., & Soliman, A. M. (2014). Investigation of concrete exposed to dual sulfate attack. *Cement and Concrete Research, 64*, 42–53.

Ng, D. S., Paul, S. C., Anggraini, V., Kong, S. Y., Qureshi, T. S., Rodriguez, C. R., & Šavija, B. (2020). Influence of SiO_2, TiO_2 and Fe_2O_3 nanoparticles on the properties of fly ash blended cement mortars. *Construction and Building Materials, 258*, 119627.

Niaki, M. H., Fereidoon, A., & Ahangari, M. G. (2018). Experimental study on the mechanical and thermal properties of basalt fiber and nanoclay reinforced polymer concrete. *Composite Structures, 191*, 231–238.

Nikbin, I. M., Mehdipour, S., Dezhampanah, S., Mohammadi, R., Mohebbi, R., Moghadam, H. H., & Sadrmomtazi, A. (2020). Effect of high temperature on mechanical and gamma ray shielding properties of concrete containing nano-TiO_2. *Radiation Physics and Chemistry, 174*, 108967.

Niroumand, H., Zain, M. F. M., & Alhosseini, S. N. (2013). The influence of nano-clays on compressive strength of earth bricks as sustainable materials. *Procedia-Social and Behavioural Sciences, 89*, 862–865.

Norhasri, M. M., Hamidah, M. S., & Fadzil, A. M. (2017). Applications of using nano material in concrete: A review. *Construction and Building Materials, 133*, 91–97.

Norhasri, M. M., Hamidah, M. S., & Fadzil, A. M. (2019). Inclusion of nano metaclayed as additive in ultra-high performance concrete (UHPC). *Construction and Building Materials, 201*, 590–598.

Pan, X., Shi, C., Zhang, J., Jia, L., & Chong, L. (2018). Effect of inorganic surface treatment on surface hardness and carbonation of cement-based materials. *Cement and Concrete Composites, 90*, 218–224.

Pan, Z., He, L., Qiu, L., Korayem, A. H., Li, G., Zhu, J. W., . . . Wang, M. C. (2015). Mechanical properties and microstructure of a graphene oxide–cement composite. *Cement and Concrete Composites, 58*, 140–147.

Park, S.-K., Kim, J.-H. J., Nam, J.-W., Phan, H. D., & Kim, J.-K. (2009). Development of anti-fungal mortar and concrete using Zeolite and Zeocarbon microcapsules. *Cement and Concrete Composites, 31*(7), 447−453.

Peyvandi, A., Sbia, L. A., Soroushian, P., & Sobolev, K. (2013). Effect of the cementitious paste density on the performance efficiency of carbon nanofiber in concrete nanocomposite. *Construction and Building Materials, 48*, 265−269.

Pivinskii, Y. E. (2007). Nanodisperse silica and some aspects of nanotechnologies in the field of silicate materials science. Part 1. *Refractories and Industrial Ceramics, 48*(6), 408−417.

Praveenkumar, T. R., Vijayalakshmi, M. M., & Meddah, M. S. (2019). Strengths and durability performances of blended cement concrete with TiO_2 nanoparticles and rice husk ash. *Construction and Building Materials, 217*, 343−351.

Qian, S. Z., Zhou, J., & Schlangen, E. (2010). Influence of curing condition and precracking time on the self-healing behavior of engineered cementitious composites. *Cement and Concrete Composites, 32*(9), 686−693.

Qing, Y., Zenan, Z., Deyu, K., & Rongshen, C. (2007). Influence of nano-SiO_2 addition on properties of hardened cement paste as compared with silica fume. *Construction and Building Materials, 21*(3), 539−545.

Qureshi, T. S., & Panesar, D. K. (2019). Impact of graphene oxide and highly reduced graphene oxide on cement-based composites. *Construction and Building Materials, 206*, 71−83.

Rajasegar, M., & Kumaar, C. M. (2020). Hybrid effect of poly vinyl alcohol, expansive minerals, nano-silica and rice husk ash on the self-healing ability of concrete. *Materials Today: Proceedings*.

Rashad, A. M. (2013). A synopsis about the effect of nano-Al_2O_3, nano-Fe_2O_3, nano-Fe_3O_4 and nano-clay on some properties of cementitious materials—A short guide for Civil Engineer. *Materials & Design (1980−2015), 52*, 143−157.

Rasouli, D., Dintcheva, N. T., Faezipour, M., La Mantia, F. P., Farahani, M. R. M., & Tajvidi, M. (2016). Effect of nano zinc oxide as UV stabilizer on the weathering performance of wood-polyethylene composite. *Polymer Degradation and Stability, 133*, 85−91.

Reddy, V. G. P., Krishna, B. M., Tadepalli, T., & Kumar, P. R. (2020). Image-based deterioration assessment of concrete. *Materials Today: Proceedings, 32*, 788−796.

Richardson, B. (2002). *Defects and deterioration in buildings: A practical guide to the science and technology of material failure*. Routledge.

Richardson, I. G. (1999). The nature of CSH in hardened cements. *Cement and Concrete Research, 29*(8), 1131−1147.

Rivadeneyra, M. A., Delgado, R., Párraga, J., Ramos-Cormenzana, A., & Delgado, G. (2006). Precipitation of minerals by 22 species of moderately halophilic bacteria in artificial marine salts media: Influence of salt concentration. *Folia Microbiologica, 51*(5), 445−453.

Rodrigues, R., Gaboreau, S., Gance, J., Ignatiadis, I., & Betelu, S. (2020). Reinforced concrete structures: A review of corrosion mechanisms and advances in electrical methods for corrosion monitoring. *Construction and Building Materials, 269*, 121240.

Rojas, D. F. H., Pineda-Gómez, P., & Guapacha, J. F. (2020). Effect of silica nanoparticles on the mechanical and physical properties of fibercement boards. *Journal of Building Engineering, 31*, 101332.

Rosendo, F. R., Pinto, L. I., de Lima, I. S., Trigueiro, P., Honório, L. M. D. C., Fonseca, M. G., & Osajima, J. A. (2020). Antimicrobial efficacy of building material based on ZnO/palygorskite against Gram-negative and Gram-positive bacteria. *Applied Clay Science, 188*, 105499.

Ruan, Y., Han, B., Yu, X., Zhang, W., & Wang, D. (2018). Carbon nanotubes reinforced reactive powder concrete. *Composites Part A: Applied Science and Manufacturing, 112*, 371−382.

Safiuddin, M. (2017). Concrete damage in field conditions and protective sealer and coating systems. *Coatings, 7*(7), 90.

Şahmaran, M., Keskin, S. B., Ozerkan, G., & Yaman, I. O. (2008). Self-healing of mechanically-loaded self-consolidating concretes with high volumes of fly ash. *Cement and Concrete Composites, 30*(10), 872–879.

Sakr, M.R., Bassuoni, M.T., & Ghazy, A. (2020). Durability of concrete superficially treated with nano-silica and silane/nano-clay coatings. Transportation Research Record, 0361198120953160.

Salemi, N., Behfarnia, K., & Zaree, S. A. (2014). Effect of nanoparticles on frost durability of concrete. *Asian Journal of Civil Engineering (Building and Housing), 15*(3), 411–420.

Salgiya, D., Jain, U., & Tongiya, S. (2019). General building defects and cracks. *International Journal of Research in Engineering, Science and Management, 2*(10), 628–631.

Salmasi, F., & Mostofinejad, D. (2020). Investigating the effects of bacterial activity on compressive strength and durability of natural lightweight aggregate concrete reinforced with steel fibers. *Construction and Building Materials, 251*, 119032.

Salthammer, T., & Fuhrmann, F. (2007). Photocatalytic surface reactions on indoor wall paint. *Environmental Science & Technology, 41*(18), 6573–6578.

Samani, A. K., & Attard, M. M. (2012). A stress–strain model for uniaxial and confined concrete under compression. *Engineering Structures, 41*, 335–349.

Sánchez, M., Faria, P., Ferrara, L., Horszczaruk, E., Jonkers, H. M., Kwiecień, A., ... Zając, B. (2018). External treatments for the preventive repair of existing constructions: A review. *Construction and Building Materials, 193*, 435–452.

Sanchez-Silva, M., & Rosowsky, D. V. (2008). Biodeterioration of construction materials: State of the art and future challenges. *Journal of Materials in Civil Engineering, 20*(5), 352–365.

Santucci, R., Meunier, O., Ott, M., Herrmann, F., Freyd, A., & de Blay, E. (2007). Fungic contamination of residence: 10 years assessment of analyses. *Revue Francaise D Allergologie Et D Immunologie Clinique, 47*(6), 402–408.

Sasmal, S., Bhuvaneshwari, B., & Iyer, N. R. (2013). Can carbon nanotubes make wonders in civil/structural engineering? *Progress in Nanotechnology and Nanomaterials, 2*(4), 117–129.

Sastry, K.G.K., Sahitya, P., & Ravitheja, A. (2020). Influence of nano TiO_2 on strength and durability properties of geopolymer concrete. Materials Today: Proceedings, https://doi.org/10.1016/j.matpr.20.0320.139.

Scherer, G. W. (2004). Stress from crystallization of salt. *Cement and Concrete Research, 34*(9), 1613–1624.

Schifano, E., Cavallini, D., De Bellis, G., Bracciale, M. P., Felici, A. C., Santarelli, M. L., & Uccelletti, D. (2020). Antibacterial effect of zinc oxide-based nanomaterials on environmental biodeteriogens affecting historical buildings. *Nanomaterials, 10*(2), 335.

Scrivener, K. L., & Kirkpatrick, R. J. (2008). Innovation in use and research on cementitious material. *Cement and Concrete Research, 38*(2), 128–136.

Sedaghatdoost, A., & Behfarnia, K. (2018). Mechanical properties of Portland cement mortar containing multi-walled carbon nanotubes at elevated temperatures. *Construction and Building Materials, 176*, 482–489.

Seifan, M., & Berenjian, A. (2020). Bio self-healing nanoconcretes. In *Smart nanoconcretes and cement-based materials* (pp. 547–558). Elsevier.

Seifan, M., Ebrahiminezhad, A., Ghasemi, Y., & Berenjian, A. (2019). Microbial calcium carbonate precipitation with high affinity to fill the concrete pore space: Nanobiotechnological approach. *Bioprocess and Biosystems Engineering, 42*(1), 37–46.

Seifan, M., Ebrahiminezhad, A., Ghasemi, Y., Samani, A. K., & Berenjian, A. (2018a). Amine-modified magnetic iron oxide nanoparticle as a promising carrier for application in bio self-healing concrete. *Applied Microbiology and Biotechnology, 102*(1), 175–184.

Seifan, M., Ebrahiminezhad, A., Ghasemi, Y., Samani, A. K., & Berenjian, A. (2018b). The role of magnetic iron oxide nanoparticles in the bacterially induced calcium carbonate precipitation. *Applied Microbiology and Biotechnology, 102*(8), 3595–3606.

Seifan, M., Samani, A. K., & Berenjian, A. (2016). Bioconcrete: Next generation of self-healing concrete. *Applied Microbiology and Biotechnology, 100*(6), 2591−2602.

Seil, J. T., & Webster, T. J. (2012). Antimicrobial applications of nanotechnology: Methods and literature. *International Journal of Nanomedicine, 7*, 2767.

Senff, L., Labrincha, J. A., Ferreira, V. M., Hotza, D., & Repette, W. L. (2009). Effect of nano-silica on rheology and fresh properties of cement pastes and mortars. *Construction and Building Materials, 23*(7), 2487−2491.

Shaheen, N., Khushnood, R. A., Khaliq, W., Murtaza, H., Iqbal, R., & Khan, M. H. (2019). Synthesis and characterization of bio-immobilized nano/micro inert and reactive additives for feasibility investigation in self-healing concrete. *Construction and Building Materials, 226*, 492−506.

Shao, Q., Zheng, K., Zhou, X., Zhou, J., & Zeng, X. (2019). Enhancement of nano-alumina on long-term strength of Portland cement and the relation to its influences on compositional and microstructural aspects. *Cement and Concrete Composites, 98*, 39−48.

Shi, H., Liu, F., Yang, L., & Han, E. (2008). Characterization of protective performance of epoxy reinforced with nanometer-sized TiO_2 and SiO_2. *Progress in Organic Coatings, 62*(4), 359−368.

Shiny, K. S., Sundararaj, R., Mamatha, N., & Lingappa, B. (2019). A new approach to wood protection: Preliminary study of biologically synthesized copper oxide nanoparticle formulation as an environmental friendly wood protectant against decay fungi and termites. *Maderas. Ciencia y Tecnología, 21*(3), 347−356.

Shu, H., Yang, M., Liu, Q., & Luo, M. (2020). Study of TiO_2-modified sol coating material in the protection of stone-built cultural heritage. *Coatings, 10*(2), 179.

Shukla, R. K., Sharma, V., Pandey, A. K., Singh, S., Sultana, S., & Dhawan, A. (2011). ROS-mediated genotoxicity induced by titanium dioxide nanoparticles in human epidermal cells. *Toxicology In Vitro, 25*(1), 231−241.

Sierra-Fernandez, A., Gomez-Villalba, L. S., De la Rosa-García, S. C., Gomez-Cornelio, S., Quintana, P., Rabanal, M. E., & Fort, R. (2018). Inorganic nanomaterials for the consolidation and antifungal protection of stone heritage. In *Advanced materials for the conservation of stone* (pp. 125−149). Cham: Springer.

Sillu, D., Kaushik, Y., & Agnihotri, S. (2020). Immobilization of enzymes onto silica-based nanomaterials for bioprocess applications. In *Immobilization strategies* (pp. 399−434). Singapore: Springer.

Sillu, D., & Agnihotri, S. (2019). Cellulase immobilization onto magnetic halloysite nanotubes: Enhanced enzyme activity and stability with high cellulose saccharification. *ACS Sustainable Chemistry & Engineering, 8*(2), 900−913.

Singh, H., & Gupta, R. (2020). Cellulose fiber as bacteria-carrier in mortar: Self-healing quantification using UPV. *Journal of Building Engineering, 28*, 101090.

Singh, N., Kalra, M., & Saxena, S. (2017). Nanoscience of cement and concrete. *Materials Today: Proceedings, 4*(4), 5478−5487.

Singh, T. (2014). A review of nanomaterials in civil engineering works. *International Journal of Structural and Civil Engineering Research, 3*, 31−35.

Snehal, K., Das, B. B., & Akanksha, M. (2020). Early age, hydration, mechanical and microstructure properties of nano-silica blended cementitious composites. *Construction and Building Materials, 233*, 117212.

Snoeck, D., & De Belie, N. (2019). Autogenous healing in strain-hardening cementitious materials with and without superabsorbent polymers: An 8-year study. *Frontiers in Materials, 6*, 48.

Soudki, K. A. (2001). Concrete problems and repair techniques. Waterloo, Ontario, Department of Civil Engineering. <https://concretequality.files.wordpress.com/2010/07/concrete-problems-andrepair-technique.pdf> Accessed on 26.01.18.

Stabnikov, V., Chu, J., Myo, A. N., & Ivanov, V. (2013). Immobilization of sand dust and associated pollutants using bioaggregation. *Water, Air, & Soil Pollution, 224*(9), 1631.

Stynoski, P., Mondal, P., & Marsh, C. (2015). Effects of silica additives on fracture properties of carbon nanotube and carbon fibre reinforced Portland cement mortar. *Cement and Concrete Composites, 55*, 232–240.

Suchorzewski, J., Prieto, M., Mueller, U., & Malaga, K. (2019). Damage and stress detection (self-sensing) in concrete with multi-walled carbon nanotubes. *Multidisciplinary Digital Publishing Institute Proceedings, 34*(1), 17.

Sun, X., Jiang, G., Bond, P. L., Keller, J., & Yuan, Z. (2015). A novel and simple treatment for control of sulfide induced sewer concrete corrosion using free nitrous acid. *Water Research, 70*, 279–287.

Supit, S. W. M., & Shaikh, F. U. A. (2015). Durability properties of high volume fly ash concrete containing nano-silica. *Materials and Structures, 48*(8), 2431–2445.

Tai, C. Y., & Chen, F. B. (1998). Polymorphism of $CaCO_3$, precipitated in a constant-composition environment. *AIChE Journal, 44*(8), 1790–1798.

Takahashi, Y., Hirata, T., Shimizu, H., Ozaki, T., & Fortin, D. (2007). A rare earth element signature of bacteria in natural waters? *Chemical Geology, 244*(3–4), 569–583.

Tamimi, A., Hassan, N. M., Fattah, K., & Talachi, A. (2016). Performance of cementitious materials produced by incorporating surface treated multiwall carbon nanotubes and silica fume. *Construction and Building Materials, 114*, 934–945.

Thomas, M. (2011). The effect of supplementary cementing materials on alkali-silica reaction: A review. *Cement and Concrete Research, 41*(12), 1224–1231.

Thompson, J. L., Silsbee, M. R., Gill, P. M., & Scheetz, B. E. (1997). Characterization of silicate sealers on concrete. *Cement and Concrete Research, 27*(10), 1561–1567.

Tong, T., Fan, Z., Liu, Q., Wang, S., Tan, S., & Yu, Q. (2016). Investigation of the effects of graphene and graphene oxide nanoplatelets on the micro-and macro-properties of cementitious materials. *Construction and Building Materials, 106*, 102–114.

Tragazikis, I. K, Dassios, K. G., Dalla, P. T., Exarchos, D. A., & Matikas, T. E. (2019). Acoustic emission investigation of the effect of graphene on the fracture behavior of cement mortars. *Engineering Fracture Mechanics, 210*, 444–451.

Troiano, F., Gulotta, D., Balloi, A., Polo, A., Toniolo, L., Lombardi, E., . . . Cappitelli, F. (2013). Successful combination of chemical and biological treatments for the cleaning of stone artworks. *International Biodeterioration & Biodegradation, 85*, 294–304.

Valentini, F., Diamanti, A., Carbone, M., Bauer, E. M., & Palleschi, G. (2012). New cleaning strategies based on carbon nanomaterials applied to the deteriorated marble surfaces: A comparative study with enzyme-based treatments. *Applied Surface Science, 258*(16), 5965–5980.

Van Tittelboom, K., De Belie, N., De Muynck, W., & Verstraete, W. (2010). Use of bacteria to repair cracks in concrete. *Cement and Concrete Research, 40*(1), 157–166.

Veltri, S., Palermo, A. M., De Filpo, G., & Xu, F. (2019). Subsurface treatment of TiO_2 nanoparticles for limestone: Prolonged surface photocatalytic biocidal activities. *Building and Environment, 149*, 655–661.

Vera-Agullo, J., Chozas-Ligero, V., Portillo-Rico, D., García-Casas, M. J., Gutiérrez-Martínez, A., Mieres-Royo, J. M., & Grávalos-Moreno, J. (2009). Mortar and concrete reinforced with nanomaterials. In *Nanotechnology in construction* (3, pp. 383–388). Berlin, Heidelberg: Springer.

Verdier, T., Coutand, M., Bertron, A., & Roques, C. (2014). A review of indoor microbial growth across building materials and sampling and analysis methods. *Building and Environment, 80*, 136–149.

Vipulanandan, C., & Mohammed, A. (2015). Smart cement modified with iron oxide nanoparticles to enhance the piezoresistive behavior and compressive strength for oil well applications. *Smart Materials and Structures, 24*(12), 125020.

Wang, J., Ersan, Y. C., Boon, N., & De Belie, N. (2016). Application of microorganisms in concrete: A promising sustainable strategy to improve concrete durability. *Applied Microbiology and Biotechnology, 100*(7), 2993–3007.

Wang, J., Mignon, A., Snoeck, D., Wiktor, V., Van Vliergerghe, S., Boon, N., & De Belie, N. (2015). Application of modified-alginate encapsulated carbonate producing bacteria in concrete: A promising strategy for crack self-healing. *Frontiers in Microbiology, 6*, 1088.

Wang, J., Van Tittelboom, K., De Belie, N., & Verstraete, W. (2012). Use of silica gel or polyurethane immobilized bacteria for self-healing concrete. *Construction and Building Materials, 26*(1), 532–540.

Wang, J. Y., Soens, H., Verstraete, W., & De Belie, N. (2014). Self-healing concrete by use of microencapsulated bacterial spores. *Cement and Concrete Research, 56*, 139–152.

Wang, M., & Yao, H. (2020). Comparison study on the adsorption behavior of chemically functionalized graphene oxide and graphene oxide on cement. *Materials, 13*(15), 3274.

Wang, S., Lim, J. L. G., & Tan, K. H. (2020). Performance of lightweight cementitious composite incorporating carbon nanofibers. *Cement and Concrete Composites, 109*, 103561.

Wang, W. C. (2017). Compressive strength and thermal conductivity of concrete with nanoclay under various high-temperatures. *Construction and Building Materials, 147*, 305–311.

Wang, X., Xing, F., Zhang, M., Han, N., & Qian, Z. (2013). Experimental study on cementitious composites embedded with organic microcapsules. *Materials, 6*(9), 4064–4081.

Warscheid, T., & Braams, J. (2000). Biodeterioration of stone: A review. *International Biodeterioration & Biodegradation, 46*(4), 343–368.

Webster, A., & May, E. (2006). Bioremediation of weathered-building stone surfaces. *Trends in Biotechnology, 24*(6), 255–260.

Wei, S., Cui, H., Jiang, Z., Liu, H., He, H., & Fang, N. (2015). Biomineralization processes of calcite induced by bacteria isolated from marine sediments. *Brazilian Journal of Microbiology, 46*(2), 455–464.

Weiner, S., & Dove, P. M. (2003). An overview of biomineralization processes and the problem of the vital effect. *Reviews in Mineralogy and Geochemistry, 54*(1), 1–29.

Wiktor, V., & Jonkers, H. M. (2015). Field performance of bacteria-based repair system: Pilot study in a parking garage. *Case Studies in Construction Materials, 2*, 11–17.

Wu, M., Hu, X., Zhang, Q., Xue, D., & Zhao, Y. (2019). Growth environment optimization for inducing bacterial mineralization and its application in concrete healing. *Construction and Building Materials, 209*, 631–643.

Wu, M., Johannesson, B., & Geiker, M. (2012). A review: Self-healing in cementitious materials and engineered cementitious composite as a self-healing material. *Construction and Building Materials, 28*(1), 571–583.

Wu, W., Xiao, X. H., Zhang, S. F., Peng, T. C., Zhou, J., Ren, F., & Jiang, C. Z. (2010). Synthesis and magnetic properties of maghemite (γ-Fe_2O_3) short-nanotubes. *Nanoscale Research Letters, 5*(9), 1474–1479.

Xia, T., Kovochich, M., Liong, M., Madler, L., Gilbert, B., Shi, H., & Nel, A. E. (2008). Comparison of the mechanism of toxicity of zinc oxide and cerium oxide nanoparticles based on dissolution and oxidative stress properties. *ACS Nano, 2*(10), 2121–2134.

Yamanaka, T., Aso, I., Togashi, S., Tanigawa, M., Shoji, K., Watanabe, T., & Suzuki, H. (2002). Corrosion by bacteria of concrete in sewerage systems and inhibitory effects of formates on their growth. *Water Research, 36*(10), 2636–2642.

Yang, Z., Gao, Y., Mu, S., Chang, H., Sun, W., & Jiang, J. (2019). Improving the chloride-binding capacity of cement paste by adding nano-Al_2O_3. *Construction and Building Materials, 195*, 415–422.

Yoo, D. Y., Kim, S., Kim, M. J., Kim, D., & Shin, H. O. (2019). Self-healing capability of asphalt concrete with carbon-based materials. *Journal of Materials Research and Technology, 8*(1), 827–839.

Yousefi, A., Allahverdi, A., & Hejazi, P. (2013). Effective dispersion of nano-TiO_2 powder for enhancement of photocatalytic properties in cement mixes. *Construction and Building Material, 41*, 224–230.

Zarzuela, R., Moreno-Garrido, I., Blasco, J., Gil, M. A., & Mosquera, M. J. (2018). Evaluation of the effectiveness of CuONPs/SiO$_2$-based treatments for building stones against the growth of phototrophic microorganisms. *Construction and Building Materials, 187*, 501−509.

Zhan, B. J., Xuan, D. X., & Poon, C. S. (2019). The effect of nanoalumina on early hydration and mechanical properties of cement pastes. *Construction and Building Materials, 202*, 169−176.

Zhan, Y., Xie, J., Wu, Y., & Wang, Y. (2020). Synergetic effect of Nano-ZnO and trinidad lake asphalt for antiaging properties of SBS-modified asphalt. *Advances in Civil Engineering, 2020*.

Zhang, L., De Schryver, P., De Gusseme, B., De Muynck, W., Boon, N., & Verstraete, W. (2008). Chemical and biological technologies for hydrogen sulfide emission control in sewer systems: A review. *Water Research, 42*(1−2), 1−12.

Zhang, M. H., & Li, H. (2011). Pore structure and chloride permeability of concrete containing nano-particles for pavement. *Construction and Building Materials, 25*(2), 608−616.

Zhang, P., Li, Q., Chen, Y., Shi, Y., & Ling, Y. F. (2019). Durability of steel fiber-reinforced concrete containing SiO$_2$ nano-particles. *Materials, 12*(13), 2184.

Zhang, R., Cheng, X., Hou, P., & Ye, Z. (2015). Influences of nano-TiO$_2$ on the properties of cement-based materials: Hydration and drying shrinkage. *Construction and Building Materials, 81*, 35−41.

Zheng, W., Chen, W. G., Feng, T., Li, W. Q., Liu, X. T., Dong, L. L., & Fu, Y. Q. (2020). Enhancing chloride ion penetration resistance into concrete by using graphene oxide reinforced waterborne epoxy coating. *Progress in Organic Coatings, 138*, 105389.

Zhu, C., Zhang, H., Xu, G., & Wu, C. (2018). Investigation of the aging behaviors of multi-dimensional nanomaterials modified different bitumens by Fourier transform infrared spectroscopy. *Construction and Building Materials, 167*, 536−542.

Zhu, T., Lu, X., & Dittrich, M. (2017). Calcification on mortar by live and UV-killed biofilm-forming cyanobacterial Gloeocapsa PCC73106. *Construction and Building Materials, 146*, 43−53.

Zhu, T., Paulo, C., Merroun, M. L., & Dittrich, M. (2015). Potential application of biomineralization by Synechococcus PCC8806 for concrete restoration. *Ecological Engineering, 82*, 459−468.

Zhu, W., Bartos, P. J., & Porro, A. (2004). Application of nanotechnology in construction. *Materials and Structures, 37*(9), 649−658.

Zhu, Y., Murali, S., Cai, W., Li, X., Suk, J. W., Potts, J. R., & Ruoff, R. S. (2010). Graphene and graphene oxide: Synthesis, properties, and applications. *Advanced Materials, 22*(35), 3906−3924.

Zielecka, M., Bujnowska, E., Kępska, B., Wenda, M., & Piotrowska, M. (2011). Antimicrobial additives for architectural paints and impregnates. *Progress in Organic Coatings, 72*(1−2), 193−201.

Zohhadi, N., Aich, N., Matta, F., Saleh, N. B., & Ziehl, P. (2015). Graphene nanoreinforcement for cement composites. In *Nanotechnology in construction* (pp. 265−270). Cham: Springer.

19

Biodegradation of micropollutants

Sarmad Ahmad Qamar[1], Adeel Ahmad Hassan[2], Komal Rizwan[3], Tahir Rasheed[2], Muhammad Bilal[4], Tuan Anh Nguyen[5], Hafiz M.N. Iqbal[6]

[1]INSTITUTE OF ORGANIC AND POLYMERIC MATERIALS, NATIONAL TAIPEI UNIVERSITY OF TECHNOLOGY, TAIPEI, TAIWAN [2]SCHOOL OF CHEMISTRY AND CHEMICAL ENGINEERING, SHANGHAI JIAO TONG UNIVERSITY, SHANGHAI, P.R. CHINA [3]DEPARTMENT OF CHEMISTRY, UNIVERSITY OF SAHIWAL, SAHIWAL, PAKISTAN [4]SCHOOL OF LIFE SCIENCE AND FOOD ENGINEERING, HUAIYIN INSTITUTE OF TECHNOLOGY, HUAI'AN, P.R. CHINA [5]INSTITUTE FOR TROPICAL TECHNOLOGY, VIETNAM ACADEMY OF SCIENCE AND TECHNOLOGY, HANOI, VIETNAM [6]TECNOLOGICO DE MONTERREY, SCHOOL OF ENGINEERING AND SCIENCES, MONTERREY, MEXICO

19.1 Introduction

The contamination of soil and water bodies through persistent environmental pollutants have attracted the attention of modern researchers (Mishra et al., 2020). These pollutants have their growing concentration every passing year (Lin et al., 2020). Such persistent environmental pollutants mainly consist of pharmaceuticals, pesticides, phthalates, polycyclic aromatics, chlorinated phenolic compounds, organics, and some metal ions (Bhatt, Bhatt, Huang, Lin, & Chen, 2020; Kurwadkar, 2019). Together with other previously reported pollutants, there are some emerging contaminants reported recently with their understudied effects. Due to the worldwide usage of chemicals, urban, industrial, and agricultural lands are going under high risk (Agency, 2007). Moreover, reported that persistent environmental pollutants comprise pesticides, chlorinated compounds, aromatic hydrocarbons, and antibiotics as 90% of their composition. Basically, these pollutants are transferring directly to the environment from different sources such as pesticides spraying on agricultural land, antibiotics from medical wastes, industrial effluents, and through wastewater treatment (Bhatt, Gangola et al., 2019; Bhatt, Pal, Bhandari, & Barh, 2019; Bhatt, Sethi, et al., 2020).

According to the European Environmental Agency, there are approximately 3000 suspected sites in the United States and 3 million contaminated points in Europe (Agency, 2007). However, other developing countries such as India, China, Nigeria, Pakistan, Ethiopia, Vietnam, Indonesia, and South Africa have more contaminated sites than those of the developed nations due to the high applications of these chemicals. The concentration of these chemicals share dependencies vary from one region to another (Bilal & Iqbal, 2020). Although many persistent environmental pollutants present in soil and water bodies in small concentrations but the toxic concentration among them varies from chemical to chemical (Bhatt, Huang, Zhang, Sharma, & Chen, 2020; Pang et al., 2020). The water residual concentrations of pesticides in the US, China, and Europe were reported from 0.6 to 921 ng/L. Because they are highly toxic to living organisms, even in traces, they are known as micropollutants. Moreover, pesticides are categorized as organochlorines, organophosphates, carbamates, and pyrethroids, varied in their threshold. Such pesticides cause irreparable damage to the environment due to nonbiodegradable and persistent in nature (Zhang, Wang, Dong, & Lv, 2020).

Pesticides applied to agricultural soil and water bodies, accumulate there, and may result their penetration into food chain which leads to severe health problems in humans and animals. Long term exposure may cause serious diseases such as cancer, endocrine disruption, asthma, and fetal deaths (Gholami-Borujeni et al., 2011). Pesticides are stored into food chains and cause severe damage in life forms. Moreover, the synthetic hormonal medicines like progestins in water system has been studied for bioconcentration within fish which may cause abnormalities in development (Steele et al., 2013). Such hazardous chemicals gather in human tissues and cause adverse health effects. For example, pesticides in high concentration led to cardiovascular issues, endocrine system, liver dysfunction, and hypertension. Together with human beings, amphibians and birds are also affected by such chemicals, such as dichlorodiphenyltrichloroethane (DDT) accumulate into birds through food chains which have been reported to lead reduction in bird populations. Among humans, children, persistent environmental pollutants, and agricultural workers suffer through the harmful effects of such chemicals. Although persistent chemicals enter through inhalation, ingestion or via skin penetrations, many persistent environmental pollutants are exposed through contaminated food (Jayaraj, Megha, & Sreedev, 2016). According to surveys, patients who are suffering from carcinogenic diseases have high concentrations of such pollutants in their bloodstreams compared with normal patients (Ejaz, Akram, Lim, Lee, & Hussain, 2004). Many studies have been reported that pesticide accumulation is one of the key causes of lymphoma and cancers of brain, breast, testes, and ovaries (Singh, Kang, Mulchandani, & Chen, 2008). Xenobiotics are applied to control pests and crop growth, but their adverse effects on plants and animals can never be ignored. Therefore less utilization of such pollutants may be helpful in controlling environmental pollutants-related diseases globally. However, persistent environmental chemicals related pollution is one of the major environmental challenges worldwide (Perreault, De Faria, & Elimelech, 2015).

Different chemical, physical, and biological methods have been reported for the degradation of these toxic chemicals (Wu et al., 2017). Persistent environmental pollutant's degradation is mainly dependent on their half-life ($t_{1/2}$), physiochemical environment, and degradation kinetics of chemicals (Sinkkonen & Paasivirta, 2000). Physiochemical methods include sedimentation, coagulation, centrifugation, hydrolysis, desorption, and photodegradation. However, such degradation methodologies have limited effectiveness and high cost. On the other hand, biological methods are ecofriendly and cost-effective for remediation of such persistent chemicals (Bhatt, Bhatt, et al., 2020). Basically, microorganisms produce enzymes used for the degradation of toxic chemicals through bond cleavage and converting into inorganic chemicals through the mineralization process (Bhatt, Huang, et al., 2020; Bhatt, Sethi, et al., 2020). However, biotransformation can be used for a process in which slow or slight conversion of one chemical form to another takes place. Microorganisms are effective for the transformation of toxic chemicals into less or nontoxic chemicals. In order to upgrade the desirable features of microbial strains and effective degradation of persistent pollutants, microbial engineering was introduced, in which different genetic tools are applied to modify the microbial strains and meet the modern requirements of bioremediation (Bhatt, Huang et al., 2020). Different microbial strains, that is, *Dehalococcoides, Burkholderia, Pseudomonas, Comamonas, Achromobacter, Alcaligenes, Rhodococcus,* and *Sphingomonas* have been designed to improve bioremediation of persistent environmental pollutants (Bilal & Iqbal, 2020). Numerous tools of genetic engineering have been developed to build catabolic routes for the degradation of hazardous chemicals. Applying microbial strains for the degradation of persistent environmental pollutants has been termed as bioremediation. Moreover, designing of microbial strains for pollutant degradation purpose using advanced technologies also considers as bioremediation (Dvořák, Nikel, Damborský, & de Lorenzo, 2017).

The principal purpose of introduction of microbial engineering for bioremediation is to build new cells and design new enzymes and pathways for degradation. Such bioremediation method provides an effective, ecofriendlier, and feasible technology for the mineralization of persistent environmental pollutants (Ramos et al., 2011). For example, 1,2,3-trichloropropane can be remediated by applying a combination of protein and metabolic engineering (Janssen & Stucki, 2020). Commonly, genetic engineering approaches have been successfully reported for the degradation of organic pollutants and dyes (Kumar et al., 2020). Likewise, *Pseudomonas* species can be applied to degrade polyethylene, polyvinyl chloride, polystyrene, and other plastic polymers with different rates of degradation (Wilkes & Aristilde, 2017). Recent advancements in synthetic biology can be measured by the engineering of microbial consortia (Chang et al., 2014). Microbial consortium performance is manifold better than that of microbial monoculture. Microbial consortia speed-up the degradation process of persistent environmental pollutants. The major reason of this robustness is that different microorganisms work for the degradation of the environmental pollutants, however, each microbe works for degradation for a limited period. Overall, engineering microbial strains are emerged as an effective pollutant degrader than any of the wild-type strains (Jeandet et al., 2018). In this chapter, we mainly focus on the engineered microbial strains for the degradation of persistent environmental pollutants. Moreover, this chapter covers the traditional and advanced techniques applied for the engineering of microbial strains and to improve the efficiency of such microbial strains for the degradation of persistent environmental pollutants.

19.2 Advanced physicochemical treatment approaches for pollutants degradation

With the increase in large-scale applications of synthetic materials, their wastes are needed to develop new methodologies for the degradation of these contaminants (Sousa, Ribeiro,

Barbosa, Pereira, & Silva, 2018). As accumulation of environmental pollutants have adverse effects on human health therefore the scientific community has been working on the development of efficient, ecofriendly, and cost-effective techniques for the removal of these pollutants (Jain, Yadav, Joshi, & Kodgire, 2019; Qamar, Ashiq, Jahangeer, Riasat, & Bilal, 2020). Several different advancements have been recorded for degradation purposes in the past few years. Different techniques such as photocatalytic fuel cells, photocatalysis, and nanoremediation techniques have been reported to seek remedies for these problems. However, physical, thermal, chemical, and biological strategies have been highly adopted to remove the contamination of soil and water (Zhan, Huang, Lin, Bhatt, & Chen, 2020).

19.3 Photocatalysis

Photocatalysis is a method in which light is used to degrade environmental pollutants (Colmenares & Xu, 2016). According to International Union of Pure and Applied Chemistry, photocatalysis is a light-dependent reaction (Dionysiou, Puma, Ye, Schneider, & Bahnemann, 2016). In this process, protons are used to excite the photocatalyst, which results in the formation of charged species, used to speed-up the oxidation–reduction reactions (Gaya & Abdullah, 2008). Superoxide (O^{-2}) and hydroxy free radical are used as photocatalysts and generated by using different metal oxides such as Fe_2O_3, ZnO, ZnS, $BiTiO_3$, etc. through photocatalysis, however, their activation can be performed through light radiation (Khaki, Shafeeyan, Raman, & Daud, 2017). These semiconductor materials possess valence and conduction bands that are separated by band gap energy (E_g). Excitation of photocatalyst leads transference of electrons from valance band (filled) to conduction bands (vacant). The energy required for this shift comes on excitation which is exactly equal to the energy gap between the bands. However, electron-hole pair interacts with the hydroxyl ions or oxygen molecule or with water molecules in order to generate extremely reactive species, such as superoxide anions (O^{-2}) or hydroxy-based radicals. These species work as an oxidizing agent and cause the oxidation of persistent environmental pollutants through oxidation reactions (Koe, Lee, Chong, Pang, & Sim, 2020).

Together with advantages, photocatalysis process has a drawback of longer band gap which requires a plenty of energy to overcome such energy gap, faulty potential for migration, high potential for electrons-holes recombination and poor capacity for light absorption (Huang et al., 2017). With the help of doping, the broader energy gap can be reduced by furnishing greater value of dipole moment, which results a greater number of electrons can be shifted from valence band to conduction band (Chiu, Chang, Chen, Sone, & Hsu, 2019). On the application of doping agent, photocatalyst develops greater dipole moment to vary the kinetics of electron shift which results greater number of electrons can shift from valence band to conduction band on the cost of low energy of band gap. There are numerous dopants can be applied for TiO_2 as photocatalyst in order to improve its response and minimize the electron-hole recombination chances. However, this results an increased degradation of persistent environmental pollutants because of their efficient consumption of electron-hole recombination and available number of photoexcited holes (Saggioro et al., 2011).

19.4 Photocatalytic fuel cells

Various reports have been published explaining the phenomena of production and electron transference at microbial anode disturb the performance of fuel cell system (Choi, 2015; Jiang et al., 2020). In order to fix microbial anode can be replaced by a high performance photoanode of TiO_2 having an efficient photogenerated electrons transfer (Gan, Gan, Clark, Su, & Zhang, 2012). In photocatalytic cells, microbial cells have reported to function as catalyst used for oxidation of environmental pollutants, and electrons are generated which are further utilized to produce electricity and hydrogen at cathode (Jiang et al., 2017). Thus photocatalytic cells form a degrading system for persistent pollutants at photoanode. Overall, such cells mainly comprise of a Pt cathode, a TiO_2 anode, a substrate, a reactor, and an electrolyte. TiO_2 and ZnO are effectively employed as photoanodes for the degradation of persistent environmental pollutants and for the generation of electricity together with the preparation of hydrogen gas at the cathode (Regonini & Clemens, 2015).

19.5 Sonochemical methods

Recently, sonochemical techniques have been appeared as an effective technology for the removal of persistent environmental pollutants. Sonochemistry is a branch of chemistry that studies the chemical reactions take place by the induction of sound in solution (Mason, 2012). Sound waves results high vapor temperature and pyrolytic degradation of environmental pollutants at bubble-water interface. Generally, ultrasound is employed for the destruction or simulate the removal of liquid-phase contaminants (Bremner, Burgess, & Chand, 2011). In the conservative approach, environmental pollutants were treated either by the decomposition process or through oxidation process. Ultrasounds compose of a consecutive set of compression and rarefaction, wherein compression generates positive pressure, however, rarefaction generates negative pressure (Xiao, Wei, Chen, & Weavers, 2014). Such pressure amplifies the tensile strength of the liquid at one point which results the generation of cavitational bubbles. These bubbles get energy from the sound waves and approach to burstable stage which further results high temperature of 5000K and pressure 1000 atm and the environment turns oxidative due to the formation of hydroxyl, hydroperoxyl higher reactive radicals, which are applied for the pyrolytic degradation of persistent environmental pollutants (Durán, Monteagudo, Sanmartín, & García-Díaz, 2013). Leong, Ashokkumar, and Kentish (2011) introduced sono-photocatalytic oxidative approach for the degradation of 96 antipyrine in aqueous solution. Likewise, sonolysis and photolysis were employed for the wastewater treatment for environmental pollutants caused by the pharmaceutical industry (Durán et al., 2013). For an efficient removal of tinidazole from aqueous solution, H_2O_2, and ultrasound were employed (Rahmani et al., 2014).

19.6 Nanoremediation

Nanoremediation is highly efficient and cost-effective approach, applied for the treatment of persistent environmental pollutants using nanoparticles (Rajan, 2011). In this technique, nanoparticles are applied as catalysts to oxidize, reduce, or sorption of contaminants. For this purpose, nanozeolites, nanotubes, metal oxides, and nanofibers are applied to treat hazardous chemicals (Aguilar-Pérez et al., 2021; Aguilar-Pérez, Heya, Parra-Saldívar, & Iqbal, 2020). High reactivity, high surface area to mass ratio, rapid mobility, and cost-effectiveness makes nanoremediation technique as unique for the treatment of organic and inorganic pollutants (Tosco, Papini, Viggi, & Sethi, 2014). As nanoparticles are feasible to be applied both for in situ and ex situ, therefore they can be utilized as catalysts for wider range of toxic chemicals, for example, inorganic anions, pesticides, nitrates, and heavy metal ion (Karn, Kuiken, & Otto, 2009). For the remediation of those heavy metals, which show higher standard negative redox potential and can easily be adsorbed to the iron hydroxide shell (Li & Zhang, 2007).

Natural/synthetic goethite ZVI, carbon nanotubes, and TiO_2 nanomembranes have been utilized for the remediation of environment (Sayan et al., 2013). In this regard, bimetallic nanoparticles have also successfully played their role for cleaning-up the environmental pollutants from the water system (Koutsospyros et al., 2012). Carbon nanotubes have also been applied for the remediation of toxic heavy metals e, g Pb^{+2}, Cr^{+3}, and Zn^{+2} and metalloids such as volatile organic, and biological impurities, dioxins, and arsenic compounds (Li et al., 2005; Rao, Sivasankar, & Sadasivam, 2009). Carbon nanotubes have been further employed for the absorption of synthetic contaminants from the water bodies and have an excellent binding capacity for numerous functional entities (Savage & Diallo, 2005). The time required for nanoremediation is remarkably less than those of conventional remediation techniques. Zhang (2003) investigated that TCE level reduced by 99% upon treatment with nZVI together with considerable reduction in time required for remediation even from years to few days.

19.7 Biosensors for environmental pollutants detection

Biosensors are used to monitor biological processes. The signals which are generated by the biosensors are read by microelectronic segments and then processed accordingly (Purohit, 2003). Biosensors can be successfully applied for the monitoring of environment for kinds of different pollutants. Tsekenis et al. (2015) introduced luminescence nanosensors using 20 modified blue-emissive upconversion for the detection of organophosphate pesticides. Such biosensors have advanced enzymatic and electrochemical sensing efficiencies. For this purpose, two component regulatory systems help to integrate the regulatory and sensory mechanisms. Such regulatory systems comprise of histidine kinase as sensor and regulatory unit to modulate the gene expression (Ravikumar, Baylon, Park, & Choi, 2017). They are further applied to detect the signals generated from different organic contaminants and heavy metals. They are designed for the detection of optical, fluorescence, electrochemical, enzymatic, piezoelectric, and thermal-based pollutants. To date, nanozymes and nanodetectors are found one of the most promising for the detection and treatment of pollutants of environment (Bidmanova et al., 2016).

19.8 Biotechnological approaches for micropollutants degradation

For bioremediation of the environmental pollutants, different biotechnological methods have been reported, and highly efficient omics methods are the basis of these techniques. Related to bioremediation of environmental pollutants a lot of data has been produced. To find the degradation capacity of microbes, various tools related to bioinformatics could be employed to study proteins and genetic materials. Microbes have been employed to enhance the quality of environment (Chen, Mulchandani, & Deshusses, 2005). Polluted places may be treated through employing different methodologies as system biology, genetic engineering, CRISPR/Cas, synthetic biology, immobilization of enzyme, and microbial electrochemistry. These biotechnological methods have been continuously developing and optimizing for better performance (Qamar, Asgher, & Bilal, 2020). These technologies have play significant role toward environmental sustainability by engineering different strains of microbes which can produce attractive and useful products/molecules from environmental pollutants is striking method to encourage the economy (Cregut, Bedas, Durand, & Thouand, 2013). Recycling of environmental pollutants is not possible in major degradation frameworks of biodegradation by utilization of engineered bacteria, which are particular to perform this task in short time, enhance the working possibility. In comparison of conventional remediation approaches, the biotechnological methods are noninvasive and cost friendly. In polluted environment the efficacy of biotechnology has been validated to degrade the environmental pollutants (Singh, Sharma, & Santal, 2016) Nanoscience, omics, findings, and innovations of genes, synthesis of new materials are technologies which enhances and improved the efficiency of bioremediation for cleaning of contaminated environment (Tripathi et al., 2017).

19.9 Microbial electrochemical system

Combination of microbial system along with the electrochemical processes is known as microbial electrochemical system (MES). Basic principles in MESs are shown in Fig. 19–1 (Wang & Ren, 2013). Simultaneously the release of reducing equivalents and conversion of released electrons to electricity from synthetic products are carried by MES (Wang, Cai, Zhou, Zhang, & Chen, 2012). Pollutants are reduced/oxidized to create nontoxic products or intermediate through series of redox reactions which are usually mediated by the bacterial metabolism. MES has several advantages in comparison to different other bioremediation methodologies as: (1) consumption of the energy is less, (2) this results in energy or electricity production which may supply power to environmental systems, (3) production of nonharmful, nontoxic products, and (4) it has ability of redox reactions and can be employed to many toxic synthetic products (Wang et al., 2012). MES is comprised of two chambers as cathode and anode, which are separated through the ion-exchange membrane (IEM) (Wang & Ren, 2013). Anode chamber is placed where microbes are present, and they may exist in form of biofilm or planktonic state. Substrate oxidation is carried out by microbial system,

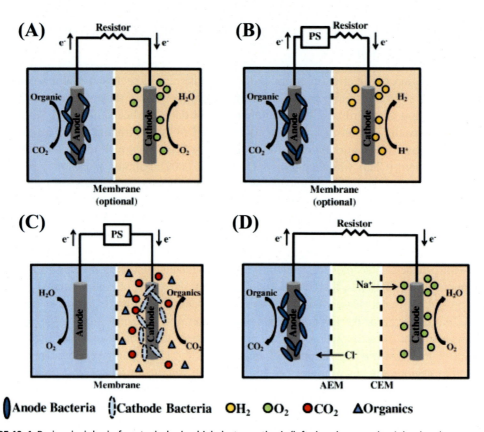

FIGURE 19–1 Basic principles in four typical microbial electrosynthesis (left chamber: anode; right chamber: cathode). (A) Electricity generation in air-cathode microbial fuel cells; (B) hydrogen generation with external power supply in microbial electrolysis cells; (C) chemical production by microbial electrosynthesis; (D) middle chamber desalination in microbial desalination cells (Wang & Ren, 2013).

which results in production of various end-products as protons, electrons, and various other secondary metabolites. Produced electrons gathered at anode and transported to the cathode through the external circuit. Although, in IEM the protons moved from anode to cathode chamber to fulfill the whole circuit on being reduced through the electrons at cathode. To produce the valuable products as hydrogen (H_2), alcohol (R-OH), methane (CH_4), and acids (RCOOH), MES carries the water oxidation at anode to produce electrons and protons in existence of external electrical potential. This consequently results in either hydrogen or electrical energy production. In some cases extra power is supplied for carrying out the redox reaction (Wang, Luo, Fallgren, Jin, & Ren, 2015).

For the treatment of wastewater, the MES has been employed (Nancharaiah & Lens, 2015). Various studies has been conducted to report the employment of MES for bioremediation of soil because soil has electrochemical characteristics (Wang et al., 2012; Wang, Yi, Cao, Fang, & Li, 2017). Various methods have been designed for the utilization of MES in remediation of soil as (1) oxidation process holds at the anode as products like petroleum behave as source of electron donors and degradation occurs during oxidation–reduction reactions (Wang et al., 2017); (2) anodic reduction: synthetic products behave as electron acceptors in reducing atmosphere and get degraded (Wang et al., 2017); (3) sorption process: products absorption upon electrodes or on electrode existing biofilm; (4) electrokinetic energy: in electric field influence, particles of soil go hydrological and physicochemical variations (Wang, Song et al., 2016); (5) cathode alkalization: utilization of protons at cathode where pH enhances during O_2 reduction (Zheng et al., 2014); and (6) cathode reduction: azo compounds, metals, halogen-based organics attain electrons that were anode produced (Wang, Deng, & Zhao, 2016; Wang, Song et al., 2016). A bioelectrochemical system has been designed and this system has been known as bioelectric well, and this can be used in polluted groundwater sources or wells for applications (Palma, Daghio, Franzetti, Petrangeli Papini, & Aulenta, 2018).

To efficiently remediate the atrazine (pesticide) polluted soil, a technique called bioelectro-venting has been invented (Domínguez-Garay, Quejigo, Dörfler, Schroll, & Esteve-Núñez, 2018). In short time period many hydrocarbons has been degraded by utilization of the anodic oxidation (Lu, Huggins, Jin, Zuo, & Ren, 2014). During a research, it was reported that in 120 days the bioremediation efficiency was around 82.1%–89.7%, demonstrated for total petroleum (Lu, Yazdi, et al., 2014). Toxic metals such as copper, cadmium, and lead reportedly may be transferred and gathered in regions of cathodes in electric filed influence, which is driven through soil MES. After the 108 and 143 days, the lead (44%) and cadmium (31%) extracted from region of anode, respectively. Around 200 mg/kg copper migrated to cathodic region in 56 days in single chambered MES (Wang, Song et al., 2016).

19.10 Enzyme-assisted remediation of micropollutants

The utilization of microbial-degradative enzymes for on-site bioremediation of polluted places is an effective and efficient approach (Abhilash et al., 2013). Various reports are available in literature that at contaminated sites, the degradative enzymes have significant capability to convert the toxic organic molecules to nonharmful, nontoxic components. For instance, the peroxidases, monooxygenases which are flavin dependent, cytochrome P450 are described to have capability to catalyze different important reactions as redox reactions, epoxides production and hydrolysis which facilitates detoxification process of synthetic products (Arslan, Imran, Khan, & Afzal, 2017). Various enzymes as nitroreductases, lignin peroxidases, tyrosinases, dehalogenases, phosphotriesterase, quinone reductases, manganese peroxidases, proteases, and laccases have significant potential to degrade environmental pollutants (Nagata, Ohtsubo, & Tsuda, 2015; Qamar et al., 2020; Restaino, Borzacchiello, Ilaria, Fedele, & Schiraldi, 2016).

Laccases are enzymes which are significantly known for their degradative efficiency for different carcinogenic contaminants like benzo[a]pyrene (95%), anthracene (80%), perylene

(73%), phenanthrene (72%), fluoranthene (70%), pyrene (66%), and fluorine (54%) (Farnet, Gil, Ruaudel, Chevremont, & Ferre, 2009; Rao, Scelza, Acevedo, Diez, & Gianfreda, 2014). The enzyme manganese peroxidases have significant potential to degrade phenanthrene (7%–30%), anthracene (32%–100%), benzothiophene (>90%), pyrene (11%–65%), and fluoranthene (25%) (Rao et al., 2014). Along with degradation of synthetic products, the enzymes also have capability increase the soil nutritional values by cycling of nutrients (Tripathi et al., 2017). Many technologies for cleaning purpose of environment have been established through enzyme technology advancements. Through engineering of enzymes, the enzymatic features as their stereoselectivity, pH, temperature, efficacy, and specificity to substrate can be enhanced. Different technologies as random-priming (Shao, Zhao, Giver, & Arnold, 1998), shuffling of DNA (Kuchner & Arnold, 1997), extension (staggered) methods (Zhao, Giver, Shao, Affholter, & Arnold, 1998), and site-directed mutagenesis (Ju & Parales, 2006) may be utilized to modify the mechanisms and enzymes features.

Through the subunits exchanging, sequencing of genes, the appropriate required features of enzymes can be attained as chimeric enzymes can be produced that are known to keep high features in comparison to its original forms (Beil, Mason, Timmis, & Pieper, 1998). Into the peptides, different binding locations can be incorporated, and then various cofactors and molecules may bind to enzyme and enhance the catalytic potential of enzyme. For bioremediation of different contaminants, this method can be easily employed (Pazirandeh, Wells, & Ryan, 1998). The following steps are involved in engineering of enzymes as modification selection in protein and this may be carried out through randomization, rational design, execution of modification by mutagenesis, and assessment/evaluation of enzyme (modified) for enhanced features and assessing the modified enzymes for improved characteristics by different screening/selection methodologies (Fig. 19–2) (Kazlauskas & Bornscheuer, 2009). For bioremediation of toxic heavy metals and radionuclides, enzymes engineering have been carried out to generate highly selective and catalytically efficient enzymes (Dhanya, 2014). By utilization of nitrobenzene 1,2-dioxygenase, the oxidation of 2,6-di-nitrotoluene can be improved 2.5 fold (Ju & Parales, 2006).

19.11 Immobilized enzymes for micropollutants degradation

Immobilization of enzyme is useful method to acquire reusable, active, and potentially stable enzymes by fixing them on surface of solids (Sastre, Reis, & Netto, 2020). Enzymes are potential catalysts that can be employed either in immobilized of free state. Enzymes in free state can be altered easily, greatly sensitive to agents which can denature them, prone to attack by protease, and they are difficult to reuse so their on-site applications are limited (Khan, Akhtar, & Husain, 2006). For bioremediation purpose, immobilization of degradative enzymes is a fascinating substitute of free enzymes. Attachment of enzyme (either in free form or soluble state) to multiple supports provide enhancement in the activity and stability of enzymes, which is known as enzyme immobilization (Ahmad & Sardar, 2015). Great yields are obtained at the low-cost through immobilized enzymes in

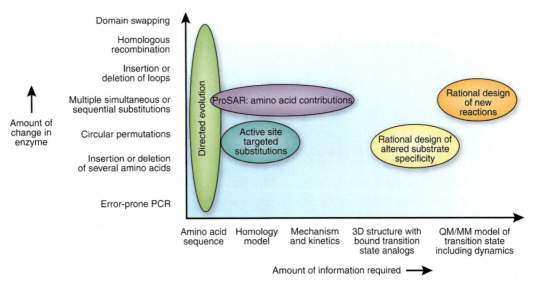

FIGURE 19–2 Protein engineering methods differ widely based on the degree that the change in enzyme, and the amount of information available for rational design. ProSAR, or protein structure–activity relationship, is described in the text. *3D*, three-dimensional; *QM/MM*, quantum mechanics or molecular mechanics (Kazlauskas & Bornscheuer, 2009).

environmental pollutants bioremediation, so they can be employed for remediation of contaminants at large scale (Corici et al., 2016). Many advantages are associated with immobilization as stability for longer time span, excellent efficiency, decrease in cost, simple and easy recovery, and enzyme recycling for industrial purposes (Sheldon, 2007). Immobilization improves the range of pH and temperature in which enzymes works also enhance the immobilized enzyme stability toward heat. In cell-free medium the intracellular enzyme stability also can be improved through immobilization (Sheldon, 2007). The features of enzymes are enhanced greatly through the process of enzyme immobilization, although alterations in configuration and potential activity of enzymes are not required (Shaheen, Asgher, Hussain, & Bhatti, 2017).

There are different methods of enzyme immobilization that are employed as entrapment crosslinking, affinity tag-binding, encapsulation, and adsorption binding (Sirisha, Jain, & Jain, 2016). Support employed for immobilizing the enzyme must have following features as huge surface area, cost-effective, not restrictive substrate. Employment of immobilized enzyme for degradation of the harmful xenobiotic molecules made this technique very significant (Mehta, Bhardwaj, Bhardwaj, Kim, & Deep, 2016). Different studies have been reported regarding enhanced biodegradation of synthetic products through the immobilization. Through the electrostatic interaction, manganese peroxidase has been immobilized to unique nanoclay and nanomaterial (Acevedo et al., 2010). Nanoclay immobilized manganese peroxidase exhibited enhanced stability in wider array of pH and temperature and long storage than that of free manganese peroxidase. Catalytic efficacy of manganese peroxidase on

PAH has been not affected negatively by immobilization in solution, which showed that the immobilization have not changed the configuration (Acevedo et al., 2010). In comparison of free enzymes, the immobilized *Momordica charantia* (bitter gourd) peroxidase exhibited improved textile dyes decolorization (Akhtar, Khan, & Husain, 2005).

Efficiency of immobilized and free enzymes from the *Arthrobacter* specie HB-5 has been determined for atrazine pesticide removal from polluted soil and same catalytic potential was noted by both kinds of enzymes for degradation of atrazine while wide range of pH was demonstrated for immobilized enzyme (Ma et al., 2011). The immobilization of horse-radish peroxidase in Ca-alginate beads has been demonstrated and discoloration of acid orange-7 (75%) and acid blue-25 (84%) dye was obtained by this strategy (Gholami-Borujeni et al., 2011). Immobilized horse-radish peroxidase enzyme discolored approximately 88% an anthraquinone dye, acid violet-109 in only 35 min (Šekuljica et al., 2015). Lignin peroxidase which was attained from fungal strains has been immobilized on the carbon nanotubes (Oliveira, da Luz, Kasuya, Ladeira, & Junior, 2018). The immobilized lignin peroxidase on carbon nanotubes efficiently discolored Remazol brilliant blue-R dye and also showed greater specific potential and catalytic activity and low K_m values in comparison of the free enzymes. By employing *Coriolopsis gallica*-laccase immobilized on mesoporous nanoframework, dichlorophen pesticide was degraded and potential decrease in toxicity of pesticide was obtained after oxidation through immobilized-laccase (Fig. 19–3) (Vidal-Limon, García Suárez, Arellano-García, Contreras, & Aguila, 2018).

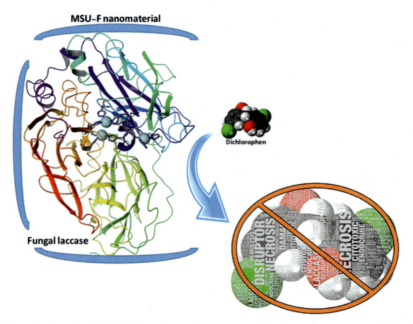

FIGURE 19–3 Degradation of pesticide dichlorophen by laccase immobilized on nanoporous materials (Vidal-Limon et al., 2018).

19.12 Nanozymes

Nanoparticle based enzyme-mimics are called as nanozymes or artificial enzymes (Castillo et al., 2021) which have a wide range of applications (Fig. 19–4) (Shin, Park, & Kim, 2015). These have capability to catalyze various reactions and reaction kinetics are followed similar to those of enzymes (Gao & Yan, 2016). These nanozymes exhibit great stability and are economically feasible (Xie, Zhang, Wang, Zheng, & Huang, 2012). Different nanoparticles are reported to mimic various enzymes as esterase, peroxidase, catalase, ferroxidase, superoxide-dismutase, phosphatase, etc. In the nanozymes, the active binding sites are absent and only particular substrate can bind and give chemical reaction (Wang, Hu, & Wei, 2016). For detection and removal of various environmental pollutants, dyes, and lignin based waste, the nanozymes can be employed (Liang et al., 2017). Phosphatase mimicking CeO_2/ g-Fe_2O_3 nanocomposites has been designed and employed for biodegradation of the parathion methyl (organophosphate-based pesticide) (Shin et al., 2015). The magnetic Fe_3O_4 nanoparticles have capability to mimic peroxidases and these have been employed for remediation of rhodamine, methylene blue, and phenol. The technologies based on implementation of nanozymes proved extremely efficient, and economical (Wei & Wang, 2008). Carbon-based nanomaterials may be utilized as nanozymes because they exhibit intrinsic enzyme type potential in various chemical reactions (Bird & Hopkins, 2016).

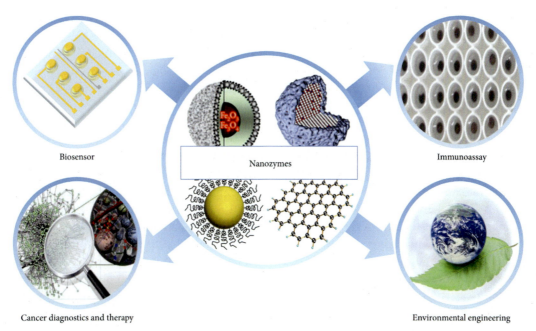

FIGURE 19–4 A wide range of applications in the field of nanozymes. The images of nanomaterials in the center ring represent (clockwise from top *left*) Fe_3O_4 nanoparticle, CeO_2 nanoparticle, graphene oxide, and Au-nanoparticle, which are the typical nanomaterials exhibiting enzyme-like activities (Shin et al., 2015).

19.13 Metabolic engineering approaches for pollutants degradation

The metabolic engineering is proceeded by different modes such as selection of appropriate genes and enhancing their expression in indigenous microbes to improve the biodegradation processes. The optimization of regulatory and genetic processes is achieved through metabolic engineering approaches to use them for the degradation of environmental pollutants (Curran & Alper, 2012). The physiologies and metabolisms of microbial strains are comprehended by the metabolic engineers through different biobased systems and "omics" techniques. The bioremediation of environmental pollutants is considered appealing by the fusion of metabolic engineering with advanced systems biology. The complex biological systems and their correlation with the boarder ecosystem could be studied using the systems biology. The engineered microbes modify their degradation processes due the genomic level modifications provided by the metabolic engineering. Moreover, the biodegradation of environmental pollutants within surroundings could be the result of modified processes (Dangi, Sharma, Hill, & Shukla, 2019; Sharma, Dangi, & Shukla, 2018). During degradation mechanism, the chemical palette of fungal and bacteria cells is increased by application of these techniques (Bartley, Kim, Medley, & Sauro, 2017).

The bioremediation potential of environmental pollutants could also be achieved by another branch known as metabolomics. Different approaches have been carried to attain the entire group of metabolites during degradation studied by the previous studies (Bonifay, Aydin, Aktas, Sunner, & Suflita, 2016; Klassen et al., 2017). The metabolites generated during the bioremediation of environmental pollutants were studied by different techniques such as nuclear magnetic resonance, direct-injection mass spectrometry, high-performance liquid chromatography, gas chromatography-mass spectroscopy, and Fourier transform-infrared spectroscopy (Chakraborty & Das, 2017). The *Sinorhizobium* sp. C4 was used to study the bioremediation of phenanthrene through comparative metabolomics (Beale, Karpe, & Ahmed, 2016). The degradation of succinate-associated metal ions and biphenyl of *Pseudomonas pseudoalcaligenes* KF707 were studied by the metabolomic processes (Booth, Weljie, & Turner, 2015). The coexpression and construction of several genes are possible now days due to the quick improvements of molecular and systems biology tools (Bhatt, Gangola et al., 2019; Bhatt, Pal et al., 2019). The large category of environmental pollutants from the surroundings are decomposed by the cascade of genes controlled through synthetic and molecular biology tools (Wargacki et al., 2012). Since the beginning of genomic scale tools, their extensively utilization for the metabolic engineering of various degradation pathways have been investigated. Each step of environmental pollutants degradation is assisted due to the accessibility of whole-genome sequences for microbes and the information derived from genome behaves as catalyst for the metabolic engineering activities (Henry et al., 2010).

The microbial metabolic processes are comprehended by various genome-scale models of organisms like *Pseudomonas putida*, *Escherichia coli*, and *Saccharomyces cerevisiae* (Reed, Vo, Schilling, & Palsson, 2003). In response to particular molecular alternations, these models are widely employed to analyze and to compare the metabolic fluxes (Curran & Alper, 2012). *Pseudomonas* utilize the intermediate metabolites to penetrate either the Entner–Doudoroff or carboxylic acid pathway. A large group of environmental pollutants are degraded by *Pseudomonas* family, belonging to the class of heterotrophic bacteria (Nikel & de Lorenzo, 2013). The biofilm of the *P. putida* KT2440 enhances the process of degradation and genetic engineering cause the recombinant *P. putida* KT2440 favorable for the degradation of 1-chlorobutane (Benedetti, de Lorenzo, & Nikel, 2016). The bioconversion of hydrocortisone into cortisone was favorable by the strain *Bacillus megaterium*. The high yield of cortisone was obtained by using protein engineering from the different mixture constituents found in *B. megaterium* (Konig, Hartz, Bernhardt, & Hannemann, 2019). The CRISPR/Cas9 genome editing approach was also applied to engineer the metabolic pathways for the degradation of environmental pollutants. The CRISPR/Cas9 genome editing tool was also used to engineer *Rhodococcus ruber* to biotransform the pathways for acrylamide (Fig. 19–5) (Liang, Jiao, Wang, Yu, & Shen, 2020). The utility of the CRISPR/cas9 genome editing is enhanced due to their vast utilization on the fungal and bacterial strains (Liang et al., 2020). Different substances such as dyes, antibiotics, hydrocarbons, and pesticides have been reported to biodegrade by various microbial resources (Zhang et al., 2016). Different conditions such as the prohibited capabilities of a single-organism degradation approach, long-time duration of degradation, and the complex structure of pollutant are responsible for the disabilities of microbial metabolic pathways for toxic chemicals degradation (Wu et al., 2017).

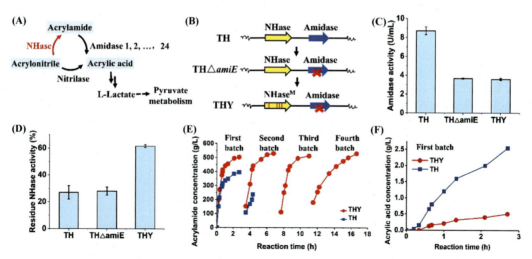

FIGURE 19–5 Applying CRISPR/Cas9-based recombineering to engineer *Rhodococcus ruber* for the bioproduction of acrylamide (Liang et al., 2020).

19.14 Invention of novel genes involved in bioremediation

The bacteria have revealed some remarkable survival mechanisms integrated into their genome as the response of the toxic metals and these calamities are subjected by the production of diverse group of enzymes and proteins (Johnsen, Wick, & Harms, 2005). The genetic and metabolic engineering is used in microorganisms to harvest the products and to provide particular characteristics to the host cells. For the removal and subsequent detoxification of pollutants in microorganisms, scientists are investigating the chances of finding new alternative schemes for microorganisms. The toxic heavy metals are converted into less toxic states due to the alternation into the oxidation state of heavy metals by the introduction of enzymes encoding genes. To experience the excellent bioremediation, the mercuric reductase was inserted into another bacterium for bacterial merA encoding (Dash & Das, 2015). Therefore we can observe that the new processes of heavy metal detoxification can be attained by the insertion of modified genes (Elbanna, Hassan, Khider, & Mandour, 2010).

The utilization of silicomodes can also be considered as another technique for bioremediation purpose. Recently, it has become possible to attain the information of any object from single origin due to the invention of the computer software. The information about toxicity, characteristics, pathways of degradation and appearance of various compounds are now reachable by the different databases. Khan and coworkers performed an amazing comparison of utilization of such computational tools to carry bioremediation. To analyze the toxicity of various compounds, they listed the various tools in their study, and also explained knowledge on the toxicity of various compounds by the description of databases (Khan, Sajid, & Cameotra, 2013). Moreover, the sources of programs were also used to check the environmental degradability of compounds. The two excellent explained programs were Biochemical Network Integrated Computational Explorer (BNICE), and Biodegradability Evaluation and Simulation System (BESS). The genes useful for the bioremediation purposes and the detailed study of interactions between various biomolecules in silico and to deduce novel proteins are now feasible with the aid of database knowledge and biological prediction software (Karp et al., 2005; Letunic et al., 2006; Von Mering et al., 2007).

19.15 Enhanced bioremediation via metabolic engineering processes

Mostly due to the suboptimal degradation pathways, the most pollutants have high maintenance in the surroundings (Timmis & Pieper, 1999). Many biological pathways are studied for the conversion of inorganic toxic metals through different enzymes during the process of organic pollutant degradation. Microorganism association in microbial bioremediation of toxic pollutants from the surrounding by the different extracellular and intracellular incidents have been vastly investigated. The toxic metals are decomposed into less toxic forms due to the synthesis of extracellular and intracellular enzymes by the resistant bacteria in response to the presence of toxic metals in the environment. The rate limiting step is also present in each enzymatic pathway. The bioremediation possibility can also be enhanced due to the manipulation of rate limiting step. The engineering of microorganisms has allowed the transformation of particular metal contaminates due to the modern technology of genetic engineering (Fig. 19−6) (Jaiswal & Shukla, 2020). These genetically modified organisms (GMOs) can be used for *in situ* removal of metal pollutants due to possibility of synthesizing genes combination which are not naturally present. The most common procedures consist of modification of present gene sequences, pathway switching, and engineering with a single gene or operon.

The microbial laccases can efficiently degrade and mineralized several hazardous pollutants such as chlorinated hydroxyl biphenyl, chlorophenol, nonylphenol, and bisphenol A (Fukuda et al., 2001; Schultz, Jonas, Hammer, & Schauer, 2001; Tsutsumi, Haneda, & Nishida, 2001). The enzymatic coupling was used for the detoxification of 2,4,6-trinitrotoluene (TNT) by Nyanhongo, Couto, and Guebitz (2006). They developed the model for isolated product to analyze the coupling of reduced TNT to organic soil materials. For instance, the basis for the identical TNT enzymatic coupling detoxification by the conversion of 2,4-diamino-6-nitrotoluene (2,4-DANT) to guaiac. The aerobic and anaerobic environments favored the catechol and soil guaiac coupling applied for TNT detoxification (Thiele, Fernandes, & Bollag, 2002). During the anaerobic phase, the microorganisms have been used to reduce the TNT from cow manure. The 74% catechol and 25% humic acid, coupling was acquired by the *Trametes villosa* based laccase reliable for the oxidative coupling in the aerobic phase.

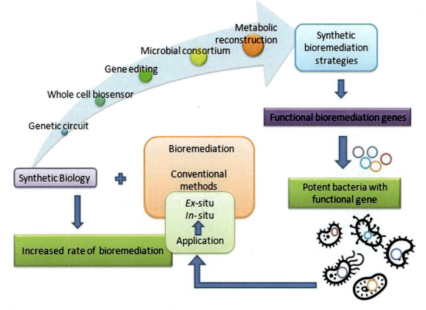

FIGURE 19–6 Genetic engineering technologies for efficient bioremediation of environmental pollutants. *Reprinted from Jaiswal and Shukla, (2020). Alternative strategies for microbial remediation of pollutants via synthetic biology. Frontiers in Microbiology, 11 Article distributed under CC-BY license.*

The phenolic humic materials were utilized for the successful synthesis of 4-amino-2, 6-dinitrotoluene (4-ADNT) and TNT metabolites (Wang, Thiele, & Bollag, 2002). The polymerization and oligomerization supported the laccase-mediated bioremediation (Cabana et al., 2007). The azo dyes such as methyl orange was found to be removed by the coupling reaction of laccase from aromatic amines with catechol (Zille, Munteanu, Gübitz, & Cavaco-Paulo, 2005). The food industrial wastewater has also been treated through the laccase in the coupling reactions (Minussi, Pastore, & Durán, 2002).

The bacteria, fungi, and yeast secrete enzymes to process microbial degradation of environmental pollutants. The polymeric chain cleavage occurs in early stage of biodegradation for the conversion of polymers into the smaller portions (Asgher, Bhatti, Ashraf, & Legge, 2008). In next step, mineralization occurs on the smaller mono/oligomers after their transportation to the cells. Various substances are also generated during this step such as the adenosine triphosphate, water, salts, CO_2, CH_4, N_2, H_2, and minerals. The enzymes are responsible for the process of biodegradation despite of their chemical nature and reliance on the environment. Enzymes are suitable to the environment due to their biological catalytic property and ability to boost the rate of chemical reaction (Bilal, Adeel, Rasheed, Zhao, & Iqbal, 2019). The molecular weight ranging from several thousand to several million g/mol has observed in different proteins like polypeptides. The presence of active site is another important feature which is responsible for the interaction between enzymes and substrate. Some enzymes are effective for the series of substrates while the other are impactful only for particular type of substrate (Qamar et al., 2020). The enzymatic activity is also supported by some other substances such as the vitamins, ATP, and metal ions. Therefore different mechanisms of catalysis have been explored due to the great variety of enzymes. Enzyme hydrolysis and oxidation are two specific steps for the degradation mechanism. Many enzymes are also involved in the ester bond cleavage for the polymer degradation through hydrolysis. Different active sites are considered for the different types of enzymes. The splitting of ester linkage is possible by the action of esterase enzyme (Siracusa, 2019). Therefore the isolation of effective genes and the improvement of their pollutant degradation potential has emerged as an interesting area of R&D.

The implantation requires the framework of synthetic genetic circuit. The recombinant DNA advisory committee recommended *P. putida* as a HVB (host vector biosafety) strain. It is also considered as safe to release in surrounding due to its recommendations as generally recognized as safe (GRAS). *P. putida* exhibit an ability to tolerate the extreme conditions of pH, temperature, toxins, solvents, oxidative stress, and osmotics, and these features enable its ideal candidature for next generation of synthetic biology framework. Furthermore, the low nutrient requirements and flexible metabolism are also particular features of *P. putida* (Gong et al., 2018; Khraisheh, Al-Ghouti, & AlMomani, 2020). Therefore the environmental bioremediation applications could be achieved by the utilization of such organism as the microbial biosorbent model (Tanveer et al., 2018). The degradation of persistent compounds is also possible by *P. putida* synthetic genetic circuit due to the creation of design including the promoter genes and expression of the gene (Adams, 2016). The bioremediation pathways can be developed by the synthetic promoters of *P. putida*, designed through the incorporation of genome with reporter system considered as an extension of synthetic biology.

The cell growth and response to external factors like pH, temperature, light, and oxygen can be controlled by the cellular system of microbial cells that also act as beneficial for them (Tropel & Van Der Meer, 2004). The amounts of various existed persistent compounds will be responded by the outside surroundings of microbes residing the polluted sites (Ray, Panjikar, & Anand, 2018; Antonacci & Scognamiglio, 2020).

The cell biosensors are catching the focus of modern research works due to their ability to analyze the presence, type, and biodegradation potential of xenobiotic compounds such as pharmaceutical residues, pesticides, paraffin, PAHs, and PCBs (Wynn, Deo, & Daunert, 2017; Heng, Ooi, Mori, & Futra, 2018; Patel et al., 2019). The detection of specific contaminants via transducer exhibits the color signal by the reporter proteins acting as microbe (Zhang & Liu, 2016). An enhanced contact must be existing between microbe and contaminant to aim the biosensor for the identification and bioremediation (Dhar, Roy, & Nigam, 2019). This would aid to encode the genes for utilizing the recalcitrant compounds as substrate and to bacterium to maintain their cellular pathways in response to the outside environment (Bilal & Iqbal, 2019; Skinder, Uqab, & Ganai, 2020). The genetic circuits are improved against the outside environmental toxins. Therefore the schemes of synthetic biology are accessible for the removal of specific toxic compound (Tay, Nguyen, & Joshi, 2017). The cell responds to the surroundings alternations by the action of regulatory system of the environmental changes. The response regulator and histidine kinase are the main components of a prokaryotic TCRS. There is an integral protein formed by the histidine kinase sensor domain with an extracellular loop present. There is also a high conserved domain, part of HK act as a transmitter domain in the last cytoplasmic transmembrane. Therefore the phosphate is added to the conserved histidine due to ability of HK to detect the external surrounding changes. The phosphorylation of aspartate residues is also carried out by the regulation of RR due to HK. Therefore the activation of gene expression is conducted because of the promoter binding (Ravikumar et al., 2017).

19.16 Electrochemical and microbial treatment of dye-containing wastewaters

Bacteria are not able to degrade majority of azo dyes of traditional aerobic wastewater treatment plants, however, the sewage sewer line physically absorb the azo dyes about 40%–80% (Pagga & Taeger, 1994). The separation of bacteria with aerobic azo reductase is complicated and the aforementioned problem is corelated with the observed issues during this isolation. Therefore different modern physicochemical treatments are required for textile wastewater because of the inefficiency of traditional modes toward aerobic sewage-treatment systems for the decolorization of effluents containing azo dyes (Schönberger, 1997). The effluents of the textile industry are quite variable due to the utilization of structurally diverse dyes within a short duration. Therefore for the treatment of textile wastewater, a large number of unspecified processes are required. Even though, for the treatment of synthetic dyes, a stable operation of continuous fungal bioreactors has been studied. However, the issues have not been resolved such as the control of biomass, nature of synthetic dyes, and large-scale applications of white-rot fungi for the removal of dyes from textile wastewater (Elisangela et al., 2009). To overcome these problems, the immobilized enzymes are being utilized (Ji, Nguyen, Hou, Hai, & Chen, 2017). The oxidation of a large number of pollutants is possible by lactase enzyme that has received the amazing limelight. The decolorization of azo dyes in sewage-treatment systems is also achieved by the bacteria responsible for the anaerobic reduction of the azo bond. The anaerobic treatment of azo compounds is not applicable for the amines and they do not metabolize easily (Brown & Hamburger 1987).

The degradation of azo dyes can be achieved by the combination of an anaerobic with an aerobic system indicated by the large number of studies (Libra, Borchert, Vigelahn, & Storm, 2004; Mohanty, Dafale, & Rao, 2006). The production of colorless toxic amines is resulted by the reductive cleavage of the azo bond under anaerobic conditions which is considered as the nonparticular and extracellular process (Van der Zee & Villaverde, 2005). However, the anerobic process is considered during the biodegradation of aromatic amines. Application of this technique was firstly considered for the Mordant Yellow 3 and sulfonated azo dye (Haug et al., 1991). From that time for various anaerobic/aerobic systems, distinct reactor configurations have been studied such as the rotating biological contactors for aerobic treatment, anaerobic baffled reactors for anaerobic processes, activated sludge, and anaerobic high-rate reactors, such as up flow anaerobic sludge blanket, fixed film, and rotating biological contactors (Dos Santos, Cervantes, & Van Lier, 2007). The 70%–95% decolorization was noticed in anaerobic regions in case of simultaneous treatment systems (Pearce, Lloyd, & Guthrie, 2003). A complete decolorization of the dyes can be achieved in anaerobic stage for the continuous anaerobic/aerobic systems supplied with the low dye concentrations to the anaerobic stage, and substrate mixtures with biochemical oxygen demand (BOD), and chemical oxygen demand (COD). The treatment of wastewater from a dyeing factory on a laboratory scale was also demonstrated with identical results (Zaoyan et al., 1992).

The removal of synthetic dyes from the aqueous solution has been attained through different techniques such as filtration, electrocoagulation, ozonation, photocatalysis, membrane filtration, adsorption, coagulation, photolysis, sonolysis, biodegradation, and wet-land treatment. The details discussion is present along with the pros and cons of dye removal methods (Martínez-Huitle & Brillas, 2009). The processes are divided into two categories, degradative (chemical and biological) methods and separative (physical and physicochemical) methods based on the principal mechanisms behind the removal of dyes. Mostly the separation process is used for the removal of dyes, but there are some cons of this method such as the concentrated dye solution in membrane processes, disposal of dye-containing sludge as in coagulation process and dye adsorbents. As compared to this, series of chemical degradation methods are used for the complex compounds. The mineralization of organic pollutants utilizes the modern oxidation processes for the advance degradation procedure until the complete oxidation degree. These ways synthesize the final products from the various ions (following heteroatoms presents at the starting organic pollutants) along with the carbon dioxide and water. During recent years, the modern oxidation ways have attained the outstanding focus for the excellent degradation of dyes apart from the various degradation techniques (Martínez-Huitle & Brillas, 2009; Oturan & Aaron, 2014).

These procedures include the synthesis of powerful reactive oxidants primarily the hydroxyl radical with redox potential of 2.8 V and is considered as the second highest oxidizing agent after the fluorine. These radicals utilize three different ways including hydroxylation (electrophilic addition to π systems), redox reaction (electron transfer) and dehydrogenation (H-atom abstraction) to attack the organic pollutants after their in situ production. The mitigation of organic pollutants including the synthetic dyes is also feasible by the electrochemical advanced oxidation processes that have gained much interest. The diffusion rate of organic pollutants depends on the active sites of anode and applied current density, production rate, and extent depend on the nature (catalytic activity) of anode material (Panizza & Cerisola, 2009; Miled, Said, & Roudesli, 2010). There is no use of external reagents for the synthesis of hydroxyl radicals which is considered as the greatest benefit of this procedure. The external addition of one of the reagents (H_2O_2 or ferrous iron) is responsible for the synthesis of hydroxyl radicals based on the Fenton chemistry including in situ electrochemical generation in case of indirect EAOPs. The use of a sufficient electrode material is required for the H_2O_2 in situ production at the cathode. The reaction between the external addition of H_2O_2/ferrous iron or their synthesis is responsible for the formation of hydroxyl radicals (Brillas, Sirés, & Oturan, 2009). In contrast to the conventional treatment techniques, EAOPs have many benefits. The reduction in energy consumption and process efficiency of wastewater could be improved. There are also some benefits such as the amenability of automation, higher pollutant removal ability, versatile nature, and functional safety (Zhou, Liu, Jiao, Wang, & Tan, 2011).

19.17 Conclusions

With the growing usage of synthetic chemicals in agricultural, industrial, and medical sectors, a large number of organic and inorganic pollutants are exposed to the environment. The concentration of such chemicals in soil and water bodies varies from one place to another. Therefore it is highly recommended to degrade these environmental pollutants using sustainable and ecofriendly approach. Conventional approaches through microbial remediation are found to be inefficient because only a few microbes can be applied for the treatment of contaminated sites. Therefore the development of modern techniques on the basis of microbial engineering can improve the remediation of synthetic chemicals both in situ and ex situ. Microbial engineering is highly efficient for the remediation of organic and environmental pollutants in the environment. Further investigations to get scientific understanding of environmental pollutants and their removal using effective technology are still under processing. Harnessing metabolism seems to be sustainable to reduce the contaminants present in soil and water bodies. Detailed understanding and research work on microbial technology is extremely vital to introduce new strategies for the management and bioremediation of persistent environmental pollutants.

References

Abhilash, P., Dubey, R. K., Tripathi, V., Srivastava, P., Verma, J. P., & Singh, H. (2013). Remediation and management of POPs-contaminated soils in a warming climate: Challenges and perspectives. *Environmental Science and Pollution Research, 20*(8), 5879–5885.

Acevedo, F., Pizzul, L., González, M., Cea, M., Gianfreda, L., & Diez, M. (2010). Degradation of polycyclic aromatic hydrocarbons by free and nanoclay-immobilized manganese peroxidase from *Anthracophyllum* discolor. *Chemosphere, 80*(3), 271–278.

Adams, B.L. (2016). The next generation of synthetic biology chassis: Moving synthetic biology from the laboratory to the field.

Agency, E.E. (2007). Progress in management of contaminated sites (CSI 015). Europe Environmental Assessment Agency, Kongan, 6DK-1050, Denmark.

Aguilar-Pérez, K. M., Heya, M. S., Parra-Saldívar, R., & Iqbal, H. M. N. (2020). Nano-biomaterials in-focus as sensing/detection cues for environmental pollutants. *Case Studies in Chemical and Environmental Engineering, 2*, 100055.

Aguilar-Pérez, K. M., Avilés-Castrillo, J. I., Ruiz-Pulido, G., Medina, D. I., Parra-Saldivar, R., & Iqbal, H. M. N. (2021). Nanoadsorbents in focus for the remediation of environmentally-related contaminants with rising toxicity concerns. *Science of The Total Environment, 779*, 146465.

Ahmad, R., & Sardar, M. (2015). Enzyme immobilization: An overview on nanoparticles as immobilization matrix. *Biochemistry and Analytical Biochemistry, 4*(2), 1.

Akhtar, S., Khan, A. A., & Husain, Q. (2005). Potential of immobilized bitter gourd (*Momordica charantia*) peroxidases in the decolorization and removal of textile dyes from polluted wastewater and dyeing effluent. *Chemosphere, 60*(3), 291–301.

Antonacci, A., & Scognamiglio, V. (2020). Biotechnological advances in the design of algae-based biosensors. *Trends in Biotechnology, 38*(3), 334–347.

Arslan, M., Imran, A., Khan, Q. M., & Afzal, M. (2017). Plant–bacteria partnerships for the remediation of persistent organic pollutants. *Environmental Science and Pollution Research, 24*(5), 4322–4336.

Asgher, M., Bhatti, H. N., Ashraf, M., & Legge, R. L. (2008). Recent developments in biodegradation of industrial pollutants by white rot fungi and their enzyme system. *Biodegradation, 19*(6), 771.

Bartley, B. A., Kim, K., Medley, J. K., & Sauro, H. M. (2017). Synthetic biology: Engineering living systems from biophysical principles. *Biophysical Journal, 112*(6), 1050–1058.

Beale, D. J., Karpe, A. V., & Ahmed, W. (2016). *Beyond metabolomics: A review of multi-omics-based approaches. Microbial metabolomics: Applications in clinical, environmental, and industrial microbiology* (pp. 289–312). Cham: Springer.

Beil, S., Mason, J. R., Timmis, K. N., & Pieper, D. H. (1998). Identification of chlorobenzene dioxygenase sequence elements involved in dechlorination of 1, 2, 4, 5-tetrachlorobenzene. *Journal of Bacteriology, 180*(21), 5520–5528.

Benedetti, I., de Lorenzo, V., & Nikel, P. I. (2016). Genetic programming of catalytic *Pseudomonas putida* biofilms for boosting biodegradation of haloalkanes. *Metabolic Engineering, 33*, 109–118.

Bhatt, P., Bhatt, K., Huang, Y., Lin, Z., & Chen, S. (2020). Esterase is a powerful tool for the biodegradation of pyrethroid insecticides. *Chemosphere, 244*, 125507.

Bhatt, P., Gangola, S., Chaudhary, P., Khati, P., Kumar, G., Sharma, A., & Srivastava, A. (2019). Pesticide induced up-regulation of esterase and aldehyde dehydrogenase in indigenous *Bacillus* spp. *Bioremediation Journal, 23*(1), 42–52.

Bhatt, P., Huang, Y., Zhang, W., Sharma, A., & Chen, S. (2020). Enhanced cypermethrin degradation kinetics and metabolic pathway in *Bacillus thuringiensis* strain SG4. *Microorganisms, 8*(2), 223.

Bhatt, P., Pal, K., Bhandari, G., & Barh, A. (2019). Modelling of the methyl halide biodegradation in bacteria and its effect on environmental systems. *Pesticide Biochemistry and Physiology, 158*, 88−100.

Bhatt, P., Sethi, K., Gangola, S., Bhandari, G., Verma, A., Adnan, M., ... Chaube, S. (2020). Modeling and simulation of atrazine biodegradation in bacteria and its effect in other living systems. *Journal of Biomolecular Structure and Dynamics, 12*, 1−11.

Bidmanova, S., Kotlanova, M., Rataj, T., Damborsky, J., Trtilek, M., & Prokop, Z. (2016). Fluorescence-based biosensor for monitoring of environmental pollutants: From concept to field application. *Biosensors and Bioelectronics, 84*, 97−105.

Bilal, M., Adeel, M., Rasheed, T., Zhao, Y., & Iqbal, H. M. (2019). Emerging contaminants of high concern and their enzyme-assisted biodegradation—A review. *Environment International, 124*, 336−353.

Bilal, M., & Iqbal, H. M. (2019). Microbial-derived biosensors for monitoring environmental contaminants: Recent advances and future outlook. *Process Safety and Environmental Protection, 124*, 8−17.

Bilal, M., & Iqbal, H. M. (2020). Microbial bioremediation as a robust process to mitigate pollutants of environmental concern. *Case Studies in Chemical and Environmental Engineering, 2*, 100011.

Bird, H. Z., & Hopkins, R. Z. (2016). Nanomaterials for selective superoxide dismutation. *Reactive Oxygen Species, 1*(1), 59−64.

Bonifay, V., Aydin, E., Aktas, D. F., Sunner, J., & Suflita, J. M. (2016). *Metabolic profiling and metabolomic procedures for investigating the biodegradation of hydrocarbons. Hydrocarbon and lipid microbiology protocols* (pp. 111−161). Berlin, Heidelberg: Springer.

Booth, S. C., Weljie, A. M., & Turner, R. J. (2015). Metabolomics reveals differences of metal toxicity in cultures of *Pseudomonas pseudoalcaligenes* KF707 grown on different carbon sources. *Frontiers in Microbiology, 6*, 827.

Brillas, E., Sirés, I., & Oturan, M. A. (2009). Electro-Fenton process and related electrochemical technologies based on Fenton's reaction chemistry. *Chemical Reviews, 109*(12), 6570−6631.

Brown, D., & Hamburger, B. (1987). The degradation of dyestuffs: Part III-Investigations of their ultimate degradability. *Chemosphere, 16*(7), 1539−1553.

Cabana, H., Jiwan, J. L. H., Rozenberg, R., Elisashvili, V., Penninckx, M., Agathos, S. N., & Jones, J. P. (2007). Elimination of endocrine disrupting chemicals nonylphenol and bisphenol A and personal care product ingredient triclosan using enzyme preparation from the white rot fungus *Coriolopsis polyzona*. *Chemosphere, 67*(4), 770−778.

Castillo, N. E. T., Melchor-Martínez, E. M., Sierra, J. S. O., Ramírez-Torres, N. M., Sosa-Hernández, J. E., Iqbal, H. M. N., & Parra-Saldívar, R. (2021). Enzyme mimics in-focus: Redefining the catalytic attributes of artificial enzymes for renewable energy production. *International Journal of Biological Macromolecules, 179*, 80−89.

Chakraborty, J., & Das, S. (2017). Application of spectroscopic techniques for monitoring microbial diversity and bioremediation. *Applied Spectroscopy Reviews, 52*(1), 1−38.

Chang, D., Chen, T., Liu, H., Xi, Y., Qing, C., Xie, Q., & Frost, R. L. (2014). A new approach to prepare ZVI and its application in removal of Cr (VI) from aqueous solution. *Chemical Engineering Journal, 244*, 264−272.

Chen, W., Mulchandani, A., & Deshusses, M. A. (2005). Environmental biotechnology: Challenges and opportunities for chemical engineers. *AIChE Journal, 51*(3), 690−695.

Chiu, Y.-H., Chang, T.-F. M., Chen, C.-Y., Sone, M., & Hsu, Y.-J. (2019). Mechanistic insights into photodegradation of organic dyes using heterostructure photocatalysts. *Catalysts, 9*(5), 430.

Choi, S. (2015). Microscale microbial fuel cells: Advances and challenges. *Biosensors and Bioelectronics, 69*, 8−25.

Colmenares, J. C., & Xu, Y.-J. (2016). *Heterogeneous photocatalysis: Green chemistry and sustainable technology*. Springer.

Corici, L., Ferrario, V., Pellis, A., Ebert, C., Lotteria, S., Cantone, S., ... Gardossi, L. (2016). Large scale applications of immobilized enzymes call for sustainable and inexpensive solutions: Rice husks as renewable alternatives to fossil-based organic resins. *RSC Advances, 6*(68), 63256−63270.

Cregut, M., Bedas, M., Durand, M.-J., & Thouand, G. (2013). New insights into polyurethane biodegradation and realistic prospects for the development of a sustainable waste recycling process. *Biotechnology Advances, 31*(8), 1634−1647.

Curran, K. A., & Alper, H. S. (2012). Expanding the chemical palate of cells by combining systems biology and metabolic engineering. *Metabolic Engineering, 14*(4), 289−297.

Dangi, A. K., Sharma, B., Hill, R. T., & Shukla, P. (2019). Bioremediation through microbes: Systems biology and metabolic engineering approach. *Critical Reviews in Biotechnology, 39*(1), 79−98.

Dash, H. R., & Das, S. (2015). Bioremediation of inorganic mercury through volatilization and biosorption by transgenic *Bacillus cereus* BW-03 (pPW-05). *International Biodeterioration & Biodegradation, 103*, 179−185.

Dhanya, M. (2014). Advances in microbial biodegradation of chlorpyrifos. *Journal of Environmental Research and Development, 9*(1), 232−240.

Dhar, D., Roy, S., & Nigam, V. K. (2019). *Advances in protein/enzyme-based biosensors for the detection of pharmaceutical contaminants in the environment. Tools, techniques and protocols for monitoring environmental contaminants* (pp. 207−229). Elsevier.

Dionysiou, D. D., Puma, G. L., Ye, J., Schneider, J., & Bahnemann, D. (2016). *Photocatalysis: Applications*. Royal Society of Chemistry.

Domínguez-Garay, A., Quejigo, J. R., Dörfler, U., Schroll, R., & Esteve-Núñez, A. (2018). Bioelectroventing: An electrochemical-assisted bioremediation strategy for cleaning-up atrazine-polluted soils. *Microbial Biotechnology, 11*(1), 50−62.

Dos Santos, A. B., Cervantes, F. J., & Van Lier, J. B. (2007). Review paper on current technologies for decolourisation of textile wastewaters: Perspectives for anaerobic biotechnology. *Bioresource Technology, 98*(12), 2369−2385.

Durán, A., Monteagudo, J., Sanmartín, I., & García-Díaz, A. (2013). Sonophotocatalytic mineralization of antipyrine in aqueous solution. *Applied Catalysis B: Environmental, 138*, 318−325.

Dvořák, P., Nikel, P. I., Damborský, J., & de Lorenzo, V. (2017). Bioremediation 3.0: Engineering pollutant-removing bacteria in the times of systemic biology. *Biotechnology Advances, 35*(7), 845−866.

Ejaz, S., Akram, W., Lim, C. W., Lee, J. J., & Hussain, I. (2004). Endocrine disrupting pesticides: A leading cause of cancer among rural people in Pakistan. *Experimental Oncology, 26*(2), 98−105.

Elbanna, K., Hassan, G., Khider, M., & Mandour, R. (2010). Safe biodegradation of textile azo dyes by newly isolated lactic acid bacteria and detection of plasmids associated with degradation. *Journal of Bioremedediation & Biodegradation, 1*, 112.

Elisangela, F., Andrea, Z., Fabio, D. G., de Menezes Cristiano, R., Regina, D. L., & Artur, C. P. (2009). Biodegradation of textile azo dyes by a facultative *Staphylococcus arlettae* strain VN-11 using a sequential microaerophilic/aerobic process. *International Biodeterioration & Biodegradation, 63*(3), 280−288.

Farnet, A., Gil, G., Ruaudel, F., Chevremont, A., & Ferre, E. (2009). Polycyclic aromatic hydrocarbon transformation with laccases of a white-rot fungus isolated from a Mediterranean schlerophyllous litter. *Geoderma, 149*(3−4), 267−271.

Fukuda, T., Uchida, H., Takashima, Y., Uwajima, T., Kawabata, T., & Suzuki, M. (2001). Degradation of bisphenol A by purified laccase from *Trametes villosa*. *Biochemical and Biophysical Research Communications, 284*(3), 704−706.

Gan, Y. X., Gan, B. J., Clark, E., Su, L., & Zhang, L. (2012). Converting environmentally hazardous materials into clean energy using a novel nanostructured photoelectrochemical fuel cell. *Materials Research Bulletin, 47*(9), 2380−2388.

Gao, L., & Yan, X. (2016). Nanozymes: An emerging field bridging nanotechnology and biology. *Science China Life Sciences, 59*(4), 400–402.

Gaya, U. I., & Abdullah, A. H. (2008). Heterogeneous photocatalytic degradation of organic contaminants over titanium dioxide: A review of fundamentals, progress and problems. *Journal of Photochemistry and Photobiology C: Photochemistry Reviews, 9*(1), 1–12.

Gholami-Borujeni, F., Mahvi, A. H., Naseri, S., Faramarzi, M. A., Nabizadeh, R., & Alimohammadi, M. (2011). Application of immobilized horseradish peroxidase for removal and detoxification of azo dye from aqueous solution. *Research Journal of Chemistry and Environment, 15*(2), 217–222.

Gong, T., Xu, X., Dang, Y., Kong, A., Wu, Y., Liang, P., . . . Yang, C. (2018). An engineered *Pseudomonas putida* can simultaneously degrade organophosphates, pyrethroids and carbamates. *Science of the Total Environment, 628*, 1258–1265.

Haug, W., Schmidt, A., Nörtemann, B., Hempel, D. C., Stolz, A., & Knackmuss, H. J. (1991). Mineralization of the sulfonated azo dye Mordant Yellow 3 by a 6-aminonaphthalene-2-sulfonate-degrading bacterial consortium. *Applied and Environmental Microbiology, 57*(11), 3144–3149.

Bremner, D. H., Burgess, A. E., & Chand, R. (2011). The chemistry of ultrasonic degradation of organic compounds. *Current Organic Chemistry, 15*(2), 168–177.

Heng, L. Y., Ooi, L., Mori, I. C., & Futra, D. (2018). *Environmental toxicity and evaluation. Environmental risk analysis for Asian-oriented, risk-based watershed management* (pp. 71–94). Singapore: Springer.

Henry, C. S., DeJongh, M., Best, A. A., Frybarger, P. M., Linsay, B., & Stevens, R. L. (2010). High-throughput generation, optimization and analysis of genome-scale metabolic models. *Nature Biotechnology, 28*(9), 977–982.

Huang, Z., Gao, Z., Gao, S., Wang, Q., Wang, Z., Huang, B., & Dai, Y. (2017). Facile synthesis of S-doped reduced TiO_{2-x} with enhanced visible-light photocatalytic performance. *Chinese Journal of Catalysis, 38*(5), 821–830.

Jain, M., Yadav, P., Joshi, A., & Kodgire, P. (2019). Advances in detection of hazardous organophosphorus compounds using organophosphorus hydrolase based biosensors. *Critical Reviews in Toxicology, 49*(5), 387–410.

Jaiswal, S., & Shukla, P. (2020). Alternative strategies for microbial remediation of pollutants via synthetic biology. *Frontiers in Microbiology, 11*, 808.

Janssen, D. B., & Stucki, G. (2020). Perspectives of genetically engineered microbes for groundwater bioremediation. *Environmental Science: Processes & Impacts, 22*(3), 487–499.

Jayaraj, R., Megha, P., & Sreedev, P. (2016). Organochlorine pesticides, their toxic effects on living organisms and their fate in the environment. *Interdisciplinary Toxicology, 9*(3–4), 90–100.

Jeandet, P., Sobarzo-Sánchez, E., Clément, C., Nabavi, S. F., Habtemariam, S., Nabavi, S. M., & Cordelier, S. (2018). Engineering stilbene metabolic pathways in microbial cells. *Biotechnology Advances, 36*(8), 2264–2283.

Ji, C., Nguyen, L. N., Hou, J., Hai, F. I., & Chen, V. (2017). Direct immobilization of laccase on titania nanoparticles from crude enzyme extracts of *P. ostreatus* culture for micro-pollutant degradation. *Separation and Purification Technology, 178*, 215–223.

Jiang, C., Yang, Q., Wang, D., Zhong, Y., Chen, F., Li, X., . . . Shang, M. (2017). Simultaneous perchlorate and nitrate removal coupled with electricity generation in autotrophic denitrifying biocathode microbial fuel cell. *Chemical Engineering Journal, 308*, 783–790.

Jiang, W., Gao, Q., Zhang, L., Wang, H., Zhang, M., Liu, X., . . . Qiu, J. (2020). Identification of the key amino acid sites of the carbofuran hydrolase CehA from a newly isolated carbofuran-degrading strain Sphingbium sp. CFD-1. *Ecotoxicology and Environmental Safety, 189*, 109938.

Johnsen, A.R., Wick, L.Y., and Harms, H. (2005). Environmental Pollution (Barking, Essex: 1987), 133, 71–84.

Ju, K.-S., & Parales, R. E. (2006). Control of substrate specificity by active-site residues in nitrobenzene dioxygenase. *Applied and Environmental Microbiology, 72*(3), 1817–1824.

Karn, B., Kuiken, T., & Otto, M. (2009). Nanotechnology and in situ remediation: A review of the benefits and potential risks. *Environmental Health Perspectives, 117*(12), 1813−1831.

Karp, P. D., Ouzounis, C. A., Moore-Kochlacs, C., Goldovsky, L., Kaipa, P., Ahrén, D., . . . López-Bigas, N. (2005). Expansion of the BioCyc collection of pathway/genome databases to 160 genomes. *Nucleic Acids Research, 33*(19), 6083−6089.

Kazlauskas, R. J., & Bornscheuer, U. T. (2009). Finding better protein engineering strategies. *Nature Chemical Biology, 5*(8), 526−529.

Khaki, M. R. D., Shafeeyan, M. S., Raman, A. A. A., & Daud, W. M. A. W. (2017). Application of doped photocatalysts for organic pollutant degradation—A review. *Journal of Environmental Management, 198*, 78−94.

Khan, A. A., Akhtar, S., & Husain, Q. (2006). Direct immobilization of polyphenol oxidases on Celite 545 from ammonium sulphate fractionated proteins of potato (*Solanum tuberosum*). *Journal of Molecular Catalysis B: Enzymatic, 40*(1−2), 58−63.

Khan, F., Sajid, M., & Cameotra, S. S. (2013). In silico approach for the bioremediation of toxic pollutants. *Journal of Petroleum & Environmental Biotechnology, 4*(161), 2.

Khraisheh, M., Al-Ghouti, M. A., & AlMomani, F. (2020). *P. putida* as biosorbent for the remediation of cobalt and phenol from industrial waste wastewaters. *Environmental Technology & Innovation, 20*, 101148.

Klassen, A., Faccio, A. T., Canuto, G. A. B., da Cruz, P. L. R., Ribeiro, H. C., Tavares, M. F. M., & Sussulini, A. (2017). Metabolomics: Definitions and significance in systems biology. *Metabolomics: From Fundamentals to Clinical Applications, 965*, 3−17.

Koe, W. S., Lee, J. W., Chong, W. C., Pang, Y. L., & Sim, L. C. (2020). An overview of photocatalytic degradation: Photocatalysts, mechanisms, and development of photocatalytic membrane. *Environmental Science and Pollution Research, 27*(3), 2522−2565.

Konig, L., Hartz, P., Bernhardt, R., & Hannemann, F. (2019). High-yield C11-oxidation of hydrocortisone by establishment of an efficient whole-cell system in *Bacillus megaterium*. *Metabolic Engineering, 55*, 59−67.

Koutsospyros, A., Pavlov, J., Fawcett, J., Strickland, D., Smolinski, B., & Braida, W. (2012). Degradation of high energetic and insensitive munitions compounds by Fe/Cu bimetal reduction. *Journal of Hazardous Materials, 219*, 75−81.

Kuchner, O., & Arnold, F. H. (1997). Directed evolution of enzyme catalysts. *Trends in Biotechnology, 15*(12), 523−530.

Kumar, A., Kumar, A., Singh, R., Singh, R., Pandey, S., Rai, A., . . . Rahul, B. (2020). *Genetically engineered bacteria for the degradation of dye and other organic compounds. Abatement of environmental pollutants* (pp. 331−350). Elsevier.

Kurwadkar, S. (2019). Occurrence and distribution of organic and inorganic pollutants in groundwater. *Water Environment Research, 91*(10), 1001−1008.

Leong, T., Ashokkumar, M., & Kentish, S. (2011). The fundamentals of power ultrasound—A review.

Letunic, I., Copley, R. R., Pils, B., Pinkert, S., Schultz, J., & Bork, P. (2006). SMART 5: Domains in the context of genomes and networks. *Nucleic Acids Research, 34*(suppl_1), D257−D260.

Li, X.-Q., & Zhang, W.-X. (2007). Sequestration of metal cations with zerovalent iron nanoparticles a study with high resolution X-ray photoelectron spectroscopy (HR-XPS). *The Journal of Physical Chemistry C, 111*(19), 6939−6946.

Li, Y.-H., Di, Z., Ding, J., Wu, D., Luan, Z., & Zhu, Y. (2005). Adsorption thermodynamic, kinetic and desorption studies of Pb2+ on carbon nanotubes. *Water Research, 39*(4), 605−609.

Liang, H., Lin, F., Zhang, Z., Liu, B., Jiang, S., Yuan, Q., & Liu, J. (2017). Multicopper laccase mimicking nanozymes with nucleotides as ligands. *ACS Applied Materials & Interfaces, 9*(2), 1352−1360.

Liang, Y., Jiao, S., Wang, M., Yu, H., & Shen, Z. (2020). A CRISPR/Cas9-based genome editing system for *Rhodococcus ruber* TH. *Metabolic Engineering, 57*, 13−22.

Libra, J. A., Borchert, M., Vigelahn, L., & Storm, T. (2004). Two stage biological treatment of a diazo reactive textile dye and the fate of the dye metabolites. *Chemosphere, 56*(2), 167–180.

Lin, Z., Zhang, W., Pang, S., Huang, Y., Mishra, S., Bhatt, P., & Chen, S. (2020). Current approaches to and future perspectives on methomyl degradation in contaminated soil/water environments. *Molecules (Basel, Switzerland), 25*(3), 738.

Lu, L., Huggins, T., Jin, S., Zuo, Y., & Ren, Z. J. (2014). Microbial metabolism and community structure in response to bioelectrochemically enhanced remediation of petroleum hydrocarbon-contaminated soil. *Environmental Science & Technology, 48*(7), 4021–4029.

Lu, L., Yazdi, H., Jin, S., Zuo, Y., Fallgren, P. H., & Ren, Z. J. (2014). Enhanced bioremediation of hydrocarbon-contaminated soil using pilot-scale bioelectrochemical systems. *Journal of Hazardous Materials, 274*, 8–15.

Ma, T., Zhu, L., Wang, J., Wang, J., Xie, H., Su, J., . . . Shao, B. (2011). Enhancement of atrazine degradation by crude and immobilized enzymes in two agricultural soils. *Environmental Earth Sciences, 64*(3), 861–867.

Martínez-Huitle, C. A., & Brillas, E. (2009). Decontamination of wastewaters containing synthetic organic dyes by electrochemical methods: A general review. *Applied Catalysis B: Environmental, 87*(3–4), 105–145.

Mason, T.J. (2012). Trends in sonochemistry and ultrasonic processing. In *Proceedings of the AIP conference*.

Mehta, J., Bhardwaj, N., Bhardwaj, S. K., Kim, K.-H., & Deep, A. (2016). Recent advances in enzyme immobilization techniques: Metal-organic frameworks as novel substrates. *Coordination Chemistry Reviews, 322*, 30–40.

Miled, W., Said, A. H., & Roudesli, S. (2010). Decolorization of high polluted textile wastewater by indirect electrochemical oxidation process. *Journal of Textile and Apparel, Technology and Management, 6*(3).

Minussi, R. C., Pastore, G. M., & Durán, N. (2002). Potential applications of laccase in the food industry. *Trends in Food Science & Technology, 13*(6–7), 205–216.

Mishra, S., Zhang, W., Lin, Z., Pang, S., Huang, Y., Bhatt, P., & Chen, S. (2020). Carbofuran toxicity and its microbial degradation in contaminated environments. *Chemosphere, 259*, 127419.

Mohanty, S., Dafale, N., & Rao, N. N. (2006). Microbial decolorization of reactive black-5 in a two-stage anaerobic–aerobic reactor using acclimatized activated textile sludge. *Biodegradation, 17*(5), 403–413.

Nagata, Y., Ohtsubo, Y., & Tsuda, M. (2015). Properties and biotechnological applications of natural and engineered haloalkane dehalogenases. *Applied Microbiology and Biotechnology, 99*(23), 9865–9881.

Nancharaiah, Y. V., & Lens, P. N. (2015). Selenium biomineralization for biotechnological applications. *Trends in Biotechnology, 33*(6), 323–330.

Nikel, P. I., & de Lorenzo, V. (2013). Engineering an anaerobic metabolic regime in *Pseudomonas putida* KT2440 for the anoxic biodegradation of 1, 3-dichloroprop-1-ene. *Metabolic Engineering, 15*, 98–112.

Nyanhongo, G. S., Couto, S. R., & Guebitz, G. M. (2006). Coupling of 2, 4, 6-trinitrotoluene (TNT) metabolites onto humic monomers by a new laccase from Trametes modesta. *Chemosphere, 64*(3), 359–370.

Oliveira, S. F., da Luz, J. M. R., Kasuya, M. C. M., Ladeira, L. O., & Junior, A. C. (2018). Enzymatic extract containing lignin peroxidase immobilized on carbon nanotubes: Potential biocatalyst in dye decolourization. *Saudi Journal of Biological Sciences, 25*(4), 651–659.

Oturan, M. A., & Aaron, J. J. (2014). Advanced oxidation processes in water/wastewater treatment: Principles and applications. A review. *Critical Reviews in Environmental Science and Technology, 44*(23), 2577–2641.

Pagga, U., & Taeger, K. (1994). Development of a method for adsorption of dyestuffs on activated sludge. *Water Research, 28*(5), 1051–1057.

Palma, E., Daghio, M., Franzetti, A., Petrangeli Papini, M., & Aulenta, F. (2018). The bioelectric well: A novel approach for in situ treatment of hydrocarbon-contaminated groundwater. *Microbial Biotechnology, 11*(1), 112–118.

Pang, S., Lin, Z., Zhang, W., Mishra, S., Bhatt, P., & Chen, S. (2020). Insights into the microbial degradation and biochemical mechanisms of neonicotinoids. *Frontiers in Microbiology, 11*, 868.

Panizza, M., & Cerisola, G. (2009). Direct and mediated anodic oxidation of organic pollutants. *Chemical Reviews, 109*(12), 6541–6569.

Patel, R., Zaveri, P., Mukherjee, A., Agarwal, P. K., More, P., & Munshi, N. S. (2019). Development of fluorescent protein-based biosensing strains: A new tool for the detection of aromatic hydrocarbon pollutants in the environment. *Ecotoxicology and Environmental Safety, 182*, 109450.

Pazirandeh, M., Wells, B. M., & Ryan, R. L. (1998). Development of bacterium-based heavy metal biosorbents: Enhanced uptake of cadmium and mercury by *Escherichia coli* expressing a metal binding motif. *Applied and Environmental Microbiology, 64*(10), 4068–4072.

Pearce, C. I., Lloyd, J. R., & Guthrie, J. T. (2003). The removal of colour from textile wastewater using whole bacterial cells: A review. *Dyes and Pigments, 58*(3), 179–196.

Perreault, F., De Faria, A. F., & Elimelech, M. (2015). Environmental applications of graphene-based nanomaterials. *Chemical Society Reviews, 44*(16), 5861–5896.

Purohit, H. (2003). Biosensors as molecular tools for use in bioremediation. *Journal of Cleaner Production, 11*(3), 293–301.

Qamar, S. A., Asgher, M., & Bilal, M. (2020). Immobilization of alkaline protease from *Bacillus brevis* using Ca-alginate entrapment strategy for improved catalytic stability, silver recovery, and dehairing potentialities. *Catalysis Letters, 150*, 3572–3583.

Qamar, S. A., Ashiq, M., Jahangeer, M., Riasat, A., & Bilal, M. (2020). Chitosan-based hybrid materials as adsorbents for textile dyes—A review. *Case Studies in Chemical and Environmental Engineering, 2*, 100021.

Rahmani, H., Gholami, M., Mahvi, A., Alimohammadi, M., Azarian, G., Esrafili, A., & Farzadkia, M. (2014). Tinidazole removal from aqueous solution by sonolysis in the presence of hydrogen peroxide. *Bulletin of Environmental Contamination and Toxicology, 92*(3), 341–346.

Rajan, C. (2011). Nanotechnology in groundwater remediation. *International Journal of Environmental Science and Development, 2*(3), 182.

Ramos, J.-L., Marqués, S., van Dillewijn, P., Espinosa-Urgel, M., Segura, A., Duque, E., & Roca, A. (2011). Laboratory research aimed at closing the gaps in microbial bioremediation. *Trends in Biotechnology, 29*(12), 641–647.

Rao, A. N., Sivasankar, B., & Sadasivam, V. (2009). Kinetic studies on the photocatalytic degradation of Direct Yellow 12 in the presence of ZnO catalyst. *Journal of Molecular Catalysis A: Chemical, 306*(1–2), 77–81.

Rao, M., Scelza, R., Acevedo, F., Diez, M., & Gianfreda, L. (2014). Enzymes as useful tools for environmental purposes. *Chemosphere, 107*, 145–162.

Ravikumar, S., Baylon, M. G., Park, S. J., & Choi, J.-I. (2017). Engineered microbial biosensors based on bacterial two-component systems as synthetic biotechnology platforms in bioremediation and biorefinery. *Microbial Cell Factories, 16*(1), 1–10.

Ray, S., Panjikar, S., & Anand, R. (2018). Design of protein-based biosensors for selective detection of benzene groups of pollutants. *ACS Sensors, 3*(9), 1632–1638.

Reed, J. L., Vo, T. D., Schilling, C. H., & Palsson, B. O. (2003). An expanded genome-scale model of *Escherichia coli* K-12 (i JR904 GSM/GPR). *Genome Biology, 4*(9), 1–12.

Regonini, D., & Clemens, F. (2015). Anodized TiO_2 nanotubes: Effect of anodizing time on film length, morphology and photoelectrochemical properties. *Materials Letters, 142*, 97–101.

Restaino, O. F., Borzacchiello, M. G., Ilaria, S., Fedele, L., & Schiraldi, C. (2016). Biotechnological process design for the production and purification of three recombinant thermophilic phosphotriesterases. *New Biotechnology* (33), S15.

Saggioro, E. M., Oliveira, A. S., Pavesi, T., Maia, C. G., Ferreira, L. F. V., & Moreira, J. C. (2011). Use of titanium dioxide photocatalysis on the remediation of model textile wastewaters containing azo dyes. *Molecules (Basel, Switzerland), 16*(12), 10370−10386.

Sastre, D. E., Reis, E. A., & Netto, C. G. M. (2020). Strategies to rationalize enzyme immobilization procedures. *Methods in Enzymology, 630*, 81−110.

Savage, N., & Diallo, M. S. (2005). Nanomaterials and water purification: Opportunities and challenges. *Journal of Nanoparticle Research, 7*(4−5), 331−342.

Sayan, B., Indranil, S., Aniruddha, M., Dhrubajyoti, C., Uday, C., & Debashis, C. (2013). Role of nanotechnology in water treatment and purification: Potential applications and implications. *International Journal of Chemical Science and Technology, 3*(3), 59.

Schönberger, H. (1997). Modern concepts for reducing the wastewater load in the textile processing. Treatment of wastewaters from textile processing. *Schriftenreihe Biologische Abwasserreinigung, 9*, 13−24.

Schultz, A., Jonas, U., Hammer, E., & Schauer, F. (2001). Dehalogenation of chlorinated hydroxybiphenyls by fungal laccase. *Applied and Environmental Microbiology, 67*(9), 4377−4381.

Šekuljica, N. Ž., Prlainović, N. Ž., Stefanović, A. B., Žuža, M. G., Čičkarić, D. Z., Mijin, D. Ž., & Knežević-Jugović, Z. D. (2015). Decolorization of anthraquinonic dyes from textile effluent using horseradish peroxidase: Optimization and kinetic study. *The Scientific World Journal, 2015*.

Shaheen, R., Asgher, M., Hussain, F., & Bhatti, H. N. (2017). Immobilized lignin peroxidase from *Ganoderma lucidum* IBL-05 with improved dye decolorization and cytotoxicity reduction properties. *International Journal of Biological Macromolecules, 103*, 57−64.

Shao, Z., Zhao, H., Giver, L., & Arnold, F. H. (1998). Random-priming in vitro recombination: An effective tool for directed evolution. *Nucleic Acids Research, 26*(2), 681−683.

Sharma, B., Dangi, A. K., & Shukla, P. (2018). Contemporary enzyme based technologies for bioremediation: A review. *Journal of Environmental Management, 210*, 10−22.

Sheldon, R. A. (2007). Enzyme immobilization: The quest for optimum performance. *Advanced Synthesis & Catalysis, 349*(8−9), 1289−1307.

Shin, H. Y., Park, T. J., & Kim, M. I. (2015). Recent research trends and future prospects in nanozymes. *Journal of Nanomaterials, 2015*.

Singh, N., Sharma, J. K., & Santal, A. R. (2016). *Biotechnological approaches to remediate soil and water using plant−microbe interactions. Phytoremediation* (pp. 131−152). Springer.

Singh, S., Kang, S. H., Mulchandani, A., & Chen, W. (2008). Bioremediation: Environmental clean-up through pathway engineering. *Current Opinion in Biotechnology, 19*(5), 437−444.

Sinkkonen, S., & Paasivirta, J. (2000). Degradation half-life times of PCDDs, PCDFs and PCBs for environmental fate modeling. *Chemosphere, 40*(9−11), 943−949.

Siracusa, V. (2019). Microbial degradation of synthetic biopolymers waste. *Polymers, 11*(6), 1066.

Sirisha, V. L., Jain, A., & Jain, A. (2016). Enzyme immobilization: An overview on methods, support material, and applications of immobilized enzymes. *Advances in Food and Nutrition Research, 79*, 179−211.

Skinder, B. M., Uqab, B., & Ganai, B. A. (2020). *Bioremediation: A sustainable and emerging tool for restoration of polluted aquatic ecosystem. Fresh water pollution dynamics and remediation* (pp. 143−165). Singapore: Springer.

Sousa, J. C., Ribeiro, A. R., Barbosa, M. O., Pereira, M. F. R., & Silva, A. M. (2018). A review on environmental monitoring of water organic pollutants identified by EU guidelines. *Journal of Hazardous Materials, 344*, 146−162.

Steele, W. B., IV, Garcia, S. N., Huggett, D. B., Venables, B. J., Barnes, S. E., III, & La Point, T. W. (2013). Tissue-specific bioconcentration of the synthetic steroid hormone medroxyprogesterone acetate in the common carp (*Cyprinus carpio*). *Environmental Toxicology and Pharmacology, 36*(3), 1120−1126.

Tanveer, T., Shaheen, K., Parveen, S., Misbah, Z. T., Babar, M. M., & Gul, A. (2018). *Omics-based bioengineering in environmental biotechnology. Omics technologies and bio-engineering* (pp. 353–364). Academic Press.

Tay, P. K. R., Nguyen, P. Q., & Joshi, N. S. (2017). A synthetic circuit for mercury bioremediation using self-assembling functional amyloids. *ACS Synthetic Biology*, 6(10), 1841–1850.

Thiele, S., Fernandes, E., & Bollag, J. M. (2002). Enzymatic transformation and binding of labeled 2, 4, 6-trinitrotoluene to humic substances during an anaerobic/aerobic incubation. *Journal of Environmental Quality*, 31(2), 437–444.

Timmis, K. N., & Pieper, D. H. (1999). Bacteria designed for bioremediation. *Trends in Biotechnology*, 17(5), 201–204.

Tosco, T., Papini, M. P., Viggi, C. C., & Sethi, R. (2014). Nanoscale zerovalent iron particles for groundwater remediation: A review. *Journal of Cleaner Production*, 77, 10–21.

Tripathi, V., Edrisi, S. A., Chen, B., Gupta, V. K., Vilu, R., Gathergood, N., & Abhilash, P. (2017). Biotechnological advances for restoring degraded land for sustainable development. *Trends in Biotechnology*, 35(9), 847–859.

Tropel, D., & Van Der Meer, J. R. (2004). Bacterial transcriptional regulators for degradation pathways of aromatic compounds. *Microbiology and Molecular Biology Reviews*, 68(3), 474–500.

Tsekenis, G., Filippidou, M., Chatzipetrou, M., Tsouti, V., Zergioti, I., & Chatzandroulis, S. (2015). Heavy metal ion detection using a capacitive micromechanical biosensor array for environmental monitoring. *Sensors and Actuators B: Chemical*, 208, 628–635.

Tsutsumi, Y., Haneda, T., & Nishida, T. (2001). Removal of estrogenic activities of bisphenol A and nonylphenol by oxidative enzymes from lignin-degrading basidiomycetes. *Chemosphere*, 42(3), 271–276.

Van der Zee, F. P., & Villaverde, S. (2005). Combined anaerobic–aerobic treatment of azo dyes—A short review of bioreactor studies. *Water Research*, 39(8), 1425–1440.

Vidal-Limon, A., García Suárez, P. C. N., Arellano-García, E., Contreras, O. E., & Aguila, S. A. (2018). Enhanced degradation of pesticide dichlorophen by laccase immobilized on nanoporous materials: A cytotoxic and molecular simulation investigation. *Bioconjugate Chemistry*, 29(4), 1073–1080.

Von Mering, C., Jensen, L. J., Kuhn, M., Chaffron, S., Doerks, T., Krüger, B., ... Bork, P. (2007). STRING 7—Recent developments in the integration and prediction of protein interactions. *Nucleic Acids Research*, 35(suppl_1), D358–D362.

Wang, C., Deng, H., & Zhao, F. (2016). The remediation of chromium (VI)-contaminated soils using microbial fuel cells. *Soil and Sediment Contamination: An International Journal*, 25(1), 1–12.

Wang, C. J., Thiele, S., & Bollag, J. M. (2002). Interaction of 2, 4, 6-trinitrotoluene (TNT) and 4-amino-2, 6-dinitrotoluene with humic monomers in the presence of oxidative enzymes. *Archives of Environmental Contamination and Toxicology*, 42(1), 1–8.

Wang, H., Luo, H., Fallgren, P. H., Jin, S., & Ren, Z. J. (2015). Bioelectrochemical system platform for sustainable environmental remediation and energy generation. *Biotechnology Advances*, 33(3–4), 317–334.

Wang, H., & Ren, Z. J. (2013). A comprehensive review of microbial electrochemical systems as a platform technology. *Biotechnology Advances*, 31(8), 1796–1807.

Wang, H., Song, H., Yu, R., Cao, X., Fang, Z., & Li, X. (2016). New process for copper migration by bioelectricity generation in soil microbial fuel cells. *Environmental Science and Pollution Research*, 23(13), 13147–13154.

Wang, H., Yi, S., Cao, X., Fang, Z., & Li, X. (2017). Reductive dechlorination of hexachlorobenzene subjected to several conditions in a bioelectrochemical system. *Ecotoxicology and Environmental Safety*, 139, 172–178.

Wang, X., Cai, Z., Zhou, Q., Zhang, Z., & Chen, C. (2012). Bioelectrochemical stimulation of petroleum hydrocarbon degradation in saline soil using U-tube microbial fuel cells. *Biotechnology and Bioengineering*, 109(2), 426–433.

Wang, X., Hu, Y., & Wei, H. (2016). Nanozymes in bionanotechnology: From sensing to therapeutics and beyond. *Inorganic Chemistry Frontiers, 3*(1), 41−60.

Wargacki, A. J., Leonard, E., Win, M. N., Regitsky, D. D., Santos, C. N. S., Kim, P. B., ... Yoshikuni, Y. (2012). An engineered microbial platform for direct biofuel production from brown macroalgae. *Science (New York, N.Y.), 335*(6066), 308−313.

Wei, H., & Wang, E. (2008). Fe_3O_4 magnetic nanoparticles as peroxidase mimetics and their applications in H_2O_2 and glucose detection. *Analytical Chemistry, 80*(6), 2250−2254.

Wilkes, R. A., & Aristilde, L. (2017). Degradation and metabolism of synthetic plastics and associated products by *Pseudomonas* sp.: Capabilities and challenges. *Journal of Applied Microbiology, 123*(3), 582−593.

Wu, M., Li, W., Dick, W. A., Ye, X., Chen, K., Kost, D., & Chen, L. (2017). Bioremediation of hydrocarbon degradation in a petroleum-contaminated soil and microbial population and activity determination. *Chemosphere, 169*, 124−130.

Wynn, D., Deo, S., & Daunert, S. (2017). Engineering rugged field assays to detect hazardous chemicals using spore-based bacterial biosensors. *Methods in Enzymology, 589*, 51−85.

Xiao, R., Wei, Z., Chen, D., & Weavers, L. K. (2014). Kinetics and mechanism of sonochemical degradation of pharmaceuticals in municipal wastewater. *Environmental Science & Technology, 48*(16), 9675−9683.

Xie, J., Zhang, X., Wang, H., Zheng, H., & Huang, Y. (2012). Analytical and environmental applications of nanoparticles as enzyme mimetics. *TrAC Trends in Analytical Chemistry, 39*, 114−129.

Zaoyan, Y., Ke, S., Guangliang, S., Fan, Y., Jinshan, D., & Huanian, M. (1992). Anaerobic−aerobic treatment of a dye wastewater by combination of RBC with activated sludge. *Water Science and Technology, 26*(9−11), 2093−2096.

Zhan, H., Huang, Y., Lin, Z., Bhatt, P., & Chen, S. (2020). New insights into the microbial degradation and catalytic mechanism of synthetic pyrethroids. *Environmental Research, 182*, 109138.

Zhang, D., & Liu, Q. (2016). Biosensors and bioelectronics on smartphone for portable biochemical detection. *Biosensors and Bioelectronics, 75*, 273−284.

Zhang, H., Zhang, S., He, F., Qin, X., Zhang, X., & Yang, Y. (2016). Characterization of a manganese peroxidase from white-rot fungus *Trametes* sp. 48424 with strong ability of degrading different types of dyes and polycyclic aromatic hydrocarbons. *Journal of Hazardous Materials, 320*, 265−277.

Zhang, W.-X. (2003). Nanoscale iron particles for environmental remediation: An overview. *Journal of Nanoparticle Research, 5*(3−4), 323−332.

Zhang, X., Wang, J., Dong, X.-X., & Lv, Y.-K. (2020). Functionalized metal-organic frameworks for photocatalytic degradation of organic pollutants in environment. *Chemosphere, 242*, 125144.

Zhao, H., Giver, L., Shao, Z., Affholter, J. A., & Arnold, F. H. (1998). Molecular evolution by staggered extension process (StEP) in vitro recombination. *Nature Biotechnology, 16*(3), 258−261.

Zheng, Y., Wang, C., Zheng, Z.-Y., Che, J., Xiao, Y., Yang, Z.-H., & Zhao, F. (2014). Ameliorating acidic soil using bioelectrochemistry systems. *RSC Advances, 4*(107), 62544−62549.

Zhou, M., Liu, L., Jiao, Y., Wang, Q., & Tan, Q. (2011). Treatment of high-salinity reverse osmosis concentrate by electrochemical oxidation on BDD and DSA electrodes. *Desalination, 277*(1−3), 201−206.

Zille, A., Munteanu, F. D., Gübitz, G. M., & Cavaco-Paulo, A. (2005). Laccase kinetics of degradation and coupling reactions. *Journal of Molecular Catalysis B: Enzymatic, 33*(1−2), 23−28.

20

Microbial degradation of environmental pollutants

Hamza Rafeeq[1], Sarmad Ahmad Qamar[2], Tuan Anh Nguyen[3], Muhammad Bilal[4], Hafiz M.N. Iqbal[5]

[1]DEPARTMENT OF BIOCHEMISTRY, UNIVERSITY OF AGRICULTURE, FAISALABAD, PAKISTAN [2]INSTITUTE OF ORGANIC AND POLYMERIC MATERIALS, NATIONAL TAIPEI UNIVERSITY OF TECHNOLOGY, TAIPEI, TAIWAN [3]INSTITUTE FOR TROPICAL TECHNOLOGY, VIETNAM ACADEMY OF SCIENCE AND TECHNOLOGY, HANOI, VIETNAM [4]SCHOOL OF LIFE SCIENCE AND FOOD ENGINEERING, HUAIYIN INSTITUTE OF TECHNOLOGY, HUAI'AN, P.R. CHINA [5]TECNOLOGICO DE MONTERREY, SCHOOL OF ENGINEERING AND SCIENCES, MONTERREY, MEXICO

20.1 Introduction

It has been a growing concern regarding increased disposal of industrial wastes to the environment, the growth of industrialization, and exploitation of natural resources mainly land and water pollution. Contamination with hazardous heavy metals and chemical substances in soil, soil-water, sediment, and the air is therefore a major threat to the environment because of the absence of technologies to break down these hazardous substances into less-toxic forms, which has a long-term impact on the ecosystem (Aguilar-Pérez et al., 2021; Aguilar-Pérez, Heya, Parra-Saldívar, & Iqba, 2020). The necessity to restore these natural resources has led to the designing of novel technologies, which prioritize the removal of pollutants, and not traditional waste due to their capacity to join the food chain (Asha & Sandeep, 2013). Due to growing human activities such as metallic mining and smelting, agriculture, waste management industries, and biomedical sectors, there may be adverse effects if resulting pollutants could not be managed well, causing the accumulation of heavy metals substances, for example, silver, cadmium, gold, mercury, cobalt, lead, arsenic, selenium, nickel, and zinc. Plants need certain metallic compounds for their proper growth and functioning in very limited amounts. However, due to the revolution in the industry, several kinds of metals are continuously increasing in soil and water, exhibit an alarming threat to human life and the aquatic biota environment. Excavation and solidification/stabilization are conventional methods for the site contaminated with heavy metals and are suitable for contamination control, however, not suitable for permanent removal of heavy metals (Seh-Bardan, Othman, Wahid, Husin, & Sadegh-Zadeh, 2012).

Several advancements in bioremediation technologies have been made over the past two decades, with the ultimate aim of effectively recycling the environmental pollutants at a very low cost, in an environmentally friendly manner. Researchers have created and modeled multiple biosanitation methods but as a result of their existing form of the pollutants, there is no process of bioremediation which function as the silver bullet, for contamination recycling. Indigenous microbial strains found in a contaminated ecosystem are the major obstacles of bioremediation and biodegradation of substances causing pollution in the environment (Khan, Swapna, Hameeda, & Reddy, 2015). One of the major advantages of bioremediation is that while compared with both physical and chemical regeneration approaches in environmental and cost-saving characteristics, bioremediation with an emphasis on one of the procedures has been presented with many strong meanings of degradation. In certain conditions, the term used for bioremediation is synonymously replaced with biodegradation, the former being a mechanism applicable to bioremediation. In this study, bioremediation is characterized as a method utilizing biological processes to reduce pollutant concentrations in an uncertain state such as detoxification, degradation, transformation, and mineralization.

Pollutants are mainly eliminated depending on the nature of the contaminant, including compound of chlorination, agrochemicals, chlorinated gases of the greenhouse, dyes, hydrocarbons, heavy metals, plastics, nuclear waste, and household waste disposal. Criteria is considered for the selection of any bioremediation technique based on the nature of the pollutant, and depth level of pollution, the type of environment, placement, cost, and policy of environment (Mazzeo, 2013; Smith et al., 2015). The elimination of the hazardous effects of many pollutants by biodegradation as a biotechnological process involves the application of microorganisms. The terms "biodegradation" and "bioremediation" can be alternatively used in environmental biotechnology. Microbes are important tools for the elimination of pollutants in sediment, water, and soil, mostly because of their benefit rather than other protocols of remedial procedures. Microorganisms have been restored to the natural environment for pollution prevention (Demnerová et al., 2005). The present chapter was aimed to review the various functions of microbes to bioremediation which provide the background relative to the identification of deficiencies in a specific area, which is currently a growing research and development (R&D) concern, as microbes can address environmental hazards in a promising and environmentally friendly way.

20.2 Bioremediation: ecological relation between microorganisms

Microorganisms interact very dynamically with other organisms as part of the largest biological group. In most cases, each microbe cannot metabolize the energy made by itself. Cometabolism is the primary mechanism for the degradation of hazardous pollutants (Mekuto, Ntwampe, & Mudumbi, 2018). In addition, synergy among pollutant degradation microorganisms is popular. The metabolites of one microbe cause another to live, which results in the mutual nutritional requirements of each other. The whole process should then begin to calculate the rate of environmental pollution of the contaminants. In addition, there is a competitive interaction within microorganisms (Dikshitulu, Baltzis, Lewandowski, & Pavlou, 1993). Toxins degradation in competitive environments is not required from time to time, even though some microbial strains degradation occurs in the ambient environment. The explanation is that other organisms can accumulate desirable factors that contribute to rapid growth and reproduction and subsequent growth of the species, which is dominant in the ecosystem, thereby inhibiting the organism involved in degradation. As a result, clarification of these microorganisms' population systems makes it possible to predict the environmental destination of toxins, to plan various bioremediation technologies in conjunction with various environmental pollutants, and to increase the declined population of novel microbial species. Degradation of pollutants by genetically engineered microorganisms using various remediation methodologies have been represented in Fig. 20–1.

20.3 Herbicides, pesticides, and fertilizers as a product of agriculture

Insect repellents are often used in farming for the prevention, management, and maintenance of crops with low yield losses (Damalas & Eleftherohorinos, 2011). Present agricultural

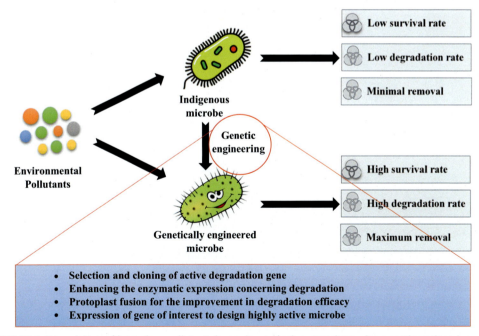

FIGURE 20–1 Degradation of hazardous pollutants by engineered microorganisms using genetic engineering approaches.

activity, however, has terminated the widespread usage of pesticides and chemical fertilizers (Li, 2018), which causes severe contamination in different media and detrimental effects on human health (Fig. 20–2) (Liu, Bilal, Duan, & Iqbal, 2019). Several strategies for physicochemical degradation and remediation of contaminated sites have been developed, in part or in full, for pesticide remediation. However, the use of these techniques has limited due to their high costs and the huge waste output. A variable range of microorganisms, for example, fungi and bacteria, have demonstrated a high ability to break down into harmless atmospheric products through a variable range of screening sources and insulation (Rayu, Nielsen, Nazaries, & Singh, 2017). Bioremediation also requires biostimulation, involving a variety of medium nutrients to support biosorption and microbial strain degradation (Wang et al., 2017). In addition, endogenous microorganisms typically exhibit limited biodegradability of various pesticide pollutants. Building highly successful and genetically modified microorganisms significantly increase the performance of degrader polluter and the ability of bacteria which responds broader environment (Yuanfan et al., 2010). A variety of different genes have been identified in different species that have the ability to degrade the pesticide, which has provided the opportunity to build genetically engineered microorganisms (Neumann et al., 2004). Potential microbial agents for various types of pesticide remediation have been summarized in Table 20–1.

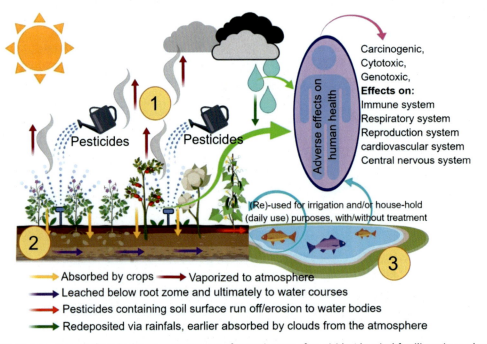

FIGURE 20–2 Pesticide drift and adverse consequences of excessive use of pesticides/chemical fertilizers in modern agriculture practices that can lead to contamination of different media including (1) air, (2) land, and (3) water (Liu, Bilal, Duan, & Iqbal, 2019).

Table 20–1 Potential microbial agents for pesticide remediation, as reported by few recent studies.

Microorganisms	Compound	References
Staphylococcus and *Bacillus* spp.	Endosulfan	Mohamed, El-Hussein, El-Siddig, and Osman (2011)
Enterobacter spp.	Chlorpyrifos	Niti, Sunita, Kamlesh, and Rakesh (2013)
Pseudomonas putida, *Arthrobacter* spp., *Acinetobacter* spp.	RIDOMIL MZ 68 MG, WP 76, Decis 2.5 EC, Malathion	Hussaini, Shaker, and Iqbal (2013), Monica et al. (2016)
Enterobacter spp., *Acinetobacter* spp., *Photobacterium* and *Pseudomonas* spp.	Methyl Parathion and Chlorpyrifos	Karanth, Liu, Olivier, and Pope (2004)

20.4 Dyestuff-based hazardous pollutants

Several industries regularly use different forms of synthetic dyeing, such as textile, detergents, leather, printers, cosmetics, food, and drug industries. Approximately, 17%−20% of total dyes are used in textile, with other industries, that is, tanning, paper, plastics, food preparation, cosmetics, and pharmaceuticals degenerate during the wastewater treatment processes (Bilal et al., 2017a; Bilal, Rasheed, Iqbal, & Yan, 2018; Chatha, Asgher, & Iqbal, 2017; Qamar, Ashiq, Jahangeer, Riasat, & Bilal, 2020). It has been demonstrated that synthetic dyes and their intermediate and degrading metabolites contain highly poisonous, carcinogenic, mutagenic, and teratogenic flavors (Bilal et al., 2016a; Bilal, Iqbal, Hu, & Zhang, 2016b; Bilal, Iqbal, Hu, Wang, & Zhang, 2017b; Bilal et al., 2018; Ito, Adachi, Yamanashi, & Shimada, 2016). Because of the high COD value, dyes decrease the clarity and solubility of oxygen water, possibly interfering with aquatic biota production and ecology equilibrium (Bilal et al., 2016a). In view of the vast array of undesired effects on the ecosystem of excessive decomposition and coloring of aqueous resources, it offers a broad opportunity for the development of certain professional approaches to the treatment of water elimination. However, conventional physical or chemical controls often find that the total rejection of various dyes with toxic pollutants is inadequate (Ahmad et al., 2015).

The aim of demanding research is to implement new, highly effective, and green-based biological strategies and to achieve a high level of efficiency in the elimination of pollutants to protect ecosystems (Khan & Husain, 2019). Biological treatment is, therefore, takes an exceptional position when it comes to rehabilitating wastewater as a substitute for conventional treatment strategies, due to its excellent coloring, cost-effectiveness, little generation of wastewater, and environmental friendliness (Han, Wang, Shi, & Qian, 2000). These microorganisms can greatly improve wastewater discoloration, which is commonly used in bioaugmentation systems because of multiple degrading impacts as a result of some target teething genes overexpression (Liang et al., 2018). Some important microbial species involved in the bioremediation processes have been described in Table 20−2.

Table 20–2 The most dominating microorganisms in the involvement of bioremediation processes.

Microorganisms	Substances	References
NAP1,2 & 4 strain of *B. subtilis*	Oil paints	Phulpoto, Qazi, Mangi, Ahmed, and Kanhar (2016)
M. roridum	Residual dyes	Jasińska, Różalska, Bernat, Paraszkiewicz, and Długoński (2012), Jasińska, Paraszkiewicz, Sip, and Długoński (2015)
Pynocoporus and *Phanerochaete*	Residual dyes	Yan, Niu, Chen, Chen, and Irbis (2014)
P. ochrochloron	Industrial dyes	Shedbalkar and Jadhav (2011)
Listeria denitrificans	Dyes of Azo from textile	(Hassan, Alam, & Anwar, 2013)
Pseudomonas aeruginosa, and *Bacillus* spp.	Remazol black (textile dye), Sulphonated diazo dye, and reactive red HE8B and RNB dyes	Shah (2013), Das, Mishra, and Verma (2016), Patel and Gupte (2016)
Exiguobacterium aurantiacums, *Acinetobacter baumannii*, *Exiguobacterium indicum*, and *Bacillus cereus*	Azo dyes	Kumar, Chaurasia, and Kumar (2016)
Klebsiella oxytoca, *Bacillus firmus*, and *Staphylococcus aureus*	Vat dyes	Adebajo, Balogun, & Akintokun, 2017

20.5 Potentially toxic heavy metals

There are various kinds of heavy metals, having a high atomic number, for example, 22–92 in all groups of the periodic table. The complete well-being of a living organism is based on certain metals such as Zn, Mn, V, Ni, Cr, Pb, Cu, Se, etc. Major chemical toxins are heavy metals including zinc, mercury, and cadmium. The trace amount of these metals, even in soil and water, can cause severe health issues for all the species. The key cause of agricultural production due to the deposition of heavy metals into soils was opposing effects on the quality of food (marketability and safety), growth of crops (phytotoxicity), and environmental health. High concentrations of heavy metals are toxic to higher life forms by nature, as their pollution causes biomagnification and damages the quality of the soil and the crops produced (Adhikari, Manna, Singh, & Wanjari, 2004). Human health can be affected by the accumulation of heavy metals in the environment (Fig. 20–3). The primary route of exposure to these components of the human population is through the absorption of these metals. In many countries, this issue increasingly attracting public and government agencies, especially in developed countries. the intakes of heavy metals by citizens across the food chain were recorded.

In recent years, heavy metal pollution in large areas has been a significant concern. Which includes mostly industrial waste incinerators, exhaust waste for vehicles, contaminants of metallurgy, and urban compost usage, chemicals, sewage drainage, and fertilizers. Industrial urban wastewater effluent containing relatively much quantity of these heavy

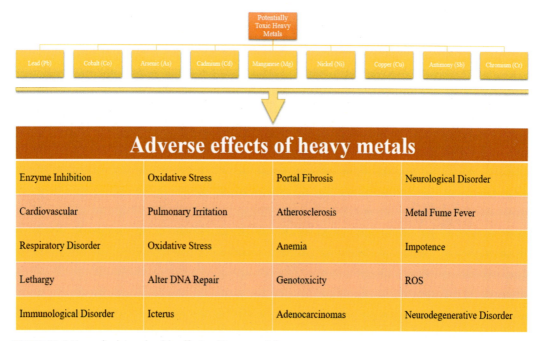

FIGURE 20–3 Several adverse health effects of heavy metals.

metals (Aguilar-Pérez, Avilés-Castrillo, & Ruiz-Pulido, 2020) (Damodaran, Suresh, & Mohan, 2011). Specific microbial enzymes may reduce or oxidize certain metals. Microbial metabolism generates oxidation and metal reduction products such as hydrogen, oxygen, and H_2O_2. Microbial metabolites could also mediate the solubilization and precipitation of metals. Aerobic oxidation of the organic or inorganic acid (nitric and sulfuric acid) microbial development will facilitate the dissolution of metal chelates (Girma, 2015).

20.6 Petroleum and aromatic compounds

Petroleum biodegradation is a complex process that depends on its nature and quantity. These hydrocarbons can be classified into four classes: saturated, aromatic, asphaltic (phenols, fatty acids, ketones, esters, and porphyria), and resins (pyridine, quinoline, carbazole, sulphoxide, and amide) (Colwell, Walker, & Cooney, 1977). Aromatic compound-degrading pathways are shown in Fig. 20–4 (Zhao et al., 2017b). One of the principal factors restricting the biodegradation of oil residues in the natural system is inadequate access for microorganisms. The hydrocarbons in petroleum bind to the soil and are not easy to degrade or extract. Hydrocarbons are differently prone to microbial attack. It is generally classified as linear alkanes > branched alkanes > small aromatics > cyclical alkanes, and hydrocarbons are vulnerable to microbial degradation. Certain compounds, such as polycyclic aromatic

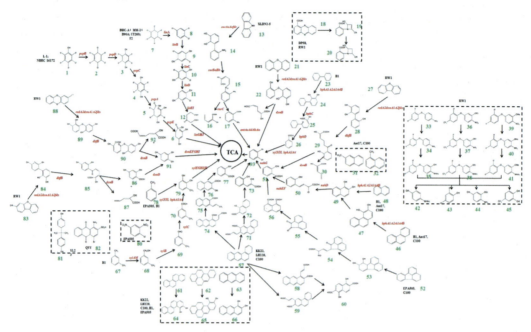

FIGURE 20–4 Aromatic compound-degrading pathways (Zhao et al., 2017b).

hydrocarbons of high molecular weight, may not be degraded as a result of the microbial attack (Atlas & Bragg, 2009). In the oxidation of gasoline, bacteria are the most active agents which act as key contaminants. Some hydrocarbon bacteria are known to feed themselves. The bacterial generations *Gordonia, Brevibacterium, Aeromicrobium, Dietzia, Burkholderia,* and *Mycobacterium* have shown that these are potential organisms of degradation extracted from the soil polluted by petroleum (Chaillan et al., 2004).

20.7 Polychlorinated biphenyls

Bacteria that utilize biphenyl was first reported in 1970 when dioxygenase attacked the biphenyl and converted it into benzoic acid. *Pseudomonas pseudoalcaligenes* KF707 can be used in the degradation of various polychlorinated biphenyls. Since that time many types of bacteria that are involved in the degradation of biphenyls are isolated and characterize (Robertson & Hansen, 2015). In the production of the PCBs, the direct chlorination of the biphenyl generate a mixture of complex, having 9–10 chlorines, and about 210 compounds (congeners) can be produced by applying this process and tri-hexa-chlorinated to hexa-chlorinated PCBs are mostly used commercially (Grimm et al., 2015). Aroclor is the trademark of major PCBs and a four-digit number code is used to identify the product of Aroclor in which the very first two digits used for the carbon of the biphenyl which is 12 in number and the second two digits show the amount of the chlorine present in the sample by weight

(Ampleman et al., 2015). The global production of PCBs is about 1.2 million tons, and these have various uses in the field of the heat exchanger, pesticides, transformer oil, hydraulic acid, and plasticizer due to both chemical and physical stability, and so on. For many years, PCBs have been released into the environment and it became a major issue as a pollutant throughout the globe. It can store in the biota because it shows lyophilic activities (Fernandez-Gonzalez, Yebra-Pimentel, Martinez-Carballo, & Simal-Gandara, 2015).

Scientists first isolated the two strains of *Achromobacter*, which is able to degrade various PCBs by converting into the chlorobenzoic acid (Vitku et al., 2016), then the degradation of the 31 different biphenyls by strain which utilizes the biphenyl and biodegradability of the PCBs depend on the substitution of the chlorine molecules such as by changing the position or the number of chlorine molecule also degradation of the PCBs depends on the strains (Li, Wang, Wang, Wang, & Zhao, 2016). Gram-negative bacteria as well as Gram-positive bacterial strains could degrade a large number of PCBs congeners, such as *Rhodococcus*. Biodegradation of PCBs can be done with anaerobic dechlorination of PCBs but using this process of dechlorination with pure strains met with limited success (Jacobson et al., 2017). PCBs are chiefly made by a class of aromatic chlorinated hydrocarbons. Due to excellent heat resistance and insulation capacity, PCBs are broadly applied as additives in heat-insulating media (Muller et al., 2017). However, because of the resistance properties, heavy accumulation characteristic, potential carcinogenic properties, and widespread contamination, PCBs source less destruction to human health and the environment (Aldhafiri, Mahmoud, Al-Sarawi, & Ismail, 2018).

There are two methods by which the PCBs biodegradation have been studied, by an aerobic microorganism which is helpful in oxidative degradation, and anaerobic method in which dechlorination of biphenyl carried out (Bogdal et al., 2017). The oxidative biodegradation of PCBs have been performed by various types of Gram-negative microorganisms, such as *Alcaligenes, Burkholderia, Pseudomonas, Achromobacter, Sphingomonas, Acinetobacter, Comamonas,* and *Ralstonia*, and Gram-positive microorganism, such as *Corynebacterium, Rhodococcus,* and *Bacillus* (Carro, García, Ignacio, & Mouteira, 2018). These microorganisms use biphenyls as energy and carbon source and catabolize the higher PCBs while using the enzymes of biphenyls (Wu et al., 2018). There are various laboratories that studied the degradation of PCBs and they draw the general concept for the PCBs biodegradability, (1) the larger the number of chlorines minimum will be the biodegradability; (2) more than two chlorine rings in the PCBs are degraded with difficulty as one chlorine ring can be degraded easily; (3) if the position of chlorine ring is the orthosubstituted such as on the 2−6 or 2−2 double chlorine rings present on the same position degraded poorly; and (4) degradation of PCBs depend on the strain of microorganisms and position on which strain is an attack on the PCBs (Jafarabadi et al., 2019).

The metabolism of PCBs through 2, 3-dioxygenation and 3, 4-dioxygenation is dependent on the PCBs chlorine substitution, and show a higher degree of degradation of various kind of PCBs congeners. While on the other side, 2, 3-dioxygenation using *P. pseudoalcaligenes* KF707 showed a very narrow range of biodegradation of PCBs (Rude et al., 2019). The strains that are used in the biodegradation were categorized into two groups with the ability to degrade17 PCBs. Various strains show the wide range of ability of degradation of PCBs, but showed weaker activation toward the double para substitution with the specificity of LB400-type designated but on the other hand, a strain which showing the narrow range of biodegradation but shows the maximum activity toward the double parasubstitution congeners and this is designated with the KF707-type of specificity (Curtis et al., 2019).

The analysis of various products of dioxygenated with several dichlorinated to penta-chlorinated biphenyl formation by various kinds of biphenyl dioxygenase which depend on the dioxygenation yield and regiospecificity on the different pattern of both non-oxidized and oxidized rings of chlorine. Many strains show the capability of PCBs biodegradation, showing various novelties (Saktrakulkla et al., 2020). A bacterial strain namely SK-3 which can be easily grown on 2, 3, and 4-chlorobiphenyl, respectively, and 4-chlorobenzoic acid and chloroacetate utilization as the energy and carbon source. Chloride ion production is the result of 4-chlorobenzoate growth (Tahir et al., 2020). Another strain of bacteria which is capable to grow in aerobic condition on the ortho-substitute which can use carbon as the energy source. In this phase of growth on 2, 2-dichlorobiphenyl, and other one position 2, 4-chlorobiphenyl the production of 2 and 4 chlorobenzoate, respectively (Donat-Vargas et al., 2020). A Gram-positive bacterial strain *Rhodococcus sp.* RHA1, transferred the 45 components into 62 KANECLOR PCBs mixture such as 200 and 400 including the trichloro, tetrachloro, and pentachlorobiphenyl. It was noticed that genes for the degradation are present in this strain. Cyanobacteria, yeast, and fungi can also metabolize the low level of biphenyl into the compounds of monohydroxy and dihydroxy. *Phanerochaete* spp., are effective for the degradation of PCBs was reported, and its ability was described as it can degrade the PCBs with various positions of chlorine ring such as the ortho, meta, and para, which indicate the nonspecificity for the various position substitutions (Raffetti, Donat-Vargas, Mentasti, Chinotti, & Donato, 2020).

Some of the biphenyl enzymes effectively degrade the PCBs with cometabolism. Enzymes of biphenyl such as the dioxygenase, which involve in the initial catalysis of 2−3-oxygenation and convert it into the compound of 2,3-dihydrodiol, this another enzyme dihydrodiol hydrogenase which dehydrogenated the 2,3-dihydrodiol into 2,3-dihydroxybiphenyl, which is responsible for the cleavage of the ring, which was dihyroxylated and produce 2-hydroxy-6-oxo-6-phenylhexa-2−4-dienoic acid and the cleavage take place on meta position of the yellowish compound. Hydrolase enzyme hydrolyzes the yellow compound and converts it into 2-hydroxypentachloro-2,4-dienoate and benzoic acid (Kang et al., 2020). Most of the bacteria which utilize the biphenyl cannot degrade any further product of chlorobenzoic acid. The intermediated of degradation, for example, compounds of dihydroxy and cleavage of meta ring of yellow compound accumulated as the end product, which is dead, and these all depend on the congeners of PCBs (Xiang, Xing, Liu, Qin, & Huang, 2020). There are various bacterial alternate to the above ones which is *B. xenovorans* LB400, which is involved in the degradation of PCBs congeners with the position of chlorine ring 2, 5-chlorophenyl while treated with 3,4-oxygenation in which the end product accumulate as compounds of 3,4-dihydrodiol.

It was investigated in more detail that the LB400 strain was used for regiospecificity of various PCBs and *Rhodococcus globerulus* P6 converted 2,4,6-trichlorobiphenyl to compound dihydroxy and accumulate as the end products compound of trihydroxy (Verma & Rani, 2020). *Sphingobium yanoikuyae* B1 strain has been reported to have a positive effect on the decomposition of a broad range of heterocyclic and monocyclic PAHs (naphthalene, phenanthrene, chrysene, biphenyl, anthracene, benz[a]anthracene, benzo[a]pyrene, dibenzo-p-dioxin, m-p xylene, toluene, and carbazole) as individual source (Adam et al., 2014; Waigi, Kang, Goikavi, Ling, & Gao, 2015; Zhao et al., 2017a). The dioxygenase enzymes are transcribed by the gene bphA1fA2fA3A4, which is type-IV, three-component dioxygenase, that carries out the hydrogenation and dihydroxylation of the respective benzene ring. Moreover, various other gene types are found to be involved in PAHs degradation, and the enzymes concerned with downstream processes of biphenyl have also been investigated (Bae et al., 2003; Muangchinda et al., 2015).

20.8 Phenazines

There are some bacteria that act as the biological control agents, that is, *Pseudomonas* spp. are some of them by which metabolites are secreted in the environment and suppress the disease in the rhizosphere. Phenazine is the subclass of these metabolites, which are secondary metabolites of nitrifying bacteria, with broad antibiotic properties and broad applications as biological control agents to a fungal plant pathogen as reported by Peng et al. (2018b). Phenazine and its derived compounds have received much scientific attention as a major compound for possible applications as an environmental sensor, microbial fuel cell design, and their excellent antimalarial, antiparasitic, antitumoral activities, and anticancer prodrug (Yue et al., 2018). Due to excellent activities against various kinds of cell types, it has been studied for their beneficiary physiological role for its producers. In the agricultural sectors, the phenazines productions such as the phenazine-1 carboxamides and phenazine 1, carboxylic acid protect the plant from the infection of fungus, which is a very pathogenic, dry-land place of cereal rhizospheres, in which phenazine accumulate and have a time of their half-life about three to four days (Jin et al., 2020).

On the other hand, phenazine also active in the other type of environment with the context of clinical aspects, for example, the patients suffering from cystic fibrosis or lungs disorder, but the turnover of these substance has not been measured yet in the systematized way (Karmegham et al., 2020). Due to having a very short half-life in the rhizosphere, which indicates the active mechanism of removal because it is unclear yet that what actually they are (Sun et al., 2020). The various phenazine with a natural compound with nitrogen and the heterocyclic ring has been described. The various species of *Pseudomonas* produce different kinds of phenazine with variation in properties and PCN, pyocyanin, and PCN are the culture of *Pseudomonas aeruginosa*. The best cursor for the phenazine are the PCA by which other kinds of phenazine can be derived, (phzS and phzM) are the enzymes that are involved in the production of PYO and also modify the PCA, PhxH act on the PCN and convert it into the PCA. PhzS act on the *P. aeruginosa* and the production of 1-hydroxyphenazine take place (Patel, Raju, Haldar, & Chatterjee, 2020). There are many varieties of the organism by which phenazine with benefits can be produced. The phenazine from the *P. aeruginosa* is helpful in the survival without oxygen, development of biofilm, signaling, and acquisition of iron, but the redox properties of the phenazine are very harmful to the various bacteria and eukaryotic organisms, which have a close association with *Pseudomonas* spp. (Guo et al., 2020).

The toxicity and the change in the phenazine depend on the change in environment, such as *Caenorhabditis elegans* which is not much sensitive again PYO at acidic pH, then PCA but in alkaline condition, it is totally opposite. Reactive oxygen species can be produced by the toxicity of phenazine which causes the hindrance in the movement of electron in the electron transport chain (Bollinger, Thies, Katzke, & Jaeger, 2020). The induction of ROS is useful for the prevention of phenazine toxic effect and capacity to transform or degrade phenazines such as PCN and PCA. A recent study conducted on the changes or transformation of phenazines was conducted in which diffusion of these phenazines with colonies of *Aspergillus fumigatus* and *P. aeruginosa* resulted in various transformation but not the phenazine removal (Sharma et al., 2020). The degradation signals for the quorum sensing and lactone acyl-homoserine has been described. In addition to the additional chemical modification, the process of turnover is considered when phenazine fate in the different environments is seeking. The degradation or alter capacity for the phenazine elaborated by some microorganisms, which belongs from the *Pseudomonas* spp. in the natural community, and the genes, which are responsible for this activity are unknown yet (DeBritto et al., 2020).

The phenazine involves in the environmental chemistry and community of microbes, furthermore, much of the information is known about the physiological function, regulation, and biosynthesis of phenazines, and very little information is known about the phenazine degradation. The microorganism with the name *S. wittichii* DP58 has been studied for its involvement in the degradation of PCA, but the genes responsible for this catalysis have not been identified (Hane et al., 2021). Here some of the genes which cause the degradation of phenazine are the member of a microbial species complex of *Mycobacterium*. With the identification of these genes which was conserved with the degradation of phenazines in mycobacteria not only divers the phylogenetic activity of the microbial species such as the *Actinobacteria* and *Proteobacterium* (Khokhar, Jadoun, Arif, & Jabin, 2021). It also suggests that enzymes responsible for catalyzing these reactions may varied with different microbes. *Mycobacterium* spp., ubiquitous, and *Pseudomonas* spp. can be isolated from the same kind of environments such as the crude oil, soil, and the patients suffering from cystic fibrosis. It may not be wondered that one organism has the capacity to utilize the product of another organism. Fig. 20–5 shows the steps in phenazine degradation by the strain B1 (Zhao et al., 2017a).

Rhodococcus and *M. fortuitum* species shared the cluster of various genes for degradation of PCA and not involved in the degradation of PYO. JVH1 strain of *Rhodococcus* was unable to degrade to PYO, having the same homology with these genes and it is suggested that PYO and PCA shared the intermediate of degradation (Silva et al., 2021). *P. aeruginosa*, PCA

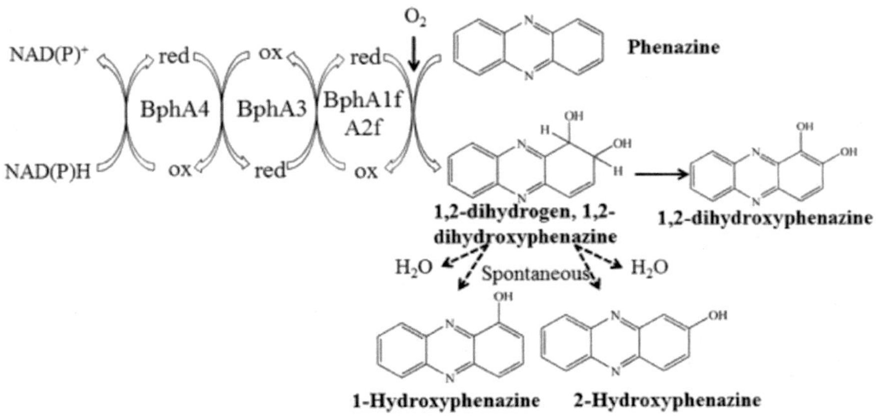

FIGURE 20–5 Initial steps in phenazine degradation by the strain B1 (Zhao et al., 2017a).

produce PYO by the actions of phzM and phzS. There is another possibility by which the DKN1213 and CT6 in first degrade the PYO to PCA which produces more mRNA for the specific genes of PCA and other one possibility is that both of PYO and PCA convert into the intermediate of each other and combine degraded by the action of various MFORT_16334, MFORT_16269, MFORT_16349, and MFORT_16319, all subsets of these proteins with 162 number of amino acids are enough for the degradation of PYO, and for degradation of PCN, amidase is required. In the action of amidase the very first catalysis takes place on PCN and is converted into the PCA and the remaining breakdown further takes place (Feng et al., 2018). The degradation of phenazine may affect the growth of phenazine-producing organisms in the natural or artificial environment (Zhang et al., 2020). Each phenazine derived from the *Pseudomonas* has different chemical and redox properties and has a great impact on the producers physiology, for example, in *P. aeruginosa*, PYO act as biofilm maturation and ferric ion reduced by the PCA by facilitating the acquisition of iron and toxicity of phenazine may benefits the species of Pseudomonas by inhibiting the competing factors of it.

20.9 Conclusion

Bioremediation provides environmentally friendly technologies for the mitigation of hazardous pollutants by promoting biobased degradation methodologies. Therefore, by the development of microbial communities and understanding, and the possible mechanisms with pollutants and in the natural environment, increasing the scientific understanding of the microbial genetics to enhance the microbial degradation abilities, leading research of the designing of novel bioremediation technologies with cost-effective perspective, and devoting sites that are set aside for long-term research, these possibilities increase significant advancements. Undoubtedly, bioremediation technologies provide novel ways with "greener" themes. Moreover, this technology provides effective cheater methodologies to treat polluted soil and groundwater. The degradation rate of unwanted materials or pollutants has been studied in comparison to biological pollutants removal and insufficient distribution with necessary nutrients. Due to microbial growth kinetics, the biological degradation in the natural environment is not very successful, leading toward less popularity. Bioremediation can only prove to be effective when environmental parameters are adjusted according to microbial growth and activity kinetics. Biological mitigation strategies have been applied in various sectors globally with varying levels of success. Chiefly, the benefits of bioremediation are higher, which is evident by the number of literature studies that are used with this technology and its improving popularity. Normally, various microbial strains have been studied from different places and their degradation ability could effectively use.

Acknowledgments

The listed author(s) are thankful to their representative universities for providing the literature services.

Conflicts of Interest

The listed author(s) declare that no competing, conflicting, and financial interests exist in this work.

References

Adam, I. K., Rein, A., Miltner, A., Fulgencio, A. C., Trapp, S., & Kästner, M. (2014). Experimental results and integrated modeling of bacterial growth on an insoluble hydrophobic substrate (phenanthrene). *Environmental Science & Technology*, 48, 8717–8726.

Adebajo, S., Balogun, S., & Akintokun, A. (2017). Decolourization of vat dyes by bacterial isolates recovered from local textile mills in Southwest, Nigeria. *Microbiology Research Journal International*, 18, 1–8.

Adhikari, T., Manna, M., Singh, M., & Wanjari, R. (2004). Bioremediation measure to minimize heavy metals accumulation in soils and crops irrigated with city effluent. *Journal of Food, Agriculture and Environment*, 2, 266–270.

Aguilar-Pérez, K. M., Avilés-Castrillo, J. I., & Ruiz-Pulido, G. (2020). Nano-sorbent materials for pharmaceutical-based wastewater effluents-An overview. *Case Studies in Chemical and Environmental Engineering*, 2, 100028.

Aguilar-Pérez, K. M., Heya, M. S., Parra-Saldívar, R., & Iqba,, H. M. N. (2020). Nano-biomaterials in-focus as sensing/detection cues for environmental pollutants. *Case Studies in Chemical and Environmental Engineering*, 2, 100055.

Aguilar-Pérez, K. M., Avilés-Castrillo, J. I., Ruiz-Pulido, G., Medina, D. I., Parra-Saldivar, R., & Iqbal, H. M. N. (2021). Nanoadsorbents in focus for the remediation of environmentally-related contaminants with rising toxicity concerns. *Science of The Total Environment*, 779, 146465.

Ahmad, A., Mohd-Setapar, S. H., Chuong, C. S., Khatoon, A., Wani, W. A., Kumar, R., & Rafatullah, M. (2015). Recent advances in new generation dye removal technologies: Novel search for approaches to reprocess wastewater. *RSC Advances*, 5, 30801–30818.

Aldhafiri, S., Mahmoud, H., Al-Sarawi, M., & Ismail, W. A. (2018). Natural attenuation potential of polychlorinated biphenyl-polluted marine sediments. *Polish Journal of Microbiology*, 67, 37–48.

Ampleman, M. D., Martinez, A., DeWall, J., Rawn, D. F., Hornbuckle, K. C., & Thorne, P. S. (2015). Inhalation and dietary exposure to PCBs in urban and rural cohorts via congener-specific measurements. *Environmental Science & Technology, 49*(2), 1156−1164.

Asha, L., & Sandeep, R. (2013). Review on bioremediation-potential tool for removing environmental pollution. *International Journal of Basic and Applied Chemical Sciences, 3*, 21−33.

Atlas, R., & Bragg, J. (2009). Bioremediation of marine oil spills: When and when not−the Exxon Valdez experience. *Microbial Biotechnology, 2*, 213−221.

Bae, M., Sul, W. J., Koh, S.-C., Lee, J. H., Zylstra, G. J., Kim, Y. M., & Kim, E. (2003). Implication of two glutathione S-transferases in the optimal metabolism of m-toluate by *Sphingomonas* yanoikuyae B1. *Antonie Van Leeuwenhoek, 84*, 25−30.

Bilal, M., Asgher, M., Parra-Saldivar, R., Hu, H., Wang, W., Zhang, X., & Iqbal, H. M. (2017a). Immobilized ligninolytic enzymes: An innovative and environmental responsive technology to tackle dye-based industrial pollutants—A review. *Science of the Total Environment, 576*, 646−659.

Bilal, M., Iqbal, H. M., Hu, H., Wang, W., & Zhang, X. (2017b). Enhanced bio-catalytic performance and dye degradation potential of chitosan-encapsulated horseradish peroxidase in a packed bed reactor system. *Science of the Total Environment, 575*, 1352−1360.

Bilal, M., Iqbal, H. M., Shah, S. Z. H., Hu, H., Wang, W., & Zhang, X. (2016a). Horseradish peroxidase-assisted approach to decolorize and detoxify dye pollutants in a packed bed bioreactor. *Journal of Environmental Management, 183*, 836−842.

Bilal, M., Iqbal, M., Hu, H., & Zhang, X. (2016b). Mutagenicity, cytotoxicity and phytotoxicity evaluation of biodegraded textile effluent by fungal ligninolytic enzymes. *Water Science and Technology, 73*, 2332−2344.

Bilal, M., Rasheed, T., Iqbal, H. M., & Yan, Y. (2018). Peroxidases-assisted removal of environmentally-related hazardous pollutants with reference to the reaction mechanisms of industrial dyes. *Science of the Total Environment, 644*, 1−13.

Bogdal, C., Niggeler, N., Glüge, J., Diefenbacher, P. S., Wächter, D., & Hungerbühler, K. (2017). Temporal trends of chlorinated paraffins and polychlorinated biphenyls in Swiss soils. *Environmental Pollution, 220*, 891−899.

Bollinger, A., Thies, S., Katzke, N., & Jaeger, K. E. (2020). The biotechnological potential of marine bacteria in the novel lineage of *Pseudomonas pertucinogena*. *Microbial Biotechnology, 13*(1), 19−31.

Carro, N., García, I., Ignacio, M., & Mouteira, A. (2018). Polychlorinated dibenzo-P-dioxins and dibenzofurans (PCDD/Fs) and dioxin-like polychlorinated biphenyls (dl-PCBS) in bivalve mollusk from Galician Rías (NW, SPAIN). *Chemosphere, 197*, 782−792.

Chaillan, F., Le Flèche, A., Bury, E., Phantavong, Y.-h, Grimont, P., Saliot, A., & Oudot, J. (2004). Identification and biodegradation potential of tropical aerobic hydrocarbon-degrading microorganisms. *Research in Microbiology, 155*, 587−595.

Chatha, S. A. S., Asgher, M., & Iqbal, H. M. (2017). Enzyme-based solutions for textile processing and dye contaminant biodegradation—A review. *Environmental Science and Pollution Research, 24*, 14005−14018.

Colwell, R. R., Walker, J. D., & Cooney, J. J. (1977). Ecological aspects of microbial degradation of petroleum in the marine environment. *CRC Critical Reviews in Microbiology, 5*, 423−445.

Curtis, S. W., Terrell, M. L., Jacobson, M. H., Cobb, D. O., Jiang, V. S., Neblett, M. F., ... Marcus, M. (2019). Thyroid hormone levels associate with exposure to polychlorinated biphenyls and polybrominated biphenyls in adults exposed as children. *Environmental Health, 18*(1), 75.

Damalas, C. A., & Eleftherohorinos, I. G. (2011). Pesticide exposure, safety issues, and risk assessment indicators. *International Journal of Environmental Research and Public Health, 8*, 1402−1419.

Damodaran, D., Suresh, G. & Mohan, R. (2011). Bioremediation of soil by removing heavy metals using *Saccharomyces cerevisiae*. In *2nd international conference on environmental science and technology*, Singapore.

Das, A., Mishra, S., & Verma, V. K. (2016). Enhanced biodecolorization of textile dye remazol navy blue using an isolated bacterial strain *Bacillus pumilus* HKG212 under improved culture conditions. *Journal of Biochemical Technology*, 6, 962−969.

DeBritto, S., Gajbar, T. D., Satapute, P., Sundaram, L., Lakshmikantha, R. Y., Jogaiah, S., & Ito, S. I. (2020). Isolation and characterization of nutrient dependent pyocyanin from *Pseudomonas aeruginosa* and its dye and agrochemical properties. *Scientific Reports*, 10(1), 1−12.

Demnerová, K., Mackova, M., Speváková, V., Beranova, K., Kochánková, L., Lovecká, P., . . . Macek, T. (2005). Two approaches to biological decontamination of groundwater and soil polluted by aromatics— Characterization of microbial populations. *International Microbiology*, 8, 205−211.

Dikshitulu, S., Baltzis, B., Lewandowski, G., & Pavlou, S. (1993). Competition between two microbial populations in a sequencing fed-batch reactor: Theory, experimental verification, and implications for waste treatment applications. *Biotechnology and Bioengineering*, 42, 643−656.

Donat-Vargas, C., Moreno-Franco, B., Laclaustra, M., Sandoval-Insausti, H., Jarauta, E., & Guallar-Castillon, P. (2020). Exposure to dietary polychlorinated biphenyls and dioxins, and its relationship with subclinical coronary atherosclerosis: The Aragon Workers' Health Study. *Environment International*, 136, 105433.

Feng, J., Qian, Y., Wang, Z., Wang, X., Xu, S., Chen, K., & Ouyang, P. (2018). Enhancing the performance of *Escherichia coli*-inoculated microbial fuel cells by introduction of the phenazine-1-carboxylic acid pathway. *Journal of Biotechnology*, 275, 1−6.

Fernandez-Gonzalez, R., Yebra-Pimentel, I., Martinez-Carballo, E., & Simal-Gandara, J. (2015). A critical review about human exposure to polychlorinated dibenzo-p-dioxins (PCDDs), polychlorinated dibenzofurans (PCDFs) and polychlorinated biphenyls (PCBs) through foods. *Critical Reviews in Food Science and Nutrition*, 55(11), 1590−1617.

Girma, G. (2015). Microbial bioremediation of some heavy metals in soils: An updated review. *Egyptian Academic Journal of Biological Sciences, G. Microbiology*, 7, 29−45.

Grimm, F. A., Hu, D., Kania-Korwel, I., Lehmler, H. J., Ludewig, G., Hornbuckle, K. C., . . . Robertson, L. W. (2015). Metabolism and metabolites of polychlorinated biphenyls. *Critical Reviews in Toxicology*, 45(3), 245−272.

Guo, S., Wang, Y., Bilal, M., Hu, H., Wang, W., & Zhang, X. (2020). Microbial synthesis of antibacterial phenazine-1, 6-dicarboxylic acid and the role of PhzG in *Pseudomonas chlororaphis* GP72AN. *Journal of Agricultural and Food Chemistry*, 68(8), 2373−2380.

Han, L.-P., Wang, J.-L., Shi, H.-C., & Qian, Y. (2000). Bioaugmentation: A new strategy for removal of recalcitrant compounds in wastewater—A case study of quinoline. *Journal of Environmental Sciences (China)*, 12, 22−25.

Hane, M., Wijaya, H. C., Nyon, Y. A., Sakihama, Y., Hashimoto, M., Matsuura, H., & Hashidoko, Y. (2021). Phenazine-1-carboxylic acid (PCA) produced by Paraburkholderia phenazinium CK-PC1 aids postgermination growth of Xyris complanata seedlings with germination induced by *Penicillium rolfsii* Y-1. *Bioscience, Biotechnology, and Biochemistry*, 85(1), 77−84.

Hussaini, S. Z., Shaker, M., & Iqbal, M. A. (2013). Isolation of bacterial for degradation of selected pesticides. *Advances in Bioresearch*, 4, 82−85.

Hassan, M. M., Alam, M. Z., & Anwar, M. N. (2013). Biodegradation of textile azo dyes by bacteria isolated from dyeing industry effluent. *Int Res J Biol Sci*, 2(8), 27−31.

Ito, T., Adachi, Y., Yamanashi, Y., & Shimada, Y. (2016). Long−term natural remediation process in textile dye−polluted river sediment driven by bacterial community changes. *Water Research*, 100, 458−465.

Jacobson, M. H., Darrow, L. A., Barr, D. B., Howards, P. P., Lyles, R. H., Terrell, M. L., . . . Marcus, M. (2017). Serum polybrominated biphenyls (PBBs) and polychlorinated biphenyls (PCBs) and thyroid function among Michigan adults several decades after the 1973−1974 PBB contamination of livestock feed. *Environmental Health Perspectives*, 125(9), 097020.

Jafarabadi, A. R., Bakhtiari, A. R., Mitra, S., Maisano, M., Cappello, T., & Jadot, C. (2019). First polychlorinated biphenyls (PCBs) monitoring in seawater, surface sediments and marine fish communities of the Persian Gulf: Distribution, levels, congener profile and health risk assessment. *Environmental Pollution, 253*, 78–88.

Jasińska, A., Paraszkiewicz, K., Sip, A., & Długoński, J. (2015). Malachite green decolorization by the filamentous fungus *Myrothecium roridum*—Mechanistic study and process optimization. *Bioresource Technology, 194*, 43–48.

Jasińska, A., Różalska, S., Bernat, P., Paraszkiewicz, K., & Długoński, J. (2012). Malachite green decolorization by non-basidiomycete filamentous fungi of *Penicillium pinophilum* and *Myrothecium roridum*. *International Biodeterioration & Biodegradation, 73*, 33–40.

Jin, Z. J., Zhou, L., Sun, S., Cui, Y., Song, K., Zhang, X., & He, Y. W. (2020). Identification of a strong quorum sensing-and thermo-regulated promoter for the biosynthesis of a new metabolite pesticide phenazine-1-carboxamide in *Pseudomonas* strain PA1201. *ACS Synthetic Biology, 9*(7), 1802–1812.

Kang, Y., Cao, S., Yan, F., Qin, N., Wang, B., Zhang, Y., . . . Duan, X. (2020). Health risks and source identification of dietary exposure to indicator polychlorinated biphenyls (PCBs) in Lanzhou, China. *Environmental Geochemistry and Health, 42*(2), 681–692.

Karanth, S., Liu, J., Olivier, K., Jr, & Pope, C. (2004). Interactive toxicity of the organophosphorus insecticides chlorpyrifos and methyl parathion in adult rats. *Toxicology and Applied Pharmacology, 196*(2), 183–190.

Karmegham, N., Vellasamy, S., Natesan, B., Sharma, M. P., Al Farraj, D. A., & Elshikh, M. S. (2020). Characterization of antifungal metabolite phenazine from rice rhizosphere fluorescent pseudomonads (FPs) and their effect on sheath blight of rice. *Saudi Journal of Biological Sciences, 27*(12), 3313–3326.

Khan, M. Y., Swapna, T., Hameeda, B., & Reddy, G. (2015). Bioremediation of heavy metals using biosurfactants. *Advances in Biodegradation and Bioremediation of Industrial Waste*, 381.

Khan, N., & Husain, Q. (2019). Continuous degradation of Direct Red 23 by calcium pectate–bound Ziziphus mauritiana peroxidase: Identification of metabolites and degradation routes. *Environmental Science and Pollution Research, 26*, 3517–3529.

Khokhar, D., Jadoun, S., Arif, R., & Jabin, S. (2021). Tuning the spectral, thermal and morphological properties of poly (o-phenylenediamine-co-vaniline). *Materials Research Innovations*, 1–11. Available from https://doi.org/10.1080/14328917.2020.1870330.

Kumar, S., Chaurasia, P., & Kumar, A. (2016). Isolation and characterization of microbial strains from textile industry effluents of Bhilwara, India: Analysis with bioremediation. *Journal of Chemical and Pharmaceutical Research, 8*, 143–150.

Li, Q. L., Wang, L. L., Wang, X., Wang, M. L., & Zhao, R. S. (2016). Magnetic metal-organic nanotubes: An adsorbent for magnetic solid-phase extraction of polychlorinated biphenyls from environmental and biological samples. *Journal of Chromatography A, 1449*, 39–47.

Li, Z. (2018). Health risk characterization of maximum legal exposures for persistent organic pollutant (POP) pesticides in residential soil: An analysis. *Journal of Environmental Management, 205*, 163–173.

Liang, Y., Hou, J., Liu, Y., Luo, Y., Tang, J., Cheng, J. J., & Daroch, M. (2018). Textile dye decolorizing *Synechococcus* PCC7942 engineered with CotA laccase. *Frontiers in Bioengineering and Biotechnology, 6*, 95.

Liu, L., Bilal, M., Duan, X., & Iqbal, H. M. N. (2019). Mitigation of environmental pollution by genetically engineered bacteria—Current challenges and future perspectives. *Science of The Total Environment, 667*, 444–454.

Mazzeo, D.E.C. (2013). Avaliação da viabilidade do lodo de esgoto como recondicionante de solos agrícolas, após processo de atenuação natural, por meio de diferentes bioensaios.

Mekuto, L., Ntwampe, S. K. O., & Mudumbi, J. B. N. (2018). Microbial communities associated with the co-metabolism of free cyanide and thiocyanate under alkaline conditions,. *3 Biotech, 8*, 93.

Mohamed, A. T., El-Hussein, A. A., El-Siddig, M. A., & Osman, A. G. (2011). Degradation of oxyfluorfen herbicide by soil microorganisms biodegradation of herbicides. *Biotechnology, 10*, 274–279.

Monica, P., Darwin, R. O., Manjunatha, B., Zúñiga, J. J., Diego, R., Bryan, R. B., ... Maddela, N. R. (2016). Evaluation of various pesticides-degrading pure bacterial cultures isolated from pesticide-contaminated soils in Ecuador. *African Journal of Biotechnology, 15*, 2224–2233.

Muangchinda, C., Chavanich, S., Viyakarn, V., Watanabe, K., Imura, S., Vangnai, A., & Pinyakong, O. (2015). Abundance and diversity of functional genes involved in the degradation of aromatic hydrocarbons in Antarctic soils and sediments around Syowa Station. *Environmental Science and Pollution Research, 22*, 4725–4735.

Muller, M. H. B., Polder, A., Brynildsrud, O. B., Karimi, M., Lie, E., Manyilizu, W. B., & Lyche, J. L. (2017). Organochlorine pesticides (OCPs) and polychlorinated biphenyls (PCBs) in human breast milk and associated health risks to nursing infants in Northern Tanzania. *Environmental Research, 154*, 425–434.

Neumann, G., Teras, R., Monson, L., Kivisaar, M., Schauer, F., & Heipieper, H. J. (2004). Simultaneous degradation of atrazine and phenol by *Pseudomonas* sp. strain ADP: Effects of toxicity and adaptation. *Applied and Environmental Microbiology, 70*, 1907–1912.

Niti, C., Sunita, S., Kamlesh, K., & Rakesh, K. (2013). Bioremediation: An emerging technology for remediation of pesticides. *Research Journal of Chemistry and Environment, 17*, 4.

Patel, N. P., Raju, M., Haldar, S., & Chatterjee, P. B. (2020). Characterization of phenazine-1-carboxylic acid by Klebsiella sp. NP-C49 from the coral environment in Gulf of Kutch, India. *Archives of Microbiology, 202*(2), 351–359.

Patel, Y., & Gupte, A. (2016). Evaluation of bioremediation potential of isolated bacterial culture YPAG-9 (*Pseudomonas aeruginosa*) for decolorization of sulfonated di-azodye Reactive red HE8B under optimized culture conditions. *International Journal of Current Microbiology and Applied Sciences, 5*, 258–272.

Peng, H., Ouyang, Y., Bilal, M., Wang, W., Hu, H., & Zhang, X. (2018b). Identification, synthesis and regulatory function of the N-acylated homoserine lactone signals produced by *Pseudomonas chlororaphis* HT66. *Microbial Cell Factories, 17*(9).

Phulpoto, A., Qazi, M., Mangi, S., Ahmed, S., & Kanhar, N. (2016). Biodegradation of oil-based paint by *Bacillus* species monocultures isolated from the paint warehouses. *International Journal of Environmental Science and Technology, 13*, 125–134.

Qamar, S. A., Ashiq, M., Jahangeer, M., Riasat, A., & Bilal, M. (2020). Chitosan-based hybrid materials as adsorbents for textile dyes—A review. *Case Studies in Chemical and Environmental Engineering, 2*, 100021.

Raffetti, E., Donat-Vargas, C., Mentasti, S., Chinotti, A., & Donato, F. (2020). Association between exposure to polychlorinated biphenyls and risk of hypertension: A systematic review and meta-analysis. *Chemosphere*, 126984.

Rayu, S., Nielsen, U. N., Nazaries, L., & Singh, B. K. (2017). Isolation and molecular characterization of novel chlorpyrifos and 3, 5, 6-trichloro-2-pyridinol-degrading bacteria from sugarcane farm soils. *Frontiers in Microbiology, 8*, 518.

Robertson, L. W., & Hansen, L. G. (Eds.), (2015). *PCBs: Recent advances in environmental toxicology and health effects*. University Press of Kentucky.

Rude, K. M., Pusceddu, M. M., Keogh, C. E., Sladek, J. A., Rabasa, G., Miller, E. N., ... Gareau, M. G. (2019). Developmental exposure to polychlorinated biphenyls (PCBs) in the maternal diet causes host-microbe defects in weanling offspring mice. *Environmental Pollution, 253*, 708–721.

Saktrakulkla, P., Lan, T., Hua, J., Marek, R. F., Thorne, P. S., & Hornbuckle, K. C. (2020). Polychlorinated biphenyls in food. *Environmental Science & Technology, 54*(18), 11443–11452.

Seh-Bardan, B. J., Othman, R., Wahid, S. A., Husin, A., & Sadegh-Zadeh, F. (2012). Bioleaching of heavy metals from mine tailings by *Aspergillus fumigatus*. *Bioremediation Journal, 16*, 57–65.

Shah, M. P. (2013). Microbial degradation of textile dye (Remazol Black B) by *Bacillus* spp. ETL-2012. *Journal of Applied & Environmental Microbiology, 1*, 6–11.

Sharma, M., Nandy, A., Taylor, N., Venkatesan, S. V., Kollath, V. O., Karan, K., & Gieg, L. M. (2020). Bioelectrochemical remediation of phenanthrene in a microbial fuel cell using an anaerobic consortium enriched from a hydrocarbon-contaminated site. *Journal of Hazardous Materials, 389*, 121845.

Shedbalkar, U., & Jadhav, J. P. (2011). Detoxification of malachite green and textile industrial effluent by *Penicillium ochrochloron*. *Biotechnology and Bioprocess Engineering, 16*, 196.

Silva, A. F., Monteiro, M., Resende, D., Braga, S. S., Coimbra, M. A., Silva, A., & Cardoso, S. M. (2021). Inclusion complex of resveratrol with γ-cyclodextrin as a functional ingredient for lemon juices. *Foods, 10*(1), 16.

Smith, E., Thavamani, P., Ramadass, K., Naidu, R., Srivastava, P., & Megharaj, M. (2015). Remediation trials for hydrocarbon-contaminated soils in arid environments: Evaluation of bioslurry and biopiling techniques. *International Biodeterioration & Biodegradation, 101*, 56–65.

Sun, T., Liu, C., Wang, J., Nian, Q., Feng, Y., Zhang, Y., ... Chen, J. (2020). A phenazine anode for high-performance aqueous rechargeable batteries in a wide temperature range. *Nano Research, 13*(3), 676–683.

Tahir, E., Cordier, S., Courtemanche, Y., Forget-Dubois, N., Desrochers-Couture, M., Bélanger, R. E., ... Muckle, G. (2020). Effects of polychlorinated biphenyls exposure on physical growth from birth to childhood and adolescence: A prospective cohort study. *Environmental Research, 189*, 109924.

Verma, M. L., & Rani, V. (2020). Biosensors for toxic metals, polychlorinated biphenyls, biological oxygen demand, endocrine disruptors, hormones, dioxin, phenolic and organophosphorus compounds: A review. *Environmental Chemistry Letters*, 1–10.

Vitku, J., Heracek, J., Sosvorova, L., Hampl, R., Chlupacova, T., Hill, M., ... Starka, L. (2016). Associations of bisphenol A and polychlorinated biphenyls with spermatogenesis and steroidogenesis in two biological fluids from men attending an infertility clinic. *Environment International, 89*, 166–173.

Waigi, M. G., Kang, F., Goikavi, C., Ling, W., & Gao, Y. (2015). Phenanthrene biodegradation by sphingomonads and its application in the contaminated soils and sediments: A review. *International Biodeterioration & Biodegradation, 104*, 333–349.

Wang, C., Zhou, Z., Liu, H., Li, J., Wang, Y., & Xu, H. (2017). Application of acclimated sewage sludge as a bio-augmentation/bio-stimulation strategy for remediating chlorpyrifos contamination in soil with/without cadmium. *Science of the Total Environment, 579*, 657–666.

Wu, W. L., Deng, X. L., Zhou, S. J., Liang, H., Yang, X. F., Wen, J., ... Zou, F. (2018). Levels, congener profiles, and dietary intake assessment of polychlorinated dibenzo-p-dioxins/dibenzofurans and dioxin-like polychlorinated biphenyls in beef, freshwater fish, and pork marketed in Guangdong Province, China. *Science of the Total Environment, 615*, 412–421.

Xiang, Y., Xing, Z., Liu, J., Qin, W., & Huang, X. (2020). Recent advances in the biodegradation of polychlorinated biphenyls. *World Journal of Microbiology and Biotechnology, 36*(10), 1–10.

Yan, J., Niu, J., Chen, D., Chen, Y., & Irbis, C. (2014). Screening of Trametes strains for efficient decolorization of malachite green at high temperatures and ionic concentrations. *International Biodeterioration & Biodegradation, 87*, 109–115.

Yuanfan, H., Jin, Z., Qing, H., Qian, W., Jiandong, J., & Shunpeng, L. (2010). Characterization of a fenpropathrin-degrading strain and construction of a genetically engineered microorganism for simultaneous degradation of methyl parathion and fenpropathrin. *Journal of Environmental Management, 91*, 2295–2300.

Yue, S. J., Bilal, M., Guo, S. Q., Hu, H. B., Wang, W., & Zhang, X. H. (2018). Enhanced trans-2, 3-dihydro-3-hydroxyanthranilic acid production by pH control and glycerol feeding strategies in engineered *Pseudomonas chlororaphis* GP72. *Journal of Chemical Technology & Biotechnology, 93*(6), 1618–1626.

Zhang, Y. M., Fang, H., Zhu, W., He, J. X., Yao, H., Wei, T. B., ... Qu, W. J. (2020). Ratiometric fluorescent sensor based oxazolo-phenazine derivatives for detect hypochlorite via oxidation reaction and its application in environmental samples. *Dyes and Pigments, 172*, 107765.

Zhao, Q., Bilal, M., Yue, S., Hu, H., Wang, W., & Zhang, X. (2017a). Identification of biphenyl 2, 3-dioxygenase and its catabolic role for phenazine degradation in Sphingobium yanoikuyae B1. *Journal of Environmental Management, 204*, 494–501.

Zhao, Q., Yue, S., Bilal, M., Hu, H., Wang, W., & Zhang, X. (2017b). Comparative genomic analysis of 26 Sphingomonas and Sphingobium strains: Dissemination of bioremediation capabilities, biodegradation potential and horizontal gene transfer. *Science of the Total Environment, 609*, 1238–1247.

Further reading

Das, N., & Chandran, P. (2011). Microbial degradation of petroleum hydrocarbon contaminants: An overview. *Biotechnology Research International, 2011*.

Joyce, L. E., Aguirre, J. D., Angeles-Boza, A. M., Chouai, A., Fu, P. K.-L., Dunbar, K. R., & Turro, C. (2010). Photophysical properties, DNA photocleavage, and photocytotoxicity of a series of dppn dirhodium (II, II) complexes. *Inorganic Chemistry, 49*, 5371–5376.

Selvakumar, S., Manivasagan, R., & Chinnappan, K. (2013). Biodegradation and decolourization of textile dye wastewater using *Ganoderma lucidum*. *3 Biotech, 3*, 71–79.

Shumkova, E., Egorova, D., Boronnikova, S., & Plotnikova, E. (2015). Polymorphism of the bphA genes in bacteria destructing biphenyl/chlorinated biphenils. *Molecular Biology, 49*, 569–580.

Zhou, Y., Wei, J., Shao, N., & Wei, D. (2013). Construction of a genetically engineered microorganism for phenanthrene biodegradation. *Journal of Basic Microbiology, 53*, 188–194.

21

Metal oxide nanoparticles for environmental remediation

Roberta Anjos de Jesus[1], Geovânia Cordeiro de Assis[2], Rodrigo José de Oliveira[3], Muhammad Bilal[4], Ram Naresh Bharagava[5], Hafiz M.N. Iqbal[6], Luiz Fernando Romanholo Ferreira[1], Renan Tavares Figueiredo[1]

[1]INSTITUTE OF TECHNOLOGY AND RESEARCH (ITP), TIRADENTES UNIVERSITY (UNIT), ARACAJU, BRAZIL [2]CHEMICAL CATALYSIS AND REACTIVITY GROUP, INSTITUTE OF CHEMISTRY AND BIOTECHNOLOGY, FEDERAL UNIVERSITY OF ALAGOAS, MACEIÓ, BRAZIL [3]DEPARTMENT OF CHEMISTRY, STATE UNIVERSITY PARAÍBA, CAMPINA GRANDE, BRAZIL [4]SCHOOL OF LIFE SCIENCE AND FOOD ENGINEERING, HUAIYIN INSTITUTE OF TECHNOLOGY, HUAI'AN, P.R. CHINA [5]LABORATORY FOR BIOREMEDIATION AND METAGENOMICS RESEARCH (LBMR), DEPARTMENT OF MICROBIOLOGY (DM), BABASAHEB BHIMRAO AMBEDKAR UNIVERSITY (A CENTRAL UNIVERSITY), LUCKNOW, INDIA [6]TECNOLOGICO DE MONTERREY, SCHOOL OF ENGINEERING AND SCIENCES, MONTERREY, MEXICO

21.1 Introduction

Plastic materials have been very successful due to their low cost and weight, as well as their versatility and durability, making them present in almost all materials used in human everyday life (Rajmohan, Ramya, Raja Viswanathan, & Varjani, 2019; Rhein & Schmid, 2020; Rodrigues et al., 2019). Plastics use has increased 20-fold since 1964 and is expected to double by 2035 (Velis, 2014). The degradation of polymers has become a serious problem to be addressed because its accumulation is increasing day by day. Many degradation methods have been carried out but were effective only under certain conditions (Kulkarni & Dasari, 2018). The harmful effects to the environment caused by plastics in recent years have focused on the need for attention and awareness of plastic degradation. Synthetic plastics, such as polyethylene (PE), are widely used in packaging and other industrial materials and agricultural applications (Esmaeili, Pourbabaee, Alikhani, Shabani, & Esmaeili, 2013), however, its final treatment is not carried out correctly, and this influences an accumulation in the environment and, consequently, pollution in general.

Plastics are lighter and cost less compared to alternative materials. However, if not properly disposed or recycled, they can persist for long periods in the environment and can also degrade into small pieces that are worrying, called microplastics (MPs) (Wright & Kelly, 2017). MPs are very small particles of plastic material (usually smaller than 5 mm or much smaller, including nanoplastics) (Picó & Barceló, 2019). They are considered "secondary MPs" when they originate from the physical, chemical, or biological degradation of larger plastics. "Primary MPs," on the other hand, can be found as components of cleaning and hygiene products, cosmetics, paints, detergents, etc. Regardless of their origin, the vast majority of MPs end up in the environment (Auta, Emenike, & Fauziah, 2017). Once released into the environment, they can be accumulated by animals, including fish and shellfish, and, consequently, consumed as food by consumers.

Animal research has shown that some plastics can pass from the airways or gastrointestinal tract into the blood or lymphatic system, spreading and accumulating in other organs (Barboza, Vieira, Branco, Carvalho, & Guilhermino, 2018; Lönnstedt & Eklöv, 2016). The possibility of plastic making this journey likely depends on size, shape, type, and countless other characteristics. Once incorporated, these plastics can potentially cause inflammation or leaching.

Plastics are a class of very stable compounds and do not have easy degradation. Therefore, currently, environmental pollution caused by synthetic plastics is recognized as a major environmental, social, and economic problem (Poornima et al., 2015). Biodegradable plastics are of great interest to the scientific community. Thus, scientists are investigating an "environmentally friendly" way to replace synthetic plastics with bioplastics or to biodegrade plastics from fossil sources, due to the inability to delete them from everyday life (Song, Murphy, Narayan, & Davies, 2009). Several processes have been/are being studied and improved to investigate the ability to degrade polymers as microbial processes, for example (Satlewal, Soni, Zaidi, Shouche, & Reeta, 2008), and enzymatic (Scherer, Fuller, Lenz, & Goodwin, 1999), where this degradation usually occurs photolytically and chemically (Zheng, Yanful, & Bassi, 2005). Changes in molecular mass and temperature, among others, are essential for the results of degradation (Ahmann & Dorgan, 2007). An area that is receiving attention is the biodegradation of polymers with metal oxide nanoparticles (Dyshlyuk et al., 2020). Nanoparticles are considered to be influential in the profile of bacterial growth and have been incorporated to improve the degradation time of polymers, in addition to maintaining and improving important materials properties (Bhatia, 2013; Dyshlyuk et al., 2020).

Nanotechnology is a growing field that provides sustainable development at the atomic level (Girigoswami, 2018). Currently, nanoparticles are receiving enormous attention from industry and academia due to their potential in the development of new, highly active materials (Bhatia, 2013; Jeevanandam, Barhoum, Chan, Dufresne, & Danquah, 2018; Yoganandham Suman, Li, & Pei, 2020). Based on composition and morphology, there are several classes of nanoparticles, such as metallic nanoparticles (for example, Au, Pd, Pt) and metal oxides (for example, TiO_2), but also including organics like fullerenes, carbon nanotubes (CNTs) and cellulose, and starch nanocrystals (Wassel, El-Naggar, & Shoueir, 2020).

Research point that around 1814 products are being sold globally, containing nanoparticles that can be released into the environment (Vance et al., 2015). Exposure of ecosystems to nanomaterials is also expected to increase with nanomaterials use increasing (Gottschalk, Ort, Scholz, & Nowack, 2011; Yoganandham Suman et al., 2020).

21.2 Fundamentals of biodegradation of organic materials

One of the main processes that determine the fate of organic materials in the environment is degradation. Under environmental conditions, the degradation of organic materials is motivated by the coexistence of biotic and abiotic factors. The biotic system includes microorganisms. These, especially bacteria and fungi, have a wide prominence in the process due to their abundance, diversity of species, ability to adapt to a wide variety of environmental conditions, among other characteristics. In contrast, abiotic factors such as pH, temperature, humidity, salinity, and presence or absence of oxygen stimulate the benign growth of the microbial population and also influence their metabolic activity in degradation (Kumar & Maiti, 2016).

Numerous studies have been devoted to the search for alternative ways to degrade organic materials as photodegradation process (Griffini, Brambilla, Levi, Del Zoppo, & Turri, 2013; Hao et al., 2019), thermal degradation (Wiecinska, 2016), chemical degradation (Nabi, Sofi, Rashid, Ingole, & Bhat, 2020), and biodegradation (Milionis et al., 2019; Smagin, Sadovnikova, Vasenev, & Smagina, 2018). In contrast to the various degradation processes, biodegradation has received wide prominence because it is considered unique since the end result is often the complete conversion of the organic substance in its original chemical species or in other inorganic portions (for example, CO_2 and H_2O).

In this context, emerging organic pollutants (EOPs), especially polymers, have been the focus of many studies for presenting potential hydrolyzable ester bonds, which are converted into H_2O and CO_2 in aerobic conditions and CH_4 is produced in anaerobic condition by the action of microorganisms (Katarzyna Leja, 2010; Kumar & Maiti, 2016; Luckachan & Pillai, 2011).

21.2.1 Occurrence and environmental impact of emerging organic pollutants

Rapid population growth, agricultural, and industrial activities result in increased production and diversification of synthetic chemicals that cause various environmental impacts (Bernhardt, Rosi, & Gessner, 2017; Peng et al., 2018). These chemicals are substances released into the environment for which there are currently no regulations and are collectively called EOPs (Pal, Gin, Lin, & Reinhard, 2010).

With rapid advances in analytical techniques, new EOPs were discovered at an ever-faster rate (Castro-Perez & Prakash, 2020; Gama, Melchert, Paixão, & Rocha, 2019) in waters and effluents. They are mainly organic compounds present as pharmaceutical and personal care products, hormones, food additives, pesticides, plasticizers, wood preservatives, disinfectants, surfactants, flame retardants, and other organic compounds generated in the environment mainly by human activities.

EOPs have been attracting great attention since many are persistent in the environment, causing environmental impacts and risks to the health of living beings. According to EUROSTAT data in 2016, more than 36% of the total production of chemicals between 2007 and 2016 is represented by compounds harmful to the environment (Bolinius, Sobek, Löf, & Undeman, 2018). Therefore, many attempts have been made to clarify concerns about the release of EOPs into the environment and to promote related measures to avoid ecological risks.

In the last decade, there have been many studies on the occurrence of EOPs. The main sources and routes of EOPs are wastewater effluents from treatment plants; septic tanks; hospital effluents; livestock activities; underground storage of industrial waste, as well as indirectly through the groundwater and surface water exchange process (Kurwadkar, 2019; Lapworth, Baran, Stuart, & Ward, 2012). Benedetti et al. (Magi, Di Carro, Mirasole, & Benedetti, 2018) detected EOPs (five pharmaceutical products, two perfluorinated compounds, and caffeine) in drinking water treatment plants in northwest Italy, although in low concentration these pollutants are harmful to health and the environment (Magi et al., 2018). Montagner et al. (2018) detected in the Brazilian city with the highest population density 58 EOPs (9 hormones, 14 pharmaceuticals and personal care products, 8 industrial compounds, 17 pesticides, and 10 illicit drugs) in 708 samples from raw and treated sewage, surface waters, and underground and potable. Among these EOPs, only 22 compounds were within the acceptable limits for drinking water criteria (Montagner et al., 2018).

With the increasing advent of technology, EOPs, especially petroleum plastics, are the most inevitable part of our daily needs. Global plastics production has increased exponentially since 1950, with more than 311 million tons produced to date (Law, 2017). According to The American Chemistry Council, seven commodity thermoplastics account for 85% of the total demand for plastics for use in almost every sector of the market.

Plastics are a class of synthetic organic polymers composed of long, chain-like molecules with high average molecular weight (Law, 2017). Their properties depend on chemical structures where side chains play an important role (Bhatia, 2013).

During the conversion of resin to a product, a wide variety of additives are added to improve the performance and appearance of the plastic (including fillers, dyes, plasticizers, flame retardants, UV stabilizers, and antimicrobial agents, among others) (Hermabessiere et al., 2017; Law, 2017). As a result, there is a class of materials with highly versatile and desirable properties (including strength, durability, lightweight, thermal and electrical insulation, and barrier features) that can take various forms (such as adhesives, foams, fibers, and rigid or flexible solids) (Akindoyo et al., 2016). Although these additives provide plastics with beneficial properties for their applications, the transformations of these additives make plastics nondegradable and dangerous to health in some cases (Priyanka & Archana, 2011).

Although recycling is an environmentally attractive solution, there are restrictions in the process because a very small part of the plastic can be recycled and the rest goes to landfill sites (Eili, Shameli, Ibrahim, & Wan Yunus, 2012). Landfills are rarely satisfactory because incineration causes the generation of highly toxic smoke that is released into the environment, causing air pollution and consequently soil and surface waters.

Table 21.1 Estimated decomposition time for some commonly used plastics (Villegas Aguilar, 2018).

Plastic	Decomposition time (years)
Fishing line	±600
Plastic bottles	±500
Plastic cutlery	±400
Lighter	±100
Plastic glass	70–80
Plastic bag	±60
Balloon	±2

The vast majority of monomers used in the manufacture of plastics, such as ethylene and propylene, are derived from fossil hydrocarbons. None of the commonly used plastics are biodegradable (Geyer, Jambeck, & Law, 2017). PE waste is usually disposed of in landfills or water bodies as waste material for decomposition/degradation. These synthetic plastics are added at the rate of 25 million/year (Bhatia, 2013) and cause increasing impacts on the environment (da Costa, Santos, Duarte, & Rocha-Santos, 2016). As a result, they accumulate rather than decompose (Barnes, Galgani, Thompson, & Barlaz, 2009). For example, polyolefins that are used in packaging and materials of this nature due to their resistance to peroxidation, water, and microorganisms, are durable during use. Table 21.1 shows the estimated decomposition times for some plastics.

Thus, the current concern of scientists is to biodegrade petroleum plastics or generate new materials (bioplastics) that replace current plastics with similar physical properties and that are biodegradable, due to the inability to exclude them from everyday life.

21.2.2 Current overview of biodegradable materials

Biodegradation is defined as a biochemical process that involves the hydrolytic cleavage of chemical substances and electron transfer processes in redox reactions under certain environmental conditions caused by microorganisms following a defined path. The biodegradation process can be in two ways: (1) aerobic and (2) anaerobic (Kumar & Maiti, 2016; Lebrero et al., 2019; Nair & Laurencin, 2007).

$$\text{Polymer} + O_2 \rightarrow CO_2 + H_2O + \text{biomass} + \text{residue}_{(s)} \tag{1}$$

$$\text{Polymer} \rightarrow CO_2 + CH_4 + H_2O + \text{biomass} + \text{residue}_{(s)} \tag{2}$$

If oxygen is present, aerobic biodegradation occurs and carbon dioxide is produced. If there is no oxygen, anaerobic degradation occurs and methane is produced. Biodegradation is expected to be the main loss mechanism for most chemicals released into the environment as it results in the production of biomass, carbon dioxide, and methane that are considered green. The materials that are being degraded by this mechanism are known as biodegradable materials.

In this context, biodegradable polymers (bioplastics), such as polybutylene succinate (PBS), polyhydroxyalkanoates (PHA), polylactic acid (PLA), polycaprolactone (PCL) have been commercially synthesized in recent years as a sustainable alternative to replace the synthetic plastics that are recognized as the main environmental pollutants of solid waste (Lackner, 2015; Wróblewska-Krepsztul et al., 2018; Zhong, Godwin, Jin, & Xiao, 2020).

Bioplastics are classified into two groups according to the Japan BioPlastics Association (Kumar & Maiti, 2016), a country that is a leader in the area of Biotechnology: (1) biodegradable plastics that refers to synthetic biodegradable polymers with an average molecular weight of at least 1000 Da (this classification includes chemically modified starch and biodegradable polymers based on polyamine acids); and (2) biomass-based plastics that are produced chemically or biochemically using renewable organic products as materials.

Bioplastics can be obtained in three ways: (1) extracted directly from biomass, such as polysaccharides (chitosan, starch, and cellulose), proteins (gluten and soy), and lipids; (2) obtained from monomers derived from biomass that require chemical substances for their conversion (for example, the lactic acid monomer intermediate, obtained from the corn starch fermentation stock, requires a chemical transformation for its conversion into PLA); and finally, (3) microbial such as PHA, PBS, and their derivatives produced by natural microorganisms or genetically modified organisms that accumulate intracellularly in the form of storage granules (Kumar & Maiti, 2016).

The susceptibility of polymeric materials to biodegradation does not depend on the starting material of a specific polymer but is related to its structure and chemical composition. Among bioplastics, polyesters have been extensively investigated due to their potentially hydrolyzable ester bonds that can be broken down to their respective monomer or oligomer units in bioactive environments (Lim, Raku, & Tokiwa, 2005; Mierzwa-Hersztek, Gondek, & Kopeć, 2019).

An overview of polyester bioplastics with similar and comparable characteristics to synthetic polymers and which have an environmentally friendly role in the world leads to PLA and PHA.

PLA is derived from 100% renewable resources (e.g., corn, beet, and starch) and offers great promise in many commodity applications due to its low cost and improved properties. PLA is obtained from the monomer lactic acid, which is produced from the microorganism-catalyzed fermentation of sugar or starch (Lackner, 2015). PLA has been widely accepted as a biodegradable polymer for packaging materials due to its rigidity, transparency, processability, and biocompatibility (Kumar & Maiti, 2016; Rasal, Janorkar, & Hirt, 2010; Zhong et al., 2020).

Another promising material applied in the packaging, medicine, and agriculture sector is PHA, which is synthesized from renewable resources (corn, sugar cane, cellulose, chitin, and others) from microbial fermentation. A commercial application is 3-hydroxybutyrate-co-3-hydroxy-valerate (PHBV) marketed under the name of Biopol (Kumar & Maiti, 2016; Zhong et al., 2020).

It is important to mention the incorporation of clays in nanocomposites. The clays most commonly used in the field of nanocomposites belong to the 2:1 layer silicate family, also called 2:1 phyllosilicate [montmorillonite (MMT), saponite, LAPONITE]. Its structure consists of layers composed of two silicon atoms coordinated tetrahedrally, fused to an octahedral sheet shared by the edge of aluminum or magnesium hydroxide (Bordes, Pollet, & Averous, 2009). MMT has been reported as a favorable approach to improve the characteristics of materials based on biopolymers (de Moraes, Müller, & Laurindo, 2012). However, according to Müller, Laurindo, and Yamashita (2012), these improvements are strongly linked to the nature of the clay (hydrophilic or hydrophobic). MMT is a low cost, hydrophilic, and ecological nanoparticle with a high specific area (Müller et al., 2012). Nanoclays have already been demonstrated to be introduced in natural rubber or polypropylene to acquire nanocomposites with better conductivity or mechanical performance (Levchenko et al., 2011). It has been shown in the literature that MMT can also potentially be used to control the release of the antimicrobial agents. According (Tunç & Duman, 2011), the amount of release of antimicrobial agent from synthesized films can be controlled with MMT concentration and temperature. These films have potential as active materials for food packaging in the food industry (Tunç & Duman, 2011).

The preparation of bionanocomposites, using low percentages of inorganic fillers, is one of the ways to improve some of the properties of biodegradable polymers, such as thermal, mechanical, and oxidative barriers. Bentonite is one of the most widely used lamellar silicates containing or clay-mineral montamilonite. This is because it is environmentally friendly and available in large quantities at a relatively low cost. (Júnior et al., 2019). Júnior et al. (2019), also investigated vermiculite clay.

Currently, bioplastics have important roles in the contemporary world due to their applications to environmental safety, biomedical implementation, packaging, among others. However, the main limitations for large-scale production of bioplastics are related to the high costs of production and difficulty of recovery through the fermentation process.

21.2.3 Factors that affect the biodegradation of polymers

Biodegradable materials decompose in the environment, where it is discarded, in the approximate period of one year, through natural biological processes, in nontoxic carbonaceous soil, water, and carbon dioxide (Katarzyna Leja, 2010; Zee, Stoutjesdijk, Heijden, & Wit, 1995). Biotic and abiotic factors, together with the chemical and physical characteristics of polymers, affect biodegradability (see Fig. 21−1; Mierzwa-Hersztek et al., 2019; Siracusa, 2019).

Abiotic parameters (temperature, pH, humidity, salinity, presence or absence of oxygen, sunlight, and photooxidation, among others) not only affect the polymers to be degraded but also influence biotic factors. The biotic factors that can cause biodegradation of polymers include, among others, microbial diversity (bacteria, fungi, among others), biosurfactants produced by microbes for fixation on the surface, and potential action of enzymatic activity, among others (Kumar & Maiti, 2016).

The chemical and physical characteristics of polymers affect the extent of biodegradation. The chemical structure (responsible for the stability of the carboxyl or hydroxyl functional group, reactivity, and hydrophilicity) is the most important factor that affects the biodegradability of polymeric materials (Katarzyna Leja, 2010; Park & Xanthos, 2009;

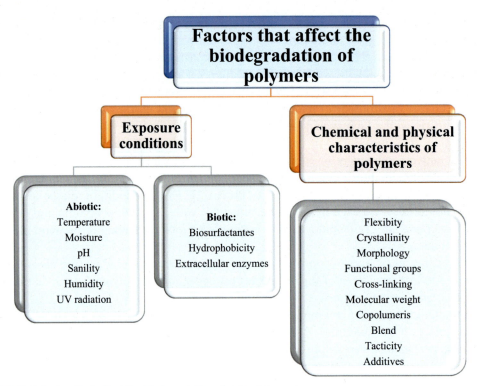

FIGURE 21–1 Factors that affect the biodegradation of polymers.

Tokiwa & Calabia, 2006). Other important factors for determining the kinetics and biodegradation mechanism of polymers are influenced by the surface morphology (surface area, shape, and size), intrinsic properties (monomeric composition, type of enantiomer, cross-linking, molecular mass and its distribution), high-order configurations (glass transition and melting temperature, among others), and processing conditions (crystallinity ratio and amorphous region, copolymerization, choice of additives and organic/inorganic fillers) typical of each class of polymers (Kumar & Maiti, 2016; Park & Xanthos, 2009; Tokiwa & Calabia, 2006).

The lower the molecular weight of the polymers, the greater the facility for biodegradation, as this way the molecules can be assimilated by microorganisms without having to perform extracorporeal digestion of the substrate. Thus, aromatic polyesters and polyurethanes have a lower rate of biodegradation compared to aliphatic polyesters. These, in turn, are more susceptible to biodegradation due to their lower molecular weight monomeric composition (Katarzyna Leja, 2010; Kumar & Maiti, 2016; Madhavan Nampoothiri, Nair, & John, 2010; Tokiwa & Calabia, 2006).

The crystallinity fraction and amorphous domain of polymers play a crucial role in determining the rate of biodegradability. A polymer with many crystalline portions has less biodegradability, because in these parts the diffusion of water is less, making it difficult for the enzymes to access restricted to the amorphous part of the polymer. Besides that, the crystalline structure the polymer chains are organized; consequently, the crystalline polymers are more resistant to degradation whereas in the amorphous domains the polymer chains are weakly compacted and, therefore, more susceptible to undergo microbial attacks and enzymatic catalysis (Kumar & Maiti, 2016). Jenkins and Harrison (Jenkins & Harrison, 2008) studied the degradation of PCL in a lipase solution evaluating the effect of molecular weight and degree of crystallinity separately. They observed that an increase in molecular weight or degree of crystallinity reduced the rate of degradation (Jenkins & Harrison, 2008).

Although the surface area is a relevant factor in the biodegradation of polymers, the effects of this factor on the rates of biodegradation are less frequently reported in the literature. Kunioka, Ninomiya, and Funabashi (2006) studied the biodegradation rate of PLA in powder form. They observed that smaller particles (0–125 μm) degrade more quickly than particles with larger sizes ($\geq$125 μm) (Kunioka et al., 2006).

Under controlled conditions on a laboratory scale, Chinaglia, Tosin, and Degli-Innocenti (2018) investigated the effect of the surface area of different polymers on the behavior of biodegradation. The microorganisms secrete enzymes that break the polymer down into smaller molecular blocks, hydroxy acids, which are used as a carbon source for the growth of these microorganisms. They observed a higher rate of biodegradation in polymers with a high surface area, since the higher, the greater the amount of enzyme per polymeric surface area (Chinaglia et al., 2018).

In general, the high melting temperature reduces biodegradation of the polymer due to interactions between polymer chains that affect the value of the heat of fusion (ΔH) and the internal rotational energies corresponding to the rigidity (or flexibility) of the polymer molecules (Kumar & Maiti, 2016; Tokiwa, Calabia, Ugwu, & Aiba, 2009). Tokiwa and Calabia observed that low molecular weight PLA and its copolymers are hydrolyzed by lipase due to the greater diffusion of microorganisms in the biodegradation medium, but poly(glycolic acid) (PGA) and high molecular weight PLA polymer are not easily hydrolyzable and are resistant to enzymatic degradation due to their high crystallinity, high temperature of fusion, and greater diffusion difficulties for biodegradation agents (Tokiwa & Calabia, 2006).

21.3 Performance of metal oxide nanoparticles in the biodegradation of organic matter

The preparation of conventional mixtures or composites using inorganic or natural fillers, respectively, are among the routes to improve some of the properties of biodegradable polymers. Thermal stability, gas barrier properties, resistance, and low viscosity, are among the properties that can be achieved by these multiphase systems (Samantaray et al., 2020). The incorporation of nanomaterials in biodegradable polymers holds strong promise in the design of ecological composites for different applications (Armentano et al., 2018). Such composites are called nanocomposites (Giannelis, 1996). These new composites are significant due to their nanoscale dispersion, even with a very low level of incorporation of nanomaterials ($\leq$5% by weight), resulting in materials with a high surface area (Sinharay & Bousmina, 2005).

The development of nanotechnology and its application in various sectors has been growing in recent years (Denet et al., 2020). In recent decades, nanotechnology has attracted intense research interest in the field of environmental remediation (Cai, Zhao, Yu, Rong, & Zhang, 2019; Kabir, Kumar, Kim, Yip, & Sohn, 2018). Nanoparticles are defined by the world federation of national standardization bodies, the International Organization for Standardization (ISO), as nanoobjects with all external dimensions at the nanoscale, where the lengths of the longest and shortest axes of the nanoobjects do not differ significantly (ISO/TS 80004-2:2015). Although the nanoscale basically varies from 1 to 100 nm (Liu, Jiang, et al., 2020; Mohajerani et al., 2019), the nanoparticles can be classified into three size ranges: greater than 500 nm, between 100 and 500 nm and between 1 and 100 nm (Commission, 2010). Regarding size and size distribution, nanoparticles can exhibit intensive size-related properties (Nam & Luong, 2019). The nanoparticles mainly include metal-based nanometric particles (metallic oxides and nanoparticles containing metals), clay minerals, graphene, activated carbon, and CNTs (Jeevanandam et al., 2018; Lei, Sun, Tsang, & Lin, 2018).

Currently, many metallic nanomaterials are being developed, such as titanium, copper, zinc, gold, silver, magnesium, and so forth (Salem & Fouda, 2020). The nanoparticles are being widely used for various purposes, such as biomedical applications, industrial production, solar fuel batteries, and energy storage, as well as applied in cosmetics and in the textile sector. It is known that the phases, sizes, and morphologies of nanomaterials significantly affect their properties and possible uses. Consequently, your personalized syntheses must be taken into account (Atta, Al-Lohedan, El-Saeed, Al-Shafey, & Wahby, 2017; Dubchak, Ogar, Mietelski, & Turnau, 2010; Honig & Spałek, 1998; Wassel et al., 2020). Polymeric matrices incorporated with nanoparticles often exhibit mechanical properties and several other notably improved properties when compared to pure polymers. The improvements generally include a higher modulus of elasticity, both in the solid and in the molten state, greater resistance and thermal stability, less permeability to gas, and mainly greater biodegradability. The main reason for these enhanced properties in nanocomposites is the stronger interfacial interaction between the polymeric matrix and the nanoparticles (Bordes et al., 2009; Sinharay & Bousmina, 2005).

The presence of nanoparticles influences the growth capacity of microbes that are considered polymer degraders (Pathak & Kumar, 2017). In addition, the incorporation of nanoparticles in the polymeric matrix or the manufacture of nanocomposites overcomes the deficiencies of biodegradable polymers, such as fragility and poor thermal and mechanical properties (Kumar & Maiti, 2016). An of the great advantages of nanoparticles is the ability to adjust the rate of biodegradation (it is possible to increase and decrease the rate compared to that of pure polymer), depending on the need. Thus, the chemical, physical, and biological properties of biodegradable polymers can be modified and controlled for sustainable applications (Kumar & Maiti, 2016). Fig. 21–2 illustrates the different sources of microplastic particles, in addition to other pollutants (household, industrial, urban, and waste products containing plastic particles) and the possible routes by which these particles are released into the environment. It also reveals the potential for the dissemination of microparticles the environment and its impact of contamination on water, soil, and air. The illustration also shows the role of nanoparticles in the biodegradation of these contaminants.

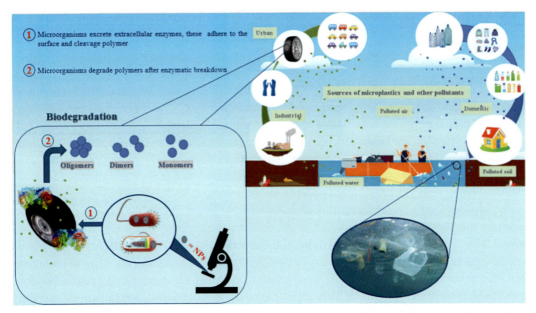

FIGURE 21–2 Sources of microplastic particles (domestic, industrial, urban, and waste products containing plastic particles) and the possible routes by which these particles are released into the environment. It also reveals the potential for biodegradation.

As the size of a solid particle decreases, the number of atoms that make up the particle becomes small (Sharma et al., 2019). In this state, fundamental physical properties, such as the melting point, can change considerably. As a result, nanoparticles show electromagnetic or physicochemical properties completely different from their properties when they are not in the nanometric order, although they are made of the same materials (Mourdikoudis, Pallares, & Thanh, 2018). The definition of nanoparticles differs depending on the materials, fields, and applications in question. In the strictest sense, they are particles that have at least a dimension less than 100 nm are considered, where the physical properties of the solid materials themselves would change dramatically. On the other hand, as already described in the literature, particles in other ranges can also be called nanoparticles (Jeevanandam et al., 2018; Mohajerani et al., 2019). Various types of nanoparticles, produced by different methods, are applied as a raw material in different fields (Khan, Saeed, & Khan, 2019).

In particular, in recent years, the literature has witnessed impressive progress in investigations into polymeric matrices a support for anchoring metallic nanoparticles. This is due to several factors, including polymer chains are flexible and can contain a variety of functional groups capable of efficiently immobilizing nanoparticles and their precursors by dispersive bonds or van der Waals. The chemical structure of the polymer also contributes to preventing particle aggregation (Abushrida et al., 2020).

Some studies have already demonstrated the influence of metallic nanoparticles incorporated in polymers. According to Liu, Li, Yuan, and Yang (2020), Liu, Jiang, et al. (2020), the nanoparticles act as a nucleating agent and, thus, control the polymeric matrix that further controls the rate of biodegradation. Silicate nanocomposites in polylactide layers (PLA-30B) showed considerable improvement in biodegradation compared to the pure polymeric matrix.

21.3.1 Metal oxides nanoparticles

Metal oxide nanoparticles have been used successfully as an adsorbent, reducer, oxidizer, and catalyst to remove various heavy metals (Siddiqui, Naushad, & Chaudhry, 2019; Zou et al., 2016), organic contaminants and other inorganic pollutants from aqueous solutions (Lu, Dong, Fan, Zuo, & Li, 2017; Santhosh et al., 2016). Metal oxide nanoparticles can also be used to remedy soil contaminated by organic contaminants and heavy metals (Li et al., 2016; Liu, Jiang, et al., 2020).

The synthesis and characterization of nanoparticles is an important area of research in materials chemistry. Especially relevant is the use of nanoparticles in nanotechnology applications. These applications generally depend on specific physical characteristics of the nanoparticles, such as size, morphology, and crystallinity, which are influenced by the method used to produce the particles (Diamandescu, Mihaila-Tarabasanu, Popescu-Pogrion, Totovina, & Bibicu, 1999; Kandori, Yamamoto, Yasukawa, & Ishikawa, 2002). Among the various methods studied, each offers a relatively different route to produce metal oxide nanoparticles of specific size and morphology (Xu & Teja, 2008). Some of the methods used to synthesize nanoparticles are discussed below.

21.3.2 Synthesis and postmodification of metal oxide nanoparticles

Coprecipitation method: The coprecipitation method has often been reported to prepare nanoparticles. It is a simple yet powerful method to obtain nanoparticles with adjustable components and various functions, avoiding sophisticated synthesis and unwanted additives (Kapuria, Sharma, Kumar, & Koner, 2018; Wang et al., 2020). In this method, the synthesis process is dominated by brief nucleation and then the growth of nuclei on the surface of the crystal by diffusion of the solute. In the coprecipitation method, a metal hydroxide is formed by converting the salt precursor into the aqueous medium with the addition of sodium hydroxide or ammonium hydroxide. To obtain the nanoparticles, the chloride salts formed are washed and the hydroxide heated. This method does not allow the control and size distribution, being a disadvantage. However, it can be adjusted by changing factors such as ionic strength, pH, temperature, and salt used during the reaction process (Schwarzer & Peukert, 2004).

Sol-gel method: The sol-gel method is relatively more convenient and cheaper for the synthesis of nanoparticles compared to the other methods reported (Aly, Abed Alrahim Mohammed, Al-Meer, Elsaid, & Barakat, 2016; Charmforoushan, Roknabadi, Shahtahmassebi, & Malaekeh-Nikouei, 2020; Wang, Zhang, Yang, & Sun, 2008). It is a method that involves condensation and hydroxylation of precursor molecules, it is often considered a wet method. Factors such as temperature, pH of the gel, agitation, and concentration of precursors are responsible for the regulation of hydrolysis and condensation reactions (Ennas et al., 1998). In this method it is possible to control the morphology, for example, an amorphous nanostructure or monodisperse phase can be formed. In addition, the sol-gel method offers several advantages for the doping process, such as the lower process temperature allows to avoid any unwanted doping phase. In this method it is also possible to work with low concentrations of metallic precursors (traces), facilitating insertion into the oxide structure. Another advantage of the method is the homogeneous distribution of metallic cations (Ba-Abbad, Takriff, Benamor, & Mohammad, 2016; Bel Hadj Tahar, Ban, Ohya, & Takahashi, 1998). TiO_2 and ZnO are examples of metallic nanoparticles that were synthesized by this method (Raghunath & Perumal, 2017).

Pechini method: The Pechini method, also known as the polymeric precursor method, includes the formation of a polymeric resin by combining cations, hydroxycarboxylic acids (typically citric acid) and polyalcohols (ethylene glycol). In this method, a polymeric resin is obtained by means of chelation and polyesterification processes, then the viscous resin is dried and calcined to produce pure oxide nanostructures. The Pechini method uses metal salts as a source of cations instead of metal alkoxides, which is independent of the control of the hydrolysis rate (Nishio, Seki, Thongrueng, Watanabe, & Tsuchiya, 1999). Homogeneous distribution of metal cations, good reproducibility, size and morphology control, synthesis of compounds controlled by stoichiometry at low process temperature are the main advantages of this method (Dimesso, 2018; Mersian, Alizadeh, & Hadi, 2018). Nanoparticles of iron(III) oxide, TiO_2, and zirconia were obtained from this method and their properties were investigated (Aflaki & Davar, 2016; Mashreghi & Ghasemi, 2015; Vargas, Diosa, & Mosquera, 2019).

Hydrothermal method: Metal nanoparticles using this method are generally obtained in aqueous media, using autoclaves under constant pressure and temperature to control the size and shape of the nanoparticles. Compared with other methods, nanoparticles prepared from the hydrothermal method have advantages such as high particle purity, good dispersity, use of relatively low temperatures, and uniform particle size (Chang et al., 2020; Yang et al., 2019). As demonstrated by Wang, Xie, Yan, and Duan (2011), using the hydrothermal method and changing certain reaction conditions, it is also possible to obtain materials with different morphologies (Wang et al., 2011). Common examples of metal oxide nanoparticles effectively synthesized by the hydrothermal method are TiO_2, ZnO, and CeO_2 nanoparticles (Goto, Shin, Yokoi, Cho, & Sekino, 2020; Lee, Cheng, & Lee, 2020; Tok, Boey, Dong, & Sun, 2007).

Wet-chemical method: Among the synthesis methods, the wet-chemical method is more popular for large-scale synthesis, due to the ease of processing and cost-effectiveness (Li, Hong, Wang, Yu, & Qi, 2009). However, it has certain limitations, such as the formation of larger grains, in addition to causing coalescence. The coalescence process produces interconnected irregularly shaped grains. This nonuniform distribution and a larger grain size affect the physicochemical properties of the synthesized nanoparticles (Arote et al., 2019). It involves a mixing metal precursor in ultrapure water, then stirring for predetermined times, and then heating and centrifuging. Fe_3O_4 and ZnO nanoparticles were prepared using this method (Wu, Zheng, & Wu, 2005).

Other methods employed in the synthesis of nanoparticles include sonochemical method (Hassanjani-Roshan, Vaezi, Shokuhfar, & Rajabali, 2011), green synthesis (Abboud et al., 2014), electrochemical (Pandey, Merwyn, Agarwal, Tripathi, & Pant, 2012), microwave

(Roy & Bhattacharya, 2012), microemulsion (Uskoković & Drofenik, 2005), thermal decomposition (Navaladian, Viswanathan, Viswanath, & Varadarajan, 2007), and solvothermal methods (Fernández-García & Rodriguez, 2011).

It is important to point out that all the properties of the synthesized nanoparticles, such as crystallinity, surface area, morphology, uniform distribution of metallic cations, optical and electronic, magnetic properties, among others, are confirmed from detailed investigations carried out by numerous characterization techniques (see Fig. 21–3) (Fasiku, John Owonubi, Malima, Hassan, & Revaprasadu, 2020).

Thus, these characterizations presented in Fig. 21–1, in addition to other existing ones, contribute strongly to the evaluation of more assertive applications based on theoretical and experimental foundations of the techniques. Currently, the preparation and characterization of transition metal oxide nanoparticles and their application as nanocatalysts have become an exciting and emerging research area. Unlike the same bulk materials, metal oxide nanoparticles such as iron oxide, zinc oxide, nickel oxide, titanium oxide, vanadium oxide, and copper oxide have important and specific optical, thermal and physicochemical properties due to its shape, surface area, and size (Benhammada et al., 2020; Herlekar, Barve, & Kumar, 2014).

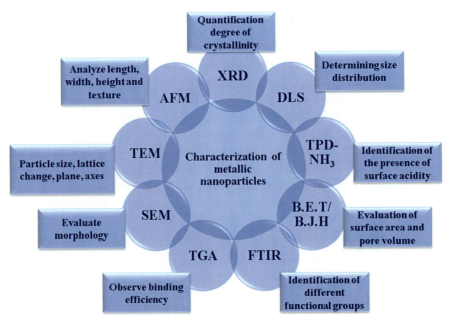

FIGURE 21–3 Summary of characterization techniques used in investigating the properties of nanoparticles. *Adapted from Fasiku, V. O., John Owonubi, S., Malima, N. M., Hassan, D., & Revaprasadu, N. (2020). Metal oxide nanoparticles: A welcome development for targeting bacteria. In Antibiotic materials in healthcare (pp. 261–286). https://doi.org/10.1016/B978-0-12-820054-4.00015-X.*

21.3.3 Metal oxide nanoparticles as catalysts for environmental remediation

Fe_3O_4: Among these metal oxide nanoparticles, iron oxide is of great interest and has many outstanding characteristics such as super and paramagnetic, high reactivity, biodegradability, availability, and biocompatibility, among others (Jagathesan & Rajiv, 2018; Mahdavi, Namvar, Ahmad, & Mohamad, 2013). Iron oxide is a transition metal that has different stoichiometric and crystalline structures, including wustite (FeO), hematite (α-Fe_2O_3), maghemite (ν-Fe_2O_3), and magnetite (Fe_3O_4). In all phases, hematite (α-Fe_2O_3) is the most stable state of iron oxide under environmental conditions, also known as ferric oxide, has a blood-red nanoscale color (Cornell & Schwertmann, 2003). In crystallizes in the space group R_3c with a hexagonal or rhombohedron crystallographic system, in which iron cations occupy the octahedral interstitial sites (Klotz, Le Godec, Strässle, & Stuhr, 2008). Hematite exhibits symmetry in C_3v (Sivula, Le Formal, & Grätzel, 2011) and there are two different FeO bond lengths (Cornell & Schwertmann, 2003; Mishra & Chun, 2015). Stimulated by the promising applications of iron oxides and by the new chemical and physical properties of nanoscale materials, considerable efforts were made in the synthesis of α-Fe_2O_3 nanostructured materials with different morphologies. So far, a variety of nanostructured α-Fe_2O_3 materials in various geometric morphologies have been successfully manufactured, such as nanoparticles (Woo et al., 2004) nanotubes (Chen, Xu, Li, & Gou, 2005), nanocubes (Zheng et al., 2006), and spheres (Gou, Wang, Park, Liu, & Yang, 2008; Zhang et al., 2009).

The literature has addressed that the use of nanoparticles based on iron oxide, such as magnetite, can accelerate the biodegradation of organic matter (Cruz Viggi et al., 2014). Besides, research has shown that magnetite can also promote the biodegradation of organic pollutants (Aulenta, Rossetti, Amalfitano, Majone, & Tandoi, 2013). With that, Yang et al. (2017) observed that the addition of magnetite in digested anaerobic sludge accelerated the biodegradation of ciprofloxacin during anaerobic digestion (Yang et al., 2017). According to the results observed by Aulenta et al. (2013), the application of nanoparticles is seen as a promising strategy to increase and favor interactions in the biodegradation process.

V_2O_5: V_2O_5 is the most promising metal oxide in the transition metal oxide series. Concerning its orthorhombic layered structure, it is a very important material for energy storage such as lithium-ion batteries, due to its excellent specific capacity. V_2O_5 is one of the most abundant transition metal oxides in the earth's crust, a low-cost, nontoxic, and highly ecological material. It is known that the electrochemical performance of size nanoparticles is greater than that of other particles of the same size (Dhoundiyal, Das, & Bhatnagar, 2020; Menezes et al., 2009; Pan et al., 2015). In the transition metal oxide series, V_2O_5 gains more attention due to its special physical and chemical properties and makes this compound more applicable as we advance in the area of nanosized particles (Yan et al., 2009). Vanadium has a series of oxidation states, which are stable for a specific composition range (VO, VO_2, V_2O_3, V_6O_{11}, and V_2O_5). Among these compositions, V_2O_5 has the highest oxidation state of vanadium (Dhoundiyal et al., 2020; Kumar & Maiti, 2016). Vanadium pentoxide is also known for its structural and electronic phase transition, which exhibits the transition

from metal to insulator and metal to semiconductor (Bahlawane & Lenoble, 2014). The properties of V_2O_5 include n-type conductivity, infrared reflectance and transmittance, 2.38 eV band gap, magnetic susceptibility, variable oxidation states, high specific capacity, high energy density, electrical resistivity, and so on. (Kim et al., 2000). Due to these excellent properties, V_2O_5, as well as the materials derived from it, are widely used as cathode materials in batteries, thermoelectric devices, storage medium, sensors, switching devices, electro-optical device, catalysis, field-effect transistors (Dhoundiyal et al., 2020; Singh et al., 2018). Vanadium oxide nanostructures have attracted a lot of attention due to their potential to improve electrochemical, optical, and thermochromic applications (Warwick & Binions, 2014). This property resulted in the extensive investigation of vanadium oxides in electrical switches and various sensors (Chou et al., 2008; Le, Kang, & Kim, 2019).

TiO₂: Titanium dioxide (TiO_2) nanoparticles have been extensively investigated as catalysts for the photodegradation of organic compounds due to their unique physical and chemical properties (Cho, Choi, Lee, Hyeon, & Lee, 2001; Nasikhudin, Diantoro, Kusumaatmaja, & Triyana, 2018). TiO_2 is one of the metal oxides with antibacterial activity. Its effectiveness as an antibacterial agent depends on several factors related to the characteristics of the TiO_2 particles used, including morphology, level of dispersion, and distribution and composition of the anatase and rutile crystalline phases (Sievers, Pollo, Corção, & Medeiros Cardozo, 2020). Four main phases of TiO_2 crystal were explored, namely: anatase, rutile, brookite, and TiO_2 (B). Rutile is a thermodynamically less stable phase, but anatase is the most stable phase at the nanoscale (Cargnello, Gordon, & Murray, 2014; Chen & Caruso, 2013; Pan, Wang, & Bahnemann, 2014). Anatase is the polymorph widely studied, showing superior photochemical performances due to the high mobility, affinity, and transmittance of electrons by visible light (Jiang et al., 2018). Anatase and rutile are the most common phases of TiO_2 (Qian et al., 2019). The brookite phase of TiO_2 has rarely been studied due to the challenges of obtaining the pure phase (Chen & Mao, 2007). In contrast to the other three natural polymorphic forms, the artificial TiO_2 (B) has four TiO_6^{2-} octahedron sharing edges with an open channel parallel to the b axis that lies between the axial oxygen (Hu, Yu, Gao, Lin, & Lou, 2015; Marchand, Brohan, & Tournoux, 1980; Motakef-Kazemi, Shojaosadati, & Morsali, 2014). Several studies have shown that TiO_2 (B) shows photocatalytic activity comparable to TiO_2 anatase, and has been investigated in this regard (Liu et al., 2012). Properties such as high stability, anticorrosion, and excellent photocatalyst, led to the abundant obtaining and wide use of TiO_2 (Jiang Et Al., 2018; Riu, Maroto, & Rius, 2006). The physicochemical properties of titanium oxide nanoparticles are mainly affected by their shape, size, surface characteristics, optical and electronic properties, and internal structure. It has been reported that a decrease in the size of titanium oxide nanoparticles leads to harmful effects on human health, therefore raising considerable concerns (Shah, Shah, Hussain, & Khan, 2017). Approximately 70% of the total pigment production volume in the world is based on titanium oxide nanoparticles, so TiO_2 is among the five most-used nanoparticles in consumer products (Shi, Magaye, Castranova, & Zhao, 2013). Common physical-chemical methods used in the preparation of titanium oxide nanoparticles require high temperature, pressure, and toxic chemicals that limit their production and potential application in the health field (Khan et al., 2019).

Its application is found in several sectors, such as coatings/paints, toothpaste, sunscreens, pharmaceuticals, and medicines (Baranowska-Wójcik, Szwajgier, Oleszczuk, & Winiarska-Mieczan, 2020). Other applications include self-cleaning systems such as windows, textiles, and car anticlog mirrors (Isaifan et al., 2017). In the field of nanomedicine, some potential applications of titanium oxide nanoparticles as agents for the treatment of tumors, imaging tools, and nanotherapeutics have been reported (Bogdan, Pławińska-Czarnak, & Zarzyńska, 2017). Also, under irradiation with UV light, titanium oxide nanoparticles exhibited antibacterial and antimicrobial properties (López de Dicastillo, Guerrero Correa, Martínez, Streitt, & José Galotto, 2020).

Antimicrobial titanium (TiO_2) nanocomposites have been actively investigated in recent years (Kubacka et al., 2015), this, due to the semiconductor nature of TiO_2 that can be activated in the presence of UV or visible radiation and generate reactive oxygen species (Jalvo, Faraldos, Bahamonde, & Rosal, 2017). The inclusion of titanium dioxide in composites can also be a very effective way of providing a biodegradation function for biodegradable polymers (Ando, Kawasaki, Yamano, Uegaki, & Nakayama, 2015). Asghar et al. (2011), studied the degradation of PE using TiO_2 nanoparticles. The results revealed that the PE-TiO_2 composition has the potential to degrade PE residues without any secondary pollution. Asghar et al. (2011), still demonstrated a general tendency for a degradation reaction (Asghar et al., 2011). Soitong and Wongsaenmai (2017) also manufactured photodegradable composites based on PE incorporated with TiO_2 and poly(ethylene oxide) (PEO), and the results showed that the composite presented an efficiency superior to the pure polymer, attributing the best result to the presence of TiO_2 (Soitong & Wongsaenmai, 2017).

ZnO: Zinc oxide is an important material with applications in several industrial sectors (Mohammadi, Aliofkhazraei, Hasanpoor, & Chipara, 2018), due to its specific chemical and physical properties, such as high thermal stability (Cheng, Xin, Leng, Yu, & Wang, 2008), mechanics (Bacaksiz et al., 2008) and chemistry (Auffan, Rose, Wiesner, & Bottero, 2009). ZnO is a direct gap metallic oxide (3.37 eV at room temperature), is classified as a semiconductor II–VI, with excellent optical and electrical properties (Liu, Li, et al., 2020). Because it has wide gap energy and great bonding energy at room temperature, it is a good candidate for transparent conductive oxide materials, which is also influenced by its high conductivity, good optical transmittance, and low-cost manufacturing (Lei, Wu, Hsu, & Lee, 2012). In addition to the properties mentioned above, ZnO has some particular properties, such as low toxicity, biocompatibility, and biodegradability (Huang, Xia, Cao, & Zeng, 2008; Sapnar et al., 2012) making it an interesting material for applications in biomedicine and ecological systems for the purpose of biodegradation (Sapnar et al., 2012; Yi, Wang, & Park, 2005).

In this context, Capelezzo et al. (2018), investigated the insertion of ZnO nanoparticles in a biodegradable polymer, resulting in a biodegradable polymeric film with antimicrobial properties (Capelezzo et al., 2018). In the same line of investigation, Agustin and Padmawijaya (2016), made bioplastics with an antimicrobial capacity to contain or inhibit the growth of pathogenic microorganisms (Agustin & Padmawijaya, 2017).

SiO_2: The advancement of nanotechnology has led to the production of nanostructured silica (SiO_2), which has been widely used in engineering. Silica particles extracted from

natural resources contain metallic impurities and are not favorable for advanced scientific and industrial applications (Rahman & Padavettan, 2012). Thus, the focus is on synthetic silica (colloidal silica, silica gels, pyrogenic silica and precipitated silica), which is pure and produced mainly in amorphous powder forms compared to natural mineral silica (quartz, tridymite, cristobalite) which is in crystalline forms (Vansant, Voort, & Vrancken, 1995). Nanoparticles of silicon dioxide (SiO_2), crystalline (quartz), or amorphous (silica), draw particular attention due to their excellent physical-chemical, mechanical, optical, and electrical properties (Ahkam, Khan, Iqbal, Murtaza, & Khan, 2019), what makes this material a strong candidate for applications in several areas, such as photonics, microelectronics, thin films, electrical/thermal insulators, has also been widely applied in catalysis, pH sensors, pharmaceutical products, paint industry, etc. (Islam et al., 2018; Lam, Lee, & Mohamed, 2009; Stöber, Fink, & Bohn, 1968). The multifunctional properties of SiO_2 nanoparticles can be adjusted by doping or inducing point defects from any continuous imperfect SiO_2 lattice including oxygen and silicon vacancies (Ahkam et al., 2019; Skuja, 1998).

Silica nanoparticles have also been investigated in polymer biodegradation studies (Pathak & Kumar, 2017). Pathak and Kumar (2017), observed the efficiency of biodegradation of low-density PE (LDPE) in microbial cultures in the presence of SiO_2 nanoparticles and their synergistic effect on morphometric parameters. The study showed the bacteria-nanoparticle interactions that significantly influence the degradation of the polymer. Nanoparticles act as enhancers of the ability of microorganisms to biodegrade (Pathak & Kumar, 2017).

21.4 Inference and future prospects

Degradation has become a serious problem to be addressed, since petroleum-derived polymers have been widely used by contemporary society. The recycling potential remains largely untapped due to the high cost and low quality of recycled plastics, with very low overall recycling rates representing only 6% of the total plastics that demand recycling (Narancic & O'Connor, 2019).

A notable characteristic of synthetic polymers is their high resistance to biodegradation, leading to their overaccumulation due to the less effective measures for their management in the environment. Microbial degradation proved to be efficient in reducing these pollutants (Jacquin et al., 2019; Urbanek, Rymowicz, & Mirończuk, 2018). Plastic-degrading bacteria isolated from mangroves and landfills were able to use PE and LDPE, respectively, as a carbon and energy source (Esmaeili et al., 2013; Kathiresan, 2003). However, the rates of degradation of synthetic plastics are still relatively slow (Devi et al., 2015).

In this context, NPs are also being explored for their potential to increase the rate of biodegradation of organic pollutants and the measures were designed to reduce environmental risks and human health. In Chapter 5, Nanophotocatalysts for Biodegradation of Materials, it was reported that NPs act on the mechanism of the growth rate of microorganisms (Bhatia, 2013; Cada, Muyot, Sison, & B, 2019). In the literature, the use of iron-based NPs for biodegradation studies of organic compounds are frequently reported (Cada et al. 2019; Kapri, Zaidi, Satlewal, & Goel, 2010; Kuang, Zhou, Chen, Megharaj, & Naidu, 2013; Yang et al., 2017).

However, other NPs have not been explored much. Therefore, this field of study has the potential to be explored with other documented NPs that influence the bacterial growth profile. In addition, the mechanisms of degradation are also not known in great detail and, consequently, constitute a challenge in this field of research resulting in a new direction for studies related to biodegradation, to make the area solid and useful at the commercial level. Therefore, the implications of NPs have wide applications as enhancers of the biodegradation of synthetic plastics and can act as a benign solution.

21.5 Acknowledgments

The authors would like to thank the Institute of Technology and Research at Tiradentes University (ITP-UNIT), State University of Paraíba (UEPB), the Paraíba State Research Support Foundation (FAPESQ) Call 05/2018, the Graduate and Research Incentive Program (PROPESQ) 2017 UEPB, the Foundation of Support to Research and Technological Innovation of the State of Sergipe [FAPITEC/SE], the Coordination for the Improvement of Higher Education Personnel (CAPES) under the Finance Code 001 and the National Council for Scientific and Technological Development (CNPq) for financial support (process no 315405/2018-0; 421147/2016-4; 315018/2018-6).

References

Abboud, Y., Saffaj, T., Chagraoui, A., El Bouari, A., Brouzi, K., Tanane, O., & Ihssane, B. (2014). Biosynthesis, characterization and antimicrobial activity of copper oxide nanoparticles (CONPs) produced using brown alga extract (*Bifurcaria bifurcata*). *Applied Nanoscience*, 4(5), 571−576. Available from https://doi.org/10.1007/s13204-013-0233-x.

Abushrida, A., Elhuni, I., Taresco, V., Marciani, L., Stolnik, S., & Garnett, M. C. (2020). A simple and efficient method for polymer coating of iron oxide nanoparticles. *Journal of Drug Delivery Science and Technology*, 55, 101460. Available from https://doi.org/10.1016/j.jddst.2019.101460.

Aflaki, M., & Davar, F. (2016). Synthesis, luminescence and photocatalyst properties of zirconia nanosheets by modified Pechini method. *Journal of Molecular Liquids*, 221, 1071−1079. Available from https://doi.org/10.1016/j.molliq.2016.06.067.

Agustin, Y. E., & Padmawijaya, K. S. (2017). Effect of glycerol and zinc oxide addition on antibacterial activity of biodegradable bioplastics from chitosan-kepok banana peel starch. *IOP Conference Series: Materials Science and Engineering*, 223, 012046. Available from https://doi.org/10.1088/1757-899X/223/1/012046.

Ahkam, Q. M., Khan, E. U., Iqbal, J., Murtaza, A., & Khan, M. T. (2019). Synthesis and characterization of cobalt-doped SiO_2 nanoparticles. *Physica B: Condensed Matter*, 572, 161−167. Available from https://doi.org/10.1016/j.physb.2019.07.044.

Ahmann, D., & Dorgan, J. R. (2007). Bioengineering for pollution prevention through development of bio-based energy and materials state of the science report. *Industrial Biotechnology*, 3(3), 218−259. Available from https://doi.org/10.1089/ind.2007.3.218.

Akindoyo, J. O., Beg, M. D. H., Ghazali, S., Islam, M. R., Jeyaratnam, N., & Yuvaraj, A. R. (2016). Polyurethane types, synthesis and applications—A review. *RSC Advances*, 6(115), 114453−114482. Available from https://doi.org/10.1039/C6RA14525F.

Aly, I. H. M., Abed Alrahim Mohammed, L., Al-Meer, S., Elsaid, K., & Barakat, N. A. M. (2016). Preparation and characterization of wollastonite/titanium oxide nanofiber bioceramic composite as a future implant material. *Ceramics International, 42*(10), 11525–11534. Available from https://doi.org/10.1016/j.ceramint.2016.02.060.

Ando, H., Kawasaki, N., Yamano, N., Uegaki, K., & Nakayama, A. (2015). Biodegradation of a poly(ε-caprolactone-co-l-lactide)−visible-light-sensitive TiO_2 composite with an on/off biodegradation function. *Polymer Degradation and Stability, 114*, 65–71. Available from https://doi.org/10.1016/j.polymdegradstab.2015.02.003.

Armentano, I., Puglia, D., Luzi, F., Arciola, C., Morena, F., Martino, S., & Torre, L. (2018). Nanocomposites based on biodegradable polymers. *Materials, 11*(5), 795. Available from https://doi.org/10.3390/ma11050795.

Arote, S. A., Pathan, A. S., Hase, Y. V., Bardapurkar, P. P., Gapale, D. L., & Palve, B. M. (2019). Investigations on synthesis, characterization and humidity sensing properties of ZnO and $ZnO-ZrO_2$ composite nanoparticles prepared by ultrasonic assisted wet chemical method. *Ultrasonics Sonochemistry, 55*, 313–321. Available from https://doi.org/10.1016/j.ultsonch.2019.01.012.

Asghar, W., Qazi, I. A., Ilyas, H., Khan, A. A., Awan, M. A., & Rizwan Aslam, M. (2011). Comparative solid phase photocatalytic degradation of polythene films with doped and undoped TiO_2 nanoparticles. *Journal of Nanomaterials, 2011*, 1–8. Available from https://doi.org/10.1155/2011/461930.

Atta, A. M., Al-Lohedan, H. A., El-Saeed, A. M., Al-Shafey, H. I., & Wahby, M. (2017). Salt-controlled self-healing nanogel composite embedded with epoxy as environmentally friendly organic coating. *Journal of Coatings Technology and Research, 14*(5), 1225–1236. Available from https://doi.org/10.1007/s11998-017-9917-6.

Auffan, M., Rose, J., Wiesner, M. R., & Bottero, J.-Y. (2009). Chemical stability of metallic nanoparticles: A parameter controlling their potential cellular toxicity in vitro. *Environmental Pollution, 157*(4), 1127–1133. Available from https://doi.org/10.1016/j.envpol.2008.10.002.

Aulenta, F., Rossetti, S., Amalfitano, S., Majone, M., & Tandoi, V. (2013). Conductive magnetite nanoparticles accelerate the microbial reductive dechlorination of trichloroethene by promoting interspecies electron transfer processes. *ChemSusChem, 6*(3), 433–436. Available from https://doi.org/10.1002/cssc.201200748.

Auta, H. S., Emenike, C., & Fauziah, S. (2017). Distribution and importance of microplastics in the marine environment: A review of the sources, fate, effects, and potential solutions. *Environment International, 102*, 165–176. Available from https://doi.org/10.1016/j.envint.2017.02.013.

Ba-Abbad, M. M., Takriff, M. S., Benamor, A., & Mohammad, A. W. (2016). Synthesis and characterisation of Co^{2+} incorporated ZnO nanoparticles prepared through a sol-gel method. *Advanced Powder Technology, 27*(6), 2439–2447. Available from https://doi.org/10.1016/j.apt.2016.08.009.

Bacaksiz, E., Parlak, M., Tomakin, M., Özçelik, A., Karakız, M., & Altunbaş, M. (2008). The effects of zinc nitrate, zinc acetate and zinc chloride precursors on investigation of structural and optical properties of ZnO thin films. *Journal of Alloys and Compounds, 466*(1−2), 447–450. Available from https://doi.org/10.1016/j.jallcom.2007.11.061.

Bahlawane, N., & Lenoble, D. (2014). Vanadium oxide compounds: Structure, properties, and growth from the gas phase. *Chemical Vapor Deposition, 20*(7−8−9), 299–311. Available from https://doi.org/10.1002/cvde.201400057.

Baranowska-Wójcik, E., Szwajgier, D., Oleszczuk, P., & Winiarska-Mieczan, A. (2020). Effects of titanium dioxide nanoparticles exposure on human health-a review. *Biological Trace Element Research, 193*(1), 118–129. Available from https://doi.org/10.1007/s12011-019-01706-6.

Barboza, L. G. A., Vieira, L. R., Branco, V., Carvalho, C., & Guilhermino, L. (2018). Microplastics increase mercury bioconcentration in gills and bioaccumulation in the liver, and cause oxidative stress and damage in *Dicentrarchus labrax* juveniles. *Scientific Reports, 8*(1), 15655. Available from https://doi.org/10.1038/s41598-018-34125-z.

Barnes, D. K. A., Galgani, F., Thompson, R. C., & Barlaz, M. (2009). Accumulation and fragmentation of plastic debris in global environments. *Philosophical Transactions of the Royal Society B: Biological Sciences, 364*(1526), 1985–1998. Available from https://doi.org/10.1098/rstb.2008.0205.

Bel Hadj Tahar, R., Ban, T., Ohya, Y., & Takahashi, Y. (1998). Tin doped indium oxide thin films: electrical properties. *Journal of Applied Physics, 83*(5), 2631−2645. Available from https://doi.org/10.1063/1.367025.

Benhammada, A., Trache, D., Kesraoui, M., Tarchoun, A. F., Chelouche, S., & Mezroua, A. (2020). Synthesis and characterization of α-Fe_2O_3 nanoparticles from different precursors and their catalytic effect on the thermal decomposition of nitrocellulose. *Thermochimica Acta, 686*, 178570. Available from https://doi.org/10.1016/j.tca.2020.178570.

Bernhardt, E. S., Rosi, E. J., & Gessner, M. O. (2017). Synthetic chemicals as agents of global change. *Frontiers in Ecology and the Environment, 15*(2), 84−90. Available from https://doi.org/10.1002/fee.1450.

Bhatia, M. (2013). Implicating nanoparticles as potential biodegradation enhancers: A review. *Journal of Nanomedicine & Nanotechnology, 04*(04). Available from https://doi.org/10.4172/2157-7439.1000175.

Bogdan, J., Pławińska-Czarnak, J., & Zarzyńska, J. (2017). Nanoparticles of titanium and zinc oxides as novel agents in tumor treatment: A review. *Nanoscale Research Letters, 12*(1), 225. Available from https://doi.org/10.1186/s11671-017-2007-y.

Bolinius, D. J., Sobek, A., Löf, M. F., & Undeman, E. (2018). Evaluating the consumption of chemical products and articles as proxies for diffuse emissions to the environment. *Environmental Science: Processes & Impacts, 20*(10), 1427−1440. Available from https://doi.org/10.1039/C8EM00270C.

Bordes, P., Pollet, E., & Averous, L. (2009). Nano-biocomposites: Biodegradable polyester/nanoclay systems. *Progress in Polymer Science, 34*(2), 125−155. Available from https://doi.org/10.1016/j.progpolymsci.2008.10.002.

Cada, E. J. G., Muyot, M. L. C., Sison, J. M. C., & B, R. Q. (2019). Enhanced in vitro biodegradation of low-density polyethylene using alkaliphilic bacterial consortium supplemented with iron oxide nanoparticles. *Philippine Science Letters, 12*, 55−69.

Cai, C., Zhao, M., Yu, Z., Rong, H., & Zhang, C. (2019). Utilization of nanomaterials for in-situ remediation of heavy metal(loid) contaminated sediments: A review. *Science of the Total Environment, 662*, 205−217. Available from https://doi.org/10.1016/j.scitotenv.2019.01.180.

Capelezzo, A. P., Mohr, L. C., Godoy, J. S., Bellei, A. S., Silva, L. L., Martins, M. A. P. M., & Mello, J. M. M. (2018). Addition of zinc oxide nanoparticles in biodegradable polymer and evaluation of its antimicrobial activity. *Materials Science Forum, 930*, 230−235. Available from https://doi.org/10.4028/www.scientific.net/MSF.930.230.

Cargnello, M., Gordon, T. R., & Murray, C. B. (2014). Solution-phase synthesis of titanium dioxide nanoparticles and nanocrystals. *Chemical Reviews, 114*(19), 9319−9345. Available from https://doi.org/10.1021/cr500170p.

Castro-Perez, J., & Prakash, C., (2020). Recent advances in mass spectrometric and other analytical techniques for the identification of drug metabolites. In *Identification and quantification of drugs, metabolites, drug metabolizing enzymes, and transporters* (pp. 39−71). Elsevier. Available from https://doi.org/10.1016/B978-0-12-820018-6.00002-8

Chang, T.-H., Lu, Y.-C., Yang, M.-J., Huang, J.-W., Linda Chang, P.-F., & Hsueh, H.-Y. (2020). Multibranched flower-like ZnO particles from eco-friendly hydrothermal synthesis as green antimicrobials in agriculture. *Journal of Cleaner Production, 262*, 121342. Available from https://doi.org/10.1016/j.jclepro.2020.121342.

Charmforoushan, A., Roknabadi, M. R., Shahtahmassebi, N., & Malaekeh-Nikouei, B. (2020). Low temperature facile synthesis of pseudowollastonite nanoparticles by the surfactant-assisted sol-gel method. *Materials Chemistry and Physics, 243*, 122629. Available from https://doi.org/10.1016/j.matchemphys.2020.122629.

Chen, D., & Caruso, R. A. (2013). Recent progress in the synthesis of spherical titania nanostructures and their applications. *Advanced Functional Materials, 23*(11), 1356−1374. Available from https://doi.org/10.1002/adfm.201201880.

Chen, J., Xu, L., Li, W., & Gou, X. (2005). Fe_2O_3 nanotubes in gas sensor and lithium-ion battery applications. *Advanced Materials, 17*(5), 582−586. Available from https://doi.org/10.1002/adma.200401101.

Chen, X., & Mao, S. S. (2007). Titanium dioxide nanomaterials: Synthesis, properties, modifications, and applications. *ChemInform, 38*(41). Available from https://doi.org/10.1002/chin.200741216.

Cheng, C., Xin, R., Leng, Y., Yu, D., & Wang, N. (2008). Chemical stability of ZnO nanostructures in simulated physiological environments and its application in determining polar directions. *Inorganic Chemistry*, *47*(17), 7868−7873. Available from https://doi.org/10.1021/ic8005234.

Chinaglia, S., Tosin, M., & Degli-Innocenti, F. (2018). Biodegradation rate of biodegradable plastics at molecular level. *Polymer Degradation and Stability*, *147*, 237−244. Available from https://doi.org/10.1016/j.polymdegradstab.2017.12.011.

Cho, Y., Choi, W., Lee, C.-H., Hyeon, T., & Lee, H.-I. (2001). Visible light-induced degradation of carbon tetrachloride on dye-sensitized TiO_2. *Environmental Science & Technology*, *35*(5), 966−970. Available from https://doi.org/10.1021/es001245e.

Chou, S.-L., Wang, J.-Z., Sun, J.-Z., Wexler, D., Forsyth, M., Liu, H.-K., & Dou, S.-X. (2008). High capacity, safety, and enhanced cyclability of lithium metal battery using a V_2O_5 nanomaterial cathode and room temperature ionic liquid electrolyte. *Chemistry of Materials*, *20*(22), 7044−7051. Available from https://doi.org/10.1021/cm801468q.

Commission, E. (2010). *Scientific basis for the definition of the term "Nanomaterial" European Commission, Scientific Committee on Emerging and Newly Identified Health Risks (SCENHR)*. Brussels.

Cornell, R.M., & Schwertmann, U. (2003). *The iron oxides: Structure, properties, reactions, occurences and uses* (2nd ed.). Wiley-VCH Verlag GmbH & Co. KGaA. Available from https://doi.org/10.1002/3527602097. ISBN:9783527302741.

Cruz Viggi, C., Rossetti, S., Fazi, S., Paiano, P., Majone, M., & Aulenta, F. (2014). Magnetite particles triggering a faster and more robust syntrophic pathway of methanogenic propionate degradation. *Environmental Science & Technology*, *48*(13), 7536−7543. Available from https://doi.org/10.1021/es5016789.

da Costa, J. P., Santos, P. S. M., Duarte, A. C., & Rocha-Santos, T. (2016). (Nano)plastics in the environment—Sources, fates and effects. *Science of the Total Environment*, *566−567*, 15−26. Available from https://doi.org/10.1016/j.scitotenv.2016.05.041.

de Moraes, J. O., Müller, C. M. O., & Laurindo, J. B. (2012). Influence of the simultaneous addition of bentonite and cellulose fibers on the mechanical and barrier properties of starch composite-films. *Food Science and Technology International*, *18*(1), 35−45. Available from https://doi.org/10.1177/1082013211427622.

Denet, E., Espina-Benitez, M. B., Pitault, I., Pollet, T., Blaha, D., Bolzinger, M.-A., & Briançon, S. (2020). Metal oxide nanoparticles for the decontamination of toxic chemical and biological compounds. *International Journal of Pharmaceutics*, *583*, 119373. Available from https://doi.org/10.1016/j.ijpharm.2020.119373.

Devi, R., Kannan, V., Natarajan, K., Nivas, D., Kannan, K., Chandru, S., & Antony, A. (2015). The role of microbes in plastic degradation. In *Environmental waste management* (pp. 341−370). CRC Press. Available from https://doi.org/10.1201/b19243-13

Dhoundiyal, H., Das, P., & Bhatnagar, M. C. (2020). Synthesis of V_2O_5 nanostructures and electrical transport properties of V_2O_5 nanoparticle. *Materials Today: Proceedings*, *26*, 2830−2832. Available from https://doi.org/10.1016/j.matpr.2020.02.589.

Diamandescu, L., Mihaila-Tarabasanu, D., Popescu-Pogrion, N., Totovina, A., & Bibicu, I. (1999). Hydrothermal synthesis and characterization of some polycrystalline α-iron oxides. *Ceramics International*, *25*(8), 689−692. Available from https://doi.org/10.1016/S0272-8842(99)00002-4.

Dimesso, L., (2018). Pechini processes: An alternate approach of the sol-gel method, preparation, properties, and applications. In *Handbook of sol-gel science and technology* (pp. 1067−1088). Springer Nature. Available from https://doi.org/10.1007/978-3-319-32101-1_123

Dubchak, S., Ogar, A., Mietelski, J. W., & Turnau, K. (2010). Influence of silver and titanium nanoparticles on arbuscular mycorrhiza colonization and accumulation of radiocaesium in *Helianthus annuus*. *Spanish Journal of Agricultural Research*, *8*(S1), 103. Available from https://doi.org/10.5424/sjar/201008S1-1228.

Dyshlyuk, L., Babich, O., Ivanova, S., Vasilchenco, N., Prosekov, A., & Sukhikh, S. (2020). Suspensions of metal nanoparticles as a basis for protection of internal surfaces of building structures from biodegradation. *Case Studies in Construction Materials*, *12*, e00319. Available from https://doi.org/10.1016/j.cscm.2019.e00319.

Eili, M., Shameli, K., Ibrahim, N. A., & Wan Yunus, W. M. Z. (2012). Degradability enhancement of poly(lactic acid) by stearate-Zn$_3$Al LDH nanolayers. *International Journal of Molecular Sciences, 13*(7), 7938−7951. Available from https://doi.org/10.3390/ijms13077938.

Ennas, G., Musinu, A., Piccaluga, G., Zedda, D., Gatteschi, D., Sangregorio, C., & Spano, G. (1998). Characterization of iron oxide nanoparticles in an Fe$_2$O$_3$-SiO$_2$ composite prepared by a sol − gel method. *Chemistry of Materials, 10*(2), 495−502. Available from https://doi.org/10.1021/cm970400u.

Esmaeili, A., Pourbabaee, A. A., Alikhani, H. A., Shabani, F., & Esmaeili, E. (2013). Biodegradation of low-density polyethylene (LDPE) by mixed culture of *Lysinibacillus xylanilyticus* and *Aspergillus niger* in soil. *PLoS One, 8*(9), e71720. Available from https://doi.org/10.1371/journal.pone.0071720.

Fasiku, V.O., John Owonubi, S., Malima, N.M., Hassan, D., & Revaprasadu, N. (2020). Metal oxide nanoparticles: A welcome development for targeting bacteria. In *Antibiotic materials in healthcare* (pp. 261−286). Elsevier. Available from https://doi.org/10.1016/B978-0-12-820054-4.00015-X

Fernández-García, M., & Rodriguez, J.A. (2011). Metal oxide nanoparticles. In *Encyclopedia of inorganic and bioinorganic chemistry*. John Wiley & Sons. Available from https://doi.org/10.1002/9781119951438.eibc0331

Gama, M. R., Melchert, W. R., Paixão, T. R. L. C., & Rocha, F. R. P. (2019). An overview of the Brazilian contributions to Green Analytical Chemistry. *Anais Da Academia Brasileira de Ciências, 91*(Suppl. 1). Available from https://doi.org/10.1590/0001-3765201920180294.

Geyer, R., Jambeck, J. R., & Law, K. L. (2017). Production, use, and fate of all plastics ever made. *Science Advances, 3*(7), e1700782. Available from https://doi.org/10.1126/sciadv.1700782.

Giannelis, E. P. (1996). Polymer layered silicate nanocomposites. *Advanced Materials, 8*(1), 29−35.

Girigoswami, K. (2018). *Toxicity of metal oxide nanoparticles*. https://doi.org/10.1007/978-3-319-72041-8_7

Goto, T., Shin, J., Yokoi, T., Cho, S. H., & Sekino, T. (2020). Photocatalytic properties and controlled morphologies of TiO$_2$-modified hydroxyapatite synthesized by the urea-assisted hydrothermal method. *Powder Technology, 373*, 468−475. Available from https://doi.org/10.1016/j.powtec.2020.06.062.

Gottschalk, F., Ort, C., Scholz, R. W., & Nowack, B. (2011). Engineered nanomaterials in rivers—Exposure scenarios for Switzerland at high spatial and temporal resolution. *Environmental Pollution, 159*(12), 3439−3445. Available from https://doi.org/10.1016/j.envpol.2011.08.023.

Gou, X., Wang, G., Park, J., Liu, H., & Yang, J. (2008). Monodisperse hematite porous nanospheres: Synthesis, characterization, and applications for gas sensors. *Nanotechnology, 19*(12), 125606. Available from https://doi.org/10.1088/0957-4484/19/12/125606.

Griffini, G., Brambilla, L., Levi, M., Del Zoppo, M., & Turri, S. (2013). Photo-degradation of a perylene-based organic luminescent solar concentrator: Molecular aspects and device implications. *Solar Energy Materials and Solar Cells, 111*, 41−48. Available from https://doi.org/10.1016/j.solmat.2012.12.021.

Hao, Z., Guo, C., Lv, J., Zhang, Y., Zhang, Y., & Xu, J. (2019). Kinetic and mechanistic study of sulfadimidine photodegradation under simulated sunlight irradiation. *Environmental Sciences Europe, 31*(1), 40. Available from https://doi.org/10.1186/s12302-019-0223-z.

Hassanjani-Roshan, A., Vaezi, M. R., Shokuhfar, A., & Rajabali, Z. (2011). Synthesis of iron oxide nanoparticles via sonochemical method and their characterization. *Particuology, 9*(1), 95−99. Available from https://doi.org/10.1016/j.partic.2010.05.013.

Herlekar, M., Barve, S., & Kumar, R. (2014). Plant-mediated green synthesis of iron nanoparticles. *Journal of Nanoparticles, 2014*, 1−9. Available from https://doi.org/10.1155/2014/140614.

Hermabessiere, L., Dehaut, A., Paul-Pont, I., Lacroix, C., Jezequel, R., Soudant, P., & Duflos, G. (2017). Occurrence and effects of plastic additives on marine environments and organisms: A review. *Chemosphere, 182*, 781−793. Available from https://doi.org/10.1016/j.chemosphere.2017.05.096.

Honig, J. M., & Spałek, J. (1998). Electronic properties of NiS$_{2-x}$Se$_x$ single crystals: From magnetic Mott − Hubbard insulators to normal metals. *Chemistry of Materials, 10*(10), 2910−2929. Available from https://doi.org/10.1021/cm9803509.

Hu, H., Yu, L., Gao, X., Lin, Z., & Lou, X. W. (David) (2015). Hierarchical tubular structures constructed from ultrathin TiO$_2$(B) nanosheets for highly reversible lithium storage. *Energy & Environmental Science, 8*(5), 1480–1483. Available from https://doi.org/10.1039/C5EE00101C.

Huang, J., Xia, C., Cao, L., & Zeng, X. (2008). Facile microwave hydrothermal synthesis of zinc oxide one-dimensional nanostructure with three-dimensional morphology. *Materials Science and Engineering: B, 150*(3), 187–193. Available from https://doi.org/10.1016/j.mseb.2008.05.014.

Isaifan, R. J., Samara, A., Suwaileh, W., Johnson, D., Yiming, W., Abdallah, A. A., & Aïssa, B. (2017). Improved self-cleaning properties of an efficient and easy to scale up TiO$_2$ thin films prepared by adsorptive self-assembly. *Scientific Reports, 7*(1), 9466. Available from https://doi.org/10.1038/s41598-017-07826-0.

Islam, S., Bakhtiar, H., Aziz, M. S. B. A., Duralim, M. B., Riaz, S., Naseem, S., & Osman, S. S. (2018). CR incorporation in mesoporous silica matrix for fiber optic pH sensing. *Sensors and Actuators A: Physical, 280*, 429–436. Available from https://doi.org/10.1016/j.sna.2018.08.016.

Jacquin, J., Cheng, J., Odobel, C., Pandin, C., Conan, P., Pujo-Pay, M., & Ghiglione, J.-F. (2019). Microbial ecotoxicology of marine plastic debris: A review on colonization and biodegradation by the "Plastisphere". *Frontiers in Microbiology, 10*. Available from https://doi.org/10.3389/fmicb.2019.00865.

Jagathesan, G., & Rajiv, P. (2018). Biosynthesis and characterization of iron oxide nanoparticles using *Eichhornia crassipes* leaf extract and assessing their antibacterial activity. *Biocatalysis and Agricultural Biotechnology, 13*, 90–94. Available from https://doi.org/10.1016/j.bcab.2017.11.014.

Jalvo, B., Faraldos, M., Bahamonde, A., & Rosal, R. (2017). Antimicrobial and antibiofilm efficacy of self-cleaning surfaces functionalized by TiO$_2$ photocatalytic nanoparticles against *Staphylococcus aureus* and *Pseudomonas putida*. *Journal of Hazardous Materials, 340*, 160–170. Available from https://doi.org/10.1016/j.jhazmat.2017.07.005.

Jeevanandam, J., Barhoum, A., Chan, Y. S., Dufresne, A., & Danquah, M. K. (2018). Review on nanoparticles and nanostructured materials: History, sources, toxicity and regulations. *Beilstein Journal of Nanotechnology, 9*, 1050–1074. Available from https://doi.org/10.3762/bjnano.9.98.

Jenkins, M. J., & Harrison, K. L. (2008). The effect of crystalline morphology on the degradation of polycaprolactone in a solution of phosphate buffer and lipase. *Polymers for Advanced Technologies, 19*(12), 1901–1906. Available from https://doi.org/10.1002/pat.1227.

Jiang, X., Manawan, M., Feng, T., Qian, R., Zhao, T., Zhou, G., & Pan, J. H. (2018). Anatase and rutile in evonik aeroxide P25: Heterojunctioned or individual nanoparticles? *Catalysis Today, 300*, 12–17. Available from https://doi.org/10.1016/j.cattod.2017.06.010.

Júnior, R. M. S., de Oliveira, T. A., Araque, L. M., Alves, T. S., de Carvalho, L. H., & Barbosa, R. (2019). Thermal behavior of biodegradable bionanocomposites: Influence of bentonite and vermiculite clays. *Journal of Materials Research and Technology, 8*(3), 3234–3243. Available from https://doi.org/10.1016/j.jmrt.2019.05.011.

Kabir, E., Kumar, V., Kim, K.-H., Yip, A. C. K., & Sohn, J. R. (2018). Environmental impacts of nanomaterials. *Journal of Environmental Management, 225*, 261–271. Available from https://doi.org/10.1016/j.jenvman.2018.07.087.

Kandori, K., Yamamoto, N., Yasukawa, A., & Ishikawa, T. (2002). Preparation and characterization of disk-shaped hematite particles by a forced hydrolysis reaction in the presence of polyvinyl alcohol. *Physical Chemistry Chemical Physics, 4*(24), 6116–6122. Available from https://doi.org/10.1039/b206095g.

Kapri, A., Zaidi, M. G. H., Satlewal, A., & Goel, R. (2010). SPION-accelerated biodegradation of low-density polyethylene by indigenous microbial consortium. *International Biodeterioration & Biodegradation, 64*(3), 238–244. Available from https://doi.org/10.1016/j.ibiod.2010.02.002.

Kapuria, N., Sharma, V., Kumar, P., & Koner, A. L. (2018). Exploration of dynamic self-assembly mediated nanoparticle formation using perylenemonoimide–pyrene conjugate: A tool towards single-component white-light emission. *Journal of Materials Chemistry C, 6*(42), 11328–11335. Available from https://doi.org/10.1039/C8TC03730B.

Katarzyna Leja, G. L. (2010). Polymers biodegradation and biodegradablepolymers—A review. *Polish Journal of Environmental Studies*, *19*(2), 255–266.

Kathiresan, K. (2003). Polythene and plastic-degrading microbes in an Indian mangrove soil. *Revista de Biologia Tropical Journal*, *51*, 629–633.

Khan, I., Saeed, K., & Khan, I. (2019). Nanoparticles: Properties, applications and toxicities. *Arabian Journal of Chemistry*, *12*(7), 908–931. Available from https://doi.org/10.1016/j.arabjc.2017.05.011.

Kim, G. T., Muster, J., Krstic, V., Park, J. G., Park, Y. W., Roth, S., & Burghard, M. (2000). Field-effect transistor made of individual V_2O_5 nanofibers. *Applied Physics Letters*, *76*(14), 1875–1877. Available from https://doi.org/10.1063/1.126197.

Klotz, S., Le Godec, Y., Strässle, T., & Stuhr, U. (2008). The $\alpha - \gamma - \varepsilon$ triple point of iron investigated by high pressure–high temperature neutron scattering. *Applied Physics Letters*, *93*(9), 091904. Available from https://doi.org/10.1063/1.2976128.

Kuang, Y., Zhou, Y., Chen, Z., Megharaj, M., & Naidu, R. (2013). Impact of Fe and Ni/Fe nanoparticles on biodegradation of phenol by the strain *Bacillus fusiformis* (BFN) at various pH values. *Bioresource Technology*, *136*, 588–594. Available from https://doi.org/10.1016/j.biortech.2013.03.018.

Kubacka, A., Diez, M. S., Rojo, D., Bargiela, R., Ciordia, S., Zapico, I., & Ferrer, M. (2015). Understanding the antimicrobial mechanism of TiO_2-based nanocomposite films in a pathogenic bacterium. *Scientific Reports*, *4*(1), 4134. Available from https://doi.org/10.1038/srep04134.

Kulkarni, A., & Dasari, H. (2018). Current status of methods used in degradation of polymers: A review. *MATEC Web of Conferences*, *144*, 02023. Available from https://doi.org/10.1051/matecconf/201814402023.

Kumar, S., & Maiti, P. (2016). Controlled biodegradation of polymers using nanoparticles and its application. *RSC Advances*, *6*(72), 67449–67480. Available from https://doi.org/10.1039/C6RA08641A.

Kunioka, M., Ninomiya, F., & Funabashi, M. (2006). Biodegradation of poly(lactic acid) powders proposed as the reference test materials for the international standard of biodegradation evaluation methods. *Polymer Degradation and Stability*, *91*(9), 1919–1928. Available from https://doi.org/10.1016/j.polymdegradstab.2006.03.003.

Kurwadkar, S. (2019). Occurrence and distribution of organic and inorganic pollutants in groundwater. *Water Environment Research*, *91*(10), 1001–1008. Available from https://doi.org/10.1002/wer.1166.

Lackner, M. (2015). Bioplastics. In *Kirk-Othmer encyclopedia of chemical technology* (pp. 1–41). John Wiley & Sons. Available from https://doi.org/10.1002/0471238961.koe00006

Lam, M. K., Lee, K. T., & Mohamed, A. R. (2009). Sulfated tin oxide as solid superacid catalyst for transesterification of waste cooking oil: An optimization study. *Applied Catalysis B: Environmental*, *93*(1–2), 134–139. Available from https://doi.org/10.1016/j.apcatb.2009.09.022.

Lapworth, D. J., Baran, N., Stuart, M. E., & Ward, R. S. (2012). Emerging organic contaminants in groundwater: A review of sources, fate and occurrence. *Environmental Pollution*, *163*, 287–303. Available from https://doi.org/10.1016/j.envpol.2011.12.034.

Law, K. L. (2017). Plastics in the marine environment. *Annual Review of Marine Science*, *9*(1), 205–229. Available from https://doi.org/10.1146/annurev-marine-010816-060409.

Le, T. K., Kang, M., & Kim, S. W. (2019). Morphology engineering, room-temperature photoluminescence behavior, and sunlight photocatalytic activity of V_2O_5 nanostructures. *Materials Characterization*, *153*, 52–59. Available from https://doi.org/10.1016/j.matchar.2019.04.046.

Lebrero, R., Osvaldo, D.F., Pérez, V., Cantera, S., Estrada, J.M., & Muñoz, R. (2019). Biological treatment of gas pollutants in partitioning bioreactors. In Huerta-Ochoa, S., Castillo-Araiza, C. O., Quijano, G. (Eds.), *Advances in Chemical Engineeringi* (vol. 54, pp. 239-274). Academic Press. Available from https://doi.org/10.1016/bs.ache.2018.12.003

Lee, H.-Y., Cheng, C.-Y., & Lee, C.-T. (2020). Bottom gate thin-film transistors using parallelly lateral ZnO nanorods grown by hydrothermal method. *Materials Science in Semiconductor Processing*, *119*, 105223. Available from https://doi.org/10.1016/j.mssp.2020.105223.

Lei, C., Sun, Y., Tsang, D. C. W., & Lin, D. (2018). Environmental transformations and ecological effects of iron-based nanoparticles. *Environmental Pollution, 232*, 10−30. Available from https://doi.org/10.1016/j.envpol.2017.09.052.

Lei, P.-H., Wu, H.-M., Hsu, C.-M., & Lee, Y.-C. (2012). Zinc oxide (ZnO) grown on sapphire substrate using dual-plasma-enhanced metal organic vapor deposition (DPEMOCVD) and its application. *Applied Surface Science, 261*, 857−862. Available from https://doi.org/10.1016/j.apsusc.2012.07.163.

Levchenko, V., Mamunya, Y., Boiteux, G., Lebovka, M., Alcouffe, P., Seytre, G., & Lebedev, E. (2011). Influence of organo-clay on electrical and mechanical properties of PP/MWCNT/OC nanocomposites. *European Polymer Journal, 47*(7), 1351−1360. Available from https://doi.org/10.1016/j.eurpolymj.2011.03.012.

Li, C., Hong, G., Wang, P., Yu, D., & Qi, L. (2009). Wet chemical approaches to patterned arrays of well-aligned ZnO nanopillars assisted by monolayer colloidal crystals. *Chemistry of Materials, 21*(5), 891−897. Available from https://doi.org/10.1021/cm802839u.

Li, L., Hu, J., Shi, X., Fan, M., Luo, J., & Wei, X. (2016). Nanoscale zero-valent metals: A review of synthesis, characterization, and applications to environmental remediation. *Environmental Science and Pollution Research, 23*(18), 17880−17900. Available from https://doi.org/10.1007/s11356-016-6626-0.

Lim, H.-A., Raku, T., & Tokiwa, Y. (2005). Hydrolysis of polyesters by serine proteases. *Biotechnology Letters, 27*(7), 459−464. Available from https://doi.org/10.1007/s10529-005-2217-8.

Liu, J., Jiang, J., Meng, Y., Aihemaiti, A., Xu, Y., Xiang, H., & Chen, X. (2020). Preparation, environmental application and prospect of biochar-supported metal nanoparticles: A review. *Journal of Hazardous Materials, 388*, 122026. Available from https://doi.org/10.1016/j.jhazmat.2020.122026.

Liu, S., Jia, H., Han, L., Wang, J., Gao, P., Xu, D., & Che, S. (2012). Nanosheet-constructed porous TiO_2-B for advanced lithium ion batteries. *Advanced Materials, 24*(24), 3201−3204. Available from https://doi.org/10.1002/adma.201201036.

Liu, Z., Li, L., Yuan, X., & Yang, P. (2020). Study on photoelectric properties of Si supported ZnO. *Journal of Alloys and Compounds, 843*, 155909. Available from https://doi.org/10.1016/j.jallcom.2020.155909.

Lönnstedt, O. M., & Eklöv, P. (2016). Environmentally relevant concentrations of microplastic particles influence larval fish ecology. *Science (New York, N.Y.), 352*(6290), 1213−1216. Available from https://doi.org/10.1126/science.aad8828.

López de Dicastillo, C., Guerrero Correa, M., Martínez, F.B., Streitt, C., & José Galotto, M. (2020). Antimicrobial effect of titanium dioxide nanoparticles. In *Antimicrobial resistance*. IntechOpen. Available from https://doi.org/10.5772/intechopen.90891

Lu, H., Dong, H., Fan, W., Zuo, J., & Li, X. (2017). Aging and behavior of functional TiO_2 nanoparticles in aqueous environment. *Journal of Hazardous Materials, 325*, 113−119. Available from https://doi.org/10.1016/j.jhazmat.2016.11.013.

Luckachan, G. E., & Pillai, C. K. S. (2011). Biodegradable polymers- A review on recent trends and emerging perspectives. *Journal of Polymers and the Environment, 19*(3), 637−676. Available from https://doi.org/10.1007/s10924-011-0317-1.

Madhavan Nampoothiri, K., Nair, N. R., & John, R. P. (2010). An overview of the recent developments in polylactide (PLA) research. *Bioresource Technology, 101*(22), 8493−8501. Available from https://doi.org/10.1016/j.biortech.2010.05.092.

Magi, E., Di Carro, M., Mirasole, C., & Benedetti, B. (2018). Combining passive sampling and tandem mass spectrometry for the determination of pharmaceuticals and other emerging pollutants in drinking water. *Microchemical Journal, 136*, 56−60. Available from https://doi.org/10.1016/j.microc.2016.10.029.

Mahdavi, M., Namvar, F., Ahmad, M., & Mohamad, R. (2013). Green biosynthesis and characterization of magnetic iron oxide (Fe_3O_4) nanoparticles using seaweed (*Sargassum muticum*) aqueous extract. *Molecules (Basel, Switzerland), 18*(5), 5954−5964. Available from https://doi.org/10.3390/molecules18055954.

Marchand, R., Brohan, L., & Tournoux, M. (1980). TiO$_2$(B) a new form of titanium dioxide and the potassium octatitanate K$_2$Ti$_8$O$_{17}$. *Materials Research Bulletin, 15*(8), 1129–1133. Available from https://doi.org/10.1016/0025-5408(80)90076-8.

Mashreghi, A., & Ghasemi, M. (2015). Investigating the effect of molar ratio between TiO$_2$ nanoparticles and titanium alkoxide in Pechini based TiO$_2$ paste on photovoltaic performance of dye-sensitized solar cells. *Renewable Energy, 75*, 481–488. Available from https://doi.org/10.1016/j.renene.2014.10.033.

Menezes, W. G., Reis, D. M., Benedetti, T. M., Oliveira, M. M., Soares, J. F., Torresi, R. M., & Zarbin, A. J. G. (2009). V$_2$O$_5$ nanoparticles obtained from a synthetic bariandite-like vanadium oxide: Synthesis, characterization and electrochemical behavior in an ionic liquid. *Journal of Colloid and Interface Science, 337*(2), 586–593. Available from https://doi.org/10.1016/j.jcis.2009.05.050.

Mersian, H., Alizadeh, M., & Hadi, N. (2018). Synthesis of zirconium doped copper oxide (CuO) nanoparticles by the Pechini route and investigation of their structural and antibacterial properties. *Ceramics International, 44*(16), 20399–20408. Available from https://doi.org/10.1016/j.ceramint.2018.08.033.

Mierzwa-Hersztek, M., Gondek, K., & Kopeć, M. (2019). Degradation of polyethylene and biocomponent-derived polymer materials: An overview. *Journal of Polymers and the Environment, 27*(3), 600–611. Available from https://doi.org/10.1007/s10924-019-01368-4.

Milionis, A., Sharma, C. S., Hopf, R., Uggowitzer, M., Bayer, I. S., & Poulikakos, D. (2019). Engineering fully organic and biodegradable superhydrophobic materials. *Advanced Materials Interfaces, 6*(1), 1801202. Available from https://doi.org/10.1002/admi.201801202.

Mishra, M., & Chun, D.-M. (2015). α-Fe$_2$O$_3$ as a photocatalytic material: A review. *Applied Catalysis A: General, 498*, 126–141. Available from https://doi.org/10.1016/j.apcata.2015.03.023.

Mohajerani., Burnett., Smith., Kurmus., Milas., Arulrajah., & Kadir, A. (2019). Nanoparticles in construction materials and other applications, and implications of nanoparticle use. *Materials, 12*(19), 3052. Available from https://doi.org/10.3390/ma12193052.

Mohammadi, E., Aliofkhazraei, M., Hasanpoor, M., & Chipara, M. (2018). Hierarchical and complex ZnO nanostructures by microwave-assisted synthesis: Morphologies, growth mechanism and classification. *Critical Reviews in Solid State and Materials Sciences, 43*(6), 475–541. Available from https://doi.org/10.1080/10408436.2017.1397501.

Montagner, C., Sodré, F., Acayaba, R., Vidal, C., Campestrini, I., Locatelli, M., ... Jardim, W. (2018). Ten years-snapshot of the occurrence of emerging contaminants in drinking, surface and ground waters and wastewaters from São Paulo State, Brazil. *Journal of the Brazilian Chemical Society*. Available from https://doi.org/10.21577/0103-5053.20180232.

Motakef-Kazemi, N., Shojaosadati, S. A., & Morsali, A. (2014). In situ synthesis of a drug-loaded MOF at room temperature. *Microporous and Mesoporous Materials, 186*, 73–79. Available from https://doi.org/10.1016/j.micromeso.2013.11.036.

Mourdikoudis, S., Pallares, R. M., & Thanh, N. T. K. (2018). Characterization techniques for nanoparticles: Comparison and complementarity upon studying nanoparticle properties. *Nanoscale, 10*(27), 12871–12934. Available from https://doi.org/10.1039/C8NR02278J.

Müller, C. M. O., Laurindo, J. B., & Yamashita, F. (2012). Composites of thermoplastic starch and nanoclays produced by extrusion and thermopressing. *Carbohydrate Polymers, 89*(2), 504–510. Available from https://doi.org/10.1016/j.carbpol.2012.03.035.

Nabi, S., Sofi, F. A., Rashid, N., Ingole, P. P., & Bhat, M. A. (2020). Au-nanoparticle loaded nickel-copper bimetallic MOF: An excellent catalyst for chemical degradation of Rhodamine B. *Inorganic Chemistry Communications, 117*, 107949. Available from https://doi.org/10.1016/j.inoche.2020.107949.

Nair, L. S., & Laurencin, C. T. (2007). Biodegradable polymers as biomaterials. *Progress in Polymer Science, 32*(8–9), 762–798. Available from https://doi.org/10.1016/j.progpolymsci.2007.05.017.

Nam, N.H., & Luong, N.H. (2019). Nanoparticles: Synthesis and applications. In *Materials for biomedical engineering* (pp. 211–240). Elsevier. Available from https://doi.org/10.1016/B978-0-08-102814-8.00008-1

Narancic, T., & O'Connor, K. E. (2019). Plastic waste as a global challenge: Are biodegradable plastics the answer to the plastic waste problem? *Microbiology (Reading, England), 165*(2), 129−137. Available from https://doi.org/10.1099/mic.0.000749.

Nasikhudin., Diantoro, M., Kusumaatmaja, A., & Triyana, K. (2018). Study on photocatalytic properties of TiO_2 nanoparticle in various pH condition. *Journal of Physics: Conference Series, 1011*, 012069. Available from https://doi.org/10.1088/1742-6596/1011/1/012069.

Navaladian, S., Viswanathan, B., Viswanath, R. P., & Varadarajan, T. K. (2007). Thermal decomposition as route for silver nanoparticles. *Nanoscale Research Letters, 2*(1), 44−48. Available from https://doi.org/10.1007/s11671-006-9028-2.

Nishio, K., Seki, N., Thongrueng, J., Watanabe, Y., & Tsuchiya, T. (1999). Preparation and properties of highly oriented $Sr_{0.3}Ba_{0.7}Nb_2O_6$ thin films by a sol-gel process. *Journal of Sol-Gel Science and Technology, 16*, 37−45.

Pal, A., Gin, K. Y.-H., Lin, A. Y.-C., & Reinhard, M. (2010). Impacts of emerging organic contaminants on freshwater resources: Review of recent occurrences, sources, fate and effects. *Science of The Total Environment, 408*(24), 6062−6069. Available from https://doi.org/10.1016/j.scitotenv.2010.09.026.

Pan, J., Li, M., Luo, Y. Y., Wu, H., Zhong, L., Wang, Q., & Li, G. H. (2015). Synthesis and SERS activity of V2O5 nanoparticles. *Applied Surface Science, 333*, 34−38. Available from https://doi.org/10.1016/j.apsusc.2015.01.242.

Pan, J. H., Wang, Q., & Bahnemann, D. W. (2014). Hydrous TiO_2 spheres: An excellent platform for the rational design of mesoporous anatase spheres for photoelectrochemical applications. *Catalysis Today, 230*, 197−204. Available from https://doi.org/10.1016/j.cattod.2013.08.007.

Pandey, P., Merwyn, S., Agarwal, G. S., Tripathi, B. K., & Pant, S. C. (2012). Electrochemical synthesis of multi-armed CuO nanoparticles and their remarkable bactericidal potential against waterborne bacteria. *Journal of Nanoparticle Research, 14*(1), 709. Available from https://doi.org/10.1007/s11051-011-0709-0.

Park, K. I., & Xanthos, M. (2009). A study on the degradation of polylactic acid in the presence of phosphonium ionic liquids. *Polymer Degradation and Stability, 94*(5), 834−844. Available from https://doi.org/10.1016/j.polymdegradstab.2009.01.030.

Pathak, V. M., & Kumar, N. (2017). Implications of SiO_2 nanoparticles for in vitro biodegradation of low-density polyethylene with potential isolates of Bacillus, Pseudomonas, and their synergistic effect on *Vigna mungo* growth. *Energy, Ecology and Environment, 2*(6), 418−427. Available from https://doi.org/10.1007/s40974-017-0068-5.

Peng, Y., Fang, W., Krauss, M., Brack, W., Wang, Z., Li, F., & Zhang, X. (2018). Screening hundreds of emerging organic pollutants (EOPs) in surface water from the Yangtze River Delta (YRD): Occurrence, distribution, ecological risk. *Environmental Pollution, 241*, 484−493. Available from https://doi.org/10.1016/j.envpol.2018.05.061.

Picó, Y., & Barceló, D. (2019). Analysis and prevention of microplastics pollution in water: Current perspectives and future directions. *ACS Omega, 4*(4), 6709−6719. Available from https://doi.org/10.1021/acsomega.9b00222.

Poornima, P., Ey., Swati, P., Harshita., Manimita., Shraddha., ... Tiwari, A. (2015). Noparticles accelerated in-vitro biodegradation of LDPE: A review. *Advances in Applied Science Research, 6*.

Priyanka, N., & Archana, T. (2011). Biodegradability of polythene and plastic by the help of microorganism: A way for brighter future. *Journal of Environmental & Analytical Toxicology, 1*(02). Available from https://doi.org/10.4172/2161-0525.1000111.

Qian, R., Zong, H., Schneider, J., Zhou, G., Zhao, T., Li, Y., & Pan, J. H. (2019). Charge carrier trapping, recombination and transfer during TiO_2 photocatalysis: An overview. *Catalysis Today, 335*, 78−90. Available from https://doi.org/10.1016/j.cattod.2018.10.053.

Raghunath, A., & Perumal, E. (2017). Metal oxide nanoparticles as antimicrobial agents: A promise for the future. *International Journal of Antimicrobial Agents, 49*(2), 137−152. Available from https://doi.org/10.1016/j.ijantimicag.2016.11.011.

Rahman, I. A., & Padavettan, V. (2012). Synthesis of silica nanoparticles by sol-gel: Size-dependent properties, surface modification, and applications in silica-polymer nanocomposites—A review. *Journal of Nanomaterials, 2012*, 1–15. Available from https://doi.org/10.1155/2012/132424.

Rajmohan, K. V. S., Ramya, C., Raja Viswanathan, M., & Varjani, S. (2019). Plastic pollutants: Effective waste management for pollution control and abatement. *Current Opinion in Environmental Science & Health, 12*, 72–84. Available from https://doi.org/10.1016/j.coesh.2019.08.006.

Rasal, R. M., Janorkar, A. V., & Hirt, D. E. (2010). Poly(lactic acid) modifications. *Progress in Polymer Science, 35*(3), 338–356. Available from https://doi.org/10.1016/j.progpolymsci.2009.12.003.

Rhein, S., & Schmid, M. (2020). Consumers' awareness of plastic packaging: More than just environmental concerns. *Resources, Conservation and Recycling, 162*, 105063. Available from https://doi.org/10.1016/j.resconrec.2020.105063.

Riu, J., Maroto, A., & Rius, F. (2006). Nanosensors in environmental analysis. *Talanta, 69*(2), 288–301. Available from https://doi.org/10.1016/j.talanta.2005.09.045.

Rodrigues, M. O., Abrantes, N., Gonçalves, F. J. M., Nogueira, H., Marques, J. C., & Gonçalves, A. M. M. (2019). Impacts of plastic products used in daily life on the environment and human health: What is known? *Environmental Toxicology and Pharmacology, 72*, 103239. Available from https://doi.org/10.1016/j.etap.2019.103239.

Roy, A., & Bhattacharya, J. (2012). Microwave-assisted synthesis and characterization of Cas nanoparticles. *International Journal of Nanoscience, 11*(05), 1250027. Available from https://doi.org/10.1142/S0219581X12500275.

Salem, S. S., & Fouda, A. (2020). Green synthesis of metallic nanoparticles and their prospective biotechnological applications: An overview. *Biological Trace Element Research*. Available from https://doi.org/10.1007/s12011-020-02138-3.

Samantaray, P. K., Little, A., Haddleton, D. M., McNally, T., Tan, B., Sun, Z., & Wan, C. (2020). Poly(glycolic acid) (PGA): A versatile building block expanding high performance and sustainable bioplastic applications. *Green Chemistry, 22*(13), 4055–4081. Available from https://doi.org/10.1039/D0GC01394C.

Santhosh, C., Velmurugan, V., Jacob, G., Jeong, S. K., Grace, A. N., & Bhatnagar, A. (2016). Role of nanomaterials in water treatment applications: A review. *Chemical Engineering Journal, 306*, 1116–1137. Available from https://doi.org/10.1016/j.cej.2016.08.053.

Sapnar, K. B., Ghule, L. A., Bankar, A., Zinjarde, S., Bhoraskar, V. N., Garadkar, K. M., & Dhole, S. D. (2012). Antimicrobial activity of 6.5 MeV electron-irradiated ZnO nanoparticles synthesized by microwave-assisted method. *International Journal of Green Nanotechnology, 4*(4), 477–483. Available from https://doi.org/10.1080/19430892.2012.738162.

Satlewal, A., Soni, R., Zaidi, M., Shouche, Y., & Reeta, G. (2008). Comparative biodegradation of HDPE and LDPE using an indigenously developed microbial consortium. *Journal of Microbiolog and Biotechnology, 18*(3), 477–482.

Scherer, T. M., Fuller, R. C., Lenz, R. W., & Goodwin, S. (1999). Hydrolase activity of an extracellular depolymerase from *Aspergillus fumigatus* with bacterial and synthetic polyesters. *Polymer Degradation and Stability, 64*(2), 267–275. Available from https://doi.org/10.1016/S0141-3910(98)00201-8.

Schwarzer, H.-C., & Peukert, W. (2004). Tailoring particle size through nanoparticle precipitation. *Chemical Engineering Communications, 191*(4), 580–606. Available from https://doi.org/10.1080/00986440490270106.

Shah, S. N. A., Shah, Z., Hussain, M., & Khan, M. (2017). Hazardous effects of titanium dioxide nanoparticles in ecosystem. *Bioinorganic Chemistry and Applications, 2017*, 1–12. Available from https://doi.org/10.1155/2017/4101735.

Sharma, G., Kumar, A., Sharma, S., Naushad, M., Prakash Dwivedi, R., ALOthman, Z. A., & Mola, G. T. (2019). Novel development of nanoparticles to bimetallic nanoparticles and their composites: A review. *Journal of King Saud University—Science, 31*(2), 257–269. Available from https://doi.org/10.1016/j.jksus.2017.06.012.

Shi, H., Magaye, R., Castranova, V., & Zhao, J. (2013). Titanium dioxide nanoparticles: A review of current toxicological data. *Particle and Fibre Toxicology, 10*(1), 15. Available from https://doi.org/10.1186/1743-8977-10-15.

Siddiqui, S. I., Naushad, M., & Chaudhry, S. A. (2019). Promising prospects of nanomaterials for arsenic water remediation: A comprehensive review. *Process Safety and Environmental Protection, 126*, 60−97. Available from https://doi.org/10.1016/j.psep.2019.03.037.

Sievers, N. V., Pollo, L. D., Corção, G., & Medeiros Cardozo, N. S. (2020). In situ synthesis of nanosized TiO_2 in polypropylene solution for the production of films with antibacterial activity. *Materials Chemistry and Physics, 246*, 122824. Available from https://doi.org/10.1016/j.matchemphys.2020.122824.

Singh, N., Umar, A., Singh, N., Fouad, H., Alothman, O. Y., & Haque, F. Z. (2018). Highly sensitive optical ammonia gas sensor based on Sn Doped V_2O_5 Nanoparticles. *Materials Research Bulletin, 108*, 266−274. Available from https://doi.org/10.1016/j.materresbull.2018.09.008.

Sinharay, S., & Bousmina, M. (2005). Biodegradable polymers and their layered silicate nanocomposites: In greening the 21st century materials world. *Progress in Materials Science, 50*(8), 962−1079. Available from https://doi.org/10.1016/j.pmatsci.2005.05.002.

Siracusa, V. (2019). Microbial degradation of synthetic biopolymers waste. *Polymers, 11*(6), 1066. Available from https://doi.org/10.3390/polym11061066.

Sivula, K., Le Formal, F., & Grätzel, M. (2011). Solar water splitting: Progress using hematite (α-Fe_2O_3) photoelectrodes. *ChemSusChem, 4*(4), 432−449. Available from https://doi.org/10.1002/cssc.201000416.

Skuja, L. (1998). Optically active oxygen-deficiency-related centers in amorphous silicon dioxide. *Journal of Non-Crystalline Solids, 239*(1−3), 16−48. Available from https://doi.org/10.1016/S0022-3093(98)00720-0.

Smagin, A., Sadovnikova, N., Vasenev, V., & Smagina, M. (2018). Biodegradation of some organic materials in soils and soil constructions: Experiments, modeling and prevention. *Materials, 11*(10), 1889. Available from https://doi.org/10.3390/ma11101889.

Soitong, T., & Wongsaenmai, S. (2017). Photo-oxidative degradation polyethylene containing titanium dioxide and poly(ethylene oxide). *Key Engineering Materials, 751*, 796−800. Available from https://doi.org/10.4028/www.scientific.net/KEM.751.796.

Song, J. H., Murphy, R. J., Narayan, R., & Davies, G. B. H. (2009). Biodegradable and compostable alternatives to conventional plastics. *Philosophical Transactions of the Royal Society B: Biological Sciences, 364*(1526), 2127−2139. Available from https://doi.org/10.1098/rstb.2008.0289.

Stöber, W., Fink, A., & Bohn, E. (1968). Controlled growth of monodisperse silica spheres in the micron size range. *Journal of Colloid and Interface Science, 26*(1), 62−69. Available from https://doi.org/10.1016/0021-9797(68)90272-5.

Tok, A. I. Y., Boey, F. Y. C., Dong, Z., & Sun, X. L. (2007). Hydrothermal synthesis of CeO_2 nano-particles. *Journal of Materials Processing Technology, 190*(1−3), 217−222. Available from https://doi.org/10.1016/j.jmatprotec.2007.02.042.

Tokiwa, Y., & Calabia, B. P. (2006). Biodegradability and biodegradation of poly(lactide). *Applied Microbiology and Biotechnology, 72*(2), 244−251. Available from https://doi.org/10.1007/s00253-006-0488-1.

Tokiwa, Y., Calabia, B., Ugwu, C., & Aiba, S. (2009). Biodegradability of plastics. *International Journal of Molecular Sciences, 10*(9), 3722−3742. Available from https://doi.org/10.3390/ijms10093722.

Tunç, S., & Duman, O. (2011). Preparation of active antimicrobial methyl cellulose/carvacrol/montmorillonite nanocomposite films and investigation of carvacrol release. *LWT—Food Science and Technology, 44*(2), 465−472. Available from https://doi.org/10.1016/j.lwt.2010.08.018.

Urbanek, A. K., Rymowicz, W., & Mirończuk, A. M. (2018). Degradation of plastics and plastic-degrading bacteria in cold marine habitats. *Applied Microbiology and Biotechnology, 102*(18), 7669−7678. Available from https://doi.org/10.1007/s00253-018-9195-y.

Uskoković, V., & Drofenik, M. (2005). Synthesis of materials within reverse micelles. *Surface Review and Letters, 12*(02), 239−277. Available from https://doi.org/10.1142/S0218625X05007001.

Vance, M. E., Kuiken, T., Vejerano, E. P., McGinnis, S. P., Hochella, M. F., Rejeski, D., & Hull, M. S. (2015). Nanotechnology in the real world: Redeveloping the nanomaterial consumer products inventory. *Beilstein Journal of Nanotechnology, 6*, 1769–1780. Available from https://doi.org/10.3762/bjnano.6.181.

Vansant, E.F., Voort, P.V.D., & Vrancken, K.C. (1995). *Characterization and chemical modification of the silica surface* (1st ed., vol. 93). ISBN: 9780080528953, Imprint: Elsevier Science, 553 pp.

Vargas, M. A., Diosa, J. E., & Mosquera, E. (2019). Data on study of hematite nanoparticles obtained from Iron(III) oxide by the Pechini method. *Data in Brief, 25*, 104183. Available from https://doi.org/10.1016/j.dib.2019.104183.

Velis, C. (2014). *Global recycling markets—plastic waste: A story for one player—China.* Report prepared by FUELogy and formatted by D-waste on behalf of International Solid Waste Association - Globalisation and Waste Management Task Force. Vienna: ISWA.

Villegas Aguilar, P. (2018). How will we solve our plastic waste problems? *The Belize Times, 6*(14), 1–4.

Wang, H., Xie, J., Yan, K., & Duan, M. (2011). Growth mechanism of different morphologies of ZnO crystals prepared by hydrothermal method. *Journal of Materials Science & Technology, 27*(2), 153–158. Available from https://doi.org/10.1016/S1005-0302(11)60041-8.

Wang, H., Zhang, Q., Yang, H., & Sun, H. (2008). Synthesis and microwave dielectric properties of $CaSiO_3$ nanopowder by the sol–gel process. *Ceramics International, 34*(6), 1405–1408. Available from https://doi.org/10.1016/j.ceramint.2007.05.001.

Wang, J., Cheng, Y., Peng, R., Cui, Q., Luo, Y., & Li, L. (2020). Co-precipitation method to prepare molecularly imprinted fluorescent polymer nanoparticles for paracetamol sensing. *Colloids and Surfaces A: Physicochemical and Engineering Aspects, 587*, 124342. Available from https://doi.org/10.1016/j.colsurfa.2019.124342.

Warwick, M. E. A., & Binions, R. (2014). Advances in thermochromic vanadium dioxide films. *Journal of Materials Chemistry A, 2*(10), 3275–3292. Available from https://doi.org/10.1039/C3TA14124A.

Wassel, A. R., El-Naggar, M. E., & Shoueir, K. (2020). Recent advances in polymer/metal/metal oxide hybrid nanostructures for catalytic applications: A review. *Journal of Environmental Chemical Engineering, 8*(5), 104175. Available from https://doi.org/10.1016/j.jece.2020.104175.

Wiecinska, P. (2016). Thermal degradation of organic additives used in colloidal shaping of ceramics investigated by the coupled DTA/TG/MS analysis. *Journal of Thermal Analysis and Calorimetry, 123*(2), 1419–1430. Available from https://doi.org/10.1007/s10973-015-5075-1.

Woo, K., Hong, J., Choi, S., Lee, H.-W., Ahn, J.-P., Kim, C. S., & Lee, S. W. (2004). Easy synthesis and magnetic properties of iron oxide nanoparticles. *Chemistry of Materials, 16*(14), 2814–2818. Available from https://doi.org/10.1021/cm049552x.

Wright, S. L., & Kelly, F. J. (2017). Plastic and human health: A micro issue? *Environmental Science & Technology, 51*(12), 6634–6647. Available from https://doi.org/10.1021/acs.est.7b00423.

Wróblewska-Krepsztul, J., Rydzkowski, T., Borowski, G., Szczypiński, M., Klepka, T., & Thakur, V. K. (2018). Recent progress in biodegradable polymers and nanocomposite-based packaging materials for sustainable environment. *International Journal of Polymer Analysis and Characterization, 23*(4), 383–395. Available from https://doi.org/10.1080/1023666X.2018.1455382.

Wu, X., Zheng, L., & Wu, D. (2005). Fabrication of superhydrophobic surfaces from microstructured ZnO-based surfaces via a wet-chemical route. *Langmuir: The ACS Journal of Surfaces and Colloids, 21*(7), 2665–2667. Available from https://doi.org/10.1021/la050275y.

Xu, C., & Teja, A. S. (2008). Continuous hydrothermal synthesis of iron oxide and PVA-protected iron oxide nanoparticles. *The Journal of Supercritical Fluids, 44*(1), 85–91. Available from https://doi.org/10.1016/j.supflu.2007.09.033.

Yan, B., Liao, L., You, Y., Xu, X., Zheng, Z., Shen, Z., & Yu, T. (2009). Single-crystalline V_2O_5 ultralong nanoribbon waveguides. *Advanced Materials, 21*(23), 2436–2440. Available from https://doi.org/10.1002/adma.200803684.

Yang, Q., Gao, X., Feng, R., Li, M., Zhang, J., Zhang, Q., & Tan, Y. (2019). MoO_3-SnO_2 catalyst prepared by hydrothermal synthesis method for dimethyl ether catalytic oxidation. *Journal of Fuel Chemistry and Technology*, *47*(8), 934–941. Available from https://doi.org/10.1016/S1872-5813(19)30038-6.

Yang, Z., Xu, X., Dai, M., Wang, L., Shi, X., & Guo, R. (2017). Accelerated ciprofloxacin biodegradation in the presence of magnetite nanoparticles. *Chemosphere*, *188*, 168–173. Available from https://doi.org/10.1016/j.chemosphere.2017.08.159.

Yi, G.-C., Wang, C., & Park, W. I. L. (2005). ZnO nanorods: Synthesis, characterization and applications. *Semiconductor Science and Technology*, *20*(4), S22–S34. Available from https://doi.org/10.1088/0268-1242/20/4/003.

Yoganandham Suman, T., Li, W.-G., & Pei, D.-S. (2020). Toxicity of metal oxide nanoparticles. In *Nanotoxicity* (pp. 107–123). https://doi.org/10.1016/B978-0-12-819943-5.00005-1

Zee, M., Stoutjesdijk, J. H., Heijden, P. A. A. W., & Wit, D. (1995). Structure-biodegradation relationships of polymeric materials. 1. Effect of degree of oxidation on biodegradability of carbohydrate polymers. *Journal of Environmental Polymer Degradation*, *3*(4), 235–242. Available from https://doi.org/10.1007/BF02068678.

Zhang, F., Yang, H., Xie, X., Li, L., Zhang, L., Yu, J., & Liu, B. (2009). Controlled synthesis and gas-sensing properties of hollow sea urchin-like α-Fe_2O_3 nanostructures and α-Fe_2O_3 nanocubes. *Sensors and Actuators B: Chemical*, *141*(2), 381–389. Available from https://doi.org/10.1016/j.snb.2009.06.049.

Zheng, Y., Cheng, Y., Wang, Y., Bao, F., Zhou, L., Wei, X., & Zheng, Q. (2006). Quasicubic α-Fe_2O_3 nanoparticles with excellent catalytic performance. *The Journal of Physical Chemistry. B*, *110*(7), 3093–3097. Available from https://doi.org/10.1021/jp056617q.

Zheng, Y., Yanful, E. K., & Bassi, A. S. (2005). A review of plastic waste biodegradation. *Critical Reviews in Biotechnology*, *25*(4), 243–250. Available from https://doi.org/10.1080/07388550500346359.

Zhong, Y., Godwin, P., Jin, Y., & Xiao, H. (2020). Biodegradable polymers and green-based antimicrobial packaging materials: A mini-review. *Advanced Industrial and Engineering Polymer Research*, *3*(1)), 27–35. Available from https://doi.org/10.1016/j.aiepr.2019.11.002.

Zou, Y., Wang, X., Khan, A., Wang, P., Liu, Y., Alsaedi, A., & Wang, X. (2016). Environmental remediation and application of nanoscale zero-valent iron and its composites for the removal of heavy metal ions: A review. *Environmental Science & Technology*, *50*(14), 7290–7304. Available from https://doi.org/10.1021/acs.est.6b01897.

22

Metal-organic framework for removal of environmental contaminants

Adnan Khan[1], Sumeet Malik[1], Nisar Ali[2], Xiaoyan Gao[2], Yong Yang[2], Muhammad Bilal[3]

[1]INSTITUTE OF CHEMICAL SCIENCES, UNIVERSITY OF PESHAWAR, PESHAWAR, PAKISTAN [2]KEY LABORATORY FOR PALYGORSKITE SCIENCE AND APPLIED TECHNOLOGY OF JIANGSU PROVINCE, NATIONAL AND LOCAL JOINT ENGINEERING RESEARCH CENTER FOR DEEP UTILIZATION TECHNOLOGY OF ROCK-SALT RESOURCE, FACULTY OF CHEMICAL ENGINEERING, HUAIYIN INSTITUTE OF TECHNOLOGY, HUAI'AN, P.R. CHINA [3]SCHOOL OF LIFE SCIENCE AND FOOD ENGINEERING, HUAIYIN INSTITUTE OF TECHNOLOGY, HUAI'AN, P.R. CHINA

Abbreviations

BTC (1,3,5-tricarboxylic acid)
BDC (benzene dicarboxilicacid)
AMCA (Amino Aluminium Citric Acid)
MIL (Materials Institut Lavoisiers)
TNS (TiO_2 nanosheets)
TCP (2,4,6-trichlorophenol)
IRA (Ion Exchange Resin Amberlite)
GNP (Gold nanoparticles)
COF (covalent organic framework)

22.1 Introduction

To meet the demands of the ever-growing world population, industrial sectors are escalating production very fast. The setup of these new industries is playing a key role in achieving higher standards of living (Nawaz, Khan, Ali, Ali, & Bilal, 2020). Alongside this, rush in industrialization is also damaging the environment. The pollutants emerging out from these industries are directly targeting living organisms. Water containing worn-out chemicals coming out of industries may cause the death of aquatic animals and plants. In humans, they are causing serious

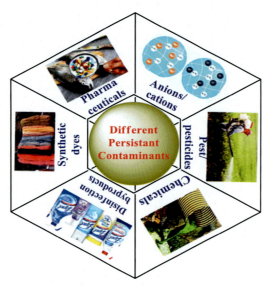

FIGURE 22–1 Representation of various emerging pollutants.

ailments (Ali, Uddin, Khan, et al., 2020; Ali, Ahmad, et al., 2020). Most commonly observed pollutants include dyes (Aziz et al., 2020; Yang, 2021), pharmaceuticals (Barczak, Dobrowolski, Borowski, & Giannakoudakis, 2020), metal ions (Khan et al., 2019; Ali, Khan, Malik, et al., 2020), acids (Khan et al., 2020), arsenic (Ali, Azeem, et al., 2020), and so forth (Fig. 22–1). These contaminants ultimately become the cause of various abnormalities like skin allergies, nausea, stomach problems, kidney failures, liver disorders, gene mutations, cancer, and so forth (Ali et al., 2019; Hanif-ur-Rehman, Shah, Khan, & Ali, 2020).

To overcome these problems, various technologies have been followed for the removal of these contaminants. These technologies include biological, physical, and chemical techniques like coagulation/flocculation (Bruno, Campo, Giustra, De Marchis, & Di Bella, 2020), precipitation (Reyes-Serrano, López-Alejo, Hernández-Cortázar, & Elizalde, 2020), membrane filtration (Chen, Zheng, Dai, Wu, & Wang, 2020; Chen, Xing, Han, Su, Li, & Lu, 2020), photolysis (Song, Liu, Guo, & Wang 2020), electrochemical degradation (Sartaj et al., 2020), advanced oxidation processes (Giwa et al., 2020), ozonation (Chávez, Quiñones, Rey, Beltrán, & Álvarez, 2020), photocatalytic degradation (Khan et al., 2016(Khan, 2021)), and sorption (Ali, Khan, Nawaz, et al., 2020). These processes have been frequently used and have provided fruitful results. But certain limitations are also associated with these processes like production of secondary pollutants, high energy, and cost consumption, time-consuming, etc. To avoid such problems, attention has been fixated on employing photocatalytic degradation and sorption processes (Ali, Bilal, Khan, Ali, Khan, et al., 2020; Ali, Bilal, Khan, Ali, Yang, et al., 2020). These processes have proven to be efficient and economically affordable strategies.

Various kinds of materials have been used as catalysts (Khan, 2021) and sorbents depending upon the ease of availability, stability, and efficiency. Recently, metal-organic

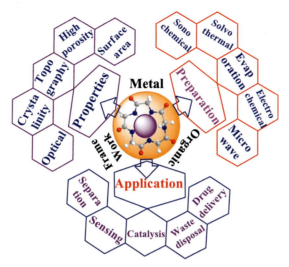

FIGURE 22–2 Preparation, properties, and applications of metal-organic frameworks.

frameworks (MOFs), also known as porous coordination polymers, have replaced the conventional agents of contaminant removal due to their advanced properties (Kumar, Bansal, Kim, & Kwon, 2018). The MOFs are very delicately designed, highly porous structures. The crystalline lattice of MOFs consists of central metal ions/clusters surrounded by multidentate-bridged organic ligands. The construction of MOFs is done via treatment under hydrothermal conditions, which provides the products with a highly porous structure (Shi et al., 2018). MOFs have proved to be very useful structures due to their highly porous molecular structure, stable nature, large surface areas, and being more susceptible to functionalization (Ansari et al., 2019). They have found applications in the fields of luminescent sensors (Zhang et al., 2018), gas storage (Duan et al., 2018), drug delivery (Lazaro & Forgan, 2019), separation (Du et al., 2019), photocatalysis (Zeng et al., 2019), and sorption (Jiang et al., 2019; Fig. 22−2).

The design of MOFs provides a highly porous structure incorporated with a well-defined framework of functionalities, thereby providing a competent material, that is, being used in multiple fields. The most important element in the structure of the MOFs is the presence of organic ligands and their arrangement. The organic ligands arranged differently from material to material not only give a variety of MOF structures but also account for the porosity of the MOFs (Rojas & Horcajada, 2020). The organic linkers provide functional groups inside the pores of the MOFs making them susceptible to the required applications. The activity of the MOFs can further be enhanced by functionalizing them with suitable functional groups. These unique features of MOFs have made them excellent and potent candidates as catalysts or sorbent material in the removal of various contaminants (Rasheed et al., 2020). In the present chapter, the deployment of MOFs and their derivatives in wastewater remediation will be discussed in detail.

22.2 Designing and properties of metal-organic frameworks

MOFs have provided better performance than the typical microporous structures like zeolites, due to the flexibility in their infrastructure, hold over their design, and functionalization of the pores (Liu et al., 2017). While considering MOFs with respect to their applications as catalysts or sorbent materials, a smaller size with a specific surface area tend to be excellent candidates. But the smaller size may also cause problems such as aggregation of particles, thereby reducing the surface energy (Saraci, Quezada-Novoa, Donnarumma, & Howarth, 2020). In case of photocatalysis, a well-defined crystalline structure is required to obtain the best results. Based on one's requirements, different forms of MOFs have been designed, including hollow structures, one-dimensional (Li et al., 2016), ultrathin two-dimensional (2D) structures (He et al., 2018), combined three-dimensional (3D) structures (Bible et al., 2018), and so forth.

One of the most noticeable properties of the MOFs is their stability toward the medium. The inability of the MOFs to sustain their structure' when interacting with other structures like solvents may limit their usage. This property of the MOFs is strongly dependent on the choice of metallic ions/clusters and the coordinated organic ligands (Cuadrado-Collados et al., 2020). In most cases, the MOFs tend to leach out at their metal localities due to instability in aqueous solution or other organic solvents they are dissolved in. This property of the MOFs sometimes tends to lower their applications in various fields. The metal centers of the MOFs tend to oxidatively leach out due to various reasons, mainly due to the weak coordination among the metals and those of the associated organic linkers (Liu, Vikrant, Kim, Kumar, & Kailasa, 2020). The active sites of the MOFs may be blocked when dissolved in an aqueous medium, thus limiting their activity as sorbent materials. In an aqueous medium, the water molecules interact with a compact structure of the MOF.

An illustrative example is the MOF-5, the oxygen atom of the water molecule replaces one of the ZnO_4 tetrahedron, which leads to ligand release from the MOF structure (De Toni, Jonchiere, Pullumbi, Coudert, & Fuchs, 2012). The release of the ligand from the MOF structure upon water treatment occurs through the ligand-displacement mechanism. This discussion shows the instability of most MOF structures in water media limiting their applications (Chen & Wu, 2019). To overcome these issues, certain steps are being taken by researchers. One of such techniques is the linker's modification to develop a hydrophobic structure, minimizing the chances of structural collapse by interaction with an aqueous medium. Another possible step could be changing the metal center, which ultimately enhances the metal-ligand bond. The introduction of additional groups that may act as protective agents for the central metal atom is yet another approach for the prevention of MOF structure from collapsing (Tan et al., 2015). These approaches may help to enhance the stability of MOFs, thereby escalating their deployment in various fields. Other high-profile properties of MOF structures include their adjustable functionalities, highly porous structure, pore aperture, crystalline nature, high surface area, and larger pore volumes (Guo et al., 2020). These properties make MOFs an excellent choice to be used as photocatalysts or sorbent materials in the removal of various contaminants, leading to a green environment.

22.3 Synthetic pathways

The morphological survey of MOFs shows that they consist of uniformly arranged infinite organic linkers as the repeating units with inorganic nodes. MOFs have been specified to have a highly porous structure (90% free volume) and great surface areas (above 6000 m²/g) (Elrasheedy, Nady, Bassyouni, & El-Shazly, 2019). There is a general concept that the MOFs are manufactured by building covalent interactions among the metal-centers and the coordinated organic ligands (Fig. 22-3; Van Vleet, Weng, Li, & Schmidt, 2018). The most commonly followed strategy for the synthesis of MOFs is the solvothermal process (Gao, Huang, Lin, Tong, & Zhang, 2016). In this process, one-pot self-assembly is designed in reaction between the metal-containing salts and those of the organic linkers takes place at an elevated temperature (up to 250°C). The high temperature and pressure are very important factors that cause the single crystalline MOF production in shorter reaction time.

Another important feature of the MOFs obtained through the solvothermal procedure is their complexity rather than the ones obtained at room temperatures. This process tends to provide a high-quality product thus frequently followed (McKinstry et al., 2016). Another common strategy to be followed for the production of MOFs is the microwave-assisted MOF growth. This strategy has also proved to be very helpful toward the production of desired quality products (Albuquerque & Herman, 2017). The ionothermal process is also being followed, which works on the principle of using ionic liquid as solvent as well as a template

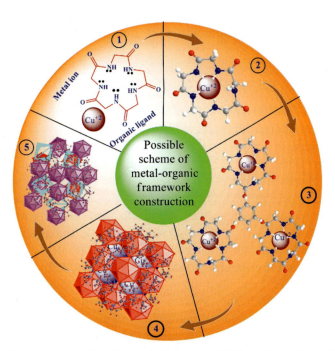

FIGURE 22-3 Construction mechanism of metal-organic frameworks (Li, Wang, Yuan, Zhang, & Chew, 2020).

(Zhang, Xu, & Jiao, 2016). Recently, metal-organic polyhedrals are being synthesized by following the stepwise approaches with a high controlling degree.

During the last decade, steps are being taken toward a "green" approach for the development of the MOF structures (Ye et al., 2017). The term "green" refers to such a process that follows a nonhazardous pathway, producing low-cost, renewable, and recyclable products. Some of the most important aspects to be considered while synthesizing MOFs through a green approach is the environmental and human hazardous effects should be minimum and have high product. The reactants should be nontoxic, and the synthesis routes of the product should not alter the properties of the final product. Low-energy and cost-requiring methods should be followed, products should be biodegradable, and the reactants should be naturally occurring biological components (Julien, Mottillo, & Friščić, 2017). The green strategies have greatly been employed for the production of MOFs.

22.4 Applications of metal-organic framework in environmental remediation

Taking into consideration the useful properties of the MOFs, as discussed earlier in detail, they have been utilized in a number of fields. The contribution of MOFs in the wastewater remediation fields cannot be overlooked. The removal of different contaminants based on MOFs has been frequently studied by following different mechanisms (Li et al., 2020). The most commonly followed mechanism is photocatalytic degradation or sorption. The removal of contaminants by MOFs through photocatalytic degradation as well as sorption is discussed in detail as follows.

22.4.1 Photocatalytic removal of contaminants using metal-organic frameworks

The general concept of photocatalytic degradation is the breakdown of hazardous contaminants into simpler nontoxic entities by making use of a catalyst in the presence of a light source (Zhang, Wang, Dong, & Lv, 2020). The process is based on the in situ production of radicals ($\bullet O_2-$ & $\bullet OH$) which degrade the contaminates into simpler and less toxic compounds (H_2O and CO_2) (Khan et al., 2016). This mechanism requires a suitable photocatalyst, which must have certain properties like the ability to absorb light, proper conduction band, and valence band positions with respect to the H_2O redox potential, charge-separation ability, and efficiency in electron-hole pair transfer to inhibit recombination of the holes and electrons, and also resistive nature in the electrolytic medium (Saeed et al., 2018).

Many of the substances have been used as successful photocatalysts (Fosso-Kankeu, Waanders, & Geldenhuys, 2015). The problem associated with such photocatalysts is their higher energy bandgaps, which reduces the light absorption ability, affecting photocatalytic efficiency. The employment of various metal sulfides has also been studied in the removal of contaminants, but the problem associated with the sulfides is their photocorrosive ability due to oxidative sulfide ions presence (Pathania et al., 2016). To avoid such problems, MOFs have grabbed the attention of researchers, due to their unique properties like porosity control, choice of metal, and organic ligands reaching out for the better light absorption property, great surface areas, optical properties, and so forth (Younis et al., 2020). Xie et al. (2019) constructed a heterostructure photocatalyst with MIL-53(Fe) and silver phosphate (Ag_3PO_4) through in situ precipitation method. The prepared composite Ag_3PO_4/MIL-53(Fe) was used for the photocatalytic degradation of antibiotic contaminants like tetracycline (TC), oxytetracycline (OTC), chlortetracycline (CTC), and deoxy tetracycline (DCL). The prepared photocatalyst showed an excellent degradation efficiency of 93.72% (TC), 90.12% (OTC), 85.54% (CTC), and 91.74% (DCL) in the presence of visible light in 1 h time. The mechanism for the degradation of the contaminants was also deeply studied and confirmed that the $^{\bullet}O_2-$, $^{\bullet}OH^-$, and h^+ radicals were the main agents during the degradation process.

Mahmoodi, Keshavarzi, Oveisi, Rahimi, and Hayati (2019), Mahmoodi, Taghizadeh, Taghizadeh, and Abdi (2019) studied the preparation of NH_2-MIL-125(Ti) magnetized with $CoFe_2O_4$ nanoparticles through the hydrothermal process. Further Ag/AgCl was deposited onto the preparation material through the in situ method and photoreduction method. The obtained Ag/AgCl@CFNMT material was used for the degradation of rhodamine B dye. The results showed the highest degradation efficiency of 89% for the selected dye. Islam et al. (2017) designed a novel visible-light-active hybrid nanocomposite, BiOI/MIL-88B(Fe) through the precipitation method. The obtained nanocomposite was utilized for the photocatalytic degradation of rhodamine B, phenol, and ciprofloxacin. The good removal efficiency by the prepared nanocomposite could be attributed to the improved charge carrier separation and reduced recombination (Fig. 22–4). The degradation efficiency of up to 88% was obtained for the removal of contaminants.

Mahmoodi and Abdi (2019) also studied the photocatalytic efficiency of the metal-organic framework (MOF-199) for the degradation of basic blue 41. The photocatalytic efficiency of 99% was achieved for the MOF-199. The operating conditions like pH, time, catalyst dosage, and so forth were also optimized. Pu et al. (2017) studied the efficiency of MOF MIL-53(Fe) for the

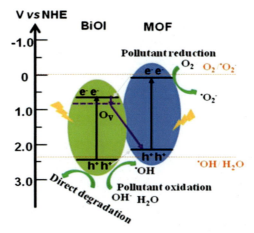

FIGURE 22–4 Mechanism of photocatalysis of BiOI/metal-organic framework (Islam et al., 2017).

degradation of orange G dye. The results showed a removal percentage of 98% after 120 minutes. Also, good stability of the prepared catalyst was obtained that showed 94.3% efficiency after the fifth cycle. Table 22–1 summarizes MOFs as photocatalysts in the removal of various contaminants.

Table 22–1 Metal-organic frameworks as photocatalysts in the removal of various contaminants.

MOF	Contaminant	Catalytic efficiency	Studied conditions pH	Time (min)	Cycles	References
MIL-53(Fe)	Acid orange 7	100%	3	90	5	Gao, Li, Li, Yao, and Zhang (2017)
[Zn(L)(H$_2$O)]·H$_2$O (**1**) (H$_2$L = 4-(pyridine-4-yl) phthalic acid)	Rhodamine B	98.5%		120	5	Dong, Shi, Li, and Wang (2019)
	Methyl orange	83.8%				
TiO$_2$ sheets/Cu-biphenylamine (TNS/Cu(BA))						
TNS/Cu(BA)	Methyl orange	99%		120		Khan, Mutahir, Wang, Lei, and Xia (2018)
	Rhodamine B	99%		160		
	Phenol	68%		180		
(2,4,6-trichlorophenol)	TCP	53%		180		
Zeolitic imidazole framework (ZIF) materials, ZIF-9 and ZIF-12	Paranitrophenol	90%	No effect	60	3	Ren et al. (2018)
ZIF-8/Fe$_2$O$_3$ composite nanofibers	Reactive red 198	94%				Mahmoodi, Keshavarzi, et al. (2019), Mahmoodi, Taghizadeh, et al. (2019)
Fe-based MOF, FeBTC (BTC = 1,3,5-tricarboxylic acid) modified with Amberlite IRA-200	Rhodamine B	99%		60		Araya et al. (2017)
TiO$_2$@NH$_2$-UiO-66 composites	Styrene	99%		600		Yao et al. (2018)
Gold nanosheets GNPs@MOF-76 (Ce) (MOF-76(1a))	p-nitrophenol	96.3		40	3	Singh et al. (2020)
ZnO/Ni$_{0.9}$Zn$_{0.1}$O-82	Methylene blue	97.4%		60		Zhong et al. (2020)
Fe$_3$O$_4$@MOF$_{UIO-66}$@COF (Covalent Organic Framework)	Malachite green	100%	12	40	5	Zheng, Yao, and Xu (2020)
	Congo red	98%	2	120		
NH$_2$-MIL-101(Fe)@CuCoNi composite	Methylene blue	99%		120	5	Chen, Zheng, et al. (2020), Chen, Xing, et al. (2020)
	Crystal violet	93%				
MOF/CCAC-5	Rhodamine B	100%				Sun et al. (2020)
	Tetracycline	98%				

MOF, Metal-organic frameworks.

22.4.2 Sorptive removal of contaminants using metal-organic frameworks

Another primary mechanism followed by the MOF structures for the removal of contaminants is the sorption process. The general idea of sorption is that the sorbent provides a binding surface for the contaminants and get attached either physically or through chemical interactive forces, thence removed from the medium. A brief look into the sorptive mechanism for the removal of contaminants shows that different interactions could be build up between the sorbent and sorbate. These interactions include electrostatic interaction, hydrogen bonding, pi-complexation, hydrophobicity, acid-base, etc. These mechanistic pathways could be followed depending upon the surface chemistry of the MOF structure and the nature of the contaminants. Zhao, Krishnaraj, Jena, Poelman, and Van Der Voort (2018) studied the removal of organic dyes methylene blue and rhodamine B onto the surface of a novel anionic framework $\{[Me_2NH_2]_{0.5}[In_{0.5}L_{0.5}] \cdot xDMF\}n(\mathbf{1})$. The MOF was prepared through a solvothermal process and had 1D channels along the b-axis. The brief insight of the removal of the cationic dyes by the prepared MOF showed that the sorption follows the ion-exchange process due to the intrinsic anionic framework of the sorbent. The results showed a sorption capacity of 76.8% and 11.2% for the Methyl Blue (MB) and Rhodamine B (RhB) dye, respectively. The difference in the sorption capacity for both cationic dyes was explained based on the size difference. MB being smaller in size, can accommodate into the MOF network, while RhB being larger in size, could not fit in.

Yang et al. (2018) also studied the preparation of a 3D reduced graphene oxide (rGO)/zeolitic imidazolate framework-67 (ZIF-67) aerogel, 3D rGO/ZIF-67 aerogel by the in-situ assembly method (Fig. 22–5). The prepared porous hydrogel was used in the successful removal of crystal violet and methyl orange (MO) with an exceptional sorption capacity of 1714.2 and 426.3 mg/g respectively. The greater sorption capacity for the anionic MO dye may be attributed to the porosity of ZIF-67 polyhedrons and the electrostatic attraction between MO and ZIF-67. Wu et al. (2019) studied the preparation of a Cu(II)-based metal-organic gel by the one-step mixing method. The prepared 3D nanoporous gel was utilized for the efficient removal of neutral red dye. The results showed a sorption capacity of 650.32 mg/g for the removal of the dye. The highest removal efficiency of 80% was attributed to the electrostatic interaction builds up between the MOF and the dye molecules as well as the intraparticles diffusion due to the nanoporous structure.

Tian et al. (2019) prepared a water-stable cationic Fe-based MOF, CPM-97-Fe, and employed it for the removal of a couple of anionic as well as cationic dyes. The obtained data were further analyzed by performing kinetics, and isotherm model studies. The dyes used as target contaminants were Reactive Brilliant Red X-3B, Acid orange 7 (AO7), Congo Red (CR), Methylene Blue (MB), RhB, Croscein Scarlet 3B (CS3B), New Coccine (NC), Acid Black 1 (AB1) with obtained sorption capacities of CR (831 mg/g), X-3B (648 mg/g), AO7 (502 mg/g), MB (380 mg/g), CS3B (356 mg/g), AB1 (325 mg/g), RhB (306 mg/g), NC (157 mg/g). The possible mechanism for the removal of these dyes included electrostatic interaction, ion exchange, p-p stacking interaction, as well as pore filling. The sorption of contaminants using MOFs is shown in Table 22–2.

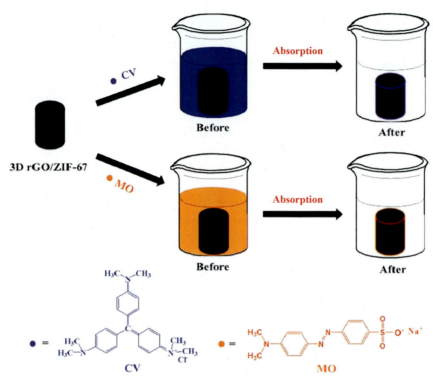

FIGURE 22–5 Schematic explanation of methyl orange and crystal violet and methyl orange by 3D reduced graphene oxide/zeolitic imidazolate framework-67 metal-organic framework (Yang et al., 2018).

Table 22–2 Metal-organic frameworks as sorbents in the removal of various contaminants.

MOF	Contaminant	Sorptive capacity	Studied conditions		Cycles	References
			pH	Time (min)		
NiCu-BTC	Methylene blue	798 mg/g	7	240		Abd El Salam and Zaki (2018)
BDC-Zn MOF	Methyl orange	2100 mg/g		60		Herrera, Reyes, Flores, Martínez, and Villanueva (2020)
Fe_3O_4@AMCA-MIL-53(Al) nanocomposite	Methylene blue	1.02 mmol/g	9	260		Alqadami, Naushad, Alothman, and Ahamad (2018)
	Malachite green	0.90 mmol/g	7			

(Continued)

Table 22-2 (Continued)

MOF	Contaminant	Sorptive capacity	Studied conditions pH	Studied conditions Time (min)	Cycles	References
MOFs, 1 and 2	Methyl orange	81.08%		1440		Ezugwu et al. (2018)
	Congo red	98.65%				
Fe-BDC MOF has	Methylene blue	94.75%		1440		
MXene and metal-organic framework	Methylene blue	140 mg/g				Arora et al. (2019)
	Acid blue 80	200 mg/g				
Ce(III)-doped UiO-67	Methylene blue	398.9 mg/g		80		Yang et al. (2020)
3D isostructural Co^{2+}/Cd^{2+} metal-doped metal-organic frameworks named compounds 1–5	Cadmium	90%	No effect			Chen et al. (2019)
3D porous anionic MOF, {[(CH_3)$_2NH_2$][Zn_6(l3-OH)(l4-O)(NSBPDC)5(H_2O)$_2$]_DMF_12H_2O}n (Zn-1) (H_2NSBPDC = 6-nitro-2,20-sulfone-4,40-dicarboxylic acid)	Methylene blue	94%		3300		Cui, Zhang, Ren, Cheng, and Gao (2019)
(NH_2-MIL-125(Ti))	Methylene blue	91.3%	7	120		Fan, Zhang, Qin, Li, and Qi (2018)

MOF, Metal-organic frameworks.

22.5 Conclusion

The rise in environmental pollution due to industrial effluents has become one of the greatest concerns for researchers. During the last few decades, steps are being taken for the removal of contaminants from wastewater bodies. This quest for the removal of contaminants through a more advanced and green approach has led to the utilization of MOFs. The MOFs possess unique properties like specific surface area, high porous structure, advanced coordination chemistry, greater pore volumes, and so forth. These properties of MOFs make them susceptible to deployment as photocatalysts and sorbent material for a variety of contaminants. The presented chapter has covered the design, synthesis, and applications of MOFs toward environmental remediation briefly.

References

Abd El Salam, H. M., & Zaki, T. (2018). Removal of hazardous cationic organic dyes from water using nickel-based metal-organic frameworks. *Inorganica Chimica Acta, 471*, 203–210.

Albuquerque, G. H., & Herman, G. S. (2017). Chemically modulated microwave-assisted synthesis of MOF-74 (Ni) and preparation of metal−organic framework-matrix based membranes for removal of metal ions from aqueous media. *Crystal Growth & Design, 17*(1), 156−162.

Ali, N., Ahmad, S., Khan, A., Khan, S., Bilal, M., Ud Din, S., ... Khan, H. (2020). Selenide-chitosan as high-performance nanophotocatalyst for accelerated degradation of pollutants. *Chemistry−An Asian Journal, 15*(17), 2660−2673.

Ali, N., Azeem, S., Khan, A., Khan, H., Kamal, T., & Asiri, A. M. (2020). Experimental studies on removal of arsenites from industrial effluents using tridodecylamine supported liquid membrane. *Environmental Science and Pollution Research, 27*, 1−12.

Ali, N., Bilal, M., Khan, A., Ali, F., Khan, H., Khan, H. A., ... Iqbal, H. M. (2020). Understanding the hierarchical assemblies and oil/water separation applications of metal-organic frameworks. *Journal of Molecular Liquids, 318*, 114273.

Ali, N., Bilal, M., Khan, A., Ali, F., Yang, Y., Khan, M., ... Iqbal, H. M. (2020). Dynamics of oil-water interface demulsification using multifunctional magnetic hybrid and assembly materials. *Journal of Molecular Liquids, 312*, 113434.

Ali, N., Khan, A., Malik, S., Badshah, S., Bilal, M., & Iqbal, H. M. (2020). Chitosan-based green sorbent material for cations removal from an aqueous environment. *Journal of Environmental Chemical Engineering, 8*, 104064.

Ali, N., Khan, A., Nawaz, S., Bilal, M., Malik, S., Badshah, S., & Iqbal, H. M. (2020). Characterization and deployment of surface-engineered chitosan-triethylenetetramine nanocomposite hybrid nano-adsorbent for divalent cations decontamination. *International Journal of Biological Macromolecules, 152*, 663−671.

Ali, N., Uddin, S., Khan, A., Khan, S., Khan, S., Ali, N., ... Bilal, M. (2020). Regenerable chitosan-bismuth cobalt selenide hybrid microspheres for mitigation of organic pollutants in an aqueous environment. *International Journal of Biological Macromolecules, 161*, 1305−1317.

Ali, N., Zada, A., Zahid, M., Ismail, A., Rafiq, M., Riaz, A., & Khan, A. (2019). Enhanced photodegradation of methylene blue with alkaline and transition-metal ferrite nanophotocatalysts under direct sun light irradiation. *Journal of the Chinese Chemical Society, 66*(4), 402−408.

Alqadami, A. A., Naushad, M., Alothman, Z. A., & Ahamad, T. (2018). Adsorptive performance of MOF nanocomposite for methylene blue and malachite green dyes: Kinetics, isotherm and mechanism. *Journal of Environmental Management, 223*, 29−36.

Ansari, A., Siddiqui, V.U., Khan, I., Akram, M.K., Ahmad, W., Siddiqi, A.K., & Asiri, A.M. (2019). Metal-organic-frameworks (MOFs) for industrial wastewater treatment. In: *Metal-organic framework composites*, vol. 1, pp. 1−28.

Araya, T., Chen, C. C., Jia, M. K., Johnson, D., Li, R., & Huang, Y. P. (2017). Selective degradation of organic dyes by a resin modified Fe-based metal-organic framework under visible light irradiation. *Optical Materials, 64*, 512−523.

Arora, C., Soni, S., Sahu, S., Mittal, J., Kumar, P., & Bajpai, P. K. (2019). Iron based metal organic framework for efficient removal of methylene blue dye from industrial waste. *Journal of Molecular Liquids, 284*, 343−352.

Aziz, A., Ali, N., Khan, A., Bilal, M., Malik, S., Ali, N., & Khan, H. (2020). Chitosan-zinc sulfide nanoparticles, characterization and their photocatalytic degradation efficiency for azo dyes. *International Journal of Biological Macromolecules, 153*, 502−512.

Barczak, M., Dobrowolski, R., Borowski, P., & Giannakoudakis, D. A. (2020). Pyridine-, thiol-and amine-functionalized mesoporous silicas for adsorptive removal of pharmaceuticals. *Microporous and Mesoporous Materials, 299*, 110132.

Bible, M., Sefa, M., Fedchak, J. A., Scherschligt, J., Natarajan, B., Ahmed, Z., & Hartings, M. R. (2018). 3D-printed acrylonitrile butadiene styrene-metal organic framework composite materials and their gas storage properties. *3D Printing and Additive Manufacturing, 5*(1), 63−72.

Bruno, P., Campo, R., Giustra, M. G., De Marchis, M., & Di Bella, G. (2020). Bench scale continuous coagulation-flocculation of saline industrial wastewater contaminated by hydrocarbons. *Journal of Water Process Engineering, 34*, 101156.

Chávez, A. M., Quiñones, D. H., Rey, A., Beltrán, F. J., & Álvarez, P. M. (2020). Simulated solar photocatalytic ozonation of contaminants of emerging concern and effluent organic matter in secondary effluents by a reusable magnetic catalyst. *Chemical Engineering Journal, 398*, 125642.

Chen, J., Xing, Z., Han, J., Su, M., Li, Y., & Lu, A. (2020). Enhanced degradation of dyes by Cu-Co-Ni nanoparticles loaded on amino-modified octahedral metal−organic framework. *Journal of Alloys and Compounds, 834*, 155106.

Chen, K., & Wu, C. D. (2019). Transformation of metal-organic frameworks into stable organic frameworks with inherited skeletons and catalytic properties. *Angewandte Chemie International Edition, 58*(24), 8119−8123.

Chen, M., Zheng, J., Dai, R., Wu, Z., & Wang, Z. (2020). Preferential removal of 2, 4-dichlorophenoxyacetic acid from contaminated waters using an electrocatalytic ceramic membrane filtration system: Mechanisms and implications. *Chemical Engineering Journal, 387*, 124132.

Chen, S., Wang, C. J., Liu, D. N., Zhu, Z. X., Qian, Y. Y., Luo, D., & Wang, Y. Y. (2019). Selective uptake of cationic organic dyes in a series of isostructural Co^{2+}/Cd^{2+} metal-doped metal−organic frameworks. *Journal of Solid State Chemistry, 270*, 180−186.

Cuadrado-Collados, C., Mouchaham, G., Daemen, L., Cheng, Y., Ramirez-Cuesta, A., Aggarwal, H., ... Silvestre-Albero, J. (2020). Quest for an optimal methane hydrate formation in the pores of hydrolytically stable metal−organic frameworks. *Journal of the American Chemical Society, 142*(31), 13391−13397.

Cui, Y. Y., Zhang, J., Ren, L. L., Cheng, A. L., & Gao, E. Q. (2019). A functional anionic metal−organic framework for selective adsorption and separation of organic dyes. *Polyhedron, 161*, 71−77.

De Toni, M., Jonchiere, R., Pullumbi, P., Coudert, F. X., & Fuchs, A. H. (2012). How can a hydrophobic MOF be water-unstable? Insight into the hydration mechanism of IRMOFs. *ChemPhysChem, 13*(15), 3497−3503.

Dong, J. P., Shi, Z. Z., Li, B., & Wang, L. Y. (2019). Synthesis of a novel 2D zinc (II) metal−organic framework for photocatalytic degradation of organic dyes in water. *Dalton Transactions, 48*(47), 17626−17632.

Du, J., Zhang, C., Pu, H., Li, Y., Jin, S., Tan, L., ... Dong, L. (2019). HKUST-1 MOFs decorated 3D copper foam with superhydrophobicity/superoleophilicity for durable oil/water separation. *Colloids and Surfaces A: Physicochemical and Engineering Aspects, 573*, 222−229.

Duan, C., Zhang, H., Li, F., Xiao, J., Luo, S., & Xi, H. (2018). Hierarchically porous metal−organic frameworks: Rapid synthesis and enhanced gas storage. *Soft Matter, 14*(47), 9589−9598.

Elrasheedy, A., Nady, N., Bassyouni, M., & El-Shazly, A. (2019). Metal organic framework based polymer mixed matrix membranes: Review on applications in water purification. *Membranes, 9*(7), 88.

Ezugwu, C. I., Asraf, M. A., Li, X., Liu, S., Kao, C. M., Zhuiykov, S., & Verpoort, F. (2018). Cationic nickel metal-organic frameworks for adsorption of negatively charged dye molecules. *Data in Brief, 18*, 1952−1961.

Fan, Y. H., Zhang, S. W., Qin, S. B., Li, X. S., & Qi, S. H. (2018). An enhanced adsorption of organic dyes onto NH_2 functionalization titanium-based metal-organic frameworks and the mechanism investigation. *Microporous and Mesoporous Materials, 263*, 120−127.

Fosso-Kankeu, E., Waanders, F., & Geldenhuys, M. (2015, November). Photocatalytic degradation of dyes using TiO_2 nanoparticles of different shapes. In *Proceedings of the seventh international conference on latest trends in engineering and technology*.

Gao, J., Huang, C., Lin, Y., Tong, P., & Zhang, L. (2016). In situ solvothermal synthesis of metal−organic framework coated fiber for highly sensitive solid-phase microextraction of polycyclic aromatic hydrocarbons. *Journal of Chromatography A, 1436*, 1−8.

Gao, Y., Li, S., Li, Y., Yao, L., & Zhang, H. (2017). Accelerated photocatalytic degradation of organic pollutant over metal-organic framework MIL-53 (Fe) under visible LED light mediated by persulfate. *Applied Catalysis B: Environmental, 202*, 165−174.

Giwa, A., Yusuf, A., Balogun, H. A., Sambudi, N. S., Bilad, M. R., Adeyemi, I., ... Curcio, S. (2020). Recent advances in advanced oxidation processes for removal of contaminants from water: A comprehensive review. *Process Safety and Environmental Protection, 146*, 220−256.

Guo, X., Zhu, N., Wang, S. P., Li, G., Bai, F. Q., Li, Y., ... Feng, S. (2020). Stimuli-responsive luminescent properties of tetraphenylethene-based strontium and cobalt metal−organic frameworks. *Angewandte Chemie, 132*(44), 19884−19889.

Hanif-ur-Rehman., Shah, A. H. A., Khan, A., & Ali, N. (2020). In situ transport and electrodeposition of Ag(I) on stainless steel electrode across carrier mediated supported liquid membrane. *Desalination and Water Treatment, 205*, 189−197.

He, T., Ni, B., Zhang, S., Gong, Y., Wang, H., Gu, L., ... Wang, X. (2018). Ultrathin 2D zirconium metal−organic framework nanosheets: Preparation and application in photocatalysis. *Small, 14*(16), 1703929.

Herrera, L. Á. A., Reyes, P. K. C., Flores, A. M. H., Martínez, L. T., & Villanueva, J. M. R. (2020). BDC-Zn MOF sensitization by MO/MB adsorption for photocatalytic hydrogen evolution under solar light. *Materials Science in Semiconductor Processing, 109*, 104950.

Islam, M. J., Kim, H. K., Reddy, D. A., Kim, Y., Ma, R., Baek, H., ... Kim, T. K. (2017). Hierarchical BiOI nanostructures supported on a metal organic framework as efficient photocatalysts for degradation of organic pollutants in water. *Dalton Transactions, 46*(18), 6013−6023.

Jiang, D., Chen, M., Wang, H., Zeng, G., Huang, D., Cheng, M., ... Wang, Z. (2019). The application of different typological and structural MOFs-based materials for the dyes adsorption. *Coordination Chemistry Reviews, 380*, 471−483.

Julien, P. A., Mottillo, C., & Friščić, T. (2017). Metal−organic frameworks meet scalable and sustainable synthesis. *Green Chemistry, 19*(12), 2729−2747.

Khan, A., Ali, N., Bilal, M., MaliK, S., Badshah, S., & Iqbal, H. (2019). Engineering functionalized chitosan-based sorbent material: characterization and sorption of toxic elements. *Applied Sciences, 23*(9), 5138−5141.

Khan, H., Gul, K., Ara, B., Khan, A., Ali, N., Ali, N., & Bilal, M. (2020). Adsorptive removal of acrylic acid from the aqueous environment using raw and chemically modified alumina: Batch adsorption, kinetic, equilibrium and thermodynamic studies. *Journal of Environmental Chemical Engineering, 8*(4), 103927.

Khan, H., Khalil, A. K., Khan, A., Saeed, K., & Ali, N. (2016). Photocatalytic degradation of bromophenol blue in aqueous medium using chitosan conjugated magnetic nanoparticles. *Korean Journal of Chemical Engineering, 33*(10), 2802−2807.

Khan, M., Khan, A., Khan, H., Ali, N., Sartaj, S, Malik, S., ... Bilal, M. (2021). Development and characterization of regenerable chitosan-coated nickel selenide nano-photocatalytic system for decontamination of toxic azo dyes. *International Journal of Biological Macromolecules, 182*, 866−878.

Khan, M. A., Mutahir, S., Wang, F., Lei, W., & Xia, M. (2018). Sensitization of TiO_2 nanosheets with Cu−biphenylamine framework to enhance photocatalytic degradation performance of toxic organic contaminants: Synthesis, mechanism and kinetic studies. *Nanotechnology, 29*(37), 375605.

Khan, S., Khan, A., Ali, N., Ahmad, S., Ahmad, W., Malik, S., ... Bilal, M. (2021). Degradation of Congo red dye using ternary metal selenide-chitosan microspheres as robust and reusable catalysts. *Environmental Technology & Innovation, 22*, 101402.

Kumar, P., Bansal, V., Kim, K. H., & Kwon, E. E. (2018). Metal-organic frameworks (MOFs) as futuristic options for wastewater treatment. *Journal of Industrial and Engineering Chemistry, 62*, 130−145.

Lazaro, I. A., & Forgan, R. S. (2019). Application of zirconium MOFs in drug delivery and biomedicine. *Coordination Chemistry Reviews, 380*, 230−259.

Li, J., Wang, H., Yuan, X., Zhang, J., & Chew, J. W. (2020). Metal-organic framework membranes for wastewater treatment and water regeneration. *Coordination Chemistry Reviews, 404*, 213116.

Li, X., Wang, Z., Zhang, B., Rykov, A. I., Ahmed, M. A., & Wang, J. (2016). Fe$_x$Co$_{3-x}$O$_4$ nanocages derived from nanoscale metal−organic frameworks for removal of bisphenol A by activation of peroxymonosulfate. *Applied Catalysis B: Environmental, 181*, 788−799.

Liu, B., Vikrant, K., Kim, K. H., Kumar, V., & Kailasa, S. K. (2020). Critical role of water stability in metal−organic frameworks and advanced modification strategies for the extension of their applicability. *Environmental Science: Nano, 7*(5), 1319−1347.

Liu, X., Zhou, Y., Zhang, J., Tang, L., Luo, L., & Zeng, G. (2017). Iron containing metal−organic frameworks: Structure, synthesis, and applications in environmental remediation. *ACS Applied Materials & Interfaces, 9*(24), 20255−20275.

Mahmoodi, N. M., & Abdi, J. (2019). Nanoporous metal-organic framework (MOF-199): Synthesis, characterization and photocatalytic degradation of Basic Blue 41. *Microchemical Journal, 144*, 436−442.

Mahmoodi, N. M., Keshavarzi, S., Oveisi, M., Rahimi, S., & Hayati, B. (2019). Metal-organic framework (ZIF-8)/inorganic nanofiber (Fe$_2$O$_3$) nanocomposite: Green synthesis and photocatalytic degradation using LED irradiation. *Journal of Molecular Liquids, 291*, 111333.

Mahmoodi, N. M., Taghizadeh, A., Taghizadeh, M., & Abdi, J. (2019). In situ deposition of Ag/AgCl on the surface of magnetic metal-organic framework nanocomposite and its application for the visible-light photocatalytic degradation of Rhodamine dye. *Journal of Hazardous Materials, 378*, 120741.

McKinstry, C., Cathcart, R. J., Cussen, E. J., Fletcher, A. J., Patwardhan, S. V., & Sefcik, J. (2016). Scalable continuous solvothermal synthesis of metal organic framework (MOF-5) crystals. *Chemical Engineering Journal, 285*, 718−725.

Nawaz, A., Khan, A., Ali, N., Ali, N., & Bilal, M. (2020). Fabrication and characterization of new ternary ferrites-chitosan nanocomposite for solar-light driven photocatalytic degradation of a model textile dye. *Environmental Technology & Innovation, 20*, 101079.

Pathania, D., Gupta, D., Ala'a, H., Sharma, G., Kumar, A., Naushad, M., . . . Alshehri, S. M. (2016). Photocatalytic degradation of highly toxic dyes using chitosan-g-poly (acrylamide)/ZnS in presence of solar irradiation. *Journal of Photochemistry and Photobiology A: Chemistry, 329*, 61−68.

Pu, M., Ma, Y., Wan, J., Wang, Y., Wang, J., & Brusseau, M. L. (2017). Activation performance and mechanism of a novel heterogeneous persulfate catalyst: Metal−organic framework MIL-53 (Fe) with Fe II/Fe III mixed-valence coordinatively unsaturated iron center. *Catalysis Science & Technology, 7*(5), 1129−1140.

Rasheed, T., Bilal, M., Hassan, A. A., Nabeel, F., Bharagava, R. N., Ferreira, L. F. R., . . . Iqbal, H. M. (2020). Environmental threatening concern and efficient removal of pharmaceutically active compounds using metal-organic frameworks as adsorbents. *Environmental Research, 185*, 109436.

Ren, W., Gao, J., Lei, C., Xie, Y., Cai, Y., Ni, Q., & Yao, J. (2018). Recyclable metal-organic framework/cellulose aerogels for activating peroxymonosulfate to degrade organic pollutants. *Chemical Engineering Journal, 349*, 766−774.

Reyes-Serrano, A., López-Alejo, J. E., Hernández-Cortázar, M. A., & Elizalde, I. (2020). Removing contaminants from tannery wastewater by chemical precipitation using CaO and Ca(OH)$_2$. *Chinese Journal of Chemical Engineering, 28*(4), 1107−1111.

Rojas, S., & Horcajada, P. (2020). Metal−organic frameworks for the removal of emerging organic contaminants in water. *Chemical Reviews, 120*(16), 8378−8415.

Saeed, K., Sadiq, M., Khan, I., Ullah, S., Ali, N., & Khan, A. (2018). Synthesis, characterization, and photocatalytic application of Pd/ZrO$_2$ and Pt/ZrO$_2$. *Applied Water Science, 8*(2), 60.

Saraci, F., Quezada-Novoa, V., Donnarumma, P. R., & Howarth, A. J. (2020). Rare-earth metal−organic frameworks: From structure to applications. *Chemical Society Reviews, 49*(22), 7949−7977.

Sartaj, S., Ali, N., Khan, A., Malik, S., Bilal, M., Khan, M., . . . Khan, S. (2020). Performance evaluation of photolytic and electrochemical oxidation processes for enhanced degradation of food dyes laden wastewater. *Water Science and Technology, 81*(5), 971−984.

Shi, Z., Xu, C., Guan, H., Li, L., Fan, L., Wang, Y., ... Zhang, R. (2018). Magnetic metal organic frameworks (MOFs) composite for removal of lead and malachite green in wastewater. *Colloids and Surfaces A: Physicochemical and Engineering Aspects, 539*, 382−390.

Singh, K., Kukkar, D., Singh, R., Kukkar, P., Bajaj, N., Singh, J., ... Kim, K. H. (2020). In situ green synthesis of Au/Ag nanostructures on a metal-organic framework surface for photocatalytic reduction of p-nitrophenol. *Journal of Industrial and Engineering Chemistry, 81*, 196−205.

Song, C., Liu, H. Y., Guo, S., & Wang, S. G. (2020). Photolysis mechanisms of tetracycline under UV irradiation in simulated aquatic environment surrounding limestone. *Chemosphere, 244*, 125582.

Sun, Z., Wu, X., Qu, K., Huang, Z., Liu, S., Dong, M., & Guo, Z. (2020). Bimetallic metal-organic frameworks anchored corncob-derived porous carbon photocatalysts for synergistic degradation of organic pollutants. *Chemosphere, 259*, 127389.

Tan, K., Nijem, N., Gao, Y., Zuluaga, S., Li, J., Thonhauser, T., & Chabal, Y. J. (2015). Water interactions in metal organic frameworks. *CrystEngComm, 17*(2), 247−260.

Tian, S., Xu, S., Liu, J., He, C., Xiong, Y., & Feng, P. (2019). Highly efficient removal of both cationic and anionic dyes from wastewater with a water-stable and eco-friendly Fe-MOF via host-guest encapsulation. *Journal of Cleaner Production, 239*, 117767.

Van Vleet, M. J., Weng, T., Li, X., & Schmidt, J. R. (2018). In situ, time-resolved, and mechanistic studies of metal−organic framework nucleation and growth. *Chemical Reviews, 118*(7), 3681−3721.

Wu, Q., He, L., Jiang, Z. W., Li, Y., Zhao, T. T., Li, Y. H., ... Li, Y. F. (2019). One-step synthesis of Cu(II) metal−organic gel as recyclable material for rapid, efficient and size selective cationic dyes adsorption. *Journal of Environmental Sciences, 86*, 203−212.

Xie, L., Yang, Z., Xiong, W., Zhou, Y., Cao, J., Peng, Y., ... Zhang, Y. (2019). Construction of MIL-53 (Fe) metal-organic framework modified by silver phosphate nanoparticles as a novel Z-scheme photocatalyst: Visible-light photocatalytic performance and mechanism investigation. *Applied Surface Science, 465*, 103−115.

Yang, J. M., Yang, B. C., Zhang, Y., Yang, R. N., Ji, S. S., Wang, Q., ... Zhang, R. Z. (2020). Rapid adsorptive removal of cationic and anionic dyes from aqueous solution by a Ce (III)-doped Zr-based metal−organic framework. *Microporous and Mesoporous Materials, 292*, 109764.

Yang, Q., Lu, R., Ren, S., Chen, C., Chen, Z., & Yang, X. (2018). Three dimensional reduced graphene oxide/ZIF-67 aerogel: Effective removal cationic and anionic dyes from water. *Chemical Engineering Journal, 348*, 202−211.

Yang, Y., Ali, N., Khan, A., Khan, S., Khan, S., Khan, H., ... Bilal, M. (2021). Chitosan-capped ternary metal selenide nanocatalysts for efficient degradation of Congo red dye in sunlight irradiation. *International Journal of Biological Macromolecules, 167*, 169−181.

Yao, P., Liu, H., Wang, D., Chen, J., Li, G., & An, T. (2018). Enhanced visible-light photocatalytic activity to volatile organic compounds degradation and deactivation resistance mechanism of titania confined inside a metal-organic framework. *Journal of Colloid and Interface Science, 522*, 174−182.

Ye, G., Zhang, D., Li, X., Leng, K., Zhang, W., Ma, J., ... Ma, S. (2017). Boosting catalytic performance of metal−organic framework by increasing the defects via a facile and green approach. *ACS Applied Materials & Interfaces, 9*(40), 34937−34943.

Younis, S. A., Kwon, E. E., Qasim, M., Kim, K. H., Kim, T., Kukkar, D., ... Ali, I. (2020). Metal-organic framework as a photocatalyst: Progress in modulation strategies and environmental/energy applications. *Progress in Energy and Combustion Science, 81*, 100870.

Zeng, T., Wang, L., Feng, L., Xu, H., Cheng, Q., & Pan, Z. (2019). Two novel organic phosphorous-based MOFs: Synthesis, characterization and photocatalytic properties. *Dalton Transactions, 48*(2), 523−534.

Zhang, X., Wang, J., Dong, X. X., & Lv, Y. K. (2020). Functionalized metal-organic frameworks for photocatalytic degradation of organic pollutants in environment. *Chemosphere, 242*, 125144.

Zhang, Y., Yuan, S., Day, G., Wang, X., Yang, X., & Zhou, H. C. (2018). Luminescent sensors based on metal-organic frameworks. *Coordination Chemistry Reviews*, 354, 28–45.

Zhang, Z. H., Xu, L., & Jiao, H. (2016). Ionothermal synthesis, structures, properties of cobalt-1, 4-benzenedicarboxylate metal–organic frameworks. *Journal of Solid State Chemistry*, 238, 217–222.

Zhao, S. N., Krishnaraj, C., Jena, H. S., Poelman, D., & Van Der Voort, P. (2018). An anionic metal-organic framework as a platform for charge-and size-dependent selective removal of cationic dyes. *Dyes and Pigments*, 156, 332–337.

Zheng, M., Yao, C., & Xu, Y. (2020). Fe_3O_4 nanoparticles decorated with UIO-66 metal–organic framework particles and encapsulated in a triazine-based covalent organic framework matrix for photodegradation of anionic dyes. *ACS Applied Nano Materials*, 3(11), 11307–11314.

Zhong, M., Qu, S. Y., Zhao, K., Fei, P., Wei, M. M., Yang, H., & Su, B. (2020).). Bimetallic metal-organic framework derived $ZnO/Ni_{0.9}Zn_{0.1}O$ nanocomposites for improved photocatalytic degradation of organic dyes. *ChemistrySelect*, 5(6), 1858–1864.

23

Effects of zeolite-based nanoparticles on the biodegradation of organic materials

Farooq Sher[1], Abu Hazafa[2,3], Tazien Rashid[4], Muhammad Bilal[5], Fatima Zafar[3,6], Zahid Mushtaq[2], Zaka Un Nisa[3,7]

[1]DEPARTMENT OF ENGINEERING, SCHOOL OF SCIENCE AND TECHNOLOGY, NOTTINGHAM TRENT UNIVERSITY, NOTTINGHAM, UNITED KINGDOM [2]DEPARTMENT OF BIOCHEMISTRY, UNIVERSITY OF AGRICULTURE, FAISALABAD, PAKISTAN [3]INTERNATIONAL SOCIETY OF ENGINEERING SCIENCE AND TECHNOLOGY, COVENTRY, UNITED KINGDOM [4]DEPARTMENT OF CHEMICAL ENGINEERING, NFC INSTITUTE OF ENGINEERING AND FERTILIZER RESEARCH, FAISALABAD, PAKISTAN [5]SCHOOL OF LIFE SCIENCE AND FOOD ENGINEERING, HUAIYIN INSTITUTE OF TECHNOLOGY, HUAI'AN, P.R. CHINA [6]SCHOOL OF BIOCHEMISTRY AND BIOTECHNOLOGY, UNIVERSITY OF THE PUNJAB, LAHORE, PAKISTAN [7]FACULTY OF MEDICINE, QUAID-I-AZAM UNIVERSITY, ISLAMABAD, PAKISTAN

23.1 Introduction

The textile industry is one of the major industries worldwide. One of the highest consumers of primary water, this sector is growing proportionally with the increasing demand for textile products worldwide. This is also causing an increase in water demand, and consequently increased effluent discharge. The major operations involved in the textile industry are spinning (twisting of fibers to form thread), weaving (arranging two different sets of threads perpendicular to each other to form fabric), and finishing. Finishing steps might contain several elements, including washing, bleaching, stabilizing, and dyeing operations (Dotto et al., 2019; Holkar et al., 2016; Wasti, 2018). There are several classes of dyes that are produced commercially. Among all of them, azo dyes (one of the most studied groups of commercial dyes, these dyes contain azo group) and anthraquinone dyes (one of the oldest dyes, this dye is the second largest used dye commercially; this dye offers good fastness but is tinctorialy weak) are major classes of commercial dyes. Among these, azo dye holds up to 70% market share of all organic dyes (Chequer et al., 2013).

Major constituents of textile effluent are color, total dissolved solids (TDS), chemical oxygen demand (COD), turbidity, and pH. The presence of dyes in the effluent affects esthetic value of water, and also possess severe environmental and health threats. They tend to adversely affect aquatic life and are also carcinogenic for human beings (Balapure, Madamwar, & Bhatt, 2017; Wijetunga, Li, & Jian, 2010). Several prominent azo dyes are degradable into amines in the intestinal environment, which are proven carcinogenic. Several unregulated azo dyes are also found to form aromatic amines with severe mutagenic tendencies (Brüschweiler & Merlot, 2017). Mutagenic threats are not limited to azo family dyes. Anthraquinone dyes, such as Disperse Blue 3 is also found to have severe toxic effects after a series of bacterial, algal, and protozoan experiments (Novotny et al., 2006). Schneider, Hafner, and Jäger (2004) confirmed 14 dyes to be mutagenic and suspected 16 other dyes products to have mutagenic potential after an extensive review of data from literature and dye manufacturers. The authors also pointed out a poor investigation of several dyes product that may lead to widespread commercial use of genotoxic dye products. Genotoxic capability of color index (CI) Disperse Blue 291 was confirmed about testing on human cells by Tsuboy et al. (2007). Azo dyes were reported to decrease the germination frequency of *Nigella sativa* seeds even at low concentration of 2.5 mM (Kumbhakar et al., 2018).

A conventional textile effluent treatment plant (ETP) includes equalization, primary settling, coagulation, carbon filtration, and biological treatment. While industrially established techniques for the removal of pH, chemical oxygen demand (COD), and biological oxygen demand (BOD) exists, the removal of coloring ingredients in wastewater poses a major challenge (Gebrati et al., 2019; Shoukat, Khan, & Jamal, 2019). According to current industrial practices, color is removed by the carbon filtration/adsorption, or by coagulation, although other chemical and biological options also exists. Both of these processes result in the removal of color from water and the concentration of that color in another phase, such as colored sludge in case of coagulation, which then poses a disposal/treatment problem of its own (Dotto et al., 2019).

Due to low biodegradability, dyes cannot be remediated by conventional biological treatment. Among different alternatives, advanced oxidation processes (AOPs) have also been suggested for textile dye remediation. AOP is the set of oxidation processes that rely on the oxidation of organic contaminants by hydroxyl radicals. AOPs have the potential to offer complete or satisfactory degradation of textile dyes and other contaminants, unlike physical removal, although they might incur higher costs (Giannakis, Lin, & Ghanbari, 2020; Khatri et al., 2018). Hydroxyl radical (.OH) is a charge-less form of hydroxyl ion, known to man in several capacities, from cleaning agent for harmful substances in the atmosphere to cause oxidative damage to cell contents, including DNA in humans. It is unstable and takes up an electron to form a hydroxyl ion, making it a very reactive oxidizing agent. Due to its instability, it is generated *in situ* for applications (Oturan & Aaron, 2014; Solomon, Kiflie, & Van, 2020).

Due to high reactivity and efficient remediation of contaminants, Fenton's reagent is being focused upon as the AOP of choice. Controversy exists among the scientific community about the exact reaction pathway and the powerful oxidative species formed, but following is generally the most agreed upon reaction mechanism [see Eqs. (23.1) and (23.2)] (Danish et al., 2017; Tasneem, Sarker, & Uddin, 2020).

$$H_2O_2 + Fe^{2+} \rightarrow H^{\cdot} + OH^{\cdot} + Fe^{3+} \tag{23.1}$$

$$Fe^{3+} + H_2O_2 \rightarrow H^+ + HO_2^{\cdot} + Fe^{2+} \tag{23.2}$$

The reactive oxygen species produced as a result, oxidizes the target contaminants. The use of conventional oxidants like hydrogen peroxide can pose a splashing hazard. Recently, sodium percarbonate (SPC) has been suggested as a novel source of H_2O_2 (Danish et al., 2017). During the past few decades, applications of engineered nanomaterials has been increased due to their large surface area, high efficiency, high accuracy, and environmentally friendly behaviors (Kamali et al., 2019). However, for organic waste and dye degradation from water, various nonmetallic and metallic compounds have been synthesized. Studies showed that nanotechnology possesses remarkable results for wastewater treatment (Tahir, Sagir, & Shahzad, 2019; Yao et al., 2019). In this regard, a chemical agent that has grabbed the attention of scientific community, especially in the last few decades due to superior remediation characteristics, is nanozerovalent iron (nZVI). Iron is a transition metal that has the most common oxidation states of Fe^{2+} and Fe^{3+}. When these oxidation states, usually Fe^{2+} are reduced by provision of two electrons, zerovalent iron, such as Fe^0 is formed. Zerovalent iron tends to convert to Fe^{2+}, displaying exceptional reductive tendency in the process, which can be used beneficially. Zerovalent iron particles with a size under 100 μm are called nZVI. Due to smaller particle size, nZVI have higher surface area and display superior reactivity then ZVI (Karam, Zaher, & Mahmoud, 2020; Raji et al., 2020; Rashid et al., 2020).

nZVI has a high surface area, due to which it can be effectively used as an adsorbent for the removal of several contaminants. nZVI can also be used as a reducing agent for chemical remediation of contaminants. Zerovalent iron can also be used to ferrous ion generation and consequent production of oxidizing species (Danish et al., 2017), according to Haber-Weiss process [see Eqs. (23.3)–(23.7)]:

$$Fe^0 + 2H^+ \rightarrow Fe^{2+} + H_2 \tag{23.3}$$

$$Fe^0 + H_2O_2 + 2H^+ \rightarrow Fe^{2+} + 2H_2O \tag{23.4}$$

$$H_2O_2 + Fe^{2+} \rightarrow OH^{\cdot} + OH^{-} + Fe^{3+} \tag{23.5}$$

$$H_2O_2 + Fe^{3+} \rightarrow H^+ + Fe^{2+} + HO_2^{\cdot} \tag{23.6}$$

$$Fe^{2+} + Fe^0 \rightarrow 3Fe^{2+} \tag{23.7}$$

Zeolites are inorganic aluminosilicate minerals that have a crystalline framework structure. There are over 40 naturally occurring and more than a hundred synthetic zeolites (Othmer and Kirk). They mostly contain aluminum (Al) and silicon (Si) and hence are often characterized by their Si/Al ratio, along with their unit cell type. Several other elements such as oxygen, calcium, iron, magnesium, and sodium are also present within the crystalline lattice. Presence of these elements, especially cations like sodium, and zeolite's ability to exchange these cations with other ions in aqueous matrix make them a viable ion exchange agent (Jha and Singh, 2011). Zeolites are also characterized based on their silicon content. If Si/Al ratio is up to 2, then zeolite is termed as low silica zeolite (Amalcime). If Si/Al ratio lies between 2 and 5, then zeolite is the intermediate silica content (Faujasite) and those zeolites having Si/Al ratio more than 5 are known as high silica content zeolite (Cliniptilolite).

The aggregation of nZVI can be substantially decreased by impregnating them with support material. This improves the availability of active sites for reaction. In several cases, functional groups present in the support material are also reported to improve the pollutant remediation process. Some of the desired characteristics of material to be used as nZVI support are (1) it should be low cost; (2) it should aid in the process of environmental remediation; and (3) it should not decrease the reactivity of nZVI.

Several materials have been suggested in the literature as nZVI support material, such as carbon nanotubes (CNTs), graphene, biochar, and clays such as bentonite. Natural zeolite has also been suggested for this role, as it fulfills all the above requirements, being cheap, adsorptive, and as an already established catalytic support. The present chapter examined the mutagenic dye remediation potential of nanoscaled nZVI in a Fenton-like role, using sodium percarbonate (SPC) as the source of reactive advance oxidative species, for both synthetic dye solution and actual textile effluents.

The textile sector, being a major user of water, calls for a technically and financially feasible solution for remediation of its effluent. While a variety of physical, chemical, or biological methods are in practice or have been suggested for textile dye remediation, there is room for improvement in such methods on both environmental and commercial fronts. The nanoscaled nZVI/SPC Fenton process, explored in the present study, can offer the following benefits:

1. Faster redox cycling of Fe^{3+} to Fe^{2+}, increasing the hydroxyl radical generation as well.
2. Distribution of nanoiron particles on support medium provides improved reactivity, due to higher surface area of nanoscale iron, and decreased aggregation due to support.
3. Generation of H_2 gas during oxidation of zerovalent iron to Fe^{2+} state, which might be collected and used in an energy provision capacity in the future, either directly or in a fuel cell.

$$Fe^0 + 2H^+ \rightarrow Fe^{2+} + H_2$$

23.2 Textile effluent composition

The textile industry is a major primary water user, utilizing up to 350 L/kg of textile product formed, and waste around 10–50% textile dyes. Textile effluent generated in Pakistan is roughly estimated to be around 1.15×10^{12} L/year. The mutagenic potential of textile dyes has forced the scientific community to characterize textile effluent. Azo and anthraquinone dyes are two most used dyes in textiles and are used for almost every fiber type. A synthetic effluent composition of textile industry is presented in Table 23–1 (Soares, Silva, & Arcy, 2016).

Textile processing adds a multitude of chemical agents, at each step of processing, in the effluent produced as a result. In the earlier steps, such as sizing, polymers like starch, or carboxymethyl cellulose are applied on yarn surface to increase its strength and prevent its breakage during the steps ahead. When these polymers are removed (desizing), they are added to the effluent. In the scouring process, hydrophilicity of fabric is increased by removing waxes or dirt from the textile. This process requires high pH as well as an increase in temperature in some cases. Effluent leaving this step contains alkalis like NaOH, wetting agents, and detergents. Scoured textile is fed to the bleaching step where the natural color and odor of textile fiber material is removed by the application of bleaching agents like

Table 23–1 Synthetic effluent characteristics for various types of organic and synthetic effluents.

Parameter	Unit	Polyester	Cotton	Polyester-cotton
pH	pH Scale	7.40	10.10	8.40
Conductivity	mS/cm	3.70	5.0	4.50
COD	mg O_2/L	2530	1112	1450
BOD_5	mg O_2/L	700	350	440
BOD_5/COD	–	0.28	0.31	0.30
Dissolved organic carbon	mg C/L	659	254	354
Biodegradability	%	64	89	81
Absorbance at 254 nm	–	4.26	0.95	2.00
Color DFZ $_{436\ nm}$	m^{-1}	3.50	21.50	16.70
Color DFZ $_{525\ nm}$	m^{-1}	1.70	20.40	17.50
Color DFZ $_{620\ nm}$	m^{-1}	1.50	1.20	0.20
Color Pt-Co scale	mg/L	75	260	230
Color at 1:40 dilution	–	Not visible	Visible	Visible
Chloride	mg C/L	42	1900	1470
Sulphate	mg SO_4^{2-}/L	934	292	137
Total dissolved nitrogen	mg N/L	11	9	10
Nitrate	mg N-NO_3^-/L	7	7	7
Nitrite	mg N-NO_2^-/L	4	2	3
Ammonia	mg N-NH_4^+/L	<0.50	<0.50	<0.50
Phosphate	mg P-PO_4^{3-}/L	6	<0.50	<0.50
Sodium	mg Na^+/L	1723	1998	1772
Magnesium	mg Mg^{2+}/L	21	16	17
Calcium	mg Ca^{2+}/L	150	73	91
Total suspended solids	mg TSS/L	38	141	98
Volatile suspended solids	mg VSS/L	36	103	83

Source: This information is taken from Soares, P. A., Silva, T. F. C. V., & Arcy, A. R. (2016). Assessment of AOPs as a polishing step in the decolourisation of bio-treated textile wastewater: Technical and economic considerations. *Journal of Photochemistry and Photobiology A: Chemistry, 317*, 26–38.

sodium hypochlorite and hydrogen peroxide. These added oxidants also aid in the removal of other pollutants (Mani, Chowdhary, & Bharagava, 2019; Saleemi et al., 2020).

Dye impart color to fabric owing to the part of molecule called a chromophore. Textile dyeing usually requires a large amount of water. Most of dyes are applied to fiber after dissolving them in aqueous media. Dyeing and finishing steps might also include the addition of pH adjusters, biocides, and fabric conditions such as starch in the case of cotton textile products (Cinperi et al., 2019). Common features of almost every textile effluent is the presence of high COD content, low biodegradability, pH in alkaline range (with usual exception of polyester effluent), and considerable presence of color (Wang et al., 2019).

23.3 Conventional methods for dye remediation

Wastewater treatment plants, either municipal or industrial, consists of a number of steps, often starting with preliminary treatment. Preliminary treatment serves to regulate the input flow to effluent treatment plant (ETP), as well as remove any coarser objects, if any. Primary treatment removes particles large enough to settle down by gravity in clarifier tanks as sludge, removing a substantial chunk of COD content as well (Yin et al., 2019; Zhou, Zhou, & Ma, 2020). After that, secondary treatment steps include some form of biological treatment, varying from anaerobic to aerobic treatment, as well as attached to suspended biological media. This step, followed by a clarifier to remove the biological sludge generated in this step, removes sufficiently biodegradable content of the effluent being treated (Zhou et al., 2020).

However, the layout of an ETP is subjected to change, often depending upon the effluent being treated. Certain processes including adsorption, membrane separation, coagulation, and nutrient removals, such as nitrogen and phosphorus, might be added as compulsory or polishing steps (Kiran et al., 2019). Due to BOD/COD values between 0.06 and 0.35, the biodegradation of textile wastewater is considered low, as a BOD_5/COD ratio less than 0.4 indicates low biodegradability (Bilińska, Gmurek, & Ledakowicz, 2016). Conventional treatment options, such as coagulation, adsorption, filtration, and biological treatment are inefficient or have other demerits since dyes are designed to resist degradation and remain stable (Sarayu and Sandhya, 2012).

Globally, a popular technique for decolorization among textile ETP is the application of activated sludge process, as the biological treatment option, along with polymeric coagulants to remove color. More than 90% color removal is achieved by a combination of both these steps. However, coagulation of color also yields dyes in the flocculation sludge (Gebrati et al., 2019; Haddad et al., 2018). The solution considered by the research community for this issue is the combination of a biological treatment step with another biological, chemical, or physical treatment step in order to improve the dye removal efficiency. The effluent containing azo dyes and reactive dyes were individually treated by sequential anaerobic/aerobic treatment. Toxicity in the case of azo dye effluent increased more than before. However, reactive dye toxicity was effectively reduced by aerobic treatment (da Silva et al., 2012).

Synthetic dye mixture, comprising of 70% cotton dyeing wastewater, synthesized from Procion Yellow H-EXL gran and Procion Deep Red H-EXL gran dyes and 30% Polyester dyeing wastewater, synthesized from Dianix Blue K-FBL and Dianix Orange K3G was subjected to biological and photochemical treatments, but the color was still visible after treatment and subsequent dilution (Soares et al., 2016). While adsorption shows comparatively more promise, generally acidic pH (around 3) is required for favorable performance (Yagub et al., 2014). The sole use of electrochemical degradation leads to poor organic carbon and turbidity removal, according to Aquino et al. (2016). They also pointed out that electrochemical treatment of real effluent is only feasible in the case of high conducting solutions; hence, requiring the addition of a salt, such as $Na_2SO_4^-$. While coagulation causes the problem of high sludge generation (Malik et al., 2017).

23.4 Advanced oxidation processes for dye remediation

AOP are the oxidation process that utilizes reactive oxygen species (such as hydroxyl radical) to oxidize the target contaminants. AOP includes processes like ozonation, electrochemical oxidation, photocatalysis, and Fenton processes (Khatri et al., 2018; Sathya, Nithya, & Balasubramanian, 2019). Major oxidants utilized for AOP are ozone, UV, and H_2O_2. Oxidation is the reaction in which a chemical species takes electron(s) from another chemical species, changing both species chemically in the process. The donating participant of reaction is oxidized while receiving one is reduced. Like reduction, environmental contaminants can also be oxidized to convert them into another form, the one that might have lesser environmental and health concerns (Fig. 23–1; Pourgholi et al., 2018).

Such resilient contaminants that do not react with ordinary oxidants can often be made to react by AOP. AOPs work by *in situ* generation of highly oxidative species, particularly hydroxyl radicals. Hydroxyl radical is highly unstable and reactive and tend to accept an election to form hydroxide ion, leading to its high oxidation potential. Hydroxyl radical has the second-highest oxidation potential (2.8 eV), behind fluorine (3 eV). Fluorine is not suitable for the remediation role, due to its gaseous and poisonous nature, leaving *in situ* production of hydroxyl radical as the best option in AOP (Zazou et al., 2019).

Photocatalysis is another AOP considered for textile dye remediation in literature. Materials including, but not limited to titanium oxide, zinc oxide, vanadium oxide, tungsten oxide, and indium oxide are being considered in the catalytic role. Remediation of dyes including Reactive Black 5, Reactive Red 45, Methylene Blue, Procion Yellow H-EXL, and Acid Violet 7 has been demonstrated using photocatalysis (Aguedach et al., 2008; Barakat, 2011; Krishnakumar and Swaminathan, 2011; Peternel et al., 2007). Basturk and Karatas (Basturk and Karatas, 2015) reported 99% decolourization efficiency using UV/H_2O_2 and determined the reaction kinetics as pseudo-first-order. However, the process was highly dependent on pH and performs optimum at pH 3. Also, H_2O_2 costs $390 to 500 per ton (Zhu and Logan, 2013).

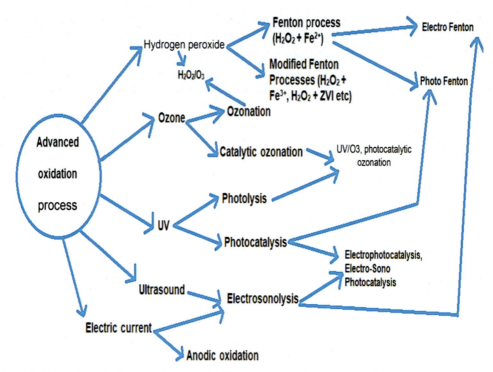

FIGURE 23–1 The schematic diagram of advanced oxidation processes for dye removal.

Ozonation and catalytic ozonation involve the use of ozone or ozone in combination with a catalyst for contaminant removal. Dye such as Congo Red (Elahmadi, Ben Salah, & Gadri, 2009), Reactive black 5 (Chu et al., 2007), Acid Red 14 (Gao et al., 2012), as well as dye bath effluents (Sevimli and Sarikaya, 2005) have been treated via ozonation. Manganese, iron, granular-activated carbon, and carbon nanotubes (CNTs) have been tried in combination with ozone to enhance dye removal. Ozone offers more redox potential than hydrogen peroxide but is twice as expensive (Asghar, Abdul Raman, & Daud, 2015).

23.4.1 Fenton process for dye remediation

Fenton reagent is an attractive alternative due to its ability to produce hydroxyl radical *in situ*. Fenton reagent is hydrogen peroxide activated by ferrous (Fe^{2+}) ion. Iron sources include iron powder, as well as iron salts like ferrous sulfate and ferrous chloride. Both these salts are most commonly found in their greenish hydrated form. Hydrogen peroxide serves as the source of reactive hydroxyl species. It is a mild blueish liquid that can decompose on contact with light, with its concentrated form, along with silver-based decomposition catalyst, acting as a propellant in rockets and other projectiles. It is also used as a bleaching agent and oxidizer (Çiner, 2018; Kaur, Sangal, & Kushwaha, 2019).

A common theme in the lab-scale experimental setup of textile dye remediation via the Fenton process is the addition of liquid hydrogen peroxide and powder iron salts in a stirred vessel (both constant and varying RPMs has been reported) along with an aqueous solution of target dye. Most of the synthetic dye solutions that have been worked on are less than 100 mg/L, as the solution above this concentration tends to deviate from the Beer-Lambert curve, which can give problematic results on UV-Vis spectrophotometry. Actual effluents are also diluted, sometimes up to an order of 10^6 to get reasonable absorption values on the UV-Vis spectrophotometer. The preferred pH range in such studies is 3–4 (Malakootian et al., 2019; Patel et al., 2019).

Remediation of Alician blue methyl Pyridium dye by activated carbon impregnated with iron in a fixed bed reactor resulted in up to 96.7% removal of dye; however, adsorption efficiency was limited due to large size of dye molecule (Duarte et al., 2013). Karthikeyan et al. (2011) reported optimum treatment of real textile effluent with homo and heterogeneous Fenton process, utilizing mesoporous activated carbon and Fenton's reagent. The optimum pH was limited to 3.5. Similar results were obtained in the removal of Reactive Black 5 (Meriç, Kaptan, & Ölmez, 2004).

23.5 Nanozerovalent iron

nZVI is an iron that has been reduced from its higher oxidation state to a zerovalent state. It can reduce other substances by reverting to its preferred Fe^{2+} state. The structure of an nZVI particle has been described in the literature by a core-shell model. Iron has a high reactive affinity toward oxygen and moisture (as evident by corrosion of large iron-based structures). This reactive tendency is enhanced manyfold on the nanoscale due to its small size. While a coarse size piece of iron might take months to corrode, nanosize iron particles corrode within minutes (Nigam Ahuja et al., 2020; Peng, Affam, & Chung, 2020).

The products formed by the reactive interaction of iron with atmospheric moisture and oxygen includes an exhaustive list of iron oxides and hydroxides. The most common one occurring is Fe_2O_3 (the oxide which gives rust its characteristic red color). Freshly formed nZVI, if left bare to atmospheric oxygen, gets rusted in the same way. However, when ethanol is used as part of the synthesis liquid media of nZVI in borohydride reduction media or coated on the particles after their synthesis, the surface atoms of iron react with atmospheric oxygen and moisture to form an oxide/hydroxide shell, predominantly consisting of FeOOH (Eljamal et al., 2018; Leybo et al., 2019).

As soon as the nZVI particle is synthesized, it forms the shell (see Fig. 23–2), which encloses the unreacted Fe^0. This FeOOH shell offer adsorption of target contaminant, but is also an electron transfer inhibitor, causing a decrease in the reductive ability of nZVI particle. The shell is reported to be 2–4 nm thick (Mu et al., 2017). The composition of oxide shell is also reported to be dependent upon the method of synthesis, may also include FeO, Fe_2O_3, and Fe_3O_4. Although, there is a lack of consensus upon the exact shell composition and reaction mechanism of the formation of shell oxides/hydroxides.

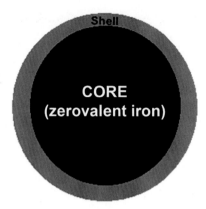

FIGURE 23-2 The schematic diagram of core-shell model of nanozerovalent iron.

The contaminant can also travel in direction of concentration gradient, from shell surface to zerovalent core, where they got reduced. The Fe^{2+} produced as a result travels in the opposite direction, from core to shell surface, where it oxidizes to Fe^{3+}. If nZVI particle is aged longer, the FeOOH layer grows denser, which decreases the shell porosity, adversely affecting the contaminant mass transfer (Karam et al., 2020; Zhao et al., 2020).

23.5.1 Nanozerovalent iron for textile dye remediation

nZVI has been suggested in the literature for dye remediation in a variety of potential roles. nZVI itself degrades the target chromophores via a reductive pathway; however, it may also be used as a precursor to *in situ* chemical oxidation. Encouraging results have been obtained over decades of experimentation. Some of the works are reviewed ahead (Rashid et al., 2020). nZVI was reported to remove up to 90% azo dye in combination with hydrogen peroxide at an optimum dose of 60–80 mg/L nanoparticles with 300–400 mg/L of the oxidant; however, with insufficient COD removal. A 2.8 times increase was required to bring a considerable decrease in COD levels (Yu et al., 2014). Shu and coworkers reported 98.9% removal of Acid black 24 dye in a 100 ppm concentration solution. Optimum pH was reported to be in the range of 4–9, with the removal to be taking place within half an hour. Dye uptake rate was 609.4 mg dye/gm nZVI, with total organic carbon content decreasing up to more than half of initial value. pH in the acidic range was confirmed to improve dye degradation by nZVI in a study of Reactive Black 5 and Reactive Red 198 removals. nZVI dosage increase also had a positive effect on dye removal (Shu et al., 2007).

Congo red, a proven carcinogenic azo dye, was remediated using nZVI. Remediation performance was reported to be significantly better than microscale zerovalent iron. An increase in nZVI dosage also increases the dye remediation; a behavior shared with remediation processes of other dyes as well (Shih and Tso, 2012). Ozone, Fe^{2+}, and nZVI were utilized in the pretreatment of actual filtered textile effluent to increase its biodegradability for the subsequent biological operations. Enhanced BOD/COD was 0.61, and color removal was reported to be 87% after treatment, as compared to only 33% of untreated effluent (Malik et al., 2018).

Ultrasonication assisted removal of toxic remazol black 133 was performed up to 80% within 15 min within the optimum pH range of 4–10. One gm of nZVI removed approximately 749 mg of dye, with fourier transform infrared spectroscopy (FT-IR) confirming the breakdown of an azo group into an amino group (Dutta et al., 2016a). Similar to other dye removal studies by nZVI, the removal efficiency increased with increasing nanoparticle dosage. Dutta et al. (2016b) also demonstrated the removal of anthraquinone dye, Reactive Blue MR, by using nZVI. The optimum pH range was reported as 8–12. However, the remediation process was negatively affected in an extreme pH range. 1 gm of nZVI removed approximately 2.2 gm of dye, which is much high than those reported for most of the azo dyes. Degradation products were confirmed to be amine, which is carcinogenic but has better biodegradability. An innovative approach was adopted by coupling the nZVI with UV/H_2O_2 process in remediation studies of Acid black 24. This resulted in a very rapid (within 10 min) removal of color and complete mineralization of nZVI content (Shu, Chang, & Chang, 2009). nZVI was also successfully used for removal of fluorescent dye acridine orange, up to 90.7% removal using 0.5 gm/0.1 L dye solution (Prema et al., 2011).

Somewhat of a dilemma is faced while utilizing the nZVI for remediation of the largest commercially used dye, azo group. Azo group dyes, as stated earlier, are resilient to conventional biodegradation. They can be degraded reduction using nZVI. However, the reduction of azo (N=N) yields amine (R-NH_2) as the degradation products, which have well documented carcinogenic concerns. However, these amines offer enhanced biodegradability, allowing the treatment of the effluent via a subsequent biological polishing step (Ken and Sinha, 2020; Nandi, De, & Haldar, 2019).

23.6 Nanoparticle aggregation

Nanoparticles, which are smaller than 100 nm, such as nZVI offer several benefits due to their small size, such as high surface area and reactivity. However, Nanoparticles are also vulnerable to aggregation, as small particles with high surface area attempt to join together to form larger stable particles. This highly affects their surface area and reactivity, as well as fate and transport (Hotze, Phenrat, & Lowry, 2010). nZVI particles aggregate due to two major forces, Van der Waal attraction, and magnetic attraction (Fig. 23–3). Iron nanoparticles can retain their aggregated state even in the liquid matrix and sometimes despite vigorous stirring. nZVI aggregation is reported to take place in two regimes (Phenrat et al., 2007).

1. In the first regime, rapid aggregation of zerovalent iron nanoparticles takes place, which leads to the formation of microscale aggregates. Particles are reported to increase up to approximately 5.5 times in size, in just 3.5 min, which indicates the rapid pace of aggregation. Rate of aggregation increases with increasing particle size.
2. These aggregate particles start to arrange themselves in chains after one minute. The aggregate particle chains formed during the first 30 min then engage in gelation activity, where these aggregate chains arrange into a network. These chains are reported to be up to 50 μm in length. Gelation is observed on particle concentrations as low as 60 ppm.

FIGURE 23–3 Representation of filtered aggregated nanozerovalent iron particles.

One of the major roles for which nZVI is being considered is the particulate environmental remediation agent that is, injected into an underground water reservoir for in situ degradation of contaminants. However, the ability to aggregate together also affects the transport of nZVI particles in underground water bodies. For effective pollutant remediation, particles should disperse well in the underground water body. The aggregation tendency is also anticipated to lead to the deposition of nanoparticles on underground formation materials, which can result in the clogging of formation pores. This will affect the porosity of underground formations (Duan, Dong, & Zhang, 2018; Gao et al., 2018; Garcia et al., 2020).

To deal with this challenge, a solution in the form of supporting nZVI on some support/carrier material has been suggested. This support/carrier material offers physical hindrance between molecular/magnetic attractions among nanoparticles, decreasing agglomeration of particles, which results in more available surface area, which leads to better reactive performance. Supported nZVI particles can also disperse better in aqueous media. Supported nZVI particles also show negligible aggregation, but the attraction among particles is much weaker and particles can be dispersed with minimal disturbance (Chen et al., 2017; Wu et al., 2019).

23.7 Support material for nanozerovalent iron

The support material for nZVI should offer the following benefits, including low cost, aid in the process of environmental remediation, and increased reactivity of nZVI. Several supports have been considered in literature (Adusei-Gyamfi and Acha, 2016) including, but not limited to, alumina, chitosan, exfoliated graphite, bentonite, rectorite, graphene oxide, mesoporous silica, activated carbon, clinoptilolite, CNTs, biochar, pumice, mesoporous carbon, sepiolite, montmorillonite, vermiculite, kaolin, coral, and oyster shells (Ezzatahmadi et al., 2017; Ishag et al., 2020; Liu and Wang, 2019; Zhou et al., 2019). A variety of conventional adsorbents, such as mesoporous silica (Hartmann, Kullmann, & Keller, 2010), activated carbon (Navalon, Alvaro, & Garcia, 2010), and novel materials including CNTs (Deng, Wen, & Wang, 2012) and reduced graphene oxide (r-GO) (Ahmad et al., 2015) have been suggested in this role. A selected few are described here.

Natural alumina supported nZVI was used to remediate Cu ions from aqueous solution by Karabelli et al. (2011), and to remove rare-earth elements by Yesiller, Eroğlu, and Shahwan (2013). Alumina supported iron nanoparticles displayed more than 80% removal of rare-earth metals. Graphene oxide supported nanoiron particles, in presence of sodium persulfate, can effectively remediate chlorinated organics, as reported by Ahmad et al. (2015). However, removal efficiency decreases with an increase in the initial concentration of chlorinated organic, and pH.

Activated carbon, the highly porous form of charcoal is well known for its excellent adsorptive capacity. There have been several successful attempts to apply the AC and nZVI composite for removal of a variety of different contaminants, such as heavy metals (Fu et al., 2013), phenol (Messele et al., 2014), acrylonitrile (Xiao et al., 2015), nitrate, and phosphate (Khalil et al., 2017) to name a few. CNTs are a cylindrical allotrope of carbon, with sizes in nanoscale, and characteristic high surface area, conductivity, and strength. CNT/nZVI has been utilized for the removal of Cr(6) (Lv et al., 2011), and selenite (Sheng et al., 2016). Zhou and coworkers demonstrated 95% removal of 2,4-dichlorophenol via bimetallic (Pd/Fe) nanoparticles supported on multiwalled CNT, in 1 min.

K10 bentonite supported nanoiron particles were investigated as an adsorbent for cobalt ions (Shahwan et al., 2010). The addition of bentonite resulted in a decrease in the aggregation of nanoiron particles. Adsorption increased with increasing pH and effective cobalt ion removal was reported up to eight successive uses. Mesoporous silica is a mesoporous form of silicon dioxide, with pores up to 50 nm. Both of its major types, Santa Barbara Amorphous-15 (SBA-15) and Mobil Composition of Matter-41 (MCM-41) have been explored as effective support material for nZVI. Petala et al. utilized iron nanoparticles up to 80 nm in size supported on MCM-41 for remediation of Cr(6) (Petala et al., 2013). Tang and group also reported up to 80% removal of p-nitrophenol by SBA-15 supported nZVI (Tang et al., 2015).

Due to its high surface area, biochar has also been suggested as an effective support material for nanoiron particles. Biochar supported nZVI has been used for activation of persulfate (Yan et al., 2015) and hydrogen peroxide (Yan et al., 2017) for environmental remediation. The effect of oxygen-containing functional groups in biochar toward promoting the generation of SO_4^- radical has also been suggested in the case of persulfate activation. Nano zerovalent iron/biochar (nZVI/BC) has been utilized in Cr(6) removal from soil (Su et al., 2016), from water (Shang et al., 2017), dye removal (Quan et al., 2014), and trichloroethylene (Dong et al., 2017). Chitosan fibers used as support for nZVI particles of size up to 100 nm have also been explored by Horzum et al. (2013) as a sorbent for the removal of inorganic arsenic. Arsenic removal declines in the alkaline range of pH.

Due to their abundance and cheapness, attention has also turned toward clay minerals, such as montmorillonite, rectorite, and various zeolites. Bhowmick et al. (2014) demonstrated the application of Mt-nZVI for the removal of arsenic from water and reported decreasing aggregation of nanoparticles due to effective dispersion on a clay surface. Danish and coworkers utilized the natural zeolite as support for nZVI and bimetallic catalysts in an advanced oxidation process for the removal of chlorinated organic solvents (Danish et al., 2017).

23.7.1 Support mediums for dye remediation

Impregnation of nZVI on some suitable support material was suggested as a remedy to combat the nanoparticle aggregation issue. This has been reported to significantly improve the nZVI performance. A wide variety of materials have been demonstrated to be used as support material for nZVI. Luo and coworkers demonstrated the removal of orange II dye using rectorite, a natural clay, as support for nZVI and reported complete degradation of dye in 10 min, while only 35% dye was removed by using unsupported nZVI (Luo et al., 2013).

91.87% removal of methyl orange in 10 min was reported using bentonite supported nanoiron/palladium composite under a pseudo-first-order kinetic model (Wang et al., 2013). However, degradation efficiency did not exceed 93.75%. A similar study showed more than 95% removal of Acid Violet Red B was reported in 9 min at a high concentration of 800 mg/L, using similar bentonite supported nZVI. This study also confirmed the pseudo-first-order kinetic model of the nZVI (Lin et al., 2014).

Jin et al. (2015) suggested the use of nZVI and kaolin in equal mass ratios, in a remediation study of Direct Fast Black G dye and reported 99.8% removal in 1 h, even in the alkaline range of 9.4. Seventy eight percent of the decolourization was obtained in the first 10 min. A comprehensive comparative study of different clay materials as support for nZVI and their application in dye remediation was done by Kerkez and coworkers (Kerkez et al., 2014). They found bentonite to be a better option (92.7% dye removal) than native clay (92.1%) and kaoline (91.6%) for remediation of an industrial azo dye. Almost twice kaoline nanocomposite was required to achieve the above-stated decolourization then bentonite supported nZVI.

A decrease in aggregation due to support of nZVI on 3-dimensional graphene (3-DG) led to up to five times increase in the degradation rate of orange IV azo dye (Wang et al., 2015). 3-DG was synthesized upon a Ni skeleton by chemical vapor deposition. Ninety-eight percent dye was removed after 3 h operation, with significant improvements being observed on decreasing pH and increasing temperature. Biochar was used as a support material for nZVI in the removal process of methyl orange (Han et al., 2015). Biochar offered improved dispersion and reactivity of nanoparticles, as well as enhanced adsorption capacity, resulting in 98.5% dye removal at a composite dose of as low as 600 mg/L, at optimum pH of 4, within 10 min. Biochar supported nZVI also showed signs of reactivity for as long as 30 days.

nZVI particles dispersed on r-GO sheet were synthesized and were utilized in the removal of Rhodamine B; 86.4% removal was reported within 20 min. Nanoparticle dispersion on graphene oxide sheet decreased aggregation; however, some aggregation persisted even after the deposition of nZVI upon r-GO sheet (Shi et al., 2017). nZVI supported on clinoptilolite was synthesized and utilized in the removal of methyl orange and methylene blue dyes by Nairat et al. (2015). Dispersion of nZVI on the coffin-like clinoptilolite significantly increased the dye removal performance. Chain-like agglomeration behavior was reduced due to the provision of physical barriers by supporting clinoptilolite particles. Synthesized cation exchange resin, already in industrial practice as a water softening agent via ion exchange, was also demonstrated as a support material for nZVI by its application in the removal of Acid Blue 113 dye (Shu et al., 2010).

In a choice of suitable support material for nZVI, one of the most important factors to consider will be the ease and cost of access in a case supported nZVI find their place in the commercially applied wastewater and underground water remediation technologies. Materials like CNT and graphene still need significant attention to bring their costs in a feasible range for massive industrial-scale production. A much more reasonable option is either the naturally available materials like clays (bentonite), aluminosilicates (zeolites), or the materials that can be synthesized by comparatively cheaper and easily available ingredients, such as activated carbon, biochar, or synthesized zeolites (Lazaratou, Vayenas, & Papoulis, 2020; Xu et al., 2019).

Instead of using $FeSO_4^-.7H_2O$ or $FeCl_3$ as the source of electron-donating iron in Fenton or modified Fenton processes, nZVI is also being considered to activate the oxidant for a generation of reactive oxygen species. nZVI, both supported or unsupported, has been utilized in Fenton or modified Fenton roles in several environmental remediation studies. nZVI was used in combination with H_2O_2 for degradation of polyvinyl alcohol, with a reported efficiency of 94% in about 60 s. The pH range of less than 3–4 was validated to be optimum for the Fenton process. Performance of freshly prepared nZVI was also reported to be far superior to that of manufactured by a commercial manufacturer (1% removal) (Lin and Hsu, 2018).

$nZVI/H_2O_2$ system was utilized in the absence and in presence of UV and ultrasound to demonstrate the removal of methyl tertiary butyl ether and $nZVI/H_2O_2$ system reinforced with ultrasonication was reported to be the best set of technologies for remediation of contaminant. Significant removal was obtained within 15 min (Samaei, Maleknia, & Azhdarpoor, 2016). Application of zerovalent iron nanoparticles and hydrogen peroxide-based Fenton system even extends up to wastes of hazardous nature. Samadi, Esfahani, and Naddafi (2013) demonstrated the remediation of a hospital waste landfill leachate in Iran and reported COD removal of 48.67% and 53% color removal at nZVI dosage of 2.5 g/L. A similar system is also reported to be effective in the case of other complicated targets, such as pharmaceutical components (Zhang et al., 2017).

23.8 Future recommendations

Despite their superior reactivity, engineering knowledge on nZVI remains far from complete, and nZVI remains a domain with several important potential research areas.

1. Use of different support materials together for nZVI instead of using a single material to optimize cost and performance, such as blending of minerals containing functional groups complementary to remediation process and low-cost adsorbents

such as agricultural waste-derived activated carbon or other such material. It will also help to lower the dependence upon nonrenewable resources, such as aluminosilicates.
2. While supporting nZVI particles on some carrier material reduces their aggregation behavior, magnetic attraction remains high, to some extent. This magnetic attraction of iron nanoparticles can be utilized in the recovery of nZVI particles from underground aquifers after a remediation job.
3. The borohydride synthesis method results in the evolution of hydrogen gas. Up to the writing of this thesis, there is almost a complete lack of studies regarding the determination of the potential of this hydrogen gas as a possible energy source.
4. Environmental fate and probability of nZVI contributes to iron overload in humans as well as flora and fauna being influenced by the aquafer/effluent treated by nZVI. Excess iron can cause several eye disorders (such as retinitis or choroiditis), hypothyroidism, and can accelerate Alzheimer's and Parkinson's disease. Detailed studies on a lab, pilot, and field-scale are required to ensure that this nanotechnology does not suffer the same fate as dichloro-diphenyl-trichloroethane (DDT) or genetically modified food.
5. Determination and validation of optimum shell thickness of nZVI, and development of techniques to control this oxide shell is critical to establish the nZVI as a wide-scale remediation option since nZVI particles with no or very thin oxide shell can react with water and corrode away even before reaching the target contaminant, while too thick a shell can heavily reduce the reactivity of nZVI.
6. The detailed and comprehensive investigation of products are formed from azo dye degradation via an advance oxidative pathway.
7. Comparative analysis of underground injection and treat-and-pump techniques for underground water remediation via Supported nZVI/Fenton system on basis of cost and efficiency (Ken and Sinha, 2020; Rashid et al., 2020; Wu et al., 2020).

23.9 Conclusions

The present chapter was aimed at exploring the textile dye remediation potential of nZVI supported on different mediums in a Fenton-like system for activation of SPC, which served as a source of hydroxyl radical generation. According to the accumulated data, it is stated that both actual affluent and synthetic textile effluent were decolorized using this nZVI/SPC system. Based on emerging evidence, we observed the maximum decoloration of 95—97% in the case of azo dye. The continuous stay of nZVI in its liquid phase presents a dilemma. Synthesis, transport, and application of nZVI in the same aqueous phase result in maximum reactivity, which also causes nZVI to exhaust sooner. Zerovalent iron is effectively distributed on the medium support, which significantly decreases the aggregation of nanoparticles. Owing to cheap and safer reagents and ingredients and significant decolorization within a short time, nZVI/SPC is anticipated to be practically feasible for industrial effluent treatment, as well as ground and surface water treatment. Although, the literature revealed that nZVI showed better dye removal results, but some points need to be updated to improve the cost and efficiency in the near future. However, it is recommended that more study is required for the utilization of nZVI/SPC in the Fenton role coupled with other AOPs, such as photolysis or sono-AOP to improve degradation performance and reduce time/reagents required.

References

Adusei-Gyamfi, J., & Acha, V. (2016). Carriers for nano zerovalent iron (nZVI): Synthesis, application and efficiency. *RSC Advances, 6*, 91025–91044.

Aguedach, A., et al. (2008). Influence of ionic strength in the adsorption and during photocatalysis of reactive black 5 azo dye on TiO_2 coated on non woven paper with SiO_2 as a binder. *Journal of Hazardous Materials, 150*, 250–256.

Ahmad, A., et al. (2015). Efficient degradation of trichloroethylene in water using persulfate activated by reduced graphene oxide-iron nanocomposite. *Environmental Science and Pollution Research, 22*, 17876–17885.

Aquino, J. M., et al. (2016). Combined coagulation and electrochemical process to treat and detoxify a real textile effluent. *Water, Air, & Soil Pollution, 227*.

Asghar, A., Abdul Raman, A. A., & Daud, W. M. A. W. (2015). Advanced oxidation processes for in-situ production of hydrogen peroxide/hydroxyl radical for textile wastewater treatment: A review. *Journal of Cleaner Production, 87*, 826–838.

Balapure, K., Madamwar, D., & Bhatt, N. (2017). Mineralization of reactive azo dyes present in simulated textile waste water using down flow microaerophilic fixed film bioreactor. *Bioresource Technology, 175*, 1–7.

Barakat, M. A. (2011). Adsorption and photodegradation of Procion yellow H-EXL dye in textile wastewater over TiO_2 suspension. *Journal of Hydro-environment Research, 5*, 137–142.

Basturk, E., & Karatas, M. (2015). Decolorization of antraquinone dye Reactive Blue 181 solution by UV/H_2O_2 process. *Journal of Photochemistry and Photobiology A: Chemistry, 299*, 67–72.

Bhowmick, S., et al. (2014). Montmorillonite-supported nanoscale zero-valent iron for removal of arsenic from aqueous solution: Kinetics and mechanism. *Chemical Engineering Journal, 243*, 14–23.

Bilińska, L., Gmurek, M., & Ledakowicz, S. (2016). Comparison between industrial and simulated textile wastewater treatment by AOPs – Biodegradability, toxicity and cost assessment. *Chemical Engineering Journal, 306*, 550–559.

Brüschweiler, B. J., & Merlot, C. (2017). Azo dyes in clothing textiles can be cleaved into a series of mutagenic aromatic amines which are not regulated yet. *Regulatory Toxicology and Pharmacology, 88*, 214–226.

Chen, X., et al. (2017). Review on nano zerovalent iron (nZVI): From modification to environmental applications. In *Proceedings of the IOP conference series: Earth and environmental science*. IOP Publishing.

Chequer, F. M. D., et al. (2013). Textile dyes: Dyeing process and environmental impact. In *Eco-friendly textile dyeing and finishing*, Intechopen, p. 27.

Chu, L.-B., et al. (2007). Enhanced ozonation of simulated dyestuff wastewater by microbubbles. *Chemosphere, 68*, 1854–1860.

Çiner, F. (2018). Application of Fenton reagent and adsorption as advanced treatment processes for removal of Maxilon Red GRL. *Global NEST Journal, 20*, 1–6.

Cinperi, N. C., et al. (2019). Treatment of woolen textile wastewater using membrane bioreactor, nanofiltration and reverse osmosis for reuse in production processes. *Journal of Cleaner Production, 223*, 837–848.

da Silva, M. E. R., et al. (2012). Sequential anaerobic/aerobic treatment of dye-containing wastewaters: Colour and COD removals, and ecotoxicity tests. *Applied Biochemistry and Biotechnology, 166*(4), 1057–1069.

Danish, M., et al. (2017). An efficient catalytic degradation of trichloroethene in a percarbonate system catalyzed by ultra-fine heterogeneous zeolite supported zero valent iron-nickel bimetallic composite. *Applied Catalysis A: General, 531*, 177–186.

Deng, J., Wen, X., & Wang, Q. (2012). Solvothermal in situ synthesis of Fe_3O_4-multi-walled carbon nanotubes with enhanced heterogeneous Fenton-like activity. *Materials Research Bulletin, 47*, 3369–3376.

Dong, H., et al. (2017). Removal of trichloroethylene by biochar supported nanoscale zero-valent iron in aqueous solution. *Separation and Purification Technology, 188*, 188–196.

Dotto, J., et al. (2019). Performance of different coagulants in the coagulation/flocculation process of textile wastewater. *Journal of Cleaner Production, 208*, 656–665.

Duan, R., Dong, Y., & Zhang, Q. (2018). Characteristics of aggregate size distribution of nanoscale zero-valent iron in aqueous suspensions and Its effect on transport process in porous media. *Water, 10*(6), 670.

Duarte, F., et al. (2013). Treatment of textile effluents by the heterogeneous Fenton process in a continuous packed-bed reactor using Fe/activated carbon as catalyst. *Chemical Engineering Journal, 232*, 34–41.

Dutta, S., et al. (2016a). Modified synthesis of nanoscale zero-valent iron and its ultrasound-assisted reactivity study on a reactive dye and textile industry effluents. *Desalination and Water Treatment, 57*, 19321–29332.

Dutta, S., et al. (2016b). Rapid reductive degradation of azo and anthraquinone dyes by nanoscale zero-valent iron. *Environmental Technology & Innovation, 5*, 176–187.

Elahmadi, M. F., Ben Salah, N., & Gadri, A. (2009). Treatment of aqueous wastes contaminated with Congo Red dye by electrochemical oxidation and ozonation processes. *Journal of Hazardous Materials, 168*, 1163–1169.

Eljamal, R., et al. (2018). Improvement of the chemical synthesis efficiency of nano-scale zero-valent iron particles. *Journal of Environmental Chemical Engineering, 6*(4), 4727–4735.

Ezzatahmadi, N., et al. (2017). Clay-supported nanoscale zero-valent iron composite materials for the remediation of contaminated aqueous solutions: A review. *Chemical Engineering Journal, 312*, 336–350.

Fu, F., et al. (2013). Removal of Cr(VI) from wastewater by supported nanoscale zero-valent iron on granular activated carbon. *Desalination and Water Treatment, 51*, 2680–2686.

Gao, M., et al. (2012). Ozonation of azo dye Acid Red 14 in a microporous tube-in-tube microchannel reactor: Decolorization and mechanism. *Chemosphere, 89*, 190–197.

Gao, Y.-q, et al. (2018). Ultrasound-assisted heterogeneous activation of persulfate by nano zero-valent iron (nZVI) for the propranolol degradation in water. *Ultrasonics Sonochemistry, 49*, 33–40.

Garcia, A. N., et al. (2020). Fate and transport of sulfidated nano zerovalent iron (S-nZVI): A field study. *Water Research, 170*, 115319.

Gebrati, L., et al. (2019). Inhibiting effect of textile wastewater on the activity of sludge from the biological treatment process of the activated sludge plant. *Saudi Journal of Biological Sciences, 26*(7), 1753–1757.

Giannakis, S., Lin, K.-Y. A., & Ghanbari, F. (2020). A review of the recent advances on the treatment of industrial wastewaters by Sulfate Radical-based Advanced Oxidation Processes (SR-AOPs). *Chemical Engineering Journal*, 127083.

Haddad, M., et al. (2018). Reduction of adsorbed dyes content in the discharged sludge coming from an industrial textile wastewater treatment plant using aerobic activated sludge process. *Journal of Environmental Management, 223*, 936–946.

Han, L., et al. (2015). Biochar supported nanoscale iron particles for the efficient removal of Methyl Orange dye in aqueous solutions. *PLoS One, 10*.

Hartmann, M., Kullmann, S., & Keller, H. (2010). Wastewater treatment with heterogeneous Fenton-type catalysts based on porous materials. *Journal of Materials Chemistry, 20*, 9002–9017.

Holkar, C. R., et al. (2016). A critical review on textile wastewater treatments: Possible approaches. *Journal of Environmental Management, 182*, 351–366.

Horzum, N., et al. (2013). Chitosan fiber-supported zero-valent iron nanoparticles as a novel sorbent for sequestration of inorganic arsenic. *RSC Advances, 3*, 7828–7837.

Hotze, E. M., Phenrat, T., & Lowry, G. V. (2010). Nanoparticle aggregation: Challenges to understanding transport and reactivity in the environment. *Journal of Environmental Quality Abstract - Special Submissions, 39*, 1909–1924.

Ishag, A., et al. (2020). Environmental application of emerging zero-valent iron-based materials on removal of radionuclides from the wastewater: A review. *Environmental Research*, 109855.

Jha, B., & Singh, D. N. (2011). A review on synthesis, characterization and industrial applications of flyash zeolites. *Journal of Materials Education, 33*(1–2), 71.

Jin, X., et al. (2015). Synthesis of kaolin supported nanoscale zero-valent iron and its degradation mechanism of Direct Fast Black G in aqueous solution. *Materials Research Bulletin, 61*, 433–438.

Kamali, M., et al. (2019). Sustainability criteria for assessing nanotechnology applicability in industrial wastewater treatment: Current status and future outlook. *Environment International, 125*, 261–276.

Karabelli, D., et al. (2011). Preparation and characterization of alumina-supported iron nanoparticles and its application for the removal of aqueous Cu^{2+} ions. *Chemical Engineering Journal, 168*, 979–984.

Karam, A., Zaher, K., & Mahmoud, A. S. (2020). Comparative studies of using nano zerovalent iron, activated carbon, and green synthesized nano zerovalent iron for textile wastewater color removal using artificial intelligence, regression analysis, adsorption isotherm, and kinetic studies. *Air, Soil Water Research, 13*, 1178622120908273.

Karthikeyan, S., et al. (2011). Treatment of textile wastewater by homogeneous and heterogeneous Fenton oxidation processes. *Desalination, 281*, 438–445.

Kaur, P., Sangal, V., & Kushwaha, J. (2019). Parametric study of electro-Fenton treatment for real textile wastewater, disposal study and its cost analysis. *International Journal of Environmental Science Technology, 16*(2), 801–810.

Ken, D. S., & Sinha, A. (2020). Recent developments in surface modification of nano zero-valent iron (nZVI): Remediation, toxicity and environmental impacts. *Environmental Nanotechnology, Monitoring Management, 14*, 100344.

Kerkez, D. V., et al. (2014). Three different clay-supported nanoscale zero-valent iron materials for industrial azo dye degradation: A comparative study. *Journal of the Taiwan Institute of Chemical Engineers, 45*, 2451–2461.

Khalil, A. M. E., et al. (2017). Optimized nano-scale zero-valent iron supported on treated activated carbon for enhanced nitrate and phosphate removal from water. *Chemical Engineering Journal, 309*, 349–365.

Khatri, J., et al. (2018). Advanced oxidation processes based on zero-valent aluminium for treating textile wastewater. *Chemical Engineering Journal, 348*, 67–73.

Kiran, S., et al. (2019). Advanced approaches for remediation of textile wastewater: A comparative study. In *Advanced functional textiles polymers: Fabrication, processing applications*, p. 201–264.

Krishnakumar, B., & Swaminathan, M. (2011). Influence of operational parameters on photocatalytic degradation of a genotoxic azo dye Acid Violet 7 in aqueous ZnO suspensions. *Spectrochimica Acta Part A: Molecular and Biomolecular Spectroscopy, 81*, 739–744.

Kumbhakar, D. V., et al. (2018). Assessment of cytotoxicity and cellular apoptosis induced by azo-dyes (methyl orange and malachite green) and heavy metals (cadmium and lead) using *Nigella sativa L.* (black cumin). *Cytologia, 83*(3), 331–336.

Lazaratou, C., Vayenas, D., & Papoulis, D. (2020). The role of clays, clay minerals and clay-based materials for nitrate removal from water systems: A review. *Applied Clay Science, 185*, 105377.

Leybo, D., et al. (2019). Effect of initial salt composition on physicochemical and structural characteristics of zero-valent iron nanopowders obtained by borohydride reduction. *Processes, 7*(10), 769.

Lin, C.-C., & Hsu, S.-T. (2018). Performance of $nZVI/H_2O_2$ process in degrading polyvinyl alcohol in aqueous solutions. *Separation and Purification Technology, 203*, 111–116.

Lin, Y., et al. (2014). Decoloration of acid violet red B by bentonite-supported nanoscale zero-valent iron: Reactivity, characterization, kinetics and reaction pathway. *Applied Clay Science, 93–94*, 56–61.

Liu, Y., & Wang, J. (2019). Reduction of nitrate by zero valent iron (ZVI)-based materials: A review. *Science of the Total Environment, 671*, 388–403.

Luo, S., et al. (2013). Synthesis of reactive nanoscale zero valent iron using rectorite supports and its application for Orange II removal. *Chemical Engineering Journal, 223*, 1–7.

Lv, X., et al. (2011). Removal of chromium(VI) from wastewater by nanoscale zero-valent iron particles supported on multiwalled carbon nanotubes. *Chemosphere, 85*, 1204–1209.

Malakootian, M., et al. (2019). Biogenic silver nanoparticles/hydrogen peroxide/ozone: Efficient degradation of reactive blue 19. *BioNanoScience*, 1–8.

Malik, S. N., et al. (2017). Comparison of coagulation, ozone and ferrate treatment processes for color, COD and toxicity removal from complex textile wastewater. *Water Science and Technology, 76*, 1001–1010.

Malik, S. N., et al. (2018). Catalytic ozone pretreatment of complex textile effluent using Fe^{2+} and zero valent iron nanoparticles. *Journal of Hazardous Materials, 357*, 363–375.

Mani, S., Chowdhary, P., & Bharagava, R. N. (2019). Textile wastewater dyes: Toxicity profile and treatment approaches. In *Emerging and eco-friendly approaches for waste management*. Springer, p. 219–244.

Meriç, S., Kaptan, D., & Ölmez, T. (2004). Color and COD removal from wastewater containing Reactive Black 5 using Fenton's oxidation process. *Chemosphere, 54*, 435–441.

Messele, S. A., et al. (2014). Catalytic wet peroxide oxidation of phenol using nanoscale zero-valent iron supported on activated carbon. *Desalination and Water Treatment, 57*, 5155–5164.

Mu, Y., et al. (2017). Iron oxide shell mediated environmental remediation properties of nano zero-valent iron. *Environmental Science: Nano, 4*, 27–45.

Nairat, M., et al. (2015). Incorporation of iron nanoparticles into clinoptilolite and its application for the removal of cationic and anionic dyes. *Journal of Industrial and Engineering Chemistry, 21*, 1143–1151.

Nandi, T., De, A., & Haldar, S. (2019). Detection of different pollutant azo dyes in wastewater using diethylene triaminepentacetic acid (DTPA) stabilized nano scale zero valent iron. *Materials Today: Proceedings, 11*, A1–A7.

Navalon, S., Alvaro, M., & Garcia, H. (2010). Heterogeneous Fenton catalysts based on clays, silicas and zeolites. *Applied Catalysis B: Environmental, 99*, 1–26.

Nigam Ahuja, N., et al. (2020). Synthesis and characterization of zero valent iron nanoparticles for textile wastewater treatment. *Pollution, 6*(4), 773–783.

Novotny, C., et al. (2006). Comparative use of bacterial, algal and protozoan tests to study toxicity of azo- and anthraquinone dyes. *Chemosphere, 63*.

Othmer, D. & Kirk, R.E. (2007). Molecular sieves, In Kirk, R.E., & Othmer, D. *Encyclopedia of chemical technology*. Wiley. (pp. 443–445).

Oturan, M., & Aaron, J.-J. (2014). Advanced oxidation processes in water/wastewater treatment: Principles and applications. A review. *Critical Reviews in Environmental Science and Technology, 44*, 577–2641.

Patel, D., et al. (2019). *Fenton process combined with coagulation for the treatment of textile wastewater*.

Peng, B., Affam, A., & W. Chung. (2020). Nano zero valent iron and ozonation for selected recalcitrant wastewater. In *Proceedings of the IOP conference series: Earth and environmental science*. IOP Publishing.

Petala, E., et al. (2013). Nanoscale zero-valent iron supported on mesoporous silica: Characterization and reactivity for Cr(VI) removal from aqueous solution. *Journal of Hazardous Materials, 261*, 295–306.

Peternel, I. T., et al. (2007). Comparative study of UV/TiO_2, UV/ZnO and photo-Fenton processes for the organic reactive dye degradation in aqueous solution. *Journal of Hazardous Materials, 148*, 477–484.

Phenrat, T., et al. (2007). Aggregation and sedimentation of aqueous nanoscale zerovalent iron dispersions. *Environmental Science and Technology, 41*, 284–290.

Pourgholi, M., et al. (2018). Removal of dye and COD from textile wastewater using AOP (UV/O_3, UV/H_2O_2, O_3/H_2O_2, and UV/H_2O_2/O_3). *Journal of Environmental Health Sustainable Development*, 621–629.

Prema, P., et al. (2011). Color removal efficiency of dyes using nanozerovalent iron treatment. *Toxicological & Environmental Chemistry, 93*, 1908–1917.

Quan, G., et al. (2014). Nanoscale zero-valent iron supported on biochar: Characterization and reactivity for degradation of acid orange 7 from aqueous solution. *Water, Air, & Soil Pollution, 225*.

Raji, M., et al. (2020). Nano zero-valent iron on activated carbon cloth support as Fenton-like catalyst for efficient color and COD removal from melanoidin wastewater. *Chemosphere, 263*, 127945.

Rashid, T., et al. (2020). Formulation of Zeolite-supported nano-metallic catalyst and its application in textile effluent treatment. *Journal of Environmental Chemical Engineering*, 104023.

Saleemi, S., et al. (2020). A green and cost-effective approach to reutilize the effluent from bleaching process. *Journal of Applied Emerging Sciences, 9*(2), 85–90.

Samadi, M. T., Esfahani, K., & Naddafi, K. (2013). Comparison the efficacy of Fenton and "nZVI + H_2O_2" processes in municipal solid waste landfill leachate treatment. *International Journal of Environmental Research, 7*, 187–194.

Samaei, M. R., Maleknia, H., & Azhdarpoor, A. (2016). A comparative study of removal of methyl tertiary-butyl ether (MTBE) from aquatic environments through advanced oxidation methods of H_2O_2/nZVI, H_2O_2/nZVI/ultrasound, and H_2O_2/nZVI/UV. *Desalination and Water Treatment, 57*, 21417–21427.

Sarayu, K., & Sandhya, S. (2012). Current technologies for biological treatment of textile wastewater—a review. *Applied Biochemistry and Biotechnology, 167*(3), 645–661.

Sathya, U., Nithya, M., & Balasubramanian, N. (2019). Evaluation of advanced oxidation processes (AOPs) integrated membrane bioreactor (MBR) for the real textile wastewater treatment. *Journal of Environmental Management, 246*, 768–775.

Schneider, K., Hafner, C., & Jäger, I. (2004). Mutagenicity of textile dye products. *Journal of Applied Toxicology, 24*, 83–91.

Sevimli, M. F., & Sarikaya, H. Z. (2005). Effect of some operational parameters on the decolorization of textile effluents and dye solutions by ozonation. *Environmental Technology, 26*, 135–144.

Shahwan, T., et al. (2010). Synthesis and characterization of bentonite/iron nanoparticles and their application as adsorbent of cobalt ions. *Applied Clay Science, 47*, 257–262.

Shang, J., et al. (2017). Removal of chromium (VI) from water using nanoscale zerovalent iron particles supported on herb-residue biochar. *Journal of Environmental Management, 197*, 331–337.

Sheng, G., et al. (2016). Enhanced sequestration of selenite in water by nanoscale zero valent iron immobilization on carbon nanotubes by a combined batch, XPS and XAFS investigation. *Carbon, 99*, 123–130.

Shi, X., et al. (2017). Optimizing the removal of rhodamine B in aqueous solutions by reduced graphene oxide-supported nanoscale zerovalent iron (nZVI/rGO) using an artificial neural network-genetic algorithm (ANN-GA). *Nanomaterials, 7*.

Shih, Y.-H., & Tso, C.-P. (2012). Fast decolorization of azo-dye congo red with zerovalent iron nanoparticles and sequential mineralization with a Fenton reaction. *Environmental Engineering Science, 29*.

Shoukat, R., Khan, S. J., & Jamal, Y. (2019). Hybrid anaerobic-aerobic biological treatment for real textile wastewater. *Journal of Water Process Engineering, 29*, 100804.

Shu, H.-Y., et al. (2007). Reduction of an azo dye Acid Black 24 solution using synthesized nanoscale zerovalent iron particles. *Journal of Colloid and Interface Science, 314*, 89–97.

Shu, H.-Y., et al. (2010). Using resin supported nano zero-valent iron particles for decoloration of Acid Blue 113 azo dye solution. *Journal of Hazardous Materials, 184*, 499–505.

Shu, H.-Y., Chang, M.-C., & Chang, C.-C. (2009). Integration of nanosized zero-valent iron particles addition with UV/H_2O_2 process for purification of azo dye Acid Black 24 solution. *Journal of Hazardous Materials, 167*, 1178–1184.

Soares, P. A., Silva, T. F. C. V., & Arcy, A. R. (2016). Assessment of AOPs as a polishing step in the decolourisation of bio-treated textile wastewater: Technical and economic considerations. *Journal of Photochemistry and Photobiology A: Chemistry, 317*, 26–38.

Solomon, D., Kiflie, Z., & Van Hulle, S. (2020). Integration of sequencing batch reactor and homo-catalytic advanced oxidation processes for the treatment of textile wastewater. *Nanotechnology for Environmental Engineering, 5*(1), 1–13.

Su, H., et al. (2016). Stabilisation of nanoscale zero-valent iron with biochar for enhanced transport and in-situ remediation of hexavalent chromium in soil. *Environmental Pollution, 214*, 94–100.

Tahir, M. B., Sagir, M., & Shahzad, K. (2019). Removal of acetylsalicylate and methyl-theobromine from aqueous environment using nano-photocatalyst WO_3-TiO_2@g-C_3N_4 composite. *Journal of Hazardous Materials, 363*, 205–213.

Tang, L., et al. (2015). Rapid reductive degradation of aqueous p-nitrophenol using nanoscale zero-valent iron particles immobilized on mesoporous silica with enhanced antioxidation effect. *Applied Surface Science, 333*, 220–228.

Tasneem, A., Sarker, P., & Uddin, M. K. (2020). Comparative efficacy of coagulation-flocculation and advanced oxidation process (AOP: Fenton) for textile wastewater treatment. *Current Journal of Applied Science Technology*, 41–51.

Tsuboy, M. S., et al. (2007). Genotoxic, mutagenic and cytotoxic effects of the commercial dye CI Disperse Blue 291 in the human hepatic cell line HepG2. *Toxicology In Vitro, 21*, 650–1655.

Wang, T., et al. (2013). Functional clay supported bimetallic nZVI/Pd nanoparticles used for removal of methyl orange from aqueous solution. *Journal of Hazardous Materials, 262*, 819–825.

Wang, W., et al. (2015). Iron nanoparticles decoration onto three-dimensional graphene for rapid and efficient degradation of azo dye. *Journal of Hazardous Materials, 299*, 50–58.

Wang, W.-L., et al. (2019). Advanced treatment of bio-treated dyeing and finishing wastewater using ozone-biological activated carbon: A study on the synergistic effects. *Chemical Engineering Journal, 359*, 168–175.

Wasti, E. (2018). *Pakistan economic survey 2017–18*. Islamabad: Finance Division, Government of Pakistan.

Wijetunga, S., Li, X.-F., & Jian, C. (2010). Effect of organic load on decolourization of textile wastewater containing acid dyes in upflow anaerobic sludge blanket reactor. *Journal of Hazardous Materials, 177*(1–3).

Wu, J., et al. (2019). Degradation of sulfamethazine by persulfate activated with organo-montmorillonite supported nano-zero valent iron. *Chemical Engineering Journal, 361*, 99–108.

Wu, Y., et al. (2020). Zero-valent iron-based technologies for removal of heavy metal (loid)s and organic pollutants from the aquatic environment: Recent advances and perspectives. *Journal of Cleaner Production, 277*, 123478.

Xiao, J., et al. (2015). Characterization of nanoscale zero-valent iron supported on granular activated carbon and its application in removal of acrylonitrile from aqueous solution. *Journal of the Taiwan Institute of Chemical Engineers, 55*, 152–158.

Xu, C., et al. (2019). Immobilization of heavy metals in vegetable-growing soils using nano zero-valent iron modified attapulgite clay. *Science of the Total Environment, 686*, 476–483.

Yagub, M. T., et al. (2014). Dye and its removal from aqueous solution by adsorption: A review. *Advances in Colloid and Interface Science, 209*, 172–184.

Yan, J., et al. (2015). Biochar supported nanoscale zerovalent iron composite used as persulfate activator for removing trichloroethylene. *Bioresource Technology, 175*, 269–274.

Yan, J., et al. (2017). Enhanced Fenton-like degradation of trichloroethylene by hydrogen peroxide activated with nanoscale zero valent iron loaded on biochar. *Scientific Reports, 7*.

Yao, T., et al. (2019). Preparation of reduced graphene oxide nanosheet/FexOy/nitrogen-doped carbon layer aerogel as photo-Fenton catalyst with enhanced degradation activity and reusability. *Journal of Hazardous Materials, 362*, 62–71.

Yesiller, S. U., Eroğlu, A. E., & Shahwan, T. (2013). Removal of aqueous rare earth elements (REEs) using nano-iron based materials. *Journal of Industrial and Engineering Chemistry, 19*, 898–907.

Yin, H., et al. (2019). Textile wastewater treatment for water reuse: A case study. *Processes, 7*(1), 34.

Yu, R.-F., et al. (2014). Monitoring of ORP, pH and DO in heterogeneous Fenton oxidation using nZVI as a catalyst for the treatment of azo-dye textile wastewater. *Journal of the Taiwan Institute of Chemical Engineers, 45*, 947–954.

Zazou, H., et al. (2019). Treatment of textile industry wastewater by electrocoagulation coupled with electrochemical advanced oxidation process. *Journal of Water Process Engineering, 28*, 214–221.

Zhang, W., et al. (2017). Removal of norfloxacin using coupled synthesized nanoscale zero-valent iron (nZVI) with H_2O_2 system: Optimization of operating conditions and degradation pathway. *Separation and Purification Technology, 172*, 158–167.

Zhao, S., et al. (2020). Enhanced removal of Cr(VI) from wastewater by nanoscale zero valent iron supported on layered double hydroxides. *Journal of Porous Materials, 27*(6), 1701–1710.

Zhou, H., Zhou, L., & Ma, K. (2020). Microfiber from textile dyeing and printing wastewater of a typical industrial park in China: Occurrence, removal and release. *Science of the Total Environment, 739*, 140329.

Zhou, Y., et al. (2019). Applications of nanoscale zero-valent iron and its composites to the removal of antibiotics: A review. *Journal of Materials Science*, 1–18.

Zhu, X., & Logan, B. E. (2013). Using single-chamber microbial fuel cells as renewable power sources of electro-Fenton reactors for organic pollutant treatment. *Journal of Hazardous Materials, 252-253*, 198–203.

24

Biodegradation of environmental pollutants using horseradish peroxidase

Hamza Rafeeq[1], Sarmad Ahmad Qamar[2], Syed Zakir Hussain Shah[3], Syed Salman Ashraf[4], Muhammad Bilal[5], Tuan Anh Nguyen[6], Hafiz M.N. Iqbal[7]

[1]DEPARTMENT OF BIOCHEMISTRY, RIPHAH INTERNATIONAL UNIVERSITY, FAISALABAD, PAKISTAN [2]INSTITUTE OF ORGANIC AND POLYMERIC MATERIALS, NATIONAL TAIPEI UNIVERSITY OF TECHNOLOGY, TAIPEI, TAIWAN [3]DEPARTMENT OF ZOOLOGY, UNIVERSITY OF GUJRAT, GUJRAT, PAKISTAN [4]DEPARTMENT OF CHEMISTRY, COLLEGE OF ARTS AND SCIENCES, KHALIFA UNIVERSITY, ABU DHABI, UNITED ARAB EMIRATES [5]SCHOOL OF LIFE SCIENCE AND FOOD ENGINEERING, HUAIYIN INSTITUTE OF TECHNOLOGY, HUAI'AN, P.R. CHINA [6]INSTITUTE FOR TROPICAL TECHNOLOGY, VIETNAM ACADEMY OF SCIENCE AND TECHNOLOGY, HANOI, VIETNAM [7]TECNOLOGICO DE MONTERREY, SCHOOL OF ENGINEERING AND SCIENCES, MONTERREY, MEXICO

24.1 Introduction

Enzymes are well-characterized biocatalysts, which play highly specific and efficient role in the various industrial sectors, and many processes of different product formation (Baumer et al., 2018). There are variety of different enzymes in nature, horseradish peroxidase (HRP) is one of them which have extensive applications in biocatalysis fields due to its high specificity, bioactivity, and selectivity. Like other enzymes, the HRP exhibit short lifespan with higher bioprocess costs, low recyclability, and stability that highly affect its wide-range applications (Kamble, Srinivasan, & Singh, 2019). Therefore the immobilization of enzymes is important to tackle the aforementioned issues and to improve catalytic properties of HRPs (Frey, Hayashi, & Buller, 2019). Furthermore, the immobilization increases the catalytic activity of enzyme, for example, high activity, specificity, selectivity, and stability against environmental changes, that is, pH and temperature (Basso & Serban, 2019). Due to oxidoreductases in nature, HRP can catalyze both kinds of oxidation and reduction reactions with peroxidases, and variety of different other compounds as well (Jacobowitz, Doyle, & Weng, 2019).

HRPs contain globular molecule with a predominantly α-helical secondary structure except for one short β-sheet region (Fig. 24–1).

In the structure of HRP, heme protein is present with molecular weight from 30–150 kDa, and a highly specific prosthetic group, such as iron (III) and ferriprotoporphyrin (IX) (Bilal, Rasheed, Zhao, & Iqbal, 2019). The term "peroxidase" represents a specific set of enzymes such as the sodium-dehydrogenase peroxidase, glutathione peroxidases, and iodide peroxidases as well as it also implies on other general enzymes of peroxidase with nonspecific nature. In living beings, especially in animals, plants, and microbes, the very important role and activity of peroxidase was not known until the 1990s (Li, Chen, Xu, & Pan, 2019). Lignification is also done by these enzymes and gives the direction for plants tissue recovery mechanisms, either damaged bodily or infected.

The oxidation of peroxides and the variety of different substrates performed by the peroxidases are involved in other forms of intermediate enzymes (He, Li, Wu et al., 2019). In the first reaction, Co-I is the resultant compound after the oxidation of ferric enzyme takes place by hydrogen peroxide, which is highly unstable. In the (Co-I) the structure of heme (Fe) is present IV = O porphyrin with π-cationic radical, and the oxidation of substrate result in the formation of second compound, namely Co-II, is produced, which is involved in the releasing of free radicals (Shi, Xu, & Gu, 2019). Consequently, Co-II molecule takes part in the regeneration of iron III state, which is achieved by the reduction of second molecule of substrate which results in the production of other radicals (Stoyanovsky et al., 2019). The schematic overview of HRP intermediate states have been shown in Fig. 24–2 (Krainer & Glieder, 2015).

By the advancement in technologies, nanostructured materials have introduced a new material class for enzyme immobilization, which have the unique set of characteristics for various complement factors, which indicate the efficiency of biocatalytic molecules,

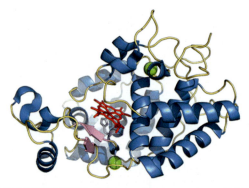

FIGURE 24–1 Structure of HRP C1A (PDB ID 1H5A). Helices and loops are shown in blue and yellow, respectively, with one short β-sheet region, represented in *pink* color. The two calcium (Ca^{+2}) ions are shown as *green* spheres. The heme group is shown in *red* and lies between the distal and the proximal domain; the proximal His170 residue (*light blue*) coordinates to the heme iron. *Reprinted from Krainer, F.W., & Glieder, A. (2015). An updated view on horseradish peroxidases: Recombinant production and biotechnological applications.* Applied Microbiology and Biotechnology, 99*(4), 1611–1625 with permission under the terms of the Creative Commons Attribution License.*

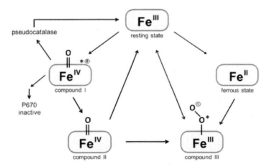

FIGURE 24-2 Schematic overview of horseradish peroxidase intermediate states. *Reprinted from Krainer, F.W., & Glieder, A. (2015). An updated view on horseradish peroxidases: Recombinant production and biotechnological applications.* Applied Microbiology and Biotechnology, 99*(4), 1611–1625 with permission under the terms of the Creative Commons Attribution License.*

including surface area, enzymes loadings, and resistance in mass transfer (Haida & Hakiman, 2019). Many of the nanoscale materials have been reported for the enzyme immobilization purposes, for example, carbon nanodots (NDs), carbon nanotubes (CNTs), magnetic nanoparticles (MNPs), graphene, gold nanoparticles, nanofibrous membranes, silicon nanoparticles, and nanodiamonds, etc. (Jiang et al., 2019). These aforementioned nanomaterials exhibit the best activity rather than the conventional method and make the introduce with stability of enzymes by providing the finest microenvironment (Qamar, Asgher, & Bilal, 2020).

The effective and novel usage of peroxides include various products of phenols, which involve the wastewater treatment, peroxide contamination removal, and the removal of food debris as well as the removal of industrial wastes and the production of value-added chemical and phenolic compounds (Bilal, Rasheed, et al., 2019). The high production of peroxides at industrial scales was performed by the tubers of HRP, however, other spices of recombinant peroxidases show the finest characteristics and have been reported for the production of peroxidases with ability to reduce the electron donor mechanism of few substrates makes it very productive in various applications of industrial and analytical fields (Kumar et al., 2019).

In ELISA, these enzymes are widely used for the production of antibodies by conjugation and also to use with minimum concentration for chromogenic product yield (Bilal & Iqbal, 2019a). Hydrogen peroxides (H_2O_2) is produced when peroxides combine with different polyenzymatic systems used for the blood glucose sensing due to oxidative nature. The effective utilization and distribution of these genetically engineered enzymes as immobilized catalysts have become very popular today by the advancements in the "green" biotechnology (Kelliher et al., 2019). The proper optimization of enzymes and degradation by helping enzyme offers the ideal and novel methodologies to reduce the environmental pollution. As a result of increased human operation in oil reservoirs, volatile farming activities, rapid industrialization, and environmental contamination has increased from the previous few decades (Basu, Green, Cheng, & Craik, 2019). Contaminants caused by their toxicity include heavy metals, nuclear waste, pesticides, greenhouse gases, petroleum hydrocarbon units, industrial wastes, paint industries waste, textile waste, metal coating sectors, preservation of wood, and paper and pulp industries are using huge amount of hazardous chemicals and dyes which may vary from the 0.1 to the 3800 mg/L, and they are making high contribution of phenols in the environment.

The presence of phenol derivatives in the matrices of water, and other material containing phenolic compounds reflected as the major pollutants in the environment, which even in very low concentration cause hindrance to their routine usage (Paquette & Lin, 2019). Various methodologies and techniques have been established for the removal and degradation of the phenolic compounds from the environment (An et al., 2019). The detoxification of toxic dyes and treatments of different pollutants have made possible by HRP enzyme. Most of the applications in the H_2O_2 presence with or without ABTS redox mediators have focused on the quantification and detection of phenolic compounds and various contaminants related to these compounds (Zhan et al., 2019). Considering the aforementioned points, the characteristics of biocatalytic degradation and detection of different pollutants by use of ideal materials and adsorbents have become the focus of modern research (Qamar, Ashiq, Jahangeer, Riasat, & Bilal, 2020).

Applications and immobilization of HRP enzyme have not been covered properly to address rising environmental contamination issues. In the advancement of biocatalysis field, the aforementioned considerations have been carefully applied in the environmental degradation of various hazardous pollutants and contaminants (Zeng, Zhou, et al., 2019). In this chapter, we aimed to discuss different nanomaterials, for example, carbon nanodots, graphene-oxide, CNTs, nanofibers, Fe_3O_4, and MNPs nanodiamonds, metal−organic frameworks, nanogels, and ordered mesoporous silica particles, which act as support materials for the immobilization of HRP (Lin et al., 2019). In each of the given section, structure and functions of nanomaterial are properly discussed, and the suitable examples of tailored HRP with industrial applications are given in detail. This chapter aimed to cover future trends and challenges for future immobilization of enzymes, which empowered the aspects of biocatalyst development, their applications in environment pollution control, and in biosensor designing.

24.2 Carbon nanotubes, carbon nanoonions, and carbon nanodots for horseradish peroxidase immobilization

The matrix of single-walled carbon nanotube (SWCNT) for the immobilization for HRP have exhibited excellent efficacy for electron transference acceleration (Asmat, Anwer, & Husain, 2019). The electrostatic force of attraction and nanoorientation by the covalent bonding and the crosslinking of glutaraldehyde are the most applicable methodologies for HRP-immobilization on the different nanomaterials (Vineh, Saboury, Poostchi, Rashidi, & Parivar, 2018). The techniques used for HRP-immobilization purposes has various challenges toward wide ranges of applications depending upon the support matrices (Alshawafi et al., 2018). Furthermore, on the CNTs the noncovalent interaction and using robust π-π interaction between the biopolymers is an effective approach for the immobilization of enzymes (Su, Zhou, Long, & Li, 2018).

For H_2O_2-sensing purpose, two methods were purposed, i.e., the use of SWCNT treated with acid for the immobilization of HRP, and for the electrochemical test acid treatment of cut-SWCNT for the fabrication purpose successively placed on the electrodes with carbon-based screen printing (Bilal, Rasheed, et al., 2018). In the chemistry of pyrene crosslinking, cut-SWCNT carbon printed electrode was used for the immobilization of HRP by Waifalkar, Chougale, Kollu, Patil, and Patil (2018). The biosensor used for the immobilization of HRP showed the great reproducibility, stability, and minimum contamination interference. The composite of maize tassel and multiwalled CNT has been used for the adsorptive immobilization of HRP (Moyo, Chikazaza, Nyamunda, & Guyo, 2013).

The SWCNTs with carboxyl functionalized adsorptive immobilization of HRP has been revealed by Zhang, Lü, Shao, and He (2014). Recently, various entities with less mediator HRP-immobilization employed by using the nanotube of peptides (Alarfaj, El-Tohamy, & Oraby, 2018). The hydrothermal reaction was used by Zhang, Dong, Zhao, Wang, and Meng (2019); Zhang, Wells, et al. (2019); and Zhang, Yao, Sun, and Tay (2019) for the development and characterize of Fe_2O_3 nanocomposites along with magnetic iron (III) oxide CNTs, and use their utilization as a support matrices for the immobilization of HRPs (Xie et al, 2019). The magnetic sphere formation was carried out with size about 340 nm on the CNTs walls. At optimized conditions, bioactivity of immobilized HRP shows more durability in varied temperature and pH values as compared with the native enzymes (Yu et al., 2019). The long-lasting efficiency, storage, and easy recovery could be observed by the nanobiocatalyst, immobilized after the completion of six repeated reactions (Zeng, Wei, et al., 2019). The solubility and biocompatibility shown by the carbon nanoonions (CNOs) and CNTs is high, which are insoluble in the organic as well as inorganic solution and exhibit several different reactions such as cycloaddition, cyclopropanation, addition of radical, and oxidation of different mineral acids (Wang et al., 2019). The less toxicity and biocompatibility have been demonstrated by modified CNOs. In the field of biological imaging and clean up environmental applications, CNOs can be used (Zhu et al., 2019).

The elemental biosensing was purposed so that CNOs can be used as support matrix in this process of pollutants detection (Bilal & Iqbal, 2019b). Due to surface area stability and electric properties, CNOs show high performance in systems rather than the systems having no CNOs as a support material. This theme is based on the potential utilization of CNOs as a carrier and support material in the enzyme's immobilization of different kinds, that is, alkaline phosphatase, HRP, and glucose oxidase (Bilal, Adeel, Rasheed, Zhao, & Iqbal, 2019). For the completion of oxidation process of CNOs, which are creating carboxylic groups on the shell of graphite and making them able to create a link with enzyme's amino group in the carbodiimide presence, which acts as a coupling reagent. The modified CNOs can be loaded on 0.5–1 mg of protein, which was revealed by thermogravimetric analysis (Asif, 2021). The immobilized enzymes exhibit the stability of storage without any change in pH and temperature.

Carbon dots are another type of nanomaterial namely with size ranges <10 nm, which are novel and intriguing carbon nanomaterial. Due to having its fascinating and intriguing properties it is widely being used in the research fields which shows the finest biocompatibility, optical robustness, chemical inertness, and various other electrochemical properties (Sohail, 2021). Carbon dots are actually pseudospherical discrete type of nanoparticles composed of amorphous or nanocrystalline cores. Its characteristics can be changed by changing the structure which strongly depends on fabrication modes and precursor used in the preparation.

Furthermore, due to the exceptional luminance and electrochemical activities carbon dots show high density and electron mobility, which makes them a valuable electrocatalytic and electrochemical nanomaterial (Jia et al., 2019; Lin et al., 2017). The various surface modification of carbon dots by different functional moiety, for example, amino and carboxylic group, these kind of nanomaterial possess very suitable characteristics for the functioning with various kinds of organic, polymeric, inorganic, and biological agents for the development of biosensor and enzymes coupling agents, as a precursor such as chitosan biopolymer and silver nitrate in the one step method of hydrothermal functionalized amino carbon dots (NH_2-CDs) can be synthesizes (Luo et al, 2018). The size of fabricated amino carbon dot studied was about 2.8 nm, with high crystallinity and stable size distribution in the presence of silver nitrate as catalyst/precursor (Elsayed et al., 2021). The HRP-immobilization on these amino carbon dots showed finest stability and biocatalytic activity. From the previous literature, the given Table 24−1 concluded NDs, CNTs, nanofibers (NFs), and NOs based material for support matrices used for the immobilization of HRP with improved biocatalytic activity (Mohamed, Al-Harbi, Almulaiky, Ibrahim, & El-Shishtawy, 2017).

24.3 Graphene and its derivatives for horseradish peroxidase immobilization

In the nanomaterial sciences graphene have gained much scientific interest in both potential application and fundamental sciences, due to having excellent characteristics including two-dimensional (2D) structural morphology, good chemical, and electrical characteristic and its various byproducts with fascinating surface area (Oh et al., 2016). Such unique morphological characteristics of graphene and its byproducts enable graphene with excellent characteristics desired for designing and synthesis of nanomaterial (Bilal, Asgher, Iqbal, Hu, & Zhang, 2016). When compared with conventional soft or hard materials, the graphene and its byproducts show high class properties. Graphene with its 2D and flexible structure enable it enhanced surface are for the nanomaterial growth and the nucleation, particularly helpful for the 2D nanomaterial, which cannot made by the traditional methods (Rehman, Bhatti, Bilal, & Asgher, 2016). Graphene oxide (GO) is very special type among graphene derivative compounds, which provide the useful models for the specific morphology formation and also provide the surface area for the growth of various kinds of nanomaterials. In addition, GO acts a constructional material for the formation of three-dimensional (3D) graphene with varied structures which may join with other kind of nanomaterial and result with classified 3D based nanomaterial (Zou et al., 2016).

Table 24–1 Carbon nanotubes, nanodots, nanoonions and nanofibers for the immobilization of horseradish peroxidase and resultant enhanced biocatalytic characteristics.

Support material	Type of immobilization	Biocatalyst with improved properties	References
Nanocapsules	Encapsulation	Strikingly increased enzymatic performance, including catalytic efficiency, thermostability, environmental tolerance and efficiency of biodegradation.	Liu et al. (2020)
ZIF-67(Co)/MWCNT	Entrapment	Improved substrate affinityEnhanced sensitivity for H_2O_2 detection and increased sensitivity.	Liu et al. (2019)
APTES-Fe_3O_4/MWCNTs	Covalent bonding	The immobilized enzyme was more stable at wide pH and temperature values compared with free enzymes.Strong effect when stored for a long time and conveniently isolated by an external magnet. Presented strong reuse potential after 6 months, maintaining 65% of its initial operation.	Zhang, Dong, et al. (2019); Zhang, Wells, et al. (2019); Zhang, Yao, et al. (2019)
Fe_3O_4/polyacrylonitrile magnetic nanofibers	Covalent linkage	The MNFs-conjugated HRP exhibited no modification in pH-optima and presented high catalytic efficiency relative to the free counterpart. The optimum HRP-coupled MNFs obtained a notable efficacy of phenol removal and was reusable for five repeated cycles with over 50% of their initial degradation remaining.	Almulaiky et al. (2019)
Pyrene modified acid-treated SWCNTs	Crosslinking	The manufactured HRP-immobilized biosensor demonstrated marked stability, good reproducibility, and limited contamination interference.	Şahin (2020)
Amino-functionalized carbon dots	Covalent attachment	Good biocatalytic activity and stability were demonstrated by the immobilization of HRP on these NH2-CDs.	Su et al. (2018)
Polyacrylamide and poly (vinyl alcohol) bi-component (PVA-PAAm) nanofibers	Covalent linkage	The immobilized enzyme revealed an optimal temperature for the free enzyme of 50°C rather than 45°C. High thermal and storage stability and reuse performance were also seen for 25 consecutive reaction cycles with retention of more than 50% of catalytic activity.	Temoçin, Inal, Gökgöz, and Yiğitoğlu (2018)
Carbon nanoonions	Covalent linkage	Without altering temperatures and pH-optima values, nanoimmobilized enzymes showed improved storage stability.	Sok and Fragoso (2018)

(Continued)

Table 24-1 (Continued)

Support material	Type of immobilization	Biocatalyst with improved properties	References
Nanodiamond-incorporated polymethyl methacrylate nanofibers scaffold	Covalent attachment	After 10 consecutive reuse cycles, the immobilized nanobiocatalyst revealed more than 60% of its primary operation. After the immobilization process, the optimum temperature of the soluble enzyme (30°C) was increased to 40°C. The immobilized HRP successfully catalyzed the oxidation of the substrate and demonstrated greater stability than that of the free equivalent to the denaturing effects of metal ions, isopropanol, urea, heptane, and butanol.	Alshawafi et al. (2018)
MWCNT/cordierite nanocomposite	Physical adsorption	Improved stability of the adsorbed derivative's catalytic, thermal, acid-base and storage relative to the free enzyme. It demonstrated outstanding recyclability as well.	Li, Cheng et al. (2017)
Titanate nanowires	Covalent attachment	Immobilized HRP is less likely to denature bio-electrocatalytic activity and maintain it. The evolved biosensor displayed up to 30 days of storage stability, maintaining 91% of its operation.	Nicolini, Ferraz, and de Resende (2016)
PI/MWCNTs nanofibers	Covalent bonding	The immobilized biocatalyst reported an improvement in retention activity of 2.38%–12.50% through the blending of MWCNTs relative to pristine nanofibers polyimides.	Zhang, Xu, Jin, Wu, and Xu (2014)
Chitosan–halloysite hybrid-nanotubes	Crosslinking	Maximum loading capacity of enzymes. Excellent data reliability, maintaining maximum operation after 35 days of storage, although over 70% of its initial activity was lost by the free HRP.	Zhai, Sun, Zhao, Gong, and Wang (2013)
Chitosan-wrapped SWCNTs	—	Immobilized HRP for nitric oxide reduction maintained high catalytic activity. The biosensor based on HRP-SWCNT-chitosan showed a strong linear detection range and fast response time (less than 6 s).	Jiang et al. (2009)

PI, Polyimide; *HRP*, Horseradish peroxidase; *MWCNT*, multiwalled carbon nanotubes; *SWCNTs*, single walled carbon nanotubes.

It has been investigated that the graphene templates can easily be vanished by burning in air that evades the usage of destructive base or acid which decrease the production cost (Bilal, Rasheed, et al., 2019). Except these, in many other matters without the use of other graphene processes can be removed and act as active and efficient component, in which minimum chances of impurities as well as to get maximum efficiency from the template. In the view of graphene versatility, formation simplicity, and large number of materials can be fabricated using the synthesis of graphene in which metal oxides, metal particles, sulfides metals, and organic semiconductors are included (Alimba & Faggio, 2019). These kinds of nanomaterials along with controlled shape, configuration, and size, show the high potential for countless storage and energy conversion applications, for example, solar cell, supera capacitors, lithium-ion batteries, and photocatalysis. Consequently, the material with graphene template shows the maximum interest in practical and theoretical point-of-view (Hajian et al, 2019).

The molecules of graphene are single atomic thickness and lattice of honeycomb with configuration of sp^2-hybridized carbon (Kashefi, Borghei, & Mahmoodi, 2019). Due to great mechanical steadiness, surface area, and the finest thermal, optical, and electrical features, graphene has gained great interest in materials chemistry. Physicists and chemists are dealing with the development of applications in wide range of area, for example, drug delivery, biosensing, hydrogen storage, super capacitors, nanobiocatalysis, nanoelectronics, solar cells, and environmental protection applications (Savun & Gineste, 2019). Owing to the aforementioned unique characteristics, graphene is considered as the good carrier of support biomolecules and enzymes immobilization (Loibner, Hagauer, Schwantzer, Berghold, & Zatloukal, 2019). Large surface area of graphene help to increase the enzymatic loading, and stability of the graphene make the possibility for reuse of immobilize enzyme. Due to the excellent electrical conductivity, the enzyme couple with graphene use in the application of biosensing (MacNevin et al., 2019). Being graphene precursor, GO also increases importance in the field of material science and its desirable matrix for biomolecular immobilization due to finest characteristics with larger surface area with high thermal and mechanical power (Magesa et al., 2019). Furthermore, GO exhibits multiple oxygen on the surface area and the edge, involving epoxy, carboxyl and hydroxyl carrying attachment for enzymes. The use of GO biocatalyst has shown the varied ranges of activities related to the biocatalysis, and different approaches used to design nanocarriers of GO (Madec et al., 2019).

GO lack functional group which is attained by the reduction of GO to certain level (Bai, Xu, & Zhang, 2020). The hydrophobic adsorption take place on the reduced GO, and the immobilization of various enzymes could be carried out by the crosslinking, covalent attachment, and adsorption through electrostatic attraction, for example, the HRP attachment on the surface of graphene formed by the interaction between p-p and other hydrophobic forces (Hwang et al., 2020). Although the immobilization of HRP takes place on the surface of reduced GO and GO by the interaction of van der Waals (electrostatic interaction) and hydrogen bonding due to the presence of large number of oxygen moieties on the surface (Ge et al., 2020). The properties of immobilized biocatalyst are driven up by the material originated from the graphene family. The deactivation of the enzymes and biomolecules, having the oxidation functionality of reduced GO have capability to remove free radicals of dithiocyanate and hydroxyl molecule (Akbarzadeh, Soheili, Hosseinifard, & Gholami, 2020). Furthermore, reduced GO can enhance the stability for enzymes by acting as agent having potential for quenching radicals. Reduced GO and GO-based material which use in the immobilization of HRP and their biocatalytic with improved features have been represented in Table 24–2.

Table 24–2 Graphene and its derivatives (graphene oxide and reduced graphene oxide)-based support materials for Horse radish peroxidase-immobilization, and their improved biocatalytic features.

Enzyme support	Immobilization type	Catalytic operation	Reference
Graphene oxide (GO)/ magnetic chitosan beads	Crosslinking	The biocatalyst engineered retained up to 10 consecutive cycles of 90% of its original catalytic performance. The optimum temperature was increased from 50°C to 55°C following the immobilization process. The immobilized HRP retained its activity by 93.72% and 60.97% respectively after 30 and 60 days of storage at 4°C.	Sahu and Raichur (2019)
Multiarmed magnetic graphene oxide composite	Covalent attachment	In comparison to the free biocatalyst, the reliability of activity, storage, and temperature after HRP-immobilization has been enhanced. After continuous reuse for eight times, the magnetic GO biocomposite immobilized enzyme maintained more than 68% of its primary activity.	Rong et al. (2019)
RGO-NH$_2$-NPs	Physical adsorption	Dramatic improvement in the catalytic efficacy, and k_{cat}/K_m values. The catalytic activity increased to 120 and 133-folds for 1.5 and 2 mg/mL of nanoparticles. The free and immobilized nanobiocatalytic system showed over 40% and 70% of initial activities after 2 h incubation at 50°C, respectively. It also provided excellent reuse capability that retained nearly 60% of activity after 10 repeated batches.	Vineh et al. (2018)
Functionalized reduced graphene oxide	Covalent binding	Immobilization resulted in catalytic constant and catalytic efficiency improvements of 6.5 and 8.5-times, respectively. Immobilized HRP showed improved recycling ability after 10 successive reuse cycles, retaining approximately 70% of primary catalytic activity. At 40°C, over 90% of the activity was preserved by the carrier supported HRP, while 40% of the activity was lost after 2 h of incubation in the case of free enzymes. After 35 days of storage, it showed greater than 97% of the initial activity.	Vineh et al. (2018)
AuNPs/S-RGO	Covalent attachment	Faster response, adequate stability of storage, inexpensive reproducibility and repeatability that are satisfactory and outstanding selectivity.	Vilian et al. (2017)
Porous graphene	—	Excellent electrochemical performance toward H_2O_2 Low determination limit High efficiency	Liu et al. (2020)
Graphene oxide/Fe$_3$O$_4$	Covalent attachment	Immobilized HRP exhibited better thermal stability than soluble enzyme. The immobilized nanobiocatalyst was effectively isolated, recovered and recycled from the complex reaction mixture using an external magnetic field.	Chang, Huang, Ding, and Tang (2016)

(Continued)

Table 24–2 (Continued)

Enzyme support	Immobilization type	Catalytic operation	Reference
Superparamagnetic Fe_3O_4/graphene oxide nanocomposite	Covalent attachment	Improved storage stability and tolerance compared to free enzymes for changes in pH and temperature. Easy recovery using an external magnetic field 66% of its original operation after four cycles was maintained by Immobilized HRP.	Chang et al. (2015)
Porous Co_3O_4 nanosheets and reduced graphene oxide composite	–	Good performance Large linear range and incredibly low limit of detection	Bai, Liu, Sun, and Gao (2016)
Poly-L histidine(P-L-His) modified reduced graphene oxide	Physical entrapment	Immobilized HRP showed outstanding electrocatalytic activity to reduce H_2O_2. Rapid H_2O_2 response along with good reproducibility and stability.	Vilian and Chen (2014)

HRP, Horse radish peroxidase.

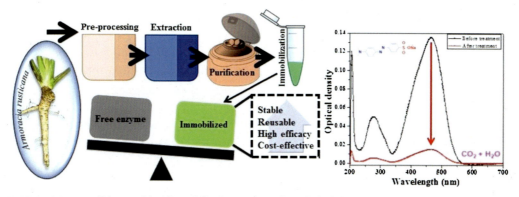

FIGURE 24–3 Horseradish peroxidase immobilized onto the polyvinyl alcohol-alginate beads and its methyl orange degradation potential (Bilal et al., 2017).

The catalytic ability and stability characteristics of HRP was assess through physical enzyme immobilization on the amino-functionalized reduced GO-NPs of 70 nm in size, exposed to compare with enzyme present in the free form (Mohammadian & Faridbod, 2018). The activity of enzymes and assays show that catalytic constant and Kcat values increase more than six-folds, with seven-folds improvement in catalytic and kcat/km values (Zhao, Qiu, et al., 2016; Zhao, Zheng, Kong, Xia, & Qu, 2016). After the immobilization of HRP, increases in biocatalytic activity with 120–130 folds for the 1 and 2 mg/mL NPs. Both immobilized and free HRP enzyme showed the similar pH-optima at 7.0. After the incubation for 2 h at 45°C–50°C, the immobilized and free form of nanobiocatalyst shows 70% and 40% of its actual activity, respectively (Fig. 24–3) (Bilal, Asgher, Iqbal, Hu, & Zhang, 2017; Bilal, Rasheed, et al., 2017). It also has shown with the good capability of reuse with around 60% activity after repetition of 10 cycles. Circular dichroism spectroscopy is used for the analysis of secondary structure which shows less content of the α-helical after the immobilization of HRP (Pitzalis, Monduzzi, & Salis, 2017).

HRP-immobilization on the activated reduced graphene was performed by covalent binding and the size of reduced GO was about 60 nm and using glutaraldehyde as crosslinking agent. The immobilization of HRP increases 6.3–8.3 times the catalytic efficiency and catalytic constant, respectively (Vilian et al., 2017). The contents of α-helix reduce from 15% to 10% after immobilization process. After the 10 repeated cycles it retains the 70% ability of primary activity of biocatalysis when HRP was immobilized (Radhakrishnan & Kim, 2015). The immobilized HRP with carrier supported which preserved about 90% of its activity at 40°C. After 2 h of incubation period and the free enzyme reduced its 40% activity (Lee et al., 2017). But after the 35 days of storage, the highest starting activity can be seen with more than 97% efficiency. There was another kind of support material for the immobilization of HRP such as multiarmed polyethylene-glycol (PEG) which GO polymer-based biocomposite (GO@Fe_3O_4@6arm-PEGNH_2) was firstly used. The amino group from the presence of 6arm-PEG-NH_2, the capacity of HRP loading becomes maximum as much as 185.36 mg/g on surface area of new nanomaterial (Aldhahri et al., 2020). When compared with free enzyme, storage, operational, and thermostability were enhanced after HRP-immobilization. The magnetic GO composites with enzyme immobilized, kept its primary activity more than 68% after eight time of reuse continually (Kergaravat, Fabiano, Soutullo, & Hernández, 2021).

The strategy of reverse coprecipitation was used for the immobilization through covalent interaction by using of 1-ethyl-3-(3-dimethylamino propyl) carbodiimide (EDC), which is a new kind of crosslinker and preparation of GO/Fe_3O_4 NPs also takes place by the ultrasonic-assisted method (Roccella et al, 2020). The optimized conditions for the different type of chemical reaction such as the ratio of phenol to PEG, pH, H_2O_2 initial concentration, and immobilized biocatalyst (Matsubara & Takemura, 2020). The effective kind of enzyme immobilization on GO/Fe_3O_4 were recognized for interaction occurrence between the functioning biomolecules of support material and entities with EDC. The great resistance for the varied in temperature can be seen by the HRP immobilized with GO/Fe_3O_4. The nanobiocatalysts, which are immobilized results with high recovery, ready separation, and recyclability using externally magnetic field, which is present in the complex of reaction mixture (Dose et al., 2019). By using the glutaraldehyde as crosslinking agent, the HRP covalent attachment on the immobilization matrix, hybrid microsphere of GO-coated Fe_2O_3/chitosan were produced (Cen, Liu, Xue, & Zheng, 2019). This procedure achieved high immobilization efficacy of about 80% and maintain its original performance of catalysis about 90% after the consecutively repeated 12 cycles.

The enhance activity of HRP recognized peroxide like molecule, conjugated with GO-Fe_2O_3. Fe_2O_3-GO composites show the great impact on the immobilization of HRP compared to other matrices of support for nanomaterial (Dalkıran, 2020). The coating of GO on the magnetic chitosan microspheres allows the GO to distribute well on beads surface of chitosan resulting the equal distribution of GO on the all catalysis beads site (Tao, Ma, Yang, & Hao, 2020). The immobilization of antibodies and enzyme on the surface of graphene make more suitable by using the AuNPs which increase the surface area for the attachment of biomolecules. AuNPs are completely coated on the matrix of graphene surface which provide it microenvironment, making the surface of graphene more favorable for the attachment of antibodies and enzymes. It is also helpful in the transfer of electron between the electrode and biomolecules (Banach-Wiśniewska, Tomaszewski, Hellal, & Ziembińska-Buczyńska, 2021). It is recently reported that the use of GO-coated with AuNPs for the detection of various harmful and clinical pollutants in the water samples such as AuNPs-decorated reduced GO, AuNPs anchored reduced GO, AuNPs-coated graphene nanocomposites, and AuNPs-assembled GO composite has recently been used for the immobilization of HRP fabrication, which is applicable in different electrochemical sensors. Being a source of sulfur as well as reducing agent modified thiol-reduced GO is used (Kim, Kwon, & Kim, 2021). The sulfur doped reduced graphene oxide (S-rGO) surface coated with the AuNPs and composites of S-rGO/AuNPs used for the immobilization of HRP. The good storage, simple synthesis, quick response, reproducibility and repeatability, and stability can be seen by using the electrode of S-rGO/ AuNPs/HRP. The identification of bisphenol A in tomato juice and in milk sample shows the exceptional selectivity (Anna, Mariusz, Hellal, & Aleksandra, 2021).

24.4 Magnetic nanoparticles for horseradish peroxidase immobilization

NPs are known as extraordinary support material for catalytic processes for enzyme attachment with unique characteristics of immobilization, involving high capacity for enzyme loading, minimum restriction in diffusion, specific surface area, and mechanical steadiness (Verma, Kumar, Das, Randhawa, & Chamundeeswari, 2020). The magnetic properties of the Fe_3O_4 NPs can be lost by the oxidation in air. Furthermore, the centrifugation and filtration process are very difficult for the recovery of NPs. By applying the material in the matrix of enzyme mobilization having the properties of ferromagnetics, which are very useful in the easy retravel and conjugated biomolecular separation from the reactor by applying the external magnetic field. By applying this process, the support of magnet avoids the requisites different process such as the filtration, decantation, and centrifugation (Xu et al., 2020). Another kind of biomaterial named Fe_3O_4 MNPs, which also work as support material for the immobilization of enzymes with excellent features of superparamagnetic with no hemolytic and genotoxicity activities. These characteristics are helpful in the HRP-immobilization on the basic Fe_3O_4 MNPs (Esmi, Nematian, Salehi, Khodadadi, & Dalai, 2021). The immobilized HRP with magnetic nanoparticle shows great characteristic of thermoresistant other than the enzyme present in free form and shows more than 50% original activity after repetition of 10 cycles (Ogunsona, Muthuraj, Ojogbo, Valerio, & Mekonnen, 2020). When compared with free form of enzyme, it shows maximum effectiveness for the oxidation of substrate and are more stable while treating with organic solvents, that is, isopropanol and triton X-100 (Yi, Dai, Zhao, & Si, 2017).

Coprecipitation is a convenient method by which MNP with average size of 43 nm can be synthesized and combined on the cryogel chitosan and vinyl alcohol for the HRP immobilization by using the covalent attachment (De Witte, Wagner, Fratila-Apachitei, Zadpoor, & Peppas, 2020). The microbar of cryogel with immobilized HRP is active material use to identify the molecules of H_2O_2 by using the oxidation of o-dianisidine and H_2O_2, and in result brown color produce which were relative to concentration of H_2O_2 (Sengel, Sahiner, Aktas, & Sahiner, 2017). The immobilization of nanobiocatalyst kept the maximum activity of the enzymes with 10 successive reactions which develop a system of chromogenic nanobiocatalyst with high efficiency by HRP immobilization on the silica-coated magnetic nanoparticle and make it resistance for oxidation in air and combination (Oktay, Demir, & Kayaman-Apohan, 2020). The fabrication of nanobiocatalysts was performed through the reversely immobilizing on magnetic nanoparticle coated with silica and HRP fabrication done through the electrostatic interaction. The capacity of HRP protein loading on Si-MNPs reaches to 5 μg/mg of modified silica nanoparticles (Erol, Cebeci, Köse, & Köse, 2017). The greatest results could be observed by immobilized HRP when compared with the free form of enzyme, and rapid chromogenic degradation of tetramethyl take place. The concentration of glutaraldehyde is about 300 μL and 7 h time for agitation the effective attachment of the HRP take place on Fe_3O_4 surface (Dzhakasheva & Lieberzeit, 2017). The separation by magnetic force for immobilized HRP on the Fe_3O_4 make it reusable in the various repeated cycles. It was investigated that the material associated with Fe_3O_4-MNPs can be easily modified for their surface functionalization by creating a linkage of enzyme with functional moieties (Georgiev, Trzebicka, Kostova, & Petrov, 2017).

The large number of reactive oxygens with functional features and larger surface area of GO makes it fascinating material for immobilization of enzymes (Carbone, Gorton, & Antiochia, 2015). By using EDC as a crosslinker of biocomposite, the immobilization of HRP and fabrication of GO/Fe_3O_4 NPs can be done with the process of ultrasonic driven reversed coprecipitation. High efficacy shown by the nanobiocatalytic immobilization systems for removal of phenolic compound take place, which may occur after repeated cycles in aqueous media by using the action of magnetic field for GO/Fe_3O_4 conjugated-HRP. It was noted that after the completion of four cycles for the recognition of polymer that in return cover the enzymes active site and minimize its catalytic activity to 40% of its original activity (Xie et al., 2016). The unique properties of ND such as larger surface area with many reactive oxygen and different functional groups on the surface, make it more interesting candidate for material scientists. These aforementioned properties of ND make it suitable support material for immobilization of enzymes as well as offers the finest choice for nanosupporting material and also for nanostructure of metal oxide (Malekmohammadi, Hadadzadeh, Farrokhpour, & Amirghofran, 2018).

Due to having the inherently magnetic features, the nanostructure based on the Fe_3O_4-ND attain great importance for the immobilization of enzymes. However, using these nanocomposites is very challenging because of the surface area modification which has been studied for increasing the magnetic features of nanocomposites based on the Fe_3O_4, that is, MNP of functionalized APTES use in the effective aminoacylase immobilization (Feng, Yu, Li, Mo, & Li, 2016). For HRP-immobilization using the MNP coated with silica, covalent bonding of assisted polyglycerol was also used for the synthesis nanocomposites of Fe_3O_4/ND for the immobilization of HRP (Kim, Choi, Kim, Wi, & Bae, 2014).

The interaction between the nanocomposite's functional groups and HRP is useful for the conjugation of enzyme on the surface layer of polyglycerol on the nanocomposites Fe_3O_4/ND. When compare with free form of enzyme, the fascinating characteristic can be seen by the immobilized enzyme such as the greatest activity at different temperature, pH, and storage duration (Kazenwadel, Franzreb, & Rapp, 2015). The increase in the reuse activity was more than 75% to its original activity after repeating six cycle which proves itself as magnetic nanobiocatalyst with good recycling ability. Best biocatalytic performance for HRP was observed using support material of magnetic and nanomagnetic nanostructure (Mohamad, Marzuki, Buang, Huyop, & Wahab, 2015).

24.5 Magnetic electrospun nanofibers for horseradish peroxidase immobilization

In comparison with different other nanoscale materials, electrospun nanofibers give various kind of benefits, for example, great surface area, porous structure, high selective material, reusability, and simple recovery (Huang, Wang, Xue, & Mao, 2018). Nanofibers are fabricated by using the method of electrospinning. For the development of the controllable and uniform structure fibers, different types of polymeric and organic materials are used in the process of electrospinning. The composite nanofibers with organic and inorganic hybrids with double characteristics of organic/inorganic materials are also developed by using the process of electrospinning, that's why use in the sensing of chemical and toxic substance absorbance (Bilal, Zhao, Rasheed, & Iqbal, 2018). The fabrication of semiconducting PbS NPs in which polynanofibers are present was performed by electrospinning. The addition of gold nanorods (AuNRs) in the polymeric solution was helpful in the synthesis of AuNRs/PVA a mat based on nanofiber which enhanced the surface and scatter the Raman substrate with sensitivity (Safarik et al., 2018). The practical method demonstrated by the electrospinning for the synthesis of hybrid organic-organic NFs composites which shows the high functional and potential characteristics (Bilal, Barceló, & Iqbal, 2020).

For degradation of phenol immobilization of HRP, and the fabrication of new type of Fe_3O_4/polyacrylonitrile (PAN) MNFs, the electrospinning has been used. The saturation point of MNFs after magnetic induction was about to 19.03 emu/g and 200–400 nm is the average size of the fabricated magnetic nanofibers. When compare with the free enzyme the HRP-conjugated-MNFs show no variation in pH, and also offers the best performance in catalytic activity (Melo et al., 2020). The combination loading of MNFs immobilized HRP with 40% Fe_3O_4 NPs show minimum binding, and maximum activity due to combing effect of magnet with MNPs. The finest MNFs immobilized HRP having the efficiency of phenol removal and easily usable for almost five cycles and 50% remain behind after the initial degradation (Almulaiky & Al-Harbi, 2019). The HRP-immobilization highly catalyze the oxidation of substrate and show greatest stability for denaturation effect of urea, butanol, isopropanol, metal ions, and heptane as compared to free enzyme. The high resistance toward photolytic cleavage from trypsin and have many applications in the field of bioremediation, biomedical, and biosensor (Courth et al., 2020).

24.6 Metal–organic frameworks for horseradish peroxidase immobilization

Several metal–organic frameworks (MOFs) have been constructed that possess larger pore size with excellent stability for the entrapment of enzyme. [PCN-333(Al)] MOFs with stability and large size of pore which is about 5.5×4.2 nm and proved to be the ideal candidate for enzyme immobilization purposes (Eldin, 2016). MOFs [i.e., PCN-333(Al)] having water resistance in larger cavities and finest surface area was selected for encapsulation and adsorption of HRP and cholesterol oxide (ChOx), respectively. The characteristic features of PCN-33/HRP and ChOx as-prepared showed the greater surface area and various number of cages mesoporous offers the fascinating loading of these enzymes. Without any modification in the chemical structure the absorbed ChOx and encapsulated HRP show the good performance of biocatalysis (Han et al., 2019). The biocatalyst immobilized on the MOFs exhibit highly stable temperature and pH variation, organic solvent and proteolytic digestion. Magnetic nanoparticle can also be used in the synthesis of MOFs and product of these with result of magnetically responsive composite which was used in dynamic localization, activate heating, and catalysis. A method is invented for the synthesis of magnetically active form of MOFs and recyclable based system for biocatalysis (Ricco, Santa Maria, Hanel, & Bhattacharya, 2020). Herein, both of the MNPs and HRP encapsulation was performed in the crystal framework-8 of zeolitic imidazole (ZIF-8). Combination of both of these protein and inorganic-NPs induces the faster assembly of ZIF-8. It was reported that MNPs inclusions enhance the catalytic activity of biocomposite based on ZIF-8, and its combine effects showed five times more activity then the complex of both these present in free form, and can be easily reutilized and recovered in many cycles. Overall, MNPs and HRP combination utilization and porous MOFs offers the many features towards biocatalytic application (Carraro et al., 2020). Mesoporous and MOFs-based immobilization of HRP and their catalytic properties are given in Table 24–3.

24.7 Mesoporous silica for horseradish peroxidase immobilization

There are many nanomaterial uses in the field of biotechnology, and mesoporous silica is one of them, which is being used in nanomedicine, inorganic catalysis, adsorption, and biocatalysis due to its fascinating structural and textural characteristics including the volume and size of pore, surface area, and the particle size (Gui et al., 2018). Many researchers are working on the application of the mesoporous silica for the immobilization of enzymes and other features of larger biomolecules. Many of the enzymes in which the immobilization take place on the mesoporous silica for the enhancement in catalytic activities have performed such as the xylanase, lipase, laccase, cytochrome C, subtilisin, chloroperoxidase, and lysosome. A system developed by Pitzalis et al. (2017) in which bienzymatic catalysis take place by applying glucose oxidase and HRP on the surface of SBA-15 silica mesoporous, and it conserved the remarkable activity after the completion of 14 repeated cycles (Valldeperas et al., 2019).

Table 24–3 Metal−organic frameworks and mesoporous silica as support materials for the immobilization of horseradish peroxidase, and their improved biocatalytic performance.

Immobilization assistance	Type of immobilization	Improved biocatalytic properties	References
Metal−organic frameworks (MOFs) [i.e., PCN-333(Al)]	Encapsulation	Without chemical modifications the encapsulated horse radish peroxidase (HRP) had exceptional biocatalytic performance. The immobilized MOF bioconjugates demonstrated high stability against variations in pH and temperature, proteolytic digestion, and organic solvents.	Zhao, Qiu, et al. (2016); Zhao, Zheng, et al. (2016)
Magnetic zeolitic imidazolate framework 8	Encapsulation	The ZIF-8-based magnetic biocomposite improved enzymatic activity. The composite bioactivity of HRP/MNP@ZIF-8 was five times higher than the composite of HRP@ZIF-8.In multiple cycles it can easily be recovered and reused.	Ricco et al. (2020)
Modified mesoporous silica nanoparticles	Cu-driven 1,3-dipolar cycloaddition reaction	Cow-fixed enzymes were highly stable in the mesopore and demonstrated significant biocatalytic performance for the hydrolyzing and colorimetric reaction of 2-methoxy phenol in several repeated cycles.	Gößl et al. (2019)
SBA-15 mesoporous silica	Covalent attachment	A marked activity after 14 recycling was preserved by the wet GOx/HRP@SBA-15 biocatalyst. It showed that phenolic compounds can be oxidized.	Pitzalis et al. (2017)

Furthermore, the biocatalytic activity of mesoporous silica-immobilized enzymes showed the great ability to oxidize phenolic compounds, for example, caffeic acid and ferulic acid. A process developed by Goud et al. (2019) described the synthesis of NPs with modified mesoporous silica with massive number of pores that enhance the enzyme loading capacity. Sp-carbonic anhydrase and sp-HRP, which is acetylene functionalized enzymes, is placed or conjugated with porous silica NPs by using a reaction of cycloaddition. The enzyme present in the pores of silica is attached covalently and show the great performance in biocatalysis after the repetition of various cycles for 4 nitrophenyl acetate (LP-MSN-CA) hydrolysis, and other reaction of 2 methoxy phenol (LP-MSN-HRP) (Sneha, Beulah, & Murthy, 2019).

24.8 Horseradish peroxidase for environmental applications

Much scientific effort has been devoted to the development of catalyst/biocatalyst-based systems for tailored environmental applications, that is, mitigation of pollutants, detection of hazardous metallic wastes, treatment of wastewater matrices, and the bioconversion of lignocellulosic waste to high-value compounds (Fig. 24−4). Designing environmentally friendly processes for industrial sectors offers various advantages including least chemical

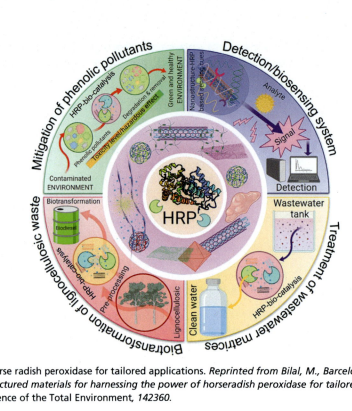

FIGURE 24–4 Horse radish peroxidase for tailored applications. *Reprinted from Bilal, M., Barceló, D., & Iqbal, H. M. (2020). Nanostructured materials for harnessing the power of horseradish peroxidase for tailored environmental applications.* Science of the Total Environment, *142360.*

utilization, energy saving, and mild operational conditions. The synthesis of "greener" processes for efficient degradation of hazardous pollutants such as industrial colorants and dyes, toxic chemicals, and gases, have become emerging research area now a days. For this purpose, HRP and HRP-based bio and nanobiocatalytic systems are being investigated for tailored environmental applications (Fig. 24–4). The explanation of industrial and biotechnological applications of HRP-based systems have been individually explained in the following sections.

24.8.1 Biosensing system of horseradish peroxidase

Many types of pollutants are present in environment such as the phenolic compound with aromatic amines are known as major organic pollutant, which are generally releasing from the water bodies into the environment (Liang et al., 2020). The phenolic compounds highly present in the effluents of the industries in the form of wastewater, plastics, and paint industries, refining units of petroleum, sector of metal coating, pulp and textile unit and preservation of wood industries with in different range of concentration that may vary from 0.1 to 3900 mg/mL. The great consumption and contribution of phenolic compounds in environments indicates the presence of phenol derivatives in matrices of water. Hence, the pollutant carrying the phenol ring considered as the major environment pollutant and are very harmful for the living organism as well. The very minute concentration in the wastewater can create hindrance to use or reuse (Boruah, Biswas, & Deb, 2019). HRP is class of peroxidases which is highly used in the treatment of toxic dyes and pollutants. Most of the HRP applications aimed for the mitigation and detection of phenol pollutants and various other contaminants related to these compounds, with or without ABTS redox mediators' availability of H_2O_2. The usage of HRP in the contamination treatment in which halogenated phenols are included, such as the 4-bromophenol, 2, 4-dichlorophenol, and phenolic acids for example, benzoquinones, and gallic acid.

Various applications have been planned for the optimization of HRP catalyzed phenol elimination from the water matrices. Modification in the HRP catalytic features, in both engineered and native forms, and cost reduction in the treatment, has made possible by different techniques for example HRP-immobilization on different matrices/carrier for support, HRP-assisted membranes and the HRP-assisted bioreactor design. It was reported that consistency between experimental and theoretical data of peroxidases and halogenated phenol to recognize the potential of bioremediation (Wu, Jiao, Ma, & Peng, 2020) By following the HRP in treatment or catalyst, different efficiency for removal about 59.18% and 93.86% were recorded for the target of halogenated phenols, such as 2—4-dichlorophenol and 4-bromophenol. The results for halogenated phenol reorganized with experimental and computational methods have shown the interactions between HRP and phenols. The biodegradation of phenolic compounds can be enhanced by using nanoencapsulation of HRP. For example, the efficiency for removal of BPA and phenols enhance up to 3.5 and 7 fold, respectively (Alias et al., 2019).

In broad terms the analytical system, used for the detection of various signals from the biological molecules such as the aptamers, cells, enzymes, antibodies, and others, on the surface of electrode is called the biosensor system (Ali, Kharbash, & Kim, 2020). However, various kinds of multiple features of surface layer biosensing electrodes can be seen, such as sensitivity, selectivity, and reproducibility. Until now many different types of biosensing systems by applying different materials and enzymes have been reported for various application in the biotechnological sectors, for example detection of pollutant, heavy metals, bioprocessing, and clinical diagnosis, etc. in the monitoring of environment with reference the biosensor assisted with HRP shows not only specific detection of pollution but also indicates the damaging effects on the biological system such as cytotoxicity and genotoxicity, and thus it helps to monitor the pollutants present in the environment such as the soil, air, and water, having similar importance for all living organism and also for human health (Dai & Liu, 2019). Mostly SiO_2-SPAABs-HRP and GO are used on the antibody detection layer and also in antibody immobilization. For the detection of biomarker and protein molecule brushes loaded with HRP, silica poly-acrylic acid (SiO_2SPAABs) an immunosensor with ultrasensitivity have been developed (Zhao, Qiu, et al., 2016; Zhao, Zheng, et al., 2016). Similarly, there are many types of HRP-based biosensor develop for the advanced detection of heavy metals which is toxic in nature such as the mercury, lead, and cadmium metal ions. Another type of biosensor based on the Pt/PANI-coPDTDA/HRP used in the application of pollutant detection from the environment, for example, lead, cadmium, and mercury (Sall, Diaw, Gningue-Sall, Efremova Aaron, & Aaron, 2020).

24.8.2 Wastewater treatment and biodegradation of dyes

According to different perspectives of environment the wastewater treatment matrices is very difficult and therefore necessary steps should be taken for the current insecurity of environment. The problem of contaminated water has directly linked to streams of wastewater which highly depends on proper regulation, social awareness, fascinating features for treatment, and ecological conditions (Crini & Lichtfouse, 2019). Due to having the complexity in the structure and nature of contaminant present in wastewater, it is very difficult to arrange all mention circumstances. Hence, there is a need to develop a highly efficient treatment system. The contamination of water matrices is also due to the high wastewater contents, for example, leather and textile industry that release their contaminated wastewater which having the full toxic effluents in water bodies (Zhang, Dong, et al., 2019; Zhang, Wells, et al., 2019; Zhang, Yao, et al., 2019). The contamination from the industrial wastes contain the dyes having serious impact on the humans, aquatic life, and all ecosystem. During the washing and dyeing process in the industries about 10%–25% of the dyes lost (Sadh, Duhan, & Duhan, 2018). Furthermore, consumption of various dyes, basic dye about 2% and 50% of reactive type dyes are eluted from the industrial waste which eventually lead to water surface and pollution reached in the ground water. According to estimation about 280,000 tons of the dyes are lost contain toxic material from the worldwide industry throughout the year (Zahan, Othman, & Muster, 2018). Up until now, different process for the treatment of chemical, physical, biological, and physiochemical advancement processes are applied in various industries. Hence, every treatment has some of the disadvantages and advantages with their own mechanism. Significantly, use of detoxification of dyes and biocatalytic degradation based on HRP which is environmentally friendly choice from system of biological treatment (Na & Lee, 2017).

24.8.3 Biotransformation of lignocellulosic wastes

The conversion of nonrenewable sources into the renewable carbon-based organic source is very difficult to maintain the sustainability issues related to the ecological, economic, and environmental issues. For the renewable resources the profitable utilization lignocellulosic biomass showed the good choice for both industrial and public domains for the production of ecofriendly biofuels, functional materials, and other value-added chemicals (Asgher, Urooj, Qamar, & Khalid, 2020; Asgher, Qamar, Bilal, & Iqbal, 2020; Asgher, Arshad, Qamar, & Khalid, 2020). Nonetheless, but the minimum utilization of the lignocellulosic in the industrial scale due to having the complex and difficult nature of lignocellulosic material (Asgher, Rani, Khalid, Qamar, & Bilal, 2021). The lignocellulose biomass is excellent resource for renewable carbon which is used in the production of industrial-scale chemical, functional materials, and biofuels (Asgher, Afzal, Qamar, & Khalid, 2020; Saldarriaga-Hernández et al., 2020). Usually lignocellulosic waste materials related to agriculture are commonly burned in the fields. The burning of lignocellulosic wastes in agriculture is very uneconomical process and produce unhealthy impact on the environment and produce hazardous gases, smoke, and fumes. Some of the reports show the contamination of smoke in environment is due to the burning of crops in nearest regions in recent times, instead of the exploitation of biomass, which is moving toward the replacement of fossil fuels with coupling system of biomass resources (Asgher, Nasir, Khalid, & Qamar, 2020; Bjelic, Hočevar, Grilc, Novak, & Likozar, 2020).

Current focus of research is related to utilization of biomass for the production of different fuels and chemicals that some of them have same properties as petroleum products. Biomass valorization into chemical and fuels is a very active research area in all over the world. When compared with various fossils carbon source, the process of biomass as an increase ratio of H/O and consequently need of less reaction's steps for the production of various chemical with easier artificial method. The upper starting part represents the natural lignin with sinapyl alcohol, pcoumaryl alcohol, and coniferyl alcohol units. The middle part represents the unique action mechanisms of ligninolytic enzymes, that is, laccase, lignin peroxidase, and manganese peroxidase as models. The last part shows various monolignin products that can be obtained after multiple steps involved in the ligninolysis processes.

24.9 Conclusion

In conclusion, immobilized HRP on/in strong nanostructured cues, for example, CNOs, GO, CNTs, rGO, NDs, NFs, MOFs, Fe_3O_4-ND, and MNPs, offers improved and inclusive catalytic function for different multifunctional applications. Thus they indicate the high utilization of these immobilized catalysts in biotechnological and industrial sectors, for example, mitigation and detection of environment pollutants, multi-feature system of biosensing, having a good reproducibility, selectivity, and sensitivity, and various other biocatalysis systems used in industrial sectors. Whenever enzymes such as the HRP are immobilized on nanostructure support/material matrices, their stability and activity are increased due to owing efficient confinement and greater surface area of the enzyme within nanostructure which is biologically engineered. During recent few years, high developments are made in the production of nanostructure carrier or support, but some of the difficulties have not been resolved yet, which need subsequent consideration by the R&D sectors in the future. From the perspective of multifunctional unique properties, the complete mechanism of synthesis and nanostructure coordination with proper support clearly predicted, for example, functional and structural coordination between the nanostructure material and enzyme is not clear yet studied and should be investigate properly for the better performance of enzyme-based biosensors, different enzymes with multiple functions such as the specificity, selectivity, detection, and reproducibility within a wide range for the target various molecules and other substance. Now, multienzymes can address or retain the resistance or/and issue of inhibition with biosensor-enzymatic systems. Furthermore, the cascade system in the multienzyme can be seen in which product of one reaction may start the other one reaction so that why cause the resistance or inhibition by its own products.

Acknowledgment

Generous support from Khalifa University (CIRA-2020-046) is graciously acknowledged. Consejo Nacional de Ciencia y Tecnología (MX) is also thankfully acknowledged for partially supporting this work under Sistema Nacional de Investigadores (SNI) program awarded to Hafiz M.N. Iqbal (CVU: 735340).

Conflicts of Interest

The listed author(s) declared that they have no competing financial interest or personal relationship that could have appeared to influence the work reported in this chapter.

References

Akbarzadeh, E., Soheili, H. Z., Hosseinifard, M., & Gholami, M. R. (2020). Preparation and characterization of novel Ag_3VO_4/Cu-MOF/rGO heterojunction for photocatalytic degradation of organic pollutants. *Materials Research Bulletin, 121*, 110621.

Alarfaj, N. A., El-Tohamy, M. F., & Oraby, H. F. (2018). CA 19-9 pancreatic tumor marker fluorescence immunosensing detection via immobilized carbon quantum dots conjugated gold nanocomposite. *International Journal of Molecular Sciences, 19*(4), 1162.

Aldhahri, M., Almulaiky, Y. Q., El-Shishtawy, R. M., Al-Shawafi, W. M., Salah, N., Alshahrie, A., & Alzahrani, H. A. (2020). Ultra-thin 2D CuO nanosheet for HRP immobilization supported by encapsulation in a polymer matrix: Characterization and dye degradation. *Catalysis Letters, 151*, 232–246.

Ali, A. A., Kharbash, R., & Kim, Y. (2020). Chemo-and biosensing applications of spiropyran and its derivatives—A review. *Analytica Chimica Acta, 1110*, 199–223.

Alias, C., Benassi, L., Bertazzi, L., Sorlini, S., Volta, M., & Gelatti, U. (2019). Environmental exposure and health effects in a highly polluted area of Northern Italy: A narrative review. *Environmental Science and Pollution Research, 26*(5), 4555–4569.

Alimba, C. G., & Faggio, C. (2019). Microplastics in the marine environment: Current trends in environmental pollution and mechanisms of toxicological profile. *Environmental Toxicology and Pharmacology, 68*, 61–74.

Almulaiky, Y. Q., & Al-Harbi, S. A. (2019). A novel peroxidase from Arabian balsam (*Commiphora gileadensis*) stems: Its purification, characterization and immobilization on a carboxymethylcellulose/Fe_3O_4 magnetic hybrid material. *International Journal of Biological Macromolecules, 133*, 767–774.

Almulaiky, Y. Q., El-Shishtawy, R. M., Aldhahri, M., Mohamed, S. A., Afifi, M., Abdulaal, W. H., & Mahyoub, J. A. (2019). Amidrazone modified acrylic fabric activated with cyanuric chloride: A novel and efficient support for horseradish peroxidase immobilization and phenol removal. *International Journal of Biological Macromolecules, 140*, 949–958.

Alshawafi, W. M., Aldhahri, M., Almulaiky, Y. Q., Salah, N., Moselhy, S. S., Ibrahim, I. H., ... Mohamed, S. A. (2018). Immobilization of horseradish peroxidase on PMMA nanofibers incorporated with nanodiamond. *Artificial Cells, Nanomedicine, and Biotechnology, 46*(Suppl. 3), S973–S981.

An, Y., Xing, H., Zhang, Y., Jia, P., Gu, X., & Teng, X. (2019). The evaluation of potential immunotoxicity induced by environmental pollutant ammonia in broilers. *Poultry Science, 98*(8), 3165–3175.

Anna, B. W., Mariusz, T., Hellal, M. S., & Aleksandra, Z. B. (2021). Effect of biomass immobilization and reduced graphene oxide on the microbial community changes and nitrogen removal at low temperatures. *Scientific Reports (Nature Publisher Group), 11*(1).

Asgher, M., Afzal, M., Qamar, S. A., & Khalid, N. (2020). Optimization of biosurfactant production from chemically mutated strain of *Bacillus subtilis* using waste automobile oil as low-cost substrate. *Environmental Sustainability*, 3(4), 405–413.

Asgher, M., Arshad, S., Qamar, S. A., & Khalid, N. (2020). Improved biosurfactant production from *Aspergillus niger* through chemical mutagenesis: Characterization and RSM optimization. *SN Applied Sciences*, 2(5), 1–11.

Asgher, M., Nasir, I., Khalid, N., & Qamar, S. A. (2020). Development of biocomposites based on bacterial cellulose reinforced delignified rice husk-PVA plasticized with glycerol. *Journal of Polymer Research*, 27(11), 1–11.

Asgher, M., Qamar, S. A., Bilal, M., & Iqbal, H. M. (2020). Bio-based active food packaging materials: Sustainable alternative to conventional petrochemical-based packaging materials. *Food Research International*, 137, 109625.

Asgher, M., Rani, A., Khalid, N., Qamar, S. A., & Bilal, M. (2021). Bioconversion of sugarcane molasses waste to high-value exopolysaccharides by engineered *Bacillus licheniformis*. *Case Studies in Chemical and Environmental Engineering*, 3, 100084.

Asgher, M., Urooj, Y., Qamar, S. A., & Khalid, N. (2020). Improved exopolysaccharide production from *Bacillus licheniformis* MS3: Optimization and structural/functional characterization. *International Journal of Biological Macromolecules*, 151, 984–992.

Asif, H. M. (2021). A review on natural antioxidants in foods and new insights on animal body compounds, role, production and future perspectives. *Saudi Journal of Medical and Pharmaceutical Science*, 7(1), 20–25.

Asmat, S., Anwer, A. H., & Husain, Q. (2019). Immobilization of lipase onto novel constructed polydopamine grafted multiwalled carbon nanotube impregnated with magnetic cobalt and its application in synthesis of fruit flavours. *International Journal of Biological Macromolecules*, 140, 484–495.

Bai, Y., Liu, M., Sun, J., & Gao, L. (2016). Fabrication of Ni-Co binary oxide/reduced graphene oxide composite with high capacitance and cyclicity as efficient electrode for supercapacitors. *Ionics*, 22(4), 535–544.

Bai, Y., Xu, T., & Zhang, X. (2020). Graphene-based biosensors for detection of biomarkers. *Micromachines*, 11(1), 60.

Banach-Wiśniewska, A., Tomaszewski, M., Hellal, M. S., & Ziembińska-Buczyńska, A. (2021). Effect of biomass immobilization and reduced graphene oxide on the microbial community changes and nitrogen removal at low temperatures. *Scientific Reports*, 11(1), 1–12.

Basso, A., & Serban, S. (2019). Industrial applications of immobilized enzymes—A review. *Molecular Catalysis*, 479, 110607.

Basu, K., Green, E. M., Cheng, Y., & Craik, C. S. (2019). Why recombinant antibodies—Benefits and applications. *Current Opinion in Biotechnology*, 60, 153–158.

Baumer, J. D., Valério, A., de Souza, S. M. G. U., Erzinger, G. S., Furigo, A., Jr, & de Souza, A. A. U. (2018). Toxicity of enzymatically decolored textile dyes solution by horseradish peroxidase. *Journal of Hazardous Materials*, 360, 82–88.

Bilal, M., Adeel, M., Rasheed, T., Zhao, Y., & Iqbal, H. M. (2019). Emerging contaminants of high concern and their enzyme-assisted biodegradation—A review. *Environment International*, 124, 336–353.

Bilal, M., Asgher, M., Iqbal, H. M., Hu, H., & Zhang, X. (2017). Biotransformation of lignocellulosic materials into value-added products—A review. *International Journal of Biological Macromolecules*, 98, 447–458.

Bilal, M., Asgher, M., Iqbal, M., Hu, H., & Zhang, X. (2016). Chitosan beads immobilized manganese peroxidase catalytic potential for detoxification and decolorization of textile effluent. *International Journal of Biological Macromolecules*, 89, 181–189.

Bilal, M., Barceló, D., & Iqbal, H. M. (2020). Nanostructured materials for harnessing the power of horseradish peroxidase for tailored environmental applications. *Science of the Total Environment*, 749, 142360.

Bilal, M., & Iqbal, H. M. (2019a). Naturally-derived biopolymers: Potential platforms for enzyme immobilization. *International Journal of Biological Macromolecules*, 130, 462–482.

Bilal, M., & Iqbal, H. M. (2019b). Persistence and impact of steroidal estrogens on the environment and their laccase-assisted removal. *Science of the Total Environment, 690,* 447−459.

Bilal, M., Rasheed, T., Iqbal, H. M., Hu, H., Wang, W., & Zhang, X. (2017). Novel characteristics of horseradish peroxidase immobilized onto the polyvinyl alcohol-alginate beads and its methyl orange degradation potential. *International Journal of Biological Macromolecules, 105,* 328−335.

Bilal, M., Rasheed, T., Iqbal, H. M., Hu, H., Wang, W., & Zhang, X. (2018). Horseradish peroxidase immobilization by copolymerization into cross-linked polyacrylamide gel and its dye degradation and detoxification potential. *International Journal of Biological Macromolecules, 113,* 983−990.

Bilal, M., Rasheed, T., Zhao, Y., & Iqbal, H. M. (2019). Agarose-chitosan hydrogel-immobilized horseradish peroxidase with sustainable bio-catalytic and dye degradation properties. *International Journal of Biological Macromolecules, 124,* 742−749.

Bilal, M., Zhao, Y., Rasheed, T., & Iqbal, H. M. (2018). Magnetic nanoparticles as versatile carriers for enzymes immobilization: A review. *International Journal of Biological Macromolecules, 120,* 2530−2544.

Bjelic, A., Hočevar, B., Grilc, M., Novak, U., & Likozar, B. (2020). A review of sustainable lignocellulose biorefining applying (natural) deep eutectic solvents (DESs) for separations, catalysis and enzymatic biotransformation processes. *Reviews in Chemical Engineering,* 1 (ahead-of-print).

Boruah, B. S., Biswas, R., & Deb, P. (2019). A green colorimetric approach towards detection of arsenic (III): A pervasive environmental pollutant. *Optics & Laser Technology, 111,* 825−829.

Carbone, M., Gorton, L., & Antiochia, R. (2015). An overview of the latest graphene-based sensors for glucose detection: The effects of graphene defects. *Electroanalysis, 27*(1), 16−31.

Carraro, F., Williams, J. D., Linares-Moreau, M., Parise, C., Liang, W., Amenitsch, H., ... Falcaro, P. (2020). Continuous-flow synthesis of ZIF-8 biocomposites with tunable particle size. *Angewandte Chemie, 132*(21), 8200−8204.

Cen, Y. K., Liu, Y. X., Xue, Y. P., & Zheng, Y. G. (2019). Immobilization of enzymes in/on membranes and their applications. *Advanced Synthesis & Catalysis, 361*(24), 5500−5515.

Chang, Q., Huang, J., Ding, Y., & Tang, H. (2016). Catalytic oxidation of phenol and 2, 4-dichlorophenol by using horseradish peroxidase immobilized on graphene oxide/Fe_3O_4. *Molecules (Basel, Switzerland), 21*(8), 1044.

Chang, Q., Jiang, G., Tang, H., Li, N., Huang, J., & Wu, L. (2015). Enzymatic removal of chlorophenols using horseradish peroxidase immobilized on superparamagnetic Fe_3O_4/graphene oxide nanocomposite. *Chinese Journal of Catalysis, 36*(7), 961−968.

Courth, K., Binsch, M., Ali, W., Ingenbosch, K., Zorn, H., Hoffmann-Jacobsen, K., ... Opwis, K. (2021). Immobilization of peroxidase on textile carrier materials and their application in the bleaching of colored whey. *Journal of Dairy Science, 104,* 1548−1559.

Crini, G., & Lichtfouse, E. (2019). Advantages and disadvantages of techniques used for wastewater treatment. *Environmental Chemistry Letters, 17*(1), 145−155.

Dai, Y., & Liu, C. C. (2019). Recent advances on electrochemical biosensing strategies toward universal point-of-care systems. *Angewandte Chemie, 131*(36), 12483−12496.

Dalkıran, B. (2020). Amperometric determination of heavy metal using an HRP inhibition biosensor based on ITO nanoparticles-ruthenium (III) hexamine trichloride composite: Central composite design optimization. *Bioelectrochemistry (Amsterdam, Netherlands), 135,* 107569.

De Witte, T. M., Wagner, A. M., Fratila-Apachitei, L. E., Zadpoor, A. A., & Peppas, N. A. (2020). Immobilization of nanocarriers within a porous chitosan scaffold for the sustained delivery of growth factors in bone tissue engineering applications. *Journal of Biomedical Materials Research. Part A, 108*(5), 1122−1135.

Dose, G., Roccella, S., Richou, M., Gallay, F., Visca, E., Greuner, H., ... You, J. H. (2019). Ultrasonic analysis of tungsten monoblock divertor mock-ups after high heat flux test. *Fusion Engineering and Design, 146,* 870−873.

Dzhakasheva, M. A., & Lieberzeit, P. A. (2017). Immobilization of an exoenzyme complex to a polyvinyl alcohol cryogel. *Eurasian Journal of Applied Biotechnology* (1), 59–67.

Eldin, M. M. (2016). Enzyme immobilization: Nanopolymers for enzyme immobilization applications. *Energy, 16*, 18.

Elsayed, M. H., Jayakumar, J., Abdellah, M., Mansoure, T. H., Zheng, K., Elewa, A. M., ... Chou, H. H. (2021). Visible-light-driven hydrogen evolution using nitrogen-doped carbon quantum dot-implanted polymer dots as metal-free photocatalysts. *Applied Catalysis B: Environmental, 283*, 119659.

Erol, K., Cebeci, B., Köse, K., & Köse, D. A. (2017). *Investigation of immobilization and kinetic properties of catalase enzyme on poly (HEMA-GMA) cryogel.*

Esmi, F., Nematian, T., Salehi, Z., Khodadadi, A. A., & Dalai, A. K. (2021). Amine and aldehyde functionalized mesoporous silica on magnetic nanoparticles for enhanced lipase immobilization, biodiesel production, and facile separation. *Fuel, 291*, 120126.

Feng, J., Yu, S., Li, J., Mo, T., & Li, P. (2016). Enhancement of the catalytic activity and stability of immobilized aminoacylase using modified magnetic Fe_3O_4 nanoparticles. *Chemical Engineering Journal, 286*, 216–222.

Frey, R., Hayashi, T., & Buller, R. M. (2019). Directed evolution of carbon–hydrogen bond activating enzymes. *Current Opinion in Biotechnology, 60*, 29–38.

Ge, L., Zhang, M., Wang, R., Li, N., Zhang, L., Liu, S., & Jiao, T. (2020). Fabrication of CS/GA/RGO/Pd composite hydrogels for highly efficient catalytic reduction of organic pollutants. *RSC Advances, 10*(26), 15091–15097.

Georgiev, G. L., Trzebicka, B., Kostova, B., & Petrov, P. D. (2017). Super-macroporous dextran cryogels via UV-induced crosslinking: Synthesis and characterization. *Polymer International, 66*(9), 1306–1311.

Gößl, D., Singer, H., Chiu, H. Y., Schmidt, A., Lichtnecker, M., Engelke, H., & Bein, T. (2019). Highly active enzymes immobilized in large pore colloidal mesoporous silica nanoparticles. *New Journal of Chemistry, 43*(4), 1671–1680.

Goud, K. Y., Kumar, V. S., Hayat, A., Gobi, K. V., Song, H., Kim, K. H., & Marty, J. L. (2019). A highly sensitive electrochemical immunosensor for zearalenone using screen-printed disposable electrodes. *Journal of Electroanalytical Chemistry, 832*, 336–342.

Gui, B., Meng, Y., Xie, Y., Du, K., Sue, A. C. H., & Wang, C. (2018). Immobilizing organic-based molecular switches into metal–organic frameworks: A promising strategy for switching in solid state. *Macromolecular Rapid Communications, 39*(1), 1700388.

Haida, Z., & Hakiman, M. (2019). A comprehensive review on the determination of enzymatic assay and nonenzymatic antioxidant activities. *Food Science & Nutrition, 7*, 1555–1563.

Hajian, R., Balderston, S., Tran, T., DeBoer, T., Etienne, J., Sandhu, M., ... Aran, K. (2019). Detection of unamplified target genes via CRISPR–Cas9 immobilized on a graphene field-effect transistor. *Nature Biomedical Engineering, 3*(6), 427–437.

Han, J., Cai, Y., Wang, Y., Gu, L., Li, C., Mao, Y., ... Ni, L. (2019). Synergetic effect of Ni^{2+} and 5-acrylamidobenzoboroxole functional groups anchoring on magnetic nanoparticles for enhanced immobilization of horseradish peroxidase. *Enzyme and Microbial Technology, 120*, 136–143.

He, L., Li, Y., Wu, Q., Wang, D. M., Li, C. M., Huang, C. Z., & Li, Y. F. (2019). Ru (III)-based metal–organic gels: Intrinsic horseradish and NADH peroxidase-mimicking nanozyme. *ACS Applied Materials & Interfaces, 11*, 29158–29166.

Huang, W. C., Wang, W., Xue, C., & Mao, X. (2018). Effective enzyme immobilization onto a magnetic chitin nanofiber composite. *ACS Sustainable Chemistry & Engineering, 6*(7), 8118–8124.

Hwang, M. T., Heiranian, M., Kim, Y., You, S., Leem, J., Taqieddin, A., ... Bashir, R. (2020). Ultrasensitive detection of nucleic acids using deformed graphene channel field effect biosensors. *Nature Communications, 11*(1), 1–11.

Jacobowitz, J. R., Doyle, W. C., & Weng, J.-K. (2019). PRX9 and PRX40 are extensin peroxidases essential for maintaining tapetum and microspore cell wall integrity during *Arabidopsis* anther development. *The Plant Cell*, *31*, 848–861.

Jia, H., Wang, Z., Yuan, T., Yuan, F., Li, X., Li, Y., . . . Yang, S. (2019). Electroluminescent warm white light-emitting diodes based on passivation enabled bright red bandgap emission carbon quantum dots. *Advanced Science*, *6*(13), 1900397.

Jiang, D., Ni, D., Rosenkrans, Z. T., Huang, P., Yan, X., & Cai, W. (2019). Nanozyme: New horizons for responsive biomedical applications. *Chemical Society Reviews*, *48*, 3683–3704.

Jiang, H. L., Kim, Y. K., Arote, R., Jere, D., Quan, J. S., Yu, J. H., . . . Cho, C. S. (2009). Mannosylated chitosan-graft-polyethylenimine as a gene carrier for Raw 264.7 cell targeting. *International Journal of Pharmaceutics*, *375*(1–2), 133–139.

Kamble, A., Srinivasan, S., & Singh, H. (2019). In-silico bioprospecting: Finding better enzymes. *Molecular Biotechnology*, *61*, 53–59.

Kashefi, S., Borghei, S. M., & Mahmoodi, N. M. (2019). Covalently immobilized laccase onto graphene oxide nanosheets: Preparation, characterization, and biodegradation of azo dyes in colored wastewater. *Journal of Molecular Liquids*, *276*, 153–162.

Kazenwadel, F., Franzreb, M., & Rapp, B. E. (2015). Synthetic enzyme supercomplexes: Co-immobilization of enzyme cascades. *Analytical Methods*, *7*(10), 4030–4037.

Kelliher, T., Starr, D., Su, X., Tang, G., Chen, Z., Carter, J., . . . Que, Q. (2019). One-step genome editing of elite crop germplasm during haploid induction. *Nature Biotechnology*, *37*(3), 287–292.

Kergaravat, S. V., Fabiano, S. N., Soutullo, A. R., & Hernández, S. R. (2021). Comparison of the performance analytical of two glyphosate electrochemical screening methods based on peroxidase enzyme inhibition. *Microchemical Journal*, *160*, 105654.

Kim, K. H., Choi, I. S., Kim, H. M., Wi, S. G., & Bae, H. J. (2014). Bioethanol production from the nutrient stress-induced microalga *Chlorella vulgaris* by enzymatic hydrolysis and immobilized yeast fermentation. *Bioresource Technology*, *153*, 47–54.

Kim, S. W., Kwon, S., & Kim, Y. K. (2021). Graphene oxide derivatives and their nanohybrid structures for laser desorption/ionization time-of-flight mass spectrometry analysis of small molecules. *Nanomaterials*, *11*(2), 288.

Krainer, F. W., & Glieder, A. (2015). An updated view on horseradish peroxidases: Recombinant production and biotechnological applications. *Applied Microbiology and Biotechnology*, *99*(4), 1611–1625.

Kumar, A., Park, G. D., Patel, S. K., Kondaveeti, S., Otari, S., Anwar, M. Z., . . . Lee, J. K. (2019). SiO_2 microparticles with carbon nanotube-derived mesopores as an efficient support for enzyme immobilization. *Chemical Engineering Journal*, *359*, 1252–1264.

Lee, S. X., Lim, H. N., Ibrahim, I., Jamil, A., Pandikumar, A., & Huang, N. M. (2017). Horseradish peroxidase-labeled silver/reduced graphene oxide thin film-modified screen-printed electrode for detection of carcinoembryonic antigen. *Biosensors and Bioelectronics*, *89*, 673–680.

Li, J., Chen, X., Xu, D., & Pan, K. (2019). Immobilization of horseradish peroxidase on electrospun magnetic nanofibers for phenol removal. *Ecotoxicology and Environmental Safety*, *170*, 716–721.

Li, Z. L., Cheng, L., Zhang, L. W., Liu, W., Ma, W. Q., & Liu, L. (2017). Preparation of a novel multi-walled-carbon-nanotube/cordierite composite support and its immobilization effect on horseradish peroxidase. *Process Safety and Environmental Protection*, *107*, 463–467.

Liang, J., Zulkifli, M. Y., Choy, S., Li, Y., Gao, M., Kong, B., . . . Liang, K. (2020). Metal–organic framework–plant nanobiohybrids as living sensors for on-site environmental pollutant detection. *Environmental Science & Technology*, *54*(18), 11356–11364.

Lin, L. S., Huang, T., Song, J., Ou, X. Y., Wang, Z., Deng, H., . . . Chen, X. (2019). Synthesis of copper peroxide nanodots for H_2O_2 self-supplying chemodynamic therapy. *Journal of the American Chemical Society*, *141*(25), 9937–9945.

Lin, S., Lin, C., He, M., Yuan, R., Zhang, Y., Zhou, Y., ... Liang, X. (2017). Solvatochromism of bright carbon dots with tunable long-wavelength emission from green to red and their application as solid-state materials for warm WLEDs. *RSC Advances, 7*(66), 41552−41560.

Liu, C., Xiao, N., Li, H., Dong, Q., Wang, Y., Li, H., ... Qiu, J. (2020). Nitrogen-doped soft carbon frameworks built of well-interconnected nanocapsules enabling a superior potassium-ion batteries anode. *Chemical Engineering Journal, 382*, 121759.

Liu, M., Xiao, X., Zhao, S., Saremi-Yarahmadi, S., Chen, M., Zheng, J., ... Chen, L. (2019). ZIF-67 derived Co@ CNTs nanoparticles: Remarkably improved hydrogen storage properties of MgH_2 and synergetic catalysis mechanism. *International Journal of Hydrogen Energy, 44*(2), 1059−1069.

Loibner, M., Hagauer, S., Schwantzer, G., Berghold, A., & Zatloukal, K. (2019). Limiting factors for wearing personal protective equipment (PPE) in a health care environment evaluated in a randomised study. *PLoS One, 14*(1), e0210775.

Luo, D., Yang, W., Wang, Z., Sadhanala, A., Hu, Q., Su, R., ... Zhu, R. (2018). Enhanced photovoltage for inverted planar heterojunction perovskite solar cells. *Science (New York, N.Y.), 360*(6396), 1442−1446.

MacNevin, C. J., Watanabe, T., Weitzman, M., Gulyani, A., Fuehrer, S., Pinkin, N. K., ... Hahn, K. M. (2019). Membrane-permeant, environment-sensitive dyes generate biosensors within living cells. *Journal of the American Chemical Society, 141*(18), 7275−7282.

Madec, M., Hébrard, L., Kammerer, J. B., Bonament, A., Rosati, E., & Lallement, C. (2019). Multiphysics simulation of biosensors involving 3D biological reaction−diffusion phenomena in a standard circuit EDA environment. *IEEE Transactions on Circuits and Systems I: Regular Papers, 66*(6), 2188−2197.

Magesa, F., Wu, Y., Tian, Y., Vianney, J. M., Buza, J., He, Q., & Tan, Y. (2019). Graphene and graphene like 2D graphitic carbon nitride: Electrochemical detection of food colorants and toxic substances in environment. *Trends in Environmental Analytical Chemistry, 23*, e00064.

Malekmohammadi, S., Hadadzadeh, H., Farrokhpour, H., & Amirghofran, Z. (2018). Immobilization of gold nanoparticles on folate-conjugated dendritic mesoporous silica-coated reduced graphene oxide nanosheets: A new nanoplatform for curcumin pH-controlled and targeted delivery. *Soft Matter, 14*(12), 2400−2410.

Matsubara, T., & Takemura, K. (2020). Containerless bioorganic reactions in a floating droplet by levitation technique using an ultrasonic wave. *Advanced Science, 8*, 2002780.

Melo, M. N., Pereira, F. M., Rocha, M. A., Ribeiro, J. G., Diz, F. M., Monteiro, W. F., ... Fricks, A. T. (2020). Immobilization and characterization of horseradish peroxidase into chitosan and chitosan/PEG nanoparticles: A comparative study. *Process Biochemistry, 98*, 160−171.

Mohamad, N. R., Marzuki, N. H. C., Buang, N. A., Huyop, F., & Wahab, R. A. (2015). An overview of technologies for immobilization of enzymes and surface analysis techniques for immobilized enzymes. *Biotechnology & Biotechnological Equipment, 29*(2), 205−220.

Mohamed, S. A., Al-Harbi, M. H., Almulaiky, Y. Q., Ibrahim, I. H., & El-Shishtawy, R. M. (2017). Immobilization of horseradish peroxidase on Fe_3O_4 magnetic nanoparticles. *Electronic Journal of Biotechnology, 27*, 84−90.

Mohammadian, N., & Faridbod, F. (2018). ALS genosensing using DNA-hybridization electrochemical biosensor based on label-free immobilization of ssDNA on Sm_2O_3 NPs-rGO/PANI composite. *Sensors and Actuators B: Chemical, 275*, 432−438.

Moyo, M., Chikazaza, L., Nyamunda, B. C., & Guyo, U. (2013). Adsorption batch studies on the removal of Pb (II) using maize tassel based activated carbon. *Journal of Chemistry, 2013*.

Na, S. Y., & Lee, Y. (2017). Elimination of trace organic contaminants during enhanced wastewater treatment with horseradish peroxidase/hydrogen peroxide (HRP/H_2O_2) catalytic process. *Catalysis Today, 282*, 86−94.

Nicolini, J. V., Ferraz, H. C., & de Resende, N. S. (2016). Immobilization of horseradish peroxidase on titanate nanowires for biosensing application. *Journal of Applied Electrochemistry, 46*(1), 17−25.

Ogunsona, E. O., Muthuraj, R., Ojogbo, E., Valerio, O., & Mekonnen, T. H. (2020). Engineered nanomaterials for antimicrobial applications: A review. *Applied Materials Today, 18*, 100473.

Oh, E., Liu, R., Nel, A., Gemill, K. B., Bilal, M., Cohen, Y., & Medintz, I. L. (2016). Meta-analysis of cellular toxicity for cadmium-containing quantum dots. *Nature Nanotechnology, 11*(5), 479.

Oktay, B., Demir, S., & Kayaman-Apohan, N. (2020). Immobilization of pectinase on polyethyleneimine based support via spontaneous amino-yne click reaction. *Food and Bioproducts Processing, 122*, 159–168.

Paquette, S., & Lin, J. C. (2019). Outpatient telemedicine program in vascular surgery reduces patient travel time, cost, and environmental pollutant emissions. *Annals of Vascular Surgery, 59*, 167–172.

Pitzalis, F., Monduzzi, M., & Salis, A. (2017). A bienzymatic biocatalyst constituted by glucose oxidase and Horseradish peroxidase immobilized on ordered mesoporous silica. *Microporous and Mesoporous Materials, 241*, 145–154.

Qamar, S. A., Asgher, M., & Bilal, M. (2020). Immobilization of alkaline protease from *Bacillus brevis* using Ca-alginate entrapment strategy for improved catalytic stability, silver recovery, and dehairing potentialities. *Catalysis Letters, 150*, 3572–3583.

Qamar, S. A., Ashiq, M., Jahangeer, M., Riasat, A., & Bilal, M. (2020). Chitosan-based hybrid materials as adsorbents for textile dyes–A review. *Case Studies in Chemical and Environmental Engineering, 2*, 100021.

Radhakrishnan, S., & Kim, S. J. (2015). An enzymatic biosensor for hydrogen peroxide based on one-pot preparation of CeO_2-reduced graphene oxide nanocomposite. *RSC Advances, 5*(17), 12937–12943.

Rehman, S., Bhatti, H. N., Bilal, M., & Asgher, M. (2016). Cross-linked enzyme aggregates (CLEAs) of *Pencilluim notatum* lipase enzyme with improved activity, stability and reusability characteristics. *International Journal of Biological Macromolecules, 91*, 1161–1169.

Ricco, A. J., Santa Maria, S. R., Hanel, R. P., & Bhattacharya, S. (2020). BioSentinel: A 6U nanosatellite for deep-space biological science. *IEEE Aerospace and Electronic Systems Magazine, 35*(3), 6–18.

Roccella, S., Dose, G., Barrett, T., Cacciotti, E., Dupont, L., Gallay, F., ... You, J. H. (2020). Ultrasonic test results before and after high heat flux testing on W-monoblock mock-ups of EU-DEMO vertical target. *Fusion Engineering and Design, 160*, 111886.

Rong, J., Zhou, Z., Wang, Y., Han, J., Li, C., Zhang, W., & Ni, L. (2019). Immobilization of horseradish peroxidase on multi-armed magnetic graphene oxide composite: Improvement of loading amount and catalytic activity. *Food Technology and Biotechnology, 57*(2), 260–271.

Sadh, P. K., Duhan, S., & Duhan, J. S. (2018). Agro-industrial wastes and their utilization using solid state fermentation: A review. *Bioresources and Bioprocessing, 5*(1), 1–15.

Safarik, I., Pospiskova, K., Baldikova, E., Savva, I., Vekas, L., Marinica, O., ... Krasia-Christoforou, T. (2018). Fabrication and bioapplications of magnetically modified chitosan-based electrospun nanofibers. *Electrospinning, 2*(1), 29–39.

Şahin, S. (2020). A simple and sensitive hydrogen peroxide detection with horseradish peroxidase immobilized on pyrene modified acid-treated single-walled carbon nanotubes. *Journal of Chemical Technology & Biotechnology, 95*(4), 1093–1099.

Sahu, M., & Raichur, A. M. (2019). Toughening of high performance tetrafunctional epoxy with poly (allyl amine) grafted graphene oxide. *Composites Part B: Engineering, 168*, 15–24.

Saldarriaga-Hernández, S., Velasco-Ayala, C., Flores, P. L. I., de Jesús Rostro-Alanis, M., Parra-Saldívar, R., Iqbal, H. M., & Carrillo-Nieves, D. (2020). Biotransformation of lignocellulosic biomass into industrially relevant products with the aid of fungi-derived lignocellulolytic enzymes. *International Journal of Biological Macromolecules, 161*, 1099–1116.

Sall, M. L., Diaw, A. K. D., Gningue-Sall, D., Efremova Aaron, S., & Aaron, J. J. (2020). Toxic heavy metals: Impact on the environment and human health, and treatment with conducting organic polymers, a review. *Environmental Science and Pollution Research, 27*, 29927–29942.

Savun, B., & Gineste, C. (2019). From protection to persecution: Threat environment and refugee scapegoating. *Journal of Peace Research, 56*(1), 88–102.

Sengel, S. B., Sahiner, M., Aktas, N., & Sahiner, N. (2017). Halloysite-carboxymethyl cellulose cryogel composite from natural sources. *Applied Clay Science, 140*, 66–74.

Shi, M.-Y., Xu, M., & Gu, Z.-Y. (2019). Copper-based two-dimensional metal-organic framework nanosheets as horseradish peroxidase mimics for glucose fluorescence sensing. *Analytica Chimica Acta, 1079*, 164–170.

Sneha, H. P., Beulah, K. C., & Murthy, P. S. (2019). *Enzyme immobilization methods and applications in the food industry. Enzymes in food biotechnology* (pp. 645–658). Academic Press.

Sohail, M. (2021). Effects of different types of microbes on blood cells, current perspectives and future directions. *Saudi Journal of Medical and Pharmaceutical Sciences, 7*(1), 1–6.

Sok, V., & Fragoso, A. (2018). Preparation and characterization of alkaline phosphatase, horseradish peroxidase, and glucose oxidase conjugates with carboxylated carbon nano-onions. *Preparative Biochemistry and Biotechnology, 48*(2), 136–143.

Stoyanovsky, D. A., Tyurina, Y. Y., Shrivastava, I., Bahar, I., Tyurin, V. A., Protchenko, O., . . . Kagan, V. E. (2019). Iron catalysis of lipid peroxidation in ferroptosis: Regulated enzymatic or random free radical reaction? *Free Radical Biology and Medicine, 133*, 153–161.

Su, Y., Zhou, X., Long, Y., & Li, W. (2018). Immobilization of horseradish peroxidase on amino-functionalized carbon dots for the sensitive detection of hydrogen peroxide. *Microchimica Acta, 185*(2), 1–8.

Tao, E., Ma, D., Yang, S., & Hao, X. (2020). Graphene oxide-montmorillonite/sodium alginate aerogel beads for selective adsorption of methylene blue in wastewater. *Journal of Alloys and Compounds, 832*, 154833.

Temoçin, Z., Inal, M., Gökgöz, M., & Yiğitoğlu, M. (2018). Immobilization of horseradish peroxidase on electrospun poly (vinyl alcohol)−polyacrylamide blend nanofiber membrane and its use in the conversion of phenol. *Polymer Bulletin, 75*(5), 1843–1865.

Valldeperas, M., Salis, A., Barauskas, J., Tiberg, F., Arnebrant, T., Razumas, V., . . . Nylander, T. (2019). Enzyme encapsulation in nanostructured self-assembled structures: Toward biofunctional supramolecular assemblies. *Current Opinion in Colloid & Interface Science, 44*, 130–142.

Verma, M. L., Kumar, S., Das, A., Randhawa, J. S., & Chamundeeswari, M. (2020). Chitin and chitosan-based support materials for enzyme immobilization and biotechnological applications. *Environmental Chemistry Letters, 18*(2), 315–323.

Vilian, A. E., & Chen, S. M. (2014). Simple approach for the immobilization of horseradish peroxidase on poly-L-histidine modified reduced graphene oxide for amperometric determination of dopamine and H_2O_2. *RSC Advances, 4*(99), 55867–55876.

Vilian, A. E., Choe, S. R., Giribabu, K., Jang, S. C., Roh, C., Huh, Y. S., & Han, Y. K. (2017). Pd nanospheres decorated reduced graphene oxide with multi-functions: Highly efficient catalytic reduction and ultrasensitive sensing of hazardous 4-nitrophenol pollutant. *Journal of Hazardous Materials, 333*, 54–62.

Vineh, M. B., Saboury, A. A., Poostchi, A. A., Rashidi, A. M., & Parivar, K. (2018). Stability and activity improvement of horseradish peroxidase by covalent immobilization on functionalized reduced graphene oxide and biodegradation of high phenol concentration. *International Journal of Biological Macromolecules, 106*, 1314–1322.

Waifalkar, P. P., Chougale, A. D., Kollu, P., Patil, P. S., & Patil, P. B. (2018). Magnetic nanoparticle decorated graphene based electrochemical nanobiosensor for H_2O_2 sensing using HRP. *Colloids and Surfaces B: Biointerfaces, 167*, 425–431.

Wang, J., Xiong, X., Takahashi, E., Zhang, L., Li, L., & Liu, X. (2019). Oxidation state of arc mantle revealed by partitioning of V, Sc, and Ti between mantle minerals and basaltic melts. *Journal of Geophysical Research: Solid Earth, 124*(5), 4617–4638.

Wu, Q., Jiao, S., Ma, M., & Peng, S. (2020). Microbial fuel cell system: A promising technology for pollutant removal and environmental remediation. *Environmental Science and Pollution Research, 27*(7), 6749–6764.

Xie, C., Lu, X., Han, L., Xu, J., Wang, Z., Jiang, L., . . . Tang, Y. (2016). Biomimetic mineralized hierarchical graphene oxide/chitosan scaffolds with adsorbability for immobilization of nanoparticles for biomedical applications. *ACS Applied Materials & Interfaces, 8*(3), 1707–1717.

Xie, X., Luo, P., Han, J., Chen, T., Wang, Y., Cai, Y., & Liu, Q. (2019). Horseradish peroxidase immobilized on the magnetic composite microspheres for high catalytic ability and operational stability. *Enzyme and Microbial Technology, 122*, 26–35.

Xu, Y., Shi, X., Hua, R., Zhang, R., Yao, Y., Zhao, B., . . . Lu, G. (2020). Remarkably catalytic activity in reduction of 4-nitrophenol and methylene blue by Fe_3O_4@ COF supported noble metal nanoparticles. *Applied Catalysis B: Environmental, 260*, 118142.

Yi, S., Dai, F., Zhao, C., & Si, Y. (2017). A reverse micelle strategy for fabricating magnetic lipase-immobilized nanoparticles with robust enzymatic activity. *Scientific Reports, 7*(1), 1–9.

Yu, B., Cheng, H., Zhuang, W., Zhu, C., Wu, J., Niu, H., . . . Ying, H. (2019). Stability and repeatability improvement of horseradish peroxidase by immobilization on amino-functionalized bacterial cellulose. *Process Biochemistry, 79*, 40–48.

Zahan, Z., Othman, M. Z., & Muster, T. H. (2018). Anaerobic digestion/co-digestion kinetic potentials of different agro-industrial wastes: A comparative batch study for C/N optimisation. *Waste Management, 71*, 663–674.

Zeng, D., Zhou, T., Ong, W. J., Wu, M., Duan, X., Xu, W., . . . Peng, D. L. (2019). Sub-5 nm ultra-fine FeP nanodots as efficient co-catalysts modified porous $g-C_3N_4$ for precious-metal-free photocatalytic hydrogen evolution under visible light. *ACS Applied Materials & Interfaces, 11*(6), 5651–5660.

Zeng, K., Wei, D., Zhang, Z., Meng, H., Huang, Z., & Zhang, X. (2019). Enhanced competitive immunomagnetic beads assay with gold nanoparticles and carbon nanotube-assisted multiple enzyme probes. *Sensors and Actuators B: Chemical, 292*, 196–202.

Zhai, C., Sun, X., Zhao, W., Gong, Z., & Wang, X. (2013). Acetylcholinesterase biosensor based on chitosan/prussian blue/multiwall carbon nanotubes/hollow gold nanospheres nanocomposite film by one-stepelectrodeposition. *Biosensors and Bioelectronics, 42*, 124–130.

Zhan, J., Liang, Y., Liu, D., Ma, X., Li, P., Zhai, W., . . . Wang, P. (2019). Pectin reduces environmental pollutant-induced obesity in mice through regulating gut microbiota: A case study of p, p′-DDE. *Environment International, 130*, 104861.

Zhang, J., Lü, F., Shao, L., & He, P. (2014). The use of biochar-amended composting to improve the humification and degradation of sewage sludge. *Bioresource Technology, 168*, 252–258.

Zhang, M. H., Dong, H., Zhao, L., Wang, D. X., & Meng, D. (2019). A review on Fenton process for organic wastewater treatment based on optimization perspective. *Science of the Total Environment, 670*, 110–121.

Zhang, S., Yao, L., Sun, A., & Tay, Y. (2019). Deep learning based recommender system: A survey and new perspectives. *ACM Computing Surveys (CSUR), 52*(1), 1–38.

Zhang, T., Xu, X. L., Jin, Y. N., Wu, J., & Xu, Z. K. (2014). Immobilization of horseradish peroxidase (HRP) on polyimide nanofibers blending with carbon nanotubes. *Journal of Molecular Catalysis B: Enzymatic, 106*, 56–62.

Zhang, W. C., Wells, J. M., Chow, K. H., Huang, H., Yuan, M., Saxena, T., . . . Slack, F. J. (2019). miR-147b-mediated TCA cycle dysfunction and pseudohypoxia initiate drug tolerance to EGFR inhibitors in lung adenocarcinoma. *Nature Metabolism, 1*(4), 460–474.

Zhao, L., Qiu, G., Anderson, C. W., Meng, B., Wang, D., Shang, L., & Feng, X. (2016). Mercury methylation in rice paddies and its possible controlling factors in the Hg mining area, Guizhou province, Southwest China. *Environmental Pollution, 215*, 1–9.

Zhao, Y., Zheng, Y., Kong, R., Xia, L., & Qu, F. (2016). Ultrasensitive electrochemical immunosensor based on horseradish peroxidase (HRP)-loaded silica-poly (acrylic acid) brushes for protein biomarker detection. *Biosensors and Bioelectronics*, *75*, 383–388.

Zhu, A., Guo, Y., Liu, G., Song, M., Liang, Y., Cai, Y., & Yin, Y. (2019). Hydroxyl radical formation upon dark oxidation of reduced iron minerals: Effects of iron species and environmental factors. *Chinese Chemical Letters*, *30*(12), 2241–2244.

Zou, B., Pu, Q., Bilal, M., Weng, Q., Zhai, L., & Nichol, J. E. (2016). High-resolution satellite mapping of fine particulates based on geographically weighted regression. *IEEE Geoscience and Remote Sensing Letters*, *13*(4), 495–499.

25

Nanobiodegradation of pharmaceutical pollutants

Tahir Rasheed[1], Komal Rizwan[2], Sameera Shafi[3], Muhammad Bilal[4]

[1]SCHOOL OF CHEMISTRY AND CHEMICAL ENGINEERING, SHANGHAI JIAO TONG UNIVERSITY, SHANGHAI, P.R. CHINA [2]DEPARTMENT OF CHEMISTRY, UNIVERSITY OF SAHIWAL, SAHIWAL, PAKISTAN [3]INSTITUTE OF CHEMISTRY, THE ISLAMIA UNIVERSITY OF BAHAWALPUR, BAHAWALNAGAR CAMPUS, BAHAWALNAGAR, PAKISTAN [4]SCHOOL OF LIFE SCIENCE AND FOOD ENGINEERING, HUAIYIN INSTITUTE OF TECHNOLOGY, HUAI'AN, P.R. CHINA

25.1 Introduction

Great attention has been drawn for the last 15 years toward pharmaceuticals because of their bioactive chemical role in the environment (Kümmerer, 2009). They are acting as a major pollutant in water bodies because their regularization is under process to some extent or not regulated yet because the authorized frameworks are under process. Continuous introduction of pharmaceuticals affecting the environment and spreading in particular areas with their small but specific amount (Kolpin et al., 2002), which can cause water pollution by spoiling the quality of water, which ultimately affects human health (Sirés & Brillas, 2012; Yuan, Hu, Hu, Qu, & Yang, 2009) and the ecosystem due to bad impact of drinking water. It can be observed in Fig. 25–1 that traces of antibiotics can penetrate and affect the surface of ground and water, respectively.

Various reasons are acting collectively for soil and water body contamination including the type of antibiotics and amount, time for contact, perseverance period, a place for their removal, and deposit (it can be in the air, water, or soil). While there are many other significant routes for the contamination of the environment, which include the cooccurrences of a group of antibiotics with different biologically active pollutants, it has been observed that in the previous two decades, a continuous chain of antibiotics remains at the microlevel causing water and soil pollution. Their continuous and uncontrolled introduction to the environment has become a major threat to the ecosystem (Boxall, 2004; Carvalho & Santos, 2016). Due to the severity of antibiotic traces into the water and soil bodies, this issue should be addressed in environmental science and engineering to overcome the pollution. For this purpose, pharmaceuticals and their multiple antibodies should be removed effectively from the wastewater right before discharging them. There must be some major efforts on a priority

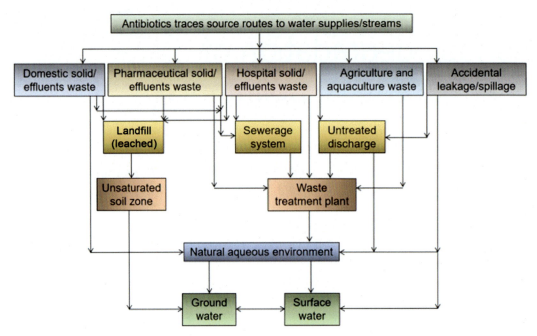

FIGURE 25–1 Possible routes of pharmaceutical compounds and antibiotics traces contamination to surface and ground water bodies (Bilal, Ashraf, Barceló, & Iqbal, 2019).

basis to overcome this issue (Daughton & Ternes, 1999; Kemper, 2008; Khetan & Collins, 2007; Kümmerer, 2003).

25.2 Environmental and ecological risks

Some specific pharmaceuticals have been designed to attach the targeted animal and human receptors. These compounds could interact with the same receptor in a different organism. A huge number of organisms from the lower phylum contain receptors the same as in the upper phylum. Prokaryotic cells that are possibly affected by antibiotics in a particular mechanism, such as synthesis nucleic acid, cell envelope and inhibition of protein synthesis play a significant role. These compounds are specifically prepared for the treatment of bacterial infection in the animal as well as in humans (Fig. 25–2) (Bilal, Mehmood, Rasheed, & Iqbal, 2020).

So, these molecules are prepared to limit or secure mammalian cells. Nontherapeutic exposure of those particular molecules is supposed to be most unfavorable affect the microorganisms of the environment than aquatic vertebrates as in fish (Le Page, Gunnarsson, Snape, & Tyler, 2017). Therefore a large number of organisms are in danger due to the toxicity of pharmaceutical compounds. This toxic effect has been observed in both micro and macroorganisms. Furthermore, a community of microbes, antibiotic-resistant bacteria,

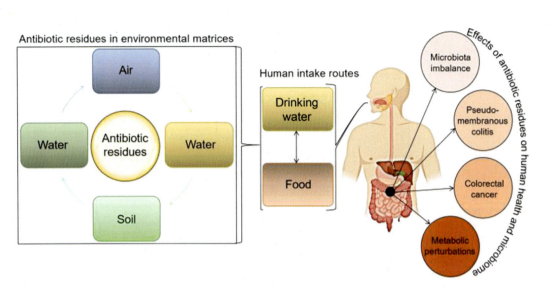

FIGURE 25–2 Adverse effects of antibiotics trace in different environmental media. *Adopted from Bilal, M., Mehmood, S., Rasheed, T., & Iqbal, H. M. N. (2020). Antibiotics traces in the aquatic environment: Persistence and adverse environmental impact. Current Opinion in Environmental Science & Health, 13, 68–74. https://doi.org/10.1016/j.coesh.2019.11.005, with permission from Elsevier. Copyright (2019) Elsevier B.V.*

altered gene expression, protein abnormalities, working of enzymes, and malformation growth are observed in rats, fish, and frogs (Johnning et al., 2013; Li et al., 2009, 2010; Marathe et al., 2013; Rutgersson et al., 2014; Zhang et al., 2011).

Antibiotics such as *trimethoprim* and *oxytetracycline* bring toxic effects in cyanobacteria *Anabaena flosaque*, green alga *Pseudokirchneriella subcapitata*, and *Daphnia magna* (Kolar et al., 2014). The population of insects in dung and aquatic invertebrates are affected by ivermectin, whereas male fish face a feminizing effect caused by ethinylestradiol (Gross-Sorokin, Roast, & Brighty, 2006; Suarez, Lifschitz, Sallovitz, & Lanusse, 2003). Injurious indirect toxic effects of pharmaceuticals are also observed by means of food chains on the various organisms. For example, kidney failure and visceral gout are serious diseases caused by diclofenac, which leads to high mortality of about 5%–86% on the Asian white-backed vulture in the subcontinent (Indian) (Oaks et al., 2004). Diclofenac bioaccumulation in vulture eats dead animals infected with diclofenac. Gammaruspulex, shrimp found in freshwater across Europe, show a free feeding priority for leaves. These leaves were not affected by oxytetracycline and sulfadiazine (Hahn & Schulz, 2007).

Chiral pharmaceuticals pretend enantiospecific toxic effects on various organisms. Human response is different for the toxicity and pharmaceutical effectiveness of enantiomers. For example, (R)-enantiomer is 100 times less active than (S)-enantiomer of ibuprofen (Kasprzyk-Hordern, 2010). Thus toxic response is different by aquatic environmental organisms. The growth of *D. magna* shows less resistance in the existence of r-atenolol than

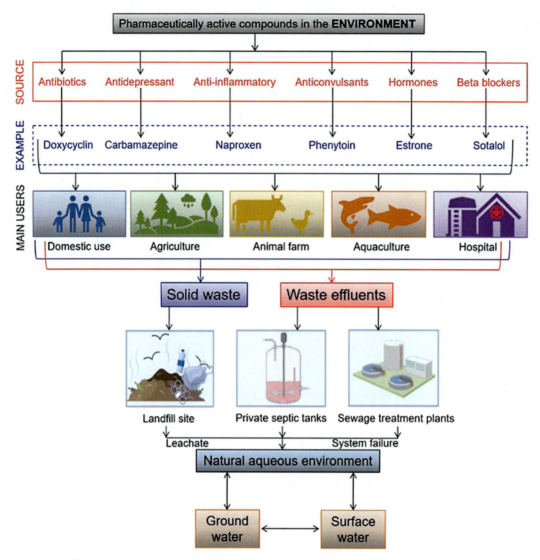

FIGURE 25–3 Pharmaceutically compounds and their transformation ways to water streams (Rasheed, Bilal et al., 2020).

s-atenolol (De Andrés, Castañeda, & Ríos, 2009). Contrarily, fluoxetine has no stereospecific toxic effect on *D. magna* (De Andrés et al., 2009). Few studies present an aquatic environment affected by the toxicity of individual enantiomers. Most of the studies on the toxicological effects of chiral pharmaceuticals make use of racemates. This shows that this area needs further intensive scale research. The transformation of Pharmaceutically active entities to our water supplies has been presented in Fig. 25–3.

25.3 Pharmaceutical removal methods

Stable environment persistency of various pharmaceuticals is an important property for humans. Persistent is improved by various functions such as fluoro-, chloro-, and nitroattached with aromatic rings, because of electron withdrawal property of these molecular structures oppose oxidation and rings remain to stabilize. This environmental persistence is reflected in ciprofloxacin antibiotics, which are present in high concentration all over the world, due to the existence of fluorine (Metcalfe, Miao, Hua, Letcher, & Servos, 2004). Clearly, advanced removal techniques are compulsory to eliminate the pharmaceuticals from wastewater and drinking water. Biological, chemical, physical, and thermal methods applied for pharmaceutical remediation. Biological methods contain trickling filters, activated sludge, anaerobic digestion, waste stabilization ponds, and aerated lagoons. Adsorption (Ahmed, Zhou, Ngo, & Guo, 2015) dialysis, evaporation, electrodialysis, flocculation, filtration, sedimentation, reverse osmosis, and stream stripping are some physical treatments, while reduction, neutralization, ion exchange, calcination, and precipitation are chemical methods. Thermal treatment contains pyrolysis and incineration (Homem & Santos, 2011; Onesios, Jim, & Bouwer, 2009). Many physical methods change the phase of pharmaceuticals from liquid to solid. In biochemical treatment, pharmaceuticals form new products such as degradation or metabolites. Mineralization can also be possible. Now summarized all methods related to aqueous media for the removal of pharmaceuticals. This section has been divided into the following different subcategories involving thermal, physical, biological, and chemical methods.

25.3.1 Biological treatment methods

Microbes have a very important and vital role in the ecosystem because they are helping to degrade xenobiotic and organic matters. The pharmaceutical's degradation taking place with the help of environmental biodegradation especially when both sewage treatment plants (STPs) and wastewater treatment plants (WWTPs) are not worked. This pharmaceutical degradation with microorganism by cometabolic or metabolic biodegradation processes occur in the presence of other compounds. *Ganoderma lucidum* and *Trametes versicolor*, bacteria, and *Chlorella sorokiniana* are the fungal strains that are known to be pharmaceuticals biodegrade (Caracciolo, Topp, & Grenni, 2015; Dawas-Massalha, Gur-Reznik, Lerman, Sabbah, & Dosoretz, 2014; Escapa, Coimbra, Paniagua, García, & Otero, 2015; Vasiliadou et al., 2016). Clofibric acid is degraded by heterotrophic bacteria to α-hydroxyisobutyric and lactic acid as well as 4-chlorophenol (Cruz-Morató et al., 2013). Salicylic acid biodegradation was carried out in just 8 h with *Pseudomonas putida* (L. Zhang, Hu, Zhu, Zhou, & Chen, 2013). Moreover, five pharmaceuticals degraded via comebolic pathway by the use of Ammonia-oxidizing bacteria and this process ammonia monooxygen used as a catalyzed during ammonia starvation. In another work, carbamazepine degradation was reported by Pseudomonas in just 20 days (Li et al., 2013).

Environmental enzymes bioremediation has used biocatalysis as the most effective biotreatment method for the last decade (Alcalde, Ferrer, Plou, & Ballesteros, 2006) Reductases, laccases, dehalogenases, bacterial mono- or dioxygenases, dehalogenases, and cytochrome P450 monooxygenases are the famous enzymes used for biocatalysis (Asif et al., 2018). But these biocatalysis applications still restricted due to the lack of a reactor system which might be flush out these enzymes with effluent. Some of the advances in biocatalysis have been summarized below (Alcalde et al., 2006). An electron-donating functional group such as alkyl(-R), alkoxy(-OR), hydroxyl(-OH), and amine(-NH$_2$) improve vulnerability to electrophilic attack through various enzymes as amide(-CONR$_2$), oxygenase, nitro(-NO$_2$), and halogen(-X) functional group frequently reduced the ability of enzymatic biotransformation (Tadkaew, Hai, McDonald, Khan, & Nghiem, 2011; Yang et al., 2013). Electron withdrawing and denoting groups play an important part in the oxidative reactions of phenolic-substrates in the case of laccases enzymes, along with diclofenac contains carboxylic group work as an electron-withdrawing group which slows down its biotransformation and an aromatic ring as an electron denoting group (Rodríguez-Delgado et al., 2016).

25.3.2 Constructed wetlands

Constructed wetlands are based on the aquatic plant-based system that provides an efficient and reliable wastewater treatment. A long history existed of treatment of water with simple operation, low cost, environmentally friendly, easy meatiness, and capability to work as an alternative of WWTP (Gersberci & Man, 1986; Kadlec & Wallace, 2008). The naturally build wetlands work as a "living filters" in the improvement of water quality by playing as a transitional zone in between water and land (Dordio, Palace, & Pinto, 2008; Sundaravadivel & Vigneswaran, 2001). These are helping to the removal of contaminants such as phosphates, metal ions, pharmaceuticals, and pesticides along with biological oxygen demand (Matamoros & Bayona, 2006). The biggest problem with this constructed wetland is its lack of complete study of removal density of remediated compound, and these wetland terms as the "Black boxes" due to the concentration of influent and effluent measured in their performance test (Scheytt, Mersmann, Lindstädt, & Heberer, 2005). These wetlands usually exhibit similar performance as STPs and WWTPs, and in these constructed wetlands approximately 70% removal of pharmaceuticals including paracetamol, salicylic acid, and tetracycline are obtained in these constructed wetlands (Hijosa-Valsero, Matamoros, Martín-Villacorta, Bécares, & Bayona, 2010; Li, Zhu, Ng, & Tan, 2014; Matamoros & Bayona, 2006; Matamoros, Caselles-Osorio, García, & Bayona, 2008; Zhang, Gersberg, Ng, & Tan, 2014). Sedimentation, sorption, volatilization, photogenerations, plant uptake, and degradation of microbial are all these chemical and biological process occurs in the aquatic plant-based system. Some studies show the removal of contamination but still lack of complete detailed of sorption, plant uptake, biological degradation and complete mechanism involved in the photodegradation, which are totally dependent on the design and operational factor of constructed wetland. The loading mode, soil mixture, bed depth, number, type of vegetation and species, and the configuration of wetlands contribute to the pharmaceutical degradation in wetland. Many pharmaceuticals work as an ionizer in water, and this ionic interaction acts as a significant role in the sorption. Sorption can also be improved by an electrostatic interaction between positively charged functional groups and negatively charged soil surface.

25.3.3 Removal of pharmaceuticals using nanofiltration

The nanofiltration process has gained more than 90% pharmaceutical removal from wastewater. A perfect mechanistic pathway is not easy to understand due to many procedures taking place at the same time. Steric hindrance (sieving), electrostatic effects and adsorption take place at the same time. Hence different compounds are removed in variable amounts. Removal is dependent on the membrane properties (permeability, hydrophobicity, surface charges, and pore size), the operating parameters (membrane potential, transmembrane pressure, feed water quality, and flux rate), and physicochemical properties (solubility, diffusibility, molecular size, polarity, hydrophobicity, and charge); (Verlicchi, Galletti, Petrovic, & Barceló, 2010). The refusal ability of negatively charged molecules is more than neutral molecules. In the case of carbadox and sulfamethoxazole nanofiltration, the refusal ability is raised from 90% to 95%. It is because at pH > 5.6 both drugs get a negative charge which gives rise to membrane charge repulsion.

25.3.4 Advanced oxidation methods

25.3.4.1 Photocatalysis

Degradation by natural or artificial light is called photocatalysis (Fig. 25–4). Direct and indirect, two forms of photochemical conversions were introduced. UV light is directly absorbed by the compound in direct photolysis and is decomposed. However, radicals and reactive oxidation states are formed by catalysts or photosensitization by indirect photolysis. It follows decomposition by chemical reaction. Photolysis depends on intensity and frequency of radiations, solution composition, hydrogen peroxide, singlet oxygen, ozone availability and formation, absorption spectra of the compound, and quantum yield (Kümmerer, 2009). Ciprofloxacin, oxytetracycline, and sulfamethoxazole photolysis by using polychromatic UV is pH-dependent. Oxytetracycline and ciprofloxacin degradation is increased and sulfamethoxazole degradation is decreased, due to pH rise from 5 to 7 (Avisar, Lester, & Mamane, 2010). Naproxen, ibuprofen, diclofenac, and clofibric acid photodegradation show a change in river

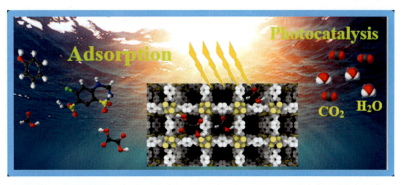

FIGURE 25–4 Schematic representation of photocatalytic process in pollutants removal of from wastewater (Rasheed, Hassan, Bilal, Hussain, & Rizwan, 2020).

water, ultrapure water, and in both systems with added 1% isopropyl alcohol (Packer, Werner, Latch, McNeill, & Arnold, 2003). One percent isopropanol enhanced degradation rates (Packer et al., 2003). This enhancement was due to the production of different radicals from isopropanol photoreduction (Packer et al., 2003). It is essential to have photocatalysis or indirect photolysis for pharmaceuticals photodegradation. For example, metronidazole degraded 6%–12% by direct photolysis, but 58%–67% increased by UV/H_2O_2 system (Shemer, Kunukcu, & Linden, 2006). Photocatalyst TiO_2 (Andreozzi, Caprio, Insola, & Marotta, 1999), a semiconductor oxide used for contaminant oxidation is being applied for pharmaceuticals degradation. Valence shell electron-photo excitation activates catalyst TiO_2 and generates electron-hole pair. At the particle surface, hydroxyl radicals are produced from water because holes have high oxidation potential (Andreozzi et al., 1999). Degradation by photocatalyst TiO_2 of different pharmaceuticals gave high removal amounts and enough mineralization (Zhao et al., 2013). In sunlight condition and simulated light, TiO_2 photooxidation removed oxytetracycline completely with 90% mineralization. This degradation was supposed to travel through different pathways: (1) photocatalytic reduction by UV/ Visible light, (2) photocatalytic reduction directly, and (3) photosensitized oxidation and reduction induced by visible light. Solution pH affects overall reaction and at pH 4.4 highest activity was obtained. Photocatalysis of carbamazepine with ZnO_2 (Haroune, Salaun, Ménard, Legault, & Bellenger, 2014), TiO_2, and N-doped TiO_2 was also affected by pH (Haroune et al., 2014). TiO_2 catalyzed photodegradation of oxytetracycline was inhibited by phosphate ions but bicarbonate, sulfate, nitrate, chloride, and ammonium did not show any change (Pereira et al., 2013). Under solar simulated irradiations, photocatalyst $BiOCl_{0.875}Br_{0.125}$ was used on bezafibrate, propranolol, ibuprofen, and carbamazepine (Lester et al., 2014). Archived degradation order was propranolol > bezafibrate > ibuprofen > carbamazepine, and kinetics also had the same order. Carbamazepine and propranolo were removed by photocatalysis but bezafibrate and ibuprofen were degraded by catalyst adsorption (Lester et al., 2014). Photocatalysis of moxifloxacins TiO_2/UV takes place with carbonyl group oxidation, hydroxylation, and ring-opening of diazobicyclo-substituents to relatively more hydrophilic compounds (Van Doorslaer, Dewulf, De Maerschalk, Van Langenhove, & Demeestere, 2015). Organic matter as humic acid influences the pharmaceutical's photocatalysis process. In the presence of humic acid, ibuprofen photocatalytic titanium dioxide decomposition is reduced from 80%–30% (Zakersalehi, Nadagouda, & Choi, 2013). TiO_2 membrane photo activity was enhanced to double after N-doping during carbamazepine decomposition (Horovitz et al., 2016). Carbamazepine photocatalytic decomposition in the presence of N-doped TiO_2 is reduced by 40% by an increase in pH due to the use of carbonates (100 mg/L as $CaCO_3$) (Avisar et al., 2013). Several aromatic and aliphatic byproducts were produced by photocatalytic degradation of clofibric acid by TiO_2 (Doll & Frimmel, 2004, 2005).

25.3.4.2 Fenton and photo-Fenton methods

In the 1890s, Fenton's reagent was introduced (Fenton, 1894). In solution, it contains hydrogen peroxide and ferrous ions, a strong oxidizing agent, and works in both homogenous and heterogeneous parameters. If conditions are heterogeneous, reactions take place by catalyst kept stationary on the heterogeneous matrix (Gan, Lau, & Ng, 2009). Hydroxyl radical production drives oxidation (Britto & Rangel, 2008). Photo-Fenton process activated by sunlight oxidized completely diclofenac in 60 min, dissolved organic carbon fully vanished with total mineralization in 100 min (Pérez-Estrada et al., 2005). However, at low pH level diclofenac precipitates, and in the homogeneous phase. Its decomposition occurs and affected by redissolution-precipitation equilibrium. Eighteen formed intermediates were identified by high performance liquid chromatography(HPLC) and gas chromatography—mass spectrometry (GC-MS) attached to time of flight mass spectrometry (TOF-MS), intermediates were formed by two theorized routes (Pérez-Estrada et al., 2005). The main path depends on the phenylacetic acid molecule and initial hydroxylation at position C-4 and quinine imine formed later followed by multistep decomposition by oxidation, hydroxylation, and decarboxylation. In another route temporary biphenyl amino part goes through C-N bond breakage by oxidation (Pérez-Estrada et al., 2005). During the photo-Fenton oxidation process of the ibuprofen, it is degraded providing depletion at ~40% TOC elimination totally (Méndez-Arriagaet al., 2011). Fe^{2+} forms complex with −COOH moiety of ibuprofen for degradation. Steps that dominate are decarboxylation and hydroxylation and 10 byproducts are formed. Photo-Fenton and Fenton process are not used for wastewater having high carbonate, nitrate, bicarbonate, chlorise, and organic matter amount, all of them block hydroxyl radicals (Méndez-Arriagaet al., 2011).

25.4 Agricultural byproducts and biosorbents

Large amount of biosorbents are produced around the world and they are sustainable. So, it is beneficial to use wastes as adsorbents. Biosorbents like biosolids, chitin, natural oxides, and agrowaste are being used as adsorbent for removal of pharmaceuticals. Adsorption potential of biosorbents are less as compare to various other adsorbents, but are beneficial due to low cost and high natural abundance (de Andrade, Oliveira, da Silva, & Vieira, 2018). Some biosorbents for the removal of pharmaceuticals are discussed below. Vegetable wastes (grape stalk, cork and yohimbe bark) were used for the pharmaceutical's removal from wastewater. An excellent removal was shown by grape stalk, gave Langmuir capability of 2.18 mg/g (Villaescusa, Fiol, Poch, Bianchi, & Bazzicalupi, 2011). For sodium diclofenac removal Isabel grape bagasse was applied (Antunes et al., 2012). Langmuir monolayer potential reported was approximately 77 mg/g and adsorption were independent of initial pharmaceutical amount. It is a reversible adsorption and bagasse regeneration is done by simple washing with H_2O, it removes almost 20% of pharmaceuticals. Surface area of Isabel grape bagasse is 2 m^2/g, but adsorption promoted by presence of functional moieties like amides, esters, alcohols, carboxyl and acids. Organic reach municipal biosolids can also be applied for pharmaceuticals removal. For removal of ciprofloxacin from waste water fulvic avid and humic acid containing biowastes were effectively applied (Carmosini & Lee, 2009). It was assumed that ciprofloxacin adsorption followed cation-exchange mechanisms and ciprofloxacin adsorption was competed by K^+ ions (Carmosini & Lee, 2009). Al-Fe-oxides were reported for ciprofloxacin sorption and with increase in ionic strength sorption decreased

(Gu & Karthikeyan, 2005). On activated sludge high Langmuir capacity of 90.9 mg/g was found for oxytetracycline. Main sorption mechanism is cation exchange (Song et al., 2014). Metal ion and pH play very important role when oxytetracycline was adsorbed on activated sludge (Song et al., 2014). At pH 5.5, maximum adsorption took place and Cd^{+2}, Ca^{+2}, Mg^{+2}, K^+, and Na^+ reduced capacity by interfering in oxytetracycline sorption. Monovalent cation has less effect on sorption than bivalent cations. However sorption is increased by Cu^{+2} due to chelation (Song et al., 2014). For 17β-estradiol adsorption, bone char adsorbent was produced from waste cattle bones (Liu et al., 2012). This char gave Langmuir monolayer adsorption capability of 10 mg/g and was regenerated easily by water/ethyl alcohol mixture (Zhang, Qiao, Zhao, & Wang, 2011). For adsorption of tetracycline, the waste tire powder and its char was applied (Lian, Song, Liu, Zhu, & Xing, 2013). A hydrophobic aromatic compound, naphthalene was used for comparison as model adsorbate. Tetracycline showed very less sorption on waste tire powder than naphthalene. But on pyrolyzed tire char adsorption was increased (under nitrogen atmosphere, pyrolyzed at 200°C, 400°C, 600°C, and 800°C, 120 min residence time) versus tire powder (Lian et al., 2013). Over a pH range (2–12), on adsorbents, Cu^{+2} and tetracycline facilitated their combine sorption. Graphite surface of char's works as π-π electron donor and it has essential part in adsorption of the tetracycline by mechanism which is π-π electron donor acceptor (Lian et al., 2013). With increase in temperature tetracycline sorption also increases. Adsorption is also increased by increase in char's pyrolysis temperature. Chars produced at temperature of 600°C and 800°C reported high sorption capacity as compared to char produced at 200°C and 400°C temperature (Lian et al., 2013).

25.5 Resins and metal oxide-based adsorbents (metal organic frameworks)

Salicylic acid was removed by employing the Filtrasorb F400, Sephabeads SP206, and Sephabeads SP207 and adsorption capabilities were 351, 82, and 45 mg/g at 20°C, respectively (Otero, Grande, & Rodrigues, 2004). On all three adsorbents adsorption of salicylic acid decreased with increase in temperature. Adsorption data fits well by Langmuir isotherms, shows adsorption was monolayer. Nonporous and mesoporous silicas ($Si-NP_8$, $Si-P_{700}$) and aluminas ($Al-P_{242}$, $Al-NP_{37}$) showed ofloxacin adsorption potentials of 1.5, 0.80, 1.2, and 1.4 μmol/m² from water, respectively (Goyne, Chorover, Kubicki, Zimmerman, & Brantley, 2005). Nonporous and mesoporous have same capacities and not much role is played by surface area (Goyne et al., 2005). For adsorbing ketoprofen, dilcofenac, carbamazepine, ibuprofen, and clofibric acid mesoporous silica (SBA-15) was employed (Bui & Choi, 2009). Due to presence of basic and acidic groups pharmaceutical pK_a are different. In pharmaceutical sorption their pK_a values has important character. With decrease in adsorbate pK_a the Freundlich adsorption capacities (KF_a) adsorption capacities decrease in the following order; ibuprofen ($pK_a = 4.91$, $K_F = 1.50$) > ketoprofen, ($pK_a = 4.45$, $K_F = 1.09$) > diclofenac ($pK_a = 4.45$. $K_F = 0.72$) > clofibric acid ($pK_a = 2.84$, $K_F = 0.56$), But this order is not followed by carbamazepine (pK_a = 13.90, K_F = 1.10). At pH 3 maximum adsorption showed by ketoprofen, ibuprofen, diclofenac, clofibric acid, and carbamazepine and they showed decrease with pH increase (Bui & Choi, 2009). The surface of silica is positively charged at pH < 4, it increases hydrogen bonding which is for silica (SBA-15) main mechanism of adsorption (Bui & Choi, 2009).

Ciprofloxacin adsorption increases by functionalization of SBA-15 by sulfonic acid using H_2SO_4 and H_2O_2 (Gao et al., 2015). Pharmaceutical sorption increases by change in adsorbent dose. Sulfamethoxazole, trimethoprim, diclofenac, carbamazepine, and ketoprofen adsorption was increased by increasing adsorbent dose. Iopromide, gemfibrozil, acetaminophen, clofibric acid, estrone, and atenolol were not affected by dose on SBA-15 (Bui, Pham, Le, & Choi, 2013). Adsorption of carbamazepine and diclofenac was observed on mercapto-functionalized HMS (M-HMS), amine-functionalized HMS (A-HMS), hexagonal mesoporous silicates (HMS), mesoporous silicates (SBA-15 and MCM-41) and powder activated carbon (PAC) (Suriyanon, Punyapalakul, & Ngamcharussrivichai, 2013). For carbamazepine and diclofenac similar rate order (PAC > M-HMS > SBA-15 > MCM-41 > HMS > A-HMS) was obtained. Adsorbent surface is directly proportional to this order.

SBA-15 functionalized with amino-Fe(III) were used for chlortetracycline, oxytetracycline and tetracycline adsorption (Zhang, Lan, Liu, & Qu, 2015). In dominant mechanism tetracyclines formed complex with Fe(III). Oxytetracycline, chlortetracycline, and tetracycline showed maximum adsorption at pH 4.4, 5.6, and 5.2, respectively. In adsorption process no role was played by ionic strength. Self-assembled compounds built from metal ions and have specific angle coordination with organic spacer compounds so that they can form crystalline arrays are called metal organic frameworks (MOFs) (Mon, Bruno, Ferrando-Soria, Armentano, & Pardo, 2018). They have very high surface area and are porous as well. Naproxen and clofibric were adsorbed on (MIL-101 and MIL-101-Fe). Adsorption capacity is correlated directly with surface area of adsorbents. The order is MIL-101 >MIL-101-Fe > AC (Hasan, Jeon, & Jhung, 2012). Naproxen adsorption increased with low initial concentration and low pH (Hasan et al., 2012). These MOFs were decorated with basic (—NH) and acidic groups (—SO_3H). MOF, MIL-101(ED-MIL-101) was decorated with basic ethylenediamine. It gave highest adsorption for both pharmaceuticals. Adsorption decreased when same MOF was decorated with acidic function that MOF is aminomethane sulfonic MIL-101 (AMSA-MIL-101) (Hasan, Choi, and Jhung, 2013). Magnetic microspheres (Q100), which are hyper-cross-linked show tetracycline sorption faster than commercial resin (NDA150, XAD-4) and magnetic resin (Q80) (Zhou et al., 2012). Aqueous phase stable MOFs were used to adsorb sulfasalazine and furosemide (Cychosz & Matzger, 2010). Most stable form of this study was MIL-100(Cr). It showed good sorption for both sulfasalazine and furosemide. Pharmaceuticals and MIL-100(Cr) showed cooperative bonding. Zr-based MOF (UiO-66) and its basic (-NH_2) and acidic (-SO_3H) decorated forms both removed diclofenac nicely (Hasan, Khan, & Jhung, 2016). Both parent and modified MOF had fast absorption than AC. It was due to smaller pore size of MOFs. Due to π-π stacking and electrostatic interaction (<pH 5.5) sorption capacities of both materials were higher than AC. Specially (-SO_3H) modified UiO-66 increased capacity and adsorption kinetics. It was due to formation of acid-base interaction between diclofenac NH and MOF (-SO_3H) groups. For NH_2-UiO-66 this trend is reverse (Hasan

et al., 2016). MAF-4, MAF-5 and MAF-6, a series of metal azolate compound were prepared and they were pyrolyzed in to porous carbons (CDM-4, CDM-5, CDM-6, respectively) respectively, than were applied for pollutant and pharmaceuticals sorption from water (CDM means carbon-derived MOFs) (Bhadra & Jhung, 2017). Among those tested CDM-6 was the best adsorbent, including AC, for the adsorbates bisphenol A, clofibric acid, oxybenzone, diclofenac sodium and salicyclic acid. Presumed adsorption mode was hydrogen bonding. CDMs were very efficient and recyclable, especially CDM-6 (Bhadra & Jhung, 2017).

To remove atenolol from water a bio-MOF-derived carbon (BMDC) was recently used (Bhadra and Jhung, 2018). Among all adsorbents atenolol with highest capacity (552 mg/g) was removed by BMDC-12h (12 h pyrolysis) (Bhadra and Jhung, 2018). For atenolol adsorption electrostatic interactions were mainly responsible. Adsorbent was regenerated by solvent washings. Triclosan was removed from aquatic media by using -NH- and −COOH and modified UiO-66 (Sarker, Song, & Jhung, 2018). Better removal capacity (189 mg/g) was shown by functionalized MOF as compared to UiO-66-NH_2, pristine UiO-66 and commercial AC. It was due to hydrogen bonding between triclosan and UiO-66-NH-CO-COOH and this form is four cycle reusable (Sarker, Song, et al., 2018).

In an Al-based MOF, a nitrogen-doped porous carbon CDIL@AlPCP was prepared by direct ionic liquid carbonization. CDIL@AlPCP is carbon derived from ionic liquid (IL)-loaded Al-based metalorganic frameworks. It effectively removed personal products (acetoaminophen, para-chloro-metaxylenol, and triclosan) and different pharmaceuticals from the aqueous phase. Hydrogen bonding is main reason for remarkable efficiency. Q_{max} value of CDIL(0.5)@AlPCP for triclosan and p-chloro-m-xylenol were 326 and 338 mg/g respectively (Sarker, An, Yoo, & Jhung, 2018). By using MOF-5 recently ciprofloxacin was removed from aqueous phase. MOF-5 gave a higher adsorption capacity (98 mg/g) as compared to AC (65 mg/g) due to electrostatic interactions between MOF-5 and ciprofloxacin (Gadipelly, Marathe, & Rathod, 2018). The adsorption capacities of naproxen, ketoprofen, indomethacin, ibuprofen and furosemide on UiO-66 surface were shown highest as compared to carbon nanotubes, AC, resins, graphenes, zeolites and other MOFs (Sarker, Song, et al., 2018). MOFs use is growing worldwide. Their acid-base, electrostatic, p-complexation, hydrophobic interaction, H-bonding, and metal sites coordination and framework has important role for pharmaceutical adsorption. The schematic Illustration has been presented in Fig. 25−5.

25.6 Nanomaterials

For pharmaceuticals adsorption, nanomaterials have been used. Nanotubes and xerogels at 400°C and 900°C were modified using nitric acid. For ciprofloxacin removal, these modified adsorbents showed second-order kinetics (Carabineiro, Thavorn-Amornsri, Pereira, & Figueiredo, 2011). They showed better adsorption capacities for both AC (Carabineiro et al., 2011) and unmodified precursors. At different concentrations and pH's multiwall carbon nanotubes (MWCNT) were used for sulfadimethoxine and sulfapyridine 854. The pK_a, pH, and ionic concentration of sulfadimethoxine ($pK_{a1} = 2.4$ and $pK_{a2} = 6.0$) and sulfapyridine ($pK_{a1} = 2.3$ and $pK_{a2} = 8.4$) influence their sorption efficiently. Adsorption is more affected

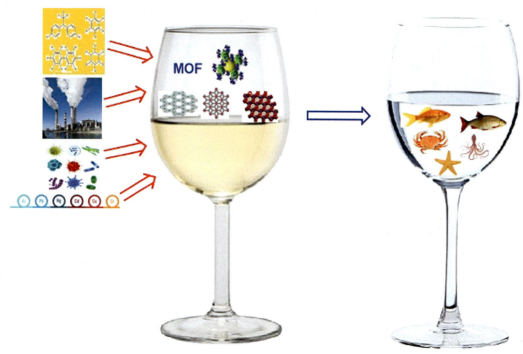

FIGURE 25–5 Schematic diagram representing metal organic frameworks role for removing pollutants from wastewater (Rasheed, Hassan et al., 2020).

by enhancement in calcium chloride concentration as compared to sodium chloride. MWCNT showed lower maximum adsorption capacities as compared to other adsorbents, but MWCNT was found very efficient after per unit surface area data normalization (Xia, Li, Zhu, Zhou, & Yang, 2013). Graphene was developed in 2004 and since then it is being used in many applications, one of them is water treatment (Lin, Xu, & Li, 2013). For doxycycline, oxytetracycline, chlortetracycline, and tetracycline adsorption graphene oxide functionalized magnetic particles were used. Ionic concentration and pH had no systemic effect on change in adsorption. The π-π interaction are reasonably important in adsorption and leads to highest removal for tetracycline (Lin et al., 2013). Molybdenum disulphide, a graphene liked structure at pH 6 has capacity of 556 mg/g is an efficient adsorbent of doxycycline (Chao et al., 2014). It is observed that composite of calcium oxide/biocomposite fiber graphene oxide has ciprofloxacin adsorption from 18 to 39 mg/g. Adsorption increases with loading increase of graphene oxide from 0% to 6% (Wu, Li, & Hong, 2013).

25.7 Conclusion

In this chapter, we outline various pharmaceutical removal methods, such as biological treatment, constructed wetlands, nanofiltration, and advanced oxidation methods, including photocatalysis, Fenton and photo-Fenton methods, agricultural byproducts and biosorbents, resins and metal oxide-based adsorbents, and nanostructured materials for the abatement of pharmaceutical contamination in the environmental niches. Pharmaceutical contaminants are highly hazardous and compromising the ecosystem integrity and ecofriendliness. They also cause a number of acute and chronic disorders and largely affect flora and fauna. Due to indiscriminate and excessive usage of pharmaceutics, exploration of treatment methods for effective removal of pharmaceutical contaminants from environment has gained importance. Nanomaterials utilization has surged in recent years to mitigate pharmaceutical contaminants due to their unique attributes, such as large surface area, high porosity, and high affinity for organic and inorganic molecules along with great catalytic characteristics. Adsorptive applicability of nanomaterials are rendered highly valuable for effective removal of pharmaceutical contaminants, and this process does not lead to the formation of new toxic materials, which are produced during other removal treatments. However, establishing these practices on the industrial scale is still in its infancy.

References

Ahmed, M. B., Zhou, J. L., Ngo, H. H., & Guo, W. (2015). Adsorptive removal of antibiotics from water and wastewater: Progress and challenges. *Science of the Total Environment, 532*, 112–126.

Alcalde, M., Ferrer, M., Plou, F. J., & Ballesteros, A. (2006). Environmental biocatalysis: From remediation with enzymes to novel green processes. *Trends in Biotechnology, 24*(6), 281–287.

Andreozzi, R., Caprio, V., Insola, A., & Marotta, R. (1999). Advanced oxidation processes (AOP) for water purification and recovery. *Catalysis Today, 53*(1), 51–59.

Antunes, M., Esteves, V. I., Guégan, R., Crespo, J. S., Fernandes, A. N., & Giovanela, M. (2012). Removal of diclofenac sodium from aqueous solution by Isabel grape bagasse. *Chemical Engineering Journal, 192*, 114–121.

Asif, M. B., Hai, F. I., Kang, J., Van De Merwe, J. P., Leusch, F. D., Price, W. E., & Nghiem, L. D. (2018). Biocatalytic degradation of pharmaceuticals, personal care products, industrial chemicals, steroid hormones and pesticides in a membrane distillation-enzymatic bioreactor. *Bioresource Technology, 247*, 528–536.

Avisar, D., Horovitz, I., Lozzi, L., Ruggieri, F., Baker, M., Abel, M.-L., & Mamane, H. (2013). Impact of water quality on removal of carbamazepine in natural waters by N-doped TiO2 photo-catalytic thin film surfaces. *Journal of Hazardous Materials, 244*, 463–471.

Avisar, D., Lester, Y., & Mamane, H. (2010). pH induced polychromatic UV treatment for the removal of a mixture of SMX, OTC and CIP from water. *Journal of Hazardous Materials, 175*(1–3), 1068–1074.

Bhadra, B. N., & Jhung, S. H. (2017). A remarkable adsorbent for removal of contaminants of emerging concern from water: Porous carbon derived from metal azolate framework-6. *Journal of Hazardous Materials, 340*, 179–188.

Bilal, M., Ashraf, S. S., Barceló, D., & Iqbal, H. M. (2019). Biocatalytic degradation/redefining "removal" fate of pharmaceutically active compounds and antibiotics in the aquatic environment. *Science of the Total Environment, 691*, 1190–1211.

Bilal, M., Mehmood, S., Rasheed, T., & Iqbal, H. M. N. (2020). Antibiotics traces in the aquatic environment: Persistence and adverse environmental impact. *Current Opinion in Environmental Science & Health, 13*, 68–74. Available from https://doi.org/10.1016/j.coesh.2019.11.005.

Boxall, A. B. (2004). The environmental side effects of medication: How are human and veterinary medicines in soils and water bodies affecting human and environmental health? *EMBO Reports, 5*(12), 1110–1116.

Britto, J. M., & Rangel, M. d. C. (2008). Processos avançados de oxidação de compostos fenólicos em efluentes industriais. *Química Nova, 31*(1), 114–122.

Bui, T. X., & Choi, H. (2009). Adsorptive removal of selected pharmaceuticals by mesoporous silica SBA-15. *Journal of Hazardous Materials, 168*(2–3), 602–608.

Bui, T. X., Pham, V. H., Le, S. T., & Choi, H. (2013). Adsorption of pharmaceuticals onto trimethylsilylated mesoporous SBA-15. *Journal of Hazardous Materials, 254*, 345–353.

Carabineiro, S., Thavorn-Amornsri, T., Pereira, M., & Figueiredo, J. (2011). Adsorption of ciprofloxacin on surface-modified carbon materials. *Water Research, 45*(15), 4583–4591.

Caracciolo, A. B., Topp, E., & Grenni, P. (2015). Pharmaceuticals in the environment: Biodegradation and effects on natural microbial communities. A review. *Journal of Pharmaceutical and Biomedical Analysis, 106*, 25–36.

Carmosini, N., & Lee, L. S. (2009). Ciprofloxacin sorption by dissolved organic carbon from reference and bio-waste materials. *Chemosphere, 77*(6), 813–820.

Carvalho, I. T., & Santos, L. (2016). Antibiotics in the aquatic environments: A review of the European scenario. *Environment International, 94*, 736–757.

Chao, Y., Zhu, W., Wu, X., Hou, F., Xun, S., Wu, P., & Li, H. (2014). Application of graphene-like layered molybdenum disulfide and its excellent adsorption behavior for doxycycline antibiotic. *Chemical Engineering Journal, 243*, 60–67.

Cruz-Morató, C., Ferrando-Climent, L., Rodriguez-Mozaz, S., Barceló, D., Marco-Urrea, E., Vicent, T., & Sarrà, M. (2013). Degradation of pharmaceuticals in non-sterile urban wastewater by *Trametes versicolor* in a fluidized bed bioreactor. *Water Research, 47*(14), 5200–5210.

Cychosz, K. A., & Matzger, A. J. (2010). Water stability of microporous coordination polymers and the adsorption of pharmaceuticals from water. *Langmuir, 26*(22), 17198–17202.

Daughton, C. G., & Ternes, T. A. (1999). Pharmaceuticals and personal care products in the environment: Agents of subtle change? *Environmental Health Perspectives, 107*(Suppl. 6), 907–938.

Dawas-Massalha, A., Gur-Reznik, S., Lerman, S., Sabbah, I., & Dosoretz, C. G. (2014). Co-metabolic oxidation of pharmaceutical compounds by a nitrifying bacterial enrichment. *Bioresource Technology, 167*, 336–342.

de Andrade, J. L. R., Oliveira, M. F., da Silva, M. G., & Vieira, M. G. (2018). Adsorption of pharmaceuticals from water and wastewater using nonconventional low-cost materials: A review. *Industrial & Engineering Chemistry Research, 57*(9), 3103–3127.

De Andrés, F., Castañeda, G., & Ríos, Á. (2009). Use of toxicity assays for enantiomeric discrimination of pharmaceutical substances. *Chirality: The Pharmacological, Biological, and Chemical Consequences of Molecular Asymmetry, 21*(8), 751–759.

Dordio, A., Palace, A., & Pinto, A.P. (2008). *Wetlands: Water living filters?*

Doll, T. E., & Frimmel, F. H. (2004). Kinetic study of photocatalytic degradation of carbamazepine, clofibric acid, iomeprol and iopromide assisted by different TiO2 materials—Determination of intermediates and reaction pathways. *Water Research, 38*(4), 955–964.

Doll, T. E., & Frimmel, F. H. (2005). Photocatalytic degradation of carbamazepine, clofibric acid and iomeprol with P25 and Hombikat UV100 in the presence of natural organic matter (NOM) and other organic water constituents. *Water Research, 39*(2–3), 403–411.

Escapa, C., Coimbra, R., Paniagua, S., García, A., & Otero, M. (2015). Nutrients and pharmaceuticals removal from wastewater by culture and harvesting of *Chlorella sorokiniana*. *Bioresource Technology, 185*, 276–284.

Fenton, H. (1894). LXXIII.—Oxidation of tartaric acid in presence of iron. Journal of the Chemical Society, Transactions, 65, 899–910.

Gersberci, R., & Man, G. (1986). *Role of aquatic plants in wastewater treatment by artificial wetlands.*

Gadipelly, C. R., Marathe, K. V., & Rathod, V. K. (2018). Effective adsorption of ciprofloxacin hydrochloride from aqueous solutions using metal-organic framework. *Separation Science and Technology, 53*(17), 2826–2832.

Gan, S., Lau, E., & Ng, H. (2009). Remediation of soils contaminated with polycyclic aromatic hydrocarbons (PAHs). *Journal of Hazardous Materials, 172*(2–3), 532–549.

Gao, J., Lu, Y., Zhang, X., Chen, J., Xu, S., Li, X., & Tan, F. (2015). Elucidating the electrostatic interaction of sulfonic acid functionalized SBA-15 for ciprofloxain adsorption. *Applied Surface Science, 349*, 224–229.

Goyne, K. W., Chorover, J., Kubicki, J. D., Zimmerman, A. R., & Brantley, S. L. (2005). Sorption of the antibiotic ofloxacin to mesoporous and nonporous alumina and silica. *Journal of Colloid and Interface Science, 283*(1), 160–170.

Gross-Sorokin, M. Y., Roast, S. D., & Brighty, G. C. (2006). Assessment of feminization of male fish in English rivers by the Environment Agency of England and Wales. *Environmental Health Perspectives, 114*(Suppl. 1), 147–151.

Gu, C., & Karthikeyan, K. (2005). Sorption of the antimicrobial ciprofloxacin to aluminum and iron hydrous oxides. *Environmental Science & Technology, 39*(23), 9166–9173.

Hahn, T., & Schulz, R. (2007). Indirect effects of antibiotics in the aquatic environment: A laboratory study on detritivore food selection behavior. *Human and Ecological Risk Assessment, 13*(3), 535–542.

Haroune, L., Salaun, M., Ménard, A., Legault, C. Y., & Bellenger, J.-P. (2014). Photocatalytic degradation of carbamazepine and three derivatives using TiO2 and ZnO: Effect of pH, ionic strength, and natural organic matter. *Science of the Total Environment, 475*, 16–22.

Hasan, Z., Jeon, J., & Jhung, S. H. (2012). Adsorptive removal of naproxen and clofibric acid from water using metal-organic frameworks. *Journal of Hazardous Materials, 209*, 151–157.

Hasan, Z., Khan, N. A., & Jhung, S. H. (2016). Adsorptive removal of diclofenac sodium from water with Zr-based metal–organic frameworks. *Chemical Engineering Journal, 284*, 1406–1413.

Hijosa-Valsero, M., Matamoros, V., Martín-Villacorta, J., Bécares, E., & Bayona, J. M. (2010). Assessment of full-scale natural systems for the removal of PPCPs from wastewater in small communities. *Water Research, 44*(5), 1429–1439.

Homem, V., & Santos, L. (2011). Degradation and removal methods of antibiotics from aqueous matrices—A review. *Journal of Environmental Management, 92*(10), 2304–2347.

Horovitz, I., Avisar, D., Baker, M. A., Grilli, R., Lozzi, L., Di Camillo, D., & Mamane, H. (2016). Carbamazepine degradation using a N-doped TiO2 coated photocatalytic membrane reactor: Influence of physical parameters. *Journal of Hazardous Materials, 310*, 98–107.

Johnning, A., Moore, E. R., Svensson-Stadler, L., Shouche, Y. S., Larsson, D. J., & Kristiansson, E. (2013). Acquired genetic mechanisms of a multiresistant bacterium isolated from a treatment plant receiving wastewater from antibiotic production. *Applied and Environmental Microbiology, 79*(23), 7256–7263.

Kadlec, R. H., & Wallace, S. (2008). *Treatment wetlands.* CRC Press.

Kasprzyk-Hordern, B. (2010). Pharmacologically active compounds in the environment and their chirality. *Chemical Society Reviews, 39*(11), 4466–4503.

Kemper, N. (2008). Veterinary antibiotics in the aquatic and terrestrial environment. *Ecological Indicators, 8*(1), 1–13.

Khetan, S. K., & Collins, T. J. (2007). Human pharmaceuticals in the aquatic environment: A challenge to green chemistry. *Chemical Reviews, 107*(6), 2319–2364.

Kolar, B., Arnuš, L., Jeretin, B., Gutmaher, A., Drobne, D., & Durjava, M. K. (2014). The toxic effect of oxytetracycline and trimethoprim in the aquatic environment. *Chemosphere, 115*, 75–80.

Kolpin, D. W., Furlong, E. T., Meyer, M. T., Thurman, E. M., Zaugg, S. D., Barber, L. B., & Buxton, H. T. (2002). Pharmaceuticals, hormones, and other organic wastewater contaminants in US streams, 1999 − 2000: A national reconnaissance. *Environmental Science & Technology*, 36(6), 1202−1211.

Kümmerer, K. (2003). Significance of antibiotics in the environment. *Journal of Antimicrobial Chemotherapy*, 52(1), 5−7.

Kümmerer, K. (2009). The presence of pharmaceuticals in the environment due to human use–present knowledge and future challenges. *Journal of Environmental Management*, 90(8), 2354−2366.

Le Page, G., Gunnarsson, L., Snape, J., & Tyler, C. R. (2017). Integrating human and environmental health in antibiotic risk assessment: A critical analysis of protection goals, species sensitivity and antimicrobial resistance. *Environment International*, 109, 155−169.

Lester, Y., Avisar, D., Gnayem, H., Sasson, Y., Shavit, M., & Mamane, H. (2014). Demonstrating a new BiOCl 0.875 Br 0.125 photocatalyst to degrade pharmaceuticals under solar irradiation. *Water, Air, & Soil Pollution*, 225(9), 2132.

Li, A., Cai, R., Cui, D., Qiu, T., Pang, C., Yang, J., . . . Ren, N. (2013). Characterization and biodegradation kinetics of a new cold-adapted carbamazepine-degrading bacterium, *Pseudomonas* sp. CBZ-4. *Journal of Environmental Sciences*, 25(11), 2281−2290.

Li, D., Yang, M., Hu, J., Zhang, J., Liu, R., Gu, X., . . . Wang, Z. (2009). Antibiotic-resistance profile in environmental bacteria isolated from penicillin production wastewater treatment plant and the receiving river. *Environmental Microbiology*, 11(6), 1506−1517.

Li, D., Yu, T., Zhang, Y., Yang, M., Li, Z., Liu, M., & Qi, R. (2010). Antibiotic resistance characteristics of environmental bacteria from an oxytetracycline production wastewater treatment plant and the receiving river. *Applied and Environmental Microbiology*, 76(11), 3444−3451.

Li, Y., Zhu, G., Ng, W. J., & Tan, S. K. (2014). A review on removing pharmaceutical contaminants from wastewater by constructed wetlands: Design, performance and mechanism. *Science of the Total Environment*, 468, 908−932.

Lian, F., Song, Z., Liu, Z., Zhu, L., & Xing, B. (2013). Mechanistic understanding of tetracycline sorption on waste tire powder and its chars as affected by Cu^{2+} and pH. *Environmental Pollution*, 178, 264−270.

Lin, Y., Xu, S., & Li, J. (2013). Fast and highly efficient tetracyclines removal from environmental waters by graphene oxide functionalized magnetic particles. *Chemical Engineering Journal*, 225, 679−685.

Liu, P., Liu, W.-J., Jiang, H., Chen, J.-J., Li, W.-W., & Yu, H.-Q. (2012). Modification of bio-char derived from fast pyrolysis of biomass and its application in removal of tetracycline from aqueous solution. *Bioresource Technology*, 121, 235−240.

Marathe, N. P., Regina, V. R., Walujkar, S. A., Charan, S. S., Moore, E. R., Larsson, D. J., & Shouche, Y. S. (2013). A treatment plant receiving waste water from multiple bulk drug manufacturers is a reservoir for highly multi-drug resistant integron-bearing bacteria. *PLoS One*, 8(10).

Matamoros, V., & Bayona, J. M. (2006). Elimination of pharmaceuticals and personal care products in subsurface flow constructed wetlands. *Environmental Science & Technology*, 40(18), 5811−5816.

Matamoros, V., Caselles-Osorio, A., García, J., & Bayona, J. M. (2008). Behaviour of pharmaceutical products and biodegradation intermediates in horizontal subsurface flow constructed wetland. A microcosm experiment. *Science of the Total Environment*, 394(1), 171−176.

Méndez-Arriaga, F., Otsu, T., Oyama, T., Gimenez, J., Esplugas, S., Hidaka, H., & Serpone, N. (2011). Photooxidation of the antidepressant drug Fluoxetine (Prozac®) in aqueous media by hybrid catalytic/ozonation processes. *Water Research*, 45(9), 2782−2794.

Metcalfe, C., Miao, X.-S., Hua, W., Letcher, R., & Servos, M. (2004). *Pharmaceuticals in the Canadian environment* (pp. 67−90). Springer.

Mon, M., Bruno, R., Ferrando-Soria, J., Armentano, D., & Pardo, E. (2018). Metal−organic framework technologies for water remediation: Towards a sustainable ecosystem. *Journal of Materials Chemistry A*, 6(12), 4912−4947.

Oaks, J. L., Gilbert, M., Virani, M. Z., Watson, R. T., Meteyer, C. U., Rideout, B. A., ... Arshad, M. (2004). Diclofenac residues as the cause of vulture population decline in Pakistan. *Nature, 427*(6975), 630–633.

Onesios, K. M., Jim, T. Y., & Bouwer, E. J. (2009). Biodegradation and removal of pharmaceuticals and personal care products in treatment systems: A review. *Biodegradation, 20*(4), 441–466.

Otero, M., Grande, C. A., & Rodrigues, A. E. (2004). Adsorption of salicylic acid onto polymeric adsorbents and activated charcoal. *Reactive and Functional Polymers, 60*, 203–213.

Packer, J. L., Werner, J. J., Latch, D. E., McNeill, K., & Arnold, W. A. (2003). Photochemical fate of pharmaceuticals in the environment: Naproxen, diclofenac, clofibric acid, and ibuprofen. *Aquatic Sciences, 65*(4), 342–351.

Pereira, J. H., Reis, A. C., Queirós, D., Nunes, O. C., Borges, M. T., Vilar, V. J., & Boaventura, R. A. (2013). Insights into solar TiO2-assisted photocatalytic oxidation of two antibiotics employed in aquatic animal production, oxolinic acid and oxytetracycline. *Science of the Total Environment, 463*, 274–283.

Pérez-Estrada, L. A., Malato, S., Gernjak, W., Agüera, A., Thurman, E. M., Ferrer, I., & Fernández-Alba, A. R. (2005). Photo-Fenton degradation of diclofenac: Identification of main intermediates and degradation pathway. *Environmental Science & Technology, 39*(21), 8300–8306.

Rasheed, T., Bilal, M., Hassan, A. A., Nabeel, F., Bharagava, R. N., Romanholo Ferreira, L. F., ... Iqbal, H. M. N. (2020). Environmental threatening concern and efficient removal of pharmaceutically active compounds using metal-organic frameworks as adsorbents. *Environmental Research, 185*, 109436. Available from https://doi.org/10.1016/j.envres.2020.109436.

Rasheed, T., Hassan, A. A., Bilal, M., Hussain, T., & Rizwan, K. (2020). Metal-organic frameworks based adsorbents: A review from removal perspective of various environmental contaminants from wastewater. *Chemosphere, 259*, 127369. Available from https://doi.org/10.1016/j.chemosphere.2020.127369.

Rodríguez-Delgado, M., Orona-Navar, C., García-Morales, R., Hernandez-Luna, C., Parra, R., Mahlknecht, J., & Ornelas-Soto, N. (2016). Biotransformation kinetics of pharmaceutical and industrial micropollutants in groundwaters by a laccase cocktail from *Pycnoporus sanguineus* CS43 fungi. *International Biodeterioration & Biodegradation, 108*, 34–41.

Rutgersson, C., Fick, J., Marathe, N., Kristiansson, E., Janzon, A., Angelin, M., ... Larsson, D. J. (2014). Fluoroquinolones and qnr genes in sediment, water, soil, and human fecal flora in an environment polluted by manufacturing discharges. *Environmental Science & Technology, 48*(14), 7825–7832.

Sarker, M., An, H. J., Yoo, D. K., & Jhung, S. H. (2018). Nitrogen-doped porous carbon from ionic liquid@ Al-metal-organic framework: A prominent adsorbent for purification of both aqueous and non-aqueous solutions. *Chemical Engineering Journal, 338*, 107–116.

Sarker, M., Song, J. Y., & Jhung, S. H. (2018). Carboxylic-acid-functionalized UiO-66-NH2: A promising adsorbent for both aqueous-and non-aqueous-phase adsorptions. *Chemical Engineering Journal, 331*, 124–131.

Scheytt, T., Mersmann, P., Lindstädt, R., & Heberer, T. (2005). Determination of sorption coefficients of pharmaceutically active substances carbamazepine, diclofenac, and ibuprofen, in sandy sediments. *Chemosphere, 60*(2), 245–253.

Shemer, H., Kunukcu, Y. K., & Linden, K. G. (2006). Degradation of the pharmaceutical metronidazole via UV, Fenton and photo-Fenton processes. *Chemosphere, 63*(2), 269–276.

Sirés, I., & Brillas, E. (2012). Remediation of water pollution caused by pharmaceutical residues based on electrochemical separation and degradation technologies: A review. *Environment International, 40*, 212–229.

Song, X., Liu, D., Zhang, G., Frigon, M., Meng, X., & Li, K. (2014). Adsorption mechanisms and the effect of oxytetracycline on activated sludge. *Bioresource Technology, 151*, 428–431.

Suarez, V., Lifschitz, A., Sallovitz, J., & Lanusse, C. (2003). Effects of ivermectin and doramectin faecal residues on the invertebrate colonization of cattle dung. *Journal of Applied Entomology, 127*(8), 481–488.

Sundaravadivel, M., & Vigneswaran, S. (2001). Constructed wetlands for wastewater treatment. *Critical Reviews in Environmental Science and Technology, 31*(4), 351–409.

Suriyanon, N., Punyapalakul, P., & Ngamcharussrivichai, C. (2013). Mechanistic study of diclofenac and carbamazepine adsorption on functionalized silica-based porous materials. *Chemical Engineering Journal, 214*, 208–218.

Tadkaew, N., Hai, F. I., McDonald, J. A., Khan, S. J., & Nghiem, L. D. (2011). Removal of trace organics by MBR treatment: The role of molecular properties. *Water Research, 45*(8), 2439–2451.

Van Doorslaer, X., Dewulf, J., De Maerschalk, J., Van Langenhove, H., & Demeestere, K. (2015). Heterogeneous photocatalysis of moxifloxacin in hospital effluent: Effect of selected matrix constituents. *Chemical Engineering Journal, 261*, 9–16.

Vasiliadou, I., Sánchez-Vázquez, R., Molina, R., Martínez, F., Melero, J., Bautista, L., . . . Morales, G. (2016). Biological removal of pharmaceutical compounds using white-rot fungi with concomitant FAME production of the residual biomass. *Journal of Environmental Management, 180*, 228–237.

Verlicchi, P., Galletti, A., Petrovic, M., & Barceló, D. (2010). Hospital effluents as a source of emerging pollutants: An overview of micropollutants and sustainable treatment options. *Journal of Hydrology, 389*(3–4), 416–428.

Villaescusa, I., Fiol, N., Poch, J., Bianchi, A., & Bazzicalupi, C. (2011). Mechanism of paracetamol removal by vegetable wastes: The contribution of π–π interactions, hydrogen bonding and hydrophobic effect. *Desalination, 270*(1–3), 135–142.

Wu, Q., Li, Z., & Hong, H. (2013). Adsorption of the quinolone antibiotic nalidixic acid onto montmorillonite and kaolinite. *Applied Clay Science, 74*, 66–73.

Xia, M., Li, A., Zhu, Z., Zhou, Q., & Yang, W. (2013). Factors influencing antibiotics adsorption onto engineered adsorbents. *Journal of Environmental Sciences, 25*(7), 1291–1299.

Yang, S., Hai, F. I., Nghiem, L. D., Nguyen, L. N., Roddick, F., & Price, W. E. (2013). Removal of bisphenol A and diclofenac by a novel fungal membrane bioreactor operated under non-sterile conditions. *International Biodeterioration & Biodegradation, 85*, 483–490.

Yuan, F., Hu, C., Hu, X., Qu, J., & Yang, M. (2009). Degradation of selected pharmaceuticals in aqueous solution with UV and UV/H_2O_2. *Water Research, 43*(6), 1766–1774.

Zakersalehi, A., Nadagouda, M., & Choi, H. (2013). Suppressing NOM access to controlled porous TiO2 particles enhances the decomposition of target water contaminants. *Catalysis Communications, 41*, 79–82.

Zhang, C.-L., Qiao, G.-L., Zhao, F., & Wang, Y. (2011). Thermodynamic and kinetic parameters of ciprofloxacin adsorption onto modified coal fly ash from aqueous solution. *Journal of Molecular Liquids, 163*(1), 53–56.

Zhang, D., Gersberg, R. M., Ng, W. J., & Tan, S. K. (2014). Removal of pharmaceuticals and personal care products in aquatic plant-based systems: A review. *Environmental Pollution, 184*, 620–639.

Zhang, L., Hu, J., Zhu, R., Zhou, Q., & Chen, J. (2013). Degradation of paracetamol by pure bacterial cultures and their microbial consortium. *Applied Microbiology and Biotechnology, 97*(8), 3687–3698.

Zhang, L., Song, X., Liu, X., Yang, L., Pan, F., & Lv, J. (2011). Studies on the removal of tetracycline by multiwalled carbon nanotubes. *Chemical Engineering Journal, 178*, 26–33.

Zhang, Z., Lan, H., Liu, H., & Qu, J. (2015). Removal of tetracycline antibiotics from aqueous solution by amino-Fe (III) functionalized SBA15. *Colloids and Surfaces A: Physicochemical and Engineering Aspects, 471*, 133–138.

Zhao, C., Pelaez, M., Duan, X., Deng, H., O'Shea, K., Fatta-Kassinos, D., & Dionysiou, D. D. (2013). Role of pH on photolytic and photocatalytic degradation of antibiotic oxytetracycline in aqueous solution under visible/solar light: Kinetics and mechanism studies. *Applied Catalysis B: Environmental, 134*, 83–92.

Zhou, Q., Li, Z., Shuang, C., Li, A., Zhang, M., & Wang, M. (2012). Efficient removal of tetracycline by reusable magnetic microspheres with a high surface area. *Chemical Engineering Journal, 210*, 350–356.

26

Nanobioremediation of insecticides and herbicides

Ammar Ali[1,2], Zaheer Ahmed[1,2], Rizwana Maqbool[1,2], Khurram Shahzad[3], Zahid Hussain Shah[4], Muhammad Zargham Ali[4], Hameed Alsamadany[5], Muhammad Bilal[6]

[1]DEPARTMENT OF PLANT BREEDING AND GENETICS, UNIVERSITY OF AGRICULTURE FAISALABAD, FAISALABAD, PAKISTAN [2]CENTER FOR ADVANCED STUDIES IN AGRICULTURE AND FOOD SECURITY, UNIVERSITY OF AGRICULTURE FAISALABAD, FAISALABAD, PAKISTAN [3]DEPARTMENT OF PLANT BREEDING AND GENETICS, UNIVERSITY OF HARIPUR, HARIPUR, PAKISTAN [4]DEPARTMENT OF PLANT BREEDING AND GENETICS, PIR MEHR ALI SHAH ARID AGRICULTURE UNIVERSITY, RAWALPINDI, PAKISTAN [5]DEPARTMENT OF BIOLOGICAL SCIENCES, KING ABDULAZIZ UNIVERSITY, JEDDAH, SAUDI ARABIA [6]SCHOOL OF LIFE SCIENCE AND FOOD ENGINEERING, HUAIYIN INSTITUTE OF TECHNOLOGY, HUAI'AN, P.R. CHINA

26.1 Nanobioremediation

Nanobioremediation is an advanced, rapidly developing innovation in which biologically synthesized nanoparticles are utilized to eliminate contaminants from the atmosphere. These nanoparticles are comprised of unique chemical, biochemical, and physical properties and thus have received extensive consideration from scientists around the globe from various areas of natural sciences as well as for bioremediation. Naturally formed nanoparticles had been seen as noteworthy in altering and detoxifying toxins which ruin the atmosphere. Additionally, these do have the potential to clean up the atmosphere at cheap price and to reduce harmful byproducts (Sinha, Mehrotra, Srivastava, Srivastava, & Singh, 2020).

The utilization of nanomaterials as nanoenabled sensors, nanopesticides, or nanofertilizers in agriculture to enhance crop yields is getting increased attention. Engineered nanomaterials (ENMs) could improve crop yield by affecting soil supply of fertilizer nutrients and plant uptake. Such materials can kill crop diseases through a number of processes, including the fabrication of reactive oxygen species (ROS), which work directly on pathogens. ENMs might also indirectly inhibit disease by boosting plant nutrition and the pathways to crop protection. Proficient utilization of ENMs can supplement or swap conventional pesticides and fertilizers, thus minimizing the influence of agricultural approaches on the environment.

The ENMs utilized as fertilizers and pesticides will be addressed to allow the practical usage of agricultural nanotechnology to accomplish world food safety (Adisa et al., 2019).

Nanotechnology's use in the production of pesticides is relatively recent and in the early stages of growth. Its research is targeted at reducing the indiscriminate usage of synthetic pesticides and ensuring their healthy use. Numerous researchers are exploring the ability of nanotechnology, in particular the nanoencapsulation method for the distribution of pesticides. Also observed was a thorough examination of different nanoencapsulation techniques and materials, application effectiveness, and present drift in research. The current chapter centered on creating developed nanoencapsulated pesticides that had slow discharge properties with improved stability, solubility, and permeability. Such properties were obtained principally by either shielding the nanoencapsulated active components from early decomposition or by growing their use in disease management over a prolonged period. The formulation of nanoencapsulated pesticide can lessen the usage of pesticides and human contact to them, which is naturally helpful to the defense of crops.

Nonetheless, deficiency of information regarding the fusion process and reduction of a cost−benefit study of nanoencapsulation equipment impeded their use within the production of pesticides. Further analysis of the actions of these products and their final environmental consequences would help to create a legislative mechanism for their commercialization (Nuruzzaman, Rahman, Liu, & Naidu, 2016).

Pesticides are unusual to water pollutants because they are intentionally utilized to combat pests in crop growing and public health (Brown, 1978). Pesticidal water waste is a serious concern in developed nations. The origins of pesticide in H_2O (water) are garbage from farm lands, human waste, and pesticide-treated orchards (Savage, 2009). These consist of numerous groups such as fungicides, insecticides, rodenticides, and herbicides. Pesticides are harmful to multiple unwanted animals like humans, owing to their lack of sufficient precision.

Due to their extensive application, they seep into ground and surface (H_2O) water, therefore is also present in water consumed by humans. These remain in the atmosphere and present a significant health risk mainly to humans. Among the possible consequences correlated with this treatment, inherent disruption had significant health ramifications for lung cancer activation, non-Hodgkin's lymphoma, leukemia, and pancreatic and prostate cancer (Ito et al., 1996). Permitted thresholds are being updated, and the acceptable thresholds are projected to exceed molecular rates over the coming years owing to growing knowledge regarding the dangers involved with drinking water pollution. Despite the public's adverse view, pesticides will also be used for several decades to maintain the supply of food to the increasing world populace (Wang & Liu, 2007). Therefore designing innovative techniques that are able to eliminating pesticide only at minimum amounts is critical.

Pesticide reduction approaches used earlier include photocatalysis, biodegradation, adsorption and isolation of membranes. Such strategies are also disadvantageous owing to their time use or cost-effectiveness. The effect of nanotechnology is becoming rapidly visible in every field of science, as well as environmental research and care (Manimegalai, Kumar, & Sharma, 2011). Nanoparticles are generally known as groups of atoms 1−100 nm in dimension (Al-Warthan, Kholoud, El-Nour, Eftaiha, & Ammar, 2010).

Nanoparticles possess different or enhanced properties relative to larger bulk substance particles, and these novel properties are obtained from the variance of unique features like the size of the particle, ionic environment, morphology, and distribution. Nanoparticles have a high surface to volume ratio, along with a smaller size of the particle (Gurunathan et al., 2009). Metallic nanoparticles like silver, gold, copper, and zerovalent iron had a remarkable catalytic role in mineralization of inorganic and organic pollutants and halocarbons. The membrane plays an essential role in water treatment systems because processes with low energy consumption are tremendously required (Kasher, 2009).

However, the membrane technique utilized as a tool for the dispersal of the nanoparticle has significant energy benefits, when opposed to other traditional approaches it often decreases the solvent need for membrane cleaning and production costs. Nonetheless, not many experiments on the cellulose acetate membrane (CAM) have been performed as help for silver nanoparticles (AgNPs) in the pesticidal mineralization in soil. In this context, a comprehensive analysis was performed for the estimation of mineralization ability of CAM-supported silver nanoparticles (Balamurugan, Sundarrajan, & Ramakrishna, 2011) (Fig. 26–1).

One of the critical issues confronting agriculture globally today is the need to monitor and cut the heavy usage of agrochemicals, for instance, insecticides, fungicides, and herbicides. It is well known that such compounds could adversely affect the atmosphere and culture as well as induce resistance in target species (Mall, Larsen, & Martin, 2018). In fact, large amounts of pesticides are potentially lost during use because of volatilization, photolysis, and oxidation; less than 0.1% of the pesticides efficiently implemented effective against the targeted species (Liang et al., 2017). Nanotechnology aims to include a means of minimizing the damaging effects of pesticides on human health and the environment as it can entail technologies that permit the controlled emission of active compounds, thus increasing safety and production of chemicals while decreasing the required quantities for use in the field.

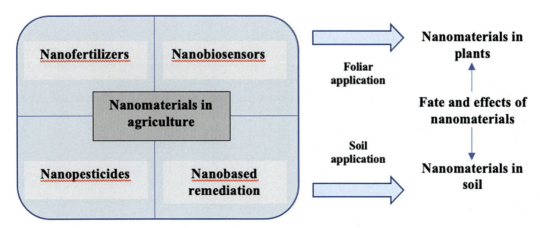

FIGURE 26–1 Application of nanotechnology in agriculture (Usman et al., 2020).

Managed systems of release are innovations that have drawn commercial, scientific, and global attention in the last few years. These innovations are utilized in fields like cosmetics, engineering, food, farming, and medicine. Given their usage in the environmental and agricultural fields, a study utilizing the Scopus database (given the duration from 2009 to 2019) found that merely 6% of nanoformulations work related to the agroindustry region, 77% of which concerned controlled release of nanoformulations, while just 23% included stimulus-receptive materials. Given stimulus-responsive items, pH was the stimulus most commonly observed (37%), accompanied by photographic and the thermal stimulation (17 and 27 pc correspondingly). Together, redox, hormones and various forms of triggers accounted for 20% of the sample (Guo et al., 2015).

26.2 Fundamentals of nanobioremediation technologies

The industry produces about 10 million tons of hazardous chemical substances annually. These substances can further react to chemicals after release, such as polychlorinated dibenzofurans or polychlorinated dibenzo-p-dioxins, that are byproducts of some chemical reactions involving chlorine (Avio, Gorbi, & Regoli, 2017).

The chemical and physical characteristics of these chemical compounds are highly diverse, and their cytotoxic activity and specific interactions with abiotic and biotic environmental effects, that is, microorganisms, plants, wind livestock, rocks, organic matter, water, and so forth have hindered effective application of remediation techniques (Jeon, Murugesan, Baldrian, Schmidt, & Chang, 2016). The integration of NPs and NMs by biotechnologies may include a step-alter in remediation capacities, reduce intermediate processes, and increase the level of degradation (Kang, 2014).

Biological therapies have been influential in comparison to physical and chemical methods for remediating contaminated areas, owing to their low price and broad variety of implementations. Bioremediation requires, among others, bioabsorption, biotransformation, microbial stabilization, and bioaccumulation. Such systems include plants and certain microorganisms like bacteria and fungi and variations of them. NMs had been combined with biological practiced in recent years to improve and facilitate the elimination of harmful substances from the atmosphere. Cecchin et al. employed the word nanobioremediation for procedures in which toxins are extracted by NPs microorganisms or plants. In fact, it called certain forms of procedures due to the essence of the individual used for contaminant remediation. More precisely, they named the methods phytonanoremediation, zoonanoremediation, and microbial nanoremediation (El-Ramady et al., 2017).

In either case, a proper relationship between NPs and living things is necessary because bioremediation make use of living organisms to remedy the polluted atmosphere. Many elements are of critical significance within this framework. For instance, it is recognized that nanonutrition, nanoparticle size, and nanotoxicity can have an impact on living organisms, and this can have an effect on the entire bioremediation process. In addition, the discovery of nanoparticles and living organisms is a concern and a field of study with scope for advanced research in terms of the short- and long-term impact of the synergistic application of biotechnology and nanomaterials, the downstream influence of nanoparticles and nanomaterials on microorganisms and trophic movement of nanomaterials in the food chain impacts human health (Wang et al., 2011).

Batch studies have shown an interaction between microorganisms and nanoparticles for the depletion of certain contaminants. However, knowledge is still lacking on the synergistic effects of nanoparticles and biotechnology during the nanobioremediation method and how these shared technologies react to contaminants of a varied nature. It should also be remembered that no data for safety on the long-standing utilization nanoparticles with microorganisms have been produced, to the best of our knowledge. Bionanoparticles have various benefits over metallic nanoparticles, for instance their biodegradability, which has less effect on the atmosphere. Existing nanotechnologies may be used in water, air, or soil remediation processes, but more cost-efficient production techniques should be developed (Fosso-Kankeu, Mulaba-Bafubiandi, & Mishra, 2014).

The regulatory system is important regarding concerns about the utilization of such materials. Scientists may contribute to understanding the interactions between NMs and biobased innovations during practices of remediation under erratic atmospheric conditions and thus provide arguments for better regulation. Finally, nanobioremediation can make an enormous contribution to sustainability, as it presents environmental benefits and is inexpensive as compared to other innovations; and even more, a spectrum of applications of nanomaterials, combined with biological treatments, has shown high efficiency in contaminant degradation that creates new opportunities to tackle environmental challenges. (Vázquez-Núñez, Molina-Guerrero, Peña-Castro, Fernández-Luqueño, & de la Rosa-Álvarez, 2020) (Fig. 26–2).

26.3 Nanobioremediation of insecticides and herbicides

Pesticides are designed to monitor and kill out most of the weeds and pests. They are categorized into various groups based on type of pest attacking the plant and also their origin. For agricultural fields, industrial pesticides, for instance, fungicides, herbicides and insecticides, are widely used. Nevertheless, the improper use of such agrochemicals has detrimental environmental consequences, such as decreased insect pollinator population, a threat to the habitat of birds and the endangered species. When consumed, the chemical pesticides after prolonged exposure often cause numerous health problems, for instance skin, hair, and issues related to the nervous system and cancer. In the past, different techniques have been developed to reduce the pesticide content, which is based on biological degradation, membrane filtration and surface adsorption. Some of the disadvantages of these strategies are delayed and slow response, less sensitivity, and precision. Recently, nanotechnology has emerged as a method to help in pesticide sensing and remediation. It will concentrate on using this technology for pesticide identification, degradation, and elimination. Nanomaterials were categorized into nanocomposites, nanotubes, and nanoparticles, which are widely used for pesticide identification, removal, and degradation. It also concentrated

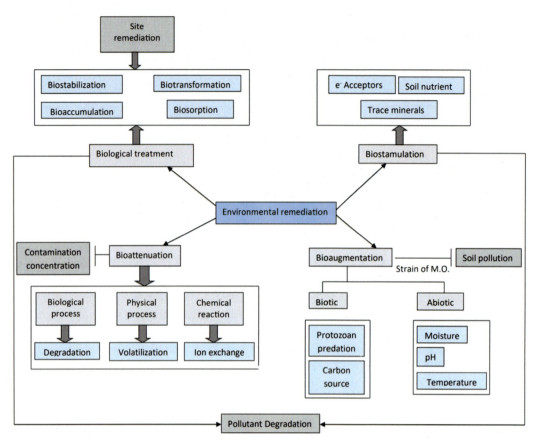

FIGURE 26–2 Working principle of environmental remediation.

on the chemistry responsible for the pesticide sensing and remedying using nanomaterials, nanoparticles of different kinds, viz. nanoparticles of metal, metal oxide nanoparticles, and bimetallic nanoparticles; halloysite nanotubes (HNT), and carbon nanotubes (CNT) were used for pesticide identification, elimination and degradation. Furthermore, numerous enzyme-based biosensors were also summarized for pesticide detection (Vázquez-Núñez et al., 2020) (Fig. 26–3).

Nanotechnology-based methods for identification, removal of toxic pesticides and degradation have gained universal attention in modern times. These methods in the identification and degradation of pesticides are considered to be extremely accurate. Various nanomaterial forms, viz. nanotubes, nanocomposites and nanoparticles have, through multiple methodologies, contributed to significant work in this field (Liu, Yuan, Yue, Zheng, & Tang, 2008). The nanoscale size, physicochemical properties, high specificity, and high surface to volume ratio of target allow the pesticides to be detected in a particular way. Various nanomaterials possess several applications to classify pesticides in various sample matrices when producing

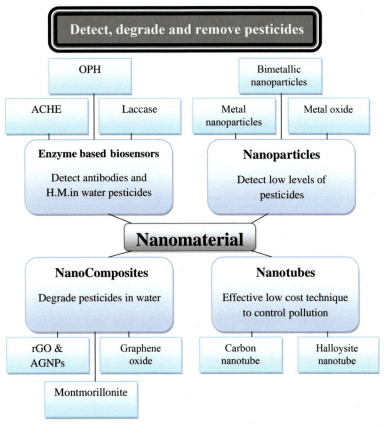

FIGURE 26–3 Chemistry involved in pesticide sensing and remedying.

biosensors. Such nanomaterials were also used for their degradation, as well to the pesticide sensing. The specific behavior of the surface and explicit surface area of nanomaterials play a significant role during the degradation of pesticide (Aragay, Pons, & Merkoçi, 2011).

Pesticides are extensively used materials that kill and control pests and also help in controlling the weed population. Too much utilization of agrochemicals has posed a menace to the endangered species, destroyed bird's habitats, and decreased populations of pollinators. These can cause numerous skin disorders as well as nervous system and eye-associated health problems. According to the World Health Organization, approximately 3 million employees in farms exhibit symptoms of poisoning by pesticide every year.

Different forms of nanomaterials discussed in this analysis were used to detect, degrade, and extract pesticides from various matrices, for instance vegetables, fruits and water. Nanoparticles like AgNPs, AuNPs NPs, and SiO_2, in addition to HNT and CNT have been utilized for detection of low concentrations pesticide. Various kinds of surface alterations have improved the specificity and sensitivity of these pesticide sensing nanomaterials. AgNPs and graphene oxide nanocomposite, nanoparticles such as ZnONPs, TiO_2 NPs, Fe/NiNPs, FeNPs, and nanotubes (HNTs) showed quick degradation of a variety of pesticides found in water, with effectiveness in all of the cases exceeding 90%. The mechanisms behind the degradation of pesticides by various nanomaterials are photocatalysis, dechlorination, and catalytic reduce. CNTs, metal oxide NPs, HNTs and nanocomposites were used for the extraction of these agrochemicals through magnetic separation or adsorption from different matrices. ZnO NPs, CNTs, and HNTs demonstrated efficiencies in the removal of 99%−100%. Modified biosensors based on enzymes with diverse nanomaterials have established sensitivity for different pesticides, rapid response, and high specificity. Cholinesterase inhibition and lacquer in the incidence of pesticides is the system responsible for the sensing of these agrochemicals (Bapat, Labade, Chaudhari, & Zinjarde, 2016). A broad surface area having tunable chemistry, rapid response sensitivity, high selectivity, and small scale are some benefits that biosensors and nanotechnology-based materials deliver over traditional practices. Yet they do have some disadvantages. Regardless of their ultrasmall scale, nanoparticles cannot easily be retrieved after they are dispersed in the water even a major problem is the toxicity of CNTs, metal oxide and metal NPs. Chemicals used for the alteration of the surface of different nanomaterials often exhibit toxic effects when exposed (Luckham & Brennan, 2010).

26.4 Nanomaterials for remediation and sensing of pesticide

In recent times, different kinds of nanomaterials, for example, nanotubes, nanoparticles, and nanocomposites, have been used for sensing and remediating pesticides. Various nanomaterials used for degradation, identification, and elimination of different pesticides are described (Fig. 26−4).

26.5 Nanoparticles

Nanoparticles are particles having a nanoscale size of about 1−100 nm that are manufactured by two methods: (1) bottom-up and (2) top-down. The top-down approach includes

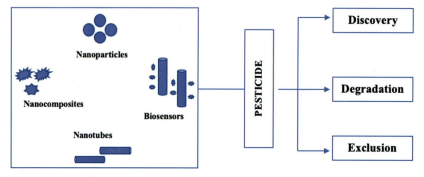

FIGURE 26–4 Nanotechnology-based materials for pesticides (Rawtani, Khatri, Tyagi, & Pandey, 2018)

breaking down bulk materials into nanoscale to get nanoparticles, whereas bottom-up process entails the piling up of molecules and atoms of material for manufacturing of nanoparticles (Pandey, Rawtani, & Agrawal, 2016). Compared with their bulk counterparts, these particles demonstrate unique physical, chemical and biological properties. These NP properties may be due to their small scale, distinctive form and high surface area. These unique properties have in recent years allowed using myriad nanoparticles for the identification, removal and degradation of various kinds of pesticides. Metal NPs, metal oxide NPs, and bimetallic NPs are the key types of NPs that various researchers have used to identify and degrade pesticides (Zhang et al., 2008).

With the development in nanomaterials and nanotechnology, their part in providing creative and efficient solutions to environmental challenges is more appropriate than ever (Seqqat, Blaney, Quesada, Kumar, & Cumbal, 2019). Because of their distinctive physicochemical properties, nanomaterials have attracted researchers from around the globe in various environmental sciences, especially in bioremediation (Tripathi, Sanjeevi, Anuradha, Chauhan, & Rathoure, 2018). They possess extensive unique properties in comparison to the bulk content and also possess distinctive visible properties for the reason that their size is small to hold the electrons, are more reactive, and also cause quantum effects (Prashanthi, Sundaram, Jeyaseelan, & Kaliannan, 2017). They have large surface area per unit, possess the ability to penetrate the polluted sites easily and also display plasmon resonance. The surface functionality of the nanoparticles can be adjusted to enhance their selectivity for the extraction of the sample. Nanoparticles may be composed by using various chemical and physical methods, but their high expense, the usage of dangerous chemicals, and the creation of toxic byproducts have provided the upper edge to the synthesis of biogenic nanoparticles (Satyanarayana & Reddy, 2018). Several studies report a variety of biogenic work papers on the development of many nanoparticles for instance Cu, Ag, Zn, Au, and Fe utilizing microorganisms (yeasts, microbes, actinomycetes, plant extracts, fungi, and algae) that are evolving as nanofabricants (Mashrai, Khanam, & Aljawfi, 2017). Biosynthesis of nanoparticles, with different biological units being used in the method, is a rapidly growing field of study in green nanotechnology (Singh et al., 2018). Nanoparticles biosynthesis is a bottom-up approach where oxidation or reduction is main reaction. Reduction of metallic compounds to their particular nanoparticles is generally because of the existence of plant phytochemicals or microbial enzymes producing antioxidants or through properties. With some optimization of the process parameters, nanoparticles of ideal sizes, and morphologies can be acquired cleanly and efficiently (Iravani, 2014). Numerous research groups in the past have concentrated on using nanotechnology as a more accurate, cost-efficient, and effective remediation method (Martins & Mata, 2015). Nanomaterials have been commonly used to combat the tiny quantities of radioactive pollutants found in food, water, and soil. Ideally, they can track, recognize, and delete these. Nanomaterials can be processed to decompose them and eliminate a specific pollutant even though they are found in microlevels. These nanomaterials may also be designed to create incredibly compact, reliable and flexible pollution-monitoring devices: nanosensors that can later communicate with and decompose contaminants to less harmful organisms after identification (Mohamed, 2017).

A wide surface to volume ratio of the nanoparticles to the bulk materials permits them a to serve as a possible tool that can also be used to control these processing processes that not only minimize the amount of required material, but also use less hazardous preliminary material and eliminate harmful waste output (Sharma, Ojha, Bharadwaj, Pathak, & Sharma, 2015).

26.6 Bimetallic nanoparticles

Bimetallic nanoparticles are the synthesis of atoms within a single nanoparticle from two separate metals. These forms of nanoparticles have drawn the interest of a lot of researchers worldwide because of several unusual and interesting properties that can be extracted from the mixture of two metals. These characteristics evolve in bimetallic nanoparticles owing to synergistic impact when metals are mixed (Zaleska-Medynska, Marchelek, Diak, & Grabowska, 2016). The composition and behavior of these nanoparticles depend on metal atoms being dispersed within nanoparticle. The color of colloidal nanoparticle samples is primarily determined by the form, scale, and distribution of the atoms of metal inside the nanoparticle. In recent years, depletion of different toxins by removal from bimetallic nanoparticles has attracted attention (Duan & Wang, 2013).

Bimetallic nanoparticles have also been used for killing a variety of pesticides, in particular Ni/Fe nanoparticles. Ni/Fe bimetallic nanoparticles are used to kill the profenofos as a catalyst. Particulate nanoscale zerovalent iron (nZVI) has acted as a means of reduction for degradation of pesticides. Ni preserved the surface of the nZVI particles against degradation and also helped to improve the reaction rate. The factor impact viz. The research examined the preliminary pH, concentration levels of profenofos, and sum of Ni/Fe nanoparticles on degradation cycle. At 1.4 mg/L concentration level of profenofos, pH 5.12 and 13.83 g/L concentration of catalyst maximum rate of removal (94.51%) was attained (Mansouriieh, Sohrabi, & Khosravi, 2015) (Table 26−1).

Even Fe/Ni NPs are used for sulfentrazone dechlorination. Total dechlorination was reported at concentrations between NPs between pH 4 and 1 g/L. In 30 min, the pesticide was observed 100% dechlorination. The catalyst active sites tended to reduce the chemical by dechlorination. The dechlorination drug toxicity assay was conducted on the *Daphnia similis*, which came out to be less harmful as compared to the pesticide (Nascimento, Lopes, Cruz, Silva, & Lima, 2016). Fe/Ni NPs, with nZVI as an iron source, is used for the degradation of 4-chlorophenol. The radicals of superoxide produced during catalytic reaction have been responsible for direct dechlorination of pesticides. Phenol and chloride ions were the degradation products found by GC-Ms. The research helped in providing an efficient mechanism for pesticides degradation via radicals of superoxide produced by a bimetallic system with Ni/nZVI particles (Shen, Mu, Wang, Ai, & Zhang, 2017).

The latest research has used green-synthesized Cu/Ag bimetallic nanoparticles as nanocatalyst for degradation of the chlorpyrifos found in water. After reducing chlorpyrifos, products found for degradation were trichloro-pyridinol and diethyl-thiophosphate. This

Table 26–1 Nanocrystalline metal oxides, used to adsorb pesticides (Taghizade Firozjaee et al., 2018).

Nanocrystalli-ne metal oxides	Modifier	Target pesticides	Adsorption amount	Reference
Fe_3O_4	Polystyrene	Organochlorine Pesticides	The adsorption potential of 10.2, 24.7, 21.3, and 33.5 mg/g, respectively, was measured for dieldrin, aldrin, endrin, and lindane.	Lan, Cheng, and Zhao (2014)
MgO and Al_2O_3	Activated carbon	Diazinon	ACNFs comprising metal oxide had a median initial adsorption rate of 19.36 μL/min in diazinon.	Behnam, Morshed, Tavanai, and Ghiaci (2013)
Al_2O_3	Cerium oxide	Dimethyl methylphosphonate (DMMP)	The DMMP adsorption was a simple 775 μg/g at 25°C.	Mitchell, Sheinker, Cox, Gatimu, and Tesfamichael (2004)
Zinc oxide	Chitosan	Permethrin	At room temperature and pH 7, 0.5 g of bionanocomposite will extract 99% of the pesticide from permethrin solution (25 mL, 0.1 mg/L).	Dehaghi, Rahmanifar, Moradi, and Azar (2014)

ACNFs, Activated carbon nano fibers.

research helped in establishing an environmentally safe means for production of bimetallic nanoparticles that can be used to cleanse water from pollution with pesticides (Rosbero & Camacho, 2017).

26.7 Nanocomposites

A nanocomposite is a combination of multiple components with measurements within nanometer diameter. The goal of manufacturing nanocomposites is to merge the properties of a variety of materials to create a new nanomaterial with improved physical and chemical properties and improvements. The nanocomposite characteristics vary greatly from those of the source materials. According to conventional composite materials, the nanocomposites show large surface area and high surface to volume ratio (Kamigaito, 1991). Such composite materials expanded interest in the environmental field, particularly for pesticides remediation from numerous samples. The key explanation for this is owing to the complex and special properties that are provided by nanomaterials, which aid in the remediation of pesticide.

A widely used phenoxy based herbicide; montmorillonite is used for elimination of MCPA (2-methyl-4-chlorophenoxyacetic acid). Montmorillonite was used for building nanocomposite with a cationic polymer, hexadimethrine. The process behind the elimination of the pesticide was electrostatic contact of anionic pesticide with the polymer ammonium group found on clay stone. Nanocomposite has proven to be an excellent adsorbent for removal of anionic pesticides (Gámiz, Hermosín, Cornejo, & Celis, 2015).

Graphene oxide is commonly used for pesticide remediation processing of nanocomposites along with various metal oxides and metal nanoparticles. The explanation behind

Graphene oxide's extreme adsorption actions against various pesticides is a close association of the organic contaminants with an aromatic chain of graphene (Zhang et al., 2015). Reduced graphene oxide (rGO) is used for the preparation of nanocomposite for processing triazine pesticides with Fe_3O_4 NPs. The electrostatic contact between pesticide and nanocomposite helped in successfully adsorbing the pesticide and then extracting it. The nanocomposite had an adsorption capacity of 93.61% for pesticides. Simple recovery from the reaction mixtures of these magnetic nanocomposites by adding external magnetic fields was an additional benefit over their wide and limited surface area (Boruah, Sharma, Hussain, & Das, 2017).

The nanocomposites of reduced graphene oxide and silver nanoparticles have been used for elimination and degradation of dichlorodiphenyldichloroethylene organochlorine, endosulfan, and chlorpyrifos pesticides. The AgNP mediated pesticides dehalogenation and the adsorption of damaged goods on rGO was the simple two-stage process behind the removal and degradation of these pesticides. The strong adsorption potential and reusability of this nanocomposite encourages its usage for remediation of multiple pesticides that are found in soil and water (Koushik, Gupta, Maliyekkal, & Pradeep, 2016). For the reduction of pesticides of organochlorine from honey, rGO was utilized in another test to produce nanocomposite having iron oxide/β-cyclodextrin. The process behind extracting the pesticide included its nanocomposite adsorption, focused on the isolation of magnetic solid material. The prepared nanocomposite helped identify very small pesticide concentrations up to levels of ppt and sub-ppt (Mahpishanian & Sereshti, 2017).

The main benefit of nanocomposites are shared properties of various materials, which considerably enhance the degradation and sensing of several pesticides. Replacement and inclusion of various nanomaterials significantly increase the likelihood of utilizing such nanocomposite structures in specific applications to remediate much smaller concentrations of pesticides.

26.8 Nanotubes

Nanotube is a thin, tubular, and porous nanomaterial with lengths ranging from nanometer to millimeter, whereas the diameter typically varies from nanometers. The high aspect ratio, broad surface region, enables nanotubes to be used as strong adsorbents for various substances. Various forms of compounds, such as proteins, nucleic acids, narcotics, and hormones have been bound to the surface and stored for different uses inside the lumen of such nanotubes. The HNTs and CNTs have gained considerable interest in the area of environmental remediation owing to their peculiar properties and wide surface area. Probability of surface alteration of nanotubes has allowed their usage from different types of samples for remediation and sensing of various pesticides.

The poisonous dyes and heavy metals were found utilizing CNT-based nano- and biosensors. CNTs were also used for the production of nanofiltration membranes for desalination applications. Such nanomaterials were also commonly used for remediation and identification of pesticides from various samples (Street, Sustich, Duncan, & Savage, 2014).

26.9 Biosensors for pesticide detection

A sensor is a systematic instrument that offers details on the concentration of multiple analytes in the ambient area, whether it is a liquid or gas. When analyte identification is dependent on a biological portion like proteins, antibodies, nucleic acids, and enzymes, the sensor is called a biosensor. Using a transducer, the indication produced after the biological element's contact by the analyte is transformed into the more easily evaluable types. Biosensors have been commonly utilized in numerous sectors like the food and fermentation, biomedical, defense, and medical industries. These were also commonly used in aqueous applications to identify heavy metallic substances and antibiotics (Saidur, Aziz, & Basirun, 2017).

Different biosensors based on enzymes exercised in conjunction by various nanomaterials in the area of pesticide detection. The nanomaterials utilized in the biosensors assist in enhancing the analyte reaction, selectivity and sensitivity. The most widely used enzymes in biosensors for pesticide identification are AChE (acetylcholinesterase), laccase, and OPH (organophosphate hydrolase). The decreased catalytic intervention of such enzymes in the existence of pesticides that acts as enzyme inhibitors is the fundamental mechanism behind the detection of pesticides by the enzyme-based biosensor.

26.10 Enzyme-responsive systems

Because of their excellent selectivity and specificity in response to internal biological stimuli, enzymatic-responsive materials have attracted significant attention in the sector of drug delivery. Nevertheless, a study using enzymatic-responsive products for the agrochemical development is still in the early phase. Enzymes play an essential role in both metabolic and biological practices, and the use of enzymes as activators for the intelligent distribution of agrochemical has several benefits, like precision, consistency along with performance, requiring precise chemical reactions under mild environmental conditions (Hu, Katti, & Gu, 2014).

Under such conditions, enzyme-immune polymers can interrelate with a biological system, resulting in reactions that are detectable by signal enhancement. The development of nanomaterials that can be caused by enzymes has taken an interest in a number of fields lately. The level of expression of different enzymes can be utilized as a catalyst which results in a site-specific, enzyme-mediated reaction of nanomaterial and controlled release of the active compounds (Ghizal, Fatima, & Srivastava, 2014) (Fig. 26–5).

To accomplish effective insect management, certain enzymes may be used as mechanisms for releasing active ingredients. Specific attention was paid to enzymes found in the midgut larva and salivary glands and insects, in clay, and generated by phytopathogenic fungus. Salivary glands and the insect midgut produce primarily carbohydrases and proteases (Akbar & Sharma, 2017). The most commonly occurring enzymes in soil include catalase, dehydrogenase, phosphatase, alkaline, and urease. Phytopathogenic fungi often discharge

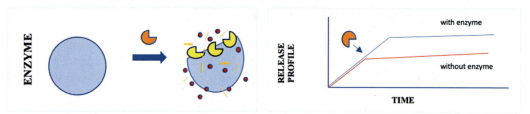

FIGURE 26–5 Release profile of active substances in reaction to the absence or presence of enzymes from fungi or insects (Camara et al., 2019).

enzymes, for instance cellulases and pectinases, that are responsible for the breakdown of plant cell walls (Fan et al., 2017).

26.11 Photoresponsive methods

Photoresponsive nanoparticles enable the spatiotemporal regulation through light irradiation of the discharge of active particles. These nanoparticles can attract light from the UV, visible, and infrared ranges of different wavelengths. As a consequence, light-responsive nanomaterials have potential uses in the agriculture sector, as the excessive sunlight radiation could cause the release of the active agents from charged particles (Esser-Kahn, Odom, Sottos, White, & Moore, 2011). The features of polymer structure that can be altered by light stimulation include optical chirality, charge, polarity, conjugation, amphiphilicity, and conformation among others. These molecular adjustments contribute to macroscopic enhancements in polymer composites including shape, optical properties, wet ability, solubility, conductivity, and adhesion (Hoogenboom, 2014). In reaction to light, the extraction of active ingredients from these nanostructured polymers starts immediately after nanocarrier surface modifications by irradiation at a defined wavelength (Bertrand & Gohy, 2017).

The integration of photoactive particles like azobenzene, orthonitrobenzyl, spyropyrane, and coumarin into polymer-based substances will provide light-controlled structures. Such molecules serve as the activated agitators of light, which induce the escape of enclosed substances that can happen either through polymerization or interaction between both the small molecule and polymer, or through polarization changes in the polymer. O-nitrobenzyl and coumarin classes consisting of polymers can be split into small molecules, although types of spyropyrane and azobenzene polymers can be formed reversibly in the existence of other wavelengths of light. Light-controlled release may also function as an on and off process where light activation activates the release of active substances. In contrast, the system may be reversed, and release interrupted in the deficiency of light or at a particular light intensity (Bruneau et al., 2019).

Use of nanotechnology in field agriculture has developed rapidly over recent years. Several manufacturing firms are already establishing collaborations with research institutions to create viable nanopesticides. Another hotspot in agricultural nanotechnology is production in stimulus-responsive nanoformulations capable of preserving stabilization of active particles under environmental circumstances, transmitting active component to the end, diminishing environmental dispersion and prolonging biological operation. While stimulus-responsive methods are already well established in the medical sector, their implementations in agricultural areas are still restricted. The sector of pesticides also needs continuing systemic work to produce better formulations of pesticides that are environmentally friendly, selective, controlled release. Effective systems which react to internal biostimulus pose a major constraint in this regard. While stimulus-responsive attributes could diminish the early deterioration of pesticides, boost their potency, and reduce collateral consequences on nontarget organisms, as mentioned in this study, several disadvantages that obstruct their large scale uses. It is essential to build nanoformulations focused on green nanotechnology that provide low price, easy methods, and controlled-discharge functionality to ensure the secure utilization of nanopesticides. In summary, it is highly promising to use smart delivery nanopesticides as an effective tool for sustainable agricultural development (Table 26–2).

The most critical obstacles to be addressed to expand marketing of nanobased pesticides can be identified as (1) creation of systematic methods for accurate risk–benefit assessment; (2) developments in the understanding of the relation between the nanomaterials and pathogens or plants; (3) establishing methods for monitoring nanopesticides in the atmosphere and evaluating their effect on human health and food protection; (4) absence of a consistent and systematic concept of nanosubstances; (5) establishment and enforcement of foreign legislation to ensure the secure production and use of stimulant-responsive nanopesticides; and (6) absence of a regional network for efficient cooperation between the private and public entities connected in nanobased drug creation (Parisi, Vigani, & Rodríguez-Cerezo, 2015). It is evident that to resolve these concerns, more in-depth work would be required before nanobased goods are completely commercialized. It is especially necessary that all possible involved parties, including governments, universities, nongovernmental organizations, research institutes, businesses, and consumers will engage (Kah, Kookana, Gogos, & Bucheli, 2018).

Table 26–2 A list of studies and their implementation on nanopesticides.

Carrier system	Agent	Purpose	Method	Reference
Chitosan	Imazapyr and imazapic	Cytotoxicity evaluation	Encapsulation	Maruyama et al. (2016)
Silica	Pyridoxine, pentoxifylline, and piracetam	Perfused brain tissue	Suspension	Jampilek et al. (2015)
Alginate	Imidacloprid	Sucking pest (leafhoppers), Cytotoxicity	Emulsion	Kumar, Bhanjana, Sharma, Sidhu, and Dilbaghi (2014)
Carboxymethyl chitosan	Methomyl	Control discharge for longer period of time	Encapsulation	Sun et al. (2014)
Tripolyphosphate/ chitosan	Paraquat	Lesser geno- and cytotoxicity	Encapsulation	Grillo et al. (2014)
Wheat gluten	Etho-fumesate	Reduction in diffusivity	Extrusion	Chevillard, Angellier-Coussy, Guillard, Gontard, and Gastaldi (2012)

26.12 Conclusion

Nanobioremediation is an advanced, rapidly developing innovation in which biologically synthesized nanoparticles are utilized to eliminate contaminants from the atmosphere. Nanoparticles have the potential to clean up the atmosphere at a cheap price and to reduce harmful byproducts. The utilization of nanomaterials as nanoenabled sensors, nanopesticides, or nanofertilizers in agriculture to enhance crop yields is getting increased attention. Use of nanotechnology in the agriculture field has developed rapidly over recent years. Several manufacturing firms are already establishing collaborations with research institutions to create viable nanopesticides. One of the critical issues confronting agriculture globally today is the need to monitor and cut the heavy usage of agrochemicals, for instance, insecticides, fungicides, and herbicides. It is well known that such compounds could adversely affect the atmosphere and culture as well as induce resistance in target species. Nanotechnology aims to include a means of minimizing the damaging effects of pesticides on human health and the environment, as it can entail technologies that permit the controlled emission of active compounds, thus increasing safety and production of chemicals while decreasing the required quantities for use in the field. Another hotspot in agricultural nanotechnology is the production in stimulus-responsive nanoformulations capable of preserving stabilization of active particles under environmental circumstances, transmitting active component to the end, diminishing environmental dispersion and prolonging biological operation. This chapter will cover different forms of nanomaterials that were used to detect, degrade, and extract pesticides from various matrices, for instance, vegetables, fruit, and water.

Conflict of interest

The authors declare no conflict of interest.

References

Adisa, I. O., Pullagurala, V. L. R., Peralta-Videa, J. R., Dimkpa, C. O., Elmer, W. H., Gardea-Torresdey, J. L., & White, J. C. (2019). Recent advances in nano-enabled fertilizers and pesticides: A critical review of mechanisms of action. *Environmental Science: Nano, 6*(7), 2002–2030.

Akbar, S. M., & Sharma, H. C. (2017). Alkaline serine proteases from *Helicoverpa armigera*: Potential candidates for industrial applications. *Archives of Insect Biochemistry and Physiology, 94*(1), e21367.

Al-Warthan, A., Kholoud, M. M., El-Nour, A., Eftaiha, A., & Ammar, R. A. A. (2010). Synthesis and applications of silver nanoparticles. *Arabian Journal of Chemistry, 3*, 135–140.

Aragay, G., Pons, J., & Merkoçi, A. (2011). Recent trends in macro-, micro-, and nanomaterial-based tools and strategies for heavy-metal detection. *Chemical Reviews, 111*(5), 3433–3458.

Avio, C. G., Gorbi, S., & Regoli, F. (2017). Plastics and microplastics in the oceans: From emerging pollutants to emerged threat. *Marine Environmental Research, 128*, 2–11.

Balamurugan, R., Sundarrajan, S., & Ramakrishna, S. (2011). Recent trends in nanofibrous membranes and their suitability for air and water filtrations. *Membranes, 1*(3), 232–248.

Bapat, G., Labade, C., Chaudhari, A., & Zinjarde, S. (2016). Silica nanoparticle based techniques for extraction, detection, and degradation of pesticides. *Advances in Colloid and Interface Science, 237*, 1–14.

Behnam, R., Morshed, M., Tavanai, H., & Ghiaci, M. (2013). Destructive adsorption of diazinon pesticide by activated carbon nanofibers containing Al_2O_3 and MgO nanoparticles. *Bulletin of Environmental Contamination and Toxicology, 91*(4), 475−480.

Bertrand, O., & Gohy, J. F. (2017). Photo-responsive polymers: Synthesis and applications. *Polymer Chemistry, 8*(1), 52−73.

Boruah, P. K., Sharma, B., Hussain, N., & Das, M. R. (2017). Magnetically recoverable Fe_3O_4/graphene nanocomposite towards efficient removal of triazine pesticides from aqueous solution: Investigation of the adsorption phenomenon and specific ion effect. *Chemosphere, 168*, 1058−1067.

Brown, A. W. A. (1978). *Ecology of pesticides*. John Wiley & Sons.

Bruneau, M., Bennici, S., Brendle, J., Dutournie, P., Limousy, L., & Pluchon, S. (2019). Systems for stimuli-controlled release: Materials and applications. *Journal of Controlled Release, 294*, 355−371.

Camara, M. C., Campos, E. V. R., Monteiro, R. A., Santo Pereira, A. D. E., de Freitas Proença, P. L., & Fraceto, L. F. (2019). Development of stimuli-responsive nano-based pesticides: Emerging opportunities for agriculture. *Journal of Nanobiotechnology, 17*(1), 100.

Chevillard, A., Angellier-Coussy, H., Guillard, V., Gontard, N., & Gastaldi, E. (2012). Controlling pesticide release via structuring agropolymer and nanoclays based materials. *Journal of Hazardous Materials, 205*, 32−39.

Dehaghi, S. M., Rahmanifar, B., Moradi, A. M., & Azar, P. A. (2014). Removal of permethrin pesticide from water by chitosan−zinc oxide nanoparticles composite as an adsorbent. *Journal of Saudi Chemical Society, 18*(4), 348−355.

Duan, S., & Wang, R. (2013). Bimetallic nanostructures with magnetic and noble metals and their physicochemical applications. *Progress in Natural Science: Materials International, 23*(2), 113−126.

El-Ramady, H., Alshaal, T., Abowaly, M., Abdalla, N., Taha, H. S., Al-Saeedi, A. H., ... Sztrik, A. (2017). *Nanoremediation for sustainable crop production,* . *Nanoscience in food and agriculture* (5, pp. 335−363). Cham: Springer.

Esser-Kahn, A. P., Odom, S. A., Sottos, N. R., White, S. R., & Moore, J. S. (2011). Triggered release from polymer capsules. *Macromolecules, 44*(14), 5539−5553.

Fan, C., Guo, M., Liang, Y., Dong, H., Ding, G., Zhang, W., ... Cao, Y. (2017). Pectin-conjugated silica microcapsules as dual-responsive carriers for increasing the stability and antimicrobial efficacy of kasugamycin. *Carbohydrate Polymers, 172*, 322−331.

Fosso-Kankeu, E., Mulaba-Bafubiandi, A.F. and Mishra, A.K. (2014). Prospects for immobilization of microbial sorbents on carbon nanotubes for biosorption: Bioremediation of heavy metals polluted water. In Application of *nanotechnology in water research* (p. 39). Wiley.

Gámiz, B., Hermosín, M. C., Cornejo, J., & Celis, R. (2015). Hexadimethrine-montmorillonite nanocomposite: Characterization and application as a pesticide adsorbent. *Applied Surface Science, 332*, 606−613.

Ghizal, R., Fatima, G. R., & Srivastava, S. (2014). Smart polymers and their applications. *International Journal of Engineering Technology, Management and Applied Sciences, 2*(4), 104−115.

Grillo, R., Pereira, A. E., Nishisaka, C. S., De Lima, R., Oehlke, K., Greiner, R., & Fraceto, L. F. (2014). Chitosan/tripolyphosphate nanoparticles loaded with paraquat herbicide: An environmentally safer alternative for weed control. *Journal of Hazardous Materials, 278*, 163−171.

Guo, M., Zhang, W., Ding, G., Guo, D., Zhu, J., Wang, B., ... Cao, Y. (2015). Preparation and characterization of enzyme-responsive emamectin benzoate microcapsules based on a copolymer matrix of silica−epichlorohydrin−carboxymethylcellulose. *RSC Advances, 5*(113), 93170−93179.

Gurunathan, S., Kalishwaralal, K., Vaidyanathan, R., Venkataraman, D., Pandian, S. R. K., Muniyandi, J., ... Eom, S. H. (2009). Biosynthesis, purification and characterization of silver nanoparticles using *Escherichia coli*. *Colloids and Surfaces B: Biointerfaces, 74*(1), 328−335.

Hoogenboom, R. (2014). Smart polymers and their applications. *Biopolymers, 408*, 436.

Hu, Q., Katti, P. S., & Gu, Z. (2014). Enzyme-responsive nanomaterials for controlled drug delivery. *Nanoscale, 6*(21), 12273–12286.

Iravani, S. (2014). Bacteria in nanoparticle synthesis: Current status and future prospects. *International Scholarly Research Notices, 2014*.

Ito, N., Hagiwara, A., Tamano, S., Futacuchiá, M., Imaida, K., & Shirai, T. (1996). Effects of pesticide mixtures at the acceptable daily intake levels on rat carcinogenesis. *Food and Chemical Toxicology, 34*(11–12), 1091–1096.

Jampilek, J., Zaruba, K., Oravec, M., Kunes, M., Babula, P., Ulbrich, P., . . . Suchy, P. (2015). Preparation of silica nanoparticles loaded with nootropics and their in vivo permeation through blood-brain barrier. *BioMed Research International, 2015*.

Jeon, J. R., Murugesan, K., Baldrian, P., Schmidt, S., & Chang, Y. S. (2016). Aerobic bacterial catabolism of persistent organic pollutants—Potential impact of biotic and abiotic interaction. *Current Opinion in Biotechnology, 38*, 71–78.

Kah, M., Kookana, R. S., Gogos, A., & Bucheli, T. D. (2018). A critical evaluation of nanopesticides and nanofertilizers against their conventional analogues. *Nature Nanotechnology, 13*(8), 677–684.

Kamigaito, O. (1991). What can be improved by nanometer composites? *Journal of the Japan Society of Powder and Powder Metallurgy, 38*(3), 315–321.

Kang, J. W. (2014). Removing environmental organic pollutants with bioremediation and phytoremediation. *Biotechnology Letters, 36*(6), 1129–1139.

Kasher, R. (2009). Membrane-based water treatment technologies: Recent achievements, and new challenges for a chemist. *Bulletin of the Israel Chemical Society, 24*, 10–18.

Koushik, D., Gupta, S. S., Maliyekkal, S. M., & Pradeep, T. (2016). Rapid dehalogenation of pesticides and organics at the interface of reduced graphene oxide–silver nanocomposite. *Journal of Hazardous Materials, 308*, 192–198.

Kumar, S., Bhanjana, G., Sharma, A., Sidhu, M. C., & Dilbaghi, N. (2014). Synthesis, characterization and on field evaluation of pesticide loaded sodium alginate nanoparticles. *Carbohydrate Polymers, 101*, 1061–1067.

Lan, J., Cheng, Y., & Zhao, Z. (2014). Effective organochlorine pesticides removal from aqueous systems by magnetic nanospheres coated with polystyrene. *Journal of Wuhan University of Technology-Material Science (Ed.), 29*(1), 168–173.

Liang, Y., Guo, M., Fan, C., Dong, H., Ding, G., Zhang, W., . . . Cao, Y. (2017). Development of novel urease-responsive pendimethalin microcapsules using silica-IPTS-PEI as controlled release carrier materials. *ACS Sustainable Chemistry & Engineering, 5*(6), 4802–4810.

Liu, S., Yuan, L., Yue, X., Zheng, Z., & Tang, Z. (2008). Recent advances in nanosensors for organophosphate pesticide detection. *Advanced Powder Technology, 19*(5), 419–441.

Luckham, R. E., & Brennan, J. D. (2010). Bioactive paper dipstick sensors for acetylcholinesterase inhibitors based on sol–gel/enzyme/gold nanoparticle composites. *Analyst, 135*(8), 2028–2035.

Mahpishanian, S., & Sereshti, H. (2017). One-step green synthesis of β-cyclodextrin/iron oxide-reduced graphene oxide nanocomposite with high supramolecular recognition capability: Application for vortex-assisted magnetic solid phase extraction of organochlorine pesticides residue from honey samples. *Journal of Chromatography A, 1485*, 32–43.

Mall, D., Larsen, A. E., & Martin, E. A. (2018). Investigating the (mis) match between natural pest control knowledge and the intensity of pesticide use. *Insects, 9*(1), 2.

Manimegalai, G., Kumar, S. S., & Sharma, C. (2011). Pesticide mineralization in water using silver nanoparticles. *International Journal of Chemical Sciences, 9*(3), 1463–1471.

Mansouriieh, N., Sohrabi, M. R., & Khosravi, M. (2015). Optimization of profenofos organophosphorus pesticide degradation by zero-valent bimetallic nanoparticles using response surface methodology. *Arabian Journal of Chemistry, 12*(8), 2524–2532.

Martins, A., & Mata, T. (2015). Nanotechnology and sustainability—Current status and future challenges. *Book: Life Cycle Analysis of Nanoparticles*, 271–306.

Maruyama, C. R., Guilger, M., Pascoli, M., Bileshy-José, N., Abhilash, P. C., Fraceto, L. F., & De Lima, R. (2016). Nanoparticles based on chitosan as carriers for the combined herbicides imazapic and imazapyr. *Scientific Reports, 6*, 19768.

Mashrai, A., Khanam, H., & Aljawfi, R. N. (2017). Biological synthesis of ZnO nanoparticles using *C. albicans* and studying their catalytic performance in the synthesis of steroidal pyrazolines. *Arabian Journal of Chemistry, 10*, S1530–S1536.

Mitchell, M. B., Sheinker, V. N., Cox, W. W., Gatimu, E. N., & Tesfamichael, A. B. (2004). The room temperature decomposition mechanism of dimethyl methylphosphonate (DMMP) on alumina-supported cerium oxide − Participation of nano-sized cerium oxide domains. *The Journal of Physical Chemistry B, 108*(5), 1634–1645.

Mohamed, E. F. (2017). Nanotechnology: Future of environmental air pollution control. *Environmental Management and Sustainable Development, 6*(2), 429–454.

Nascimento, M. A., Lopes, R. P., Cruz, J. C., Silva, A. A., & Lima, C. F. (2016). Sulfentrazone dechlorination by iron-nickel bimetallic nanoparticles. *Environmental Pollution, 211*, 406–413.

Nuruzzaman, M. D., Rahman, M. M., Liu, Y., & Naidu, R. (2016). Nanoencapsulation, nano-guard for pesticides: A new window for safe application. *Journal of Agricultural and Food Chemistry, 64*(7), 1447–1483.

Pandey, G., Rawtani, D., & Agrawal, Y. K. (2016). *Aspects of nanoelectronics in materials development. Nanoelectronics and materials development.* IntechOpen.

Parisi, C., Vigani, M., & Rodríguez-Cerezo, E. (2015). Agricultural nanotechnologies: What are the current possibilities? *Nano Today, 10*(2), 124–127.

Prashanthi, M., Sundaram, R., Jeyaseelan, A., & Kaliannan, T. (Eds.), (2017). *Bioremediation and sustainable technologies for cleaner environment.* Cham: Springer.

Rawtani, D., Khatri, N., Tyagi, S., & Pandey, G. (2018). Nanotechnology-based recent approaches for sensing and remediation of pesticides. *Journal of Environmental Management, 206*, 749–762.

Rosbero, T. M. S., & Camacho, D. H. (2017). Green preparation and characterization of tentacle-like silver/copper nanoparticles for catalytic degradation of toxic chlorpyrifos in water. *Journal of Environmental Chemical Engineering, 5*(3), 2524–2532.

Saidur, M. R., Aziz, A. A., & Basirun, W. J. (2017). Recent advances in DNA-based electrochemical biosensors for heavy metal ion detection: A review. *Biosensors and Bioelectronics, 90*, 125–139.

Satyanarayana, T., & Reddy, S. S. (2018). A review on chemical and physical synthesis methods of nanomaterials. *International Journal for Research in Applied Science & Engineering Technology, 6*, 2885–2889.

Savage, N. F. (Ed.), (2009). *Nanotechnology applications for clean water* (p.589). Norwich, NY: William Andrew.

Seqqat, R., Blaney, L., Quesada, D., Kumar, B., & Cumbal, L. (2019). *Nanoparticles for environment, engineering, and nanomedicine.*

Sharma, N., Ojha, H., Bharadwaj, A., Pathak, D. P., & Sharma, R. K. (2015). Preparation and catalytic applications of nanomaterials: A review. *RSC Advances, 5*(66), 53381–53403.

Shen, W., Mu, Y., Wang, B., Ai, Z., & Zhang, L. (2017). Enhanced aerobic degradation of 4-chlorophenol with iron-nickel nanoparticles. *Applied Surface Science, 393*, 316–324.

Singh, J., Dutta, T., Kim, K. H., Rawat, M., Samddar, P., & Kumar, P. (2018). 'Green'synthesis of metals and their oxide nanoparticles: Applications for environmental remediation. *Journal of Nanobiotechnology, 16*(1), 84.

Sinha, S., Mehrotra, T., Srivastava, A., Srivastava, A., & Singh, R. (2020). *Nanobioremediation technologies for potential application in environmental cleanup,* . Environmental biotechnology (2, pp. 53–73). Cham: Springer.

Street, A., Sustich, R., Duncan, J., & Savage, N. (Eds.), (2014). *Nanotechnology applications for clean water: Solutions for improving water quality*. William Andrew.

Sun, C., Shu, K., Wang, W., Ye, Z., Liu, T., Gao, Y., ... Yin, Y. (2014). Encapsulation and controlled release of hydrophilic pesticide in shell cross-linked nanocapsules containing aqueous core. *International Journal of Pharmaceutics, 463*(1), 108–114.

Taghizade Firozjaee, T., Mehrdadi, N., Baghdadi, M., & Nabi Bidhendi, G. R. (2018). Application of nanotechnology in pesticides removal from aqueous solutions—A review. *International Journal of Nanoscience and Nanotechnology, 14*(1), 43–56.

Tripathi, S., Sanjeevi, R., Anuradha, J., Chauhan, D. S., & Rathoure, A. K. (2018). *Nano-bioremediation: Nanotechnology and bioremediation. Biostimulation remediation technologies for groundwater contaminants* (pp. 202–219). IGI Global.

Usman, M., Farooq, M., Wakeel, A., Nawaz, A., Cheema, S. A., ur Rehman., ... Sanaullah, M. (2020). Nanotechnology in agriculture: Current status, challenges and future opportunities. *Science of the Total Environment, 721*, 137778.

Vázquez-Núñez, E., Molina-Guerrero, C. E., Peña-Castro, J. M., Fernández-Luqueño, F., & de la Rosa-Álvarez, M. (2020). Use of nanotechnology for the bioremediation of contaminants: A review. *Processes, 8*(7), 826.

Wang, C. J., & Liu, Z. Q. (2007). Foliar uptake of pesticides—Present status and future challenge. *Pesticide Biochemistry and Physiology, 87*(1), 1–8.

Wang, Y., Morin, G., Ona-Nguema, G., Juillot, F., Calas, G., & Brown, G. E., Jr (2011). Distinctive arsenic (V) trapping modes by magnetite nanoparticles induced by different sorption processes. *Environmental Science & Technology, 45*(17), 7258–7266.

Zaleska-Medynska, A., Marchelek, M., Diak, M., & Grabowska, E. (2016). Noble metal-based bimetallic nanoparticles: The effect of the structure on the optical, catalytic and photocatalytic properties. *Advances in Colloid and Interface Science, 229*, 80–107.

Zhang, C., Zhang, R. Z., Ma, Y. Q., Guan, W. B., Wu, X. L., Liu, X., ... Pan, C. P. (2015). Preparation of cellulose/graphene composite and its applications for triazine pesticides adsorption from water. *ACS Sustainable Chemistry & Engineering, 3*(3), 396–405.

Zhang, S. P., Shan, L. G., Tian, Z. R., Zheng, Y., Shi, L. Y., & Zhang, D. S. (2008). Study of enzyme biosensor based on carbon nanotubes modified electrode for detection of pesticides residue. *Chinese Chemical Letters, 19*(5), 592–594.

27

Microbial-induced corrosion of metals with presence of nanoparticles

Mohammad Tabish[1], Ayesha Zarin[2], Muhammad Uzair Malik[1], Muhammad Abubaker Khan[3], Jingmao Zhao[1], Ghulam Yasin[3,4]

[1]STATE KEY LABORATORY OF ELECTROCHEMICAL PROCESS AND TECHNOLOGY FOR MATERIALS, COLLEGE OF MATERIALS SCIENCE AND ENGINEERING, BEIJING UNIVERSITY OF CHEMICAL TECHNOLOGY, BEIJING, P.R. CHINA [2]INSTITUTE OF CHEMICAL SCIENCES, BAHAUDDIN ZAKARIYA UNIVERSITY, MULTAN, PAKISTAN [3]SCHOOL OF PHYSICS AND OPTICAL ENGINEERING, SHENZHEN UNIVERSITY, SHENZHEN, P.R. CHINA [4]INSTITUTE FOR ADVANCED STUDY, SHENZHEN UNIVERSITY, SHENZHEN, P.R. CHINA

27.1 Introduction

Microbial-induced corrosion (MIC) is the degradation of different metallic materials due to the influence of microorganisms. MIC involves different kinds of methods through which corrosion is directly or indirectly involved due to reactive microbes. It was founded over 100 years ago (Gaines, 1910) and indicated in 1963 that mildly corrosive soils containing microbes can cause MIC and enhance the corrosion rates (Chaker, 1989), though the importance of MIC was not broadly identified at this time. Now it is famous because of the enlarged awareness between practitioners and researchers in different engineering sectors, particularly in the oil and gas fields. As we know, the presence of microorganisms increases the rate of corrosion by several orders (Beech & Coutinho, 2003). According to Flemming et al. (Flemming, 1996; Flemming & Schaule, 1994), microbial-induced corrosion rates are 20% higher than corrosion costs, which consume billions of dollars per annum in the United States and over 50% of failure in pipelines have been identified due to MIC (Booth, 1964; Rajasekar et al., 2007). In several fields, MIC destroys systems and equipment particularly in oil and gas (Videla, 2002), marine environments (Aktas et al., 2017), medical devices (Widmer, 2001), water systems (Rhoads, Pruden, & Edwards, 2017), nuclear waste storage facilities (Forte Giacobone, Rodriguez, Burkart, & Pizarro, 2011), and aviation fuel systems (Dai et al., 2016). Other corrosion processes have also been involved with this such as pitting (Campaignolle & Crolet, 1997; Little, Ray, & Pope, 2000; Mansfeld & Little, 1991), crevice

(Forte Giacobone et al., 2011), underdeposit corrosion (Wang & Melchers, 2017), and stress corrosion cracking (Wu et al., 2015). MIC tends to the deterioration of several materials. In many industries, piping materials usually consist of carbon steel. Pipeline failures contribute to major economic losses and environmental damage. It was strongly believed that the 2006 Trans-Alaska Pipeline leaked due to MIC. That leakage caused major economic losses and ecological issues (Jacobson, 2007). Stainless steels (SS) are not protected from MIC pitting corrosion (Manam et al., 2017; Zhou et al., 2018). MIC or biocorrosion is a well-known research area but it took many years to determine the impact of particular microorganisms on their specific environment, the interrelationship efficiency of natural microbial populations, and the physical and chemical characteristics of biofilms (Jia, Unsal, Xu, Lekbach, & Gu, 2019). Biofilms consist of many kinds of microorganisms including bacteria, fungi, and protists that may cause rusting to the surface of metals, alloys, or ceramics such as magnesium (Ahmadkhaniha et al., 2016), aluminum (Dai et al., 2016), zinc (Ilhan-Sungur, Unsal-Istek, & Cansever, 2015), and even concrete (Harbulakova, Estokova, Stevulova, Luptáková, & Foraiova, 2013; Li et al., 2018). MIC is a mechanism influenced graphically by reactions in the presence of microorganisms in the substrate, at the substrate or electrolyte, making it difficult to assess, predict, and mitigate. However, in the corrosion control process MIC faces some threats (Sowards, Williamson, Weeks, McColskey, & Spear, 2014). Most of the microorganisms involved in rusting methods in the form of attachment, colonization, alteration of local environment after the adsorption of organic molecules (proteins, polysaccharides, fatty acids, lipids, etc.) on the surfaces (Little, Wagner, & Mansfeld, 1992). The bacterial attachment is the most substantial step in MIC (Bairi, George, & Mudali, 2012; Moradi, Song, Yang, Jiang, & He, 2014; Yuan & Pehkonen, 2009). So, it is most significant step to recognize the various MIC causing microbes and to alleviate the different MIC mechanisms to protect the materials. For the development of MIC resistance, more than one method can be used to protect the metal substrate, for example, the use of the mechanical or chemical cleaning, corrosion inhibitors or a biocides that are in the form of coating or nanoparticles (NPs) incorporated coatings which can provide a resistant barrier to micro/nanofouling and corrosive species (Anjum, Ali, Khan, Zhao, & Yasin, 2020; Jabbar et al., 2017; Khan et al., 2020; Makhlouf, 2014; Tabish, Yasin et al., 2021; Yasin, Anjum et al., 2020; Yasin, Arif, Nizam et al., 2018; Yasin, Arif, Shakeel et al., 2018; Yasin, Khan et al., 2018; Yasin, Mehtab, Shakeel, Mushtaq et al., 2020). The coated NPs are adsorbed on the surface of substrate to obtain productive results. These techniques are not only beneficial for corrosion protection as well as these methods also improve the other properties of the coatings. The main reason to use NPs is having a little momentum and they can be easily coated or it is possible to improve the wear resistance of metals. Moreover, NP of inorganic materials enhance the barrier properties and also provide resistance against weather degradations as well as self-cleaning properties (Martins et al., 2006; Turhan, Weiser, Jha, & Virtanen, 2011; Zubillaga et al., 2009). So, corrosion protective coatings are also maintained by NPs (Hu & Dong, 1998). Thus, this chapter draws the reader's attention toward the NPs-based various coatings for the improvements in MIC protection.

27.2 Corrosive microbes

The causative bacteria has been dedicated to a great deal of research on metal corrosion (Chen, Howdyshell, Howdyshell, & Ju, 2014; Lou et al., 2016; Moradi, Duan, Ashassi-Sorkhabi, & Luan, 2011), such as sulfate-reducing bacteria (SRB), nitrate-reducing bacteria (NRB), iron-oxidizing bacteria (FeOB), iron-reducing bacteria (FeRB), and acid-producing bacteria (APB) (Dai et al., 2016). Archaea becomes critical in extreme conditions, such as reservoirs, where the temperature can easily exceed 70°C or above (Stetter et al., 1993). Fungi have become an important part of MIC investigations particularly in damp and warm climates such as Southeast Asia (Qu et al., 2015). Organic acids are the source of corrosion for steel because these organic acids are produced by bacteria and fungi. It was notified that these acids have been reported to cause corrosion of other metals, including magnesium alloy and zinc (Jia et al., 2019; Juzeliūnas et al., 2007; Qu et al., 2017).

27.2.1 Sulfate-reducing bacteria

In MIC literature, SRB is the most studied bacteria because of its practical importance (Enning & Garrelfs, 2014; Gu, Jia, Unsal, & Xu, 2019; Venzlaff et al., 2013). The valence range of sulfur elements is from -2 to $+6$. In addition to sulfate, SRB can also be used for terminal electron acceptors by consuming sulfur compounds with a valence higher than -2. Bisulfite (HSO_3^-), thiosulfate ($S_2O_3^{2-}$), and elemental sulfur are used as terminal electron acceptors (Thauer, Stackebrandt, & Hamilton, 2007). SRB are rigid anaerobes (that do not require oxygen to grow), but they can endure for a certain period without growth when exposed to oxygen (Barton & Tomei, 1995). SRB refers to live bottom inhabitant either they are found in the open-air system or mixed-culture biofilms. SRB is required typically anaerobic system for its growth (Dannenberg, Kroder, Dilling, & Cypionka, 1992; Thauer et al., 2007). Sulfates have been exploited as the terminal electron acceptor for SRB respiration. Organic compounds for example volatile fatty acids are mostly utilized as electron donors. Molecular hydrogen (H_2) can be used as an electron donor in hydrogenated-positive SRB. For SRB metabolism the energy is provided by the electron exchange and this electron exchange occurs by the redox reaction (Thauer et al., 2007; Xu, Li, Song, & Gu, 2013). Some SRBs can even grow autotrophically on CO_2 and H_2, much like methanogens; and some SRB easily use saturated and unsaturated hydrocarbons that are freely available in oil pipelines (Novelli, 1944; Xu & Gu, 2014).

27.2.2 Nitrate-reducing bacteria

In the petroleum and gas industries, nitrate is often injected into NRB growth in order to suppress SRB growth by competitive sulfates reduction by SRB in mitigating reservoir souring (Fida, Chen, Okpala, & Voordouw, 2016; Gieg, Jack, & Foght, 2011). However, the iron oxidation due to NRB is thermodynamically more feasible and shows an adverse effect than iron oxidation with SRB (Xu et al., 2013). During a seven-day lab test, *Bacillus licheniformis* biofilm has shown more corrosion rate in comparison with sulfate-reducing *Desulfovibrio vulgaris* biofilm on steel (Xu et al., 2013). Nitrate-reducing *Pseudomonas aeruginosa* corrosion was reported on 304 stainless steel as well (Jia, Yang, Xu, & Gu, 2017). Hence, the injection of nitrate should be properly measured so that nitrate does not enter into pipelines. Unpolluted seawater does not contain nitrate, unlike sulfate. Therefore nitrate is introduced by agricultural runoff into soils or water systems. In current years, researchers have realized that NRB and SRB both are important in soil MIC investigations (Wan et al., 2018).

27.2.3 Acid-producing bacteria

APB have the ability to produce organic acids that are responsible for the MIC and lower the pH value under their biofilms (Xu, Li, & Gu, 2016). The pH in the bulk fluid is higher than that is beneath a biofilm. According to Vroom et al. (1999) that the pH value must be different by 2 or even higher values for two neighboring sites within the same biofilm. The proton attack is thermodynamically auspicious under low pH with iron oxidation. In the fermentation of organic substrate most of the heterotrophic bacterials discharge organic acids. The types and quantities of acids formed depend on the substrate molecules and the type of microorganisms. Organic acids may increase the tendency toward corrosion. When acid is trapped between the metal and biofilm the effect of the metabolites is very intensive. From *Clostridium aceticum* we can form acetic acid and from sulfur-oxidizing bacteria we can form sulfuric acid, for example, such as *Thiobacillus thioxiduns*, are noticeable contributors toward corrosion. In addition, organic acids of the Krebs cycle can prevent the formation of oxide film hence promote the electrochemical oxidation of different metals (Little et al., 1992).

27.2.4 Metal oxidizing microbes

FeOB is known for the cause of MIC (Liu, Gu, Asif, Zhang, & Liu, 2017). The occurrence of FeOB has been recognized in numerous atmospheres which also include pipelines (Maeda et al., 1999) and water tanks (Kobrin, 1976). FeOB metabolism can deposit iron hydroxides and promote steel corrosion (Eqs. 27.1 and 27.2; Iverson, 1987), especially pitting corrosion (Moradi et al., 2011; Starosvetsky, Starosvetsky, Pokroy, Hilel, & Armon, 2008). *Acidithiobacillus ferrooxidans* (Kelly & Wood, 2000) is a famous FeOB that has potent involvement in metal oxidation activities (Bevilaqua et al., 2009; Van den Eynde, Paradelo, & Monterroso, 2009), especially under acidic conditions. The ferrous (Fe^{2+}) is the only source of energy for CO_2 reduction from pH 2.0 to 4.5 (Eq. 27.3) (Leathen, Kinsel, & Braley, 1956). For FeOB growth, the energy was generated from the oxidation of ferrous ions into ferric ions (Wang, Ju, Castaneda, Cheng, & Zhang Newby, 2014). This moves the equilibrium of iron oxidation in a forward direction.

$$Fe^{2+} + 2(OH)^- \rightarrow Fe(OH)_2 \tag{27.1}$$

$$Fe(OH)_2 + \frac{1}{2}H_2O + \frac{1}{4}O_2 \rightarrow Fe(OH)_3 \tag{27.2}$$

$$24Fe^{2+} + 6CO_2 + 24H^+ \rightarrow 24Fe^{3+} + C_6H_{12}O_6 + 6H_2O \tag{27.3}$$

27.2.5 Fungi

Fungi are ubiquitous and are eukaryotic microorganisms. In MIC, fungi are not well studied. However, they can act as a prominent factor in damp and hot environments. Fungi are concerned with MIC of numerous metals containing carbon steel, stainless steel, copper, and aluminum (Binkauskienė, Bučinskienė, & Lugauskas, 2017; Cojocaru et al., 2016). *Aspergillus niger* (a facultative filamentous fungus) promotes MIC pitting corrosion in magnesium alloy (Qu et al., 2015), whereas the microorganism fungus *Hormoconis (Cladosporium) resinae* that is found in aviation fuel is responsible for the corrosion of Al fuel tanks (Salvarezza, De Mele, & Videla, 1983; Videla, Guiamet, DoValle, & Reinoso, 1988; Ayllon & Rosales, 1994). In natural surroundings, oxygen is absorbed by fungal biofilms and supports anaerobic microbes like SRB to live under the biofilm (Usher, Kaksonen, Cole, & Marney, 2014). It can also damage hydrocarbons that produce organic acids (Little, Staehle, & Davis, 2001).

27.3 Metal-based nanoparticles

Nanotechnology is a noteworthy part that has revealed numerous applications in optoelectronics, agriculture, medicine, textiles, etc. (Dasgupta et al., 2015; Mehtab et al., 2019; Nanda & Majeed 2014; Tabish, Malik et al., 2021; Yasin, Arif, Mehtab, Lu et al., 2020) because of the exceptional properties of nanomaterials. NP have many exceptional properties that include optical, magnetic, physiochemical, and mechanical properties, which make NP suitable for many applications. Because of the importance of nanotechnology in the field of science and technology many countries doing advanced research in that field. Richard Feynman gives the first concept of nanotechnology in his talk that "There's Plenty of Room at the Bottom" (The title of the classic talk at the yearly conference of the American Physical Society, the California Institute of Technology in 1959; Feynman, 1959). The commonly used method for the fabrication of NP includes chemical, physical, and biological methods. The chemical method shows the best results in the formation of NP because a large number of NP can produce in a short time with good control on particle size and distribution. (Alkilany, Bani Yaseen, & Kailani, 2015; He, Protesescu, Caputo, Krumeich, & Kovalenko, 2015).

With the alteration of concentration of reacting chemicals and by managing the reaction conditions, various types of nanoparticle shapes can be formed using chemical methods. Besides all of these advantages, chemical methods also have some really bad effects on the environment while dealing with toxic chemicals and form hazardous wastes. Similarly, many physical methods can also be used for the production of metal NP. The commonly used physical methods are laser ablation or cluster beam deposition, sputter deposition, microwave-assisted synthesis, and so forth. But the physical method also contains the involvement of high temperature, pressure, and radiation (Alzahrani, Sharfalddin, & Alamodi, 2015; Dzido, Markowski, Małachowska-Jutsz, Prusik, & Jarzębski, 2015). Therefore biogenic synthesis is also a source of big interest for scientists. The development of experimental procedures for the production of NP of definite size and shape is an essential improvement of nanotechnology (Durán et al., 2010; Vala, Shah, & Patel, 2014). Hence, that is why researchers are predicting biological systems that can work as effective systems for the production of many different metal NP (Kar, Murmu, Saha, Tandon, & Acharya, 2014). Biological methods contain various types of microorganisms such as bacteria, fungi, algae, cyanobacteria, actinomycetes, myxobacteria, and plants, which are being proficiently used for the formation of both NP like intracellular and extracellular (Adil et al., 2015; Ahmed, Ahmad, Swami, & Ikram, 2016; Chen et al., 2014; Patel, Berthold, Puranik, & Gantar, 2015; Singh, Shedbalkar, Wadhwani, & Chopade, 2015). Due to the higher rate of success, microorganisms are gaining importance in the formation of NP (Adil et al., 2015; Chen et al., 2014; Patel et al., 2015; Singh et al., 2015; Shelar & Chavan, 2014). The most important microorganism used for the formation of NP is fungi because they are universally distributed in nature and play a vital role in the production of metal NP (Kar et al., 2014; Rai, Ribeiro, Mattoso, & Durán, 2015).

NPs are responsible for the fast development in the field of nanotechnology. Their special size-oriented properties make these materials essential and superior in many zones (Ali, Hira, Zafar, ul Haq, & Hussain, 2016; Yasin, Arif, Mehtab, Shakeel, Khan, etal., 2020). In recent times, various NPs have been exploited for the protection of biocorrosion and the prevention of biofilms' growth such as copper NP (CuNPs) (Zarasvand & Rai, 2016), silver NP (AgNPs) (Khowdiary, El-Henawy, Shawky, Sameeh, & Negm, 2017), zinc oxide NP (ZnONPs) (Karbowniczek et al., 2017), titanium oxide nanoparticle (TiO_2NPs) (Ma, Shi, Di, Yao, & Liu, 2009; Nguyen et al., 2021), iron NP (FeNPs) (Singh, Singh, & Bala, 2020), silica NP (SiO_2NPs) (Hench & Paschall, 1973) and gold NP (AuNPs) (Gopinath et al., 2016).

27.3.1 Copper nanoparticles

Copper has been used as an economical and active constituent in animal tissues, sterilizing liquids and environments, so, CuNPs can be an important antimicrobial agent (Mahmoodi, Elmi, & Hallaj-Nezhadi, 2018). Because of the size and surface area, CuNPs have exceptional physical and chemical properties. The properties of CuNPs as an antibacterial agent have been extensively examined (Chatterjee, Chakraborty, & Basu, 2014; Guo et al., 2017; Vimbela, Ngo, Fraze, Yang, & Stout, 2017). Alloys contain 65% or above copper inhibit bacteria within 90 min (George, 2005). It is difficult to fully understand the mechanism behind the intercalation of antibacterial activity of metal oxide (Huh and Kwon, 2011). It is supposed that CuNPs affect carboxylic groups and amine groups of peptidoglycans, which are the major constituents of the bacterial cell wall, which causes the death of bacteria due to membrane tear (Chatterjee et al., 2014; Mahmoodi et al., 2018; Vimbela et al., 2017).

Kalajahi, Rasekh, Yazdian, Neshati, and Taghavi (2020) utilized copper NP doped carbon quantum dots (Cu/CQDs)-based nanohybrid as an inhibitor for MIC. The antibacterial efficacy was determined by a dose-response test. It was confirmed that 50 ppm is the optimum concentration of Cu/CQDs to inhibit the SRB and the optimum time is 24 h, the reduction in the SRB population is high (the SRB growth is 10^1 cells/mL). Besides, the electrochemical measurement of MIC was examined on X60 steel in the mixture of sterilized seawater (SSW) with 50 ppm Cu/QDs and 1.0×10^5 cell/mL of SRB as well as in the individual environments. The results showed an increase in corrosion protection due to the Cu/QDs. It has shown that an increased charge transfer resistance (R_{ct}) value is obtained after fitting the curves, as well as the corresponding scanning electron microscope (SEM) morphology in the given environments, demonstrated changes of the surface after incubation for 15 days to understand the effect of the Cu/QDs as shown in Fig. 27–1. After 15 days of incubation, the Cu/QDs formed a homogenous layer on the immersed surface by preventing the accessibility of aggressive agents toward the surface. Similarly, the polarization test revealed a decrease in the corrosion current density for Cu/QDs-SRB system.

Ituen, Ekemini, Yuanhua, Li, and Singh (2020) studied the effect of extract of tangerine (Citrus reticulata) peels (ETP) and ETP mediated Cu NP (ETP-CuNPs) for antimicrobial corrosion protection application. The ETP-CuNPs were utilized as a corrosion inhibitor in the Desulfovibrio sp and 1 M HCl acid environments. They studied the efficacy of ETP and ETP-CuNPs at different temperatures that is, 303K and 333K on X80 steel. The produced ETP-CuNPs size is 54 – 72 nm, which was confirmed by Transmission electron microscopy

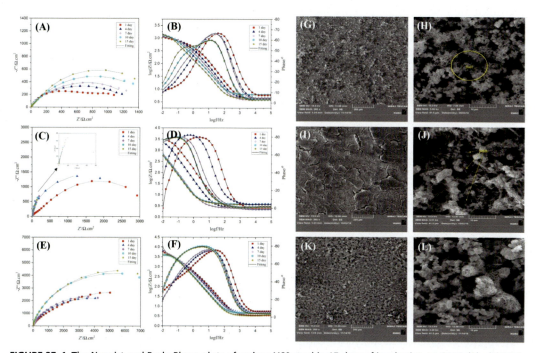

FIGURE 27–1 The Nyquist and Bode-Phase plots of carbon X60 steel in 15 days of incubation, at 30°C: (A)–(B) SSW, (C)–(D) SRB; (E)–(F) Cu/CQDs-SRB. Field emission scanning electron microscopy (FESEM) analysis of corrosion and biofilm products on X60 steel coupons after 15 days, (G)–(H) SSW, (I)–(J) SRB, and (K)–(L) Cu/CQD-SRB (Kalajahi et al., 2020). *SRB, sulfate-reducing bacteria; SSW, sterilized seawater.*

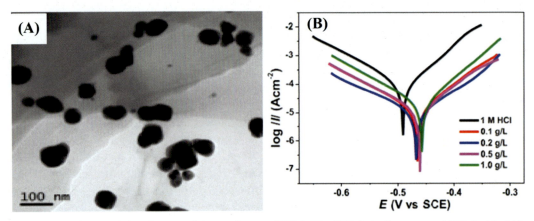

FIGURE 27-2 (A) Transmission electron microscopy image of ETP-CuNPs, (B) Polarization curves of X80 steel in 1 M HCl with or without different concentrations of ETP-CuNPs (Ituen et al., 2020).

with round-shaped particulates and having no agglomeration as shown in Fig. 27-2A. The three-log reduction in *Desulfovibio* sp. population was caused by 1.96 mg/L minimum inhibitory concentration of ETP-CuNPs because of the more electron active sites and showed more inhibition than ETP in both aggressive environments. The polarization curves displayed the corrosion inhibition performance on X80 steel with different concentrations of ETP-CuNPs in 1 M HCl as shown in Fig. 27-2B.

Tabesh, Salimijazi, Kharaziha, Mahmoudi, and Hejazi (2019) employed the electrophoretic deposition method to consume the Cu NP in chitosan-based nanocomposite coatings for the antibacterial and corrosion resistance of the 316 L stainless steel substrate. They reported the concentration effect of CuNPs in the chitosan-copper nanocomposite coatings. The results concluded that the CuNPs with a particle size of 11 ± 6 nm showed excellent corrosion resistance properties in simulated body fluid (SBF) with the current density value of 20 $\mu A/cm^2$ as well as exhibited lethal influence on both positive and negative gram bacteria. In another study, it was found that copper oxide NP (CuONPs) unveiled good biocorrosion protection ability with a minimum inhibitory concentration of 100 $\mu g/mL$ (Zarasvand & Rai, 2016).

27.3.2 Silver nanoparticles

AgNPs have exhibited improved antimicrobial properties related to both the silver ion reservoir and direct interaction with microorganisms (Tamayo et al., 2014), placing them as promising candidates for their use as biocorrosion inhibitors. González et al. (2019) synthesized silanol-based hybrid polymers doped with AgNPs and encapsulated by SiO_2 using a sol-gel technique, which was used for the biocorrosion protection of 2024T3 Al-alloy in *P. aeruginosa*. The AgNPs comprised the size of about 30 nm and showed higher antibacterial properties. Moreover, the resultant coating displayed an increase in hydrophobicity and

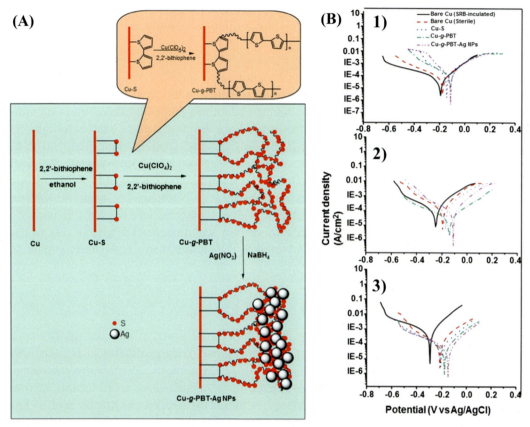

FIGURE 27–3 (A) Schematic illustration of the 2,2′-bithiophen grafting polymerization process followed by successive immobilization of AgNPs on a premodified copper surface; (B) polarization curves of the bare C, Cu–S, Cu-g-PBT and Cu-g-PBT-AgNP coatings after inoculation in the *Desulfovibrio desulfuricans* for (1) 5, (2) 14, and (3) 30 days (Wan et al., 2010).

an impedance measurement confirmed the effective corrosion protection against *P. aeruginosa*. As reported elsewhere, AgNPs incorporated with bithiophene polymer (PBT) were considered as the inhibitor for the microbial corrosion and bacterial adhesion as shown in Fig. 27–3A. Electrochemical impedance spectroscopy (EIS) and Tafel polarization was employed to validate the corrosion and adhesion of the SRB (*Desulfovibrio desulfuricans*) on the functionalized Cu surface. The experimental results evinced that the Cu-g-PBT-AgNP had significant inhibition over prolonged time. The corrosion protection efficacy was about 98% as compared to bare Cu substrate after 30 days of immersion as shown in Fig. 27–3B (Wan, Yuan, Neoh, & Kang, 2010).

San Keskin, Yaylaci, Durgun, Deniz, and Nazır (2021) proposed a biosynthesis method of AgNPs via *Lysinibacillus* sp. NOSK bacteria. They considered the green AgNPs with spherical shape and 42 nm size as a corrosion inhibitor for the copper surface in the marine environment affected by a halophilic bacterium, *Heterixalus variabilis*. In recent times, AgNPs were also reported as the antifungal activity for *Aspergillus* species (Bocate et al., 2019).

Plant extract functionalized AgNPs were used as efficient mild steel (MS1010) inhibitor against *Bacillus thuringiensis EN2*, from the cooling tower. The EIS and surface study proved that the AgNPs obstruct the formation of biofilms on the steel surface with the efficiency of 77% as compared to just plant extract (inhibition efficiency 52%) It was demonstrated that AgNPs formed a barrier protective layer on the surface which inhibit the bacterial action and adhesion thereby protect from corrosion (Narenkumar et al., 2018). In another study, a cationic surfactant modified AgNPs were investigated in SRB (*Desulfomonas pigra*) serial dilution method. The nanomaterials were able to kill all the SRB cells even with high concentrations (Aiad, El-Sukkary, Soliman, El-Awady, & Shaban, 2014). In a similar study, the AgNPs modified by thiol surfactant was used to evaluate the antibacterial activity in SRB. The antibacterial activity was improved might be due to the formation of reaction products that are involved in the biofilm distraction (Azzam, Sami, & Kandile, 2012).

27.3.3 Zinc oxide nanoparticles

ZnONPs have been projected as an impactful antibacterial agent in different applications. ZnONPs is a significant ceramic material that has good biocompatibility to human cells (Feng, Wei, Shuai, & Peng, 2014; Mousa et al., 2018). Furthermore, ZnONPs show attractive antimicrobial properties, which can firmly inhibit microbial adhesion, reproduction, biofilm formation (Karbowniczek et al., 2017). Mousa et al. (2018) applied poly(lactic acid) (PLA) embedded ZnONPs coating on AZ31 Mg alloy and found that the coating had good biocorrosion resistance and antibacterial properties.

However, the ZnONPs have a very harmful ecological effect owing to their toxic nature (Rasool & Lee, 2016). To resolve this problem, the NPs are intermixed with some antimicrobial polymers, which not only decrease the cytotoxicity and environmental impact but also increase the stability and performance of the composite coating without influencing their functional properties. For instance, Rasheed et al. (2019) suggested that chitosan-ZnONPs nanocomposite (CZNC) was consumed for the inhibition of mixed SRBs culture. The CZNC synthesized morphology structure is shown in Fig. 27–4A and B in which ZnONPs are intermixed with the chitosan matrix. The inhibition efficiency of CZNC in SRB was concentration-dependent and at 250 μg/mL CZNC-10(with 10% ZnONPs) offered more than 73% inhibition of sulfate reduction (Fig. 27–4C). The microbial inhibition performance of CZNC-10 was examined on the carbon steel substrate. The EIS analysis evinced the charge transfer resistance (R_{ct}) value of CZNC-10 was increased 3.2 times than the control sample with the maximum inhibition capability of 74%.

In another study, the multifunctional polypyrrole/zinc oxide (Ppy/ZnO)-based composite coatings were synthesized using cyclic voltammetry method targeting to improve the antimicrobial induced corrosion, biocompatibility, and antibacterial activity of Mg alloy. The SEM

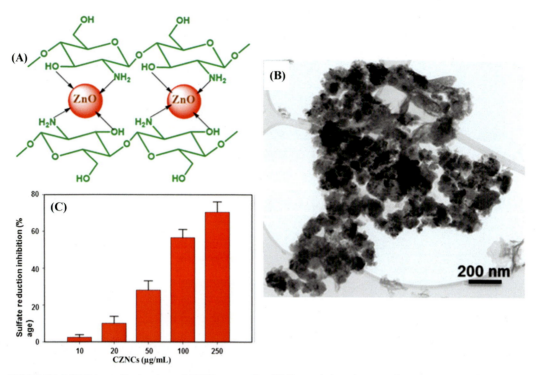

FIGURE 27–4 (A) Schematic structure of CZNCs composite, (B) Transmission electron microscopy image of CZNCs composite, (C) Different concentration of CZNCs (0–250 μg/mL) for SRBs sludge biomass (1000 mg volatile suspended solids) in injecting seawater at 35°C (Rasheed et al., 2019), *CZNCs, chitosan-ZnONPs nanocomposites; SRB, sulfate-reducing bacteria.*

micrographs confirmed the bare Mg surface, Ppy, and Ppy/ZnO coating surface as shown in Fig. 27–5A. The electrochemical results (Fig. 27–5B) showed that the Ppy/ZnO significantly reduced corrosion rates in SBF and representing the corrosion current density value of 2.4×10^{-6} μA/cm² as well as the $R_{ct} = 1.4 \times 10^4$ Ω cm². The authors suggested that the Mg alloy coated with Ppy/ZnO had shown promising cytocompatibility and osteogenic properties. Fig. 27–5C showed the SEM micrograph of the corroded surface after immersion for 15 days (Guo et al., 2020).

27.3.4 Titanium oxide nanoparticle

TiO$_2$NPs have shown special antibacterial properties and are appropriate for the preparation of antibacterial products. In artificial seawater and SRB culture media, the antimicrobial corrosion efficiency of TiO$_2$NPs in amorphous Ni-P-Cr composite coating has been measured. The passivation effect was observed when the Ni−P−Cr/TiO$_2$ nanocomposite coating was immersed in the control and SRB culture medium (Ma et al., 2009). The photocatalytic bactericidal property has been inducted into the nanocomposite

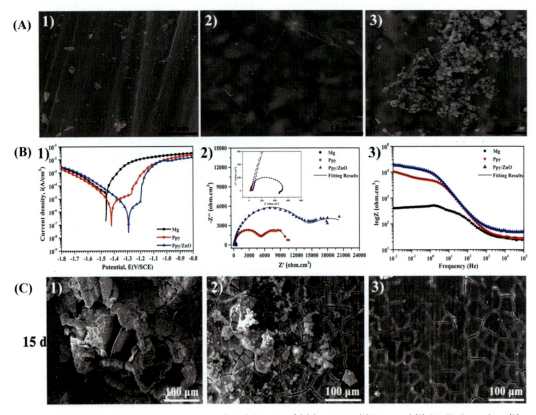

FIGURE 27–5 (A) Scanning electron microscope (SEM) images of (1) bare Mg, (2) Ppy, and (3) Ppy/ZnO coating; (B) electrochemical measurement: (1) potentiodynamic polarization, (2) Nyquist, and (3) Bode-impedance of the coatings in simulated body fluid electrolyte; (C) SEM images of (1) bare Mg, (2) Ppy, and (3) Ppy/ZnO coating after immersion for 15 days in the electrolyte (Guo et al., 2020).

coatings using Ce-doped TiO_2NPs coatings on 304 stainless steel in the SRB environment. The Ce-doped TiO_2 coating exhibited excellent photocatalytic bactericidal performance (95%) than pure TiO_2 film (85%) (Wang, Wang, Hong, & Yin, 2010). In another study, Zaveri, McEwen, Karpagavalli, and Zhou (2010) investigated the bioimplant Ti–6Al–4V alloy coated with TiO_2NPs in three different SBF solutions. The different electrochemical analyses revealed that the TiO_2NPs (50–100 nm) coated Ti–6Al–4V surface showed enhanced corrosion resistance properties due to an increase in the preexisting oxide layer. In a similar work, Höhn and Virtanen (2015) studied the biocompatibility and microbial corrosion properties of Ti–6Al–4V alloy followed by the TiO_2NPs coating by spin coating technique. The effect of the coating on the metal ion release, corrosion behavior, and biomimetic apatite was investigated in Dulbecco's Modified Eagle's Medium (DMEM) at 37.5°C with 5% CO_2 content in the incubator. The micrographs of TiO_2NPs coating on the surface as shown in Fig. 27–6A and B. The bare Ti–6Al–4V alloy contains Al metal and can release Al from its surface. The Al

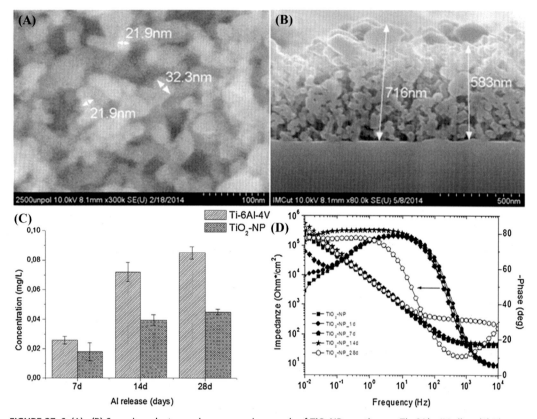

FIGURE 27–6 (A)–(B) Scanning electron microscope micrograph of TiO$_2$NPs coating on Ti–6Al–4V alloy, (C) The release concentration of Al via ICP-OES analysis of the electrolyte after immersion of samples over different time, and (D) Bode-plot of Ti–6Al–4V alloy with TiO$_2$NPs coating (Höhn & Virtanen, 2015).

release was determined from the Ti–6Al–4V alloy with or without TiO$_2$NPs by ICP-OES analysis as shown in Fig. 27–6C. It was found that there was less amount of Al release as compared to bare Ti–6Al–4V alloy after immersion in DMEM for 28 days. The impedance value demonstrated the good corrosion resistance properties in DMEM electrolyte for a prolonged time (Fig. 27–6D).

27.3.5 Iron nanoparticles

FeNPs are one of the most advanced and used biotechnological and microbiological nanomaterials owing to their exceptional physicochemical properties. Fe$_3$O$_4$NPs are applied as a corrosion inhibitor for the protection of iron from iron-corroding bacteria (ICB) such as *Halanaerobium sp.* Das, Kerkar, Meena, and Mishra (2017) studied the toxicological effect of Fe$_3$O$_4$NPs against ICB and observed the antimicrobial protection ability. The co-precipitation method was used to synthesize the NPs which was confirmed by TEM as shown in

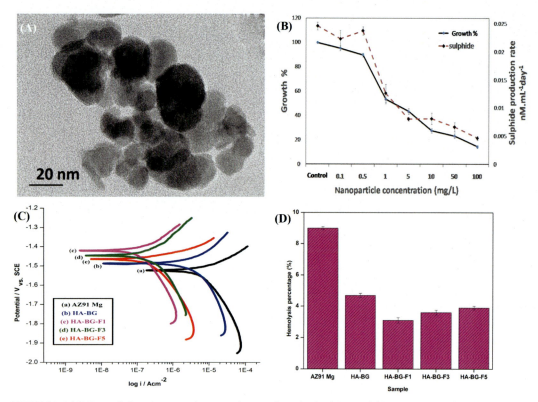

FIGURE 27–7 (A) Transmission electron microscopy image of synthesized FeNPs, (B) Iron nanoparticles concentration versus iron-corroding bacteria and sulfide production rate plot (Das et al., 2017), (C) Polarization curves of AZ91 with or without HA-BG-Fe coating, and (D) Hemeolysis ratio plot of HA-BG-Fe coating with different concentration of Fe_3O_4NPs (Singh et al., 2020).

Fig. 27–7A. The growth efficiency of ICB *Halanaerobium sp.* strain L4, was evaluated using different concentration of Fe_3O_4NPs as shown in Fig. 27–7B, it was confirmed from the growth curve that by increasing the amount of NPs the growth was reduced. Furthermore, using 100 mg/L Fe_3O_4NPs, the sulfide production rate was decreased to 11.8%, which thereby minimize the corrosion rate. Recently, an electrophoretic deposition procedure was utilized to develop Fe_3O_4NPs-1% incorporated hydroxyapatite-bioglass-chitosan nanocomposite (HA-BG-Fe1%) coatings on AZ91 Mg alloy. The polarization curves (Fig. 27–7C) revealed the insertion of Fe_3O_4NPs reduced the i_{corr} value to 398 μA/cm², also the hemocompatibility ratio is <5% (Fig. 27–7D), which confirmed the potential materials for implant applications (Singh et al., 2020).

27.3.6 Silica nanoparticles

Silica (SiO_2) is the most common source, which is nontoxic and extremely biocompatible. Hench and Paschall (1973) demonstrated that silica plays a major role in the osteogenic

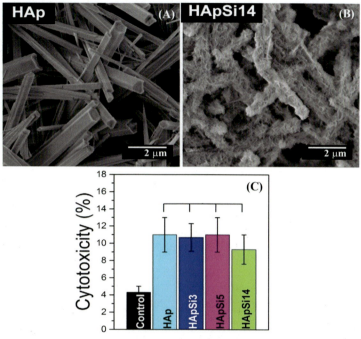

FIGURE 27–8 (A)–(B) Scanning electron microscope images of HAp and Hap-SiO$_2$ coatings, and (C) percentage cytotoxicity plot of different concentrations of HAp-Si coatings (Sutha et al., 2013).

ability and bioactivity of bioactive glass (biomaterials containing 46%–60% SiO$_2$). SiO$_2$NPs have not only a higher surface area but also have shown silanol (Si-OH) groups when in hydrated form and behave like crystal nuclei for apatites which thereby affect the bioactivity. Bartkowiak et al. (2018) prepared hydroxyapatite (HAp)-based composite coatings enriched with varying concentrations of SiO$_2$NPs through the hydrothermal process on titanium substrate. SiO$_2$NPs were considered as a bioactive agent for the HAp coatings with unique osseointegration properties. SEM observation confirmed the formation of bone-like apatite for Hap-SiO$_2$ coatings as compared to HAp after immersed in SBF for four days as shown in Fig. 27–8A and B. The cytotoxicity outcomes (Fig. 27–8C) evinced the biocompatibility of HAp-SiO$_2$ coatings for titanium implants. In another study, different concentrations of SiO$_2$NPs doped HAp-based chitosan (SH/CTS) composite coatings were developed on stainless steel (SS) substrate through a facile ultrasonication process. The enhanced corrosion resistance behavior in SBF revealed the long-term biocompatibility of SH/CTS on SS (Sutha, Kavitha, Karunakaran, & Rajendran, 2013). Xu, Sun, Jiang, Munroe, and Xie (2018) fabricated MoO$_3$-SiO$_2$ nanocomposite coating using the sputter-deposition method on 316 L SS. The intermixed nanocrystalline MoO$_3$ and amorphous SiO$_2$ coating with the thickness of 25 μm showed the greater antimicrobial property and impede the growth of SRB owing to the creation Mo(V)-S complexes. The electrochemical test corroborated that the 316 L SS exhibited significant resistance to biocorrosion when exposed to Postgate's C seawater.

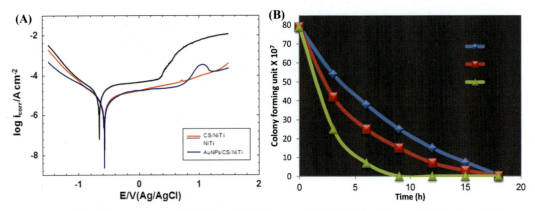

FIGURE 27–9 (A) Polarization curves of NiTi alloy with chitosan and AuNPs/CS coatings, and (B) kinetics of NiTi alloys for the antibacterial activity toward *Staphylococcus aureus* (Ahmed et al., 2014).

27.3.7 Gold nanoparticles

AuNPs are widely being used as antibacterial (Annamalai, Christina, Sudha, Kalpana, & Lakshmi, 2013), biomedical (Castañeda, Merkoçi, Pumera, & Alegret, 2007), antibiofilm (Apte et al., 2013), antifungal (Jayaseelan, Ramkumar, Rahuman, & Perumal, 2013), entomological, and parasitological applications (Jeyalalitha, Murugan, & Madhiyazhagan, 2013). Recently, owing to its inexpensive, simple, and ecofriendly protocols, the botanical synthesis of AuNPs has gained rising popularity. For example, Gopinath et al. (2016) synthesized AuNPs from the Gloriosa superba leaf extract and analyzed the antibacterial and antibiofilm activities. The 20 nm-sized AuNPs verified no toxic actions and good antibacterial characteristics. Ahmed, Fadl-allah, El-Bagoury, and El-Rab (2014) constructed the AuNPs incorporated chitosan (AuNPs/CS) composite coating by electrodeposition method on NiTi alloy. The coating reduced the Ni release from the surface by 20-fold than the uncoated NiTi alloy in Hank's solution after soaking for 4 days. Furthermore, the protection efficiency verified that the AuNPs/CS coating delivered 99.2% protection to the NiTi alloy in Hank's solution at 37°C with the corrosion current density of 0.01 $\mu A/cm^2$ as shown in Fig. 27–9A. Moreover, the AuNPs could kill 99% of the *Staphylococcus aureus* and offered rapid bacterial suppression rate (Fig. 27–9B) which affirmed the promising material for the antimicrobial application.

27.4 Conclusion

The use of nanotechnology is deemed a developing strategy to protect the metals and alloys against various microbial corrosives for many engineering applications. NP are considered as the most ecofriendly alternative biocides to protect different metals such as steel, Al, Mg, and Ti from MIC. Several synthesis methods have been developed for the formulation of NPs like, mechanical, physical, chemical, and biological, but their basic mechanism for corrosion resistance is quite similar to react with microbes and suppress them. Accordingly, different coatings techniques along with NPs have been utilized, that is, inorganic or organic nanocomposite coatings, which are responsible for the protection. Although NPs-based biocides have shown significant inhibition, however, certain limitations which are high-energy demand, low dispersibility, high cost, and/or inapplicability under some industrial conditions, which are still under consideration.

References

Adil, S. F., Assal, M. E., Khan, M., Al-Warthan, A., Siddiqui, M. R. H., & Liz-Marzán, L. M. (2015). Biogenic synthesis of metallic nanoparticles and prospects toward green chemistry. *Dalton Transactions*, *44*(21), 9709−9717.

Ahmadkhaniha, D., Järvenpää, A., Jaskari, M., Sohi, M. H., Zarei-Hanzaki, A., Fedel, M., . . . Karjalainen, L. P. (2016). Microstructural modification of pure Mg for improving mechanical and biocorrosion properties. *Journal of the Mechanical Behavior of Biomedical Materials*, *61*, 360−370.

Ahmed, R. A., Fadl-allah, S. A., El-Bagoury, N., & El-Rab, S. M. F. G. (2014). Improvement of corrosion resistance and antibacterial effect of NiTi orthopedic materials by chitosan and gold nanoparticles. *Applied Surface Science*, *292*, 390−399. Available from https://doi.org/10.1016/j.apsusc.2013.11.150.

Ahmed, S., Ahmad, M., Swami, B. L., & Ikram, S. (2016). A review on plants extract mediated synthesis of silver nanoparticles for antimicrobial applications: A green expertise. *Journal of Advanced Research*, *7*(1), 17−28.

Aiad, I., El-Sukkary, M. M., Soliman, E. A., El-Awady, M. Y., & Shaban, S. M. (2014). In situ and green synthesis of silver nanoparticles and their biological activity. *Journal of Industrial and Engineering Chemistry*, *20*(5), 3430−3439. Available from https://doi.org/10.1016/j.jiec.2013.12.031.

Aktas, D. F., Sorrell, K. R., Duncan, K. E., Wawrik, B., Callaghan, A. V., & Suflita, J. M. (2017). Anaerobic hydrocarbon biodegradation and biocorrosion of carbon steel in marine environments: The impact of different ultra low sulfur diesels and bioaugmentation. *International Biodeterioration & Biodegradation*, *118*, 45−56. Available from https://doi.org/10.1016/j.ibiod.2016.12.013.

Ali, A., Hira Zafar, M. Z., ul Haq, I., . . . Hussain, A. (2016). Synthesis, characterization, applications, and challenges of iron oxide nanoparticles. *Nanotechnology, Science and Applications*, *9*, 49.

Alkilany, A. M., Bani Yaseen, A. I., & Kailani, M. H. (2015). Synthesis of monodispersed gold nanoparticles with exceptional colloidal stability with grafted polyethylene glycol-g-polyvinyl alcohol. *Journal of Nanomaterials*, *2015*, 712359.

Alzahrani, E., Sharfalddin, A., & Alamodi, M. (2015). Microwave-hydrothermal synthesis of ferric oxide doped with cobalt. *Advances in Nanoparticles*, *4*(02), 53.

Anjum, M. J., Ali, H., Khan, W. Q., Zhao, J., & Yasin, G. (2020). *Metal/metal oxide nanoparticles as corrosion inhibitors. Corrosion protection at the nanoscale* (pp. 181−201). Elsevier.

Annamalai, A., Christina, V. L. P., Sudha, D., Kalpana, M., & Lakshmi, P. T. V. (2013). Green synthesis, characterization and antimicrobial activity of Au NPs using *Euphorbia hirta* L. leaf extract. *Colloids and Surfaces B: Biointerfaces*, *108*, 60−65. Available from https://doi.org/10.1016/j.colsurfb.2013.02.012.

Apte, M., Girme, G., Nair, R., Bankar, A., Ravi Kumar, A., & Zinjarde, S. (2013). Melanin mediated synthesis of gold nanoparticles by *Yarrowia* lipolytica. *Materials Letters*, *95*, 149−152. Available from https://doi.org/10.1016/j.matlet.2012.12.087.

Ayllon, E. S., & Rosales, B. M. (1994). Electrochemical test for predicting microbiologically influenced corrosion of aluminum and AA 7005 alloy. *Corrosion*, *50*(8), 571−575.

Azzam, E. M. S., Sami, R. M., & Kandile, N. G. (2012). Activity inhibition of sulfate reducing bacteria using some cationic thiol surfactants and their nanostructures. *American Journal of Biochemistry*, *2*(3), 29−35.

Bairi, L. R., George, R. P., & Mudali, U. K. (2012). Microbially induced corrosion of D9 stainless steel−zirconium metal waste form alloy under simulated geological repository environment. *Corrosion Science, 61,* 19−27.

Bartkowiak, A., Suchanek, K., Menaszek, E., Szaraniec, B., Lekki, J., Perzanowski, M., & Marszałek, M. (2018). Biological effect of hydrothermally synthesized silica nanoparticles within crystalline hydroxyapatite coatings for titanium implants. *Materials Science and Engineering: C, 92,* 88−95. Available from https://doi.org/10.1016/j.msec.2018.06.043.

Barton, L. L., & Tomei, F. A. (1995). Characteristics and activities of sulfate-reducing bacteria. *Sulfate-reducing bacteria* (pp. 1−32). Springer.

Beech, I. B., & Coutinho, C. (2003). Biofilms on corroding materials. In P. Lens, et al. (Eds.), *Biofilms in medicine industry and environmental biotechnology: Characteristics, analysis and control.* London: IWA Publishing.

Bevilaqua, D., Acciari, H. A., Arena, F. A., Benedetti, A. V., Fugivara, C. S., Filho, G. T., & Júnior, O. G. (2009). Utilization of electrochemical impedance spectroscopy for monitoring bornite (Cu_5FeS_4) oxidation by *Acidithiobacillus ferrooxidans*. *Minerals Engineering, 22*(3), 254−262. Available from https://doi.org/10.1016/j.mineng.2008.07.010.

Binkauskienė, E., Bučinskienė, D., & Lugauskas, A. (2017). *Long-time corrosion of metals and profiles of fungi on their surface in outdoor environments in Lithuania. Mycoremediation and environmental sustainability* (pp. 105−117). Springer.

Bocate, K. P., Reis, G. F., de Souza, P. C., Junior, A. G. O., Durán, N., Nakazato, G., . . . Panagio, L. A. (2019). Antifungal activity of silver nanoparticles and simvastatin against toxigenic species of Aspergillus. *International Journal of Food Microbiology, 291,* 79−86.

Booth, G. H. (1964). Sulphur bacteria in relation to corrosion. *Journal of Applied Bacteriology, 27*(1), 174−181. Available from https://doi.org/10.1111/j.1365-2672.1964.tb04825.x.

Campaignolle, X., & Crolet, J. L. (1997). Method for studying stabilization of localized corrosion on carbon steel by sulfate-reducing bacteria. *Corrosion, 53*(6), 440−447.

Castañeda, M. T., Merkoçi, A., Pumera, M., & Alegret, S. (2007). Electrochemical genosensors for biomedical applications based on gold nanoparticles. *Biosensors and Bioelectronics, 22*(9), 1961−1967. Available from https://doi.org/10.1016/j.bios.2006.08.031.

Chaker, V. (1989). *Effects of soil characteristics on corrosion.* ASTM International.

Chatterjee, A. K., Chakraborty, R., & Basu, T. (2014). Mechanism of antibacterial activity of copper nanoparticles. *Nanotechnology, 25*(13), 135101.

Chen, G., Yi, B., Zeng, G., Niu, Q., Yan, M., Chen, A., . . . Zhang, Q. (2014). Facile green extracellular biosynthesis of CdS quantum dots by white rot fungus *Phanerochaete chrysosporium*. *Colloids and Surfaces B: Biointerfaces, 117,* 199−205.

Chen, Y., Howdyshell, R., Howdyshell, S., & Ju, L.-K. (2014). Characterizing pitting corrosion caused by a long-term starving sulfate-reducing bacterium surviving on carbon steel and effects of surface roughness. *Corrosion, 70*(8), 767−780.

Cojocaru, A., Prioteasa, P., Szatmari, I., Radu, E., Udrea, O., & Visan, T. (2016). EIS study on biocorrosion of some steels and copper in Czapek Dox medium containing *Aspergillus niger* fungus. *Revista de Chimie, 67,* 1264−1270.

Dai, X., Wang, H., Ju, L.-K., Cheng, G., Cong, H., & Newby, B.-mZ. (2016). Corrosion of aluminum alloy 2024 caused by *Aspergillus niger*. *International Biodeterioration & Biodegradation, 115,* 1−10. Available from https://doi.org/10.1016/j.ibiod.2016.07.009.

Dannenberg, S., Kroder, M., Dilling, W., & Cypionka, H. (1992). Oxidation of H_2, organic compounds and inorganic sulfur compounds coupled to reduction of O_2 or nitrate by sulfate-reducing bacteria. *Archives of Microbiology, 158*(2), 93−99.

Das, K. R., Kerkar, S., Meena, Y., & Mishra, S. (2017). Effects of iron nanoparticles on iron-corroding bacteria. *3 Biotech, 7*(6), 385. Available from https://doi.org/10.1007/s13205-017-1018-9.

Dasgupta, N., Ranjan, S., Mundekkad, D., Ramalingam, C., Shanker, R., & Kumar, A. (2015). Nanotechnology in agro-food: From field to plate. *Food Research International, 69*, 381–400.

Durán, N., Marcato, P. D., Conti, R. D., Alves, O. L., Costa, F., & Brocchi, M. (2010). Potential use of silver nanoparticles on pathogenic bacteria, their toxicity and possible mechanisms of action. *Journal of the Brazilian Chemical Society, 21*(6), 949–959.

Dzido, G., Markowski, P., Małachowska-Jutsz, A., Prusik, K., & Jarzębski, A. B. (2015). Rapid continuous microwave-assisted synthesis of silver nanoparticles to achieve very high productivity and full yield: From mechanistic study to optimal fabrication strategy. *Journal of Nanoparticle Research, 17*(1), 1–15.

Enning, D., & Garrelfs, J. (2014). Corrosion of iron by sulfate-reducing bacteria: New views of an old problem. *Applied and Environmental Microbiology, 80*(4), 1226–1236.

Feng, P., Wei, P., Shuai, C., & Peng, S. (2014). Characterization of mechanical and biological properties of 3-D scaffolds reinforced with zinc oxide for bone tissue engineering. *PLoS One, 9*(1), e87755.

Feynman, R. (1959). *There is plenty of room at the bottom: An invitation to enter a new field of physics.*

Fida, T. T., Chen, C., Okpala, G., & Voordouw, G. (2016). Implications of limited thermophilicity of nitrite reduction for control of sulfide production in oil reservoirs. *Applied and Environmental Microbiology, 82*(14), 4190–4199.

H.C. Flemming. (1996). *Biofouling and microbiologically influenced corrosion (MIC)-An economical and technical overview. Microbial Deterioration of Materials.*

Flemming, H. C., & Schaule, G. (1994). Microbial deterioration of materials-biofilm and biofouling: Biofouling. *Mikrobielle Werkstoffzerstoerung-Biofilm und Biofouling: Biofouling, Werkstoffe und Korrosion;(Germany), 45*(1).

Forte Giacobone, A. F., Rodriguez, S. A., Burkart, A. L., & Pizarro, R. A. (2011). Microbiological induced corrosion of AA 6061 nuclear alloy in highly diluted media by *Bacillus cereus* RE 10. *International Biodeterioration & Biodegradation, 65*(8), 1161–1168. Available from https://doi.org/10.1016/j.ibiod.2011.08.012.

Gaines, R. H. (1910). Bacterial activity as a corrosive influence in the soil. *Journal of Industrial & Engineering Chemistry, 2*(4), 128–130. Available from https://doi.org/10.1021/ie50016a003.

George, M. W. (2005). Nanoscience, nanotechnology, and chemistry. *Small (Weinheim an der Bergstrasse, Germany), 1*, 172–179.

Gieg, L. M., Jack, T. R., & Foght, J. M. (2011). Biological souring and mitigation in oil reservoirs. *Applied Microbiology and Biotechnology, 92*(2), 263–282. Available from https://doi.org/10.1007/s00253-011-3542-6.

González, E. A., Leiva, N., Vejar, N., Sancy, M., Gulppi, M., Azócar, M. I., ... Páez, M. A. (2019). Sol–gel coatings doped with encapsulated silver nanoparticles: Inhibition of biocorrosion on 2024-T_3 aluminum alloy promoted by *Pseudomonas aeruginosa*. *Journal of Materials Research and Technology, 8*(2), 1809–1818. Available from https://doi.org/10.1016/j.jmrt.2018.12.011.

Gopinath, K., Kumaraguru, S., Bhakyaraj, K., Mohan, S., Venkatesh, K. S., Esakkirajan, M., ... Arumugam, A. (2016). Green synthesis of silver, gold and silver/gold bimetallic nanoparticles using the *Gloriosa superba* leaf extract and their antibacterial and antibiofilm activities. *Microbial Pathogenesis, 101*, 1–11. Available from https://doi.org/10.1016/j.micpath.2016.10.011.

Gu, T., Jia, R., Unsal, T., & Xu, D. (2019). Toward a better understanding of microbiologically influenced corrosion caused by sulfate reducing bacteria. *Journal of Materials Science & Technology, 35*(4), 631–636. Available from https://doi.org/10.1016/j.jmst.2018.10.026.

Guo, J., Gao, S.-H., Lu, J., Bond, P. L., Verstraete, W., & Yuan, Z. (2017). Copper oxide nanoparticles induce lysogenic bacteriophage and metal-resistance genes in *Pseudomonas aeruginosa* PAO1. *ACS Applied Materials & Interfaces, 9*(27), 22298–22307.

Guo, Y., Jia, S., Qiao, L., Su, Y., Gu, R., Li, G., & Lian, J. (2020). A multifunctional polypyrrole/zinc oxide composite coating on biodegradable magnesium alloys for orthopedic implants. *Colloids and Surfaces B: Biointerfaces, 194*, 111186. Available from https://doi.org/10.1016/j.colsurfb.2020.111186.

Harbulakova, V. O., Estokova, A., Stevulova, N., Luptáková, A., & Foraiova, K. (2013). Current trends in investigation of concrete biodeterioration. *Procedia Engineering, 65*, 346–351.

He, M., Protesescu, L., Caputo, R., Krumeich, F., & Kovalenko, M. V. (2015). A general synthesis strategy for monodisperse metallic and metalloid nanoparticles (In, Ga, Bi, Sb, Zn, Cu, Sn, and their alloys) via in situ formed metal long-chain amides. *Chemistry of Materials, 27*(2), 635–647.

Hench, L. L., & Paschall, H. A. (1973). Direct chemical bond of bioactive glass-ceramic materials to bone and muscle. *Journal of Biomedical Materials Research, 7*(3), 25–42.

Höhn, S., & Virtanen, S. (2015). Biocorrosion of TiO_2 nanoparticle coating of Ti–6Al–4V in DMEM under specific in vitro conditions. *Applied Surface Science, 329*, 356–362. Available from https://doi.org/10.1016/j.apsusc.2014.12.114.

Hu, Z. S., & Dong, J. X. (1998). Study on antiwear and reducing friction additive of nanometer titanium oxide. *Wear, 216*(1), 92–96. Available from https://doi.org/10.1016/S0043-1648(97)00252-4.

Huh, A. J., & Kwon, Y. J. (2011). "Nanoantibiotics": A new paradigm for treating infectious diseases using nanomaterials in the antibiotics resistant era. *Journal of Controlled Release, 156*(2), 128–145.

Ilhan-Sungur, E., Unsal-Istek, T., & Cansever, N. (2015). Microbiologically influenced corrosion of galvanized steel by *Desulfovibrio* sp. and *Desulfosporosinus* sp. in the presence of Ag–Cu ions. *Materials Chemistry and Physics, 162*, 839–851.

Ituen, E., Ekemini, E., Yuanhua, L., Li, R., & Singh, A. (2020). Mitigation of microbial biodeterioration and acid corrosion of pipework steel using Citrus reticulata peels extract mediated copper nanoparticles composite. *International Biodeterioration & Biodegradation, 149*, 104935. Available from https://doi.org/10.1016/j.ibiod.2020.104935.

Iverson, W. P. (1987). Microbial corrosion of metals. In A. I. Laskin (Ed.), *Advances in applied microbiology* (pp. 1–36). Academic Press.

Jabbar, A., Yasin, G., Khan, W. Q., Anwar, M. Y., Korai, R. M., Nizam, M. N., & Muhyodin, G. (2017). Electrochemical deposition of nickel graphene composite coatings: Effect of deposition temperature on its surface morphology and corrosion resistance. *RSC Advances, 7*(49), 31100–31109.

Jacobson, G. A. (2007). Corrosion at Prudhoe Bay: A lesson on the line. *Materials Performance, 46*(8).

Jayaseelan, C., Ramkumar, R., Rahuman, A. A., & Perumal, P. (2013). Green synthesis of gold nanoparticles using seed aqueous extract of *Abelmoschus esculentus* and its antifungal activity. *Industrial Crops and Products, 45*, 423–429. Available from https://doi.org/10.1016/j.indcrop.2012.12.019.

Jeyalalitha, T., Murugan, K., & Madhiyazhagan, P. (2013). Bioefficacy of plant-mediated gold nanoparticles and *Anthocepholus cadamba* on filarial vector, *Culex quinquefasciatus* (Insecta: Diptera: Culicidae). *Parasitology Research, 112*(3), 1053–1063.

Jia, R., Unsal, T., Xu, D., Lekbach, Y., & Gu, T. (2019). Microbiologically influenced corrosion and current mitigation strategies: A state of the art review. *International Biodeterioration & Biodegradation, 137*, 42–58. Available from https://doi.org/10.1016/j.ibiod.2018.11.007.

Jia, R., Yang, D., Xu, D., & Gu, T. (2017). Anaerobic corrosion of 304 stainless steel caused by the *Pseudomonas aeruginosa* biofilm. *Frontiers in Microbiology, 8*, 2335.

Juzeliūnas, E., Ramanauskas, R., Lugauskas, A., Leinartas, K., Samulevičienė, M., Sudavičius, A., & Juškėnas, R. (2007). Microbially influenced corrosion of zinc and aluminium—Two-year subjection to influence of *Aspergillus niger*. *Corrosion Science, 49*(11), 4098–4112. Available from https://doi.org/10.1016/j.corsci.2007.05.004.

Kalajahi, S. T., Rasekh, B., Yazdian, F., Neshati, J., & Taghavi, L. (2020). Green mitigation of microbial corrosion by copper nanoparticles doped carbon quantum dots nanohybrid. *Environmental Science and Pollution Research, 27*(32), 40537–40551. Available from https://doi.org/10.1007/s11356-020-10043-4.

Kar, P. K., Murmu, S., Saha, S., Tandon, V., & Acharya, K. (2014). Anthelmintic efficacy of gold nanoparticles derived from a phytopathogenic fungus, *Nigrospora oryzae*. *PLoS One, 9*(1), e84693.

Karbowniczek, J., Cordero-Arias, L., Virtanen, S., Misra, S. K., Valsami-Jones, E., Tuchscherr, L., ... Boccaccini, A. R. (2017). Electrophoretic deposition of organic/inorganic composite coatings containing ZnO nanoparticles exhibiting antibacterial properties. *Materials Science and Engineering: C, 77*, 780–789. Available from https://doi.org/10.1016/j.msec.2017.03.180.

Kelly, D. P., & Wood, A. P. (2000). Reclassification of some species of *Thiobacillus* to the newly designated genera *Acidithiobacillus* gen. nov., *Halothiobacillus* gen. nov. and *Thermithiobacillus* gen. nov. *International Journal of Systematic and Evolutionary Microbiology, 50*(2), 511–516.

Khan, M. A., Wang, Y., Anjum, M. J., Yasin, G., Malik, A., Nazeer, F., ... Zhang, H. (2020). Effect of heat treatment on the precipitate behaviour, corrosion resistance and high temperature tensile properties of 7055 aluminum alloy synthesis by novel spray deposited followed by hot extrusion. *Vacuum, 174*, 109185.

Khowdiary, M. M., El-Henawy, A. A., Shawky, A. M., Sameeh, M. Y., & Negm, N. A. (2017). Synthesis, characterization and biocidal efficiency of quaternary ammonium polymers silver nanohybrids against sulfate reducing bacteria. *Journal of Molecular Liquids, 230*, 163–168. Available from https://doi.org/10.1016/j.molliq.2017.01.022.

Kobrin, G. (1976). Corrosion by microbiological organisms in natural waters. *Materials Performance (MP), 15*(7).

Leathen, W. W., Kinsel, N. A., & Braley, S. A., Sr (1956). *Ferrobacillus ferrooxidans*: A chemosynthetic autotrophic bacterium. *Journal of Bacteriology, 72*(5), 700.

Li, Y., Xu, D., Chen, C., Li, X., Jia, R., Zhang, D., ... Gu, T. (2018). Anaerobic microbiologically influenced corrosion mechanisms interpreted using bioenergetics and bioelectrochemistry: A review. *Journal of Materials Science & Technology, 34*(10), 1713–1718.

Little, B., Staehle, R., & Davis, R. (2001). Fungal influenced corrosion of post-tensioned cables. *International Biodeterioration & Biodegradation, 47*(2), 71–77. Available from https://doi.org/10.1016/S0964-8305(01)00039-7.

Little, B., Wagner, P., & Mansfeld, F. (1992). An overview of microbiologically influenced corrosion. *Electrochimica Acta, 37*(12), 2185–2194.

Little, B. J., Ray, R. I., & Pope, R. K. (2000). Relationship between corrosion and the biological sulfur cycle: A review. *Corrosion, 56*(4), 433–443.

Liu, H., Gu, T., Asif, M., Zhang, G., & Liu, H. (2017). The corrosion behavior and mechanism of carbon steel induced by extracellular polymeric substances of iron-oxidizing bacteria. *Corrosion Science, 114*, 102–111. Available from https://doi.org/10.1016/j.corsci.2016.10.025.

Lou, Y., Lin, L., Xu, D., Zhao, S., Yang, C., Liu, J., ... Yang, K. (2016). Antibacterial ability of a novel Cu-bearing 2205 duplex stainless steel against *Pseudomonas aeruginosa* biofilm in artificial seawater. *International Biodeterioration & Biodegradation, 110*, 199–205. Available from https://doi.org/10.1016/j.ibiod.2016.03.026.

Ma, J., Shi, Y., Di, J., Yao, Z., & Liu, H. (2009). Effect of TiO_2 nanoparticles on anticorrosion property in amorphous Ni-P-Cr composite coating in artificial seawater and microbial environment. *Materials and Corrosion, 60*(4), 274–279. Available from https://doi.org/10.1002/maco.200805066.

Maeda, T., Negishi, A., Komoto, H., Oshima, Y., Kamimura, K., & Sugio, T. (1999). Isolation of iron-oxidizing bacteria from corroded concretes of sewage treatment plants. *Journal of Bioscience and Bioengineering, 88*(3), 300–305. Available from https://doi.org/10.1016/S1389-1723(00)80013-4.

Mahmoodi, S., Elmi, A., & Hallaj-Nezhadi, S. (2018). Copper nanoparticles as antibacterial agents. *Journal of Molecular Pharmaceutics and Organic Process Research, 6*(1), 1–7.

Makhlouf, A. S. H. (2014). Techniques for synthesizing and applying smart coatings for material protection. In A. S. H. Makhlouf (Ed.), *Handbook of smart coatings for materials protection* (pp. 56–74). Woodhead Publishing.

Manam, N. S., Harun, W. S. W., Shri, D. N. A., Ghani, S. A. C., Kurniawan, T., Ismail, M. H., & Ibrahim, M. H. I. (2017). Study of corrosion in biocompatible metals for implants: A review. *Journal of Alloys and Compounds, 701*, 698–715.

Mansfeld, F., & Little, B. (1991). A technical review of electrochemical techniques applied to microbiologically influenced corrosion. *Corrosion Science*, *32*(3), 247–272. Available from https://doi.org/10.1016/0010-938X(91)90072-W.

Martins, J. I., Costa, S. C., Bazzaoui, M., Gonçalves, G., Fortunato, E., & Martins, R. (2006). Electrodeposition of polypyrrole on aluminium in aqueous tartaric solution. *Electrochimica Acta*, *51*(26), 5802–5810. Available from https://doi.org/10.1016/j.electacta.2006.03.015.

Mehtab, T., Yasin, G., Arif, M., Shakeel, M., Korai, R. M., Nadeem, M., ... Lu, X. (2019). Metal-organic frameworks for energy storage devices: Batteries and supercapacitors. *Journal of Energy Storage*, *21*, 632–646. Available from https://doi.org/10.1016/j.est.2018.12.025.

Moradi, M., Duan, J., Ashassi-Sorkhabi, H., & Luan, X. (2011). De-alloying of 316 stainless steel in the presence of a mixture of metal-oxidizing bacteria. *Corrosion Science*, *53*(12), 4282–4290. Available from https://doi.org/10.1016/j.corsci.2011.08.043.

Moradi, M., Song, Z., Yang, L., Jiang, J., & He, J. (2014). Effect of marine *Pseudoalteromonas* sp. on the microstructure and corrosion behaviour of 2205 duplex stainless steel. *Corrosion Science*, *84*, 103–112.

Mousa, H. M., Abdal-hay, A., Bartnikowski, M., Mohamed, I. M. A., Yasin, A. S., Ivanovski, S., ... Kim, C. S. (2018). A multifunctional zinc oxide/poly(lactic acid) nanocomposite layer coated on magnesium alloys for controlled degradation and antibacterial function. *ACS Biomaterials Science & Engineering*, *4*(6), 2169–2180. Available from https://doi.org/10.1021/acsbiomaterials.8b00277.

Nanda, A., & Majeed, S. (2014). Enhanced antibacterial efficacy of biosynthesized AgNPs from *Penicillium glabrum* (MTCC1985) pooled with different drugs. *International Journal of PharmTech Research*, *6*, 217–223.

Narenkumar, J., Parthipan, P., Madhavan, J., Murugan, K., Marpu, S. B., Suresh, A. K., & Rajasekar, A. (2018). Bioengineered silver nanoparticles as potent anti-corrosive inhibitor for mild steel in cooling towers. *Environmental Science and Pollution Research*, *25*(6), 5412–5420.

Nguyen, V. T., Tabish, M., Yasin, G., Bilal, M., Nguyen, T. H., Van, C. P., ... Nguyen, T. A. (2021). A facile strategy for the construction of TiO_2/Ag nanohybrid-based polyethylene nanocomposite for antimicrobial applications. *Nano-Structures & Nano-Objects*, *25*, 100671. Available from https://doi.org/10.1016/j.nanoso.2021.100671.

Novelli, G. D. (1944). Assimilation of petroleum hydrocarbons by sulfate-reducing bacteria. *Journal of Bacteriology*, *7*, 47–48.

Patel, V., Berthold, D., Puranik, P., & Gantar, M. (2015). Screening of cyanobacteria and microalgae for their ability to synthesize silver nanoparticles with antibacterial activity. *Biotechnology Reports*, *5*, 112–119.

Qu, Q., Li, S., Li, L., Zuo, L., Ran, X., Qu, Y., & Zhu, B. (2017). Adsorption and corrosion behaviour of Trichoderma harzianum for AZ31B magnesium alloy in artificial seawater. *Corrosion Science*, *118*, 12–23. Available from https://doi.org/10.1016/j.corsci.2017.01.005.

Qu, Q., Wang, L., Li, L., He, Y., Yang, M., & Ding, Z. (2015). Effect of the fungus, *Aspergillus niger*, on the corrosion behaviour of AZ31B magnesium alloy in artificial seawater. *Corrosion Science*, *98*, 249–259.

Rai, M., Ribeiro, C., Mattoso, L., & Duran, N. (2015). *Nanotechnologies in food and agriculture*. Springer.

Rajasekar, A., Ganesh Babu, T., Karutha Pandian, S., Maruthamuthu, S., Palaniswamy, N., & Rajendran, A. (2007). Biodegradation and corrosion behavior of manganese oxidizer *Bacillus cereus* ACE4 in diesel transporting pipeline. *Corrosion Science*, *49*(6), 2694–2710. Available from https://doi.org/10.1016/j.corsci.2006.12.004.

Rasheed, P. A., Jabbar, K. A., Rasool, K., Pandey, R. P., Sliem, M. H., Helal, M., ... Mahmoud, K. A. (2019). Controlling the biocorrosion of sulfate-reducing bacteria (SRB) on carbon steel using ZnO/chitosan nanocomposite as an eco-friendly biocide. *Corrosion Science*, *148*, 397–406. Available from https://doi.org/10.1016/j.corsci.2018.12.028.

Rasool, K., & Lee, D. S. (2016). Effect of ZnO nanoparticles on biodegradation and biotransformation of co-substrate and sulphonated azo dye in anaerobic biological sulfate reduction processes. *International*

Biodeterioration & Biodegradation, 109, 150−156. Available from https://doi.org/10.1016/j.ibiod.2016.01.015.

Rhoads, W. J., Pruden, A., & Edwards, M. A. (2017). Interactive effects of corrosion, copper, and chloramines on *Legionella* and mycobacteria in hot water plumbing. *Environmental Science & Technology, 51*(12), 7065−7075. Available from https://doi.org/10.1021/acs.est.6b05616.

Salvarezza, R. C., De Mele, M. F. L., & Videla, H. A. (1983). Mechanisms of the microbial corrosion of aluminum alloys. *Corrosion, 39*(1), 26−32.

San Keskin, N. O., Yaylaci, E., Durgun, S. G., Deniz, F., & Nazır, H. (2021). Anticorrosive properties of green silver nanoparticles to prevent microbiologically influenced corrosion on copper in the marine environment. *Journal of Marine Science and Application*. Available from https://doi.org/10.1007/s11804-020-00188-6.

Shelar, G. B., & Chavan, A. M. (2014). Fungus-mediated biosynthesis of silver nanoparticles and its antibacterial activity. *Archives of Appied Science Research, 6*, 111−114.

Singh, R., Shedbalkar, U. U., Wadhwani, S. A., & Chopade, B. A. (2015). Bacteriagenic silver nanoparticles: Synthesis, mechanism, and applications. *Applied Microbiology and Biotechnology, 99*(11), 4579−4593.

Singh, S., Singh, G., & Bala, N. (2020). Electrophoretic deposition of Fe_3O_4 nanoparticles incorporated hydroxyapatite-bioglass-chitosan nanocomposite coating on AZ91 Mg alloy. *Materials Today Communications*, 101870. Available from https://doi.org/10.1016/j.mtcomm.2020.101870.

Sowards, J. W., Williamson, C. H. D., Weeks, T. S., McColskey, J. D., & Spear, J. R. (2014). The effect of *Acetobacter* sp. and a sulfate-reducing bacterial consortium from ethanol fuel environments on fatigue crack propagation in pipeline and storage tank steels. *Corrosion Science, 79*, 128−138.

Starosvetsky, J., Starosvetsky, D., Pokroy, B., Hilel, T., & Armon, R. (2008). Electrochemical behaviour of stainless steels in media containing iron-oxidizing bacteria (IOB) by corrosion process modeling. *Corrosion Science, 50*(2), 540−547. Available from https://doi.org/10.1016/j.corsci.2007.07.008.

Stetter, K. O., Huber, R., Blöchl, E., Kurr, M., Eden, R. D., Fielder, M., . . . Vance, I. (1993). Hyperthermophilic archaea are thriving in deep North Sea and Alaskan oil reservoirs. *Nature, 365*(6448), 743−745.

Sutha, S., Kavitha, K., Karunakaran, G., & Rajendran, V. (2013). In-vitro bioactivity, biocorrosion and antibacterial activity of silicon integrated hydroxyapatite/chitosan composite coating on 316L stainless steel implants. *Materials Science and Engineering: C, 33*(7), 4046−4054. Available from https://doi.org/10.1016/j.msec.2013.05.047.

Tabesh, E., Salimijazi, H. R., Kharaziha, M., Mahmoudi, M., & Hejazi, M. (2019). Development of an in-situ chitosan-copper nanoparticle coating by electrophoretic deposition. *Surface and Coatings Technology, 364*, 239−247.

Tabish, M., Malik, M. U., Khan, M. A., Yasin, G., Asif, H. M., Anjum, M. J., . . . Nazir, M. T. (2021). Construction of NiCo/graphene nanocomposite coating with bulges-like morphology for enhanced mechanical properties and corrosion resistance performance. *Journal of Alloys and Compounds, 867*, 159138. Available from https://doi.org/10.1016/j.jallcom.2021.159138.

Tabish, M., Yasin, G., Anjum, M. J., Malik, M. U., Zhao, J., Yang, Q., . . . Khan, W. Q. (2021). Reviewing the current status of layered double hydroxide-based smart nanocontainers for corrosion inhibiting applications. *Journal of Materials Research and Technology, 10*, 390−421. Available from https://doi.org/10.1016/j.jmrt.2020.12.025.

Tamayo, L. A., Zapata, P. A., Vejar, N. D., Azócar, M. I., Gulppi, M. A., Zhou, X., . . . Páez, M. A. (2014). Release of silver and copper nanoparticles from polyethylene nanocomposites and their penetration into Listeria monocytogenes. *Materials Science and Engineering: C, 40*, 24−31. Available from https://doi.org/10.1016/j.msec.2014.03.037.

Thauer, R. K., Stackebrandt, E., & Hamilton, W. A. (2007). Energy metabolism and phylogenetic diversity of sulphate-reducing bacteria. *Sulphate-Reducing Bacteria*, 1−38.

Turhan, M. C., Weiser, M., Jha, H., & Virtanen, S. (2011). Optimization of electrochemical polymerization parameters of polypyrrole on Mg–Al alloy (AZ91D) electrodes and corrosion performance. *Electrochimica Acta, 56*(15), 5347–5354. Available from https://doi.org/10.1016/j.electacta.2011.03.120.

Usher, K. M., Kaksonen, A. H., Cole, I., & Marney, D. (2014). Critical review: Microbially influenced corrosion of buried carbon steel pipes. *International Biodeterioration & Biodegradation, 93*, 84–106.

Vala, A. K., Shah, S., & Patel, R. (2014). Biogenesis of silver nanoparticles by marine-derived fungus *Aspergillus flavus* from Bhavnagar Coast, Gulf of Khambhat, India. *Journal of Marine Biology & Oceanography, 3*(1), 1–3.

Van den Eynde, V. C., Paradelo, R., & Monterroso, C. (2009). Passivation techniques to prevent corrosion of iron sulphides in roofing slates. *Corrosion Science, 51*(10), 2387–2392. Available from https://doi.org/10.1016/j.corsci.2009.06.025.

Venzlaff, H., Enning, D., Srinivasan, J., Mayrhofer, K. J. J., Hassel, A. W., Widdel, F., & Stratmann, M. (2013). Accelerated cathodic reaction in microbial corrosion of iron due to direct electron uptake by sulfate-reducing bacteria. *Corrosion Science, 66*, 88–96. Available from https://doi.org/10.1016/j.corsci.2012.09.006.

Videla, H. A. (2002). Prevention and control of biocorrosion. *International Biodeterioration & Biodegradation, 49*(4), 259–270. Available from https://doi.org/10.1016/S0964-8305(02)00053-7.

Videla, H. A., Guiamet, P. S., DoValle, S., & Reinoso, E. H. (1988). *Effects of fungal and bacterial contaminants of kerosene fuels on the corrosion of storage and distribution systems.* Houston, TX: National Assoc. of Corrosion Engineers.

Vimbela, G. V., Ngo, S. M., Fraze, C., Yang, L., & Stout, D. A. (2017). Antibacterial properties and toxicity from metallic nanomaterials. *International Journal of Nanomedicine, 12*, 3941.

Vroom, J. M., De Grauw, K. J., Gerritsen, H. C., Bradshaw, D. J., Marsh, P. D., Watson, G. K., ... Allison, C. (1999). Depth penetration and detection of pH gradients in biofilms by two-photon excitation microscopy. *Applied and Environmental Microbiology, 65*(8), 3502–3511.

Wan, D., Yuan, S., Neoh, K. G., & Kang, E. T. (2010). Surface functionalization of copper via oxidative graft polymerization of 2,2′-bithiophene and immobilization of silver nanoparticles for combating biocorrosion. *ACS Applied Materials & Interfaces, 2*(6), 1653–1662. Available from https://doi.org/10.1021/am100186n.

Wan, H., Song, D., Zhang, D., Du, C., Xu, D., Liu, Z., ... Li, X. (2018). Corrosion effect of *Bacillus cereus* on X80 pipeline steel in a Beijing soil environment. *Bioelectrochemistry (Amsterdam, Netherlands), 121*, 18–26.

Wang, H., Ju, L.-K., Castaneda, H., Cheng, G., & Zhang Newby, B.-m (2014). Corrosion of carbon steel C1010 in the presence of iron oxidizing bacteria *Acidithiobacillus ferrooxidans*. *Corrosion Science, 89*, 250–257. Available from https://doi.org/10.1016/j.corsci.2014.09.005.

Wang, H., Wang, Z., Hong, H., & Yin, Y. (2010). Preparation of cerium-doped TiO_2 film on 304 stainless steel and its bactericidal effect in the presence of sulfate-reducing bacteria (SRB). *Materials Chemistry and Physics, 124*(1), 791–794. Available from https://doi.org/10.1016/j.matchemphys.2010.07.063.

Wang, X., & Melchers, R. E. (2017). Corrosion of carbon steel in presence of mixed deposits under stagnant seawater conditions. *Journal of Loss Prevention in the Process Industries, 45*, 29–42. Available from https://doi.org/10.1016/j.jlp.2016.11.013.

Widmer, A. F. (2001). New developments in diagnosis and treatment of infection in orthopedic implants. *Clinical Infectious Diseases, 33*(Suppl._2), S94–S106.

Wu, T., Yan, M., Zeng, D., Xu, J., Sun, C., Yu, C., & Ke, W. (2015). Stress corrosion cracking of X80 steel in the presence of sulfate-reducing bacteria. *Journal of Materials Science & Technology, 31*(4), 413–422.

Xu, D., & Gu, T. (2014). Carbon source starvation triggered more aggressive corrosion against carbon steel by the *Desulfovibrio vulgaris* biofilm. *International Biodeterioration & Biodegradation, 91*, 74–81. Available from https://doi.org/10.1016/j.ibiod.2014.03.014.

Xu, D., Li, Y., & Gu, T. (2016). Mechanistic modeling of biocorrosion caused by biofilms of sulfate reducing bacteria and acid producing bacteria. *Bioelectrochemistry (Amsterdam, Netherlands), 110*, 52−58. Available from https://doi.org/10.1016/j.bioelechem.2016.03.003.

Xu, D., Li, Y., Song, F., & Gu, T. (2013). Laboratory investigation of microbiologically influenced corrosion of C1018 carbon steel by nitrate reducing bacterium *Bacillus licheniformis*. *Corrosion Science, 77*, 385−390. Available from https://doi.org/10.1016/j.corsci.2013.07.044.

Xu, J., Sun, T. T., Jiang, S., Munroe, P., & Xie, Z.-H. (2018). Antimicrobial and biocorrosion-resistant MoO_3-SiO_2 nanocomposite coating prepared by double cathode glow discharge technique. *Applied Surface Science, 447*, 500−511. Available from https://doi.org/10.1016/j.apsusc.2018.04.026.

Yasin, G., Anjum, M. J., Malik, M. U., Khan, M. A., Khan, W. Q., Arif, M., ... Tabish, M. (2020). Revealing the erosion-corrosion performance of sphere-shaped morphology of nickel matrix nanocomposite strengthened with reduced graphene oxide nanoplatelets. *Diamond and Related Materials, 104*, 107763.

Yasin, G., Arif, M., Mehtab, T., Lu, X., Yu, D., Muhammad, N., ... Song, H. (2020). Understanding and suppression strategies toward stable Li metal anode for safe lithium batteries. *Energy Storage Materials, 25*, 644−678.

Yasin, G., Arif, M., Mehtab, T., Shakeel, M., Khan, M. A., & Khan, W. Q. (2020). *Metallic nanocomposite coatings. Corrosion protection at the nanoscale* (pp. 245−274). Elsevier.

Yasin, G., Arif, M., Mehtab, T., Shakeel, M., Mushtaq, M. A., Kumar, A., ... Song, H. (2020). A novel strategy for the synthesis of hard carbon spheres encapsulated with graphene networks as a low-cost and large-scalable anode material for fast sodium storage with an ultralong cycle life. *Inorganic Chemistry Frontiers, 7*(2), 402−410.

Yasin, G., Arif, M., Nizam, M. N., Shakeel, M., Khan, M. A., Khan, W. Q., ... Zuo, Y. (2018). Effect of surfactant concentration in electrolyte on the fabrication and properties of nickel-graphene nanocomposite coating synthesized by electrochemical co-deposition. *RSC Advances, 8*(36), 20039−20047.

Yasin, G., Arif, M., Shakeel, M., Dun, Y., Zuo, Y., Khan, W. Q., ... Nadeem, M. (2018). Exploring the nickel--graphene nanocomposite coatings for superior corrosion resistance: Manipulating the effect of deposition current density on its morphology, mechanical properties, and erosion-corrosion performance. *Advanced Engineering Materials, 20*(7), 1701166.

Yasin, G., Khan, M. A., Arif, M., Shakeel, M., Hassan, T. M., Khan, W. Q., ... Zuo, Y. (2018). Synthesis of spheres-like Ni/graphene nanocomposite as an efficient anti-corrosive coating; effect of graphene content on its morphology and mechanical properties. *Journal of Alloys and Compounds, 755*, 79−88.

Yuan, S. J., & Pehkonen, S. O. (2009). AFM study of microbial colonization and its deleterious effect on 304 stainless steel by *Pseudomonas* NCIMB 2021 and *Desulfovibrio desulfuricans* in simulated seawater. *Corrosion Science, 51*(6), 1372−1385.

Zarasvand, K. A., & Rai, V. R. (2016). Inhibition of a sulfate reducing bacterium, *Desulfovibrio marinisediminis* GSR3, by biosynthesized copper oxide nanoparticles. *3 Biotech, 6*(1), 84.

Zaveri, N., McEwen, G. D., Karpagavalli, R., & Zhou, A. (2010). Biocorrosion studies of TiO_2 nanoparticle-coated Ti−6Al−4V implant in simulated biofluids. *Journal of Nanoparticle Research, 12*(5), 1609−1623. Available from https://doi.org/10.1007/s11051-009-9699-6.

Zhou, E., Li, H., Yang, C., Wang, J., Xu, D., Zhang, D., & Gu, T. (2018). Accelerated corrosion of 2304 duplex stainless steel by marine *Pseudomonas aeruginosa* biofilm. *International Biodeterioration & Biodegradation, 127*, 1−9.

Zubillaga, O., Cano, F. J., Azkarate, I., Molchan, I. S., Thompson, G. E., & Skeldon, P. (2009). Anodic films containing polyaniline and nanoparticles for corrosion protection of AA2024T3 aluminium alloy. *Surface and Coatings Technology, 203*(10), 1494−1501. Available from https://doi.org/10.1016/j.surfcoat.2008.11.023.

Index

Note: Page numbers followed by "*f*" and "*t*" refer to figures and tables, respectively

A

Abiotic reactions, 376–378
Acacia catechu. *See* Khair (*Acacia catechu*)
Acacia spp., 262
Acaricides, 66
Acetic acid, 397–398
Acetylcholinesterase (AChE), 667
Achromobacter, 478–479, 517
Acid Violet 7, 585
Acid-producing bacteria (APB), 677–678
Acidic anhydride (AA), 232
Acidithiobacillus thiooxidans, 410–411
Acinetobacter spp., 4–5, 451–452, 513t, 517
Activated carbon, 591
Admixtures, 369
Adsorption, 9
Advanced oxidation methods, 641–643
   Fenton and photo-Fenton methods, 642–643
   photocatalysis, 641–642
Advanced oxidation processes (AOPs), 87, 580
   for dye remediation, 585–587
      Fenton process, 586–587
   effective factors in, 93–97
      direct photolysis, 96
      dosage of photocatalyst, 96–97
      effect of H_2O_2, 95–96
      pH of solution, 93–95
   for water and wastewater treatment, 87–90
      carbon-based structure, 88–89
      metal oxide, 87–88
      MOF based structure, 89–90
Advanced physicochemical treatment approaches for pollutants degradation, 479–480
Aerobic biodegradation, 247, 327
Aeromicrobium, 515–516
Agar, 200–201
Aggregate estimation respirometric framework, 228
Agricultural/agriculture
   byproducts, 643–644
   herbicides, pesticides, and fertilizers as product of, 511–512
Agrochemical compounds, 153
Alcaligenes, 478–479, 517
Algae, 343
   based nanoparticle biosynthesis, 40–41
   degradation activity by, 253–254
Alginate, 199–200
Aliphatic polyesters, 176–178
Alkali-carbonate reaction (ACR), 408–409
Alkali-silica reaction (ASR), 408–409
Alternaria, 410–411
Alumina (Al_2O_3), 33, 436
   nanoparticles, 293
Aluminum (Al), 581–582
American Society for Testing and Materials (ASTM), 203
Amine functionalized MWNTs (a-MWNTs), 188–189, 229–230
Amine-functionalized HMS (A-HMS), 645
3-aminopropyltriethoxy silane (APTES), 457
Amphiphilic polyurethane (APU), 19
Amylopectin, 199
Amylose, 199
Anabaena flosaque, 637
Anaerobic biodegradation, 247–248, 327
Anthraquinone dyes, 579–580
Antibiotic(s), 637
   biodegradation of, 22–23
   contaminants, 566–567
Antimicrobial
   agents in concrete, 383
   titanium nanocomposites, 545
Antrodia, 283–284
Aquatic ecosystems, 67
Aqueous emulsions, 348–349

Arabidopsis thaliana, 280
Aromatic compounds, petroleum and, 515–516
Aromatic polyesters, 176–178
Arsenic, 509
Arthrobacter spp., 4–5, 513t
Aspergillus, 4–5, 252, 410–411
    A. flavus, 291
    A. fumigatus, 42–43
    A. nidulans, 40
    A. niger, 40, 190–191, 291, 351, 422–423, 679
    A. sydowii, 190–191
    A. versicolor, 352–353
Assimilation stage, 327
Assimilatory chemical biodeterioration, 396
Astragalus sinicus, 280
Atomic force microscopy, 227
Attrition, 71–72
AuNPs incorporated chitosan (AuNPs/CS), 690
Aureobasidium pullulans, 40
Azadirachta indica, 432–433

B

Bacillus spp., 4–5, 513t, 517
    B. brevis, 37–40
    B. cereus, 37–40, 47
    B. cohnii, 451–452
    B. fusiformis, 76, 162–163
    B. licheniformis, 451–452, 677–678
    B. magaterium, 451–452
    B. pasteurii, 451–452
    B. pseudofirmus, 451–452
    B. sphaericus, 10–11, 451–452
    B. subtilis, 349, 429–430
Bacteria, 277
    degradation activity by, 251–252
    nanoparticles biosynthesis, 37–40
    nanoparticles synthesized from, 13
    proliferation, 422–423
Bacterial polymer, 215
Basidiomycetes, 283–284
Bassia latifolia. See Illupai (*Bassia latifolia*)
Bentonite nano/microparticles (BNMPs), 455
Benzo[ghi]perylene (BghiP), 163–164
β-cyclodextrin (β-CD), 113–115
β-tricalcium phosphate (TCP), 106–107

Bimetallic nanoparticles, 659–660, 664–665
Bio-MOF-derived carbon (BMDC), 646
Bioactive glass (BG), 113–115
Bioaugmentation, 4–5
Bioceramics, 124–126
Biochars (BCs), 155, 582
Biochemical Network Integrated Computational Explorer (BNICE), 492
Biocides, 66, 384–385
Biocompatible molecules, 106–107
Biocompatible polyester, 215
Bioconcept of calcite, 451
Bioconcrete and limitations, 451–454
Biocorrosion of metals, 1
Biodegradability, 176–178
Biodegradability Evaluation and Simulation System (BESS), 492
Biodegradable materials, 106–107, 533–535
Biodegradable nanoparticles
    applications, 145–148
        in imaging technology, 146–147
        in industries, material sciences, and electronics, 147–148
    chitosan, 138–140
    dendrimers, 140
    lipid-based nanoparticles, 140–141
    PCL, 138
    PLGA, 138
    types, 137–141
Biodegradable plastics (BPs), 176–178, 214–216, 222, 247–248, 325, 327, 530
Biodegradable polyester, 215
Biodegradable polymers (BDPs), 217–218, 240, 534
    broad-spectrum perspectives and status of, 214–215
    challenges and benefits of, 215–217
Biodegradation, 1, 176–178, 217–218, 221–226, 240, 275, 323, 344, 510, 533. *See also* Degradation; Nanobiodegradation
    of antibiotics and personal care products, 22–23
    biological synthesis of nanoparticles, 12–14
    bioremediation of hydrophobic contaminants, 18–20

chemical means of protecting wood from, 284–287
composting, 222–223
of dyes, 622
enzyme-encapsulated nanoparticles in, 142–144
general considerations of biodegradation, 74
of materials, 141
mechanistic pathway of enzymatic degradation, 223–225
microbial degradation, 223
microorganisms' role in, 276–278
nanobioremediation, 15–16
nanomaterials effects on biodegradation behavior, 109–126
   calcium phosphates effects on biodegradation behavior, 124–126
   CBNs effects on biodegradation behavior, 116–118
   nano zinc-oxide effects on biodegradation behavior, 109–113
   nanogold effects on biodegradation behavior, 115–116
   nanosilver effects on biodegradation behavior, 113–115
   nanotitanium dioxide effects on biodegradation behavior, 118–123
nanomaterials role in bioremediation, 16
nanoparticles as enhancers of biodegradation, 74–76
   iron nanoparticles, 75–76
   silver nanoparticles, 76
   titanium nanoparticles, 76
nanoparticles role, 74–76
   in biodegradation, 11–12
at nanoscale level, 249–251
of natural plastic, 330–331
of organic materials, 531–537
of phenolic compounds, 21–22
of plastics, 246–249, 325–326
   aerobic biodegradation, 247
   anaerobic biodegradation, 247–248
   biodegradation standards for plastics, 248–249
   factors affecting biodegradation of plastics, 249
remediation of pollutants using nanotechnology, 16–18
sol-gel technology to protect marble from, 350–361
of synthetic dyes by nanoparticles, 20–21
of synthetic plastic, 327–328
of thermo set plastics, 333
types, 326–327
   assimilation, 327
   biodeterioration, 326
   biofragmentation, 326–327
Biodeposition, 451
Biodeterioration, 1, 326, 390–391
Bioelectro-venting, 485
Biofilm(s), 381–382, 675–676
   formation, 342–343
   technology, 162
Biofragmentation, 326–327
Biogenic deterioration
   of building materials, 409–410
   in buildings and other cementitious materials, 406–411
Biogenic uraninite, 19–20
Bioglasses, 124–126
Biological deterioration
   classification of, 393–396
      chemical classification, 395–396
      physical or mechanical classification, 393–394
      soiling (esthetic)/fouling classification, 394–395
   of concrete, 391–393
   control concrete biodeterioration, 398–399
Biological methods, 478–479, 679–680
Biological synthesis of nanoparticles, 12–14
Biological therapies, 658
Biological treatment methods, 639–640
Biologically controlled mineralization (BCM), 450
Biologically induced mineralization (BIM), 450
Biologically influenced mineralization (BIFM), 450
Biomass-based plastics, 214–215
Biomedical applications of nanoparticle, 72
Biomimetics nanotheranostics, 146–147

Bionanotechnology, 31–32
Biophysical destruction, 339
Bioplastics, 180–181, 214–215, 217–218, 534
    enzymatic degradation of, 245–246
Bioreducing agents, in nanoparticle biosynthesis, 41–42
Bioremediation (BRM), 4–5, 11, 510–511, 658. *See also* Environmental remediation
    dendrimers in, 17
    of environmental pollutants, 490
    of hydrophobic contaminants, 18–20
        bioremediation of heavy metals, 20
        soil bioremediation, 19
        uranium bioremediation, 19–20
    invention of novel genes involved in, 492
    nanocrystals and carbon nanotubes in, 18
    of radioactive wastes, 10–11
    role of nanomaterials in, 16
    single-enzyme nanoparticles in, 18
    technologies, 510
Biosensing system of HRP, 620–621
Biosensors
    for environmental pollutants detection, 482
    for pesticide detection, 667
Biosorbents, 643–644
Biostability
    of epoxysiloxane coatings, 358–361
    laboratory testing of epoxysiloxane coatings for, 353–354
Biostimulation, 4–5
Biosynthesis pathways, 42–44
    extracellular biosynthesis, 42–43
    intracellular biosynthesis, 43–44
Biotechnology
    biotechnological approaches for micropollutants degradation, 483
    reduction strategies for waste management of plastics by using, 246
Biotic reactions, 376–378
Biotransformation of lignocellulosic wastes, 622–623
Biphasic calcium phosphate (BCP), 116–117
Blended cements, 369
Blended plastic biodegradation, 332
Bone tissue regeneration (BTR), 116–117
Boron-containing substances, 285
Bottom-up approach, 69–71, 662–663. *See also* Top-down approach
    chemical reduction, 70
    electrochemical method, 71
    green synthesis, 71
    laser ablation, 70–71
    microemulsion, 70
    microwave, 70
Brevibacillus borstelensis, 193–194
Brevibacterium, 515–516
    B. casei, 43–44
2-bromoethanesulfonate (BES), 75
Brown alga, 199–200
Brown rot fungus (*Poria placenta*), 291
Burkholderia, 478–479, 515–517
Butyric acid, 397–398

C
Cadmium, 509
Caenorhabditis elegans, 520
Calcium aluminate cement (CAC), 378–379
Calcium carbonate, 450–451
Calcium phosphate (Ca-P), 124–126
    effects on biodegradation behavior, 124–126
Calcium silicate hydrate (CSH), 413–414
Cancer
    cancer-based nanodrug delivery, 139–140
    nanotheranostics, 146–147
Candida, 4–5
    C. albicans, 32–33
Capillary electrophoresis–mass spectroscopy, 200
Carbamates, 285, 477–478
Carbon
    carbon-based fullerene materials, 116–117
    carbon-based structure, 88–89
    dots, 608
Carbon dioxide (CO_2), 397–398
    evolution, 334
Carbon materials (CMs), 155
Carbon nanodots (NDs), 604–605
    for HRP immobilization, 606–608, 609t
Carbon nanoonions (CNOs), 607
    for HRP immobilization, 606–608, 609t

Carbon nanotubes (CNTs), 2–3, 32–33, 69, 108, 144, 155, 441–444, 482, 530, 582, 604–605, 659–660
  in bioremediation, 18
  for HRP immobilization, 606–608, 609t
Carbon nitrides (CN), 93
Carbon-based nanomaterials (CBNs), 116–117, 441–446. See also Inorganic nanomaterials
  CNTs, 441–444
  effects on biodegradation behavior, 116–118
  graphene-based nanomaterials, 444–446
  nanoclay, 446–450
Carbon-based nanoparticles, 32–33, 36–37, 69, 155–157. See also Metal-based nanoparticles
  CNTs, 155
  fullerene nanoparticles, 156–157
  graphene based, 155–156
Carbonation of concrete, 380
Carboxymethyl chitin nanoparticles, 139–140
Casuarina
  C. equisatifolia, 264
  C. junghuhniana, 264
Catalysis, 72
Catalytic ozonation, 586
Causative bacteria, 677
Cell biosensors, 495
Cellular metabolism, 181–182, 221
Cellular toxicity, 3–4
Cellulose, 159, 186–188, 198–199, 530
Cellulose acetate membrane (CAM), 657
  plasmid, 279
Cement-based materials, 427–429
  antimicrobial agents in concrete, 383
  biocide, 384–385
  biofilms, 381–382
  changing redox conditions, 383
  chemical removal of sulfide, 384
  concrete materials deterioration, 378–380
  corrosion and deterioration, 371
  generation of sulfuric acid, 376–378
  inhibiting activities of sulfate reducing bacteria, 383
  measures against concrete biodeterioration, 380–381
  MICD of, 371–376
  microbiologically-induced corrosion in sewer structures, 376
  minimizing sulfide in sewer environment, 383
  nanomaterials for minimizing microbial attack, 386
  other measures, 384
  parameters and concrete deterioration, 370f
CEN. See European Committee for Standardization (CEN)
Centaurea virgata, 20
Ceramic nanoparticles, 33, 68
Chemical classification of biodeterioration, 395–396
Chemical fertilizers, 511–512
Chemical induced deterioration, 408–409
Chemical methods of stone protection, 345
Chemical recycling, 239
Chemical reduction, 70
Chemical removal of sulfide, 384
Chemical sulfate attack, 427–429
Chitosan, 137–140, 139f, 200
Chitosan-TPP nanoparticles, 292–293
Chitosan-ZnONPs nanocomposite (CZNC), 684
Chlorella sorokiniana, 639
Chloride ingress, 438–439
5-chloro-2-methyl-4-isothiazol-3-one (Kathon), 286
Chlortetracycline (CTC), 566–567
Cholesterol oxide (ChOx), 618
Chromophoric motifs, 219
Ciprofloxacin (CIP), 22, 75, 645
Citrates, 285
Citrus reticulata, 432–433
Civil engineering
  nanomaterials types in, 418–450
    carbon-based nanomaterials, 441–446
    inorganic nanomaterials, 419–441
Cladosporium, 410–411
  C. cladosporioides, 351, 423–424
  C. herbarum, 352–353
Clay nanoparticles, 166–167
Clays, 582
Clostridium aceticum, 678

Cobalt (Co), 509
   Co-doping, 92
   Co-I molecule, 604
   Co-II molecule, 604
Collagen, 199
Colonization on concrete, 396–397
Comamonas, 478–479, 517
   C. acidovorans, 188
   C. testosteroni, 279–280
Cometabolism, 510–511
Commercial polymers, 186–188
Composting, 182–183, 222–223, 323
Computed tomography, 146–147
Concrete(s), 369–370, 374, 390, 406–407
   antimicrobial agents in, 383
   biological deterioration of, 391–393
   classification of biological deterioration, 393–396
   control concrete biodeterioration, 398–399
   conventional, 390
   deterioration in, 390–391
   materials deterioration, 378–380
   measures against concrete biodeterioration, 380–381
   microorganisms
      and chain on concrete, 397
      and colonization on concrete, 396–397
      microorganism-induced deterioration mechanism process, 397–398
Condensation polymers, 186–188
Conduction band (CB), 85–86
Congo red, 588
Coniophora, 283–284
Constructed wetlands, 640
Construction materials technology, 412–413
Contaminants, 510
   photocatalytic removal of, 566–568
   sorptive removal of, 569–570
Control composting test, 335–336
Conventional concrete, 390
Conventional methods for dye remediation, 584–585
Copolymerization, 93
Copper (Cu), 432
   naphthenates, 285

Copper nanoparticles (CuNPs), 432, 680–682
Copper NP carbon quantum dots (Cu/CQDs), 680–681
Copper oxide (CuO), 432–433
Copper oxide NP (CuONPs), 432–433, 682
Coprecipitation method, 540, 615–616
Corrosion, 370
   and deterioration of cement-based materials, 371, 372t
   of reinforced steel rebar, 408–409
Corrosive microbes, 677–679
   APB, 678
   fungi, 679
   metal oxidizing microbes, 678
   NRB, 677–678
   SRB, 677
Corynebacterium, 517
Crystal framework-8 of zeolitic imidazole (ZIF-8), 618
Cyamopsis tetragonalobus, 200
Cyanobacteria, 343
Cyanobacterial nanoparticle biosynthesis, 37–40
Cyclodextrins (CD), 18
Cylindrobasidium, 283–284
Cystoseira trinodis, 40–41

D

Daphnia magna, 213–214, 637
Decabromodiphenyl ethane, 153
Dechlorane plus, 153
Degradation, 106–107
   activity
      by algae, 253–254
      by bacteria, 251–252
      by fungi, 252–253
      using insects, 254–255
      using worms, 255
   of polymers, 218–225, 529
      biodegradation, 221–225
      hydrolysis, 220
      photodegradation, 219
      thermal degradation, 219–220
      weathering, 218–219
Dehalococcoides, 478–479
Dehalogenases, 485

Dendrimers, 140, 140f
  in bioremediation, 17
Deoxy tetracycline (DCL), 566–567
Deoxyribonucleic acid (DNA), 118
Depolymerization, 181–182, 221
Desmodesmus, 40–41
Desulfovibrio vulgaris, 677–678
Deterioration in concrete, 390–391
  MICD, 391
  microbiological-induced deterioration, 390–391
Detonation nanodiamond (DND), 352–353
2,4-diamino-6-nitrotoluene (2,4-DANT), 493–494
4,5-dichloro-2-n-octyl-4-isothiazol-3-one (DCOIT), 286
Dichlorodiphenyltrichloroethane, 153
Didecyldimethylammonium tetrafluoroborate (DBF), 285
Dietzia, 515–516
Differential scanning calorimetric analysis (DSC analysis), 202
Direct photolysis, 96
Disperse Blue 3, 579–580
Dissimilatory chemical biodeterioration, 396
Doping method, 90
Dosage of photocatalyst, 96–97
Doxorubicin, 68
Drug delivery systems, 47
Dulbecco's Modified Eagle's Medium (DMEM), 685–687
Durability, 369
Dye remediation
  advanced oxidation processes for, 585–587
  conventional methods for, 584–585
Dye-containing wastewaters
  electrochemical and microbial treatment of, 495–497
Dyes biodegradation, 622
Dyestuff-based hazardous pollutants, 513

E

Ecoconcretes, 425–427
Ecological risks, 636–638
Effluent treatment plant (ETP), 580
Eleagnum angustifolia, 20
Electrochemical method, 71

Electrochemical of dye-containing wastewaters, 495–497
Elemental analyzer/isotope ratio mass spectrometry, 203
ELISA, 605–606
Emerging organic pollutants (EOPs), 531
  occurrence and environmental impact of, 531–533
Emerging pollutants (EPs), 61–62
  occurrence of, 62–67
Energy capture, 72–73
Engineered nanomaterials (ENMs), 655–656
Engineered nanoscale materials, 1
Engineered timber. *See* Softwood
Engineering-constructive methods, 344
Enhanced bioremediation via metabolic engineering processes, 492–495
Enterobacter spp., 513t
Environmental contamination, 62
Environmental enzymes bioremediation, 639–640
Environmental pollutants, 477
  biosensors for environmental pollutants detection, 482
  microbial degradation of environmental pollutants
    bioremediation, 510–511
    dyestuff-based hazardous pollutants, 513
    herbicides, pesticides, and fertilizers as product of agriculture, 511–512
    petroleum and aromatic compounds, 515–516
    phenazines, 519–521
    polychlorinated biphenyls, 516–519
    potentially toxic heavy metals, 514–515
Environmental pollution, 9
Environmental remediation
  MOF applications in, 566–570
  photocatalytic removal of contaminants using MOF, 566–568
  sorptive removal of contaminants using MOF, 569–570
Environmental reorganization, 15–16
Environmental risks, 636–638
Environmentally friendly fungicidal additive, 352–353

Enzymatic degradation
  of bioplastics, 245–246
  and mechanism, 184
  mechanistic pathway of, 223–225
  microbes and, 144–145
Enzymatic method to degrade polyhydroxyalkanoate, 331–332
  catalytic domain, 331
  linker region, 331–332
  subdomain, 331
Enzyme(s), 18, 603
  enzyme-assisted remediation of micropollutants, 485–486
  enzyme-encapsulated nanoparticles
    applications of biodegradable nanoparticles, 145–148
    in biodegradation, 142–144
    biodegradation of materials, 141
    different types of biodegradable nanoparticles, 137–141
    enzyme-encapsulated nanoparticles in biodegradation, 142–144
    microbes and enzymatic degradation, 144–145
  enzyme-responsive systems, 667–668
  immobilization, 165–166
EPA. See US Environmental Protection Agency (EPA)
Epoxy-titanate sols, 351
Epoxysiloxane
  coatings for biostability, 353–354, 358–361
  synthesis of epoxysiloxane sols, 352
Escherichia coli, 32–33, 47, 490–491
Ester type PU (ES-PU), 188
Ether type PU (ET-PU), 188
1-ethyl-3-(3-dimethylaminopropyl) carbodiimide (EDC), 115–116
Ethylene vinyl alcohol, 333
ETP mediated Cu NP (ETP-CuNPs), 681–682
Eucalyptus (*Eucalyptus* spp.), 262
Euphorbia tirucalli, 12–13
European beech sapwood (*Fagus sylvatica* L.), 291
European Committee for Standardization (CEN), 203
European Environmental Agency, 477–478

Exiguobacterium mexicanum, 451–452
Exopolysaccharides (EPS), 41–42
Extracellular biosynthesis, 42–43
Extracellular depolymerase, 326
Extracellular matrix (ECM), 106–107

F
Fagus sylvatica L. *See* European beech sapwood (*Fagus sylvatica* L.)
Fenton method, 642–643
Fenton process for dye remediation, 586–587
Fenton's reagent, 580–581
Fertilizers as product of agriculture, 511–512
Field tests of friendly wood protection coatings, 295–297
Filler dimension effect on biodegradation, 232–233
Filtration, 9
Fire retardant, 153
Flame absorption atomic method, 20
Flavobacterium, 451–452
Fluorescence imaging, 146–147
Fluorine-doped hydroxyapatite (FHAp), 124–126
Forest, 263
Forming clear zone, 334
Fossil resources, plastics from, 188–195
Fourier transform infrared spectroscopy (FTIR spectroscopy), 202
Freshwater, plastic wastes in, 242
Fuelwood, 271–272
Fullerenes, 32–33, 69, 156–157, 530
  nanoparticles, 156–157
Functionalized MWCNT, 229
Fungi, 40, 679
  degradation activity by, 252–253
  nanoparticles synthesized from, 14
  proliferation, 422–423
Fungicidal carbamates, 285–286
Fungicides, 66
Furniture, 270–271
Fusarium, 4–5
  F. moniliforme, 181–182
  F. oxysporum, 40
  F. solani, 329

G

Ganoderma lucidum, 639
Gas chromatography–mass spectrometry (GC-MS), 222
Gasification, 178
Gel permeation chromatography (GPC), 190–191, 202, 252
Gelatine, 199
Gelidium amansii, 40–41
Generally recognized as safe (GRAS), 494–495
Genetic engineering, 478–479
Genetically engineered microorganisms, degradation by, 279–280
Geobacter sulfurreducens, 41–42
Gloeophyllum, 283–284
Gluconacetobacter, 37–40
Glutathione peroxidases, 604
Gold, 509
　metal NPs, 15–16
Gold nanoparticles (AuNPs), 604–605, 680, 690
Gordonia, 515–516
Gram-negative microorganisms, 517
Gram-positive microorganism, 517
Graphene, 2–3, 141, 444–445, 582, 604–605
　and derivatives for HRP immobilization, 608–615, 612t
　graphene based nanoparticles, 155–156
　graphene-based nanomaterials, 444–446
　graphene-based photocatalysts, 89
Graphene family nanomaterials (GFN), 444–445
Graphene nanosheets (GNS), 444–445
Graphene oxide (GO), 116–117, 444–445, 608, 611, 665–666
Graphene quantum dots (GQDs), 116–117
Graphitic carbon nitride, 89
Gravimetric determination of weight loss, 201
Gravimetric estimation respirometric frameworks, 228
Green approach, 566
Green biosynthesis of plant-based NPs, 12–13
Green biotechnology, 605–606
Green synthesis, 71, 289–290
Growth regulators, 66
Guar gum (GG), 200

Gundelia tournefortii, 20
Gypsum crust, 341, 342f

H

Halanaerobium sp., 687–688
Halloysite nanotubes (HNTs), 446–447, 659–660
Halomonas, 451–452
　H. elongata, 37–40
Hardwood timber, 261
Hazardous pollutants, 493–494
Heavy metals, 666
　bioremediation of, 20
Helminthosporium, 410–411
Heme protein, 604
Herbicides, 66
　nanobioremediation of, 659–662
　as product of agriculture, 511–512
Heterotrophs, 343
Hexagonal mesoporous silicates (HMS), 645
High-density polyethylene (HDPE), 242–243
High-strength concrete (HSC), 429–430
HIV-based nanodrug delivery, 139–140
Hormonema dematioides, 351
Horseradish peroxidase (HRP), 2–3, 3f, 603
　C1A structure, 604f
　carbon nanotubes, nanoonions, and nanodots, 606–608, 609t
　for environmental applications, 619–623
　　biosensing system of HRP, 620–621
　　biotransformation of lignocellulosic wastes, 622–623
　　wastewater treatment and biodegradation of dyes, 622
　graphene and derivatives for HRP immobilization, 608–615, 612t
　intermediate states, 605f
　magnetic electrospun nanofibers for HRP immobilization, 617
　magnetic nanoparticles for HRP immobilization, 615–617
　mesoporous silica for HRP immobilization, 618–619
　metal–organic frameworks for HRP immobilization, 618
Human contamination with EPs, 66

Hyaluronic acid (HA), 113–115
Hybrid carbon nanotubes (HCNTs), 18
Hybrid coatings based on wax and silicones, 295–297
Hydrated biogenic uraninite, 19–20
Hydrogen peroxides (H_2O_2), 605–606
Hydrolysis, 220
Hydrolyzable compounds, 350
Hydrolyzable ester bonds of biodegradable plastics, 214
Hydrophilic PGA, 138
Hydrophobic contaminants, bioremediation of, 18–20
Hydrophobic impregnation, 405–406
Hydrophobization, 287
Hydrothermal method, 541
Hydroxyapatite (HAp), 688–689
3-hydroxybutyrate-co-3-hydroxy-valerate (PHBV), 534
Hydroxyl radical (. OH), 580
Hypocrea lixii, 422–423

I

Illupai (*Bassia latifolia*), 266–267
Imaging technology, biodegradable nanoparticles in, 146–147
Immobilization of enzymes, 603
Immobilized enzymes for micropollutants degradation, 486–488
Immune system, 3–4
Incineration, 239, 243–244
Indian forest types, 261
Indigenous microbial strains, 4–5
Induced biomineralization, 340
Industrial plastic waste, 325
Industrialization, 9, 561–562
Inorganic copper compounds, 285
Inorganic nanomaterials, 419–441. *See also* Carbon-based nanomaterials
    nanoalumina, 436–439
    nanocopper, 432–433
    nanoiron oxide, 439–441
    nanosilica, 424–430
    nanosilver, 430–432
    nanotitanium dioxide, 419–424
    nanozinc, 433–436

Inorganic NPs, 13
Insect repellents, 511–512
Insecticides, 66
    nanobioremediation of, 659–662
Insects, 277
    degradation activity using, 254–255
Internalization of nanoparticles, 34–36
International Organization for Standardization (ISO), 203, 538
Intimate coupling, 160–161
Intracellular biosynthesis, 43–44
Intracellular depolymerase, 326
Iodide peroxidises, 604
3-iodo-2-propynylbutyl carbamate (IPBC), 285–286
Ion exchange membrane (IEM), 483–484
Ionothermal process, 565–566
Iron, 581
    iron-based technology, 15–16
Iron nanoparticles (FeNPs), 75–76, 680, 687–688
Iron oxide
    magnetic nanoparticles, 157–159
    NPs, 72
Iron oxidizing bacteria (FeOB), 677–679
Iron-corroding bacteria (ICB), 687–688
Iron-oxide nano/microparticles (INMPs), 455
Iron-reducing bacteria (FeRB), 677
Iron(II, II) oxide (Fe_3O_4), 457, 543
Isothiazolones, 285–286

K

K10 bentonite, 591
Khair (*Acacia catechu*), 262

L

Laccases, 157–158, 485–486
Lacquers, 431–432
Lactic acid, 228, 397–398
Lactobacillus plantarum, 37–40
Land, plastic wastes on, 241–242
Landfilling, 239, 243
Lantana camara, 432–433
Laponite, 166–167, 534–535
Laser ablation, 70–71
Laser cleaning, 345

Leaching of highly toxic chemicals from plastic wastes, 241
Lead, 509
Lentinus tigrinus, 190–191
Leucaena leucocephala. See Subabul (*Leucaena leucocephala*)
Lignification, 604
Lignin, 159
Lignin peroxidase (LiP), 144, 485
Lignocellulosic wastes, biotransformation of, 622–623
Lime technology, 349
Lipases, 224, 226–227
Lipid based nanoparticles, 33, 68, 140–141
    solid lipid nanoparticle, 141*f*
Liposome-based NPs, 31–32
Long-chain hydrocarbons, 15–16
Low-density polyethylene (LDPE), 242–243
Low-toxic biocides to protect wood from wood destroying fungi, 297–306
    evaluation of effectiveness of biocidal compositions
        for protecting wood from xylotrophic fungi in moist chamber, 302–306
        for protecting wood from xylotrophic fungi on nutrient medium, 297–302

M

Magnetic electrospun nanofibers for HRP immobilization, 617
Magnetic metals, 10–11
Magnetic nanoparticles (MNPs), 604–605
    for HRP immobilization, 615–617
Magnetic nanotheranostics, 146–147
Magnetic resonance imaging (MRI), 146–147
Manganese peroxidases, 164, 485
Marine borers, 277
Marine ecosystems, plastic wastes in, 242
Marrubium vulgare, 20
Mealworm (*Tenebrio molitor*), 255
Mechanical induced deterioration, 407–408
Mechanical recycling, 239
Melia dubia, 264
Membrane, 9
Mercapto-functionalized HMS (M-HMS), 645

Mercury, 509
Mesenchymal stem cells (MSCs), 124–126
Mesoporous silica for HRP immobilization, 618–619
Mesoporous silica nanoparticles (MSNs), 113–115
Mesorhizobium huakuii, 280
Metabolic engineering approaches
    enhanced bioremediation via, 492–495
    for pollutants degradation, 490–491
Metabolomics, 490
Metal clusters, 89–90
Metal dioxides, 285
Metal doping, 90
Metal organic frameworks (MOFs), 89–90, 165–166, 562–563, 618, 644–646
    applications in environmental remediation, 566–570
    based structure, 89–90
    designing and properties of, 564
    for HRP immobilization, 618
    nanoparticles, 165–166
    preparation, properties, and applications of, 563*f*
    representation of emerging pollutants, 562*f*
    synthetic pathways, 565–566
Metal oxide, 87–88
    metal oxide-based adsorbents, 644–646
Metal oxide nanoparticles, 540, 659–660
    as catalysts for environmental remediation, 543–546
    fundamentals of biodegradation of organic materials, 531–537
        current overview of biodegradable materials, 533–535
        factors affecting biodegradation of polymers, 535–537
        occurrence and environmental impact of EOPs, 531–533
    inference and future prospects, 546–547
    metal oxide nanoparticles performance in organic matter biodegradation, 537–546
    performance in organic matter biodegradation, 537–546
    synthesis and postmodification of, 540–542

Metal oxidizing microbes, 678
Metal-based nanoparticles, 146–147, 679–690.
    See also Carbon-based nanoparticles
  copper nanoparticles, 680–682
  gold nanoparticles, 690
  iron nanoparticles, 687–688
  manganese peroxidase, 164
  nickel/iron nanoparticles, 162–163
  silica nanoparticles, 688–689
  silver nanoparticles, 164–165, 682–684
  titanium oxide nanoparticle, 685–687
  zinc oxide nanoparticles, 163–164, 684–685
Metallic nanoparticles, 13, 32, 68, 539–540, 657
Methylene Blue, 585
Microbes, 41, 451–452, 510, 639
  and enzymatic degradation, 144–145
  waste plastics degradation using, 251–254
Microbial biomineralization, 450–451
Microbial degradation, 183–184, 223
  of plastics, 201–203
Microbial electrochemical system (MES), 483–485
Microbial engineering for bioremediation, 479
Microbial inoculum, 222–223
Microbial mediated nanoparticles biosynthesis, mechanisms of, 41
Microbial mediated synthesis of nanoparticles, 37–41
  algae based nanoparticle biosynthesis, 40–41
  bacterial and cyanobacterial nanoparticle biosynthesis, 37–40
  yeast and mold nanoparticle biosynthesis, 40
Microbial methods, 4–5
Microbial treatment of dye-containing wastewaters, 495–497
Microbial-induced calcium carbonate precipitation (MICCP), 406
Microbial/microbiological-induced corrosion (MIC), 370, 390, 675–676
  corrosive microbes, 677–679
  metal-based nanoparticles, 679–690
  in sewer structures, 376
Microbial/microbiological-influenced concrete deterioration (MICD), 370–371, 391
  of cement-based materials, 371–376
    classification of microbes in microbial-induced corrosion, 375t
Microbiological-influenced metal deterioration (MIMD), 391
Microbiological(ly)-induced deterioration, 390–391
  and protection of outdoor stone monuments
    methods to counter stone biodegradation, 344–349
    sol-gel technology to protect marble, 350–361
    stone destruction under influence of microorganisms, 339–344
Microcystins (MCs), 67
Microemulsion, 70
Microorganisms, 74, 180–182, 246–247, 371–374, 510, 679–680
  and chain on concrete, 397
  and colonization on concrete, 396–397
  ecological relation between, 510–511
  microorganism-induced deterioration mechanism process, 397–398
  role in biodegradation, 276–278
    bacteria, 277
    insects, 277
    marine borers, 277
    mold and stain fungi, 277
    shipworms, 277
    termites, 278
    wood lice, 277
  stone destruction under influence of, 339–344
Micropatterning, 71
Microplastics (MPs), 529–530
Micropollutants
  advanced physicochemical treatment approaches for pollutants degradation, 479–480
  biodegradation of, 477–479
  biosensors for environmental pollutants detection, 482
  biotechnological approaches for micropollutants degradation, 483
  electrochemical and microbial treatment of dye-containing wastewaters, 495–497

enhanced bioremediation via metabolic engineering processes, 492–495
enzyme-assisted remediation of micropollutants, 485–486
immobilized enzymes for micropollutants degradation, 486–488
invention of novel genes involved in bioremediation, 492
MES, 483–485
metabolic engineering approaches for pollutants degradation, 490–491
nanoremediation, 481–482
nanozymes, 489
photocatalysis, 480
photocatalytic fuel cells, 481
sonochemical methods, 481
Microscopy observations of surface, 202
Microwave, 70
Modified creazote oils, 285
Mold fungi, 277
Mold nanoparticle biosynthesis, 40
Molecular allotropes of carbon. *See* Fullerenes
Molecular biology, 279
Molluscicides, 66
Montmorillonite (MMT), 534–535
Mucor circinelloides, 422–423
Multiwalled carbon nanotubes (MWCNTs), 18, 116–117, 442, 646–647
Mycobacterium, 515–516
  M. fortuitum, 520–521
Myconanotechnology, 40

N

NAH plasmid, 279
Nano zinc-oxide (nZnO), 109–111
  effects on biodegradation behavior, 109–113
Nanoalumina, 436–439
Nanobarium titanate, 165
Nanobiodegradation. *See also* Biodegradation
  degradation activity
    using insects, 254–255
    using worms, 255
  enzymatic degradation of bioplastics, 245–246
  impacts of plastic wastes accumulation, 241–242
  plastic wastes in freshwater and marine ecosystems, 242
  plastic wastes on land, 241–242
  at nanoscale level, 249–251
  photooxidative degradation of waste plastics, 244
  plastic waste
    disposal methods, 243–244
    pollution and environmental impacts, 240–241
  of plastics, 246–249
  reduction strategies for waste management of plastics by using biotechnology, 246
  thermal degradation of waste plastics, 245
  types of plastics targeted, 242–243
  waste plastics degradation using microbes, 251–254
Nanobioremediation (NBR), 2–3, 15–16, 655–658
  bimetallic nanoparticles, 664–665
  biosensors for pesticide detection, 667
  enzyme-responsive systems, 667–668
  fundamentals of nanobioremediation technologies, 658–659
  nanobioremediation of insecticides and herbicides, 659–662
  nanocomposites, 665–666
  nanomaterials for remediation and sensing of pesticide, 662
  nanoparticles, 662–664
  nanotubes, 666
  photoresponsive methods, 668–669
Nanobiosensors, 47
Nanobiotechnology, 31–32
Nanoclays, 166–167, 446–450
Nanocoated concrete, 429
Nanocomposites, 665–666
Nanocopper, 432–433
Nanocrystals, 147–148
  in bioremediation, 18
Nanodiamonds, 604–605
Nanodispersed titanium dioxide, 347–348
Nanoenabled sensors, 655–656
Nanoengineered self-healing concrete, 454–458
  applications of nanoengineered bio concrete, 456*t*

Nanofertilizer, 655–656
Nanofibers, 108
Nanofibrous membranes, 604–605
Nanofiltration, pharmaceuticals removal using, 641
Nanogold effects on biodegradation behavior, 115–116
Nanohydroxyapatite (nHAp), 113–115
Nanoiron and derivatives, 17
Nanoiron oxide, 439–441
Nanomaterials (NMs), 10–12, 31–32, 108, 141, 369, 646–647, 655–656, 663–664
   application in, 44–47
      industrial applications of nanomaterials, 48t
      tissue engineering, 108–109
      wound healing, 109
   bioconcrete and limitations, 451–454
   bioreducing agents' role in nanoparticle biosynthesis, 41–42
   biosynthesis pathways, 42–44
   challenges
      and future prospects, 458
      pertaining to applications of, 47–49
   contemporary defects and biogenic deterioration, 406–411
   current practices to overcome degradation, 412
   different nanomaterials effects on biodegradation behavior, 109–126
   mechanism of action, 34–37
      generation of reactive oxygen species, 36
      leakage of cellular components in response to nanoparticles exposure, 36–37
      translocation and internalization of nanoparticles, 34–36
   mechanisms of microbial mediated nanoparticles biosynthesis, 41
   microbial biomineralization, 450–451
   microbial mediated synthesis of nanoparticles, 37–41
   for minimizing microbial attack, 386
   for mitigating deterioration, 412–418
   nanoengineered self-healing concrete, 454–458
   nanomaterial-microbe interaction and mechanism, 33–37
   nanoparticle biosynthesis by using microorganisms, 45t
   for remediation and sensing of pesticide, 662
   responses to, 37
      nanomaterial-microbe interactions, 37
   role in bioremediation, 16
   types
      in civil engineering, 418–450
      of nanoparticles, 32–33
Nanomedicine, 46
Nanomembranes, 86
Nanometer, 406
Nanoparticles (NPs), 9–10, 12, 31–32, 62, 106–107, 154, 348, 406, 538, 655–657, 662–664, 675–676, 679–680. *See also* Enzyme-encapsulated nanoparticles
   aggregation, 589–590
      support mediums for dye remediation, 592–593
   applications, 72–73
      biomedical applications, 72
      catalysis, 72
      energy capture, 72–73
   bimetallic, 664–665
   biodegradability of polymers and impact of, 228–230
   biological synthesis of, 12–14
   chemistry, 67–74
   as enhancers for biodegradation
      carbon-based nanoparticles, 155–157
      clay nanoparticles, 166–167
      iron oxide magnetic nanoparticles, 157–159
      metal-based nanoparticles, 162–165
      MOF nanoparticles, 165–166
      nanobarium titanate, 165
      titanium oxide–based nanoparticles, 160–162
   microbial mediated synthesis of, 37–41
   nanoparticle-based materials
      future prospects, 76–77
      nanoparticle chemistry, 67–74
      nanoparticles' role in biodegradation, 74–76
      NORMAN list of emerging substances, 63t
      occurrence of emerging pollutants, 62–67

nanoparticle-modified smart cement composites, 439–440
preparation, 69–72, 69f
  bottom-up approach, 70–71
  top-down approach, 71–72
roles in biodegradation, 11–12
synthesized from bacteria, 13
synthesized from fungi and yeast, 14
synthesized from plants, 12–13
synthetic dyes biodegradation by, 20–21
toxicity, 73–74
types, 32–33
  carbon-based nanoparticles, 32–33
  ceramic nanoparticles, 33
  lipid based nanoparticles, 33
  metallic nanoparticles, 32
  polymeric nanoparticles, 33
Nanopesticides, 655–656
Nanophotocatalysis, 85–86
  advanced oxidation processes for water and wastewater treatment, 87–90
  effective factors in advanced oxidation processes, 93–97
  mechanisms, 97–98
  methods of improving photocatalytic efficiency, 90–93
    co-doping, 92
    metal doping, 90
    morphology control, 92
    semiconductor coupling, 91
    surface sensitization by organic ligands, 90–91
  process, 86f
Nanoporous scaffolds, 108
Nanoremediation, 481–482
Nanoscale surfaces, 108
Nanosilica, 424–430
Nanosilver (nAg), 113–115, 430–432
  effects on biodegradation behavior, 113–115
Nanosorbents, 86
Nanostructured materials, 2–3, 604–605
Nanotechnology (NT), 9–11, 31, 108, 137, 346–347, 530, 538, 656, 679
  nanotechnology-based methods, 660–661
  nanotechnology-derived nanomaterials, 31
  remediation of pollutants using, 16–18
  for wood protection, 289–295
Nanotitanium dioxide (TiO_2), 419–424
  effects on biodegradation behavior, 118–123
Nanotubes, 666
Nanozerovalent iron (nZVI), 17, 581, 587–589
  support material for, 590–593
  for textile dye remediation, 588–589
Nanozinc, 433–436
Nanozymes, 489
Natural goethite ZVI, 482
Natural plastic biodegradation, 330–331
  PHB, 330–331
Natural zeolite, 582
Nematicides, 66
Neolentinus, 283–284
Nickel, 509
  nickel/iron nanoparticles, 162–163
Nicotinamide adenine dinucleotide (NADH), 41–42
Nicotinamide adenine dinucleotide phosphate (NADPH), 41–42
Nigella sativa, 579–580
Nitrate-reducing bacteria (NRB), 376, 677–678
Nitroreductases, 485
Noaea mucronata, 20
Noble metallic NPs, 15–16
Nonbiodegradable plastics, 180–181, 325
NORMAN network, 62, 67
Nylon, 328–329

O

OCT plasmid, 279
One-pot self-assembly, 565
Ordinary Portland cement (OPC), 369
Organic compounds, 85
Organic ligands, surface sensitization by, 90–91
Organic linkers, 89–90
Organic materials
  biodegradation of, 531–537
  inference and future prospects, 167
  nanoparticles as enhancers for biodegradation, 155–167
Organic pollutant compounds, 153–154
Organic treatments, 405–406

Organic-inorganic epoxysiloxane, 351
Organo montmorillonite (OMMT), 230
Organochlorines, 15–16, 477–478
Organophosphates, 477–478
Oxalic acid, 341
Oxidoreductases, 603
Oxygen consumption, 334
Oxytetracycline (OTC), 566–567
Ozonation, 586

P
p-nitrophenol-nanosilica (PNP-NS), 429–430
Paints, 431–432
Partially biomass-based plastics, 214–215
PBS-co-hexane succinate (PBS-co-HS), 225
Pechini method, 541
Penicillium, 410–411
    P. chrysogenum, 291, 423–424
    P. fellutanum, 42–43
    P. spinulosum, 352–353
Pentachlorophenol (PCP), 16
Peroxidase, 604
Peroxides, 18
Personal care products, 67
    biodegradation of, 22–23
Pesticides, 478, 511–512, 656, 659–661
    biosensors for pesticide detection, 667
    nanomaterials for remediation and sensing of, 662
    as product of agriculture, 511–512
Petroleum and aromatic compounds, 515–516
Phanerochaete chrysosporium, 144, 190–191, 198–199
Pharmaceutical pollutants
    agricultural byproducts and biosorbents, 643–644
    environmental and ecological risks, 636–638
    nanobiodegradation of, 635–636
    nanomaterials, 646–647
    pharmaceutical removal methods, 639–643
        advanced oxidation methods, 641–643
        biological treatment methods, 639–640
        constructed wetlands, 640
        pharmaceuticals removal using nanofiltration, 641
    resins and metal oxide-based adsorbents, 644–646
Pharmaceutical removal methods, 639–643
    advanced oxidation methods, 641–643
    biological treatment methods, 639–640
    constructed wetlands, 640
    pharmaceuticals removal using nanofiltration, 641
Pharmaceuticals and personal care products (PPCPs), 22–23
Phenanthrene (PHEN), 18–19
Phenazines, 519–521
Phenol derivatives, 606
Phenolic compounds (PCs), 21–22
    biodegradation of, 21–22
Phlebiopsis, 283–284
Phormidium fragile, 37–40
Phosphotriesterase, 485
Photo-Fenton method, 642–643
Photoacoustic imaging, 146–147
Photobacterium, 513*t*
Photobiodegradation plastics, 333
Photocatalysis, 85, 480, 585, 641–642
    mechanisms, 97–98
Photocatalyst, dosage of, 96–97
Photocatalytic degradation, 566
Photocatalytic fuel cells, 481
Photocatalytic removal of contaminants using MOF, 566–568
Photodegradation, 219
Photooxidation, 178
Photooxidative degradation of waste plastics, 244
Photoresponsive methods, 668–669
Photosensitizing substances, 350
Photothermal therapy, 146–147
Phototrophic organisms, 343
Phototrophs, 343
Physical induced deterioration, 407
Physiochemical methods, 478–479
Phytochelatin synthase (PCS), 280
Phytoremediation, 15–16
Pinus sylvestris L. *See* Scots pine (*Pinus sylvestris* L.)
Plants
    nanoparticles synthesized from, 12–13
    protection products, 66

Plasmids, 279
Plastic industrial waste material
　biodegradation, 323
　　biodegradable plastics, 325–327
　　biodegradation of synthetic plastic, 327–328
　　chemical formation of polyester, 328
　　commonly used plastics, 324
　　industrial plastic waste, 325
　　industries, 324–325
　　nonbiodegradable plastics, 325
　　of blended plastic, 332
　　of natural plastic, 330–331
　　of thermo set plastics, 333
　　plastic, 323–324
　　polyester, 328
　　polymer production, 324
　　structure of polyester, 328–330
　　types of biodegradation, 326–327
　enzymatic method to degrade polyhydroxyalkanoate, 331–332
　nylon and polyethylene, 328–329
　other degrading polymers, 332–333
　　ethylene vinyl alcohol, 333
　　photobiodegradation plastic, 333
　standard testing methods
　　carbon dioxide evolution and oxygen consumption, 334
　　control composting test, 335–336
　　experimental observation, 334
　　forming clear zone, 334
　　radiolabeling, 334
Plastic waste
　biodegradation of plastic-based waste materials
　　biodegradability of various polymers and impact of nanoparticle, 228–230
　　broad-spectrum perspectives and status of biodegradable polymers, 214–215
　　different fillers and influence on biodegradation, 231–232
　　different kinds of degradation of polymers, 218–225
　　filler dimension effect on biodegradation, 232–233
　　influence of factors on polymers biodegradability, 226–227
　　techniques and methodologies employed for polymer degradation, 227–228
　disposal methods, 176–178, 239, 243–244
　　incineration, 243–244
　　landfilling, 243
　　recycling, 244
　pollution and environmental impacts, 240–241
　　leaching of highly toxic chemicals from plastic wastes, 241
Plastics, 175–176, 239, 323–324, 529–530
　biodegradation, 181–188
　　composting, 182–183
　　enzymatic degradation and mechanism, 184
　　factors influencing biodegradability of polymers, 184–186
　　microbial degradation, 183–184
　　synthetic and other approaches to, 186–188
　design properties, 180
　degradability properties, 180–181
　determination techniques for microbial degradation of, 201–203
　　differential scanning calorimetric analysis, 202
　　Fourier transform infrared spectroscopy, 202
　　gel permeation chromatography, 202
　　gravimetric determination of weight loss, 201
　　microscopy observations of surface, 202
　　other recently reported analytical techniques, 203
　　radiolabeling, 203
　　standard methods, 203
　　thermogravimetrical analysis, 201–202
　from fossil resources
　　polybutylene succinate, 190
　　polycaprolactone, 188–189
　　polyethylene, 191–193
　　polypropiolactone, 189–190
　　polypropylene, 193–194
　　polystyrene, 195
　　polyurethane, 188
　　polyvinyl chloride, 190–191
　future prospects, 203–204

Plastics (*Continued*)
  identifications and classifications, 179–181
  materials, 529
  methods in biodegradation of, 178*f*
  from renewable resources, 195–201
  technologies for solid waste management, 182*f*
  thermal properties, 179–180
    thermoplastics properties, 180
    thermosetting polymers, 180
  waste management of, 178–179
Plectonema boryanum, 43–44
Plesiophthalmus davidis, 254–255
Pleurotus ostreatus, 295
Poisonous dyes, 666
Pollutants, 9, 510
Pollutants remediation using nanotechnology, 16–18
  dendrimers in bioremediation, 17
  nanocrystals and carbon nanotubes in bioremediation, 18
  nanoiron and derivatives, 17
  single-enzyme nanoparticles in bioremediation, 18
Poly 3-hydroxybutyrate (P3HB), 116–117
Poly-D-L-lactide-co-glycolide (PLGA), 106–107, 137–138, 139*f*
Poly(acrylamide-co-acrylamidoglycolic acid)/guar gum (PAAG), 113–115
Poly(ethylene terephthalate), 213
Poly(ethylene) glycol altered urethane acrylate (PMUA), 18–19
Poly(hydroxyalkanoates) (PHAs), 186–188, 195–197, 534
  enzymatic method to degrade, 331–332
Poly(lactic acid) and copolymers, 197–198
Poly(propylene fumarate) (PPF), 106–107
Polyacrylonitrile (PAN), 617
Polybrominated diphenyl ethers, 153
Polybutylene succinate (PBS), 190, 534
Polycaprolactone (PCL), 137–138, 188–189, 328–329, 534
Polychlorinated biphenyls (PCBs), 18–19, 153, 516–519
Polycyclic aromatic hydrocarbons (PAHs), 18–19, 158–159, 159*f*
Polydimethylsiloxane (PDMS), 293
Polyester, 328
  chemical formation of, 328
  structure of, 328–330
    PCL, 328–329
    PUR, 329–330
Polyethylene (PE), 175–176, 191, 213, 242–243, 529
Polyethylene, 192–193, 328–329
Polyethylene glycols (PEG), 146–147
Polyethylene tetrachloride (PET), 242–243
Polygalactomannan, 200
Polyglycolic acid (PGA), 138
Polyhydroxybutyrate (PHB), 330–331
Polylactic acid (PLA), 138, 232, 329, 534
Polymeric nanoparticles, 33, 68
Polymers, 159, 176–178, 239, 324, 326
  coatings, 421–422
  degradation, 143–144, 218–225
    influence of factors on polymers biodegradability, 226–227
    techniques and methodologies employed for, 227–228
  factors affecting biodegradation of, 535–537
  factors influencing biodegradability of, 184–186
  nanoparticles, 141
  polymer-based nanomaterials, 137
  production, 324
Polynuclear aromatic hydrocarbons, 19
Polypropiolactone (PPL), 189–190
Polypropylene (PP), 175–176, 193–194, 213, 242–243
Polypropylene fiber (PPF), 437–438
Polypyrrole-polyaniline (PPy-PANI), 16–17
Polypyrrole/zinc oxide (Ppy/ZnO), 684–685
Polystyrene (PS), 175–176, 195, 213, 242–243
Polyurethane (PUR), 188, 329–330
Polyurethane acrylate anionomer (UAA), 19
Polyvinyl butyral (PVB), 292
Polyvinyl chloride (PVC), 175–176, 190–191, 213, 242–243
Pongamia pinnata, 432–433

Poplar (*Poplus* spp), 262
Poria placenta. *See* Brown rot fungus (*Poria placenta*)
Positron emission tomography (PET), 146–147
Potentially toxic heavy metals, 514–515
Precipitation, 9
Primary MPs, 529–530
Pristine MWNTs (*p*-MWNTs), 188–189, 229–230
Procion Yellow H-EXL, 585
Proteases, 485
Protective epoxysiloxane coatings, structure and composition of, 354–358
Protein data bank (PDB), 145
Pseudokirchneriella subcapitata, 637
Pseudomonas, 4–5, 478–479, 513t, 517
    P. aeruginosa, 36–37
    P. delafieldii, 16
    P. fluorescens, 21
    P. pseudoalcaligenes KF707, 516–517
    P. putida, 181–182, 279, 490–491, 513t
    P. stutzeri, 181–182
Pterocarpus santalinus. *See* Red sander (*Pterocarpus santalinus*)
Pyrethroids, 477–478
Pyrolysis, 71, 178

Q

Q10 coenzyme, 32–33
Quaternary ammonium compounds (QACs), 285–286
Quaternized chitosan oligomers, 292–293
Quinone reductases, 485

R

Radiolabeling, 203, 334
Radiotherapy, 146–147
Ralstonia, 517
    R. eutropha AE104, 280
Rapid chloride permeability tests (RCPT), 429
Reactive Black 5, 585
Reactive oxygen species (ROS), 32–33, 655–656
    generation of, 36
Reactive Red 45, 585
Recycling of waste plastic, 244
Red sander (*Pterocarpus santalinus*), 268

Reduced graphene oxide (rGO), 444–445, 590–591, 665–666
Reduction strategies for plastics waste management, 246
    combined approaches, 246
Reflectometric interference spectroscopy, 203
Reinforced concrete (RC), 408–409
Relative humidity (RH), 380
Remediation, 11
    methods, 66
    nanomaterials for, 662
    of pollutants using nanotechnology, 16–18
Renewable resources
    others polymers
        agar, 200–201
        alginate, 199–200
        cellulose, 198–199
        chitosan, 200
        gelatine, 199
        guar gum, 200
        starch, 199
    plastics from, 195–201
    poly(hydroxyalkanoate), 195–197
    poly(lactic acid) and copolymers, 197–198
Repellents, 66
Research and development (R&D), 510
Reseda lutea, 20
Resins, 324, 644–646
Reverse coprecipitation, 614
Rhamnus virgata (RV), 12–13
Rhizopus
    R. arrhizus, 228–229
    R. oryzae NS5, 253
    R. stolonifer, 40
Rhodococcus, 478–479, 517, 520–521
    R. globerulus P6, 519
    R. jostii, 76
Rhodonia placenta, 290
Rhodopseudomonas
    R. capsulata, 42–43
    R. palustris, 280
Ring-opening polymerization, 229
Rodenticides, 66
Rosewood, 265–266

S

Saccharomyces cerevisiae, 40, 490−491
Saccharophagus degradans, 200−201
Sal (*Shorea robusta*), 262, 267
Salicylic acid, 644−645
Salmonella typhi, 37−40
Sandblasting method, 345
Saponite, 534−535
Sargassum
    S. crassifolium, 40−41
    S. ilicifolium, 40−41
    S. muticum, 40−41
    S. wightii, 42−43
Scanning electron microscopy (SEM), 227, 353, 427, 680−681
Scariola orientalis, 20
Scenedesmus, 40−41
Scots pine (*Pinus sylvestris* L.), 291
Secondary building units (SBUs), 89−90
Secondary MPs, 529−530
Selenium, 509
Semiconductor
    coupling, 91
    NPs, 10−11, 68−69
Sensor, 667
Serpula, 283−284
    S. lacrymans, 295
Sewer environment, minimizing sulfide in, 383
Sewer structures, microbiologically-induced corrosion in, 376
Shipworms, 277
Shish kebab effect, 436−437
Shorea robusta. See Sal (*Shorea robusta*)
Silica (SiO_2), 33, 545−546, 688−689
    silica-based nanomaterials, 2−3
Silica nanoparticles (SiO_2NPs), 680, 688−689
Silicomodes, 492
Silicon (Si), 581−582
    nanoparticles, 604−605
Silicones, hybrid coatings based on wax and, 295−297
Silver, 509
    nanomaterials, 47
Silver nanoparticles (AgNPs), 12−13, 15−16, 76, 164−165, 290, 657, 682−684

Single-enzyme nanoparticles in bioremediation, 18
Single-walled carbon nanotubes (SWCNTs), 18, 606−607
Sinorhizobium sp. C4, 490
Small-angle X-ray scattering (SAXS), 353, 356−357
Sodium percarbonate (SPC), 581
Softwood, 263
    timber, 261
Soil
    bacteria, 144
    bioremediation, 19
    contamination, 477, 635−636
    microorganisms, 325
Soiling (esthetic)/fouling classification, 394−395
Sol-gel method, 294, 540−541
    materials and methods
        climatic tests of protective coating experimental samples in Antarctica, 354
        environmentally friendly fungicidal additive, 352−353
        laboratory testing of epoxysiloxane coatings for biostability, 353−354
        methods for studying microstructure of coatings, 353
        synthesis of epoxysiloxane sols, 352
    to protect marble, 350−361
    to protect wood from wood destroying fungi
        hybrid coatings based on wax and silicones, 295−297
        low-toxic biocides to protect wood from wood destroying fungi, 297−306
    results
        biostability of epoxysiloxane coatings, 358−361
        structure and composition of protective epoxysiloxane coatings, 354−358
Sono-photocatalytic oxidative approach, 481
Sonochemical methods, 481
Sorptive removal of contaminants using MOF, 569−570
Sparassis latifolia, 144
Sphingobium, 4−5
    S. yanoikuyae, 157
    B1 strain, 519

Sphingomonas, 4–5, 478–479, 517
Stachybotrys, 410–411
Stain fungi, 277
Stainless steels (SS), 675–676, 688–689
Staphylococcus, 513t
    S. aureus, 37–40, 47, 430–431, 690
Starch, 186–188, 199
    nanocrystals, 530
Sterilized seawater (SSW), 680–681
Stone biodegradation, methods to counter, 344–349
Stone destruction under influence of microorganisms, 339–344
Streptomyces coelicolor, 200–201
Styrene, 213–214
Subabul (*Leucaena leucocephala*), 262
Sulfamides, 285
Sulfate-reducing bacteria (SRB), 376, 677
    inhibiting activities of, 383
Sulfatization, 341
Sulfide
    chemical removal of, 384
    minimizing sulfide in sewer environment, 383
Sulfonamides, 286
Sulfuric acid, generation of, 376–378
Superhydrophobic coatings, 288
Superworm (*Zophobas atratus*), 255
Surface functionality of NPs, 74
Surface sensitization by organic ligands, 90–91
Synergy, 510–511
Synthesized organic compounds, 153
Synthetic dyes
    biodegradation by nanoparticles, 20–21
    mixture, 585
Synthetic goethite ZVI, 482
Synthetic plastics, 214
    biodegradation, 327–328

T

Tapinella, 283–284
Teak (*Tectona grandis*), 262, 265
Tenebrio
    T. molitor, 254
    T. obscurus, 254

Tenebrio molitor. See Mealworm (*Tenebrio molitor*)
Termites, 278
Tetracycline (TC), 162, 566–567
Tetraethoxysilane (TEOS), 295
Tetraselmis, 40–41
Textile
    effluent composition, 582–584
    industry, 579, 582–583
    nanozerovalent iron for textile dye remediation, 588–589
Thermal degradation
    of polymers, 219–220
    of waste plastics, 245
Thermo set plastics biodegradation, 333
Thermogravimetric analysis (TGA), 115–116, 190–191, 201–202
Thermophysical methods, 344
Thermoplastics, 324
Thermosets, 324
Thermosetting polymers, 180
Thiobacillus
    T. intermedius, 410–411
    T. neapolitanus, 410–411
    T. novellus, 410–411
    T. thiooxidans, 397
    T. thioparus, 410–411
    T. thioxiduns, 678
Three-dimensional graphene (3D grapheme), 608
Timber, 261
Timber industry-based waste materials, biodegradation of
    annual consumption of wood, 273
    background, 262
    biodegradation, 275
    consumption at wood-based industry, 270–272
        fuelwood, 271–272
        furniture, 270–271
    degradation by genetically engineered microorganisms, 279–280
    factors affecting microbial degradation, 278–279
        biological factors, 278–279

Timber industry-based waste materials, biodegradation of (*Continued*)
  environmental factors, 279
  plantation and cultivation of timber, 264–268
    illupai, 266–267
    red sander, 268
    rosewood, 265–266
    sal, 267
    teak, 265
  production of timber, 268–270
  value chain of timber-based industry, 273–274
  waste problem, 274–275
  wood from tree, 262–264
Tissue engineering, 106–107
  nanomaterials application in, 108–109
Titanium nanoparticles, 76
Titanium oxide (TiO_2), 33, 160, 544
  nanomembranes, 482
Titanium oxide nanoparticle (TiO_2NPs), 160–162, 680, 685–687
Top-down approach, 69–72, 662–663. *See also* Bottom-up approach
  attrition, 71–72
  micropatterning, 71
  pyrolysis, 71
Toxicity of NPs, 73–74
Toxins degradation, 510–511
Trabusiella guamensis, 144
Traditional bioremediation, 2–3
Trametes
  T. versicolor, 639
  T. villosa, 493–494
Translocation of nanoparticles, 34–36
Treatment processes, 85
Trema micrantha, 262–263
Triazoles, 285
Tributyl phosphate (TBP), 297
2,4,5-trichlorophenol, 160–161
Trichoderma, 4–5
  T. reesei, 198–199
Triethanolamine (TEOA), 93
2,4,6-trinitrotoluene (TNT), 493–494
Tripolyphosphate (TPP), 292–293
Trypsin, 18

Two-dimensional structural morphology (2D structural morphology), 608
Tyrosinases, 485

U

Ubiquinone, 32–33
Ulocladium
  U. chartarum, 352–353
  U. ilicis, 291
UltracidTM, 161
Ultrahigh performance concrete (UHPC), 424–425
Ultrasonic pulse velocity (UPV), 419–421
Ultrasonication, 589
Ultrasound, 481
  ultrasound-based insulin delivery system, 139–140
Ultraviolet (UV), 253, 410–411
  degradation, 178
  UV-visible spectroscopy, 12–13
Uranium bioremediation, 19–20
Urbanization, 9
US Environmental Protection Agency (EPA), 213
US Food and Drug Administration, 47

V

Valance band (VB), 85–86
Value chain of timber-based industry, 273–274
Vanadium(V) oxide(V_2O_5), 543–544

W

Waste plastics, 179, 240–241
  degradation using microbes, 251–254
    activity by algae, 253–254
    activity by bacteria, 251–252
    activity by fungi, 252–253
Wastes, 9
  management of plastics, 178–179
  wood, 274–275
Wastewater treatment, 622
  AOPs for, 87–90
  plants, 584
Water body contamination, 635–636
Water treatment, AOPs for, 87–90
Weathering, 218–219
Weissella viridescens, 21

Wet-chemical method, 541
White-rot fungus (*Trametes versicolor*), 284, 291
Wood
   biodegradation under influence of wood-destroying fungi, 283–284
   field tests of friendly wood protection coatings, 295–297
   hydrophobization, 287
   lice, 277
   products
      carbon impact equation, 276
      chemical means of protecting wood from biodegradation, 284–287
      field tests of friendly wood protection coatings in climatic conditions, 295–297
      nanotechnology for wood protection, 289–295
      sol-gel technology to protect wood from wood destroying fungi, 295–306
      surface hydrophobization, 287–289
      wood biodegradation under influence of wood-destroying fungi, 283–284
   from tree, 262–264
Worms, degradation activity using, 255
Wound healing, nanomaterials application in, 109

X

Xanthomonas, 4–5
Xenobiotic compounds, 3–4
XYL plasmid, 279
Xylotrophic fungi
   biocidal compositions for protecting wood from xylotrophic fungi
      in moist chamber, 302–306
      on nutrient medium, 297–302

Y

Yeast nanoparticles
   biosynthesis, 40
   nanoparticles synthesized from yeast, 14

Z

Zeolites, 581–582
Zero dimension (0D), 10–11
Zinc, 509
Zinc oxide (ZnO), 433, 545
Zinc oxide nanoparticles (ZnONPs), 163–164, 680, 684–685
Zirconia (ZrO_2), 33
Zophobas atratus. See Superworm (*Zophobas atratus*)

Printed in the United States
by Baker & Taylor Publisher Services